Practical Plant Nematology

2nd Edition

Practical Plant Nematology

2nd Edition

Edited by

Rosa H. Manzanilla-López

Centro de Desarrollo de Productos Bióticos, Instituto Politécnico Nacional, Yautepec, Morelos, Mexico

David J. Hunt

CABI, Silwood Park, Buckhurst Road, Ascot, Berkshire SL5 7TA, UK

Nahum Marbán-Mendoza

Universidad Autónoma Chapingo, Postgrado en Protección Vegetal, Chapingo, Estado de México, Mexico

and

Said Infante Gil

Editorial del Colegio de Postgraduados, Montecillo, Texcoco, Estado de México, Mexico

CABI

CABI is a trading name of CAB International

CABI
Nosworthy Way
Wallingford
Oxfordshire OX10 8DE
UK

Tel: +44 (0)1491 832111
E-mail: info@cabi.org
Website: www.cabi.org

CABI
200 Portland Street
Boston
MA 02114
USA

Tel: +1 (617)682-9015
E-mail: cabi-nao@cabi.org

A catalogue record for this book is available from the British Library, London, UK.

A catalogue record for this book is available from the British Library, London, UK.

ISBN-13: 9781836990390 (hardback)
9781836990406 (OA ePDF)
9781836990413 (OA ePub)

DOI: 10.1079/9781836990413.0000

Commissioning Editor: Rebecca Stubbs
Editorial Assistant: Emma McCann
Production Editor: James Bishop

Typeset by Straive, Pondicherry, India
Printed and bound by CPI Group (UK) Ltd, Croydon, CR0 4YY

Contents

Contributors

Antonio Archidona-Yuste
Instituto de Agricultura Sostenible/CSIC, Apdo. 4084, 14080 Cordoba, Spain; antonio.archidona@ias.csic.es; antonioj.archidonayuste@gmail.com
Wim Bert
Department of Biology, Nematology Unit, Ghent University, Ledeganckstraat 35, 9000 Ghent, Belgium; wim.bert@ugent.be
Anselmo J. Cabrera-Hidalgo
TecNM-Tlatlauquitepec (ITSTL), Carretera Federal Amozoc-Nautla km 122+600 Almoloni, Tlatlauquitepec 73907, Puebla, Mexico; anselmo.cabrera@tlatlauquitepec.tecnm.mx; ajcabrerahidalgo@gmail.com
Juvenil E. Cares
University of Brasilia, Institute of Biological Sciences, Department of Plant Pathology, Brasilia, DF, Brazil; cares.j28@gmail.com
Pablo Castillo
Instituto de Agricultura Sostenible/CSIC, Apdo. 4084, 14080 Cordoba, Spain; ag1cascp@uco.es; p.castillo@csic.es
Senyu Chen
Southern Research and Outreach Center, 35838 120 St, Waseca, MN 56093, USA; chenx099@umn.edu
Ignacio Cid del Prado-Vera
Colegio de Postgraduados, Montecillo, Texcoco, CP 56230, Edo. de Mexico, Mexico; i.cid@colpos.mx; iciddelpradovera@gmail.com
Danny Coyne
IITA, Kasarani, P.O. Box 30772-00100, Nairobi, Kenya; d.coyne@cgiar.org
Wilfrida Decraemer
Department of Biology, Nematology Unit, Ghent University, Ledeganckstraat 35, 9000 Ghent, Belgium; wilfrida.decraemer@ugent.be
Johan Desaeger
University of Florida, Gulf Coast Research and Education Center, 14625 CR 672, Wimauma, Florida 33598, USA; jad@ufl.edu
Don W. Dickson
Entomology and Nematology Department, University of Florida, P.O. Box 110620, Gainesville, FL 32611-0620, USA; dwd@ufl.edu

Howard Ferris
Department of Entomology and Nematology, University of California, Davis, CA 95616, USA; hferris@ucdavis.edu
Mariella M. Finetti-Sialer
National Research Council, Institute of Biosciences and Bioresources (IBBR-CNR), Via Amendola 165/A, 70126 Bari, Italy; mariella.finetti@cnr.it
Javier Franco Ponce
Consultant Nematologist, Calle Los Cerezos 338, Dpto. 103, Surco, Lima, Peru; javierfrancoponce@gmail.com
Robin M. Giblin-Davis
Fort Lauderdale Research and Education Center, University of Florida, 3505 College Avenue, Davie, FL 33314-7798, USA; giblin@ufl.edu
Cynthia Gleason
Department of Plant Pathology, Washington State University, Pullman, WA 99164, USA; cynthia.gleason@wsu.edu
Vilmar Gonzaga
Embrapa Dairy Cattle, Juiz de Fora, MG, Brazil; vilmar@cenargen.embrapa.br
Deepak Haarith Rajendran
Entomology and Nematology Department, University of Florida, Gainesville, FL USA; deepakrajendran@ufl.edu
Alemayehu Habteweld
USDA ARS, Mycology Nematology Genetic Diversity and Biology Laboratory, Building 010A, Room 118, BARC-West, 10300 Baltimore Avenue, Beltsville, MD 20705, USA; Alemayehu.Habteweld@usda.gov
Zafar Handoo
USDA ARS, Mycology Nematology Genetic Diversity and Biology Laboratory, Building 010A, Room 118, BARC-West, 10300 Baltimore Avenue, Beltsville, MD 20705, USA; Zafar.Handoo@usda.gov
Solveig Haukeland
International Centre for Insect Physiology and Ecology (*icipe*), Kasarani, Nairobi, Kenya, P.O. Box 30772-00100; Norwegian Institute of Bioeconomy Research, Norway; shaukeland@icipe.org
Sue Hockland
Independent Plant Nematology Consultant, Harrogate, UK; Department of Plant Protection, Sand Hutton, York, YO41 1LZ, UK; sue.hockland@btinternet.com
Oleksandr Holovachov
Department of Invertebrate Zoology, Swedish Museum of Natural History, Box 50007, SE-104 05 Stockholm, Sweden; oleksandr.holovachov@nrm.se
David J. Hunt
CABI, Silwood Park, Buckhurst Road, Ascot, Berkshire SL5 7TA, UK; tabanids.bite@gmail.com
Renato N. Inserra
Florida Department of Agriculture and Consumer Services, Division of Plant Industry/Nematology Section, P.O. Box 147100, Gainesville, FL 32614-7100, USA; inserra@doacs.state.fl.us
Natsumi Kanzaki
Kansai Research Center, FFPRI, 68 Nagaikyutaroh, Momoyama, Fushimi, Kyoto, 612-0855 Japan; natsumikanzaki@gmail.com
James Kisaakye
International Institute of Tropical Agriculture (IITA), Kasarani, Nairobi, Kenya, P.O. Box 30772-00100; jkisaakye@icipe.org
Rosa H. Manzanilla-López
Centro de Desarrollo de Productos Bióticos, Instituto Politécnico Nacional, Carretera Yautepec-Jojutla km 85, San Isidro, CP 62739, Morelos, Mexico. Present address: 16 Coleswood Road, Harpenden, AL5 1EQ, UK; rosa.manzanilla@gmail.com

Nahum Marbán-Mendoza
Universidad Autónoma Chapingo, Departamento de Parasitología Agrícola, Chapingo, Estado de México, CP 56230, Mexico; nmarbanm@yahoo.com.mx
Juan E. Palomares-Rius
Instituto de Agricultura Sostenible/CSIC, Apdo. 4084, 14080 Cordoba, Spain; juapariu@hotmail.com
Hugo H. Mejía-Madrid
Facultad de Ciencias and ENES Juriquilla, UNAM, Mexico; hugo_mejia_madrid@ciencias.unam.mx; hugoharlan@gmail.com
Walter Peraza
Laboratorio de Nematología, Escuela de Ciencias Agrarias, Universidad Nacional, AP 86-3000, Heredia, Costa Rica; aesquive@una.ac.cr
Roland N. Perry
School of Health, Medicine and Life Sciences, University of Hertfordshire, AL10 9AB, UK; r.perry2@herts.ac.uk
Stephen J. Powers
Stats Powers Ltd, South Petherton, Somerset, TA13 5HP, UK; powerssj@gmail.com
Thomas O. Powers
University of Nebraska, Lincoln, NE 68583-0722, USA; tpowers@unlnotes.unl.edu
Juliano V.G. Prendi
Universidade de Brasília, Brazil; julianoprendi@gmail.com
James A. Price
CMS, The James Hutton Institute, Invergowrie, Dundee, UK; james.price@hutton.ac.uk
Thomas Prior
Department of Plant Protection, Sand Hutton, York, YO41 1LZ, UK; thomas.prior@fera.co.uk
Patrick Quénéhervé
IDR (Institute of Research for Development), UMR 186 Résistance des Plantes aux Bioagresseurs, Pôle de Recherche Agro- environnementale de la Martinique, BP 214-97285 Le Lamentin Cedex 2, Martinique (French West Indies); patrick.queneherve@gmail.com
Ian T. Riley
School of Agriculture, Food and Wine, Adelaide University, SA 5005, Australia; ian.t.riley@uni.sydney.edu.au; ian.t.riley@gmail.com
Sara Sánchez-Moreno
Department of Biodiversity and Evolutionary Biology, National Museum of Natural Sciences (CSIC), Madrid 28006, Spain; sarasm@inia.csic.es
Jason Stanley
Florida Department of Agriculture and Consumer Services, Division of Plant Industry, P.O. Box 147100, Gainesville, FL 32614-7100, USA; stanlej@doacs.state.fl.us
Richard Sikora
Bonn University, Germany; rsikora@uni.bonn.de
Sergei A. Subbotin
Plant Pest Diagnostics Center, California Department of Food and Agriculture, 3294 Meadowview Road, Sacramento, CA 95832-1448, USA; Center of Parasitology of A.N. Severtsov Institute of Ecology and Evolution, Russian Academy of Sciences, Moscow 117071, Russia; sergei.a.subbotin@gmail.com
Renata Tenente
SHIN, Q1 2 CONJ 8 Cs 6, Brasília, DF, Brazil; renatatenente2@gmail.com
Esther van den Berg
National Collection of Nematodes, Biosystematics Division, ARC-Plant Protection Research Institute, Private Bag X134, Queenswood 0121, South Africa

Ingrid Varela-Benavides
Doctorado en Ciencias Naturales para el Desarrollo, Instituto Tecnológico de Costa Rica, Costa Rica; invarela@itcr.ac.cr
Edgar Villar-Luna
CIIDIR Michoacán, Unidad Michoacán, Justo Sierra Ote. 28, Centro, 59510, Jiquilpan de Juárez, Michoacán, Mexico; edgar_villarled@yahoo.com; evillarl@ipn.mx
Koon-Hui Wang
University of Hawaii, USA; koonhui@hawaii.edu
Tristan T. Watson
Agricultural Center (LSU AgCenter), FL, USA; TWatson@AgCenter.LSU.edu; TWatson@gcenter.lsu.edu

Preface

The first edition of this book was published in 2012 and the need for a revised edition became apparent, particularly because of advances in certain major areas such as molecular methodologies and the impact of molecular systematics on nematode taxonomy and the species concept.

Nematodes are among the more numerous and speciose invertebrate organisms. They can be found virtually in any habitat and niche on earth, either as free-living animals or in a variety of interactions with other organisms, including diverse types of symbiotic associations, parasitism of humans and other animals (both vertebrate and invertebrate) and plants. The economic importance of plant-parasitic nematodes has long been recognized and our knowledge of this group is continuously expanding, although there is still much to learn.

Due to their microscopic size, plant-parasitic nematodes are commonly called 'the hidden enemy'. This cryptic nature, unfortunately, accounts for the reduced attention paid to them – they may even be excluded from major crop protection and breeding programmes until it is too late. Yield losses attributed to nematodes can be difficult to assess due to several factors, including lack of expertise in the area (especially taxonomy) for making an accurate diagnosis, and lack of nematology courses and training in basic nematology as part of crop protection, research and extension programme curricula. We can add to this scenario the great complexity of evaluating and putting quantifiable figures to yield losses in multicrop systems, especially in low-input subsistence farming conditions in the tropics, in comparison to major high-value crops and high-tech (and high-input) agricultural systems commonly found in temperate countries. It has been estimated that plant-parasitic nematodes are responsible for annual losses of billions of US dollars on commodity, staple and industrial crops worldwide. With global changes in temperature, it is likely that this figure will increase as nematode pests are able to thrive in previously unsuitable areas and increase their reproductive rate, thus potentially becoming an even greater threat to crop production systems.

Although a number of specialized monographs on nematodes have been published since the first edition of this book, there remains a need for an introductory text aimed at students of agronomy, biology, extension, phytosanitation and other professionals involved in crop protection activities and plant disease diagnostics. The second edition of this book, as with the first, is intended to address this need by covering, at an introductory level, the various aspects, including taxonomy, that lie at the foundations of practical plant nematology and therefore provide a practical overview of this area of crop protection and phytopathology. Since the first edition, much progress has been made in the field of molecular studies and this area has been extensively updated to provide an overview of the current status of this field.

The book is divided into three parts. The first part covers the biology, morphology, taxonomy and practical aspects of symptomatology, sampling, preparation and identification of nematodes using both classical and molecular approaches. This section is also intended to provide guidelines for those wishing to begin working with plant-parasitic nematodes.

The second part of the book deals with the taxonomy, diversity and bionomics of the most economically important plant-parasitic nematode groups. Keys to major taxa and genera are provided and are followed by sections that summarize and illustrate the diagnostic features of major plant-parasitic genera and species. The importance of soil nematodes to ecosystems and agroecosystems is increasingly recognized in soil health and soil web studies, areas where nematodes are key players. The inclusion of a chapter dedicated to the diagnostics of the major groups of free-living soil nematodes is therefore important, this chapter also serving to provide a more complete picture of the highly diverse soil nematode fauna.

The third part of the book deals with the ecological (e.g., nematodes as bioindicators), biochemical and molecular processes involved in plant-nematode interactions, and with the chemical and non-chemical methods used to manage nematodes as part of an integrated pest management approach. There is a new chapter on integrated nematode management. Statutory measures dealing with quarantine issues and knowledge dissemination (farmer field schools and knowledge transfer) aspects are also included to demonstrate the need for a more holistic approach when dealing with plant-parasitic nematodes. Finally, a statistics chapter outlining the planning and analysis of experiments is provided, this being an area where many students frequently require advice and support.

Within the confines of a single volume, it is not possible to cover every nematode group and topic in a comprehensive manner. This book aims to provide an accessible introduction to the basic principles at a level that will hopefully be sufficient for the majority of users. For those who wish to pursue a particular topic further, additional information sources are indicated wherever possible.

The Editors
September 2025

Dedication

To past and present teachers of nematology and crop protection.

Acknowledgements

We would like to thank the chapter contributors and the many colleagues who generously supplied photographs and figures to illustrate the book.

The Editors
September 2025

1 Biology and Host-parasite Interactions

James A. Price[1]* and Roland N. Perry[2]

[1]*CMS, The James Hutton Institute, Invergowrie, Dundee, UK;* [2]*School of Health, Medicine and Life Sciences, University of Hertfordshire, Hatfield, UK*

Abstract

This chapter serves as an introduction to plant-parasitic nematodes and their importance as pests of cultivated plants or crops. The chapter introduces the biology, life cycle (embryogenesis and juvenile hatching), reproduction (amphimixis, automixis and apomixis), thermal requirements, survival strategies, including quiescence and diapause, host-parasite interactions, nematode attraction to plants (semiochemicals, including allelochemicals and pheromones), nematode feeding-habits (i.e., sedentary endoparasitic, migratory endoparasitic, semi-endoparasitic and ectoparasitic), plant penetration, plant-nematode co-evolution, and control and management, including physical, chemical, biological and cultural methods plus resistance and tolerance to nematode attack. The chapter ends with a short summary of the potential for future management strategies.

1.1. Introduction

Plant-parasitic nematodes have enormous economic and social impacts, especially in countries where nematodes may cause total crop loss. All crop plants are parasitized by one or more species of nematodes, and in addition to the detrimental effects on the growth and yield of the crops, the damage caused, especially to root crops (such as carrots), can render the produce unmarketable and eliminate income. Most plant-parasitic nematodes are microscopic in size (less than 1 mm in length and between 15 and 20 μm in diameter). Thus, even though they are major pests, most plant-parasitic nematodes are not visible to the naked eye, and it is often difficult to convince growers and advisers that the crop problems are caused by nematodes. Education and outreach seminars are an essential component of enhancing awareness of nematode problems (see Chapter 22, this volume) and are a necessary prelude to accurate identification and effective and targeted control or management strategies (see section 1.8). It should also be remembered that plant-parasitic nematodes are only one component of a multifaceted ecosystem, and their interactions with other pathogens also need to be evaluated when assessing disease complexes.

*E-mail: james.price@hutton.ac.uk

A revision of Perry, R.N. and Tovar-Soto, A. (2012) Biology and host-parasite interactions. In: Manzanilla-López, R.H. and Marbán-Mendoza, N. (eds) *Practical Plant Nematology*. Mundi-Prensa and Biblioteca Básica de Agricultura, Montecillo, Mexico.

(eds R.H. Manzanilla-López, D.J. Hunt, N. Marbán-Mendoza and S. Infante Gil)
DOI: 10.1079/9781836990413.0001

Plant-parasitic nematodes have a needle-like structure, termed stylet, onchiostyle or odontostylet in different taxonomic groups, which is used to pierce and disrupt plant tissue and, during feeding, to introduce nematode secretions and to withdraw plant cell contents (Gheysen and Jones, 2024). Although it is always difficult to put a framework onto a large, biologically disparate group, plant-parasitic nematodes can be divided broadly into four categories based on their feeding strategy (Fig. 1.1). **Sedentary endoparasitic** nematodes, such as the root-knot nematodes, *Meloidogyne* spp., and the cyst nematodes, *Heterodera* and *Globodera* spp., enter host roots, set up a feeding site within the root tissue and feed internally, remaining sessile at the feeding site. This group has attracted considerable research interest, and *Meloidogyne* is the most economically important genus worldwide (Jones *et al.*, 2013). **Migratory endoparasitic** nematodes, such as *Pratylenchus* spp. and *Radopholus* spp., invade plant hosts but have no fixed feeding site within the plant and do not become sessile. These nematodes move around the plant tissue and feed from numerous cells, causing considerable damage, and can move in and out of several plants. There are also those that feed on the aerial parts of plants (stems, leaves, buds and seeds), such as species of *Aphelenchoides*. **Semi-endoparasitic** nematodes, such as *Rotylenchulus* and *Tylenchulus*, become sessile after penetrating the root but only the anterior part of the nematode penetrates and remains embedded in the root tissue, the rest of the body being outside the root in the soil. **Ectoparasitic nematodes** remain outside the roots to feed externally, puncturing individual

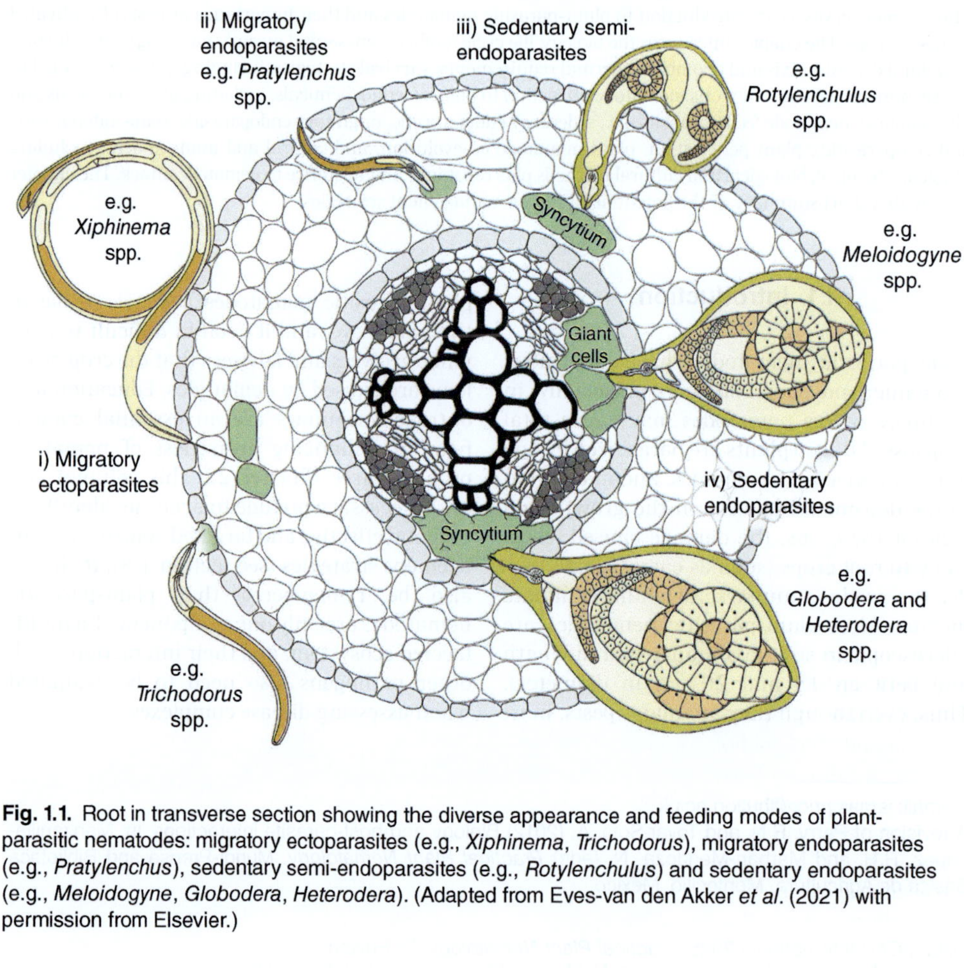

Fig. 1.1. Root in transverse section showing the diverse appearance and feeding modes of plant-parasitic nematodes: migratory ectoparasites (e.g., *Xiphinema*, *Trichodorus*), migratory endoparasites (e.g., *Pratylenchus*), sedentary semi-endoparasites (e.g., *Rotylenchulus*) and sedentary endoparasites (e.g., *Meloidogyne*, *Globodera*, *Heterodera*). (Adapted from Eves-van den Akker *et al.* (2021) with permission from Elsevier.)

cells with their stylet, extracting food and then withdrawing the stylet to move on to another cell to repeat the process. Some ectoparasitic species, for example, some obese forms of *Paratylenchus*, become permanently attached to the root by the deeply embedded stylet. Some nematodes are difficult to categorize. Genera, such as *Hoplolaimus* or *Helicotylenchus*, may be either semi-endoparasitic or migratory ecto-endoparasitic, depending on the host. The so-called 'false' root-knot nematode, *Nacobbus aberrans*, has two main methods of developing to adulthood (Manzanilla-López *et al.*, 2002): in the first strategy, the second-stage juvenile (J2) penetrates the root, and development may then occur inside or outside the root, with repeated invasion and emigration until the adult stage is reached; in the second strategy, once the J2 has invaded, development occurs solely within the root. In both cases, the mature female becomes sedentary at a feeding site in the root.

The conserved morphology of nematodes and the absence of extensive fossil records make discussion of the origins of plant parasitism speculative and inconclusive. Information has been obtained from molecular phylogenies based on the small subunit ribosomal RNA (SSU RNA), which demonstrates that plant parasitism has arisen independently on at least three separate occasions (see Subbotin and Moens, 2024, for discussion). The rRNA phylogenies also support convergent evolution of sedentary endoparasitism and feeding site establishment by root-knot and cyst nematodes (see section 1.4), rather than the theory that the two groups shared a common ancestor. Evolutionary reconstruction using SSU is based on a single gene. More recently, multigene phylogeny, using sequences of genes from nematode transcriptome and genome projects, gave better supported relationships for some clades of plant-parasitic nematodes (Smythe *et al.*, 2019; Ahmed *et al.*, 2022).

The co-evolution of plants and plant-parasitic nematodes has resulted in remarkable synchrony of the host and parasite life cycles, which enhances the chances of nematode infection, survival and reproduction. In this introductory chapter, we discuss basic biology to introduce the links between nematode and host life cycles and where and how synchrony is achieved. Aspects of the biology of nematodes that enhance survival and dispersal will be presented, especially as they impinge on control and management efficiency. Genomic and transcriptomic sequences, expressed sequence tag libraries and tools for functional analysis, such as RNAi, underpin functional genomic studies; the section on genomics (see section 1.5) will introduce this aspect. Knowledge of the biology of nematodes is clearly an important basis for effective damage prevention, and the current methods of management and control will be summarized.

1.2. Hatching and Eggshells

Since the previous edition of this book, our understanding of nematode eggshells and hatching has significantly progressed. There is little variation in the average size of nematode eggs (55-133 μm in length and 17-79 μm in width; Bird and Bird, 1991). To align different branches of nematology, naming conventions for plant-parasitic nematode eggshells should be updated. This requires the old tri- and tetralaminar eggshell models to be modernized to the currently proposed hexalaminar model (Bond and Huffman, 2023), summarized in Figure 1.2. Much of our current eggshell knowledge is based on *Caenorhabditis elegans* and, given the molecular and structural similarities between cyst nematode and *C. elegans* eggshells, it is reasonable to assume that there will be significant evolutionary parallelism for this crucial development stage. Comparative identification of molecular components between nematode eggshells is severely lacking with many similarities between eggshell structures being drawn from electron microscopy images.

In amphimixis (see section 1.6), following sperm penetration of the oocyte, the activated oocyte undergoes two meiotic divisions in which the eggshell begins to form. During gamete fusion and formation of the zygote, the egg passes through the uterus. Rapid remodelling of the cell surface occurs quickly to prevent polyspermy. This outermost eggshell layer, the *pellicula ovi* (formerly vitelline membrane), could initiate from either the oocyte or the maternal uterine layer as the oocyte passes through the oviduct. Chitin synthesis follows setting a slower block to polyspermy at chitin binding domains underneath the *pellicula ovi*. This chitin layer thickens, providing the eggshell with rigidity and offering physical protection to the developing embryo.

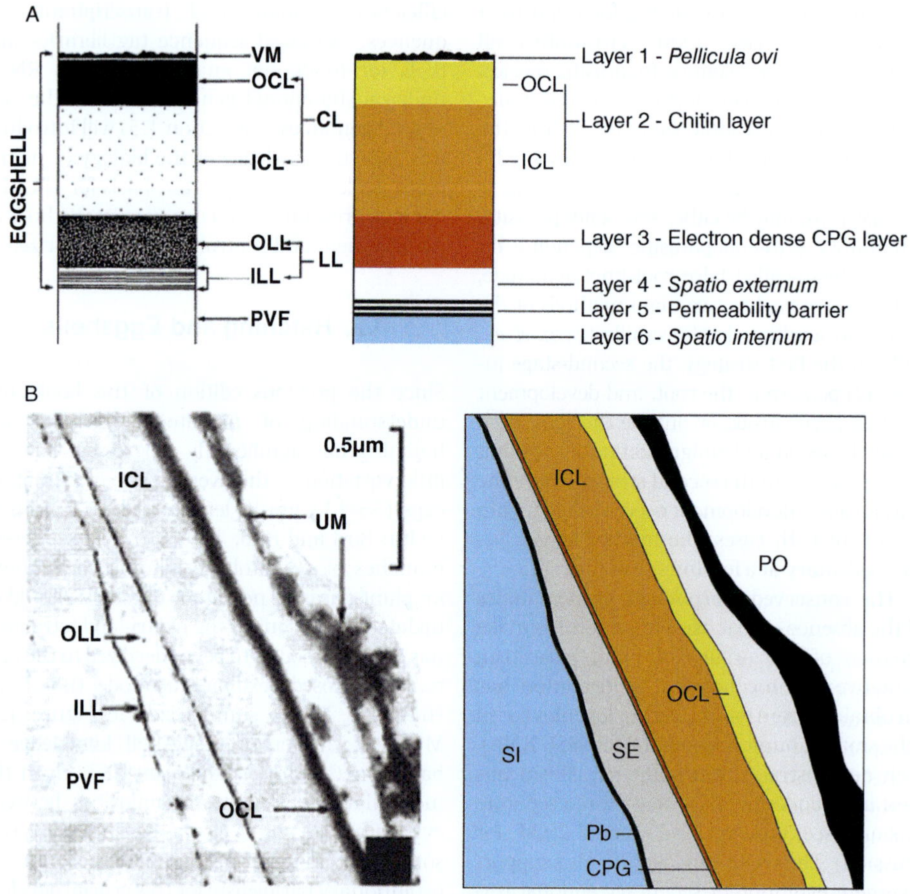

Fig. 1.2. A (left): Eggshell schematic modified from Perry and Trett (1986) showing layers through a generalized tylenchid eggshell (not to scale); (right) Tylenchid eggshell layer schematic coloured and renamed following Bond and Huffman (2023). B (left): Transmission electron microscopy image of *Heterodera glycines* (Perry and Trett, 1986); (right) interpretive tracing of transmission electron microscopy image coloured to match eggshell layers in A. Key: VM = vitelline membrane; OCL = outer chitinous layer; ICL = inner chitinous layer; CL = chitinous layer; OLL = outer lipid layer; ILL = inner lipid layer; LL = lipid layer; PVF = perivitelline fluid; UM = uterine material; PO = *pellicula ovi*; CPG = chondroitin proteoglycan layer; SE = *spatio externum*; Pb = permeability barrier; SI = *spatio internum*. (Courtesy James Price.)

The final electron-dense layer of the rigid eggshell wall (layers 1-3) is formed of lipids and chondroitin proteoglycans (CPG) acting as a molecular sieve. The exact components and function of this layer remain to be clarified; however, CPG have now been identified in both *C. elegans* and *Globodera rostochiensis* (Olson *et al.*, 2012; Price *et al.*, 2023). Beneath the CPG layer is the *spatio externum* (previously extra-embryonic matrix), a fluid-filled space separating the rigid eggshell wall and permeability barrier, preventing cytokinesis failures. The permeability barrier (layer 5) chemically separates the developing embryo from the outside environment. This barrier retains the cryoprotectant trehalose in the *spatio internum* (previously peri-embryonic space) and, in cyst nematodes, provides an environment supportive of longevity in diapause before hatching in the presence of a suitable host. This lipid layer is the main permeability barrier of the eggshell and makes the egg resistant to chemicals, including non-fumigant nematicides.

In each egg, the embryo develops to a first-stage juvenile (J1), which, in some longidorids and trichodorids, hatches. However, in most species of plant-parasitic nematodes, the J1 moults within the egg to become the J2. Physiological adaptations, such as different states of dormancy, are important for survival and persistence in the absence of a host and are frequently associated with the unhatched juveniles (Perry, 1989, 2002; see section 1.6).

Soil type is an important abiotic factor for soil-dwelling stages of nematodes. In general, coarse-textured soils favour hatching and subsequent invasion of roots by plant-parasitic nematodes, providing suitable conditions for aeration and nematode migration. Maximum hatching of plant-parasitic nematodes usually occurs when soil water content is at field capacity, whilst drought and water logging inhibit hatching. Given suitable environmental conditions, such as appropriate temperature, oxygen availability and moisture levels, and an absence of physiological barriers such as diapause (see section 1.6), hatching of most species occurs without requiring specific cues emanating from the food source. However, in some species, co-evolution of host and parasite has resulted in a sophisticated relationship whereby the nematode does not hatch unless stimulated by diffusates from host roots (host root diffusates; also termed root exudates or root leachates), which contain hatch-stimulating chemicals (hatching factors; HF). Hatching in response to host root diffusates has the advantage of ensuring that the nematodes hatch and leave the protection of the egg when host roots are close by; thus, synchrony of host availability and nematode hatch is advantageous for nematode survival.

The overt response to hatch stimulation is eclosion (or hatch from the egg), but detailed analysis of the hatching response of a few species of parasitic nematodes that depend on host stimulation for hatch has demonstrated that a sequence of inter-related events occurs between stimulation and eclosion. The hatching process can be divided into three phases: changes in the eggshell permeability, metabolic activation of the juvenile and eclosion. The chronological order of the first two phases differs between genera. For example, in *Meloidogyne* spp., activation of the juvenile appears to occur first and causes eggshell changes; in others, such as *Globodera* spp., alteration of eggshell permeability characteristics appears to be a necessary pre-requisite for activation of the juvenile (Perry, 2002) (Fig. 1.3). The agents for initiation of these responses vary between species and genera of nematodes but have been studied most extensively in species of root-knot and cyst nematodes. Hatching and survival attributes of these species are associated with the 'packaging' of eggs into ecological units (Perry and Moens, 2011). Female root-knot nematodes lay eggs in a gelatinous matrix, which comprises an irregular meshwork of glycoprotein material. The gelatinous matrix surrounds the eggs and retains them in a package called an egg mass. With cyst nematodes, the death of the mature female is followed by polyphenol oxidase tanning of the cuticle, resulting in a hard, brown cyst. Egg masses and cysts can each contain several hundred eggs. Where conditions are especially conducive for plant growth and nematode development, so-called 'giant cysts' may be produced with large numbers of eggs; however, this 'gigantism' is not a heritable trait (Price *et al.*, 2024a).

Hatching of *Meloidogyne* spp. is, in general, temperature dependent, and hatching occurs when temperatures are favourable without the need for stimulus from root diffusates. However, there are exceptions. For example, a proportion of the unhatched J2 of *M. hapla*, *M. triticoryzae* and *M. chitwoodi* is dependent on root diffusates for hatching, especially in later generations during a host growing season (Gaur *et al.*, 2000; Perry and Wesemael, 2008). Although a few other species from other groups (e.g., *Rotylenchulus reniformis*) hatch in response to host root diffusates, this phenomenon is most common among the cyst nematodes but even in this group reliance on host stimulation for hatch varies. *G. rostochiensis* and *Globodera pallida*, have a very restricted host range and are almost completely dependent on host diffusates for hatch, whereas *Heterodera schachtii*, for example, has a very wide host range (some 218 plant species, including many weeds) and hatches well in water (Perry, 2002). Cyst nematodes can be classified into three broad categories based on the hatching of J2 in water: (i) low-water hatch (*G. rostochiensis*, *G. pallida*, *H. cruciferae*, *H. carotae*, *H. goettingiana*, *H. humuli*); (ii) moderate-water hatch (*H. trifolii*, *H. galeopsidis*, *H. glycines*) and (iii) high-water hatch (*H. schachtii*, *H. avenae*). In general, dependence on root diffusates is

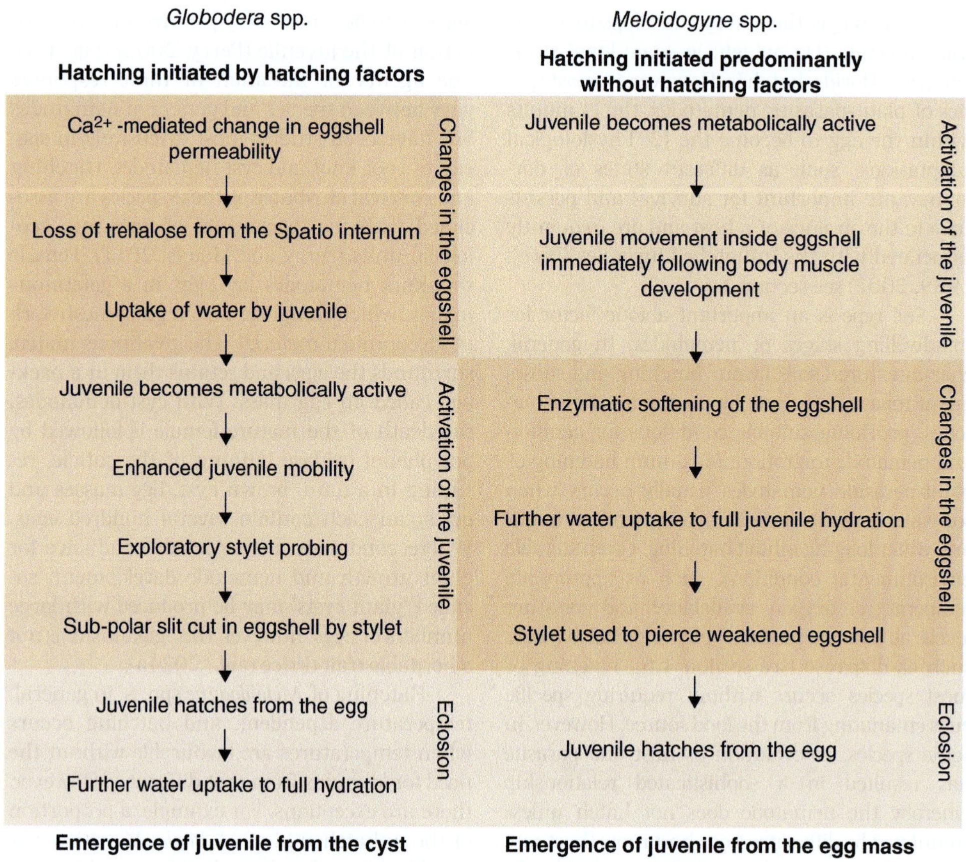

Fig. 1.3. Flow diagram displaying differences in chronological order of hatching phases for *Globodera* spp. and *Meloidogyne* spp. second-stage juveniles (Modified from Perry *et al.*, 2024.)

a survival mechanism, with high dependence related to a narrow host range and low dependence related to a broad host range. *Heterodera avenae* may appear to be an exception to this, as this species has a large hatch in water but a relatively narrow host range; however, the hosts are very common. There is evidence that the hatch response of *Globodera* spp. to host root diffusate varies with the potato cultivar used (Deliopoulous *et al.*, 2011). Although the variation could be attributed to differences in plant growth rather than differences in HF, Devine and Jones (2001) showed genetic segregation of HF in hybrid potato crosses.

The dependence of *G. rostochiensis* and *G. pallida* on a plant-derived hatching stimulus is an obvious control target, with the aim of inducing hatch in the absence of a host plant and thus causing the nematodes to die of starvation, the so-called 'suicide hatch'. Ngala *et al.* (2024) termed this strategy 'lure and starve' and examined *in situ* hatching of *G. pallida* and *H. carotae* following exogenous inundative applications of their respective host root diffusates in-field experiments and found hatching induction of up to 83% for *G. pallida* and 54% for *H. carotae*. Such a control approach depends, among other aspects, on production of sufficient root diffusate, timing of application and ensuring the diffusates penetrate the soil to the level where cysts accumulate.

Greater understanding of HF could allow a synthetic alternative to be developed. The first HF to be purified was the terpenoid, glycinoeclepin A, which induced hatching of *H. glycines*, followed by purification of glycinoeclepins B and C.

The potato steroid glycoalkaloids α-solanine and α-chaconine initiate the hatching of *G. rostochiensis* and *G. pallida*. Solanoeclepin A, a hatching stimulant for *Globodera* spp., has been successfully synthesized (Tanino *et al.*, 2011; Guerrieri *et al.*, 2021). However, its widespread use remains prohibitively expensive. The related triterpenoid HF solanoeclepin B is less active than solanoeclepin A but is metabolized by soil organisms to solanoeclepin A (Shimizu *et al.*, 2023). HF stimulate hatching of *H. glycines* and *G. rostochiensis* at concentrations as low as 10^{-10} g ml^{-1}. While total chemical synthesis for some HF has been achieved, there has been no successful control strategy using analogues or natural compounds of the HF to induce hatch in the field.

Host root diffusates induce a cascade of inter-related changes leading to eclosion, and the sequence of events has been discussed in detail by Jones *et al.* (1998), Perry (2002) and Perry *et al.* (2024). Unhatched J2 of *Globodera* and *Heterodera* spp. are surrounded by perivitelline fluid, which contains trehalose. Trehalose generates an osmotic pressure that reduces the water content of the J2 and inhibits movement because the turgor pressure is insufficient to antagonize the longitudinal musculature. For hatching to occur, the osmotic pressure needs to be removed. In *G. rostochiensis* and some other species, this is achieved by an HF-induced change in permeability of the inner lipoprotein membranes of the eggshell. In both *G. rostochiensis* and *G. pallida*, a 5 min exposure to host diffusate is sufficient to stimulate hatch (Perry and Beane, 1982), suggesting the involvement of a receptor-ligand interaction between the HF and the eggshell lipoprotein membrane. An annexin identified in eggshells of *G. rostochiensis* has been shown to respond directly to HF, altering lipid binding properties in response to host exposure (Price *et al.*, 2023). The change in eggshell permeability enables trehalose to leave the egg, allowing an influx of water and subsequent rehydration of the J2 to a water content commensurate with movement. The presence of trehalose in eggs containing unhatched, viable J2 has been used as the basis for a viability assay for *G. rostochiensis* and *G. pallida* (van den Elsen *et al.*, 2012; Ebrahimi *et al.*, 2015).

The eggshell of *G. rostochiensis* remains rigid during the hatching process, and there is no evidence of enzyme involvement. However, enzymes have been implicated in softening of the eggshell prior to eclosion in other species, including *Xiphinema diversicaudatum*, *Aphelenchus avenae* and *M. incognita*. Rehydration of the J2 of *G. rostochiensis* is accompanied by increased metabolic activity due in part to removal of osmotic pressure and hydration and in part to direct stimulation of the J2 by root diffusate. Changes in gene expression of *G. rostochiensis* J2 appear to occur during or immediately after the hatching process (Jones *et al.*, 1997), but more work is needed on the molecular aspects of the hatching response.

The J2 of *Globodera* spp. uses its stylet to cut a regular series of perforations in the subpolar region of the eggshell, and the J2 hatches through the resulting slit. Once hatched, the J2 then leaves the cyst via either of the natural openings of the cyst, of the fenestral region or of the neck where the female's head has broken away. The second-stage juvenile of *D. dipsaci* also cuts a slit in the eggshell, but the stylet thrusts are more random, and the J2 uses its head to force open the slit. For *Pratylenchus penetrans* and *H. avenae*, a single stylet thrust penetrates the eggshell, and the head extends this into a tear. Once the J2 are released into the soil, they begin to search for a suitable host. During this soil phase, they do not feed and depend on stored energy reserves; thus, they need to locate and commence feeding without delay. Under optimal conditions for movement, J2 of *G. rostochiensis* must locate a host root and set up a feeding site within 6-11 days of hatching, otherwise it will exhaust its energy reserves and die.

1.3. Attraction to Plants

Chemicals that cause interactions between organisms are called **semiochemicals**; these include **allelochemicals**, which mediate interspecific responses, such as nematodes responding to diffusate from host roots, and **pheromones**, which mediate intraspecific responses, such as male nematodes responding to sex pheromones from female nematodes. Around actively growing roots, there exist several gradients of volatile and non-volatile compounds, including amino acids, ions, pH, temperature and CO_2. It is evident that nematodes use their chemosensory sensilla, primarily the amphids, to orientate towards the roots using at least some of these gradients (Curtis *et al.*, 2009; Perry and Holden-Dye, 2024). The ability to orientate towards stimuli from plant roots enhances the chances of host location and reduces the time without food (Perry, 1997). Evaluating the reality of the attractiveness or otherwise of an individual compound is difficult. Information is usually based on *in vitro* behavioural studies, often using agar plate bioassays to evaluate horizontal migration, which bear little if any resemblance to the situation in the soil. The use of pluronic gel, micro-moulded substrates, 'pipette bulb' assay and olfactometers provides more informative analyses of nematode movement and attraction in three-dimensional environments (Perry, 2024). Although care must always be exercised in extrapolating from *in vitro* assays to the field situation, some generalizations can be made, and certain compounds are strongly implicated in orientating nematodes to the roots.

Gradients can be separated into three types (Perry, 2005): **long-distance attractants**, which enable nematodes to move to the root area; **short-distance attractants**, which enable the nematode to orientate to individual roots, and **local attractants**, which are used by endoparasitic nematodes to locate the preferred invasion site. There is clear experimental evidence that CO_2 is a long-distance attractant, although *M. hapla*, *M. incognita* and *M. javanica* mixed with pluronic gel did not show any sign of attraction when CO_2 was delivered to a concentric circle on the gel surface (Wang *et al.*, 2009). With cyst nematodes, such as *Globodera* spp., it is apparent that the J2 responds to host root diffusate, and the evidence is persuasive that diffusate contains chemicals that constitute short-distance attractants (Perry, 1997; Rolfe *et al.*, 2000). Diffusates from the roots of the host plant, potato, increased the activity of the infective J2 of *G. rostochiensis* and also attracted them to the roots. Devine and Jones (2002) showed that the chemicals in potato root diffusate responsible for stimulating hatch differ from those responsible for attracting the J2 to the root. Electrophysiological analysis of sensory responses (Perry, 2001) demonstrated that J2 of *G. rostochiensis* responded to potato root diffusate but not to root diffusate from the non-host sugar beet, thus indicating that response to diffusates may be host specific. Pudasaini *et al.* (2006) found that the migration of *P. penetrans* towards a host depends on both the initial distance between the nematode and the host and the nature of the host. These authors considered that the attractiveness of the host to *P. penetrans* seems to be correlated with its efficiency as a host; the attractiveness of hosts also declines with age.

The orientation of J2 of cyst and root-knot nematodes to the preferred invasion site, the root tip, is well established but the active factors that constitute the 'local attractants' are unknown. The nematodes may orientate to an electrical potential gradient at the elongation zone of the root tip, but the relative importance of electrical and chemical attractants for root tip location has not been evaluated; in addition, the elevated temperature at the zone of root elongation may influence nematode perception.

Blocking sensory perception so that the nematodes are unable to orientate to roots and thus exhaust their food reserves and die is an attractive control option but may be difficult to achieve. Interfering with chemoreception by blocking specific nematode amphidial secretions using monoclonal and polyclonal antibodies retarded the movement of J2 of *M. javanica* (Stewart *et al.*, 1993) and *G. pallida* (Fioretti *et al.*, 2002) and significantly reduced infection of plants. However, responses were not permanently blocked as, after a period of between 0.5 and 1.5 h, turnover of sensilla secretions presumably 'unblocked' the amphids. A similar effect occurred with the bionematicide, DiTera®, where disruption of sensory perception was reversible (Twomey *et al.*, 2002). Lectins bind to amphids of *Meloidogyne* spp., and treatment of infected soil with lectins significantly reduced root galling of *M. incognita* in tomato, probably by blocking the nematode chemoreceptors and interfering with host finding (Marbán-Mendoza *et al.*, 1987).

1.4. Penetration and Feeding

The preferred invasion site for endoparasitic root-knot and cyst nematodes is behind the root tip in the zone of elongation. When a root is encountered, its surface is explored for a suitable penetration site, and at this stage, plant chemicals in the rhizosphere can influence nematode behaviour (Curtis *et al.*, 2009). Second-stage juveniles of cyst nematodes, with their more robust stylets, cut cortical cell walls and migrate through cells until they reach the differentiating vascular cylinder. This intracellular migration causes considerable damage to host tissue, resulting in necrosis from the invasion point to the feeding site. By contrast, root-knot nematodes migrate intercellularly. After invasion, J2 of root-knot nematodes move towards the root tip until they reach the root apex, where they turn around, avoiding the barrier of the endodermis, and migrate back up the root until they reach a site near the vascular cylinder. Migration of both cyst and root-knot nematodes through root tissue is facilitated by cell wall-degrading enzymes from the subventral glands, which are secreted through the stylet. These enzymes weaken or break down cell walls and include cellulases, pectate lyases, and, in root-knot nematodes, xylanase and polygalacturonase (Gheysen and Jones, 2024). These enzymes are not present in non-parasitic nematodes or other invertebrates, and it is likely that they were acquired by horizontal gene transfer from bacteria to plant-parasitic nematodes.

For all species of plant-parasitic nematodes, feeding depends on the use of the hollow stylet, which is inserted into a cell in order to extract its contents. The convergent evolution of cyst nematodes, root-knot nematodes, and *Nacobbus* has resulted in the same outcome: feeding on nematode-induced nutrient sinks. However, the method of achieving the end result and the feeding sites themselves differ. A cyst nematode selects a cell, becomes sessile, and a multinucleate feeding site, termed a syncytium, is formed by gradual incorporation of hundreds of adjacent cells as the intervening cell walls disintegrate. The mature sedentary female of *N. aberrans* also sets up a syncytium. Root-knot nematodes become sessile and induce the formation of several binucleate cells followed by mitosis that does not cause cell division. These nuclei further divide, several large multinucleate giant cells result, and cells surrounding them also enlarge to form a gall or root-knot.

In both cyst and root-knot nematodes, secretions from the dorsal and ventral pharyngeal glands (Fig. 1.4) play a central role in the induction and maintenance of the feeding sites. The secretions are crucial for facilitating movement to the feeding site and modulating the host defence response (Hassan *et al.*, 2010; see Chapter 18, this volume). The proteins and other molecules involved are collectively termed nematode 'effectors'.

1.5. Genomics

With the progression of sequencing technology, the methodologies for DNA preparation have also evolved. What was once accomplished through single-pass sequencing of cloned cDNA from expressed sequencing tags (EST) has been replaced with more routine, and accessible, long-read, whole-genome sequencing. Rapid advances in the availability of genomic and transcriptomic data have consequently helped develop various aspects of research on practical plant nematology.

Lengthy molecular approaches can now be initially replaced with widely available bioinformatic pipelines for quick gene discovery. This has helped to further our understanding of host-nematode interactions. Direct analysis of transcriptomic sequences assists the identification of nematode effector proteins, expression of which can be localized using *in situ* hybridization in the laboratory. Effector gene function can consequently be characterized through increasingly accessible RNA interference (RNAi) techniques (Gheysen and Jones, 2024). Characterization of the changes nematodes induces in their host and how effector proteins provoke these modifications offers routes for novel management strategies. Similarly, advances in genomic data availability improve understanding of plant-parasitic nematode evolution. Inferring taxonomy and phylogeny from genomic data has now become routine as nematology moves away from simpler gene-based phylogenetics to genome-based phylogenomics (Subbotin and Moens, 2024). Recognizing the evolutionary relationships between pathogens, how parasitism evolved, and the host-parasite relationship could be used to infer how plant-parasitic nematodes will continue to evolve and how control strategies must adapt accordingly.

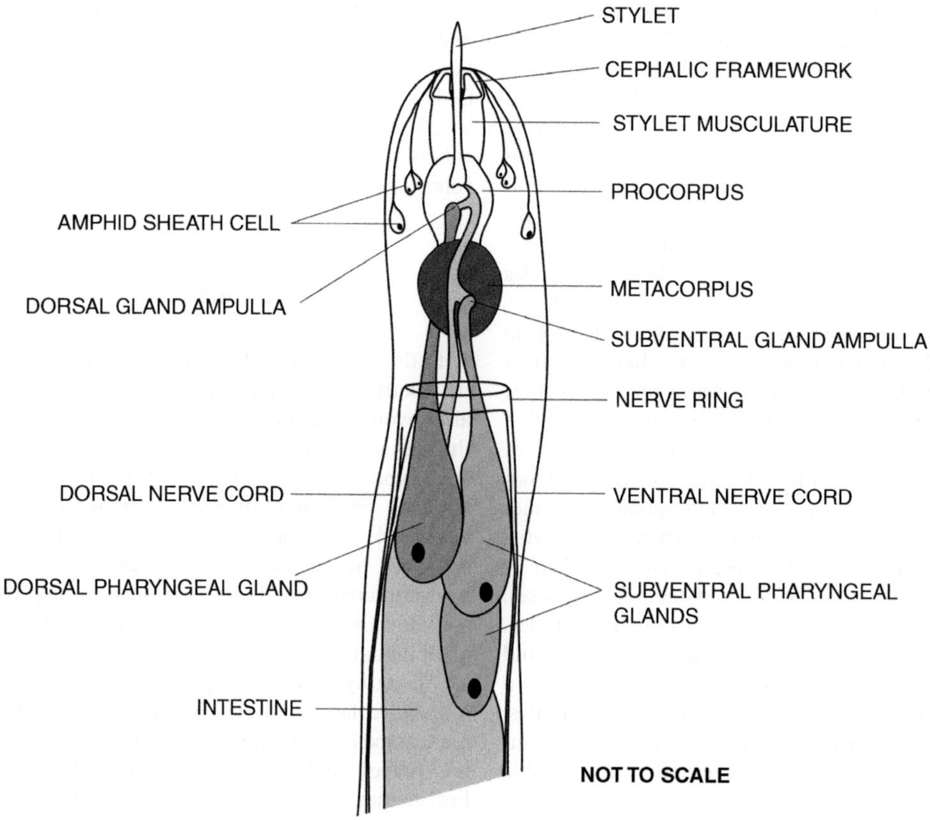

Fig. 1.4. Digital drawing of generalized plant-parasitic nematode showing relative positions of structures. (Courtesy James Price.)

Advances in genomic data availability can also be applied to diagnostics, localizing where control strategies need to be applied and reducing pathogen dispersal through transport of contaminated plant or soil material. Diagnostic regions of DNA are currently being exploited for the identification of potato cyst nematode (PCN) in Scottish seed potato-growing land. Use of species-specific probes in TaqMan qPCR assays allows quantification and speciation of approximately 17,000 samples annually (Reid *et al.*, 2015). By only permitting seed to be produced on PCN-free land, potatoes are eligible for export in agreement with European Union (EU) directives. Diagnostic markers have similarly been used with rapid loop-mediated isothermal amplification (LAMP) techniques, which, when combined with lateral flow devices, allow production of in-field detection for several root-knot and cyst nematode species (Niu *et al.*, 2011; Peng *et al.*, 2017; Yao *et al.*, 2021).

1.6. Development and Multiplication

Once feeding commences, the nematode can continue development. The majority of nematodes have four juvenile stages before development to the adult; thus, there are four moults where the old cuticle is replaced by a new cuticle. Exceptions include certain species of *Xiphinema*, which have only three moults and three juvenile stages. During the moulting process, the cuticle is either shed completely or, as in *Meloidogyne*, partially absorbed. Nematodes increase in size between moults, not during the moulting process. The moult from fourth-stage juvenile (J4) to adult of the fungal feeding nematode, *Aphelenchoides hamatus*, took 12-13 h to complete. Initial changes included the loss of knobs and shaft of the stylet as the head retracted away from the old conus. The nematode's volume and water content decreased, enabling it to retract within and away from the old cuticle. During the early period, the nematodes were quiescent; later on, they began to retract and bend their heads followed by movement of the entire body. The adult then expanded rapidly to burst the old cuticle. Shrinkage and subsequent expansion of the nematodes is thought to be due to water loss controlled by the nematodes, followed by rapid water re-entry (Wright and Perry, 1991).

After the final moult to adult, reproduction occurs. Nematodes exhibit a variety of sexual and asexual reproductive methods, and most information is available on *Meloidogyne* (reviewed by Chitwood and Perry, 2009). Sexual reproduction (**amphimixis**) can occur in species where there are two separate sexes, where the haploid (n) male (spermatocytes) and female (oocytes) gametes fuse to form the zygote and restore the diploid ($2n$) complement of chromosomes. In *Anguina tritici* and *Ditylenchus dipsaci*, for example, female nematodes are the homogametic sex (genetically XX) and males are heterogametic (sometimes XO, usually XY), and the sex ratio is determined genetically. In genera such as *Globodera*, *Heterodera* and *Meloidogyne*, sex chromosomes are absent, and the sex ratio may be environmentally influenced. In many genera, including *Heterodera*, *Pratylenchus* and *Radopholus*, the female can be fertilized by several males, thus enhancing the genetic diversity of the offspring.

In species of nematodes that reproduce asexually, males are absent or occur only rarely. There are two main types of parthenogenesis, meiotic parthenogenesis (**automixis**) and mitotic parthenogenesis (**apomixis**). In meiotic parthenogenesis, there is a first meiotic division in the oocytes, although there are variations between species. This meiotic division allows some genetic re-organization, even though the diploid chromosome number is restored by self-fertilization. *Meloidogyne hapla* race A exhibits facultative meiotic parthenogenesis, oogenesis and spermatogenesis proceeding as in amphimictic species to yield one haploid nucleus and two polar bodies per oocyte. Although parthenogenesis predominates, amphimixis can occur if sperm are present. However, if the sperm are not present, the egg pronucleus recombines with the second (haploid) polar body to restore the diploid state. In monosexual populations of *Aphelenchus avenae*, females produce only female progeny, and reproduction is by obligate meiotic parthenogenesis. Meiosis produces only one polar body, with the egg nucleus having the $2n$ chromosome number that then develops into the zygote.

Mitotic parthenogenesis, which occurs in several species of *Meloidogyne*, is the most common method of asexual reproduction and is always obligate; the only division is mitosis, and the oocytes retain the diploid chromosome number. This would seem to prevent any genetic reorganization, except for that resulting from mutations. Frequently, mitotic parthenogenesis is associated with polyploidy, which may increase the likelihood of mutation. Several of the most widespread and economically important species of *Meloidogyne* are obligate mitotic parthenogens. Populations of the same *Meloidogyne* species may differ in mode of reproduction; for example, 29 of 32 populations of *M. hapla* reproduced by facultative meiotic parthenogenesis, the others by mitotic parthenogenesis (Triantaphyllou, 1966).

In hermaphrodites, both egg and sperm are produced in the same individual. Usually, the sperm is produced first and is stored in the spermatheca, then the gonad produces oocytes, which are fertilized by the sperm until the sperm supply is exhausted. Hermaphroditism is a common method of reproduction among free-living nematodes but is relatively rare in plant-parasitic nematodes, being found in some members of the Criconematoidea and in *Radopholus similis* and species of *Paratrichodorus*.

1.7. Temperature Requirements and Survival

Most species of nematodes are adapted to specific environmental temperature ranges, and the distribution of species is related to temperature preferences (Trudgill *et al.*, 2005). Within the preferred temperature ranges, the rate of development of a given species slows down as the environmental temperature decreases, and the low temperature at which development ceases is called the **base temperature**. As temperature increases within the range, the rate of development increases up to a **temperature optimum**, which is ideal for development. Above the temperature optimum, the rate of development decreases and eventually ceases at the **temperature maximum**.

The worldwide distribution of *Meloidogyne* provides a good illustration of variation within a genus of preferred temperature ranges in relation to development. *Meloidogyne chitwoodi*, for example, exhibits a base temperature of 4°C, compared with ca 8, 10 and 14°C for *M. hapla*, *M. incognita* and *M. javanica*, respectively. In relation to temperature requirements, root-knot nematodes can be divided into two distinct groups of species, **thermophils** and **cryophils**, which can be separated by their ability to survive lipid-phase transitions that occur at 10°C. *Meloidogyne chitwoodi*, *M. hapla* and probably *M. naasi* are cryophils and able to survive soil temperatures below 10°C; *M. arenaria*, *M. javanica* and *M. exigua* are thermophils and have limited or no survival at temperatures below 10°C.

For obligate parasitic species, there are situations where persistence of a population requires survival of the infective stages outside the host (Perry and Moens, 2011). This may occur when environmental conditions, such as temperature extremes and dehydration, exist that are not commensurate with life cycle progression or when the host is not available, requiring the nematode to survive in the absence of food. Nematodes are poikilothermic organisms and lack the ability to modify their micro-environment if it becomes unfavourable or move away over long distances. To overcome adverse environmental conditions during growth and development, a resting stage (or a resting period) has evolved in the life cycle of many nematodes. Plant-parasitic nematodes can survive unfavourable conditions by entering dormant states, which include **quiescence**, **diapause** and **dauer** formation.

Quiescent states are induced by unfavourable environmental conditions and are readily reversible when favourable conditions return; facultative quiescence is a response to sudden environmental changes, and obligate quiescence is life stage-specific. Either form may be induced by desiccation (anhydrobiosis), high salt concentration (osmobiosis), lack of oxygen (anoxybiosis), low temperature (cryobiosis) and high temperature (thermobiosis). Desiccation survival has attracted the most interest, partly because the ability to survive dryness is essential for the distribution of infective stages via wind and dry plant material, such as seeds, and partly because of the spectacular abilities of some nematodes to survive desiccation for periods considerably in excess of the normal life cycle. Some of the most astonishing survival and dispersal attributes are found in species of nematodes that live within aerial parts of plants, especially stems, leaves and seeds, the best documented of which are the abilities of J2 of *Anguina tritici* and J4 of *Ditylenchus dipsaci* (Fig. 1.5) to survive in a dormant, desiccated state for many years (Moens and Perry, 2009). For example, *D. dipsaci* J4 have been recorded as surviving for more than 20 years in the dry state, yet the whole life cycle of the nematode in favourable environmental conditions takes only about 4 weeks from egg to egg-laying adult. Nematode anhydrobiotes can be grouped into those that depend on environmental factors to control water loss and those, such as *D. dipsaci* J4, that have intrinsic abilities to control water loss; Perry and Moens (2011) termed these two groups **external dehydration strategists** and **innate dehydration strategists**, respectively. Both groups require controlled drying in order to survive: the first group to prolong the time to lethal low water content and the second group to enable biochemical changes to take place to facilitate long-term survival.

Diapause in cyst and root-knot nematodes is essentially a strategy to overcome cyclic long-term conditions, such as seasonal conditions and/or the absence of the host, that are not conducive to hatch and infection. Diapause is categorized in two distinct types. In facultative diapause, exogenous environmental factors are presumed to induce this form of dormancy, which is terminated either by endogenous factors, whose rate of operation is dependent on prevailing physical environmental conditions, or by the presence of the host. Obligate diapause is initiated presumably by endogenous factors (which may or may not be modulated by the host) and is terminated by predetermined periods of specific environmental conditions (Womersley *et al.*, 1998; Nyczepir and Thomas, 2009).

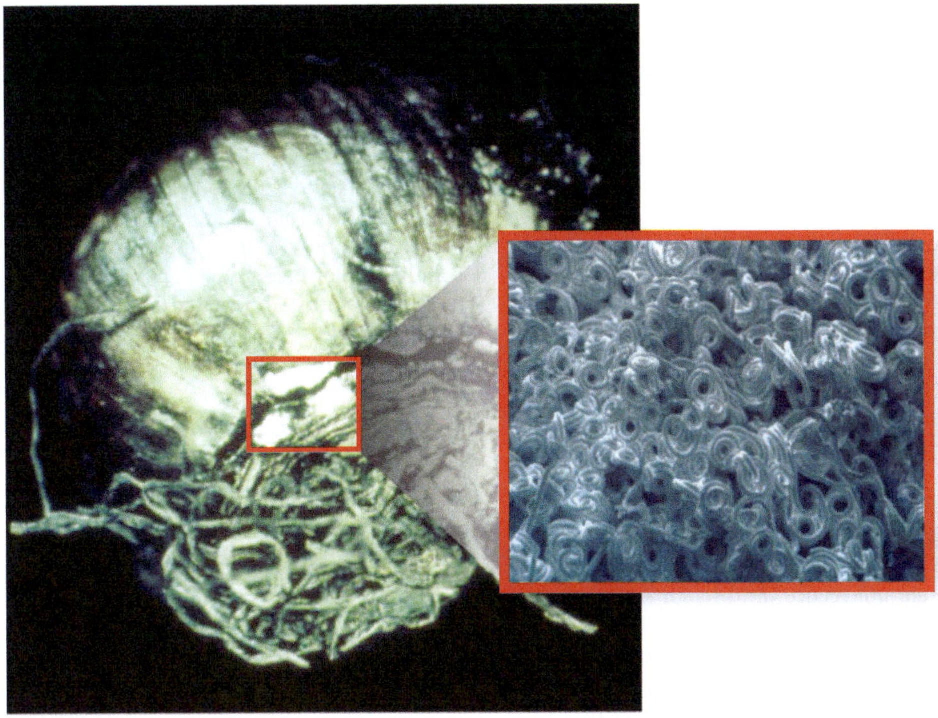

Fig. 1.5. Narcissus bulb with accumulation of desiccated *Ditylenchus dipsaci* fourth-stage juveniles (J4) in a mass termed eelworm wool, and an inset showing a transmission electron microscope image of the eelworm wool with coiled, clumped, desiccated J4. (From Moens and Perry, 2009; both images courtesy of Roland N. Perry.)

The duration of obligate diapause is affected by the photoperiod experienced by the infected plant, with unhatched J2 from plants grown under continuous light showing no obligate diapause.

The percentage of unhatched J2 that enter diapause varies from less than 10% for the predominantly tropical *M. arenaria* to 94% for *M. naasi*, which has a temperate distribution. Obligate diapause is a necessary part of the life cycle of *M. naasi* and involves the effects of temperature. *Meloidogyne naasi* only produces one generation of eggs by the end of the host plant growing season, all of which are protected by a gelatinous matrix until the following spring and must be subjected to a distinct three-phase temperature regime if hatch is to be successful and synchrony with the growing season of the hosts is to be maintained. Embryonic development (during late summer-autumn) occurs at 20°C, chilling (winter) at 5–10°C, and hatch (spring) at around 20°C. Some J2 require a second winter period before hatching is successful, indicating different temporal levels of diapause within the egg population, which are initiated by endogenous factors long before the onset of winter temperatures. Temperature is not the only factor influencing the induction and duration of diapause. In *M. incognita*, there is evidence that the number of eggs entering diapause is influenced by the age of the female. This may indicate age-induced physiological stress as the controlling mechanism, although the effect of changes in host physiology on the female is likely to be an important component (Womersley *et al.*, 1998). Diapause habits can be changed in favourable environments. In Kenya, persistently warmer soils, more consistent photoperiods and the constant availability of a host mean that *G. rostochiensis* can complete up to three life cycles in less than a year compared to one per year across Europe (Mwangi *et al.*, 2021).

Some species of cyst nematodes that do not enter diapause are ready to hatch soon after egg formation and completion of embryogenesis, but a specific chemical stimulus from the host root is required before eclosion can take place (see section 1.2). In other cyst nematode species, a proportion of unhatched J2 enter diapause after embryogenesis and do not hatch, even though environmental conditions, including the presence of host root diffusate, are favourable. In these instances, a time period is necessary before diapause ends, and in temperate regions this is often associated with prevention of hatch until the winter season has been completed. It is a method of ensuring that unhatched J2 overwinter and do not hatch if an atypical period of favourable conditions occurs during the winter period. In species dependent on host root diffusates to stimulate hatch, diapause can be terminated and immediately be followed by a period of obligate quiescence until hatch stimulation by diffusate occurs (Perry, 1999). Molecular studies have shown differences in gene expression changes between quiescence and diapause in unhatched J2 of *G. rostochiensis* after hydration or exposure to root diffusate (Palomares-Rius *et al.*, 2016).

The dauer is a morphologically, behaviourally and physiologically adapted alternative juvenile stage for surviving unfavourable conditions. Dauer has been defined by Vlaar *et al.* (2021) as a state in which the juvenile is non-feeding, essentially non-ageing, and has a conserved dafachronic acid biosynthetic pathway that produces (a chemical analogue of) dafachronic acid. Under adverse conditions, *D. dipsaci* development stops at the J4 stage and, compared with J4 in a population feeding and developing under ideal conditions, the survival form of J4 has more lipid reserves and shows a propensity to aggregate (Perry and Moens, 2011), properties that reflect the dauer state. However, evidence for dauer pathways in plant-parasitic nematodes has come mainly from bioinformatic studies and comparison with the extensive work on *C. elegans* (Vlaar *et al.*, 2021). Kikuchi *et al.* (2007) identified nine homologues for *daf* (dauer formation) *C. elegans* genes. *Meloidogyne hapla* has 14 orthologues of *C. elegans daf* genes (Abad *et al.*, 2008; Abad and Opperman, 2009) but not the *daf-28* orthologue, which plays a key role in the signal transduction pathway. Zhang *et al.* (2022) found that the signal transduction pathway accompanying the moult from *B. xylophilus* dauers was quite different to that of *C. elegans*. It appears that basic development mechanisms are conserved but signalling is not. As yet, there is insufficient information to be able to link individual *daf* genes to specific survival traits (Elling *et al.*, 2007; Vlaar *et al.*, 2021) but with the increased availability of genomic information, the occurrence of dauers in plant-parasitic nematodes is likely to be substantiated.

1.8. Control and Management

The terms **control** and **management** have different meanings in relation to reducing nematode populations below economically damaging density levels. Control refers to the use of an individual practice, whereas management refers to a combination of control practices, called integrated pest management (IPM) (see Chapters 19 and 20, this volume). As pointed out by Nyczepir and Thomas (2009), the choice and combination of control strategies in IPM systems depend on the value and characteristics of the crop(s), the various options that can be employed, and information available on relevant biotic and abiotic components. Gaps in knowledge and uncertainties inherent in IPM systems mean that growers are more likely to use a control practice rather than management strategies to suppress plant-parasitic nematodes.

The objective of nematode control and management is to prevent significant loss of yield and quality in vulnerable crop plants and, in the longer term, to keep nematode populations below economically damaging levels. The control approaches may be classified broadly as: physical, chemical, biological, cultural and resistance.

1.8.1. Physical Control

Nematodes in soil may be killed by heating the soil water to a lethal temperature. This may be done for small areas by electrically heating the soil or by passing steam through it, although both approaches consume a significant amount of energy. An alternative, and more frequently used, procedure is to cover the soil with one or two layers of polyethylene film and allow the sun to raise the soil temperature. This is known as soil solarization (Gaur and Perry, 1991), and it is effective in hot climates to treat the topsoil but is usually ineffective at depths greater than approximately 20 cm. Infested plant tissue (roots, seeds, bulbs, corms, tubers and plantlets) may be freed of nematodes by dipping in water at a carefully chosen and controlled temperature, long enough to kill viable nematodes. There is often a risk of plant damage from hot water treatment, especially of non-dormant tissues, and for these, dipping in cold solutions or suspensions of nematicides may be preferable. Some nematodes may be controlled by flooding the land for several weeks or months, but other species seem to be unaffected by this practice (Whitehead, 1998). The advantage of controlling nematodes by physical methods is the absence of the residues that are frequently associated with chemical control methods.

1.8.2. Chemical Control

Most plant-parasitic nematodes spend a part of their life cycle in the soil, and to obtain effective control, the soil has to be disinfested. Chemicals that kill nematodes are termed **nematicides**. Compounds that provide sub-lethal dosages that disrupt nematode behaviour, are termed **nematistats**. However, the term nematicide will be used as the umbrella term in this section.

Chemical control was the mainstay for reducing populations of plant-parasitic nematode (PPN) in most moderate-to-high-value crops in intensive production systems. However, environmental and human health concerns have steadily reduced the availability of such control options, starting with the suspension by the US Environmental Protection Agency of the use of 1,2-dibromo-3-chloropropane (DBCP) as a soil fumigant for all crops. Numerous other nematicides have suffered a similar fate or have been voluntarily withdrawn; for example, methyl bromide has been withdrawn from use in the USA and Western Europe due to its role in ozone depletion, and within the EU, 1,3-dichloropropene (1,3-D) has been omitted from the list of authorized nematicides.

The second half of the 20th century saw the development of the organophosphates such as fenamiphos, ethoprophos and fosthiazate, together with the carbamate compounds carbofuran, aldicarb and oxamyl. Although still in use in many countries around the world, their relatively high acute mammalian toxicity means that their use is limited; across the EU in 2023, only fosthiazate was still registered. Recent new fumigant nematicides include fluensulfone, fluoroalkenyl, fluopyram, fluzaindolizine and cyclobutrifluram (Back *et al.*, 2024). Where the application of nematistats is timed correctly, juveniles can be halted in the soil. This reduction of the initial population infecting and damaging the crop allows yields to be sustained under heavy infections, mimicking a natural host tolerance to nematode infections (see section 1.8.5).

Nematicides are an important component of nematode control and management programmes, including IPM; the focus of chemical control is now on the development of new nematicides with improved efficacy, and much more favourable mammalian and ecotoxicity profiles (Desaeger *et al.*, 2020; Sikora *et al.*, 2021; see Chapter 20, this volume). The use of chemicals becomes economic when other methods are too costly or difficult to apply, or when a method such as rotation is inadequate because it does not suppress the nematode populations sufficiently to enable the grower to plant the host more frequently in the rotation. Chemicals are also very useful in high-value horticultural crops and may be essential for producing nematode-free produce in export crops.

1.8.3. Biological Control

Many natural enemies – such as virus, rickettsiae, fungi, predatory nematodes, springtails and mites – attack plant-parasitic nematodes in soil and reduce their populations (Viaene *et al.*, 2024). It is important to determine the nature and extent of such constraints on nematode multiplication to establish whether these enemies can be exploited to reduce damage and increase crop yields. Some general principles for biological control are evident, and there are well-documented cases where effective biological control has been established, principally in cryptic habitats (see Chapter 21, this volume). However, it is increasingly evident that biological control strategies alone are unlikely to be effective at the field scale; at best, their use will probably be as part of an IPM system and, even then, will depend on effective mass production methods, which, for most potential control organisms, remain to be developed. The most promising agents for the control of plant-parasitic nematodes are bacteria and fungi, and many studies have been conducted with the fungi *Paecilomyces lilacinus* and *Pochonia chlamydosporia* and bacteria belonging to *Pasteuria* spp.

Paecilomyces lilacinus has been used to control *G. pallida*, *N. aberrans* and some *Meloidogyne* spp. (Hallmann *et al.*, 2009). *Pochonia chlamydosporia* is a facultative parasite distributed in soils that suppresses populations of plant-parasitic nematodes and has shown potential as a parasite of eggs of sedentary endoparasitic nematodes. Studies using native strains in cryptic habitats (e.g., glasshouses, small plots or containers) have provided some promising results (Manzanilla-López and Lopez-Llorca, 2017; Manzanilla-López *et al.*, 2009), but effective control on a larger scale has not been demonstrated.

P. penetrans is the most widespread species of *Pasteuria* and mainly parasitizes *Meloidogyne* spp.; *P. thornei* and *P. nishizawae* parasitize *Pratylenchus* spp., and *Heterodera* and *Globodera* spp., respectively (Davies, 2009). The bacterial dome-shaped spores attach to the cuticle of the nematode, and when the nematode enters the root, the spores germinate and the germ tubes penetrate the cuticle, enter the body of the nematode, and eventually produce sporangia and endospores. The bacteria destroy the reproductive capacity of the nematode. When the plant and nematodes decompose, millions of spores are released into the soil, and these are resistant to environmental fluctuations. Using *Pasteuria* spp. as a control option has two main disadvantages that have hampered development: the host specificity of strains of *Pasteuria* species, and the obligate parasitism and hence difficulty for mass production. Pasteuria Bioscience Inc. (USA) (now Syngenta) achieved *in vitro* culture of *Pasteuria* spp., and products have been used successfully in niche markets against *Belonolaimus longicaudatus* on turf (Econem®) and against *H. glycines* (ClarivaTM pn). Such usage paves the way for field-scale applications, but continuing research about the specificity of the populations towards different nematode species is needed before the potential of *Pasteuria* spp. as a field control option can be properly evaluated.

1.8.4. Cultural Control

Cultural control is a term that includes measures that limit population increase of plant-parasitic nematodes. The measures are primarily: (i) crop rotation, (ii) trap crops, (iii) incorporation of organic amendments, (iv) crop phenology and (v) sanitation. Most of these measures are more applicable to annual rather than perennial crops (Nyczepir and Thomas, 2009).

Crop rotation is one of the most widely used practices in developed countries. This practice was carried out by farmers before they recognized its importance as a measure to control populations of plant-parasitic nematodes. To be effective, it should include plants that are unfavourable as hosts for nematodes. The level of control is based on the degree of resistance of plants used in the rotation and the time before susceptible crops are planted again. The disadvantages of crop rotation are that the non-host or resistant plants may be of limited or no commercial value, and some nematodes, such as some species of *Meloidogyne*, have a wide host range so selection of appropriate crops for control is difficult. Thus, in addition to knowledge of the biology and host range of the nematode(s) to be controlled, correct identification of species and, in some cases, even races/pathotypes is essential.

The use of trap crops involves the planting of fast-growing plants that are highly susceptible to the species of nematode to be controlled. Once the plants have achieved some growth and the nematodes have invaded, the crop is removed, thus destroying nematodes that have penetrated the plant tissues. The disadvantage of trap cropping is that timing of the harvest is essential to ensure that nematode invasion but not reproduction has occurred. Also, the crop may have limited commercial value, which means that the field is essentially out of commercial production for a period; this economic loss has to be weighed against the economic benefits of controlling nematode damage on subsequent crops.

Incorporation of organic matter into soil is a long-standing practice used by farmers to control plant-parasitic nematodes. A variety of organic matter has been used, including animal waste, slurries, compost and crop residues, all of which have been shown to reduce populations of plant-parasitic nematodes, such as *M. incognita, M. javanica, G. rostochiensis, Tylenchulus semipenetrans* and *N. aberrans*. These amendments help control parasitic nematode populations through various mechanisms, which include the stimulation and increase of nematode antagonists, the release of compounds with nematicidal properties, interference with the physiology of nematodes, and induction of physical and chemical properties in the soil; incorporation of nutrient-rich organic matter also stimulates microorganisms, resulting in increased ammonia production, which is nematicidal. The use of amendments containing chitin stimulates chitin-degrading microorganisms, which destroy the eggshells of nematodes. Thoden *et al.* (2011) consider that a significant part of the positive effects of organic amendments on crop yield is due to the proliferation of non-pathogenic, free-living nematodes and their overall positive effects on soil microbial populations, organic matter decomposition, nutrient availability, plant morphology and ecosystem stability. The drawback of the addition of organic matter is that although plant growth may be enhanced, thus enabling plants to withstand nematode invasion better, the nematode population may actually increase. The term biofumigation has been used in relation to organic amendments that release volatile substances and has especially been the term of choice for the incorporation of cruciferous plants and others that release the nematicidal compound isothiocyanate (Thoden *et al.*, 2011).

Allelopathy is the ability of an organism to produce chemicals that are toxic to other organisms, and some cover crops release substances that are lethal to nematodes. Marigold, *Tagetes* spp., can be used as such a cover crop (Krueger *et al.*, 2010) because it produces a toxic substance called alpha-terthienyl (Gommers and Bakker, 1988). Marigold can suppress 14 genera of plant-parasitic nematodes but is most effective against *Pratylenchus* spp. and *Meloidogyne* spp. African marigolds, *Tagetes erecta* and French marigolds, *T. patula*, are the most commonly used species, but *T. patula* appears to be most effective against the widest range of nematodes (Krueger *et al.*, 2010). However, there are drawbacks to using *Tagetes* for nematode management. Planting large areas with marigold could be an expensive treatment that may not be cost-effective. Also, alpha-terthienyl is only released by active, living marigold roots, because it is inactivated on exposure to near-UV light; thus, there is no benefit in treating a field with marigold extracts of homogenized plant parts (Ploeg, 2000).

Knowledge of crop phenology can prevent the establishment of a particular species of plant-parasitic nematode or avoid the formation of a new generation by modifying harvest times. An example of this technique is the early harvest of potatoes in Ayrshire, UK, before temperatures are favourable for hatch (Hominick, 1979) or early harvest in South America in order to prevent invasion of a second generation of nematodes (Yépez, 1972).

Sanitation is an important component of nematode management, comprising a set of practices aimed at eliminating or reducing the amount of inoculum present in a plant-growing area, as well as preventing the spread of nematodes to healthy plants. Sanitation includes the removal and destruction of infected plant material, inspection and use of certified plant material and cleaning of farm equipment.

1.8.5. Resistant Plants

The use of resistant cultivars provides the ideal method for suppressing growth and reproduction of plant-parasitic nematodes and has the advantages of reducing or eliminating periods of rotation, being environmentally benign, and not requiring special application techniques. There are numerous examples of effective resistance to plant-parasitic nematodes (Starr and Mercer, 2009; Claudius-Cole, 2024), a classic one being the use of the *Mi* gene that confers resistance to three species of *Meloidogyne* (*M. arenaria, M. incognita* and *M. javanica*), which has been introgressed into several tomato varieties (Williamson and Roberts, 2009). However, use of host plant resistance for control of *Meloidogyne* is limited because only a small number of plant species are resistant to this nematode, and there are many crops for which appropriate resistance loci have not been identified; thus, bioengineering is important to design effective and durable forms of crops resistant to *Meloidogyne* (Atkinson *et al.*, 2009).

Resistance and **tolerance** are distinct traits. Endoparasitic species can still invade and damage the roots of resistant varieties and, under high initial population pressure, resistant varieties can exhibit the same symptoms as non-resistant ones, including reduced yield. Tolerant varieties can withstand root damage caused by nematode invasion, continuing regular water and nutrient uptake and maintaining good yields despite infection. However, without resistance, tolerant varieties will enable nematode populations to multiply.

Deploying resistant genotypes has a potential disadvantage as more virulent field populations of nematodes may be selected, making some resistant varieties redundant. This is particularly visible in PCN, where repeated exposure to resistant genotypes has selected for a virulent population of *G. pallida* in Germany (Mwangi *et al.*, 2019). Selection for virulence is specific to the source of resistance that the population has been exposed to (Varypatakis *et al.*, 2019). However, additive effects in population reduction can be seen when stacking multiple sources of resistance into one genotype, reducing the likelihood of selecting for resistance-breaking populations and providing an option for controlling populations with increased virulence (Price *et al.*, 2024b).

1.9. Conclusions and Future Directions

There is limited information regarding many aspects of nematode biology and host-parasite interaction. However, the urgent need to find environmentally acceptable methods of management has provided the impetus to increase understanding of plant-nematode interactions. This drive has resulted in advances in genomic and transcriptomic dataset availability. Consequentially, this has developed greater understanding of key aspects of life cycle biology. Using this information to understand more about plant-parasitic nematode eggshells and hatching will help develop novel chemicals that can be targetted specifically against the pathogen. These data can also be exploited to recognize how biological controls can be effectively utilized. Cultural changes are also beginning to impact how we perceive plant-parasitic nematode control. Increased discussions around the use of genetically modified or edited crops often focus on the benefit of introducing pathogen control into current, agronomically favourable genotypes (see Chapter 18, this volume). The challenge now will be to translate this information into viable, environmentally benign control or management options.

1.10. References

Abad, P. and Opperman, C.H. (2009) The complete sequence of the genomes of *Meloidogyne incognita* and *Meloidogyne hapla*. In: Perry, R.N., Moens, M. and Starr, J.L. (eds), *Root-knot Nematodes*. CAB International, Wallingford, UK, pp. 363–376. DOI: 10.1079/9781845934927.0363.

Abad, P., Gouzy, J., Aury, J.M., Castagnone-Sereno, P., Danchin, E.G., *et al.* (2008) Genome sequence of the metazoan plant-parasitic nematode *Meloidogyne incognita*. *Nature Biotechnology* 26, 909–915. DOI: 10.1038/nbt.1482.

Atkinson, H.J., Urwin, P.E. and Hussey, R.S. (2009) Plant biotechnology and control. In: Perry, R.N., Moens, M. and Starr, J.L. (eds), *Root-knot Nematodes*. CAB International, Wallingford, UK, pp. 338–362. DOI: 10.1079/9781845934927.0338.

Ahmed, M., Roberts, N.G., Adediran, F., Smythe, A.B., Kocot, K.M., *et al.* (2022) Phylogenomic analysis of the Phylum Nematoda: conflicts and congruences with morphology, 18S rRNA, and mitogenomes. *Frontiers in Ecology Evolution* 9, 769565. DOI: 10.3389/fevo.2021.769565.

Back, M.A., Hare, M.C. and Thoden, T.C. (2024) Chemical control of nematodes. In: Perry, R.N., Moens, M. and Jones, J.T. (eds), *Plant Nematology*, 3rd edn. CAB International, Wallingford, UK, pp. 514–538.

Bird, A.F. and Bird, D.J. (1991) *The Structure of Nematodes.* Academic Press, New York and London.

Bond, A.T. and Huffman, D.G. (2023) Nematode eggshells: a new anatomical and terminological framework, with a critical review of relevant literature and suggested guidelines for the interpretation and reporting of eggshell imagery. *Parasite* 30, 6. DOI: 10.1051/parasite/2023007.

Chitwood, D.J. and Perry, R.N. (2009) Reproduction, physiology and biochemistry. In: Perry, R.N., Moens, M. and Starr, J.L. (eds), *Root-knot Nematodes*. CAB International, Wallingford, UK, pp. 182–200.

Claudius-Cole, A.O. (2024) Nematode resistance in crops. In: Perry, R.N., Moens, M. and Jones, J.T. (eds), *Plant Nematology*, 3rd edn. CAB International, Wallingford, UK, pp. 472–497.

Curtis, R.H.C., Robinson, A.F. and Perry, R.N. (2009) Hatch and host location. In: Perry, R.N., Moens, M. and Starr, J.L. (eds), *Root-knot Nematodes*. CAB International, Wallingford, UK, pp. 139–162.

Davies, K.G. (2009) Understanding the interactions between an obligate hyperparasite bacterium, *Pasteuria penetrans*, and its obligate plant-parasitic nematode host, *Meloidogyne* spp. *Advances in Parasitology* 68, 211–245. DOI: 10.1016/S0065-308X(08)00609-X.

Deliopoulous, T., Jones, P.W. and Haydock, P.P.J. (2011) Variation in *in vitro* hatch of potato cyst nematodes in response to different potato cultivars inoculated with isolates of arbuscular mycorrhizal fungi. *Nematology* 13, 661–672. DOI: 10.1163/138855410X538144.

Desaeger, J., Wram, C. and Zasada, I. (2020) New reduced-risk agricultural nematicides – rationale and review. *Journal of Nematology* 52, e2020-91. DOI: 10.21307/jofnem-2020-091.

Devine, K.J. and Jones, P.W. (2001) Potato cyst nematode hatching activity and hatching factors in inter-specific *Solanum* hybrids. *Nematology* 3, 141–149. DOI: 10.1163/156854101750236268.

Devine, K.J. and Jones, P.W. (2002) Investigations into the chemoattraction of the potato cyst nematodes *Globodera rostochiensis* and *G. pallida* towards fractionated potato root leachate. *Nematology* 5, 65–75. DOI: 10.1163/156854102765216704.

Ebrahimi, N., Viaene, N. and Moens, M. (2015) Optimizing trehalose-based quantification of live eggs in potato cyst nematodes (*Globodera rostochiensis* and *G. pallida*). *Plant Disease* 99, 947–953. DOI: 10.1094/pdis-09-14-0940-re.

Elling, A.A., Mitreva, M., Recknor, J., Gai, X., Martin, J., *et al.* (2007) Divergent evolution of arrested development in the dauer stage of *Caenorhabditis elegans* and the infective stage of *Heterodera glycines*. *Genome Biology* 8, R211. DOI: 10.1186/gb-2007-8-10-r211.

Eves-van den Akker, E.S., Stojiiković, B. and Gheysen, G. (2021) Recent applications of biotechnological approaches to elucidate the biology of plant-nematode interactions. *Current Opinion in Biotechnology* 70, 122–130. DOI: 10.1016/j.copbio.2021.03.008.

Fioretti, L., Porter, A., Haydock, P.J. and Curtis, R. (2002) Monoclonal antibodies reactive with secreted-excreted products from the amphids and the cuticle surface of *Globodera pallida* affect nematode movement and delay invasion of potato roots. *International Journal for Parasitology* 32, 1709–1718. DOI: 10.1016/s0020-7519(02)00178-9.

Gaur, H.S. and Perry, R.N. (1991) The use of soil solarization for control of plant parasitic nematodes. *Nematological Abstracts* 60, 153–167.

Gaur, H.S., Beane, J. and Perry, R.N. (2000) The influence of root diffusate, host age and water regimes on hatching of the root-knot nematode, *Meloidogyne triticoryzae*. *Nematology* 2, 191–199. DOI: 10.1163/156854100509088.

Gheysen, G. and Jones, J.T. (2024) Molecular aspects of plant-nematode interactions. In: Perry, R.N., Moens, M. and Jones, J.T. (eds), *Plant Nematology*, 3rd edn. CAB International, Wallingford, UK, pp. 292–315.

Gommers, F.J. and Bakker, J. (1988) Physiological diseases induced by plant responses or products. In: Poinar Jr, G.O. and Jansson, H.-B. (eds), *Diseases of Nematodes. Vol. I*. CRC Press Inc., Boca Raton, Florida, pp. 3–22.

Guerrieri, A., Floková, K., Vlaar, L.E., Schilder, M.L., Kramer, G., *et al.* (2021) UPLC-MS/MS analysis and biological activity of the potato cyst nematode hatching stimulant, solanoeclepin A, in the root exudate of *Solanum* spp. *Planta* 254, 1–13. DOI: 10.1007/s00425-021-03766-2.

Hallmann, J., Davies, K.G. and Sikora, R. (2009) Biological control using microbial pathogens, endophytes and antagonists. In: Perry, R.N., Moens, M. and Starr, J.L. (eds), *Root-knot Nematodes*. CAB International, Wallingford, UK, pp. 380–411.

Hassan, S., Behm, C.A. and Mathesius, U. (2010) Effectors of plant parasitic nematodes that re-program root cell development. *Functional Plant Biology* 37, 933–942. DOI: 10.1071/FP10112.

Hominick, W.M. (1979) Selection for hatching at low temperatures in *Globodera rostochiensis* by continuous cultivation of early potatoes. *Nematologica* 25, 322–332. DOI: 10.1163/187529279X00055.

Jones, J.T., Robertson, L., Perry, R.N. and Robertson, W.M. (1997) Changes in gene expression during stimulation and hatching of the potato cyst nematode *Globodera rostochiensis. Parasitology* 114, 309–315. DOI: 10.1017/S0031182096008451.

Jones, J.T., Haegeman, A., Danchin, E.G.J., Gaur, H.S., Helder, J., *et al.* (2013) Top 10 plant-parasitic nematodes. *Molecular Plant Pathology* 14, 946–961. DOI: 10.1111/mpp.12057.

Jones, P.W., Tylka, G.L. and Perry, R.N. (1998) Hatching. In: Perry, R.N. and Wright, D.J. (eds), *The Physiology and Biochemistry of Free-living and Plant-parasitic Nematodes*. CABI Publishing, Wallingford, UK, pp. 181–212.

Kikuchi, T., Aikawa, T., Kosaka, H., Pritchard, L., Ogura, N., *et al.* (2007) Expressed sequence tag (EST) analysis of the pine wood nematodes *Bursaphelenchus xylophilus* and *B. mucronatus*. *Molecular and Biochemical Parasitology* 155, 9–17. DOI: 10.1016/j.molbiopara.2007.05.002.

Krueger, R., Dover, K.E., McSorley, R. and Wang, K.-H. (2010) *Marigolds (*Tagetes *spp.) for nematode management.* Florida Cooperative Extension Service, Institute of Food and Agricultural Sciences, University of Florida, Florida.

Manzanilla-López, R.H. and Lopez-Llorca, L.V. (eds) (2017) *Perspectives in Sustainable Nematode Management Through* Pochonia chlamydosporia *Applications for Root and Rhizosphere Health*. Springer, Cham, Switzerland. DOI: 10.1007/978-3-319-59224-4.

Manzanilla-López, R.H., Costilla, M.A., Doucet, M., Franco, J., Inserra, R.N., *et al.* (2002) The genus *Nacobbus* Thorne and Allen, 1944 (Nematoda: Pratylenchidae): systematics, distribution, biology and management. *Nematropica* 32, 149–227.

Manzanilla-López, R.H., Atkins, S.D., Clark, I.M., Kerry, B.R. and Hirsch, P.R. (2009) Measuring abundance, diversity and parasitic ability in the populations of the nematophagous fungus *Pochonia chlamydosporia* var. *chlamydosporia*. *Biocontrol Science and Technology* 19, 391–406. DOI: 10.1080/09583150902783793.

Marbán-Mendoza, N., Jeyaprakash, A., Jansson, H.-B., Damon, H.A. and Zuckerman, B.M. (1987) Control of root-knot nematode on tomato by lectins. *Journal of Nematology* 19, 31–335.

Moens, M. and Perry, R.N. (2009) Migratory plant endoparasitic nematodes: a group rich in contrasts and diversity. *Annual Review of Phytopathology* 47, 313–332. DOI: 10.1146/annurev-phyto-080508-081846.

Mwangi, J.M., Niere, B., Finckh, M.R., Krüssel, S. and Kiewnick, S. (2019) Reproduction and life history traits of a resistance breaking *Globodera pallida* population. *Journal of Nematology* 51, e2019-79. DOI: 10.21307/jofnem-2019-079.

Mwangi, J.M., Mwangi, G.N., Finckh, M.R. and Kiewnick, S. (2021) Biology, pathotype, and virulence of *Globodera rostochiensis* populations from Kenya. *Journal of Nematology* 53, e2021-03. DOI: 10.21307/jofnem-2021-003.

Ngala, B., Dewaegeneire, P., Robilliard, E., Mariette, N., Manceau, F., *et al.* (2024) Lure and starve: host root exudates to suppress field populations of cyst nematodes. *Applied Soil Ecology* 201, 105490. DOI: 10.1016/j.apsoil.2024.105490.

Niu, J., Guo, Q., Jian, H., Chen, C., Yang, D., *et al.* (2011) Rapid detection of *Meloidogyne* spp. by LAMP assay in soil and roots. *Crop Protection*. 30(8). DOI: 10.1016/j.cropro.2011.03.028.

Nyczepir, A.P. and Thomas, S.H. (2009) Current and future management strategies in intensive crop production systems. In: Perry, R.N., Moens, M. and Starr, J.L. (eds), *Root-knot Nematodes*. CAB International, Wallingford, UK, pp. 412–443.

Olson, S.K., Greenan, G., Desai, A., Müller-Reichert, T. and Oegema, K. (2012) Hierarchical assembly of the eggshell and permeability barrier in *C. elegans*. *The Journal of Cell Biology* 198, 731–748. DOI: 10.1083/jcb.201206008.

Palomares-Rius, J.E., Hedley, P., Cock, P.J.A., Morris, J.A., Jones, J.T., *et al.* (2016) Gene expression changes in diapause or quiescent potato cyst nematode, *Globodera pallida*, eggs after hydration or exposure to tomato root diffusate. *PeerJ* 1654. DOI: 10.7717/peerj.1654.

Peng, H., Long, H., Huang, W., Liu, J., Cui, J., *et al.* (2017) Rapid, simple and direct detection of *Meloidogyne hapla* from infected root galls using loop-mediated isothermal amplification combined with FTA technology. *Scientific Reports* 7, 44853. DOI: 10.1038/srep44853.

Perry, R.N. (1989) Dormancy and hatching of nematode eggs. *Parasitology Today* 5, 377–383. DOI: 10.1017/S0031182000084626.

Perry, R.N. (1997) Plant signals in nematode hatching and attraction. In: Grundler, F., Ohl, S. and Fenoll, C. (eds), *Cellular and Molecular Aspects of Plant-nematode Interactions*. Kluwer Academic Publishers, Dortrecht, Netherlands, pp. 38–50.

Perry, R.N. (1999) Desiccation survival of parasitic nematodes. *Parasitology* 119(2001), S19–30. DOI: 10.1017/S0031182000084626.

Perry, R.N. (2001) Analysis of the sensory responses of parasitic nematodes using electrophysiology. *International Journal for Parasitology* 31, 908–917. DOI: 10.1016/s0020-7519(01)00198-9.

Perry, R.N. (2002) Hatching. In: Lee, D.L. (ed), *The Biology of Nematodes*. Taylor and Francis, London, UK, pp. 147–169.

Perry, R.N. (2005) An evaluation of types of attractants enabling plant-parasitic nematodes to locate plant roots. *Russian Journal of Nematology* 13, 83–88.

Perry, R.N. (2024) Detection of the effects of root exudates (diffusates) on nematode hatching and attraction. In: Molinari, S. (ed), *Plant-nematode Interactions. Methods in Molecular Biology*. Humana, New York, New York, pp. 271–289. DOI: 10.1007/978-1-0716-3638-1_8.

Perry, R.N. and Beane, J. (1982) The effect of brief exposures to potato root diffusate on the hatching of *Globodera rostochiensis*. *Revue de Nématologie* 5, 221–224.

Perry, R.N. and Holden-Dye, L. (2024) Behaviour and sensory perception. In: Perry, R.N., Moens, M. and Jones, J.T. (eds), *Plant Nematology*, 3rd edn. CAB International, Wallingford, UK, pp. 262–291.

Perry, R.N. and Moens, M. (2011) Survival of parasitic nematodes outside the host. In: Perry, R.N. and Wharton, D.A. (eds), *Molecular and Physiological Basis of Nematode Survival*. CAB International, Wallingford, UK, pp. 1–27.

Perry, R.N. and Trett, M.W. (1986) Ultrastructure of the eggshell of *Heterodera schachtii* and *H. glycines* (Nematoda: Tylenchida). *Revue de Nématologie* 9, pp. 399–403.

Perry, R.N. and Wesemael, W. (2008) Host plant effects on hatching of root-knot nematodes. *Russian Journal of Nematology* 16, 1–5.

Perry, R.N., Wright, D.J., Chitwood, D.J. and Price, J.A. (2024) Reproduction, physiology and biochemistry. In: Perry, R.N., Moens, M. and Jones, J.T. (eds), *Plant Nematology*, 3rd edn. CAB International, Wallingford, UK, pp. 231–261.

Ploeg, A. (2000) Effects of amending soil with *Tagetes patula* cv. Single Gold on *Meloidogyne incognita* infestation on tomato. *Nematology* 2, 489–493. DOI: 10.1163/156854100509394.

Price, J.A., Ali, M.F., Major, L.L., Smith, T.K. and Jones, J.T. (2023) An eggshell-localised annexin plays a key role in the coordination of the life cycle of a plant-parasitic nematode with its host. *PLoS Pathogens* 19(2), e1011147. DOI: 10.1371/journal.ppat.1011147.

Price, J.A., Hunt, K., Nyagah, N., Orage, C., Mwaura, O., *et al.* (2024a) Characterisation of 'giant' cysts of the potato cyst nematode *Globodera rostochiensis* present in potato fields in Kenya. *Nematology* 26 (Suppl. 2), 1–7. DOI: 10.1163/15685411-bja10339.

Price, J.A., Preedy, K., Young, V., Todd, D. and Blok, V.C. (2024b) Stacking host resistance genes to control *Globodera pallida* populations with different virulence. *European Journal of Plant Pathology* 168, 373–381. DOI: 10.1007/s10658-023-02761-5.

Pudasaini, M.P., Schomaker, C.H., Been, T.H. and Moens, M. (2006) Vertical distribution of the plant-parasitic nematode, *Pratylenchus penetrans*, under four field crops. *Phytopathology* 96, 226–233. DOI: 10.1094/PHYTO-96-0226.

Reid, A., Evans, F., Mulholland, V., Cole, Y. and Pickup, J. (2015) High-throughput diagnosis of potato cyst nematodes in soil samples. In: Lacomme, C. (ed), *Plant Pathology: Techniques and Protocols*. Humana Press, New York, New York, pp. 137–148. DOI 10.1007/978-1-4939-2620-6_11.

Rolfe, R.N., Barrett, J. and Perry, R.N. (2000) Analysis of chemosensory responses of second stage juveniles of *Globodera rostochiensis* using electrophysiological techniques. *Nematology* 2, 523–533. DOI: 10.1163/156854100509448.

Shimizu, K., Akiyama, R., Okamura, Y., Ogawa, C., Masuda, Y., *et al.* (2023) Solanoeclepin B, a hatching factor for potato cyst nematode. *Science Advances* 9, 4166. DOI: 10.1126/sciadv.adf4166.

Sikora, R.A., Desaeger, J. and Molendijk, L.P.G. (eds) (2021) *Integrated Nematode Management: State-of-the-art and Visions for the Future*. CAB International, Wallingford, UK.

Smythe, A., Holovachov, O. and Kocot, K. (2019) Improved phylogenomic sampling of free-living nematodes enhances resolution of higher-level nematode phylogeny. *BMC Evolutionary Biology* 19, 121. DOI: 10.1186/s12862-019-1444-x.

Starr, J.L. and Mercer, C.F. (2009) Development of resistant varieties. In: Perry, R.N., Moens, M. and Starr, J.L. (eds), *Root-knot Nematodes*. CAB International, Wallingford, UK, pp. 326–337.

Stewart, G.R., Perry, R.N. and Wright, D.J. (1993) Studies on the amphid specific glycoprotein gp32 in different life cycle stages of *Meloidogyne* species. *Parasitology* 107, 573–578. DOI: 10.1017/S0031182000068165.

Subbotin, S.A. and Moens, M. (2024) Molecular Systematics. In: Perry, R.N., Moens, M. and Jones, J.T. (eds), *Plant Nematology*, 3rd edn. CAB International, Wallingford, UK, pp. 41–73.

Tanino, K., Takahashi, M., Tomata, Y., Tokura, H., Uehara, T., *et al.* (2011) Total synthesis of solanoeclepin A. *Nature Chemistry* 3, 484–488. DOI: 10.1038/nchem.1044.

Thoden, T.C., Korthals, G.W. and Termorshuizen, A.J. (2011) Organic amendments and their influences on plant-parasitic and free-living nematodes: a promising method for nematode management? *Nematology* 13, 133–153. DOI: 10.1163/138855410X541834.

Triantaphyllou, A.C. (1966) Polyploidy and reproductive patterns in the root-knot nematode *Meloidogyne hapla*. *Journal of Morphology* 118, 403–413. DOI: 10.1002/jmor.1051180308.

Trudgill, D.L., Honek, A., Li, D. and van Straalen, N.M. (2005) Thermal time – concepts and utility. *Annals of Applied Biology* 146, 1–14. DOI: 10.1111/j.1744-7348.2005.04088.x.

Twomey, U., Rolfe, R.N., Warrior, P. and Perry, R.N. (2002) Effects of the biological nematicide, DiTera®, on movement and sensory responses of second-stage juveniles of *Globodera rostochiensis*, and stylet activity of *G. rostochiensis* and fourth-stage juveniles of *Ditylenchus dipsaci*. *Nematology* 4, 909–915. DOI: 10.1163/156854102321122520.

van den Elsen, S., Ave, M., Schoenmakers, N., Landeweert, R., Bakker, J., *et al.* (2012) A rapid, sensitive, and cost-effective assay to estimate viability of potato cyst nematodes. *Phytopathology* 102, 140–146. DOI: 10.1094/PHYTO-02-11-0051.

Varypatakis, K., Jones, J.T. and Blok, V.C. (2019) Screening of populations of *Globodera pallida* selected for increased virulence on several potato varieties. *Nematology*, 21, 995–998. DOI: 10.1163/15685411-00003283.

Viaene, N., Coyne, D.L. and Davies, K.G. (2024) Biological and cultural management. In: Perry, R.N., Moens, M. and Jones, J.T. (eds), *Plant Nematology*, 3rd edn. CAB International, Wallingford, UK, pp. 443–471.

Vlaar, L.E., Bertran, A., Rahimi, M., Dong, L., Kammenga, J.E., *et al.* (2021) On the role of dauer in the adaptation of nematodes to a parasitic lifestyle. *Parasites Vectors* 14, 554. DOI: 10.1186/s13071-021-04953-6.

Wang, C., Bruening, G. and Williamson, V.M. (2009) Determination of preferred pH for root-knot nematode aggregation using Pluronic F-127 Gel. *Journal of Chemical Ecology* 35, 1242-1251. DOI: 10.1007/s10886-009-9703-8.

Whitehead, A.G. (1998) *Plant Nematode Control*. CABI Publishing, Wallingford, UK.

Williamson, V.M. and Roberts, P.A. (2009) Mechanisms and genetics of resistance. In: Perry, R.N., Moens, M. and Starr, J.L. (eds), *Root-knot Nematodes*. CAB International, Wallingford, UK, pp. 301–325.

Wright, K. and Perry, R.N. (1991) Moulting of *Aphelenchoides hamatus*, with especial reference to the formation of the stomatostyle. *Revue de Nématologie* 14, 497–504.

Womersley, C.Z., Wharton, D.A. and Higa, L.M. (1998) Survival biology. In: Perry, R.N. and Wright, D.J. (eds), *The Physiology and Biochemistry of Free-living and Plant-parasitic Nematodes*. CABI Publishing, Wallingford, UK, pp. 271–302.

Yao, K., Peng, D., Jiang, C., Zhao, W., Li, G., *et al.* (2021) Rapid and visual detection of *Heterodera schachtii* using recombinase polymerase amplification combined with Cas12a-mediated technology. *International Journal of Molecular Sciences* 22, 12577. DOI: 10.3390/ijms222212577.

Yépez, G.T. (1972). *Los Nematodos enemigos de la Agricultura*. Universidad Central de Venezuela. Facultad de Agronomía, Instituto de Zoología Agrícola. Maracay, Venezuela.

Zhang, W., Li, Y., Liu, Z., Li, D., Wen, X., *et al.* (2022) Transcriptome analysis of dauer moulting of a plant parasitic nematode, *Bursaphelenchus xylophilus* promoted by pine volatile β-Pinene. *Agronomy* 12(9), 2114. DOI: 10.3390/agronomy12092114.

2 General Morphology and Morphometrics of Plant-parasitic Nematodes

David J. Hunt[1]* and Juan E. Palomares-Rius[2]
[1]*CABI, Silwood Park, Buckhurst Road, Ascot, Berkshire, UK;* [2]*Institute for Sustainable Agriculture-CSIC, Av. Menéndez Pidal SN, Córdoba, Spain*

Abstract
The chapter presents an introduction to nematode diversity, their ecological role (including habitat and niche characteristics) and their free-living and parasitic habits. The main section is concerned with the general morphology of the major organ systems of nematodes, which are described and illustrated in detail, particular attention being paid to the anatomical features commonly used in diagnostic characterisation and taxonomic identification. A section on nematode morphometrics touches on some of the problems with this classical approach and defines the abbreviations most commonly used in nematode characterisation. A simple dichotomous key to the major groups of soil nematodes is included.

2.1. Introduction

Nematodes are currently regarded as being the most numerous multicellular animals on the planet. Their diversity is still widely disputed, with some estimates of the number of extant species reaching into the millions. They occupy niches in just about every biotype and exist in unimaginably large numbers. Although essentially tubular in structure, their basic form has been adapted into an astonishingly wide variety of shapes and sizes.

Nematodes are lower invertebrate animals. While most species are free-living in marine, freshwater or soil environments, many are parasites of various types of plants and animals, including man, and may be of considerable agricultural, clinical and veterinary importance. Nematodes occur in soils and sediments, at the bottom of lakes and rivers and at enormous depths in the oceans; some species can withstand temperatures constantly below freezing point while others live in the waters of hot springs. They are even found in the Dry Valleys of Antarctica and in deserts, where they spend most of their life in an anhydrobiotic state, reducing their activity to periods when suitable conditions (mainly the presence of liquid water) return.

The feeding habits of nematodes are extremely varied, with many species being microphagous

*E-mail: tabanids.bite@gmail.com

A revision of Hunt, D.J. and Palomares-Rius, J.E. (2012) General morphology and morphometrics of plant-parasitic nematodes. In: Manzanilla-López, R.H. and Marbán-Mendoza, N. (eds) *Practical Plant Nematology*. Mundi-Prensa and Biblioteca Básica de Agricultura, Montecillo, Mexico.

(eds R.H. Manzanilla-López, D.J. Hunt, N. Marbán-Mendoza and S. Infante Gil)
DOI: 10.1079/9781836990413.0002

and feeding on micro-organisms, while others are saprophagous, feeding on dead and decaying organic matter, or herbivorous, obtaining their nourishment by parasitizing plants. Yet other species are carnivores or predators, obtaining their food from other animals, either alive or dead. Many species parasitize, or are associated with, other invertebrates, those parasitizing vertebrates often being particularly injurious. It has been estimated that a single hectare of soil from arable land may contain as many as 6,000,000,000 nematodes, while marine beach sand may contain 3,000,000,000 per hectare. A single wheat gall typically contains around 11,000-18,000 specimens of *Anguina tritici*, but as many as 90,000 have been recorded.

By 1930, some 4500 species of nematode had been described. This rose to 9000 by 1950, and the present-day number of known species of nematodes is probably in excess of 30,000, over half of which are parasites. The estimated number of existing species ranges from 500,000 to several million, although the only certainty is that we have no accurate idea of their total number. New molecular technologies, such as the extensive sequencing of DNA samples, are helping to determine this biodiversity, at least in terms of molecular characterization. As the vast majority of nematode species are not yet known to science, the taxonomy and phylogeny of this group is still in its early stages, although much progress has been made in the last decades using increasingly powerful molecular tools.

Soil nematodes are essentially aquatic organisms, as they require, at the very least, a thin film of water to be active. However, very few species can actually swim – they cannot generate a sufficiently rapid wave motion to counter gravity and, hence, sink in free water. As is often the case with nematodes, there are always exceptions to the rule – some species of plant-parasitic nematodes are adapted to aquatic environments and are able to swim in an effective manner. Examples include *Ditylenchus angustus* and *Aphelenchoides besseyi*, both of which are adapted to flooded rice fields, while the microbivorous *Turbatrix aceti* swims actively in vinegar and often congregates in dense swarms near the surface of the liquid. The majority of soil nematodes, however, are adapted to moving through the soil interstices, the thin film of water that surrounds the soil particles, serving to lubricate their passage. As the nematodes depend on the system of pores within the soil to facilitate movement, it comes as no surprise to learn that fine-textured soils, such as clays, tend to have fewer large nematodes than coarse-grained soils, such as sands or sandy-loams, although this effect may be ameliorated in soils that are undisturbed by cultivation and have hence acquired an extensive system of pores developed, for example, by the activity of plant roots growing through the soil.

The account that follows is based on that of Hunt and Palomares-Rius (2012), with the major morphological characteristics of plant-parasitic nematodes being discussed, contrasted and illustrated. The literature sources have been updated where appropriate. Little attention is given to the morphology of non-parasitic nematodes, with the exception of some basic comparisons of stoma type, as this aspect is covered in greater detail in Chapter 16. The commonly used measurements and ratios used in morphometrics are also discussed. The treatment, albeit often fairly simplistic, is intended to offer a pragmatic approach to identification rather than provide a detailed treatise on body form and organ systems. For those wishing to explore further details, the web link maintained by the nematology group at the University of Nebraska-Lincoln (http://nematode.unl.edu/) and the Department of Entomology and Nematology at the University of California (http://nemaplex.ucdavis.edu/) provides much information. The classic work by Chitwood and Chitwood (1959) is still relevant in many respects, if somewhat daunting to the novice. The pictorial guides by Mai and Lyon (1975) and Mai *et al.* (1996) are excellent and profusely illustrated guides to plant-parasitic genera. The book by Bird and Bird (1991) can be recommended, and there are several subsequent publications that deal with the morphology of plant-parasitic nematodes in varying degrees of detail (e.g., Hooper and Evans, 1993, Hunt *et al.*, 2003, 2018; Peña-Santiago, 2020; Decraemer and Hunt, 2024), while Geraert (2006) monographs the morphology of the Tylenchida and provides detailed appraisals of the Tylenchidae, Criconematidae, Dolichodoridae, Pratylenchidae (Geraert, 2008, 2010, 2011, 2013) and Tylenchulidae (Ghaderi *et al.*, 2016). The morphology of more specialized taxa, such as cyst nematodes (Subbotin *et al.*, 2010;

Perry *et al.*, 2018), root-knot nematodes (e.g., Eisenback and Hunt, 2009; Subbotin *et al.*, 2021) and sheath nematodes (Chitambar and Subbotin, 2014), should also be consulted, but older monographs – such as De Coninck (1965), Maggenti (1981), Siddiqi (1986, 2000), Nickle (1991), Hunt (1993) and Decraemer (1995) – still contain much useful information and are recommended as valuable resources. Additional overviews of the general morphology of many nematode structures and organs, even in 3D, are in the web dedicated to *C. elegans* and other nematodes (https://wormatlas.org/index.html). Some new technologies, such as transmission electron microscopy (TEM), new fixation methods and the use of powerful computers, can be used for nematode anatomical reconstruction in 3D (https://wormatlas.org/index.html). However, these techniques have barely been explored in plant-parasitic nematodes.

2.2. Morphology

The following text should be used in conjunction with Figures 2.1-2.13. Figures 2.1-2.4 show general body form and typical morphology using line drawings, light micrographs and scanning electron micrographs. Figure 2.5 deals with the body cuticle; Figure 2.6 shows the labial region and anterior sensilla; stoma and pharyngeal form are shown in Figures 2.7-2.11; Figure 2.12 focuses on the female genital system, while Figure 2.13 deals with the male reproductive

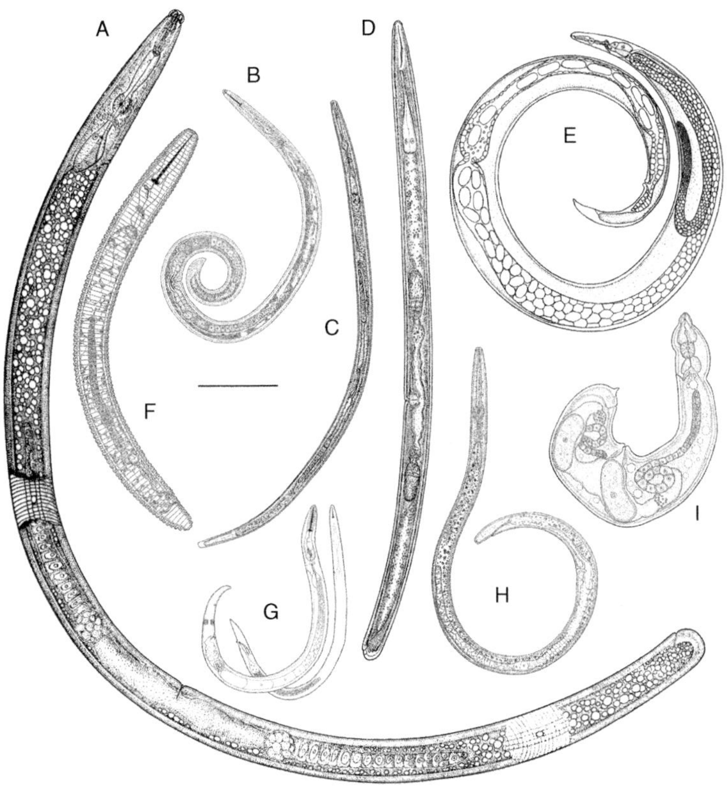

Fig. 2.1. Relative body size and range in form of mature females of some common plant-parasitic nematodes. A: *Hoplolaimus galeatus*; B: *Helicotylenchus dihystera*; C: *Tylenchorhynchus annulatus*; D: *Trichodorus primitivus*; E: *Anguina tritici* (note that, because of its length of 3-5 mm, this nematode is shown at a different scale to the others); F: *Criconemoides xenoplax*; G: *Paratylenchus bukowinensis*; H: *Aphelenchus avenae*; I: *Rotylenchulus reniformis*. Figure digitally composed from drawings taken from *CIH Descriptions of Plant-parasitic Nematodes*. (Scale bar: A-D, F-I = 100 μm; E = 250 μm.)

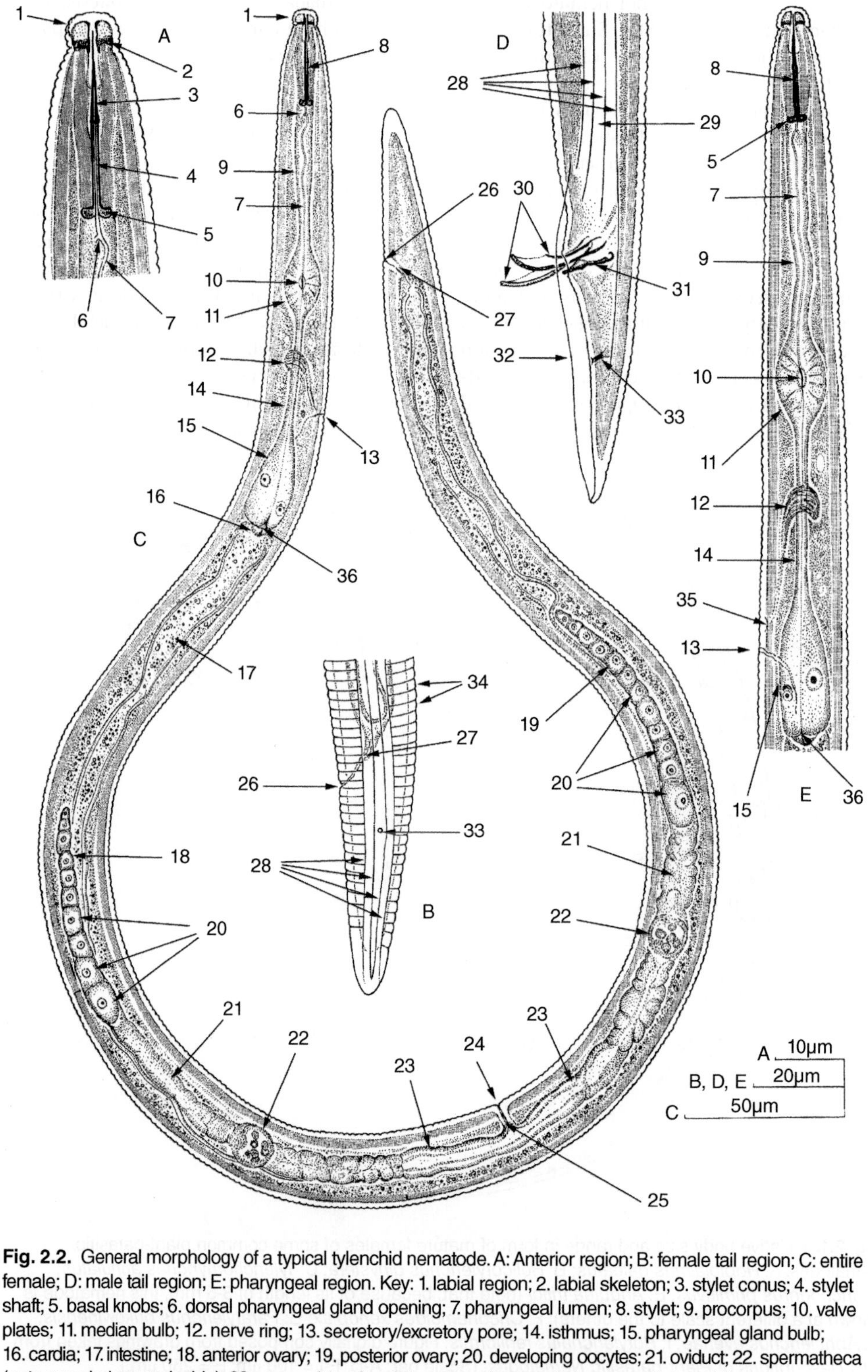

Fig. 2.2. General morphology of a typical tylenchid nematode. A: Anterior region; B: female tail region; C: entire female; D: male tail region; E: pharyngeal region. Key: 1. labial region; 2. labial skeleton; 3. stylet conus; 4. stylet shaft; 5. basal knobs; 6. dorsal pharyngeal gland opening; 7. pharyngeal lumen; 8. stylet; 9. procorpus; 10. valve plates; 11. median bulb; 12. nerve ring; 13. secretory/excretory pore; 14. isthmus; 15. pharyngeal gland bulb; 16. cardia; 17. intestine; 18. anterior ovary; 19. posterior ovary; 20. developing oocytes; 21. oviduct; 22. spermatheca (note rounded sperm inside); 23. uterus; 24. vulva; 25. vagina; 26. anus; 27. rectum; 28. lateral lines or incisures; 29. lateral field; 30. spicules; 31. gubernaculum; 32. bursa; 33. phasmid; 34. annules; 35. hemizonid; 36. pharyngo-intestinal junction. (Modified after Siddiqi (1972), courtesy CABI.)

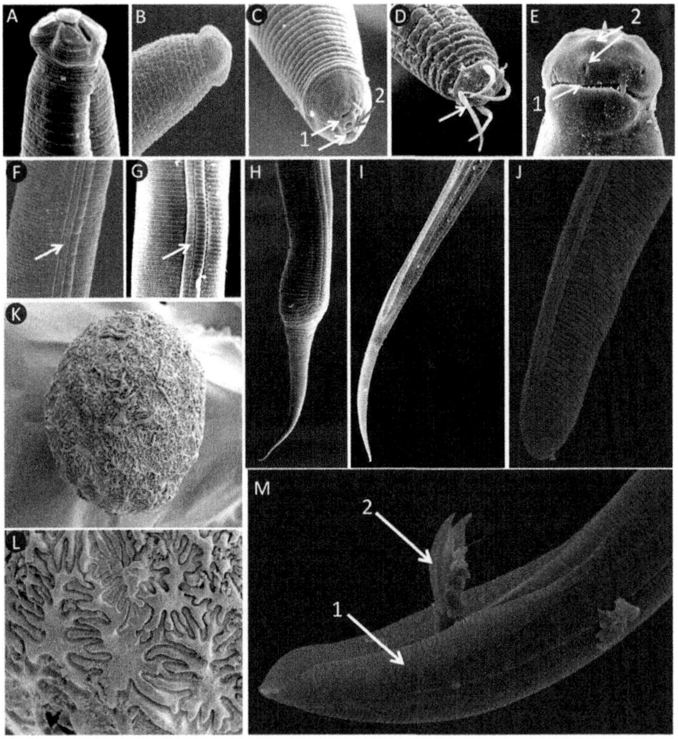

Fig. 2.3. SEM micrographs. A: *Dolichodorus*; B: *Tylenchorhynchus*; C: second-stage *Sphaeronema alni* (1 = amphids, 2 = oral aperture); D: *Eutylenchus excretorius* showing cephalic setae; E: *Paralongidorus litoralis* (1 = amphidial aperture, 2 = labial papillae); F: *Tylenchorhynchus* showing lateral fields (arrow); G: *Dolichodorus* showing lateral fields (arrow); H: *Dolichodorus* sp. tail; I: *Eutylenchus excretorius* tail; J: *Tylenchorhynchus* tail; K, L: cystoid and ornamentation of *Meloidoderita kirjanovae*; M: male of *Tylenchorhynchus* (1 = bursa, 2 = spicules).

system. In the text, major features or characters are indicated using bold font.

2.2.1. General Form

Soil nematodes are typically **vermiform** (worm-like) and comprise a more or less cylindrical body that tapers to varying degrees towards the anterior and posterior extremities. This typical shape may, however, be greatly modified, as seen in the obese females of semi-endoparasitic or endoparasitic nematodes, such as cyst, root-knot and certain pratylenchids and criconematids, and in unusual morphotypes, as found in many criconematids (Fig. 2.1). Body length varies from a few hundred microns to over 9 m in *Placentonema gigantissima*, a parasite of the sperm whale. Plant-parasitic nematodes rarely exceed 1.5 mm in size (and many are much smaller), although the largest, members of *Paralongidorus*, can reach a length of over 12 mm. The body is usually smooth, only rarely bearing 'appendages', as seen in the Bunonematidae, a group of fascinating bacterial feeders of rather bizarre appearance.

Nematodes may be thought of as tubes, with the internal organs also being tubular or derived therefrom (Fig. 2.2). The anterior extremity is often, and erroneously, called the **head**, although more accurately it should be referred to as the **labial region**. It contains the mouth opening, or **oral aperture**, which opens to the **stoma** or **buccal cavity**. The stoma is followed by the **pharynx** or **oesophagus** (both terms are in use), and this region is often referred to as the **neck** or **cervical region**. The region posterior

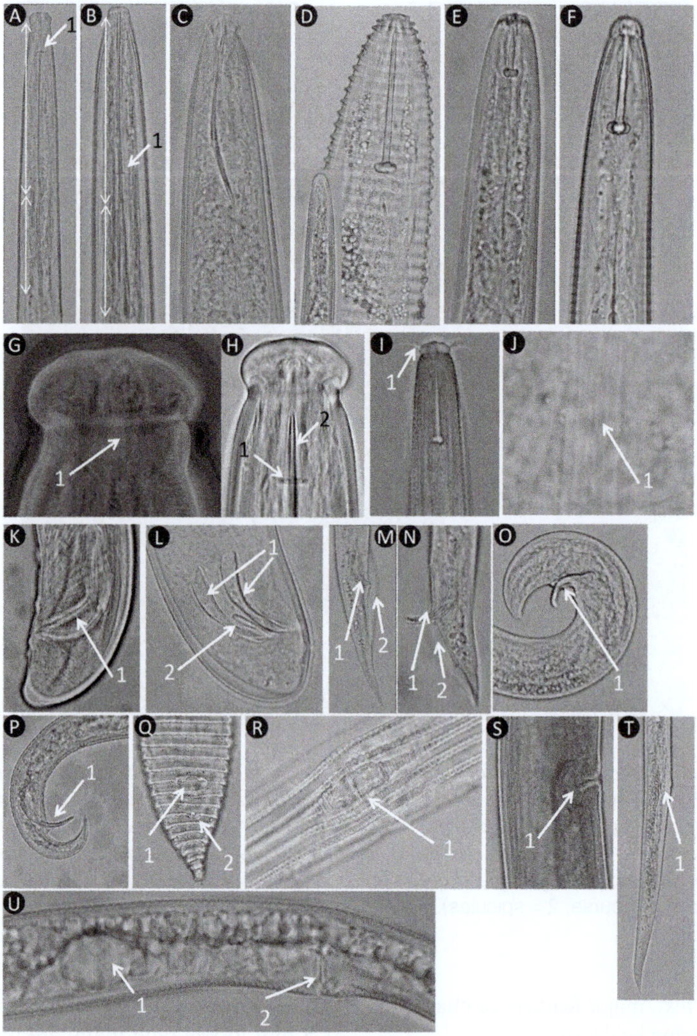

Fig. 2.4. Light micrographs. A, B: Anterior region of *Longidorus* and *Xiphinema*, respectively, showing 1: guide ring and different stylet parts, odontostylet (above), odontophore (below); C: *Trichodorus* showing onchiostyle, D: *Criconemoides annulatiformis*, male without stylet is shown inset, E: *Pratylenchus*; F: *Rotylenchus eximius*; G: *Paralongidorus litoralis* showing amphids; H: *Paralongidorus litoralis* showing 1: guiding ring and 2: stylet; I: *Eutylenchus excretorius* showing cephalic setae; J: *Ditylenchus gigas* showing lateral fields; K: *Paralongidorus* male tail showing spicules; L: *Trichodorus* male tail showing 1: spicules and 2: gubernaculum; M: *Ditylenchus gigas* showing 1: spicules and 2: bursa; N: *Pratylenchus* spp. showing 1: spicules and 2: bursa; O: *Schistonchus caprifici* showing characteristic shape of Aphelenchoidea spicule; P: *Criconemoides annulatiformis* male tail showing spicules; Q: *Criconemoides annulatiformis* female tail showing 1: vulva and 2: anus; R: *Eutylenchus excretorius* female showing vulva; S: *Xiphinema* showing vulva; T: *Ditylenchus gigas* female tail showing vulva; U: *Pratylenchus* showing 1: spermatheca and 2: vulva.

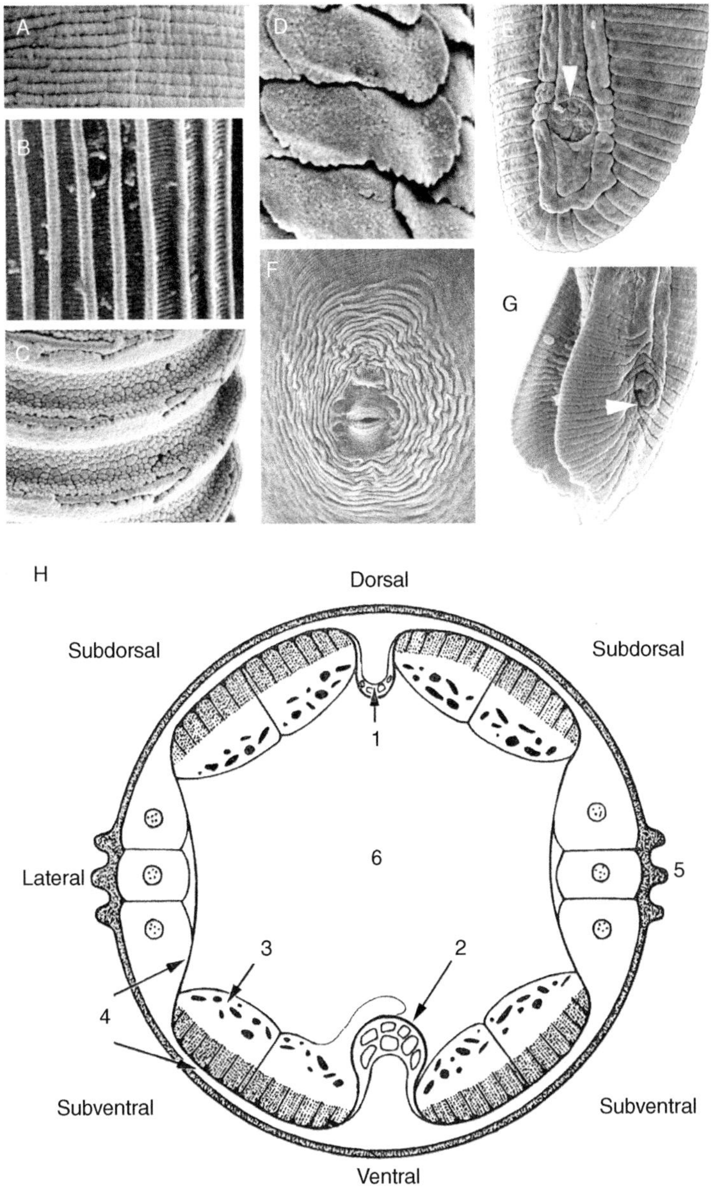

Fig. 2.5. A-F. Surface structure of the body cuticle. A: Transverse striae, SEM (Trichodoridae); B: longitudinal ridges (Actinolaimidae, Dorylaimida) (after Vinciguerra and Claus, 2000; courtesy Universidad de Jaén); C, D: *Criconema paradoxiger*; C: external cuticular layer in female; D: scales in juvenile (after Decraemer and Geraert, 1992; courtesy Brill); E: lateral field with incisures, areolation (arrow) and scutellum (arrowhead) in *Scutellonema*; F: perineal pattern in *Meloidogyne* (after Siddiqi, 1986; courtesy CABI); G: caudal alae and scutellum (arrowhead) in *Scutellonema* male; H: transverse section of nematode showing sectors, lateral fields, somatic musculature, pseudocoelom and chords (diagrammatic; internal organs omitted). Key: 1. dorsal epidermal chord; 2. ventral chord with ventral nerve; 3. muscle cell; 4. basal lamina; 5. lateral fields with ridges or alae; 6. pseudocoelom (based on Bird and Bird, 1991). (Figure adapted from Decraemer and Hunt (2006), courtesy CABI.)

to the anus or cloacal aperture is called the **tail**. The tail may be very short (even absent in the adult female of some endoparasitic species), although it can also be extremely long and filiform. Tail shape is extremely variable in nematodes and is a useful characteristic for identification purposes. Its shape may differ between juveniles and adults or between the sexes. The tail serves for locomotion and anchorage during copulation and may assist in the exit of the juvenile from the eggshell.

The body is encased in a flexible **cuticle**, which is pierced by apertures for the digestive, secretory/excretory and genital systems, by various sensory structures such as papillae or sensilla of various types, and the amphids. Internal to the cuticle is the **hypodermis** and **somatic musculature**. These components form the body wall that surrounds the **pseudocoelomic cavity** (embryologically, nematodes do not possess a true coelom as the cavity is not lined by its own epithelium), and it is within this pseudocoelomic fluid-filled cavity that the tubular organs, such as the digestive and reproductive systems, are suspended.

Nematodes typically display a bilateral symmetry although the symmetry of the cephalic region and/or stoma may be **hexaradiate** or **triradiate**. The orientation of a nematode and the relative position of morphological structures within it are referred to by certain terms. Viewing the nematode body as a transverse section, the three primary positions are **dorsal** (equivalent to the back of the animal), **ventral** (=the belly) and **lateral** (left and right sides). The sector between the dorsal and the lateral axes is termed **subdorsal**; that between the lateral and the ventral axes is often referred to as **subventral**, although this term is imprecise in terms of its etymology. The axis halfway between the dorsal and the lateral axes is known as **dorsolateral** and that between the lateral and the ventral axes is **ventro-lateral**. Other commonly used terms include **distal** and **proximal**, although unfortunately these are often used with opposite meanings! These terms should be interpreted with respect to the organ being so described, distal being taken to refer to that part of the organ furthest from the midpoint or external opening, while proximal refers to a point closer to the middle or external opening. In the case of the female genital tract, for example, the structures adjacent to the external aperture, the vulva, are proximal, whereas those furthest away, such as the ovary, are distal, regardless of whether the ovary itself is outstretched or reflexed.

When nematodes die in an unstressed state (such as proper heat relaxation), they usually assume a typical form or **habitus**. This is often curved or more spiral in form – rarely do they die absolutely straight. The inner surface of the curve or spiral is almost invariably ventral and, as this is also the surface that bears the external apertures such as the secretory/excretory pore, vulva, anus or cloacal aperture, even a straight nematode can be correctly orientated through accurate observation. Of course, if nematodes are improperly killed or fixed, for example, by dunking them in cold fixative and thus subjecting them to a slow death, the body is likely to distort and twist, thus making observation and orientation of the various structures much more difficult. The habitus of a properly killed nematode can provide a useful hint of its identity; for example, *Helicotylenchus* species often tend to die in a spiral form, whereas *Pratylenchus* species tend to die in a straight to slightly ventrally arcuate form, *Radopholus* dies more or less straight, and males of *Aphelenchoides* assume an unmistakable walking-stick shape and so on. Nematodes are difficult enough to identify as it is, so avoid making the matter worse by poor fixation, which results in highly distorted and twisted animals.

2.2.2. Body Cuticle

A multilayered cuticle is almost invariably present but may be absent in some specialized parasitic nematodes inhabiting the haemocoel of insects and mites in which the absorption of food is through the general body surface, for example, *Fergusobia*. The structure of the nematode cuticle varies widely across the Nematoda, may be extremely complex and may differ in structure and number of constituent layers between the various stages of a nematode or between different regions of the same nematode (Decraemer *et al.*, 2003; Decraemer and Hunt, 2024). For the pragmatic agenda of this chapter, however, it is not necessary to go into cuticle structure and ultrastructure in great detail, aspects of which can be found in other sources such as Siddiqi (2000), Decraemer *et al.* (2003) and Decraemer and Hunt (2024). The cuticle acts like a protective skin and covers the external surfaces of the nematode, invaginating at orifices, such as the oral aperture, **amphids**, **phasmids**, secretory-excretory (S-E) pore, vulva, anus and cloacal aperture, and forming a lining to the **cheilostom** (or lip cavity), **amphidial fovea**, **terminal canal**, proximal duct of the S-E system, ***pars distalis vaginae***, **cloaca** and **rectum**. In the absence of a skeleton and circular muscles, the cuticle apparently functions as an antagonistic system, preventing radial deformation of the body when the longitudinal muscles contract during locomotion. The cuticle also has a protective role and, due to its semi-permeable nature, is involved in secretion-excretion and gaseous exchange and so on.

The body cuticle maybe smooth but is often marked by transverse and/or longitudinal grooves or striae. The transverse striae may be very fine and superficial or be rather deeper and wider apart, thereby delimiting the **annules** (sing. **annule** but also referred to as **annuli** or **annulus**, respectively). The fine transverse striae are present in Longidoridae and Trichodoridae, but they are indistinct and difficult to observe by light microscopy, with the cuticle appearing smooth. When both transverse and longitudinal striae are present all over the body and intersect so that the cuticle is divided into blocks or tiles, it is said to be **tessellated**. Apart from striae, longitudinal elevations or ridges may also be present. These may be restricted to the lateral regions, where they form the **lateral fields**, or occur around the circumference of the body. **Alae** are thickened wing-like extensions of cuticle which are often found laterally or sublaterally on the body but may also be localized in the male caudal region, when they form the **bursa**. Other cuticular outgrowths, including transverse or longitudinal flaps, occur at the female genital opening, where they may cover the vulva or stand guard on either side. More elaborate, and often spectacular cuticular ornamentation, such as setae, papillae, tubercles, warts, spines, bands, plates and pores, may also be present. In some plant-parasitic nematodes, such cuticular ornamentations are important diagnostic features, especially in the heavily annulated forms of Criconematidae, where they exist in a bewildering variety of patterns and shapes.

In all nematodes, the lateral body cuticle is thickened and modified to form the **lateral chords**. Externally, each lateral chord may be developed to form the **lateral fields**, one on each side of the body. The lateral fields, which are not developed in all nematode groups, may be used to recognize the exact lateral orientation of the body, although in some species (including males of cyst nematodes and root-knot nematodes), they may spiral around the body, which is itself twisted. In Tylenchomorpha, these are marked by **longitudinal lines** or **incisures** and may be elevated to form **longitudinal ridges** or **bands**. The lateral fields appear to act like a series of pleats and allow the diameter of the nematode to increase during growth, accumulation of food reserves and/or presence of eggs, for example. The ridges may be interrupted by the transverse striae, in which case, the lateral field is described as being **areolated**. Such areolations may be restricted, often to the anterior region of the body, or occur along its full length. The number of longitudinal lines or incisures is of taxonomic importance, but, as their number decreases towards the extremities, particularly so in the anterior region, they should ideally be counted in the mid-body area, where their number is more stable. The number of longitudinal ridges is often confused with the number of lines or incisures, but there is always one ridge less than there are lines so, for example, a nematode with three ridges would have four lines; one ridge, two lines and so on. The lateral field may show fusions of the transverse or longitudinal striae, and these are known as **anastomoses**. Some nematodes lack lateral fields altogether, this being most notable in the Longidoridae and Trichodoridae, but also occurring in Criconematidae juveniles and females and obese females of certain genera (*Heterodera*, *Sphaeronema*), whereas in others it is present as a simple groove (*Belonolaimus*, some *Hemicycliophora* spp.) or as a zig-zag line (some criconematids). Specialized cuticular formations may occur at or around the vulva and anus, a well-known example being the **perineal patterns** of mature female root-knot nematodes.

2.2.3. Labial Region

In nematodes, the basic pattern of the labial region (also known as the cephalic region or 'head') consists of six lips arranged around a central mouth opening or oral aperture (Fig. 2.6). Of these six lips, two are subdorsal, two lateral and two subventral. The basic pattern can be heavily modified, with adjacent lips fusing to form three lips, for example, one being dorsal and the other two ventro-sublateral. In other species, the lateral lips may be reduced or absent and, in extreme cases, all six lips may fuse. In Tylenchomorpha, the anterior end shows an amalgamated, usually hexagonal, labial region and so lip-like differentiations, when present, are better referred to as lip sectors or lip areas. Loof and De Grisse (1974) introduced the term **pseudolips** for the six areas around the oral opening of some Criconematidae. The area around the oral aperture may be differentiated into an **oral disc** and/or a **labial disc**. In many groups of Tylenchomorpha, the cuticle of the labial region is invaginated to form a **prestoma**, with the **stoma** itself being located at a deeper level.

In Tylenchomorpha, the labial region is internally supported by a **cephalic** or **labial framework** or **skeleton**. This may be well developed and heavily cuticularized or weak and difficult to discern under light microscopy. The framework is hexaradiate and typically comprises a series of six finger-like ribs that arch towards the oral aperture, with these ribs being

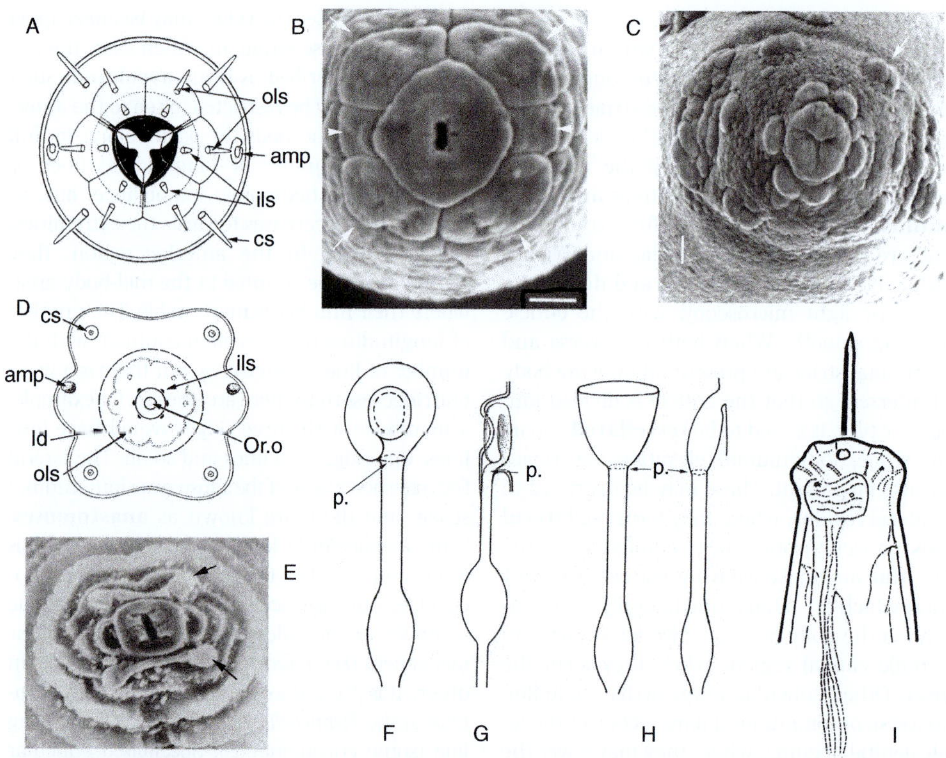

Fig. 2.6. Labial region and anterior sensilla. A: Basic scheme (amp: amphid, cs: cephalic sensillum, ils: inner labial sensillum, ols: outer labial sensillum) (after de Coninck, 1965); B: *Scutellonema* with six lip areas (arrows); *en face* view; C: *Paratrichodorus* in *en face* view with outer labial and cephalic papillae in a single circlet (arrow); D: arrangement in *Aphelenchoides* (amp: amphid, cs: cephalic sensillum, ils: inner labial sensillum, ld: labial disc, ols: outer labial sensillum, Or.o: oral opening); E: *Criconemoides*, *en face* view showing pseudolips (arrows) (after Van den Berg and De Waele, 1989). Amphid structure. F, G: Spiral amphid, lateral and ventral views (p = pore); H: pocket-like amphidial fovea in lateral (left) and ventral (right) view (p = pore); I: *Xiphidorus*, pouch-like amphidial fovea with pore-like opening. (Figure adapted from Decraemer and Hunt (2006), courtesy CABI.)

connected basally by short circumferential extensions, two per finger-like rib, that form the basal framework, a cuticular structure that defines the base of the labial region and differentiates it from the rest of the body. Additional extensions of the finger-like ribs extending into the body proper may also be present, and such 'roots' enhance the strength and rigidity of the labial region. The degree of cuticularization and complexity of form reflects the feeding habit of the nematode (=functional morphology). Those feeding on thin-walled cells, such as fungi or root hairs, have a weak skeleton, as only weak muscles are required to protrude the delicate stylet typical of such nematodes, while nematodes with a robust stylet need more powerful protractor muscles, and hence functional morphology demands that the supporting framework is stronger so as to provide adequate insertion points for the muscles and to prevent the labial region from deforming under compression forces when the muscles contract. The labial region can be **continuous** with the body contour (i.e., the contour of the labial region flows seamlessly into that of the body), or more or less **offset** from the rest of the body – by either a depression or a constriction – or be broader than the adjoining body and, therefore, **expanded**. In Tylenchomorpha, the true height of the labial region is not always easy to establish, as it may be offset at a different level to the basal framework. The labial region itself varies from low and flattened to high and conical in shape – valuable taxonomic characters that should not be overlooked – and may be smooth or bear transverse striae that divide it into annules. Longitudinal striae may also be present in some genera, with the striae dividing one or more annules into blocks.

The labial region bears an array of anterior sense organs, the **sensilla** (singular: **sensillum**), each composed of a neuronal and non-neuronal section formed by two epidermal cells, the socket cell and the sheath cell (Coomans, 1979). In nematodes, there are primitively 12 labial sensilla and four cephalic sensilla arranged in three circlets so that there are six inner labial sensilla, six outer labial sensilla and four cephalic sensilla. This is the well-known 6 + 6 + 4 pattern. Two chemoreceptor sense organs, the **amphids**, are primitively located clearly posterior to the three circlets of anterior sensilla, but in more derived forms, such as in Tylenchomorpha, they have migrated anteriad onto the labial region. Each lip bears an inner and an outer sensillum on its radial axis, with this hexaradial pattern being maintained when lip number is secondarily reduced. The four cephalic sensilla are bilaterally arranged (two laterodorsal, two lateroventral) and represent the anteriormost somatic sensilla. In the plant-parasitic Longidoridae and Trichodoridae, the cephalic sensilla have migrated onto the lip region and are close to the outer labial sensilla, thereby forming a single circlet or 6 + (6 + 4) pattern. In Tylenchomorpha, the anterior sensilla are arranged in three circlets, but, because of the small size of the labial region, the two posterior circlets are located close together (Fig. 2.6).

In general, the six inner labial sensilla protrude from the surrounding cuticle via a terminal pore on top of a **papilla** (plural: **papillae**[*]). In a number of plant- and animal-parasitic nematodes, the inner labial sensilla either have pore-like openings around the oral opening or inside the prestoma (*Pratylenchus*), or lack pores entirely, with the receptors ending blind in the cuticle of the oral disc (*Hemicycliophora*). Inner labial sensilla that open to the environment are chemoreceptive, whereas those that are covered or embedded in the cuticle are mechano-receptive. In most tylenchs, the inner labial sensilla possess two ciliary receptors and have a combined chemo- and mechano-receptive function. In *Longidorus*, four such receptors can be found, but there are only two or three in *Trichodorus*.

The outer labial sensilla may protrude via papillae or **setae**, but in many plant- and animal-parasitic nematodes, they end in simple pores or are embedded in the cuticle, with the cuticle above each termination possibly showing a slight depression. The lateral outer labial sensilla are often reduced, a phenomenon that may be related to the development of the **amphids** (*Meloidogyne*). The cephalic sensilla are submedian in position and usually protrude from the surrounding cuticle as setae or papillae with a terminal pore. In many plant-parasitic nematodes, they are embedded in the cuticle.

The main constituent parts of an amphid are the **aperture**, the **fovea**, the **canalis** and the **fusus** or **sensillar pouch**. In plant-parasitic nematodes, the fovea is an invagination of the cuticle, forming a pocket connected to the exterior through an aperture. The amphidial aperture – typically located laterally but sometimes shifted dorsad – varies in form. For example, in Trichodoridae, it is a postlabial transverse slit, whereas in Longidoridae, it is of variable shape and size, being either a postlabial transverse slit of varying width, or a minute pore. In Tylenchomorpha, the amphid openings have migrated onto the labial region and are usually close to the oral opening, their slit-like to oval openings being greatly reduced in size. The fovea varies in size and shape and may be completely or partially filled with a gelatinous substance (*corpus gelatum*) secreted by the amphidial gland. In Longidoridae, the fovea varies in shape, ranging from stirrup- or goblet-shaped to a pouch, which may be bilobed. In tylenchs, there is no sharp demarcation between the fovea and the canalis, and the sensillar pouch may be located as far posterior as the stylet knobs. The amphids are the largest of the chemoreceptors and possess a much greater number of receptors than other sensilla. In Tylenchomorpha, lip patterns and arrangement of the anterior sensilla and amphids are important diagnostic features in identification and for the analysis of relationships, while in Longidoridae, the position of the amphids and fovea shape are of diagnostic value.

2.2.4. Secretory/Excretory System

The **S-E system** is often referred to as the 'excretory system' on morphological grounds, but physiological evidence is indicative of both a secretory and osmoregulatory function – hence the expanded term now in vogue. The S-E system may be glandular, in which case it generally consists of a **renette cell** or ventral gland connected to a ventral pore by a duct, the terminal part of which is lined by cuticle; or tubulo-glandular, a more complex system consisting of an H-shaped cell. In Tylenchomorpha, the S-E system is of the tubulo-glandular type and is asymmetrical with a single renette cell situated laterally or latero-ventrally and usually posterior to the pharynx. The S-E duct leads to the **S-E pore**, which opens ventrally, usually posterior to the nerve ring, although exceptionally close to the cephalic region or as far posterior as the vulval region (*Tylenchulus*). In the latter case, the renette cell is enlarged and produces a gelatinous matrix into which the eggs are deposited. In Trichodoridae, on the other hand, the S-E system is not developed, and only a ventral pore and a very short duct are present, while in Longidoridae, it is not clear whether an S-E system exists.

2.2.5. Nervous System

The most prominent feature of the central nervous system is the **nerve ring**, which usually encircles the isthmus of the pharynx (=**circumpharyngeal**), rarely the intestine (=**circumintestinal**). It is connected to several ganglia, with longitudinal nerves running anteriorly towards the anterior sensilla in the cephalic region, while posteriorly, four large nerves, the largest being the ventral nerve, run through each of the four epidermal chords. Four smaller nerves, two laterodorsal and two lateroventral, run adjacent. The cell bodies of the anterior sensilla nerves are usually located in six separate ganglia anterior to the nerve ring. The amphidial nerves have an indirect connection with the nerve ring, with their cell bodies being located in the paired lateral ganglia. Of the posterior longitudinal nerves, the lateral nerve contains a few ganglia, including the lumbar ganglia in the tail region. The ventral nerve has a chain of ganglia, the anteriormost being the retrovesicular ganglion, whereas the dorsal and submedian nerves lack posterior ganglia. Posterior to the nerve ring, the longitudinal nerves are connected to each other by commissures. The presence and position of some of these commissures are of taxonomic importance, with the latero-ventral commissure being known as the **hemizonid**. This is readily visible as a refractive, often lenticular, body near the S-E pore. Another commissure, although smaller, lies at a short distance from the hemizonid and is referred to as the **hemizonion**. In the tail region, **caudalids** represent the paired anal-lumbar commissure, which links the preanal ganglion to the lumbar ganglion.

The peripheral nervous system consists of a lattice-like network of nerves that connect setae and papillae. It coordinates impulses from these somatic sense organs, both with each other and with the central nervous system. This system has mainly been described from free-living marine taxa. The innervation of **deirids** (lateral cervical papillae common in Tylenchomorpha and usually found in the vicinity of the nerve ring) is from the lateral nerves of the lateral ganglia. Other somatic sense organs common in Tylenchomorpha (although they may be lacking, as in criconematids) are the **postdeirids** and **phasmids**, which are usually more or less opposite one another and situated on or near the tail, although they may migrate anteriad to varying degrees, as in some hoplolaimids. Phasmids have the basic structure of a sensillum, with the ciliary receptor being in open connection with the environment via a pore. This pore may be provided with a plug, probably secreted by the sheath cell, and which may be small, as in many Tylenchomorpha, or large and shield-like, as in some hoplolaimids, in which case it is referred to as a **scutellum** (plural = **scutellae**). Although the pharynx possesses its own nervous system, it is also connected to the nerve ring.

2.2.6. Digestive System

With the colonization of new habitats and niches during evolution, the wide diversity of available food sources and methods of ingestion has resulted in significant variation in the structure of the nematode digestive system. As a result, the form of the anterior section, the **pharynx** (also commonly, if erroneously, referred to as the **oesophagus** as it is lined by the external cuticle), together with its anterior-most section, the **stoma**, is of fundamental importance in identification, especially at higher levels of classification (Figs 2.7-2.10). The pharynx offers the added bonus that, even under the relatively low power of the stereomicroscope, its structure and form is usually readily visible and, together with stoma form, facilitates a simple and reliable decision process as to whether a nematode is a plant parasite or not and to which major group (or even family) it belongs.

The pharynx is primarily a pumping organ that actively sucks food into the intestine, although in some cases, it is merely a non-muscular tube serving as a conduit for liquid food pushed into the body by, for example, the turgor pressure of the contents of plant cells. The digestive system consists of three regions: the **stomodeum**, **mesenteron** and **proctodeum**. Only the mesenteron or mid-gut is of endodermal origin, the stomodeum is of mixed ecto-mesodermal origin, while the proctodeum and rectum are formed from ectoderm. The stomodeum region is often referred to as the **neck**, **pharyngeal region** or **cervical region**.

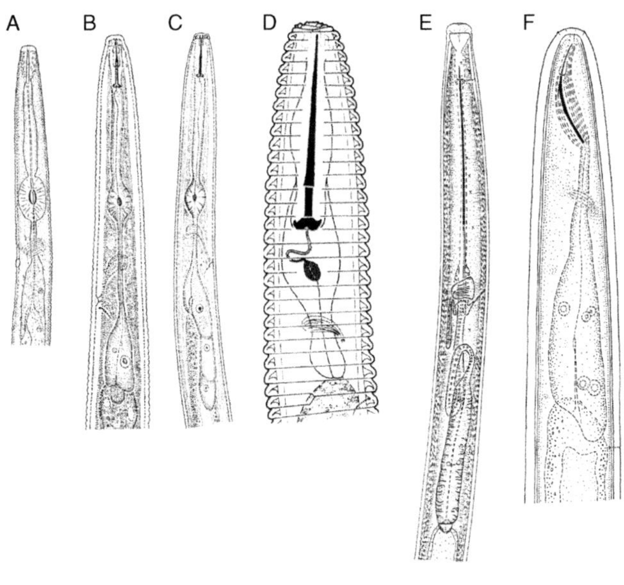

Fig. 2.7. Pharyngeal forms in plant-parasitic nematodes. A: Aphelenchid (*Aphelenchus*); B: Tylenchid (*Tylenchorhynchus*); C: Tylenchid (*Pratylenchus*); D: Tylenchid (*Criconemoides*); E: Longidorid (*Paralongidorus*); F: Trichodorid (*Paratrichodorus*). Drawings not to scale and from various sources.

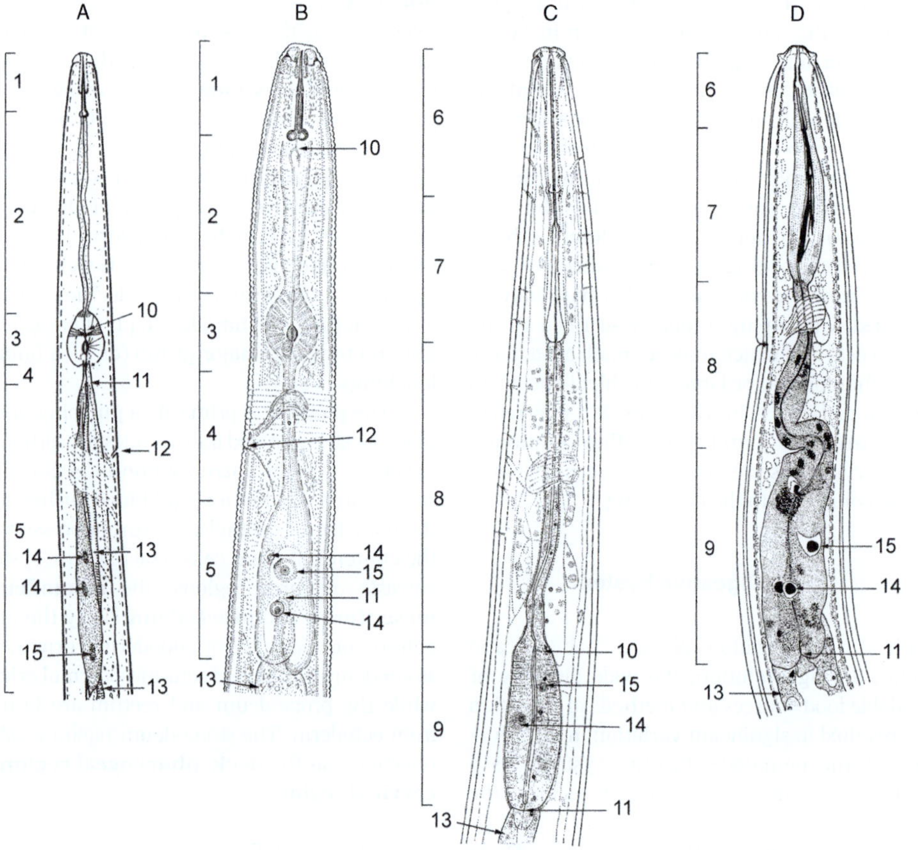

Fig. 2.8. Detailed anatomy of pharyngeal region of plant-parasitic nematode taxa. A: *Aphelenchoides* (Aphelenchoididae); B: *Pratylenchoides* (Merliniinae); C: *Paraxiphidorus* (Longidoridae); D: *Paratrichodorus* (Trichodoridae). Key: 1. stylet; 2. procorpus; 3. metacorpus; 4. isthmus; 5. pharyngeal glands (overlapping intestine in A, abutting intestine in B); 6. cheilostom; 7. pharyngostom; 8. narrow anterior region of pharynx; 9. pharyngeal bulb; 10. dorsal pharyngeal gland orifice; 11. pharyngo-intestinal junction; 12. secretory/excretory pore; 13. intestine; 14. ventro-sublateral pharyngeal gland nuclei; 15. dorsal pharyngeal gland nucleus. A, after Shepherd *et al.* (1980); B, after Siddiqi (1986); C, after Decraemer and De Waele (1981); D, after Decraemer *et al.* (1998). (Figure adapted from Decraemer and Hunt (2006), courtesy CABI.)

Stomodeum

The **stomodeum** comprises the mouth opening or **stoma** (itself formed from the **cheilostom** and **pharyngostom**), **pharynx** and **cardia**. The **oral aperture**, which is usually round or oval in shape, is mostly situated terminally. The cheilostom or lip cavity is hexaradiate or triradiate and is lined by body cuticle. The cuticle lining may be sclerotized to varying degrees and is supported by simple rods. Originally, the cheilostom was delimited by the posterior extent of the lips, but in derived conditions, it may extend further posteriorly, as in Longidoridae, where the cheilostom represents the stoma *sensu stricto*, extending from the oral opening as far posterior as the guiding ring. In Tylenchomorpha, the cheilostom largely lines the cavity of the cephalic framework through which the **spear** or **stylet** moves. The pharyngostom is triradially symmetrical (appearing Y-shaped in apical

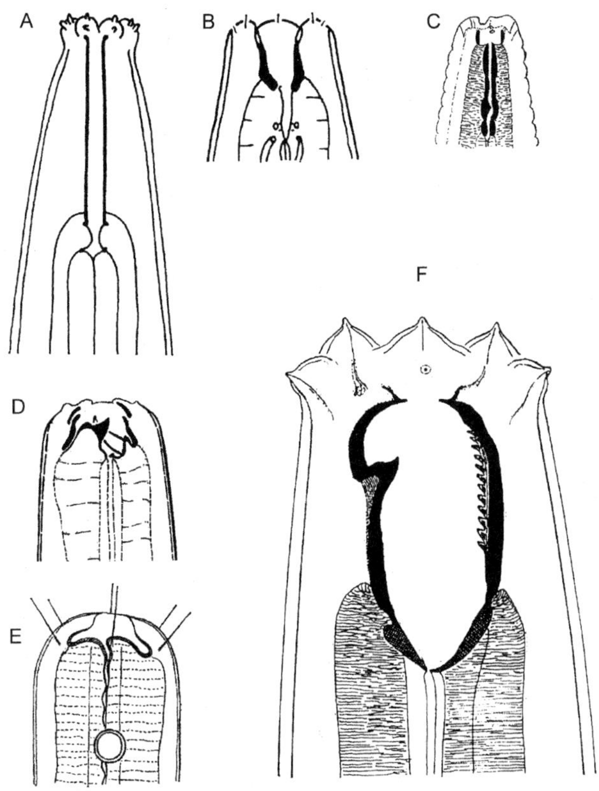

Fig. 2.9. Variation in stoma form – those lacking a mouth spear. A: *Rhabditis*; B: *Panagrolaimus*, C: *Cephalobus*; D: Diplogastrid; E: Monhysterid; F: Mononchid. Drawings not to scale and from various sources. Note that the stoma is usually tubular, cup shaped or barrel shaped and may be simple or armed with teeth and other cuticularizations.

view), as is the pharynx, of which it is a specialized part. Its structure reflects the method of feeding and the food source. The stoma may lack a spear and, depending on feeding habit, be of variable form (Fig. 2.9), but in plant-parasitic nematodes, some entomophilic and some predatory nematodes, the pharyngostom possesses a hollow, axial, protrusible, spear or stylet. In Trichodoridae, the 'spear' is actually a dorsal tooth (the **onchium**), with food flowing around, not through, it as seen in the aforementioned groups (Figs 2.10, 2.11). The cuticularized spear or stylet in the majority of plant-parasitic nematodes is either a hollow **stomatostyle** (Tylenchomorpha) or **odontostyle** (Dorylaimida, Longidoridae) (Figs 2.10, 2.11). The form and size of the spear are important diagnostic characteristics and, together with the form of the pharynx, greatly assist the identification process.

The stomatostyle is formed from parts of the stoma that have become fused. It consists of three parts: a **conus** with a ventral aperture, a more or less cylindrical **shaft** and a posterior region that may enlarge to form three **basal knobs**, the size and shape of which vary depending on functional requirements as they act as attachment points for the stylet protractor muscles (Figs 2.10, 2.11B). Tylenchids have no stylet retractor muscles, with stylet retraction occurring from the resulting tension in the alimentary tract and body pressure once the protractor muscles have relaxed.

The dorylaimid spear, or **odontostylet**, is made up of two parts: an anterior **odontostyle**,

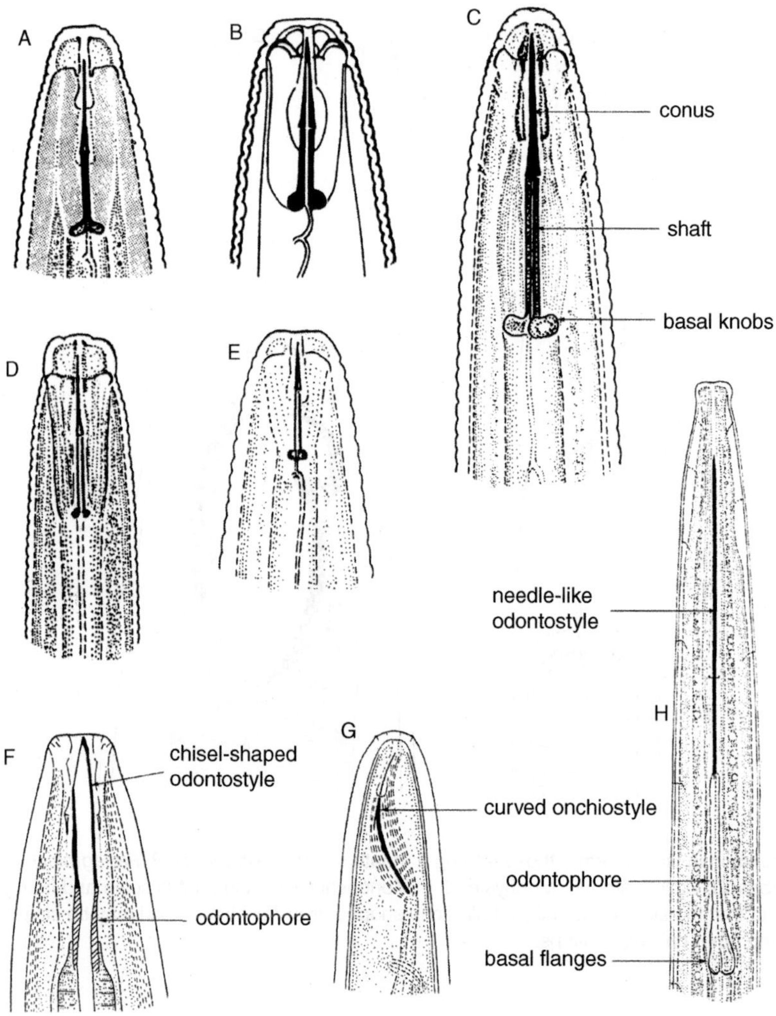

Fig. 2.10. Variation in stoma form – those with a mouth spear. A: *Tylenchorhynchus*; B: *Pratylenchus*; C: *Helicotylenchus*; D: *Aphelenchoides*; E: *Ditylenchus*; F: omnivorous dorylaimid (chisel-shaped odontostyle); G: *Trichodorus* (curved onchiostyle); H: plant-parasitic dorylaim with needle-like odontostyle (*Xiphinema*). Note variations in strength and length of spear, these being related to feeding habit. A-E have a tripartite spear, typically comprising conus, shaft and basal knobs; F-H have a bipartite spear with a basal extension. Drawings not to scale and from various sources.

a hollow, needle-like structure that developed from a single ventral tooth and which has a dorsal aperture anteriorly and a furcate or simple posterior end, and a posterior stylet extension or **odontophore**, a modification of the anterior pharyngeal region, which acts as a supporting structure for the odontostyle and an attachment point for the protractor muscles (Figs 2.10, 2.11A). The base of the odontophore may be enlarged to form three **flanges** for enhanced insertion of the odontostylet protractor muscles, and this development is associated with the ability to feed on woody roots that require greater force to penetrate – more powerful muscles need stronger insertion points. The odontostyle passes through a **guiding ring**, a cuticularized structure that

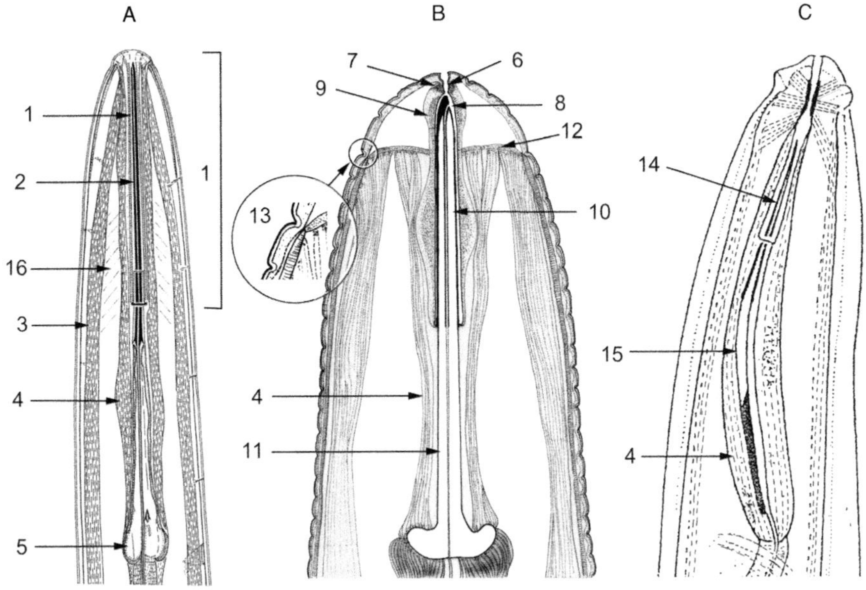

Fig. 2.11. Detailed anatomy of stoma region and types of feeding apparatus in plant-parasitic nematodes. A: Odontostyle and odontophore (Longidoridae); B: stomatostylet (Tylenchomorpha); C: onchiostyle (Trichodoridae). Key: 1. cheilostom; 2. odontostyle; 3. somatic muscles; 4. stylet protractor muscles; 5. odontophore with flanges; 6. Prestoma; 7. thickening of cuticle around prestoma; 8. stylet opening; 9. stoma; 10. stylet conus; 11. stylet shaft and knobs; 12, 13. basal labial framework; 14. onchium; 15. onchiophore; 16. dilatores buccae. A, after Coomans (1985); B, after Endo (1980); C, after Maafi and Decraemer (2002). (Figure adapted from Decraemer and Hunt (2006), courtesy CABI.)

represents the junction between the pharyngeal and the labial cuticles. Longidorids possess eight **odontostyle protractor muscles** (typical of dorylaims), but true retractor muscles are poorly developed or absent. **Pharyngeal retractor muscles** at the level of the narrow part of the pharynx assist in retracting the odontostyle. Because of the extreme elongation of the cheilostom, *Xiphinema* species possess a special set of ***dilatores buccae*** **muscles**, which counteract compression of the cheilostom wall on odontostyle protraction. In Longidoridae, the odontostyle is long, attenuated and needle-like. When it is protruded, the anterior stomatal lining folds along the odontostyle, thereby forming a **guiding sheath**, with the inner cuticular lining being separated from the rest of the wall (except at the level of the guiding ring and at its posterior end) by a fluid-filled cavity, which acts as a hydrostatic skeleton. In *Longidorus*, for example, the hydrostatic tissue forms four compensation sacs to regulate the pressure on odontostyle protraction. The odontostyle is secreted by a cell in the ventro-sublateral sector of the pharynx, whereas the odontophore, together with the guiding sheath and the pharyngeal lining, is derived from pharyngeal tissue.

The ventrally curved **onchiostyle** found in Trichodoridae consists of an anterior pick-like tooth or **onchium**, which has a solid tip, and a posterior support or extension, the **onchiophore**, which is formed by the thickened and sclerotized dorsal lumen wall of the pharyngostom (Figs 2.10, 2.11C). The onchiostyle is protruded by contraction of protractor muscles in the pharyngostom. There are no retractor muscles. As the onchiostyle is solid, ingested food flows around it rather than through a lumen, as in the stomatostyle of the Tylenchomorpha and the odontostyle of the Longidoridae. Thus, the feeding spear in plant-parasitic nematodes has evolved independently on at least three occasions (Tylenchomorpha, Longidoridae, Trichodoridae).

The pharynx, which is lined with cuticle, contains radial muscle fibres and typically has a triradiate or Y-shaped lumen orientated such that there are two subdorsal radii and one ventral radius, that is, the bottom leg of the Y is ventral. In Tylenchomorpha, the pharynx is typically **tripartite**, with the anterior part, the **corpus**, being subdivided into an anterior, muscular, **procorpus**, usually cylindroid in form, and a larger diameter, more robust and muscular **metacorpus** located posteriorly (Figs 2.7, 2.8A, B). The metacorpus of most plant-parasitic tylenchs contains a cuticularized triradiate **valve apparatus**, the lumen of which can be opened by powerful musculature, thereby facilitating a vigorous pumping action. In aphelenchs, the metacorpus is massive, typically filling the body cavity at that point and being markedly offset anteriorly and posteriorly from the remainder of the pharynx. It is quite distinct from that of most tylenchs and serves as a major diagnostic characteristic that is conveniently visible under the stereomicroscope. In most tylenchs, on the other hand, the metacorpus is fusiform, tapering gradually into the anterior and posterior regions of the pharynx and not filling the body cavity. As always, beware that there are exceptions, with the metacorpus of *Hoplolaimus* being singularly robust and more or less offset, while in most criconematids, the procorpus and metacorpus are amalgamated into a massive structure with enlarged valve plates, a typical and very obvious feature for this group. The corpus is followed by a non-muscular **isthmus** (which may be very short or absent) and three **pharyngeal glands** arranged either within a terminal bulb (**postcorpus**) or as a lobe or lobes overlapping the intestine ventrally, dorsally or predominantly laterally. The glands are secretory in nature and are involved in digestion processes. In tylenchs, the **dorsal pharyngeal gland** and its nucleus is most anterior and is followed by the two **subventral gland nuclei**, but in aphelenchs, the subventral nuclei and their glands are anteriormost, with the dorsal gland and nucleus most posterior. Nematodes that produce a lot of salivary secretions have better developed glands, for example, as in J2 root-knot and cyst nematodes, where the secretions stimulate the plant root to form the complex trophic system, which is essential for these sedentary endoparasites to complete their life cycle. The arrangement of the pharyngeal glands is of taxonomic and phylogenetic importance. From the dorsal and subventral glands arise separate ducts, which extend forward and form an **ampulla** before opening into the lumen of the pharynx through cuticularized branched tubes. Within the Tylenchomorpha, the position of the **dorsal gland orifice** is of taxonomic importance and differentiates the Aphelenchoidea, where it is in the metacorpus and located slightly anterior to the valve plates, from the other taxa of the infra-order, where it is in the procorpus and often just a short distance posterior to the stylet base.

In Longidoridae, the pharynx is of rather different form, being bottle-shaped or **bipartite**. The pharynx continues posterior to the pharyngostom as a narrow flexible tube followed by an offset, muscular and glandular bulb that is usually cylindroid in form and may be heavily muscularized. The narrow anterior section may be coiled back on itself, particularly so in forms with a very long odontostylet, with the coiled section extending to allow the spear to be fully protracted (Figs 2.7, 2.8C). The expanded section of the pharynx, or terminal bulb, typically contains five pharyngeal glands, one dorsal and four subventral, and their respective orifi, with the cuticular lining of the bulb lumen being reinforced by six longitudinal platelets.

In Trichodoridae, the pharynx is broadly similar in shape to that of dorylaims and longidorids, except that the narrow anterior section only gradually expands into a largely glandular basal bulb (i.e., it is not distinctly offset from the anterior section), which contains five gland nuclei and associated orifi (Figs 2.7, 2.8D). The **pharyngo-intestinal junction** may be directly abutting or characterized by various types of overlap (a ventral overlap by the pharyngeal glands, a dorsal overlap by the intestine over the pharynx or both types of overlap together).

The pharyngo-intestinal junction or **cardia** (sometimes known as the cardiac valve) exists in many different types. In Tylenchomorpha, it usually consists of two cells of the pharynx (in some cases, it is tricellular) that differ in size and position in relation to the anterior intestine, whereas in Trichodoridae, it is only weakly developed.

Mesenteron

The **intestine** is entirely endodermal in origin. It is a simple, single-layered tube, either cellular or syncytial (criconematids), with or without a clear lumen lined with microvilli. The function of the intestine is mainly one of absorption (with its lumen being bordered with microvilli), storage and secretion of proteins and enzymes. In advanced root parasites, such as *Meloidogyne*, the intestine becomes a storage organ, losing connection with the anus while its thin wall lacks microvilli. Anteriorly, the intestine may be differentiated into a ventricular part that may overlap the pharynx (Trichodoridae). Posteriorly, the intestine may be differentiated into a **prerectum**, itself separated from the intestine by a valve-like structure formed by columnar cells (Longidoridae).

Proctodeum

The **rectum** is of ectodermal origin and is lined with the body cuticle. It is a very simple short tube, which is apparently formed from a few cells and opens to the exterior *via* the **anus**. The anal aperture may be a simple pore or a crescentic slit. The junction between the rectum and the prerectum is controlled by an H-shaped sphincter muscle. Defecation is facilitated by the *dilator ani* muscle. In criconematids, the rectum and anus are poorly developed and are probably non-functional. In *Meloidogyne*, six large **rectal glands** produce a **gelatinous matrix** in which the eggs are deposited. In Mononchida and Dorylaimida, males possess rectal glands consisting of 3-5 pairs of cells located dorso-laterally or laterally from the posterior intestine (prerectum). Caudally or dorso-caudally from the spicules, the ducts of these rectal glands form loops that run anteriorly between the spicules and open via the dorsal wall of the **cloaca** (Coomans and Loof, 1986).

2.2.7. Reproductive System

Nematodes display a range of reproductive strategies and, although the majority of nematodes have separate sexes (**dioecious**) and are **gonochoristic** (individuals are either male or female), bisexual individuals (**hermaphrodites**) occur in some species. Gonochoristic species reproduce by **amphimixis** or cross-fertilization, whereas uniparental reproduction (**autotoky**) takes the form of either **parthenogenesis**, where development occurs through females producing female offspring (i.e., without fertilization), or by self-fertilization in hermaphrodites, where male and female gametes are produced in the same individual (**automixis**).

Female reproductive system

As with other organ systems, the female genital tract is tubular in form and is divided into a number of more or less clear divisions, each with a different function (Fig. 2.12). There are usually two genital tracts, a system which is known as **didelphic** (i.e., having two uteri). If both tracts are opposed, that is, extend in opposite directions, then the system is also said to be **amphidelphic** or **amphididelphic**. There may be a tendency to lose one of the genital tracts, either entirely (**monodelphic**) or partially (**pseudomonodelphic**); in the latter case, the reduced tract is non-functional, although it may still retain recognizable parts, including ovarian cells. The reduced genital tract, which is usually the posterior one, may be atrophied to such an extent that it is entirely absent or represented only by an undifferentiated **post-vulval** or **post-uterine sac**, the length of which varies considerably, being very short, as in Tylenchidae and criconematids, or very long, as in some *Aphelenchoides*, for example. Strictly speaking, the terms ‘post-vulval sac’ and ‘post-uterine sac’, although often used indiscriminately, are not necessarily interchangeable, as in species where the vagina slopes anteriad, the post-uterine sac can actually be pre-vulval, at least in part. In addition, in those species of *Xiphinema* where the anterior ovary regresses, the remnant may be referred to as a **pre-vulval** or **pre-uterine** sac. A genital system with one branch or, rarely, two branches (e.g., as seen in the highly specialized root-knot and cyst nematodes) running anteriorly is described as **prodelphic**, whereas if the branch runs posteriorly (some *Xiphinema* spp.), it is **opisthodelphic**. Combining these terms allows the state of the genital system to be accurately conveyed, the most frequent combinations

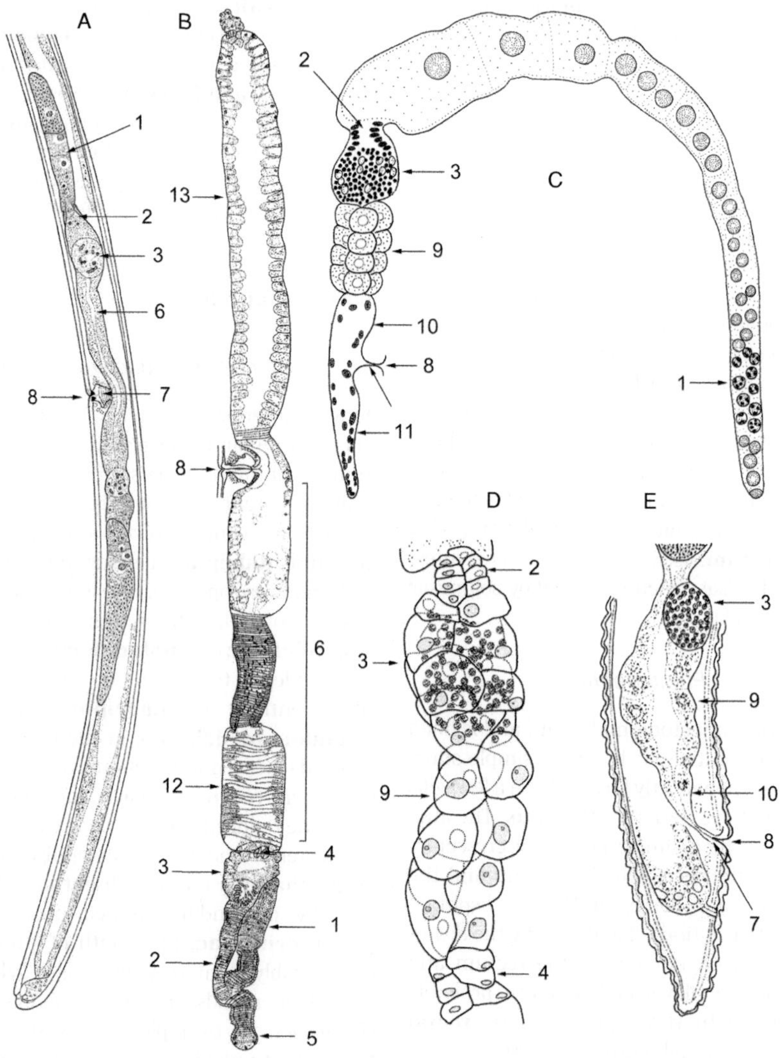

Fig. 2.12. Female reproductive system in plant-parasitic taxa. A: Didelphic-amphidelphic system (*Trichodorus*, Trichodoridae); B: pseudomonodelphic system (*Xiphinema surinamense*, Longidoridae); C-E: monodelphic system. C: Entire female reproductive system of *Pratylenchus coffeae* (Pratylenchidae); D: detail of oviduct-uterus region in *Rotylenchus goodeyi* (Tylenchoidea); E: posterior body region (*Hemicriconemoides minor*, Criconematoidea). Key: 1. ovary; 2. oviduct; 3. spermatheca; 4. sphincter/valve; 5. ovarial sac; 6. uterus; 7. Vagina;, 8. vulva; 9. crustaformeria; 10. uterine sac; 11. post-vulval uterine sac; 12. uterus pars dilatata; 13. reduced anterior branch; 14. sperm. A, after Decraemer (1991); B, after Decraemer *et al.* (1998); C, after Román and Hirschmann (1969); D, after Bert *et al.* (2003); E, after Decraemer and Geraert (1992). (Figure adapted from Decraemer and Hunt (2006), courtesy CABI.)

being **amphi-didelphic**, **mono-prodelphic** and **mono-opisthodelphic**, although in obese nematodes with a terminal or subterminal vulva, the term **di-prodelphic** is applicable. The terms **monogonic** and **digonic** refer, respectively, to the presence of one or two ovaries. The ovary or ovaries may be outstretched, that is, continuing the line of the genital branch away from the vulva; or reflexed, in which case they turn back towards the vulva. The ovary or ovaries are always **outstretched** in Tylenchomorpha and always **reflexed** in Longidoridae and Trichodoridae. Exceptionally, as in Tylenchomorpha species with obese females, the ovaries may show multiple flexures or coiling due to the excessive growth of the genital system in a confined space. A double flexure may also be present, particularly in aphelenchs, with the first being posteriad some distance posterior to the pharynx and the second anteriad so as to resume the directional axis of the remainder of the genital tract. In immature females of *Rotylenchulus* and obese females of *Verutus*, for example, both genital tracts are flexed in this manner. The genital system opens to the exterior *via* a mid-ventral **vulva**. The vulva may be a transverse or longitudinal slit, may be pore-like or show a more oval structure. In many didelphic systems, the vulva is located at about 50% of the entire body length from the anterior end, although it may migrate more anteriad or posteriad in monodelphic conditions, or be virtually subterminal, particularly in those with obese females such as root-knot and cyst nematodes.

Each genital branch consists of an **ovary** (=gonad) and a **gonoduct**. The gonoduct is formed from the **oviduct** and **uterus** and may contain one or two **sphincters** (valves) and a **spermatheca**. The spermatheca, the shape of which can be an important taxonomic characteristic, is either a specialized part of the **oviduct** or the **uterus**. In amphimictic species, it usually contains sperm, thereby confirming to the observer that males are present, even if only female nematodes have been recovered in the sample. The spermatheca may be present or absent, and in parthenogenetic species, it is small and empty.

The **ovary** is usually a tubular structure and can be divided into three main zones. The blind end, or **cap cell**, usually functions as a germinal zone and is followed by the growth zone and ripening zone. In some taxa, the **oocytes** produced by the cap cell are connected to a central protoplasmic core or **rachis**. In species without a rachis, the oocytes are interconnected by protoplasmic bridges. The developing oocytes may be arranged in one or more rows.

The **oviduct** may show great variation in structure and development and is considered to be of fundamental importance in nematode systematics (Geraert, 1983). In Tylenchomorpha, the oviduct is generally formed by two rows of few cells (e.g., three or four cells in Tylenchoidea, Criconematoidea and two cells in Aphelenchoididae). The number of cells corresponds to major taxa and is of diagnostic and phylogenetic importance (Bert *et al.*, 2008). In Trichodoridae, the oviduct consists of two cells, while in Longidoridae, it consists largely of flattened disc-like cells around a collapsed lumen and a wider part that acts as a **spermatheca**. The number of spermatheca cells and their shape and spatial arrangement are of taxonomic importance in the Tylenchomorpha (Bert *et al.*, 2003).

The **uterus** is usually subdivided into sections that may include a glandular, a muscular and a non-muscular portion, although in *Paratrichodorus*, it is a simple tube. In Tylenchomorpha, the uterus *sensu stricto* is restricted to the eggshell-producing region of the gonoduct, the **crustaformeria**, of which the number and arrangement of cells (e.g., **tricolumella** or **quadricolumella** for columns of three and four cells, respectively) are of taxonomic importance. In Longidoridae, the uterus may vary from very short and simple, to very long and complex with local uterine differentiations such as the **Z-organ** or **pseudo Z-organ** (*Xiphinema*). Part of the uterus or uteri opposite the vagina may be differentiated into a muscular **ovejector** to assist with egg laying. Various sets of muscles attach to the vagina, some dilating the lumen while others constrict or suspend it in the body, thereby preventing it from prolapsing during egg laying. This process is also assisted by the vulval dilators, which connect the vulva to the lateral body wall. Eggs may be laid singly or stuck together in masses in a **gelatinous matrix** secreted by the female. Such **egg masses** are associated with species where the females swell and become sedentary, although some obese genera retain all the eggs within the body, with the female cuticle tanning on death to form a tough **cyst**. **Egg masses** and **cysts** serve to protect the vulnerable eggs. Most nematode eggs are morphologically very similar, being ellipsoidal in shape with a transparent shell and of similar size, irrespective of the size of the adult. The developing juvenile is folded several times within the egg.

Male reproductive system

The male genital system may have a single testis (**monorchic**) or two testes (**diorchic**) (Fig. 2.13).

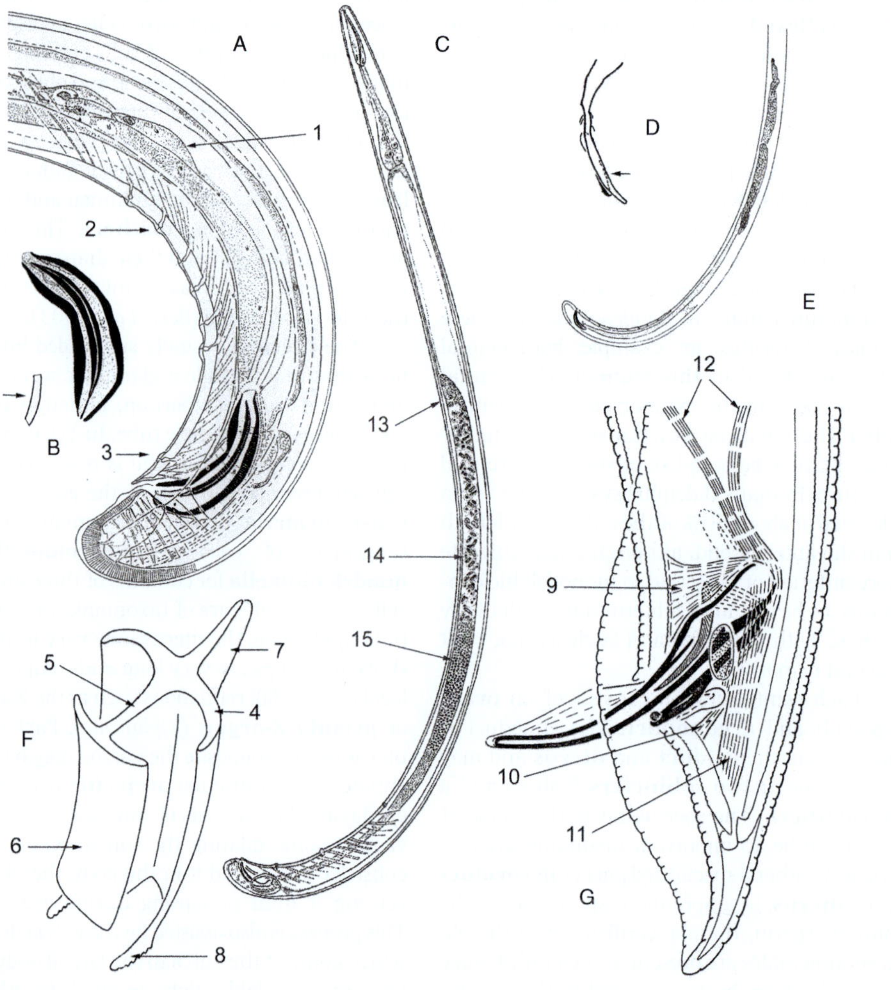

Fig. 2.13. Male reproductive system in plant-parasitic taxa. A: Posterior body region (*Paraxiphidorus*); B: spicule and guiding piece or crura (arrow); C: monorchic system (*Trichodorus*); D: spicule with velum (arrow); E: diorchic system (*Xiphinema*); F: gubernaculum; G: muscles and copulatory apparatus of *Hoplolaimus*. Key: 1. ejaculatory gland; 2. mid-ventral supplement; 3. adanal supplement; 4. corpus; 5. cuneus; 6. capitulum; 7. apophysis or apodeme; 8. crura; 9. anterior spicule protractor; 10. gubernaculum protractor; 11. posterior spicule protractor; 12. spicule retractor muscles; 13. germinal zone of testis; 14. vesicular seminalis; 15. vas deferens. A, B, after Decraemer *et al.* (1998); C, after Decraemer (1991); D, after Luc and Coomans (1992); F, after Maggenti (1980); G, after Coomans (1962). (Figure adapted from Decraemer and Hunt (2006), courtesy CABI.)

Tylenchomorpha and Trichodoridae males are monorchic; those of Longidoridae are diorchic, with the posterior testis being reflexed. *Meloidogyne* males may have two testes due to female to male sex reversal during the development phase. In diorchic males, the testes join proximally and continue as a single posteriorly directed ***vas deferens***. In many monorchic males, the testis is first followed by a storage sac for spermatocytes, the ***vesicula seminalis*** or **seminal vesicle**, which then continues as the *vas deferens*. Sperm are continuously produced in most plant-parasitic taxa, but in criconematids, they are produced before the final moult. The testis then appear degenerate, although well-developed sperm fill the gonoduct. Nematode sperm cells are non-flagellate and non-ciliate and show amoeboid motility. They exhibit considerable diversity in size and shape, often being unusually large in aphelenchs and *Ditylenchus*, for example. In some cases, as in Trichodoridae, sperm form can be used taxonomically.

The rectum and vas deferens open into a shared cavity lined by cuticle, the **cloaca**. The copulatory apparatus is located within this cavity and generally consists of two, equally sized (**isometric**) and similarly shaped (**isomorphic**) cuticularized, tubular structures called **spicules**. These are protrusible and, in some species, may be **dimorphic** or partly fused, reduced to a single spicule or, rarely, absent. Each spicule contains sensilla with one or two dendrites and dendritic process or receptor enclosed in a channel leading to a pore(s) near the spicule tip. In several taxa, the spicules bear ornamentations such as **striae**, **bristles**, a ventral **velum** or subventral **vela** (**flanges**), the presence and form of which are of taxonomic importance. Spicule morphology among aphelenchs, tylenchs, dorylaims and trichodorids is quite distinctive and serves as a useful taxonomic characteristic. The **gubernaculum** is a cuticular thickening of the dorsal wall of the **spicular pouch** and acts as a guide during spicule protrusion. It can range from very simple to complex in structure and may be protrusible. The copulatory apparatus typically functions by means of sets of protractor and retractor muscles. There are protractor and retractor muscles for each spicule as well as protractor and retractor muscles for the gubernaculum. In Trichodoridae, the spicule protractor muscles form a capsule of suspensor muscles, which is not directly attached to the spicules. The spicules are forced out by increased turgor pressure when the fluid-filled capsule is compressed and 'sucked back' in when the muscles relax. Various accessory genital structures – such as the **bursa** (**=caudal alae**), a cuticular extension around the genital area and the tail, **pre-** and **postcloacal supplements** (organs related to reproduction) or **genital papillae** – may also be present in the caudal region. In Rhabditomorpha, **bursal rays**, which have a sensory and strengthening function, are distinctive, being embedded in the bursal wing, often numerous and of great taxonomic importance. A few rays also support the bursa in certain Tylenchomorpha, such as *Aphelenchus*. The arrangement of the pre- and postcloacal supplements in Longidoridae and Trichodoridae or genital papillae in Rhabditomorpha are of taxonomic importance. Paired genital papillae (**hypoptygma**) may also be present on the postcloacal lip, as in Merliniinae (Tylenchomorpha).

2.3. Morphometrics

As a supplement to morphological characters, measurements of various structures and some associated ratios have long been used in nematode diagnostics. De Man (1884) proposed a number of basic ratios that could be used to diagnose various taxa and these, together with later additions, are often referred to as the de Man ratios or indices. Cobb (1920) proposed what became known as the 'Cobb formula' for expressing various morphometric parameters, although this has long fallen from use. Figure 2.14 illustrates how to measure the basic morphometric parameters for a typical plant-parasitic nematode.

As nematodes are often morphologically conserved, measurements and ratios of body parts may be used to supplement nematode identification procedures, although their importance needs to be interpreted with care as some are more variable than others, perhaps because of allometric growth patterns or body-distorting events such as gonad maturity, presence of retained eggs and so on. The commonly used morphometrics are usually referred to by standard abbreviations and those most often utilized are listed below.

It is important to note that structures such as body length, tail length and so on, are measured along their midline, whether curved or not, while parameters such as body width (or diameter) are measured at 90° to the longitudinal body axis. Another important proviso is that stylet length refers to the *actual* length of the structure, that is, from the base of the shaft or knobs to the tip of the conus, not from the stylet base to the top of the labial extremity, as the stylet tip itself may be several microns posterior to this point or, in some cases (e.g., *Xiphinema*, *Longidorus*), protruded substantially beyond.

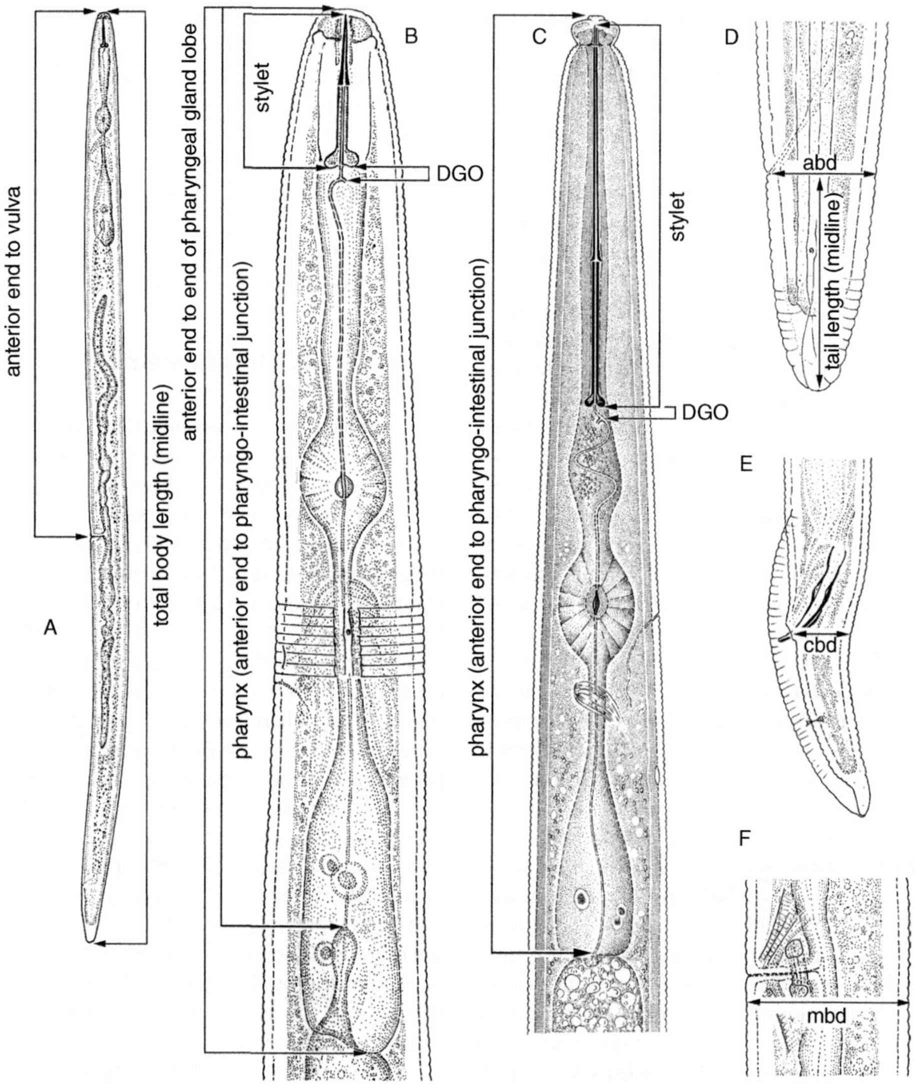

Fig. 2.14. How to take the basic measurements of a nematode. A: Entire nematode showing how to measure total body length along the mid-line from anterior end to tail tip, and distance of vulval aperture from the anterior end, also measured along the mid-line; B, C: pharyngeal region showing how to measure stylet length, distance of dorsal gland orifice (DGO) from basal knobs, pharynx length (anterior end to pharyngo-intestinal junction) and anterior end to tip of pharyngeal gland lobes (this is not the same as the true pharynx length but is usually taken in species where the glands overlap the intestine, the pharyngo-intestinal junction itself often being obscure); D: female tail region showing how to measure tail length and anal body diameter (abd) by extrapolating a line at 90° to the longitudinal axis; E: male tail region showing corresponding measurement for body diameter at the cloacal aperture (cbd). Spicule length is measured from tip to tip along the curved median line; F: mid-body region showing how to take the maximum body diameter (mbd), again at 90° to the longitudinal body axis. (Figure digitally compiled from line drawings by Orton Williams (1974) and Siddiqi (1974), courtesy CABI.)

Spicule length should be measured along the curved median line of the organ, not as a straight line from one end to the other. The latter measurement, sometimes known as the chord, is often taken because it is easier to calculate with an eyepiece micrometre; however, it usually bears little relationship to actual spicule length, particularly when the spicules are strongly curved.

Listed below are some of the most common morphometric abbreviations:

L	=	Total body length (anterior end to tail tip as measured along body axis).
L'	=	Body length, as measured along body axis, from anterior end to anus/cloacal aperture. Used when the tail is very long and where variations in length, whether natural or through breakage, can greatly influence ratio c.
a	=	Total body length divided by maximum body width.
b	=	Total body length divided by pharyngeal length (the pharynx, or oesophagus, is defined as being from the anterior end to the pharyngo-intestinal junction, that is, **not** to the posterior tip of overlapping gland lobes).
b'	=	Total body length divided by distance from anterior end of body to posterior end of pharyngeal glands.
c	=	Total body length divided by tail length.
c'	=	Tail length divided by anal/cloacal aperture body width.
V	=	Position of vulva from anterior end expressed as percentage of body length (note that V is defined as a percentage and so does not require a % sign). Superior figures to left or right of the V value refer to the extent of anterior and/or posterior gonads or uterine sac, also expressed as a percentage of body length.
V'	=	Position of vulva from anterior end expressed as percentage of distance from anterior end to anus.
T	=	Distance between cloacal aperture and anteriormost part of testis expressed as percentage of body length.
m	=	Length of conical part of stylet as percentage of whole stylet length.
O	=	Distance of dorsal pharyngeal gland opening posterior to stylet knobs as percentage of stylet length
MB	=	Distance of median bulb from anterior end expressed as a percentage of total pharynx/oesophagus length.
Caudal ratio A	=	Length of hyaline tail region divided by its proximal width.
Caudal ratio B	=	Length of hyaline tail region divided by its width at a point 5 μm from its terminus.

The following additional parameters are usually used in the taxonomy of criconematids:

R	=	Total number of body annules.
R_{oes}	=	Number of annules from anterior end to base of pharynx/oesophagus.
R_{ex}	=	Number of annules from anterior end to excretory pore.
R_V	=	Number of annules from posterior end to vulva.
R_{Van}	=	Number of annules from vulva to anus.
R_{an}	=	Number of annules from posterior end to anus.

This list is not complete, and other, more specialized, abbreviations may be used for particular genera or families of nematodes, including cyst nematodes (see, e.g., Hooper, 1986; Subbotin *et al.*, 2010). It goes without saying that unusual abbreviations should be defined by the author(s) within the publication in which they appear. Some parameters, depending on the genus studied, are more important than others, and information pertaining to these can be obtained from monographs on the group under consideration.

2.4. Hints for Identifying Nematodes

The most important point when identifying nematodes is experience. With sufficient experience, even the most badly fixed and distorted nematode may rapidly be identified, at least to genus. So, the best way to identify nematodes is to assemble some basic knowledge of morphology and then gain experience by looking at different types of nematodes. Start off with freshly extracted nematodes from soil or plant tissues (the latter point could also help in genus differentiation) that have been heat-relaxed (gentle warming to about 60°C) in a small drop of water on a slide – the features will be so much clearer than after fixation as long as care is taken to avoid overheating or even boiling the animals (in such cases, the body contents will appear blocky or vacuolated, and the organs will be distorted or even wrenched from their attachment points). Get into the habit of assembling pertinent information by looking at a nematode according to a protocol. For example, first ascertain the relative size (length and width, presence of relatively few but very large annules, etc.) and shape of the nematode, supplementing this with how it moves if you are fortunate enough to be looking at live material. Remember that most plant-parasitic nematodes move in a rather slow and sinuous fashion (most – not all, however), whereas bacterial feeders, such as *Rhabditis*, tend to thrash around with rapid flexures of the body. Predators move rather fast and purposefully, but usually with an undulate motion quite different from that of rhabditids. Examine the gut contents. Plant parasites and many other nematodes have greyish intestinal contents, whereas algal-feeding dorylaims and the like may have the gut contents that are tinted greenish or brownish. What shape is the anterior extremity or labial region? Is it low and flattened or barely rounded, high and conical or expanded, for example? Is the labial skeleton very weak (=invisible), or does it show moderate or heavy development? The next step is to see what sort of buccal cavity the nematode has. Ask yourself whether it has a spear or not and, if not, whether the stoma is tubular or cup- or barrel-shaped? Be careful as the tubular stoma of some rhabditids may look like a spear at low power – and some plant parasites actually lack a spear in certain stages (e.g., degenerate males of *Radopholus* and criconematids, J4 females of *Paratylenchus*). If a spear is present, what shape is it? Stomatostyle (comprising conus, shaft and probably basal knobs of varying development), odontostyle (clearly tubular with a chisel-shaped aperture or long, attenuated and needle-like) or onchiostyle (curved and bristle-like)? Couple this information with the type of pharynx. Is it bipartite or tripartite; if tripartite, are there valve plates present in the metacorpus region or are they restricted to the basal bulb (some nematodes, such as pseudodiplogastrids, actually have valve plates in both bulbs)? By this stage, you should have a good idea whether you have a plant parasite or not and to which major group it belongs. If you have a tylenchomorph, check to see if the pharyngeal glands overlap the intestine (and in which direction) or abut the intestine more or less at right angles in the form of a bulb. Gland overlap can often be ascertained using the fact that the pharyngeal tissues are rather transparent whereas those of the intestine are more grey and denser in nature. Hence, if the junction between light (pharyngeal) tissue and darker (intestinal) tissue is at right angles to the body axis, then the glands are non-overlapping, that is, they abut the intestine; if the junction is at an angle, however, there is an overlap, the direction of which can be ascertained by reference to the direction of slope compared to the dorsal or ventral body surface (the excretory pore, vulva and anus or cloaca all open on the ventral surface, which is also usually the inside of the curve in properly heat-relaxed material). Also check the metacorpus – is it very robust, almost filling the body cavity and offset from the rest of the pharynx anteriorly and posteriorly (as in aphelenchs); or is it more fusiform, gradually swelling from the anterior pharynx and then tapering into the isthmus (as in most tylenchs)? Of course, the main diagnostic feature distinguishing aphelenchs from tylenchs is the location of the dorsal pharyngeal gland opening, but this is difficult to see at the best of times, let alone under a stereomicroscope, so take the percentage shot and use the metacorpus character, always remembering, of course, that nematodes invariably have exceptions to the rule. Having got the stoma and pharynx characteristics established, check whether it is a female, male or juvenile by looking for the vulval aperture on the ventral surface or the presence of spicules, probably a bursa too. If female, is the vulva median or displaced anteriorly or posteriorly? How many genital tracts are there (it is often easier to look for and count the ovaries, as the oocytes are closely packed and have denser cytoplasm than the remainder of the genital tract)? Remember that in tylenchomorphs, the ovaries are outstretched (pointed end directed towards either body extremity), whereas in longidorids and trichodorids (also all rhabditids, other dorylaims and many other soil nematodes), they are reflexed back towards the vulva (pointed ends directed towards the vulva). If the specimen is a male, what shape are the spicules – they differ markedly between aphelenchs (except for *Aphelenchus avenae*, which are tylench-like), tylenchs, longidorids and trichodorids. Is there a bursa present? If so, does it extend to the tail tip or is it restricted to the cloacal aperture region? Finally, what is the tail shape and how long is it relative to the width at anus or cloacal aperture? All of these features can often be seen in fresh material under the stereomicroscope, particularly by the experienced observer, or at low power with a compound microscope. Experience and consistent technique are key requirements to interpreting nematode structure in order. Asking the correct questions allows further progress along the identification pathway (see Goodey, 1959). A dichotomous approach is advocated – if it has character A, then this should lead you to a look for the presence of, say, character C or D; if character B is detected, then check for character E or F. With experience, this process becomes so rapid that the user is not even aware of it, and an immediate genus name springs to mind from the briefest examination.

Some general 'pointers' to vermiform nematodes are listed below, but be aware that exceptions abound:

- Spear present in buccal cavity: aphelenchs, tylenchs, dorylaims (including longidorids), trichodorids (although the 'spear' is actually a tooth).
- Spear absent in buccal cavity: rhabditids, araeolaimids, cephalobids, diplogastrids, mononchids.
- Spear with conus, shaft and, usually, knobs or basal swellings: aphelenchs (but *A. avenae* has a smooth base to the shaft), tylenchs.
- Spear tubular with a chisel-shaped aperture: most dorylaims apart from longidorids.
- Spear long, attenuated and needle-like: longidorids, exceptionally some other dorylaims.
- Spear curved and bristle-like: trichodorids.
- Pharynx tripartite: aphelenchs, tylenchs and many other microbivorous soil nematodes.
- Pharynx bipartite: most dorylaims (some are more complex), trichodorids.
- Pharyngeal glands non-overlapping: *Tylenchus* group, most anguinids, *Tylenchorhynchus* group, all criconematids, all longidorids, most trichodorids.
- Pharyngeal glands overlapping: most of the remaining tylenchomorphs, including *Aphelenchoides*, *Aphelenchus*, *Belonolaimus*, *Pratylenchus*, *Helicotylenchus*, *Radopholus*, *Hoplolaimus*, *Meloidogyne*, *Heterodera*, *Globodera*, some anguinids, some trichodorids.
- Vulva posterior, genital tract monoprodelphic: *Tylenchus* group, anguinids, some hoplolaimids (rare), all criconematids, some trichodorids.
- Vulva anterior, genital tract opisthodelphic: some *Xiphinema*.
- Vulva median, monoprodelphic: *Trophurus*.
- Vulva median, amphididelphic: most other nematodes.

2.4.1. Simple key to Major Groups of Soil Nematodes

1. Spear absent 2
Spear present 3
2. Stoma very large, barrel-shaped, armed with teeth and/or denticles .... Mononchida
Stoma not as above .... microbivorous/predatory soil nematodes
3. Pharynx tripartite, spear with conus, shaft and, usually, basal knobs 4
Pharynx bipartite, spear chisel-shaped, needle-like or curved and bristle-like 5

4. Metacorpus robust, filling body cavity, offset .... Aphelenchoidea
Metacorpus fusiform, not filling body cavity .... tylenchs
5. Spear chisel-shaped or needle-like, basal pharynx strongly offset from anterior section .... dorylaims 6
Spear curved, bristle-like .... Trichodoridae
6. Spear needle-like .... Longidoridae
Spear chisel-shaped .... other dorylaims

Note

* Note the singular/plural form: sensillum/sensilla, but papilla/papillae.

2.5. References

Bert, W., Karssen, G., Van Driessche, R. and Geraert, E. (2003) The cellular structure of the female reproductive system within the Heteroderinae and Meloidogyninae (Nematoda). *Nematology* 4, 953–964. DOI: 10.1163/156854102321122575.

Bert, W., Leliaert, F., Vierstraete, A.R., Vanfleteren, J.R. and Borgonie, G. (2008) Molecular phylogeny of the Tylenchina and evolution of the female gonoduct (Nematoda: Rhabditida). *Molecular Phylogenetics and Evolution* 48, 728–744.

Bird, A.F. and Bird, D.J. (1991) *The Structure of Nematodes*. Academic Press, New York, New York.

Chitambar, J.J. and Subbotin, S.A. (2014) *Systematics of the Sheath Nematodes of the Superfamily Hemicycliophoroidea. Nematology Monographs and Perspectives, 10* (Series eds: Hunt, D.J. and Perry, R.N.). Brill, Leiden, Netherlands.

Chitwood, B.G. and Chitwood, M.B. (1959) *An Introduction to Nematology*. Monumental Press, Baltimore, Maryland.

Cobb, N.A. (1920) One hundred new nemas. (Type species of 100 new genera). *Contributions to a Science of Nematology* 9, 217–343.

Coomans, A. (1979) The anterior sensilla of nematodes. *Revue de Nématologie* 2, 259–283.

Coomans, A. and Loof, P.A.A. (1986) Observations on the glands of the male reproductive system in dorylaims and its phylogenetic importance. *Révue de Nematologie* 9, 261–265.

De Coninck, L. (1965). Classe des Nématodes – Généralités. In: Grassé, P. (ed), *Traité de Zoologie*. Masson, Paris, France.

Decraemer, W. (1995) *The Family Trichodoridae: Stubby Root and Virus Vector Nematodes*. Kluwer, Dordrecht, Netherlands.

Decraemer, W. and Hunt, D.J. (2024) Structure and classification. In: Perry, R.N., Moens, M. and Jones, J.T. (eds), *Plant Nematology*, 3rd edn. CABI Publishing, Wallingford, UK, pp. 3–40.

Decraemer, W., Karanastasi, E., Brown, D.J.F. and Backeljau, T. (2003) Review of the ultrastructure of the nematode body cuticle and its phylogenetic interpretation. *Biological Reviews* 78, 465–510.

Eisenback, J.D. and Hunt, D.J. (2009) General morphology. In: Perry, R.N., Moens, M. and Starr, J.L. (eds), *Root-knot Nematodes*. CAB International, Wallingford, UK, pp. 18–54.

Geraert, E. (1983) The use of the female reproductive system in nematode systematics. In: Stone, A.R., Platt, H.M. and Khalil, L.F. (eds) *Concepts in Nematode Systematics*. Academic Press, London, pp. 73–84.

Geraert, E. (2006) *Functional and detailed morphology of the Tylenchida (Nematoda). Nematology Monographs and Perspectives, 4* (Series eds: Hunt, D.J. and Perry, R.N.). Brill, Leiden, Netherlands.

Geraert, E. (2008) *The Tylenchidae of the World. Identification of the Family Tylenchidae (Nematoda)*. Academia Press, Ghent, Belgium.

Geraert, E. (2010) *The Criconematidae of the World. Identification of the Family Criconematidae (Nematoda: Tylenchida)*. Academia Press, Ghent, Belgium.

Geraert, E. (2011) *The Dolichodoridae of the World. Identification of the Family Dolichodoridae (Nematoda: Tylenchida)*. Academia Press, Ghent, Belgium.

Geraert, E. (2013) *The Pratylenchidae of the World. Identification of the Family Pratylenchidae (Nematoda: Tylenchida)*. Academia Press, Ghent, Belgium.

Ghaderi, R., Geraert, E. and Karegar, A. (2016) *The Tylenchulidae of the World*. Academia Press, Ghent, Belgium.

Goodey, J.B. (1959) Data to be considered, observed and, where possible, reported upon when presenting descriptions of new species. *Nematologica* 4, 211–216.

Hooper, D.J. (1986) Drawing and measuring nematodes. In: Southey, J.F (ed), *Laboratory Methods for Work with Plant and Soil Nematodes*, 6th edn. Reference Book 402. HMSO, London, UK, pp. 87–94.

Hooper, D.J. and Evans, K. (1993) Extraction, identification and control of plant parasitic nematodes. In: Evans, K., Trudgill, D.L. and Webster, J.M. (eds), *Plant Parasitic Nematodes in Temperate Agriculture*. CAB International, Wallingford, UK, pp. 1–59.

Hunt, D.J. (1993) *Aphelenchida, Longidoridae and Trichodoridae: their Systematics and Bionomics*. CABI Publishing, Wallingford, UK.

Hunt, D.J. and Palomares-Rius, J.E. (2012) General morphology and morphometrics of plant-parasitic nematodes. In: Manzanilla-López, R.H. and Marbán-Mendoza, N. (eds), *Practical Plant Nematology*. Colegio de Postgraduados and Mundi-Prensa, Biblioteca Básica de Agricultura, Jalisco, Mexico, pp. 25–64.

Hunt, D.J., Luc, M. and Manzanilla-López, R.H. (2003) Morphology. In: Luc. M., Sikora, R.A. and Bridge, J. (eds), *Plant Parasitic Nematodes in Tropical and Subtropical Agriculture*, 2nd edn. CAB International, Wallingford, UK, pp. 11–52.

Hunt, D.J., Palomares-Rius, J.E. and Manzanilla-López, R.H. (2018). Identification, morphology and biology of plant parasitic nematodes. In: Sikora, R.A., Coyne, D., Hallmann, J. and Timper, P. (eds), *Plant Parasitic Nematodes in Subtropical and Tropical Agriculture*, 3rd edn. CAB International, Wallingford, UK, pp. 20–61.

Loof, P.A.A. and De Grisse, A.T. (1974) Taxonomic and nomenclatural observations on the genus *Criconemella* De Grisse & Loof, 1965 *sensu* Luc & Raski, 1981 (Criconematidae). *Mededelingen van de Faculteit Landbouwwetenschappen, Rijksuniversiteit Gent* 54/1, 53–74.

Mai, W.F. and Lyon, H.H. (1975) *Pictorial Key to Genera of Plant-parasitic Nematodes*. Cornell University Press, Ithaca, New York.

Mai, W.F., Mullin, P.G., Lyon, H.H. and Loeffler, K. (1996) *Plant-parasitic Nematodes: a Pictorial Key to Genera*. Cornell University Press, Ithaca, New York.

Maggenti, A.R. (1981) *General Nematology*. Springer, Dordrecht, Netherlands.

de Man, J.G. (1884) *Die frei in der reinen Werde und im süssen Wasser lebenden Nematoden der niederländischen Fauna*. Leiden, Netherlands, pp. 1–206.

Nickle, W.R. (ed) (1991) *Manual of Agricultural Helminthology*. Marcel Dekker, New York, New York.

Peña-Santiago, R. (2020). *Morphology and Bionomics of Dorylaims (Nematoda: Dorylaimida)*. In: *Nematology Monographs and Perspectives, 13* (Series eds: Hunt, D.J. and Perry, R.N.). Brill, Leiden, Netherlands.

Perry, R.N., Moens, M. and Jones, J.T. (eds) (2018) *Cyst Nematodes*. CAB International, Wallingford, UK.

Siddiqi, M.R. (1986) *Tylenchida Parasites of Plants and Insects*. CAB, Farnham, UK.

Siddiqi, M.R. (2000) *Tylenchida Parasites of Plants and Insects*, 2nd edn. CABI Publishing, Wallingford, UK.

Subbotin, S.A., Mundo-Ocampo, M. and Baldwin, J.R. (2010) *Systematics of Cyst Nematodes (Nematoda: Heteroderinae). Nematology Monographs and Perspectives, 8A, 8B*. (Series eds: Hunt, D.J. and Perry, R.N.). Brill, Leiden, Netherlands.

Subbotin, S.A., Palomares-Rius, J.E. and Castillo, P. (2021). *Root-knot Nematodes*. In: *Nematology Monographs and Perspectives, 14* (Series eds: Hunt, D.J. and Perry, R.N.). Brill, Leiden, Netherlands.

Table 2.1. Key to the major groups of soil nematodes together with their main associated morphological characters.

Stylet present	Three-part pharynx	**Metacorpus less than 3/4 body width, dorsal pharyngeal gland outlet in procorpus. TYLENCHS**	Lip region without setae	Female greatly enlarged (pear-shaped or saccate); sedentary in roots	Pear-shaped female	One ovary	Excretory pore near nerve ring – Genera ***Sphaeronema***, ***Trophonema***	
							Excretory pore posterior to nerve ring – Several genera, ***Nacobbus***, ***Tylenchulus***	
						Two ovaries – Genus ***Rotylenchulus***		
					Female shape round with perineal pattern, root galling – Genus ***Meloidogyne***			
					Female becoming a cyst, usually not galling roots	Cyst stage vulva terminal on prominence, anus dorsal. Second-stage juvenile stylet less than 30 µm.- Genera ***Globodera*** or ***Heterodera***		
						Other genera, ***Punctodera, Cryphodera, Atalodera***, etc.		
				Female vermiform	Tail less than six anal body widths long	Procorpus and metacorpus swollen, combined into a large valvular bulb	Mature female with extra cuticle or sheath.- Genera ***Hemicycliophora***, ***Hemicriconemoides***	
							Mature female without extra cuticle or sheath	Ornamentation on cuticle and annuli – Genera ***Criconema***, ***Criconemoides***, ***Ogma***
								No prominent ornamentation – Many genera, ***Tylenchulus***, ***Paratylenchus***, ***Cacoparus***.
						Procorpus and metacorpus clearly separated	Many important plant-parasites: ***Anguina***, ***Ditylenchus***, ***Pratylenchus***, ***Rotylenchus***, ***Radopholus***, ***Hoplolaimus***, etc.	
					Tail equal to or longer than six anal body widths	Female with 2 ovaries.- Genera ***Psilenchus***, ***Brachydorus***		
						Female with 1 ovary – Genera ***Caloosia***, ***Tylodorus***, ***Filenchus***		
				Metacorpus absent or reduced (some form galls on plants) – Genera ***Nothanguina***, ***Anguina***,...				
			Lip region with setae	**No plant parasites** – Genera ***Atylenchus***, ***Eutylenchus***				
		Metacorpus as wide as body width, dorsal pharyngeal gland outlet in metacorpus APHELENCHS	a < 80, vulval flap absent, vagina normal	Tail of female cylindroid with blunt tip, 6-14 incisures in lateral fields, male with bursa and gubernaculum – Genus ***Aphelenchus***				
				Tail of female conoid and mucronate, 2-4 incisures in lateral fields, male without bursa and gubernaculum – Genus ***Aphelenchoides***				
			a > 80, vulval flap present, vagina curved	Genus ***Bursaphelenchus*** **(in part)**				
	Two-part pharynx, spear without swellings	Stylet short, curved, body short and thick, cigar shaped (0.45-1.5 mm long) Family. Trichodoridae. Genera ***Trichodorus***, ***Paratrichodorus***, ***Monotrichodorus***, etc.						
		Sclerotized basal flanges in spear, guiding ring close to basal flanges – Genus ***Xiphinema***						
		Nonsclerotized basal flanges in spear, guiding ring near lip region	Amphid opening minute, pore like – Genus ***Longidorus***					
			Amphid opening wide, slit-like – Genus ***Paralongidorus***					
		Stylet straight usually not very long. Dorylaimoidea, most of them **non plant-parasitic nematodes**						
Stylet absent	**Non-plant-parasitic nematodes**							

*Adapted from Mai and Lyon (1975).

3 Taxonomy and Systematics

Rosa H. Manzanilla-López[1]*, Hugo H. Mejía-Madrid[2] and David J. Hunt[3]
[1]*Centro de Desarrollo de Productos Bióticos, Instituto Politécnico Nacional, Yautepec, Morelos, Mexico;* [2]*Facultad de Ciencias and ENES Juriquilla, Queretaro, UNAM, Mexico;* [3]*CABI, Silwood Park, Buckhurst Road, Ascot, Berkshire, UK*

Abstract

This chapter deals with species concepts and the various approaches to the systematic and taxonomic concepts commonly used in morphological and biological-based systems of classification. Attention is given to the molecular species concept and the molecular tools used in bar codes, genomics and bioinformatics that are used in combination with classic species concepts in the quest for a stable and robust classification system. For reference, the chapter introduces the major groups of plant-parasitic nematodes under the headings Aphelenchoidea, Tylenchina, Dorylaimida and Triplonchida. A simplified systematic classification of plant-parasitic nematode taxa to genus level is provided, this system being used as the *de facto* taxonomic framework throughout the book. A glossary of commonly used nomenclatural terms is appended.

3.1. Introduction

What is a species? A common enough and harmless-sounding question, yet one that evokes much argument, discussion and, often, outright disagreement. Species are, after all, the units of diversity and the foundation stone for systematic schemes. Although pragmatic for the mathematical analysis of ecosystem biodiversity, the term 'unit' implies a rather more assured or encapsulated definition than is actually the case in the real world. So, what does the term 'species' imply? One response to the question is that 'a species is what a competent person says it is'. This sounds trite, yet actually gets straight to the point – a species is more a matter of interpretation and opinion rather than something carved in stone, that is, it is a subjective method of imposing order on variation. Species-defining parameters are not immutable: they change. For example, all root-knot nematodes were once regarded as belonging to a single species and were referred to as *Heterodera marioni*, with the taxon being based more on symptomatology than anything else until Chitwood (1949) revived the old generic name *Meloidogyne* and re-established (albeit somewhat arbitrarily) a number of previously proposed (and poorly delineated) names as being valid. Now there are over 100 nominal species, and the majority of recent proposals are

*E-mail: rosa.manzanilla@gmail.com
A revision of Manzanilla-López, R.H. and Hunt, D.J. (2012) Taxonomy and systematics. In: Manzanilla-López, R.H. and Marbán-Mendoza, N. (eds), *Practical Plant Nematology*. Mundi-Prensa and Biblioteca Básica de Agricultura, Montecillo, Mexico.

(eds R.H. Manzanilla-López, D.J. Hunt, N. Marbán-Mendoza and S. Infante Gil)
DOI: 10.1079/9781836990413.0003

dependent on molecular sequence data rather than the 'classic' techniques of morphology and morphometrics, techniques that have lost importance, even credibility, in the face of the phenotypic plasticity and cryptic speciation now known to be so prevalent in the genus.

The problem in species definition stems from the fact that, as humans, we often try to impose, for our own reasons, a definition that is too rigid for the variation, both immediately evident and cryptic, that exists in nature – organisms are not immutable but exhibit variation over geographic range and change over time, for example. In an attempt to address these difficulties, we have evolved the 'species concept', an idea that itself changes over time, with taxonomic fashion and with the advent of new tools (such as molecular methodologies), powerful and insightful tools no doubt, but ones that also carry within them similar subjective decision-making processes to those that bedevil classical taxonomy.

Given this background there are, therefore, numerous species concepts, each with a staunch band of ardent supporters, yet none can sustain a definitive claim for veracity. A concept is just that – an idea, perhaps probabilistic, but surely not defined with sufficient precision to encapsulate the entire 'identity' of an organism in a populational, rather than a unitary sense. Of course, we need to have 'species', as they form convenient operational taxonomic units (OTU) that enable us to build classification systems by placing apparently similar units together, or to study biodiversity in a given environment – in short, they are the building blocks that we use to try to create (or impose) some sort of order out of seemingly endless and potentially confusing variation.

Given this background, it would seem more plausible, if less comfortable, to view species as amorphous shades of grey, endlessly adjusting tone and dimension, rather than as sharply delineated black or white boxes – smudges, rather than cubes. Notwithstanding this, a named species is ultimately based on the holotype, a single specimen that is unchanging and exhibits no variation. Such are the conflicts inherent in taxonomy.

3.2. Morphological and Biological-based Classification

Taxonomy and systematics tend to be used as synonyms but they are different subjects. **Systematics** is the scientific study of types and diversity of the organisms and their relationships, while **taxonomy** is the theory and study of classification, including its basis, principles, procedures and rules (López Caballero and Pérez Suárez, 1999). The variation found between individuals or groups of specimens has been traditionally assigned by taxonomists and evolutionists to different categories (not necessarily equivalent) that reflect different approaches towards the classification of living organisms. The basic unit for the taxonomist is the species, which is also considered as the unit of diversity. Species identification can be based on morphological characters (e.g., morphological and morphometrical), physiological, ecological, ethological and/or other biological attributes, such as type of reproduction – either sexually (**amphimixis**) or asexually by parthenogenesis (**apomixis**) – and, more recently, on molecular data. Depending on the criteria selected, many different concepts of species can be found in the literature, and the number of species in a specific genus will likely depend on the prevalent concept of species (and/or methods to separate species) applied to the genus by taxonomical authorities (i.e., taxonomists).

Species, taxa and classification schemes are arbitrary categories created by scientists to provide temporary working frameworks to assess similarity, dissimilarity, biodiversity, relatedness and evolutionary patterns of organisms. Taxonomic traits or characters can be defined (or agreed upon) for a species or groups in **phylogenetic** (e.g., common descent based on evolutionary relatedness) or non-phylogenetic terms. Character variation can be assigned to different states of the character or trait (e.g., primitive or derived) that have different weight and taxonomical value related to their uniqueness or commonality, for example, in relation to their diagnostic use in the different taxonomical ranks within a classification system. Above the species level, the main taxonomic categories (**ranks or taxa**) include **genus**, **family**, **class**, **order** and **phylum**. Below the species rank is the **subspecies**. The basic unit for the evolutionist and the geneticist (and, by extension, the nematologist and plant pathologist) is the population or deme (i.e., a local breeding population). The deme has no taxonomic equivalent, but race has a taxonomic equivalent in the subspecies, although races are frequently placed as subordinates of a subspecies.

3.3. Historical Background

Linnaeus was the founder of the modern binomial system of nomenclature (*Systema Naturae*, 1753 for *Species Plantarum* and 1758, 10th edition of *Systema Naturae*). Linnaean species were considered immutable and monotypic (i.e., morphospecies), and species and higher taxa were based on morphological characters as seen in museums, collections and other preserved specimens. In the 19th century, Darwin and Wallace (1859) developed the theory of natural selection and descent with modification, with species being placed at the end of lines of descent (Claridge *et al.*, 1997). The unification of genetics, systematics and evolutionary biology (i.e., new synthesis, neo-Darwinism) transformed the species concepts in the 1930s and 1940s. The biological species concept (BSC), which considers species as actually or potentially interbreeding units that are reproductively isolated from other groups, was formulated by Mayr (1942, 1969), and it has become one of the most common concepts called upon to define or identify species. As in any other species concept, the BSC has limitations in its applicability; for example, it does not apply to asexual or parthenogenetic organisms, in which functional mating and fusion of gametes do not take place. It is difficult to establish boundaries between species based on the mechanism of reproduction, because BSC requires adequate reproductive isolation in order to maintain the integrities of the interacting species populations, and reproductive isolation may not always be completed (Claridge *et al.*, 1997). Besides, reproductive isolation may also occur without associated morphological or other obvious differentiation characters (i.e., **cryptic species**). Another problem is the reproductive isolation status of allopatric populations and hybrids. If reproductive isolation can be built up, it can also break down. In the latter case, two species can gradually merge into one or produce hybrid individuals. Perhaps one of the greatest problems with the BSC is that data demonstrating that a given species is unequivocally reproductively isolated from its congeners is rarely available, that is, the species is usually assumed to be isolated, often just because it looks different. The polytypic nature of biological species has long being recognized, and a series of categories – from superspecies to subspecies – has been used to allocate species variation (Claridge *et al.*, 1997).

Time scale is also an important factor to be considered in species concepts (i.e., evolutionary species). Although species are evolving lineages, having both retrospective and prospective trajectories, at any point in time, a species is determined by what has happened to it in the past, since reference to future evolutionary trends cannot be made – lineages can only be recognized in retrospect (Hull, 1997).

Some of the difficulties mentioned above have led to rejection of the BSC in favour of a more general phylogenetic species concept that defines species as the 'smallest aggregation of populations (sexual) or lineages (asexual) diagnosable by a unique combination of character states in comparable individuals' (Nixon and Wheeler, 1990 cited by Claridge *et al.*, 1997), but one of the phylogenetic species concept disadvantage lies in its application to sibling species complexes where BSC can be considered to be a more appropriate approach to reveal them (Claridge *et al.*, 1997).

Until cladistic or phylogenetic systematics was proposed by Hennig (1950, 1966), phylogenetic reconstructions were established *via* the so-called 'threefold parallelism' (López Caballero and Pérez Suárez, 1999) integrated by comparative anatomy, **ontogeny** (history of development and growth of an individual) and palaeontology (fossil record). These were the three criteria used to establish similarity or relatedness between organisms. From the evolutionary criteria, similarity can be classified into **homology** (traits inherited through a common ancestor) and **homoplasy** (traits not inherited through a common ancestor). Homoplasy can be subdivided into convergence and reversion. **Convergence** refers to the independent origin of a similar trait present in two or more unrelated species, while parallelism refers to the development of a similar trait in distinct species that have a common ancestry, and **reversion** refers to the new occurrence of the ancestral trait (or character).

According to Hennig's scheme, all taxa must be strictly monophyletic (pertaining to a single phylum or other taxon of organism). In this scheme, monophyly does not apply to species (Hull, 1997). Monophyletic groups are characterized by **synapomorphies** (i.e., shared derived characters) while species are characterized by **autapomorphies** (unique characters not shared with the sister group). Monophyly can be extended to the species level, for example, when a series of populations can exhibit a tree-line structure (i.e., monophyletic species concept). The monophyletic species concept is synchronous with respect to its grouping criterion because monophyly, in the sense of sister group relationship, has no temporal dimension (Hull, 1997).

The **diagnostic species concept** differs from the monophyletic species concept in that the diagnostic characters that define species need not be autapomorphies but may be any unique combination of primitive and derived characters (Hull, 1997).

The **phenetic species concept** is considered by Hull (1997) as the most general concept because males and females or their progeny may be placed differently when applying the OTU approach. At the other extreme, the biological and mate recognition concepts are the least general because they apply only to species that reproduce sexually. Hull (1997) also considered the diagnostic species concept as being the closest to the most operational and most applicable concept. The monophyletic species concept is as operational as the methods of cladistic analysis allow.

At present, two of the most used approaches for phylogenetic reconstruction include cladism and phenetics (and, more recently, cladistic and phylogenetic analysis of molecular and morphological data (see Chapter 6, this volume)). Cladism aims to distinguish derived homologies (**apomorphies**) from the primitive characters (**plesiomorphies**) using only the first as valid characters for grouping the taxa. Cladistic analysis includes different steps such as: taxa selection, building of a matrix of characters, search for the most parsimonious trees and evaluation of phylogenetic hypotheses. **Parsimony** is a criterion used to select the alternative phylogenetic hypothesis, and under this criterion, the preferred hypothesis minimizes the number of additional transformations of the characters (López Caballero and Pérez Suárez, 1999). The goal of cladistic analysis is the individuation of characters so that they nest perfectly (Hull, 1997). Phenetics (also known as numerical taxonomy) was developed as a consequence of the increasing availability of computers in the 1950s. Character data are analysed through one or several computer programmes that will group the **OTU** (Sokal and Sneath, 1963) according to their similarity or difference indexes to produce a dendrogram (also called **phenogram**). The latter can be transformed into a classification, but transformation can be considered as subjective in the absence of a clear guide to define what level of similarity is enough to establish the different levels in the classification. A general outline of cladistics, phenetics and evolutionary methods can be found in López Caballero and Pérez Suárez (1999).

3.4. Molecular Species

Molecular tools are being increasingly incorporated into classification studies, as shown by the growing information on nucleotide sequences of small subunit ribosomal DNA (18S rDNA) sequences, large subunit (LSU, 28S) rDNA, internally transcribed spacer (ITS) regions and mitochondrial genes from nematodes (Powers, 2004; Holterman *et al.*, 2009; see also Chapters 5 and 6, this volume). Eukaryotic rDNA is composed of tandem repeating units coding for 18S, 5.8S and 28S rRNA, separated by an intergenic spacer (IGS). The DNA sequences of the rDNA encoding 18S, 5.8S and 28S genes are more conserved than the ITS or IGS (Wendt *et al.*, 1993). The division of the DNA cistron into domains that evolve at different rates has allowed the use of this region in diagnostic and evolutionary studies of nematode species (Cherry *et al.*, 1997; Nguyen *et al.*, 2001). Most phylogenetic studies have utilized 18S and 28S and mitochondrial DNA, D2-D3 expansion region of LSU and at least four other genes (Adams *et al.*, 2009). Although molecular identification tools and research have advanced (Bogale *et al.*, 2020), the validity and practicalities of molecular approaches to taxonomy have been subjected to a variety of criticisms (Vogler and Monaghan, 2007), and there are examples in the literature that show the existing conflict in accepting molecular results even though a rigorous molecular method will allow 'molecular species' (or specific sequence recognition) of nematodes to be separated in a fast and replicable way. Nowadays, it is strongly recommended that classic new species description, along with morphometrics, drawings, pictures and written description, be accompanied (and supported) by not only molecular analyses, traditional **type** and **paratype** slides, but also a **DNA voucher** (e.g., whole organism or DNA samples). Furthermore, a set of multiple loci has been suggested for molecular phylogenetic delimitation of species in order to avoid tokogenetic (i.e., reproductive events) 'pollution' (Nadler, 2002) or an extension of the methodological approach offered by molecular biologists (Adams, 2001), where finding heritable characters, identification of fixed characters (conserved molecular traits), polarization of characters and identification of autapomorphies are suggested. In many nematode groups (including cyst and root-knot nematodes), traditional descriptive practices are becoming inadequate for the task because of the sheer number of nominal species, their relatively conserved morphology and an inadequate appreciation or knowledge of the variation component generated by the taxon. Supporting molecular data are, therefore, increasingly becoming an essential requirement when proposing a new species. Indeed, some journals will no longer publish new taxa unless they are accompanied by molecular data. That is not say that molecular methodologies will overcome all difficulties, as decisions as to which gene or genes to sequence and the significance of base pair differences remain subjective. The enhanced precision and power of molecular study, particularly in so fundamental an area, has much to commend it, however, and also opens the door to identifying all stages of an organism, regardless of the phenotype.

3.5. Bar Codes

The potential use of a sequence or sequences of DNA to identify all life forms on earth gave origin to the 'DNA barcode' concept in the early 2000s. The molecular barcode derived from single-specimen PCR and sequence of the 5′ segment of the small subunit (SSU) of the ribosomal RNA gene was used subsequently to develop a **molecular taxonomic unit** (MOTU) scheme for soil nematodes to assess biodiversity (Floyd *et al.*, 2002). The nucleotide sequence from a common gene may serve as a unique identifier for every species, and a MOTU could be assigned to named nematode species using a database of SSU sequences. However, among other factors, the low numbers of DNA nematode sequences available in gene banks that are accompanied by a binomial name of a known, taxonomically identified species, for either soil nematodes or plant-parasitic nematode species, have limited the large-scale use of this and other **metagenomic** (e.g., analysis of DNA biodiversity at any community level) approaches (see Chapter 5, this volume). It has also brought into account not only the lack of economic resources in the taxonomy of nematodes but also the need for a joint species identification approach (i.e., classic and molecular) and the problems raised by the relatively few qualified taxonomists available worldwide. It is expected that pyrosequencing platforms like the 454® (Roche) or the Illumina® (Solexa), and more recently portable third-generation sequencers (Knot *et al.*, 2020), will keep providing a robust, parallel approach for amplifying and sequencing DNA templates, the development of high-throughput platforms such as microchips, microarrays and nanochips (Finetti-Sialer and Manzanilla-López, 2011; Bogale *et al.*, 2020). A standard DNA taxonomic analysis will include broad sampling of the target taxa across their geographic range, followed by large-scale sequencing of representative samples for a DNA profile (Vogler and Monaghan, 2007), and ideally, those samples could also be linked to a nematode species name in the gene banks.

3.6. Genomics and Bioinformatics

In general terms, genomics aims to study a living organism's genome, taking into account its primary structure to understand its organization and gene functions as well as trying to establish its phylogenetic relationships with other organisms (see Chapter 6, this volume). It is encouraging that genome sequencing of plant-parasitic nematode species has accelerated thanks to new generation sequencing (NGS) technology, and many species have been sequenced, including those of major economic importance, for example, *Aphelenchoides*, *Ditylenchus*, *Heterodera*, *Globodera*, *Meloidogyne*, *Radopholus similis* and *Rotylenchulus reniformis*, among others (Xhuge *et al.*, 2023).

The availability of bioinformatics expertise and fully sequenced genomes of nematode species is opening new possibilities for the exploration of ontogeny (and phylogeny) at molecular levels by allowing the identification and comparison of genes separated by speciation (**orthologs**) or gene duplications (**paralogs**) of genes within the same lineage, their function and their switch on and off process that occur through different stages of development of the nematode.

3.7. Species Concepts in Nematology

As mentioned above, one of the most advocated species concepts in nematology is the BSC. According to this concept, it is necessary to have demonstrated that there is no reproductive strategy other than amphimixis for a taxon at the species level. Reproductive strategies (amphimixis and apomixis) are important to assess genetic variation within both populations and species of nematodes (Castagnone-Sereno, 2006; Chitwood and Perry, 2009). The BSC has been advocated as a means of refuting molecular species, but the fact remains that the vast majority of **nominal species** have been described (and almost always regarded as valid) without any cross-mating observations whatsoever – the fundamental requirement of the BSC for reproductive isolation is thus conveniently assumed, not proven. In cases where taxa usually accepted as valid are subsequently found to be capable of interbreeding, other criteria, such as geographic isolation, need for specific pollinators and so on, are then invoked as a work around, but perhaps certain groups of organisms are just less fussy about swapping DNA than we like to think.

Bickford *et al.* (2006), in a more pragmatic approach, have pointed out: 'In the interest of creating a species concept that can be applied unambiguously to molecular investigations of species limits, several authors have coined phylogenetic or genealogical species concepts that define species in terms of **monophyly** (i.e., *derived from a single common ancestor*) without explicit regard to reproduction barriers, thus side-stepping the difficulties associated with diagnosing reproductively isolated species'. Although molecular methods have been shown to facilitate species separation (or specific sequence recognition) in a rapid and repeatable manner, their acceptance is not universal, as they are not easily translated into a taxonomical trait that can be identified, measured and assigned to a recognizable **binomen** (i.e., genus and species names) within a broadly known and formal operational reference framework (e.g., systematics) and eventually to a phylogeny. This dilemma should not, however, be allowed to inhibit the potential for new approaches to help delineate species (Nadler, 2002; Manzanilla-López, 2010). It is important to bear in mind that, while Linnaean nomenclature provides evidence for the level of the taxonomic hierarchy, it does not follow that DNA-based entities are only valid if they correspond precisely to the existing species names (Vogler and Monaghan, 2007).

3.8. Classification System Stability

Siddiqi (1986) pointed out that 'The task of a taxonomist is not only to identify and describe new and known genera and species but also to try to explain how these taxa are related to each other (i.e., **phylogenetics**) and how they fit into a (**natural**) system of classification'. We must bear in mind that there are several taxonomic approaches based on different philosophical and methodological frameworks to assess relationships between taxa, and that their validation or synonymy can be subject to new ideas, informatics methodology and the interpretation of character states. According to Siddiqi (1986), classification systems may become unstable, among other reasons, because most systems lack predictability, and the characters used to specify groups will likely not remain congruent and parsimonious when new taxa and characters are discovered. There is also the conflict between 'lumpers' and 'splitters', with the first group focusing more on similarities between taxa (thus leading to a deflationary scheme in terms of the number of taxa at each rank) and the latter deriving their impetus by emphasizing differences, thereby producing a more inflationary scheme (Siddiqi, 1986).

According to Siddiqi (1986), the phylogenetic histories of Tylenchida and other orders of the Nematoda can never be correctly stated because of a lack of fossil records, and so the evolutionary pathways of these animals can only be inferred. Therefore, it is important to determine as well as possible the homologies and character states in order to indicate the possible origins and evolutionary line of the taxa concerned. Biochemical, cytogenetic, embryological, ecological and molecular taxonomy data should all be included in this exercise so as to understand the diversity of plant and soil nematodes, with the aim being to produce a more meaningful and stable classification. So far, molecular technologies have shown that an increasing number of major species of plant-parasitic nematodes are actually species complexes (e.g., Reid *et al.*, 2003; He *et al.*, 2005; Tandingan De Ley, *et al.*, 2007; Subbotin and Ryss, 2024). Determining species boundaries within such complexes requires further studies using not one but several molecular approaches and, in some cases, full genome sequencing (Finetti-Sialer and Manzanilla-López, 2011).

3.9. Outline Classification of Plant-parasitic Nematodes

Until relatively recently, morphological features were primarily used to identify and place species in different taxonomic categories (taxa), although molecular methodologies are now of increasing importance (see Holterman *et al.*, 2006, 2009; van Megen *et al.*, 2009; Ahmed and Holovachov, 2021; Qing *et al.*, 2025). It is clear, however, that delineation of nematode taxa using only some molecular characters is not always congruent with their genetic and phylogenetic (common ancestry) affiliation (Waite *et al.*, 2003).

A classification of the phylum based primarily on molecular phylogeny has been proposed (see Blaxter *et al.*, 1998; De Ley and Blaxter, 2002, 2004) in which the classes **Chromadorea** and **Enoplea** replace the former **Secernentea** and **Adenophorea**. Following this classification scheme, the order **Dorylaimida** (=subclass **Dorylaimia** in the molecular-based scheme) contains mostly omnivorous species but also plant nematodes that are vectors of viruses, and the suborder **Tylenchina** (=subclass **Chromadoria**) contains the majority of the plant-parasitic nematodes, including the most highly specialized parasites.

3.9.1. Aphelenchoidea

Aphelenchs are predominantly free-living and mycetophagous in habit, although many are insect associates, some are insect parasites and some even parasitize both plants and insects. They are typically slender, vermiform nematodes ranging 0.2-2 mm in length. They are frequently found in habitats, such as soil and decaying plant material, lichens, and in the tunnels of a wide variety of wood-boring beetles, particularly from the Scolytidae and Cerambycidae (Hunt, 1993). The most abundant and widespread species of aphelench belong to the genera *Aphelenchus* and *Aphelenchoides*. A few species of *Aphelenchoides* have evolved to parasitize higher plants, and such species are commonly known as the bud and leaf nematodes. Other important plant pathogens are found in the genus *Bursaphelenchus*, namely *B. cocophilus* and *B. xylophilus*.

The aphelenchs are relatively conserved in their morphology, with systematic schemes such as those by Hunt (1993, 2008) emphasizing differences in biology and life style as much as morphology, whereas Kanzaki and Giblin-Davis (2012) adopted a more molecular-based approach. The increasing availability of molecular data from this under-studied group is beginning to paint a clearer picture and, while largely confirming the existing major groupings such as families and subfamilies, has undermined the supposed monophyly of 'morphological genera' such as *Aphelenchoides*, *Laimaphelenchus* and *Schistonchus*. There is still much to resolve, and research is greatly hindered by the plethora of older descriptions of often poorly characterized species, including the types of important genera, including *Aphelenchoides*. A more detailed discussion of the current status of this group is presented in Chapter 7, although a root and branch revision may be required in future.

3.9.2. Tylenchina

According to the monograph by Siddiqi (1986), this taxon (which he regarded as an order – the Tylenchida) comprised 216 valid genera and 2200 valid species. In the second edition of his *magnum opus*, Siddiqi (2000) recognized 233 valid genera and 3404 valid species, a substantial increase. Andrássy (2007), however, took a more conservative approach, regarding only 211 genera as valid with a total of 2876 valid species. Opinions as to what is, or is not, a valid taxon, therefore, differ widely, their status often being subject to the individual judgement of the taxonomist compiling the list.

The Tylenchina includes representatives of all types of plant parasitism. Extreme forms of plant parasitism in the group can be represented by members of Tylenchidae feeding ectoparasitically on algae, lichen, moss and on the surface of roots (Siddiqi, 2000) and the highly evolved semi-endoparasitic, cyst-forming Heteroderidae (e.g., *Heterodera, Globodera*), species that elicit complex feeding sites in the host plant and are considered to be the pinnacle of plant parasitism.

Molecular studies on tylenchs have been quite extensive. As with the aphelenchs, molecular systems are not always congruent with the classical schemes that are primarily based on morphology, a particular example being the fact that *Radopholus* (and other genera in the Radopholinae) clade with hoplolaims rather than with the pratylenchids, their historical bedfellows.

3.9.3. Dorylaimida

This large and diverse order includes relatively few plant-parasitic nematodes, prominent among them being members of the Longidoridae. Longidorids are long to very long nematodes (1.5 to >12 mm). They can tolerate both very dry soils and waterlogged mud and are polyphagous ectoparasites of plant roots. Members of the *Longidorus* and *Paralongidorus* are often associated with softer-rooted herbaceous plants, whereas *Xiphinema* includes more robustly built species equipped with much more heavily cuticularized spears and are hence better adapted to feed deep within the tougher roots of woody plants. *Paralongidorus (sensu lato)* is known mainly from Africa, Asia, Australasia and Europe, whereas *Xiphinema* is more widespread. A number of genera are of more restricted distribution; for example, *Xiphidorus* is exclusive to South America (Hunt, 1993). Several species are known to transmit plant viruses.

Molecular techniques are increasingly being employed and probably represent the only reliable approach to the *Xiphinema americanum* group, a large and complex assemblage of nominal species so adjacent in their morphology and morphometrics that almost every new population seems to straddle several 'species' boundaries. Speciation within the *americanum* group is not just an esoteric exercise, as several species are known to transmit plant viruses (see Chapter 15). Molecular studies also indicate that the *X. americanum* group forms a distinct clade from the remaining species of the genus and, if this clade is regarded as having generic status, all other *Xiphinema* species will need to be relocated to another genus/genera.

3.9.4. Triplonchida

This order includes the family Trichodoridae. The Trichodoridae currently comprise six generally recognized genera. The two largest genera – *Trichodorus*, with over 50 species, and *Paratrichodorus*, with over 30 species – now occur worldwide, although the latter was probably introduced to South America by human means. *Allotrichodorus, Ecuadorus* and *Monotrichodorus*, on the other hand, have fewer species and have been recorded only from South and Central America (Decraemer, 1995). Trichodorid species are ectoparasites that usually aggregate near the root tip in the zone of elongation. They pierce plant cells with a long spear-like tooth (the onchiostyle), which is used like a pick to penetrate cell walls, although, unlike the spear in other groups of plant parasites, no food passes through its structure, as the tip is solid and not hollow. They mainly feed on epidermal cells and root hairs and prefer the meristematic tissues near the tip of the growing root. A number of species are known to transmit plant viruses.

As with the longidorids, molecular studies have also had an impact and resulted in the recognition of *Nanidorus* as a valid taxon rather than being regarded as a junior synonym of *Paratrichodorus* (see Duarte *et al.*, 2010; Kumari and Subbotin, 2012). Evidence for species complexes has also been found.

3.10. A Simplified Classification of Plant-parasitic Nematodes

Nematode classification, for long based largely on morphological characters, is undergoing a revolution as molecular data become increasingly pertinent in establishing phylogenetic relationships. For this reason, the higher classification is particularly fluid with, for example, proposals for infraorders and so on, and the bringing together of groups that, previously, under classical systematic systems, had been assumed to be but distantly related (De Ley and Blaxter, 2002, 2004). For plant-parasitic nematodes, the following schemes, some based purely on 'classical' morphology and others genuflexing to molecular data, include those of Siddiqi (2000), Andrássy (2007) and Decraemer and Hunt (2024) for the Tylenchina; Hunt (1993, 2008), Andrássy (2007) and Kanzaki and Giblin-Davis (2012; Chapter 7, this volume) for the Aphelenchoidea; Hunt (1993), Andrássy (2009) and Decraemer and Hunt (2024) for the Longidoridae; and Hunt (1993), Decraemer (1995), Andrássy (2009) and Decraemer and Hunt (2024) for the Trichodoridae. Other pertinent monographs dealing in part or entirely with plant-parasitic forms include, among others, Geraert (2008, 2010, 2011, 2013), Subbotin *et al.* (2010a, b; 2021), Chitambar and Subbotin (2014) and Peña-Santiago (2021). Hodda (2022a, b) provides exhaustive overviews. Following on from De Ley and Blaxter (2002, 2004), an excellent molecular overview of the Nematoda was published by van Megen *et al.* (2009). Pertinent later works include those by Ahmed and Holovachov (2021) and Qing *et al.* (2025).

It is useful to note that the various systematic ranks have different suffixes attached to the stem, thus enabling the appropriate taxonomic category to be identified without further ado:

Class	-ea
Subclass	-ia
Order	-ida
Infraorder	-omorpha
Suborder	-ina
Superfamily	-oidea
Family	-idae
Subfamily	-inae
Tribe	-ini

As an example, take the genus *Tylenchus*. The stem is *tylench-* and the root or ending is *-us*. Using the stem and combining it with the appropriate root listed above, we therefore have, in ascending order, the following potential ranks: Tylenchini, Tylenchinae, Tylenchidae, Tylenchoidea, Tylenchina, Tylenchomorpha, Tylenchida and so on. Each rank is based on the type genus, in this case, *Tylenchus*, with the genus itself having a type species that, in turn, ultimately defines the genus. In the case of *Tylenchus*, the name (or binomen, as it is a combination of a generic and a specific epithet) of the type species is *Tylenchus davainei* Bastian, 1865 (the author and date of proposal constitute the authority for the binomen), a species recovered from a mossy boulder in a stream near Falmouth, England. This taxon, therefore, serves as the type species for all the ranks listed above, including the suborder and order. The fact that the identity of the species so simply figured by Bastian is doubtful (it was probably a *Ditylenchus*) is another matter and one that has been conveniently 'tidied-up' by judicious application of the rules of nomenclature.

The following scheme is modified and updated from that of Decraemer and Hunt (2024), itself based on the higher categories proposed by De Ley and Blaxter (2004) and populated at the generic level from a blend of various authors, including Siddiqi (2000), Hunt (1993, 2008), Decraemer (1995), Andrássy (2007, 2009) and Subbotin *et al.* (2010a, b). The genera are arranged in alphabetical order under their immediate higher rank. Major plant-parasitic genera are indicated by **bold** font. Less important genera are included to demonstrate the relative diversity of the various higher taxa, and major molecular phylogenetic-based adjustments to the systematics are noted. Major synonyms (indicated by =) of the most important plant-parasitic genera are listed to avoid undue confusion as to the fate of a familiar genus. Genera are listed alphabetically under the higher category. This scheme is used throughout the book so that the reader may be comforted by a certain predictability. It is not perfect; still less will it be universally accepted. Nevertheless, it is hoped that it will serve its purpose as a pragmatic framework in which to locate the various genera and higher taxa and to gain some understanding of the complex issues involved, particularly where molecular phylogenetics and classical morphological taxonomy interact. Those wishing for either greater detail at the generic level or alternative systematic schemes are referred to exemplars by Siddiqi (1986,

2000), Andrássy (2007, 2009), Hunt (1993, 2008), Decraemer (1995) and Subbotin *et al.* (2010a, b). Van Megen *et al.* (2009) provided a phylogeny based on what were then the most recent molecular criteria, with detailed trees showing relationships between nematodes in general and within the Tylenchomorpha, Longidoridae and Trichodoridae. Qing *et al.* (2025), among others, provides a more recent overview.

PHYLUM NEMATODA POTTS, 1932
CLASS CHROMADOREA INGLIS, 1983
- Subclass Chromadoria Pearse, 1942
  - Order Rhabditida Chitwood, 1933
    - Suborder Tylenchina Thorne, 1949
      - Infraorder Tylenchomorpha De Ley & Blaxter, 2002
        - Superfamily Aphelenchoidea Fuchs, 1937[1]
          - Family Aphelenchidae Fuchs, 1937
            - Subfamily Aphelenchinae Fuchs, 1937
              - *Aphelenchus* Bastian, 1865
            - Subfamily Paraphelenchinae T. Goodey, 1951
              - *Paraphelenchus* Micoletzky, 1922
          - Family Aphelenchoididae Skarbilovich, 1947
            - Subfamily Aphelenchoidinae Skarbilovich, 1947
              - ***Aphelenchoides*** Fischer, 1894
              - *Ficophagus* Davies & Bartholomaeus, 2015 *in* Davies *et al.*, 2015
              - *Martininema* Davies & Bartholomaeus, 2015 *in* Davies *et al.*, 2015
              - *Schistonchus* Cobb, 1927
            - Subfamily Parasitaphelenchinae Rühm, 1956
              - ***Bursaphelenchus*** Fuchs, 1937
        - Superfamily Tylenchoidea Örley, 1880
          - Family Tylenchidae Örley, 1880
            - Subfamily Tylenchinae Örley, 1880
              - *Aglenchus* Andrássy, 1954
              - *Tylenchus* Bastian, 1865
              - *Coslenchus* Siddiqi, 1978
              - *Filenchus* Andrássy, 1954
            - Subfamily Boleodorinae Khan, 1964
              - *Basiria* Siddiqi, 1959
              - *Boleodorus* Thorne, 1941
            - Subfamily Thadinae Siddiqi, 1986
              - *Thada* Thorne, 1941
            - Subfamily Duosulciinae Siddiqi, 1979
              - *Duosulcius* Siddiqi, 1979
              - *Malenchus* Andrássy, 1968
              - *Miculenchus* Andrássy, 1959
              - *Ottolenchus* Husain & Khan, 1967
            - Subfamily Tanzaniinae Siddiqi, 2000
              - *Tanzanius* Siddiqi, 1991
            - Subfamily Ecphyadophorinae Skarbilovich, 1959
              - *Ecphyadophora* de Man, 1921
              - *Sigmolenchus* Gharakhani, Pourjam, Abolafia, Castillo & Pedram, 2020
            - Subfamily Ecphyadophoroidinae Siddiqi, 1986
              - *Ecphyadophoroides* Corbett, 1964
            - Subfamily Atylenchinae Skarbilovich, 1959
              - *Atylenchus* Cobb, 1913

Subfamily Eutylenchinae Siddiqi, 1986
Eutylenchus Cobb, 1913
Subfamily Tylodorinae Paramonov, 1967
Tylodorus Meagher, 1964
Subfamily Pleurotylenchinae Andrássy, 1976
Cephalenchus Goodey, 1962
Pleurotylenchus Szczygiel, 1969
Subfamily Epicharinematinae Maqbool & Shahina, 1985
Epicharinema Raski, Maggenti, Koshy & Sosamma, 1980
Gracilancea Siddiqi, 1976
Family Dolichodoridae Chitwood *in* Chitwood & Chitwood, 1950
Subfamily Dolichodorinae Chitwood *in* Chitwood & Chitwood, 1950
Dolichodorus Cobb, 1914
Neodolichodorus Andrássy, 1976
Subfamily Brachydorinae Siddiqi, 2000
Brachydorus de Guiran & Germani, 1968
Subfamily Telotylenchinae Siddiqi, 1960
Bitylenchus Filipjev, 1934
Histotylenchus Siddiqi, 1971
Neodolichorhynchus Jairajpuri & Hunt, 1984
Paratrophurus Arias, 1970
Quinisulcius Siddiqi, 1971
Sauertylenchus Sher, 1974
Telotylenchoides Siddiqi, 1971
Telotylenchus Siddiqi, 1960
Trichotylenchus Whitehead, 1960
Trophurus Loof, 1956
Tylenchorhynchus Cobb, 1913
Uliginotylenchus Siddiqi, 1971
Subfamily Meiodorinae Siddiqi, 1976
Meiodorus Siddiqi, 1976
Subfamily Macrotrophurinae Fotedar & Handoo, 1978
Macrotrophurus Loof, 1958
Subfamily Merliniinae Siddiqi, 1971
Amplimerlinius Siddiqi, 1976
Geocenamus Thorne & Malek, 1968
Macrotylenchus Sturhan, 2012
Merlinius Siddiqi, 1970
Nagelus Thorne & Malek, 1968
Paramerlinius Sturhan, 2012
Pratylenchoides Winslow, 1958
Scutylenchus Jairajpuri, 1971
Telomerlinius Siddiqi & Sturhan, 2014
Family Belonolaimidae Whitehead, 1960
Subfamily Belonolaiminae Whitehead, 1960
Belonolaimus Steiner, 1949
Carphodorus Colbran, 1965
Ibipora Monteiro & Lordello, 1977
Morulaimus Sauer, 1966
Family Hoplolaimidae Filipjev, 1934
Subfamily Hoplolaiminae Filipjev, 1934
Antarctylus Sher, 1973
Aorolaimus Sher, 1963

= *Peltamigratus* Sher, 1964
Aphasmatylenchus Sher, 1965
Helicotylenchus Steiner, 1945
= *Rotylenchoides* Whitehead, 1958
Hoplolaimus Daday, 1905
= *Basirolaimus* Shamsi, 1979
Rotylenchus Filipjev, 1936
Scutellonema Andrássy, 1958
Subfamily Rotylenchulinae Husain & Khan, 1967
Acontylus Meagher, 1968
Bilobodera Sharma & Siddiqi, 1992
Rotylenchulus Linford & Oliveira, 1940
Senegalonema Germani, Luc & Baldwin, 1984
Verutus Esser, 1981
Subfamily Heteroderinae Filipjev & Schuurmans Stekhoven, 1941
Betulodera Sturhan, 2002
Cactodera Krall & Krall, 1978
Dolichodera Mulvey & Ebsary, 1980
Globodera Skarbilovich, 1959
Heterodera Schmidt, 1871
= *Afenestrata* Baldwin & Bell, 1985
Paradolichodera Sturhan, Wouts & Subbotin, 2007
Punctodera Mulvey & Stone, 1976
Vittatidera Bernard, Handoo, Powers, Donald & Heinz, 2010
Subfamily Ataloderinae Wouts, 1973
Atalodera Wouts & Sher, 1971
Bellodera Wouts, 1985
Camelodera Krall, Shagalina & Ivanova, 1988
Cryphodera Colbran, 1966
Ekphymatodera Bernard & Mundo-Ocampo, 1989
Hylonema Luc, Taylor & Cadet, 1978
Rhizonema Cid del Prado, Lownsbery & Maggenti, 1983
Sarisodera Wouts & Sher, 1971
Subfamily Meloidoderinae Golden, 1971
Meloidodera Chitwood, Hannon & Esser, 1956
Subfamily Meloidogyninae Skarbilovich, 1959
Meloidogyne Göldi, 1887
= *Hypsoperine* Sledge & Golden 1964
= *Spartonema* Siddiqi, 2000
Subfamily Nacobboderinae Golden & Jensen, 1974
Bursadera Ivanova & Krall, 1985
Meloinema Choi & Geraert, 1974
Family Pratylenchidae Thorne, 1949
Subfamily Pratylenchinae Thorne, 1949
Apratylenchus Trinh, Waeyenberge, Nguyen, Baldwin, Karssen & Moens, 2009
Pratylenchus Filipjev, 1936
Zygotylenchus Siddiqi, 1963
Subfamily Hirschmanniellinae Fotedar & Handoo, 1978
Hirschmanniella Luc & Goodey, 1964

Subfamily Radopholinae Allen & Sher, 1967
Achlysiella Hunt, Bridge & Machon, 1989
Apratylenchoides Sher, 1973
Hoplotylus s'Jacob, 1960
= *Radopholoides* De Guiran, 1967
Radopholus Thorne, 1949
Zygradus Siddiqi, 1991
Subfamily Nacobbinae Chitwood *in* Chitwood & Chitwood, 1950
Nacobbus Thorne & Allen, 1944
Superfamily Criconematoidea Taylor, 1936
Family Criconematidae Taylor, 1936
Subfamily Criconematinae Taylor, 1936
Bakernema Wu, 1964
Criconema Hofmänner & Menzel, 1914
Lobocriconema De Grisse & Loof, 1965
Neolobocriconema Mehta & Raski, 1971
Ogma Southern, 1914
Subfamily Macroposthoniinae Skarbilovich, 1959
Criconemoides Taylor, 1936
= *Macroposthonia* de Man, 1880
= *Criconemella* De Grisse & Loof, 1965
= *Mesocriconema* Andrássy, 1965
= *Madinema* Khan, Chawla & Saha, 1976
Discocriconemella De Grisse & Loof, 1965
Xenocriconemella De Grisse & Loof, 1965
Subfamily Hemicriconemoidinae Andrássy, 1979
Hemicriconemoides Chitwood & Birchfield, 1957
Family Hemicycliophoridae Skarbilovich, 1959
Subfamily Hemicycliophorinae Skarbilovich, 1959
Hemicycliophora de Man, 1921
= *Colbranium* Andrássy, 1979
= *Aulosphora* Siddiqi, 1980
= *Loofia* Siddiqi, 1980
Subfamily Caloosiinae Siddiqi, 1980
Caloosia Siddiqi & Goodey, 1964
Hemicaloosia Ray & Das, 1978
Family Tylenchulidae Skarbilovich, 1947
Subfamily Tylenchulinae Skarbilovich, 1947
Trophotylenchulus Raski, 1957
Tylenchulus Cobb, 1913
Subfamily Sphaeronematinae Raski & Sher, 1952
Sphaeronema Raski & Sher, 1952
Subfamily Meloidoderitinae Kirjanova & Poghossian, 1973
Meloidoderita Poghossian, 1966
Subfamily Paratylenchinae Thorne, 1949
Paratylenchus Micoletzky, 1922
= *Cacopaurus* Thorne, 1943
= *Gracilacus* Raski, 1962
Tylenchocriconema Raski & Siddiqui, 1975

Superfamily Anguinoidea Nicoll, 1935

Family Anguinidae Nicoll, 1935

Afrina Brzeski, 1981

Anguina Scopoli, 1777

Cotylenchus Álvarez-Ortega & Subbotin, 2024

Diptylenchus Khan, Chawla & Seshadri, 1969

Ditylenchoides Subbotin & Ryss, 2024

Ditylenchus Filipjev, 1936

Ficotylus Davies, 2009

Halenchus Cobb & Cobb, 1933

Heteroanguina Chizhov, 1980

Indoditylenchus Sinha, Choudhury & Baqri, 1985

Litylenchus Zhao, Davies, Alexander & Riley, 2011

Mesoanguina Chizhov & Subbotin, 1985

Nothanguina Whitehead, 1959

Nothotylenchus Thorne, 1941

Orrina Brzeski, 1981

Palaeoanguina Poinar, 2011 [fossil genus]

Paraditylenchus Subbotin & Ryss, 2024

Pseudohalenchus Tarjan, 1958

Pterotylenchus Siddiqi & Lenné, 1984

Safianema Siddiqi, 1980

Subanguina Paramonov, 1967

Zeatylenchus Zhao, Davies, Alexander & Riley, 2013

CLASS ENOPLEA INGLIS, 1983

Subclass Dorylaimia Inglis, 1983

Order Dorylaimida Pearse, 1942

Suborder Dorylaimina Pearse, 1942

Superfamily Dorylaimoidea Thorne, 1935[2]

Family Longidoridae Thorne, 1935

Subfamily Longidorinae Thorne, 1935

Australodorus Coomans, Olmos, Casella & Chaves, 2004

Longidoroides Khan, Chawla & Saha, 1978

Longidorus Micoletzky, 1922

Paralongidorus Siddiqi, Hooper & Khan, 1963[3]

Paraxiphidorus Coomans & Chaves, 1995

Xiphidorus Monteiro, 1976

Subfamily Xiphinematinae Dalmasso, 1969

Xiphinema Cobb, 1913

Subclass Enoplia Pearse, 1942

Order Triplonchida Cobb, 1920

Suborder Diphtherophorina Coomans & Loof, 1970

Superfamily Diphtherophoroidea Micoletzky, 1922

Family Trichodoridae Thorne, 1935

Allotrichodorus Rodriguez-M., Sher & Siddiqi, 1978

Ecuadorus Siddiqi, 2002

Monotrichodorus Andrássy, 1976

Nanidorus Siddiqi, 1974

Paratrichodorus Siddiqi, 1974

= *Atlantadorus* Siddiqi, 1974

Trichodorus Cobb, 1913

3.11. Glossary of Some Commonly Used Nomenclatural Terms

Apud. After; in the sense of.
Binomen (pl. **binomina**). The combination of two names, generic and specific, to form the binomial scientific name of a species.
Code. Abbreviation referring to the *International Code of Zoological Nomenclature*.
Et. And.
Genus dubium (pl. ***genera dubia***). A genus, or genera, of doubtful validity.
Genus incertae sedis (pl. ***genera incertae sedis***). A genus, or genera, of uncertain systematic position.
Genus inquirendum (pl. ***genera inquirenda***). Genus or genera requiring further study to validate.
Holotype. A single specimen designated from the type series on which a scientific name rests. It may be male, female or juvenile, the choice varying from group to group, depending on the diagnostic character set.
Homonym. An identical nomen for a taxon. It may be junior or senior, depending on date of authorship; objective or subjective depending on whether it is based on the same taxon or arises from a matter of opinion, or primary or secondary, depending on whether the homonymy arose from an original combination of two names under the same genus or from a subsequent combination.
Incertae sedis. Of uncertain taxonomic position.
Lapsus calami. Literally a slip of the pen. A printer's error or inadvertent spelling mistake by the author.
Nec. Not.
New combination. The first combination of a generic name and previously published specific **name**. Usually abbreviated as 'n. comb'.
Nomen (pl. **nomina**). A name or names.
Nomen novum (pl. ***nomina nova***). New or replacement name(s).
Nomen nudum (pl. ***nomina nuda***). A name, or names, without a valid description (literally a naked name).
Objective. Factual, not based on opinion (see 'subjective').
Paratype. All the remainder of the type series once the holotype has been designated.
Partim. In part.
Sensu lato. In the widest or broad sense.
Sensu stricto. In the restricted or narrow sense.
Species dubia (pl. ***species dubiae***). Species of dubious validity.
Species incertae sedis. Species of uncertain systematic position.
Species indeterminata (pl. ***species indeterminatae***). Species where the description is too poor to allow reidentification.
Species inquirenda (pl. ***species inquirendae***). Species requiring further study in order to validate.
Subjective. Based on opinion (see 'objective').
Synonym. A taxon regarded as identical to another. It may be junior or senior, depending on date of authorship.
Taxon (pl. **taxa**). Taxonomic unit(s) in a systematic scheme.
Type series. That material designated by the author(s) as being representative of the named taxon. Ideally, it should all come from the same locality, not be a composite from different samplings.
Vel. Or.
A more extensive glossary can be found in the *International Code of Zoological Nomenclature*, either in printed copy (4th edition, 1999) or online at https://www.iczn.org/the-code/the-code-online/.

Notes

[1] This scheme only covers the plant-parasitic forms. For a complete overview of this group, see Chapter 7.
[2] Plant-parasitic forms occur in other groups (*e.g.*, *Californidorus*, *Longidorella*) although little is known about their biology or parasitic status.
[3] *Siddiqia* Khan, Chawla & Saha, 1978 is not included as a valid genus (see Peña-Santiago, 2021) because *Paralongidorus* is paraphyletic with species arranged in two subclades clustering within *Longidorus* (Mwamula *et al.*, 2020).

3.12. References

Adams, B.J. (2001) The species delimitation uncertainty principle. *Journal of Nematology* 22, 153–160.

Adams, B.J., Dillman, A.R. and Finlinson, C. (2009) Molecular taxonomy and phylogeny. In: Perry, R.N., Moens, M. and Starr, J.L. (eds), *Root-knot Nematodes*. CAB International, Wallingford, UK, pp. 119–138.

Ahmed, M. and Holovachov, O. (2021) Twenty years after De Ley and Blaxter – how far did we progress in understanding the phylogeny of the phylum Nematoda? *Animals* 11(12), 3479.

Andrássy, I. (2007) *Free-living nematodes of Hungary, II (Nematoda Errantia). Pedozoologica Hungarica, 4* (Series eds: Csuzdi, C. and Mahunka, S.). Hungarian Natural History Museum and Systematic Zoology Research Group of the Hungarian Academy of Sciences, Budapest, Hungary.

Andrássy, I. (2009) *Free-living Nematodes of Hungary (Nematoda Errantia). Volume III. Pedozoologica Hungarica, 5* (Series eds: Csuzdi, C. and Mahunka, S.). Hungarian Natural History Museum and Systematic Zoology Research Group of the Hungarian Academy of Sciences, Budapest, Hungary.

Bickford, D., Lohman, D.J., Sodhi, N.S., Ng, P.K.L., Meier, R., *et al.* (2006) Cryptic species as a window on diversity and conservation. *Trends in Ecology and Evolution* 22, 148–155.

Blaxter, M.L., De Ley, P., Garey, J.R., Liu, L.X., Scheldeman, P., *et al.* (1998) A molecular evolutionary framework for the phylum Nematoda. *Nature* 392, 71–75.

Bogale, M., Baniya, A. and Digennaro, P. (2020) Nematode identification techniques and recent advances. *Plants* 9, 1260.

Castagnone-Sereno, P. (2006) Genetic variability and adaptive evolution in parthenogenetic root-knot nematodes. *Heredity* 96, 282–289.

Cherry, T., Szalanski, A.L., Todd, T.C. and Powers, T.O. (1997) The internal transcribed spacer region of *Belonolaimus* (Nemata: Belonolaimidae). *Journal of Nematology* 29, 23–29.

Chitambar, J.J. and Subbotin, S.A. (2014) *Systematics of the Sheath Nematodes of the Superfamily Hemicycliophoroidea. Nematology Monographs and Perspectives, 10* (Series eds: Hunt, D.J. and Perry, R.N.). Brill, Leiden, Netherlands.

Chitwood, B.G. (1949) Root-knot nematodes – Part 1. A revision of the genus *Meloidogyne* Goeldi, 1887. *Proceedings of the Helminthological Society of Washington* 16, 90–104.

Chitwood, D.J. and Perry, R.N. (2009) Reproduction, physiology and biochemistry. In Perry, R.N., Moens, M. and Starr, J.L. (eds), *Root-knot Nematodes*. CAB International, Wallingford, UK, pp. 182–200.

Claridge, M.F., Dawah, H.A. and Wilson, M.R. (1997) Practical approaches to species concepts for living organisms. In: Claridge, M.F. and Wilson, M.R. (eds), *Species. The Units of Biodiversity*. The Systematics Association Special Volume Series 54. Chapman and Hall, London, UK, pp. 1–15.

Decraemer, W. (1995) *The Family Trichodoridae: Stubby Root and Virus Vector Nematodes*. Kluwer Academic Publishers, Dordrecht, Netherlands.

Decraemer, W. and Hunt, D.J. (2024) Structure and classification. In: Perry, R.N., Moens, M. and Jones, J.T. (eds), *Plant Nematology*, 3rd edn. CABI Publishing, Wallingford, UK, pp. 3–40.

De Ley, P. and Blaxter, M.L. (2002) Systematic position and phylogeny. In: Lee, D.L. (ed), *The Biology of Nematodes*. Taylor and Francis, London, New York, pp. 1–30.

De Ley, P. and Blaxter, M.L. (2004) A new system for Nematoda: combining morphological characters with molecular trees and translating clades into ranks and taxa. In: Cook, R. and Hunt, D.J. (eds), *Proceedings of the Fourth International Congress of Nematology. 8–13 June 2002. Tenerife, Spain. Nematology Monographs and Perspectives,* 2. Brill, Leiden, Netherlands, pp. 633–653.

Duarte, I.M., De Almeida, M.T.M., Brown. D.J.F., Marques. I., Neilson. R., *et al.* (2010) Phylogenetic relationships, based on SSU rDNA sequences, among the didelphic genera of the family Trichodoridae from Portugal. *Nematology* 12, 171–180.

Finetti-Sialer, M.M. and Manzanilla-López, R.H. (2011) Exploiting "-omics" and molecular approaches in plant nematology research. In: Rodríguez Herrera, R., Aguilar, C.N., Kilpatrick Simpson-Williamson, J. and Gutiérrez Sánchez, G. (eds), *Phytopathology in the Omics Era*. Research Signpost, Kerala, India, pp. 39–68.

Floyd, R., Abebe, E., Papert, A. and Blaxter, M.L. (2002) Molecular barcodes for soil nematode identification. *Molecular Ecology* 11, 839–850.

Geraert, E. (2008) *The Tylenchidae of the World. Identification of the Family Tylenchidae (Nematoda: Tylenchida)*. Academia Press, Ghent, Belgium.

Geraert, E. (2010) *The Criconematidae of the World. Identification of the Family Criconematidae (Nematoda)*. Academia Press, Ghent, Belgium.

Geraert, E. (2011) *The Dolichodoridae of the World. Identification of the Family Dolichodoridae*. Academia Press, Ghent, Belgium.

Geraert, E. (2013) *The Pratylenchidae of the World. Identification of the Family Pratylenchidae (Nematoda: Tylenchida)*. Academia Press, Ghent, Belgium.

He, Y., Subbotin, S.A., Rubstova, T.V., Lamberti, F., *et al.* (2005) A molecular phylogenetic approach to Longidoridae (Nematoda: Dorylaimida). *Nematology* 7, 111–124.

Hennig, W. (1950) *Grundzüge einer Theorie der Phylogenetischen Systematik*. Berlin, Deutscher Zentralverlab.

Hennig, W. (1966) *Phylogenetic Systematics*. University of Illinois Press, Chicago, Illinois.

Hodda, M. (2022a) Phylum Nematoda: a classification, catalogue and index of valid genera, with a census of valid species. *Zootaxa* 5114(1), 1–289. DOI: 10.11646/zootaxa.5114.1.1.

Hodda, M. (2022b) Phylum Nematoda: trends in species descriptions, the documentation of diversity, systematics, and the species concept. *Zootaxa* 5114, 290–317. DOI: 10.11646/zootaxa.5114.1.2.

Holterman, M., van der Wurf, A., van den Elsen, S., van Megen, H., *et al.* (2006) Phylum-wide analysis of SSU rDNA reveals deep phylogenetic relationships among nematodes and accelerated evolution toward crown clades. *Molecular Biology and Evolution* 23, 1792–1800.

Holterman, M., Karssen, G., van den Elsen, S., van Megen, H., Bakker, J., *et al.* (2009) Small subunit rDNA-based phylogeny of the Tylenchida sheds light on relationships among some high-impact plant-parasitic nematodes and the evolution of plant feeding. *Phytopathology* 99, 227–235.

Hull, D.L. (1997) The ideal species concept – and why we cannot get it. In: Claridge, M.F., Dawah, H.A. and Wilson, M.R. (eds), *Species. The Units of Biodiversity*. The Systematics Association Special Volume Series 54. Chapman and Hall, London, UK, pp. 357–380.

Hunt, D.J. (1993) *Aphelenchida, Longidoridae and Trichodoridae: Their Systematics and Bionomics*. CABI Publishing, Wallingford, UK.

Hunt, D.J. (2008) A checklist of the Aphelenchoidea (Nematoda: Tylenchina). *Journal of Nematode Morphology and Systematics* 10(2007), 99–135.

International Code of Zoological Nomenclature, 4th edn. (1999) International Commission on Zoological Nomenclature, London, UK.

Kanzaki, N. and Giblin-Davis, R.M. (2012) Aphelenchoidea. In: Manzanilla-López, R.H. and Marbán-Mendoza, N. (eds), *Practical Plant Nematology*. Colegio de Postgraduados and Mundi-Prensa, Biblioteca Básica de Agricultura, Jalisco, Mexico, pp. 161–208.

Knot, I.E., Zouganelis, G.D. Weedal, G.D., Wich, S.A. and Rae, R. (2020) DNA barcoding of nematodes using the MiniOn. *Frontiers in Ecology and Evolution. Section Environmental Informatics and Remote Sensing*, 8. DOI: 10.3389/fevo.2020.00100.

Kumari, S. and Subbotin, S.A. (2012). Molecular characterization and diagnostics of stubby root and virus vector nematodes of the family Trichodoridae (Nematoda: Triplonchida) using ribosomal RNA genes. *Plant Pathology* 16, 1021–1031. DOI: 10.1111/j.1365-3059.2012.02598.x.

López Caballero, E. and Pérez Suárez, G. (1999) Métodos de análisis en la reconstrucción filogenética. *Boletín SEA* 26, 45–56.

Manzanilla-López, R.H. (2010) Speciation within *Nacobbus*: consilience or controversy? *Nematology* 12, 321–334.

Mayr, E. (1942) *Systematics and the Origin of Species*. Columbia University Press, New York, New York.

Mayr, E. (1969) *Principles of Systematic Zoology*. McGraw Hill, New York, New York.

Nadler, S. (2002) Species delimitation and nematode biodiversity: phylogenies rule. *Nematology* 4, 615–625.

Nguyen, K.B., Maruniak, J. and Adams, B.J. (2001) Diagnostic and phylogenetic ability of the rDNA internal transcriptional spacer sequences of *Steinernema*. *Journal of Nematology* 33, 73–82.

Nixon, K. and Wheeler, Q.D. (1990) An amplification of the phylogenetic species concept. *Cladistics* 6, 211–223.

Peña-Santiago, R. (2021) *Morphology and Bionomics of Dorylaims (Nematoda: Dorylaimida)*. In: *Nematology Monographs and Perspectives, 13* (Series eds: Hunt, D.J. and Perry, R.N.). Brill, Leiden, Netherlands.

Powers, T. (2004) Nematode molecular diagnostics: from bands to barcodes. *Annual Review of Phytopathology* 42, 367–383.

Qing, X., Zhang, Y.M., Sun, S., Ahmed, M., Lo, W-S., *et al.* (2025) Phylogenomic insights into the evolution and origin of Nematoda. *Systematic Biology* 74, 349–358. DOI: 10.1101/2023.12.13.571554.

Reid, A., Manzanilla-López, R.H. and Hunt, D.J. (2003) *Nacobbus aberrans* (Thorne, 1935) Thorne & Allen, 1944 (Nematoda: Pratylenchidae): a nascent species complex revealed by RFLP analysis and sequencing of the ITS-rDNA region. *Nematology* 5, 441–451.

Siddiqi, M.R. (1986) *Tylenchida Parasites of Plants and Insects*. Commonwealth Agricultural Bureaux, Farnham Royal, UK.

Siddiqi, M.R. (2000) *Tylenchida Parasites of Plants and Insects*, 2nd edn. CABI Publishing, Wallingford, UK.

Sokal, R.R. and Sneath, P.H.A. (1963) *The Principles of Numerical Taxonomy*. W.H. Freeman and Co., San Francisco, California.

Subbotin, S.A. and Ryss, A.Yu. (2024) Revision of the genus *Ditylenchus* Filipjev, 1936: *Ditylenchoides* gen. n. and *Paraditylenchus* gen. n. (Nematoda: Anguinoidea). *Russian Journal of Nematology* 32, 91–102.

Subbotin, S.A., Mundo-Ocampo, M. and Baldwin, J.R. (2010a) *Systematics of Cyst Nematodes (Nematoda: Heteroderinae). Nematology Monographs and Perspectives, 8A*. (Series eds: Hunt, D.J. and Perry, R.N.). Brill, Leiden, Netherlands.

Subbotin, S.A., Mundo-Ocampo, M. and Baldwin, J.R. (2010b) *Systematics of Cyst Nematodes (Nematoda: Heteroderinae). Nematology Monographs and Perspectives, 8B* (Series eds: Hunt, D.J. and Perry, R.N.). Brill, Leiden, Netherlands.

Subbotin, S.A., Palomares-Rius, J.E. and Castillo, P. (2021) *Root-knot Nematodes. Nematology Monographs and Perspectives, 14* (Series eds: Hunt, D.J. and Perry, R.N.). Brill, Leiden, Netherlands.

Tandingan De Ley, I., Mundo-Ocampo, M., Yoder, M. and De Ley, P. (2007) Nematodes from vernal pools in the Santa Rosa Plateau Ecological Reserve, California I. *Hirschmanniella santarosae* sp. n. (Nematoda: Pratylenchidae), a cryptic sibling species of *H. pomponiensis* Abdel-Rahman & Maggenti, 1987. *Nematology* 9, 405–429.

van Megen, H., van den Elsen, S., Holterman, M., Karssen, G., Mooyman, P., *et al.* (2009) A phylogenetic tree of nematodes based on about 1200 full-length small subunit ribosomal DNA sequences. *Nematology* 11, 927–950.

Vogler, A.P. and Monaghan, M.T. (2007) Recent advances in DNA taxonomy. *Journal of Zoological Systematics and Evolutionary Research* 45, 1–10.

Waite, I.S., O'Donnell, A.G., Harrison, A., Davies, J.T., Colvan, S.R., *et al.* (2003) Design and evaluation of nematode 18S rDNA primers for PCR and denaturing gradient gel electrophoresis (DGGE) of soil community DNA. *Soil Biology and Biochemistry* 35, 1165–1173.

Wendt, K.R., Vrain, T.C. and Webster, J.M. (1993) Separation of three species of *Ditylenchus* and some host races of *D. dipsaci* by restriction fragment polymorphism. *Journal of Nematology* 25, 555–563.

Zhuge, J., Zhou, X., Zhou, L., Hu, J. and Guo, K. (2023) The plant parasitic nematode database: a comprehensive data platform for plant parasitic nematode research. *International Journal of Molecular Sciences* 24 (23), 16841. DOI: 10.3390/ijms242316841.

4 Methodology and Symptomatology

Rosa H. Manzanilla-López
Centro de Desarrollo de Productos Bióticos, Instituto Politécnico Nacional, Yautepec, Morelos, Mexico

Abstract
Isolation and identification of the plant-parasitic nematodes associated with a decline in plant health and crop yield are key elements to diagnostics. Selection of adequate control methods and their insertion into an integrated pest- or integrated nematode management programmes are determined by the identification of the etiological agent. In this chapter, common above- and below-ground symptoms of plants affected by plant-parasitic nematodes, together with the most frequently used methods for sampling and extracting nematodes from soil, plant organs and tissues, are described and illustrated. Protocols relating to the most widely used laboratory and field methods for sampling and extraction of nematodes from soil and plant tissues are included, together with nematode counting, fixation and staining of nematodes in plant organs and tissues, as well as the processing (killing, fixation, dehydration, mounting) of nematodes for permanent slide preparations.

4.1. Introduction

Isolation and identification of plant-parasitic nematodes associated with a decline in plant health and crop yield are key elements to diagnostics. Selection of adequate control methods and their insertion into integrated pest- or integrated nematode management programmes are determined by the identification of the etiological agent. In this chapter, common above- and below-ground symptoms of plants affected by plant-parasitic nematodes and the most frequently used methods for sampling and extracting nematodes from soil, plant organs and tissues are described. There are several manuals that cover these topics more extensively and the reader is referred to Southey (1986), Zuckerman *et al.* (1990), Shurtleff and Averre (2000), Coyne *et al.* (2007), Perry *et al.* (2021) and Bulletin OEPPO (2021).

4.2. Symptomatology

It is difficult to establish nematodes as the sole cause of a disease since plant symptoms are similar to nutrient deficiency and inadequate watering/irrigation. Poor or tired soils can be

*E-mail: rosa.manzanilla@gmail.com
A revision of Manzanilla-López, R.H. (2012) Methodology and symptomatology. In: Manzanilla-López, R.H. and Marbán-Mendoza, N. (eds), *Practical Plant Nematology*. Mundi-Prensa and Biblioteca Básica de Agricultura, Montecillo, Mexico.

(eds R.H. Manzanilla-López, D.J. Hunt, N. Marbán-Mendoza and S. Infante Gil)
DOI: 10.1079/9781836990413.0004

due to the build-up of populations of nematodes over time that eventually results in a reduction of root intake of nutrients and water transport, thus affecting plant health and producing visible symptoms of disease or physiological impairment. Disease symptoms caused by nematodes vary depending on plant host, organs and tissues affected. Symptoms can be subdivided into two large groups: above-ground (aerial) or below-ground (subterranean). Damage can range from localized microscopic, superficial, mechanical lesions, produced by piercing by the nematode stylet, to extensive and deeper necrotic lesions that weaken the root anchorage system of the plant (e.g., *Radopholus similis*). However, to establish nematodes as one of the possible causes of the symptoms observed, it is essential to extract and identify them from soil, plant organs and, whenever required, to apply Koch's postulates to find physical evidence of the pathogen (nematode) in association with the unhealthy plant material: (i) the suspect nematode must be found in unhealthy but not healthy plant individuals; (ii) the suspect nematode must be isolated from unhealthy plants and grown in a pure culture; (iii) inoculation of a susceptible host with the suspect nematode sourced from a pure culture, and original disease symptoms must develop afterwards; (iv) the same nematode must be re-isolated from plants inoculated under experimental conditions.

4.2.1. Field Symptoms

Nematode effects can be seen in the field as patchy areas of variable size with poorly grown, yellowing, stunted plants and bare soil areas (Figs 4.1, 4.2). In potato fields, patches are often elliptical in shape, with the largest population density in the centre of the patch (Haydock and Perry, 1988).

Fig. 4.1. Field symptoms. A: *Heterodera schachtii* on sugar beet; B: spread of *H. schachtii*; C: faba bean 'Alfred' infested by *Ditylenchus gigas*; D: red clover patch infected by *Ditylenchus dipsaci*; E: patchiness in lucerne field infected with *D. dipsaci*; F: cereal root eelworm on oats.
(A-F Copyright Rothamsted Research.)

Fig. 4.2. Field symptoms (cont.). A: *Nacobbus aberrans* on bean field (J.A. Hernández); B: *N. aberrans* infestation in tomato field (J. Cristóbal-Alejo, courtesy of *Nematropica*); C: *Meloidogyne* sp. on *Hypericum* (R.H. Manzanilla-López); D: *Heterodera* and *Meloidogyne* on golf turf (R.H. Manzanilla-López).

4.2.2. Above-ground Symptoms

Symptoms are similar to those resulting from root damage that interferes with the support of the plant and nutrient absorption. They include yellowing; stunting (i.e., reduced plant growth); wilting, swelling and distortion of aerial plant parts; early senescence and low yield (Figs 4.3-4.6).

Important above-ground symptoms to look for include:

1. Stunting and deformation of the plant, 'stem break' (e.g., *Ditylenchus dipsaci* on tobacco); stunting and swelling at the base of the stem, shortening of internodes (*D. dipsaci* on lucerne).
2. Distorted apical growth, death of terminal apex or growth zone, rolling of leaves and inflorescence (*Aphelenchoides besseyi* and *Aphelenchoides fragariae* on strawberry).
3. Changes in leaf colour: dark green to black-coloured spots of angular, cuneiform shape on leaves, inter-veinal chlorosis and necrosis in leaves (e.g., *Aphelenchoides ritzemabosi* on chrysanthemum); red pigmentation on leaves (*Heterodera cruciferae* on cabbage).
4. Galls on stem, leaves and seeds (*Anguina* spp.).
5. Twisted leaves and lesions of yellowish colour in stems and leaves, necrosis of stem bases (e.g., *D. dipsaci*).
6. Twisted leaves, perhaps with white tips (i.e., *A. besseyi* on rice).
7. Twisted panicles and empty grains (e.g., *Ditylenchus angustus* on rice).
8. Reduced foliage and die back (*Tylenchulus semipenetrans*).
9. Yellowing and collapse of leaves followed by sudden death of palms (*Bursaphelenchus cocophilus*).
10. Red necrosis in stem vascular bundles, forming a red ring in coconut palm stem (*B. cocophilus*).
11. Yellowing and wilting of needles, leading to eventual death of pine trees (*Bursaphelenchus xylophilus*).
12. Necrotic patches on pods (*D. dipsaci*)
13. Small fruits, fruit drops and malformation of fruits.

4.2.3. Underground Symptoms of Roots, Root Systems and Plant Storage Organs

Plants show a general decline due to reduced absorption capacity of the root system. Symptoms to look out for (Figs 4.7-4.12) include:

1. Reduced root systems, particularly secondary roots.
2. Excessive branching of secondary roots (*Meloidogyne hapla*, *Pratylenchus* spp., *Nacobbus*

Fig. 4.3. Above-ground symptoms (cont.). A: *Ditylenchus gigas* on field bean pods (note necrotic patches); B: *D. gigas* 'eelworm wool' on field bean (*Vicia faba*) seed in its pod; C: faba bean seeds with testa removed to show necrotic cotyledons infested with *D. gigas*. (A-C Copyright Rothamsted Research.)

Fig. 4.4. Above-ground symptoms. A: Oats infested with *Ditylenchus dipsaci* cf. non-infested resistant cv. (right); B: *Anguina graminophila* on *Agrostis* sp.; C: faba bean 'Alfred' with stems infected by *D. gigas* cf. healthy stems; D: *Aphelenchoides ritzemabosi*-infested chrysanthemum plant; E: *A. ritzemabosi* infesting chrysanthemum leaf; F: *Aphelenchoides fragariae* in leaf stained with methyl blue in lactophenol. (A-F Copyright Rothamsted Research.)

Fig. 4.5. Above-ground symptoms (cont.). A: *Anguina tritici*. Green wheat ears healthy and infested. B: Healthy wheat ear (left) and one infected by *A. tritici* (right) showing spreading glumes and black galls within (D.J. Hunt); C: *A. tritici*-infected wheat ear showing spreading glumes and black galls within (arrows) (D.J. Hunt); D: healthy grain (left) and mature galls (right) (after J. Bridge); E: young galls. (A, E Copyright Rothamsted Research.)

aberrans sensu lato), 'bearded' roots (*Heterodera carotae*, *H. schachtii*).

3. Necrosis (localized or extensive) along roots, brown to reddish in colour (*Pratylenchus* spp., *Radopholus* spp.).

4. Root galls of different size (*Meloidogyne* spp., *N. aberrans sensu lato*).

5. Swellings on roots or hooked tips (*Subanguina* spp., *Xiphinema* spp.).

6. Roots with rounded, terminal galls (*Longidorus* spp., *Hemicycliophora* spp.).

7. Accumulation of soil particles on roots, commonly known as 'dirty roots' (*T. semipenetrans*).

8. Suppression of root growth, stumpy or 'fanged' roots (*Trichodorus* spp., *Paratrichodorus* spp.).

9. Roots encrusted with cysts (*Heterodera* spp., *Globodera* spp., *Cactodera* spp., *Punctodera* spp.).

10. 'Bottle brush-like' roots (*H. cruciferae*, *Punctodera chalcoensis*).

11. Small storage roots (*H. schachtii*).

12. Scabby appearance of tuber surface, sunken lesions (*Pratylenchus scribneri*).

13. Small, off-white spots that enlarge and darken later on, woolly in texture (*Ditylenchoides destructor* on potato).

14. Cysts attached to tuber surface (*Globodera* spp.) or storage roots (*Heterodera schachtii*).

15. Tuber surface with 'pimples', wart-like protuberances (*P. penetrans*, *Meloidogyne chitwoodi*).

16. Internal rot of tubers, corms and bulbs (*D. destructor*, *Scutellonema bradys*).

Fig. 4.6. Above-ground symptoms (cont.). A: Coconut palm with red ring disease caused by *Bursaphelenchus cocophilus*; B: infested coconut trunk cut through to show characteristic 'red ring' of red-ring disease; C: 'Corchosis' (corky root, a disease complex that includes *M. incognita*, *Pratylenchus*, *Helicotylenchus*, *Fusarium* and *Trichoderma*) on coffee trees (A. Tovar Soto); D: Pine wilt disease, caused by *Bursaphelenchus xylophilus*, in Portugal (M. Mota). (A, B Copyright Rothamsted Research.)

17. Soft bulbs with concentric circles of brownish scales when cut (*D. dipsaci* on *Allium*).
18. Lesions on groundnut pods (*Pratylenchus* spp., *Criconemoides* spp.).
19. Groundnut seeds (*Arachis hypogaea*) brown in colour, shrunken or wrinkled (*Aphelenchoides arachis*).

Images of and information on a variety of symptoms caused by different nematode species can be found in Bridge and Starr (2007) and at: nemaplex.ucdavis.edu/Uppermnus/topmnu.htm.

4.2.4. Disease Complexes

Plant-parasitic nematodes can interact with other organisms, causing disease complexes (Manzanilla-López and Starr, 2009; Sikora *et al.*, 2018; Parrado and Quintanilla, 2024), with the symptoms varying according to the nematode and soil-borne pathogens involved (Fig. 4.6C).

4.2.5. Nematode Virus Vectors

Some viruses are transmitted by nematodes (see Chapter 15, this volume). General symptoms can include ring spotting, spraing arch (Fig. 4.12B), yellowing, vein banding and leaf curling. When diagnosis of a viral disease vectored by a nematode is made, it is important to try to see if the nematode vector is present as well. Virus-induced symptoms may take several years to develop in the aerial parts of the plant. However,

Fig. 4.7. Below-ground symptoms of roots, root system and plant storage organs. A: *Subanguina radicicola* on *Poa annua* (root system and galls); B: Docking disorder of sugar beet (stubby root) caused by *Trichodorus* spp.; C: *S. radicicola* on *Poa annua* (root galls); D: blueberry cuttings infected by *Trichodorus* sp. (left), compared with 'healthy' cuttings infected by *Tetylenchus joctus* (right). (A-D Copyright Rothamsted Research.)

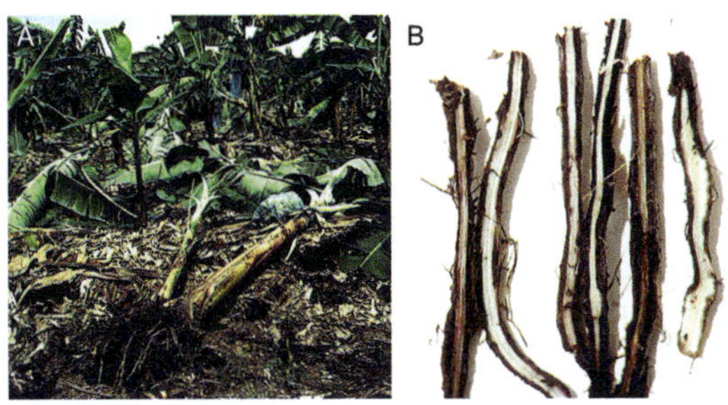

Fig. 4.8. Below-ground symptoms of roots, root system and plant storage organs. A: Banana trees uprooted or toppled as a result of root destruction caused by *Radopholus similis*; B: roots split lengthways to show infection of cortex by *R. similis.* (After J. Bridge.)

nematode-transmitted viruses can be detected in the roots during the first two years after planting (Taylor and Brown, 1997; Singh *et al.*, 2020).

4.3. Sampling

Estimating population densities with precision largely depends on accurate sampling and appropriate and consistently applied extraction procedures (see also Hallmann *et al.*, 2021; Viane *et al.*, 2021; Been *et al.*, 2024; den Nijs *et al.*, 2021). Sampling requires taking one or several samples for subsequent analysis in the laboratory to obtain information about nematode populations in the soil, associated with plants or living inside plant tissues and organs. Before sampling, it is necessary to define the objective(s): (i) **survey** (what genera of nematodes are present?); (ii) **diagnostic** (etiological agent

Fig. 4.9. Below-ground symptoms (galls). A: *Nacobbus aberrans* on beans (R.H. Manzanilla-López, courtesy of *Nematropica*); B: *N. aberrans* on potato (Copyright Rothamsted Research); C: *N. aberrans* on tomato (J. Cristóbal-Alejo, courtesy of *Nematropica*).

of a disease); (iii) **prospective** (what is the distribution of the nematodes in a determined area?); (iv) **quantitative** (quantification of one or several populations during a pre-established period of time, population dynamics, nematicide tests, etc.); (v) **quarantine/statutory** (preventing the entrance or movement of nematodes of quarantine importance into nematode-free areas); (vi) **experimental** or (vii) **commercial advisory**.

Selection of the nematode sampling method will depend on crop, crop combination, nematode species, period of sampling and soil type (den Nijs *et al.*, 2021). It is worth mentioning that soil-inhabiting nematodes are part of larger nematode communities that include a number of different species occupying different habitats, niches and life habits (e.g., free living and parasites). It is, therefore, important to consider the biology of the nematodes. For example, the presence/absence of juvenile/adult stages in soil and roots, anhydrobiotic stages, feeding habits, behaviour (i.e., vertical and horizontal migration), ecology, season of the year when active stages are found in the soil and distribution patterns (i.e., regular, random, aggregated). The typical spatial distribution of plant-parasitic nematodes is aggregated. This distribution can be favoured by some biological features, such as oviposition and number of eggs laid by individual females (e.g., cyst nematodes and root-knot nematodes), root distribution and depth, nematode dispersion through agricultural machinery, direction of ploughing, microclimate response and interactions with natural enemies. Some nematodes show a Poisson distribution due to their patchy distribution in the field. There are statistical functions to describe nematode distribution, including contagious distribution, negative binomial, Thomas, Neyman A, Polia Aeppli, and log-normal discrete (Barker and Campbell, 1981; Been *et al.*, 2024).

Fig. 4.10. Below-ground symptoms (tubers and storage organs). A: *Meloidogyne hapla* on carrots showing galls and deformities; B: *M. hapla* on carrot showing 'fingers' or 'fanged' root system; C: *Meloidogyne chitwoodi* 'mezquinos' on potato (A. Tovar-Soto); D: *Nacobbus aberrans* galls on sprouting potato tuber; E: *N. aberrans* on sugar beet. (A, D, E Copyright Rothamsted Research.)

4.3.1. Sampling Plan

Before starting to sample, it is necessary to design a plan that includes a sampling pattern, size unit (weight) and number of samples to be taken according to the area to be sampled, the season or period of sampling.

4.3.2. Sampling Patterns

In general, systematic sampling produces more reliable results than random sampling (Fig. 4.13). For annual crops with shallow roots (e.g., grasses), sampling is done at a depth of 8-12 cm (Barker, 1978, 1985) (Fig. 4.13A). In plots with a single plant (e.g., trees), samples are taken in the drip zone at 20 cm depth at five different points arranged like the points of a star. An additional five samples are taken closer to the stem or trunk, with each point positioned between the outer star points (Fig. 4.13B). In the case of perennials with deep roots, such as vine and citrus, samples are taken at two or three different depths (15, 30, 60-100 cm). For bare soils, it is recommended to take 20-40 subsamples, with each subsample being taken at 20 cm depth in a systematic, zigzag pattern (Shurtleff and Averre III, 2000) (Fig. 4.13C). In annual crops (established in rows), sampling must be conducted in the root zone at a depth of 5-12 cm. When the area to be sampled is large, a

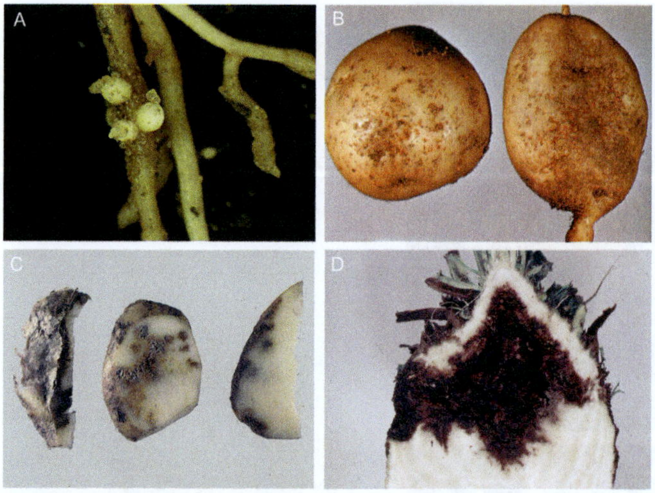

Fig. 4.11. Below-ground symptoms. A: Cysts of *Heterodera glycines* on soybean roots (R.H. Manzanilla-López); B: cysts on potato tubers; C: *Ditylenchoides destructor* on potato tuber; D: crown rot of sugar beet (longitudinal section) caused by *Ditylenchus dipsaci.* (B-D Copyright Rothamsted Research.)

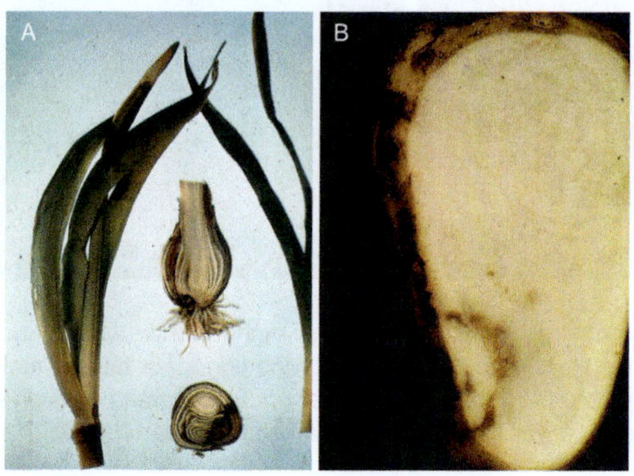

Fig. 4.12. A: Narcissus plant infested with *Ditylenchus dipsaci* (Rothamsted Research nematology picture archive.); B: Spraing arch in potato caused by Rattle virus transmitted by trichodorid nematodes (SHRI print). (A, B Copyright Rothamsted Research.)

minimum of 20-30 samples covering the whole field are collected following a zigzag pattern (Barker, 1978; Barker and Campbell, 1981). If the field plot is too large, it can be divided into smaller areas (sub-plots), taking core samples in a 'W' pattern from each sub-plot (Fig. 4.13D). The total number of samples taken per sub-plot can be combined to make a single composite sample. Soil should be mixed thoroughly but carefully to avoid damaging the nematodes, particularly the larger and more delicate forms, due to abrasion between the soil particles. It is important to bear in mind that there are specified sampling protocols for species of quarantine importance, such as cyst and root-knot nematodes (see Haydock and Perry, 1998; Duncan and Phillips, 2009; Reid *et al.*, 2010; Chen *et al.*, 2011; Pickup *et al.*, 2018).

4.3.3. Selecting Sampling Areas

Sampling areas are selected based on the presence of patches and diseased plants in the field, previous cropping history and yield reduction.

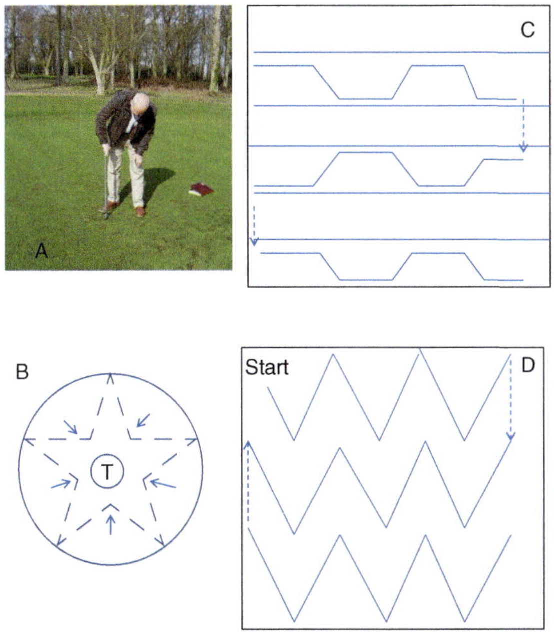

Fig. 4.13. Sampling. A: Taking core samples; B: sampling pattern within the drip line (cores taken at points of a star and points closer (arrows) to the trunk (T) of a tree or a shrub; C: sampling pattern for a field under fallow or without symptoms; D: sampling pattern of a field planted with a row crop (after Shurtleff and Averre III, 2000).

4.3.4. Number of Samples

There are several procedures to estimate a representative number of samples (Barker, 1978; Barker and Campbell, 1981; McSorley, 1987; Been *et al.*, 2024). The minimum number of samples varies according to plot size. Some authors recommend 20 or fewer for small plots (1-5 m^2), 25 samples ha^{-1} or 30 or more for large plots (>100 m^2).

4.3.5. Sampling Procedures

There are basic recommendations to follow when taking samples: (i) samples should be taken when nematodes are at detectable levels; (ii) the soil must be damp and preferably below field capacity; (iii) check for plants with visible symptoms, collect roots from the diseased plant and place them separately in a polythene bag containing a damp piece of cotton or paper; (iv) sample apparently healthy plants of the same crop growing in the vicinity of the diseased plants; (v) temperature and humidity fluctuations must be considered when taking samples since they can affect the number of nematodes. It is, therefore, not recommended to collect soil and roots from the surface layer (ca 5 cm). In areas with hot and dry summers, samples must be taken at 30-40 cm depth; (vi) take notes of relevant information regarding crop and plot history (e.g., previous crops) and (vii) samples should include soil, plant roots and other organs and tissues if necessary.

4.3.6. Sampling Equipment

A variety of tools and equipment can be used when taking samples from soil of different textures and at different depths (Fig. 4.14A). A spade, trowel or a sample corer in the shape of a tube (2.5 cm diam. × 20 cm long) are the most commonly used tools. The latter is widely used for sampling cyst nematodes, for example, and removes ca 100 cm^3 of soil per core. Garden spades can be used as well. However, there can be specifications for selecting a sampling tool, for

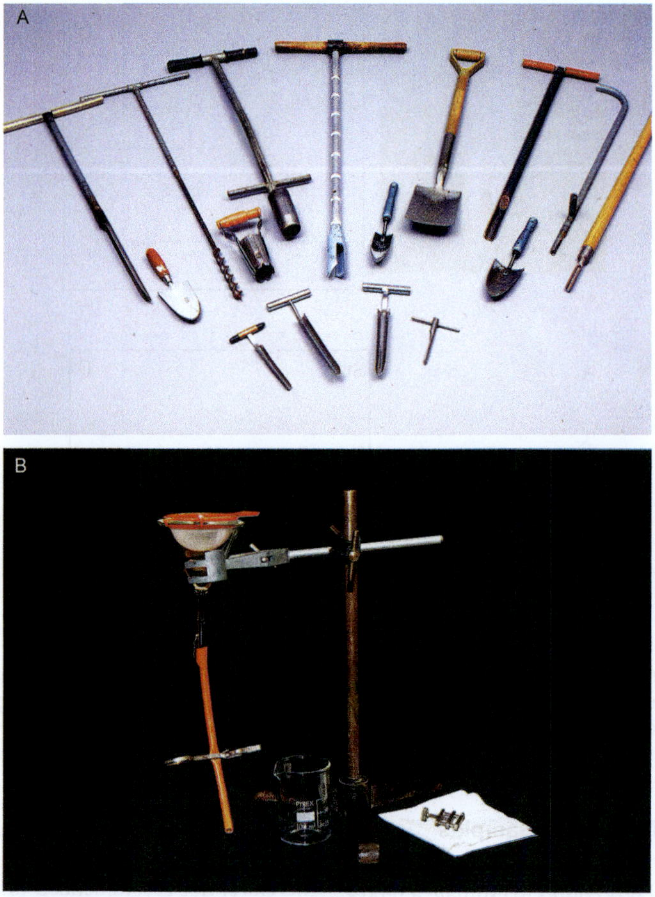

Fig. 4.14. Sampling and extraction. A: Sampling tools (Rothamsted Research nematology picture archive); B: Baermann funnel extraction filter showing funnel support, sieve, glass funnel, latex tubing, clamp, beaker. (A, B Copyright Rothamsted Research.)

example, when taking samples for routine estimation of some species, such as the potato cyst nematode, where core samples (each 4.5 ml vol) are taken using a cheese-type corer to a maximum depth of 5 cm (Haydock and Perry, 1998).

4.3.7. Soil Samples

The top 3-5 cm of soil is usually removed and discarded (except in turf and other various crops) at each sampling point. Samples (ca 50 g each) are taken at a depth of 15-30 cm (or deeper if necessary), and close to the roots. Avoid taking soil samples around dead or severely damaged plants, since nematode populations may be low. For diagnostic purposes, it is recommended to take paired samples that include 10-30 cores of soil from healthy plants adjacent to diseased plants and a similar amount from one or more planting areas. Once samples are collected, they are placed inside a strong polythene bag, sealed and labelled with water-proof ink or pencil. Labels should be attached to the outside of the bag and include information such as: (i) name and address of farmer; (ii) plot location, including global positioning system (GPS) coordinates; (iii) crop; (iv) sampling date and (v) collector. Keep duplicate records of samples and other relevant information, such as crop history and grower observations, in a field notebook.

According to Haydock and Perry (1998), the final size of the sample to be examined in the laboratory must be large enough to achieve the accuracy and sensitivity required and be derived from sufficient points to be representative of the sampled area.

4.3.8. Samples of Plant Material

Root and soil samples should be taken at the same points. Root samples consist of secondary (feeder) roots and the depth at which samples are taken will depend on the growth of the root system. Root samples should include material from plants suspected of being infected as well as samples from healthy adjacent roots. If plants have been subjected to drought and heat before sampling, root samples should be taken from the interior of the root system or from a place where they are still damp and in fresh condition. Root and soil samples should be kept in separate bags. Above-ground samples of plant material should include stems, leaves and shoots.

4.3.9. Care and Handling of Samples

Samples contain live nematodes, and they should be kept active to facilitate their extraction. Temperature and humidity affect nematode viability and, therefore, samples contained in bags should be kept cool and protected from direct sunlight and evaporation. Samples can be transported to the laboratory in a cool box, cold bag or a polystyrene box. If samples are not going to be processed immediately, they should be kept cool, with storage temperature depending on the geographical region: 4°C in northern and temperate regions, 10-16°C in warmer regions and 16-18°C in the tropics and subtropics (Shurtleff and Averre III, 2000).

4.4. Extraction Methods – Soil

Nematode extraction methods from soil, plant organs and tissues are either 'active', when the nematodes move naturally from the soil or tissues to the water interface or 'passive', when nematodes are separated and transferred to an aqueous medium using physical methods, such as sieving or elutriation.

Appropriate selection of extraction procedures depends on different features, such as parasitic habit (i.e., ectoparasitic, sedentary, migratory, semi-endoparasitic), mobility, dispersal capacity, nematode size and other characteristics that may facilitate or hinder extraction. Taxonomic studies usually require nematodes to be extracted in good physical condition and with minimum physical damage. Other important features to consider are soil type (e.g., silty, sandy, etc.), equipment and time availability to process samples. Extraction methods vary in their efficiency and ability to process several samples at a time. In order to obtain replicable results, extraction protocols should be executed with utmost care and attention, as variations or errors in technique can affect the numbers and type of nematodes extracted. The commonest extraction procedures include the Baermann funnel, various tray techniques, sieving-flotation-centrifugation, Fenwick can and mist chamber.

4.4.1. Baermann Funnel and Variants

It is a very simple and cheap method of extracting nematodes but has low efficiency, with nematodes often sticking to the sides of the equipment and/or suffering from lack of oxygen. It is only used for small soil samples (30-300 g). Second-stage juveniles, non-motile stages and sluggish nematodes (such as criconematids, longidorids and trichodorids) are recovered in low numbers, and eggs may also hatch during incubation. The basic equipment consists of a beaker (plastic or glass) filled with water on top of which is placed a plastic sieve or a wire mesh basket lined with a piece of tissue (Kleenex wipes, roll tissue) that contains the soil sample, with the latter being kept in contact with the water contained in the beaker. Alternatively, a glass funnel (10-15 cm diam.) with a 10-15 cm stem fitted with rubber or plastic tubing (8 cm long) and sealed with a spring or a screw clamp can be used instead of a beaker. The funnel is placed on a funnel support or rack, filled with water and the plastic tubing is purged to eliminate air bubbles before placing the sample on top of the funnel. Nematodes are collected 24-36 h later (Fig. 4.14B).

4.4.2. The 'Whitehead and Hemming Tray' Extraction Method

Often just referred to as the Whitehead tray, this is a simple, efficient and very convenient method for extracting active free living and plant-parasitic nematodes from soil (see also Bell and Watson, 2001), but soil samples must not be dried and ideally should be used straight from the field. If this is not possible, samples should be kept in sealed plastic bags in a refrigerator or cold room until used. The basic system can be readily modified to use locally available materials (including various plastic containers and mosquito mesh) and is cheap, if somewhat intensive in its use of space (Fig. 4.15A, B). Racking systems can be built to ameliorate the latter difficulty.

Procedure (Rothamsted Research, UK)

1. Make sure all soil aggregates in the sample are gently broken up, with all stones and large pieces of roots removed. Dry soils should be pre-soaked for several hours before processing.
2. Measure out a 200 ml representative subsample of the soil sample.
3. The extraction equipment consists of a nest of two plastic seed trays (36 × 23 cm) – an unmodified bottom tray (the collection tray) but with the base of the upper tray partially cut out and replaced by a coarse plastic mesh (this part forms the extraction filter).
4. Across the base of this top tray, lay a single ply tissue. Two pieces will be necessary to make

Fig. 4.15. Extraction methods. A: Whitehead and Hemming tray; B: soil samples on Whitehead and Hemming trays with flat-bottomed basins with plastic sieves and lids at bottom right (R.H. Manzanilla-López); C: sieves of different mesh size; D: sieves, buckets, washing bottle, measuring cylinder, beaker used in the sieving decanting method; E: plastic containers with holes and mesh; F: small sieves made from rings cut from plastic drainpipe with mesh of different sizes. (A, D-F Copyright Rothamsted Research.)

sure that the whole base and the lower part of the sides of the tray are covered, and the joint between the tissues can be sealed by dampening the overlap with water from a wash bottle or a similar device. This prevents soil particles from contaminating the tray below, thus giving a much cleaner extraction and one where the nematodes are more readily visible.

5. Sprinkle the 200 ml subsample of soil across the tissue on the top tray (the extraction filter), being careful to spread the soil evenly. Place the top tray carefully into the bottom tray and stand the assembly on a flat surface where it will not be disturbed.

6. Gently add about 1 l of water into the bottom tray. Keep adding water until its level reaches and moistens the soil (do not swamp the sample, as this leads to anaerobic conditions, which hamper nematode movement).

7. Leave the system undisturbed for 24 h (48 h if it is suspected that the sample contains *Trichodorus*).

8. Carefully remove the top tray, taking care not to break the tissue, and discard its contents.

9. Transfer half of the water in the bottom tray into a 1 l measuring cylinder via a wide-mouthed funnel. Gently sluice the remainder around the tray and add to the cylinder. Wash the tray with a little water and add this to the cylinder too.

10. Allow the sample to settle in the cylinder, leaving it undisturbed for 6 h.

11. Siphon off and discard all but the bottom 100 ml of water from the cylinder.

12. Transfer the last 100 ml into a beaker. Wash out the cylinder and add this also to the beaker.

13. Examine the sample in a counting dish under a binocular microscope.

4.4.3. Sieving

Sieving is one of the methods that can be used to get cleaner samples of nematodes. It can be used alone or in combination with other protocols such as the Baermann funnel or flotation-centrifugation. The latter technique allows the rapid processing of a larger quantity of soil samples and has a higher recovery efficiency than the Baermann funnel but requires more expensive equipment, such as sieves, a centrifuge and consumables (e.g., kaolin, $MgSO_4$, sugar). The principle of the technique is based upon the fact that the specific gravity of most nematodes is slightly greater than water, and therefore, when nematodes contained in a soil sample are resuspended in water, they will settle more slowly than the soil particles, which rapidly sink. While still in suspension, nematodes are sieved through a series of sieves of different mesh aperture in inches (60, 100, 75, 53 mesh inch^{-1}) or micron equivalent (250, 150, 200, 63, 53, 30 mesh µm^{-1}). Nematodes are collected after a process of swirling, sedimentation (1-2 min), decanting and washing through each sieve. Samples collected from each sieve can be processed both separated and combined (i.e., samples collected from different mesh aperture sieves) through centrifugation in a solution of specific gravity in which nematodes will float (Table 4.1). After centrifugation, nematodes are rinsed thoroughly to avoid distortion and plasmolysis. Non-mobile stages of nematodes are also recovered with this technique.

Sieves

Sieves are cylinders with a height of 4.5-7 cm and varying diameters and are produced in stainless steel or brass with a mesh made of stainless steel, nylon or terylene (Fig. 4.15C-F). As these are expensive, they can be readily made using polyvinyl chloride (PVC) cylinders of a suitable diameter and gluing on a fine mesh fabric (Fig. 4.15F). The fabric mesh fineness can be referred to by numbers (e.g., 10, 20, 50, 100, 200, 325, 500), with the mesh or screen aperture (number of openings) being specified in mm, inches or microns. Sieve meshes are selected according to the size of the nematode to be extracted from the sample (Table 4.2). To avoid blockage, it is advisable to wet the sieve

Table 4.1. Solute weight (g) dissolved in water to prepare 1 l of solution of the indicated specific gravity (Hooper, 1986a).

Specific gravity (20°C)	1.15	1.18	1.22
Saccarose (sugar cane)	401	484	588
$MgSO_4$ (anhydrous)	166	200	245
$MgSO_4 + 7H_2O$	339	409	503
$ZnSO_4$ (anhydrous)	156	187	229
$ZnSO_4 + 7H_2O$	279	335	410

Table 4.2. Sieve screen fineness and mesh aperture (Hooper, 1986a; Shurtleff and Averre III, 2000).

Nematode size	Genera	Meshes/ inch	Mesh size (μm)
Large	*Anguina*, *Belonolaimus*, *Dolichodorus*, *Longidorus*, *Xiphinema*, cysts	60	250
Medium	*Criconemoides*, *Paratrichodorus*, *Paratylenchus*, *Pratylenchus*, *Radopholus*	150-170	105-90
Small	Juveniles	325-350	46-45
Eggs	–	500	30

surface before and after use and to clean them with a water jet sprayed from the lower side.

Sieving-flotation-centrifugation

Soil samples (200 g) are processed by sedimentation, sieving and flotation techniques with 60-, 100-, 200-, 325- and 500-mesh inch^{-1} sieves (Jenkins, 1964) (Fig. 4.15D) followed by centrifugal flotation (Zacheo and Lamberti, 1974) with sucrose solution (1.18 sp.gr.). Modifications include repeated swirling, sedimentation and sieving procedures (three times) through 325- and 500-mesh inch^{-1} sieves.

PROCEDURE.

1. Mix soil sample thoroughly and gently break any lumps. Add 200 ml of tap water to a measuring cylinder (1000 ml). Add a sample of soil to raise the cylinder level up to 400 ml. Mix.
2. Pour the mixture through a 60-mesh sieve and collect in a bucket. Make up to 2 l.
3. Stir (swirl) the mixture and allow the suspension to settle for 60 s before decanting the supernatant through a 100-mesh sieve. Wash the sieve and collect the material retained on the sieve in a glass beaker. Refill the bucket and repeat the process using a 200-mesh sieve and collect in a beaker.
4. Refill the bucket and repeat the process twice, using a 325- followed by a 500-mesh sieve. Sievings (soil materials and nematodes) are collected in the same beaker.
5. Stir the sievings thoroughly and pour into 100 ml centrifuge tubes.
6. Add 1 g kaolin to each tube, mix, balance and centrifuge for 5 min (2900 g for vermiform nematodes; 1150 g for cyst nematode, 420 g for *Criconemoides*; Shurtleff and Averre III, 2000).
7. Discard the supernatant and resuspend the pellet in sucrose solution (450 g l^{-1} solution), mix, balance and centrifuge for 3 min at the appropriate value as listed above.
8. Pour the supernatant through a 500-mesh sieve and wash thoroughly before collecting in a beaker for examination.

Note: To estimate the relative centrifugal force (RFC), use the formula g = 0.00001118 × radius of centrifuge arm to tip of tube in centimeter × (speed in rev min^{-1})^2. Speeds used for nematode extraction are equivalent to 700-2900 g with 1800 g being commonly used (Hooper, 1986a). Avoid expressing centrifuge speed as rpm, as g depends on distance from the spindle (rotor arm length or angle) as well as rotational speed.

Immersion sieving method for extracting longidorids

This method, also known as the Flegg technique, is used to extract long and delicate nematodes, such as *Longidorus* and *Xiphinema*, and is a variant of sieving but with the top mesh submerged beneath the water surface so as to avoid damaging delicate nematodes against the exposed mesh. Due to the delicate nature of these nematodes, all of the following processing steps must be done gently and carefully. All soil samples must be fresh and kept cool and moist until they are processed.

PROCEDURE (ROTHAMSTED RESEARCH, UK).

1. Carefully break apart all soil aggregates by hand. Pull the aggregates apart rather than breaking them by exerting pressure.
2. Mix the soil gently, and measure a representative subsample of 100 ml into a beaker (250 ml, plastic Tri-pour). Add water.
3. Use a plastic washing-up bowl containing two supports on its base and with a square hole cut in the side in such a position as to provide a

controlled depth of water sufficient to cover the base of a sieve (Fig. 4.15E).

4. Carefully lower a 150-mesh sieve (105 μm aperture, 20 cm diam.) into the bowl, ensuring no air bubbles are caught beneath it.

5. Wash the soil from the 250 ml beaker into a tall-form 1 l capacity plastic beaker into which ca 200 ml of water had previously been added. Fill the beaker with water.

6. With a broad, round-edged spatula, stir the water above the soil and gently swirl it around. Gradually increase the speed of stirring to ensure that all of the soil particles are taken into suspension.

7. Stop the swirling and wait 10 s. Pour the supernatant into the submerged sieve, refill the beaker and repeat the stirring, settling and sieving.

8. Gently move the sieve from side to side under the water in the bowl before washing the sieve contents with a gentle stream of water.

9. Transfer the sieve contents into the original (clean) labelled 250 ml beaker and examine in a counting dish under a binocular microscope.

Fenwick can

This is used to extract *Heterodera*, *Punctodera*, *Cactodera* and *Globodera* cysts from dry soil (Fig. 4.16D). It can process up to 300 g of soil with a 70% cyst recovery (Shepherd, 1986) but has the disadvantage that, using dry soil, only *Globodera* cysts will remain viable after extraction.

PROCEDURE (AFTER SHEPHERD, 1986).

1. Fill the can with water and wet the collecting sieves.

2. Place the soil sample (ca 300 g) in the 1 mm aperture sieve on top of the funnel and wash the soil into the can with a strong jet of water from above.

3. Collect the organic material overflowing into the can collar on the collecting sieves (i.e., an 840 μm aperture over a 250 μm aperture sieve) placed below the outlet from the collar.

4. Wash the sample from the sieve surface using a wash bottle and collect the sample into a small container.

5. Insert a strip of filter paper around the inside of a 50 ml beaker.

6. Place the sample collected from the sieve in the beaker, add water and swirl the contents gently.

7. The water level is raised at the periphery during swirling, and the floating cysts and debris adhere to the filter paper. A drop of detergent can be added to break superficial water tension, thereby facilitating cyst adherence to the paper. The paper strip is carefully removed, allowed to dry and the cysts collected. A more complex version of the Fenwick can is an automated washing carousel (Fig. 4.17A). Soil samples are air-dried at 37°C and then processed through the carousel (Reid *et al.*, 2010), which allows the processing of 250 soil samples in 4 h (John Pickup, pers. comm.).

Extraction of cysts of Heterodera *spp. from soils*

PROCEDURE (AFTER DUNN, 1969).

1. Wash 600 cm^3 of soil through a 24-mesh sieve onto a 60-mesh sieve.

2. Transfer the residue from the 60-mesh sieve into 50 ml centrifuge tubes and suspend in a sucrose solution of specific gravity 1.23 (solutions of appropriate density made of inorganic salts can substitute sugar).

3. Centrifuge for 5 min at 1150 g and decant the supernatant onto a 60-mesh sieve. Rinse briefly to remove the sugar solution and wash into a beaker or counting dish for inspection.

Elutriators and the Trudgill column

Elutriation techniques are quantitatively better but require expensive equipment (elutriators) and large quantities of water (Fig. 4.16D). During elutriation, a stream of water flowing at a specific rate from below separates the nematodes from soil particles while maintaining the nematodes in suspension. The Trudgill column is a further refinement of the elutriation technique and can be used to extract nematodes of any kind and size, including cysts (Fig. 4.16A, B). It is useful to recover *Heterodera* females as well as white or yellow females of *Globodera* from washed roots and soil residues (Trudgill *et al.*, 1973; Shepherd, 1986).

Fig. 4.16. Extraction methods (cont.). A, B: Trudgill tower or Trudgill column elutriator; C: Fenwick can (R.H. Manzanilla-López); D: Oostenbrink elutriator. (A, B, D Copyright Rothamsted Research.)

Fluidizing column extraction of Heterodera *cysts*

This method is often referred to as the Trudgill tower or Trudgill column.

PROCEDURE (AFTER TRUDGILL *ET AL.*, 1973).

1. Measure 100 ml of water and pour it into a beaker. Take a soil sample with cysts (100-200 g), and mix it with the water.
2. Pass the soil mixture through a coarse sieve (8 mm aperture).
3. Add the sample to the Trudgill column (about one-third full of water).
4. Adjust the upward water flow (3.5 cm per second), allowing it to run for 2.5 min to mix and fluidize the suspension.
5. Catch the overflow on a bank of two sieves (60- and 100-mesh aperture)
6. Increase the speed of the upward water flow (7.0 cm sec^{-1}) for another 2.5 min.
7. Wash the sieves with water and collect them into a beaker for processing via centrifugation-flotation with sugar.

Soil bioassays

One method to assess nematode populations of plant-endoparasitic species (i.e., *Meloidogyne*, *Nacobbus*) in bare soils is to use a soil bioassay that basically consists of growing a suitable host in infested soil in order to allow galls to develop and thereby assess the level of infection. The closed bag method is used to detect *N. aberrans* in soil taken from fallow fields.

Fig. 4.17. Extraction methods (cont.). A: Automated washing carousel (courtesy of J. Pickup, SASA); B: mist chamber. (Copyright Rothamsted Research.)

PROCEDURE (AFTER ORTUÑO *ET AL.*, 2005). Homogenize the soil sample, and take three subsamples (400 g each) and transfer 200 g of soil to polythene bags (10 × 20 cm). Water the soil contained in the bag with 50 ml of water (field capacity), and sow the potato seed or a healthy chit (without pathogens). Add the remaining 200 g of soil and 50 ml of water. Close the bag by folding the top of the bag and staple it closed. Place the bag in an incubator at 25°C for 30 days in the dark. If an incubator is not available, a cardboard box lined (base and lateral sides) with about six sheets of newspaper and protected from light can be used instead and the samples evaluated at 45 days after planting. After this period, galls should be evident. If not, leave the bioassay for another 10 days. Empty the bag and the contents. Check the roots, if no galls are observed then the soil can be considered free of the nematode.

Nematode cultures

A nematode culture is a method of multiplying nematodes by enabling them to reproduce and increase in a culture medium or on a specific host under controlled conditions. To establish cultures of plant-parasitic nematodes, soil samples containing the target nematode species must be collected from the plant rhizosphere or infested plant organs in natural sites or agriculture fields. Accurate identification of the host plant is essential, as is recording the original sampling area geographical coordinates or GPS location. Conventional glasshouse culture techniques, for example, using an appropriate substrate, susceptible plants in pots or microtanks (thermostatically controlled water tanks), require considerable space, glasshouse and tool hygiene protocols, to avoid cross-contamination with nearby cultures and careful management of factors that affect nematode invasion and multiplication, including plant age, relative humidity, temperature and adequate water (see Manzanilla-López and Ehlers, 2021). For more information on the collection and inoculum preparation of different plant-parasitic nematode species, see Wesemael (2021). It is worth pointing out that root-knot nematode glasshouse cultures must be started with a single egg mass (or J2 freshly hatched from a single egg mass) and cyst nematodes cultures with one cyst to ensure that cultures contain only one nematode species and not a mixture of species or populations. Nematode species in glass culture should be accurately identified by classic and molecular methods (see Chapter 5, this volume; Bulletin OEPP, 2021; Brito, 2021; Subbotin, 2021) and checked regularly for cross-contamination. Glasshouse cultures can be also used to start nematode cultures under aseptic conditions (i.e., contamination-free).

Successful *in vitro* cultures and production of nematodes in the numbers and physical condition needed for small or large experiments, either in the laboratory, glasshouse or field, require a basic knowledge of the nematode species and host biology; nematode host culturing methods; and media composition to grow both nematode and host. Petri-dish and vitroplant culture protocols (among others) are available for different plant-parasitic nematodes species, for example, *Pratylenchus*, *Radopholus*, *Aphelenchoides* and *Ditylenchus* on carrot discs, *Bursaphelenchus* on media containing a non-sporulating strain of *Botrytis cinerea*, and so on (see Bulletin OEPP, 2021; Manzanilla-López and Ehlers, 2021).

4.5. Extraction of Nematodes from Plant Tissues

Observations to check for the presence of nematodes in plants can be done under the stereoscopic microscope at 10-50 × magnifications. To facilitate observation of nematodes, the roots, leaves or other plant organs should be gently washed in water to remove soil and debris. Samples can be cut into smaller pieces (1-2 cm long) before placing them in a Petri or Syracuse dish containing 2-4 ml of water (Shurtleff and Averre III, 2000). Plant samples can be gently torn or teased apart with dissecting needles or fine insulin syringe needles, or sliced with razor blades or scalpels. Nematodes can be recovered from seed materials (e.g., *D. dipsaci*) by dissecting candidate seeds in water under the stereomicroscope at 20×. *B. xylophilus* can be extracted by taking trunk cores and soaking in water for several days. Nematodes can also be extracted from small chips of infested wood by the Baermann funnel technique (Smith, 1997). If nematodes are not observed immediately, samples can be left for a couple of hours or overnight before being re-examined under the stereo-microscope.

4.5.1. Mist Chamber

Nematodes on plant material move to the bottom of a funnel stimulated by the water mist that is intermittently sprayed (Hooper, 1986b). This method is suitable to recover most motile nematodes, second-stage juveniles (*Meloidogyne*, *Nacobbus*), ectoparasitic and migratory endoparasitic nematodes from plant tissues (e.g., bulb tissue, leaves, stems, seeds, roots). However, it requires specialized equipment, such as spray nozzles (Fig. 4.17B). Alternatively, roots, stems or leaves, preferably chopped into small pieces (ca 1 cm) can be incubated in water contained in a covered glass jar (or a similar container, including plastic bags) for 24 h, agitating the jar contents three or four times a day. A few drops of dilute hydrogen peroxide can be added to stimulate the exit from root tissues of juvenile stages and immature females of *Nacobbus* (the peroxide helps to macerate plant tissues and also provides oxygen, thereby improving extraction efficiency). Nematodes are subsequently collected from the water and examined every 24 h. Mist chamber extraction can be continued for 7 days or more.

4.5.2. Trituration of Roots

This technique is useful for extracting eggs, second-stage juveniles and other stages of migratory and endoparasitic nematodes. Roots are cut into small segments (1-2 cm) and thoroughly mixed. One sample (4-8 g of roots) is macerated in a blender for ca 2 min (Fig. 4.18A). The macerate is then sieved, filtered through sieves (Fig. 4.18B) or processed by centrifugation-flotation (optional) (Coolen and D'Herde, 1972). A variant of this technique is the trituration of potato tubers or storage root peelings (sugarbeet) to detect *Nacobbus* third- and fourth-stage juveniles, as well as immature females and males (Costilla, 1985).

Fig. 4.18. A: Courgette tissue, triturated in a Warburg blender, and equipment for filtering nematodes from macerated plant tissues; B: flat-bottomed evaporating dish containing cotton-wool milk filter, folded for extra thickness, supported on a muslin cloth secured with an elastic band to a plastic ring. (Copyright Rothamsted Research.)

4.6. Counting Nematodes

Once nematodes are collected in a known volume of water (or fixative), aliquots from a uniform suspension of the nematodes (obtained by, e.g., swirling the contents) can be taken and transferred to a counting dish, where the nematodes are counted with the aid of a hand tally counter. There is a range of commercially available counting dishes with different volume capacity, some being suitable to fit onto the stage of a compound microscope, but others can also be custom-made in the laboratory using Perspex or, more simply, merely by cutting a grid pattern on a plastic Petri dish with a scalpel (Fig. 4.19). Counting of nematodes can be done at 10-50× under uniform illumination by transmitted light. The number of nematodes should be expressed as the number of individuals per unit of volume (e.g., 100 or 500 cm^3 of soil) and not only as the number of nematodes present in the sample.

For counting cysts collected from wet or dry soil samples, several devices have been designed (Shepherd, 1986), including some that can used on the stage of a dissecting stereo-microscope (Fig. 4.19). Cysts can be separated and collected from dry soil samples using a modified version of the entomological aspirator or pooter (Fig. 4.19A).

To estimate juvenile numbers in cysts, the cysts (ca 50) can be squashed in a drop of water on a channelled (15 mm wide × 0.05 mm deep) aluminium slide (7.5 × 2.5 mm) by rolling over the cysts with a glass rod (Reid, 1955; Turner, 1998) (Fig. 4.19). Viability can be determined by vital stains such as phloxine B and Nile Blue A, which differentiate between living and dead nematodes (Shepherd, 1986; Turner, 1998).

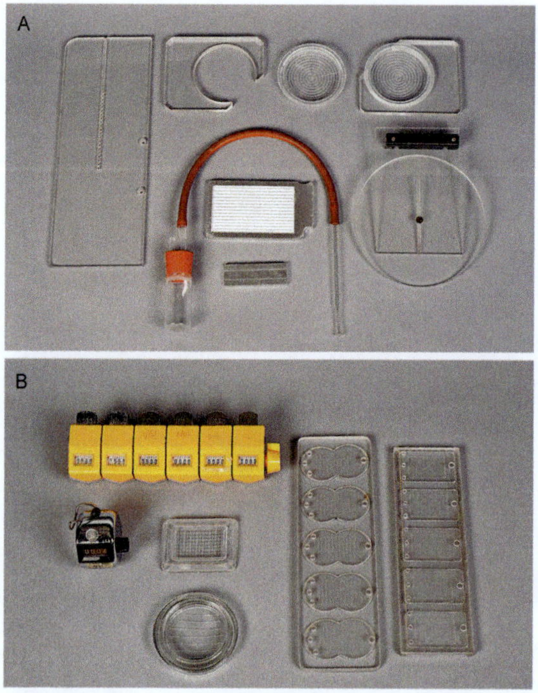

Fig. 4.19. A: Dishes for counting vermiform and cyst nematodes. Clockwise: Perspex base for rotatable Perspex stage used for recovering nematode cysts from moist debris; Doncaster dish base; Doncaster dish removed from base; Doncaster dish assembled with base, used for recovering nematode cysts from moist debris; Hesling aspirator and aluminium counting tray for recovering nematode cysts from dried debris, aluminium slide to crush cysts. B: Counting devices. Clockwise: multicounter; Fenwick multichamber counting slide with engraved graticules; Syracuse dish; rectangular counting dish; counter. (A, B Copyright Rothamsted Research.)

4.7. Fixing and Staining of Plant Organs and Tissues

To identify nematodes inside above- and below-plant organs and tissues, it is necessary to process the plant samples through fixation, staining, clearing and mounting in temporary or permanent slides (Hooper, 1986c); for more details, see Eisenback and Hunt (2021) and Bulletin OEPP (2021).

4.7.1. Root Fixation

Roots can be washed and then fixed in cold 5-10% formaldehyde, FA 4:10 or triethanolamine-formaldehyde (TAF). For histological studies, formalin acetic alcohol (FAA) is more suitable (Hooper, 1986d).

4.7.2. Staining Roots

Acid fuchsin lactoglycerol

Gently wash the roots till they are free from soil or other adhering debris; cut the roots into small pieces and plunge them into a deep beaker containing boiling lactoglycerol (equal volumes of glycerol, lactic acid and distilled water) with acid fuchsin stain (0.05%). This process should be done in a fume hood, and the glass beaker should be covered with a lid once the roots have been placed inside. Instead of a hot plate, a microwave oven can be used to boil the roots. Continue boiling for about 3 min. The time should be adjusted to suit the root material, with harder, more lignified roots requiring more time than softer roots, such as those from tomato. Several small samples can be stained in one operation by wrapping them separately in pieces of muslin cheese cloth, although each package must be labelled, perhaps using a piece of plastic plant label tied to, or enclosed within, the bag. Remove the roots and allow to cool. Wash off excess stain in running water and place in a Petri dish. Cover the roots with a mixture of water and glycerol (equal volumes of glycerol and distilled water) acidified with a few drops of lactic acid. Allow root tissue to differentiate for a time period ranging from several hours to several days. Examine in lactoglycerol (or glycerol) under the stereo-microscope. Nematodes will be stained red, whereas the rest of the roots remain largely unstained. Variants of the technique involve bleaching the roots before staining in a microwave oven (see Byrd *et al.*, 1983; Shurtleff and Averre III, 2000). Because of the potential harmful effects of acid fuchsin and phloxine B to humans, it is necessary to use protective eyewear, chemical-resistant gloves, a respirator and follow appropriate disposal procedures for chemical compounds.

Red food colour stain

An alternative to acid fuchsin and phloxine B stains is the McCormick Shilling red food colour (or similar food colour brand). A solution made of 12.5% (v/v) red food colour dissolved in distilled water is used to stain nematodes inside roots (Thies *et al.*, 2002). Root segments (1-2 cm) are placed in 1.5% NaOCl solution contained in a 200 ml beaker and agitated for 4 min, after which roots are removed, rinsed in running water (30 s) and submerged in tap water (15 min). Root segments are then plunged in the staining solution and brought to a boil for 30 s, and allowed to cool for 30 s at room temperature. Roots are rinsed with tap water and swirled for 15 s in previously heated (ca 40°C) acidified glycerol (40 ml glycerol plus five drops of 5N HCl) and allowed to cool before mounting on glass slides. Delicate seedling roots can be placed in the staining solution and left for 15 min at room temperature to avoid damage caused by the boiling stain solution. Egg masses of *Meloidogyne* spp. can be stained by submersing roots for 15 min in a 20% red food colour solution (Thies *et al.*, 2002).

4.7.3. Leaf Staining

Some chemical reagents that used to be commonly available to process and stain plant tissue samples for microscopic observation of nematodes inside leaf tissue have been banned or are difficult to acquire because of health and safety issues. Plant histological techniques should be adapted or modified, whenever possible, to replace toxic chemicals with safer ones. However, stains such as cotton blue, green malachite and acid fuchsin are still used to stain nematodes within plant tissue and can be combined with lactoglycerol to clear plant tissues prepared for semi-permanent slides. Other methods do not require chemicals at all (see section 4.5), such as the incubation of infected leaves in water for 48 h (*A. fragariae*, *A. ritzemabosi*) or soaking infected seeds (*A. besseyi*) in tap water (Xie *et al.*, 2019).

4.8. Killing, Fixation, Dehydration and Processing of Nematodes for Permanent Slides

Gentle heat (55-60°C) is recommended to kill nematodes in order to preserve internal and external structures in the best possible condition. Nematodes can be killed before fixation or be killed and fixed at the same time. Avoid over-heating nematodes (never boil them to death!), as internal structures will be disrupted, thus making identification, particularly to the species level, very difficult.

Nematodes are concentrated in a small amount of water (<20 ml) contained in a glass tube or any other glass container by either decanting or centrifuging prior to be heated directly on the flame of an alcohol lamp, on a hot plate, in an oven or microwave oven. Alternatively, the container with the nematodes can be partially immersed in a water bath (50-60°C) for a couple of minutes. This period is sufficient for the water in the container to reach the temperature required to kill the nematodes. After killing the nematodes, it is recommended to let them cool before adding a fixative. The most common fixatives are 4% formaldehyde, FA 4:1 (formaldehyde-glacial acetic acid), FAA (formaldehyde-alcohol-glacial acetic acid) and TAF (trethanolamine formaldehyde). Formaldehyde (4%) is a good fixative, especially if added immediately after FA 4:1 while TAF has the advantage of keeping the nematode appearance and body contents in very good condition and also prevents the specimens from completely desiccating due to evaporation as the triethanolamine is non-volatile (a trace of glycerol can be added to FA 4:1 and other fixatives so as to achieve the same effect – about 5% by volume should be sufficient). Nematodes are left in the fixative at least for one week. After fixation, intracellular components of tissues became insoluble, resulting in increased refraction under the microscope due to protein coagulation. To remove refringence, the fixative in tissues can be substituted by 96% ethanol by gradually dehydrating and clearing the specimens through a series of solutions made of ethanol, glycerol and distilled water (Seinhorst, 1959; Hooper, 1986a). The whole process can take 1-1.5 days (quick method) or up to several days (slow method). However, according to some authors, the slow method can give better results, particularly for more delicate nematodes or those with a thicker cuticle, such as criconematids, or more susceptible cuticle structure, such as *Hoplolaimus*.

4.8.1. Seinhorst Method for Permanent Slides

Fixed nematodes are placed from the fixative into a small container (i.e., watch glass or a micro-Syracuse) filled with 4% formaldehyde. Containers and nematodes are placed in a desiccator (Fig. 4.20) filled for ca 10% of its volume with ethanol and left until the fixative is replaced by ethanol (ca 4 h). The alcohol is then carefully removed from the container with a syringe or a Pasteur pipette under the stereoscopic microscope, before refilling the small container with 0.1 ml of Seinhorst solution I (20 parts 96% ethanol, 1 part glycerol, 79 parts distilled water). The container is then placed into a second desiccator containing dehydrated anhydrous silica gel. Samples can be left for several days in the desiccator until the alcohol and distilled water of the Seinhorst solution I evaporates (slow method). Alternatively, the small container with nematodes in Seinhorst solution I can be placed in an oven in a Petri dish, partially closed to allow slow evaporation of the ethanol at 30-40°C for 12 h (quick method). Once the ethanol has evaporated, leaving only a thin film of glycerol, Seinhorst solution II is added (5 parts glycerol and 95 parts 96% ethanol) and the container is returned to the silica gel desiccator and left until the alcohol has completely evaporated. Alternatively, the container with nematodes in Seinhorst solution II can be left in an oven for 3 h (quick method). After this period, the nematodes are ready to be mounted. However, there are many other methods available that require much less preparation time (see Ryss,

Fig. 4.20. Seinhorst method of killing, fixing and processing nematodes. This figure shows fixative, Seinhorst solutions 1 and 2, desiccator, apparatus for dropper in which FA 4:1 is heated for killing and fixing nematodes, excavated glass block. (Copyright Rothamsted Research.)

2003), one of the simplest being to place the nematodes in a drop of fixative in an excavated glass block and then filling with a 5% glycerol solution. The block may be left at room temperature or in a cool oven (about 35-40°C) and partially covered with a lid to control evaporation, at least in the initial stages. Nematodes are soon left nicely cleared and in pure glycerol, from which they can be mounted on to slides.

Although these techniques usually work well, occasionally, the nematodes may end up looking like flat ribbons. This seems to be particularly so for ring nematodes and *Hoplolaimus* and is probably due to a combination of processing being done too rapidly and the cuticle structure of the nematodes concerned. Though these ribbon-like nematodes may appear unpromising, they can usually be recovered if placed in a drop of glycerol on a slide and gently heated using a small flame from a spirit burner. Care being taken not to apply too much heat. Frequent examination under the stereo-microscope should be done to check the progress of this 're-inflation' and to avoid over-heating and browning of the specimens.

4.8.2. Fixatives

Formalin 4%	
Formaldehyde 40%	10 ml
Distilled water	90 ml
FAA	
Ethanol 96%	200 ml
Formaldehyde 40%	6 ml
Glacial acetic acid	1 ml
Distilled water	40 ml
FA 4:1	
Formaldehyde 40%	10 ml
Acetic acid	1 ml
Distilled water	89 ml
TAF	
Formalin (40% formaldehyde)	7 ml
Triethanolamine	2 ml
Distilled water	up to 100 ml
DESS	

This fixative solution, made of 20% dimethyl sulphoxide (DMSO), 0.25M disodium ethylenediaminetetraacetic acid (EDTA) (pH 7.5) and saturated NaCl, can be used instead of formalin-based fixatives to reduce chemical health hazards. The DESS solution (see section 4.9.1) preserves individual nematodes, nematode extracts or entire soil/sediment samples, which can be stored for months at room temperature (Yoder *et al.*, 2006).

Note: Always prepare the fixatives in a fume cabinet. Fixatives should be kept in a Winchester bottle properly and properly labelled (Hooper, 1986c).

4.9. Processing Specimens for DNA Extraction

If specimens are to be studied both morphologically and molecularly (see Bulletin OEPP, 2021), after killing by gentle heat some of the nematodes should be preserved for at least a week in a solution of 1M NaCl (S.A. Subbotin, pers. comm.). Nematodes can also be preserved in DESS (Yoder *et al.*, 2006).

4.9.1. Preparing DESS

1. Weigh 23.265 g of EDTA disodium (molar mass or formula weight = 372.24) to prepare a 250 ml solution. Place the salt into a beaker.
2. Add 50 ml of deionized water to the EDTA salt and stir.
3. Make a 1M NaOH solution to pH the EDTA. It will take ca 50 ml to adjust EDTA pH to 7.5.
4. The EDTA will dissolve slowly, and heating the solution at 30°C may help to speed up the process.
5. Once all the EDTA salt is dissolved, bring the volume up to 200 ml with deionized water. Then add the 20% DMSO solution (i.e., 50 ml DMSO for 250 ml). Stir for a few minutes.
6. Add NaCl until it no longer dissolves. Pour the solution into a bottle, leaving most of the salt crystals in the beaker.

4.10. Preparation of Nematode Specimens for Mounting

To make a temporary or permanent microscopic slide, nematodes (either alive or processed) can be placed in a drop of appropriate liquid mounting medium (i.e., water for live specimens, glycerol for processed specimens) sandwiched between a glass slide and a cover glass. Cover glasses should be sealed to avoid evaporation of aqueous mounting media and consequent desiccation of the nematodes (Fig. 4.21). Nematodes can be mounted in tap or distilled water for living specimens or in a suitable media for permanent slides (usually anhydrous glycerol). A solution made of 4% formaldehyde (165 ml), glycerol (15 ml) and distilled water (1500 ml) can be used as a temporary mounting medium for freshly killed specimens (I. Cid del Prado-Vera, pers. comm.). Cover glasses can be sealed with paraffin wax or nail polish, nail polish with an acrylic hardener being the best to use. To avoid flattening the nematodes, and therefore, distorting certain features, such as shape of the tail and/or cephalic region, or altering morphometric ratios based on body diameter, and so on, it is recommended to use at least three cover glass supports, arranged in a triangle, made of fibreglass or Ballotini beads, with a diameter slightly larger than that of the nematodes to be protected. Alternatively, the wax ring method can be used (Fig. 4.21A). In this method, nematodes are placed in the centre of a small drop of mounting medium surrounded by a wax ring that serves both as a seal and cover glass support. A 19 mm round cover glass is applied to the wax ring, and the slide is placed on a hotplate at 65°C for a few seconds. The wax melts, allowing the cover glass to settle down and keeping the glycerol and nematodes in the centre of the mount (Hooper, 1986c). Even though the wax acts as a support, it can be wise to provide additional supports within the glycerol drop. Care should be taken to ensure that these are pushed down through the drop so that they rest on the surface of the slide and are not floating.

4.10.1. Electron Microscopy

If there is access to electron microscope or bio-imaging facilities, a subsample of specimens should be prepared for this kind of microscopy,

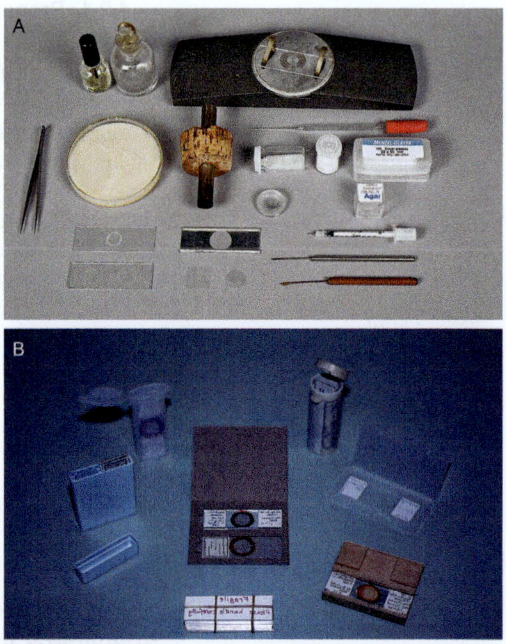

Fig. 4.21. Slide mounting materials. A: Top row: nail varnish and anhydrous glycerol bottles, ringing table for applying seal around cover glass. Middle row: fine forceps, wax, device to make wax rings, Ballotini glass beads, micro Syracuse dish, cover glasses, glass slide with a wax ring applied, excavated glass slide, aluminium double cover glass slide, square and round cover glasses, insulin syringe, eyebrow hair mounted on aluminium needle, nylon bristle mounted on wooden handling needle; B: containers for transporting permanent slides. (Copyright Rothamsted Research.)

especially for morphological studies or when describing a new species. Depending on SEM or cryo-SEM protocols, specimens may or may not require chemical fixation before being processed (see Eisenback, 1986; Carta *et al.*, 2003; Yushin *et al.*, 2021).

4.10.2. Vulval-cone Mounts of Cyst Nematodes

Dry cysts are pre-soaked overnight and then placed, using fine forceps, on a Perspex slide on the stage of a stereo-microscope. The posterior end of the cyst or vulval cone is then cut off and trimmed around the fenestral area (no more than 5-10 times the fenestral area). The vulval cones or fenestral areas are then gently cleaned with an eyebrow handling needle to avoid damage to the structures. Vulval cones are rinsed in distilled water and passed through 70, 95 and 100% ethanol to clove oil and mounted in Canada balsam or Euparal and protected with a cover glass. Ballotini glass beads or glass fibre can be used as supports to avoid crushing the vulval cones. Slides are dried in an oven (70°C). Cones can also be mounted in glycerine jelly and sealed. If cysts are too brittle, the vulval cone can be trimmed in a drop of lactoglycerol (Hooper, 1986c).

Other methods, such as staining of roots and making perineal patterns for *Meloidogyne*, are described in Chapter 11, this volume.

4.11. Health and Safety

It is important to emphasize that many of the chemicals and procedures used in nematology are potentially hazardous. All procedures should be the subject of appropriate risk analysis, and recommended procedures, including the use of designated personal protective equipment (PPE), should be rigorously implemented and followed. Hazard data sheets for the chemicals involved must be available to staff, who should be fully informed of any potential risk. Individual countries have their own legal requirements in these matters, and it is the responsibility of the senior manager and user to ensure that the appropriate legislation and procedures are followed.

4.12. References

Barker, K.R. (1978) Determining nematode population responses to control agents. In: Zehr, E. (ed), *Methods for Evaluating Plant Fungicides, Nematicides and Bactericides*. Phytopathology Society, St. Paul, Minnesota, pp. 114–125.

Barker, K.R. (1985) Sampling nematode communities. In: Barker, K.R., Carter, C.C. and Sasser, J.N. (eds), *An Advanced Treatise on* Meloidogyne *Vol. II. Methodology. International Meloidogyne Project*. North Carolina State University, Raleigh, North Carolina, pp. 3–17.

Barker, K.R. and Campbell, C.L. (1981) Sampling nematode populations. In: Zuckerman, B.M. and Rohde, R.A. (eds), *Plant Parasitic Nematodes, Vol. III*. Academic Press, London, UK, pp. 451–474.

Been, T.H., Schomaker, C.K. and Jones, J.T. (2024) Distribution patterns and sampling. In: Perry, R.N., Moens, M. and Jones, J.T. (eds), *Plant Nematology*, 3rd edn. CABI, Wallingford, UK, pp. 380–414.

Bell, N. and Watson, R. (2001) Optimising the Whitehead and Hemming tray method to extract plant parasitic and other nematodes from two soils under pasture. *Nematology* 3, 179–185. DOI: 10.1163/156854101750236312.

Bridge, J. and Starr, J.L. (2007) *Plant Nematodes of Agricultural Importance: A Color Handbook*. Manson Publishing, London, UK.

Brito, J.A. (2021) Isoelectric focusing of proteins. In: Perry, R.N., Hunt, D.J. and Subbotin, S.A. (eds), *Techniques for Work with Plant and Soil Nematodes*. CABI, Wallingford, UK, pp. 206–217.

Bulletin OEPP EPPO Bulletin PM7/148 (1). (2021) Guidelines for the management of nematode collections used for the production and maintenance of reference material. *EPPO Bulletin* 51, 507–548. DOI: 10.1111/epp.12798.

Byrd, D.W., Kirkpatrick, T. and Baker, K.R. (1983) An improved technique for clearing and staining plant tissues for detection of nematodes. *Journal of Nematology* 4, 206–212.

Carta, L.K., Wergin, W.P., Erbe, E.F. and Murphy, C.A. (2003) A comparison of low-temperature and ambient-temperature SEM for viewing nematode faces. *Journal of Nematology* 35, 78–81.

Chen, P., Liu, S.-C., Liu, H.-I., Chen, T.-W. and Chiang, K.-S. (2011) Probability of detecting nematode infestations for quarantine sampling with imperfect extraction efficacy. *Journal of Nematology* 43, 16–24.

Coolen, W.A. and D'Herde, C.J. (1972) *A Method for the Quantitative Extraction of Nematodes from Plant Tissue*. Agricultural Research Administration, State Agricultural Research Centre, Ministry of Agriculture, Ghent, Belgium.

Costilla, M.A. (1985) Un método rápido para la extracción y observación de estados juveniles de *Nacobbus aberrans* en tubérculos de papa. *Revista Industrial y Agrícola de Tucumán* 62, 163–170.

Coyne, D.L., Nicol, J.M. and Claudius-Cole, B. (2007) *Practical Plant Nematology: a Field and Laboratory Guide*. SPM-IPM Secretariat, International Institute of Tropical Agriculture (IITA), Cotonou, Benin.

den Nijs, L.J.M.F., Pickup, J. and Ehlers, R.U. (2021) Sampling. In: Perry, R.N., Hunt, D.J. and Subbotin, S.A. (eds), *Techniques for Work with Plant and Soil Nematodes*. CABI, Wallingford, UK, pp. 1–11.

Duncan, L.W. and Phillips, M.S. (2009) Sampling root-knot nematodes. In: Perry, R.N., Moens, M. and Starr, J.L. (eds), *Root-knot Nematodes*. CAB International, Wallingford, UK, pp. 275–300.

Dunn, R.A. (1969) Extraction of cysts of *Heterodera* species from soils by centrifugation in high density solutions. *Journal of Nematology* 1, 7.

Eisenback, J.D. (1986) A comparison of techniques useful for preparing nematodes for scanning electron microscopy. *Journal of Nematology* 18, 479–487.

Eisenback, J.D. and Hunt, D.J. (2021) Handling, fixing, staining and mounting nematodes. In: Perry, R.N., Hunt, D.J. and Subbotin, S.A. (eds), *Techniques for Work with Plant and Soil Nematodes*. CABI, Wallingford, UK, pp. 71–87.

Hallmann, J., Daub, M. and Wesemael, M. (2021) Estimating numbers. In: Perry, R.N., Hunt, D.J. and Subbotin, S.A. (eds), *Techniques for Work with Plant and Soil Nematodes*. CABI, Wallingford, UK, pp. 42–59.

Haydock, P.J. and Perry, J.N. (1998) The principles and practice of sampling for the detection of potato cyst nematodes. In: Marks, R.J. and Brodie, B.B. (eds), *Potato Cyst Nematodes*. CAB International, Wallingford, UK, pp. 61–74.

Hooper, D.J. (1986a) Extraction of free-living stages from soil. In: Southey, J.F. (ed), *Laboratory Methods for Work with Plant and Soil Nematodes*, 6th edn. Ministry of Agriculture Fisheries and Food, Reference Book 402, Her Majesty's Stationery Office, London, UK, pp. 5–30.

Hooper, D.J. (1986b) Extraction of nematodes from plant material. In: Southey, J.F. (ed), *Laboratory Methods for Work with Plant and Soil Nematodes*, 6th edn. Ministry of Agriculture, Fisheries and Food, Reference Book 402, Her Majesty's Stationery Office, London, UK, pp. 50–58.

Hooper, D.J. (1986c) Handling, fixing, staining and mounting nematodes. In: Southey, J.F. (ed), *Laboratory Methods for Work with Plant and Soil Nematodes*, 6th edn. Ministry of Agriculture, Fisheries and Food, Reference Book 402, Her Majesty's Stationery Office, London, UK, pp. 59–80.

Hooper, D.J. (1986d) Preserving and staining nematodes in plant tissues. In: Southey, J.F. (ed), *Laboratory Methods for Work with Plant and Soil Nematodes*, 6th edn. Ministry of Agriculture, Fisheries and Food, Reference Book 402, Her Majesty's Stationery Office, London, UK, pp. 81–85.

Jenkins, W.R. (1964) A rapid centrifugal-flotation technique for separating nematodes from soil. *Plant Disease Reporter* 48, 692.

Manzanilla-López, R.H. and Ehlers, R.-U. (2021) Culturing techniques. In: Perry, R.N., Hunt, D.J. and Subbotin, S.A. (eds), *Techniques for Work with Plant and Soil Nematodes*. CABI, Wallingford, UK, pp. 88–151.

Manzanilla-López, R.H. and Starr, J.L. (2009) Interactions with other pathogens. In: Perry, R.N., Moens, M. and Starr, J.L. (eds), *Root-knot Nematodes*. CAB International, Wallingford, UK, pp. 223–245.

McSorley, R. (1987) Extraction of nematodes and sampling methods. In: Brown, R.H. and Kerry, B.R. (eds), *Principles and Practice of Nematode Control in Crops*. Academic Press, Sydney, Australia, pp. 13–47.

Ortuño, N., Franco, J., Ramos, J., Oros, R., Main, G., *et al.* (2005) *Desarrollo del manejo integrado del nematodo rosario de la papa* Nacobbus aberrans *en Bolivia*. Proyecto Papa Andina (CIP-COSUDE), Cochabamba, Bolivia, pp. 51–52.

Parrado, L.M. and Quintanilla, M. (2024) Plant-parasitic nematode disease complexes as overlooked challenges to crop production. *Frontiers in Plant Science, Section Plant Pathogen Interactions* 15, 1439951. DOI: 10.3389/fpls.2024.1439951.

Perry, R.N., Hunt, D.J. and Subbotin, S.A. (eds) (2021) *Techniques for Work with Plant and Soil Nematodes*. CABI, Wallingford, UK.

Pickup, J., Roberts, A.M.I. and den Nijs, L.J.M.F. (2018) Quarantine, distribution patterns and sampling. In: Perry, R.N., Moens, M. and Jones, J.T. (eds), *Cyst Nematodes*. CABI, Wallingford, UK, pp. 128–157. DOI: 10.1079/9781786390837.0128.

Reid, A., Kenyon, D.M., Evans, F.F., Mulholland, V., Pickup, J., *et al.* (2010) Development of a high-throughput for the detection and species determination of PCN. In: Haydock, P. (ed), *Aspects of Applied Biologists 103, 3rd symposium on Potato Cyst Nematodes*. Association of Applied Biologists, Warwick, UK, pp. 13–16.

Reid, E. (1955) A rolling method for opening cysts of potato eelworm. *Plant Pathology* 4, 28–29.

Ryss, A.Y. (2003) Express technique to prepare collection slides of nematodes. *Zoosystematica Rossica* 11, 257–260.

Seinhorst, J.W. (1959) A rapid method for the transfer of nematodes from fixatives to anhydrous glycerin. *Nematologica* 4, 67–69.

Shepherd, A.M. (1986) Extraction and estimation of cyst nematodes. In: Southey, J.F. (ed), *Laboratory Methods for Work with Plant and Soil Nematodes*, 6th edn. Ministry of Agriculture, Fisheries and Food. Reference Book 402, Her Majesty's Stationery Office, London, UK, pp. 31–50.

Shurtleff, M.C. and Averre III, C.W. (2000) *Diagnosing Plant Diseases Caused by Nematodes*. APS Press, St. Paul, Minnesota.

Sikora, R.A., Coyne, D., Hallmann, J. and Timper, P. (eds) (2018) *Plant Parasitic Nematodes in Subtropical and Tropical Agriculture*, 3rd edn. CAB International, Wallingford, UK.

Singh, S., Awasthi, L.P., Jangre, A. and Nirmalkar, V.K. (2020) Transmission of plant viruses through soil-inhabiting nematode vectors. In: Awasthi, L.P. (ed), *Applied Plant Virology. Advances, Detection, and Antiviral Strategies*. Academic Press, London, UK, pp. 291–300. DOI: 10.1016/B978-0-12-818654-1.00022-0.

Smith, I.M., McNamara, D.G., Scott, P.R., Holderness, M. and Burger, B. (eds) (1997) *Quarantine Pests for the European Union and for the European and Mediterranean Plant Protection Organization*. CAB International, Wallingford, UK, pp. 581–591.

Southey, J.F. (ed) (1986) *Laboratory Methods for Working with Plant and Soil Nematodes*, 6th edn. Reference Book 42, Ministry of Agriculture, Fisheries and Food, Her Majesty's Stationery Office, London, UK, pp. 1–4.

Taylor, C.E. and Brown, D.J.F. (1997) *Nematode Vectors of Plant Viruses*. CAB International, Wallingford, UK.

Thies, J., Merril, S.B. and Corley, E.L. (2002) Red food coloring stain: New safer procedures for staining nematodes in roots and egg masses on root surfaces. *Journal of Nematology* 34, 179–181.

Trudgill, D.L., Evans, K. and Faulkner, G. (1973) A fluidizing column for extracting nematodes from soil. *Nematologica* 18(1972), 469–475.

Turner, S.J. (1998) Sample preparation, soil extraction and laboratory facilities for the detection of potato cyst nematodes. In: Marks, R.J. and Brodie, B.B (eds), *Potato Cyst Nematodes*. CAB International, Wallingford, UK, pp. 75–90.

Viane, N., Hallmann, J. and Teklu, M.G. (2021) Methods for nematode extraction. In: Perry, R.N., Hunt, D.J. and Subbotin, S.A. (eds), *Techniques for Work with Plant and Soil Nematodes*. CABI, Wallingford, UK, pp. 12–41.

Wesemael, W. (2021) Screening plants for resistance/susceptibility to plant-parasitic nematodes. In: Perry, R.N., Hunt, D.J. and Subbotin, S.A. (eds), *Techniques for Work with Plant and Soil Nematodes*. CABI, Wallingford, UK, pp. 60–70.

Xie, J., Yang, F., Wang, Y., Peng, Y. and Ji, H. (2019) Studies on the efficiency of different inoculation method of rice white-tip nematode, *Aphelenchoides besseyi*. *Nematology* 21, 673–678. DOI: 10.1163/15685411-00003244.

Yoder, M., De Ley, I.T., King, I., Mundo-Ocampo, M., Mann, J., *et al.* (2006) DESS: a versatile solution for preserving morphology and extractable DNA of nematodes. *Nematology* 8, 367–376.

Yushin, V.V., Claeys, M., Leunissen, J.L.M. and Zograf, J. (2021) Electron microscopy techniques. In: Perry, R.N., Hunt, D.J. and Subbotin, S.A. (eds), *Techniques for Work with Plant and Soil Nematodes*. CABI, Wallingford, UK, pp. 132–151.

Zacheo, G. and Lamberti, F. (1974) Un metodo rapido per l'estrazione di uova di nematodi dal suolo e dai tessuti vegetali. *Nematologia Mediterranea* 2, 55–59.

Zuckerman, B.M., Mai, W.F. and Krusberg, L.R. (eds) (1990) *Plant Nematology Manual*, revised edn. University of Massachusetts Agricultural Experiment Station, Amherst, Massachusetts.

5 Molecular Diagnostics of Nematodes: Some Practical Considerations

Thomas O. Powers[1] and Mariella M. Finetti-Sialer[2]*

[1]*University of Nebraska, Lincoln, Nebraska, USA;* [2]*National Research Council Institute of Biosciences and Bioresources, Bari, Italy*

Abstract

Molecular diagnostics can be broadly defined as the application of molecular techniques for the purpose of identification. This chapter focuses on practical aspects of establishing and conducting a programme in nematode molecular diagnostics. Molecular diagnostics and barcodes are introduced and supplemented by some practical considerations appropriate to select the most appropriate approaches to nematode diagnostic and identification, and to conduct a programme in nematode molecular diagnostics. Case studies in the nematode molecular diagnostics for cyst nematodes (*Globodera* spp.), root-knot nematodes (*Meloidogyne* spp.), and stem and gall nematodes (*Ditylenchus* spp.) are provided as examples.

5.1. Introduction

Molecular diagnostics can be broadly defined as the application of molecular techniques for the purpose of identification. This chapter focuses on practical aspects of establishing and conducting a programme in nematode molecular diagnostics. Collectively, previous reviews of molecular diagnostics document the rapid growth of the field and the amazing technological advances that have driven the development of diagnostic protocols (Jones *et al.*, 1997; Floyd *et al.*, 2002; Blaxter, 2004; Powers, 2004; Blok, 2005; Subbotin and Moens, 2006; De Waele and Elsen, 2007; Blok and Powers, 2009; Neilson *et al.*, 2009; Abebe *et al.*, 2011; Castagnone-Sereno *et al.*, 2011; Oliveira *et al.*, 2011; Bogale *et al.*, 2020; Subbotin, 2021). At the high end of the technological spectrum, entire terrestrial or aquatic communities of organisms are now being analysed from a single DNA extraction taken directly from the environment. Early applications of environmental DNA sequencing helped reveal the presence of innumerable soil bacteria that were formerly unknown due to their failure to grow using standard isolation through culturing. This environmental sequencing approach is now being applied to eukaryote soil communities, along with nematode populations (Ficetola *et al.*, 2024; Van Himbeeck *et al.*, 2024). Virtually unthinkable several decades ago, systematic comparisons of entire genomes

*E-mail: mariella.finetti@cnr.it

A revision of Powers, T.O. and Ramírez-Suárez, A. (2012) Molecular diagnostics of nematodes: some practical considerations. In: Manzanilla-López, R.H. and Marbán-Mendoza, N. (eds), *Practical Plant Nematology*. Mundi-Prensa and Biblioteca Básica de Agricultura Montecillo, Mexico.

 Practical Plant Nematology 2nd Edition
(eds R.H. Manzanilla-López, D.J. Hunt, N. Marbán-Mendoza and S. Infante Gil)
DOI: 10.1079/9781836990413.0005

have been made possible by high-throughput nucleotide sequencing and advances in bioinformatics. This new field of comparative biology has been termed phylogenomics. Both of these examples of technological wizardry point to an exciting future for species discovery and identification.

Molecular diagnostics, however, can be conducted at a more modest scale. Techniques that were once considered too expensive or excessively complex are now routinely conducted in laboratories with a small-scale diagnostic budget. At the centre of most molecular diagnostic approaches is polymerase chain reaction (PCR). PCR is a DNA amplification technique that produces a large amount of DNA from a minute amount of template. In the case of nematodes, a single juvenile comprising fewer than 1000 cells can be the source of multiple PCR reactions, each generating millions of copies of specific DNA sequences. The PCR process is rather simple (Fig. 5.1). The reaction basically consists of five components:

1. The DNA template that includes the region of DNA to be amplified.
2. A pair of nucleotide primers, typically short 10-20 nucleotide stretches of DNA positioned on opposite strands that delimit the region to be amplified.
3. Deoxynucleoside triphosphates that serve as the building blocks of the DNA amplification product.
4. A thermal-stable DNA polymerase that synthesizes DNA.
5. A reaction solution that includes ions and buffers necessary for an optimum environment for DNA synthesis.

Many variations of PCR have been adapted for molecular diagnostics of nematodes. For example, Oliveira *et al.* (2011) provide a table that lists 26 quantitative PCR assays for nematode species. Quantitative PCR combines the specificity of traditional PCR, with an assessment of the rate of generated amplification products, to provide an estimate of template abundance in the sample. This estimate can then be used to determine the number of target nematodes in the soil. The publications listed in Table 5.1 provide examples of molecular diagnostic assays for plant-parasitic nematodes and give an indication of the research priorities in the molecular differentiation of plant-parasitic nematode species.

5.2. Practical Considerations

Before selecting among the various diagnostic assays, there are some practical considerations that could aid investigators in choosing the most appropriate diagnostic approach.

1. What is the question? This is an important and deceptively simple question that should precede a diagnostic investigation. Not every diagnostic decision needs a molecular solution. Is *Heterodera* present in my soil sample? This is an example of a question that is easily answered with a set of soil sieves and a microscope. Molecular diagnostics should be used to address precise questions of identity. For example, plant protection officers need to know if a specific set of regulated nematode species is present in a field before issuing phytosanitary certification. This certification is required in many countries

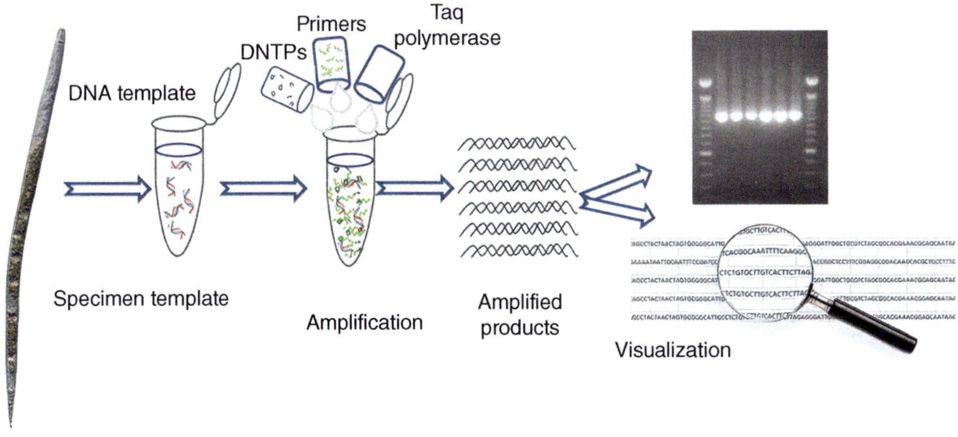

Fig. 5.1. The PCR process.

Table 5.1. Examples of studies employing DNA diagnostics for nematode identification.

Diagnostic target	PCR target	Molecular assay	Authors
Bursaphelenchus xylophilus	DNA topoisomerase I	TaqMan	Huang *et al.* (2010)
Bursaphelenchus xylophilus and *B. mucronatus*	5S rDNA	SYBR Green	Kang *et al.* (2009)
Bursaphelenchus xylophilus	Hsp70	TaqMan	Leal *et al.* (2007)
Bursaphelenchus xylophilus	gDNA	RPA-LFD	Zhou *et al.* (2022)
Bursaphelenchus spp.	ITS	Multiplex	Zhuo *et al.* (2011)
Ditylenchus destructor	ITS, D2/D3	PCR-RFLP, DNA sequence	Subbotin *et al.* (2011)
Ditylenchus destructor	rDNA-ITS	qRPA	Gao *et al.* (2021)
Ditylenchus destructor and *D. dipsaci*	ITS	Multiplex, DNA sequence	Marek *et al.* (2010)
Ditylenchus gigas	ITS, D2/D3, COI, Hsp90	PCR-RFLP DNA sequence	Vovlas *et al.* (2011)
Globodera spp.	Chorismate mutase	TaqMan	Yu *et al.* (2011)
Globodera spp.	ITS	Multiplex, DNA sequence	Skantar *et al.* (2011)
Globodera spp.	Hsp90	Multiplex, TaqMan	Madani *et al.* (2011)
Globodera spp.	ITS	PCR-RFLP	Sirca *et al.* (2010)
Globodera rostochiensis and *G. pallida*	D2/D3	DNA sequence	Douda *et al.* (2010)
Globodera rostochiensis and *G. pallida*	Cytb, ITS, D2/D3	PCR-RFLP, DNA sequence	Madani *et al.* (2010)
Globodera rostochiensis	ITS	SYBR Green	Toyota *et al.* (2008)
Globodera rostochiensis	ITS	SYBR Green	Goto *et al.* (2011)
Globodera rostochiensis and *G. artemisiae*	ITS	TaqMan	Nowaczyk *et al.* (2008)
Heterodera filipjevi and *H. avenae*	ITS	PCR-RFLP	Yan and Smiley (2010)
Heterodera glycines	ITS	SYBR Green	Goto *et al.* (2011)
Heterodera glycines	D2/D3, SCAR	Multiplex	Ou *et al.* (2008)
Heterodera schachtii	gDNA	SCAR-PCR, RPD marker	Jiang *et al.* (2021)
Heterodera schachtii	ITS	SYBR green	Madani *et al.* (2005)
Meloidogyne arenaria	SCAR	SYBR Green	Agudelo *et al.* (2011)
Meloidogyne javanica	ITS	SYBR Green	Berry *et al.* (2008)
Meloidogyne chitwoodi and *M. fallax*	ITS	TaqMan	Zijlstra and Van Hoof (2006)
Meloidogyne chitwoodi	gDNA	SSR, INDEL and SCAR primers	Hu *et al.* (2023)
Meloidogyne incognita	ITS	SYBR Green	Goto *et al.* (2011)
Pratylenchus neglectus and *P. thornei*	D3	PCR product (+/– soil assay)	Yan *et al.* (2008)
Pratylenchus penetrans	D2/D3, ITS	Multiplex	Waeyenberge *et al.* (2009)
Pratylenchus penetrans	D2-D3rDNA	COXI, PP5, B-1-4-endogluconase	Bogale *et al.* (2021)
Pratylenchus thornei	ITS	SYBR Green, soil assay	Yan *et al.* (2012)
Pratylenchus zeae	ITS	SYBR Green	Berry *et al.* (2008)
Xiphinema elongatum	ITS	SYBR Green	Berry *et al.* (2008)
Xiphinema chambersi	ITS, D2/D3, COI	PCR assay	Handoo *et al.* (2016)

to allow the movement of agricultural commodities across international borders. In some cases, the diagnostic target of the assay may be a single species such as *Meloidogyne chitwoodi*, a frequently regulated species in potato production and commerce. In this case, a simple presence/absence assay will suffice. In other cases, simultaneous identification of several species may be desirable (Hodson *et al.*, 2023). For example, in the central USA, farm managers may be concerned about specific nematode species that have the potential to interact with other plant pathogens. In potato fields, the interaction of *Pratylenchus penetrans* or *P. scribneri* with *Verticillium* species may cause a disease known as potato early dying (Riedel *et al.*, 1985). Distinguishing these two *Pratylenchus* species from the other species of *Pratylenchus* commonly found in central USA soils would improve management decisions (Fig. 5.2). For this purpose, possibly a multiplex quantitative PCR assay that recognizes both species would be preferred (Yan *et al.*, 2012).

2. Is there an adequate reference database to support conclusions based on a molecular investigation? For example, there may be PCR primer sets that allow you to generate amplified PCR products from *Meloidogyne graminicola*. However, if verified DNA sequences of *M. graminicola* do not exist within a reference database, your identification could be questioned based on the lack of comparative data. Building the comprehensive reference database is an issue that must be resolved by the nematology scientific community.

Fig. 5.2. Morphological comparison of *Pratylenchus scribneri* (A) and *P. penetrans* (B).

Quality of entries, linkage to multiple types of data and open access are important features of reliable databases. In addition to standard sequence databases like GenBank and EMBOL, several nematode-specific databases are under development. These include the *Nematode Tree of Life* (*Nema*TOL) and the Quarantine Barcode of Life (QBOL).

3. Have the necessary comparative studies been conducted to unambiguously interpret your results? This is an exceptionally important question. Whereas generation of nucleotide sequence data is a routine process, the determination of species identity based on nucleotide sequence data is anything but routine. Without the detailed underlying taxonomic studies that examine and *establish species boundaries*, a query nucleotide sequence may not have the necessary context to interpret the results. Reliability and precision in diagnostic methods are *directly related* to the quality of the corresponding *taxonomic framework*. To date, 575 complete genomes of various nematode species are deposited in the NCBI database, representing over 30 plant-parasitic nematode species. As time moves on, this database will continue to grow, providing more comprehensive and complete information, with improved genome annotation. This expansion will enhance the ability to select more appropriate and selective primer pairs for an accurate identification of nematode species and/or populations.

4. Do standard operating procedures (SOP) exist that can answer the question? Plant nematologists have not been as active as animal parasitologists in producing SOP for species diagnostics. The *European Plant Protection Organization* (EPPO) is a leader in attempting to formalize approaches for the identification of nematodes of quarantine and regulatory importance. Examples of these protocols are found at the *EPPO Quarantine Pest website* (http://www.eppo.org/QUARANTINE/quarantine.htm).

5. Is next-generation sequencing (NGS) convenient for plant-nematode identification? Given the great diversity of nematodes compared to other phyla, DNA analysis through molecular techniques and metagenomic approaches shows great promise for rapidly uncovering this hidden diversity and aiding in specimen identification. While NGS is an excellent tool for studying genomes, functional genomics, transcriptomics and gene expression (Lu *et al.*, 2020), its application to nematode identification is not without challenges. To implement this technique effectively, several practical obstacles must be addressed, beginning with sample collection, as outlined below. After obtaining a purified and representative sample, selecting the most suitable NGS platform is crucial. The platform should be practical, user-friendly and capable of providing an accessible analysis pipeline for the operator (van de Vossenberg *et al.*, 2023). Nematodes contain very small amounts of DNA, but advancements in sequencing technologies now allow us to enrich specific sequences, such as ribosomal DNA (rDNA) or mitochondrial genes useful for identification. Several platforms are available for nematode sequencing, ranging from those optimized for short-read determination (e.g., Illumina™) to those that provide long reads (e.g., Nanopore™). These approaches have been successfully applied to plant-parasitic nematodes (Ciancio *et al.*, 2018), with Nanopore technology offering the added benefit of portability, allowing sequencing to be performed both in the laboratory and in the field. However, when working with soil samples, it is important to keep in mind that total nucleic acid extraction captures DNA from both living and dead specimens. Since DNA can persist in the environment for extended periods, there is a risk of detecting inactive nematode populations at the time of sampling. Depending on the sequencing goal, and to mitigate this issue, the focus should shift to RNA molecules. Once RNA is reverse-transcribed into cDNA, it can then be used as a template for further analysis.

6. What is Plan B? If results are inconclusive, or produce an unexpected result, what is the next step? In the case of nematodes of worldwide economic and regulatory concern, there are often multiple molecular markers that have been incorporated into diagnostic protocols. A diagnostic investigation should be able to achieve a positive conclusion by more than one approach.

5.3. Basic Steps in Molecular Approaches for Nematode Identification

1. Nematode Extraction from Soil or DNA Extraction from Soil: Most approaches to nematode molecular diagnostics start with a separation and isolation of nematodes from the soil (Fig. 5.3). Standard techniques include soil sieving, flotation, centrifugation, automatic elutriation, funnels or trays. Initial amounts of soil for processing and reporting generally range from 100 to 250 g. The non-uniform distribution of nematodes in the soil suggests a sampling strategy that maximizes the potential of recovering target nematodes, especially those present in relatively low abundance. Newly published assays examining extraction of DNA directly from the soil have demonstrated promising results in nematode identification when compared with methods that start with an isolation of nematodes from the soil. Currently, the upper limit in commercially available kits for DNA extraction from the soil is 10 g. The applicability of these kits will increase with the increased capacity to extract DNA from larger amounts of soil.

2. DNA Extraction from Nematodes: When nematodes are isolated and extracted from soil or roots, sufficient amounts of DNA for PCR can be rapidly purified. One nematode provides enough DNA for PCR, although some investigators have

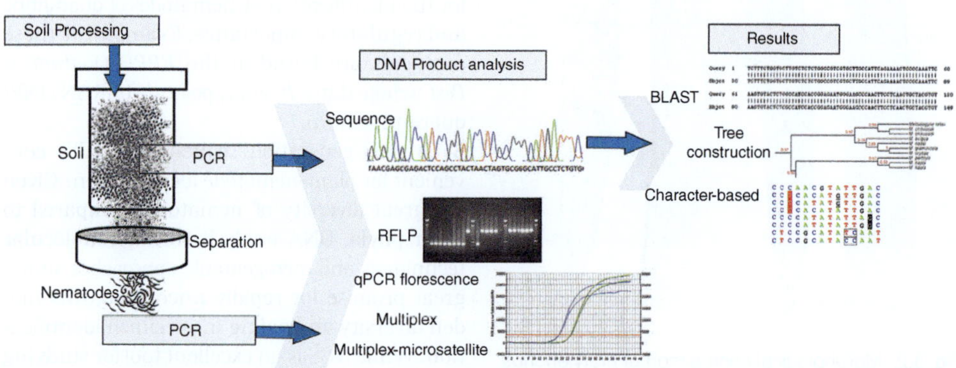

Fig. 5.3. Molecular diagnostics diagram.

pooled nematodes to obtain larger quantities of DNA. In some cases, pooling nematodes introduces the possibility of combining different nematode genotypes. The first step is usually manually disrupting the nematode cuticle, either by pressure applied by a disposable micropipette tip, fine pin or cover slip. Remember that precautions must be taken to prevent cross-contamination of samples, so all disruption tools should be sterile. Nematode substrate in PCR reaction tubes may be frozen for future amplification or a short freeze-thaw step may improve membrane disruption and release of cellular DNA. The next stage typically includes the addition of a lysis buffer and a proteinase followed by incubation and then inactivation of the proteinase. At this point, unless large numbers of nematodes were included in the extraction process, no further purification is necessary, and the sample substrate is ready to serve as template in a PCR reaction. Commercial DNA extraction kits are available, but costs of assays can be dramatically decreased using your own laboratory ingredients. It should be noted that all DNA extraction methods from nematodes are considered 'voucher eating' protocols. If it is necessary to preserve some visual record of the diagnostic target, that step needs to occur prior to DNA extraction.

3. Selection of Appropriate PCR Assay: As mentioned above, the diagnostic question together with other practical considerations will dictate the choice of the appropriate assay. When more than a single identification is required, multiplex assays may be necessary. Multiplex PCR refers to PCR reactions that contain more than a single primer set in a single reaction to amplify DNA from additional gene regions or additional species. Often, multiplex assays incorporate an internal positive control for amplification in their design. Typically, one primer pair in the reaction will amplify a product from all target species in a reaction, with additional primers designed to bind to template from only one of the potential target species. Multiplex assays can be used in a conventional PCR reaction, where amplification products are evaluated by gel electrophoresis, or in a real-time PCR reaction that monitors fluorescence associated with increasing amplification product. Real-time PCR assays are fast and do not require any post-PCR manipulation. There are two primary chemistries used in the fluorescent detection of amplified products. SYBR-Green fluorescent DNA stain incorporates directly into the amplified double-stranded DNA product. TaqMan fluorescent probes are characterized by a labelled oligonucleotide specific to the DNA target template. Probes bound to the target are displaced during the amplification process, creating a fluorescent signal. Both chemistries have been featured in diagnostic assays of nematodes. Other high-specificity probes, such as molecular beacons and Scorpion primers, have been applied for the identification of *Meloidogyne* spp. in single or multiple infections (Ciancio *et al.*, 2005; Anderson and Gleason, 2023). These probes have also been employed for the detection of grapevine fan leaf virus (GFLV) particles in its vector *Xiphinema index* (Finetti-Sialer and Ciancio, 2005). The combined use of different probes, fluorophores and primers offers the advantage of amplifying specific targets in a mixed sample within a single reaction tube. Multiplex reactions require considerable testing for the specificity of the primers and an evaluation of possible non-target interactions among primers and templates. One disadvantage of multiplex and multiplex real-time PCR assays is that they are designed to be specific for a limited number of nematodes. Simpler PCR-restriction fragment length polymorphism (PCR-RFLP) assays may be able to identify a wider range of species, but require post-PCR processing. The restriction digestions associated with PCR-RFLP approaches add time and expense to the assays, and require optimum conditions in the digestion stage to avoid partial digestions. The cost of DNA sequencing, the availability of commercial DNA services and the rapid turn-around times of these services makes DNA sequencing a feasible approach for diagnostics. DNA sequencing of amplified PCR products has the advantages of producing considerable taxonomic information in the form of polymorphic nucleotide sites within the product. Therefore, PCR primers can be designed to amplify regions of the genome that include diagnostic nucleotide sites for a large number of species. There are several additional advantages of DNA sequencing as a diagnostic approach. Each sequence generated helps build the species database, providing validation and an improved understanding of within-species genetic variability. DNA sequencing will also recover unknown or cryptic species, thereby serving an initial step in the monitoring of new or emerging pest species.

4. Selection of Diagnostic Gene Regions: The choice of a genetic marker in diagnostic assays depends on a number of characteristics. Genetic markers vary in their taxonomic resolution, ease of amplification, ease of interpretation, versatility and suitability for addressing different diagnostic questions. A quick scan of Table 5.1 shows that many nematode assays have used the internal transcribed spacer region (ITS) of the repeating array of ribosomal genes in the nuclear genome (Fig. 5.4). This gene region has the advantage of conserved sites for PCR primer binding, yet amplifies a product that generally displays considerable generic variation among species. Disadvantages include the possibility of variation among the copies of ITS within an individual, and, for methods that require comparative nucleotide sequence, the numerous insertions and deletions within the non-coding spacers complicate the process of DNA alignment. The regions that code for the large (28S) and small (18S) ribosomal genes have also been used extensively in species identification and as a means for assessing the relationships among species. Some plant-parasitic genera are relatively well-represented by ribosomal genes in public databases. One disadvantage to ribosomal genes is that the taxonomic resolution may be too coarse and not allow discrimination at the desired level of specificity. Single-copy nuclear genes are receiving increased attention as diagnostic markers. The heat shock protein gene, HSP-90, and chorismate mutase gene are two genes currently being evaluated for their diagnostic potential. Mitochondrial genes are commonly used in diagnostic applications among a wide range of organisms. The mitochondrial genome of animals is highly repeated within animal cells, presenting a robust target for PCR amplification. A single mitochondrial gene, cytochrome oxidase subunit I (COI), has been formalized as a marker for species identification by the Barcoding of Life project. Researchers in Nematology have been slow to adopt DNA barcoding with COI, in part due to the difficulty in designing PCR primers that have broad taxonomic applicability. Other concerns about mitochondrial genes as diagnostic markers involve

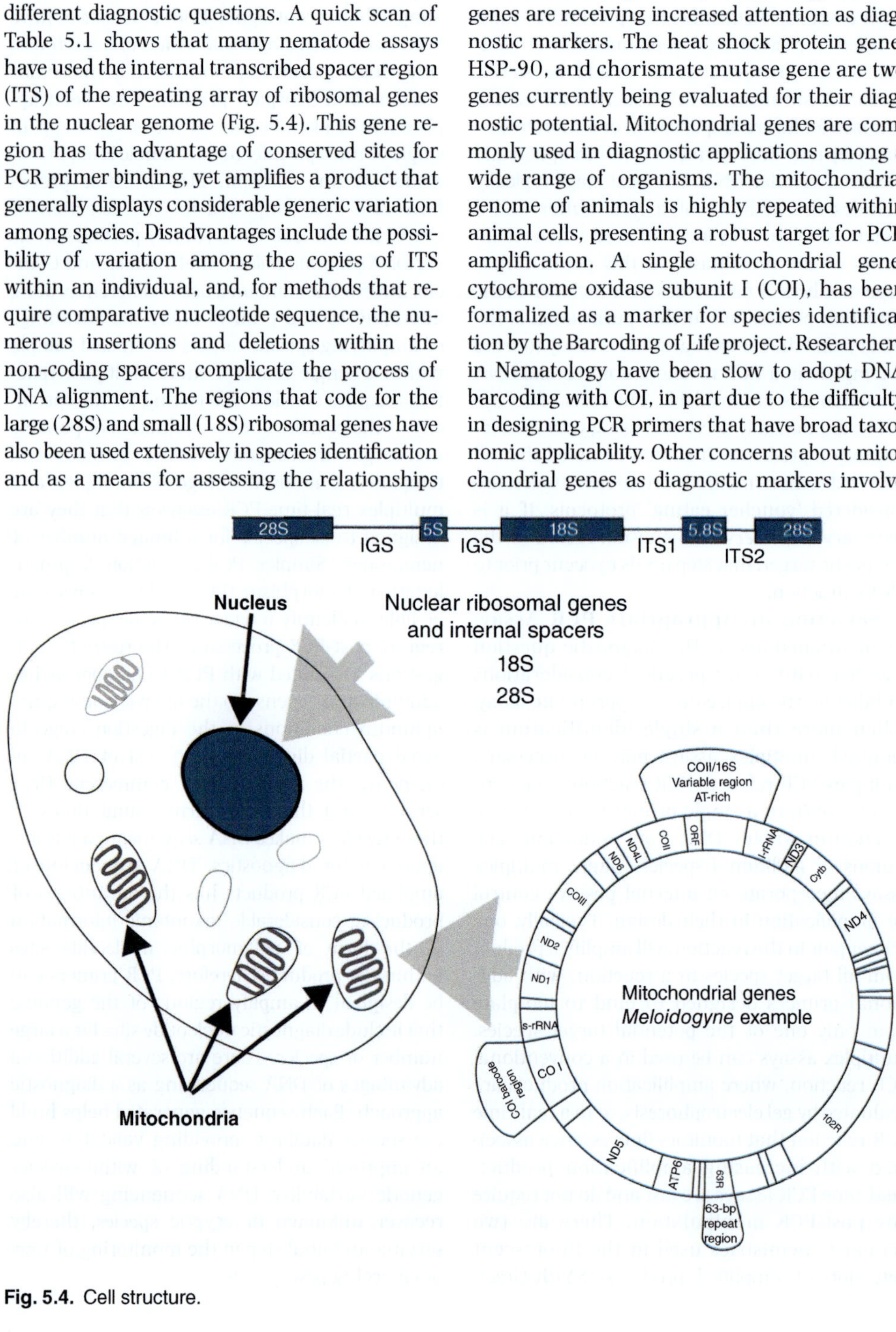

Fig. 5.4. Cell structure.

the potential of mitochondrial genotypes crossing species barriers and the possibility of a lack of congruence between gene trees and species trees. To a lesser degree, anonymous genetic markers, sequence-characterized amplified regions (SCAR) and microsatellites have been applied as the basis for nematode discrimination. It is anticipated that these regions will become more widely used as a result of increasing interest in nematode comparative genomics.

5. Interpretation of Results: Many of the diagnostic assays mentioned above require little additional interpretation from the user. In a real-time PCR approach, accumulation of amplified product implies a positive result for the diagnostic target. Similarly, in an RFLP-PCR reaction, production of amplified products and digestion fragments of the predicted sizes is considered a positive result. Target specificity is built into the assay, and confidence in the results depends on the validation steps in assay development. DNA sequence, however, may be interpreted by a variety of procedures. Comparison to other database sequences in a standard Basic Local Alignment Search Tool (BLAST) operation can locate the most similar sequences within the database. Occasionally, a query nucleotide sequence can result in ambiguous matches equidistant to two or more species in the database. These ambiguities may be clarified by applying a non-distance tree-based method, which places a query sequence in a phylogenetic context. Numerous free tree-building programmes are available online. A method that does not rely on trees or a calculation of distance, but searches for diagnostic nucleotides that are fixed for specified taxa, is called character-based diagnosis. As with all other analytic methods, character-based diagnostics depends on a thorough taxonomic identification of species boundaries prior to the diagnostic analysis.

5.4. Case Studies in Nematode Molecular Diagnostics

Case studies can help provide an indication of the state-of-the-art in nematode diagnostics. Examined in greater detail below are three groups of plant-parasitic nematodes that are actively being targeted for improved diagnostic methods.

5.4.1. *Globodera* Species

The detection and discrimination of *Globodera* species is one of the most important diagnostic issues facing nematologists. *Globodera* species differ in their host reactions, physiology, ecology and distribution. Precise identification is necessary for both management and satisfying regulatory requirements for international trade. Prior to molecular assays, identification of *Globodera* species largely relied on characteristics of the cyst stage and measurements of the infective juvenile stage. Initially, molecular assays focused on distinguishing the two potato cyst nematodes, *G. rostochiensis* and *G. pallida.* PCR-RFLP patterns of amplified ITS regions clearly demonstrated nucleotide differences detectable by restriction enzyme digestion. These nucleotide differences were subsequently elucidated by DNA sequencing. The sequence analysis led to the development of species-specific primer sets that were quickly employed in multiplex PCR reactions to generate different-sized fragments to differentiate the species by agarose gel electrophoresis. To ensure the species specificity of these primers, they were tested against other species of *Globodera* that did not reproduce on potato. Subsequent studies used the ITS sequence polymorphisms to create primers for real-time PCR assays. Now these primers need to be re-evaluated as species-specific primers, as new taxonomic studies have revealed additional genetic variation among cyst species (Skantar *et al.*, 2011; Subbotin *et al.*, 2011), including the discovery and description of *G. ellingtonae*, a new potato cyst species (Handoo *et al.*, 2012). The re-evaluation of the specificity of diagnostic primer sets is a continual process, especially during periods of rapid growth and accumulation of genetic knowledge.

Some studies have focused on the global distribution of genetic variation and the biogeographic patterns of cyst species. DNA sequencing and phylogenetic analysis of the mitochondrial gene cytochrome b has been conducted on a large number of *Globodera* isolates from the Andean mountains of South America. This geographic region is considered to be the origin of domesticated potato and the potato cyst nematodes. These studies have identified suspected localities from which European introductions of PCN occurred. They have also highlighted the genetic complexity that confronts research scientists and regulatory officials who are attempting to control and manage these nematodes.

5.4.2. *Meloidogyne* Species

There are over 100 described species of *Meloidogyne*. Many of them cause economic damage to a wide range of host plants and many appear to have been dispersed across the planet through agricultural activities. Since the adult female stage of *Meloidogyne* is embedded in root tissue and males of most economically important species are seldom seen, it is the non-descript infective second-stage juvenile that is often the diagnostic target in soil samples. PCR has made this stage available for analysis. Some of the published diagnostic approaches use techniques that can identify multiple species simultaneously. Adam *et al.* (2007) designed a molecular key for seven *Meloidogyne* species based on amplification and size polymorphism of amplified products. Depending on the size of an initial amplification of the intergenic spacer region between the 5S and the 18S ribosomal genes, a second round of amplification using different primer sets provides species-level resolution. They also demonstrated the ability to conduct as many as 15 amplifications from a single infective juvenile. Other assays target specific species. *Meloidogyne enterolobii*, formerly known as *M. mayaguensis*, is an aggressive and highly pathogenic species that appears to be spreading in tropical and subtropical climates. Recognition of this species is important because standard forms of host-plant resistances are not effective against *M. enterolobii*. The EPPO lists two SOP for identification of this species, one based on amplifications of specific-sized products of the ribosomal intergenic region and the second based on amplification of a repeated region of the mitochondrial genome. Newer assays for this species include specific PCR primers discovered through screening genomic DNA by the randomly amplified polymorphic DNA (RAPD) procedure and a multiplex analysis that incorporated PCR amplification from galled-root tissue (Hu *et al.*, 2011). For improved speed and accuracy, high-resolution melting curve analysis of amplified ribosomal intergenic is being evaluated as a high-throughput method for nematode diagnostics. Still, only a small percentage of described *Meloidogyne* species have been characterized by any form of molecular analysis. In particular, species from tropical regions constitute a conspicuous gap in knowledge in the understanding of *Meloidogyne* genetic diversity.

5.4.3. *Ditylenchus* Species

Ditylenchus species present a special challenge to diagnosticians and regulatory officials. The genus, *sensu lato* (see Chapter 14), has more than 60 species and some can be subdivided into numerous host races. The feeding of *Ditylenchus dipsaci* may include as many as 500 host species. Anhydrobiotic survival stages complicate nematode recovery, and the ubiquitous mycophagous soil forms strain efforts to assess non-target effects for specific PCR primer design. The ITS region was exploited early in the development of *Ditylenchus* species discrimination with emphasis placed on differentiation of *D. dipsaci* and *D. destructor*. Recent diagnostic results have identified seven ITS haplotypes in *D. destructor.* An analysis of *D. dipsaci* supported the description of a new species, *D. gigas*, based on a comparative examination of ITS, 28S, 18S and COI sequence. These new sequences provide an opportunity to fine-tune the multiplex assays devoted to *Ditylenchus* species discrimination. The application of COI mitochondrial sequences is notable and will help broaden that basis of *Ditylenchus* molecular identification, which is presently heavily reliant on the ITS region.

5.5. Summary

Molecular diagnosis of plant-parasitic nematodes is still in early stages of development. Continual changes and upgrades in recommended diagnostic protocols are evidence of rapid growth in the field. There are, however, urgent needs for more comprehensive analyses that examine the broad geographic range of most species. With present-day near-instantaneous transport of agricultural commodities, nematode pest species can travel the globe without restraint. It is not unrealistic to expect that future commerce will accelerate the global re-assortment of nematodes, thereby providing greater challenges to those who seek to limit the impact of emerging pest species.

5.6. References

Abebe, E., Mekete, T. and Thomas, W.K. (2011) A critique of current methods in nematode taxonomy. *African Journal of Biotechnology* 10, 312–323.

Adam, M.A.M., Phillips, M.S. and Blok, V.C. (2007) Molecular diagnostic key for identification for single juveniles of seven common and economically important species of root-knot nematode (*Meloidogyne* spp.). *Plant Pathology* 56, 190–197.

Agudelo, P., Lewis, S. and Fortnum, B. (2011) Validation of a real-time polymerase chain reaction assay for the identification of *Meloidogyne arenaria*. *Plant Disease* 95, 835–838.

Anderson, S.D. and Gleason, C.A. (2023) A molecular beacon real-time polymerase chain reaction assay for the identification of *M. chitwoodi*, *M. fallax*, and *M. minor*. *Frontiers in Plant Science* 14, 1096239.

Berry, S.D., Fargette, M., Spaull, V.W., Morand, S. and Cadet, P. (2008) Detection and quantification of root-knot nematode (*Meloidogyne javanica*), lesion nematode (*Pratylenchus zeae*) and dagger nematode (*Xiphinema elongatum*) parasites of sugarcane using real-time PCR. *Molecular and Cellular Probes* 22, 168–176.

Blaxter, M.L. (2004) The promise of a DNA taxonomy. *Philosophical Transactions of the Royal Society of London Biological Sciences* 359, 669–679.

Blok, V. (2005) Achievements in and future prospects for molecular diagnostics of plant-parasitic nematodes. *Canadian Journal of Plant Pathology* 27, 176–185.

Blok, V. and Powers, T. (2009) Biochemical and molecular identification. In: Perry, R.N., Moens, M. and Starr, J.L. (eds), *Root-knot Nematodes*. CAB International, Wallingford, UK, pp. 98–118.

Bogale, M., Baniya, A. and Digennaro, P. (2020) Nematode identification techniques and recent advances. *Plants* 9, 1260.

Bogale, M., Tadesse, B., Nuaima, R.H., Honermeier, B., Hallman, J., *et al.* (2021) Morphometric and molecular diversity among seven European isolates of *Pratylenchus penetrans*. *Plants* 10(4), 674. DOI: 10.3390/plants10040674.

Castagnone-Sereno, P., Skantar, A. and Robertson, L. (2011) Molecular tools for diagnostics. *Genomics and Molecular Genetics of Plant-Nematode Interactions* 4, 443–464.

Ciancio, A., Loffredo, A., Paradies, F., Turturo, C. and Finetti-Sialer, M. (2005) Detection of *Meloidogyne incognita* and *Pochonia chlamydosporia* by fluorogenic molecular probes. *OEPP/EPPO Bulletin* 35, 157–164.

Ciancio, A., Rosso, L.C., Colagiero, M. and Pentimone, I. (2018) Application of nanopore technology for fast sequencing of very long DNA fragments from plant-parasitic nematodes. *Abstracts 50th Annual Organization of Nematologists of Tropical America. Nematropica* 48, Abstract 6.

De Waele, D. and Elsen, A. (2007) Challenges in tropical plant nematology. *Annual Review of Phytopathology* 45, 457–485.

Douda, O., Zouhar, M., Novakova, E., Mazakova, J. and Rysanek, P. (2010) Variability of D2/D3 segment sequences of several populations and pathotypes of potato cyst nematodes (*Globodera rostochiensis*, *Globodera pallida*). *Plant Protection Science* 46, 171–180.

Ficetola, G.F., Guerrieri, A., Cantera, I. and Bonin, A. (2024) In silico assessment of 18S rDNA metabarcoding markers for the characterization of nematode communities. *PLoS One* 19, e0298905.

Finetti-Sialer, M.M. and Ciancio, A. (2005) Isolate-specific detection of grapevine fanleaf virus from *Xiphinema index* through DNA-based molecular probes. *Phytopathology* 95, 262–268.

Floyd, R., Abebe, E., Papert, A. and Blaxter, M. (2002) Molecular barcodes for soil nematode identification. *Molecular Ecology* 11, 839–850.

Gao, B., Ma, J., Li, X., Wang, R. and Chen, S. (2021) A real-time recombinase polymerase amplification assay for fast and accurate detection of *Ditylenchus destructor*. *Molecular and Cellular Probes* 61, 101788.

Goto, K., Min, Y.Y., Sato, E. and Toyota, K. (2011) A multiplex real-time PCR assay for the simultaneous quantification of the major plant-parasitic nematodes in Japan. *Nematology* 13, 713–720.

Handoo, Z.A., Carta, L.K., Skantar, A.M. and Chitwood, D.J. (2012) Description of *Globodera ellingtonae* n. sp. (Nematoda: Heteroderidae) from Oregon. *Journal of Nematology* 44, 40–57.

Handoo, Z.A., Carta, L.K., Skantar, A.M., Subbotin, S.A. and Fraedrich, S.W. (2016) Molecular and morphological characterization of *Xiphinema chambersi* population from live oak in Jekyll Island, Georgia, with comments on morphometric variation. *Journal of Nematology* 48, 20–27. DOI: 10.21307/jofnem-2017-005.

Hodson, A.K., Celayir, T. and Quiroz Alonso, A. (2023) A real-time PCR assay to detect and quantify root-knot nematodes from soil extracts. *Plant Disease* 107, 2169–2176.

Hu, M.X., Zhou, K. and Liao, J.L. (2011) Multiplex PCR for the simultaneous identification and detection of *Meloidogyne incognita*, *M. enterolobii*, and *M. javanica* using DNA extracted directly from individual galls. *The American Phytopathological Society* 101, 1270–1277.

Hu, S., Franco, J., Bali, S., Chavoshi, S., Brown, C., *et al.* (2023) Diagnostic molecular markers for identification of different races and a pathotype of Columbia root knot nematode. *Phytofrontiers*. Special Focus Issue 3(1), 199–205. DOI: 10.1094/PHYTOFR-03-22-0035-FI.

Huang, L., Ye, J., Wu, X., Xu, X., Sheng, J., *et al.* (2010) Detection of the pine wood nematode using real-time PCR assay to target the DNA topoisomerase I gene. *European Journal of Plant Pathology* 127, 89–98.

Jiang, C., Zhang, Y., Yao, K., Abdulsalam, S., Li, G., *et al.* (2021) Development of a species-specific SAR-PCR assay for direct detention of sugar beet cyst nematode (*Heterodera schachtii*) from infected roots and soil samples. *Life* 11(2), 1358. DOI: 10.3390/life11121358.

Jones, J., Phillips, S. and Armstrong, M. (1997) Molecular approaches in plant nematology. *Fundamental and Applied Nematology* 20, 1–14.

Kang, J.S., Moon, I.S, Lee, S.G., Shin, S.C. and Lee, S.H. (2009) Rapid and accurate prediction of the frequencies of *Bursaphelenchus xylophilus* and *B. mucronatus* in mixed nematode samples using real-time species-specific PCR. *Nematology* 11, 289–299.

Leal, I., Green, M., Allen, E., Humble, L. and Rott, M. (2007) Application of a real-time PCR method for the detection of pine wood nematode, *Bursaphelenchus xylophilus*, in wood samples from lodgepole pine. *Nematology* 9, 351–362.

Lu, M.R., Lai, C.K., Liao, B.Y. and Tsai, I.J. (2020) Comparative transcriptomics across nematode life cycles reveal gene expression conservation and correlated evolution in adjacent developmental stages. *Genome Biology and Evolution* 12, 1019–1030.

Madani, M., Subbotin, S.A. and Moens, M. (2005) Quantitative detection of the potato cyst nematode, *Globodera pallida*, and the beet cyst nematode, *Heterodera schachtii*, using Real-Time PCR with SYBR green I dye. *Molecular Cellular Probes* 19, 81–86. DOI: 10.1016/j.mcp.2004.09.006.

Madani, M., Subbotin, S.A., Ward, L.J., Li, X. and De Boer, S.H. (2010) Molecular characterization of Canadian populations of potato cyst nematodes, *Globodera rostochiensis* and *G. pallida* using ribosome nuclear RNA and cytochrome *b* genes. *Canadian Journal of Plant Pathology* 32, 252–263. DOI: 10.1080/07060661003740033.

Madani, M., Ward, L.J. and De Boer, S.H. (2011) Hsp90 gene, an additional target for discrimination between the potato cyst nematodes, *Globodera rostochiensis* and *G. pallida*, and the related species *G. tabacum tabacum*. *European Journal of Plant Pathology* 130, 271–285.

Marek, M., Zouhar, M., Douda, O., Mazakova, J. and Rysanek, P. (2010) Bioinformatics-assisted characterization of the ITS1-5 8S-ITS2 segments of nuclear rRNA gene clusters, and its exploitation in molecular diagnostics of European crop-parasitic nematodes of the genus *Ditylenchus*. *Plant Pathology* 59, 931–943.

Neilson, R., Donn, S., Griffiths, B., Daniell, T., Rybarczky, K., *et al.* (2009) Molecular tools for analysing nematode assemblages. In: Wilson, M. and Kakouli-Duarte, T. (eds), *Nematodes as Environmental Indicators*. CAB International, Wallingford, UK, pp. 188–207.

Nowaczyk, K., Dobosz, R., Kornobis, S. and Obrepalska-Steplowska, A. (2008) TaqMan Real-Time PCR-based approach for differentiation between *Globodera rostochiensis* (golden nematode) and *Globodera artemisiae* species. *Parasitology Research* 103, 577–581.

Oliveira, C., Monteiro, A. and Blok, V. (2011) Morphological and molecular diagnostics for plant-parasitic nematodes: working together to get the identification done. *Tropical Plant Pathology* 36, 65–73.

Ou, S., Peng, D., Liu, X., Li, Y. and Moens, M. (2008) Identification of *Heterodera glycines* using PCR with sequence characterized amplified regions (SCAR) primers. *Nematology* 10, 397–403.

Powers, T. (2004) Nematode molecular diagnostics: from bands to barcodes. *Annual Review of Phytopathology* 43, 367–385.

Riedel, R.M., Rowe, R.C. and Martin, M.J. (1985) Differential interactions of *Pratylenchus crenatus*, *P. penetrans*, and *P. scribneri* with *Verticillium dahliae* in potato early dying disease. *Phytopathology* 75, 419–422.

Sirca, S., Geric Stare, B., Strajnar, P. and Urek, G. (2010) PCR-RFLP diagnostic method for identifying *Globodera* species in Slovenia. *Phytopathologia Mediterranea* 49, 361–369.

Skantar, A.M., Handoo, Z.A., Zasada, I.A., Ingram, R.E., Carta, L.K., *et al.* (2011) Morphological and molecular characterization of *Globodera* populations from Oregon and Idaho. *Phytopathology* 101, 480–491.

Subbotin, S.A. (2021) Molecular identification of nematodes using polymerase chain reaction (PCR). In: Perry, R.N., Hunt, D.J. and Subbotin, S.A. (eds), *Techniques for Work with Plant and Soil Nematodes*. CAB International, Wallingford, UK, pp. 218–239.

Subbotin, S.A., Deimi, M., Zheng, J. and Chizhov, V. (2011) Length variation and repetitive sequences of internal transcribed spacer of ribosomal RNA gene, diagnostics and relationships of populations of potato rot nematode, *Ditylenchus destructor* Thorne, 1945 (Tylenchida: Anguinidae). *Nematology* 13, 773–785.

Subbotin, S.A. and Moens, M. (2006) Molecular taxonomy and phylogeny. In: Perry, R.N. and Moens, M. (eds), *Plant Nematology*. CAB International, Wallingford, UK, pp. 33–58.

Toyota, K., Shirakashi, T., Sato, E., Wada, S. and Min, Y.Y. (2008) Development of a real-time PCR method for the potato-cyst nematode *Globodera rostochiensis* and the root-knot nematode *Meloidogyne incognita*. *Soil Science and Plant Nutrition* 54, 72–76.

van de Vossenberg, B.T.L.H, van Kempen, T.L., van Duijnhoven, A.A.L.A.M., Warbroek, T., van der Gouw, L.P., *et al.* (2023) Development and validation of a high-throughput sequencing test for mitogenome and rDNA assembly and annotation, and its use in support of nematode identification of regulatory concern. *PhytoFrontiers* 3, 103–112.

Van Himbeeck, R., Cazzaniga, S.G., van den Elsen, S., Vrielink, J.O., Aslan, S.K., *et al.* (2024) A full-length SSU rRNA-based workflow for high-resolution monitoring of nematode communities reveals direct and indirect responses to plant-based manipulations. *Soil Biology and Biochemistry* 189, 109263.

Vovlas, N., Troccoli, A., Palomares-Rius, J., De Luca, F., Liebanas, G., *et al.* (2011) *Ditylenchus gigas* n. sp. parasitizing broad bean: a new stem nematode singled out from the *Ditylenchus dipsaci* species complex using a polyphasic approach with molecular phylogeny. *Plant Pathology* 60, 762–775.

Waeyenberge, L., Viaene, N. and Moens, M. (2009) Species-specific duplex PCR for the detection of *Pratylenchus penetrans*. *Nematology* 11, 847–857.

Yan, G. and Smiley, R. (2010) Distinguishing *Heterodera filipjevi* and *H. avenae* using polymerase chain reaction-restriction fragment length polymorphism and cyst morphology. *Phytopathology* 100, 216–224.

Yan, G., Smiley, R. and Okubara, P. (2012) Detection and qualification of *Pratylenchus thornei* in DNA extracted from soil using real-time PCR. *Phytopathology* 102, 14–22.

Yan, G., Smiley, R., Okubara, P., Skantar, A., Easley, S., *et al.* (2008) Detection and discrimination of *Pratylenchus neglectus* and *P. thornei* in DNA extracts from soil. *Plant Disease* 92, 1480–1487.

Yu, H., Chronis, D., Lu, S. and Wang, X. (2011) Chorismate mutase: an alternatively spliced parasitism gene and a diagnostic marker for three important *Globodera* nematode species. *European Journal of Plant Pathology* 129, 89–102.

Zhuo, K., Luo, M., Cui, R. and Liao, J. (2011) A multiplex one-step PCR method for the simultaneous identification of *Bursaphelenchus xylophilus*, *B. mucronatus* and *B. doui* - three species within the *xylophilus* group. *Forest Pathology* 41, 66–69.

Zhou, Q., Liu, Y., Wang, Z., Wang, H., Zhang, X., *et al.* (2022) Rapid onsite detection of the *Bursaphelenchus xylophilus* using recombinase polymerase amplification combined with lateral flow dipstick that eliminates interference from primer-dependent artifacts. *Frontiers in Plant Science* 13, 856109. DOI: 10.3389/fpls.2022.856109.

Zijlstra, C. and Van Hoof, R.A. (2006) A multiplex real-time polymerase chain reaction (TaqMan) assay for the simultaneous detection of *Meloidogyne chitwoodi* and *M. fallax*. *Phytopathology* 96, 1255–1262.

6 Molecular Phylogenetics

Sergei A. Subbotin[1,2*]
[1]*Plant Pest Diagnostics Center, California Department of Food and Agriculture, Sacramento, CA, USA;* [2]*Center of Parasitology of A.N. Severtsov Institute of Ecology and Evolution of the Russian Academy of Sciences, Moscow, Russia*

Abstract

Molecular phylogenetics is the branch of systematics that analyzes genome information, especially sequence and structure of DNA, RNA molecules and amino acid chains to assess the inference and interpretation of evolutionary relations between groups of organisms. Inferring phylogenetic trees from molecular sequence data is a fundamental method used in evolutionary and systematic studies. This chapter provides an introduction to molecular phylogenetics to assess, based on the DNA or protein sequences of the organism, the inference and interpretation of evolutionary relations between a group of organisms. The use of graphic representation of evolutionary relationships in the form of a phylogenetic tree using software packages is introduced and illustrated.

6.1. Introduction

Molecular phylogenetics is the branch of systematics that analyses genome information, especially the sequence and structure of DNA, RNA molecules and amino acid chains to assess the inference and interpretation of evolutionary relations between groups of organisms. Inferring phylogenetic trees from molecular sequence data is a fundamental method used in evolutionary and systematic studies (Hillis *et al.*, 1996; Nei and Kumar, 2000; Yang and Rannala, 2012).

Molecular data have a number of important advantages over morphological data. Firstly, molecular characters are much more abundant than morphological features, and much information on DNA or amino acid sequences can be obtained from small quantities of tissue using different technical procedures in a relatively short time period. Secondly, DNA sequences are strictly heritable entities and ideally suited to statistical analysis, whereas morphological characters can be influenced by various environmental and other factors and can barely be modelled. Thirdly, the interpretation of molecular characters, such as the assessment of homologies, is generally easier than that of morphological characters. It should also be underlined that a phylogeny reconstructed based on DNA is not always an accurate representation of the history of species.

*E-mail: sergei.a.subbotin@gmail.com

A revision of Loza-Reyes, E. (2012) Molecular phylogenetics using maximum-likelihood. In: Manzanilla-López, R.H. and Marbán-Mendoza, N. (eds), *Practical Plant Nematology*. Mundi-Prensa and Biblioteca Básica de Agricultura, Montecillo, Mexico.

 Practical Plant Nematology 2nd Edition
(eds R.H. Manzanilla-López, D.J. Hunt, N. Marbán-Mendoza and S. Infante Gil)

DOI: 10.1079/9781836990413.0006

It might happen just because not all of the genes in a genome share the same history. However, this problem can be resolved with further analysis using multi-gene reconstruction of phylogenetic relationships.

Molecular phylogenetics compares and analyses either a single or a few genes. In recent years, systematists have become increasingly interested in building phylogenetic trees using many thousands of nucleotides and genomic data. The approach attempting to infer evolutionary relationships at a genomic level is called phylogenomics. The field of phylogenomics has advanced rapidly in the past decade due to the now widespread availability of next-generation sequencing technologies.

6.2. Genes Used for Molecular Phylogenetics

To determine true evolutionary relationships between organisms, it is essential that the correct gene fragment or molecule should be chosen for a phylogenetic study. It should be universally distributed across the group chosen for study; it should be represented as orthologous, or as a single copy in the genome, not paralogous, where copies were created by a duplication event within the same genome; it must be functionally homologous in each organism; and it is critical in sequence comparisons to be able properly to align in order to identify regions of homology and heterogeneity.

The eukaryotic cell contains two different genomes: that of the nucleus and that of mitochondria. Datasets from both genomes are used for molecular phylogenetics. Mitochondria are inherited maternally, whereas the nucleus is biparental. Historically, the only nuclear genes with a high enough copy number for easy study were ribosomal genes. These genes encode ribosomal RNAs (rRNA), which form nearly two-thirds of the mass of the ribosome. Nuclear ribosomal RNA genes represent the oldest repetitive fraction universal to all eukaryotic genomes.

6.3. Phylogenetic Inference

Phylogenetic analysis is a complex field of study that embraces a variety of techniques that can be applied to a wide range of evolutionary questions. However, a complete understanding of all assumptions for each method involved in the analysis is essential for a correct interpretation of the results. A workflow of a molecular phylogenetic project could be presented as follows: (i) selection and sampling of a group of organisms, which are the target of the study, and some related organism(s) defined as outgroup taxa; (ii) choice of molecular markers; (iii) amplification, sequencing of genes and assembling of sequence data; (iv) alignment, or establishment of homology between nucleotide positions of genes; (v) construction of a phylogenetic tree or network using distance or discrete methods and making an assessment of the reliability of its branches and (vi) testing of different alternative hypotheses using trees and datasets. There are many software programmes that perform alignment, model selection, phylogenetic tree reconstruction and visualization (Table 6.1).

6.3.1. Alignment

The major step of any phylogenetic study is the construction of alignment or establishment of positional homology between nucleotides or amino acid bases that have descended from a common ancestral base. Errors incurred in this step can introduce noise and bias in the phylogenetic reconstruction. The best way to compare the homologous residues is to align sequences one on top of another in a visual display, so that, ideally, each homologous base from different sequences lines up in the same column. Each column of the multiple sequence alignment could be considered homologous. Three types of aligned pairs are distinguished: (i) *matches* (same nucleotide appears for all sequences); (ii) *mismatches* (different nucleotides were found in the same position) and (iii) *gaps* (no base in a particular position for at least for one of the sequences). A gap indicates that a deletion has occurred in one sequence or an insertion has occurred in another sequence. However, the alignment itself does not enable these mutational events to be distinguished. Therefore, insertions and deletions are sometimes collectively referred to as *indels* (Page and Holmes, 1998).

The optimal alignment is considered to be that in which the number of mismatches and gaps are minimized according to the desired criteria. The Clustal program series is one of

Table 6.1. Phylogenetic software.

Software	Website
BEAST (**B**ayesian **E**volutionary **A**nalysis **S**ampling **T**rees) is a cross-platform program for Bayesian phylogenetic analysis of molecular sequences using MCMC.	https://beast.community https://www.beast2.org
BioEdit is a free biological sequence alignment editor with an intuitive multiple-document interface with convenient features designed to make alignment and manipulation of sequences relatively easy.	https://bioedit.software.informer.com
Clustal is a series of widely used computer programs used in Bioinformatics for multiple sequence alignment.	http://www.clustal.org
FigTree is designed as a graphical viewer of phylogenetic trees and as a program for producing publication-ready figures.	http://tree.bio.ed.ac.uk/software/figtree/
ForCon is a user-friendly software tool for the conversion of nucleic acid and amino acid sequence alignments.	http://bioinformatics.psb.ugent.be/webtools/ForCon/
GARLI (**G**enetic **A**lgorithm for **R**apid **L**ikelihood **I**nference) is a program for inferring phylogenetic trees. It implements nucleotide, amino acid and codon-based models of sequence evolution, and runs on all platforms.	https://code.google.com/archive/p/garli/
Gblocks is a program that eliminates poorly aligned positions and divergent regions of an alignment of DNA or protein sequences.	https://www.biologiaevolutiva.org/jcastresana/Gblocks.html
jModelTest is a program for selecting the model of nucleotide substitution that best fits the data.	http://evomics.org/resources/software/molecular-evolution-software/modeltest
MacClade is a program for phylogenetic analysis with analytical strength in studies of character evolution. It also provides many tools for entering and editing data and many descriptive statistics as well as for producing tree diagrams and charts.	http://macclade.org
MAFFT (**M**ultiple **A**lignment using **F**ast **F**ourier **T**ransform) is a program used to create multiple sequence alignments of amino acid or nucleotide sequences.	https://mafft.cbrc.jp/alignment/software/
MEGA (**M**olecular **E**volutionary **G**enetic **A**nalysis) is a software for conducting statistical analysis of molecular evolution and for constructing phylogenetic trees.	https://www.megasoftware.net
MrBayes is a program for the Bayesian estimation of phylogeny. The program uses a Markov chain Monte Carlo (MCMC) technique to approximate the posterior probabilities of trees.	http://nbisweden.github.io/MrBayes
MUSCLE (**Mu**ltiple **S**equence **C**omparison by **L**og-**E**xpectation) is a software for multiple sequence alignment of protein and nucleotide sequences.	http://www.drive5.com/muscle/
PAUP* -(**P**hylogenetic **A**nalysis **U**sing **P**arsimony and other methods) is the most sophisticated and user-friendly program for phylogenetic analysis, with many options. It includes parsimony, distance matrix and maximum likelihood methods.	http://paup.phylosolutions.com
RAxML (**R**andomized **Ax**elerated **M**aximum **L**ikelihood) is a program for sequential and parallel maximum likelihood-based inference of large phylogenetic trees.	https://cme.h-its.org/exelixis/web/software/raxml/index.html
T-Coffee (**T**ree-based **Co**nsistency **O**bjective **F**unction **f**or Alignm**e**nt **E**valuation) is a multiple sequence alignment software using a progressive approach.	http://www.tcoffee.org
TreeView is a simple program for displaying and manipulating phylogenetic trees.	https://treeview-x.en.softonic.com

the most commonly used computerized alignment programmes using a progressive alignment approach. Clustal performs a global-multiple sequence alignment by the progressive method using a three-step process: (i) performs pair-wise alignment of all the sequences by dynamic programming; it aligns each sequence against each other giving a similarity matrix; (ii) creates a guide dendrogram using the similarity matrix and (iii) starts by aligning the two most similar sequences. Following the guide dendrogram, the next sequences are added in, aligning them to the existing alignment and inserting gaps as necessary. Multiple sequence alignment software MAFFT and MUSCLE also use the progressive alignment approach together with iterative refinement and consistency-based approaches that further enhance the accuracy and robustness of the alignments. Progressive alignment is a strictly mathematical process that is completely independent of biological reality. The use of structural components of the given molecule can significantly improve estimations of homology, thus generating a better alignment.

Any automatic alignment should be visually checked and then manually edited if necessary. As erroneously aligned sites in alignment can harm the phylogenetic signal and may have an influence on phylogenetic reconstruction, they can be removed manually or by using filtering computer programmes, for example, Gblock, which eliminates poorly aligned positions and divergent regions. The difficulty of filtering alignments, however, lies in effectively detecting unreliable columns without removing phylogenetically informative sites. These tools can also lead to a decrease in accuracy, with loss of phylogenetic signal exceeding non-phylogenetic signal or noise. It is recommended to perform phylogenetic analysis based on a series of alignments generated under different parameters (gap open and gap extension penalties) (Fig. 6.1) or algorithms, including filtered alignments, to determine how ambiguous regions in the alignment can affect the results and level of robustness for tree topology.

6.3.2. Evolutionary Models

In order to reconstruct the evolutionary tree some assumptions about the evolutionary process for the studied gene or molecules should be made. Evolutionary substitution models for nucleotides and amino acids are implemented in a different way in distance, maximum likelihood (ML) and Bayesian inference (BI). The substitution model is a description of the way sequences evolved in time by nucleotide or amino acid replacements. The nucleotide substitution process of DNA sequence can be described by a so-called 'homogeneous' Markov process that uses the Q matrix, which specifies the relative rates of change of each nucleotide along the sequences. (See equation at bottom of the page).

The *ri*'s are the rate parameters that define the rates of changes between nucleotides, and the πi's are the base frequencies. There are many models that could be derived from the Q matrix. Parameters that vary for different models of sequence evolution: (i) base frequencies (equal, unequal); (ii) number of substitution types (1-6) and (iii) equal rates or among-site rate variation.

Many models of increasing complexity have been described, both for nucleotide and amino acid sequences. In general, models of sequence evolution are built following two main approaches, either using properties calculated through the comparison of large numbers of observed sequences (empirically) or on the basis of the chemical or biological properties of DNA or amino acids (parametrically). Empirical models result in fixed parameter values that are estimated only once and then assumed to be applicable to all datasets. The modelling of amino acid replacement, such

$$
Q = \{q_{ij}\} = \begin{array}{c} \\ \\ \\ \\ \end{array}
\begin{array}{cccc} A & C & G & T \end{array}
$$

$$
Q = \{q_{ij}\} = \begin{pmatrix}
-r_2\pi_C - r_4\pi_G - r_6\pi_T & r_2\pi C & r_4\pi_G & r_6\pi_T \\
r_1\pi_A & -r_1\pi_A - r_8\pi_G - r_{10}\pi_T & r_8\pi_G & r_{10}\pi_T \\
r_3\pi_A & r_7\pi C & -r_3\pi_A - r_7\pi_C - r_{12}\pi_T & r_{12}\pi_T \\
r_5\pi_A & r_9\pi C & r_{11}\pi_G & -r_5\pi_A - r_9\pi_C - r_{11}\pi_G
\end{pmatrix}
$$

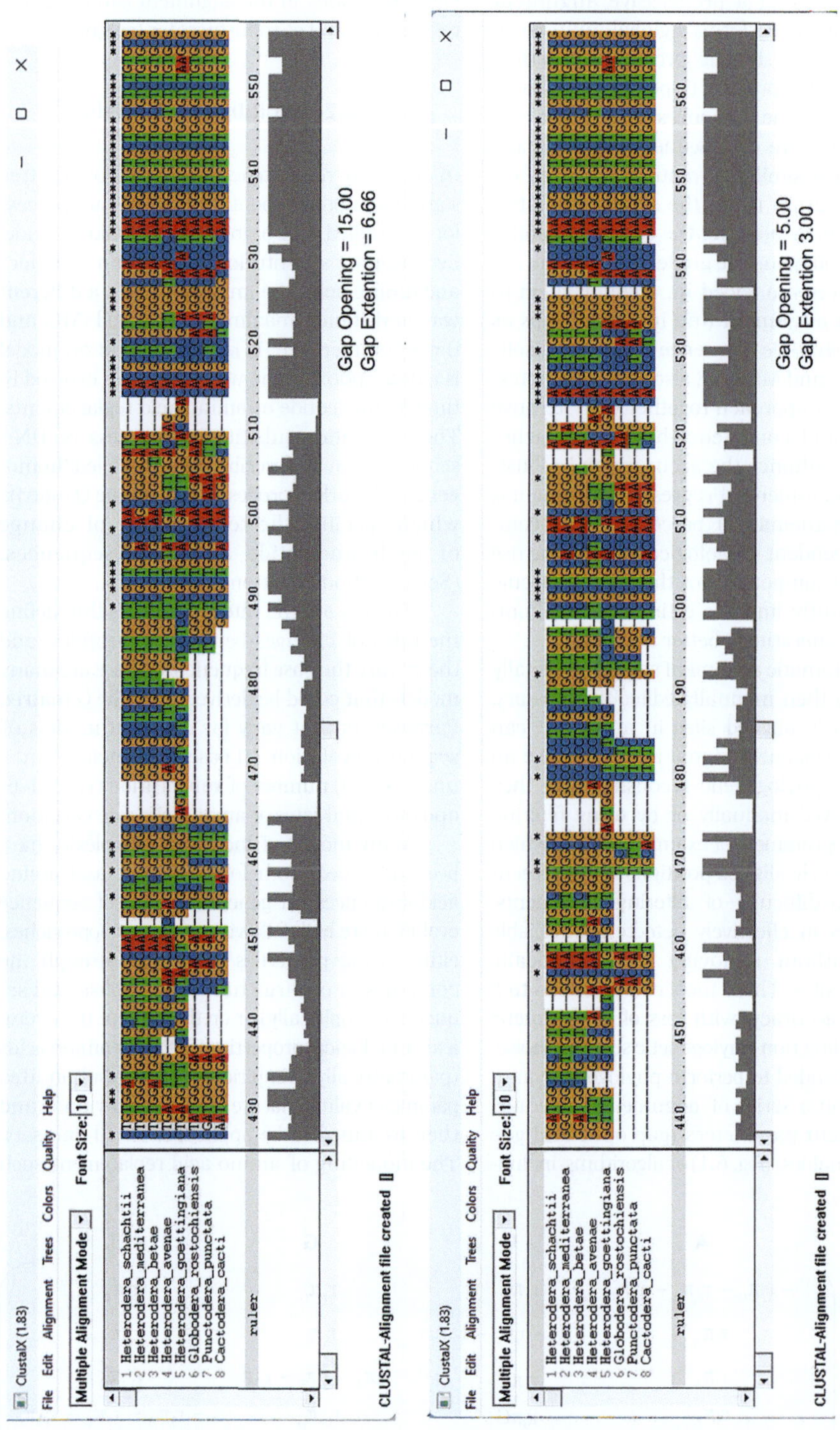

Fig. 6.1. Alignment of the ITS rRNA gene sequences of several cyst-forming nematodes generated by ClustalX under different gap opening and gap extension penalties.

as mtREV and JTT, has concentrated on this empirical approach. By contrast, parametric models allow the parameter models to be derived from the dataset in each particular analysis. The modelling of nucleotide replacement has concentrated on this parametric approach (Lemey *et al.*, 2009).

The Jukes-Cantor model (JC69) was one of the first proposed and is perhaps the simplest model of nucleotide sequence evolution. It assumes that the four bases have equal frequencies and that all substitutions are equally likely. The general time-reversible model (GTR) is the most general model where all eight free parameters of the reversible nucleotide rate Q matrix are specified. The best-fit model of evolution for the data set can be selected through statistical testing. The fit to the data of different models can be compared through likelihood ration tests (LRT) or information criteria to select the best-fit model within a set of possible ones.

Most models of DNA sequence evolutions share the following assumptions: (i) the probability of a change from nucleotide *i* to nucleotide *j* is independent and does not depend on the previous state of nucleotide *i*; (ii) substitution events are independent across sites, so that the probability of changes in one site does not affect the probability of change in another site and (iii) the process of substitution remains constant over time and relative frequencies of nucleotide are at equilibrium. These assumptions may be unrealistic in real data, although these assumptions are still made in order to make phylogenetic reconstruction more computationally feasible. Rate heterogeneity can be incorporated into ML analysis by adding a relative rate component, *r*, to the substitution probability expressions. One can accommodate the rate variation by assuming that rate r for any site is a random variable drawn from a statistical distribution. The most commonly used distribution is the gamma. The resulting model is represented by a suffix 'G', the distances are sometimes called gamma distances. A discontinuous gamma distribution is most frequently used to model this rate heterogeneity over sites in phylogenetics. It has been shown the tremendous impact it has on phylogeny estimation and molecular evolution. A single parameter (alpha or shape parameter) controls the form of the gamma distribution, making it very convenient in a phylogenetic context. When alpha has a value < 1 there is strong among-site variation present in the data set. The higher the value of alpha, the lower the heterogeneity. Sometimes a proportion of invariant sites (*I*) is also used to model rate heterogeneity over sites. Several substitution models for RNA evolution, including rRNA, have been recently proposed as an attempt to add biological realism to modelling process. The assumption that each nucleotide evolves independently is not really correct for RNA molecules. Paired-site substitution models can account for the helixes of these molecules.

A proper characterization of the process of sequence evolution is essential in molecular phylogenetic inference as phylogenetic methods tend to be less accurate or inconsistent when an incorrect model of sequence evolution is assumed. In general, model selection strategies attempt to find the appropriate level of complexity on the basis of the available data. Increasing model complexity not only improves the fit to the data but also increases the error in the estimated parameters. Therefore, the use of over-parametrized models should be avoided in order to keep estimates as precise as possible. In recent years, several statistical methods (based on hypothesis testing) have been developed for selecting best-fit models of sequence evolution for a given dataset.

The best-fit model of evolution for data set can be selected through statistical testing. The fit to the data of different models can be compared through LRT or *information criteria* to select the best-fit model within a set of possible ones.

One of the simplest ways of comparing two models is to estimate their likelihoods using the LRT statistic: $\Delta = 2\ (\ln L_1 - \ln L_0)$, where L_1 is the ML under the alternative hypothesis and L_0 is the ML under the null hypothesis. When the model compared is nested, this statistic is asymptotically distributed as χ^2 with q degrees of freedom, where q is difference in number of free parameters between the two models. The difference in log-likelihood close to zero means that the alternative hypothesis does not fit the data significantly better than the null hypothesis. Another approach for comparing models is to compare all competing models at the same time by calculating the *minimum theoretical information criterion*, or *Akaike information criterion*. The LRT tests and AIC calculation are implemented in a very popular tool called the ModelTest program. ModelTest selects the best-fit model from among several dozen competing models by performing LRT, AICs and Bayesian Information Criterion (BIC) calculations based on the likelihood scores (Lemey *et al.*, 2009).

6.3.3. Methods to Infer Phylogenetic Trees

The methods for constructing phylogenetic trees from molecular data can be grouped into two major groups: (i) *distance methods* and (ii) *discrete methods*. In distance methods, sequences are converted into a distance matrix that represents an estimate of the evolutionary distances between sequences, from which a phylogenetic tree is constructed, by considering the relationships among these distance values, which are supposed to represent distances between taxa. In contrast to distance-based methods, maximum parsimony (MP), ML method and BI maps the history of characters onto a tree. Each method requires some assumptions about evolution (Baum and Smith, 2012; Hall, 2017).

Minimum evolution method

Minimum evolution method is very useful for analysis of sequences. In this method, the sum of all branch lengths is computed for all plausible topologies, and the topology that has the smallest sum value is chosen as the best tree.

The two most popular distance methods, *UPGMA* (Unweighted Pair-Group Method with Arithmetic Mean) and *Neighbour Joining* (NJ), are both algorithmic methods, that is, they use a specific series of calculations to estimate a tree. The calculation involves manipulations of a *distance matrix* that is derived from a multiple alignment. Starting with the multiple alignment, both methods calculate for each pair of taxa the distance, or the fraction of differences, between the two sequences and write that distance to a matrix. These cluster methods construct a tree by linking the least distant pair of taxa, followed by successively more distant taxa. The NJ method is the simplest version of the minimum evolution method. This method does not examine all possible topologies, but at each stage of taxon clustering a minimum evolution principle is used. The NJ algorithm is extremely popular, because it is relatively fast and performs well when the divergence between sequences is low. One advantage of the NJ method is computational efficiency as it does not need to compare as many trees under an optimality criterion and so this method is also useful for analysing large datasets.

Maximum parsimony

MP is important and one of the earliest methods of phylogenetic inference. This method was widely used for analysing discrete morphological characters. During the late 1970s, it began to be applied to molecular data. Parsimony principle is based on the principle that among competing hypotheses, the one with the fewest assumptions should be selected (Felsenstein, 2004).

MP directly uses character states and is based on an optimality criterion, which is a rule to decide which of the alternative trees is the best. The goal of parsimony is to find the tree with the minimum total tree length, or the minimum amount of evolutionary changes, that is, the transformation of one character state to another. The better a tree fits the data, the fewer homoplasies will be required and the fewer character state changes will be required.

Several different parsimony methods have been developed for treating data sets. If all nucleotide substitutions were equally weighted, such a procedure is called *unweighted or equally weighted parsimony*. MP methods that assign different weights for various characters are named *weighted parsimony*. If transitions are ignored and only transversions are considered, such a method is called *transversion parsimony*. In the search for an MP tree, sites that have the same nucleotide for all taxa (invariable) are eliminated from the analysis, and only *variable sites* are used. Among variable sites, it distinguishes the *parsimony informative* sites and nucleotide sites that are informative for MP tree construction.

Parsimony analysis minimizes the number of characters that change in parallel or reverse. If there are many such *homoplasious* characters, which are not present in their common ancestor, then it is possible that the phylogenetic tree is an artefact of the characters that have been chosen. A slight change in characters can lead to a different tree. For this reason, it has been proposed to calculate the *consistency index* (CI) to measure the extent of homoplasy. It equals the fewest number of steps in a transformation series divided by the number of times character state arises on the tree. Lower CI indicates that many characters contradict the evolutionary tree. The

CI is also sensitive to the number of taxa in an analysis, and analyses with many taxa tend to have lower CIs than analyses with fewer taxa.

The problem of finding optimal trees under the MP criterion can be divided into two subproblems: (i) determining the tree length and (ii) searching over all possible topologies tree with minimum length. When the number of taxa is small, it is possible to evaluate each of the possible trees or to conduct an *exhaustive search*. An exhaustive search is carried out by finding each of the possible trees by using a *branch-additional algorithm*. However, if the number of trees is large, the application of this approach is impossible and a *heuristic algorithm*, a shortcut strategy, is used. *Heuristic* methods usually proceed in two steps: (i) a single tree is built by adding one taxon at a time, placing the added taxon on the branch that gives the best tree for the subset, and (ii) when a tree containing all taxa is at hand, another step tries to find a better tree by moving subtrees to other branches, retaining the new tree if it is better than the previous one.

The strength of parsimony is its simplicity and it is easy to describe and understand. As with any method, MP has its pitfalls and may even be seriously misleading. A major weakness of parsimony is that it cannot incorporate knowledge of the process of sequence evolution in tree reconstruction. Moreover, if some sequences have evolved much faster than others, homoplasies will have occurred more often among the branches leading to these sequences than others, so parsimony tends to cluster these branches together. This phenomenon is called *long-branch attraction*. This effect can be reduced by additional sampling taxa that are related to those terminating the long branches. Removing fast-evolving sites can also be helpful.

Maximum likelihood method

The ML method is very often used for sequence data and is the most appealing way to estimate phylogeny. This method is one of the standard tools of statistics and was first applied to phylogenetics several decades ago. The first algorithm for ML analysis of DNA sequence data was developed by Felsenstein (1981). ML requires three elements: a model of sequence evolution, a tree and the observed data. The tree that makes our data the most probable evolutionary outcome is the ML estimate of the phylogeny. The likelihood is not the probability that the tree is true, rather it is the probability that the tree has given rise to the data collected. This is often written as:

$$L = \Pr(D|H)$$

where Pr (D | H) is the probability of getting the data D given hypothesis H. As L is often a very small number, likelihood is often expressed as natural logarithms and referred to as *log-likelihood*. ML is also an optimality method: the best tree is the one that renders the observed sequences most likely to have evolved under the assumed evolutionary model. The likelihood of a site is the probability of the observed states at that site given all the possible combinations of states at the internal nodes of the tree (ancestral states). The likelihood of a tree is the product of the likelihoods for each site of the alignment. The tree with the highest likelihood is the best estimate of the true phylogeny.

The strong statistical foundations of likelihood-based methods probably make them the most robust way for estimating molecular phylogenies and understanding sequence evolution. However, there are also criticisms of this method related to the fact that the result may be especially dependent on the correctness of the employed model of sequence evolution. Another main obstacle of ML methods is computing time because algorithms that find the ML score must search through a multi-dimensional space of parameters to find a tree. However, the exponential increase in computer power and the development of faster algorithms have largely solved this problem.

Bayesian inference

BI methods were introduced in the 1990s and have since revolutionized the ways of analysing gene and genomic sequence datasets (Yang and Rannala, 1997). Bayesian approaches to phylogenetics are relatively new and based on a quantity called the *posterior probability* (PP) of a tree. Bayes' theorem establishes that the relationships are based on the following formula:

$$\Pr[\text{Tree} \mid \text{Data}] = \frac{\Pr[\text{Data} \mid \text{Tree}] \times \Pr[\text{Tree}]}{\Pr[\text{Data}]}$$

and is used to combine the prior probability of phylogeny (Pr[Tree]) with the likelihood (Pr[Data | Tree]) to produce a PP distribution on trees (Pr[Tree | Data]). The PP of a tree can be interpreted as the probability that the tree is correct. The PP involves a summation over all trees, for each tree, integration over all possible combinations of branch length and substitution model parameter values. It is impossible to do analytically and *Markov Chain Monte Carlo (MCMC)* is used to approximate the PP of trees. This MCMC variant involves running several independent chains simultaneously (typically four). In each generation, every chain proposes and accepts/rejects moves independently, and a swap of the states is attempted between two randomly chosen chains. States are only sampled from one of the chains, designated as the 'cold' chain (the rest are 'heated' chains). If the cold chain gets stuck in local optima (a low probability hill in the posterior density landscape), it has a chance to escape by swapping with another chain that may be on a higher hill. BI has the advantage of a strong connection with the likelihood framework and its powerful statistical foundations. Moreover, as a result of the MCMC process, there is a PP associated with each node on the inferred Bayesian tree (the fraction of times a clade occurs among the sampled trees) that can be used as a measure of support for that node.

ML and Bayesian analysis are based upon the likelihood function but there are fundamental differences in how the two methods threat parameters. Unlike ML, which searches for a single most likely tree, Bayesian MCMC searches for the 'best set of trees' in the landscape of possible trees. The initial state of the chain is a tree with a combination of branch lengths and parameters of the substitution model.

The most important aspect of MCMC BI is its computational efficiency. BI was found to perform best with a dataset containing missing data and alignment gaps. This method allows complex models of sequence evolution to be implemented, for example, to use different models for stems and loops of secondary structure of rRNA molecule for dataset. Bootstrap (BS) values and Bayesian PP both measure phylogenetic uncertainty, although they are calculated very differently. PP for clades that have been calculated from real data sets often appear to be too high and overestimate the robustness of relationships.

6.3.4. Phylogenetic Tree and Network

The result of a molecular phylogenetic analysis is expressed in phylogenetic graphs: phylogenetic tree and network (Fig. 6.2). The term 'phylogenetic graph' is used to refer to branching evolutionary histories or evolutionary relationships between genes or organisms. Phylogenetic graphs play a prominent role in modern biology as they provide a concise way to visualize evolution as descent from common ancestors. It consists of *nodes* connected by *branches* (Fig. 6.2A). The branch length usually represents the evolutionary distances between two consecutive nodes.

There are various terms referring to various parts of tree. Taxa placed at the ends of branches are called 'terminal taxa', or 'terminal nodes' ('leaves'), and are also often called 'operational taxonomic units (OTUs)' (Fig. 6.2A). The branches leading to terminal taxa are called 'terminal branches'. *Internal nodes*, or branching points, represent hypothetical ancestors and may be called 'hypothetical taxonomic units (HTUs)'.

The ancestor of all the sequences that comprise the tree is the *root* of the tree and is named as an *outgroup*. The remaining OTUs form the *ingroup*. A group of taxa that belong to the same branch is called a *clade*. A clade, or a *monophyletic group*, is a group that includes an ancestor and all descendants of that ancestor, whereas a *paraphyletic group* contains an ancestor but only some of its descendants and a *polyphyletic group* contains various organisms with no recent common ancestor (Fig. 6.3). *Sister group* or *sister taxa* are two groups on a tree with the same immediate common ancestor (Fig. 6.2A). The branching patterns, or the order of nodes, is termed the 'topology' of the tree. If three branches connect to an internal node, then the node represents a *bifurcation*, or *dichotomy* (Fig. 6.2A). A fully binary tree is called *fully resolved*. If there are more than three branches connected to an internal node, then the node represents a *multifurcation*, or *polytomy* (Fig. 6.2A). Polytomy can be distinguished as either *hard polytomy* or *soft polytomy*. Hard polytomy is meant to represent explosive radiation in which a single common ancestor gave rise to multiple distinct lineages at the same time. In contrast, soft polytomy represents unresolved topology with uncertainty in branching orders.

Often parsimony analyses find multiple trees with the same length but with a different branch order. Rather than choosing among these trees, systematists simply want to see which groups can be found in all the shortest trees. There are approaches to summarize data

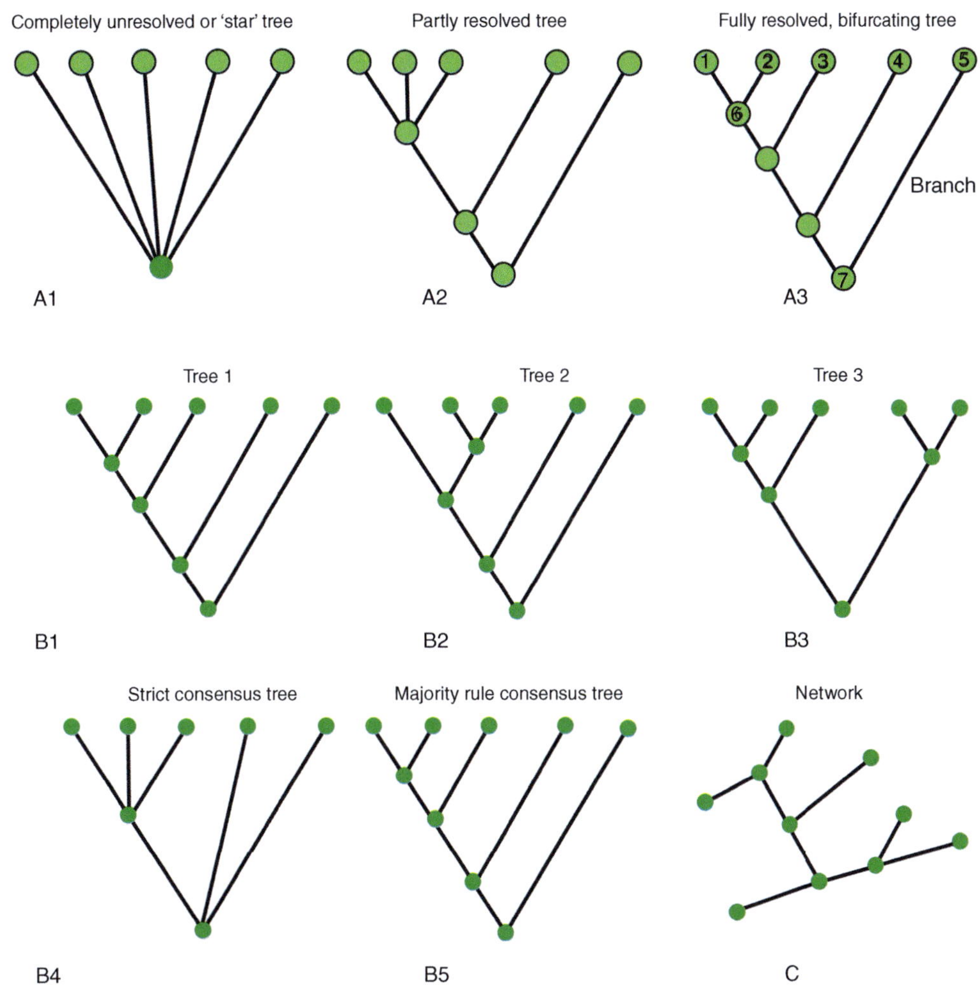

Fig. 6.2. A1, A2, A3: Phylogenetic trees with different resolution between clades (1, 2, 3, 4, 5 - terminal nodes, taxa; 1 , 2 - sister taxa; 6 - immediate common ancestor for 1 an 2 taxa; 7 – root); B1, B2, B3: three phylogenetic trees with different topology and strict consensus tree (B4) and majority rule consensus tree (B5) generated from these three trees; C: phylogenetic network.

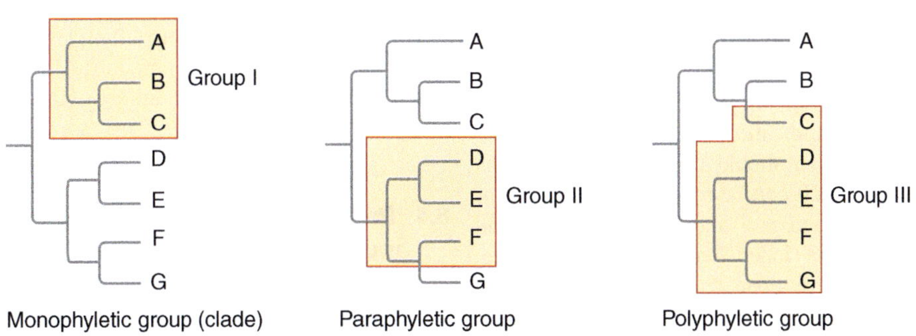

Fig. 6.3. Phylogenetic trees with different taxa grouping.

that are common to two or more trees in a single tree. The resulting tree is called a *consensus tree*. It is usually distinguished from the *strict consensus tree* by including only those groups (or splits) that occurred in all the trees being considered or the *majority consensus rule tree* including those groups (or splits) that occurred in more than half the trees being considered (Fig. 6.2B). When several interior branches of a phylogenetic tree have low statistical support values, it is often useful to produce a multifurcating tree by assuming that all interior branches have a branch length equal to 0. This multifurcating tree is called a *condensed tree*. To present phylogenetic trees there are several widely accepted rules. Branch lengths are almost always drawn to scale. BS values should be displayed as percentages and only statistical support values of 50% and higher are presented for easier understanding and comparison with other trees.

6.3.5. Supertree and Supermatrix Approaches

Multiple sequence data sets are now available for studying phylogenetic relationships of taxa. Integrating this information in studied taxa into a unified statement of evolutionary history is a major challenge in the field of systematics. There are two different strategies for this task: *supertree* and *supermatrix* approaches. The supertree approach separately analyses each gene and then uses heuristic algorithms to assemble the subtrees for individual genes into a supertree for all species. In the supermatrix approach, all character data are joined into a single data set, and a *total evidence analysis* of this large matrix is used for reconstruction of phylogeny.

6.3.6. Evaluation of the Reliability of Inferred Trees

The estimation of phylogeny should be accompanied by an indication of its confidence limit. Phylogenetic trees should always be evaluated for reliability, which can be measured as the probability that the taxa of a given clade are always members of that clade. Often the quality of trees is determined using nodal support measures, such as non-parametric bootstrapping, jack-knifing or Bayesian PP.

Perhaps, the most commonly used method is non-parametric bootstrapping. To estimate the confidence level by bootstrapping, or the BS value of clade, a series of pseudo-samples or pseudo-alignments are first generated by randomly re-sampling the sites in the original alignment with replacement. In such pseudo-alignments, some characters are not included at all, while others may be included twice or more. Secondly, for each pseudo-alignment, a tree is constructed, and the proportion of each clade among all the BS replicates is computed in a majority-rule consensus tree. If the BS and PP values of the clade obtained in the analysis are greater than 95%, the branch is considered to be statistically significant. BS value less than 70% and PP value less than 85% should be treated with caution for MP, ML, NJ and BI trees, respectively. Examples of phylogenetic trees with statistical support for branches for different nematode genera are given in Figure 6.4.

6.3.7. Testing of Evolutionary Hypothesis

It should be emphasized that the output of the phylogenetic analysis is a phylogenetic tree, which represents only one of the hypotheses of evolutionary relationships and should be carefully tested. Using different genes and different phylogenetic inference methods, different tree topologies can be obtained for the same organism group. Once phylogenetic trees are obtained from a molecular dataset using different methods, they should be compared with each other or with trees generated from other, non-molecular datasets. There are several tests that allow the evaluation of alternative hypotheses and which determine under certain assumptions and provide probability values indicating significant or non-significant differences between topologies if one tree is statistically worse than another. These include the following: Templeton test, Kishino-Hasegawa test, Shimodaira-Hasegawa test, Shimodaira Approximately Unbiased test and Bayes' factors.

6.4. Problems of Phylogenetic Analysis and General Recommendations

When working with molecular phylogeny, the researcher should remember that there are

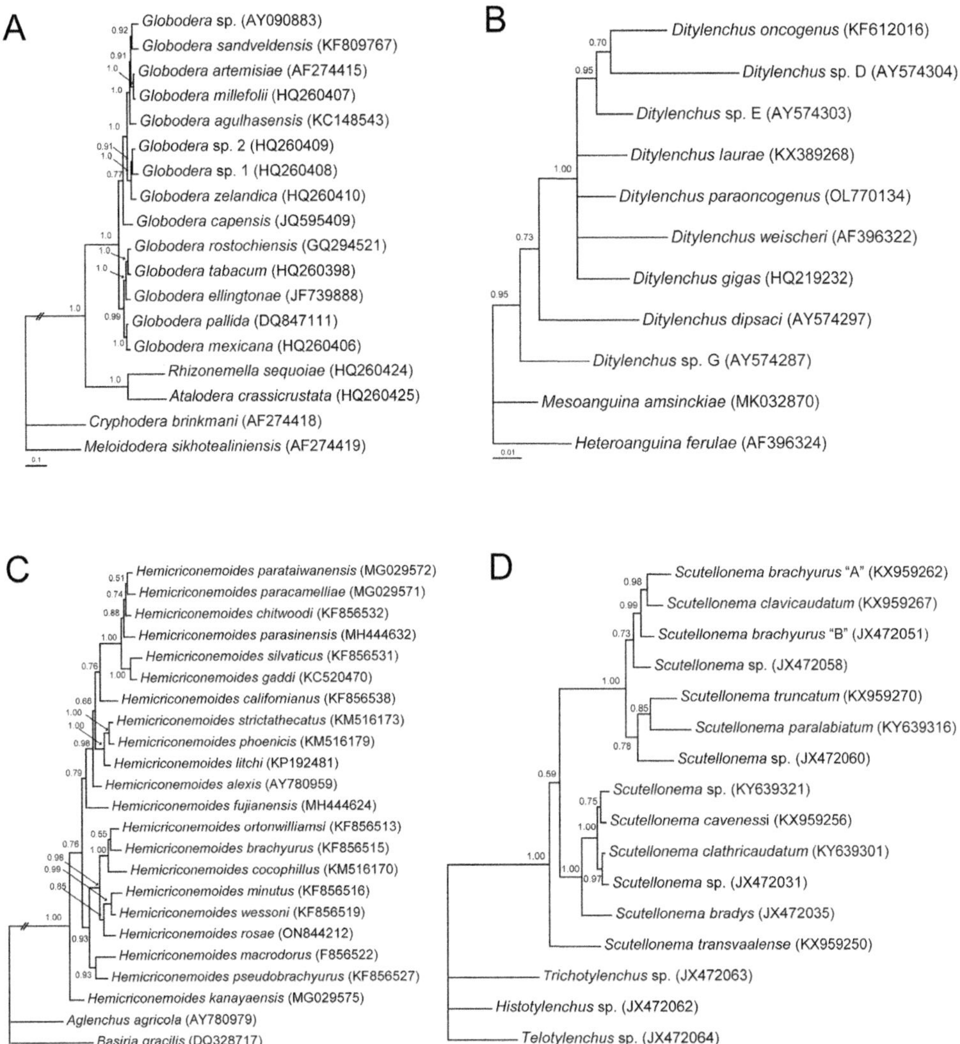

Fig. 6.4. Phylogenetic relationships of four genera as inferred from Bayesian analysis of the rRNA gene sequences under the GTR + I + G model. A: *Globodera* (ITS rRNA gene). B: *Ditylenchus* (ITS rRNA gene). C: *Hemicriconemoides* (D2-D3 of 28S rRNA gene). D: *Scutellonema* (D2-D3 of 28S rRNA gene).

many different phylogenetic algorithms and that not all of these methods are equally good for all data. Very often, data can be subject to both systematic biases and noise that affect our chances of getting the correct tree (saturation, alignment artefacts, base compositional biases, branch length or rate asymmetries leading to long-branch attractions). The quality of data and trees should always be considered with caution.

Although you need not understand all the details of the various phylogenetic methods, an understanding of the basic properties is essential for informed choice of method and interpretation of results.

Molecular systematics yields gene trees. Accurate gene trees may not be accurate organismal trees. Gene duplications, lateral transfer, introgression or recombination, incorrect inclusion of paralogous sequences in the studied dataset and incomplete lineage sorting due to retention of ancestral polymorphisms can produce mismatches between gene and organismal phylogenies. Multiple analyses using different phylogenetic algorithms help the researcher to understand the behaviour of branches in the trees. Be an intelligent user of phylogenetic software, not a mindless button-pusher!

6.5. References

Baum, D. and Smith, S. (2012) *Tree Thinking. An Introduction to Phylogenetic Biology*. Roberts Publishers, Greenwood Village, Colorado.

Felsenstein, J. (1981) Evolutionary trees from DNA sequences: a maximum likelihood approach. *Journal of Molecular Evolution* 17, 368–376.

Felsenstein, J. (2004) *Inferring Phylogenies.* Sinauer Associates, Sunderland, Massachusetts.

Hall, B.G. (2017) *Phylogenetic Trees Made Easy: A How-to Manual*, 5th edn. Sinauer Associates, Sunderland, Massachusetts.

Hillis, D.M., Moritz, C. and Mable, B.K. (1996) *Molecular Systematics*, 2nd edn. Sinauer Associates, Sunderland, Massachusetts.

Lemey, P., Salemi, M. and Vandamme A.-M. (eds) (2009) *The Phylogenetic Handbook. A Practical Approach to Phylogenetic Analysis and Hypothesis Testing*, 2nd edn. Cambridge University Press, Cambridge, UK.

Nei, M. and Kumar, S. (2000) *Molecular Evolution and Phylogenetics.* Oxford University Press, Oxford, UK.

Page, R.D.M. and Holmes, E.C. (1998) *Molecular Evolution: A Phylogenetic Approach*. Blackwell Science, Oxford, UK.

Yang, Z. and Rannala, B. (1997) Bayesian phylogenetic inference using DNA sequences: a Markov chain Monte Carlo Method. *Molecular Biology and Evolution* 14, 717–724. DOI: 10.1093/oxfordjournals.molbev.a025811.

Yang, Z. and Rannala, B. (2012) Molecular phylogenetics: principles and practice. *Nature Reviews Genetics* 13, 303–314. DOI: 10.1038/nrg3186.

7 Aphelenchoidea

Natsumi Kanzaki[1,2]* and Robin M. Giblin-Davis[1]
[1]Fort Lauderdale Research and Education Center, University of Florida/IFAS 3205 College Avenue, Davie, FL, USA; [2]Kansai Research Center, Forestry and Forest Products Research Institute, Fushimi, Kyoto, Japan

Abstract

The Aphelenchoidea is a large and diverse group whose relationships lack precision due to an abundance of old and inadequate descriptions, a lack of molecular data for many groups and the fact that they are often dismissed by nematologists, only a few species being currently recognised as important plant-parasites, most being fungivores, insect associates or predatory. The general diagnostic characters are discussed and the current taxonomic system with an outline classification for the entire superfamily provided. Morphological keys to families, subfamilies and genera are provided. Species of major economic importance, including *Aphelenchoides besseyi* (the White Tip Nematode), *A. bicaudatus*, *A. fragariae* (the Strawberry Nematode), *A. ritzemabosi* (the Chrysanthemum Foliar Nematode), *Bursaphelenchus cocophilus* (the Red Ring Nematode of coconut palms), *B. mucronatus*, *B. sexdentati*, and *B. xylophilus* (the Pine Wood Nematode), are discussed in detail. The chapter concludes with the current status of the molecular identification and systematics of aphelenchs and provides updated phylogenetic trees.

7.1. Introduction

Nematodes in the Aphelenchoidea are morphologically similar to each other based upon gross morphology; that is, most of them have a regular thread-like body, thin stylet with indistinctive basal swellings and a well-developed metacorpus (middle and muscular portion of a three-part pharynx). However, their biological traits are much more divergent when compared with their morphological similarity. They comprise plant-parasitic species, including several important plant pathogens, for example, *Bursaphelenchus xylophilus*, *Bursaphelenchus cocophilus*, *Aphelenchoides fragariae* and *Aphelenchoides besseyi*, many species of entomophilic/entomoparasitic nematodes, for example, Entaphelenchinae, Ektaphelenchinae and Acugutturinae, predators, for example, Seinurinae and mycophagous species including *Aphelenchus avenae*, which is being explored for its potential as a biological control agent for plant-pathogenic soil-inhabiting fungi. In addition, a culturable hermaphroditic *Bursaphelenchus* species, *Bursaphelenchus okinawaensis* is currently becoming an important satellite model nematode. Therefore, the taxonomy and systematics of Aphelenchoidea have been at the forefront of developing accurate identification

*E-mail: natsumikanzaki@gmail.com

A revision of Kanzaki, N. and Giblin-Davis, R.M. (2012) Aphelenchoidea. In: Manzanilla-López, R.H. and Marbán-Mendoza, N. (eds), *Practical Plant Nematology*. Mundi-Prensa and Biblioteca Básica de Agricultura, Montecillo, Mexico.

(eds R.H. Manzanilla-López, D.J. Hunt, N. Marbán-Mendoza and S. Infante Gil)
DOI: 10.1079/9781836990413.0007

systems for control and risk assessments for many species. Because of their practical importance, many of these nematodes have been studied for a long time with the result being many unclear and incomplete descriptions. Some of this is due to the unevenness of older descriptions and some to over-reliance at the light microscopic level on gross morphology of small invertebrates that display a high level of convergence in most currently used characters. This has led to confusion. For example, *Aphelenchoides* has about 200 nominal species (Gu *et al.*, 2023; Kanzaki and Tanaka, 2025) with most being so close in quantitative and qualitative macromorphology that they are difficult to discern. Add to this that type specimens and molecular data are only available for a limited number of species, and you have a recipe for a crisis in understanding the inherent identity and diversity of the members of the group, if, in fact, it is a natural grouping. In the present chapter, the current situation of the taxonomy and systematics of the superfamily are introduced, and some biological characters for each subfamily and/or genus are discussed.

7.2. General Characters of the Aphelenchoidea and Its Families

The Aphelenchoidea has two phylogenetically different families, Aphelenchidae and Aphelenchoididae. The members of the superfamily are mostly slender and thread-like nematodes with body length varying between 0.2 and 2 mm, but slightly shorter than 1 mm in an average adult nematode. Some insect parasites and plant parasites have exceptionally long, swollen or obese bodies being up to 3 mm or more in length. The cuticle is usually thin and annulated with fine transverse striations. The lateral field has been reported from many species and is assumed to be present in all species. The width and number of incisures varies among species from 2 to 12. The lip region is usually off-set, separated from the rest of the body by a weak or clear constriction. The lip region is usually composed of six equal-sized lip sectors, with an exception; that is, the two lateral sectors are narrower than the other four in *Ektaphelenchus* and *Devibursaphelenchus*. The lateral view of the lip region is rounded or roundish square, rarely spherical (only in *Anomyctus*). Annulation is present on the lip region in some groups. The stylet is thin and is separated into two parts, conus and shaft, and the stylet lumen is usually narrow, although some insect parasites and predators have a wide lumen. The base of the stylet can be variable, that is, without any knobs or swellings (most predators and insect parasites), with weak swellings (most species) or with clear basal knobs (some fungal-feeding genera). The most definitive character of Aphelenchoidea is the composition and structure of the alimentary tract, especially the structures of the procorpus and median bulb. The procorpus is a uniform cylindrical tube connecting the stylet and metacorpus. The metacorpus forms a well-developed and muscular median bulb, which works as a pump to ingest nutritional fluid from the stylet to the intestine or postcorpus. The bulb has a prominent valve apparatus (valve plates) at the central or posterior part, and all three pharyngeal gland orifices with the dorsal gland orifice anterior to the valve plate and the other two (ventrosublaterals) located posterior to the plate. The lateral view of the median bulb is typically oval, but spherical or rounded rectangular in some groups. The postcorpus is known to be composed of an isthmus and glandular basal bulb in *Paraphelenchus*, or forming a narrow isthmus in *Aphelenchus*, located immediately posterior to the median bulb, and separated from the intestine by a small pharyngo-intestinal valve. The pharyngo-intestinal junction is situated immediately posterior to the median bulb in Aphelenchoididae and the isthmus is missing. The pharyngeal gland lobes are observed in *Aphelenchus* and Aphelenchoididae. The gland lobes overlap the intestine dorsally (Aphelenchoididae) or isthmus and intestine (*Aphelenchus*). The nerve ring is located posterior to the median bulb, surrounding the isthmus (*Paraphelenchus*), isthmus and gland lobes (*Aphelenchus*) or intestine and gland lobes (Aphelenchoididae). The intestine is cellular and endodermal in origin with a distinctly lined lumen. The anus and rectum are usually present, except for several insect parasites, which do not have a functional anus and rectum and where the intestine ends in a blind sac. Genital systems of males and females are single. The female genital system is composed of an ovary, oviduct, spermatheca, crustaformeria, uterus or uterine sac, vagina/vulva and post-vulval uterine sac, but in many cases, each organ is not clearly distinctive in microscopic observations.

In some species (genera), some parts, for example, spermatheca, are not clearly separated from the other parts or are missing. A post-vulval uterine sac is usually present and conspicuous, but very small or absent in several species (genera). The oocytes in the ovary are arranged as single or double rows at the anterior part, single to multiple rows in the middle part and large cells are arranged in a single row in the posterior part. The male genital system is composed of a testis and vas deferens. The testis has sperm cells in single or multiple rows and the *vas deferens* is fused with the intestine to form a single tube at the posterior end and connected to the cloacal opening. The male spicules are paired and usually rather straight in Aphelenchidae but rosethorn shaped or a derivation of this shape in Aphelenchoididae. A straight gubernaculum is present in Aphelenchidae and missing or small and indistinctive in Aphelenchoididae. The tails of both sexes are usually short and conoid with various tail tip shapes, that is, several tube-like projections, raspberry shape (with many small knob-like projections), mucronated (one or more, up to three), rounded, pointed or with a spike-like projection. Some of the predatory species and a couple of others have filiform tails. The general morphology of aphelenchs is shown in Figure 7.1.

7.3. Current Taxonomic System

Hunt (1993; 2008) and Kanzaki and Giblin-Davis (2012) organized the taxonomic system of the superfamily and presented the genus and species catalogue. Here, the current taxonomic system of the Aphelenchoidea is introduced according to Hunt (2008) with some updates following recent revisions (e.g., Braasch, 2009; Kanzaki *et al.*, 2009b).

7.3.1. Outline Classification

Class Chromadorea Inglis, 1983
- Subclass Chromadoria Pearse, 1942
  - Order Rhabditida Chitwood, 1933
    - Suborder Tylenchina Thorne, 1949
      - Infraorder Tylenchomorpha De Ley & Blaxter, 2002
        - Superfamily Aphelenchoidea Fuchs, 1937

Family Aphelenchidae Fuchs, 1937
- Subfamily Aphelenchinae Fuchs, 1937
  - *Aphelenchus* Bastian, 1865
  - = *Isonchus* Cobb, 1913
  - = *Metaphelenchus* Steiner, 1943
- Subfamily Paraphelenchinae T. Goodey, 1951
  - *Paraphelenchus* Micoletzky, 1922
  - = *Aphelenchus* (*Paraphelenchus*) Micoletzky, 1922

Family Aphelenchoididae Skarbilovich, 1947
- Subfamily Aphelenchoidinae Skarbilovich, 1947
  - *Aphelenchoides* Fischer, 1894
  - = *Chitinoaphelenchus* Micoletzky, 1922
  - = *Pathoaphelenchus* Cobb, 1927
  - = *Asteroaphelenchoides* Drozdovsky, 1967
  - = *Paraphelenchoides* Haque, 1967
  - = *Pseudaphelenchoides* Drozdovsky, 1967
  - = *Ptychaphelenchus* Hodda, 2009
  - *Ficophagus* Davies & Bartholomaeus, 2015 *in* Davies *et al.*, 2015
  - *Laimaphelenchus* Fuchs, 1937
  - = *Ruidosaphelenchus* Laumond & Carle, 1971
  - *Martininema* Davies & Bartholomaeus, 2015 *in* Davies *et al.*, 2015

Punchaulus De Ley & Coomans, 1996
Robustodorus Andrássy, 2007
= *Megadorus* J.B. Goodey, 1960b *nec Megadorus* Linnavuori, 1959
Schistonchus Cobb, 1927
= *Aphelenchus* (*Schistonchus*) Cobb, 1927
Subfamily Anomyctinae J.B. Goodey, 1960b
Anomyctus Allen, 1940
Subfamily Tylaphelenchinae Kanzaki, Li, Lan & Giblin-Davis, 2014b
Basilaphelenchus Pedram, Kanzaki, Giblin-Davis & Pourjam, 2018
Pseudaphelenchus Kanzaki & Giblin-Davis, 2009 *in* Kanzaki *et al.*, 2009b
Tylaphelenchus Rühm, 1956
Subfamily Seinurinae Husain & Khan, 1967
Seinura Fuchs, 1931
= *Paraseinura* Timm, 1960
= *Indaphelenchus* Bajaj, Bishnoi & Dabur, 1994
Aprutides Scognamiglio, Talamé & s'Jacob, 1970
Papuaphelenchus Andrássy, 1973
Subfamily Ektaphelenchinae Paramonov, 1964
Ektaphelenchus Fuchs, 1937
= *Parasitaphelenchus* (*Ektaphelenchus*) Fuchs, 1937
= *Ectaphelenchus* Rühm, 1956
= *Cryptaphelenchoides* J.B. Goodey, 1960b
= *Caballeroides* Chaturvedi & Khera, 1977
Cryptaphelenchus Fuchs, 1937
= *Parasitaphelenchus* (*Cryptaphelenchus*) Fuchs, 1937
= *Parasitaphelenchus* (*Steineria*) Fuchs, 1937 *nec Steineria* Micoletzky, 1922
Devibursaphelenchus Kakuliya, 1967
Ektaphelenchoides Baujard, 1984
Lenisaphelenchus Kanzaki & Hamaguchi, 2020
Subfamily Acugutturinae Hunt, 1980
Acugutturus Hunt, 1980
Noctuidonema Remillet & Silvain, 1988
Setonema Poinar, 2011 [fossil genus]
Vampyronema Hunt, 1993
Subfamily Parasitaphelenchinae Rühm, 1956
Parasitaphelenchus Fuchs, 1930
Bursaphelenchus Fuchs, 1937
= *Berntsenus*[1] Massey, 1974
= *Rhadinaphelenchus* J.B. Goodey, 1960a
= *Huntaphelenchoides* Nickle, 1970b
= *Omemeea* Massey, 1971
= *Teragramia* Massey, 1974
= *Ipsaphelenchus* Lieutier & Laumond, 1978
Potensaphelenchus Gu, Liu, Abolafia & Pedram, 2021
Ruehmaphelenchus J.B. Goodey, 1963
Sheraphelenchus Nickle, 1970a
Subfamily Entaphelenchinae Nickle, 1970a
Entaphelenchus Wachek, 1955
Peraphelenchus Wachek, 1955
Praecocilenchus Poinar, 1969
Roveaphelenchus Nickle, 1970a

Genera incertae sedis:

Chalicidonema Poinar, 2011
Cretaciaphelenchoides Poinar, 2011
Oligoaphelenchoides Poinar, 1977
Palaeoaphelenchoides Poinar, 2011

7.4. Morphological Key to Genera

Hunt (1993) presented keys to families, subfamilies and genera of Aphelenchoidea. Although the taxonomic ranks were revised by Hunt (2008), the outline of the classification was not altered, thus, a simplified key to each aphelench genus based on Hunt (1993) is presented here.

7.4.1. Key to Families

1. Spicules slender, with well-developed gubernaculum. Pharynx with distinctive isthmus, nerve ring only surrounding pharynx .... Aphelenchidae
Spicules mostly robust, rosethorn shaped or derived there from. Gubernaculum absent, or if present, consisting of a small structure around distal tip of spicules. Pharynx without distinctive isthmus, nerve ring surrounding pharynx and intestine[2] .... Aphelenchoididae

7.4.2. Key to Subfamilies and Genera of Aphelenchidae

As Aphelenchinae and Paraphelenchinae are monogeneric, their respective genera key out with the subfamily.

1. Male with prominent peloderan bursa supported by four pairs of limb-like papillae. Female vulval aperture oval-shaped. Pharyngeal glands forming a long dorsal overlapping lobe .... Aphelenchinae (*Aphelenchus*)
Male lacking bursa. Vulva in form of a transverse slit. Pharyngeal glands small, retained within a non-overlapping basal bulb .... Paraphelenchinae (*Paraphelenchus*)

7.4.3. Key to Subfamilies and Genera of Aphelenchoididae

Aphelenchoididae is separated into eight subfamilies and four fossil genera (*Chalicidonema*, *Cretaciaphelenchoides*, *Oligoaphelenchoides* and *Palaeoaphelenchoides*), which could be close to Aphelenchoidinae or Parasitaphelenchinae but have not been placed in any of the subfamilies (*genera incertae sedis*) as some of them were described based upon juvenile or unidentifiable stages found in amber. A morphological key to subfamilies or unplaced genera is presented here excluding a fossil genus, *Setonema*. *Punchaulus*, which has a distinct isthmus, that is, intermediate form between Aphelenchidae and Aphelenchoididae, and is tentatively placed in the Aphelenchoidinae, can be distinguished from the other genera or subfamilies, and thus appears in the first couplet of the key. It could be treated as a *genus incertae sedis*, but further discussion will be provided elsewhere.

1. Short (*ca* 20 μm) isthmus present, with small (less than 400 μm) body, short stylet possessing well-developed basal swellings, spherical median bulb .... *Punchaulus* (tentatively placed in Aphelenchoidinae)
Isthmus absent 2

2. Cephalic (lip) region high, almost spherical, with prominent sclerotized oral disc.. Anomyctinae (= *Anomyctus*)
Cephalic (lip) region low, rounded, lacking obvious oral disc 3
3. Stylet of both sexes very long (50-180 µm) with conus comprising majority of stylet length. Ectoparasites of insects .... Acugutturinae (4)
Stylet less than 50 µm (usually 10-20 µm) long, never attenuated 7
4. Adult and juveniles not markedly swollen. Spicules of normal size (less than 30 µm) and rostrum not tubular, thorn shaped. Parasites of cockroaches ....*Acugutturus*
Adult and juveniles markedly swollen. Spicules very large (more than 38 µm) with a tubular rostrum, massive. Parasites of moths 5
5. Stylet longer than 100 µm, without prominent basal knobs. Spicules 74-109 µm long with prominent, bluntly rounded apex and complex sheaths and sclerotization of blade (= calomus/lamina complex)..*Noctuidonema*
Stylet shorter than 80 µm, with prominent basal knobs. Spicules 38-49 µm long, lacking a prominent, rounded apex and with simple blade (= calomus/lamina complex) .... *Vampyronema*
6. Three adult forms; male (with thread-like body), immature female (thread-like body), mature parasitic female (with swollen body). Endoparasitic in haemocoel of beetles ....Entaphelenchinae (8)
Mature females not swollen or endoparasitic. Only two adult forms (exceptions: *Bursaphelenchus luxuriosae* has four forms: propagative male and female and phoretic male and female, and the *Bursaphelenchus aberrans* group, i.e., *B. aberrans, B. brachycephalus, B. pityokteini* and *B. sinensis*, has three or four forms: fungal-feeding propagative male and female and predatory male [not observed in all species] and female).. 11
7. Only mature parasitic female form found in insect haemocoel. Parasites of staphylinid beetles ....*Entaphelenchus*
Three adult forms found within insect haemocoel. Parasites of silphid, curculionid and staphylinid beetles 9
8. Mature parasitic female oviparous, male corkscrew-shaped (spiral-shaped) when killed by heat. Female post-vulval uterine sac long ....*Peraphelenchus*
Mature parasitic female ovoviviparous, post-vulval uterine sac absent in mature parasitic female.. 10
9. Vulva in mature parasitic female very posteriad at about 96-97%..................... *Roveaphelenchus*
Vulva more anteriad at about 80-84% ....*Praecocilenchus*
10. Females lacking anus and rectum, or if present, indistinct and non-functional. Stylet lumen wide in most genera with exception of *Cryptaphelenchus*. Ecto- and endoparasites of insects .... Ektaphelenchinae (12)
Females with functional anus and rectum... 16
11. Body length less than 500 µm, usually even shorter. 13
Body length more than 500 µm.. 14
12. Stylet with narrow lumen. Median bulb spherical..*Cryptaphelenchus*
Stylet with wide lumen. Median bulb elongate oval. Endoparasite of insects.......... *Lenisaphelenchus*
13. Tail long, filiform in female, spicate in male .... *Ektaphelenchoides*
Tail short with bluntly, or finely rounded, terminus in both sexes 15
14. Lip cap-like with lateral lip sectors narrower than others. Small bursal flap present at male tail tip... *Devibursaphelenchus*
Lip cap-like, lateral lip sectors narrower than or equal to others. Bursal flap absent.. *Ektaphelenchus*
15. Stylet long and slender with wide lumen. Median bulb exceptionally prominent, elongate oval or rounded rectangular with valve situated in posterior half. Most species are known to be predators .... Seinurinae (17)
Stylet with narrow lumen. Median bulb oval or spherical... 19

16. Tail tip initially conoid, then filiform to a finely pointed terminus .... *Seinura*
Tail more elongate conoid, tapering to a rounded or clavate terminus; not filiform 18
17. Tail tip broadly rounded, clavate .... *Aprutides*
Tail tip finely rounded, not clavate .... *Papuaphelenchus*
18. Males with small bursa-like flap (bursal flap), or short conical tail with or without spike-like projection at tail tip. Male tail excluding spike-like projection less than 3 cloacal body diam. long .... Parasitaphelenchinae (20)
Males have conical tail lacking such a bursal flap or spike-like projection, and male tail usually longer than 3 cloacal body diam. 24
19. Post-vulval uterine sac absent. Female tail subconoid, evenly tapering to a pointed terminus. Males without small bursal flap. Condylus of spicules greatly elongated, continuing line of dorsal limb. Associated with nitidulid beetles .... *Sheraphelenchus*
Post-vulval uterine sac present 21
20. Fourth-stage juveniles endoparasitic in the insect haemocoel (body cavity). Vulva very posteriad (85-90%). Spicules partially fused. Male tail not strongly recurved when killed by heat .... *Parasitaphelenchus*
Third- (or fourth-) stage dauer juveniles ectophoretic, exceptionally endophoretic. Spicules usually separate but reported to be partially fused in some species. Male tail strongly recurved when killed by gentle heat 22
21. Vulva more anteriad at 70-80% .... *Bursaphelenchus*
Males without spike-like projection at tail tip. Females with conical tail 23
22. Male spicule mitten-shaped with conspicuous dorsal and ventral limbs. Often associated with ambrosia beetles .... *Ruehmaphelenchus*
Male spicule thorn shaped. Stylet well developed with somewhat wide lumen .... *Potensaphelenchus*
23. Both sexes with small-midsized body (less than 700 µm), stylet with small, but conspicuous basal knobs or swellings, spherical median bulb and conical tail .... Tylaphelenchinae (24)
Not as above .... Aphelenchoidinae (26)
24. Male with long bursa enveloping whole tail .... *Pseudaphelenchus*
Male without bursa or bursal flap 25
25. Stylet with posteriorly directed short rod-like basal knobs .... *Basilaphelenchus*
Stylet with small and conspicuous basal knobs .... *Tylaphelenchus*
26. Tail tip bearing four pedunculate tubercles with fringed margins, or with raspberry-shaped appendage. Vulval flap may be present .... *Laimaphelenchus*
Tail tip not as above, but may be variously mucronated or conical. Vulval flap absent 27
27. Stylet very robust, about 17 µm long with massive, rounded basal knobs. Cuticular annulations coarse .... *Robustodorus*
Stylet not as above combination of characters. Cuticular annulations fine 28
28. Stylet robust, 21-24 µm long with well-developed knobs. Associated with figs and fig wasps 29
Stylet and other characters not as above .... *Aphelenchoides*
29. Secretory-excretory pore near the lip region (mostly overlapping with the level of stylet) .... *Ficophagus*
Secretory-excretory pore posterior to stylet 30
30. Secretory-excretory pore at the level of or posterior to median bulb. Lip with labial disc .... *Schistonchus sensu stricto*
Secretory-excretory pore at the level of or anterior to median bulb. Lip without labial disc .... *Martininema*

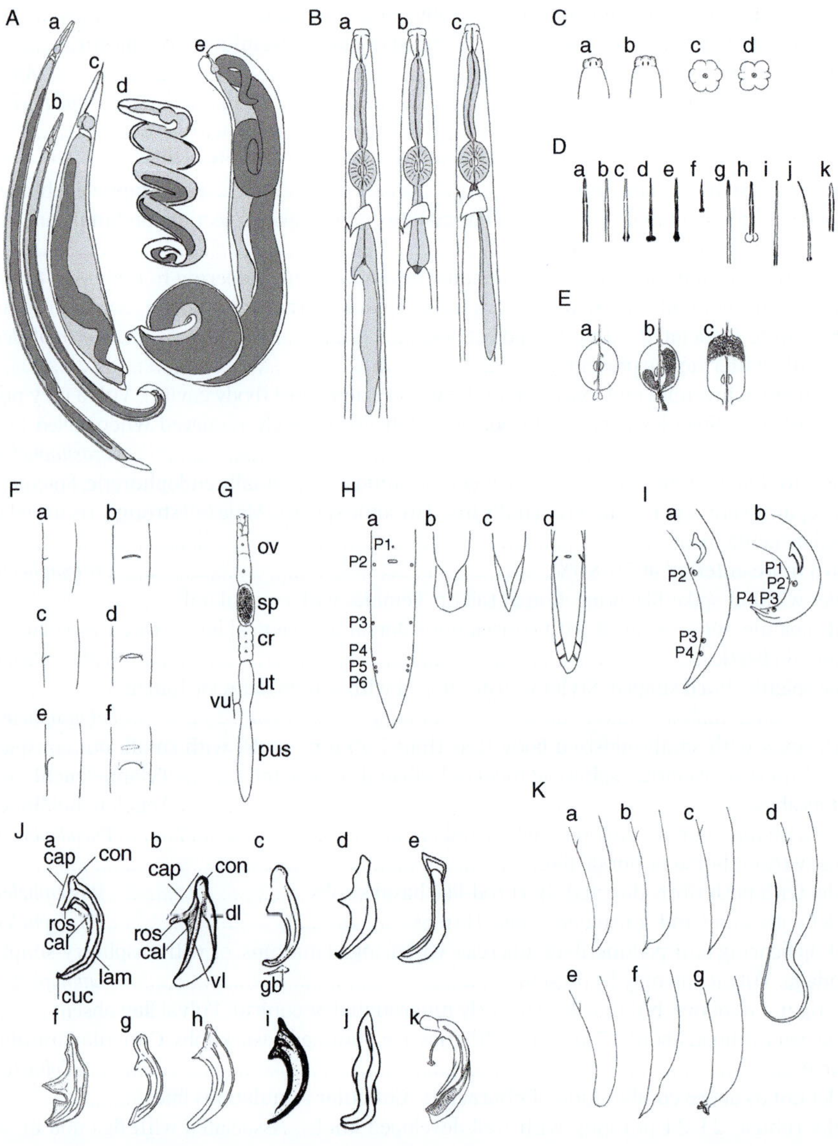

Fig. 7.1. General morphology of aphelenchs. A: Variation in body shape (a: *Bursaphelenchus* female; b: *Bursaphelenchus* male; c: *Noctuidonema* female; d: *Peraphelenchus* male; e: *Roveaphelenchus parasitic* female); B: anterior part (a: *Aphelenchus*; b: *Paraphelenchus*; c: Aphelenchoididae); C: lip region (a: rounded lip; b: squared lip; c: *en face* view of six equal-sized lips; d: *en face* view of lip which has two narrow lateral sectors); D: variation of stylet; E: variation of median bulb (metacorpus) (a: regular oval shape; b: Entaphelenchinae has well-developed pharyngeal glands; c: anterior 1/3 is glandular in many of Ektaphelenchinae and Seinurinae); F: lateral and ventral views of vulval opening of females (a, b: without flap apparatus; c, d: with short side flap; e, f: with long vulval flap); G: general morphology of female reproductive tract (ov: ovary; sp: spermatheca; cr: crustaformeria or quadricolumella; ut: uterus; vu: vulva; pus: post-vulval uterine sac; ovary and crustaformeria are usually connected by oviduct); H: ventral view of male tail (a: papillae arrangement where, P1: ventral single papilla; P2-P6: subventral paired papillae; b, c: bursal flap; d: bursa and limb-like papillae); I: lateral view of male tail; J: variation of male spicule (cap: capitulum; con: condylus; ros: rostrum; cal: calomus; lam: lamina; cuc: cucullus; dl: dorsal limb; vl: ventral limb; gb: gubernaculum or apophysis); K: variation of female tail (the variation in tail tip is observed in both females and males).

7.5. Morphological and Biological Characters of Aphelench Subfamilies

The Aphelenchoidea embraces about 30 genera with various feeding traits, for example, plant parasites, fungal feeders, insect parasites and predators. The general morphological and biological characters of these subfamilies and genera are noted here.

7.5.1. Family Aphelenchidae Fuchs, 1937

The Aphelenchidae has two genera, each being placed within a monogeneric subfamily. The most important morphological trait is the form of the pharynx and intestine, the species in the Aphelenchidae having a clearly distinct isthmus following a well-developed median bulb. Thus, the nerve ring surrounds only the pharynx, and pharynx/intestine junction (pharyngo-intestinal valve) is situated posterior to the nerve ring. The female tail is rather short and rounded or mucronated, and the presence of three to four pairs of genital papillae and a linear gubernaculum in males appears to be shared characters of the family. The generic definitions of *Aphelenchus* and *Paraphelenchus* are described below, instead of the definition of each subfamily. The members of this family are mostly soil-dwelling fungal feeders.

Subfamily Aphelenchinae Fuchs, 1937

Aphelenchus is morphologically characterized by a broad lateral field with many (usually >10, with an exception: *Aphelenchus sparsus* has 4-6) incisures, well-developed and ovoid median bulb (= metacorpus) with centrally located valve plates, pharyngeal glands in a dorsal lobe which is surrounded by the nerve ring, female vulva, lacking vulval flap and tail, rather short with bluntly rounded terminus and male tail region, with separate and slender spicules, linear gubernaculum and well-developed bursa, which is supported by three to four pairs of rib-like genital papillae.

The nematodes in this genus are reported from wide-ranging temperate, subtropical and tropical regions of the world. Males are missing in many populations, and most species are regarded as parthenogenetic. Most species are recovered from the soil environment where decayed plant materials are present and are thus regarded as mycophagous free-living nematodes. Although there are several reports about plant-parasitism and insect association, severe damage to host plants and close relationships with insects have not been reported.

Subfamily Paraphelenchinae T. Goodey, 1951

Paraphelenchus is morphologically characterized by a broad lateral field with six or eight incisures; a well-developed and ovoid median bulb (= metacorpus), with large crescentic valve plates; pharyngeal glands contained within a spatulate bulb (forming a basal bulb), which does not overlap the intestine; a female vulva lacking a vulval flap; a tail, rather short with bluntly rounded terminus with mucron; and male with separate and slender spicules, a linear gubernaculum and male tail region lacking a bursal apparatus, but with five pairs of genital papillae.

The nematodes in this genus are reported from a variety of temperate, subtropical and tropical areas of the world. The bionomics of the genus are similar to *Aphelenchus*; that is, the nematodes are found from soil and rotting plant materials and are assumed to feed on a wide range of fungi. However, its mycophagy of commercially important fungi may be stronger than that of *Aphelenchus*; some species having been reported as pests of mushroom production; for example, *Paraphelenchus myceliophthorus* was found causing damage on the cultivated mushroom, *Agaricus bisporus* (summarized by Hunt, 1993). Although the transmission mode is not clear, *Paraphelenchus* spp. are sometimes recovered from wood samples of dead pine trees in Japan (Akiba, unpubl. obs.). Some species may have a loose association with insects.

7.5.2. Family Aphelenchoididae Skarbilovich, 1947

The family has 36 genera separated into eight subfamilies and four *genera incertae sedis*. The most important morphological character is the characteristic composition of pharynx and intestine. In contrast to Aphelenchidae, members of Aphelenchoididae do not have a clear isthmus, and the pharyngo-intestinal junction – which forms a small valve – is situated just posterior to the median bulb while the pharyngeal gland lobes overlap the intestine dorsally. Therefore, the nerve ring, which is usually located a little posterior to the median bulb, surrounds both the pharyngeal glands and the intestine. The other important character concerns the male spicules and gubernaculum. The spicules of members of the Aphelenchoididae are robust, rosethorn shaped or a derivation of such, and a gubernaculum is either absent or indistinct if present. The rather narrow lateral field, which has six or fewer incisures, is also a common trait of the family.

Biological characters of each subfamily are fairly variable, that is, mycophagous free-living species, obligate and facultative plant parasites, endo- and ectoparasites of insects and predators. Many of the species are associated with insects as phoretic or parasitic species. The habitats that they occupy are also varied, many of them are found in soil and dead wood, but others are reported from the aquatic environment including fresh and seawater.

Subfamily Acugutturinae Hunt, 1980

Acugutturinae contains only three monospecific genera (= three species in total) which are ectoparasites of insects. The subfamily is morphologically characterized by its highly derived stylet, extremely long (50-60 µm in *Acugutturus*; 106-185 µm in *Noctuidonema*; 53-69 µm in *Vampyronema*) and its conus comprises more than 75% of the total length, two pairs of male genital papillae and an indistinctive, probably non-functional anus and rectum in females (Hunt, 1980, 1993; Remillet and Silvain, 1988; Anderson and Laumond, 1992).

These nematodes were mostly found in the American tropics and subtropics. *Acugutturus* is associated with cockroaches and the other two are associated with noctuid moths (Hunt, 1980; Remillet and Silvain, 1988; Anderson and Laumond, 1992). They penetrate the thin layer of the host body cuticle, for example, inter-segmental membrane, with their stylet sucking body fluids from the hosts. Thus, their body colour often assumes that of the host haemolymph. As is the case with other insect parasitic nematodes, they have not been successfully cultured (summarized by Hunt, 1993) and may not be suitable for any practical biological control applications.

Subfamily Anomyctinae J.B. Goodey, 1960b

Anomyctinae is a monotypic subfamily, containing only one genus/species, *Anomyctus xenurus*. The subfamily was previously included in the Aphelenchoidinae. However, molecular phylogenetic analyses revealed that *A. xenurus* belongs to Clade 3 of the family, a group of insect parasites and predators (e.g., Kanzaki *et al.*, 2019b). In addition, *A. xenurus* has characteristic lip and stylet and possesses a seemingly functional rectum and anus, which differentiates the subfamily from the other subfamilies belonging to the clade (Seinurinae, Acugutturinae, Entaphelenchinae and Ektaphelenchinae).

Subfamily Aphelenchoidinae Skarbilovich, 1947

This subfamily contains nine divergent genera with either weak typological definitions or serves as a home for unresolved clades included in the Aphelenchoididae. The possession of a slender stylet with a narrow lumen and basal knobs or swellings, presence of a post-vulval uterine sac in most genera, rosethorn-shaped spicules and absence of a bursa and gubernaculum could be listed as common characters, although a small gubernaculum-like attachment is reported in several species belonging to fig wasp and fig associates (*Schistonchus sensu stricto, Ficophagus* and *Martininema*) species.

The bionomics also vary among genera; that is, some *Aphelenchoides* and *Laimaphelenchus* are associated with insects, particularly bark beetles, and can be isolated from dead trees (e.g., Rühm, 1956; Massey, 1974; Poinar, 1975) while others are mostly soil-inhabiting free-living fungal feeders (e.g., Yeates *et al.*, 1993; summarized by Hunt, 1993). Plant-parasitism or facultative plant-parasitism has evolved separately in at least two subclades of *Aphelenchoides sensu lato* (one subclade involves *A. besseyi* and *Aphelenchoides ritzemabosi* while the other involves *Aphelenchoides blastophthorus* and *A. fragariae*), *Ficophagus*, *Martininema*, *Robustodorus* and *Schistonchus sensu stricto*.

The subfamily is a home for unresolved genera, most notably at least five paraphyletic groupings currently called '*Aphelenchoides*' based upon molecular phylogeny and the morphologically characteristic but currently non-sequenced genus *Punchaulus* (see De Ley and Coomans, 1996). These characteristic species, especially *Punchaulus*, which has an Aphelenchidae-like pharynx and intestine, should be reisolated and examined closely to determine their phylogenetic relationships and systematics.

Subfamily Ektaphelenchinae Paramonov, 1964

These nematodes are ectoparasites and endoparasites of insects, particularly bark beetles and some other groups of beetles (weevils and longhorn beetles). The subfamily is morphologically variable among genera; for example, *Ektaphelenchus*, *Ektaphelenchoides*, *Devibursaphelenchus* and *Lenisaphelenchus* have a stylet with a wide lumen and long (elongated oval or rounded rectangular) median bulb which has a glandular anterior one-third and valve apparatus situated at the posterior half, but *Cryptaphelenchus* has a stylet with a narrow lumen and a spherical median bulb with the valve apparatus situated at the middle (Fuchs, 1937; Rühm, 1956). The only shared trait of the subfamily appears to be the absence of a functional anus and rectum in females.

The host association and the other biological characters also differ among genera and species. *Cryptaphelenchus* is isolated from the Malpighian tubules as well as the frass and tunnel surface of bark beetles and can be cultured on fungal plates, for example, *Botrytis cinerea* and *Ophiostoma* spp. (e.g., Zhao *et al.*, 2008; Kanzaki *et al.*, 2022a). The other four genera have not been successfully cultured and are known from various organs of different species. *Ektaphelenchus* and *Devibursaphelenchus* are mostly isolated from sheath-like or cocoon-like structures on the elytra or the membranous hindwings of bark beetles. *Ektaphelenchoides* is isolated from the body cavity of longhorn beetles and the surface of maggots. Besides their insect parasitism, predation is suspected in several species of these genera. In fact, *Devibursaphelenchus alienae* was kept in laboratory culture for *ca* one year using *Acrobeloides* sp. as its food nematode (Kanzaki *et al.*, 2022b).

Although these species can parasitize insects, the impact of parasitism seems slight, almost like endophoresy, and does not affect the host performance (Kanzaki *et al.*, 2008b). Further, the parasitic species are not easily cultured. This subfamily does not seem to be suitable for any practical application, for example, biological control.

Subfamily Entaphelenchinae Nickle, 1970a

This subfamily consists of endoparasites of the haemocoel of coleopteran insects. The most conspicuous character is their shared biological traits, possession of three different adult forms, that is, male, immature free-living female and mature parasitic female. The male and immature females have small to middle-sized and normal nematode forms (thread-like body shape), but the mature parasitic female is large and sausage-shaped to obese and oviparous or ovoviviparous. Further, these genera share several morphological traits: stylet without basal knobs or swellings, very posterior female vulva, robust male spicules and absence of bursal flap of males.

They are typically obligate endoparasites of beetles belonging to Staphylinidae, Silphidae and Curculionidae. Although the detailed bionomics differ among genera, they infest the teneral adult or pupae, subsequently develop, mate and oviposit in the haemocoel of their adult hosts.

Poinar (1969) reported the possibility that *Praecocilenchus rhaphidophorus* could affect the reproduction of its host *Rhynchophorus* weevils. However, in *Rhynchophorus bilineatus* in New Britain, where *P. rhaphidophorus* occurs naturally in about 15% of populations, the weevils were still capable of reproduction but produced fewer eggs and had reduced fat bodies when compared with non-infested weevils (Bedford, 1974).

Subfamily Parasitaphelenchinae Rühm, 1956

Parasitaphelenchinae contains five genera, *Bursaphelenchus*, *Parasitaphelenchus*, *Potensaphelenchus*, *Ruehmaphelenchus* and *Sheraphelenchus*. Members of this subfamily are mostly associated with insects, particularly bark beetles (Curculionidae: Scolytinae), and other groups of beetles (weevils, longhorn beetles, ambrosia beetles and nitidulids etc.) and bees. Morphologically, the genera and species are variable and it is difficult to define the subfamily by morphology. However, several male characters, such as presence of seven or more genital papillae, rather stout spicule morphology and a short male tail, are shared among species, although there are a couple of exceptions.

The nematodes use their associated insects as their host or carrier. *Bursaphelenchus* dauers usually use the undersurface of the elytra; the tracheal system and the external reproductive organs of beetles; the oviducts, poison sac and/or Dufour's gland of bees; or inter-segmental folds or haemocoel of fig wasps for their transport. Contrastingly, third- or fourth-stage parasitic juveniles of *Parasitaphelenchus* enter the haemocoel of their host and so obtain nutrients. However, the impact of the parasitism seems moderate and these nematodes may not significantly affect the host performance. While, three other genera, *Ruehmaphelenchus*, *Sheraphelenchus* and *Potensaphelenchus* have been reported to be relatively simple phoretic associates of ambrosia beetles, nitidulid beetles and a longhorn beetle (*Spondylis buprestoides*), respectively (e.g., Massey and Hinds, 1970; Gu *et al.*, 2021). *Bursaphelenchus* is currently a paraphyletic assemblage of species with several monophyletic species groups punctuated by the genera of *Parasitaphelenchus*, *Potensaphelenchus*, *Ruehmaphelenchus* and *Sheraphelenchus*. In addition, based on its molecular phylogenetic status, *Bursaphelenchus abruptus*, a fungal-feeding nematode carried phoretically in the reproductive tracts of soil-dwelling bees, is close to *Potensaphelenchus stammeri* at the basal position of Clade 4, which could be transferred to *Potensaphelenchus* or a separate genus. Insect-vectored facultative plant-parasitism or obligate plant-parasitism has evolved at least three times out of this mostly fungal-feeding grouping, most notably with the *B. xylophilus*/*Bursaphelenchus mucronatus* group, *B. cocophilus* and the *Bursaphelenchus suri* and *Bursaphelenchus sycophilus* group.

In addition to their host/carrier usage, two interesting biological characters have been reported in *Bursaphelenchus*. One is hermaphroditism in *B. okinawaensis*, which is the first hermaphroditic species in the superfamily and is becoming an important model system for studying fungal-feeding nematodes (e.g., Shinya *et al.*, 2022). The other is the feeding dimorphism in the *B. aberrans* group of the genus. The morphotype previously treated as a different genus, *Berntsenus* was confirmed to be an alternate predatory form of the genus. So far, feeding (stomatal) dimorphism has only been reported in the Diplogastridae and these *Bursaphelenchus* spp. could be good comparative models for studying these phenomena.

Subfamily Seinurinae Husain & Khan, 1967

Most species of Seinurinae are known, or considered to be, predators inhabiting the soil environment. The subfamily is morphologically characterized by a long and slender stylet with a wide lumen, rather long (elongated oval or rounded rectangular) and prominent median bulb, with a valve apparatus located in the posterior half, male spicules with a prominent condylus and rostrum, functional anus and rectum in females and a long and elongated female tail with usually filiform, exceptionally rounded, terminus.

Their feeding mode is notable as the predators feed on many different groups of nematode prey, which they inject with pharyngeal gland secretions. The secretions immobilize or paralyse (and probably pre-digest) the prey (Hechler, 1963; Wood, 1975).

The life cycle is rather rapid, one of the juvenile stages being considered as missing from the life cycle; that is, they probably have three juvenile stages instead of four (Wood, 1975).

They could be a potential biological control agent for plant-parasitic nematodes in the soil. However, because of the difficulty in culturing, none of these nematodes has been applied practically.

Subfamily Tylaphelenchinae Kanzaki, Li, Lan & Giblin-Davis, 2014b

Tylaphelenchinae was recently established based on its molecular phylogenetic status and some typological similarity, that is, forming a monophyletic group separable from Aphelenchoidinae, and possessing a small body, rather conspicuous and characteristic stylet knobs and spherical median bulb. The subfamily contains three genera, *Tylaphelenchus*, *Pseudaphelenchus* and *Basilaphelenchus*. The former two are considered insect associates utilizing bark beetles (Rühm, 1956) and termites (Kanzaki *et al.*, 2009b, 2010) as their carriers, but the detailed insect association of *Basilaphelenchus* has not been confirmed.

7.6. Potentially Important Genera and Species for Agriculture and Forestry

Within the above-mentioned 28 aphelench genera, some of them are potentially important for agriculture and forestry as biological control agents (*Aphelenchus*) plant parasites (*Aphelenchoides*, *Ficophagus*, *Martininema* and *Schistonchus sensu stricto*) and plant pathogens (*Bursaphelenchus*).

Aphelenchus is expected to be a biological control agent for soil fungi and other species of soil nematodes. *Aphelenchoides* contains several important plant parasites; *Ficophagus*, *Martininema* and *Schistonchus sensu stricto* are all associated with the syconia of *Ficus* feeding on the florets; and *Bursaphelenchus* contains two internationally important quarantine species and several other potential pathogens. Their economic/environmental impact, bionomics and management will be discussed briefly.

7.6.1. *Aphelenchus* (Fig. 7.2)

The morphological definition of the genus is described above. The genus contains soil-inhabiting mycophagous, and facultative plant-parasitic nematodes, although severe damage has not been reported on agricultural plant species. They mostly feed on various species of soil fungi including some important plant pathogens. Therefore, one of these *Aphelenchus* species, *A. avenae* is purported to be a biological control agent for plant-pathogenic soil fungi.

A. avenae (Fig. 7.2) is a cosmopolitan species and can be isolated from all over the world. Biological traits are variable among populations. For example, Evans and Fisher (1970) reported the variation of sex ratio (= reproductive manner) among populations, and Goodey and Hooper (1965) and Jairajpuri (1968) reported variations in generation time among populations. The species may contain several cryptic species.

Although a certain level of plant-parasitism is reported for *A. avenae* (summarized by Hunt. 1993), the nematode is purported to be a biological control agent against soil-borne plant-pathogenic fungi, such as *Fusarium solani*, *Fusarium oxysporum* and *Rhizoctonia solani*. Ishibashi and Choi (1991) conducted a glasshouse experiment using cucumber seedlings to test the suppressive effects of *A. avenae* and an entomopathogenic nematode, *Steinernema carpocapsae* to *R. solani*, the root-knot nematode, *Meloidogyne incognita* and turnip moth, *Agrotis segetum*. They found that massive applications of *A. avenae* suppressed the damage by *R. solani*, and mixed application suppressed root-knot nematode damage. Ingham (1988), however, suggested that fungal-feeding nematodes may cause negative effects to the plants by feeding on mutualistic mycorrhizal fungi.

The above-mentioned advantages, that is, reducing the population of plant-pathogenic fungi with its mycophagy and interference with other plant-parasitic nematodes, outweigh the negative attributes and *A. avenae* may be an important component of an integrated pest management programme in the future.

7.6.2. *Aphelenchoides* (Figs 7.3, 7.4)

Aphelenchoides consists of about 200 nominal species that are morphologically very similar to each other (see Hunt, 1993, 2008; Shahina, 1996). Therefore, it is almost impossible to identify the species solely based on morphological observations, and the genus doubtlessly contains many synonyms and cryptic species (e.g., Subbotin *et al.*, 2020). *Aphelenchoides* is the type genus of the Aphelenchoididae and lacks conspicuous morphological characters. The key characters of the family are the posterior positioning of the female vulva at 60-75% of the body length and a lack of a vulval flap, female tail that is generally conoid with various tail tips, for example, bluntly/finely rounded, digitate or bifurcate etc., and strongly hooked male tail bearing three pairs of genital papillae but without bursa or bursal flap apparatus. The typical thorn-shaped male spicule is very similar among the species and could also be an important generic character.

Regardless of the morphological similarities among species, the genus is molecularly inferred to be polyphyletic and the biological characters are extremely variable among the species. For example, some species have been described from aquatic plants in fresh and seawater (summarized by Timm and Franklin, 1969; Bert and Borgonie, 2006); some species are associated with insects (Rühm, 1956; Massey, 1974; Poinar,

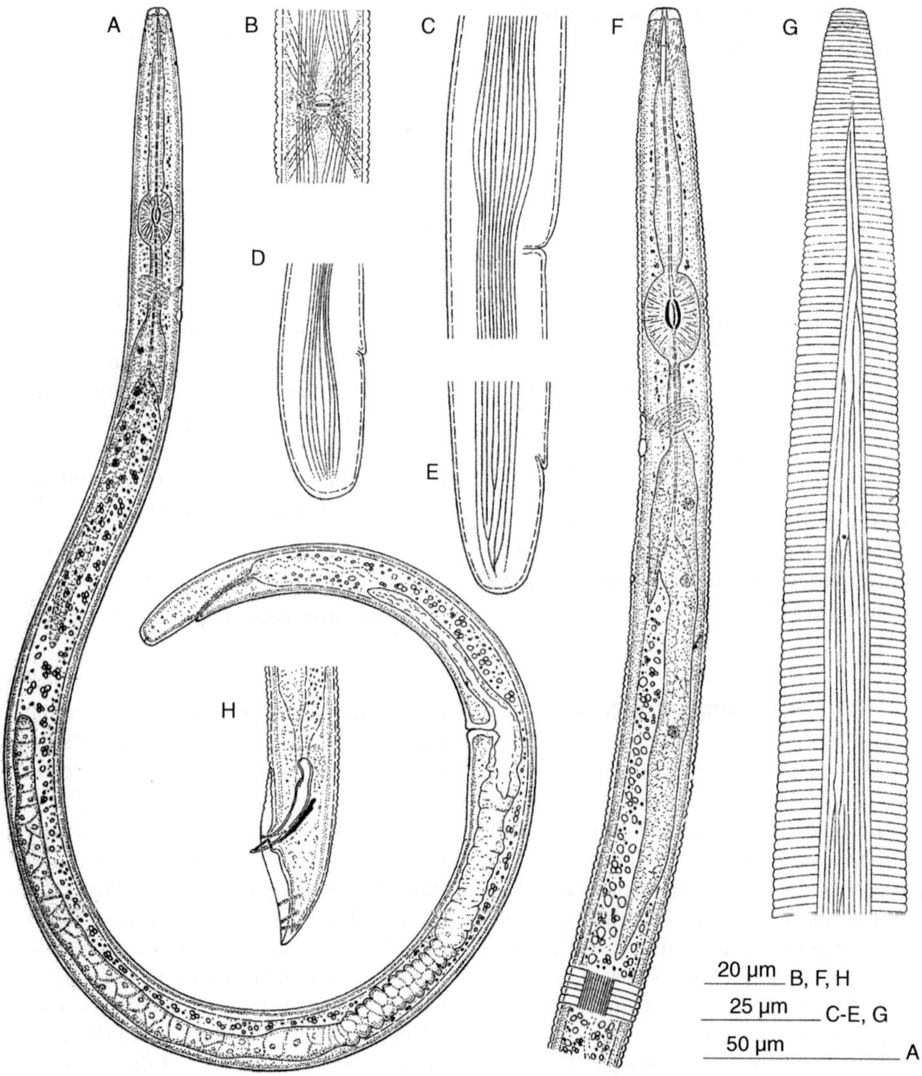

Fig. 7.2. *Aphelenchus avenae*. A: Entire female; B: female vulva, ventral view; C: vulval region, lateral view, showing multiple lateral lines; D, E: tail region; F: pharyngeal region; G: superficial view of anterior region showing lateral line development; H: male tail region showing spicules and bursa supported by ribs. (After Hooper (1974), courtesy CABI.)

1975; Kanzaki, 2006); some are soil-inhabiting fungal feeders (Hunt, 1993; Yeates *et al.*, 1993); and others are obligate or facultative plant parasites (Hunt, 1993; Yeates *et al.*, 1993).

The genus contains several important plant-parasitic species, namely, the white tip nematode of rice, *A. besseyi sensu lato*, the strawberry nematode, *A. fragariae* and chrysanthemum leaf nematode, *A. ritzemabosi*. A commonly found soil nematode, *Aphelenchoides bicaudatus* is also known as a pest for mushrooms (McLeod, 1967). More detailed information about these species is provided in *CIH Descriptions of Plant-parasitic Nematodes* (Franklin and Siddiqi, 1972; Siddiqi, 1974, 1975, 1976) and only summaries are presented here.

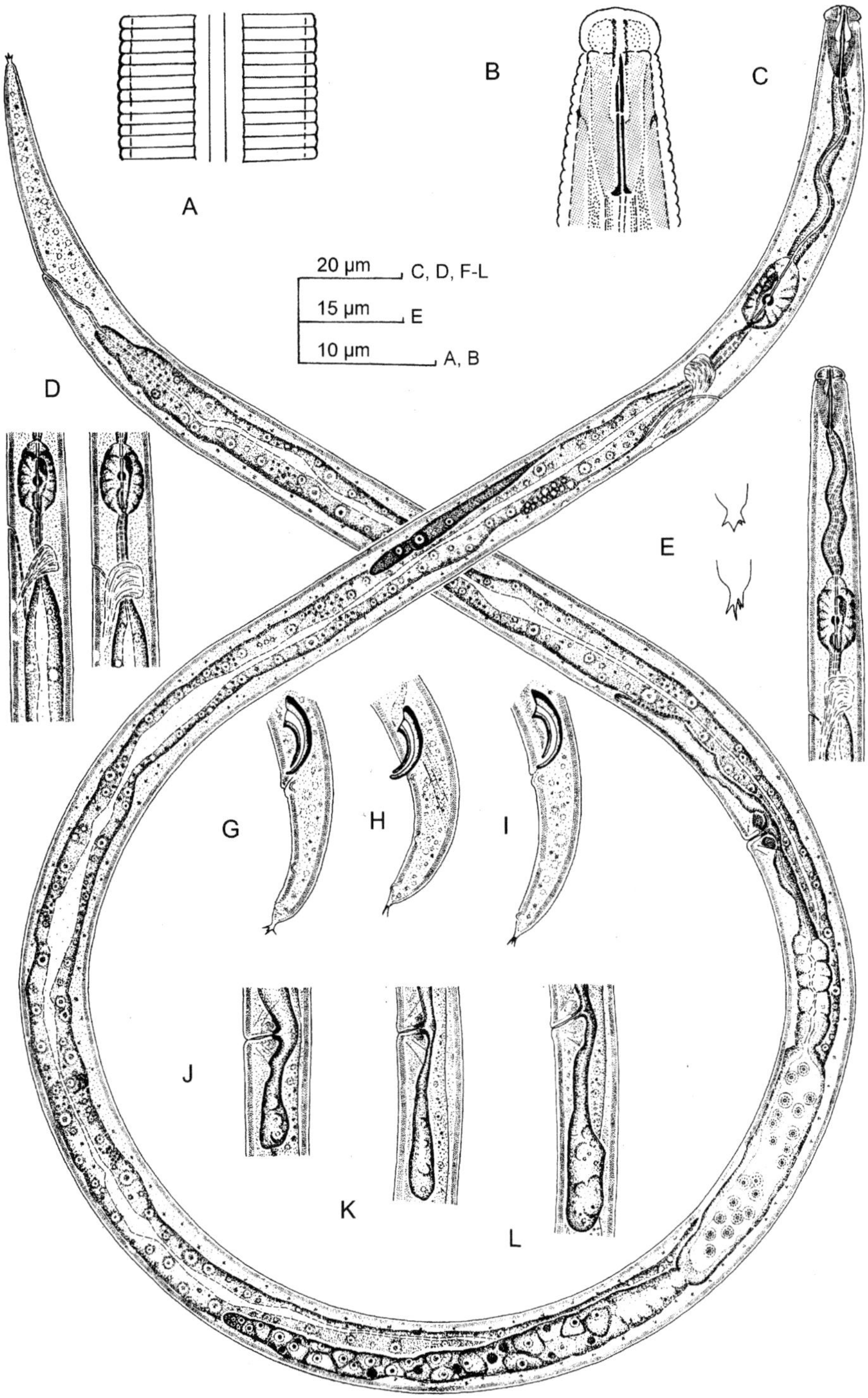

Fig. 7.3. *Aphelenchoides besseyi*. A: Lateral field showing four lines; B: anterior region and stylet; C: entire female; D: median bulb region of pharynx; E: female tail tip showing terminal mucrons; F: pharynx; G-I: male tail region showing spicules; K-M: female vulval region showing post-uterine sac development. (After Fortuner (1970) and Franklin and Siddiqi (1972), courtesy CABI.)

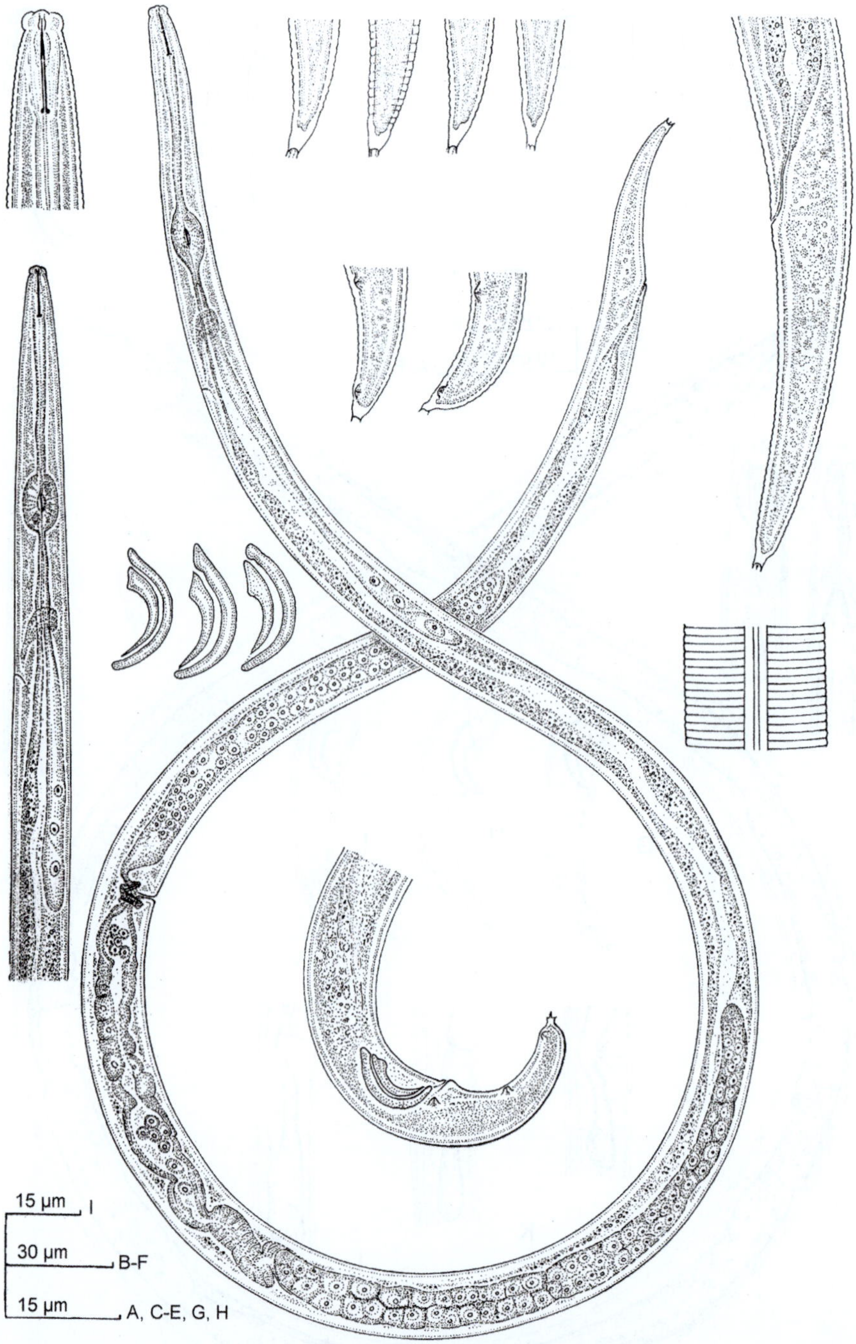

Fig. 7.4. *Aphelenchoides ritzemabosi*. A: Anterior region and stylet; B: entire female; C: female tail tips showing mucrons; D: male tail tips showing mucrons; E: female tail with terminal mucrons; F: pharynx; G: male spicules; H: lateral field showing four lateral lines; I: male tail region. (After Siddiqi (1974), courtesy CABI.)

Aphelenchoides besseyi, sensu lato (Fig. 7.3)

This species is a mycophagous/facultative plant parasite and causes the white tip disease of rice. It is widely spread throughout the rice-growing areas of the world (Fortuner and Williams, 1975), and forms a species complex (e.g., Subbotin *et al.*, 2020; Lai *et al.*, 2023). The biology, economic loss and control of *A. besseyi s. l.* are reviewed by Bridge *et al.* (2005).

The typical symptom of white tip disease in early growth is the chlorotic tips of new leaves from the leaf sheath. These chlorotic parts later dry and curl. The infected young leaves are speckled with a white splash pattern. The nematode is seed-borne and can survive in the infected seed for several years under anhydrous conditions. The infected seeds have lower fecundity, delayed germination and smaller size compared with uninfected seeds. The nematodes rapidly become active with appropriate moisture and temperature and start to ectoparasitically feed on the apical meristem of the rice plant.

For the control of white tip nematodes, heat-mediated disinfection is commonly used, that is, soaking the infected seeds in hot water at 55-61°C for 10-15 min (Vuong and Rodriguez, 1970) or chemical solutions instead of hot water (Yasu, 1968). Vuong and Rodriguez (1970) also reported that soil treatment with fensulfothion, diazinon or disulfoton can increase rice yields. Other cultural methods have been suggested as effective (Cralley, 1957).

Aphelenchoides fragariae

This species has a cosmopolitan distribution with a wide range of over 250 host-plant species from 47 families. *A. fragariae* is generally known as the strawberry nematode or the strawberry foliar nematode. Although it can feed on some fungal species and reproduce, it is usually recovered as a plant parasite, mostly infecting the above-ground parts of its hosts. As an endoparasite, *A. fragariae* enters the host plant, presumably through stomata or by cuticular penetration through the softer parts of the leaf. Its movement through, and feeding on, the mesophyll cause leaf blotches that are often delimited by large veins (Jagdale and Grewal, 2006). The nematode also can be an ectoparasite, living and feeding within the folded crown and runner buds of the strawberry plant causing small dry and brownish lesions on the expanded leaves.

Decontamination is important for controlling this nematode; that is, eliminate infected plants and replant with healthy ones. Hot water treatment is also available, and several researchers have suggested different conditions contingent upon the host, for example, 46-47°C for 10-15 min for strawberry, or 36°C for 6 h or 41°C for 1 h for lily bulbs (summarized by Siddiqi, 1975). Although chemical treatment with several different chemicals have been suggested as effective, for example, parathion, aldicarb, Systox, thiophos and methomyl, many of these soil treatment chemicals have been discontinued and are no longer available. More low-cost and environmentally sustainable control alternatives are needed for this nematode.

Aphelenchoides ritzemabosi (Fig. 7.4)

A. ritzemabosi is a widespread species that is generally known as the chrysanthemum foliar nematode. It is an obligate plant parasite with about 200 reported species of host plants. Under laboratory conditions, it was inoculated onto several species of fungi but did not propagate on any of them (Krusberg, 1961). The nematode inhabits leaves, buds, the growing points and outer layers of the stem as endo- and ectoparasites causing retardation of growth, distortion and deformation of leaves in the early stages of infection and brown spots and blotches in the later stages. When the nematode infests flower buds, they become deformed and undersized blossoms. Because the nematode is an obligate parasite it moves from diseased leaves (tissues) to healthy tissue, spreading from plant to plant through water splash (rain) and direct contact of diseased and healthy leaves.

Control is similar to that of *A. fragariae* as both species are associated with above-ground parts of plants as endo- and ectoparasites. Hot water treatment and several different chemical treatments are suggested.

Aphelenchoides bicaudatus

This widespread species is recovered from the rhizosphere of important crops, for example, corn, cotton and potato. The nematode is known as a mycophagous species that can feed on many species of plant-pathogenic and useful (edible mushroom-forming) fungi and is a potential pest of mushroom production (McLeod, 1967). Although not known to cause severe damage, it has been found associated with the production of an important mushroom, *A. bisporus*. Practical studies on controlling it have not yet been conducted, but some nematode-trapping fungi seem to hold some potential as a biocontrol agent (Cayrol *et al.*, 1972).

Another mycophagous species, *Agaricus compositus*, is also considered to be a pest in mushroom production.

7.6.3. *Schistonchus, sensu lato*

Schistonchus comprises a group of entomophilic (phoretic) and plant-parasitic nematodes. The phoretic (weakly parasitic?) nematode is carried by female fig wasps from old to new figs (syconia of *Ficus* spp.). After the wasps enter and pollinate the figs the nematodes leave the wasp and feed on flower tissues causing deformation of the flowers (Giblin-Davis *et al.*, 1995). Thus, although practical economic damage has not been reported, this genus could be regarded as a potential plant pathogen.

This genus is of scientific interest because of the tritrophic interactions between the plant, fig wasp and nematode (see Giblin-Davis *et al.*, 1995; Davies *et al.*, 2010, 2015). Although the genus '*Schistonchus*' was separated into three genera, *Schistonchus sensu stricto*, *Ficophagus* and *Martininema*, their biological characters are similar to each other, and *Ficophagus laevigata* causes similar tissue damage within its host syconia (Center *et al.*, 1999; Giblin-Davis *et al.*, 2023).

7.6.4. *Bursaphelenchus* (Figs 7.5, 7.6)

Bursaphelenchus consists of more than 100 nominal species and is divided into several species groups (Kanzaki, 2008; Braasch *et al.*, 2009). Most *Bursaphelenchus* species are phoretically associated with insects, typically bark beetles, weevils, longhorn beetles, bees, fig wasps and several other beetle groups. The genus is morphologically characterized by a strongly hooked male tail with a small bursal flap at the tail tip, and three pairs of genital papillae and a ventral single P1 papilla. The secondary loss of the bursal flap and P1 have been observed in *Bursaphelenchus kiyoharai* and *Bursaphelenchus penai* (Kanzaki *et al.*, 2011, 2014a), but they are apparently stable pleisomorphic characters with an occasional reversal.

The genus contains two lethal plant pathogens, *B. cocophilus* and *B. xylophilus*, and two other species suspected of posing potential risks to pine trees, that is, *Bursaphelenchus sexdentati* and *B. mucronatus*, and a substantial amount of molecular and morphological work has therefore been done (Ye *et al.*, 2007, Kanzaki, 2008, Ryss and Subbotin, 2017; Kanzaki and Giblin-Davis, 2018).

Bursaphelenchus cocophilus (Fig. 7.5)

Distributed from Mexico south through to the northern part of South America and the causal agent of the red ring disease (RRD) and little leaf disease on several species of palm trees. RRD is one of the most important wilt diseases of coconut and African oil palms in the neotropics causing annual losses of 10-15% (Giblin-Davis, 1993, Kanzaki and Giblin-Davis, 2016). It is chiefly vectored not only by the American palm weevil, *Rhynchophorus palmarum*, but also potentially by the weevils *Dynamis borassi* and *Metamasius hemipterus*. *B. cocophilus* is transmitted to susceptible palms during oviposition or other activities and only a few nematodes are necessary for successful transmission. In coconut palms, symptoms include typical wilt with premature coconut drop (except mature nuts) and premature senescence of progressively younger leaves, which often break at the base and hang. Stem, petiole and root cross sections often show a red ring of anthocyanin-rich pigments and these tissues usually yield large numbers of dispersal third-stage juveniles which occur intercellularly in parenchymal tissue. Nematode infestation causes irreversible wilt that kills the tree in a couple of months because of tylose formation and vascular occlusion of water-conducting xylem in the vascular bundles. Palm weevils colonize RRD trees and during feeding and tunnelling become associated with the dispersal stage of the nematodes, which appear synchronized for transmission (Giblin-Davis, 1993). RRD control has been focused on aggressive phytosanitation

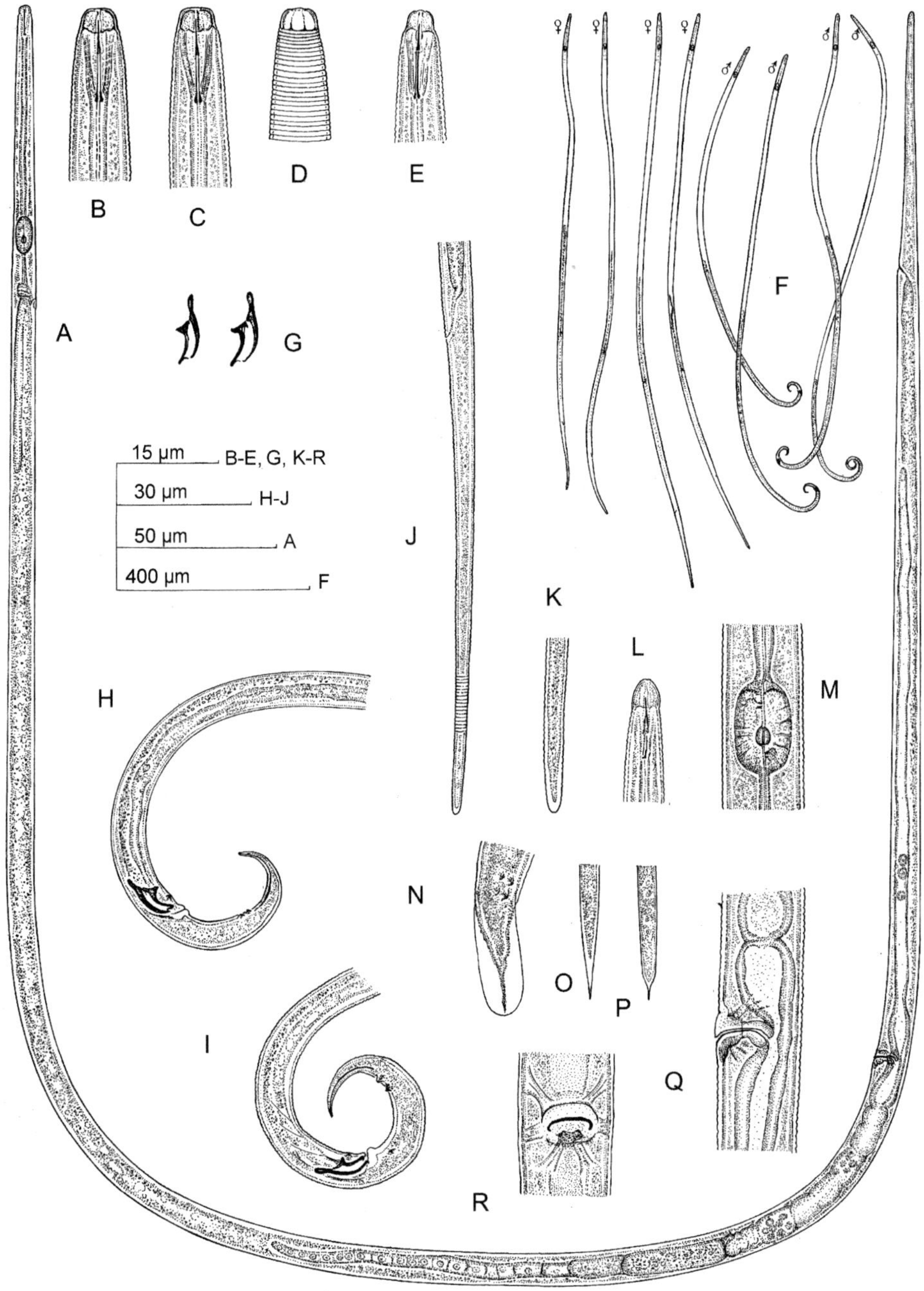

Fig. 7.5. *Bursaphelenchus cocophilus*. A: Entire female; B-D: female anterior region; E: male anterior region; F: entire worms; G: male spicules; H, I: male posterior region; J: female tail; K: female tail tip; M: median bulb; N: male tail tip showing bursal flap; O, P: juvenile tail tips; Q: female vulval region, lateral view; R: vagina in ventral view. (After Braithwaite and Siddiqi (1975), courtesy CABI.)

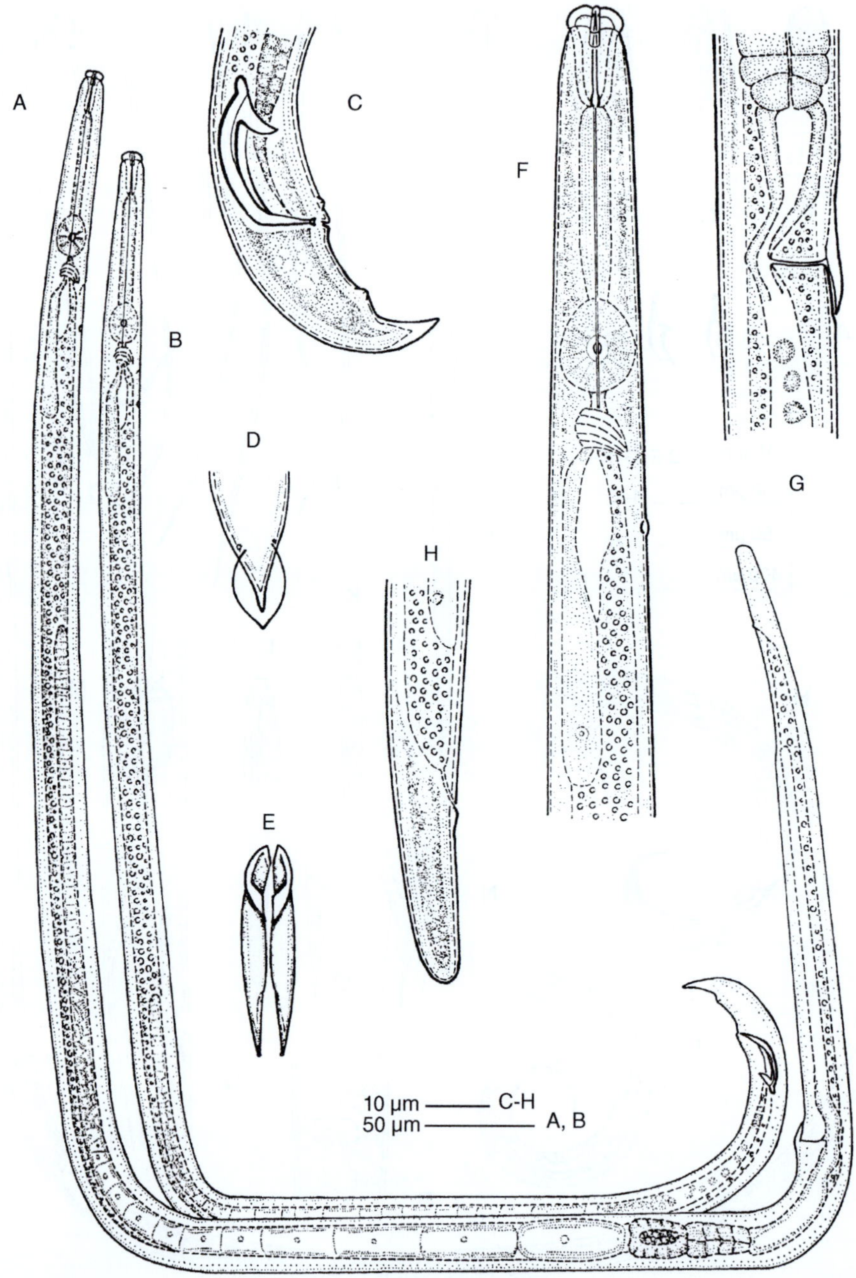

Fig. 7.6. *Bursaphelenchus xylophilus*. A: Entire female; B: entire male; C: male tail region; D: ventral view of male 'bursa'; E: ventral view of spicules; F: pharyngeal region; G: female vulval region showing flap; H: female tail showing typical rounded tip. (After Mamiya and Kiyohara (1972), courtesy *Nematologica*.)

and the use of male palm weevil aggregation pheromones for mass trapping of the weevil vectors (Griffith *et al.* 2018).

Bursaphelenchus xylophilus (Fig. 7.6)

This species is generally known as the pine wood or pine wilt nematode, the causal agent of pine wilt disease. Based upon molecular phylogenetic analysis and hybridization studies, *B. xylophilus* is considered to be native to the middle and eastern regions of North America (Nickle *et al.*, 1981; Mamiya, 1987) and was introduced into Japan at the beginning of 20th century. Since the first report, the disease has devastated the susceptible red and black pine forests in East Asia (Japan, Korea, Taiwan and Continental China). Recently, the nematode invaded the EU (Portugal and Spain) causing severe damage to native pine forests (Mota *et al.*, 1999). The annual loss of pine trees in Japan, for example, is about 0.5-1 million m^3 (= 1-2 million trees of 25 cm diameter at breast height).

The nematode has a populational mutualistic relationship with pine sawyers, the longhorn beetles belonging to the genus *Monochamus*; that is, the nematode kills the tree and so increases oviposition resources for the insect while the insect carries the nematode to enable its spread. The typical disease cycle in temperate East Asia is as follows (see Kobayashi *et al.*, 1984; Mamiya, 1984).

In the late spring to early summer (May to July), the adult beetles, which harbour the nematode dauers in their tracheal system, emerge from the dead trees and feed on the bark of pine twigs during their maturation feed. Nematodes invade trees through the wounds made by their vector beetles during this feeding, moult to the adult, move inside the tree through the resin canals and start to feed on the plant tissue. During this process, although the detailed mechanisms have not been fully clarified, susceptible pine species show a type of hypersensitive reaction and their internal tissues (water conductance, resin canal, etc.) are broken and show a systemic wilt symptom before the late summer (August to September). Thus, the disease symptom appears as a sudden death of the tree. Then, in the same season, the tree tissues collapse, and the nematodes propagate in the dead/dying tree feeding on fungi, for example, blue-staining fungi, *Ophiostoma* spp. occurring on the dead pine tissues. In the late summer, mature beetles mate and females come back to a dead/dying tree to oviposit their eggs. The eggs hatch in 10-15 days and the larvae start to feed on the dead tree; young larvae feed on inner bark and, after several moults, bore into deep wood to form the pupal chamber. Because the beetle is usually univoltine in the temperate zone, the larvae become the vectors (adults) in the following season.

The progress of disease symptoms is highly dependent on environmental conditions; that is, high temperature, drought and shading (photosynthesis inhibition) enhance the disease, while low temperature suppresses disease progression. In the cool temperate zone in Japan, the progress of disease symptoms is suppressed, and, in some cases, it takes two or more years from infection to occurrence of visual symptoms (Futai, 2003). Also in the cool Canadian temperate forest, the nematode parasitized susceptible Scots pine (*Pinus sylvestris*) trees for more than 10 years without causing any disease symptoms (Halik and Bergdahl, 1994; Bergdahl and Halik, 1998).

The nematode is mostly associated with several different groups of longhorn beetles; however, the major vectors are *Monochamus carolinensis* in North America (Rutherford and Webster, 1987), *Monochamus alternatus* in East Asia (Kobayashi *et al.*, 1984) and *Monochamus galloprovincialis* in the EU (Sousa *et al.*, 2001). The pine trees native to the middle and eastern part of North America, the natural distribution range of pathogenic *B. xylophilus*, are not susceptible to this nematode.

Bursaphelenchus mucronatus

The closest relative of *B. xylophilus* is *B. mucronatus* which is distributed through Europe, Russia and East Asia (see Ryss *et al.*, 2005; Kanzaki, 2008). The species is separated into two population groups, namely, the 'European type' and the 'Asian type' by molecular sequence analysis and some minor morphological differences, which probably justify these two groups as being different subspecies (Kanzaki *et al.*, 2008a). The life history is similar to *B. xylophilus*, that is, associated with *Monochamus* beetles in each area, and isolated from dead pine trees.

B. mucronatus is not a strong pathogen, but sometimes kills stressed host pine trees. In a glasshouse experiment Braasch (2000) demonstrated rather strong pathogenicity of the European type of *B. mucronatus* to several different European pine species. The disease symptom caused by *B. mucronatus* was similar to that caused by *B. xylophilus*, and was enhanced by high temperature and drought conditions

(Braasch, 2000; Skarmoutsos and Michalopoulos-Skarmoutsos, 2000). Kanzaki and Futai (2006) showed that the pathogenicity of the Asian type of *B. mucronatus* to potted seedlings of Japanese red pine (*Pinus densiflora*) is enhanced under conditions of shade. The pathogenicity of *B. mucronatus* is weak, and the species may not kill healthy trees. However, because the pathogenicity is enhanced by some environmental factors, the species may cause a certain level of pine mortality during times of environmental stress.

Bursaphelenchus sexdentati

This species is distributed through Europe and Russia. Although it has not been reported from Asia, the species is probably distributed in Asian pine forests because its carrier insects, *Ips* spp., are distributed in Asian countries. *B. sexdentati* is associated with *Ips sexdentatus* and several other bark beetle species (see Ryss *et al.*, 2005).

Most of the bark beetles utilize newly dead or dying trees. Therefore, the bark beetle-associated nematodes are usually saprophytic species that feed on microbes occurring on the walls of the beetle tunnel. However, *B. sexdentati* is often isolated from dead pine trees in the Mediterranean and related countries and is suspected to be related to local pine forest decline.

In an inoculation test, Skarmoutsos and Michalopoulos-Skarmoutsos (2000) demonstrated moderate pathogenicity of *B. mucronatus* and *B. sexdentati* and weak pathogenicity of *B. leoni* to European native pine species. Severe damage by this species has not been reported under natural conditions so far, although *B. sexdentati* is worth keeping an eye on as a potential pathogen.

7.7. Molecular Identification and Systematics of Aphelenchs

Molecular methods have recently been applied to a variety of nematodes and other organisms for practical diagnosis and reconstruction of taxonomic frameworks in the face of new evidence revealing convergence. Because of its economic importance in forestry and forestry products and potential plant quarantine issues, *Bursaphelenchus* has received a fair amount of research concerning sequence and primer information for molecular identification. However, the sequence information for many of the lesser known aphelench groups and species is not sufficiently characterized or available in sequence databases; that is, GenBank does not cover all aphelench subfamilies. Thus, the expansion of the database to include more sequences from this diverse grouping of nematodes is necessary to reconstruct the molecular phylogenetic framework of the superfamily.

The current molecular identification methods for *Bursaphelenchus* spp. are introduced here and, subsequently, the molecular phylogenetic relationships among aphelenchid genera and subfamily, which were analysed based on the sequence data stored in the GenBank database and the authors' original data, will be discussed.

7.7.1. Molecular Identification of *Bursaphelenchus* spp.

The morphological identification of nematode species is not easy; it requires adult nematode materials in good condition, knowledge of nematode morphology and a good microscope. Even if the nematodes are properly fixed, mounted and observed, cryptic species cannot be distinguished by morphology alone. Therefore, it is recommended to start identification using resources such as Hunt (1993, 2008), followed by comparisons to up-to-date literature and confirmation with molecular methods.

Because of the importance of accurately identifying *B. xylophilus*, molecular methods have been studied for *Bursaphelenchus* since late 1980, namely, RFLP (Abad *et al.*, 1991), dot-blot analysis (Webster *et al.*, 1990), RAPD (Braasch *et al.*, 1995), PCR-RFLP (Hoyer *et al.*, 1998; Iwahori *et al.*, 1998), species-specific PCR (Kang, 2004; Matsunaga and Togashi, 2004; Takeuchi *et al.*, 2005; Hu *et al.*, 2011), real-time PCR (Cao *et al.*, 2005; Leal *et al.*, 2007; Takeuchi and Futai, 2009; Huang *et al.*, 2010; Ye and Giblin-Davis, 2013), SCAR (Chen *et al.*, 2011) and LAMP (Kikuchi *et al.*, 2009). Within these methodologies, PCR-RFLP, species-specific PCR and LAMP are often used due to their simplicity, lower cost and reliability.

The general molecular techniques are covered in Chapter 5 and several species-specific primers and identification primers are provided below. Matsunaga and Togashi (2004) developed species-specific primer sets to distinguish *B. xylophilus*, *B. mucronatus* and their hybrid. Subsequently, Takeuchi *et al.* (2005) developed species-specific amplification using a nested-PCR technique and successfully detected nematodes from living and dead tree samples without extracting nematodes. Real-time PCR-based identification was provided by several authors. Thereafter, Kikuchi *et al.* (2009) developed a LAMP-based method that allowed detection of the nematode from dead tree tissue directly. In addition, similar LAMP methods have been developed for *B. cocophilus* (Ide *et al.*, 2017). Nevertheless, the present methodologies target nematodes, that is, detecting nematode-derived DNA, and thus, these are not very sensitive to early-stage infection. Therefore, novel methodology, for example, detecting pathogen-specific host response, is necessary for practical use.

Several phylogenetic primers for SSU, LSU and COI are shown in Figure 7.7. Within these primer sets, 18SF (= VRF1) and 28SR or VRF2 are useful for ITS-RFLP analysis.

However, the molecular identification methods are probably most needed for applications involving the detection and identification of plant-parasitic *Aphelenchoides* species as the species in this genus are morphologically very similar to each other and involve many synonyms and cryptic species (e.g., Subbotin *et al.*, 2020; Lai *et al.*, 2023).

7.7.2. Molecular Systematics of Aphelenchs

Several different genetic loci have been used for inferences of nematode phylogeny and, up to the present, the ribosomal DNA region (near full-length SSU or 18S ribosomal RNA, internal transcribed spacer including ITS1, ITS2 and 5.8S ribosomal RNA gene, D2-D3 LSU or D2-D3 expansion segments of the 28S ribosomal RNA) and mitochondrial COI have been applied to aphelenchs. Within these genes, near full-length SSU is most conservative, making it suitable for analysis among species within a rather divergent

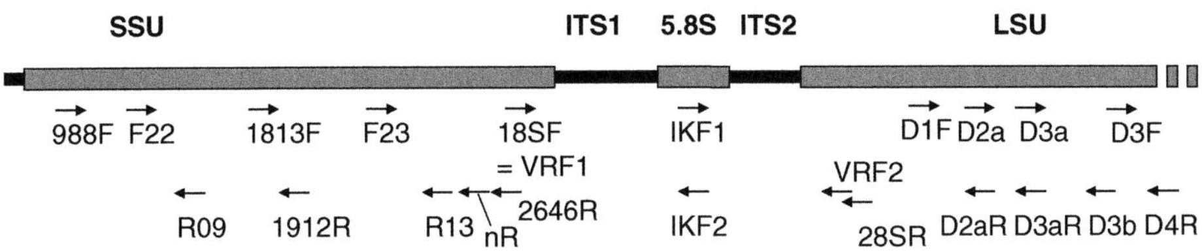

Sequences of rDNA primers

SSU988F: CTCAAAGATTAAGCCATGC
SSUF22: TCCAAGGAAGGCAGCAGGC
SSUR09: GCCTGCTGCCTTCCTTGGA
SSU1813F: CTGCGTGAGAGGTGAAAT
SSU1912R: TTTACGGTCAGAACTAGGG
SSUF23: ATTCCGATAACGAGCGAGA
SSUR13: GGGCATCACAGACCTGTTA
SSUnR: TTACGACTTTTGCCCGGTTC
SSU2646R: GCTACCTTGTTACGACTTTT

COIF1: CCTACTATGATTGGTGGTTTTG GT AAT
COIR2: GTAGCAGTAAAATAAGCACG

18SF: CGTAACAAGGTAGCTGTAG
IKF1: GGGTCGATGAAGAACGCAG
IKF2: CTGCGTTCTTCATCGACC
28SR: TTTCACTCGCCGTTACTAAGG
VRF2: TCCTCCGCTAAATGATATG
D1F: AAGGATTCCCTTAGTAACGGCGATTG
D2a: ACAAGTACCGTGAGGGAAAGTTG
D2aR: CAA CTT TCC CTC ACG GTA CTTGT
D3b: TCGGAAGGAACCAGCTACTA
D3a: GACCCGTCTTGAAACACGGA
D3aR: TCCGTGTTTCAAGACGGGTC
D3F: GAACCATCTAGTAGCTGGTTCC
D4R: GCGGTATTTGCTACTACCAYYAMGATCTGC

Fig. 7.7. Position and sequences of phylogenetic primers. Figure updated based on the previous version of this chapter (Kanzaki and Giblin-Davis, 2012) and several recent publications by the authors (e.g., Holterman *et al.*, 2006).

genus or higher ranking (Holterman *et al.*, 2006). D2-D3 LSU is also relatively conservative and therefore suitable for the analyses of species among closely related genera, within a genus or a lower taxonomic ranking. ITS and COI are rather variable and effectively distinguish closely related species or different populations within a species (e.g., Kanzaki and Futai, 2002a, b). A combined analysis also works well (e.g., Ye *et al.*, 2007).

The inferred molecular phylogenetic relationship among the species/genera belonging to Tylenchomorpha, where Aphelenchoidea belongs, and Panagrolaimomorpha, the closest relative of Tylenchomorpha, is shown in Figure 7.8. The phylogenetic tree was generated by Bayesian analysis based on near full-length SSU.

The Aphelenchoidea does not form a monophyletic clade and the Aphelenchidae and Aphelenchoididae are clearly separated from each other; that is, *A. avenae* (= Aphelenchidae) is in a basal position of the tylenchid plant/insect parasite clade, which was formerly referred to as 'Tylenchida' (see Siddiqi, 2000) and Aphelenchoididae forms a separate, well-supported clade. The Aphelenchoididae are further separated into four phylogenetic groups (Clades 1-4 in Fig. 7.8).

Clade 1 represents Tylaphelenchinae consisting of *Tylaphelenchus* and *Basilaphelenchus*. The members of the subfamily share a small body, characteristic stylet knobs and spherical median bulb. In addition, *Pseudaphelenchus* has an intermediate morphology between Aphelenchidae and the other members of the Aphelenchoididae, that is, aphelenchoidid-like pharynx and spicule and aphelenchid-like true male bursa with bursal limb-like genital papillae. However, based on the molecular tree (Fig. 7.8), *Aphelenchus* and *Pseudaphelenchus* (and other Tylaphelenchinae species) are not inferred to be sisters. Thus, the evolutionary origin of this character remains in question until detailed morphological and developmental studies are done. So far, the nominal representative of the subfamily, *Tylaphelenchus*, has not been reisolated nor molecularly characterized. Reisolation of the genus is necessary to confirm the monophyly of subfamily.

Clade 2 is separated into two distinctive subclades (Clade 2a and 2b: Fig. 7.8). Interestingly, both 2a and 2b contain '*Aphelenchoides*' and even the genus is paraphyletic within subclade 2a. The taxon '*Aphelenchoides*' doubtless needs to be re-examined and separated into several different genera. However, members of the genus are typologically similar to each other, that is, possibly morphologically convergent, or retaining the ancestral morphology, and so far, the type species of the genus, *A. kuehnii*, has not been reisolated and sequenced. Therefore, the true *Aphelenchoides* generic name, which is based on the poorly characterized type species, cannot yet be specified. Previously, *Laimaphelenchus* and *Schistonchus* were also separated into subclades 2a and 2b (see the previous version of the chapter, Kanzaki and Giblin-Davis, 2012), but some *Laimaphelenchus* spp. were transferred to *Aphelenchoides* and *Schistonchus* was subsequently separated into three independent genera, their generic status being settled.

Clade 3 consists of rather derived or specialized subfamilies, three entomoparasitic subfamilies (Ektaphelenchinae, Acugutturinae and Entaphelenchinae) and two predator subfamilies (Anomyctinae and Seinurinae). Although some genera, for example, *Seinura* and *Cryptaphelenchus*, are monophyletic, three ektaphelenchid genera, *Ektaphelenchus*, *Ektaphelenchoides* and *Devibursaphelenchus* are paraphyletic; that is, the subfamily is paraphyletic. Therefore, further comparative studies are necessary to resolve the paraphyly of Ektaphelenchinae and genera belonging to the subfamily. In addition, to date only one sequence has been deposited for Acugutturinae (*Noctuidonema* sp.) and Entaphelenchinae (*Peraphelenchus* sp.), and both are closely related to ektaphelenchid species.

Clade 4 consists of five genera, *Bursaphelenchus*, *Parasitaphelenchus*, *Potensaphelenchus*, *Ruehmaphelenchus* and *Sheraphelenchus*; that is, the clade represents the subfamily Parasitaphelenchinae. Within these five genera, the currently monotypic *Potensaphelenchus* and *B. abruptus* form a separate clade from the other species at the basal position of the clade (= subfamily), and the other three genera (*Parasitaphelenchus*, *Ruehmaphelenchus* and *Sheraphelenchus*) each form subclades within *Bursaphelenchus*.

7.7.3. Distinction between Families and Superfamilies of Aphelenchs

According to Siddiqi (2000), the Aphelenchida (= Aphelenchoidea in this chapter) is distinguished

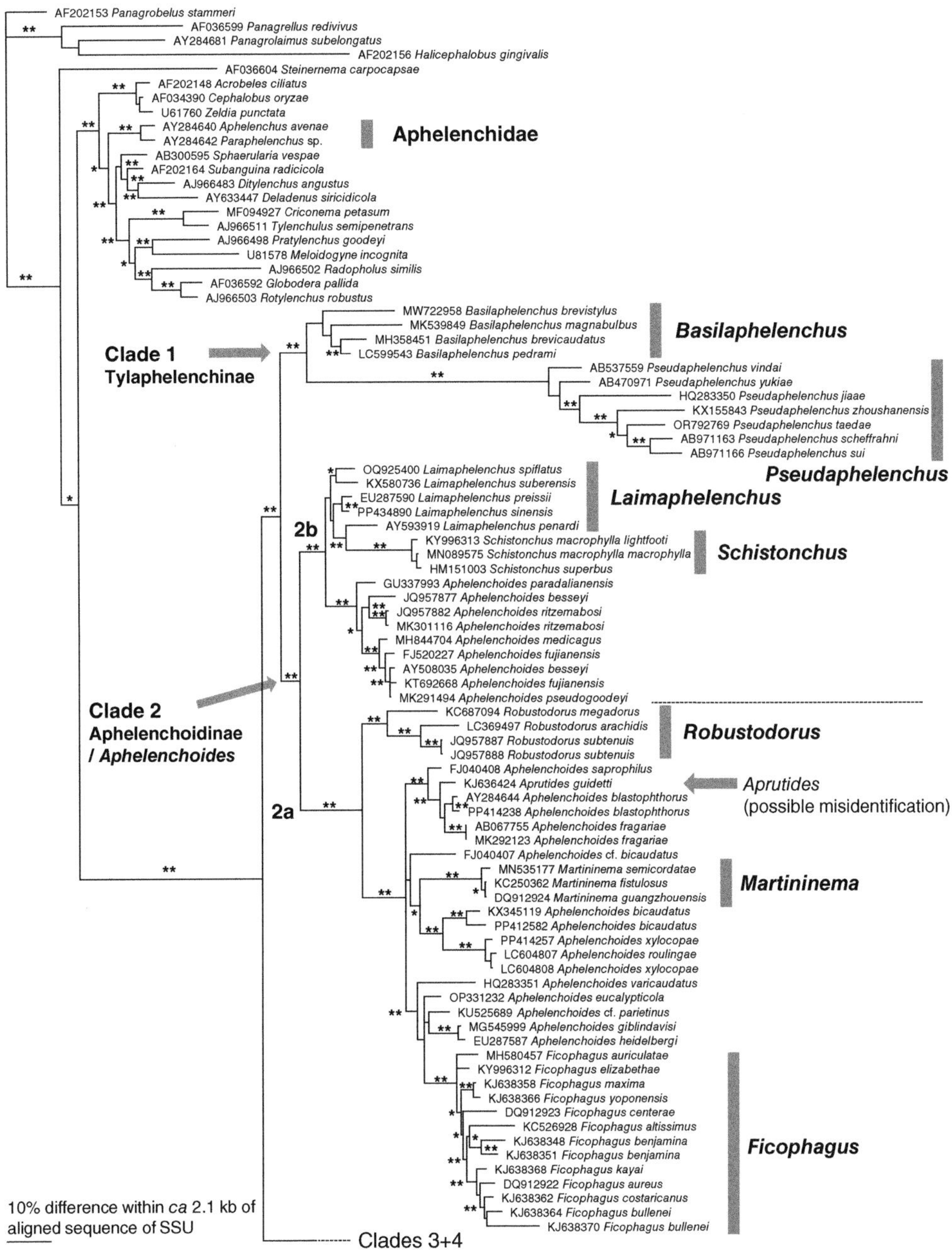

Fig. 7.8. Molecular phylogenetic relationships among *Aphelenchus* and their relatives inferred from near full-length SSU. The Bayesian tree was generated under GTR + G + I model (AIC = 108755.133; lnL = -54000.062). Analytical parameters are as follows: freqA = 0.26, freqC = 0.20, freqG = 0.27, freqT = 0.27; R(a) = 0.87, R(b) = 2.46, R(c) = 1.29, R(d) = 0.61, R(e) = 3.47, R(f) = 1.00; Pinva = 0.15; Shape = 0.57). Posterior probability values exceeding 50% are given on appropriate clades, and the nodes with 100% and 90-99% supports were indicated with ** and *, respectively.

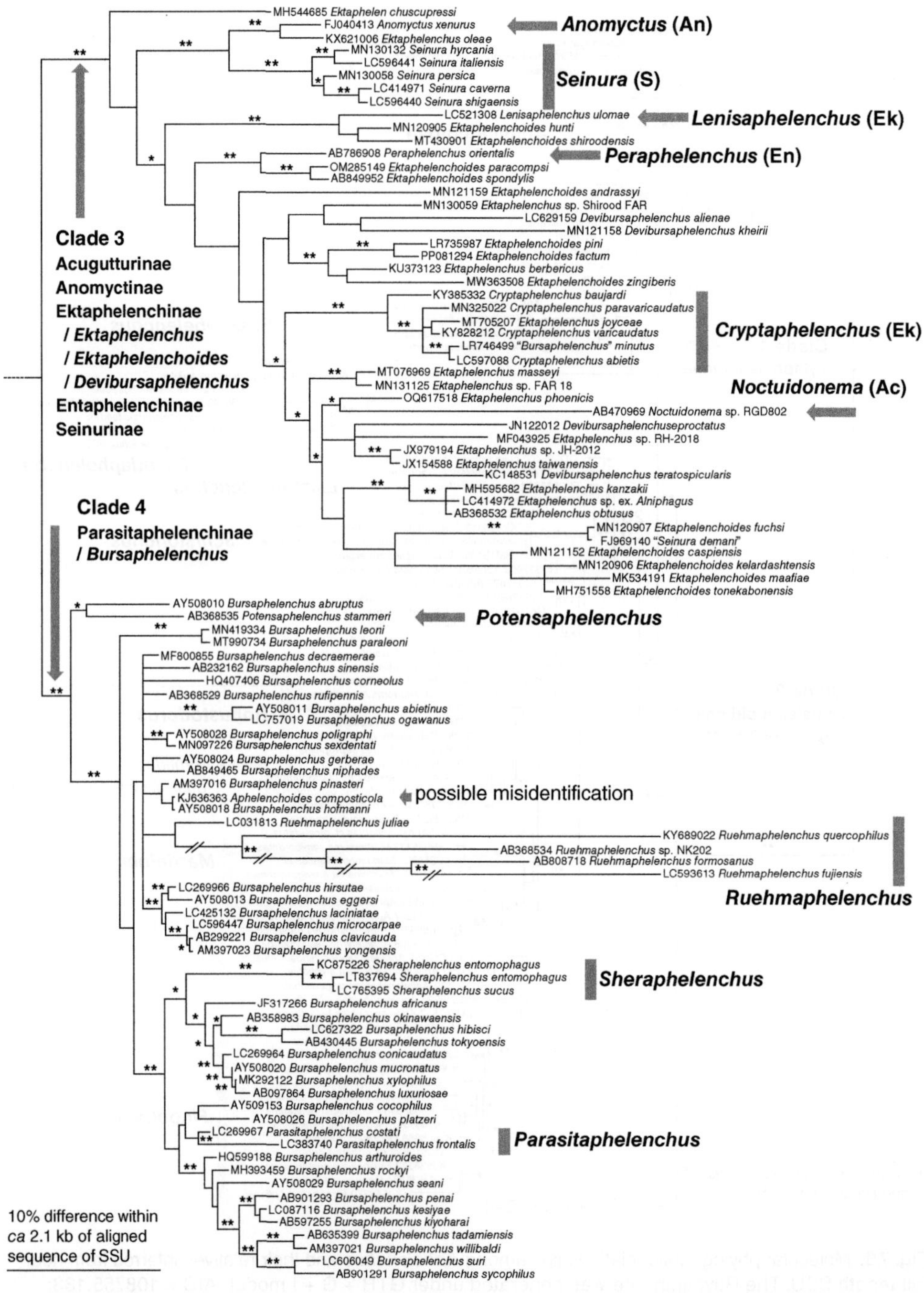

Fig. 7.8. Continued.

from Tylenchida *sensu* Siddiqi (2000) by eight major characters; (i) amphid position: laterosubdorsal *vs* lateral; (ii) origin of stylet shaft and knob: both originating from the telorhabdion and not separable *vs* metarhabdion and telorhabdion; (iii) dorsal pharyngeal gland orifice: in the muscular median bulb (metacorpus), anterior to cuticular valve apparatus *vs* in the anterior pharynx (procorpus), immediately posterior to the stylet base; (iv) metacorpus: forming a well-developed muscular median bulb with muscular valve anterior to central cuticular valve apparatus *vs* does not always form median bulb, if formed, not well developed, without the muscular valve anterior to central cuticular valve; (v) anus structure: (if present) conspicuous and forming large slit *vs* inconspicuous and forming pore-like opening; (vi) sperm: large, nucleus showing discrete chromosome *vs* usually small, nucleus not showing discrete chromosome; (vii) genital papillae or bursal limb (= rib-like papillae): present *vs* absent, and (viii) general spicule morphology: mostly thorn shaped *vs* not thorn shaped.

However, as shown above, Aphelenchidae is clustered with other tylenchids (and cephalobids), located at the basal position of tylenchid clade, and apparently apart from Aphelenchoididae; that is, these two families should be reclassified in two different superfamilies (Fig. 7.8, also see Meldal *et al.*, 2007). Based upon molecular phylogenetic analysis, the above tylenchid characters are considered to be a derived character from an Aphelenchidae-like ancestor, or the Aphelenchidae obtained Aphelenchoididae-like convergent characters as apomorphies. Therefore, the taxonomic framework of the superfamily should be reconsidered and key taxonomic characters defined according to the new framework (= Aphelenchidae + tylenchids *vs* Aphelenchoididae), which reflects the natural (evolutionary) clades.

When we compare their diagnostic characters, Aphelenchoididae is distinguished from Aphelenchidae + tylenchids by: (i) absence *vs* presence of a distinctive isthmus; (ii) thorn-shaped spicules *vs* rather blunt (not thorn-shaped) spicules; (iii) absence *vs* presence of conspicuous gubernaculum, and (iv) number of lateral incisures, ≤ 6 *vs* ≥ 6, although there are several intermediate genera/species.

The molecular phylogenetic analysis of these intermediate species/genera may be useful for evaluating the utility of these characters. For example, *Pseudaphelenchus* has an Aphelenchidae-like bursa and *Aphelenchoides*-like pharynx, and aligns molecularly with the Aphelenchoididae (Kanzaki *et al.*, 2009b, 2010). This suggests that the Aphelenchidae-like bursa is potentially a convergent character and that it cannot be applied to the family definition without a greater understanding of its developmental evolution.

Contrastingly, *Punchaulus* was described by De Ley and Coomans (1996) from Isla Santa Cruz, Galapagos, as a soil-dwelling mycophagous nematode. As stated by the authors, the genus has a distinctive isthmus and thorn-shaped spicules without a gubernaculum. They recognized these conflicting morphologies and chose Aphelenchoididae as its putative home accepting the thorn-shaped spicules and the absence of gubernaculum as the more evolutionarily relevant characters. Unfortunately, no molecular analysis has been conducted for *Punchaulus*. Thus, reisolation and molecular and morphological observation of the genus are necessary to elucidate the distinction between Aphelenchidae and Aphelenchoididae.

The morphological and structural evolutionary processes of these nematodes have not been fully clarified. However, Ragsdale *et al.* (2008, 2009, 2011) and Ragsdale and Baldwin (2010) observed the ultrastructure of the stomatostylet and metacorpus of *A. avenae*, compared it with a variety of nematode lineages and suggested the importance of cellular-level observations of structure (and development) for phylogenetic reconstruction. Ultimately, the integrated taxonomic classification of the 'Aphelenchoidea' will be reconstructed based upon a better understanding using developmental morphology and molecular phylogeny with a strong emphasis on understanding truly homologous characters.

7.7.4. Molecular Phylogeny *vs* Traditional Morphological Taxonomy in Aphelenchoididae

The Aphelenchoididae, which formed a well-supported monophyletic clade, was separated into four clades, namely, the Tylaphelenchinae (Clade 1), Aphelenchoidinae (Clade 2), five subfamilies of the insect parasites/predator clade (Clade 3) and the Parasitaphelenchinae (Clade 4) (Fig. 7.8).

The first clade, Tylaphelenchinae was discussed above. Its characteristic morphologies need to be compared with its close relatives, which have not yet been discovered and *Tylaphelenchus* spp., for morphological and molecular analysis.

The second clade was separated into two subclades and, interestingly, both of these contain *Aphelenchoides*. In the traditional taxonomy,

Aphelenchoides has typical aphelenchoidid morphology, and inhabits various environments, freshwater, seawater, soil and wood, and they have a conical tail with various tail tips, that is, rounded, pointed, mucronate, raspberry-shaped or digitate.

Aphelenchoides contains about 200 morphologically similar nominal species, many of their descriptions not being clear or distinctive by modern standards. The most difficult problem in the genus is the lack of robust morphological characters and the high degree of intraspecific variation; that is, most of the species have 'typical *Aphelenchoides*-type spicules', and female tail tip varies among individuals. Thus, it may not be easy to reconstruct the taxonomic system for the genus (see Shahina, 1996). The most likely solution will entail reisolation, brief typological re-description, for example, following OEPP/EPPO identification code, and deposition of a molecular profile for several diagnostic loci for each species. Zhuo *et al.* (2010) isolated 13 species of *Aphelenchoides* and briefly described and molecularly analysed each for their classification to species. The genus was recently synonymyzed to *Aphelenchoides* as its junior synonym (Fang *et al.*, 2024). Also, Yoshida *et al.* (2009) observed the development of *A. besseyi* and determined its very characteristic early embryogenesis and anterior-posterior axis formation pattern, which could be useful taxonomic characters. These kinds of elucidating efforts followed by molecular and morphological reconstruction are necessary for this genus (e.g., Subbotin *et al.*, 2020; Lai *et al.*, 2023). Two other genera, *Laimaphelenchus* and *Schistonchus* were previously paraphyletic groups. *Laimaphelenchus* was originally defined by four pedunculate tubercles with fringed margins on the tail tip and the presence of a vulval flap. However, the definition has been subsequently expanded, and the morphological traits now partially merge with those of *Aphelenchoides*, for example, thick mucron, a character of *Aphelenchoides*, revealed to be a raspberry-shaped extension under SEM observation, and the species was described as *Laimaphelenchus* expanding the generic definition (e.g., Zhao *et al.*, 2007). Thereafter, based on molecular phylogenetic analysis, several '*Laimaphelenchus*' species were transferred to *Aphelenchoides* (e.g., Carta *et al.*, 2016), and the monophyly of true *Laimaphelenchus* is currently supported. While, as stated above, *Schistonchus s.l.* was mostly characterized by its typical plant-parasitic form of lip and stylet and characteristic association with figs and fig wasps (Giblin-Davis *et al.*, 1995; Center *et al.*, 1999, Davies *et al.*, 2010). However, based upon detailed morphological examination and molecular phylogenetic analyses, the taxon was separated into three independent genera, *Schistonchus s.s.*, *Ficophagus* and *Martininema* (Davies *et al.*, 2015), and the paraphyly of *Schistonchus s.l.* was thus resolved.

Aphelenchoidinae has another genus, *Robustodorus* (*Punchaulus* has been discussed above). *Robustodorus* has a highly specialized stylet morphology and was previously a monotypic genus consisting of only *Robustodorus megadorus* (previously known as *Megadorus megadorus*). However, after reisolation of *R. megadorus* and *Robustodorus arachidis* (as *Aphelenchoides arachidis*) followed by morphological and molecular analysis, the genus was confirmed as a monophyletic group within Aphelenchoidinae (Clade 2a) (Ryss *et al.*, 2013; Kanzaki *et al.*, 2018a).

Ptychaphelenchus (its monotypic species, *Ptychaphelenchus eucalypticola*) was described by Hodda (2009), and molecularly characterized by Huston *et al.* (2022). Although the genus is characterized by its small bursal flaps on both sides of male mid-tail, the character is difficult to distinguish, and somewhat similar to relatively large and conspicuous P3 papillae.

Clade 3 consists of insect parasites and predators. The genera/subfamilies in this clade have variously derived or highly specialized characters. For example, Ektaphelenchinae lacks a functional anus and rectum in the female, Acugutturinae has an extremely long stylet and swollen body, Entaphelenchinae has three adult forms (parasitic and infective females and male) and Anomyctinae and Seinurinae are the only known predators in this superfamily (except for the predatory form of *Bursaphelenchus*).

The morphological similarity between Ektaphelenchinae and Seinurinae has been mentioned by several authors, for example, stylet morphology of *Ektaphelenchoides spondylis* and *Seinura musicola* (Kanzaki *et al.*, 2009a), and relatively long metacorpus is shared among several genera (species) within these two subfamilies (see Hunt, 1993, 2008). In fact, *Ektaphelenchoides* and *Seinura* can be separated only by the presence/absence of a functional rectum and anus. *Devibursaphelenchus* is also included in this clade. It is defined by the presence of a male bursal flap and was previously synonymized under *Bursaphelenchus*. However, the molecular data suggests that the bursal flap of *Devibursaphelenchus* occurred independently from *Bursaphelenchus*. Interestingly, *Noctuidonema* sp. (Acugutturinae) is also inferred as a member of this clade. This monotypic genus is known as an ectoparasite of moths (Remillet and Silvain, 1988) and it is phylogenetically close to ektaphelenchid species. Ectoparasitism of the insects could be a common (homologous) character within the Ektaphelenchinae, and Acugutturinae (at least *Noctuidonema*, because other genera in the subfamily have not been molecularly examined) seems derived from an ektaphelenchid stem ancestor. The situation of Entaphelenchinae (*Peraphelenchus*) is similar to Acugutturinae; that is, *Peraphelenchus orientalis*, the only representative of the subfamily at present, is included this clade, and close to *Ektaphelenchoides* species. Further, the first authors' recent observation (Kanzaki *et al.*, unpubl. obs.) suggests that *Entaphelenchus* is not close to *Peraphelenchus*, and the monophyly of Entaphelenchinae is not supported. Thus, to confirm their generic and subfamily status, further isolation of new and nominal species belonging to Clade 3 is necessary.

It is difficult to find a common character shared among these three subfamilies, probably because each subfamily (or genus) has acquired highly specialized morphological (= functional) and life history (= physiological) traits. As stated by previous authors (e.g., Blaxter *et al.*, 1998; Meldal *et al.*, 2007), parasitic groups radiate faster than free-living groups. For example, Clade 3 in Blaxter *et al.* (1998) contains animal parasitic nematodes and they are highly divergent in morphology and life history, although the genetic distances are rather short compared with the other free-living groups. A possible explanation for this fast radiation rate is that parasites (or other kinds of groups with highly specialized host or habitats) must adapt to their hosts' life cycle and physiological characters and, therefore, must respond with functional and physiological adaptations quickly and often. Contrastingly, the effectively free-living forms, either living as bacterial, fungal and/or perhaps some plant-parasitic feeders, are already well adapted and have either not been under evolutionary pressure to change for a long time, or convergence has occurred frequently (Blaxter *et al.*, 1998).

Currently, the number of available sequences is still insufficient to understand the molecular phylogenetic relationships among Clade 3 species/genera/subfamilies. However, based on the present phylogenetic relationship and their feeding habitats, predator groups (Anomyctinae and Seinurinae) are located at a relatively basal position of the clade, suggesting that predators occurred from a mycophagous ancestor first, and ecto- and endoparasites evolved from predators. Interestingly, within Clade 3 genera, *Cryptaphelenchus*, which is culturable using food fungi (e.g., Kanzaki *et al.*, 2022b), and typologically similar to fungal-feeding species, that is, with rather short stylets with a narrow lumen, spherical median bulb without a glandular portion and a thread-like body shape, is located at a derived position. Considering that some more basal ektaphelenchids can be cultured using prey nematodes, but they do not feed on fungi (e.g., Gu *et al.*, 2010; Kanzaki, 2014; Miraeiz *et al.*, 2017; Kanzaki *et al.*, 2018b), the clade (*Cryptaphelenchus*) re-obtained mycophagy in the form of a reversion.

Detailed morphological and biological comparisons between these subfamilies using more species are needed to recover homologous characters representative of phylogeny. The physiological characters at the genomics level may be useful to characterize this group; that is, ingestion/digestion-related genes for insect parasitism and predation may be similar, because both are feeding on invertebrates' body fluid. Kikuchi *et al.* (2007) and Karim *et al.* (2009) analysed the expressed sequence tags (EST) of *B. xylophilus* and *A. avenae* comparing their genetic composition with those of other plant-parasitic species with focus on plant-parasitic and fungal-feeding genes. Although the genomics-based phylogenetic data have not yet been published, presence/absence of specific genes may be helpful in characterizing the clade.

Lastly, Clade 4 represents the Parasitaphelenchinae, and the species/genera belonging to this clade share a P1 precloacal single papilla, although several species have secondarily lost this character. The clade mostly consists of *Bursaphelenchus* spp. and the *B. abruptus*/*P. stammeri* clade, which is closest to, or included in, this clade suggesting that the generic status of *Potensaphelenchus* is clearly supported. The closeness between *B. abruptus* and *A. stammeri* is interesting. Based upon their molecular phylogenetic position, these two species either clearly form a single genus, or a new genus should be established for *B. abruptus* because of its typological similarity to the *xylophilus* group of *Bursaphelenchus*, for example, narrow and long spicule forming trapezoidal shape, presence of terminal bursal flap, papilliform P3 and P4 in males and conspicuous vulval flap in females.

Giblin-Davis *et al.* (1993) observed the lip morphology of *B. abruptus* and suggested several differences from that of *B. xylophilus*, that is, shape and position of amphid apparatus, presence/absence of lip annulation and presence/absence of oval depression. As suggested by DeCrappeo and Giblin-Davis (2001) and Davies *et al.* (2010), lip morphology can sometimes reflect generic-level differences. These ultrastructural differences could be the key to establishing a new genus for *B. abruptus*.

Within the other four genera, *Parasitaphelenchus*, *Ruehmaphelenchus* and *Sheraphelenchus* form inner subclades as the parts of the large *Bursaphelenchus* clade, and thus, *Bursaphelenchus* becomes paraphyletic. Typologically, *Ruehmaphelenchus* is separated from the other genera by the conical male tail lacking a bursal flap, and presence of a labial disc, although the latter has not been confirmed in all species; *Sheraphelenchus* is distinguished from the others by the conical male tail with a spike-like projection and lack of a female post-vulval uterine sac, and *Parasitaphelenchus* is characterized by the presence of parasitic juvenile stage, weak curvature of male tail and mitten-shaped spicule with clear dorsal and ventral limbs (e.g., Hunt, 1993, 2008). However, recently, several authors have provided data suggestive of insect parasitism in *Bursaphelenchus* spp. (Tomalak and Filipiak, 2010; Ryss *et al.*, 2015), and some typological characters, for example, mitten-shaped spicule, are shared between *Bursaphelenchus* and *Parasitaphelenchus*. Thus, the delimitation of these two genera is becoming blurred. In addition, secondary loss of the bursal flap, a generic character of *Ruehmaphelenchus* and *Sheraphelenchus*, can be found in *Bursaphelenchus* (*B. kiyoharai* and *B. penai*) (Kanzaki *et al.*, 2011, 2014a). Therefore, following further study, either *Bursaphelenchus* should be separated into several genera according to their typological characters and molecular phylogenetic status (Ryss *et al.*, 2005; Braasch *et al.*, 2009; Ryss and Subbotin, 2017; Kanzaki and Giblin-Davis, 2018), or the clade except for the above-mentioned two species (*P. stammeri* and *B. abruptus*) should be lumped into *Parasitaphelenchus* according to the priority of the generic description.

Currently, a reasonably good number of molecular sequences have been accumulated for Aphelenchoidea. However, the sequences cover 24 of the 36 nominal genera, and some of these should be separated into multiple genera whereas others should be lumped into one. Further field collections, especially from Clade 3 subfamilies, are necessary to understand the whole picture of aphelenchoidid phylogeny.

7.8. Conclusion: Remarks for Future Work

Recent molecular systematic research has expanded our knowledge about the putative relationships among families and genera – there are many examples of morphological convergence and several examples of rapid radiation. Re-evaluation of morphological and biological characters based on molecular phylogeny will be crucial in reconstructing the taxonomic system for members of the Aphelenchoidea. We have elucidated several key questions to be resolved by more representative sampling and morphological re-examination and molecular sequencing and phylogenetic analysis.

The control and management issues for plant-parasitic members of the Aphelenchoidea will remain a challenge. Chemical control of soil-borne species is becoming increasingly difficult as the most effective chemicals are not environmental friendly. This puts emphasis on host-plant resistance, cultural control methods and biological control, which are all information intensive, justifying the need for nematologists and plant-pathologists to continue research.

Currently, the most important species in the Aphelenchoidea is probably *B. xylophilus*, which is devastating Asian pine forests and has now invaded the EU. As seen in Asian forests, once the disease has spread through a large area it is almost impossible to eradicate. Also, the nematode sometimes infests trees without manifesting any symptoms, a feature that aids in its undetected persistence and spread. The best way to control disease is through quarantine measures to prevent invasion and quick elimination at the early stage of invasion. This should serve as a warning to Mexico and Central and northern South American countries, one of the centres of divergence of the genus *Pinus* (Farjon, 1984), that have many species of potentially susceptible pine species (Futai and Furuno, 1979; Furuno *et al.*, 1993).

The Aphelenchoidea are an interesting and unique assemblage of nematodes that have radiated from fungal feeding to a variety of other feeding niches. Some aphelenchs should be good candidates for models to study evolutionary biology of feeding adaptations and/or niche expansion. Following the total genomic sequencing of *Caenorhabditis elegans* (The *C. elegans* Sequencing Consortium, 1998), many genome projects were carried out for various nematode species. Aphelenchoidea is an appropriate model for genomic analyses of fungal-feeding/plant-parasitic nematodes and the expansion into predation and insect parasitism. Several research groups have begun working on genomic sequencing of aphelenchoidid plant parasites (e.g., Jones *et al.*, 2008; Kikuchi *et al.*, 2011). Not only the practically important plant parasites but also some aphelenchoidids have become/or are becoming interesting satellite model species; for example, *B. sinensis* and its close relatives (*B. aberrans* group) show feeding (lip, stylet and pharyngeal glands) dimorphism as stomatal dimorphism/polymorphism in some diplogastrids (Kanzaki *et al.*, 2019a; Kanzaki and Giblin-Davis, 2020); *B. okinawaensis* has been confirmed as the first hermaphroditic species in the family, and is being intensively studied as a model system because of its reproductive mode (Shinya *et al.*, 2022); *B. luxuriosae* and some of its close relatives have phoretic and propagative adult forms (Kanzaki *et al.*, 2009c, 2013); and *Seinura caverna*, a second experimentally confirmed hermaphrodite in the family, can be used as a model for a predatory aphelenchoidid (Ekino *et al.*, 2023), and evolutionary studies of predation. Genomics-based physiological and evolutionary analyses will have important ramifications in aphelench research in the near future.

Notes

[1] *Berntsenus* was formerly treated as a *genus incertae sedis* (Hunt, 1993), and Hunt (2008) placed it in the subfamily Ektaphelenchinae. However, it was later confirmed to be an alternate (predatory) form of *Bursaphelenchus* (Kanzaki & Giblin-Davis, 2020).

[2] But *Punchaulus* has an intermediate morphology between Aphelenchidae and Aphelenchoididae; that is, it has rosethorn-shaped spicules without a gubernaculum, and a distinctive isthmus. De Ley and Coomans (1996) mentioned these intermediate characteristics and tentatively placed it in Aphelenchoididae.

7.9. References

Abad, P., Tares, S., Brugier, N. and de Guiran, G. (1991) Characterization of the relationships in the pinewood nematode species complex (PWNSC) (*Bursaphelenchus* spp.) using a heterologous unc-22 DNA probe from *Caenorhabditis elegans*. *Parasitology* 102, 303–308.

Allen, M.W. (1940) *Anomyctus xenurus*, a new genus and species of Tylenchoidea (Nematoda). *Proceedings of the Helminthological Society of Washington* 7, 96–98.

Anderson, R.V. and Laumond, C. (1992) *Noctuidonema daptria*, n. sp. (Nematoda: Aphelenchoididae), an ectoparasite of the moth *Lesmone porcia* (Stoll). *Journal of Nematology* 24, 16–22.

Andrássy, I. (1973) Ein Meeresrelikt und einige andere bemerkenswerte Nematodenarten aus Neuguinea. *Opuscula Zoologica (Budapest)* 12, 3–19.

Andrássy, I. (2007) Free-living Nematodes of Hungary (Nematoda errantia), II. In: Csuzdi, C. and Mahunka, S. (eds), *Pedozoologica Hungarica, 4.* Hungarian Natural History Museum and Systematic Zoology Research Group of the Hungarian Academy of Sciences, Budapest, Hungary.

Bajaj, H.K., Bishnoi, S.P. and Dabur, K.R. (1994) Description of *Indaphelenchus siddiqii* gen. n., sp. n. (Nematoda: Aphelenchida) from India. *Indian Journal of Nematology* 24, 211–213.

Bastian, H.C. (1865) Monograph on the Anguillulidae, or free nematoids, marine, land, and freshwater; with descriptions of 100 new species. *Transactions of the Linnaean Society of London* 25, 73–184.

Baujard, P. (1984) Remarques sur la sous-famille des Ektaphelenchinae Paramonov, 1964 et proposition d'*Ektaphelenchoides* n. gen. (Nematoda: Aphelenchoididae). *Revue de Nématologie* 7, 147–171.

Bedford, G.O. (1974) Parasitism of the palm weevil *Rhynchophorus bilineatus* (Montrouzier) (Coleoptera: Curculionidae) by *Praecocilenchus rhaphidophorus* (Poinar) (Nematoda: Aphelenchoidea) in New Britain. *Australian Journal of Entomology* 13, 155–156.

Bergdahl, D.R. and Halik, S. (1998) Inoculated *Pinus sylvestris* serve as long term hosts for *Bursaphelenchus xylophilus*. In: Futai, K., Togashi, K. and Ikeda, T. (eds), *Proceedings of the IUFRO Meeting on Sustainability of Pine Forests in Relation to Pine Wilt and Decline*. Shokado, Kyoto, Japan, pp. 73–78.

Bert, W. and Borgonie, G. (2006) Order Rhabditida: Suborder Tylenchina. In: Eyualem-Abebe, Traunspurger, W. and Andrássy, I. (eds), *Freshwater Nematodes: Ecology and Taxonomy*. CAB International, Wallingford, UK, pp. 648–695.

Blaxter, M.L., De Ley, P., Garey, J.R., Liu, L.X., Scheldeman, P., *et al.* (1998) A molecular evolutionary framework for the phylum Nematoda. *Nature* 392, 71–75.

Braasch, H. (2000) Influence of temperature and water supply on mortality of 3-year-old pines inoculated with *Bursaphelenchus xylophilus* and *B. mucronatus*. *Nachrichtenblatt des Deutschen Pflanzenschutzdienstes (Stuttgart)* 52, 244–249.

Braasch, H. (2009) Re-establishment of *Devibursaphelenchus* Kakulia, 1967 (Nematoda, Aphelenchoididae) and proposal for a new combination of several *Bursaphelenchus* species. *Journal of Nematode Morphology and Systematics* 12, 1–5.

Braasch, H., Burgermeister, W. and Gu, J. (2009) Revised intra-generic grouping of *Bursaphelenchus* Fuchs, 1937 (Nematoda: Aphelenchoididae). *Journal of Nematode Morphology and Systematics* 12, 65–81.

Braasch, H., Burgermeister, W. and Pastrik, K.H. (1995) Differentiation of three *Bursaphelenchus* species by means of RAPD-PCR. *Nachrichtenblatt des Deutschen Pflanzenschutzdienstes (Stuttgart)* 47, 310–314.

Bridge, J., Plowright, R.A. and Peng, D. (2005) Nematode parasites of rice. In: Luc, M., Sikora, R.A. and Bridge, J. (eds), *Plant Parasitic Nematodes in Subtropical and Tropical Agriculture*, 2nd edn. CAB International, Wallingford, UK, pp. 87–130.

Cao, A.X., Liu, X.Z., Zhu, S.F. and Lu, B.S. (2005) Detection of pinewood nematode, *Bursaphelenchus xylophilus*, using a real-time polymerase chain reaction assay. *Phytopathology* 95, 566–571.

Carta, L.K., Li, S., Skantar, A.M. and Newcombe, G. (2016) Morphological and molecular characterization of two *Aphelenchoides* endophytic in poplar leaves. *Journal of Nematology* 48, 28–33.

Cayrol, J.C., Roudeillac, P. and B'Chir, M.M. (1972) Etudes preliminaires sur les possibilites d'utilisation des champignons nematophages comme moyen de lute biologique en champignonniere. *Revue de Zoologie Agricole, et de Pathologie Végétale* 71, 118–138.

Center, B.J., Giblin-Davis, R.M., Herre, E.A. and Chung-Schickler, G.C. (1999) Histological comparisons of parasitism by *Schistonchus* spp. (Nemata: Aphelenchoididae) in Neotropical *Ficus* spp. *Journal of Nematology* 31, 393–406.

Chaturvedi, Y. and Khera, S. (1977) *Caballeroides olitorius*, gen. nov., (Nematoda: Aphelenchoididae) from the rhizosphere of jute. In: *Excerta parasitológica en memoria del Doctor Eduardo Caballero y Caballero.* University of Mexico, Mexico City, Mexico, pp. 419–422.

Chen, F.-M., Negi, S. and Ye, J.-R. (2011) A SCAR molecular marker to distinguish *Bursaphelenchus mucronatus* from the pinewood nematode, *B. xylophilus*. *Forest Pathology* 41, 376–381.

Chitwood, B.G. (1933) A revised classification of Nematoda. *Journal of Parasitology* 20, 131.

Cobb, N.A. (1913) New nematode genera found inhabiting fresh water and non-brackish soils. *Journal of the Washington Academy of Sciences* 3, 432–444.

Cobb, N.A. (1927) Note on a new nema, *Aphelenchus retusus*, with a proposed division of *Aphelenchus* into three subgenera. *Journal of Parasitology* 14, 57–58.

Cralley, E.M. (1957) The effect of seeding methods on the severity of white tip of rice. *Phytopathology* 47, 7. [Abstr.]

Davies, K., Bartholomaeus, F., Ye, W., Kanzaki, N. and Giblin-Davis, R.M. (2010) *Schistonchus* (Aphelenchoididae) from *Ficus* (Moraceae) in Australia, with description of *S. aculeata* sp. n. *Nematology* 12, 935–958.

Davies, K.A., Bartholomaeus, F., Kanzaki, N., Ye, W. and Giblin-Davis, R.M. (2015) A review of the taxonomy, phylogeny, distribution and co-evolution of *Schistonchus* (Nematoda: Aphelenchoididae) and proposal of *Ficophagus* n. gen. and *Martininema* n. gen. *Nematology* 17, 761–829.

DeCrappeo, N. and Giblin-Davis, R.M. (2001) *Schistonchus aureus* n. sp. and *Schistonchus laevigatus* n. sp. (Aphelenchoididae): associates of native Floridian *Ficus* spp. and their *Pegoscapus* pollinators (Agaonidae). *Journal of Nematology* 33, 91–103.
De Ley, P. and Blaxter, M.L. (2002) Systematic position and phylogeny. In: Lee, D.L. (ed), *The Biology of Nematodes.* Taylor and Francis, London, UK, pp. 1–30.
De Ley, P. and Coomans, A. (1996) Terrestrial nematodes of the Galapagos archipelago. 6. *Punchaulus gemellensis*, a new genus and species of Aphelenchina (Tylenchida). *Fundamental and Applied Nematology* 19, 159–165.
Drozdovsky, E.M. (1967) [Use of the characteristics of embryonic development in the classification of eelworms.] *Trudy Gel'mintologicheskoi Laboratorii* 18, 22–29.
Ekino, T., Yoshiga, T., Takeuchi-Kaneko, Y., Ichihara, Y., Kanzaki, N. and Shinya, R. (2023) Highlighting potential physical and chemical cues involved in conspecific recognition system in a predator nematode, *Seinura caverna*. *Integrative and Comparative Biology* 63, 865–876.
Evans, A.A.F. and Fisher, J.M. (1970) Some factors affecting number and size of nematodes in populations of *Aphelenchus avenae*. *Nematologica* 16, 295–304.
Fang, Y., Gu, J., Ma, X. and Li, H. (2024) Description of *Aphelenchoides yixingensis* n. sp. (Tylenchina: Aphelenchoididae) found in *Pinus massoniana* from China, and proposal for genus *Ptychaphelenchus* as a synonym of *Aphelenchoides*. *Nematology* 26, 763–778.
Farjon, A. (1984) *Pines. Drawings and Descriptions of the Genus* Pinus. Brill, Leiden, Netherlands.
Fischer, M. (1894) Über eine Clematis-krankheit. *Bericht aus dem Physiolischen Laboratorium des Landwirthschaftlichen. Instituts der Universitat Halle* 3(11), 1–11.
Fortuner, R. and Williams, K.J.O. (1975) Review of literature on *Aphelenchoides besseyi* Christie, 1942, the nematode causing 'white tip' disease in rice. *Helminthological Abstracts* B44, 1–40.
Franklin, M.T. and Siddiqi, M.R. (1972) *Aphelenchoides besseyi. CIH Descriptions of Plant-parasitic Nematodes* Set 1, No. 4. Commonwealth Agricultural Bureaux, Farnham Royal, UK.
Fuchs, A.G. (1930) Neue an Borken- und Rüsselkäfer gebundene Nematoden, halbparasitische und Wohnungseinmieter. *Zoologische Jahrbücher, Abteilung für Systematik, Ökologie und Geographie der Tiere* 59, 505–646.
Fuchs, A.G. (1931) *Seinura* gen. nov. *Zoologischer Anzeiger* 94, 226–228.
Fuchs, A.G. (1937) Neue parasitische und halbparasitische Nematoden bei Borkenkäfern und einige andere Nematoden. I. Teil. *Zoologische Jahrbücher, Abteilung für Systematik, Ökologie und Geographie der Tiere* 70, 291–380.
Furuno, T., Nakai, I., Uenaka, K. and Haya, K. (1993) [The pine wilt upon the exotic pine species introduced in Kamigamo and Shirahama Experimental Stations of Kyoto University. Various resistances among genus *Pinus* to pine wood nematode, *Bursaphelenchus xylophilus* Steiner & Buhrer.] *Reports of Kyoto University Forest* 25, 20–34.
Futai, K. (2003) Role of asymptomatic carrier trees in epidemic spread of pine wilt disease. *Journal of Forest Research* 8, 253–260.
Futai, K. and Furuno, T. (1979) [The variety of resistances among pine-species to pine wood nematode, *Bursaphelenchus lignicolus*.] *Bulletin of the Kyoto University Forests* 51, 23–36.
Giblin-Davis, R.M. (1993) Interactions of nematodes with insects. In: Khan, W. (ed), *Nematode Interactions*. Chapman and Hall, London, UK, pp. 302–344.
Giblin-Davis, R.M., Center, B.J., Nadel, H., Frank, J.H. and William, R.B. (1995) Nematodes associated with fig wasps, *Pegoscapus* spp. (Agaonidae), and syconia of native Floridian figs (*Ficus* spp.). *Journal of Nematology* 27, 1–14.
Giblin-Davis, R.M., Mundo-Ocampo, M., Baldwin, J.G., Norden, B.B. and Batra, S.W. (1993) Description of *Bursaphelenchus abruptus* n. sp. (Nemata: Aphelenchoididae), an associate of a digger bee. *Journal of Nematology* 25, 161–172.
Giblin-Davis, R.M., Kanzaki, N. and Williams, D.S. (2023) Ultrastructural study of anther parasitism of *Ficus laevigata* by *Ficophagus laevigatus* (Aphelenchoididae). *Journal of Nematology* 55(1), 20230010. Doi: 10.2478/jofnem-2023-0010.
Goodey, J.B. (1960a) *Rhadinaphelenchus cocophilus* (Cobb, 1919) n. comb., the nematode associated with "red-ring" disease of coconut. *Nematologica* 5, 98–102.
Goodey, J.B. (1960b) The classification of the Aphelenchoidea Fuchs, 1937. *Nematologica* 5, 111–126.
Goodey, J.B. (1963) *Soil and Freshwater Nematodes*, 2nd edn. Methuen and Co., London, UK.
Goodey, J.B. and Hooper, D.J. (1965) A neotype of *Aphelenchus avenae* Bastian, 1865 and the rejection of *Metaphelenchus* Steiner, 1943. *Nematologica* 11, 55–65.

Goodey, T. (1951) *Soil and Freshwater Nematodes.* Methuen and Co., London, UK.

Griffith, R., Giblin-Davis, R.M., Koshy, P.K., Sosamma, V.K. and Kanzaki, N. (2018) Nematode parasites of coconut and other palms. In: Sikora, R.A. Coyne, D., Hallmann, J. and Timper, P. (eds), *Plant Parasitic Nematodes in Subtropical and Tropical Agriculture*, 3rd edn. CAB International, Wallingford, UK, pp. 504–535.

Gu, J., Fang, Y. and Ma, X. (2023) *Aphelenchoides koreanus* n. sp. (Tylenchina: Aphelenchoididae) found in packaging wood from South Korea. *Nematology* 25, 659–668. Doi: 10.1163/15685411-bja10245.

Gu, J., Liu, L., Abolafia, J. and Pedram, M. (2021) A revision of the taxonomy of *Aphelenchoides stammeri* Körner, 1954 (Rhabditida: Aphelenchoididae) and proposal for a new genus. *Nematology* 23, 215–228.

Gu, J., Wang, J. and Zheng, J.W. (2010) *Devibursaphelenchus wangi* sp. n. (Nematoda: Ektaphelenchinae) feeding on *Aphelenchoides* sp. *Russian Journal of Nematology* 18, 49–57.

Halik, S. and Bergdahl, D.R. (1994) Long-term survival of *Bursaphelenchus xylophilus* in living *Pinus sylvestris* in an established plantation. *European Journal of Forest Pathology* 24, 357–363.

Haque, M.M. (1967) [*Paraphelenchoides capsuloplanus* n. g., n. sp. (Nematoda, Aphelenchoididae).] *Zoologichesky Zhurnal* 46, 1251–1253.

Hechler, H.C. (1963) Description, developmental biology, and feeding habitat of *Seinura tenuicaudata* (de Man) J.B. Goodey, 1960 (Nematoda: Aphelenchoididae), a nematode predator. *Proceedings of the Helminthological Society of Washington* 30, 182–195.

Hodda, M. (2009) *Ptychaphelenchus eucalypticola* gen. nov., sp. nov. (Nematoda: Panagrolaimida: Aphelenchoidoididae) from wood and bark of *Eucalyptus macrorhyncha* in Australia, with a discussion of generic relationships and a key to genera of Aphelenchoididae and Parasitaphelenchidae. *Transactions of the Royal Society of South Australia* 133, 183–194.

Holterman, M., van der Wurff, A., van den Elsen, S., van Megen, H., Holovachov, O., *et al.* (2006) Phylum-wide analysis of SSU rDNA reveals deep phylogenetic relationships among nematodes and accelerated evolution toward crown clades. *Molecular Biology and Evolution* 23, 1792–1800.

Hoyer, U., Burgermeister, W. and Braasch, H. (1998) Identification of *Bursaphelenchus* species (Nematoda, Aphelenchoididae) on the basis of amplified ribosomal DNA (ITS-RFLP). *Nachrichtenblatt des Deutschen Pflanzenschutzdienstes (Stuttgart)* 50, 273–277.

Hu, Y.Q., Kong, X.C., Wang, X.R., Zhong, T.K., Zhu, X.W., *et al.* (2011) Direct PCR-based method for detecting *Bursaphelenchus xylophilus*, the pine wood nematode in wood tissue of *Pinus massoniana*. *Forest Pathology* 41, 165–168. Doi: 10.1111/j.1439-0329.2010.00692.x

Huang, L., Ye, J.R., Wu, X.Q., Xu, X.L., Sheng, J.M., *et al.* (2010) Detection of the pine wood nematode using a real-time PCR assay to target the DNA topoisomerase I gene. *European Journal of Plant Pathology* 127, 89–98.

Hunt, D.J. (1980) *Acugutturus parasiticus* n. g., n. sp., a remarkable ectoparasitic aphelenchoid nematode from *Periplaneta americana* (L.), with proposal of Acugutturinae n. subf. *Systematic Parasitology* 1, 167–170.

Hunt, D.J. (1993) *Aphelenchida, Longidoridae and Trichodoridae: their Systematics and Bionomics.* CAB International, Wallingford, UK.

Hunt, D.J. (2008) A checklist of the Aphelenchoidea (Nematoda: Tylenchina). *Journal of Nematode Morphology and Systematics* 10(2007), 99–135.

Husain, S.I. and Khan, A.M. (1967) On the status of the genera of the superfamily Aphelenchoidea (Fuchs, 1937) Thorne, 1949 with the descriptions of six new species of nematodes from India. *Proceedings of the Helminthological Society of Washington* 34, 167–174.

Huston, D.C., Khudhir, M. and Hodda, M. (2022) Phylogenetic position of *Ptychaphelenchus eucalypticola* Hodda, 2009 within the Aphelenchoidoidea Skarbilovich, 1947 (Siddiqi, 1980) inferred from partial 18S and 28S rDNA gene sequences. *Nematology* 25, 59–76.

Ide, T., Kanzaki, N., Parra Giraldo, P.P. and Giblin-Davis, R.M. (2017) Loop-mediated isothermal amplification (LAMP) for detection of the red ring nematode, *Bursaphelenchus cocophilus* of palms. *Nematology* 19, 559–567.

Ingham, R.E. (1988) Interactions between nematodes and vesicular-arbuscular mycorrhizae. *Agriculture Ecosystems and Environment* 24, 169–182.

Inglis, W.G. (1983) An outline classification of the phylum Nematoda. *Australian Journal of Zoology* 31, 243–255.

Ishibashi, N. and Choi, D.R. (1991) Biological control of soil pests by mixed application of entomopathogenic and fungivorous nematodes. *Journal of Nematology* 23, 175–81.

Iwahori, H., Tsuda, K., Kanzaki, N., Izui, K. and Futai, K. (1998) PCR-RFLP and sequencing analysis of ribosomal DNA of *Bursaphelenchus* nematodes related to pine wilt disease. *Fundamental and Applied Nematology* 21, 655–666.

Jagdale, G.B. and Grewal, P.S. (2006) Infection behavior and overwintering survival of foliar nematodes, *Aphelenchoides fragariae*, on *Hosta*. *Journal of Nematology* 38, 130–136.

Jairajpuri, M.S. (1968) Some studies on the morphology and biology of *Aphelenchus avenae* Bastian, 1865 (Nematoda: Aphelenchoidea). *Labdev Journal of Science and Technology* 6B, 27–34.

Jones, J.T., Moens, M., Mota, M.M., Li, H. and Kikuchi, T. (2008) *Bursaphelenchus xylophilus*: opportunities in comparative genomics and molecular host-parasite interactions. *Molecular Plant Pathology* 9, 357–368.

Kakulia, G.A. (1967) [New nematode genus, *Devibursaphelenchus* n. g. (Nematoda: Aphelenchoididae).] *Soobschcheniya Akademii Nauk Gruzinskoi SSR* 47, 439–443.

Kang, J.S.K. (2004) Development of an efficient PCR-based diagnosis protocol for the identification of the pinewood nematode, *Bursaphelenchus xylophilus* (Nematoda: Aphelenchoididae). *Nematology* 6, 279–285.

Kanzaki, N. (2006) Description of *Aphelenchoides xylocopae* n. sp. (Nematoda: Aphelenchoididae), the first observed association between nematodes and carpenter bees. *Nematology* 8, 555–562.

Kanzaki, N. (2008) Taxonomy and systematics of the nematode genus *Bursaphelenchus* (Nematoda: Parasitaphelenchidae). In: Zhao, B.G., Futai, K., Sutherland, J. and Takeuchi, Y. (eds), *Pine Wilt Disease*. Springer Japan, Tokyo, Japan, pp. 44–66.

Kanzaki, N. (2014) *Ektaphelenchoides spondylis* is a predatory nematode. *Nematology* 16, 245–247.

Kanzaki, N., Aikawa, T., Maehara, N. and Matsumoto, K. (2008a) *Bursaphelenchus doui* Braasch, Gu, Burgermeister & Zhang, 2005 (Aphelenchida: Parasitaphelenchidae), an associate of *Monochamus subfasciatus* Bates (Coleoptera: Cerambycidae) and *Pinus densiflora* Sieb. & Zucc. *Nematology* 10, 69–78.

Kanzaki, N. and Futai, K. (2002a) A PCR primer set for determination of phylogenetic relationships of *Bursaphelenchus* species within the *xylophilus* group. *Nematology* 4, 35–41.

Kanzaki, N. and Futai, K. (2002b) Phylogenetic analysis of the phoretic association between *Bursaphelenchus conicaudatus* (Nematoda: Aphelenchoididae) and *Psacothea hilaris* (Coleoptera: Cerambycidae). *Nematology* 4, 759–771.

Kanzaki, N. and Futai, K. (2006) Is *Bursaphelenchus mucronatus* a weak pathogen to the Japanese red pine? *Nematology* 8, 485–489.

Kanzaki, N., Ekino, T. and Degawa, Y. (2022a) *Cryptaphelenchus abietis* n. sp. (Tylenchomorpha: Aphelenchoididae) isolated from *Cryphalus piceae* (Ratzeburg) (Coleoptera: Scolytinae) emerged from *Abies veitchii* Lindl. (Pinaceae) from Nagano, Japan. *Nematology* 24, 65–84.

Kanzaki, N., Ekino, T. and Giblin-Davis, R.M. (2019a) Feeding dimorphism in a mycophagous nematode, *Bursaphelenchus sinensis*. *Scientific Reports* 9, 13956.

Kanzaki, N., Ekino, T., Ide, T., Masuya, H. and Degawa, Y. (2018b) Three new species of parasitaphelenchids, *Parasitaphelenchus frontalis* n. sp., *P. costati* n. sp., and *Bursaphelenchus hirsutae* n. sp. (Nematoda: Aphelenchoididae), isolated from bark beetles from Japan. *Nematology* 20, 957–1005.

Kanzaki, N., Ekino, T. and Masuya, H. (2019b) *Seinura caverna* n. sp. (Tylenchomorpha: Aphelenchoididae) an androdioecious species isolated from bat guano in a calcareous cave. *Nematology* 21, 207–225.

Kanzaki, N. and Giblin-Davis, R.M. (2012) Aphelenchoidea. In: Manzanilla-López, R.H. and Marbán-Mendoza, N. (eds), *Practical Plant Nematology*. Colegio de Postgraduados and Mundi-Prensa, Biblioteca Básica de Agricultura, Jalisco, Mexico, pp. 161–208.

Kanzaki, N. and Giblin-Davis, R.M. (2016) Pine wilt and red ring, lethal plant diseases caused by insect-mediated *Bursaphelenchus* nematodes. In: Brown, J.K. (ed), *Vector-mediated Transmission of Plant Pathogens*. APS Press, St. Paul, Minnesota, 87–107.

Kanzaki, N. and Giblin-Davis, R.M. (2018) Diversity and plant pathogenicity of *Bursaphelenchus* and related nematodes in relation to their vector bionomics. *Current Forestry Reports* 4(2), 85–100.

Kanzaki, N. and Giblin-Davis, R.M. (2020) The genus *Berntsenus* Massey, 1974 is a junior synonym of *Bursaphelenchus* Fuchs, 1937. *Nematology* 22, 677–695.

Kanzaki, N., Giblin-Davis, R.M., Cardoza, Y.J., Ye, W., Raffa, K.F., *et al.* (2008b) *Bursaphelenchus rufipennis* n. sp. (Parasitaphelenchinae) and redescription of *Ektaphelenchus obtusus* (Ektaphelenchinae), associates from nematangia on the hind wings of *Dendroctonus rufipennis* (Coleoptera: Scolytidae). *Nematology* 10, 925–955.

Kanzaki, N., Giblin-Davis, R.M., Carrillo, D., Duncan, R. and Gonzalez, R. (2014a) *Bursaphelenchus penai* n. sp. (Tylenchomorpha: Aphelenchoididae), a phoretic associate of ambrosia beetles (Coleoptera: Scolytinae) from avocado in Florida. *Nematology* 16, 683–693.

Kanzaki, N., Giblin-Davis, R.M. and Center, B.J. (2009a) Description of *Ektaphelenchoides spondylis* n. sp. (Nematoda: Ektaphelenchinae) isolated from *Spondylis buprestoides* (Coleoptera: Cerambycidae) in Japan. *Nematology* 11, 181–188.

Kanzaki, N., Giblin-Davis, R.M., Herre, E.A., Scheffrahn, R.H. and Center, B.J. (2010) *Pseudaphelenchus vindai* n. sp. (Tylenchomorpha: Aphelenchoididae) associated with termites (Termitidae) in Barro Colorado Island, Panama. *Nematology* 11, 869–881.

Kanzaki, N., Giblin-Davis, R.M., Scheffrahn, R.H., Center, B.J. and Davies, K.A. (2009b) *Pseudaphelenchus yukiae* n. gen., n. sp. (Tylenchina: Aphelenchoididae) associated with *Cylindrotermes macrognathus* (Termitidae: Termitinae) in La Selva, Costa Rica. *Nematology* 11, 869–881.

Kanzaki, N. and Hamaguchi, K. (2020) *Lenisaphelenchus ulomae* n. gen., n. sp. (Rhabditida: Aphelenchoididae) isolated from the body cavity of *Uloma marseuli* Nakane (Coleoptera: Tenebrionidae) from Japan. *Nematology* 22, 961–974.

Kanzaki, N., Hamaguchi, K. and Takeuchi-Kaneko, Y. (2022b) *Devibursaphelenchus alienae* n. sp. isolated from the bark of dead *Quercus aliena* in Japan. *Nematology* 24, 205–224.

Kanzaki, N., Li, H.-F., Lan, Y.-C. and Giblin-Davis, R.M. (2014b) Description of two *Pseudaphelenchus* species (Tylenchomorpha: Aphelenchoididae) associated with Asian termites and proposal of new subfamily Tylaphelenchinae n. subfam. *Nematology* 16, 963–978.

Kanzaki, N., Maehara, N., Aikawa, T., Giblin-Davis, R.M. and Center, B.J. (2009c) The first report of a putative "entomoparasitic adult form" of *Bursaphelenchus*. *Journal of Parasitology* 95, 113–119.

Kanzaki, N., Maehara, N., Aikawa, T., Masuya, H. and Giblin-Davis, R.M. (2011) Description of *Bursaphelenchus kiyoharai* n. sp. (Tylenchina: Aphelenchoididae) with remarks on the taxonomic framework of the Parasitaphelenchinae Rühm, 1956 and Aphelenchoidinae Fuchs, 1937. *Nematology* 13. 787–804.

Kanzaki, N., Maehara, N., Aikawa, T. and Nakamura, K. (2013) An entomoparasitic adult form in *Bursaphelenchus doui* (Nematoda: Tylenchomorpha) associated with *Acalolepta fraudatrix*. *Journal of Parasitology* 99(5), 803–815.

Kanzaki, N., Shokoohi, E., Fourie, H., Swart, A., Muller, L., *et al.* (2018a) On the morphology and phylogeny of *Robustodorus* Andrássy, 2007 and two '*Aphelenchoides*' species (Nematoda: Aphelenchoidinae). *Nematology* 20, 601–615.

Kanzaki, N. and Tanaka, R. (2025). *Aphelenchoides epiphyticus* n. sp. (Tylenchomorpha Aphelenchoides) isolated from figs of Ficus septica in Japan. *Nematology* 27, 697–715.

Karim, N., Jones, J.T., Okada, H. and Kikuchi, T. (2009) Analysis of expressed sequence tags and identification of genes encoding cell-wall-degrading enzymes from the fungivorous nematode *Aphelenchus avenae*. *BMC Genomics* 10, 525.

Kikuchi, T., Aikawa, T., Kosaka, H., Pritchard, L., Ogura, N., *et al.* (2007) Expressed sequence tag (EST) analysis of the pine wood nematode *Bursaphelenchus xylophilus* and *B. mucronatus*. *Molecular and Biochemical Parasitology* 155, 9–17.

Kikuchi, T., Aikawa, T., Oeda, Y., Karim, N. and Kanzaki, N. (2009) A rapid and precise diagnostic method for detecting the pinewood nematode *Bursaphelenchus xylophilus* by loop-mediated isothermal amplification. *Phytopathology* 99, 1365–1369.

Kikuchi, T., Cotton, J.A., Dalzell, J.J., Hasegawa, K., Kanzaki, N., *et al.* (2011) Genomic insights into the origin of parasitism in the emerging plant pathogen *Bursaphelenchus xylophilus*. *PLoS Pathogens* 7, e1002219.

Kobayashi, F., Yamava, A. and Ikeda, T. (1984) The Japanese pine sawyer beetle as the vector of pine wilt. *Annual Review of Entomology* 29(1), 115–135.

Krusberg, L.R. (1961) Studies on the culturing and parasitism of plant-parasitic nematodes, in particular *Ditylenchus dipsaci* and *Aphelenchoides ritzemabosi* on alfalfa tissues. *Nematologica* 6, 181–200.

Lai, C.-K., Lee, Y.-C., Ke, H.-M., Lu, M.-R., Liu, W.-A., *et al.* (2023) The *Aphelenchoides* genomes reveal substantial horizontal gene transfers in the last common ancestor of free-living and major plant-parasitic nematodes. *Molecular Ecology Resources* 23, 905–919.

Laumond, C. and Carle, P. (1971) Nématodes associés et parasites de *Blastophagus destruens* Woll. (Col. Scolytidae). *Entomophaga* 16, 51–66.

Leal, I., Green, M., Allen, E., Humble, L. and Rott, M. (2007) Application of a real-time PCR method for the detection of pine wood nematode, *Bursaphelenchus xylophilus*, in wood samples from lodgepole pine. *Nematology* 9, 351–362.

Lieutier, F. and Laumond, C. (1978) Nématodes parasites et associés à *Ips sexdentatus* et *Ips typographus* (Coleoptera, Scolytidae) en région parisienne. *Nematologica* 24, 184–200.

Mamiya, Y. (1984) The pine wood nematode. In: Nickle, W.R. (ed), *Plant and Insect Nematodes.* Marcel Dekker, New York, New York, pp. 589–625.
Mamiya, Y. (1987) Origin of the pine wood nematode and its distribution outside the United States. In: Wingfield, M.J. (ed), *Pathogenicity of the Pine Wood Nematode.* APS Press, St. Paul, Minnesota, pp. 59–65.
Massey, C.L. (1971) *Omemeea maxbassiensis* n. gen., n. sp. (Nematoda: Aphelenchoididae) from galleries of the bark beetle *Lepersinus californicus* Sw. (Coleoptera: Scolytidae) in North Dakota. *Journal of Nematology* 3, 289–291.
Massey, C.L. (1974) *Biology and Taxonomy of Nematode Parasites and Associates of Bark Beetles in the United States.* Agriculture Handbook No. 446, USDA Forest Service, Washington, D.C.
Massey, C.L. and Hinds, T.E. (1970) Nematodes from aspen cankers in Colorado and New Mexico. *Canadian Journal of Zoology* 48, 97–108.
Matsunaga, K. and Togashi, K. (2004) A simple method for discriminating *Bursaphelenchus xylophilus* and *B. mucronatus* by species-specific polymerase chain reaction primer pairs. *Nematology* 6, 273–277.
McLeod, R.W. (1967) *Aphelenchoides bicaudatus*, a parasite of cultivated mushroom. *Nature* 214, 1163–1164.
Meldal, B.H.M., Debenham, N.J., De Ley, P., De Ley, I.T., Vanfleteren, J.R., *et al.* (2007) An improved molecular phylogeny of the Nematoda with special emphasis on marine taxa. *Molecular Phylogenetics and Evolution* 42, 622–636.
Micoletzky, H. (1922) Die freilebenden Erd- Nematoden mit besonderer Berucksichtigung der Steiermark und der Bukowina, zugleich mit einer Revision sämtlicher nicht mariner, freilebender Nematoden in Form von Genus - Beschreibunger und Bestimmungsschlüsseln. *Archiv für Naturgeschichte, Berlin* 87(1921), 1–650.
Miraeiz, E., Heydari, R., Adeldoost, Y. and Ye, W. (2017) Description of *Ektaphelenchus oleae* n. sp. (Rhabditida: Seinurinae) from Iran. *Nematology* 19, 1123–1134.
Mota, M.M., Braasch, H., Bravo, M.A., Penas, A.C., Burgermeister, W., *et al.* (1999) First report of *Bursaphelenchus xylophilus* in Portugal and in Europe. *Nematology* 7, 727–734.
Nickle, W.R. (1970a) Description of Entaphelenchidae fam. n., *Roveaphelenchus jonesi* gen. n., sp. n., and *Sheraphelenchus entomophagus* gen. n., sp. n. (Nematoda: Aphelenchoidea). *Proceedings of the Helminthological Society of Washington* 37, 105–109.
Nickle, W.R. (1970b) A taxonomic review of the genera of the Aphelenchoidea (Fuchs, 1937) Thorne, 1949 (Nematoda: Tylenchida). *Journal of Nematology* 2, 375–392.
Nickle, W.R., Golden, A.M., Mamiya, Y. and Wergin, W.P. (1981) On the taxonomy and morphology of the pine wood nematode, *Bursaphelenchus xylophilus* (Steiner & Buhrer 1934) Nickle 1971. *Journal of Nematology* 13, 385–392.
Paramonov, A.A. (1964) [*Fundamentals of Phytohelminthology. Vol. II. Taxonomy of Phytonematodes.*], Nauka, Moscow, USSR.
Pearse, A.S. (1942) *An Introduction to Parasitology.* Charles C. Thomas, Springfield, Illinois.
Pedram, M., Kanzaki, N., Giblin-Davis, R.M. and Pourjam, E. (2018) A molecular phylogenetic approach for unravelling the taxonomic status of *Basilaphelenchus persicus* n. gen., n. sp. (Aphelenchoididae: Tylaphelenchinae). *Nematology* 20, 567–582.
Poinar Jr, G.O. (1969) *Praecocilenchus rhaphidophorus* n. gen., n. sp. (Nematoda: Aphelenchoidea) parasitizing *Rhynchophorus bilineatus* (Montrouzier) (Coleoptera: Curculionidae) in New Britain. *Journal of Nematology* 1, 227–231.
Poinar Jr, G.O. (1975) *Entomogenous Nematodes.* Brill, Leiden, Netherlands.
Poinar Jr, G.O. (1977) Fossil nematodes from Mexican amber. *Nematologica* 23, 232–238.
Poinar Jr, G.O. (2011) *The Evolutionary History of Nematodes: as Revealed in Stone, Amber and Mummies. Nematology Monographs and Perspectives, 9.* Brill, Leiden, Netherlands.
Ragsdale, E.J. and Baldwin, J.G. (2010) Resolving phylogenetic incongruence to articulate homology and phenotypic evolution: a case study from Nematoda. *Proceedings of the Royal Society of London B* 277, 1299–1307.
Ragsdale, E.J., Crum, J., Ellisman, M.H. and Baldwin, J.G. (2008) Three-dimensional reconstruction of the stomatostylet and anterior epidermis in the nematode *Aphelenchus avenae* (Nematoda: Aphelenchidae) with implications for the evolution of plant parasitism. *Journal of Morphology* 269, 1181–1196.
Ragsdale, E.J., Ngo, P.T., Crum, J., Ellisman, M.H. and Baldwin, J.G. (2009) Comparative, three-dimensional anterior sensory reconstruction of *Aphelenchus avenae* (Nematoda: Tylenchomorpha). *Journal of Comparative Neurology* 517, 616–632.
Ragsdale, E.J., Ngo, P.T., Crum, J., Ellisman, M.H. and Baldwin, J.G. (2011) Reconstruction of the pharyngeal corpus of *Aphelenchus avenae* (Nematoda: Tylenchomorpha), with implications for phylogenetic congruence. *Zoological Journal of the Linnean Society* 161, 1–30.

Remillet, M. and Silvain, J.F. (1988) *Noctuidonema guyanense* n. g., n. sp. (Nematoda: Aphelenchoididae) ectoparasite of noctuids of the genus *Spodoptera* (Lepidoptera: Noctuidae). *Revue de Nématologie* 11, 21–24.

Rühm, W. (1956) Die Nematoden der Ipiden. *Parasitologische SchriftenReihe* 6, 1–435.

Rutherford, T.A. and Webster, J.M. (1987) Distribution of pine wilt disease with respect to temperature in North America, Japan, and Europe. *Canadian Journal of Forest Research* 17, 1050–1059.

Ryss, A., Vieira, P., Mota, M.M. and Kulkinich, O. (2005) A synopsis of the genus *Bursaphelenchus* Fuchs, 1937 (Aphelenchida: Parasitaphelenchidae) with keys to species. *Nematology* 7, 393–458.

Ryss, A.Y., McClure, M.A., Nischwitz, C., Dhiman, C. and Subbotin, S.A. (2013) Redescription of *Robustodorus megadorus* with molecular characterization and analysis of its phylogenetic position within the family Aphelenchoididae. *Journal of Nematology* 45, 237–252.

Ryss, A.Y., Polyanina, K., Popovichev, B.G. and Subbotin, S.A. (2015) Description of *Bursaphelenchus ulmophilus* sp. n. (Nematoda: Parasitaphelenchinae) associated with Dutch elm disease of *Ulmus glabra* Huds. in the Russian North West. *Nematology* 17, 685–703.

Ryss, A.Y. and Subbotin, S.A. (2017) [Coevolution of wood-inhabiting nematodes of the genus *Bursaphelenchus* Fuchs, 1937 with their insect vectors and plant hosts.] *Zhurnal Obshchei Biologii* 78, 32–61.

Scognamiglio, A., Talamé, M. and s'Jacob, J.J. (1970) *Aprutides martuccii* (Nematoda: Aphelenchoididae) n. g., n. sp. *Bolletino del Laboratorio di Entomologia Agraria 'Filippo Silvestri'* 28, 3–11.

Shahina, F. (1996) A diagnostic compendium of the genus *Aphelenchoides* Fischer, 1894 (Nematoda: Aphelenchida) with some new records of the group from Pakistan. *Pakistan Journal of Nematology* 14, 1–32.

Shinya, R., Sun, S., Dayi, M., Tsai, I.J., Miyama, A., *et al.* (2022) Possible stochastic sex determination in *Bursaphelenchus* nematodes. *Nature Communications* 13, 2574.

Siddiqi, M.R. (1974) *Aphelenchoides ritzemabosi. CIH Descriptions of Plant-parasitic Nematodes* Set 3, No. 32. Commonwealth Agricultural Bureaux, Farnham Royal, UK.

Siddiqi, M.R. (1975) *Aphelenchoides fragariae. CIH Descriptions of Plant-parasitic Nematodes* Set 5, No. 74. Commonwealth Agricultural Bureaux, Farnham Royal, UK.

Siddiqi, M.R. (1976) *Aphelenchoides bicaudatus. CIH Descriptions of Plant-Parasitic nematodes* Set 6, No. 84. Commonwealth Agricultural Bureaux, Farnham Royal, UK.

Siddiqi, M.R. (2000) *Tylenchida Parasites of Plants and Insects*, 2nd edn. CAB International, Wallingford, UK.

Skarbilovich, T.S. (1947) [Revision of the systematics of the nematode family Anguillulinidae Baylis and Daubney, 1926.] *Doklady Akademii Nauk SSR* 57, 307–308.

Skarmoutsos, G. and Michalopoulos-Skarmoutsos, H. (2000) Pathogenicity of *Bursaphelenchus sexdentati, Bursaphelenchus leoni* and *Bursaphelenchus hellenicus* on European pine seedlings. *Forest Pathology* 30, 149–156.

Sousa, E., Bravo, M.A., Pires, J., Naves, P., Penas, A.C., *et al.* (2001) *Bursaphelenchus xylophilus* (Nematoda; Aphelenchoididae) associated with *Monochamus galloprovincialis* (Coleoptera; Cerambycidae) in Portugal. *Nematology* 3, 89–91.

Steiner, G. (1943) New nematodes associated with a disease of the papaya in Chile. *Boletín del Departemento de Sanidad Vegetal, Santiago de Chile* 3, 95–116.

Subbotin, S.A., Oliveira, C.J., Álvarez-Ortega, S., Desaeger, J.A., Crow, W., *et al.* (2020) The taxonomic status of *Aphelenchoides besseyi* Christie, 1942 (Nematoda: Aphelenchoididae) populations from the southeastern USA, and description of *Aphelenchoides pseudobesseyi* sp. n. *Nematology* 23, 381–413. Doi: 10.1163/15685411-bja10048

Takeuchi, Y. and Futai, K. (2009) Diagnosis and quantification of the pine wood nematode, *Bursaphelenchus xylophilus* (Steiner & Buhrer), in wood of *Pinus thunbergii* with real-time PCR. *Nematological Research* 39, 9–16.

Takeuchi, Y., Kanzaki, N. and Futai, K. (2005) A nested PCR-based method for detecting the pine wood nematode, *Bursaphelenchus xylophilus*, from pine wood. *Nematology* 7, 775–782.

The *C. elegans* Sequencing Consortium. (1998) Genome sequence of the nematode *C. elegans*: a platform for investigating biology. *Science* 282, 2012–2018.

Thorne, G. (1949) On the classification of the *Tylenchida*, new order (Nematoda, Phasmidia). *Proceedings of the Helminthological Society of Washington* 16, 37–73.

Timm, R.W. (1960) *Paraseinura* (Nematoda: Aphelenchoididae), a new genus from East Pakistan. *Nematologica* 5, 171–174.

Timm, R.W. and Franklin, M.T. (1969) Two marine species of *Aphelenchoides*. *Nematologica* 15, 370–375.

Tomalak, M. and Filipiak, A. (2010) Description of *Bursaphelenchus populi* sp. n. (Nematoda: Parasitaphelenchidae), a new member of the *xylophilus* group from aspen, *Populus tremula* L., in Europe. *Nematology* 12, 399–416.

Vuong, H.H. and Rodriguez, H. (1970) Lutte contre les nématodes du riz à Madagascar (résultats d'expérimentation 1968–1969). *Agronomie Tropicale* 25, 52–66.

Wachek, F. (1955) Die entoparasitischen Tylenchiden. *Parasitologische SchriftenReihe* 3, 1–119.

Webster, J.M., Anderson, R.V., Baillie, D.L., Beckenbach, K., Curran, J., *et al.* (1990) DNA probes for differentiating isolates of the pinewood nematode species complex. *Revue de Nématologie* 13, 255–263.

Wood, F.H. (1975) Biology of *Seinura demani* (Nematoda: Aphelenchoididae). *Nematologica* 21, 49–64.

Yasu, M. (1968) Disinfection of rice seeds. *Japan Agricultural Research Quarterly* 3, 6–9.

Ye, W. and Giblin-Davis. R.M. (2013) Molecular characterization and real-time PCR assay for pine wood nematode *Bursaphelenchus xylophilus* (Nematoda: Parasitaphelenchidae). *PLoS One* 8(11), e78804:1–14.

Ye, W., Giblin-Davis, R.M., Braasch, H., Morris, K. and Thomas, W.K. (2007) Phylogenetic relationships among *Bursaphelenchus* species (Nematoda: Parasitaphelenchidae) inferred from nuclear ribosomal and mitochondrial DNA sequence data. *Molecular Phylogenetics and Evolution* 43, 1185–1197.

Yeates, G.W., Bongers, T., de Goede, R.G., Freckman, D.W. and Georgieva, S. (1993) Feeding habits in soil nematode families and genera - an outline for soil ecologists. *Journal of Nematology* 25, 315–331.

Yoshida, K., Hasegawa, K., Mochiji, N. and Miwa, J. (2009) Early embryogenesis and anterior-posterior axis formation in the white-tip nematode *Aphelenchoides besseyi* (Nematoda: Aphelenchoididae). *Journal of Nematology* 41, 17–22.

Zhao, Z.Q., Davies, K.A., Riley, I.T. and Nobbs, J.M. (2007) *Laimaphelenchus heidelbergi* sp. nov. (Nematoda: Aphelenchina) from Victoria, Australia, and emendment of the diagnosis of the genus. *Transactions of the Royal Society of South Australia* 131, 182–191.

Zhao, Z.Q, Ye, W., Giblin-Davis, R.M., Li, D., Thomas, W.K., Davies, K.A. and Riley, I. T. (2008) Morphological and molecular analysis of six aphelenchoidoids from Australian conifers and their relationship to *Bursaphelenchus* (Fuchs, 1937). *Nematology* 10, 663–678.

Zhuo, K., Cui, R., Ye, W., Luo, M., Wang, H., *et al.* (2010) Morphological and molecular characterization of *Aphelenchoides fujianensis* n. sp. (Nematoda: Aphelenchoididae) from *Pinus massoniana* in China. *Zootaxa* 2509, 39–52.

8 Tylenchidae and Dolichodoridae

Wim Bert[1]* and David J. Hunt[2]

[1]*Nematology Research Unit, Department of Biology, Ghent University, Ghent, Belgium;* [2]*CABI, Silwood Park, Buckhurst Road, Ascot, Berkshire, UK*

Abstract

An overview of the bionomics, morphology and diagnosis, together with taxonomic keys to the major groups and genera of the families Tylenchidae and Dolichodoridae is provided. The morphology of the major groups and genera is discussed and illustrated in detail and hints provided as to how to approach identification of the taxa. In the family Tylenchidae, the genera treated in greater detail include *Aglenchus, Basiria, Boleodorus, Coslenchus, Filenchus* and *Tylenchus*, whilst in the family Dolichodoridae, the major genera considered include *Belonolaimus, Dolichodorus, Merlinius, Pratylenchoides, Quinisulcius, Trophurus*, and *Tylenchorhynchus*. Each genus is diagnosed and pertinent information provided as to distribution, host range, economic importance, etc.

8.1. Introduction

Members of the Tylenchidae Örley, 1880 and Dolichodoridae Chitwood *in* Chitwood & Chitwood, 1950, both being families within the Tylenchoidea Örley, 1880, are usually abundant in most soils, including those cultivated for agricultural purposes. The nematodes themselves are usually quite distinctive, even under the low magnification of a stereomicroscope, although attribution to genera and species is often more problematic, even for a specialist.

Tylenchidae are easily characterized by their relatively weak labial skeleton, moderate stylet, non-overlapping base to the pharynx, posterior vulva, monoprodelphic female genital tract, adanal male bursa and elongate, often filiform, tail. Exceptionally, some genera/species have a more robust stylet and this in turn is associated with a stronger labial skeleton, a functional adaptation necessary for attachment and operation of the stronger protractor muscles that are required for operation of a stronger and longer stylet. There are many nominal genera of Tylenchidae, although relatively few can be regarded as true phytoparasites – most are root-hair feeders or epidermal root browsers while some are fungal feeders.

Dolichodoridae are also easily characterized, typically being more robust than Tylenchidae with

*E-mail: wim.bert@ugent.be

A revision of Hunt, D.J., Bert, W. and Siddiqi, M.R. (2012) Tylenchidae and Dolichodoridae. In: Manzanilla-López, R.H. and Marbán-Mendoza, N. (eds), *Practical Plant Nematology*. Mundi-Prensa and Biblioteca Básica de Agricultura, Montecillo, Mexico.

(eds R.H. Manzanilla-López, D.J. Hunt, N. Marbán-Mendoza and S. Infante Gil)
DOI: 10.1079/9781836990413.0008

a stronger labial skeleton, stronger stylet, more powerful pharynx but still with a non-overlapping junction with the intestine (exceptionally overlapping), median vulva, almost invariably with amphididelphic female genital tracts (exceptionally monoprodelphic), male bursa extending to or beyond the tail tip and medium conoid to cylindroid tail, often with a rounded tip of one form or another. Members of this family are migratory ectoparasites of plant roots, usually feeding on only the outer layers of the cortex although more specialist genera, such as *Belonolaimus* and *Dolichodorus*, may feed much deeper within the root tissue, often penetrating to the stele with their exceptionally long stylet. The genera within this group are diverse, systematic schemes often varying widely depending on taxonomic viewpoint and the importance attached to such features as the number of lines in the lateral fields. Whether this latter character is used for generic status or not, it is an important feature that must be ascertained in order to facilitate accurate species determination.

8.2. Tylenchidae

Tylenchidae are cosmopolitan and inhabit various terrestrial biotopes. Soil or moss samples usually contain large numbers of taxa belonging to this family (Siddiqi, 2000) and they are also known from limnic habitats (Bert and Borgonie, 2006). Despite their high diversity and numeric importance, their feeding habits are poorly known. Generally, they are considered to be associates of algae, mosses, lichens and plant roots (Siddiqi, 2000) and some are able to feed on fungi (Okada *et al.*, 2005). Although certain Tylenchidae are considered weak parasites of plants (Siddiqi, 2000), they do not cause economic loss. When identifying soil nematodes, members of the Tylenchidae, although often abundant, are all too frequently dismissed as '*Tylenchus, sensu lato*', thereby overlooking the enormous range of morphological diversity encompassed by what is seen as a relatively inaccessible group. Furthermore, although Tylenchidae may account for up to 30-40% of all nematodes collected from a study site (Yeates and Bird, 1994; Qing and Bert, 2019), the family is often left out in nematode ecology analysis because of ignorance of the correct feeding type or its presumed association with plants.

Tylenchidae fall within the Tylenchomorpha and are characterized by supposedly ancestral morphological characters, such as a filiform tail, a small and thin stylet and other weakly developed structures that are associated with feeding. Limited molecular data appear to support their early divergence within the Tylenchomorpha. However, their phylogenetic position is relatively poorly resolved which may be explained by their complex evolutionary history, including their polyphyletic nature (Qing and Bert, 2019).

The Tylenchidae family is highly diverse, comprising 44 genera and 412 species according to Qing and Bert (2019). Since then, approximately one new genus and 23 new species have been described (as of May 2024). However, diversity estimations range from 2000 to 10,000 species, indicating that 75-95% of the species remain undiscovered (Qing and Bert, 2019). Notably, more than a third of the genera are monospecific and numerous species have never been recorded since their first description. Tylenchidae are superficially morphologically minimalistic; species are often differentiated based on differences in measurements and without sufficient consideration for natural variation. Thus, diagnostic resolution is often low and many species are not well described according to present-day standards. Furthermore, many genera, especially the large constipated taxa such as *Filenchus*, are long overdue for revision. All this makes the identification of Tylenchidae an especially difficult task.

Geraert (2008) provided a comprehensive compendium of the extant species descriptions of Tylenchidae around the world, although without indulging in a taxonomic review. The book provides exhaustive keys to genus and species levels. Descriptions and morphometrics are given for more than 400 Tylenchidae species, the details of many of which are otherwise especially difficult to acquire.

The most widespread genera within the Tylenchidae, which often also comprise the most species, include *Aglenchus*, *Coslenchus*, *Basiria*, *Boleodorus*, *Filenchus* and *Tylenchus*. These six genera will be further illustrated in this chapter. The description of the genera is mainly based on Siddiqi (2000), Andrássy (2007) and Geraert (2008). For their taxonomic status, that is, generic synonyms and number of species within the genera, Geraert (2008) is followed. However,

Psilenchus and related genera are the subject of long-standing discussions concerning their position, whether within the Tylenchidae (e.g., Geraert, 2008) or in a more derived position (e.g., Siddiqi, 2000). *Psilenchus* combines morphological features of early diverging (e.g., weak stylet, elongated tail, etc.) and more derived plant parasites (e.g., phasmids on the tail and didelphic genital system). Molecular phylogenetic analyses separate *Psilenchus* and several other genera from other Tylenchidae (Bert *et al.*, 2008; Holterman *et al.*, 2009; Qing and Bert, 2019). Here we regard the didelphic 'Tylenchidae' genera (*Psilenchus* and *Atetylenchus*) as *incertae sedis* and, not being of economic importance and mostly rather rare in occurrence, do not discuss them further.

8.2.1. Family Tylenchidae Örley, 1880

(Figs 8.1-8.3)

Diagnosis (after Siddiqi, 2000): Small to medium nematodes (L = 0.3-1.3 mm), slender and vermiform, usually under 1 mm long. Cuticle finely or heavily annulated, rarely smooth. Lateral field mostly with two, three or four incisures. **Deirids normally present. Phasmid-like structures, if present, usually advulval.** Amphids vary from small oblique slits to long sinuous clefts longitudinal directed. Labial region generally elevated, rounded and annulated. Labial framework weak, **stylet usually weak and slender (6-21 µm**, few exceptions). Basal knobs or swellings small, rounded or absent; protractors divergent, not parallel to stylet.

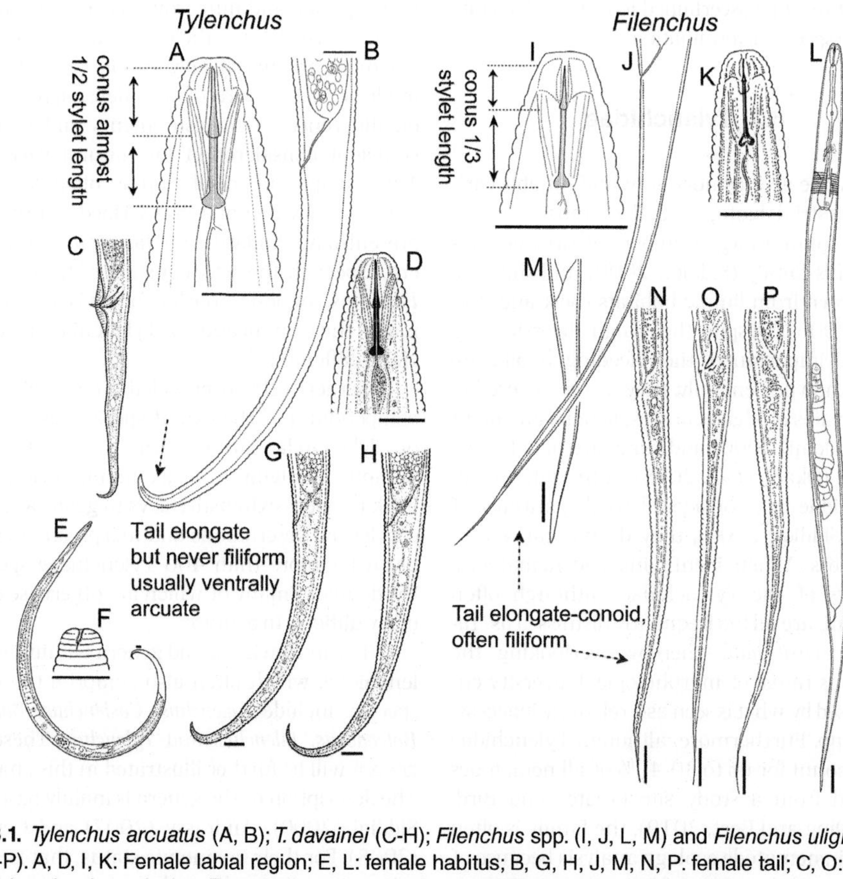

Fig. 8.1. *Tylenchus arcuatus* (A, B); *T. davainei* (C-H); *Filenchus* spp. (I, J, L, M) and *Filenchus uliginosus* (K, N-P). A, D, I, K: Female labial region; E, L: female habitus; B, G, H, J, M, N, P: female tail; C, O: male tail region showing spicules; F: cephalic region showing amphidial aperture. (Scale bars = 10 µm.) (A-H adapted from Andrássy (1977); K, N-P adapted from Siddiqi (2000), courtesy CABI.)

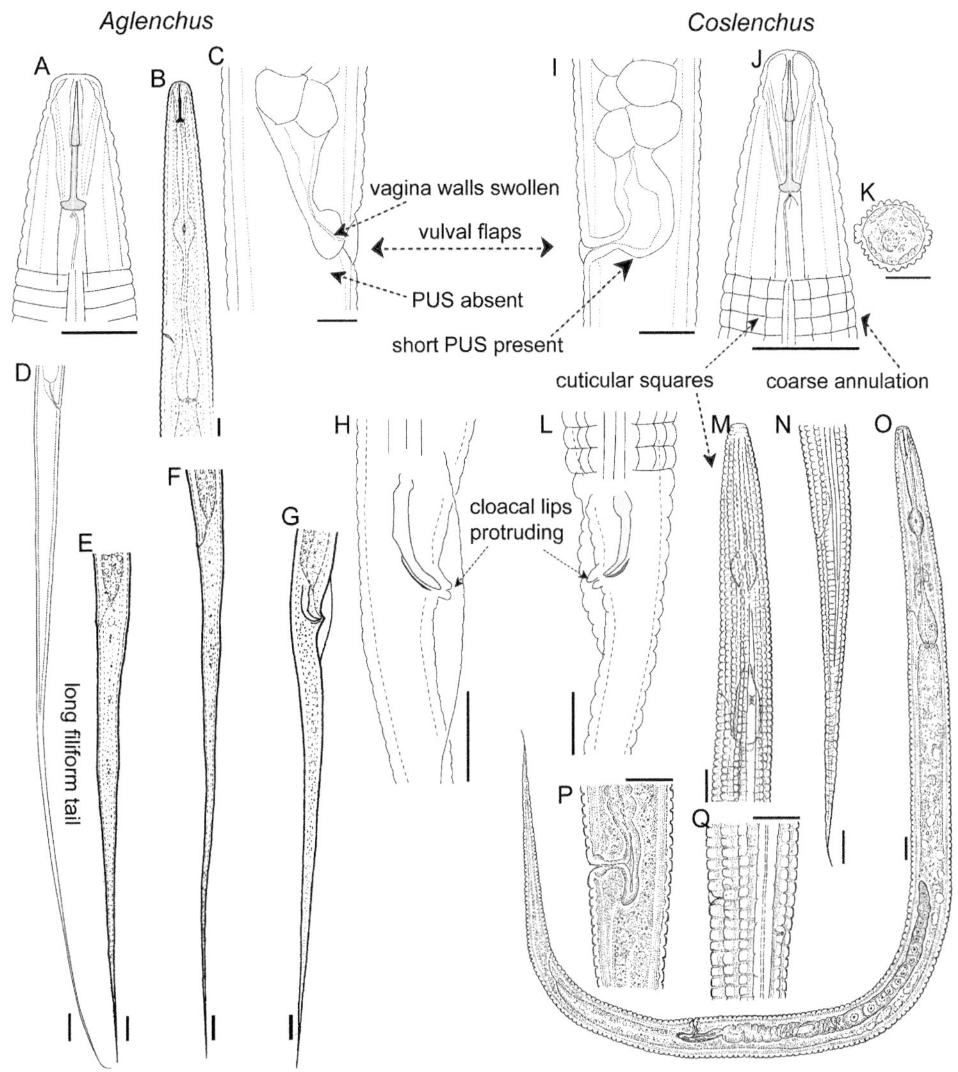

Fig. 8.2. *Aglenchus agricola* (A-H); *Coslenchus costatus* (I, J, L); *C. turkeyensis* (M-Q) and *C. multigyrus* (K). A, J: Female labial region; D, E, F, N: female tail; B, M: female pharynx; C, I, P, Q: female vulval region; G: male tail region showing spicules; H, L: male cloacal region; K: cross-section of body near middle; O: female habitus. (Scale bars = 10 μm.) (B, E-G: adapted from Andrássy (1980), courtesy *Acta Zoologica Hungarica*; K, M-Q: adapted from Siddiqi (1981), courtesy *Nematologica*.)

Pharynx divided into slender procorpus, elliptical metacorpus usually with valves, long slender isthmus followed by symmetrical pyriform glandular region offset from intestine (basal bulb). Female genital system prodelphic. **Columned uterus with four rows of cells.** Post-uterine sac typically less than one vulval body diameter long. Testis outstretched, tip rarely reflexed. **Bursa adanal, simple**, not enveloping tail, rarely absent. Sperm cells with little cytoplasm. Tail elongate-conoid to filiform, similar in both sexes.

Comments: The important characteristics of this group include: (i) weak labial skeleton; (ii) weak stylet and basal knobs; (iii) basal pharyngeal bulb not overlapping intestine; (iv) female

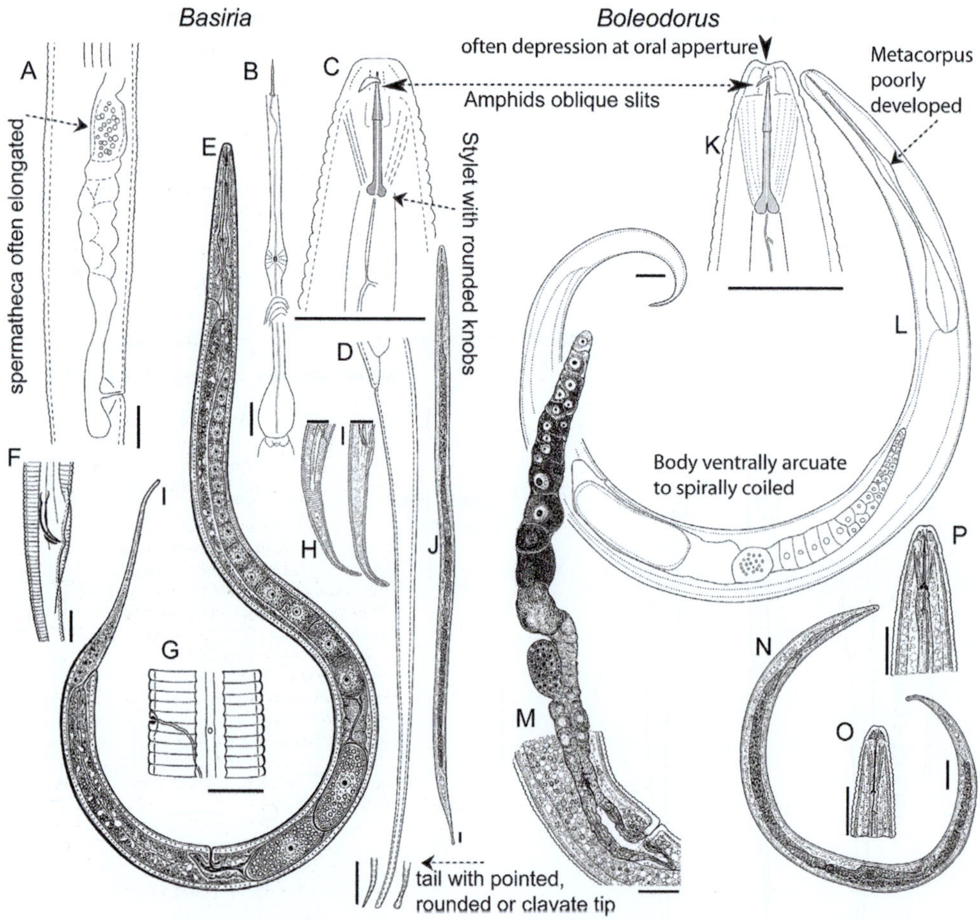

Fig. 8.3. *Basiria* spp. (A-D); *B. graminophila* (E-I); *B. citri* (J); *Boleodorus* sp. (K, L) and *B. thylactus* (M-P). A, M: Female gonoduct region; B: female pharynx; C, K, P, O: female labial region; E, J, L, N: female habitus; F: male cloacal region; G: deirid and excretory pore; H, D: female tail; I: male tail region showing spicules. (Scale bars = 10 µm.) (E-J, M-P: adapted from Siddiqi (2000), courtesy CABI.)

genital tract monodelphic, prodelphic; (v) 'phasmids', if visible, present near vulva, not usually within lateral field; (vi) tail elongate-filiform in both sexes and (vii) male bursa adanal.

The body is generally vermiform (worm-like), with the posterior portion slender and tapering sharply to pointed or rounded terminus. The vulva is positioned in the posterior section of the body, although due to the attenuated tail of some genera, the vulval ratio can be greatly affected by variations in development of, or damage to, the filamentous portion. This is resolved by using V' (dividing the length from labial region to vulva by the length from labial region to anus × 100). On death, the body can be slightly curved ventrally (*Discotylenchus*) to open C shaped (*Tylenchus*).

The pharynx is very variable in shape. The metacorpus can be either valvate or non-valvate, muscular or non-muscular, with or without median bulb. There is usually a long isthmus and basal bulb. The basal bulb can overlap the intestine, but the gland nuclei remain within the bulb. The labial region shape is variable and can be used to distinguish genera and species. The stylet is usually delicate and short (the average for the family is about 13 µm).

The genital tract is monodelphic, prodelphic with a post-uterine sac of variable length usually being present. The spermatheca can be offset or axial. The length of the post-uterine sac is important in identifying genera and species. The area around the vulva is also important (the presence or absence of cuticular flaps, angle at which the vagina exits via the vulva). The 'phasmid' is located post-median (in posterior section of the body or just posterior to the vulva) and not always within the lateral field.

The female tail shape varies between genera but is usually attenuated and tapering to a pointed terminus. The male tail is similar in shape to that of the female, except for the presence of spicules, gubernaculum and bursa. The spicules are tylenchid in shape, with a simple gubernaculum and adanal bursa. The presence of males in populations is dependent on species – some are parthenogenetic and others amphimictic.

Bionomics: Associates of algae, mosses, lichens, fungi, root hairs and plant roots, but generally not root parasites of any significance. The allocation of feeding habits is largely unknown but most likely genus- or even species-specific.

8.2.2. Systematic Scheme (commonly encountered genera are in bold)

Tylenchoidea Örley, 1880
- Tylenchidae Örley, 1880
  - Tylenchinae Örley, 1880
    - *Allotylenchus* Andrássy, 1984
    - *Cervoannulatus* Bajaj, 1998
    - *Cucullitylenchus* Huang & Raski, 1986
    - *Discotylenchus* Siddiqi, 1980
    - ***Filenchus*** Andrássy, 1954
    - *Fraglenchus* Siddiqi, 2000
    - *Gracilancea* Siddiqi, 1976
    - *Irantylenchus* Kheiri, 1972
    - *Labrys* Qing & Bert, 2018
    - *Malenchus* Andrássy, 1968
    - *Miculenchus* Andrássy, 1959
    - *Polenchus* Andrássy, 1980
    - *Sakia* Khan, 1964
    - *Silenchus* Andrássy, 2001
    - *Tanzanius* Siddiqi, 1991
    - ***Tylenchus*** Bastian, 1865
  - Atylenchinae Skarbilovich, 1959
    - ***Aglenchus*** Andrássy, 1954
    - *Atylenchus* Cobb, 1913
    - ***Coslenchus*** Siddiqi, 1978
    - *Pleurotylenchus* Szczygiel, 1969
  - Boleodorinae Khan, 1964
    - ***Basiria*** Siddiqi, 1959
    - ***Boleodorus*** Thorne, 1941
    - *Discopersicus* Yaghoubi, Pourjam, Alvarez-Ortega, Liebanas, Atighi & Pedram, 2016
    - *Neopsilenchus* Thorne & Malek, 1968
    - *Neothada* Khan, 1973
    - *Ridgellus* Siddiqi, 2000
    - *Thada* Thorne, 1941
  - Ecphyadophorinae Skarbilovich, 1959
    - *Chilenchus* Siddiqi, 2000
    - *Ecphyadophora* de Man, 1921
    - *Ecphyadophoroides* Corbett, 1964

Epicharinema Raski, Maggenti, Koshy & Sosamma, 1980
Lelenchus Andrássy, 1954
Mitranema Siddiqi, 1986
Sigmolenchus Gharakhani, Pourjam, Abolafia, Castillo & Pedram, 2020
Tenunemellus Siddiqi, 1986
Tremonema Siddiqi, 1994
Ultratenella Siddiqi, 1994
Tylodorinae Paramonov, 1967
Arboritynchus Reay, 1991
Campbellenchus Wouts, 1978
Cephalenchus Goodey, 1962
Eutylenchus Cobb, 1913
Tylodorus Meagher, 1964

8.2.3. Key to Genera of Tylenchidae, Based on Geraert (2008).

(Commonly encountered genera are in **bold**; not all genera included)

1. Stylet less than 22 µm long 2
Stylet 22-104 µm long .... *Epicharinema, Tylodorus, Arboritynchus, Campbellenchus, Gracilancea*
2. Head provided with setae .... *Atylenchus, Eutylenchus*
Head without setae 3
3. Longitudinal ridges outside lateral fields present 4
Longitudinal ridges outside lateral fields absent 6
4. Stylet conus about one-third of stylet length .... *Ecphyadophoroides, Neothada, Ridgellus*
Stylet conus about one-half of stylet length 5
5. Vulva covered by ventral flap; stylet 17-19 µm .... *Pleurotylenchus*
Vulva with lateral flaps (sometimes small); stylet usually less than 15 µm long .... ***Coslenchus***
6. Stylet conus about one-third of stylet length 7
Stylet conus about one-half of stylet length 17
7. Head high with oblique or transverse amphidial slit located laterally 8
Head variously shaped, amphidial slit longitudinal orientated or only present anteriorly (then not seen by light microscopy) 10
8. Median pharyngeal bulb non-muscular, non-valvate; body ventrally curved; female reproductive system with offset spermatheca .... ***Boleodorus***
Median pharyngeal bulb muscular, valvate; body straight or slightly curved 9
9. Stylet with basal knobs .... ***Basiria***
Stylet without basal knobs .... *Neopsilenchus*
10. Head with dome-shaped structure or small disc-like structure anteriorly .... *Cucullitylenchus, Mitranema, Discotylenchus*
Head with smooth contour 11
11. Very slender body (a = 60-180) .... *Ecphyadophora, Tenunemellus, Ultratenella, Chilenchus*
Body not extremely slender 12
12. Cuticular annuli deeply incised .... *Thada, Miculenchus, Silenchus, Malenchus*
Cuticular annuli not deeply incised 13

13. Head dorsoventrally flattened with longitudinal sinuous amphidial apertures; large amphidial pouches; body very slender .... *Lelenchus*
Head quadrangular, amphidial apertures not sinuous; amphidial pouches not distinct 14
14. Clavate stylet knobs with ventrally situated opening of pharyngeal lumen; dorsal gland opening nearly one stylet length posterior to knobs *Irantylenchus*
Stylet with rounded knobs; dorsal gland opening close to knobs 15
15. Lateral vulval membranes present .... *Fraglenchus*
Lateral vulval membranes absent 16
16. Median bulb degenerate, no valves .... *Sakia*
Median bulb slightly or well developed, with valves .... ***Filenchus***
17. Vulva with lateral flaps 18
Vulva without lateral flaps 19
18. Lateral field composed of two protruding bands; vagina thickened; post-uterine sac absent .... ***Aglenchus***
Lateral field and vagina not as above .... *Allotylenchus, Fraglenchus, Cephalenchus*
19. Pharynx without median bulb .... *Tanzanius*
Pharynx with median bulb 20
20. Lateral field and body annulations at mid-body mostly inconspicuous .... *Cervoannulatus, Polenchus*
Lateral field and body annulations distinct 21
21. Vulva with projecting vulval lips, male with elongated bursa .... *Silenchus*
Vulva plain, male with short, adcloacal bursa .... ***Tylenchus***

Genus *Tylenchus* Bastian, 1865
= *Aerotylenchus* Fotedar & Handoo, 1979
= *Areotylenchus* Fortuner, 1984
(Fig. 8.1A-H)

Diagnosis: **Body ventrally curved**, usually 0.6-1.2 mm long. Cuticle distinctly annulated. **Lateral field with four lines**, outer bands in some species areolated. Deirids usually just posterior to excretory pore level. Labial region finely annulated, not offset. Stylet relatively robust, 12-21 µm long, **conus forming about half of stylet length**, knobs distinct. Dorsal gland orifice close to knobs. Median pharyngeal bulb well developed, muscular with pronounced thickening of lumen walls. Glandular bulb offset from intestine, usually valvate and pyriform. Female reproductive system prodelphic, spermatheca a partly offset pouch, post-uterine sac spacious, about equal in length to corresponding body diameter. Males generally as common as females, bursa adanal with crenate margins. **Tails similar in both sexes, elongate but never filiform; usually clearly ventrally arcuate**, often hooked.
Type species: *Tylenchus davainei* Bastian, 1865
Comments: Geraert (2008) lists 28 species (and 30 *species inquirendae* or *incertae sedis*). Other keys to *Tylenchus* spp. are given by Andrássy (1979) and Maqbool and Shahina (1987). In the 19th century, when *Tylenchus* was erected, every species possessing a buccal spear with knobs (= a stomatostyle) was placed in *Tylenchus*. Clearly, with further enrichment of species within the genus, *Tylenchus* became very heterogeneous. Later, the genus was revised several times and new families and genera proposed. Raski and Geraert (1987) rediagnosed the genus while separating *Tylenchus* from *Filenchus*. Although *Tylenchus* is quite similar to *Filenchus*, it can be easily differentiated (see below under *Filenchus*).

Genus *Filenchus* Andrássy, 1954
= *Dactylotylenchus* Wu, 1968
= *Duosulcius* Siddiqi, 1979
= *Lambertia* Brzeski, 1977
= *Ottolenchus* Husain & Khan, 1965
= *Zanenchus* Siddiqi, 1979
(Fig. 8.1I-P)

Diagnosis: Body straight to arcuate, very small to large (L = 0.3-1.3 mm) but seldom over 1 mm.

Cuticle usually with annulation fine, sometimes coarse. Lateral field with two to four lines. **Amphidial apertures longitudinal slits or clefts extending from labial plate to lateral labial sides.** Deirids at excretory pore level. Labial sclerotization delicate. Labial region usually not offset, width at base about double its height. Stylet delicate, 6-15 µm long, **conus forming about one-third of stylet length**, knobs present. Dorsal gland orifice close to knobs. Median pharyngeal bulb is weak or well developed. Terminal bulb pyriform to elongate. Female reproductive system prodelphic, outstretched, spermatheca mostly offset, post-uterine sac absent or small to well developed, never exceeding one vulval body diameter in length. Male with small, slender spicules, gubernaculum often trough-like; bursa adanal. **Tail elongate-conoid, curved or straight, to filiform**, even hair-like in outline; tail tip curved, hooked or more or less straight.

Type species: *Filenchus vulgaris* (Brzeski, 1963) Lownsbery & Lownsbery, 1985

Comments: Geraert (2008) provides a key to 94 species. *Filenchus* is an extremely common and diverse terrestrial genus although identification of the species is often problematic. Currently, *Filenchus*, *sensu* Geraert and Raski (1987) and Geraert (2008), is a genus lacking an unambiguous definition, referred to in this context as a 'catch-all' genus and as such rejected by Siddiqi (2000). The genus *Filenchus* is more characterized by an absence of characters rather than the presence of clear, apomorphic, characters. In dichotomous keys to identify Tylenchidae, *Filenchus* is often found at the end. This clearly leads to a polyphyletic genus. However, the very limited molecular analyses in combination with morphological data do not permit corroboration of any alternative for the current generic definition (Bert *et al.*, 2010).

Filenchus is similar to *Tylenchus* but differs by the more delicate stylet with the cone forming 30-40% of the total stylet (*vs* about 50%) and the elongate, slit-like apertures extending 3-4 annuli beyond the labial plate in most *Filenchus* species (*vs* the larger pit-like amphidial opening essentially confined to the labial plate in *Tylenchus*, as seen using Scanning Electron Microscopy [SEM]).

Genus *Aglenchus* Andrássy, 1954

(Fig. 8.2A-H)

Diagnosis: Body small (L = 0.4-0.7 mm), straight to only slightly arcuate. Uniformly tapering from base of pharynx and posteriorly from vulva to the elongate, straight tail. **Cuticle relatively coarsely annulated without longitudinal markings. Lateral field composed of two protruding bands**, separated by a groove; in lateral view three or four lateral lines may appear depending on width of groove. Deirids clearly visible on central lateral band, located slightly posterior to excretory pore. Amphidial apertures restricted to labial plate in female. Stylet relatively strong (compared to *Filenchus*), 7-13 µm long, conus about 41-46% of total length. Knobs rounded, well developed. Dorsal gland orifice close to knobs. Median bulb with conspicuous valves, basal bulb elongate-pyriform. **Vulva sunken in body**, larger outer and smaller inner lips; **lateral vulval flaps present**, always well visible. **Vagina walls often markedly swollen, oblique to body axis. Post-uterine sac absent**. Distance between vulva and anus always shorter than tail. Male with distinct bursa with crenate margin. **Cloacal lips protruding, forming a small tube**. Tail in both sexes similar in form and length; straight, filiform, varying in length from 8 to 17 anal body diameters. Tail tip sharply pointed, unstriated.

Type species: *Aglenchus agricola* (de Man, 1884) Meyl, 1961

Comments: The type species is very common. Only seven other species are known. In having prominent vulval flaps, *Aglenchus* is similar to *Coslenchus* but differs in lacking cuticular ridges, having an anteriorly curved vagina and in the absence of a post-uterine sac.

Genus *Coslenchus* Siddiqi, 1978

= *Cosaglenchus* Siddiqui & Khan, 1983

= *Paktylenchus* Maqbool, 1983

(Fig. 8.2I-Q)

Diagnosis: Body small (L = 0.3-0.7 mm), straight to slightly arcuate. **Cuticle coarsely annulated with longitudinal ridges**, showing minute squares outside lateral field. Lateral field composed of two or three protruding ridges. Pore-like amphidial apertures restricted to labial plate. Labial region continuous or slightly offset, usually striated. Stylet rather short but relatively well developed, conus slightly shorter than shaft. Median pharyngeal bulb muscular, with valves.

Deirids in vicinity of excretory pore. **Vulva sunken in body, surrounded by lateral membranes. Vagina walls mostly thickened, perpendicular to body axis** or slightly anteriorly directed. Spermatheca offset, usually without sperm. Post-uterine sac present. Male usually rare, **cloacal lips protruding, forming a short, distinct, tube**, bursa crenate. Tail of both sexes similar, straight, elongate-conoid to filiform.

Type species: *Coslenchus costatus* (de Man, 1921) Siddiqi, 1978

Comments: Geraert (2008) lists 38 species. Mizukubo and Minagawa (1986) discuss the phylogenetic relationships within *Coslenchus* spp. The most recent detailed review by Brzeski (1987) considered 23 species valid.

Genus *Basiria* Siddiqi, 1959

= *Clavilenchus* Jairajpuri, 1966
= *Pseudobasiria* Jahan, 1966
= *Basiroides* Thorne & Malek, 1968
= *Duotylenchus* Saha & Khan, 1982
= *Neobasiria* Javad, 1982
= *Basirienchus* Geraert & Raski, 1986
= *Rhabdotylenchus* Xie, Feng, Li & Yin, 1994
(Fig. 8.3A-J)

Diagnosis: Body straight or slightly bent ventrally, shorter than 1 mm. Cuticle with distinct annulation, but becoming less distinct on tail. Lateral field with two, four or six lines. Labial region usually high, without annulation. **Amphidial apertures forming broad oblique slits visible on lateral region of labial region**, shape variable, sometimes in form of an inverted V. Stylet delicate, 7-15 µm, conus usually comprising 30-40% of stylet length; **knobs rounded**, small but distinct, often asymmetrical. **Orifice of dorsal pharyngeal gland often as much as one stylet length posterior to stylet base**. Median bulb usually poorly developed, muscular with refractive valve plates. Pharyngo-intestinal valve distinct, flat or conical. Deirids usually present at level of excretory pore. Female reproductive system may extend to pharynx, ovary with single row of oocytes almost along its whole length. Spermatheca often elongated, not offset. Vulval slit relatively long, transverse. Post-uterine sac less than one vulval body diameter long. Spermatozoa rounded, bursa adanal. Tail short and conical to very long and filiform; terminus pointed, rounded or clavate.

Type species: *Basiria graminophila* Siddiqi, 1959

Comments: According to Geraert (2008) the genus contains 42 species. A valuable revision of the genus is available (Karegar and Geraert, 1997a, b, c, 1998).

Genus *Boleodorus* Thorne, 1941

(Fig. 8.3K-P)

Diagnosis: Small nematodes (L = 0.3-0.7 mm). Body **ventrally arcuate to spirally coiled**. Cuticle finely annulated, lateral field with four (rarely six) lines. Labial region often elevated, conoid rounded, generally with depression at oral aperture. Amphidial apertures more or less oblique slits visible on lateral surfaces of labial region. Stylet weakly developed, **knobs flanged**. Pharynx with **weakly developed median bulb**, bulb without sclerotized valves. **Female reproductive system with short ovary**, oocytes mostly in multiple rows, short post-uterine sac. Spermatheca offset, usually with small refractive sperm, bursa weakly developed. **Tail elongate, conical, ventrally curved**, mostly with finely rounded tip.

Type species: *Boleodorus thylactus* Thorne, 1941

Comments: The genus contains 30 described species. Recently one new species has been described from Iran.

8.3. Dolichodoridae

The Dolichodoridae comprises a large number of diverse species found throughout the world, although perhaps most common in cooler, more temperate soils. All the species are ectoparasitic on the roots of plants and some species have been shown to be of economic importance on certain crops. This group has experienced a degree of flux in recent years as some taxonomists corrected what they perceived to be an inflated systematic scheme, particularly at the generic level. As a result, many of the numerous genera accepted by Siddiqi (1986, 2000), for example, were not accepted by workers such as Fortuner and Luc (1987). In particular, the character of number of incisures in the lateral field has not been accepted as being of generic significance, a point of view which results in the synonymy, for example, of *Quinisulcius* under *Tylenchorhynchus*, despite the fact that the former genus is easily recognizable from the latter under the stereomicroscope without resorting to counting the lateral lines. Excessively conservative schemes end up masking valid differences and run the risk of making the identification of species within such large aggregate genera more difficult than is necessary. In this scheme the lateral line character is maintained as a reliable and easily ascertained means of identifying groups (= genera) of similar nematodes. Personal preference, however, may dictate an alternative scheme involving several large genera, each embracing a large number of species, the number of lateral lines still being utilized, but as a means of division into smaller, more manageable, groups of species. Molecular phylogenies suggest that the Dolichodoridae have evolved separately in at least three different clades within the Tylenchomorpha and is therefore a polyphyletic assemblage (Bert *et al.*, 2008; Holterman *et al.*, 2009). Several authors therefore also consider the grouping of a diverse assemblage of taxa not justified. Recent significant taxonomic revision included the removal by Sturhan (2012) of *Pratylenchoides* Winslow, 1958 from Pratylenchidae Thorne, 1949 to the Merliniidae as earlier proposed by Ryss (1993). According to Sturhan (2012) and Siddiqi and Sturhan (2014), the Merliniidae (= Merliniinae in this chapter) consists of two subfamilies: Merliniinae comprising *Geocenamus* Thorne & Malek, 1968, *Merlinius* Siddiqi, 1970, *Paramerlinius* Sturhan, 2012, *Macrotylenchus* Sturhan, 2012, *Telomerlinius* Siddiqi & Sturhan, 2014, *Amplimerlinius* Siddiqi, 1976 and *Nagelus* Thorne & Malek, 1968; and Pratylenchoidinae Sturhan, 2012 with *Pratylenchoides*. In this account, we accept the proposed genera but place *Pratylenchoides* within the Merliniinae rather than in a separate subfamily as proposed by Sturhan (2012), and emphasize that the position of *Pratylenchoides* remains uncertain.

8.3.1. Characters Used in the Identification of Dolichodoridae

General morphology of the Dolichodoridae is illustrated in Figures 8.4-8.6 and these drawings should be used in conjunction with the information cited below.

Body size and death habitus: large-sized (1.5 mm or longer) forms include *Dolichodorus*, *Belonolaimus*, *Macrotrophurus* while medium-sized (0.7 to about 1 mm) and small-sized (less than 0.7 mm) constitute almost all the rest of the dolichodorids. The body shape is always elongate-cylindrical tapering towards both ends, it is never obese (Fig. 8.4C). The habitus is usually straight to arcuate, but is characteristically spiral in *Neodolichodorus* and some *Amplimerlinius*.

Cuticle: The cuticle is at least two layered, marked by transverse striae making coarse or fine annules (striae absent in *Macrotrophurus*). The lateral regions are thickened and marked by longitudinal incisures except in *Belonolaimus*, which has a single lateral groove on each side running from labial region to tail tip. Lateral fields may or may not be areolated, that is, traversed by the transverse striae, and the areolation may be complete or incomplete. Longitudinal striae may be present all along the body (*Tessellus* and *Scutylenchus*) or may be restricted to the anterior end.

The character of the presence of either six, five or three incisures in the lateral field has been used to distinguish Merliniinae, *Quinisulcius*, *Uliginotylenchus* and *Divittus*, respectively. *Triversus* and *Dolichodorus* were differentiated from *Tylenchorhynchus* and *Neodolichodorus*, respectively also by three *vs* four incisures. The presence of longitudinal cuticular ridges outside the lateral fields has been used to diagnose *Trilineellus*, *Mulkorhynchus*, *Neodolichorhynchus* and *Prodolichorhynchus*. Localized thickening of the cuticle at the tail terminus has been used to characterize the genera *Trophurus*, *Macrotrophurus*, *Paratrophurus* and *Amplimerlinius*.

Amphids, labial and labial sensilla, deirids and phasmids: Amphids and sensilla are best seen in SEM. Amphid apertures are longitudinal slit-like in *Dolichodorus* and pore-like or dorso-ventrally elongated in most of the other genera. They are always located near the oral opening except in *Macrotrophurus* where they are a little more posterior. Outer labial sensilla do not usually open to the surface whereas cephalic sensilla usually open to the surface in the form of papillae or pits. Deirids are present only in Merliniinae and have been reported in some species of *Quinisulcius*. Phasmids are pore-like on tail and extend into the bursa in the male, except in *Neodolichodorus* in which they are preanal. Phasmids are conspicuous and 'plugged' in some Merliniinae.

Labial region: Shape (low, high, offset, continuous, smooth or striated) and degree of sclerotization

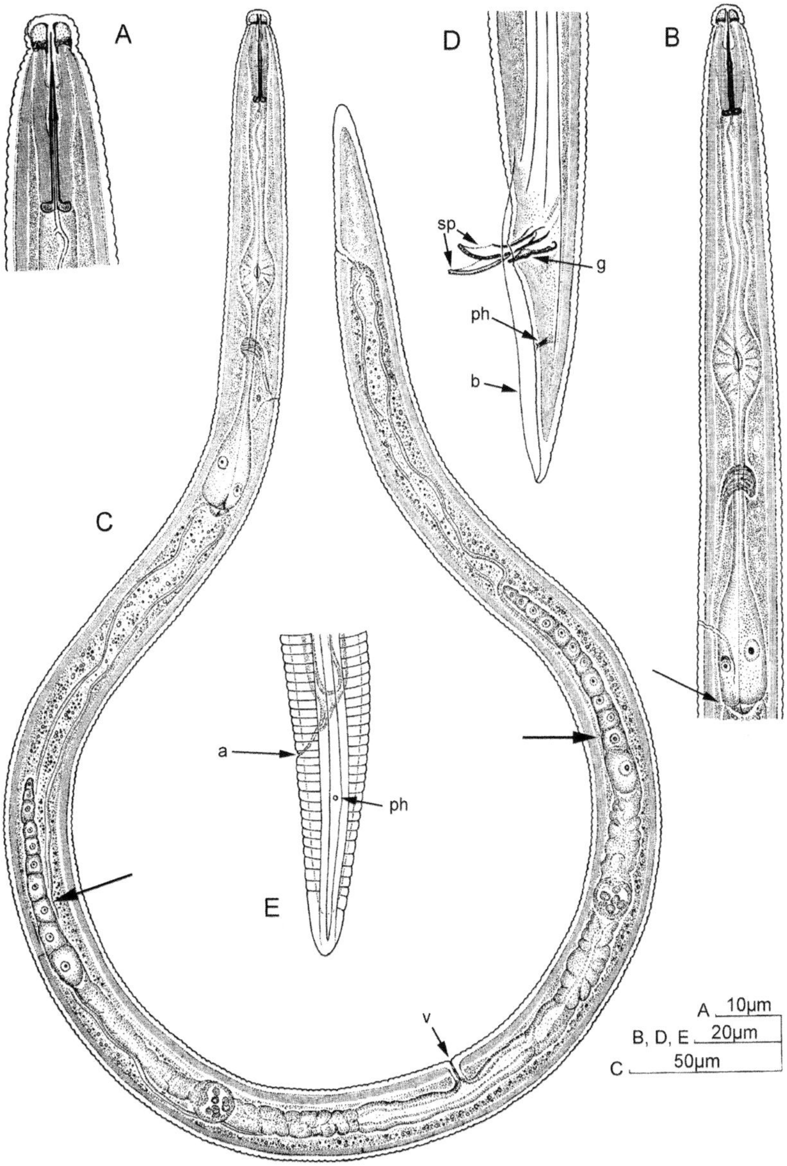

Fig. 8.4. General morphology of a typical member of the Dolichodoridae. A: Anterior region showing head shape and stylet form; B: pharynx showing non-overlapping basal bulb (arrow); C: entire female showing non-overlapping pharynx base, median vulva (v) and amphididelphic genital tracts (arrows); D: male tail region showing spicules (sp), gubernaculum (g), bursa (b) and phasmid (ph); E: female tail region showing anus (a) and phasmid (ph). Note also the four more or less parallel lines or incisures in the lateral field. (Adapted from Siddiqi (1972a), courtesy CABI.)

of the labial framework are important characters (Fig. 8.5). In *en face* it is usually circular, but maybe 2, 4 or 6 lobed. *Merlinius* has six radial grooves. Several genera do not have a distinct labial disc, but *Dolichodorus, Belonolaimus, Sauertylenchus* have a rounded labial disc.

Stylet: A short stylet (< 35 μm) with conus about as long as the shaft is a common feature, although some genera have a very long stylet (Fig. 8.5). Basal knobs are always present, their shape having diagnostic value. A long stylet with the conus much longer than the shaft is

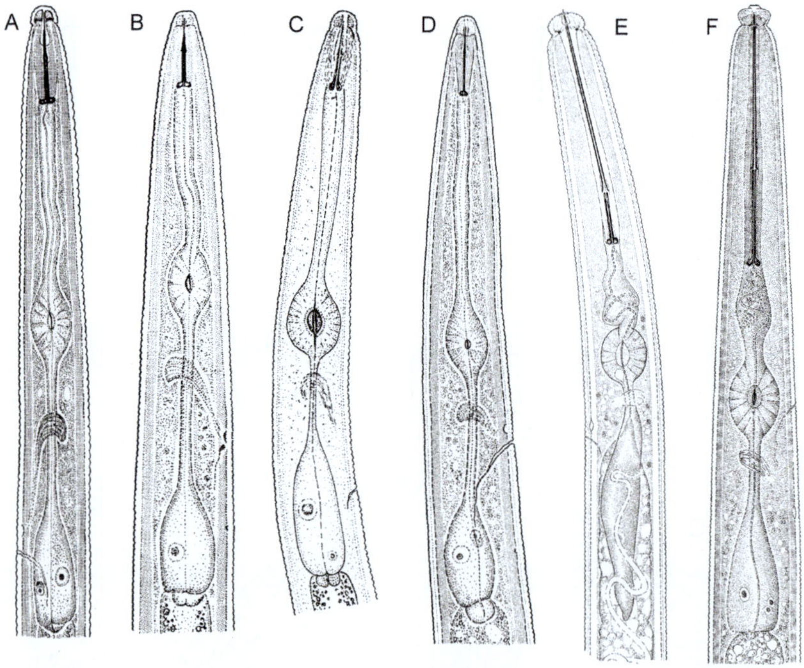

Fig. 8.5. Comparison of pharyngeal form in Dolichodoridae. A: *Tylenchorhynchus*; B: *Merlinius*; C: *Quinisulcius*; D: *Trophurus*; E: *Belonolaimus*; F: *Dolichodorus*. (Drawings not to scale and from various authors, courtesy CABI.)

found in *Dolichodorus*, *Neodolichodorus*, *Macrotrophurus* and *Hexadorus*. As a rule, a long stylet has the conus tubular until its tip and a short stylet has a conus that appears to be solid in its anterior third because the aperture is located a little behind the tip.

Pharynx: Most dolichodorids have a tylenchoid pharynx with a muscular median bulb, a slender isthmus and a glandular, non-overlapping basal bulb separated from the intestine by a cardia (Figs 8.4B, 8.5A-D, F). Belonolaiminae and some Telotylenchinae, however, have the glands lobe-like and extending over the anterior end of the intestine, the dorsal gland forming most of the lobe (Fig. 8.5E).

Intestine: A simple tube with the lumen indistinct. It extends into the tail cavity in some taxa (e.g., in *Trophurus*, Dolichodorinae), but never does so in Merliniinae. Serpentine canals or fasciculi may or may not be present (Fig. 8.6K).

Tail: One of the most important characters is the size and shape of the female tail (Fig. 8.6G-L). The male tail is always conical and enveloped by a bursa (Fig. 8.4D, 8.6A-F). The tail is relatively long when compared to the anal body diameter and this is a simple and relatively reliable method for distinguishing members of this group from the hoplolaimids.

Female reproductive system: The shape and position of vulva, presence/absence of epiptygma, sclerotization of vagina and whether the spermatheca is axial or offset and spherical or lobed when distended with sperm are all important characters. As a rule, dolichodorids are didelphic with two equally developed reproductive branches (i.e., amphididelphic), but in *Trophurus* the posterior branch is rudimentary (i.e., the genital system is pseudomonodelphic). There are always three glandular uterine cells in cross-section (= tricolumella).

Male reproductive system: The testis is single and outstretched and produces small rounded sperm. Shape and size of spicules, gubernaculum and bursa are all important characters. Dolichodorinae and Meiodorinae have a trilobed bursa with large lateral lobes. Merliniinae differs from Telotylenchinae in having a low bursa whose margins hardly project beyond the tail contour when viewed laterally. *Neodolichorhynchus* has a bursa which has a trilobed tip. The bursal tip may be flagelliform, as in species formerly placed in *Triversus*. The spicules have a broadly rounded tip

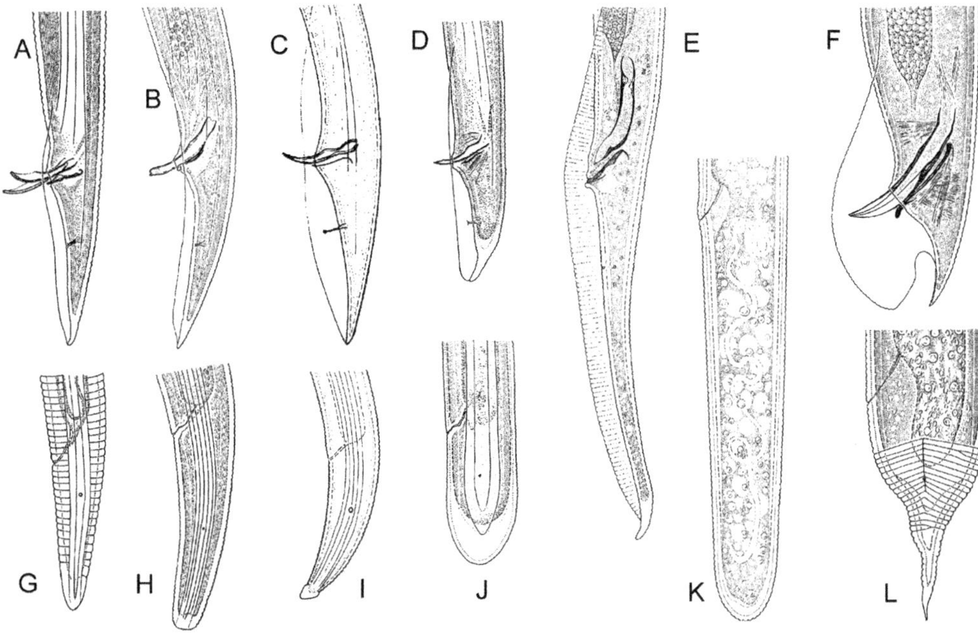

Fig. 8.6. Comparison of male and female tail shape in Dolichodoridae. Male tails, A-F. A: *Tylenchorhynchus*; B: *Merlinius*; C: *Quinisulcius*; D: *Trophurus*; E: *Belonolaimus*; F: *Dolichodorus*. Female tails, G-L. G: *Tylenchorhynchus*; H: *Merlinius*; I: *Quinisulcius*; J: *Trophurus*; K: *Belonolaimus*; L: *Dolichodorus*. (Drawings not to scale and from various authors, courtesy CABI.)

with a large central depression in Merliniinae, whereas most other dolichodorids have spicules with a narrow tip and large distal flanges. The gubernaculum is trough-like and fixed in Merliniinae and rod-like and protrusible in Telotylenchinae and many other groups. The proximal end of the gubernaculum may be rounded or pointed and directed anteriorly or posteriorly. A pair of pedunculate papillae (= hypoptygma) is present on the posterior cloacal lip in Merliniinae.

8.3.2. Family Dolichodoridae Chitwood *in* Chitwood & Chitwood, 1950

Geraert (2011) provides a valuable illustrated monograph of the valid species of Dolichodoridae, although his taxonomic scheme, particularly at the generic level, differs somewhat from that followed herein. Other good sources covering the family include Siddiqi (1986, 2000), Fortuner and Luc (1987), Luc and Fortuner (1987), Nickle (1991) and Andrássy (2007). Information as to pathogenicity may be sought in Sikora *et al.* (2018).

Diagnosis (after Siddiqi, 2000): Tylenchoidea. Small to large nematodes (L = 0.5-1.5 mm), straight, usually arcuate or more curved. Cuticle strongly annulated (except in *Macrotrophurus*). Lateral fields with 1-6 incisures. Deirids present or absent. **Phasmids pore-like, caudal in location**. Labial region annulated, round, **hexagonal or 4-lobed; labial disc indistinct** (only in SEM appearing flat rectangular or cross shaped) or distinct; with minute central, oval oral aperture surrounded by **six labial papillae or pits**; labial sensilla not on surface; amphidial apertures pore- or slit-like, at lateral margins of labial disc (except *Macrotrophurus*). Stylet well developed, knobbed; conus not much shorter than shaft. **Pharyngeal glands generally in either a basal bulb abutting intestine or dorsal gland enlarged, forming most of lobe extending over intestine; subventral glands not distinctly enlarged**. Didelphic, amphidelphic, rarely pseudo-monoprodelphic. Vulva median or submedian, transverse slit-like, rarely transversely oval; lips simple or modified. Vagina sometimes cuticularized. Spermathecae axial or offset, round, oval or two- or four-lobed. **Tail dissimilar between sexes**.

8.3.3. Systematic Scheme

(Major genera in **bold** type)

Tylenchoidea Örley, 1880

Dolichodoridae Chitwood *in* Chitwood & Chitwood, 1950

Dolichodorinae Chitwood *in* Chitwood & Chitwood, 1950

Dolichodorus Cobb, 1914

Neodolichodorus Andrássy, 1976

Brachydorinae Siddiqi, 2000

Brachydorus de Guiran & Germani, 1968

Telotylenchinae Siddiqi, 1960

Bitylenchus Filipjev, 1934

Histotylenchus Siddiqi, 1971

Neodolichorhynchus Jairajpuri & Hunt, 1984

Paratrophurus Arias, 1970

Sauertylenchus Sher, 1974

Telotylenchoides Siddiqi, 1971

Telotylenchus Siddiqi, 1960

Trichotylenchus Whitehead, 1960

Trophurus Loof, 1956

Tylenchorhynchus Cobb, 1913

Quinisulcius Siddiqi, 1971

Uliginotylenchus Siddiqi, 1971

Meiodorinae Siddiqi, 1976

Meiodorus Siddiqi, 1976

Macrotrophurinae Fotedar & Handoo, 1978

Macrotrophurus Loof, 1958

Merliniinae Siddiqi, 1971

Amplimerlinius Siddiqi, 1976

Geocenamus Thorne & Malek, 1968

Merlinius Siddiqi, 1970

Macrotylenchus Sturhan, 2012

Nagelus Thorne & Malek, 1968

Paramerlinius Sturhan, 2012

Pratylenchoides Winslow, 1958

Scutylenchus Jairajpuri, 1971

Telomerlinius Siddiqi & Sturhan, 2014

Belonolaiminae Whitehead, 1960

Belonolaimus Steiner, 1949

Carphodorus Colbran, 1965

Ibipora Monteiro & Lordello, 1977

Morulaimus Sauer, 1966

Incertae sedis

Psilenchidae Paramonov, 1967

Psilenchinae Paramonov, 1967

Psilenchus de Man, 1921

Antarctenchus Spaull, 1972

Atetylenchus Khan, 1973

8.3.4. Key to Subfamilies of Dolichodoridae

1. Female tail elongate-conoid with narrow, mucronate, spicate or finely drawn-out tip except *Neodolichodorus* in which tail is obtusely rounded, under two anal body diameters long and with terminal cuticle not abnormally thickened; bursa trilobed 2
Female tail not elongate-conoid or with narrow, mucronate, spicate or finely drawn-out tip (except rarely in some *Merlinius* and *Triversus* spp.), if under two anal body diameters long, then with terminal cuticle abnormally thickened; bursa not trilobed (except *Neodolichorhynchus* in which only terminus is trilobed) 3
2. Stylet very long (> 50 μm), conus much longer than shaft; labial disc conspicuous .... Dolichodorinae
Stylet not very long (< 40 μm), conus about as long as shaft; labial disc inconspicuous .... Meiodorinae
3. Labial region usually 4-lobed; lateral labial areas reduced; amphidial apertures longitudinal slit-like; pharyngeal gland lobe-like, extending over intestine, pharyngo-intestinal junction less than one body diameter posterior to median bulb ...... Belonolaiminae
Labial region not usually 4-lobed; lateral labial areas not reduced, amphidial apertures pore-like or transversely oval; pharyngeal glands in a bulb abutting intestine, if forming a lobe extending over intestine then pharyngo-intestinal junction more than one body diameter posterior to median bulb 4
4. Lateral field with six incisures; male usually with hypoptygma or paired papillae on posterior anal lip .... Merliniinae
Lateral field with 3-5 incisures; male without hypoptygma 5
5. Amphidial apertures conspicuous, post-labial; stylet > 80 μm long, conus much longer than the shaft .... Macrotrophurinae
Amphidial aperture inconspicuous, labial; stylet < 50 μm, conus not much longer than the shaft; deirids absent, dorsal gland abutting, rarely overlapping intestine .... Telotylenchinae
Deirids present, female tail subcylindrical, general morphology resembling *Pratylenchidae* .... Merliniinae (*Pratylenchoides*)

8.3.5. Subfamily Dolichodorinae Chitwood *in* Chitwood & Chitwood, 1950

Diagnosis (after Siddiqi, 2000): Dolichodoridae. Long nematodes (L = 1.5-3 mm). Lateral field with three or four incisures, areolated. Labial region offset, distinctly annulated, framework strongly cuticularized; **labial disc distinct; stylet very long (> 50 μm) with conus markedly longer than shaft**, knobs prominent. Intestinal fasciculi and postrectal sac present. ***Vagina vera* cuticularized.** Ovaries paired. Female and juvenile tails long, filiform to spicate or, if obtuse to mamillate, then less than two anal body diameters long and with annulated terminus. **Male tail short**, conical, **bursa trilobed** with two large lateral flap-like lobes and small terminal lobe; spicules with or without distal flanges; gubernaculum large, protrusible or fixed.
Type genus: *Dolichodorus* Cobb, 1914
Other genus: *Neodolichodorus* Andrássy, 1976

8.3.6. Key to Genera of Dolichodorinae

1. Lateral field with 3 incisures; female tail spicate or filiform .... *Dolichodorus*
Lateral field with 4 incisures; female tail obtuse or mamillate .... *Neodolichodorus*

Genus *Dolichodorus* Cobb, 1914
(Fig. 8.7)

Diagnosis (after Siddiqi, 2000): Dolichodorinae. Body straight to arcuate when relaxed. **Lateral field with three incisures**, bands completely areolated. Phasmids slightly postanal. Amphidial apertures in form of longitudinal slits. **Labial region prominently 4-lobed** with a large offset labial disc. Labial framework with massive cuticularization of basal plate. Stylet exceptionally elongate (54-170 μm), stylet guide elongate-tubular. Procorpus swollen with convoluted lumen when stylet retracted. Median and basal bulbs well developed, joined by a short, slender, isthmus. Excretory pore located well anterior to hemizonid. Vulval lips not modified. **Vaginal sclerotization symmetrical in lateral view**. Female tail convex-conoid anteriorly, then conoid-spicate to filiform. Postrectal intestinal sac only filling convex-conoid region of tail. Bursa large, trilobed. Spicules robust, with large flanges; gubernaculum robust, protrusible.

Type species: *Dolichodorus heterocephalus* Cobb, 1914

Comments: A small genus with fewer than 20 known species, mostly from the Americas. Although ectoparasitic on plant roots, particularly at the tips, the long stylet allows these nematodes to penetrate deep within the tissue, often reaching the vascular cylinder. This can result in severe stunting and root injury leading to above ground symptoms typical of root damage or drought, including wilting, chlorosis and stunting. The type

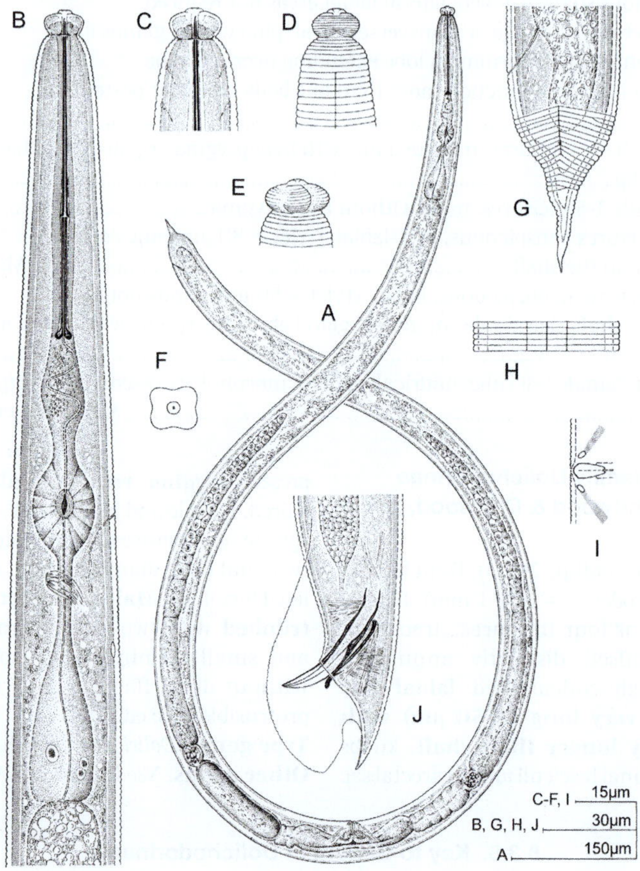

Fig. 8.7. *Dolichodorus heterocephalus*. A: Entire female; B: pharyngeal region; C: labial region; D, E: superficial view of labial region; F: *en face* view; G: female tail region. Note spicate tip; H: lateral field showing three incisures or lines; I: female vulval region; J: male tail region showing copulatory apparatus and lobed bursa. (Adapted from Orton Williams (1974a), courtesy CABI.)

species, *D. heterocephalus*, known as the awl nematode, is usually found in wet soil or around the edges of lakes, streams and ponds or in the water. Under cultivated conditions it is most likely to occur where soil moisture is high although *D. heterocephalus* does not appear to be dependent on wet conditions. Hosts include celery, sweet corn, water chestnut, bean, tomato, pepper, cranberry, lettuce, carrots, cabbage and carnation. *D. heterocephalus* has been recorded principally from the USA, where it is mainly confined to the eastern States, particularly Florida. All records of severe damage to crops have so far come from this State and, although the awl nematode can be devastating where it occurs, outbreaks are localized and it does not seem to be common or widespread enough to make it a pest of major importance (see Orton Williams, 1974a; Smart and Nguyen, 1991). The genus has also been recorded from South Africa and Zimbabwe.

8.3.7. Subfamily Telotylenchinae Siddiqi, 1960

Diagnosis (after Siddiqi, 2000): Dolichodoridae. Cuticle prominently annulated; longitudinal striae and grooves may be present. **Lateral fields each with three to five incisures. Deirids absent**. Amphidial apertures pore-like, labial in location, indistinct. Labial region annulated; labial disc indistinct (except *Sauertylenchus*); framework with light to moderate cuticularization (strong in some *Paratrophurus* spp.). **Stylet small to moderately long (11-45 µm) with conus about as long as shaft**. Distinct basal knobs. Median pharyngeal bulb well developed, glands either enclosed in a non-overlapping or abutting basal bulb or dorsal gland enlarged and forming a long lobe overlapping intestine. Intestinal fasciculi and postrectal sac present or absent. Vulva slit-like. Ovaries usually paired, opposed, outstretched, but monoprodelphic in *Trophurus*. Female tail conoid, subcylindroid, cylindroid or subclavate, rarely spicate, generally 2-6 anal body diameters long. Male tail elongate-conoid. **Bursa simple**, occasionally tip doubly indented. Spicules arcuate with large distal flanges, **tip narrowly pointed or indented**. Gubernaculum rod-like, usually protrusible or fixed. **Hypoptygma absent**.

Type genus: *Telotylenchus* Siddiqi, 1960

Other major genera:
Bitylenchus Filipjev, 1934
Histotylenchus Siddiqi, 1971
Neodolichorhynchus Jairajpuri & Hunt, 1984
Paratrophurus Arias, 1970
Quinisulcius Siddiqi, 1971
Sauertylenchus Sher, 1974
Trophurus Loof, 1956
Tylenchorhynchus Cobb, 1913
Uliginotylenchus Siddiqi, 1971

8.3.8. Key to Major Genera of Telotylenchinae

1. Pharyngeal glands extending over intestine as a long lobe 2
   Pharyngeal glands not forming a lobe, abutting intestine 3
2. Stylet conus asymmetrical; intestine extending beyond rectum into tail .... *Histotylenchus*
   Stylet conus symmetrical; intestine not extending beyond rectum into tail .... *Telotylenchus*
3. Longitudinal ridges or lamellae present outside lateral fields .... *Neodolichorhynchus*
   Longitudinal ridges or lamellae absent outside lateral fields 4
4. Lateral fields with 5 incisures .... *Quinisulcius*
   Lateral fields with 3 or 4 incisures 5
5. Lateral fields with 3 incisures, areolated; vulva in a body depression, with small epiptygma .... *Uliginotylenchus*
   Lateral fields with 3 or 4 incisures, if 3 then not areolated; vulva generally not in a body depression, without epiptygma 6

6. Labial disc conspicuous; stylet > 30 μm long .... *Sauertylenchus*
Labial disc not conspicuous; stylet < 30 μm long 7
7. Terminal cuticle of tail abnormally thickened 9
Terminal cuticle of tail not abnormally thickened 8
8. Outer bands lateral field areolated, a large postanal intestinal sac containing intestinal granules and fasciculi, cuticle at terminus thickened, gubernaculum lacking a crest .... *Bitylenchus*
Outer band not areolated, postanal intestinal sac absent or less developed, tail not pronouncedly thickened .... *Tylenchorhynchus*
9. Female genital system monoprodelphic with posterior branch reduced to a sac, median vulva .... *Trophurus*
Female genital system didelphic, median vulva .... *Paratrophurus*

Genus *Tylenchorhynchus* Cobb, 1913
= *Triversus* Sher, 1974
= *Trilineellus* Lewis & Golden, 1981
= *Divittus* Jairajpuri, 1984
= *Morasinema* Javed, 1984
= *Tessellus* Jairajpuri & Hunt, 1984
= *Macrorhynchus* Sultan, Singh & Sakhuja, 1991
(Fig. 8.8)

Diagnosis (after Siddiqi, 2000): Telotylenchinae. Body about 1 mm or less in length. Cuticle prominently annulated, may be marked by longitudinal striae but **longitudinal ridges or lamellae outside lateral fields are absent. Cuticle not abnormally thickened on tail.** Lateral field with three or four incisures, generally not areolated posterior to pharyngeal region. **Labial region without longitudinal indentations on annules**, offset or continuous, annulated or rarely smooth; labial disc indistinct, in SEM *en face* squarish, flattened; framework light to moderately cuticularized. Stylet well developed, generally 15-25 μm long; conus appearing solid anteriorly; knobs prominent. Median bulb round or oval, with distinct valve plates, usually marked off from procorpus and isthmus. Basal bulb offset from intestine or with base slightly extending over intestine. Cardia prominent. Vulva near middle of body, generally at 50-54%, lips not modified. Spermatheca round, axial. Ovaries paired, outstretched. Female tail conoid, subcylindroid, cylindroid or subclavate; terminus smooth, rarely striated, not narrowing to a point. Males generally present. Tail enveloped by a large, simple bursa. Spicules distally flanged, terminus narrow, indented or pointed. **Gubernaculum** usually **protrusible** well developed, about half as long as spicule, generally rod-like.

Type species: *Tylenchorhynchus cylindricus* Cobb, 1913

Comments: A large and diverse genus with well over 100 species currently recognized as valid in *Tylenchorhynchus*, some 20 of which form the *Bitylenchus* group, a group that is sometimes regarded as a valid genus in its own right (see Siddiqi, 2000). At present, a number of nominal genera are regarded as being junior synonyms, but opinions differ widely as to whether taxa within the assemblage deserve generic status or not. Several species enjoy a broad distribution. In the broad concept of the genus, the first character to be used when dividing the species into groups is the number of incisures in the lateral field.

These are ectoparasitic nematodes and, although they may occur in huge numbers in the rhizosphere, most species are only mild pathogens. Perhaps the most studied species is *Tylenchorhynchus annulatus* (Cassidy, 1930) Golden, 1971, a common nematode in rice fields, although it is not considered an important parasite of rice (see Bridge *et al.*, 2005). Other hosts include grasses and sugarcane. *T. annulatus* has a nearly worldwide distribution, especially in tropical and subtropical areas. It is absent from Europe.

Genus *Trophurus* Loof, 1956
= *Clavaurotylenchus* Caveness, 1958
(Fig. 8.9)

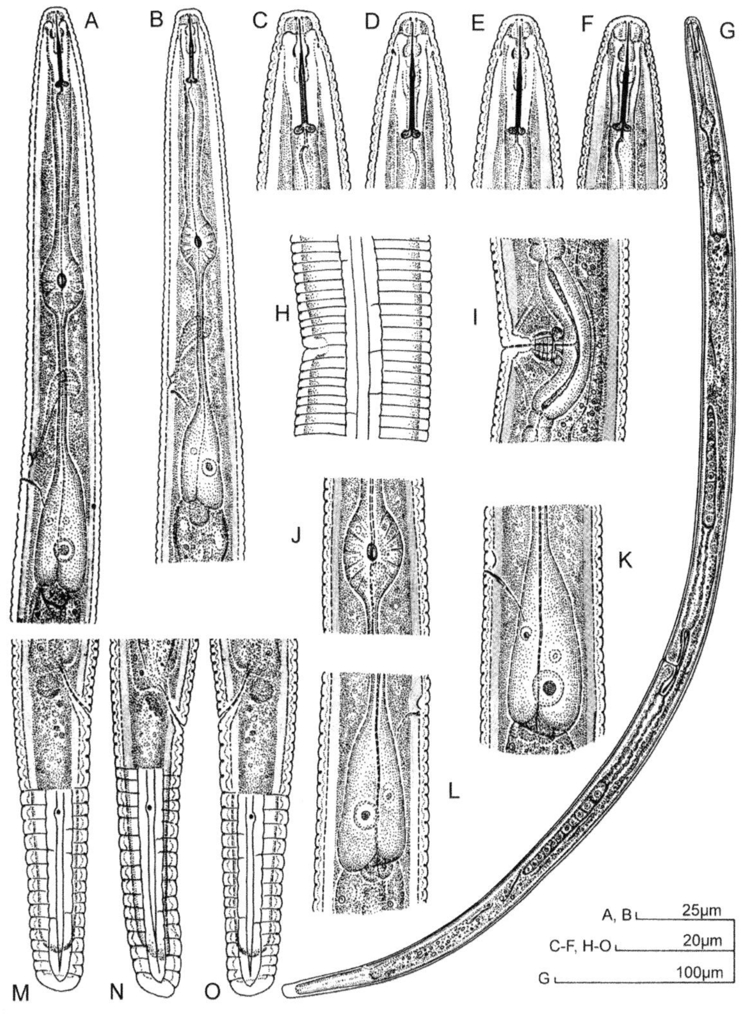

Fig. 8.8. *Tylenchorhynchus annulatus*. A: Female pharyngeal region; B: male pharyngeal region; C-F: anterior region and stylet; G: entire female; H: female vulval region, superficial view showing annulation and four lateral incisures; I: female vulval region; J: median bulb; K, L: non-overlapping pharyngeal base; M-O: female tail region. Note four incisures in lateral field and punctiform phasmid. (Adapted from Siddiqi (1976), courtesy CABI.)

Diagnosis (after Siddiqi, 2000): Telotylenchinae. Body 0.5-1.1 mm long, slender (a = 25-50). Deirids absent. Phasmids distinct, located near middle of tail. Labial region elevated, continuous, rounded or truncate. Stylet slender, 14-22 µm long with solid-appearing tip and small round knobs. Median bulb oval, muscular, located posterior to middle of pharynx. Basal bulb pyriform, offset from intestine; cardia prominent. Vulva at 53-61% of body length, lips not modified. **Ovary single**, outstretched. **Posterior branch represented by a uterine sac, often showing rudiments of the degenerate posterior ovary. Spermatheca with offset lobes. Intestine usually extending beyond rectum into tail cavity. Female tail cylindroid, subclavate or subconoid with broadly rounded terminus, 1.2-3 anal body diameters long; hyaline terminal cuticle very prominent**. Males present. Spicules slender,

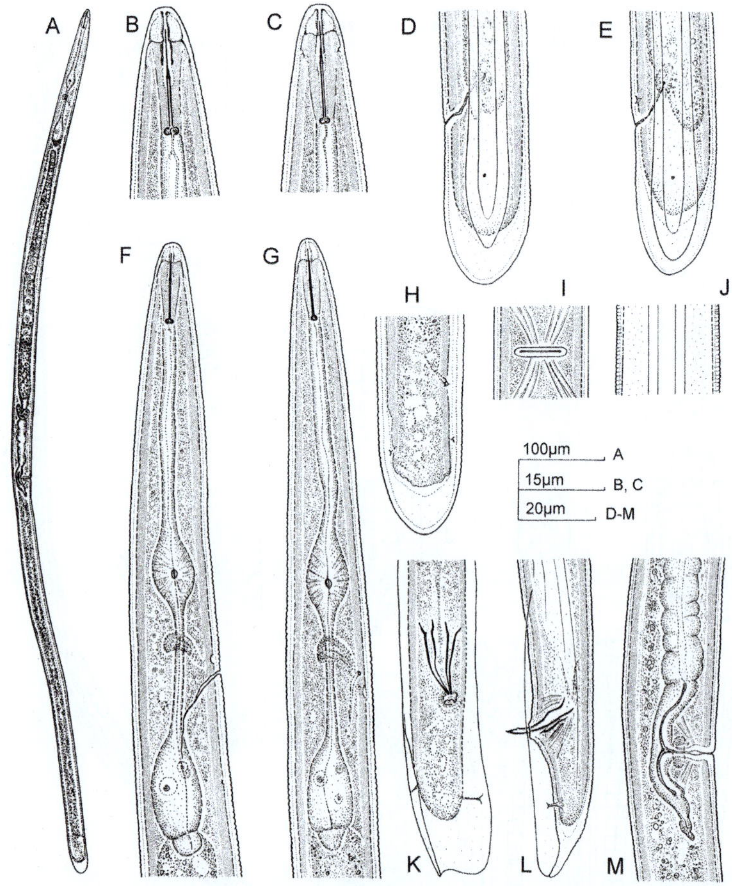

Fig. 8.9. *Trophurus imperialis*. A: Entire female showing monoprodelphic genital tract and median vulva; B, C: anterior region; D, E: female tail region (lateral); F, G: pharyngeal region with non-overlapping pharyngeal base; H: female tail (rolled); I: female vulva, ventral view; J: lateral field at mid-body; K, L: male tail region; M: female vulval region. (Adapted from Siddiqi (1973), courtesy CABI.)

about 11-25 µm long. Gubernaculum small, linear, fixed. Bursa large, enveloping tail, terminally indented in type species.

Type species: *Trophurus imperialis* Loof, 1956

Comments: About 12 species described from the Americas, Europe and Asia. Relatively rare yet distinctive nematodes and not reported to be damaging to crop plants.

Genus *Quinisulcius* Siddiqi, 1971

(Fig. 8.10)

Diagnosis (after Siddiqi, 2000): Telotylenchinae. Small nematodes (L = 0.47-0.9 mm) **strongly curved ventrad when relaxed**. Annules prominent. **Lateral field with five incisures**, smooth or areolated. Deirids absent. Labial region offset, rounded, finely annulated; labial disc indistinct; framework with light to moderate sclerotization. Stylet moderately strong, 12-24 µm long (16-19 µm long in type species); conus appearing solid anteriorly; knobs rounded, rarely cupped anteriorly. Median bulb well developed, oval, not offset from procorpus by a constriction. Basal bulb large, usually offset from intestine. Cardia prominent. Vulva at 51-59% of body length, lips not modified. Spermathecae round, axial or slightly offset, rarely functional. Ovaries paired, outstretched. Female tail conoid, usually ventrally arcuate, terminal annule enlarged, smooth or striated. **Males rare or absent**. Bursa well developed,

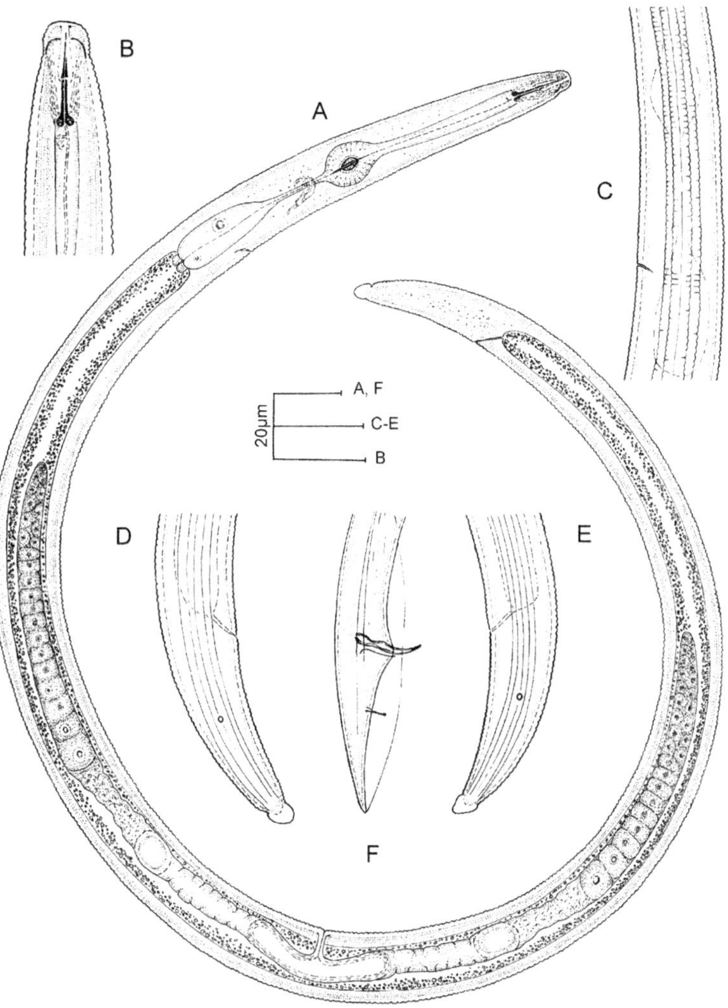

Fig. 8.10. *Quinisulcius capitatus*. A: Entire female; B: anterior region and stylet; C: areolated lateral field with five incisures; D, E: female tail region; F: male tail region. (Adapted from Jairajpuri (1985), courtesy CABI.)

simple, enveloping entire tail. Spicules arcuate, with poorly developed distal flanges, 14-24 µm long. Gubernaculum modified, with proximal end directed dorsally.

Type species: *Quinisulcius capitatus* (Allen, 1955) Siddiqi, 1971

Comments: A small genus with fewer than 20 described species and which has been regarded as a junior synonym of *Tylenchorhynchus*. However, the character of five lines in the lateral field is entirely consistent and unique to this assemblage within the family. Furthermore, *Quinisulcius* species are readily recognized as such under the stereomicroscope. As they form such a distinct group or operational taxonomic unit (OTU), the genus level (itself a rather subjective taxon) is considered appropriate herein.

8.3.9. Subfamily Merliniinae Siddiqi, 1971

Diagnosis (modified after Siddiqi, 2000; Sturhan, 2012; Hosseinvand *et al.*, 2023): Dolichodoridae.

Lateral fields with six incisures (except *Telomerlinius* and some *Pratylenchoides* spp.). **Deirids present** (except in *Scutylenchus*, *Geocenamus* and *Telomerlinius*). Phasmids usually prominent, on tail. Labial region annulated; labial disc indistinct, rarely distinct (*Geocenamus* and *Telomerlinius*). Stylet under 50 μm (except *Macrotylenchus*, up to 137 μm), with distinct round basal knobs. Median and basal bulbs well developed. **Vulva small, pore-like, transversely oval or slit-like**, usually with epiptygma. Ovaries paired. **Spermathecae 2-4 lobed**. Postrectal intestinal sac absent. Female tail conoid, cylindroid, subcylindroid or subclavate, between two and six anal body diameters long; terminal cuticle may be greatly thickened. Male tail conical, about as long as that of female. Bursa simple, moderately developed, enveloping tail. **Hypoptygma** (a pair of papillae on posterior lip of cloacal aperture) **always present. Spicules rather cylindroid** (cephalated in *Pratylenchoides*), straight to arcuate, **with distal ends broadly rounded** (pointed in *Pratylenchoides*), **notched and devoid of ventrolateral flanges or vela. Gubernaculum simple, trough-like, fixed.**

Type genus: *Merlinius* Siddiqi, 1970

Other genera:

Amplimerlinius Siddiqi, 1976
Geocenamus Thorne & Malek, 1968
Macrotylenchus Sturhan, 2012
Nagelus Thorne & Malek, 1968
Paramerlinius Sturhan, 2012
Pratylenchoides Winslow, 1958
Scutylenchus Jairajpuri, 1971
Telomerlinius Siddiqi & Sturhan, 2014

8.3.10. Key to Genera of Merliniinae (modified from Hosseinvand *et al.*, 2023)

1. Deirid present 2
Deirid absent 7
2. Generally similar to Pratylenchidae, dorsal gland nucleus lying opposite or just anterior to pharyngo-intestinal junction, female tail subcylindroid, pointed spicula .... *Pratylenchoides*
Not as above 3
3. Deirid situated in the four-incisures region of the lateral field 4
Deirid situated in the six-incisures region of the lateral field 6
4. Stylet slender and more than 90 μm, tail terminus annulated, without refractive inner cuticle layer .... *Macrotylenchus*
Stylet short and usually under 50 μm, tail terminus usually smooth, with refractive inner cuticle layer surrounding the tail tip smooth, ... 5
5. Cephalic framework heavily sclerotized, stylet 20-50 μm smooth, ... *Paramerlinius*
Cephalic framework moderate to weak, stylet less than 20 μm .... *Merlinius*
6. Cephalic region continuous with body contour, female tail approximately cylindrical, with refractive inner cuticle layer around its end .... *Amplimerlinius*
Lip region slightly offset by expansion, female tail elongate-conoid to subcylindrical without refractive inner cuticle layer around its end .... *Nagelus*
7. Lateral field with four incisures, cephalic region distinctly offset, stylet slender, basal bulb overlapping intestine .... *Telomerlinius*
Lateral field usually with six incisures, cephalic region continuous to slightly offset, stylet usually robust, basal bulb offset 8
8. Body cuticle marked by longitudinal striae or grooves over entire body .... *Scutylenchus*
Body cuticle without longitudinal striae or grooves (or rarely at anterior or posterior end of body) .... *Geocenamus*

Genus *Pratylenchoides* Winslow, 1958
Diagnosis (after Siddiqi, 2000): Merliniinae. Small nematodes 0.5-0.9 mm long, straight to arcuate on relaxation. **Generally similar to Pratylenchidae in general form**. Lateral fields with 4-6 incisures, **deirids distinct**, cuticle lacking longitudinal striae or grooves. Labial region low, continuous, hemispherical to anteriorly flattened. Pharyngeal glands mostly overlapping intestine dorsally, **dorsal gland nucleus lying opposite or just anterior to pharyngo-intestinal junction**. Amphididelphic. **Female tail subcylindroid**, terminus annulated. Male tail subcylindroid tapering to a conical hyaline tip, more than twice cloacal body diam. long. Spicula pointed, **Gubernaculum small, non-protrusible**.
Type species: *Pratylenchoides crenicauda* Winslow, 1958
Comments: About 15 other species known. Migratory endoparasites of roots, usually found in temperate climatic zones.

Genus *Merlinius* Siddiqi, 1970
(Fig. 8.11)
Diagnosis (after Siddiqi, 2000): Merliniinae. Body less than 1 mm long, straight to arcuate upon relaxation; cuticle lacking longitudinal striae or grooves; lateral fields usually not areolated posterior to pharyngeal region. Deirids distinct. Labial region continuous or slightly offset by expansion; **annules broken by six radial grooves; labial disc indistinct. Stylet usually less than 20 µm long, with conus not tubular near tip**. Median bulb located at or anterior to middle of pharynx. **Vulva closed, slit-like**; epiptygma indistinct. *Vagina vera* in lateral view irregular, not squarish. Female tail conoid to subcylindrical; terminal cuticle of normal thickness.
Type species: *Merlinius brevidens* (Allen, 1955) Siddiqi, 1970
Comments: About 25 or so species described and usually more common in temperate areas.

Genus *Amplimerlinius* Siddiqi, 1976
(Fig. 8.12)

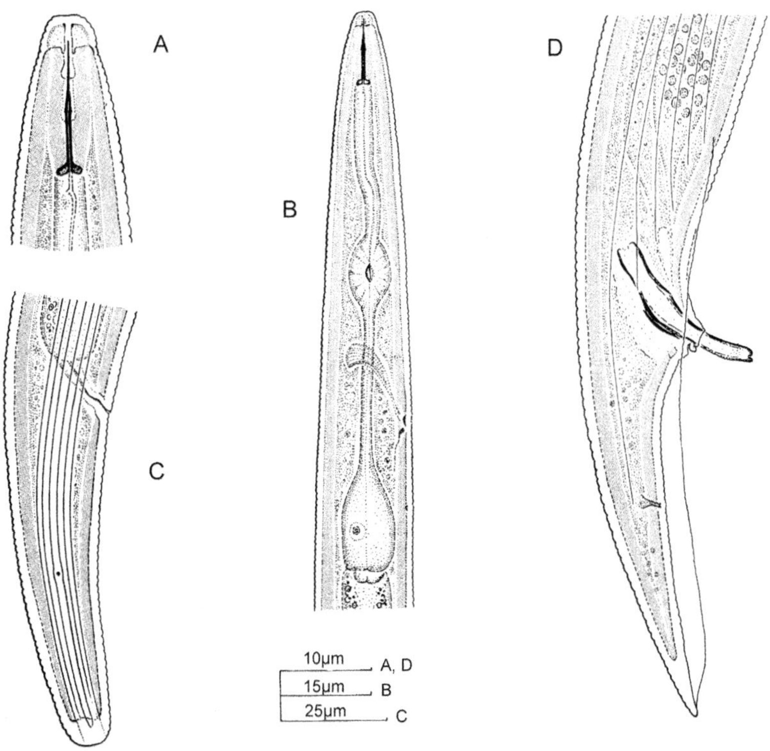

Fig. 8.11. *Merlinius brevidens*. A: Anterior region; B: pharyngeal region with non-overlapping pharyngeal base; C: female tail region; D: male tail region. (Adapted from Siddiqi (1972b), courtesy CABI.)

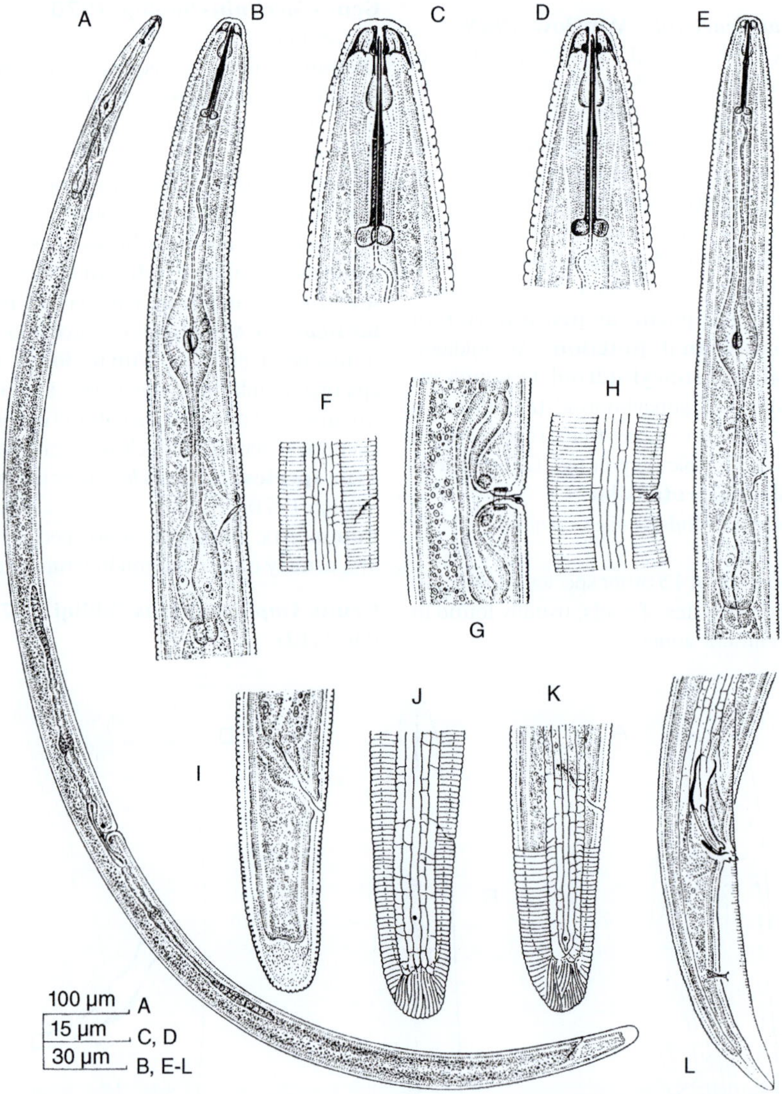

Fig. 8.12. *Amplimerlinius amplus*. A: Female; B: female pharyngeal region; C: female labial region; D: male labial region; E: male pharyngeal region; F, H: lateral field; G: vulval region; I-K: female tail; L: male tail. (After Siddiqi, courtesy CABI.)

Diagnosis (after Siddiqi, 2000): Merliniinae. Medium to large nematodes (L = 1-2 mm), arcuate to strongly curved upon relaxation. Cuticle with prominent, smooth annules. Lateral fields generally with areolation. Deirids distinct. **Labial region continuous**, minutely but distinctly annulated, **annules not broken by radial grooves**, framework heavily cuticularized, inner margins forming a large bulboid stylet guide; labial disc indistinct. **Stylet robust**, 21-42 µm long, with conus tubular up to tip; basal knobs large. Median bulb well developed, located at or posterior to middle of pharynx. Vulva open, situated in a small cavity, with epiptygma. Vagina not cuticularized. **Female tail cylindrical**, occasionally subclavate, with hemispherical, annulated terminus; **cuticle abnormally thickened at tip**. Spicules robust. Gubernaculum trough shaped.

Type species: *Amplimerlinius amplus* Siddiqi, 1976
Comments: Fewer than 20 species described.

8.3.11. Subfamily Belonolaiminae Whitehead, 1960

Diagnosis (after Siddiqi, 2000): Dolichodoridae. Medium to large nematodes (L = 1-3 mm). Lateral fields each with one, two or four incisures, areolated. Deirids absent. Phasmids on tail. Amphidial apertures obscure, probably longitudinal, cleft-like. **Labial region** offset by a constriction, rarely continuous, finely but distinctly annulated (annules 6-10), **usually 4-lobed, lateral sectors much smaller than submedians; labial disc prominent**, round to lemon shaped; six pits around oral aperture; framework lightly to heavily cuticularized, its inner margins extend as elongated, **tubular vestibulum extension**, posteriorly giving support for the attachment of most of the protractor myofibrils. **Stylet very long (62-140 μm), conus comprising 55-75% of its total length**, usually over twice shaft length; knobs rounded. Isthmus generally short. Pharyngeal glands extending over intestine laterodorsally, laterally or lateroventrally; dorsal gland longer and larger than subventrals. **Pharyngo-intestinal junction indistinct, less than one corresponding body diameter from median bulb**. Intestinal fasciculi present. **Vulva with epiptygma**. Vagina in some species cuticularized. Ovaries paired, outstretched. Spermathecae axial or slightly offset. Female tail cylindroid, subcylindroid or conoid, over 1.5 anal body diameters long. Male tail conoid; bursa enclosing tail tip. Testis single, outstretched. Spicules well developed. Gubernaculum large, modified, usually with titillae and recurved distal end.
Type genus: *Belonolaimus* Steiner, 1949
Other genera:
Carphodorus Colbran, 1965
Ibipora Monteiro & Lordello, 1977
Morulaimus Sauer, 1966

8.3.12. Key to Genera of Belonolaiminae

1. Lateral field with single groove-like incisure .... *Belonolaimus*
Lateral field with 2-4 incisures 2
2. Labial region distinctly 4-lobed .... *Ibipora*
Labial region not distinctly 4-lobed 3
3. Lateral field with only 2 inner incisures distinct; labial disc rounded .... *Carphodorus*
Lateral field with all 4 incisures distinct; labial disc lemon shaped .... *Morulaimus*

Genus *Belonolaimus* Steiner, 1949
(Fig. 8.13)

Diagnosis (after Siddiqi, 2000): Belonolaiminae. Large nematodes (L = 2-3 mm), slender (a = 50-80). **Lateral field with single incisure or groove from labial region to tail**, indenting annular contour. Phasmids pore-like, inconspicuous, located near middle of tail. Labial region large, rounded, offset, divided by longitudinal grooves into four well-separated sectors – two subdorsal, two subventral; two very small lateral pseudolips present near apex and bear inconspicuous amphidial apertures; **labial disc offset, round, conspicuous**; framework with basal plate moderately cuticularized and arches lightly cuticularized. Excretory pore located near pharyngo-intestinal junction. **Stylet very long (100-160 μm)**, with conus more than twice length of shaft. Median bulb round, isthmus short, **pharyngo-intestinal junction situated close to median bulb** and anterior to main body of glands. **Pharyngeal glands overlapping intestine lateroventrally**. Intestinal fasciculi broad, conspicuous. Vulva equatorial, with epiptygma. Spermathecae axial, slightly offset. Ovaries paired. **Tails elongate**, in female cylindroid, 3-6 anal body diameters long; with **postanal extension of intestine and annulated terminus**; in male conoid, enveloped by a low bursa. Spicules robust; slightly arcuate, flanged. Gubernaculum large, not protrusible, proximal end directed posteriorly.
Type species: *Belonolaimus gracilis* Steiner, 1949

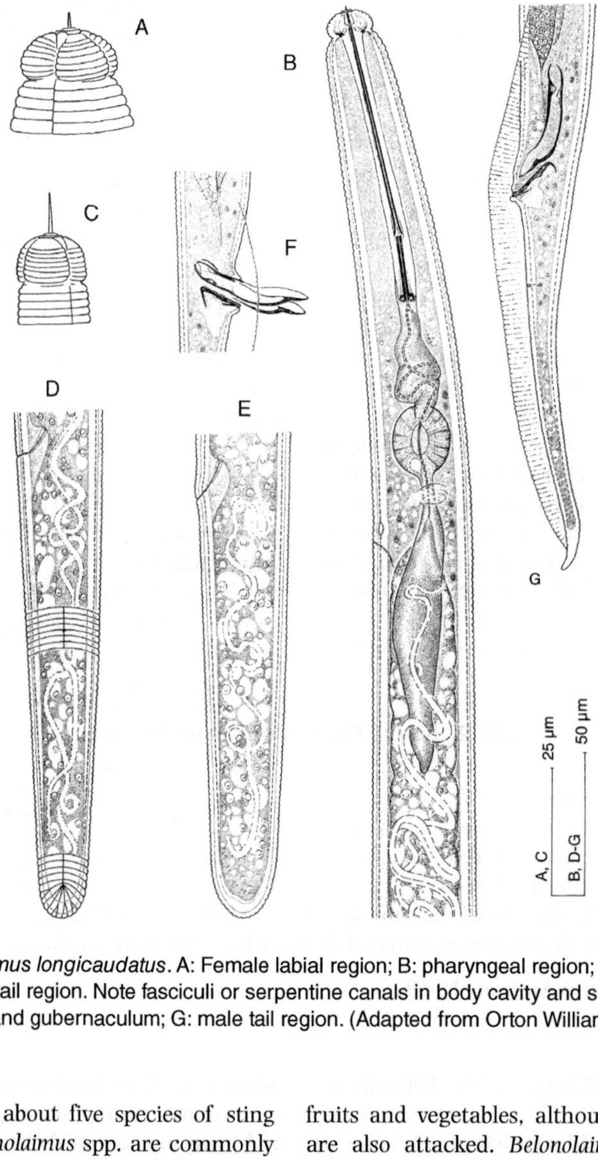

Fig. 8.13. *Belonolaimus longicaudatus*. A: Female labial region; B: pharyngeal region; C: male labial region; D, E: female tail region. Note fasciculi or serpentine canals in body cavity and single lateral line in D; F: male spicules and gubernaculum; G: male tail region. (Adapted from Orton Williams (1974b), courtesy CABI.)

Comments: Only about five species of sting nematode, as *Belonolaimus* spp. are commonly referred to, have been proposed. Having a very long stylet, these nematodes, although ectoparasitic, can feed deep within the root tissue, penetrating through the cortex to the vascular cylinder. They can cause considerable damage resulting in a reduced root system characterized by short, stubby branches with lesions and necrosis. Field symptoms include stunting, wilting under water stress, chlorosis and possibly death of the plant. The genus is restricted to the Americas but has a wide host range, including cereals, fruit trees, cotton, peanut, corn, various fruits and vegetables, although many grasses are also attacked. *Belonolaimus longicaudatus* Rau, 1985 is considered to be among the most important pests of amenity turf (e.g., golf courses) and cotton. This nematode also causes yield losses in vegetable crops, citrus, cotton and peanut, and increases the susceptibility of plants to *Fusarium* (see Sikora *et al.*, 2018). It is restricted to light sandy soils and seems to require high temperatures to thrive (see Smart and Nguyen, 1991; Davis *et al.*, 2018). It is found mainly in the south-eastern Atlantic and Gulf coastal areas of the USA, although it has also been recorded from inland regions such as

Kansas and Nebraska and has been introduced with turf grasses to California and certain islands in the Caribbean region, and also to Australia. The species has also been recorded from other countries but these records may require confirmation. The genus *Ibipora* Monteiro & Lordello, 1977 is sometimes regarded as a junior synonym of *Belonolaimus*. *Ibipora* was described from Brazil and has a similar number of species to *Belonolaimus sensu stricto*. *Ibipora* spp. have four incisures in the lateral field as opposed to a single line in *Belonolaimus*.

8.4. References

Andrássy, I. (1977) *Tylenchus davainei*. *CIH Descriptions of Plant-parasitic Nematodes*, Set 7, No. 97. Commonwealth Agricultural Bureaux, Farnham Royal, UK.

Andrássy, I. (1979) The genera and species of the family Tylenchidae Örley, 1880 (Nematoda). The genus *Tylenchus* Bastian, 1865. *Acta Zoologica Hungarica* 25, 1–33.

Andrássy, I. (1980) The genera and species of the family Tylenchidae Örley, 1880 (Nematoda). The genus *Aglenchus* (Andrássy, 1965) Meyl, 1961, *Miculenchus* Andrássy, 1959, and *Polenchus* gen. n. *Acta Zoologica Hungarica* 26, 1–20.

Andrássy, I. (1984) The genera and species of the family Tylenchidae Örley, 1880 (Nematoda). The genera *Cephalenchus* (Goodey, 1962) Golden, 1971 and *Allotylenchus* gen. n. *Acta Zoologica Hungarica* 30, 1–28.

Andrássy, I. (2007) *Free-living nematodes of Hungary, II (Nematoda Errantia)*. In: *Pedozoologica Hungarica, 4* (Series eds: Csuzdi, C. and Mahunka, S.). Hungarian Natural History Museum and Systematic Zoology Research Group of the Hungarian Academy of Sciences, Budapest, Hungary.

Bert, W. and Borgonie, G. (2006) Suborder Tylenchina. In: Eyualem, A., Andrássy, I. and Traunspurger, W. (eds), *Freshwater Nematodes: Ecology and Taxonomy*. CABI Publishing, Wallingford, UK, pp. 648–695.

Bert, W., Leliaert, F., Vierstraete, A.R., Vanfleteren, J.R. and Borgonie, G. (2008) Molecular phylogeny of the Tylenchina and evolution of the female gonoduct (Nematoda: Rhabditida). *Molecular Phylogenetics and Evolution* 48, 728–744.

Bert, W., Okada, H., Tavernier, I., Borgonie, G. and Houthoofd, W. (2010) Morphological, morphometrical and molecular characterisation of *Filenchus fungivorus* n. sp., a fungivorous nematode from Japan in a most likely polyphyletic genus (Nematoda: Tylenchina). *Nematology* 12, 235–246.

Bridge, J., Plowright, R.A. and Peng, D. (2005) Nematode parasites of rice. In: Luc, M., Sikora, R.A. and Bridge, J. (eds), *Plant Parasitic Nematodes in Subtropical and Tropical Agriculture*, 2nd edn. CABI Publishing, Wallingford, UK, pp. 87–130.

Brzeski, M.W. (1987) Taxonomic notes on *Coslenchus* Siddiqi, 1978 (Nematoda: Tylenchidae). *Annales Zoologici* 28, 417–436.

Davis, R.F., Galbieri, R. and Asmus, G.L. (2018) Nematode parasites of cotton and other tropical fibre crops. In: Sikora, R.A., Coyne, D., Hallmann, J. and Timper, P. (eds), *Plant Parasitic nematodes in Subtropical and Tropical Agriculture*, 3rd edn. CABI Publishing, Wallingford, UK, pp. 738–754.

Fortuner, R. and Luc, M. (1987) A reappraisal of Tylenchina (Nemata). 6. The family Belonolaimidae Whitehead, 1960. *Revue de Nématologie* 10, 183–202.

Geraert, E. (2008) *The Tylenchidae of the World. Identification of the Family Tylenchidae (Nematoda)*. Academia Press, Ghent, Belgium.

Geraert, E. (2011) *The Dolichodoridae of the World. Identification of the Family Dolichodoridae (Nematoda: Tylenchida)*. Academia Press, Ghent, Belgium.

Geraert, E. and Raski, D.J. (1987) A reappraisal of the Tylenchina (Nemata) 3. The family Tylenchidae Örley, 1880. *Revue de Nématologie* 19, 143–161.

Holterman, M., Karssen, G., van den Elsen, S., van Megen, H., Bakker, J., *et al.* (2009) Small subunit rDNA-based phylogeny of the Tylenchida sheds light on relationships among some high-impact plant-parasitic nematodes and the evolution of plant feeding. *Phytopathology* 99, 227–235.

Hosseinvand, M., Eskandari, A., Azizi, K., Palomares-Rius, J.E., Castillo, P., *et al.* (2023) Integrative taxonomy and phylogenetic relationships between representatives of genera within Merliniinae (Nematoda: Merliniidae), with new data on fourteen known and one unknown species. *Journal of Helminthology* 97, e89. DOI: 10.1017/S0022149X23000640.

Jairajpuri, M.S. (1985) *Quinisulcius capitatus*. *CIH Descriptions of Plant-parasitic Nematodes*, Set 8, No. 111. Commonwealth Agricultural Bureaux, Farnham Royal, UK.

Karegar, A. and Geraert, E. (1997a) The genus *Basiria* Siddiqi, 1959 (Nematoda: Tylenchidae) II. Species with four lateral lines and anterior median bulb. *Nematologica* 43, 383–406.

Karegar, A. and Geraert, E. (1997b) The genus *Basiria* Siddiqi, 1959 (Nematoda: Tylenchidae) III. Species with four lateral lines and posterior median bulb. *Nematologica* 43, 407–429.

Karegar, A. and Geraert, E. (1997c) The genus *Basiria* Siddiqi, 1959 (Nematoda: Tylenchidae). I. Introduction and species with two lateral lines. *Nematologica* 43, 327–339.

Karegar, A. and Geraert, E. (1998) The genus *Basiria* Siddiqi, 1959 (Nematoda: Tylenchidae) IV. General discussion, genus diagnosis and key to the species. *Nematologica* 44, 1–13.

Luc, M. and Fortuner, R. (1987) A reappraisal of Tylenchina (Nemata). 5. The family Dolichodoridae Chitwood, 1950. *Revue de Nématologie* 10, 177–181.

Maqbool, M.A. and Shahina, F. (1987) Description of two new and data on three known species of the genus *Tylenchus* Bastian. 1865 (Nematoda: Tylenchidae) from Pakistan with a key to the species of *Tylenchus*. *Pakistan Journal of Nematology* 5, 53–67.

Mizukubo, T. and Minagawa, N. (1986) The genus *Coslenchus* Siddiqi, 1978 (Tylenchidae: Nematoda) from Japan. II. Synonymy of *Coslenchus* over *Cosaglenchus* and *Paktylenchus* based on the phylogenic relationships and a key to the species. *Japanese Journal of Nematology* 15(1985), 14–25.

Nickle, W.R. (ed) (1991) *Manual of Agricultural Nematology*. Marcel Dekker, New York, New York.

Okada, H., Harada, H. and Kadota, I. (2005) Fungal-feeding habits of six nematode isolates in the genus *Filenchus*. *Soil Biology and Biochemistry* 37, 1113–1120.

Orton Williams, K.J. (1974a) *Dolichodorus heterocephalus. CIH Descriptions of Plant-parasitic Nematodes*, Set 4, No. 56. Commonwealth Agricultural Bureaux, Farnham Royal, UK.

Orton Williams, K.J. (1974b) *Belonolaimus longicaudatus. CIH Descriptions of Plant-Parasitic nematodes*, Set 3, No. 40. Commonwealth Agricultural Bureaux, Farnham Royal, UK.

Qing, X. and Bert, W. (2019) Family Tylenchidae (Nematoda): An overview and perspectives. *Organisms Diversity and Evolution* 19, 391–408.

Ryss, A.Y. (1993) Phylogeny of the order Tylenchida (Nematoda). *Russian Journal of Nematology* 1, 74–95.

Siddiqi, M.R. (1972a) *Tylenchorhynchus cylindricus. CIH Descriptions of Plant-parasitic Nematodes*, Set 1, No. 7. Commonwealth Agricultural Bureaux, Farnham Royal, UK.

Siddiqi, M.R. (1972b) *Merlinius brevidens. CIH Descriptions of Plant-parasitic Nematodes*, Set 1, No. 8. Commonwealth Agricultural Bureaux, Farnham Royal, UK.

Siddiqi, M.R. (1973) *Trophurus imperialis. CIH Descriptions of Plant-parasitic Nematodes*, Set 2, No. 22. Commonwealth Agricultural Bureaux, Farnham Royal, UK.

Siddiqi, M.R. (1976) *Tylenchorhynchus annulatus. CIH Descriptions of Plant-parasitic Nematodes*, Set 6, No. 85. Commonwealth Agricultural Bureaux, Farnham Royal, UK.

Siddiqi, M.R. (1981) Six new species of *Coslenchus* Siddiqi, 1978 (Nematoda; Tylenchidae). *Nematologica* 26(1980), 432–447.

Siddiqi, M.R. (1986) *Tylenchida Parasites of Plants and Insects*. Commonwealth Agricultural Bureaux, Farnham Royal, UK.

Siddiqi, M.R. (2000) *Tylenchida Parasites of Plants and Insects*, 2nd edn. CAB International, Wallingford, UK.

Siddiqi, M.R. and Sturhan D. (2014) A remarkable new nematode genus *Telomerlinius* gen. n. (Tylenchida: Merliniidae), with descriptions of two new species. *International Journal of Nematology* 24, 40–48

Sikora, R.A., Coyne, D., Hallmann, J. and Timper, P. (eds) (2018) *Plant Parasitic Nematodes in Subtropical and Tropical Agriculture*, 3rd edn. CABI Publishing, Wallingford, UK.

Smart, G.C. and Nguyen, K.B. (1991) Sting and awl nematodes. In: Nickle, W.R. (ed), *Manual of Agricultural Nematology*. Marcel Dekker, New York, New York, pp. 627–668.

Sturhan, D. (2012) Contribution to a revision of the family Merliniidae Ryss, 1998, with proposal of Pratylenchoidinae subfam. n., *Paramerlinius* gen. n., *Macrotylenchus* gen. n. and description of *M. hylophilus* sp. n. (Tylenchida). *Journal of Nematode Morphology and Systematics* 15, 127–147.

Yeates, G.W. and Bird, A.F. (1994) Some observations on the influence of agricultural practices on the nematode faunas of some south Australian soils. *Fundamental and Applied Nematology* 17, 133–145.

9 Hoplolaimidae

Esther van den Berg[1], Patrick Quénéhervé[2] and David J. Hunt[3]*
[1]National Collection of Nematodes, Biosystematics Division, ARC-Plant Protection Research Institute, Queenswood, South Africa; [2]IRD, UMR 186 Résistance des Plantes aux Bioagresseurs, Pôle de Recherche Agro-environnementale de la Martinique, Martinique (French West Indies); [3]CABI, Silwood Park, Buckhurst Road, Ascot, Berkshire, UK

Abstract
The family Hoplolaimidae, divided into the subfamilies Hoplolaiminae and Rotylenchulinae, is discussed and illustrated at the level of the major economically important genera. A simple key to the subfamilies and most important genera is provided. The morphology and bionomics of each of the major genera (*Acontylus*, *Aorolaimus*, *Aphasmatylenchus*, *Bilobodera*, *Helicotylenchus*, *Hoplolaimus*, *Rotylenchulus*, *Rotylenchus*, *Senegalonema*, *Scutellonema* and *Verutus*) are provided, supplemented with more detailed information for 18 selected species that are either of major economic importance or taxonomic importance. The chapter also covers management aspects of the most economically important species of *Hoplolaimus* spp. (lance nematodes), *Rotylenchulus* spp. (reniform nematodes), *Scutellonema bradys* (yam nematode) and *Helicotylenchus* spp. (spiral nematodes).

9.1. Introduction

The Hoplolaimidae Filipjev, 1934 includes two subfamilies – Hoplolaiminae Filipjev, 1934 and Rotylenchulinae Husain & Khan, 1967 – and contains 12 currently valid genera in which the mature females vary from vermiform to posteriorly swollen, obese or kidney-shaped. The hoplolaimids are essentially obligatory, migratory and sedentary root ectoparasites. They may feed with the anterior body embedded in a root (semi-endoparasitic) or, rarely, completely within the root (endoparasitic). Distribution is worldwide in all climatic regions with a number of species recognized as important pests of crops and implicated in causing major economic yield losses.

Reproduction usually occurs either sexually or parthenogenetically, although the female gonads may produce both ova and sperm. Females have a vulval opening situated between about 50 and 90% from the anterior end. In some groups pronounced sexual dimorphism occurs where the mature female is sedentary and much more obese than the male, the female being kidney shaped or saccate in form whereas the male always remains vermiform.

Feeding in this group occurs mainly ectoparasitically and semi-endoparasitically, but a

*E-mail: tabanids.bite@gmail.com

A revision of Van den Berg, E. and Quénéhervé, P. (2012) Hoplolaimidae. In: Manzanilla-López, R.H. and Marbán-Mendoza, N. (eds), *Practical Plant Nematology*. Mundi-Prensa and Biblioteca Básica de Agricultura, Montecillo, Mexico.

 Practical Plant Nematology 2nd Edition
(eds R.H. Manzanilla-López, D.J. Hunt, N. Marbán-Mendoza and S. Infante Gil)
DOI: 10.1079/9781836990413.0009

few species are true endoparasites. Piercing cell walls with the stylet may cause wounds that may predispose the plant to invasion by other soil pathogens, such as fungi and bacteria, which then cause further damage.

In the absence of a host, nematodes may survive in the soil or in plant residues. Provided that the environment dries slowly, some species are able to enter a reversible anhydrobiotic state (a condition where suspension of growth and development is induced by desiccation) enabling them to survive unfavourable conditions such as cold and dry seasons, temperature extremes and absence of hosts. In some genera, the eggs are the survival stage, being protected by a gelatinous matrix.

Additional information sources for the group include works such as Siddiqi (2000), Castillo and Vovlas (2005), Luc *et al.* (2005), Uzma *et al.* (2015), Sikora *et al.* (2018) and Tran *et al.* (2023). As with other groups, molecular diagnostics are of increasing importance to differentiate species and in constructing a meaningful phylogeny.

9.1.1. Key to Subfamilies and Genera

1. Mature females vermiform .... Hoplolaiminae (2)
Mature females either posteriorly swollen, obese, bilobed or kidney shaped .... Rotylenchulinae (8)
2. Phasmids enlarged to form shield-like scutella .... 3.
Phasmids punctiform, not enlarged to form scutella .... 5.
3. Scutella arranged as one posterior to vulva and one anterior, stylet massive with tulip-shaped knobs .... *Hoplolaimus*
Scutella either on tail or both posterior between vulva and anus, stylet strong but not massive, knobs not tulip-shaped .... 4.
4. Scutella more or less opposite one another and on or near tail .... *Scutellonema*
Scutella not opposite one another, usually between anus and vulva .... *Aorolaimus*
5. Pharyngeal gland overlap mostly dorsal .... 6.
Pharyngeal gland overlap mostly ventral .... 7.
6. Female tail short, about as long as anal body diam., usually bluntly rounded (many species) .... *Rotylenchus*
Female tail 2-3 times anal body diam. long, tapering, conoid (one species) .... *Antarctylus*
7. Phasmids present on or near tail .... *Helicotylenchus*
Phasmids absent .... *Aphasmatylenchus*
8. Mature female kidney shaped .... 9.
Mature female not kidney shaped .... 11.
9. Migratory vermiform immature female present, gland lobe well developed in immature female, mostly ventral or lateral, male with conoid tail, cloacal aperture not subterminal .... 10.
Migratory vermiform immature female absent, gland lobe in other stages short, mostly ventral, male with short rounded tail, cloacal aperture more or less subterminal .... *Verutus*
10. Dorsal gland orifice more than 0.5 stylet lengths posterior to knobs; immature female genital tracts double-flexed; male bursa not reaching tail tip .... *Rotylenchulus*
Dorsal gland orifice less than 0.5 stylet lengths posterior to knobs; immature female genital tracts outstretched; male bursa enclosing tail tip .... *Senegalonema*
11. Body vermiform but mature female more swollen posteriorly, vulva posterior .... *Acontylus*
Body bilobed and greatly swollen in mature female, vulva median .... *Bilobodera*

9.1.2. General Morphological Diagnosis

Labial region mostly broadly or narrowly rounded, set off or continuous, rarely conical, mostly annulated, rarely smooth; when annulated sometimes with longitudinal striae on basal labial annulus, rarely striae on other annuli. Labial framework ranges from weak to very strong. Stylet ranging from short to long and massive, with stylet knobs rounded, flattened, indented, sloping posteriorly, anchor-shaped to tulip-shaped anteriorly; cone about as long as shaft. Dorsal pharyngeal gland opening 4-20 µm from base of stylet. Median bulb well developed, usually rounded. Deirids absent. Pharyngeal glands overlapping intestine dorsally, ventrally or laterally, occasionally abutting. Lateral field without any, or with up to four incisures. Vulva situated from about mid-body to far posteriad, with two equally long genital tracts, sometimes with a double flexure, vulva rarely situated posteriorly with one genital tract degenerate or represented by a short post-vulval uterine sac. Epiptygma often present. Phasmids mostly punctiform, anterior or posterior to anus; sometimes enlarged into shield-like scutella, situated near anus, or far anterior on body, anterior and/or posterior to vulva, rarely completely absent. Tail short, normally less than two anal body diameters long, generally more curved dorsally, mostly broadly or narrowly rounded, more convex conoid on dorsal side, frequently with a small to large ventral process or mucro, rarely conical. Caudalids and cephalids sometimes present. Male with anterior end slightly less developed than female, sometimes degenerate and non-functional. Spicules and gubernaculum varying from small and weakly developed to large and strongly developed, vela may be prominent. Bursa ranging from very well developed and enveloping tail tip to not very distinct and not enveloping tail tip. Gubernaculum often with titillae.

9.2. Hoplolaiminae Filipjev, 1934

= Nemonchinae Skarbilovich, 1959
= Rotylenchoidinae Whitehead, 1958
= Aphasmatylenchinae Sher, 1965
= Rotylenchinae Golden, 1971
= Pararotylenchinae Baldwin & Bell, 1981

9.2.1. General Characters

These nematodes are easily recognized under the dissecting microscope by their medium to large size with body habitus ranging from slightly curved ventrad, C-shaped, to a complete spiral. The stylet is clearly visible and has prominent knobs. The pharyngeal gland overlapping the intestine can be long or short and either mostly dorsal or ventral in position, although occasionally there is no overlap at all, the gland then abutting the intestine. The vulva is usually located in the mid-body region but is occasionally much further posterior and is visible as a straight line in a clear area, sometimes with a slight indentation in the body contour. The body is vermiform, the tail shape varying from narrowly to broadly rounded, more curved dorsally, conical or with different variations of ventral processes or a mucro. When present, males are usually similar to females except for being slightly smaller with a less well-developed stylet and pharyngeal region and sometimes more rounded labial region. The tail is typically conoid and curved ventrad with a distinct bursa and spicules.

These nematodes are migratory ecto-, semi-endo- or endoparasites of higher plants, eggs being laid free in the soil or in plant roots. In this account, seven genera are recognized in the Hoplolaiminae: *Hoplolaimus*, *Aorolaimus*, *Rotylenchus*, *Helicotylenchus*, *Scutellonema*, *Aphasmatylenchus* and *Antarctylus*. *Peltamigratus*, although once widely recognized, is now regarded as a junior synonym of *Aorolaimus*.

9.2.2. Diagnosis

Female and male vermiform. Labial region high, rounded, conoid, truncate, sometimes more flattened anteriorly, set off to continuous, annulated or smooth. Stylet ranging from medium sized to very strong with well-developed stylet knobs that are indented, straight or tulip-like anteriorly. Pharynx overlapping intestine dorsally or ventrally, rarely abutting. Lateral field with none or up to four incisures. Vulva situated at mid-body or far posterior, usually with two equally developed genital tracts but posterior one may either be degenerate or reduced to a post-vulval uterine sac. Phasmids punctiform and near anus, or enlarged and shield-like (scutella) and situated further anterior on body, one being anterior and one posterior to the vulva, or just anterior or just posterior to vulva or, rarely, absent. Tail short, broadly to narrowly rounded, more curved dorsally, with ventral projection; rarely conoid. Male similar to female except for a slightly less developed, although still functional, stylet and pharynx.

Genus *Hoplolaimus* von Daday, 1905
= *Nemonchus* Cobb, 1913
= *Hoplolaimoides* Shakil, 1973
= *Basirolaimus* Shamsi, 1979

Diagnosis

Commonly known as lance nematodes, the genus contains over 30 species. Habitus mostly straight or C-shaped, large, 1-2 mm long. Labial region high, rounded or anteriorly flattened, set off with several annuli, in many cases basal annulus divided into blocks, rarely no longitudinal or transverse striae. Labial framework and stylet massive, basal knobs with anterior projections, almost tulip-like. Scutella present, usually one anterior and one posterior to vulva, **but some species have** both **scutella** (= phasmids) posterior **to vulva** or both anterior **to vulva.** Pharyngeal glands **with the largest overlap dorsally and laterally;** three to six gland nuclei present (those with six nuclei have been placed in *Basirolaimus*, which is regarded as a junior synonym of *Hoplolaimus* herein). **Dorsal pharyngeal gland opening usually about 3-10 µm from stylet base**. Lateral field with none, or up to four incisures, **generally areolated at level of, or anterior to scutella, sometimes with scattered areolations over entire body, rarely not areolated. Vulva situated at mid-body with two equally developed, outstretched genital tracts. Epiptygma present, single or double. Female tail** round, shorter than body diameter at anus, **annulated. When present, male similar to female except for being slightly smaller with a slightly smaller stylet. Spicules and gubernaculum well developed, bursa distinct, rounded, enveloping tail tip.**

Representative Species

Hoplolaimus galeatus (Cobb, 1913) Thorne, 1935 (Figs 9.1A-F, 9.2A-D)

Measurements (after Doucet, 1980; Fortuner, 1991; Vovlas *et al.*, 1991a)

Female: L = 1100-1940 µm; a = 22-34; b = 6.3-10.8; c = 42-82; c' = 0.6-0.7; V = 52-60; stylet = 41.5-54 µm; o = 8.2-17; anterior scutellum = 23-46%; posterior scutellum = 72-88%; tail = 17.5-36.2 µm.

Male: L = 1050-1560 µm; a = 25-32; b = 8.3-10.3; c = 28-40; c' = 1.5-1.7; stylet = 40-48 µm; o = 10-17; anterior scutellum = 29-45%; posterior scutellum = 75-89%; spicules = 40-52 µm; gubernaculum = 20-28 µm; tail = 32.3-44.2 µm.

Description

Female: Habitus ventrally arcuate. **Labial region hemispherical**, set off with **4-6 annuli**; basal annulus with **26-36** longitudinal striations. Labial framework well developed. Stylet large with tulip-shaped stylet knobs. Dorsal pharyngeal gland opening near base of stylet. **Pharyngeal glands overlapping** intestine **dorsally with three gland nuclei**. Excretory pore situated near level of pharyngo-intestinal valve. **Lateral field** with **four incisures**, usually areolated but rarely only partly so. Two enlarged **scutella located anterior and posterior to vulva**. Vulva opening transverse. Epiptygma single or double, usually conspicuous. Two outstretched genital tracts, spermathecae filled with sperm. **Tail broadly to narrowly rounded,** 10-16 annuli long; annuli on tail mostly wider than on rest of body. *Male:* **Similar to female** except for being slightly smaller and more slender. **Labial region higher** with striae on basal annulus. Spicules slightly arcuate, usually with a distinct velum. Gubernaculum with two lateral titillae protruding from cloacal orifice. Capitulum present. Bursa well developed, enveloping tail tip.

Hoplolaimus galeatus is representative of the group of species with four incisures in the lateral field and three pharyngeal gland nuclei.

Bionomics and Distribution: Widely distributed throughout the USA; also reported from Canada, Central and South America and India from a large variety of hosts such as cotton, grape, maize, turf grasses and other gramineaceous plants and pine, oak and sycamore trees. Endoparasitic on cotton causing considerable damage to cortical and vascular tissue. *H. galeatus* causes extensive root damage on sycamore trees but is unable to penetrate completely within the roots and its body partially protrudes out of the root. On pine, most of the cortex of infested roots is destroyed.

***Hoplolaimus pararobustus* (Schuurmans Stekhoven & Teunissen, 1938) Coomans, 1963 (Figs 9.1G-J, 9.3A-F)**

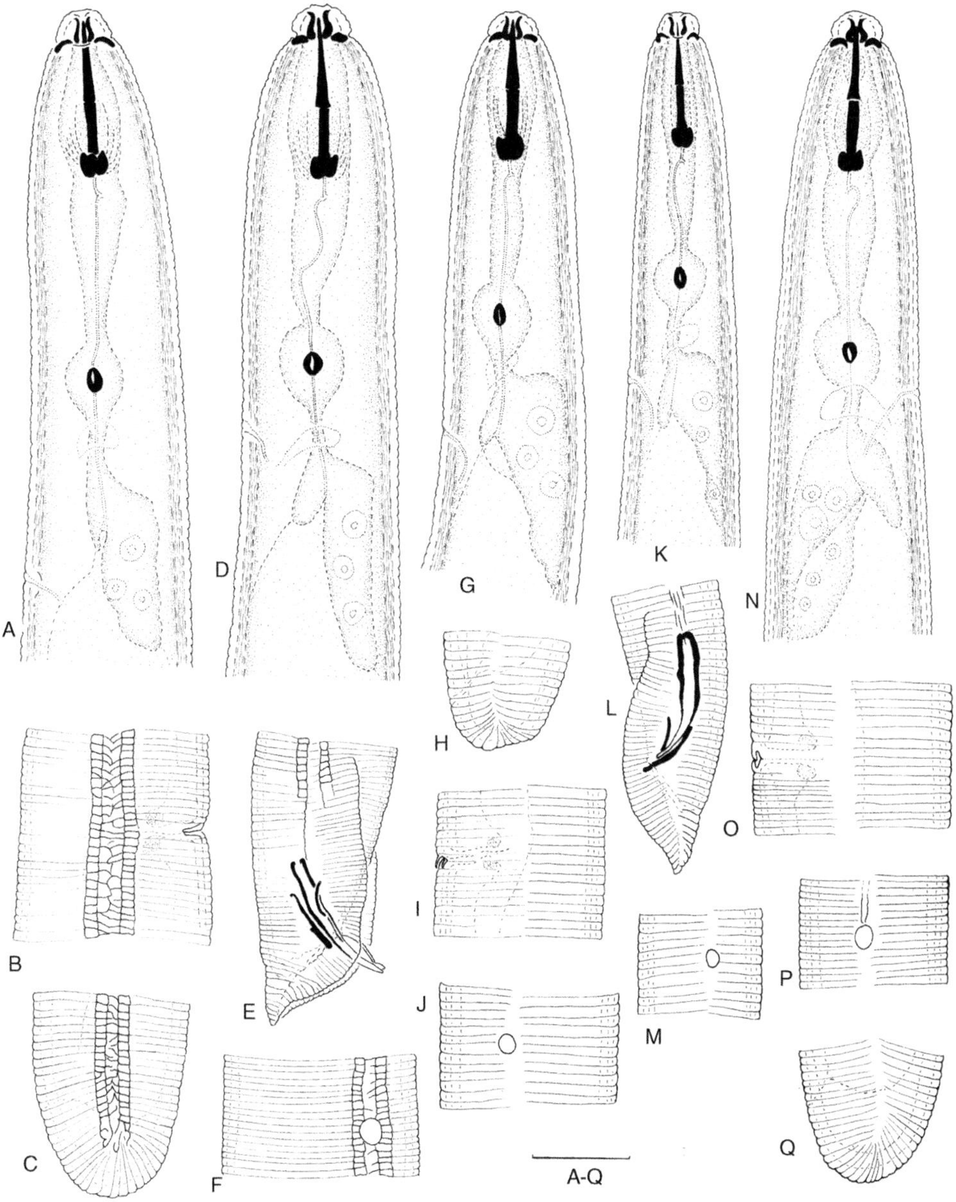

Fig. 9.1. *Hoplolaimus galeatus*. Female, A-C. A: Anterior region; B: lateral field at mid-body; C: tail region. Male, D-F. D: Anterior region; E: tail region; F: lateral field and scutellum. *Hoplolaimus pararobustus*. Female. G: Anterior region; H: tail region; I: lateral field at mid-body; J: lateral field area and scutellum. *Hoplolaimus seinhorsti*. Male, K-M. K: Anterior region; L: posterior region; M: lateral field area and scutellum. Female, N-Q. N: Anterior region; O: lateral field area at mid-body; P: lateral field area and scutellum; Q: tail region. (Scale bar = 30 μm.)

Measurements (after Fortuner, 1991; Sakwe and Geraert, 1992; Larizza *et al.*, 1998).

Female: L = 940-1800 μm; a = 20-39; b = 6-14.1; c = 40-164; c' = 0.4-0.9; V = 51-62; stylet = 37.5-49.7 μm; o = 6.8-13.4; anterior scutellum at 22-52%, posterior scutellum at 58-89% of body length; tail = 16-28 μm.

Male: L = 920-1500 μm; a = 21-37.2; b = 6.2-13.8; c = 22.2-51.9; c' = 1.4-2.1; stylet = 35-48.4 μm; o = 7.1-12.5; anterior phasmid = 23.7-50.4%;

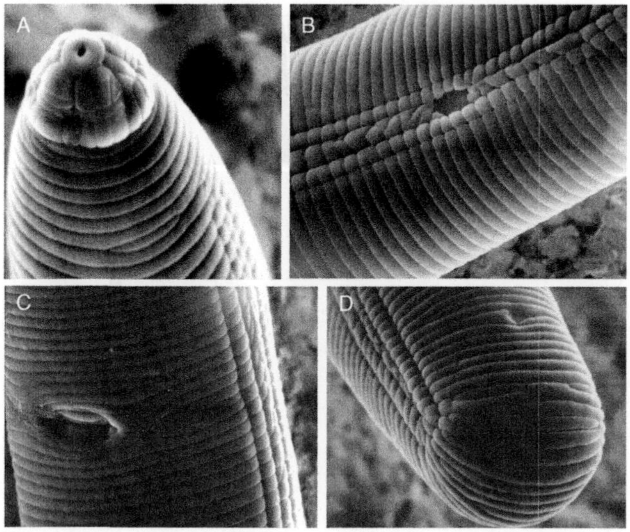

Fig. 9.2. *Hoplolaimus galeatus*. Female. A: Labial region, anterior view; B: lateral field at scutellum; C: vulval area, ventral view; D: tail region. (Photos courtesy of Dr N. Vovlas, CNR, Bari, Italy.)

posterior phasmid = 68.9-85.4%; spicules = 40-57 µm; gubernaculum = 15.4-31 µm; tail = 26.1-35.1 µm.

Description

Female: Habitus straight to slightly arcuate ventrally. **Labial region hemispherical**, set off from body with **3-6 annuli**. Longitudinal striae on annuli variable with **7-25 striae** on basal annulus. Stylet well developed with tulip-shaped stylet knobs. **Pharyngeal glands** overlapping dorsally or dorsolaterally, with **three pharyngeal gland nuclei**. Excretory pore situated from opposite median bulb to as far posterior as opposite middle of basal pharyngeal gland. **Lateral fields** represented by a **single incisure** along entire body length **or** by **none to four incomplete incisures**. Vulva a transverse slit with one epiptygma present, either anterior or posterior, frequently folded into vagina. Two outstretched genital tracts present with rounded, filled spermathecae. Two enlarged **scutella located anterior and posterior to vulva. Tail** short, **rounded to hemispherical** with 7-15 annuli. *Male*: Similar to female. **Labial region rounded, higher** than that of female with 3-5 annuli. Spicules ventrally arcuate, gubernaculum protruding from cloacal opening. **Bursa large, rounded, enveloping tail tip.**

Hoplolaimus pararobustus is representative of a group in which the lateral field is degenerate, never showing the usual complement of four incisures, and with three, not six, pharyngeal gland nuclei.

Bionomics and distribution: Found in most countries of the humid and sub-humid tropics of Africa on various crops, *viz*., coffee, banana, rice, sugarcane, yam, palm and various fruit trees and other plants. On banana, it feeds mostly endoparasitically. On coffee, it feeds semi-endoparasitically causing brown necrotic lesions on the roots. Cortex penetration results in cavities and ruptured cells. Numerous irregular brown necrotic lesions develop on the roots of infested coffee plants. Also known from grass in India and various plants in Pakistan and Sri Lanka.

***Hoplolaimus seinhorsti* Luc, 1958**
(Figs 9.1K-Q, 9.3G-I)

Measurements (after Van den Berg, 1976)

Female: L = 1000-1630 µm; a = 24.1-40; b = 5.2-14.2; c = 38-74; c' = 0.7-0.9; V = 50.5-61; stylet = 38-45 µm; tail = 22-28.7 µm; anterior scutellum at 24-46%, posterior scutellum at 68-90% of body length.

Description

Female: Habitus ranging from arcuate ventrad to an open spiral. **Labial region hemispherical**, well

Fig. 9.3. *Hoplolaimus pararobustus*. Female, A-C. A: Labial region, lateral view; B: scutellum; C: tail region. Male, D-F. D: Labial region, lateral view; E: lateral field at mid-body; F: tail region. *Hoplolaimus seinhorsti*. Female. G: Labial region, *en face* view; H: scutellum; I: tail region.

set off from body with **4-6 annuli**. Longitudinal striae variable with **8-12 longitudinal striae on basal annulus.** Labial framework well developed. Stylet massive, **stylet knobs with two blunt projections on anterior margin separated by two or three small teeth,** often indistinct. **Pharyngeal glands overlapping intestine dorsally** with **six gland nuclei**. **Lateral field with none, a single or several fine incomplete incisures.** Scutella large, one situated **anterior and one posterior to vulva**, situated on either side of body. Vulva situated at mid-body with a single or double epiptygma with two outstretched genital tracts. **Spermathecae not observed. Tail rounded** with 10-17 annuli. *Male:* Not found.

Hoplolaimus seinhorsti is representative of a group with one or no incisures in the lateral field and **six** pharyngeal gland nuclei (such species previously being placed in *Basirolaimus*).

Bionomics and distribution: Reported from tropical countries in Africa, south-east Asia, Central America and South America, it being associated with weeds and various crops including cotton, sugarcane, banana, tomato, cowpea, rice, maize and vegetables. It feeds ectoparasitically on most hosts but is endoparasitic on banana and cowpea, where it causes considerable mechanical damage to root cells. Lateral feeder roots are necrotic or missing. On cotton it was found within the cortical parenchyma where the damage may favour penetration by fungi.

Other species of note

Hoplolaimus indicus is another representative of the group that has six pharyngeal gland nuclei and predominantly a single incisure in the lateral field. It is reported to be a migratory endoparasite of rice in India and Nepal, and widespread on various other crops, fruit trees and grasses, etc. Several species are pathogenic to cotton, including *H. aegypti* in Egypt, *Hoplolaimus columbus* in the USA and Egypt and *H. indicus* in India. All species exhibit ecto- and endoparasitic associations.

H. columbus is an important endoparasite of soybean in the USA, penetrating the endodermis, pericycle and phloem. Soybeans are stunted and produce only a few pods. Other hosts are wheat, maize, millet, lima bean, common bean, watermelon, okra and numerous weeds. On cotton it is semi-endoparasitic and penetrates the cortex but not the endodermis. High populations of more than 4000/cm^3 of soil were associated with the roots of stunted and chlorotic soybean plants that produced few pods. In infested cotton fields, plants were stunted with purple leaves and the bolls were small or shedding.

Hoplolaimus diadematus, found in South America (Brazil, French Guiana) from various unidentified plants, is unique in having the basal annulus of the labial region markedly inflated and divided into blocks by 20 longitudinal striae. It also has two incisures in the lateral field.

Genus *Aorolaimus* Sher, 1963
= *Peltamigratus* Sher, 1964

Diagnosis

A genus with over 50 valid species. Female body spiral to C-shaped. **Labial region** slightly offset or continuous **with or without annuli and/or longitudinal striae**. Stylet medium sized with stylet **knobs flattened, indented or sloping** anteriorly. **Lateral field** with up to **four incisures**, not areolated except in anterior region and sometimes at scutella (phasmids). Pharyngeal glands overlapping intestine dorsally and laterally. Tail short and rounded. **Phasmids enlarged to form scutella**, **variably situated on body, not** opposite each other, anterior to anus, sometimes one anterior to and the other posterior to vulva. **Male** similar to female but **slightly smaller**, spicules and gubernaculum not massive, bursa enveloping tail, bilobed or regular.

Migratory endoparasites of root and/or tubers. Most species are polyphagous. Reproduction is either amphimictic or parthenogenetic. Largely restricted to South America and parts of Africa. Can potentially be confused with *Hoplolaimus*, but generally not so robust.

Representative species

Aorolaimus luci (Sher, 1963) Fortuner, 1987 (Figs 9.4A-G, 9.5A-I)

Measurements (after Sher, 1963a; Van den Berg and Cadet, 1991)

Female: L = 700-880 µm; a = 24-31; b = 6.1-8.3; c = 38-81.5; c' = 0.5-0.8; V = 52-58; anterior scutellum = 77-84%; posterior scutellum = 83-89%; stylet = 26-29 µm; tail = 11-13.2 µm. *Male*: L = 680-830 µm; a = 25.9-31.2; b = 5.1-8.7; c = 45-69; c' = 0.6-0.8; stylet = 25.7-29.4 µm; anterior scutellum = 79%; posterior scutellum = 86%; spicules = 27-31 µm; gubernaculum = 12-15.1 µm; tail = 11-14 µm.

Description

Female: Habitus arcuate ventrad to spiral. **Labial region** slightly set off, **rounded, slightly flattened anteriorly with none or indistinct annuli, with or without longitudinal striae**. Labial framework and stylet strongly developed with stylet knobs indented or slanting anteriorly. Dorsal pharyngeal gland opening 5.5-6.6 µm from base of stylet. Pharynx overlapping dorsally. **Vulva** situated at **mid-body** with **two** outstretched **genital tracts**. **Spermathecae filled with sperm**. Epiptygma double, protruding outwards. **Lateral field** with **four incisures**, **areolated only slightly anteriorly but not on rest of body** nor around scutella; outer two lines disappearing beyond posterior scutellum, rarely continuing to anal level; inner two lines continuing to middle of tail where they coalesce or disappear. **Tail rounded or conoid** with 8-22 annuli, distal annuli narrower. *Male:* Similar to female. **Bursa** broadly rounded posteriorly, **projecting well beyond tail terminus**; coarser, more pointed crenations along posterior margin.

Distribution: Known from the West Indies (Guadeloupe, Martinique, Trinidad), French Guiana and Colombia, associated with various crops, weeds and grass pasture. It was originally described from pineapple on Martinique.

Genus *Aphasmatylenchus* Sher, 1965

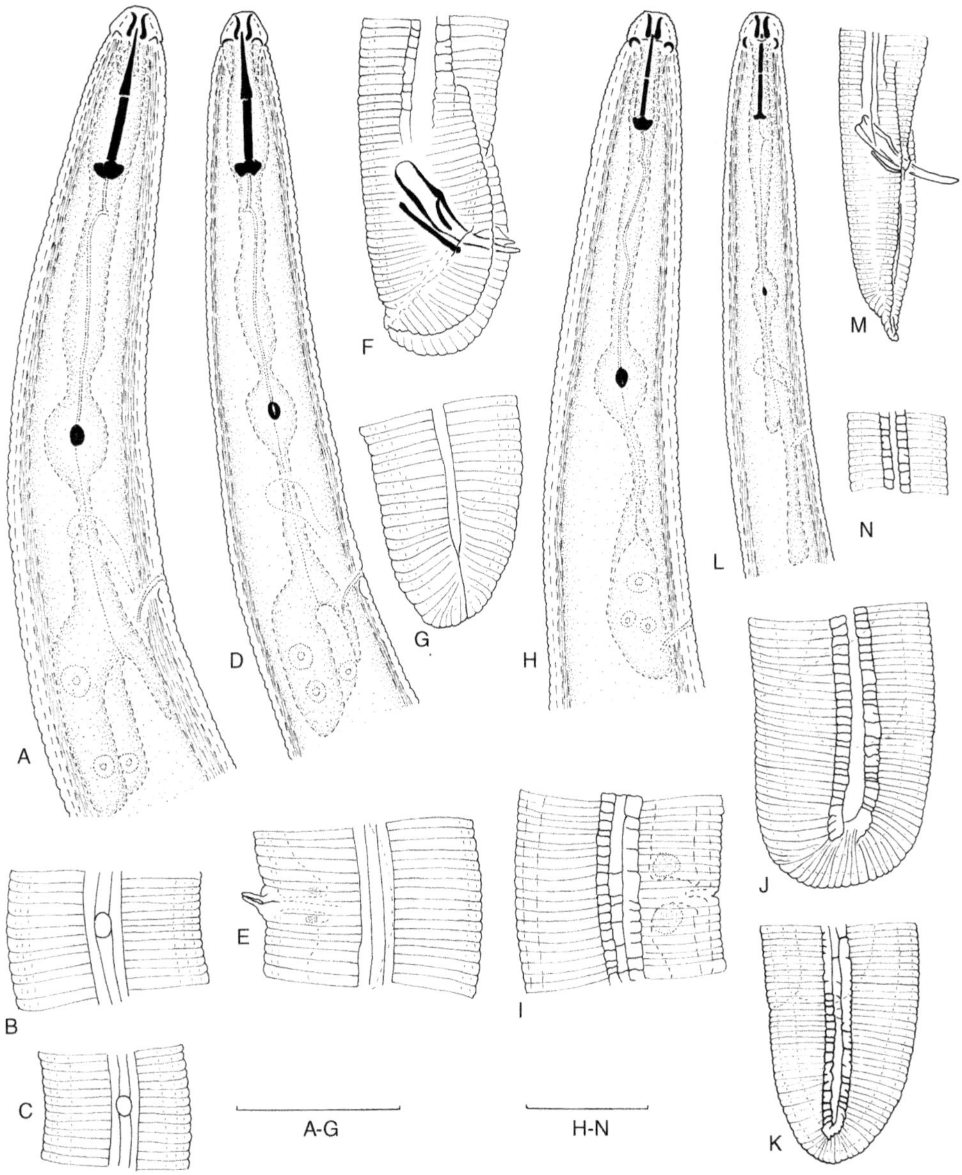

Fig. 9.4. *Aorolaimus luci.* A: Female, anterior region; B: Female scutellum; C: Male scutellum; D: Male, anterior region; E: Female, lateral field area at mid-body; F: Male, tail region; G: Female, tail region. *Aphasmatylenchus nigeriensis*. Female, H-K. H: Anterior region; I: Lateral field at mid-body; J, K: Tail region. Male, L-N. L: Anterior region; M: Tail region; N: Lateral field at mid-body. (Scale bars = 30 μm.)

Diagnosis

A small genus with four species. Female habitus spiral to C-shaped. **Labial region** slightly set off from body, **with 7-11 annuli lacking longitudinal striae. Lateral field with four incisures well or incompletely areolated**. Stylet and labial framework well developed. **Pharyngeal glands overlapping** intestine **ventrally** and **laterally**. **Vulva** situated at **mid-body** with equally developed genital tracts. Epiptygma not

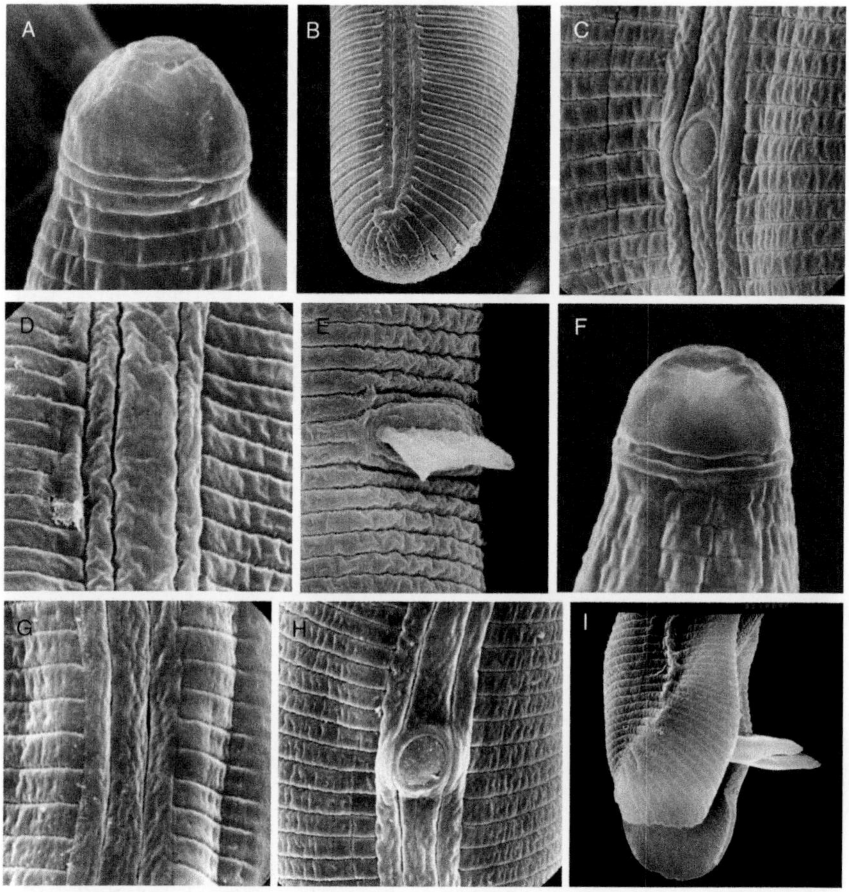

Fig. 9.5. *Aorolaimus luci*. Female, A-E. A: Labial region, lateral view; B: tail region; C: scutellum; D: lateral field at mid-body; E: epiptygma protruding from vulva. Male, F-I. F: Labial region, lateral view; G: lateral field at mid-body; H: scutellum; I: tail region.

seen. **Phasmids absent. Tail 1.5-2 anal body diameters long, more curved dorsally with rounded ends.** Male with a slight sexual dimorphism visible as a smaller anterior end. **Tail conoid, elongate with a long hyaline part. Bursa enveloping tail tip.**

Representative species

Aphasmatylenchus nigeriensis **Sher, 1965 (Figs 9.4H-N, 9.6A-G)**

Measurements (after Sher, 1965; Van den Berg and Cadet, 1991; Vovlas *et al.*, 1991b; Baujard *et al.*, 1998)

Female: L = 680-1556 μm; a = 20.7-38; b = 5-9.8; b' = 5.4-11.1; c = 20-34.9; c' = 0.8-2.4; o = 23.3-35.3; V = 49-60; stylet = 27-35.7 μm; tail = 28-70 μm.

Male: L = 600-960 μm; a = 28.1-35; b' = 4.8-7.8; c = 15-19; c' = 1.3-2.9; stylet = 23.5-27 μm; tail = 36.8-51.5 μm; spicules = 31-36 μm; gubernaculum = 11-15 μm.

Description

Female: Habitus ranging from open C to complete spiral. **Labial region rounded,** slightly set off with **7-10 annuli lacking longitudinal striations**. Labial framework well developed. **Stylet knobs flattened** to slightly **sloping posteriorly**. Pharyngeal glands overlapping ventrally and laterally. **Lateral field** with **four incisures**, **outer bands well or incompletely areolated over entire body**, **inner band occasionally slightly areolated.** Vulva situated at mid-body with two outstretched genital tracts. **Long intestinal overlap over rectum. Phasmid absent**. Tail narrowly to broadly

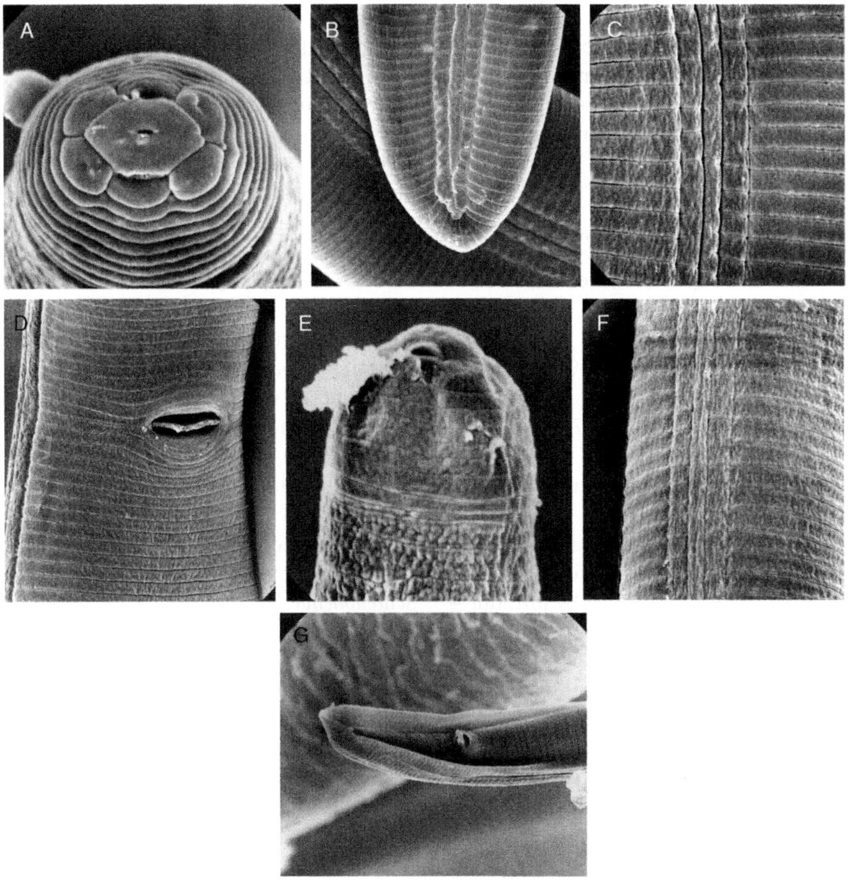

Fig. 9.6. *Aphasmatylenchus nigeriensis*. Female, A-D. A: Labial region, *en face* view; B: tail region; C: lateral field at mid-body; D: vulval area, ventral view. Male, E-G. E: Labial region, lateral view; F: lateral field at mid-body; G: tail region, ventral view.

rounded with 20-36 annuli. Fasciculi present. *Male:* Similar to female except for **less developed** stylet and pharynx. **Bursa low, enveloping tail. Tail** tapering to a **finely rounded tip. Bionomics and distribution**: Recorded in moist soils in West African countries such as Nigeria, Liberia and Ivory Coast from uncultivated land, forest, cacao and rubber trees. Also known from tropical forest in French Guiana.

Genus *Scutellonema* Andrássy, 1958

Diagnosis

A large genus with about 50 species. Small to large nematodes, habitus spiral, C-shaped or almost straight. **Labial region** set off or continuous, **rounded, rarely conical** and **usually annulated** with **basal annulus often divided into blocks** by longitudinal lines. Labial framework is sometimes strongly developed but mostly average sized together with stylet. **Stylet knobs rounded or indented** anteriorly. Pharyngeal gland with greatest **overlap dorsally** and dorsolaterally. Lateral field with four incisures, usually **areolated near scutella and anteriorly**, sometimes areolations scattered over entire body. Vulva situated at mid-body. Epiptygma present, sometimes well protruding but usually folded into vagina or covering vulval opening. **Female tail short, rounded** or dorsally **convex conoid rarely conical or with ventral projection. Phasmids** greatly enlarged to form scutella, opposite each other, **pre- or postanal. Male similar** to female with spicules and gubernaculum. **Bursa large and enveloping tail tip.**

Migratory endoparasites of roots and/or tubers. Most species are polyphagous. Reproduction is either amphimictic or parthenogenetic. This genus is mainly restricted to Africa. A recent review of the phylogeny and other aspects of the genus was by Van den Berg *et al.* (2017).

Representative species

Scutellonema brachyurus (Steiner, 1938) Andrássy, 1958
(Figs 9.7G-J, 9.8A-G)
Measurements (after Fortuner, 1991; Van den Berg, 1998)
Female: L = 500-1000 µm; a = 16.7-36; b = 4.5-10.3; c = 37.2-127.7; c' = 0.4-1; V = 52-67; stylet = 21-31 µm; tail = 7-19 µm; scutellum diameter = 3.5-5 µm.
Male: L = 526-850 µm; a = 22-33; b = 5-7.5; c = 31-58; c' = 1-2; stylet = 20-27 µm; spicules = 20-27.2 µm; gubernaculum = 9-14 µm; tail = 13-19 µm.
Description
Female: Habitus ranging from slightly arcuate ventrad to a complete spiral. **Labial region rounded, slightly set off with 3-5 annuli**; basal annulus with **6-8 longitudinal striations**. Labial framework strongly sclerotized. Stylet well developed with **stylet knobs rounded to slightly indented** anteriorly. Dorsal pharyngeal gland opening 4-7 µm from base of stylet. **Vulva** situated at **mid-body** with one or two epiptygma folded into vagina. Two outstretched genital tracts; spermathecae very rarely filled with sperm. Vulval glands small, round to oval. **Lateral field with four lines**, areolated opposite **anterior** end **and at scutella. Scutella situated from two annuli anterior to two annuli posterior to anus. Tail** varying from **narrowly to bluntly rounded**, often more so on dorsal side with 7-18 annuli; annuli slightly larger on tip. *Male*: **Very rare.** Similar to female except for a **weaker stylet** and pharyngeal area. Spicules and gubernaculum well developed. **Bursa well developed, broadly rounded and enveloping tail tip**. Scutella situated opposite to just posterior to level of cloacal aperture.
Bionomics and distribution: *Scutellonema brachyurus* has a worldwide distribution, especially in Africa and is associated with a large variety of crops and natural vegetation. It is primarily ectoparasitic on roots but instances of semi-endoparasitism and endoparasitism have been reported. It feeds ecto- and endoparasitically on African violets and semi-endoparasitically on the roots of sycamore, with only the anterior end of the body embedded in the cortex. It produced cavities in the roots of *Hippeastrum* sp. and red spider lilies. *S. brachyurus* produces lesions on the roots of tobacco, markedly reducing the root growth, top growth and height of the plants. A high density of *S. brachyurus* can cause stimulation rather than a reduction in root growth of longleaf pine. *S. brachyurus* is the commonest species of the genus in South Africa, recorded from almost every crop plant, fruit tree, ornamental and natural vegetation, and is very frequently found on tobacco in South Africa and Zimbabwe, where it causes root lesions and reduction in top weight and height. Also causes poor growth of sugarcane in Madagascar.

Scutellonema bradys (Steiner & LeHew, 1937) Andrássy, 1958
(Figs 9.7A-F, 9.9A-F)
Measurements (after Sher, 1963b; Germani *et al.*, 1985)
Female: L = 750-1190 µm; a = 20-54. 3; b = 4.8-11.5; c = 25-60; c' = 0.5-1.16; V = 36-62; stylet = 24-32 µm; tail = 27 µm; scutellum diam. = 1.5-5 µm.
Male: L = 650-1080 µm; a = 20.1-44.8; b = 5.9-11; c = 21.5-53; c' = 0.9-2.1; stylet = 20-29 µm; spicules = 23-35 µm; gubernaculum = 12-17 µm; tail = 11-16 µm.
Description
Female: Habitus straight to slightly arcuate. Labial region broadly rounded, deeply set off with 7-9 annuli; no longitudinal striations on basal annulus. Labial framework and stylet well developed with stylet knobs flattened, indented or irregular anteriorly. Pharynx overlapping dorsally or dorsolaterally. Vulva situated at mid-body with two outstretched genital tracts. Vulval slit with conspicuous internal thickenings at each end. Spermathecae filled with sperm. Lateral fields with four incisures, areolated anteriorly and opposite scutella and, rarely, partially at mid-body. Scutella situated from slightly anterior to slightly posterior to anus. Tail rounded to squarish with 13-23 annuli. *Male*: Similar to female except for sexual dimorphism. Bursa rather narrow, enveloping tail tip. Scutella usually just posterior to cloacal aperture.

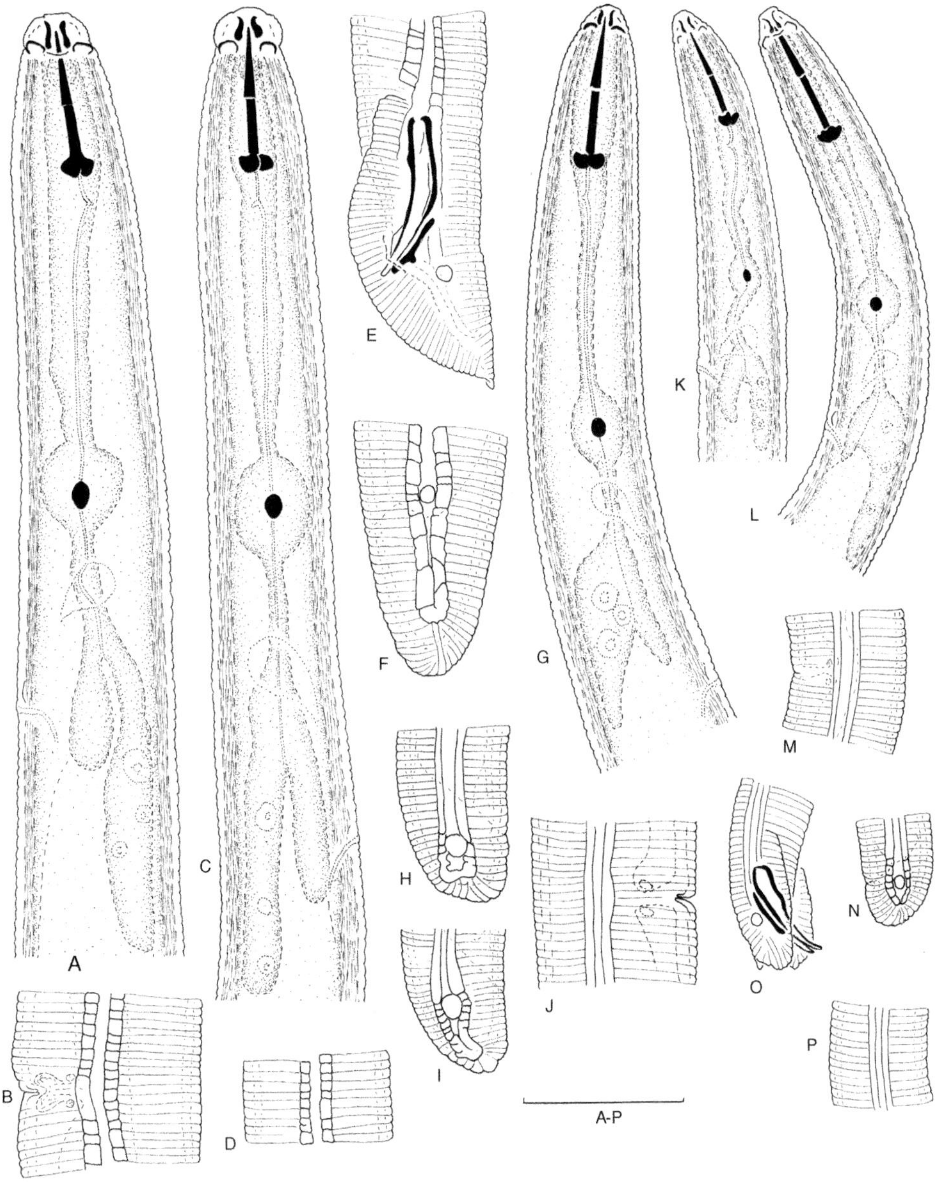

Fig. 9.7. *Scutellonema bradys*. A: Female, anterior region; B: female, lateral field at mid-body; C: male, anterior region; D: male, lateral field at mid-body; E: male, tail region; F: female, tail region. *Scutellonema brachyurus*. Female. G: Anterior region; H, I: tail regions; J: lateral field at mid-body. *Scutellonema truncatum*. K: Female, anterior region; L: male, anterior region; M: female, lateral field at mid-body; N: female, tail region; O: male, tail region; P: male lateral field at mid-body. (Scale bar = 30 μm.)

Bionomics and distribution: *Scutellonema bradys*, the yam nematode, is a very important pest of yam in Africa, Central and South America, India and southern USA where it causes the disease 'dry rot of yam'. The dissemination of this species follows the geographic dissemination of yam planting material. It is predominantly an endoparasite of yam roots and tubers causing superficial lesions that facilitate the colonization of deeper tissue by fungi and potentially leading

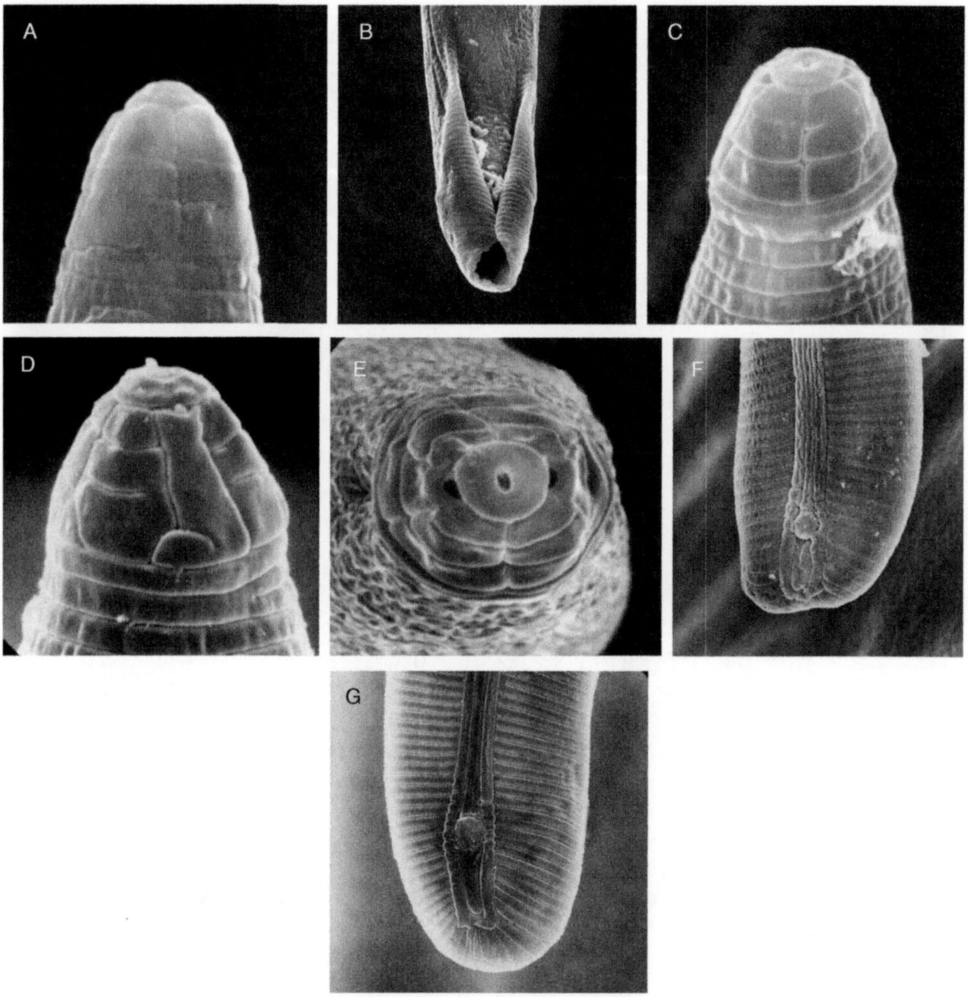

Fig. 9.8. *Scutellonema brachyurus*. Male, A, B. A: Labial region, lateral view; B: tail region; Female, C-G. C, D: Labial region, lateral view; E: labial region, *en face* view; F, G: tail region.

to complete destruction of tubers. Large populations build up in stored yam tubers and cause severe damage in yam barns. The nematodes attack the primary roots growing from the rhizomatous head of the tuber and can enter the developing tubers through their growing points. Invasion of the tubers can also occur through cracks or damaged areas in the epidermis. Eggs are laid in soil or plant tissue (roots and tubers) where they hatch. All stages seem to be infective.

Scutellonema truncatum Sher, 1963 (Figs 9.7K-P, 9.10A-F)

Measurements (after Van den Berg and Heyns, 1973; Van den Berg and Van den Oever, 1993)

Female: L = 500-800 μm; a = 18.3-34.4; b = 5.2-8.6; c = 41.7-158; c' = 0.3-0.7; V = 55-63; stylet = 21.3-29 μm; tail = 4-16.2 μm; scutellum diam. = 1.8-3.7 μm.

Male: L = 544-562 μm; a = 28-32; b = 5; c = 46-78; c' = 0.6-1; stylet = 22-23 μm; spicules = 20 μm; gubernaculum = 7-9 μm; tail = 7-12 μm.

Description

Female: Habitus spiral. **Labial region conical, blunt, not or slightly set off without annuli; six longitudinal incisures present**. Labial framework and stylet moderate with **stylet knobs flattened and indented anteriorly**. **Pharynx overlapping dorsally**. **Vulva** situated at **mid-body**. Two outstretched genital tracts

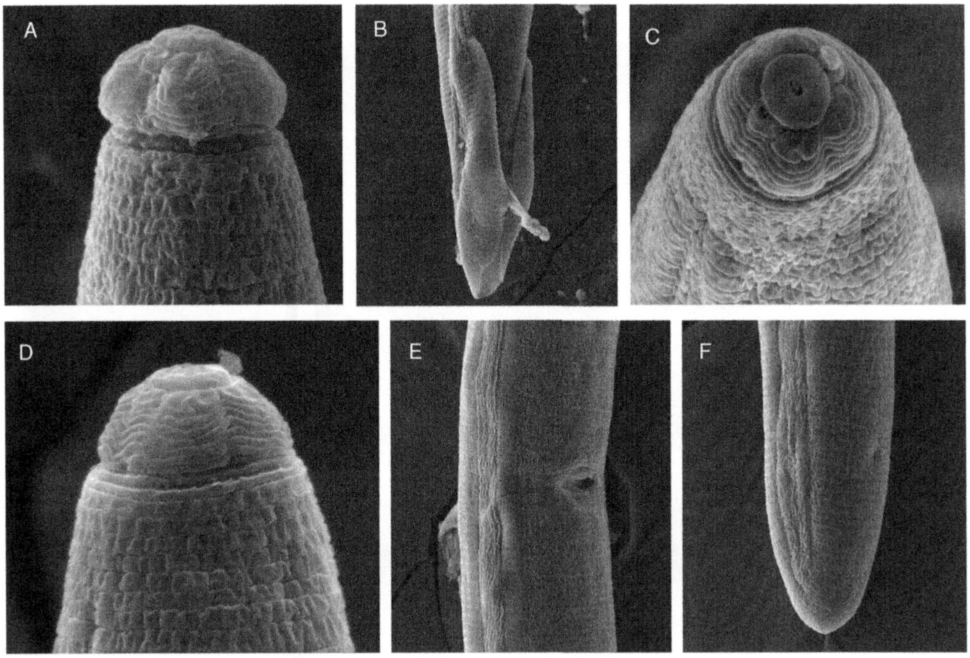

Fig. 9.9. *Scutellonema bradys*. Male, A, B. A: Labial region, lateral view; B: tail region. Female, C-F. C: Labial region, *en face* view; D: labial region, lateral view; E: vulval area, ventral view; F: tail region, with scutellum.

with **spermathecae mostly empty** but rarely filled with sperm when males are present. **Lateral field with four incisures, areolated anteriorly and at scutella.** Scutella situated in region of anus. **Tail rounded, more convex dorsally, rarely with rounded point on ventral side** with 7-16 annuli. *Male*: Similar to female. Scutella situated opposite or just posterior to cloacal aperture. Bursa enveloping tail tip. **Bionomics and distribution**: Originally described from Zimbabwe. Abundant and widespread in South Africa on grasses, *Aloe* sp., pineapple, fruit trees and indigenous vegetation and also reported from pumpkin in Mozambique and from Botswana.

Genus *Rotylenchus* Filipjev, 1936

= *Anguillulina* (*Rotylenchus*) Filipjev, 1936
= *Gottholdsteineri* Andrássy, 1958
= *Orientylus* Jairajpuri & Siddiqi, 1977
= *Calvatylus* Jairajpuri & Siddiqi, 1977
= *Interrotylenchus* Eroshenko, 1984
= *Scutellonemoides* Eroshenko, 1984
= *Varotylus* Siddiqi, 1986
= *Pararotylenchus* Baldwin & Bell, 1981
= *Plesiorotylenchus* Vovlas, Castillo & Lamberti, 1993

Diagnosis

Females moderate to long, habitus spiral or C-shaped. **Labial region set off or continuous, rounded, hemispherical, conoid or truncate, generally annulated (none or up to 10 annuli)** and **with mostly longitudinal striae** dividing some or most of the annuli into blocks, **rarely without annuli**. Labial framework and stylet moderate to robust; stylet 20-65.5 μm long with rounded to indented stylet knobs. **Pharyngeal glands overlapping dorsally** and dorsolaterally, **rarely not overlapping, mostly with three gland nuclei, but rarely up to seven present. Vulva situated from** anterior **(45%) to** well posterior to mid-body and **as far as 76%** (*Rotylenchus whiteheadi*), with **two genital tracts, rarely one, or a postvulval uterine sac** (*R. whiteheadi*). Epiptygma single or double. Female **tail, short, rounded, hemispherical**, usually more **convex dorsally, rarely conical or with a small rounded ventral projection. Phasmids usually punctiform**, located from **quite far anterior to posterior to anus. Lateral field** with **four** incisures, **mostly areolated anteriorly, rarely sparsely areolated over entire body**, or

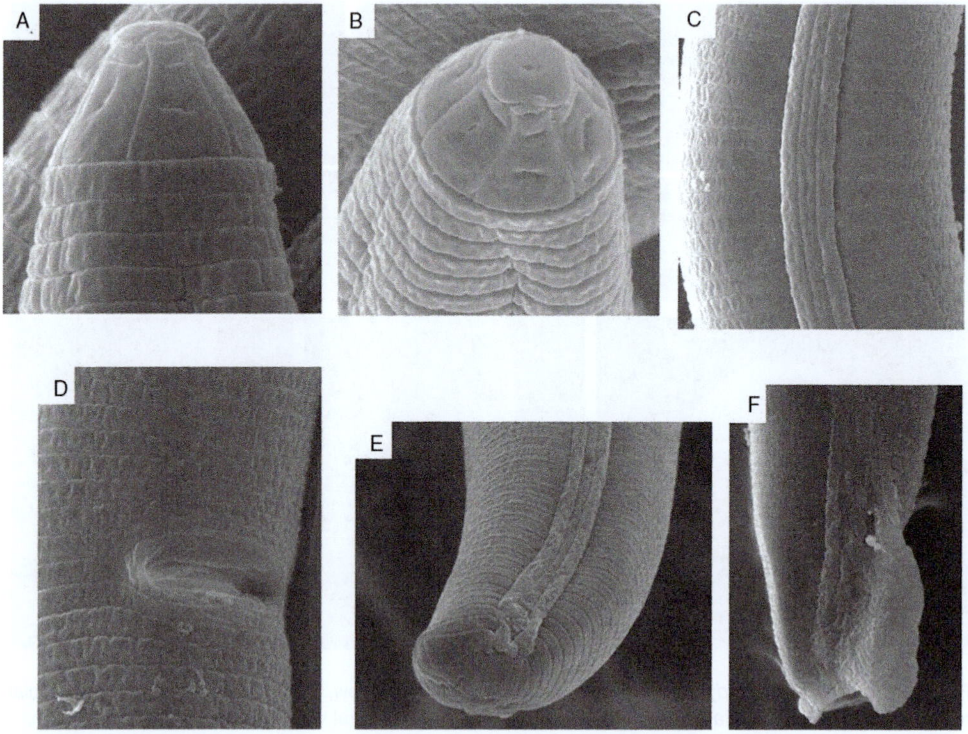

Fig. 9.10. *Scutellonema truncatum*. Female, A-E. A: Labial region, lateral view; B: labial region, *en face* view; C: lateral field at mid-body; D: vulval area, ventral view; E: tail region. Male. F: Tail region.

irregularly at mid-body or restricted to phasmid region. **Annuli outside lateral field rarely with longitudinal striations in pharyngeal region, or over entire body**. **Males**, when present, similar to female except for being **slightly smaller** and stylet slightly smaller, sometimes no sexual dimorphism present. Spicules well developed and **bursa mostly enveloping tail, rarely with a fringe-like appearance** of annuli on posterior part or projecting ventrally beyond tail tip.

Bionomics and distribution: *Rotylenchus* is a large genus, its species having a worldwide distribution and occurring on every continent, including Antarctica where *R. capensis* was found associated with moss in lithosols from inland nunataks in western Dronning Maud Land. *Rotylenchus* species are migratory ectoparasites on the root hairs and epidermal cells of the root systems of higher plants. Damage symptoms include necrotic lesions and cavities in epidermal and cortical tissue, yellowing and stunting of top growth, reduced root systems and yield loss. Semi-endoparasitism has also been reported. *Rotylenchus* species have been recorded from sugarcane from various countries including Barbados, Colombia, Cuba, Dominican Republic, Fiji, India, Japan, Malaysia, Mauritius, The Philippines and South Africa. *Rotylenchus* species also caused a significant reduction in root and top weights of *Pinus* seedlings in the savannah area of Nigeria and caused damage to fruit trees in China.

Representative species

Rotylenchus alius Van den Berg, 1986 (Figs 9.11A-G, 9.12A-C)

Measurements (after Van den Berg, 1986a, 1998)

Female: L = 490-900 μm; a = 18-33.1; b = 4.7-7; c = 24-51; c' = 0.9-2; V = 55-66; stylet = 20-31.3 μm; tail = 11-29 μm.

Male: L = 490-700 μm; a = 28-40.2; b = 5-6; c = 22-31; c' = 1.9-2.1; stylet = 17.7-23 μm; spicules = 18-24 μm; gubernaculum = 7-10 μm; tail = 19-29 μm.

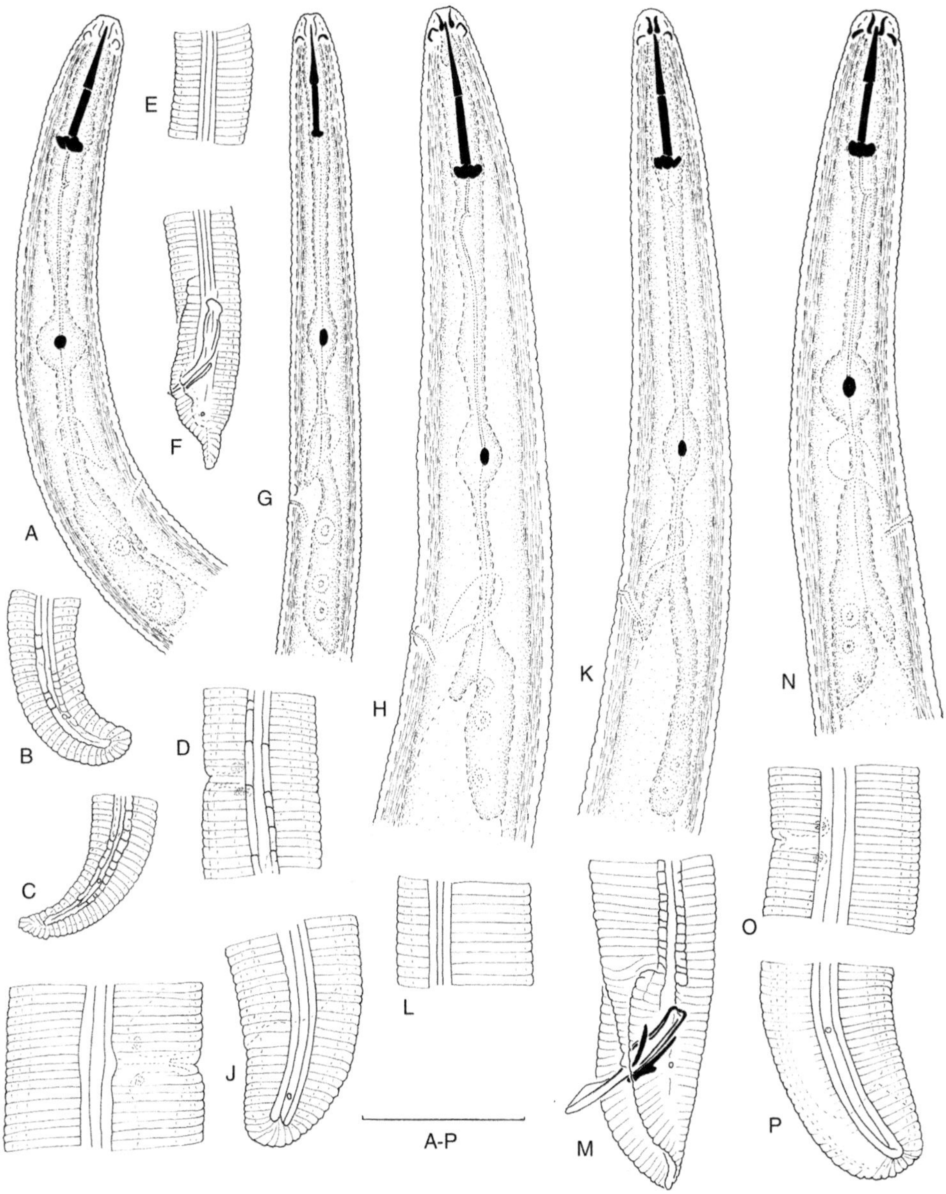

Fig. 9.11. *Rotylenchus alius*. Female, A-D. A: Anterior region; B, C: tail region; D: lateral field at mid-body. Male, E-G. E: Lateral field at mid-body; F: tail region; G: anterior region. *Rotylenchus caudaphasmidius*. Female, H-J. H: Anterior region; I: lateral field at mid-body; J: tail region. Male, K-M. K: Anterior region; L: lateral field at mid-body; M: tail region. *Rotylenchus unisexus*. Female. N: Anterior region; O: lateral field at mid-body; P: tail region. (Scale bar = 30 μm.)

Description

Female: Habitus an open spiral. **Labial region truncate, blunt anteriorly, not set off** with **4-5 annuli. Basal annulus without longitudinal striae.** Labial framework moderate. Stylet slender with stylet knobs flattened to indented anteriorly. **Pharyngeal glands short with a dorsal overlap**. Vulva situated at mid-body

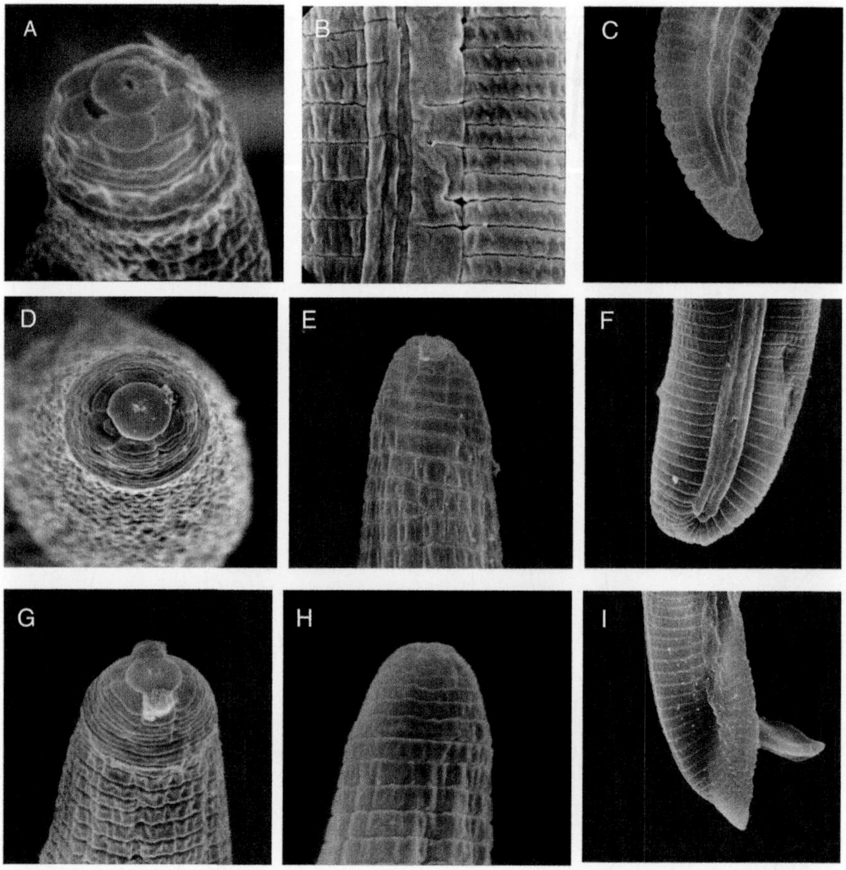

Fig. 9.12. *Rotylenchus alius*. Female. A: Labial region, *en face* view; B: lateral field at mid-body; C: tail region. *Rotylenchus caudaphasmidius*. Female, D-F. D: Labial region, *en face* view; E: labial region, lateral view; F: tail region. Male, G-I. G: Labial region, anterior view; H: labial region lateral view; I: tail region.

with two outstretched genital tracts. Spermatheca round, filled with sperm. **Lateral field with four incisures, areolated anteriorly and incompletely entire body; middle band rarely with transverse incisures**; no longitudinal striations on rest of body. **Phasmids punctiform**, situated from **three annuli anterior to four annuli posterior to anus. Tail curved ventrally with a ventral point**, with 8-15 annuli. *Male*: Similar to female. Phasmids situated anterior to cloacal aperture. **Tail ending in a long, acute tip. Bursa very narrow**, enveloping tail tip.

Bionomics and distribution: Found associated with natural vegetation in South Africa and also reported in Korea from *Musa* sp. and on barley in Iran.

***Rotylenchus caudaphasmidius* Sher, 1965 (Figs 9.11H-M, 9.12D-I)**

Measurements (after Van den Berg *et al.*, 1997; Castillo and Vovlas, 2005)

Female: L = 600-900 μm; a = 22-32; b = 4.3-7; c = 36-68; c' = 0.8-1.2; V = 55-63; stylet = 23.5-32 μm; tail = 14-20 μm.

Male: L = 600-810 μm; a = 25-36.6; b = 4.8-7.2; c = 28-42.2; c' = 1.4-1.8; stylet = 21-30 μm; spicules = 23-33 μm; gubernaculum = 8-14 μm; tail = 19-24 μm.

Description

Female: Habitus C-shaped, open 6-shaped or spiral. **Labial region narrow, rounded, not set off with 4-5 annuli; basal annulus with six indistinct longitudinal incisures.** Labial framework distinct. Stylet moderate with stylet knobs flattened to slightly concave anteriorly. Dorsal pharyngeal gland opening 6-9 µm from stylet base. **Pharyngeal glands overlapping dorsally**. **Lateral field** with **four** incisures, **areolated anteriorly**, **rarely** occasional areolations **in outer bands and opposite phasmid**. **Cuticle outside** lateral field **without longitudinal striations**. Vulva situated at mid-body with two outstretched genital tracts. Spermathecae large, round, filled with sperm. Epiptygma single or double, folded into vagina. **Phasmids punctiform, 4-8 annuli posterior to anus**. Tail rounded, more so dorsally with 12-16 annuli, some annuli more irregular on tip. *Male*: Similar to female. Habitus arcuate ventrally. No longitudinal striations on basal labial annulus. Phasmids situated from just posterior to cloacal aperture to about mid-tail.

Bionomics and distribution: Described from Peru. Also known from natural vegetation in several localities in South Africa, from cotton in Botswana, *Aspidosperma* sp. in French Guiana and forest soil in Dominica, West Indies.

Rotylenchus unisexus **Sher, 1965**
(Figs 9.11N-P, 9.13A-D)

Measurements (after Van den Berg and Heyns, 1974; Van den Berg, 1986b)

Female: L = 500-1000 µm; a = 19-34.7; b = 5.2-9; c = 30.3-105.6; c' = 0.3-1.3; V = 47-59; stylet = 19.5-29 µm; tail = 6.6-23.5 µm.

Description

Female: Habitus C-shaped to a complete spiral. **Labial region hemispherical, not or very slightly set off with 3-5 annuli**, **basal annulus** with **16-19 longitudinal striations**. Labial framework moderately developed. Stylet moderate with stylet knobs flattened to indented anteriorly. Dorsal pharyngeal gland opening 2.9-8.1 µm from stylet base. **Pharyngeal glands overlapping dorsally. Lateral field with four incisures**, **areolated anteriorly**. No longitudinal striations on rest of body. Vulva situated at mid-body with two outstretched genital tracts. Spermathecae empty. Epiptygma mostly double, folded into vagina. **Phasmids punctiform, 2-17 annuli anterior to anus. Tail narrowly to broadly rounded, rarely**

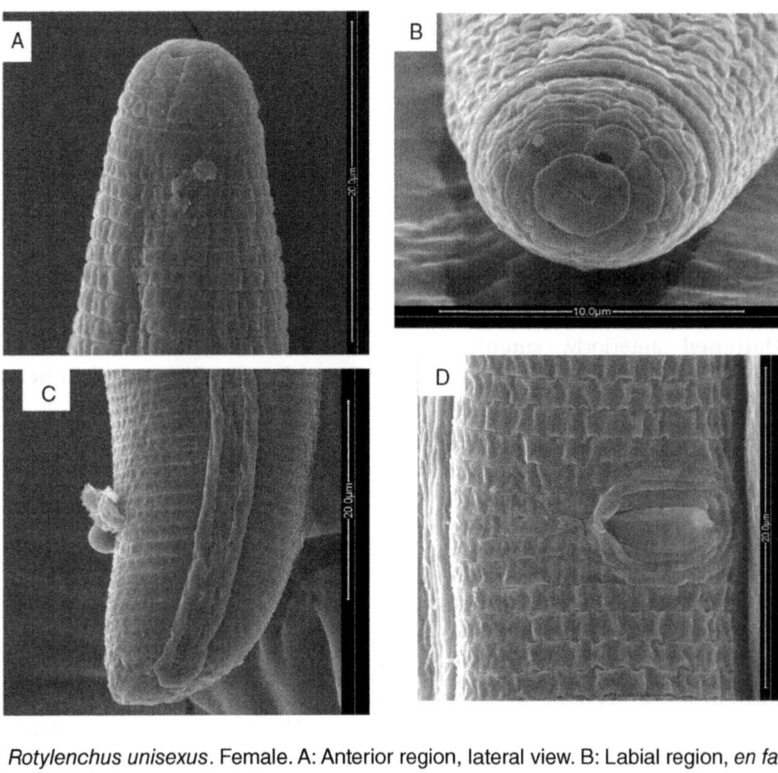

Fig. 9.13. *Rotylenchus unisexus*. Female. A: Anterior region, lateral view. B: Labial region, *en face* view. C: Tail region. D: Vulval area, ventral view.

pointed or indented with 6-15 annuli. *Male*: Not found.

Bionomics and distribution: *Rotylenchus unisexus* is widespread in Africa, *viz*., Ethiopia, Kenya, Zaire and South Africa, from the rhizosphere of various crops, grasses and reeds, ornamentals, indigenous plants and trees. Also found in New Zealand, Norfolk Island, vineyards in China and various hosts in Spain.

Other species of note

English boxwood (*Buxus sempervirens* L.) is attacked by *Rotylenchus buxophilus*; the nematodes partially embedding themselves within the roots and causing numerous tiny, brown necrotic lesions; *Rotylenchus robustus* causes stunting, yellowing, a severe reduction in the size of the root system and yield loss in carrots; and *Rotylenchus uniformis* was found to cause severe injury to peas. The latter species is also a parasite of carrots, grass, young trees and several conifers. Another parasite of carrots is *Rotylenchus laurentinus*, found along the Adriatic coast of Italy.

Genus *Helicotylenchus* Steiner, 1945
= *Rotylenchoides* Whitehead, 1958
= *Zimmermania* Shamsi, 1973

Diagnosis

Female habitus spiral to nearly straight. **Lateral field** with **four lines sparsely areolated except anteriorly. Labial region continuous to slightly set off, rounded, conical or slightly flattened** anteriorly, **smooth or annulated, lacking longitudinal striae on basal annulus.** Stylet and labial framework average sized. **Pharyngeal glands overlapping intestine ventrally** and laterally, **rarely dorsally**. Lateral field with four incisures. **Vulva** mostly situated near **mid-body**, occasionally far posteriorly. Paired **genital tracts** about **equal in length** or posterior tract reduced, or sometimes a post-vulval uterine sac present. Epiptygma present but folded into vagina. **Tail 1-2.5 anal body diameters long**, more **curved dorsally, rounded or asymmetrical, sometimes with a digitate ventral projection or mucro. Phasmids punctiform, pre- or postanal.** Male similar to female except for being **slightly smaller** and anterior part not so well developed. Spicules well developed and bursa enveloping tail.

Bionomics and distribution: *Helicotylenchus* is a large cosmopolitan genus of spiral nematodes with over 230 species. They can survive in a variety of habitats and are also capable of anhydrobiosis and are well adapted to colder conditions. Identification is hindered by the sheer number of species and phenotypic plasticity, while molecular studies indicate the occurrence of cryptic speciation (Subbotin *et al*., 2015; Mwamula *et al*., 2024, among others). Feeding causes root lesions and facilitates the entry of fungal pathogens. *Helicotylenchus stylocercus* was recovered from aerial litter found on the branches of trees in the Cloud Forest in Costa Rica, and *Helicotylenchus teres* (= *Helicotylenchus thornei*) on moss on tree trunks in Puerto Rico. The first tylenchid reported from continental Antarctica was a *Helicotylenchus* sp. *Helicotylenchus pseudorobustus* was reported from freshwater and *Helicotylenchus hydrophilus* has been reported as only occurring in or near wet environments. A number of *Helicotylenchus* spp., that is, *Helicotylenchus cavenessi*, *H. crenacauda*, *Helicotylenchus dihystera*, *Helicotylenchus erythrinae*, *Helicotylenchus microcephalus*, *Helicotylenchus mucronatus* and *Helicotylenchus multicinctus*, have been reported from aquatic vascular plants. In contrast to these reports, *Helicotylenchus* spp. have been found in areas with a mean annual rainfall of less than 100 mm. Reproduction in *Helicotylenchus* is either amphimictic or by mitotic parthenogenesis.

Representative species

Helicotylenchus brevis (Whitehead, 1958) Fortuner, 1984
(Figs 9.14A-D, 9.15A-E)

Measurements (after Marais, 2001)

Female: L = 376-594 μm; a = 16.9-27.9; b = 3.5-5; c = 29.2-107.7; c' = 0.3-0.9; V = 84-91; stylet = 25-35 μm; tail = 4-18 μm.

Male: L = 362-627 μm; a = 24.7-32.1; b = 4.4-5; c = 26.3-37.6; c' = 1.2-2; stylet = 19-27 μm; spicules = 15-28 μm; gubernaculum = 7-9 μm; tail = 11-20 μm.

Description

Female: Habitus straight to C-shaped. **Labial region flatly rounded, not set off, with 4-6 annuli**. Labial framework and stylet moderate with stylet knobs rounded, flattened or indented anteriorly. Dorsal pharyngeal gland opening 6-12 µm from base of stylet. **Pharynx with a long ventral overlap. Vulva very far posterior. Posterior genital tract reduced to a post-vulval uterine sac.** Spermatheca filled with sperm. **Lateral field with four incisures, areolated anteriorly**; outer two lines crenate posterior to vulva; **rarely incompletely areolated on tail. Phasmids punctiform situated three annuli anterior to three annuli posterior to anus. Tail short, rounded** with 4-9 annuli. *Male:* Similar to female, slightly smaller anteriorly. Lateral field areolated anteriorly and

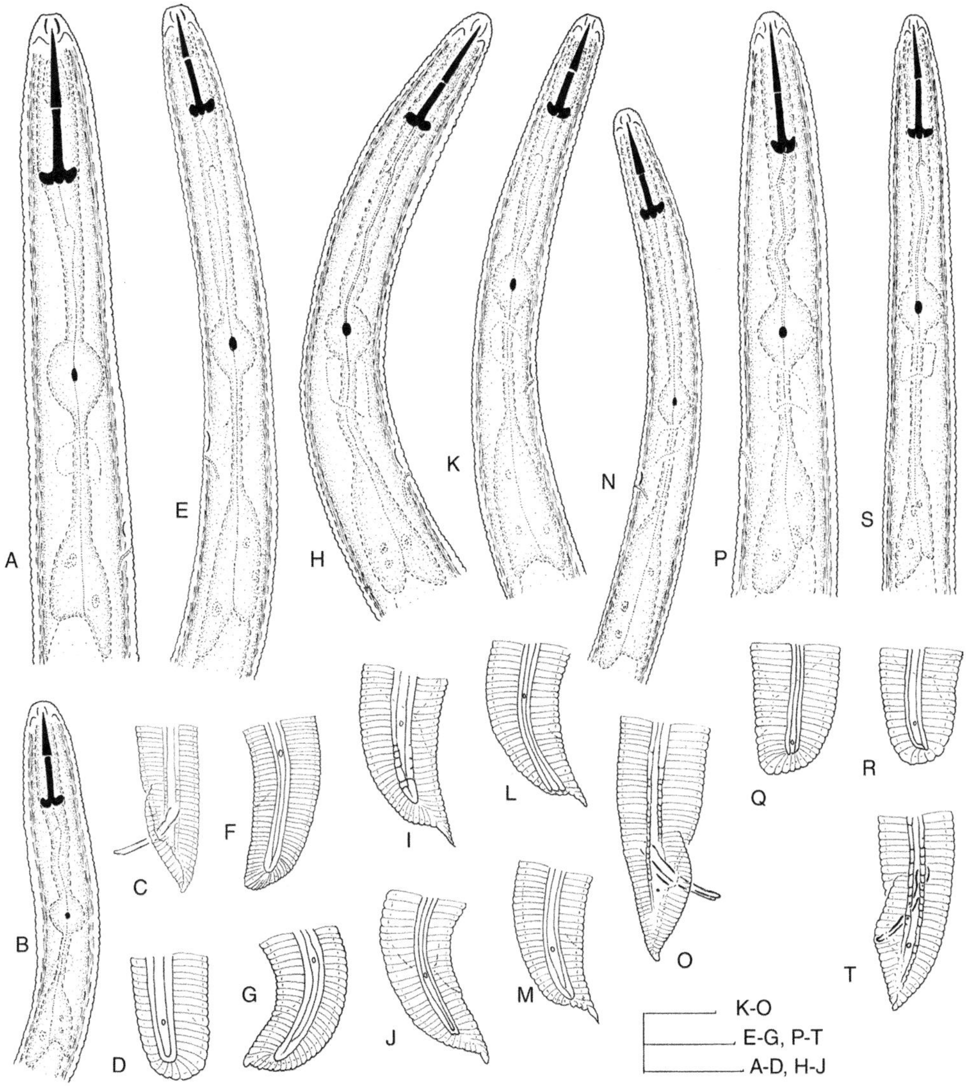

Fig. 9.14. *Helicotylenchus brevis*. A: Female, anterior region; B: male, anterior region; C: male, tail region; D: female, tail region. *Helicotylenchus dihystera*. Female. E: Anterior region; F, G: tail variations. *Helicotylenchus pseudorobustus*. Female. H: Anterior region; I, J: tail regions. *Helicotylenchus erythrinae*. Female, K-M. K: Anterior region; L, M: tail variations. Male, N, O. N: Anterior region; O: tail region. *Helicotylenchus multicinctus*. Female, P-R. P: Anterior region; Q, R: tail variations. Male, S, T. S: Anterior region; T: tail region. (Scale bar = 20 µm). (All species redrawn from Marais, 2001.)

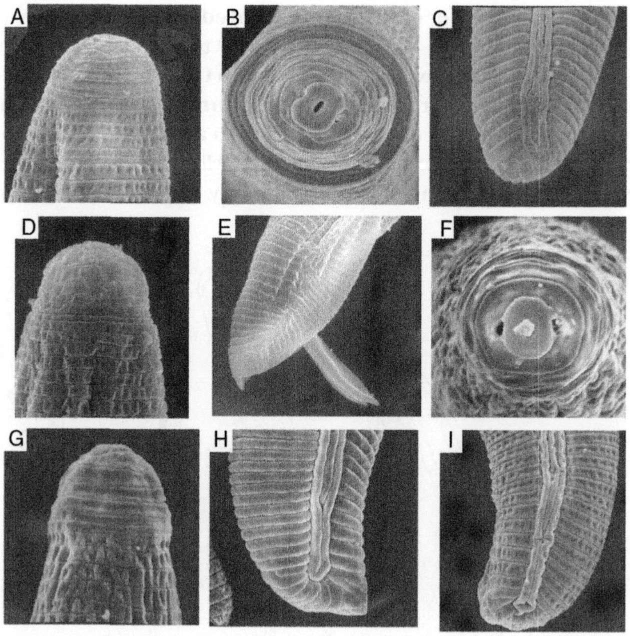

Fig. 9.15. *Helicotylenchus brevis*. Female, A-C. A: Labial region, lateral view; B: labial region *en face* view; C: tail region. Male, D, E. D: Labial region, lateral view; E: tail region. *Helicotylenchus dihystera*. Female. F: Labial region, *en face* view; G: labial region, lateral view; H, I: tail region.

opposite bursa. Tail long, conical. Phasmids opposite cloacal aperture.

Helicotylenchus brevis represents those species in the genus where the posterior genital tract is reduced, such species previously being placed in *Rotylenchoides*.

Bionomics and distribution: Described from banana, Tanzania, and since reported from South Africa from several hosts, *viz*., banana, mango, natural vegetation and bulbous plants.

***Helicotylenchus dihystera* (Cobb, 1893) Sher, 1961**

(Figs 9.14E-G, 9.15F-I)

Measurements (after Fortuner, 1991; Marais, 2001)

Female: L = 440-894 µm; a = 18.1-47; b = 3.7-6.5; c = 15.3-65; c' = 0.7-1.5; V = 57-72; stylet = 20-31 µm; tail = 12-30 µm.

Description

Female: Habitus straight, C-shaped to spiral. **Labial region rounded** to slightly **flattened** anteriorly with **4-7 annuli**. Labial framework and stylet moderate. Stylet knobs rounded, flattened, indented or sloping anteriorly. Dorsal pharyngeal gland opening 8-17 µm from stylet base. **Pharyngeal glands overlapping ventrally. Vulva situated medially to post-medially** with two outstretched genital tracts. Spermathecae empty. **Lateral field with four lines, areolated anteriorly and crenate or areolated posterior to vulva and on tail. Phasmids punctiform, situated from opposite to 19 annuli anterior to anus, very rarely posterior to anus. Tail** with 8-22 ventral annuli, **asymmetrically rounded, dorsally curved with or without projection.** If present, projection with rounded end. *Male*: Extremely rare.

Bionomics and distribution: *H. dihystera* was originally described from Australia, but has a cosmopolitan distribution with a very large host list (but note that recent molecular research suggests that, as with a number of other species, it may represent a species complex). It is an ectoparasite or a sedentary endoparasite. Typically, several cells are penetrated before feeding begins. The nematodes are partially or fully embedded in the root where they feed from a single cell. Also found on boot leaves and sheaths of rice plants. Linked with unhealthy bowling greens in Australia, reduced root systems of olive seedlings in Egypt and reduction of leaf size and plant height of guava seedlings in South Africa. Causes chlorosis, sparse root development and stunted shoots in brinjal, tomato and wheat. Interaction with *Fusarium oxysporum* f. sp. increases the incidence of leaf infections, defoliation, growth reductions and wilt in guava. In potted experiments with *H. dihystera* in sugarcane, the root cortex collapses causing stunting and malformation of the primary roots and reduction in side roots. The species increases the effects of bacterial wilt on carnation and may be of major importance in root diseases of pines that involve *Phytophthora cinnamomi*. On the other hand, where large proportions of *H. dihystera* occur within the nematode community, they have been linked to higher yields of sugarcane. It is much affected by environmental stress factors and is capable of anhydrobiosis.

Helicotylenchus erythrinae **(Zimmermann, 1904) Golden, 1956**
(Figs 9.14K-O, 9.16A-D)

Measurements (after Marais, 2001)
Female: L = 440-797 μm; a = 19.5-33.5; b = 4-6.4; c = 22.7-47.5; c' = 1-1.9; V = 59-69; stylet = 20-30 μm; tail = 11-29 μm.
Male: L = 444-737 μm; a = 24.8-38.9; b = 4.3-5.9; c = 19.1-37.5; c' = 1.2-2.4; stylet = 17.7-25 μm; spicules = 15-25 μm; gubernaculum = 5-9 μm; tail = 13-26 μm.

Description
Female: Habitus C-shaped to spiral. **Labial region rounded, not set off with 5-7 annuli**. Labial framework and stylet moderate. Stylet knobs rounded, flattened, indented or sloping posteriorly anteriorly. Dorsal pharyngeal gland outlet 4-23 μm from stylet base. **Pharynx with a long ventral overlap. Vulva situated medially to post-medially** with two outstretched genital tracts. Spermathecae filled with sperm. Epiptygma folded into vagina. **Lateral field areolated anteriorly, rarely incomplete over rest of body and posterior to vulva. Phasmids punctiform, situated four annuli posterior to 12 annuli anterior to anus. Tail** with 8-17 ventral annuli, **more curved dorsally with a rounded or pointed and annulated ventral projection. Mucro often present.** *Male*: Similar to female. Habitus straight, C-shaped to spiral. Lateral field areolated anteriorly, but incompletely areolated on posterior part of body, as much as four cloacal body diameters anterior to bursa. Phasmids situated about one cloacal body diameter anterior to cloacal aperture. **Tail with a ventrally curved, rounded or pointed ventral projection, rarely a mucro present on tail tip.**

Bionomics and distribution: *H. erythrinae* has a worldwide distribution in natural vegetation and forests in countries such as Indonesia, Philippines, Malaysia, India, New Zealand and Venezuela, occurring on a wide variety of hosts such as maize, peach, banana, avocado, coffee, taro, cassava, cacao, grass and ornamentals.

Helicotylenchus multicinctus **(Cobb, 1893) Golden, 1956**
(Figs 9.14P-T, 9.16E-H)

Measurements (after Marais, 2001)
Female: L = 385-826 μm; a = 18.6-38.9; b = 3.8-5.3; c = 32.5-67.1; c' = 0.5-1.3; V = 63-75; stylet = 21-30 μm; tail = 7-19 μm.
Male: L = 360-798 μm; a = 24.1-40.7; b = 3.8-6.9; c = 24-42.6; c' = 1-2.2; stylet = 18-25 μm; spicules = 16-25 μm; gubernaculum = 5-9 μm; tail = 12-24 μm.

Description
Female: Habitus straight, C-shaped to spiral. **Labial region rounded, not set off with 3-5 annuli.** Labial framework and stylet moderate with stylet knobs flattened, rounded to indented anteriorly. Dorsal pharyngeal gland opening 7-12 μm from stylet base. **Pharynx with long ventral overlap. Vulva situated post-median** with two outstretched genital tracts, **posterior tract reduced**. Spermatheca filled with sperm. Lateral field areolated anteriorly. **Phasmids punctiform, situated from 4 annuli posterior to 13 annuli anterior to anus.** Tail 5-15 annuli long, rounded, more curved dorsally without a ventral projection. *Male*: Similar to female except for a slightly smaller anterior region. Lateral field incompletely areolated anteriorly and opposite bursa. Tail curved ventrad with rounded projection.

Bionomics and distribution: *H. multicinctus* is, after the burrowing nematode (*Radopholus similis*) and the lesion nematode (*Pratylenchus coffeae*) probably the third most damaging nematode on banana and plantains worldwide. It is primarily an endoparasite, feeding on sub-epidermal layers of the roots and rhizomes causing typical discrete and shallow lesions, although it can also survive and complete its life cycle as an ectoparasite. Attacked plants show aboveground symptoms. Under heavy infestation, the productive life of the crop is greatly reduced. Secondary infection by fungi, such as *Rhizoctonia* and *Rhizobium*, enhances the process of root necrosis. Also associated with *Pinus elliotti* seedling decline in the USA. Recorded around grass, tropical and subtropical fruit trees, sugarcane, coffee and *Solanum* sp. in South Africa. Also associated with other plants, including grass, ornamental plants, mango, citrus, cocoa, cassava, oil palm, pineapples, maize, pawpaw, rice, yam, avocado, tea, grapevine and pine trees.

***Helicotylenchus pseudorobustus* (Steiner, 1914) Golden, 1956**
(Figs 9.14H-J, 9.16 I, J)

Measurements (after Yeates and Wouts, 1992; Marais, 2001)

Female: L = 439-930 μm; a = 19.2-44; b = 3.9-8.1; c = 20.6-61; c' = 0.8-1.8; V = 57-71; stylet = 20-30 μm; tail = 11-23 μm.

Description

Female: Habitus spiral. **Labial region rounded not set off with 4-6 annuli**. Labial framework

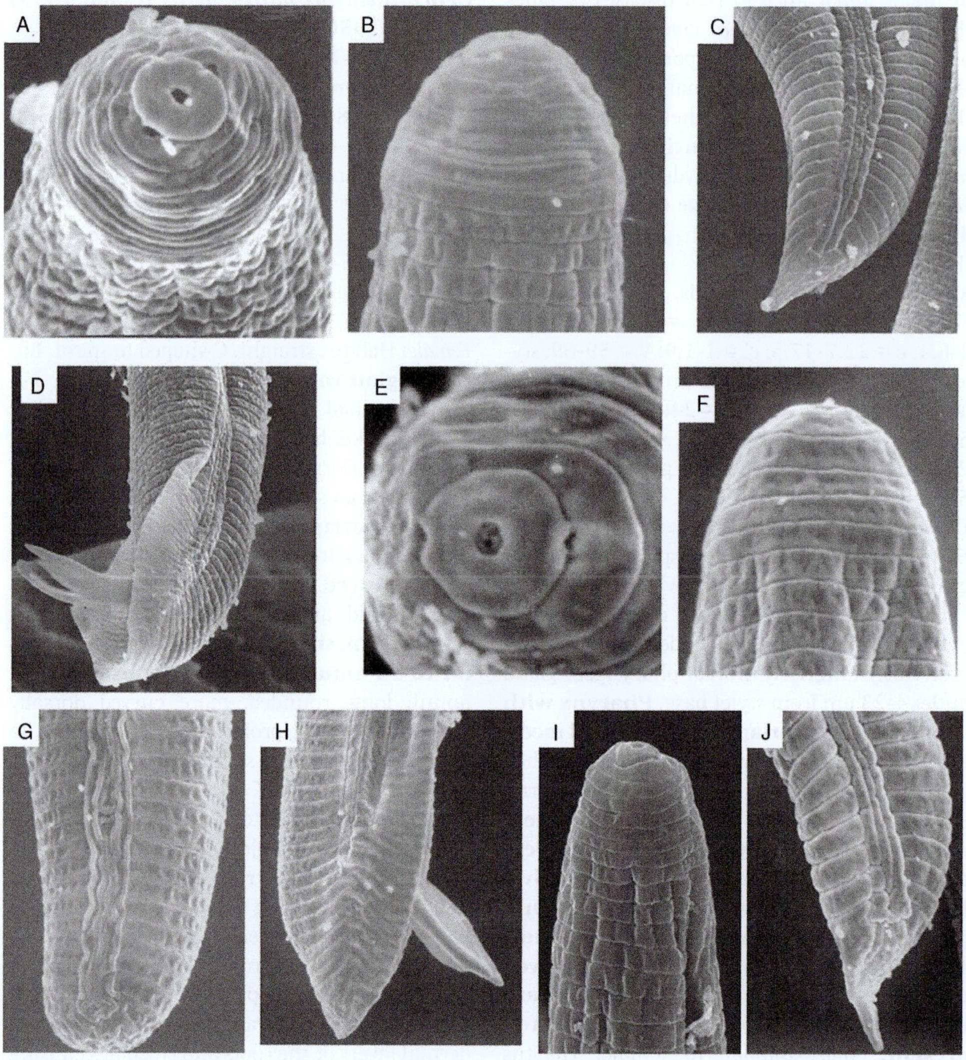

Fig. 9.16. *Helicotylenchus erythrinae*. Female, A-C. A: Labial region, *en face* view; B: labial region lateral view; C: tail region. Male. D: Tail region. *Helicotylenchus multicinctus*. Female, E-G. E: Labial region, *en face* view; F: labial region, lateral view; G: tail region. Male. H: Tail region. *Helicotylenchus pseudorobustus*. Female. I: Labial region; J: tail region.

and stylet moderate with stylet knobs rounded, flattened or indented anteriorly. Dorsal pharyngeal gland opening 8-16 μm from base of stylet. **Pharynx with long ventral overlap**. **Vulva** situated **medially to post-medially** with two outstretched genital tracts. Spermathecae empty. **Lateral field with four incisures, areolated anteriorly and crenate and areolated posterior to vulva and on tail. Phasmids punctiform, situated from two annuli posterior to 14 annuli anterior to anus. Tail** with 6-19 annuli, more **curved dorsally with a rounded to pointed ventral projection, rarely with a mucro**. *Male*: Not found.

Bionomics and distribution: *H. pseudorobustus* has a worldwide distribution in tropical and temperate countries from a wide range of host plants such as grasses, cereal crops, vegetables, sugarcane, rice, soybeans, tobacco, vineyards, ornamentals, fruit and forest trees, etc. (Subbotin *et al.*, 2015). It is semi-endoparasitic with the anterior body embedded in the root tissue and feeds mainly on pericycle, endodermal and cortical cells causing necrosis of the surrounding tissue.

Other species of note

Helicotylenchus spp. can be a problem in turf grass where they cause poor growth, yellowing and thinning of the sward with poorly developed roots and premature sloughing of cortical tissues. Pine seedlings acted as hosts to *Helicotylenchus digonicus* and the nematodes reduced the overall development of the plants. Large populations of *Helicotylenchus* spp. were also often encountered in the rhizosphere of pine trees exhibiting poor growth in Nigeria. *H. cavenessi*, described from around cassava in Nigeria, is also reported from Egypt. Recorded from around potato, temperate fruit, coffee, cotton, *Kniphofia* sp., introduced trees, *Ficus sycamorus* and from natural veld in South Africa. Large populations of *H. indicus* caused stunting of sugarcane under experimental conditions. The reduction in root and shoot weight of maize plants might be due to inhibition of nutrient uptake by the plant. Several species, including *Helicotylenchus oleae*, cause root necrosis and can affect the growth of olive trees. *Helicotylenchus varicaudatus* feed as a sedentary endoparasite. Most species are migratory ectoparasites, browsing on the roots and migrating in the soil between feeds. Their feeding causes distortion and collapse of cells. Roots become stunted and malformed with fewer lateral roots. *H. pseudorobustus* and *H. oleae* induced a single modified 'food-cell' with high metabolic activity; their feeding affected only epidermal and cortical cells with no evidence of vascular damage.

Genus *Antarctylus* Sher, 1973

A genus with only one species, *Antarctylus humus* Sher, 1973. Found in peat soil samples from Auckland Island and Macquarie Island in the sub-Antarctic. It was regarded as being most closely related to *Helicotylenchus*.

Diagnosis

Female habitus spiral shape. **Labial region rounded, tapering slightly, not set off with 6-7 fine annuli**. Stylet medium sized with knobs flattened to pointed anteriorly. **Pharyngeal glands overlapping intestine dorsally. Lateral field with four incisures. Vulva situated at mid-body** with epiptygma, single or double but inconspicuous. Two equally developed genital tracts. **Tail** tapering to a **finely rounded or pointed tip, curved ventrad** with 18-28 annuli. **Phasmids punctiform, situated from 6-11 annuli anterior to anus**. Male similar to female except for slightly less developed anterior end and pharynx. **Tail** tapering and **finely pointed with a long hyaline tip** and a **flattened**, but less distinct **bursa** enveloping tail tip. **Spicules and gubernaculum distinct but not very strong.**

9.3. Rotylenchulinae Husain & Khan, 1967

= Acontylinae Fotedar & Handoo, 1978
= Verutinae Esser, 1981

Diagnosis

Mature female habitus swollen, kidney shaped or bilobed. Labial region not as high as in Hoplolaiminae, not set off, with or without annulation. Labial sclerotization and stylet well developed in juveniles and females but atrophied in males. Male sexually dimorphic with anterior end smaller and pharynx sometimes degenerate and non-functional. Dorsal pharyngeal gland opening far from stylet base. Pharyngeal glands overlapping mostly ventrally and laterally. Genital tracts of mature female didelphic, reflexed or coiled; posterior branch rarely a post-vulval uterine sac. Tail tip of immature females and males with a long hyaline portion. Mature females with or without tail. Juvenile tails with smaller hyaline portion. Phasmids punctiform, located near anus or on tail. Bursa reduced, adanal, rarely enveloping tail. Species of the five genera (*Rotylenchulus*, *Acontylus*, *Senegalonema*, *Verutus* and *Bilobodera*), regarded as being in the Rotylenchulinae herein, are sessile, semi-endoparasites that lay eggs in a gelatinous matrix.

Genus *Rotylenchulus* Linford & Oliveira, 1940
= *Spyrotylenchus* (= *Spirotylenchus*) Lordello & Cesnik, 1958
= Leiperotylenchus Das, 1960
Diagnosis (after Jatala, 1991)
Immature female: **Vermiform, spiral to C-shaped. Labial region low rounded to high flattened, continuous, annulated or smooth.** Labial framework and stylet strong with stylet knobs mostly sloping posteriorly, rarely indented. **Dorsal pharyngeal gland opening about 0.5-2 times stylet length posterior to stylet base.** Pharyngeal **overlap** mostly **lateral. Lateral field** with **four** incisures. **Vulva posterior to mid-body, genital tract comprising two short branches with double flexures** near distal end. **Tail** elongate conoid with **prominent hyaline portion on tail tip.** *Mature female:* **Body swollen, kidney shaped** with irregular, less swollen neck and lateral fields not seen. Tail conical, rarely broadly rounded, with or without hyaline terminal portion. **Vulva behind mid-body with protruding lips, genital tracts** about equal, **greatly convoluted. Eggs deposited in a gelatinous matrix.** *Male:* **Vermiform. Stylet and pharynx vestigial.** Tail similar to immature female. **Bursa low**, difficult to see, **not reaching tail tip.** Phasmids punctiform in anterior region of tail. **Spicules and gubernaculum distinct.**
Bionomics and distribution: A small genus with about ten species and having a wide host range and a worldwide distribution. Many of the species have been reported on crops from the African region, for example, Kenya, Burundi, Nigeria, Tanzania, Mozambique, Cameroon, South Africa and Egypt. The mature female feeds in the root pericycle on syncytia and giant cells. Males do not feed. Reproduction is either amphimictic or parthenogenetic.

The vermiform immature female is the infective stage. The first moult takes place within the egg and the second-stage juvenile (J2) hatches from the egg. The males and juveniles (J2 to preadult J4) are found in the soil. The vermiform immature females penetrate the host root and induce syncytium-type trophic sites. While feeding, they mature to become sedentary and kidney shaped, usually with the anterior portion of the body embedded in the root while the enlarged posterior region protrudes from the root surface. The eggs are deposited within a gelatinous matrix secreted by the female as an external egg mass on the root. Soil particles tend to adhere to the egg masses giving them a 'dirty' appearance. The number of eggs per mass ranges from 50 to 60, exceptionally 150. Under favourable conditions, a generation is completed within 3-4 weeks.

Root galls are generally not formed but galls have been reported on pistachio. Heavy infections cause discolouration and impaired nutrient uptake resulting in stunted plants and irregular patches of poor growth in the field.

Representative species

Rotylenchulus reniformis Linford & Oliveira, 1940
(Figs 9.17I-P, 9.18E-I)
Measurements (after Jatala, 1991; Agudelo *et al.*, 2005)
Immature female: L = 340-560 µm; a = 20.3-31.9; b = 2.3-4.1; c = 10-22.8; c' = 2-4.1; stylet = 16-22 µm; V = 66-78; h = 2-12 µm.
Mature female: L = 380-520 µm; a = 4-5; V = 68-73.
Male: L = 360-525 µm; a = 22.3-33.9; b = 3.1-4.4; c = 11.3-18.3; c' = 2-4.5; stylet = 12-18 µm; spicules = 18-34 µm; gubernaculum = 7-9 µm; h = 4-9 µm.
Description
Immature female: Vermiform, migratory. Habitus open spiral to C-shape. **Labial region high, conoid, continuous with 4-6 annuli.** Labial framework and stylet well sclerotized with stylet knobs rounded anteriorly. **Dorsal pharyngeal gland outlet about one stylet length posterior to stylet base.** Pharyngeal glands overlapping mainly ventrally and laterally. **Vulva post-median with two short genital tracts with double flexures** near distal end. Tail tapering to a narrow, rounded terminus with

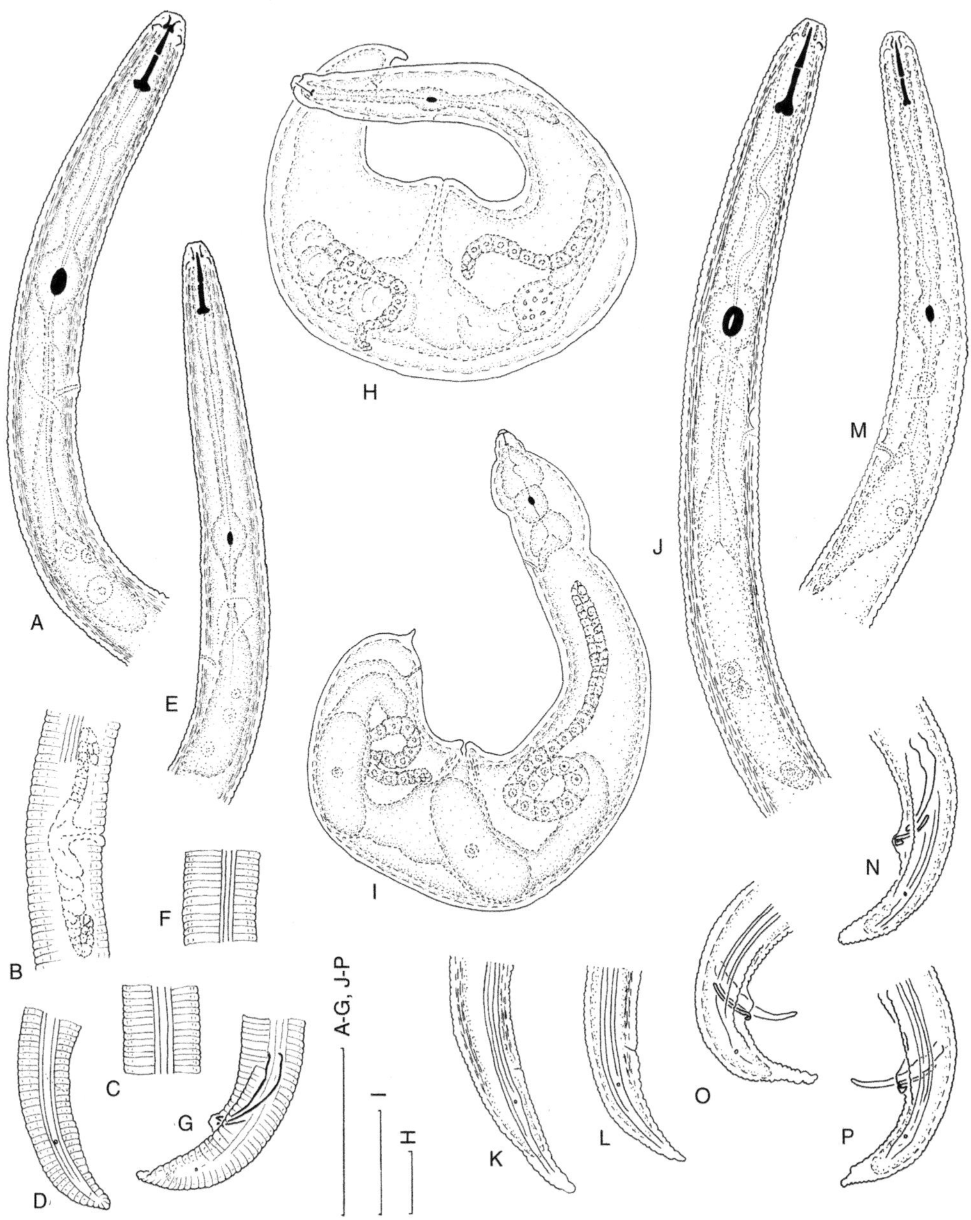

Fig. 9.17. *Rotylenchulus parvus*. Immature female, A-G. A: Anterior region; B: vulval area; C: lateral field at mid-body; D: tail region. Male, E-G. E: Anterior region; F: lateral field at mid-body; G: tail region. Mature female. H: Entire body. *Rotylenchulus reniformis*. Mature female. I: Entire body. Immature female, J-L. J: Anterior region; K, L: tail region. Male, M-P. M: Anterior region; N-P: tail region. (Scale bars: A-G, J-P = 30 μm; H = 50 μm; I = 60 μm.) (Fig. I redrawn from Dasgupta *et al.*, 1968.)

about 20-24 annuli. **Hyaline portion of tail 4-8 μm. Phasmids** punctiform, about a body width or less **behind anus.** *Mature female*: Body ventrally arcuate, **swollen, kidney shaped** with an irregular, less swollen neck. **Vulva raised** with spermathecae round to irregular, usually filled with sperm cells. **Body hemispherical beyond anus with a 5-9 μm long terminal**

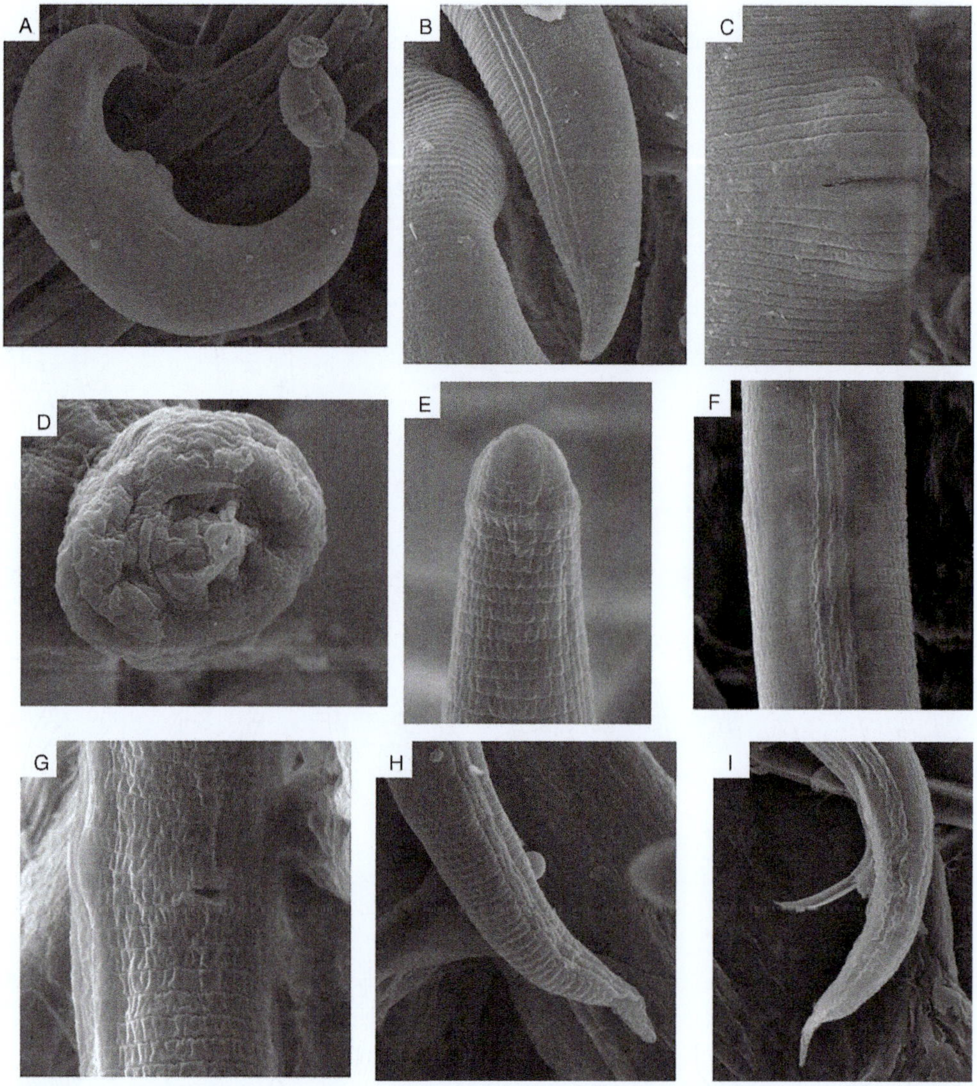

Fig. 9.18. *Rotylenchulus parvus*. Mature female. A: Entire female; B: tail region; C: vulval region; D: labial region, *en face* view. *Rotylenchulus reniformis*. Female, E-H. E: Labial region; F: lateral field; G: vulval area, ventral view; H: tail region. Male. I: Tail region.

portion. *Male*: Vermiform slightly arcuate ventrad. Anterior part less developed **with weak stylet.** Spicules arcuate ventrally with gubernaculum linear and non-protruding. **Bursa reduced and subterminal.**

Bionomics and distribution: *Rotylenchulus reniformis* is widely distributed throughout tropical and subtropical countries and attacks a large number of cultivated plants and fruit trees. It is considered one of the most injurious nematode pests in agricultural soils. It is a serious pest of sweet potato in many countries and is an important pest of mung bean in The Philippines. On pigeon pea in Jamaica, it caused yellowing of new leaves, dieback of twigs and stems and premature death of many plants; it also caused stunting and chlorosis in soybean.

This is one the most important nematode affecting vegetables, being second only to species of *Meloidogyne*. It has been recorded from many tropical and subtropical countries and has an extensive host range, including many vegetable crops. The reniform nematode can survive in moist soil in the absence of host plants for 7 months and for 6 months in dry soil. After 4 months, 84% of the nematodes were still alive. The intensity of the brinjal mosaic virus, okra yellow vein mosaic and charcoal rot caused by *Macrophomina phaseolina* was enhanced by the presence of *R. reniformis*.

R. reniformis has been implicated in severe plant damage and yield reduction of papaya in Puerto Rico and associated tree death and toppling in Trinidad. It has been reported to infect papaya on all continents and causes economic damage to passion fruit in various countries, such as Brunei, Belize, Colombia and Brazil. Serious damage and stubby root systems have been reported for *R. reniformis* on coffee and it is an important parasite of coffee in India, New Guinea, Fiji, Tonga, Western Samoa and the Ivory Coast. It is a major parasite of pineapples in the warmer regions of the world causing discolouration of leaves and poor growth and is a confirmed parasite of cotton. The symptoms include stunted growth, poorly developed roots and chlorosis.

Rotylenchulus parvus **(Williams, 1960) Sher, 1961**
(Figs 9.17A-H, 9.18A-D)

Measurements (after Van den Berg, 1978)

Immature female: L = 231-432 μm; a = 15.3-32.3; b = 2-4.4; c = 13.1-27.1; c' = 1.3-3.9; V = 56-69; stylet = 10.7-19.9 μm; tail = 10.5-27.2 μm; h = 0.8-5.2 μm.

Mature female: L = 280-380 μm; a = 3.3-4.6; V = 56-72; stylet = 9-11.5 μm.

Male: L = 280-439 μm; a = 26-39.2; b = 3.8-4; c = 11-14.6; c' = 3.1-3.6; stylet = 9.2-11.8 μm; h = 2.6-4.4 μm, spicules = 15.4-20.6 μm; gubernaculum = 5-6.6 μm; tail = 22.8-31.6 μm.

Description

Immature female: Habitus C-shaped to spiral. **Labial region broadly rounded**, almost flattened, not set off with **3-5 very fine, indistinct annuli**. Labial framework and stylet moderate, stylet knobs sloping posteriorly. **Dorsal pharyngeal gland opening 7.7-19.9 μm from stylet base. Pharynx overlapping ventrally** and laterally. Vulva situated post-median with two outstretched genital tracts each with a reflexed tip. **Lateral field with four incisures**, outer two extending almost to tail terminus, inner two terminating near phasmid. **Phasmids punctiform situated from 2 to 12 annuli posterior to anus. Tail** with 15-38 ventral annuli, conoid, ventrally arcuate, terminus varying from **sharply pointed, irregular, to more broadly rounded**, frequently with a ventral projection; **hyaline portion short to fairly long.** *Mature female*: Habitus obese, **kidney shaped**, posterior portion often twisted and crossing neck region; anterior part irregularly swollen. Tail without projection. Anus often indistinct. **Vulva prominent, post-medially with two convoluted genital tracts.** *Male*: **Rare.** Anterior part, stylet and pharynx less developed. Tail finely rounded. **Bursa reduced**, not reaching tail tip. Spicules well developed. Gubernaculum linear.

Bionomics and distribution: Described from sugarcane in Mauritius and widely distributed in many African countries. Also found in the USA, the Caribbean, Australia and India. Hosts include papaya, maize, sun hemp, cotton, various grasses, vegetables, tomato, fruit trees, cowpea and vines.

Rotylenchulus parvus is an obligate, sedentary parasite. Males are rare and reproduction is parthenogenetic, the life cycle taking 27-36 days. Eggs are laid in a gelatinous matrix deposited outside the host tissue.

Genus *Acontylus* Meagher, 1968

The only species, *Acontylus vipriensis* Meagher, 1968, was described from *Eucalyptus* and occurred at two localities in Australia. It was also associated with Black Mustard (*Brassica nigra*). The females are semi-endoparasitic with their anterior portion embedded in the root up to the level of the excretory pore. Males and juveniles have not been observed feeding. The eggs are deposited singly in a gelatinous matrix.

Diagnosis

Female body straight to slightly **ventrally arcuate, swollen in the vulval area. Labial region high, slightly set off**, trapezoid in lateral view. Labial framework and stylet well developed. **Stylet knobs sloping posteriorly. Lateral field with four incisures. Pharyngeal glands** long, **overlapping intestine dorsally** and laterally. **Lateral field with four incisures. Vulva situated far posteriorly** with anterior genital tract outstretched and **posterior one reduced to a post-vulval uterine sac.** Epiptygma not seen. **Phasmids punctiform, at**

anus level. Male vermiform. **Labial region knob-like, high, rounded, distinctly set off.** Labial framework, stylet and pharynx reduced in size. Bursa enveloping tail tip. **Juveniles with a rounded tail with a long terminal projection.**

Genus *Senegalonema* Germani, Luc & Baldwin, 1984

Only one species, *Senegalonema sorghi* Germani, Luc & Baldwin, 1984, from Senegal. Male and immature female body vermiform, C-shaped. Labial region high, not set off, not annulated. **Labial framework and stylet medium sized.** Pharyngeal glands overlapping mostly laterally, **rarely ventrally.** Lateral field **with** four incisures. Vulva at mid-body **with two genital tracts outstretched.** Mature female **semi-endoparasitic,** sessile in roots. **Body swollen posteriorly,** variously shaped to kidney shaped. **Pharyngeal glands overlapping laterally.** Vulva post-median **with two long, very convoluted genital tracts.** Phasmids enlarged. **Male with slight sexual dimorphism seen in the** slightly smaller anterior end. Pharynx functional. Bursa enveloping tail tip.

Genus *Verutus* Esser, 1981

Only two species known, *Verutus volvingentis* Esser, 1981 and *Verutus californicus* Baldwin, Bernard & Mundo-Ocampo, 1989 from the USA (*Verutus mesoangustus* was subsequently transferred to *Bilobodera* – see below). According to a molecular study of *V. volvingentis* by Subbotin *et al.* (2018), *Verutus* is phylogenetically more closely related to the Heteroderinae than the Rotylenchulinae. **Mature female reniform or sausage-shaped**, often with strong ventral curvature. **Vulva** large, situated **post-medially. Tail usually absent, anus slightly ventral to posterior terminus.** Two genital tracts, reflexed at the spermathecae and once or twice in the maturation zone. **Male vermiform, four incisures in lateral field. Tail bluntly hemispherical, annulated with cloacal aperture and protruding spicules subterminal. Bursa absent. No migratory immature females. Phasmids only seen in early-stage juveniles.**

Genus *Bilobodera* Sharma & Siddiqi, 1992

Only two species are known, the type species, *Bilobodera flexa* Sharma & Siddiqi, 1992 from India and *Bilobodera mesoangusta* (Minagawa, 1986) Sharma & Siddiqi, 1992 (originally described as *V. mesoangustus*) from Japan. **Body bilobed** with **deep dorsal depression opposite median vulva.** Two highly convoluted genital tracts. **Anus subterminal. Male vermiform, four lines in lateral field, no bursa**, tail rounded with **cloacal aperture and protruding spicules subterminal. No migratory immature females. Phasmids seen only in early-stage juveniles.**

9.4. Management of the Hoplolaimidae

Most ectoparasitic species of this family are only rarely associated with crop damage whereas the migratory endoparasites may be important plant parasites. However, the distinction between migratory ectoparasites and migratory endoparasites is not always clear as some members of this family may exhibit both types of parasitic behaviour.

Some hoplolaimids are able to cause severe damage on soybean (*H. columbus*), cotton (*H. columbus, R. reniformis*), yam (*S. bradys*), peas (*H. seinhorsti*), pineapple (*R. reniformis*) and banana (*H. multicinctus*). Other members of this family are often considered as second-class pathogens or parasites of little or no economic significance, except locally when they pullulate (e.g., *Aphasmatylenchus straturatus* on peanut in West Africa). For most species, their biological traits and pathogenicity to plants remain unknown but for a general overview of their importance in tropical agriculture, the volumes by Luc *et al.* (2005, 2013) and Sikora *et al.* (2018) are good starting points.

9.4.1. The lance nematodes (*Hoplolaimus* spp.)

The lance nematodes are relatively large robust migratory parasites that feed both endo- and ectoparasitically causing extensive root necrosis on a large variety of hosts from trees to grasses, including many important crops (coffee, fruit trees, cotton, soybean, peas and vegetables). For example, *H. galeatus* can cause extensive damage to roots and mortality to pine seedlings. Chemical control has proven relatively ineffective on lance nematodes even if effective on other species of nematodes. Carefully planned crop rotations and good control of weeds are important factors for successful management of these nematodes.

9.4.2. The reniform nematodes (*Rotylenchulus* spp.)

Among the reniform nematodes, *R. reniformis* is a sedentary semi-endoparasite that is very widely distributed in tropical and warm temperate regions. Stunted growth and yield suppression are widely observed on a large variety of crops once initial *R. reniformis* population exceed 1000 individuals/100 cm^3 soil. Management has been primarily through the use of fumigant and granular nematicide applications. Germplasm of many crops, for example, cotton, maize, soybean, tomato, potato and pineapple, have been screened for resistance to *R. reniformis* with relatively good success, even if this research effort is less extensive than for root-knot and cyst-nematode resistance. Although reniform nematodes have an extensive host range, rotations with non-host plants can provide effective management.

9.4.3. The yam nematodes (*Scutellonema bradys*)

S. bradys is well known as the yam nematode and is responsible for the 'dry rot' disease indigenous to West Africa but has subsequently been spread worldwide in yam-growing areas. It is a migratory endoparasite able to infest roots and tubers. After harvest, and in the later stages of the disease, infested tissues turn from yellow to dark brown and cracks appear in the skin of the tubers. Damage can involve 100% tuber loss in yam barns. The use of nematode-free planting material (vitroplants, hot-water treated seed tubers, vine cuttings) planted in clean soil (crop rotation with non- or poor-host plants) is the only safe management practice for this crop. Chemical control is not an option because of toxicity and cost problems.

Within this genus, another species, *Scutellonema cavenessi*, is associated with many cultivated plants in all the north-western African countries and pest management is difficult since most crops grown in rotation are susceptible hosts. Indeed, there are no known plant cultivars resistant to *S. bradys* or *S. cavenessi*.

9.4.4. The spiral nematodes (*Helicotylenchus* spp.)

Within *Helicotylenchus*, *H. dihystera* is the most cosmopolitan species recorded and has the largest host list. It is an ecto- or semi-endoparasite on the roots of many plants. Damage to crops is sometimes observed but pathogenicity remains difficult to demonstrate. However, different susceptibilities to *H. dihystera* have been observed among cultivars of sugarcane.

Second only to the burrowing nematode, *R. similis*, the banana spiral nematode *H. multicinctus* is the most widespread and abundant endoparasitic nematode on the banana crop worldwide. However, *H. multicinctus* is often regarded as the main parasitic nematode on bananas where environmental conditions are suboptimal for the crop with regard to latitude, temperature and rainfall. The management methods used for *R. similis* apply (use of banana vitroplants and appropriate fallow), although rotation break crops have not yet been clearly identified for the spiral nematode.

In conclusion, practical management methods against most of these hoplolaimids rely mainly on the use of rotation break crops as the use of chemicals has become increasingly restricted due to environmental fate, toxicology and carcinogenicity issues.

9.5. Acknowledgements

Dr Louwrens Tiedt, SEM Unit, North West University, Potchefstroom, South Africa, is thanked for the scanning electron microscope photographs. Mrs Elsa van Niekerk, ARC-PPRI, Pretoria, South Africa, is thanked for scanning the photoplates.

9.6. References

Agudelo, P., Robbins, R.T., Stewart, J.McD. and Szalanski, A.L. (2005) Intraspecific variability of *Rotylenchulus reniformis* from cotton-growing regions in the United States. *Journal of Nematology* 37, 105–114.

Baujard, P., Vovlas, N., Mounport, D. and Martiny, B. (1998) Description of *Aphasmatylenchus liberiensis* n. sp., and observations on other species of the genus *Aphasmatylenchus* Sher, 1965 (Nematoda: Hoplolaimidae). *Fundamental and Applied Nematology* 21, 129–138.

Castillo, P. and Vovlas, N. (2005) *Bionomics and Identification of the Genus* Rotylenchus *(Nematoda: Hoplolaimidae). Nematology Monographs and Perspectives, 3.* (Series eds: Hunt, D.J. and Perry, R.N.). Brill, Leiden, Netherlands.

Doucet, M. (1980) Description de deux nouveaux *Peltamigratus* et d'une population d'*Hoplolaimus galeatus* (Nematoda: Tylenchida) de la Province de Cordoba, Argentine. *Nematologica* 26, 34–46.

Fortuner, R. (1991) The Hoplolaimidae. In: Nickle, W.R. (ed), *Manual of Agricultural Nematology*. Marcel Dekker, New York, New York, pp. 669–719.

Germani, G., Baldwin, J.G., Bell, A.H. and Wu, X.Y. (1985) Revision of the genus *Scutellonema* Andrássy, 1958 (Nematoda: Tylenchida). *Revue de Nématologie* 8, 289–390.

Jatala, P. (1991) Reniform and false root-knot nematodes, *Rotylenchulus* and *Nacobbus* spp. In: Nickle, W.R. (ed), *Manual of Agricultural Nematology*. Marcel Dekker, New York, New York.

Larizza, A., Lamberti, F. and Ekanayake, H.M.R.K. (1998) The genus *Hoplolaimus* in Sri Lanka (Nematoda: Tylenchida). *Nematologia Mediterranea* 26, 79–86.

Luc, M., Sikora, R.A. and Bridge, J. (eds) (2005) *Plant Parasitic Nematodes in Subtropical and Tropical Agriculture*, 2nd edn. CABI Publishing, Wallingford, UK.

Marais, M. (2001) *A Monograph of the Genus* Helicotylenchus *Steiner, 1945 (Nemata: Hoplolaimidae).* Ph.D. Dissertation, University of Stellenbosch, Stellenbosch, South Africa.

Mwamula, A.K., Kwon, O.G., Kwon, C., Kim, Y.S., Kim, Y.H., *et al.* (2024) A revision of the phylogeny of *Helicotylenchus* Steiner, 1945 (Tylenchida: Hoplolaimidae) as inferred from ribosomal and mitochondrial DNA. *Plant Pathology Journal* 40, 171–191. DOI: 10.5423/PPJ.OA.01.2024.0013.

Sakwe, P.N. and Geraert, E. (1992) Plant parasitic nematodes from Cameroon: Belonolaimidae, Criconematidae and Hoplolaimidae (Nematoda: Tylenchida). *Medelingen van de faculteit LandbouwWetenschappen, Universiteit Gent* 57, 857–877.

Sher, S.A. (1963a) Revision of the Hoplolaiminae (Nematoda) II. *Hoplolaimus* Daday, 1905 and *Aorolaimus* n. gen. *Nematologica* 9, 267–295.

Sher, S.A. (1963b) Revision of the Hoplolaiminae (Nematode) III. *Scutellonema* Andrássy, 1958. *Nematologica* 9, 421–443.

Sher, S A. (1964) Revision of the Hoplolaiminae (Nematoda). IV. *Peltamigratus* n. gen. *Nematologica* 9, 421–443.

Sher, S.A. (1965) *Aphasmatylenchus nigeriensis* n. gen., n. sp. (Aphasmatylenchinae n. subfam.: Tylenchoidea: Nematoda) from Nigerian soil. *Proceedings of the Helminthological Society of Washington* 32, 172–176.

Siddiqi, M.R. (2000) *Tylenchida Parasites of Plants and Insects*, 2nd edn. CAB International. Wallingford, UK.

Sikora, R.A., Coyne, D., Hallmann, J. and Timper, P. (eds) (2018) *Plant Parasitic Nematodes in Subtropical and Tropical Agriculture*, 3rd edn. CABI Publishing, Wallingford, UK.

Subbotin, S.A., Vau, S. and Inserra, R.N. (2018) Molecular characterisation and phylogenetic relationship of *Verutus volvingentis* Esser, 1981 with other cystoid nematodes of the family Heteroderidae (Nematoda: Tylenchida), and some morphological details of its immature life stages. *Russian Journal of Nematology* 26, 145–157.

Subbotin, S.A., Vovlas, N., Yeates, G.W., Hallmann, J., Kiewnick, S., *et al.* (2015) Morphological and molecular characterisation of *Helicotylenchus pseudorobustus* (Steiner, 1914) Golden, 1956 and related species (Tylenchida: Hoplolaimidae) with a phylogeny of the genus. *Nematology* 17, 27–52.

Tran, V.D., Vu, V.L., Nguyen, H.T. and Trinh, Q.P. (2023) An updated species list of the genus *Rotylenchus* (Nematoda: Hoplolaimidae) and a browser-based interactive key to species identification. *Australasian Plant Pathology* 53, 79–88. DOI: 10.1007/s13313-023-00962-4.

Uzma, I., Nasira, K., Firoza, K. and Shahina, F. (2015) Review of the genus *Helicotylenchus* Steiner, 1945 (Nematoda: Hoplolaimidae) with updated diagnostic compendium. *Pakistan Journal of Nematology* 33, 115–160.

Van den Berg, E. (1976) *Hoplolaimus seinhorsti. CIH Descriptions of Plant-parasitic Nematodes*, Set 6, No. 76. Commonwealth Agricultural Bureaux, Farnham Royal, UK.

Van den Berg, E. (1978) The genus *Rotylenchulus* Linford & Oliveira, 1940 (Rotylenchulinae: Nematoda) in South Africa. *Phytophylactica* 10, 57–64.

Van den Berg, E. (1986a) Two new *Rotylenchus* species from the Cape Province with notes on some known *Rotylenchus* species (Rotylenchinae: Nematoda). *Phytophylactica* 18, 169–176.

Van den Berg, E. (1986b) One new and some known *Rotylenchus* species with a key to the South African species of the genus (Rotylenchinae: Nematoda). *Phytophylactica* 18, 195–201.

Van den Berg, E. (1998) New records and notes on known species of Hoplolaimidae (Nemata) in South Africa. *Journal of Nematode Morphology and Systematics* 1, 29–46.

Van den Berg, E. and Cadet, P. (1991) One new and some known plant parasitic nematode species from the French Caribbean (Nemata: Tylenchina). *Revue de Nématologie* 14, 389–405.

Van den Berg, E. and Heyns, J. (1973) South African Hoplolaiminae. 2. The genus *Scutellonema* Andrássy, 1958. *Phytophylactica* 5, 23–40.

Van den Berg, E. and Heyns, J. (1974) South African Hoplolaiminae. 3. The genus *Rotylenchus* Filip'ev, 1936. *Phytophylactica* 6, 165–184.

Van den Berg, E. and Van den Oever, H.A.M. (1993) Description of the male of *Scutellonema truncatum* Sher, 1963 from Mozambique (Nemata: Hoplolaimidae). *Phytophylactica* 25, 157–158.

Van den Berg, E. and Quénéhervé, P. (2012) Hoplolaimidae. In: Manzanilla-López, R.H. & Marbán-Mendoza, N. (eds) *Practical Plant Nematology*. Colegio de Postgraduados and Mundi-Prensa, Biblioteca Básica de Agricultura, Guadalajara, Jalisco, Mexico, pp. 251–297.

Van den Berg, E., Quénéhervé, P. and Tiedt, R.L. (1997) New records and descriptions of plant nematodes from rain forest soil in French Guiana, with a first description of males of *Mesocriconema surinamense* (De Grisse & Maas, 1970) Loof & De Grisse, 1989 (Tylenchida: Nemata). *Nematologica* 43, 341–354.

Van den Berg, E., Tiedt, L.R., Stanley, J.D., Inserra, R.N. and Subbotin, S.A. (2017) Characterisation of some *Scutellonema* species (Tylenchida: Hoplolaimidae) occurring in Botswana, South Africa, Costa Rica and the USA, with description of *S. clavicaudatum* sp. n. and a molecular phylogeny of the genus. *Nematology* 19, 131–173.

Vovlas, N., Castillo, P. and Gomez Barcina, A. (1991a) SEM observations on two species of *Hoplolaimus* Daday, 1905 (Nematoda: Hoplolaimidae). *Nematologia Mediterranea* 19, 305–309.

Vovlas, N., Troccoli, A. and Lamberti, F. (1991b) Morpho-anatomical observations on *Aphasmatylenchus nigeriensis* and *Scutellonema clathricaudatum* from West-Africa. *Nematologia Mediterranea* 19, 259–264.

Yeates, G.W. and Wouts, W.M. (1992) *Helicotylenchus* spp. (Nematoda: Tylenchida) from managed soils in New Zealand. *New Zealand Journal of Zoology* 19, 13–23.

10 Cyst Nematodes

Sergei A. Subbotin[1,2]* and Javier Franco[3]
[1]*Plant Pest Diagnostics Center, California Department of Food and Agriculture, Sacramento, CA, USA;* [2]*Center of Parasitology of A.N. Severtsov Institute of Ecology and Evolution of the Russian Academy of Sciences, Moscow, Russia;*
[3]*Los Cerezos 338, Lima, Peru*

Abstract

Cyst nematodes are one of the economically important parasitic nematode groups worldwide. They are especially adapted to a parasitic mode of life and have sexual dimorphism with vermiform males and swollen sedentary females. Detailed information on their morphology, life cycle and plant-host relationships are provided, in addition to bionomics, control, pathotypes, biochemical and molecular identification, phylogenetic relationships and systematics of the genera and major species. A dichotomous key to the genera is included. The genera *Betulodera, Cactodera, Dolichodera, Globodera, Heterodera, Paradolichodera, Punctodera* and *Vittatidera* are fully diagnosed and several species are considered in more detail with information on morphology, diagnostics, bionomics and distribution. Descriptions of 19 economically important species of *Heterodera* and four species of *Globodera* are given in detail.

10.1. Introduction

Cyst nematodes are one of the economically important parasitic nematode groups. These nematodes are especially adapted to a parasitic mode of life and exhibit sexual dimorphism, with vermiform males and swollen sedentary females. Cyst nematodes are taxa defined by their capacity to retain eggs inside the female body, which is transformed into a persistent tanned cyst at the completion of the female life cycle. Each embryonated egg inside the cysts contains a dormant second-stage juvenile (Fig. 10.1). Eight genera: *Heterodera* (89 species), *Globodera* (14 species), *Cactodera* (19 species), *Dolichodera* (1 species), *Paradolichodera* (1 species), *Betulodera* (1 species), *Punctodera* (5 species) and *Vittatidera* (1 species), and total of 131 valid species are presently recognized within the group (Subbotin *et al.*, 2010; Handoo and Subbotin, 2018; Singh *et al.*, 2020; Lax *et al.*, 2021).

10.2. Morphology

The morphology of cyst nematodes is rendered uniquely complex by their sexual dimorphism, parasitic habits, ontogeny and dormant stages

*E-mail: sergei.a.subbotin@gmail.com

A revision of Subbotin, S.A. and Franco, J. (2012) Cyst nematodes. In: Manzanilla-López, R.H. and Marbán-Mendoza, N. (eds), *Practical Plant Nematology*. Mundi-Prensa and Biblioteca Básica de Agricultura, Montecillo, Mexico.

(eds R.H. Manzanilla-López, D.J. Hunt, N. Marbán-Mendoza and S. Infante Gil)
DOI: 10.1079/9781836990413.0010

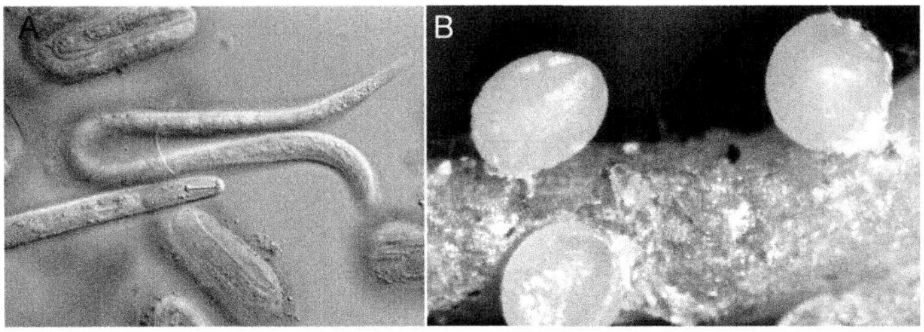

Fig. 10.1. Cyst nematodes, *Globodera pallida*. A: Second-stage juveniles and eggs; B: white females on a root. (Courtesy of U. Zunke and T. Vrain.)

(Figs 10.1, 10.2). The specialized morphology of cyst nematodes reflects a complex life history. The morphology of the egg, juvenile stages, male, female and cyst – including their features – varies among genera and species and is used for diagnostics.

10.2.1. Egg

The eggs of cyst nematodes are not readily distinguishable from those of tylenchids. Egg dimensions are often included in species descriptions, suggesting that eggs vary in size and shape among species, although these differences are not often used in diagnostics. In *Cactodera* species, light and scanning microscopy revealed distinctive textures on the surface of eggs that may be particular to certain species. For example, the eggshell of *C. cacti*, *C. thornei*, *C. eremica* and *C. milleri* is covered by punctation, whereas *C. weissi*, *C. amaranthi* and *C. salina* have smooth eggshells.

10.2.2. Second-stage Juvenile

The second-stage juveniles (J2) of cyst nematodes are vermiform, with a rounded head region and a tapering tail (Fig. 10.2). After gently killing by heating, the juvenile body becomes slightly curved. Although body length may vary among specimens and populations, average body length is sufficiently stable for species. The overall J2 body length (300-700 μm), length and shape of the tail (30-90 μm) and distinctive hyaline region (20-40 μm) are useful in species diagnosis (Wouts and Weischer, 1977; Wouts and Baldwin, 1998). Additional useful features may include stylet length, position and shape of the stylet knobs and number of incisures. The lateral incisures are informative characters since J2 of most species have four incisures, and J2 of species from the *Bifenestra*, *Sacchari* and *Cyperi* groups, *Betulodera* and *Dolichodera* have three incisures. An important diagnostic feature of the surface cuticle concerning the labial patterns of J2, and their potential for defining taxa was first noted by Stone (1975). Although many cyst nematode J2 retain this basic pattern, in many other species, this pattern is modified, often by fusion of adjacent submedial lips with one another and with the labial disc; sometimes, the submedial lips further fuse with the anterior labial annuli. The lip patterns are relatively stable and useful in diagnostics. The reproductive system in migratory newly hatched J2, including males and females, is an oval primordium located near the centre of the body and comprises four cells.

10.2.3. Male

Although rare or absent in parthenogenetic forms, males are common for most species of cyst nematodes. They may be found in soil during their life span of about 10 days, but they are more often found on roots near young females, including within the egg-sac that is attached to the female's posterior region. Often, several males are found with a single female, and both sexes can mate several times. Males are usually not used for diagnostics, as they have similar morphology and only a few species-specific characters, for example, body length, spicule tip and stylet knob structure.

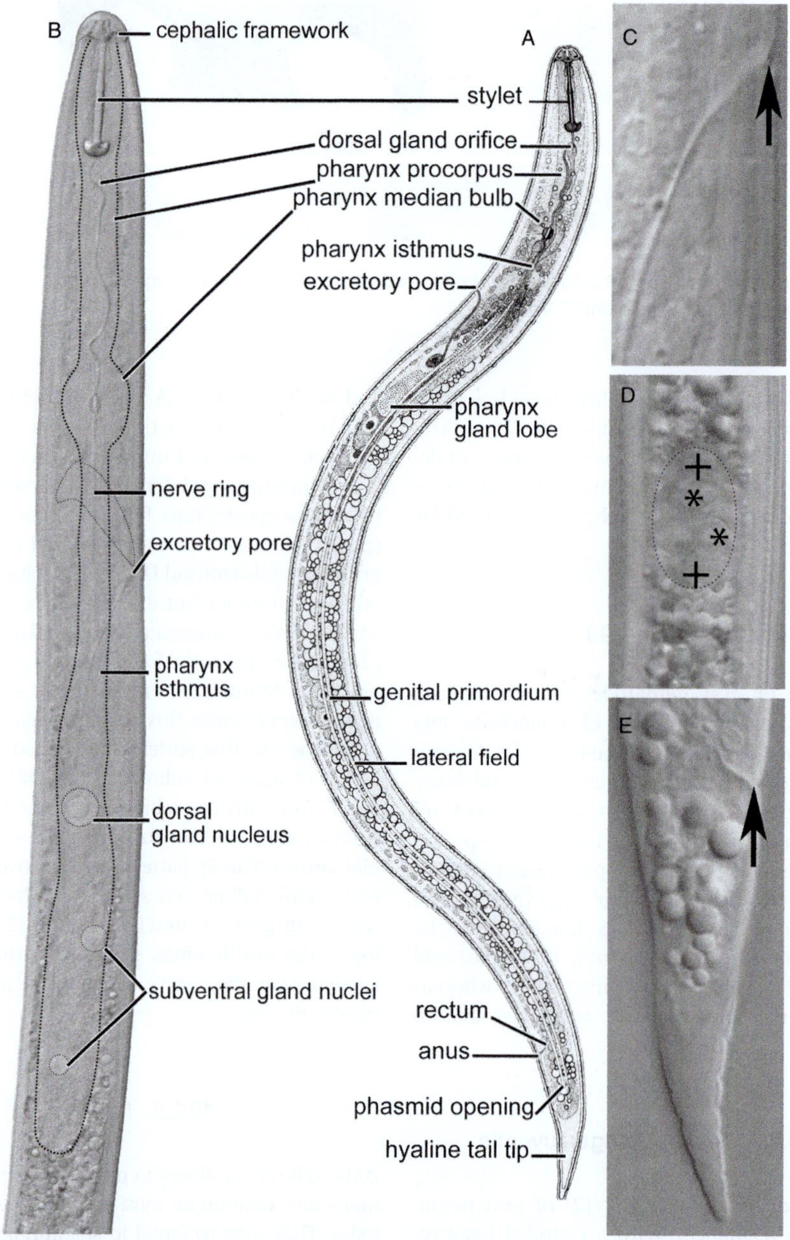

Fig. 10.2. Second-stage juveniles of cyst nematodes (*Heterodera schachtii*). A: Left lateral view (after Cobb from Thorne, 1961); B: right lateral view of anterior end, including pharyngeal region; C: excretory pore (arrow) and duct; D: genital primordium (+ = epithelial cells; ∗ = germ cells); E: tail region, including anus (arrow) and hyaline region. (After Subbotin *et al.*, 2010.)

10.2.4. Cyst

The cyst is the stage richest in diagnostic characters (Figs 10.3-10.6). Furthermore, cysts are the stage often most readily available from soil samples. Cysts may be obtained directly from infected roots or recovered from soil using a variety of techniques. All cysts include a narrow

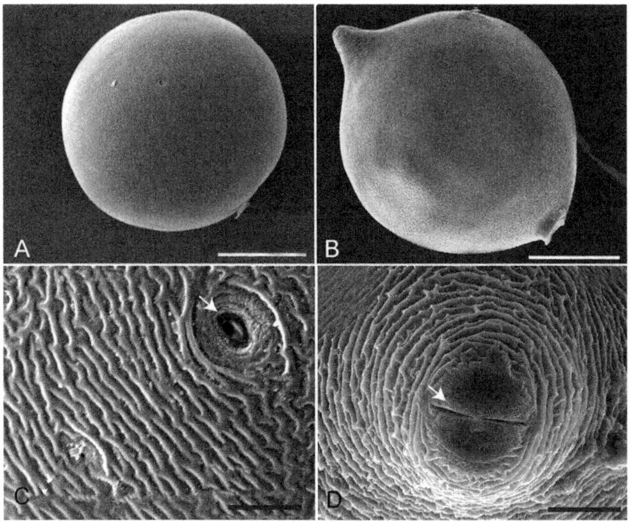

Fig. 10.3. SEM of cyst nematodes. A: Cyst of *Globodera pallida*; B: cyst of *Heterodera* zeae; C: terminal region of *Globodera pallida*; D: vulval cone of *Heterodera glycines* female. Arrows indicate vulval slit. (Scale bars: A, B = 200 μm; C = 15 μm; D = 30 μm. (Courtesy of T. Powers, USA.)

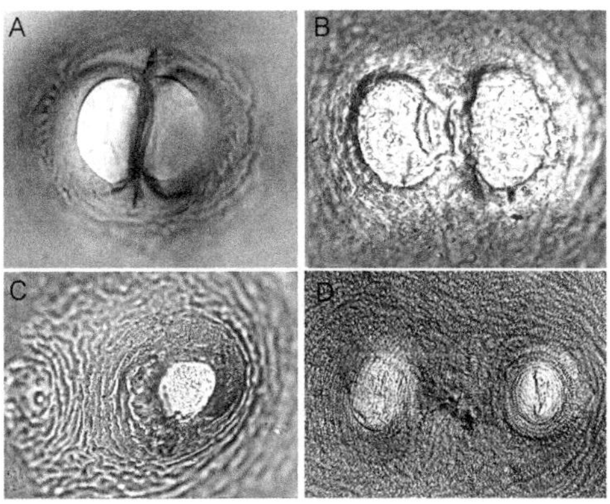

Fig. 10.4. Vulval plates. A: *Heterodera trifolii* (ambifenestrate); B: *Heterodera* sp. (bifenestrate); C: *Cactodera radicale* (circumfenestrate); D: *Punctodera punctata* (circumfenestrate). (After Subbotin *et al.*, 2010] with modifications.)

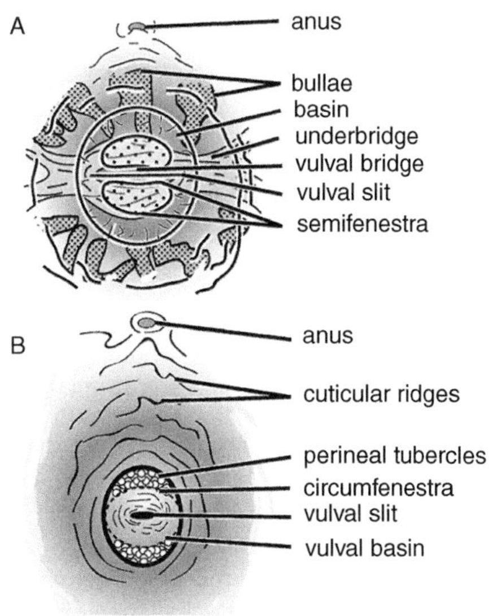

Fig. 10.5. Terminal view of tail region of young cysts (composite of species for genus) showing typical diagnostic features and associated terminology. A: Cone of *Heterodera* (ambifenestrate type); B: terminal region of *Globodera* (circumfenestrate type). (After Subbotin *et al.*, 2010.)

anterior protuberance or neck and swollen body; the cyst is swollen and varies considerably in details that may be of diagnostic importance. This includes the presence or absence of a terminal posterior protrusion, the vulval cone, which determines if the female is globose when the cone is absent, as in *Globodera, Punctodera, Dolichodera* and *Paradolichodera*, or lemon-shaped when the cone is present, as in *Heterodera, Cactodera, Betulodera* and *Vittatidera*. The formation and proportional size of the cone is affected by the relative elasticity of the body wall as the female enlarges, and this may be affected by development in relation to cuticle composition, layering

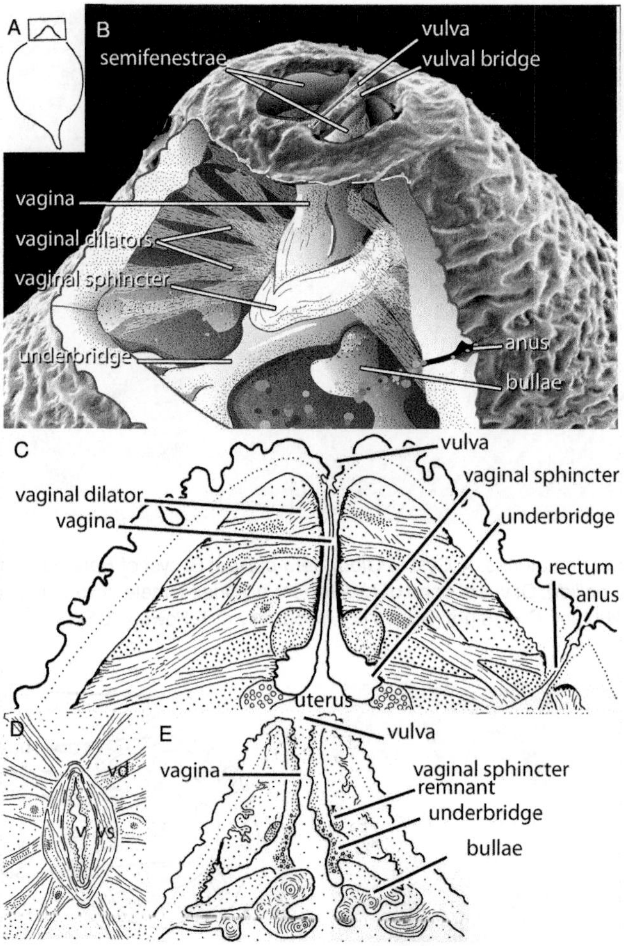

Fig. 10.6. Drawings depicting basic structures of terminal region of cyst nematodes. A: Overview of female. Box indicates terminal region as enlarged in B and C; B: ventrolateral 3D view with cutaway showing internal structures. Some features that may be exclusive to either female or cyst stages are combined to show a more complete range of possible structures; C: right lateral view of *Heterodera schachtii*; D: transverse view through vulva (v) at level of vaginal sphincter (vs) and vaginal dilators (vd) of *H. schachtii*; E: dorsal view of mature cyst. (After Subbotin *et al.*, 2010.)

and thickness, as well as attachment points and arrangement of vaginal muscles. Thus, size and shape, while reflecting underlying taxon-specific genetics, are variable because they are also subject to environmental factors, including nutrition, age, host, physical pressure and crowding. Regardless of limitations, specific trends in size of females and cysts are of some diagnostic value in combination with other characters. Overall cuticle thickness apparently varies among species; for example, a diagnostic feature for *Globodera mali* is the presence of a thin and transparent cuticle that persists in the cyst. Regardless of overall layering, the cuticle structure is distinctively thin-walled and consists of a loose mesh of fibres in a discrete area of the posterior region of the body. In this area, the cuticle ruptures at cyst maturity, forming taxon-specific apertures, the fenestrae. These apertures may allow the exit of J2 from the cyst. The absence or presence of fenestration and type of fenestration (semifenestrate, bifenestrate or circumfenestrate) is used in generic diagnostics (Figs 10.4, 10.5). Other structures of the vulval cone – such as the vulval slit, vulval bridge, underbridge and bullae – serve as important taxonomic characters (Fig. 10.6).

Taxonomic characters associated with fenestrae include the distance from the anus to the nearest edge of the fenestrae. Dividing this measurement by fenestral length gives Granek's ratio, which is useful for the differentiation of *Globodera* species.

To the outside of the cuticle, a subcrystalline layer may be present, and, while variable, its occurrence has long been recognized as somewhat taxon-specific. By definition, cysts are tanned, and the pigmentation and transitions of colour, as they mature, are of diagnostic value. Franklin (1951) was among the first to note that cyst colour at maturity may vary among species from light tan to nearly black. In *G. rostochiensis*, for example, the cysts are said to pass through a golden stage to brown as they mature, whereas in *G. pallida* they change directly to brown. Cyst colour transformations are not limited to *Globodera* and are observed in other genera. For example, females of *H. betae* change from white through yellow to brown, whereas females of *H. schachtii* do not have a distinct yellow colour.

10.3. Life Cycle

The life cycle of cyst nematodes comprises the egg, four juvenile stages and the adults (Fig. 10.7). The first-stage juvenile is limited to the egg; the J2 hatches from the egg while within the cyst or the gelatinous matrix that comprises the egg-sac. Hatching is the result of a complex interplay among external signals and the internal readiness of the J2 to emerge. Hatching of large numbers of J2 for many cyst nematode species occurs in response to host root diffusates. The J2 that hatch within the cyst leave the cyst *via* either the open fenestral region or the opening created at the neck, where the female's head has deteriorated or otherwise broken away. The J2 are then attracted towards the host root, responding to gradients of a variety of stimuli. Once at the invasion site, a J2 enters the root epidermis using repeated thrusts of its protrusible, pointed stylet. Penetration usually occurs in the zone of elongation behind the growing root tip, and the J2 migrates intracellularly and crosses the cortical tissue towards the vascular cylinder. Reaching the pericycle, the J2 carefully penetrates a single cell with its stylet and starts feeding. After establishing a feeding site, the J2 undergoes three moults to reach the adult stage.

Cyst nematodes are sexually dimorphic. Adults include sedentary obese egg-producing females and vermiform migratory males in most species. Males of amphimictic species are attracted to pheromones, which may be present in exudates secreted through the vulva of females, often into a gelatinous matrix. After mating, eggs develop and contain maturing embryos. In most species, the majority of eggs are retained

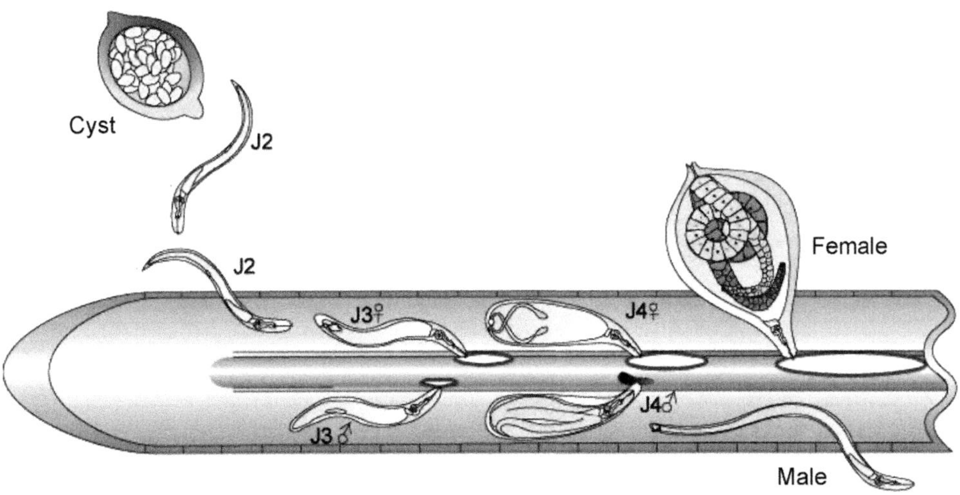

Fig. 10.7. Life cycle of an amphimictic cyst nematode. (After Lilley *et al.* (2005) with modification.)

within the female that then dies. The cuticle of the female body is tanned by a polyphenol oxidase to form the tough, leather-like, cyst wall. Females are capable of producing up to 500 to 700 eggs. Females of some species of *Heterodera* can produce a large gelatinous matrix, in which some eggs are deposited. The life cycle, from egg to egg, varies depending on the temperature and other environmental conditions but is usually completed in about 30 days. The number of generations per year also varies from one to several and depends on species. In some species, eggs within the cyst are especially persistent when conditions are not favourable for plant growth and infection. Under these conditions, they can persist in soil for many years. The combination of diapause and quiescence enables the potato cyst nematodes (PCN) *G. rostochiensis* and *G. pallida*, for example, to persist in soil for more than 20 years in the absence of a host plant and biological antagonists.

10.4. Plant-host Relationships

Cyst nematodes are highly specialized parasites infecting plants from different plant orders. Almost half of the known cyst nematode species were described from plants in the order Poales, one of the largest orders of flowering plants.

Cyst nematode parasitism is a complex process involving the interchange of information between the nematode and the plant host. After penetrating a root, the J2 injects secretions of the dorsal pharyngeal gland into this cell. These secretions mediate a massive reprogramming of cell development and induce the formation of a large, multi-nucleate feeding structure, a syncytium (Fig. 10.8), which is formed by dissolution of walls initially separating adjacent cells. The syncytium contains large, metabolically active cells acting as a nutrient sink and is characterized by multiple enlarged nuclei with nucleoli, dense cytoplasm and a proliferation of mitochondria, endoplasmic reticulum and plastids.

Cyst nematodes parasitizing roots cause remarkable but not specific symptoms. Field symptoms of severe cyst nematode infection often include reduced growth, stunting, wilting, chlorosis of leaves and reduction of quality and quantity of yield. Above-ground symptoms of damage are not unique to cyst nematodes. They could be mistaken for nutrient deficiencies, drought stress, herbicide injury or other plant diseases and often can remain undetected for several years. Symptoms of infection are expressed more fully when mineral nutrition is limited or root growth is otherwise impeded. Areas of severely affected and symptomatic plants are often round or elliptical in shape. Severely infected plants may die before flowering. Most below-ground symptoms of cyst nematode damage are also not unique. Infected roots are dwarfed or stunted. Extensive branching and local thickening of roots can also be observed. Cyst nematode infections also may make the roots more susceptible to attack by other

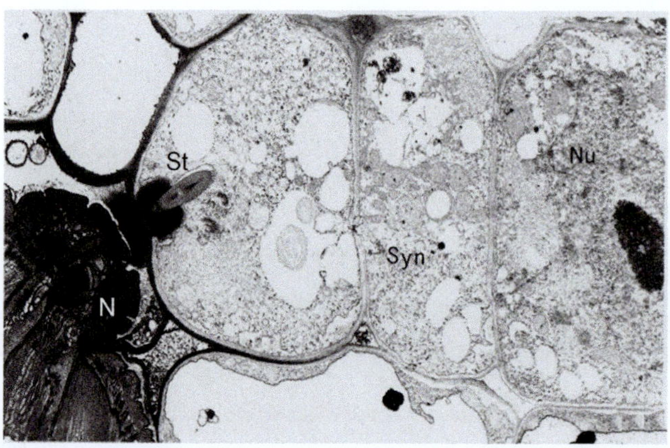

Fig. 10.8. Feeding of *Heterodera glycines* on soybean roots. An infective second-stage juvenile with stylet extended into a syncytium, N = nematode; Syn = syncytium; Nu = nucleus. (After Endo, 1978.)

soil-borne plant pathogens. Many abiotic factors, for example, fertility, pH, soil type and organic matter content, influence nematode population development and damage severity. Although worldwide losses to cyst nematodes are considerable, they are difficult to quantify, because they are often influenced by interactions with pathogenic fungi (*Rhizoctonia*, *Fusarium* and *Pythium*) and bacteria.

10.5. Control

Several strategies are used for control of cyst nematodes, and rotation with non-host crops and growing resistant cultivars are common management methods. Two or more years of using non-host crops or resistant varieties may be needed to reduce populations below damaging thresholds. Resistant varieties against several species of cyst nematodes have been developed in many research centres. For example, several cereal cultivars with resistance to *Heterodera avenae* are currently on the market. Cultivar resistance is considered one of the best methods for cereal cyst nematode (CCN) control and has been found to be successful in several countries, such as Australia, Sweden and France, on a farm scale. All of the sources of resistance reported to date against CCN feature single-gene inheritance. Six *Cre* genes for *H. avenae* resistance in wheat were derived from *Aegilops* species. Other resistance genes were derived from *Triticum aestivum* (*Cre1* and *Cre8*) and *Secale cereale* (*CreR*). Several other sources of resistance (*CreX* and *CreY*) are also reported, but their genetic control and gene designation are still unknown. Most of these resistance genes have been introgressed into hexaploid wheat. The *Cre1* gene is highly effective against populations of CCN from Europe, North Africa and North America and moderately effective or ineffective against populations in Australia and Asia.

Management of soybean cyst nematode (SCN) includes rotation with non-host crops, such as alfalfa, maize and wheat, and planting of SCN-resistant soybean cultivars. Soybean breeding lines named Peking or PI 88788 have been the major source of resistance to SCN for the last 25 years, are very effective against the common races of SCN, and the resistance is easily incorporated into new varieties. The single dominant gene, *H1* from *S. tuberosum* ssp. *andigena*, has been widely used in potato breeding programmes against *Globodera rostochiensis* pathotype Ro1 and Ro4 since its discovery in 1953. Maris Piper, which was released by the Plant Breeding Institute in Cambridge in 1966, was the first cultivar with this gene. Despite sustained efforts to breed for high levels of resistance to the white PCN *G. pallida*, none of the varieties on display have complete resistance. Moreover, H1 is ineffective against other *G. rostochiensis* pathotypes and *G. pallida*. In spite of reduced nematode reproduction, resistant cultivars do not prevent root penetration and associated stresses on the plant.

A nematicide application can provide effective control of cyst nematodes. Because of high cost, safety concerns and inconsistent results, nematicides are usually recommended when cyst nematode populations cannot be controlled by other methods. Biological control products are not widely commercially available, but cyst nematode populations in some locations might be maintained below an economic threshold by fungal (*Paecilomyces lilacinus* and *Pochonia*) and bacterial parasites (*Pasteuria*) of eggs and juveniles.

10.6. Pathotypes

A number of studies have shown intraspecific variation in relation to nematode behaviour on hosts. Host races and pathotypes have become an important part of an infraspecific classification of some cyst nematodes. They have been recognized and classified for *G. pallida*, *G. rostochiensis*, *H. avenae*, *H. australis*, *H. cajani*, *H. filipjevi*, *H. hordecalis*, *H. schachtii* and *H. zeae*. PCN are classified into pathotypes on their ability to multiply on different potato clones. However, their pathotype status is still a matter of debate. Kort *et al.* (1977) proposed the International Scheme for *Globodera* pathotypes in which, among the European populations, five pathotypes within *G. rostochiensis* and three within *G. pallida* were recognized. A second approach, the Latin American Scheme (Canto Saenz and de Scurrah, 1977), recognized among the South American populations of *G. rostochiensis* and *G. pallida*, four and seven pathotypes, respectively. The distribution of pathotypes of both species according to the Latin American Scheme for South American countries was re-analysed by Franco *et al.* (1998) (Table 10.1). Using the standard differential clones of Latin and European schemes, several pathotypes of *G. rostochiensis* and *G. pallida* have been identified in PCN populations collected in most of the South American countries. This information and the results of other local in-country studies, as in Venezuela (M. Cordero, Tachira, Venezuela, 1996, personal communication), have shown that the P4A and P5A pathotypes of *G. pallida* are the most predominant, although the P6A pathotype is also present. In the southern region of South America, the situation is complex due to the presence of both species and several pathotypes.

A race test for *H. glycines* was first developed based on variable comparative development of females on four differential soybean lines. This test was proposed by a group of nematologists and soybean breeders (Golden *et al.*, 1970; Riggs and Schmitt, 1988) to distinguish 16 races of *H. glycines*. However, the scheme was shown to describe inadequately the extensive genetic diversity of *H. glycines* populations that existed in areas of soybean production, and it was noted that the classification could be expanded to include many more groups (or types) by increasing the number of differentials. Thus, Riggs *et al.* (1981) distinguished up to 25 'races' on 12 resistant soybean lines, demonstrating that the set of four differentials used by Golden *et al.* (1970) did not account for all the variability in host compatibility of *H. glycines* (Niblack *et al.*, 2002). Niblack *et al.* (2002) proposed an HG Type test to describe population variation better and to expand the flexibility of the race classification system, with evaluation also based on the development of adult females.

10.7. Biochemical and Molecular Identification

Biochemical approaches, such as gel electrophoresis for separating protein and enzyme profiles, have shown great potential for assisting in identification of cyst nematodes. Isoelectric focusing (IEF) was used in nematology laboratories for routine diagnostics of *G. pallida* and *G. rostochiensis*. By comparison with the biochemical approaches, analysis of DNA for diagnostics has several advantages. DNA does not rely on expressed products and, therefore, is not influenced by environmental conditions or developmental stage. The DNA diagnostics, being much more sensitive than any biochemical diagnostic technique, can be used with nanograms of DNA extracted from a single specimen or even from a fragment of an individual nematode. For diagnostic purposes, various types of samples can potentially be used as total soil extracts that include a mixture of nematode species and infected plant materials. Dried, alcohol-fixed, formalin-fixed specimens or specimens embedded and mounted on permanent slide can also be used for diagnostics

The main DNA regions targeted for diagnostics of cyst nematodes are nuclear ribosomal RNA genes. These include 18S, 28S and especially the internal transcribed spacer 1 (ITS1) and internal transcribed spacer 2 (ITS2), which are situated between 18S and 5.8S and 5.8S and 28S rRNA genes, respectively. The flanking 18S and 28S rRNA gene regions contain rather conservative nucleotide sequences, and, thus, universal primers can be designed and used for amplification of the ITS regions. Genes of mitochondrial DNA (mtDNA), especially *COI*, with their relatively higher rate of mutations relative to rRNA genes, have great potential for species barcoding, diagnostics of races and populations of cyst nematodes.

PCR with species-specific primer and sequencing of rRNA or *COI* genes are presently used for diagnostics of many cyst nematode species (Subbotin *et al.*, 2010). However, the PCR and sequence techniques should be used intelligently, and the researcher or adviser must be aware of the peculiarities of each method. Where there is doubt, it is always necessary to confirm identification by several methods, including the use of traditional morphological features.

10.8. Phylogenetic Relationships

The relationship of Heteroderinae to other Tylenchina has long been the subject of speculation, and

Table 10.1. Pathotype groups of the potato cyst nematodes, *Globodera rostochiensis* and *G. pallida*.

Species and potato accession	*Globodera* species / Virulence group / European pathotypes / South American pathotypes / Ploidy, Resistance gene	*G. rostochiensis*					*G. pallida*						
	Virulence group	Ro1		Ro3		Ro5	Pa1			-	-	Pa2/3	
	European pathotypes	Ro1	Ro4	Ro2	Ro3	Ro5	Pa1		-	-	Pa2	Pa3	-
	Ploidy, Resistance gene	R1A	R1B	R2A	R3A	-	P1A	P1B	P2A	P3A	P4A	P5A	P6A
Solanum tuberosum ssp. *tuberosum*	4x, (minor)	+/-	+	+	+	+	+	+	+	+	+	+	+
S. tuberosum spp. andigena CPC 1673	4x, H1 on chromosome 5	-	-	+	+	+	+	"	"	"	+	+	"
S. kurtzianum KTT 60.21.19	2x, K1 K2 A & B	-	(+)	-	(+)	(+)	+	+	+	+	+	+	+
S. vernei GLKS 58.1642.4	2x .Quantitative	-	+	-	-	+	+	+	+	-	+	+	+
S. vernei Vt 62.33.3	2x. Quantitative		-	-	-	+	-	+	-	-	-	+	+
Ex. *S. multidissectum* hybrid P55/7	2x , 1 + polygenes H2	+	+	+	+	+	-	-/+	-	+	+	+	+
S. t. ssp. *andigena*	H3 + polygenes	+	"	"	"	"	(-)	"	"	"	(-)	(-)	"
CIP 280090.10	Quantitative	"	"	"	"	"	"	"	"	"	-	-	+
S. vernei hybrid 69.1377/94	2x, Polygenes	-	-	-	-	-	-	"	"	"	-	-	"
S. vernei hybrid 63.346/19	2x, Polygenes	-	-	-	-	-	+	"	"	"	+	+	"
S. spegazzinii	2x, Fa = H1	-	-	+	+	+	"	"	"	"	"	"	"
S. spegazzinni	2x, Fb+2 minor	+	-	+	-	-	"	"	"	"	"	"	"
	Glo1 on chromosome 7	(-)	+	+	+	+	"	"	"	"	"	"	"

+ = compatible interaction: nematode multiplication, susceptible potato;
- = incompatible interaction: nematode no multiplication, resistant potato; ( ) = partial or uncertain interaction; " = no information.

several authors have reviewed the evolution and phylogeny of cyst nematodes. New insight into the phylogeny of cyst nematodes has been provided by analysis of ribosomal RNA gene sequences. Molecular data support the hypothesis of Wouts (1985) that *Verutus* branches basal to, and as a sister to the clade of, all other heteroderids. Molecular phylogenies also support monophyly of *Heterodera*, *Cactodera*, *Punctodera*, *Globodera*, and they resolve some of the basal relationships within *Heterodera*, usually placing the *Goettingiana* group as basal clade. Molecular data also support the classification of cyst nematodes into several groups: *Afenestrata*, *Avenae*, *Bifenestrata*, *Cardiolata*, *Cyperi*, *Goettingiana*, *Humuli*, *Sacchari* and *Schachtii*.

10.9. Classification

(After Chapter 3; major genera in bold)

Order Rhabditida Chitwood, 1933
Suborder Tylenchina Thorne, 1949
Infraorder Tylenchomorpha De Ley & Blaxter, 2002
Superfamily Tylenchoidea Örley, 1880
Family Hoplolaimidae Filipjev, 1934
Subfamily Heteroderinae Filipjev & Schuurmans Stekhoven, 1941
Betulodera Sturhan, 2002
Cactodera Krall & Krall, 1978
Dolichodera Mulvey & Ebsary, 1980
Globodera Skarbilovich, 1959
Heterodera Schmidt, 1871
= *Afenestrata* Baldwin & Bell, 1985
Paradolichodera Sturhan, Wouts & Subbotin, 2007
Punctodera Mulvey & Stone, 1976
Vittatidera Bernard, Handoo, Powers, Donald & Heinz, 2010

10.9.1. Subfamily Heteroderinae

Diagnosis (after Siddiqi (2000) with modifications): Hoplolaimidae. *Mature female:* spherical, pear or lemon-shaped with a short neck, turning into a tough, hard-walled, yellowish, light to dark brown or black cyst containing eggs and juveniles; in some species, eggs may be laid in a gelatinous matrix. Cuticle surface with a zigzag or lace-like pattern, annulation absent on mature female and cyst. Vulva and anus usually close together, almost terminal. Clear, translucent vulval fenestra usually present (absent in species of *Afenestrata* group of *Heterodera*), an anal fenestra present only in *Punctodera*. *Male*: developed through metamorphosis, labial region annulated, lateral field with four or three incisures, tail short, hemispherical, without bursa. *J2*: three or four incisures in lateral field.

Type genus: *Heterodera* Schmidt, 1871

Other genera:

Betulodera Sturhan, 2002
Cactodera Krall & Krall, 1978
Dolichodera Mulvey & Ebsary, 1980
Globodera Skarbilovich, 1959
Paradolichodera Sturhan, Wouts & Subbotin, 2007
Punctodera Mulvey & Stone, 1976
Vittatidera Bernard, Handoo, Powers, Donald & Heinz, 2010

10.9.2. Identification Key to Genera of Heteroderinae

1. Cysts circumfenestrate 2
- Cysts ambifenestrate or bifenestrate, or without fenestration .... *Heterodera*

2. Cysts with terminal cone 3
- Cysts without terminal cone 4

3. Vulval slit in cysts >10 µm long, J2 with four lateral incisures.. 5
- Vulval slit in cysts 5-8 µm long, J2 with three lateral incisures.................................. *Betulodera*

4. J2 with average stylet >18 µm...*Cactodera*
- J2 with stylet = 14-17 µm .... *Vittatidera*

5. Anal region with fenestration...*Punctodera*
- Anal region without fenestration 6

6. Mature female and cyst spheroid, perineal tubercles usually present .... *Globodera*
- Mature female and cyst elongate-oval shape, perineal tubercles absent.................................. 7

7. Bullae absent in cysts, J2 with dorsal gland orifice (DGO) = 11-15 µm, c = 9-11, stylet <20 µm .... *Paradolichodera*
- Bullae present in cysts, J2 with DGO = 7-8 µm, c = 5-6, stylet = 22-24 µm .... *Dolichodera*

10.10. Genera and Major Species

10.10.1. Genus *Heterodera* Schmidt, 1871

Diagnosis: *Mature female* and *cyst*: body more or less lemon-shaped, with a posterior cone (except apparently secondarily lost in *H. bamboosi*). Vulval slit terminal or sunken into cone, anus on dorsal vulval lip. Vulval slit 6-6.8 µm in length. Cuticle surface with zigzag or lace-like pattern of ridges. Vulval fenestration ambifenestrate, bifenestrate or absent. D-layer in cuticle rudimentary or absent. Subcrystalline layer present or absent. Anus without fenestration. Bullae and underbridge present or absent. *Eggs* retained in body; in some cases, egg-sac also present. *J2* with lateral field marked by three or four incisures, phasmids punctiform or lens-like. Several distinct morphological-molecular groups: *Afenestrata*, *Avenae*, *Bifenestrata*, *Cardiolata*, *Cyperi*, *Goettingiana*, *Humuli*, *Sacchari* and *Schachtii*, are distinguished within this genus.

Type species: *Heterodera schachtii* Schmidt, 1871

Total number of species: 89

***Heterodera avenae* species complex – the Cereal Cyst Nematodes**

(Figs 10.9A, B, 10.10)

Presently, four species are collectively named as CCN: the European CCN *H. avenae* Wollenweber, 1924 the Filipjev CCN *H. filipjevi* (Madzhidov, 1981) Stelter, 1984, the Sturhan or Chinese CCN *H. sturhani* Subbotin, 2015, and the Australian CCN *H. australis* Subbotin, Rumpenhorst, Sturhan & Moens, 2002, and they are considered as major, economically important, nematode pests in cereal-growing areas. *H. avenae* and *H. filipjevi* have been found in many countries of Europe, Asia and Northern America, where they may occur often in mixed populations, whereas *H. australis* has a limited distribution and has only been reported to date in Australia, while *H. sturhani* has been reported to date in Australia and China (Subbotin *et al.*, 2018).

Young plants heavily infected by CNN are stunted, and their lower leaves are often chlorotic, forming pale green patches in the field. Mature plants have a reduced number of tillers, and the roots are shallow and have a 'bushy-knotted' appearance (Smiley *et al.*, 2017). In Europe, more than 50% of the fields in major cereal-growing areas are infected by European CCN (Rivoal and Cook, 1993), with annual yield losses reaching £3 million (Nicol and Rivoal, 2008). In the USA, annual losses of at least $3.4 million were estimated in wheat production in the states of Idaho, Oregon and Washington due to European CCN. The yield losses it causes in wheat range from 15 to 20% in Pakistan, and from 40 to 92% in wheat and from 17 to 77% in barley in Saudi Arabia. In China, yield losses of wheat crop induced by the Sturhan CCN can reach up to 70%. Hosts of CCN include many species of cereals and grasses (Williams and Siddiqi, 1972). CCN has only one generation per year, with J2 hatch from the eggs determined largely by temperature (Rivoal and Cook, 1993).

European CCN *H. avenae*

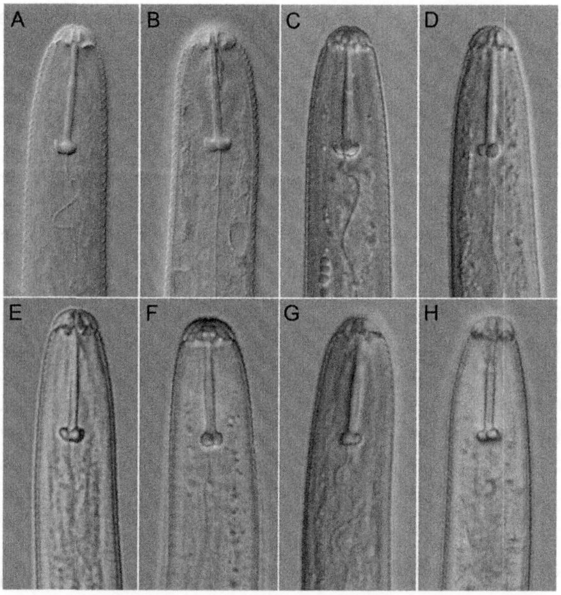

Fig. 10.9. Heads of second-stage juveniles. A: *Heterodera avenae*; B: *H. filipjevi*; C: *H. schachtii*; D: *H. cruciferae*; E: *H. humuli*; F: *Globodera rostochiensis*; G: *Betulodera* sp.; H: *Punctodera punctata*. Scale bar, A-H = 10 μm. (After Handoo and Subbotin, 2018.)

Measurements

Cysts: L = 400-1000 μm; W = 360-744 μm; L/W ratio = 0.8-1.8; fenestral length = 32-55 μm; vulval slit = 7-12 μm.

J2: L = 505-603 μm; stylet = 24-28 μm; hyaline region = 34-50 μm; tail = 52-79 μm.

Description: *Cyst:* lemon-shaped, with prominent neck and vulval cone. Subcrystalline layer conspicuous, sloughing off with formation of dark brown cyst. Bifenestrate (Fig. 10.11A), bullae prominent, crowded beneath the vulval cone. *J2:* vermiform, with a sharply pointed tail. Stylet well developed, with large, anteriorly flattened to concave basal knobs.

Remarks: European CCN *H. avenae* belongs to the *Avenae* group and to the *H. avenae* complex. It differs from the similar species *H. australis*, *H. aucklandica*, *H. sturhani*, *H. riparia*, *H. pratensis* and *H. arenaria* by morphometric characters of J2 and cysts and by sequences of the ITS-rRNA and *COI* gene. *COI* gene sequences allow the differentiation of the species of CCN from each other and from related *Heterodera* species that parasitize grasses.

***Heterodera betae* Wouts, Rumpenhorst & Sturhan, 2001 – the Yellow Beet Cyst Nematode**

The yellow beet cyst nematode was observed by Maas *et al.* (1976) during routine examination of samples collected from the southern sugar beet-growing regions of the Netherlands. It was considered a biotype of the clover cyst nematode, *H. trifolii*, and subsequently named *H. trifolii forma specialis 'betae'*, capable of parasitizing sugar beet, a crop not normally affected by the clover cyst nematode. Comparative detailed morphological and molecular analysis of several populations of the yellow beet cyst nematode with closely related species allowed Wouts *et al.* (2001) to report reliable diagnostic features and describe it as a new species, *H. betae*. The yellow beet cyst nematode reproduces on Cruciferae, Chenopodiaceae, Polygonaceae, Caryophyllaceae and a few Leguminosae, which are good hosts for *H. betae*. In natural environments, this species was found to parasitize the sea beet. Several species of *Beta* and *Brassica* species, rhubarb, radish, spinach, common chickweed, pinks, garden pea, cabbage and broad beans are hosts for yellow beet cyst nematode. In the field, the threshold level at which damage occurs varies considerably between host species. Damage occurs in sugar beet at a level of ca 500 J2 and eggs per 100 cm^3 of soil at the beginning of the season, whereas in peas and beans, the threshold is over 800 J2. The yellow beet cyst nematode reaches epidemic levels only in lighter soils. This species is reported from several European

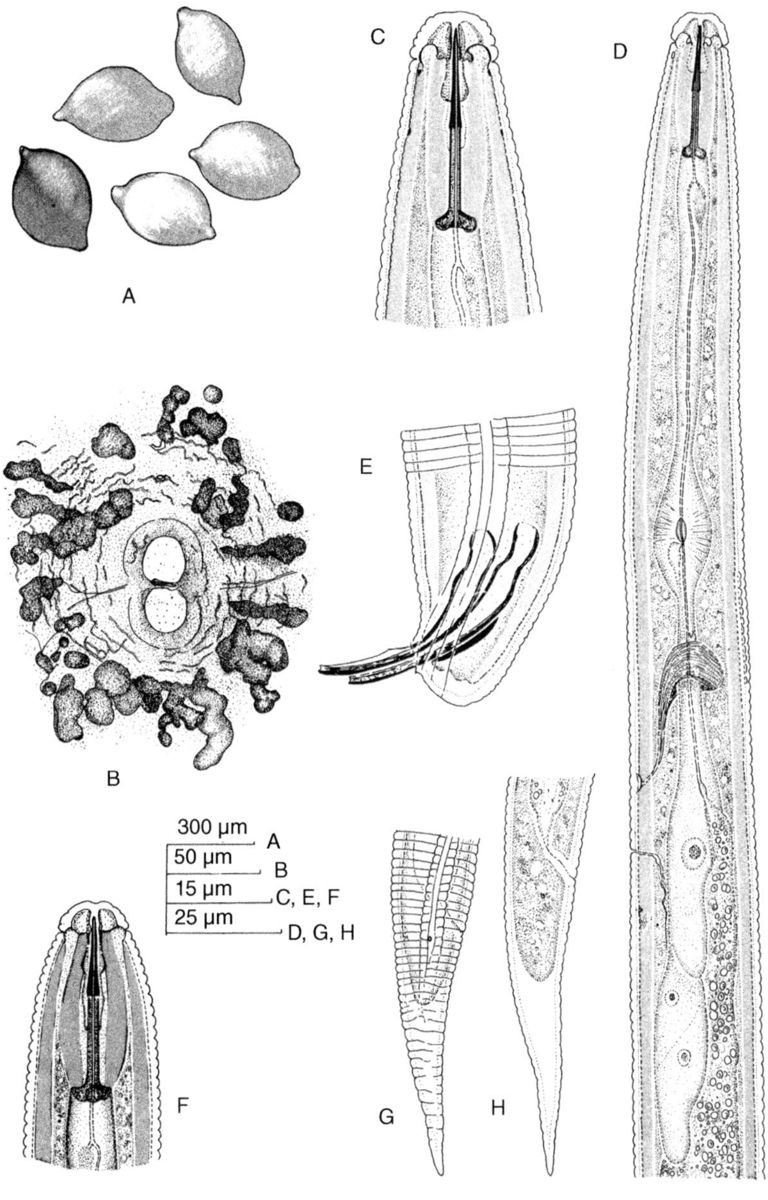

Fig. 10.10. *Heterodera avenae*. A: Cysts; B: vulval cone; C: anterior region of male; D: pharyngeal region of male; E: tail of male; F: anterior region of second-stage juvenile (J2); G, H: tail of J2. (After Williams and Siddiqi, 1972.)

countries and Korea. This nematode has 35-36 chromosomes and reproduces by mitotic parthenogenesis.

Measurements

Cysts: L = 475-1160 µm; W = 225-702 µm; L/W = 1.5-2.5; fenestral length =30.3-75 µm; vulval slit = 36-67 µm.

J2: L = 525-672 µm; stylet = 26-33 µm; hyaline region = 32-50 µm; tail = 65-84 µm.

Description: *Cyst:* lemon-shaped, often asymmetrical. Vulval cone high and wide, ambifenestrate (Fig. 10.11B), in young cysts surrounded by gelatinous matter containing up to 30 eggs. Vulval bridge narrow. Heavily pigmented bullae

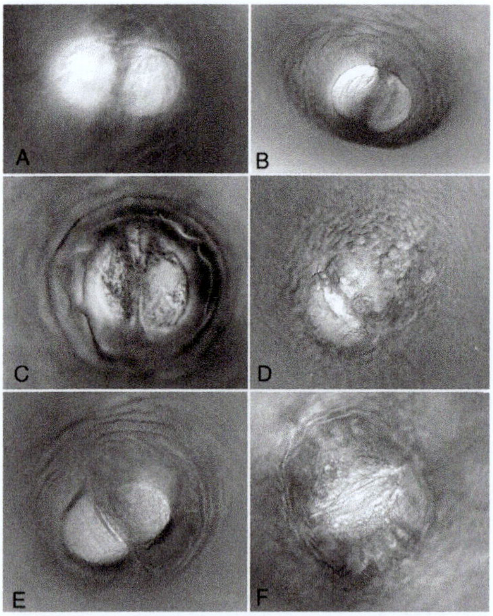

Fig. 10.11. Vulval plates. A: *Heterodera avenae*; B: *H. betae*; C: *H. cajani*; D: *H. carotae*; E: *H. ciceri*; F: *H. cruciferae*. (After Subbotin *et al.*, 2010 with modifications.)

present, mainly anterior to underbridge, globular, some with basal stem, close to cyst wall. *J2:* body slightly curved ventrally. Stylet robust, knobs heavy, deeply concave anteriorly, convex posteriorly. Tail gradually tapering to slender rounded terminus, hyaline region forming 50-60% of the total tail length.

Remarks: The yellow beet cyst nematode can be easily distinguished morphologically from the bisexual sugar beet nematode, *H. schachtii*, by the larger fenestrae, longer body, stylet and length of the hyaline part of the J2 tail. PCR-ITS-RFLP with the restriction enzymes MvaI and ScrFI generated a unique profile for this species. The *COI* sequence of *H. betae* was similar to those of *H. trifolii* and *H. galeopsidis* (Subbotin *et al.*, 2024).

Heterodera cajani Koshy, 1967 – the Pigeon Pea Cyst Nematode

The pigeon pea cyst nematode was described by Koshy (1967) from roots of pigeon pea, *Cajanus cajan* (L.) Mill., in India. Surveys indicate that *Heterodera cajani* is the most widespread cyst nematode in India and is now present in all major pigeon pea-growing regions of India and Pakistan. The damage induced by this nematode was also reported from Pakistan, Egypt and Myanmar. The list includes several dozen species of Fabaceae and Pedaliaceae. More than nine generations per year have been reported for a vegetation season. Nematode damage has been reported in many agronomic crops. This nematode is one of the major limiting factors in the production of pulses, *viz.*, pigeon pea, cowpea, green gram, black gram, which causes stunting of plants, reduction in the number of nodules on roots, with heavy crop yield losses to the tune of 30-67%.

Measurements

Cysts: L = 390-690 μm; W = 175-510 μm; L/W = 1.0-2.5; fenestral length = 21-69 μm; vulval slit = 39-60 μm.

J2: L = 324-515 μm; stylet = 20-33 μm; hyaline region = 17-40 μm; tail = 32-64 μm.

Description: *Cyst:* light to dark brown, typically lemon-shaped with a protruding neck and vulva. Subcrystalline layer present on young cysts. Ambifenestrate (Fig. 10.11C) with strong underbridge. Bullae prominent, dark brown, located beneath the underbridge. *J2:* body tapering anteriorly and posteriorly but more so posteriorly. Labial region offset, bearing four annuli. Stylet strong, well developed with anteriorly directed or rounded knobs. Tail with bluntly rounded and narrow terminus.

Remarks: *Heterodera cajani* belongs to the *Schachtii* group. It differs from several species of this group by a smaller cyst size. The nematode morphologically resembles *H. medicaginis* and *H. mediterranea* and differs from *H. medicaginis* by a longer vulval slit and from *H. mediterranea* by a longer J2 body and tail. This species could be clearly differentiated from all other *Heterodera* using sequences of the ITS-rRNA and *COI* genes.

Heterodera carotae Jones, 1950 – the Carrot Cyst Nematode

Heterodera carotae is present throughout the carrot-growing regions. It was found that only *D. carota* ssp. *sativa*, *D. carota* ssp. *carota* and *D. pulcherrimus* (Willd.), several wild Umbelliferae, such as *Torilis leptophylla*, *T. arvensis* and *T. japanica* are hosts. Up to four generations may occur in favourable growing conditions. In infected fields, irregular plant growth was observed, with the patches enlarging over time. Foliage is stunted, reddish and may dry out when infestations are heavy. Carrots are usually small, abnormally developed and have numerous radicles, which give the roots a characteristic bearded appearance, referred to as 'hairy root'. Carrot cyst nematodes have caused up to a 50% decrease in crop yield. This species was reported from several European countries, USA, Canada, Mexico and South Africa.

Measurements

Cysts: L = 218-625 μm; W = 165-500 μm; L/W ratio = 1.9-2.8; fenestral length = 27-36 μm; vulval slit = 43-51 μm.

J2: L = 375-452 μm; stylet = 22-25 μm; hyaline region = 20-31.8 μm; tail = 43.5-59 μm.

Description: *Mature cysts:* small, lemon-shaped. Bullae absent. Underbridge bifurcate, slender, unsclerotized. Fenestration indistinct, ambifenestrate (Fig. 10.11D). Vulval bridge frequently broken in older specimens. *J2:* labial region slightly offset with four indistinct annuli. Stylet knobs with concave anterior faces. The most common variation is the presence of one to three spherical refractive bodies within the tail, sometimes associated with swelling.

Remarks: *Heterodera carotae* belongs to the *Goettingiana* group. It most closely resembles *H. cruciferae* and *H. goettingiana*; from the first species, it differs on average by a longer J2 hyaline region and a longer vulval slit in cysts. From the second species, it differs by the shorter J2 tail and shorter cyst length. This species could be differentiated from other related *Heterodera* based on the *COI* gene sequence.

***Heterodera ciceri* Vovlas, Greco & Di Vito, 1985 – the Chickpea Cyst Nematode**

This nematode was found in several fields of chickpea (*Cicer arientinum* L.), where it was associated with a marked decline in plant health (Vovlas *et al.*, 1985). It is abundant and widespread in Northern Syria, several localities in Turkey, Jordan and Lebanon, where it causes severe yield loss of chickpea and lentil in fields. A host range study demonstrated that females and cysts of *H. ciceri* occurred in large numbers in the roots of chickpea, lentil, grass pea and pea. Few females were found in the roots of haricot bean and lupin (*Lupinus albus* L.), but no reproduction occurred on sugar beet (*Beta vulgaris* L.), curled dock (*Rumex crispus* L.) or *Trifolium* spp. Under field conditions, it is inferred that there is only one generation of the nematode during the growing season of spring-sown chickpea. The tolerance limit of chickpea is around 1 egg/cm^3 of soil. Yield losses of 20 and 50% can be expected in fields infected with 8 or 16 eggs/cm^3 of soil, respectively, and complete crop failure can occur in fields infested with more than 60 eggs/cm^3.

Measurements

Cysts: L = 570-930 μm; W = 350-550 μm; L/W ratio = 1.6-2.4; fenestral length = 32-52 μm; vulval slit = 32-60 μm.

J2: L = 440-585 μm; stylet = 27-30 μm; hyaline region = 31-42 μm; tail = 53-72 μm.

Description: *Cyst:* lemon-shaped with a distinct neck and a prominent cone. Cyst covered with a white subcrystalline layer. Ambifenestrate (Fig. 10.11H) with numerous prominent, dark brown bullae irregularly distributed at the periphery of the vulval cone and, in most cysts, beneath the underbridge level. Underbridge well developed with furcate ends and central thickening. *J2:* labial region hemispherical, slightly offset with three labial annuli, oral disc plate dorsoventrally elongated and two rounded lateral sectors bearing large, semilunar-shaped, amphidial apertures. Lateral field with four incisures. Tail irregularly annulated, tapering uniformly, ending rather abruptly in a finally rounded terminus.

Remarks: *Heterodera ciceri* belongs to the *Schachtii* group. It resembles *H. trifolii*, *H. rosii*, and *H. daverti* in its general morphology. It differs from *H. trifolii* by the presence of males, the different host-range (none of the *Trifolium* spp. tested were hosts of *H. ciceri*) and by the fenestral measurements. The *COI* gene sequence clearly discriminates *H. ciceri* from all other *Heterodera*.

***Heterodera cruciferae* Franklin, 1945 – the Cabbage Cyst Nematode**

Heterodera cruciferae was first reported from the United Kingdom. This species has a narrower host range than *H. schachtii* but apparently infects all species of *Brassica*, including common agricultural crops such as cabbage, Brussels sprouts, cauliflower, broccoli and radish (Baldwin and Mundo-Ocampo, 1991). The cabbage cyst nematode is considered as a pest of cabbage and could cause significant crop losses. Franklin (1951) reported several *Brassica* species as experimental hosts of *H. cruciferae*. Crop losses induced by *H. cruciferae* were reported in the yield of Brussels sprouts in California. *Heterodera cruciferae* parasitizes cool-weather or winter-grown crops; hence, the number of generations completed in a season depends on the growing period. Up to three generations might occur in Europe (Stone and Rowe, 1976). This species is distributed across many European and Asian countries, as well as in USA, Australia and northern Africa.

Measurements

Cysts: L = 355-690 µm; W = 288-571 µm; L/W ratio = 0.6-1.8; fenestral length = 22-59 µm; vulval slit = 29-55 µm.

J2: L = 333-504 µm; stylet = 20-25 µm; hyaline region = 16-30 µm; tail = 26-58 µm.

Description: *Cyst:* broad, almost spherical to lemon-shaped. Ambifenestrate (Fig. 10.11F), abullate, with very low semifenestral arches separated by a narrow vulval bridge, semifenestrae unobstructed in mature cysts, but in newly formed cysts, body wall may still be intact. *J2:* labial region offset with three or four annuli and a dorsoventrally elongated oral disc flanked by lateral lips bearing amphidial apertures. Stylet knobs: anterior face flat, rounded or slightly concave (Fig. 10.9D). Tail tapering uniformly to a fine, rounded terminus.

Remarks: *Heterodera cruciferae* belongs to the *Goettingiana* group and is closely related with *H. carotae*. The *COI* gene sequence discriminates *H. cruciferae* from all other *Heterodera*.

Heterodera elachista Ohshima, 1974 – the Japanese Cyst Nematode

This nematode species has been found to be widely distributed in Japan, mainly in upland rice fields. Based on a series of experiments, this nematode was shown to be the major factor causing continuous cropping failure of upland rice and was tentatively identified as *H. oryzae* (Ohshima, 1974). *Heterodera elachista* was also identified from a rice field in Iran and Korea, on roots of *Carex* sp., rice and corn in China and corn in Northern Italy. The yield damage induced by this nematode ranges from 7 to 19% (Bridge *et al.*, 1990). Corn plants 'Rixxer' in commercial fields at Ferrara Province, Italy, infected by the nematode showed severe decline with stunting, a patched distribution of field, and heavily infected roots with a high proliferation of short lateral roots. Greater yield losses were observed when the roots were invaded by the nematodes before tillering. In a field experiment, negative correlations between yields and final soil cyst populations were more evident in early, compared to later, sown crops. Nematodes had the most severe impact during the later stages of plant growth. This species was reported from Italy, Japan, Iran and China.

Measurements

Cysts: L = 328-586 µm; W = 229-540 µm; L/W ratio = 1.1-1.9; fenestral length = 23-50 µm; vulval slit = 26-55 µm.

J2: L = 330-525 µm; stylet = 16-25 µm; hyaline region = 25-50 µm; tail = 44-87 µm.

Description: *Cyst:* light to dark brown, spherical to lemon-shaped. Subcrystalline layer present. Ambifenestra low (Fig. 10.12A) and rounded, the width is approximately equal to the length. Vulval bridge narrow, medium-sized underbridge; cone terminus with a few large, peripheral, dark brown bullae lacking finger-like projections. *J2:* body slightly curved ventrally when killed by gentle heat. Stylet well developed with rounded or anteriorly concave knobs. Tail long, narrowly tapering to a very fine, rounded terminus.

Remarks: *Heterodera elachista* belongs to the *Cyperi* group and is morphologically close to *H. oryzae*, *H. sacchari* and *H. leuceilyma*. It differs from *H. sacchari* and *H. leuceilyma* by the absence of finger-shaped projections on the slender underbridge of the cysts and from *H. oryzae* by smaller cysts and shorter J2. *Heterodera elachista* could be

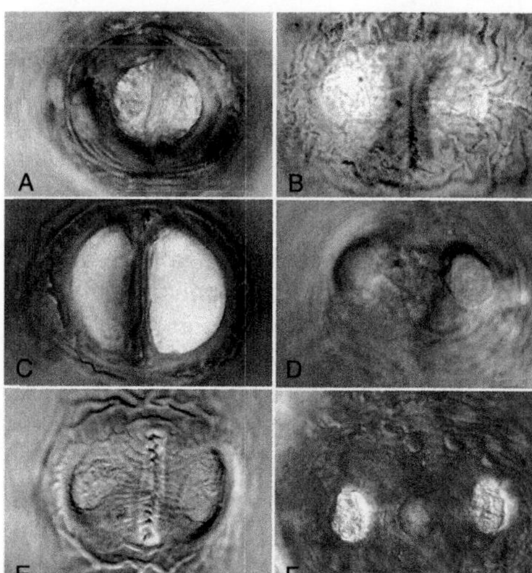

Fig. 10.12. Vulval plates. A: *Heterodera elachista*; B: *H. fici*; C: *H. glycines*; D: *H. hordecalis*; E: *H. humuli*; F: *H. latipons*. (After Subbotin *et al.*, 2010 with modifications.)

differentiated from other *Heterodera* species based on the ITS-rRNA and *COI* gene sequences.

Heterodera fici **Kirjanova, 1954 – the Fig Cyst Nematode**

The fig cyst nematode, *H. fici*, was first described from roots of plant *Ficus elastica* Roxb. growing in a pot, which was imported from Harbin, China, to Russia. Subsequent surveys revealed that this nematode is widely distributed in natural conditions in several Mediterranean, south-west countries and California, where it infects figs, and in glasshouses from many countries where it parasitizes ornamental plants from the genus *Ficus*. Under field conditions in the Mediterranean region, this nematode develops more than one generation per year. The nematode can be considered as a potential threat in fig nurseries, where fig rootstocks are often obtained from seeds. It is also noteworthy that the nematode has caused considerable damage to potted plants. Poor plant growth and leaf chlorosis have been reported in both edible and ornamental figs infected with *H. fici*. The host range of *H. fici* is restricted to the species of the genus *Ficus*, including *F. carica* L., *F. benghalensis* L., *F. lyrata* Warb, *F. australis* Willd. and *F. rubiginosa* Desf (Kirjanova and Krall, 1971).

Measurements

Cysts: L = 340-697 μm; W = 272-560 μm; L/W = 1.0-1.7; fenestral length = 42-74 μm; vulval slit = 35-56 μm.

J2: L = 328-470 μm; stylet = 20-23 μm; hyaline region = 18-33 μm; tail = 40-61 μm.

Description: *Cyst:* body light to dark brown, basically lemon-shaped, neck and vulval cone distinct. Neck protruding, curved posteriorly. Fenestra ambifenestrate, sometimes top of cone appearing bifenestrate (Fig. 10.12B). This bifenestrate appearance could be misleading unless many vulval cones are examined to reveal the true ambifenestrate vulval cone of this species. Semifenestra symmetrical, separated by vulval bridge, surrounded by well-developed basin. Bullae dome-shaped, small, scattered around the underbridge plane. Underbridge weakly developed, with furcate ends. *J2:* body vermiform; tapering at both extremities, more so posteriorly. Stylet well developed, basal knobs rounded, directed slightly anteriorly. Tail long, tapering, terminus rounded.

Remarks: *Heterodera fici* belongs to the *Humuli* group. It differs from other members of the *Humuli* group (*H. humuli*, *H. ripae*, *H. vallicola* and *H. litoralis*) by having ambifenestrate rather than bifenestrate cysts and a longer vulval slit. *Heterodera fici* could be differentiated from other *Heterodera* species based on the ITS-rRNA and *COI* gene sequences.

Heterodera glycines **Ichinohe, 1952 – the Soybean Cyst Nematode**

(Fig. 10.13)

A cyst nematode parasitizing soybean plants, *Glycine max* L., and causing 'yellow dwarf' symptoms, was recorded from Shirakawa, Fukushima Prefecture, Japan, in 1915. Ichinohe (1952) was the first to make careful morphological comparisons with other *Heterodera* species and to give a specific name and a brief description of this nematode. In Japan, yield losses have been estimated at 10-70% (Ichinohe, 1988). Presently, *H. glycines* occurs in most countries of the world where soybean is produced. In a study of losses predicted in ten soybean-producing countries, together accounting for 97% of the world crop, *H. glycines* appeared to be the most important constraint (losses of 8,969,400 t) on yield and estimated at US$1960 million (Wrather *et al.*, 2001). In these countries, total yield losses attributed to *H. glycines* were greater than those for any other pest of the crop. The SCN is widely distributed throughout the north-central United States, where different maturity groups with the same source of SCN resistance are grown. State surveys in the region report that 14-63% of the fields are infested with SCN.

Heterodera glycines has a broad host range, particularly within the family Fabaceae, but also occurs on other families. More than 66 weed species of nine families are suitable hosts. Riggs (1992) provided a list of non-fabaceous hosts comprising 63 species in 50 genera from 22 families (e.g., Boraginaceae, Capparaceae, Caryophyllaceae, Chenopodiaceae, Brassicaceae, Lamiaceae, Fabaceae, Scrophulariaceae, Solanaceae). In field conditions, *H. glycines* was also found in several other plants, including henbit (*Lamium amplexicaule*), purple deadnettle (*Lamium purpureum*), mouse-ear chickweed (*Cerastium holosteoides*) and common chickweed (*Stellaria media*) (Riggs, 1992).

Three to five generations develop during the cropping season. Optimum temperature is 23-28°C; development stops below 14°C and above 34°C. In the absence of a host, J2 and eggs in cysts may remain viable in soil for 6-8 years.

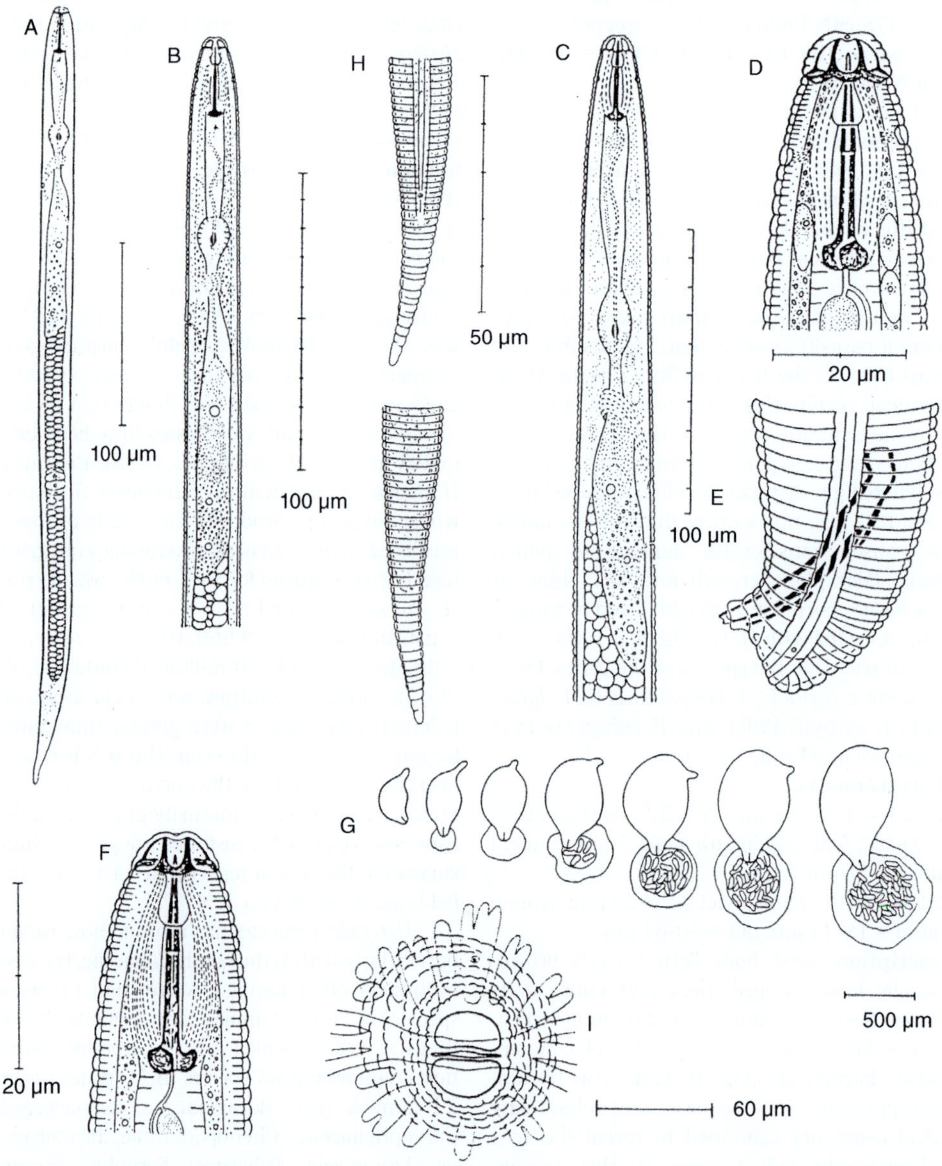

Fig. 10.13. *Heterodera glycines*. A, B, F-H: Second-stage juvenile. A: Entire body; B: pharyngeal region; H: tails; F: anterior region; C-E: male. C: Pharyngeal region; D: anterior region; E: tail; G: cysts with egg-sacs; I: vulval cone. (After Burrows and Stone, 1985.)

Heterodera glycines disturbs root growth, interferes with nodulation and causes early yellowing of soybean plants. The above-ground symptoms of damage on individual plants and appearance of infested fields are usually not sufficiently specific to allow direct identification. Infected plants are predisposed to *Fusarium* wilt. Sudden death syndrome is a soil-borne disease of soybean caused by the fungus *Fusarium solani* in association with *H. glycines*.

Measurements

Cysts: L = 340-920 µm; W = 200-688 µm; L/W ratio = 1.0-2.4; fenestral length = 35-72 µm; vulval slit = 36-60 µm.

J2: L = 345-504 μm; stylet = 21-25 μm; hyaline region = 18-36 μm; tail = 35-59 μm.

Description: *Cyst:* mainly lemon-shaped, sometimes round with a protruding neck and cone. Ambifenestrate (Fig. 10.12C), bullae prominent, located at or anterior to underbridge, extending into the vulval cone from the interior of the body wall cuticle. Shape varying from round to finger-like, round bullae differently sized, finger-like bullae of variable length and thickness. Underbridge well developed. *J2:* body vermiform with regularly annulated cuticle. Stylet robust with anteriorly protruding knobs. Tail tapering uniformly to a finely rounded terminus.

Remarks: *Heterodera glycines* belongs to the *Schachtii* group and is distinguished from similar species by a combination of morphological and morphometric characteristics. It differs from *H. schachtii* by shape of the J2 stylet knobs (slightly convex *vs* moderately or strongly concave), shorter average J2 stylet and longer average fenestral length. Although the ITS-rRNA gene sequences do not clearly differentiate this species from some *H. medicaginis* populations, the *COI* and *hsp90* gene sequences allow a clear separation of this species from all others.

Races and HG Types: The race test for *H. glycines* was first developed based on comparative development of females on four differential soybean lines and was proposed by a group of nematologists and soybean breeders (Riggs and Schmitt, 1988). The proposed system of races is based on the reaction of four host differentials for identification of 16 races of *H. glycines*. Twelve races are known to occur in the USA, three races (1, 3 and 5) are present in Japan, six in Argentina (1, 3, 5, 6, 9 and 14) and seven in China. Niblack *et al.* (2002) proposed an HG Type test to better describe population variation and to expand the flexibility of the race classification system. The HG Type system uses three of the four resistant soybean genotypes currently used as indicator hosts (Peking [= PI 548402], PI 88788 and PI 90763).

Heterodera goettingiana **Liebscher, 1892 – the Pea Cyst Nematode**

In 1890, G. Liebscher reported infection and yield loss of pea (*Pisum sativum* L.) and vetch (*Vicia sativa* L.) by nematodes identified as *H. schachtii* in fields of the Agricultural Institute at Göttingen, Germany. Two years later, he described this species as *H. goettingiana*. Several researchers have reported diseases of peas, primarily in European countries, caused by this nematode; however, little was known about the biology and pathogenicity of *H. goettingiana* until the mid-1900s (Franklin, 1951). Infected pea fields show sharply delineated patches with dwarfed, poorly branched and yellowing plants that die prematurely. Infected plants either fail to flower or flower too early. The root system is poorly developed. Development takes 3-15 weeks, depending on soil temperature and moisture as well as host species. Hosts of *H. goettingiana* include species from the family Fabaceae and the genus *Sonchus*. One or two generations occur during the growing season in the UK, and three generations may develop in southern Italy. Cysts with eggs can remain viable in the absence of a host for 12 years.

Measurements

Cysts: L = 400-780 μm; W = 310-540 μm; L/W ratio = 1.3-2.2; fenestral length = 43-71 μm; vulval slit = 43-61 μm.

J2: L = 408-519 μm; stylet = 23-26 μm; hyaline region = 27-38 μm; tail = 54-74 μm.

Description: *Cyst:* lemon-shaped with light to dark brown cyst wall. Subcrystalline layer not visible. Vulval cone ambifenestrate. In some old cysts, vulval bridge ruptured, with fenestrae joining to form a large, oval fenestrum. Bullae absent, although bullae-like structures and vulval denticles present. Underbridge weak. *J2:* body vermiform, curved ventrally after fixation. Labial region hemispherical, with two to five annuli, slightly offset from body. Lateral field with four incisures, not areolated. Stylet knobs rounded, slightly projecting anteriorly. Tail tapering uniformly to a finely rounded terminus.

Remarks: *Heterodera goettingiana* belongs to the *Goettingiana* group. It differs from several other representatives of the *Goettingiana* group (*H. cruciferae, H. carotae, H. ciceri, H. scutellariae, H. persica*), by an average longer J2 body, longer tail and longer hyaline region. The ITS-rRNA and *COI* gene sequences clearly differentiate this species from all others.

Heterodera hordecalis **Andersson, 1974 – the Barley Cyst Nematode**

In 1967, heavy infestations by a cyst nematode were found on barley in a field in Halland province, Sweden. This nematode was later described by Andersson (1974) as a new species. *Heterodera hordecalis* is reported from several

European, Asian and northern Africa countries. It also infects wheat, rye and many grasses.

Measurements

Cysts: L = 330-950 μm; W = 255-680 μm; L/W ratio = 1.0-1.5; fenestral length = 47-80 μm; vulval slit = 14-33 μm.

J2: L = 410-550 μm; stylet = 21-26 μm; hyaline region = 29-46 μm; tail = 44-60 μm.

Description: *Cyst:* body ovoid, with a distinct neck and vulval cone. Newly formed cysts with remnants of subcrystalline layer. Bifenestral (Fig. 10.12D), semifenestrae widely separated. Semifenestrae usually appearing oval in focal plane at surface but round or horseshoe-shaped when focused at mid-bridge level. In centre of bridge, a rigid, often dumbbell-like structure surrounds the anterior end of the vagina. From the narrow part of this structure, fine ridges usually extend through weaker parts of the bridge towards the semifenestrae. Underbridge extremely strong, sometimes with pronounced thickening in middle, and with ends basically bifurcate, each branch consisting of a number of cords, although often irregularly splayed. Bullae most commonly absent. *J2:* labial region slightly offset with three or four annuli. Lateral field with four incisures. Tail tip finely rounded.

Remarks: *Heterodera hordecalis* belongs to the *Avenae* group and is most similar to *H. latipons*. The most striking distinguishing features are those of the vulval slit, which is much longer in *H. hordecalis* than in *H. latipons*. The structures of the middle part of the bridge are much more rigid in *H. hordecalis* than in *H. latipons*. The stylet knobs of the J2 are more anteriorly projecting in *H. hordecalis* than in *H. latipons*, although this can be variable in the latter, and the tail and hyaline region are slightly longer in *H. hordecalis* than in *H. latipons*. The ITS-rRNA and *COI* gene sequences clearly differentiate this species from all others.

Heterodera humuli **Filipjev, 1934 – the Hop Cyst Nematode**

Cyst-forming nematode from hop plants was first recorded in 1894 in England. Filipjev (1934) gave a formal description of this species. Presently, *Heterodera humuli* was reported from many European and Asian countries, the USA, Canada, Mexico, Australia, New Zealand and South Africa. In an experimental study, the nematode significantly reduced plant height and fresh and dry weight of shoots. Infected plants showed more severe nutrient deficiency symptoms. A negative correlation has been demonstrated between the numbers of J2 in soil and cone yield in hop plantations. The growth of nematode-infested plants of a wilt-susceptible hop cultivar was significantly reduced when the plant pathogenic fungus *Verticillium albo-atrum* was also present. Hosts of the hop cyst nematode include *Humulus lupulus*, *Urtica dioica* and *U. urens*. There are several reports on experimental hosts. Only one generation is produced per year.

Measurements

Cysts: L = 290-610 μm; W = 245-450 μm; L/W ratio = 1.2-1.7; fenestral length = 49-76 μm; vulval slit = 33-43 μm.

J2: L = 336-468 μm; stylet = 21-25 μm; hyaline region = 22-30 μm; tail = 42-53 μm.

Description: *Cyst:* lemon-shaped, occasionally nearly spherical. Abullate, thin-walled, light coloured, darkening with age. Bifenestrate (Fig. 10.12E), vulval bridge broad, semifenestrae circular or subcircular, often obscured by a thin membrane with a fine fingerprint-like pattern. Underbridge slender, weak, with furcate ends. *J2:* labial region rounded, offset with two to four annuli (Fig. 10.9E). Lateral field with four incisures, reducing to three anteriorly and posteriorly, not areolated. Tail tapering, terminus often constricted and irregularly shaped.

Remarks: *Heterodera humuli* belongs to the *Humuli* group and is similar to *H. ripae* and *H. vallicola*. It differs from *H. ripae* by a longer J2 tail and hyaline region and from *H. vallicola* by more slender cyst and longer fenestra. The ITS-rRNA and *COI* gene sequences clearly differentiate this species from all others (Subbotin *et al.*, 2022).

Heterodera latipons **Franklin, 1969 – the Mediterranean CCN**

In the early 1960s, a cyst nematode similar to *H. avenae* was detected in Israel and Libya on the roots of stunted wheat and barley plants. It was morphologically studied and described by Franklin (1969) as a new species under the name *H. latipons*, based on characteristics of the Israel population. This nematode was later recorded in many European and Asian countries. *Heterodera latipons* often occurs in mixed populations and *H. avenae* in cereal cropping systems.

Hosts include barley (*Hordeum vulgare* L.), oat (*Avena sativa* L.), rye (*Secale cereale* L.), *Phalaris minor*, *P. paradoxa* and *Elytrigia repens*. *Triticum durum* Desf. was considered to be a poor host of this nematode. Yield losses as high as 50% were reported for barley in Cyprus. In Syria, the nematode causes average yield losses of 20 and 30% in barley and durum wheat, respectively, and the nematode was more damaging under water stress conditions. Moreover, damage is more severe in fields infested concomitantly with *H. latipons* and the fungus *Bipolaris sorokiniana* (Sacc.) Shoemaker, the causal agent of the common root rot and seedling blight of barley, that is, the presence of the nematode increases the aggressiveness of the fungus (Scholz, 2001). In all areas studied, *H. latipons* completed only one life cycle during the growing season (Mor *et al.*, 1992).

Measurements

Cysts: L = 300-700 μm; W = 320-560 μm; L/W ratio = 0.6-1.7; fenestral length = 52-76 μm; vulval slit = 6-11 μm.

J2: L = 401-559 μm; stylet = 22-25 μm; hyaline region = 20-36 μm; tail = 42-68 μm.

Description: *Cyst:* dark to mid-brown, covered with a white subcrystalline layer. Bifenestrate (Fig. 10.12H), semifenestrae separated by a distance greater than fenestral width, vulval slit short. Strong underbridge with pronounced thickening in middle and with ends splayed. Bullae usually absent, sometimes present at the underbridge level. *J2:* body slightly curved dorsoventrally when killed by heat. Stylet with well developed, anteriorly concave, knobs.

Remarks: *Heterodera latipons* belongs to the *Avenae* group and closely resembles *H. hordecalis* and *H. turcomanica*. These nematodes share similar circular semifenestrae separated by a distance longer than the semifenestra diameter and a rather typical underbridge but with a pronounced enlargement underlying the vulval slit. The most important differentiating character between *H. latipons* and *H. hordecalis* is the vulval slit, which is much shorter in *H. latipons*. The ITS-rRNA and *COI* gene sequences clearly differentiate this species from all others. Recently, a putatively new cyst nematode species from Syria and Turkey, related to *H. latipons*, was revealed from the analysis of *COI* and ITS-rRNA gene datasets (Subbotin *et al.*, 2018).

***Heterodera oryzae* Luc & Brizuela, 1961 – the Rice Cyst Nematode**

Heterodera oryzae was described by Luc and Brizuela (1961) from swamp rice fields in the central part of the Ivory Coast. It was also found parasitizing banana, *Musa* sp., in Senegal. In experimental tests, J2 of *H. oryzae* penetrated into the roots of *Mariscus umbellatus* and *Zea mays* and developed into numerous females. In these hosts, populations can be increased, or at least maintained. In roots of *Glycine hispida*, *Pueraria phaseoloides* and *Solanum lycopersicum*, only a few females developed. *Cyperus umbellatus* and elephant grass (*Pennisetum purpureum*) are considered as hosts for *H. oryzae*. Host yield loss induced by this nematode has not been clearly shown. It is considered to be less aggressive than other cyst nematodes infecting rice (Luc, 1986).

Measurements

Cysts: L = 310-999 μm; W = 220-690 μm; L/W ratio = 0.9-2.3; fenestral length = 19-46 μm; vulval slit = 34-65 μm.

J2: L = 350-589 μm; stylet = 16-25 μm; hyaline region = 22-45 μm; tail = 46-69 μm.

Description: *Cyst:* mature cysts dark brown to black, lemon-shaped with prominent vulval cone. Ambifenestrate (Fig. 10.14A), underbridge long with dendroid-like extremities attaching to cyst wall but without finger-like projections. Bullae present but number, size, shape and location vary, although usually small, rather spherical and scattered inside the vulval cone. *J2:* body straight or very slightly ventrally curved when heat-relaxed. Labial region with three or four annuli. Lateral field with three incisures. Tail conical, with elongated terminus and pointed tip.

Remarks: *Heterodera oryzae* belongs to the *Cyperi* group and is morphologically similar to *H. oryzicola* and *H. elachista*. It differs from *H. oryzicola* by having a larger cyst, longer J2 stylet, longer J2 hyaline region and anteriorly projecting J2 stylet knobs.

***Heterodera sacchari* Luc & Merny, 1963 – the Sugar Cane Cyst Nematode**

This species was originally described from sugar cane in the Congo, Africa (Luc and Merny, 1963). Rice (*Oryza sativa*) and sugar cane are the major field crops infected by this nematode. *Oryza glaberrima* is also infected, but cysts are smaller and contain fewer eggs. Pot experiments showed that maize, millet and sorghum, commonly cultivated in cropping systems with rice in West Africa, were poor hosts of *H. sacchari*. Infected sugar cane plants were stunted and thin, and secondary roots are less abundant than in healthy plants. Rice plants attacked by

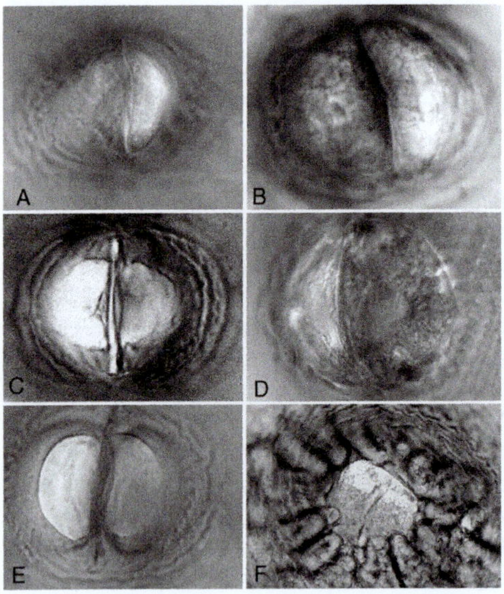

Fig. 10.14. Vulval plates. A: *Heterodera oryzae*; B: *H. sacchari*; C: *H. schachtii*; D: *H. sorghi*; E: *H. trifolii*; F: *H. zeae*. (After Subbotin *et al.*, 2010 with modifications and courtesy of U. Zunke.)

H. sacchari showed intense chlorosis, delayed and reduced tillering and reduced grain yield. The roots of attacked plants were twiggy, very necrotic and blackened. A field study conducted in 1997 at the West Africa Rice Development Association in Cote d'Ivoire showed that under low water availability in sandy soil, damage and reduced upland rice grain yield was significant (Coyne and Plowright, 2000). *Heterodera sacchari* has been found across several African countries, and also in India, Pakistan and Trinidad.

Measurements

Cysts: L = 380-1030 μm; W = 280-830 μm; L/W ratio = 1.0-2.0; fenestral length = 29-60 μm; vulval slit = 44-62 μm.

J2: L = 360-560 μm; stylet = 19-26 μm; hyaline region = 20-36 μm; tail = 46-66 μm.

Description: *Mature cyst:* brown to dark brown, lemon-shaped with prominent vulval cone and medium-sized neck. Cuticle with a lace-like pattern, numerous 'perforations' evenly but irregularly distributed on the inner layer. Cyst cone ambifenestrate (Fig. 10.14B), fenestrae rather obscurely demarcated, underbridge strongly developed with finger-like projections, few peripheral bullae. *J2:* body straight or slightly ventrally curved when heat-relaxed, slightly tapering at the anterior end, more attenuated at the posterior end. Labial region dome-shaped, with three annuli, labial framework heavily sclerotized. Lateral field composed of three lines, generally irregularly areolated. Stylet strong with knobs anteriorly concave. Tail elongated, conical, with pointed terminus.

Remarks: *Heterodera sacchari* belongs to the *Sacchari* group and to the *H. sacchari* species complex together with *H. leuceilyma* and *H. goldeni* (Tanha Maafi *et al.*, 2007). The three species are very similar in morphometrics and other morphological characters and are not easily distinguished. The ITS-rRNA and *COI* gene sequences clearly differentiate this species from all others.

Heterodera schachtii Schmidt, 1871 – the Sugar Beet Cyst Nematode

The sugar beet cyst nematode has been recognized as a plant pathogen since 1859, when it was associated with stunted and declining sugar beet in Germany. In the following years, it became recognized as a pest of great importance in beet-growing areas of several European countries. *Heterodera schachtii* is found in all major sugar beet-producing areas of the world, favouring temperate regions but apparently tolerating a broad range of climates. It is widespread in Europe, USA and Canada (Baldwin and Mundo-Ocampo, 1991). Annual yield loss in the EU countries, based upon world market sugar prices, was estimated in 1999 at up to €90 million (Müller, 1999). The optimum temperature for development is ca 25°C. In some climates, three to five generations may complete development on sugar beet in one season (Franklin, 1972). *Heterodera schachtii* was described from *Beta vulgaris* L., and it parasitizes mainly plants of the families Amaranthaceae (Caryophyllales) (many species of *Beta* and *Chenopodium*) and Brassicaceae (Brassicales) (*Brassica oleracea*, *B. napus*, *B. rapa*, *Rhaphanus sativus* and many others including diversity of common weeds) (Franklin, 1972). Some plants from Polygonaceae, Scrophulariaceae, Caryophyllaceae and Solanaceae are susceptible to nematode infection.

Measurements

Cysts: L = 480-960 μm; W = 396-696 μm; L/W ratio = 0.9-2.0; fenestral length = 28-48 μm; vulval slit = 33-54 μm.

J2: L = 400-512 μm; stylet = 23-28 μm; hyaline region = 17-33 μm; tail = 40-56 μm.

Description: *Cyst:* light to dark brown in colour. Ambifenestrate (Fig. 10.14C), within cone, remnants of vagina attached to side walls by underbridge and a number of irregularly arranged, dark brown molar-shaped bullae situated a short distance beneath the vulval bridge. *J2:* labial region offset, hemispherical, with four indistinct annuli. Stylet moderately heavy with prominent, forward-directed knobs (Fig. 10.9C). Tail acutely conical with rounded tip, distinct hyaline terminal section 1-1.25 stylet lengths long.

Remarks: *Heterodera schachtii* belongs to the *Schachtii* group and is distinguished from closely related species (*H. trifolii, H. glycines, H. betae* and others) by a combination of morphological and morphometric characteristics. PCR-ITS-RFLP profile generated by MvaI is unique for *H. schachtii*. The ITS-rRNA and *COI* gene sequences of *H. schachtii* differentiate this species from other *Heterodera*. The *COI* gene sequences distinguished southern Spanish populations from all other *H. schachtii* populations (Subbotin *et al.*, 2024).

***Heterodera sorghi* Jain, Sethi, Swarup & Srivastava, 1982 – the Sorghum Cyst Nematode**

Heterodera sorghi was first recorded and described by Jain *et al.* (1982) from the roots of sorghum in Uttar Pradesh, India. Subsequently, it was found in other states of India. Hosts of this species include maize, rice and barley. The life cycle studies of *H. sorghi* carried out on maize revealed that the nematode was able to complete its life cycle from J2 to J2 in 24 days at 28-36° C. More than four generations may be completed in a maize cropping season of 100-110 days duration. Significant reduction in the shoot length and fresh shoot and root weights of maize was observed at an initial inoculum level of 4 J2/cm^3 of soil, and, hence, this inoculum level was considered as the minimum damaging threshold level for this host.

Measurements

Cysts: L = 550-910 μm; W = 400-600 μm; L/W ratio = 1.2-1.8; fenestral length = 40-56 μm; vulval slit = 28-39 μm.

J2: L = 420-525 μm; stylet = 20-22 μm; hyaline region = 25-35 μm; tail = 42-60 μm.

Description: *Cyst:* light to dark brown, lemon-shaped with prominent neck and vulval cone. Cyst cone ambifenestrate (Fig. 10.14D), semifenestrae separated by narrow vulval bridge. Underbridge strongly developed without finger-like projections in the middle. Bullae dark brown, numerous, occurring at the level of underbridge, varying in size, shape and number. *J2:* body slightly curved ventrally when heat-relaxed. Lateral field with three incisures, not areolated. Stylet strong, cone and shaft almost equal in length, knobs with concave anterior surfaces. Tail tapering posteriorly, conical.

Remarks: *Heterodera sorghi* belongs to the *Sacchari* group and is most closely related to *H. sinensis*. It differs from *H. sinensis* by shorter average J2 body and shorter hyaline region. The ITS-rRNA gene sequence can be used to identify this species.

***Heterodera trifolii* Goffart, 1932 – the Clover Cyst Nematode**

The clover cyst nematode is an important pest of diverse agricultural crops and many pasture plants, especially parasitizing white clover, *Trifolium repens*. *Heterodera trifolii* is a cosmopolitan species found in all continents. Some authors propose considering *H. trifolii* within a conglomerate of independently evolved mitotic parthenogenetic populations comprising polyploidy and aneuploid forms, with 3n = 24-28 and 4n = 34-35 chromosomes and host races with more or less extended host ranges. Good hosts within Fabaceae for *H. trifolii* are also: *T. repens*, *Trifolium* spp., *Lotus oroboides*, *Medicago falcata*, *Melilotus officinalis*, *Melilotus albus*, *Melilotus* spp., *Kummerowia stipulacea*, *K. striata*, *Vicia villosa*, *Phaseolus vulgaris* and *Sesbania macrocarpa*. Several generations may occur during a year, so population density may increase rapidly (Mulvey, 1959). At optimal temperature of about 25°C, the life cycle is completed in 17 days (Baldwin and Mundo-Ocampo, 1991). This nematode causes reduction in yield, nitrogen fixation and persistence of this plant in pastures. Several studies have shown yield losses and growth suppression of white clover due to *H. trifolii* infection.

Measurements

Cysts: L = 360-1020 μm; W = 195-680 μm; L/W = 1.2-2.7; fenestral length = 40-80 μm; vulval slit = 39-66 μm.

J2: L = 461-678 μm; stylet = 23-31 μm; hyaline region = 27-45 μm; tail = 49-78 μm.

Description: *Cyst:* brown to dark brown, containing a few to several hundred eggs. Vulval cone distinct. Ambifenestrate cysts (Fig. 10.14E), with underbridge and bullae strong and elongated. Long vulval slit and strongly pigmented underbridge with bifurcate ends. *J2:* labial region offset, with three or four annuli, labial framework heavily sclerotized. Lateral field with four incisures, not areolated, outer incisures crenate. Stylet robust, anterior surfaces of knobs concave. Tail conoid, tapering uniformly to a finely rounded terminus.

Remarks: *Heterodera trifolii* is a member of the *H. trifolii* species complex, including *H. betae, H. lespedezae* and *H. galeopsidis*, and belongs to the *Schachtii* group. It differs from similar species *H. betae* by a shorter J2 hyaline region and shorter stylet. The ITS-rRNA and *COI* gene sequences of *H. trifolii* can differentiate this species from some related *Heterodera.*

Heterodera zeae **Koshy, Swarup & Sethi, 1971 – the corn or maize cyst nematode**

Heterodera zeae was described from corn (*Zea mays*) by Koshy *et al.* (1970) from the state of Rajasthan, India. Presently, this species has also been reported from Afghanistan, Nepal, Pakistan, China, Thailand, Egypt, Spain, Portugal, Greece and USA. It is widely distributed and economically important (Koshy and Swarup, 1971). *Heterodera zeae* is widely distributed in the maize-growing regions of India, causing substantial damage to this crop. Infected maize plants are stunted; leaf emergence is retarded, and fresh and dry weights are reduced. Suppression of maize plant growth by the Maryland population of *H. zeae* was demonstrated in tests in field microplots and in plant growth chambers. Maize growth (dry weight) and yield were suppressed by 13-73% in 4 out of 5 years in the presence of *H. zeae* (Krusberg *et al.*, 1997). However, maize yield suppression has not been demonstrated in maize fields in Maryland, where this species was found in 1981. The high soil temperatures in India and Egypt may be a primary reason why *H. zeae* is economically important to maize in those countries (Hashmi *et al.*, 1993).

Many researchers have evaluated the host range for different populations of *H. zeae*. Populations from India and the USA apparently differ in their host range, although these have not yet been compared under identical conditions.

Measurements

Cysts: L = 342-805 μm; W = 245-551 μm; L/W ratio = 1.0-2.2; fenestral length = 35-58 μm; vulval slit = 29-58 μm.

J2: L = 350-484 μm; stylet = 19-25 μm; hyaline region = 17-30 μm; tail = 32-50 μm.

Description: *Cyst:* light brown, basically lemon-shaped, cuticle thin-walled, lacking a subcrystalline layer. Thin subcrystalline layer discernible in young cysts only. Fenestra ambifenestrate (Fig. 10.14H), semifenestrae separated by a fairly wide vulval bridge, fenestral length and width variable, basin wide but generally poorly defined. Bullae prominent, immediately beneath (anteriad) the underbridge and characteristically arranged as four finger-like bullae in a distinct formation (observed in all but a few of the several hundred cysts examined). Immediately beneath finger-like bullae are a number of randomly arranged bullae. Underbridge simple, short and thin, found in all but a few of cysts examined, lacking forking at ends. *J2:* body typically vermiform, tapering at both ends. Labial region slightly offset, rounded, with low profile, with three to five annuli, labial framework moderately developed. Lateral field with four distinct lines. Stylet strongly developed with round or slightly anteriorly directed knobs. Tail short, tapering conically, with an acutely rounded terminus.

Remarks: *Heterodera zeae* differs from other species by the location of the bullae at two levels, that is, four, finger-like, bullae located immediately beneath a short, thin, underbridge and with many randomly located bullae further below. The ITS-rRNA and *COI* gene sequences of *H. zeae* differentiate this species from other *Heterodera*.

10.10.2. Genus *Globodera* Skarbilovich, 1959

Diagnosis (after Baldwin and Mundo-Ocampo, 1991 and Siddiqi, 2000): *Mature female and cyst*: spheroid, lacking terminal cone. Vulval area circumfenestrate. Vulva located in a cavity beneath outline of body, vulval slit <15 μm. No anal fenestra. Vaginal remnants, underbridge and bullae rarely present. Cuticle with distinct D-layer. All eggs retained in body, egg-sac absent. *Egg*: surface smooth. *Male*: lateral field with four lines, spicules >30 μm, distally pointed. *J2* with four incisures in lateral field. Tail conical, pointed, phasmids punctiform. *En face* pattern typically with six separate lips, sometimes with fusion of adjacent submedial lips.

Type species: *Globodera rostochiensis* (Wollenweber, 1923) Skarbilovich, 1959
Total number of species: 14
Presently, PCN include three species: *G. rostochiensis, G. pallida* and *G. ellingtonae*, which have been reported from many countries and are jointly considered to be one of the most economically important pests of potato. It is now generally accepted that PCN are native to South America (Stone, 1979), where they are the principal pests of Andean potato crops. The Peruvian and Bolivian Andes have been considered as the centres of origin of PCN. The three species of PCN have different distributions in the Andes. In South America, PCN is mainly found between 2000 and 4000 m above sea level, with the heaviest infestations between 2900 and 3800 m above sea level (Franco, 1977). Mapping of the distribution for these PCN species in some of the Andean countries showed that most Peruvian PCN populations up to north of Lake Titicaca are *G. pallida*. The distribution of *G. rostochiensis* is mainly from south of Lake Titicaca to southern Bolivia. The areas around Lake Titicaca are predominantly mixtures of both species. *Globodera ellingtonae* was found in southern Bolivia, north-west Argentina and northern Chile. The centre of origin of *G. pallida* is likely in the south of Peru between Cusco and Lake Titicaca; the south of Bolivia or north-west Argentina might indeed be considered as an ancient centre of origin for *G. rostochiensis, G. ellingtonae* and *G. tabacum*. Molecular analysis revealed that that the source of European *G. pallida* was from a single, restricted area in the south of Peru, located between the north shore of Lake Titicaca and Cusco and that of European *G. rostochiensis* was from an area in the La Paz Department of Bolivia, at the western country border, sharing Lake Titicaca with adjacent Peru (see Picard *et al.*, 2007, Plantard *et al.*, 2008, Subbotin *et al.*, 2020). *Globodera ellingtonae* presently found in the USA was likely introduced from Chile. *Globodera rostochiensis* and *G. pallida* have been introduced in many parts of the world, particularly to Europe, and also to the USA, Canada, New Zealand and numerous other countries where potatoes are grown. The recent population genetic study revealed that *G. rostochiensis* was first introduced into Europe from the Andean region of South America. Genetically less diverse populations sampled in North America, Africa, Asia and Oceania, most likely originate from Europe, which served as a secondary distribution hub for the worldwide invasion of *G. rostochiensis* (Esquibet *et al.*, 2024).

Globodera rostochiensis (Wollenweber, 1923) Skarbilovich, 1959 – the Golden Potato Cyst Nematode
(Fig. 10.15A, B, 10.16)
The golden cyst nematode, associated with potato plants, *Solanum tuberosum*, from Rostock, Germany, was first reported in 1881 and was initially considered to be *H. schachtii*, this being the only known species of cyst nematode at that time. During the early 1900s, it became more widely known throughout Europe and was formally described in 1923. In temperate regions, *G. rostochiensis* usually completes only one generation, although a second generation may be initiated but not completed – J2 hatch from first-generation eggs, but are unable to reach the adult stage (Jones, 1950; Evans, 1969). In subtropical regions, two generations might occur. Development of one generation requires 6-10 weeks. The J2 can go into diapause and remain viable for many years, with hatching continuing for 25 years or more. Heavily infected plants become yellow and stunted. Infected plants have reduced root systems, which are abnormally branched and brownish in colour. Symptoms in the field first appear in small patches. At low nematode densities, tuber sizes are reduced, whereas at higher densities both number and size of tubers can be reduced. At 8 and 64 eggs/g of soil, yield losses of about 20 and 70%, respectively, can be expected. Seinhorst (1965, 1982), Brown (1969) and Elston *et al.* (1991) proposed models that described the relationships of PCN population densities before planting with potato yield and post-harvest nematode populations. The damaging effect of PCN is determined not only by nematode density but also by such factors as cultivar, crop husbandry and environmental conditions. PCN are responsible for annual potato tuber losses of up to 9% in Europe (Evans and Stone, 1977); information on economic importance in some South American countries is little or unavailable, although yield losses in Bolivia and Peru have been estimated to be ca US$13,000,000 and US$128,000,000, respectively (Franco and Gonzalez, 2010).

Hosts include potato, *Solanum tuberosum* (Solanaceae, Solanales), tomato, *Solanum*

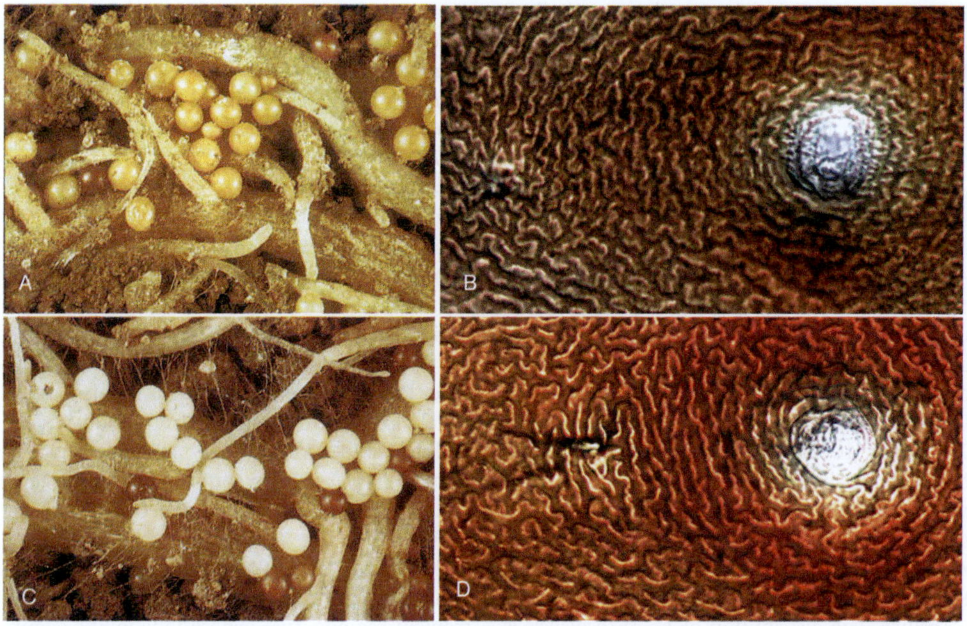

Fig. 10.15. Potato cyst nematodes. A, B: *Globodera rostochiensis*, C, D: *G. pallida*. A, C: Females and cysts on roots; B, D: Vulval plates. (A, D – courtesy of I. Cid Del Prado; B, D – courtesy of V.N. Chizhov.)

lycopersicum L., and eggplant, *S. melongena* L. Other hosts include many *Solanum* spp., *Datura* spp., *Hyoscyamus niger* L., *Nicotiana acuminata* Hook., *Physalis* spp., *Physochlaina orientalis* G. Don, *Salpiglossis* spp., *Capsicum annuum* L., *Saracha jaltomata* Schlecht.

Measurements

Cysts: L = 450-990 µm; W = 250-810 µm; L/W ratio = 0.9-1.8; fenestral diam. = 8-21 µm; number of ridges between anus and fenestra = 12-31; Granek's ratio = 1.2-9.5.

J2: L = 366-505 µm; stylet = 19-24 µm; DGO = 2.4-6.7 µm; hyaline region = 18-30 µm; tail = 36-57 µm.

Description: *Female:* colour changes from white to yellow to light golden as female matures to cyst stage. *Cyst* – brown, ovate to spherical in shape with a protruding neck, circumfenestrate (Fig. 10.15A, B), abullate. Fenestra – circular, anus conspicuous at the apex of a V-shaped subsurface cuticular mark. *J2:* body tapers at both extremities but more at posterior end. Stylet well developed, with prominent rounded knobs when viewed laterally (Fig. 10.9F). Lateral fields with four lines extending for most of body length. Tail tapering to small, rounded terminus.

Remarks: *Globodera rostochiensis* is morphologically similar to *G. pallida* and *G. tabacum*. It differs from *G. pallida* by having yellow or gold maturing females in contrast to cream-coloured ones of *G. pallida*, higher number of ridges between the vulva and the anus, larger mean for Granek's ratio, stylet knob shape, shorter average stylet length and rounded *vs* a more pointed J2 tail terminus. The ITS-rRNA and *COI* gene sequences clearly differentiate this species from all others.

Globodera pallida **Stone, 1973 – the Pale Potato Cyst Nematode** (Fig. 10.15C, D, 10.17)

The pale cyst nematode, *G. pallida*, is considered to be a major pest of potato crops in cool temperate climates. It is reported from several countries in Europe, Asia, Africa and South America. In Central and North America, *G. pallida* has been reported in Panama, USA and Canada, but in the last two countries, PCN have a rather restricted distribution, with only small infested areas because of rigorous phytosanitary regulations and seed potato certification programmes, compared to the widespread infestations found in European countries. *Globodera pallida* develops one generation for a vegetation season. This species is adapted to cool temperatures and is able to

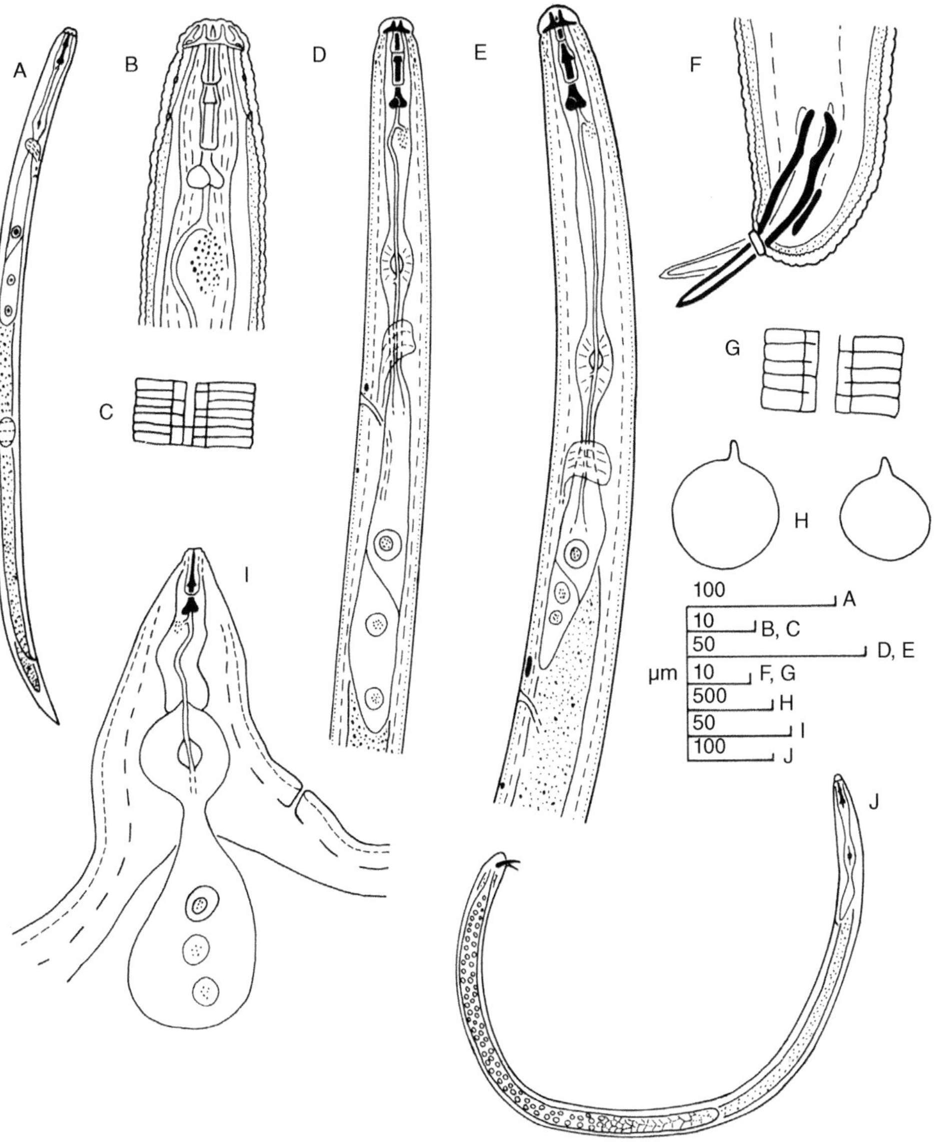

Fig. 10.16. *Globodera rostochiensis*. A-D: J2; A: entire body; B: lip region; C: lateral field; D: pharyngeal region; E-G: male; E: pharyngeal region; F: tail; G: lateral field; H: entire cysts; I: anterior region of female; J: entire male. (After Stone, 1973a.)

hatch earlier in the year and develop at 2°C cooler than *G. rostochiensis* (Langeslag *et al.*, 1982). The symptoms of attack by *G. pallida* are similar to those for *G. rostochiensis*, and the damage threshold is 1-2 eggs/g soil. Hosts includes potato, *Solanum tuberosum* (Solanaceae, Solanales), eggplant, *Solanum melongena* L., tomato, *S. lycopersicum*, and many species of *Solanum*; black henbane, *Hyoscyamus niger*.

Measurements

Cysts: L = 420-748 μm; W = 400-685 μm; fenestral diam. = 17.5-25 μm; number of ridges between anus and fenestra = 7-26; Granek's ratio = 1.2-3.6.

J2: L = 380-533 μm; stylet = 20-26 μm; DGO = 2.7-5 μm; hyaline region = 20-31 μm; tail = 31-59 μm.

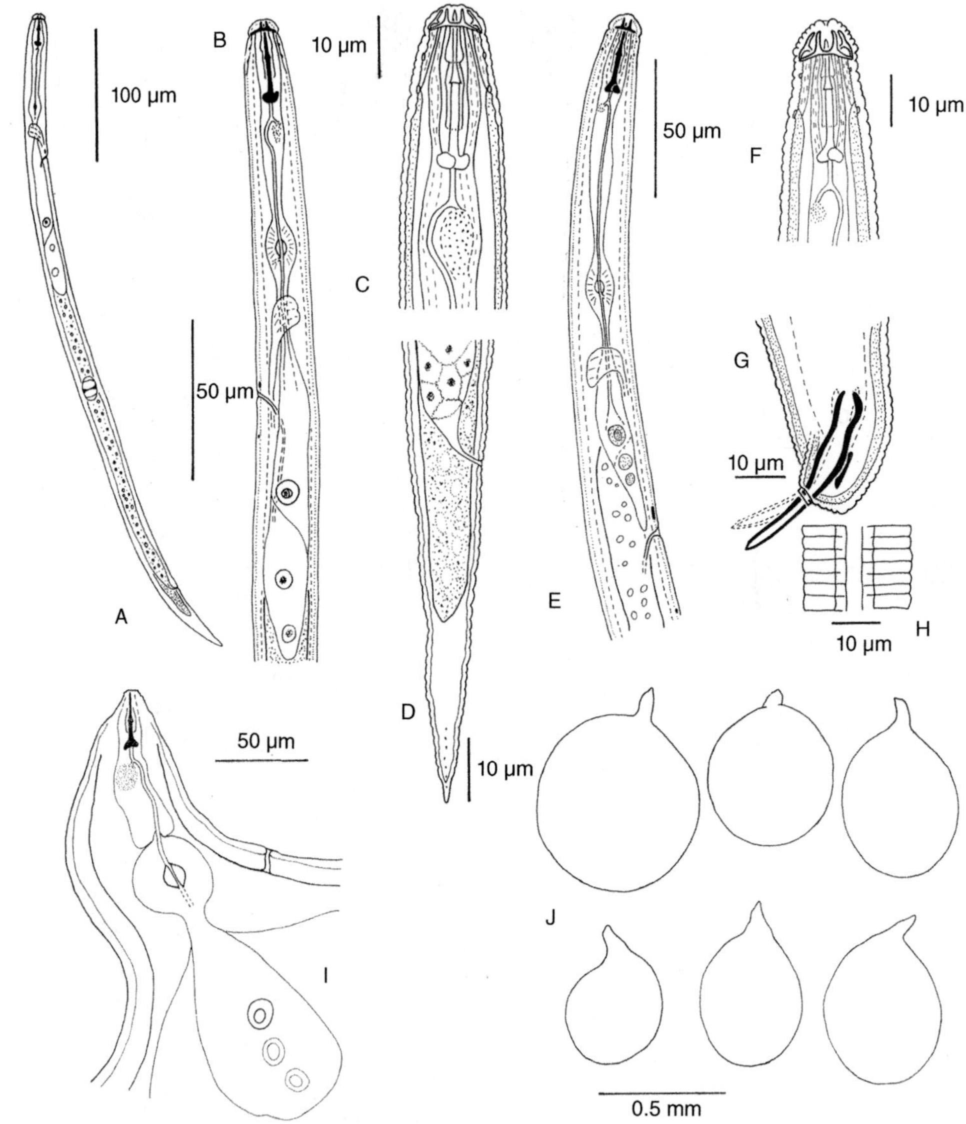

Fig. 10.17. *Globodera pallida*. A-D: Second-stage juvenile. A: Entire body; B, C: anterior region; D: tail; E-H: males. E, F: anterior region; G: tail; H: lateral field; I: anterior region of female; J: cysts. (After Stone, 1973b.)

Description: *Female:* white in colour, some populations passing, after 4-6 weeks, through a cream stage, turning glossy brown when dead. *Cyst* vulval region intact or fenestrated, with single circumfenestrate (Fig. 10.15C, D) opening occupying all or part of vulval basin, abullate. *J2:* lateral field with four incisures but with three anteriorly and posteriorly, occasionally completely areolated. Stylet well developed, basal knobs with distinct anterior projection as viewed laterally. Tail tapering uniformly with a finely rounded point, hyaline region forming about half of tail region.

Remarks: *Globodera pallida* is most closely related to *G. rostochiensis* and *G. tabacum*. It differs from *G. rostochiensis* by cream-coloured females in contrast to yellow or gold, smaller number of ridges between the vulva and the anus, smaller mean for Granek's ratio, stylet knob shape, longer stylet length, tail terminus and presence of refractive bodies on the hyaline part of the tail (usually four to seven refractive bodies *vs* absence) in J2. The ITS-rRNA and *COI* gene sequences clearly differentiate this species from all others.

***Globodera ellingtonae* Handoo, Carta, Skantar & Chitwood, 2012 – the Ellington Potato Cyst Nematode**

Globodera ellingtonae was first described from samples collected in the USA, in Oregon and Idaho, and was later reported in Argentina, Chile and Bolivia. *Globodera ellingtonae* can reproduce on potato and tomato; however, it was not considered to be a strong potato pest in experiments conducted in Oregon.

Measurements

Cysts: L = 370-860 μm; W = 320-890 μm; fenestral diam. = 12-42.5 μm; number of ridges between anus and fenestra = 8-25; Granek's ratio = 0.9-5.9.

J2: L = 365-526 μm; stylet = 19-24 μm; DGO = 6-8.5 μm; hyaline region = 19-32.5 μm; tail = 39-56 μm.

Description: *Female body:* white, turns yellow to pale brown as eggs mature, ovate to rounded or subspherical in shape with elongate, protruding neck. *Cysts*: light brown to brown in colour, spherical to subspherical. Young cysts occasionally with clustered vulval bodies of variable size and shape similar to bullae. *J2*: stylet short, well-developed, basal knobs rounded posteriorly, with a forward projection anteriorly. Lateral field consists of four incisures, with occasional areolation. Tail tapers uniformly, ending with a peg-like, finely rounded to pointed terminus.

Remarks: *Globodera ellingtonae* differs from *G. pallida* by a slightly shorter stylet mean length of females and males, the white body of the female becomes yellowish and then brown *vs* a change from white directly to brown. It differs from *G. rostochiensis* by having a lower mean number of perineal lines between the anus and the vulva and lower mean Granek's ratio. The ITS-rRNA and *COI* gene sequences clearly differentiate this species from all others.

***Globodera tabacum* (Lownsbery & Lownsbery, 1954) Skarbilovich, 1959 – the Tobacco Cyst Nematode**

Globodera tabacum is considered as a serious and important pest of shade and broadleaf tobacco. It recorded from several countries in Europe, Asia, Africa, South and North America. *Globodera tabacum* is a polytypic species containing the following subspecies: *G. tabacum tabacum* (Lownsbery and Lownsbery, 1954); *G. tabacum virginiae* (Miller and Gray, 1968) and *G. tabacum solanacearum* (Miller and Gray, 1972). All three subspecies develop on tobacco and horsenettle but otherwise differ in host preference. *Globodera tabacum* parasitizes *Nicotiana tabacum* L., horsenettle, *Solanum carolinense* L., tomato, and species of the genera: *Nicotiana*, *Solanum*, as well as *Atropa belladona* L., *Hyoscyamus niger* L., *Nicandra physalodes* (L.) Gaertn., *Capsicum annuum* L. Two or more generations usually occur. Infected tobacco plants have small root systems, and above-ground symptoms are similar to those associated with severe root-knot and lesion nematode infestations. Nematode infection is often associated with increased damage from bacterial wilt, black shank. Farmers in Virginia, USA, have recorded complete crop failures, but losses generally average 15%. A high density of nematode populations early in the growing season can reduce flue-cured tobacco yield by 25-50%, although tobacco may escape significant losses from moderate populations, especially under favourable growing conditions.

Measurements

Cysts: L = 337-937 μm; W =232-812 μm; L/W ratio = 0.9-1.5; fenestral diam. = 13-38 μm; number of ridges between anus and fenestra = 5-26; Granek's ratio = 1.4-4.9.

J2: L = 384-621 μm; stylet = 19-27 μm; DGO = 4.3-9 μm; hyaline region = 16-37 μm; tail = 34-71 μm.

Description: *Female*: body ovate to spherical with elongate neck, white, turns yellow. *Cyst:* light shiny brown, circumfenestrate, abullate. *J2*: well-developed rounded basal knobs. Terminus of tail finely rounded.

Remarks: *Globodera tabacum* differs from *G. rostochiensis* by J2 with longer mean values of body length, mean stylet and by cysts with smaller mean number of cuticular ridges. It differs from *G. mexicana* by J2, with longer mean body length, and from *G. pallida* by cysts with a smaller mean number of cuticular ridges and by J2 with longer means of body length. The ITS-rRNA and *COI* gene sequences clearly differentiate this species from all others. The subspecies can be identified based on *COI* gene sequences.

10.10.3. Genus *Betulodera* Sturhan, 2002 (Figs 10.9G, 10.18)

Diagnosis (after Sturhan, 2002): *Cyst:* lemon-shaped, pear-shaped or spheroid with insignificant, obtuse, vulval cone. Cyst wall thick, with irregular network-like pattern, D-layer absent (no punctations in inner, deeper layers of the cyst wall), subcrystalline layer heavily developed. Vulva terminal, surrounded by circumfenestration, vulval slit short (<10 μm), underbridge absent, denticles occasionally present, anus without fenestration. *Male:* body twisted, no cloacal tube, spicules with bifid distal tips, phasmid openings punctiform. *J2:* lateral field with three incisures, phasmid openings punctiform, without lens-like structure, labial region with three or four labial annuli and labial disc fused with submedial lips.

Type and only species: *Betulodera betulae* (Hirschmann & Riggs, 1969) Sturhan, 2002.

10.10.4. Genus *Cactodera* Krall & Krall, 1978

Diagnosis (after Sturhan, 2002): *Mature female and cyst*: lemon-shaped to spherical, with posterior protuberance. Vulva terminal, vulval slit <30 μm, fenestra circumfenestrate (Fig. 10.4C). Anus without fenestration. Bullae and underbridge absent, vulval denticles usually present. Cuticle with D-layer. *J2* have lateral field with four incisures, phasmid openings punctiform. Labial disc and six labial sectors present.

Type species: *Cactodera cacti* (Filipjev & Schuurmans Stekhoven, 1941) Krall & Krall, 1978

Total species number: 19

***Cactodera cacti* (Filipjev & Schuurmans Stekhoven, 1941) Krall & Krall, 1978 – the Cactus Cyst Nematode**

A cyst nematode infecting cacti, *Discocactus akkermannii* (Lindl.) Barthlott and *Cereus speciosus* K. Schumacher, both of which were expressing declining symptoms, was first recorded and described from Maartensdijk, near Utrecht, Netherlands. The cactus cyst nematode is distributed worldwide, mainly on plants of the family Cactaceae grown in glasshouses as ornamentals. The dispersal of *C. cacti* from native regions in the Americas is, beyond doubt, associated with the international trade of infested ornamental cactus plants around the world. The cactus cyst nematode has been associated with, or infecting, plants belonging to three families: Cactaceae (Caryophyllales); Umbelliferae (order Apiales): and Euphorbiaceae (order Malpighiales) (Kirjanova and Krall, 1971; Mulk, 1977; Esser, 1992). Infected plants may exhibit various symptoms, including branched roots and increased numbers of rootlets. Plants become reddish brown to yellow in colour, wilted, stunted, with reduced flower production and shortening of the flowering period. With high infection, the host may die.

Measurements

Cysts: L = 328-780 μm; W = 240-598 μm; L/W ratio = 1.1-2.0; fenestral diam. = 16-48 μm.

J2: L = 344-584 μm; stylet = 21-26 μm; hyaline region = 12-21 μm; tail = 34-60 μm.

Description: *Female:* Body lemon-shaped to almost spherical, pearly white, yellow, or golden, maturing to light brown. *Cyst*: usually lemon-shaped, but may be rounded with protruding neck and vulva, light or medium brown, sometimes reddish brown. Vulval denticles generally present, visible beneath fenestral surface. Cone tops abullate, circumfenestrate. *J2:* body vermiform, tapering anteriorly and posteriorly. Tail tapering, with hyaline region often shorter than stylet. Eggshells heavily punctate, as seen under optical microscope with oil immersion.

Remarks: *Cactodera cacti* resembles *C. weissi*, *C. acnidae*, *C. milleri* and *C. galinsogae*. It differs from *C. weissi* and *C. acnidae* in having heavily punctate eggshells in contrast to shells without visible markings, and by possessing a larger stylet in J2. It differs from J2 of *C. galinsogae* by a longer tail and from *C. milleri* by cysts with a larger fenestral diameter. The ITS-rRNA and *COI* gene sequences clearly differentiate this species from all others.

10.10.5. Genus *Dolichodera* Mulvey & Ebsary, 1980 (Fig. 10.19)

Diagnosis (after Siddiqi, 2000): *Female*: body elongate-oval, without terminal protuberance, white, swollen part 400-500 μm long, 140-270 μm wide, 2-2.8 times as long as wide, neck moderately long. Cuticle not annulated but with fine irregular striae. Vulval area terminal or just subterminal, circumfenestrate, fenestra

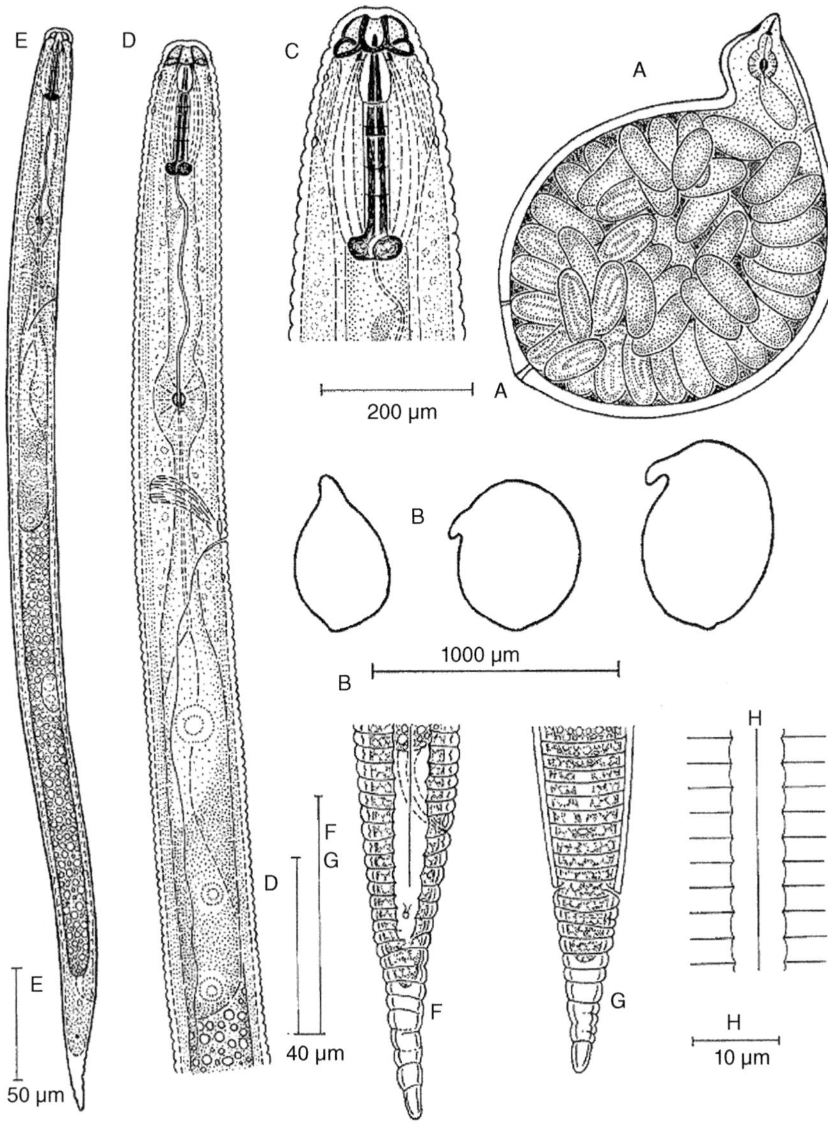

Fig. 10.18. *Betulodera betulae*. Females and cysts. A: Female filled with eggs (lateral); B: shape and size ranges of cysts. Second-stage juvenile. C: Labial region (lateral); E: full-length outline (lateral); D: pharyngeal region (lateral); F: tails (lateral); G: tails (ventral); H: lateral field. (After Hirschmann and Riggs, 1969.)

ca 20 μm in diameter, bullae present, perineal tubercles absent. Anus pore-like, lacking a fenestra, located 10-13 μm dorsal to the vulval fenestral margin. *Cyst:* light brown, elongate-oval. Several large bullae present. Perineal tubercles absent. Vulva circumfenestrate, underbridge absent. Anus lacking fenestra. *Male*: not found. *J2* with long tail (95-120 μm). Lateral field with three incisures, inner one faint. Labial region hemispherical, offset, with two annuli. Tail tip narrowly rounded. Phasmid openings lacking a lens-like ampulla, located about one anal body diameter posterior to the anus.

Type and only species: *Dolichodera fluvialis* Mulvey & Ebsary, 1980.

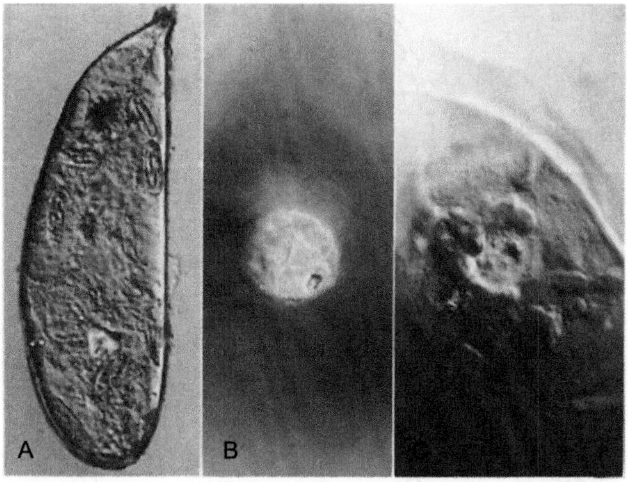

Fig. 10.19. *Dolichodera fluvialis*. Cyst. A: Whole cyst with slightly subterminal vulva fenestra; B: fenestral area, which has deteriorated to form a round aperture; C: bullae in fenestral area. (After Mulvey and Ebsary, 1980.)

10.10.6. Genus *Paradolichodera* Sturhan, Wouts & Subbotin, 2007 (Fig. 10.20)

Diagnosis (after Sturhan *et al.*, 2007): *Mature female and cyst*: elongate-ovoid, with a rounded posterior end. Cuticle transparent, with faint transverse striations on the anterior part of the body and posteriorly mostly with faint irregular ridges superimposed on distinct punctations. Cuticle turning yellowish to light brown on death, covered by a subcrystalline-like film. Eggs retained in the body, egg-sac not observed. Labial disc squarish. Stylet well developed. Vulva terminal or subterminal, vulval slit short, circumfenestrate. Anus lacking fenestration. *Male*: body not twisted, lateral field with four incisures. Cloacal tube present, spicules rounded at tip. Phasmids lacking. *J2*: long, extremely slender for family, lateral fields indistinct. Stylet short (<20 µm). DGO located at more than half stylet length posterior to the stylet base, pharyngeal glands long, filling body cavity. Tail long, slender, phasmid openings punctiform.

Type and only species: *Paradolichodera tenuissima* Sturhan, Wouts & Subbotin, 2007

10.10.7. Genus *Punctodera* Mulvey & Stone, 1976

Diagnosis (after Siddiqi, 2000): *Mature female and cyst*: Spherical, pear-shaped or ovoid, with a short, projecting neck and heavy subcrystalline layer. Cuticle reticulate, subcuticle with punctations. D-layer present. Terminal region not cone-shaped; cyst light to dark brown. Vulval slit extremely short (<5 µm), anus at a short distance from vulval fenestra. Circumfenestrate (Fig. 10.4D), fenestra surrounding vulva 16-40 µm (ca 20 µm in type species) in diameter, anus offset towards ventral margin of fenestra, an anal fenestra with shape and size similar to vulval fenestra present. Underbridge and perineal papilla-like tubercles absent. Bullae present or absent. Egg retained in the body, no egg-sac. *Male*: vermiform, less than 1.5 mm long. DGO 2.6-4.6 µm. Spicules 31-33 µm long, distally pointed. Tail less than 0.5 anal body diameter long, cloacal lips not forming a tube. *J2*: body 0.35-0.49 mm, stylet 24-26 µm and tail conical, 63-78 µm long, hyaline region in type species 38-41 µm long. Lateral field with four incisures. Pharyngeal glands filling body diameter. Phasmid openings punctiform, without a lens-like structure. Parasites of monocotyledonous plants.

Type species: *Punctodera punctata* (Thorne, 1928) Mulvey & Stone, 1976

Total number of species: 5

***Punctodera chalcoensis* Stone, Sosa Moss & Mulvey, 1976 – the Mexican Corn Cyst Nematode**

(Fig. 10.21)

The Mexican corn cyst nematode was first observed in the late 1950s in maize fields in

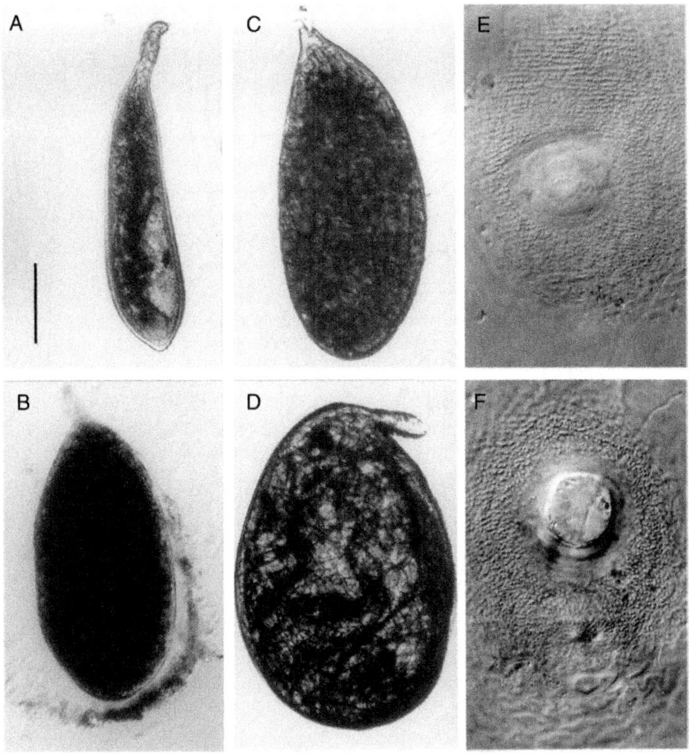

Fig. 10.20. *Paradolichodera tenuissima*. A: Young female; B: female with dissolved 'subcrystalline' layer; C: cyst filled with eggs (holotype); D: older cyst containing eggshells and a few juveniles; E: female/young cyst posterior end showing vulval slit and developing circumfenestra; F: cyst posterior end with vulval fenestra and anus (below). Scale bar: A-D = 200 µm; E, F = 20 µm. (After Sturhan *et al.*, 2007.)

Huamantla, Tlaxcala, Mexico, and was originally identified as a Mexican race of *Heterodera punctata* (Baldwin and Mundo-Ocampo, 1991). Further detailed morphological and morphometric observations revealed differences between these populations and *P. punctata* and *P. matadorensis* and led to the description of a new species, *P. chalcoensis*, by Stone *et al.* (1976). It is highly damaging to maize in Mexico, causing severe yellowing, stunting and even death of young seedlings (Baldwin and Mundo-Ocampo, 1991). Heavily attacked plants have stunted root systems with many short laterals, giving a bottle-brush effect, and the aerial parts of the plant appear unthrifty. Pathogenicity studies of *P. chalcoensis* on maize in the glasshouse showed an inverse correlation between root and top weight of mature plants and the number of nematode cysts in the inoculum.

Measurements

Cysts: L = 441 µm; W = 416 µm; L/W ratio = 1.1; fenestral diam. = 18.1-20 µm.

J2: L = 522 µm; stylet = 24.4 µm; hyaline region = 38.6 µm; tail = 66.2 µm.

Description: *Cyst*: spherical or subspherical with protruding neck, pale to dark brown, darkening with age. New cysts often retaining subcrystalline layer. Old cysts losing thin walls of vulval and anal fenestrae, younger cysts showing incomplete fenestration. Some specimens show small scattered bullae in perineal region or just anterior to vulval fenestra, but bullae may be lacking from many cysts. *J2*: vermiform body adopting a slight ventral curve on heat relaxation. Well-developed stylet with massive basal knobs, knobs rounded posteriorly, flat to shallowly concave anteriorly.

Remarks: *Punctodera chalcoensis* resembles *P. matadorensis* and *P. stonei* the most. It differs from *P. matadorensis* by the shape of the J2 stylet

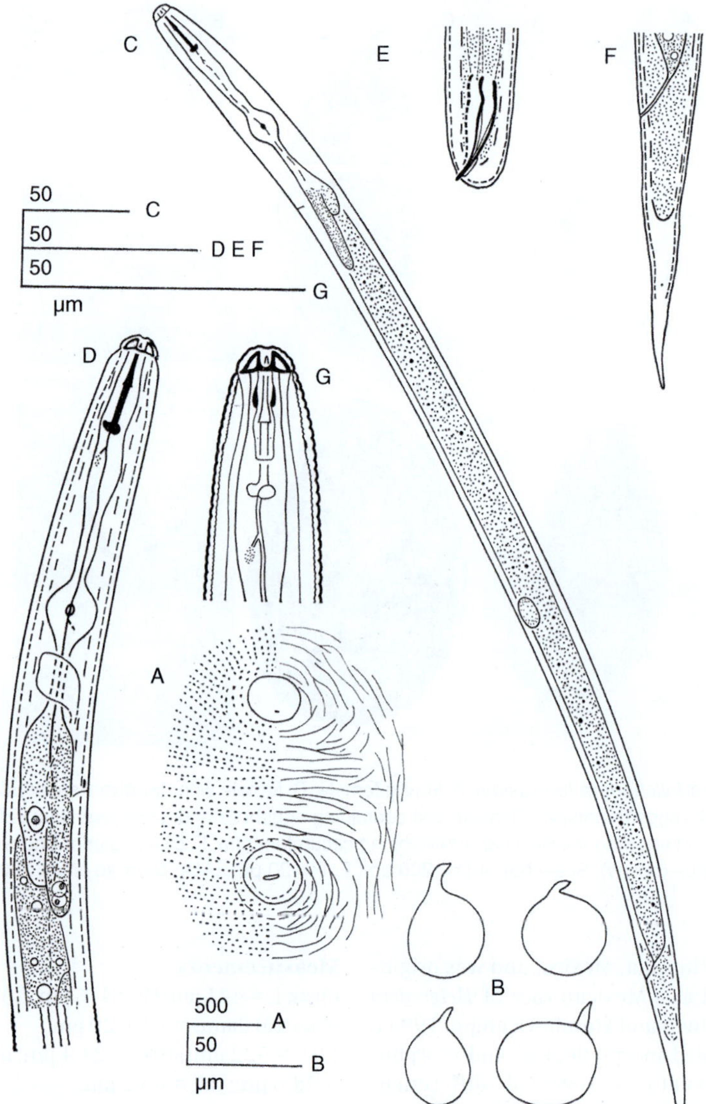

Fig. 10.21. *Punctodera chalcoensis*. A: Perineum of female; B: females; C: second-stage juvenile (J2); D: pharyngeal region of J2; F: anterior region of J2; E: male tail; F: tail of J2. (After Stone *et al*., 1976.)

knobs, which are anteriorly flat to slightly concave in *P. chalcoensis* in contrast to strongly concave. They also differ in the distribution of bullae in the cyst perineum, these being small scattered or absent in *P. chalcoensis* in contrast to being massive and always present in *P. matadorensis*. *Punctodera chalcoensis* differs from *P. stonei* by the shorter hyaline region of J2 and by its distinctive host range. The ITS-rRNA and *COI* gene sequences clearly differentiate this species from all others.

10.10.8. Genus *Vittatidera* Bernard, Handoo, Powers, Donald & Heinz, 2010

Diagnosis (after Bernard *et al.*, 2010 with modifications): *Cyst*: orange-brown to brown, lemon-shaped with short necks and vulval cone. Vulval aperture circular to rhomboid, circumfenestrate, with irregular denticle-like protuberances around the periphery of the orifice. Bullae, vulval bridge and vulval underbridge absent.

Male: variable length, stylet knobs rounded. *J2*: shows conoid tail with a narrowly rounded tip, phasmid apertures pore-like. Stylet length less than 18 μm. Stylet knobs rounded. Lateral field with four incisures. Eggshell smooth.

Type and only species: *Vittatidera zeaphila* Bernard, Handoo, Powers, Donald & Heinz, 2010.

Vittatidera zeaphila **Bernard, Handoo, Powers, Donald & Heinz, 2010**
(Fig. 10.22)

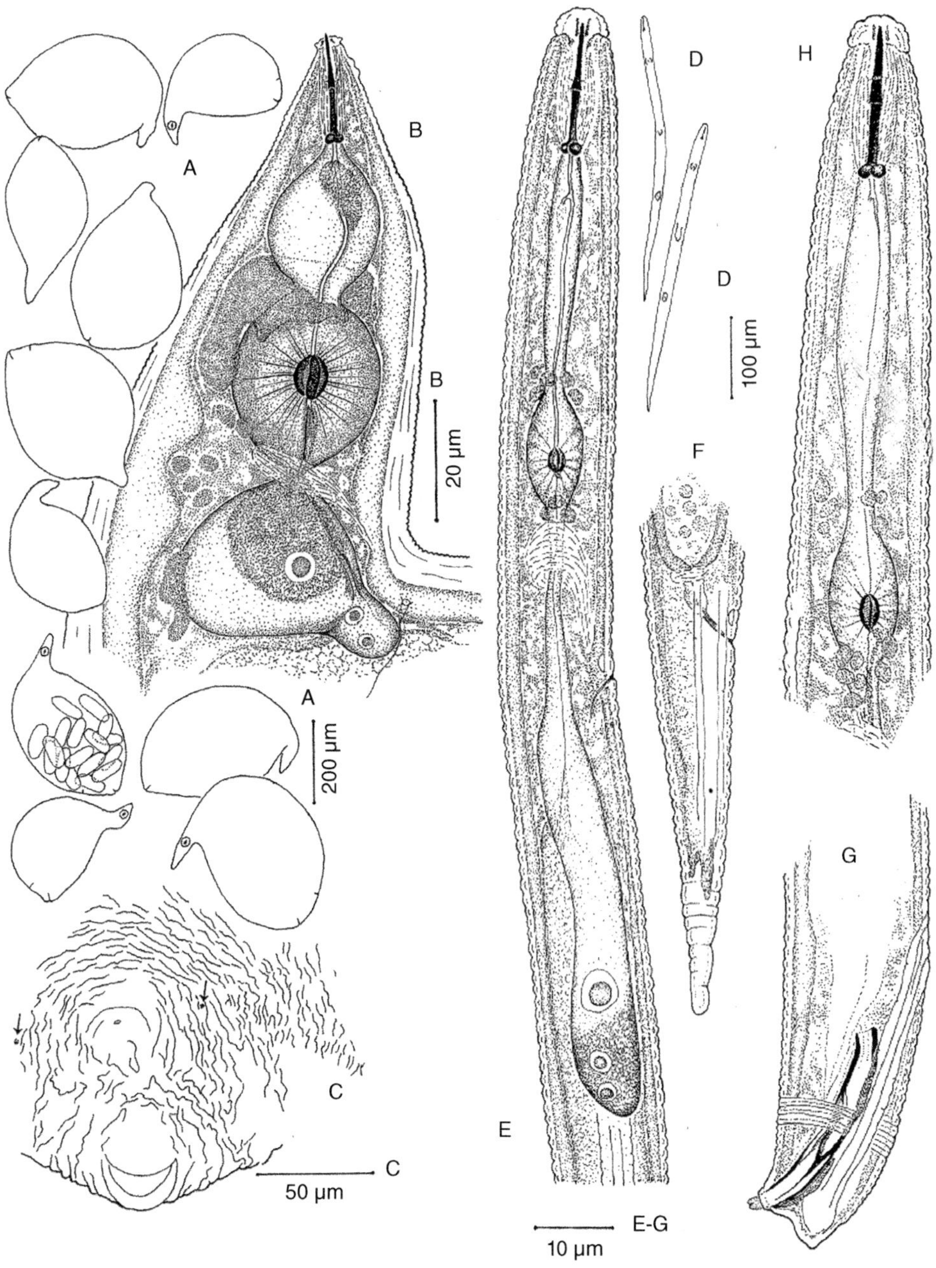

Fig. 10.22. *Vittatidera zeaphila*. A: Outline of females and cysts; B: anterior end of female; C: perianal region; D: outlines of second-stage juveniles (J2); E: anterior region of J2; F: tail of J2; H: anterior region of male; G: tail of male. (After Bernard *et al.* (2010) with modification.)

In 2006, the Missouri Extension Nematology Laboratory discovered a new species of corn cyst nematode from a sample collected in a corn field of north-western Tennessee, Obion County, USA. This nematode was considered as a representative of a new genus of the subfamily Heteroderinae. It resembles representatives of the genera *Cactodera* and *Betulodera* in having lemon-shaped cysts and circumfenestrate vulval area. Similar cyst specimens had previously been collected in 1978 from Lauderdale County, Tennessee, on goosegrass, *Eleusine indica*. *Vittatidera zeaphila* is presently found only in the USA (Bernard *et al.*, 2010).

Measurements

Cysts: L = 470 (385-608) μm; W = 287 (179-391) μm; L/W = 1.7 (1.4-2.2); fenestral diam. = 36 (28-48) μm; anus to fenestral distance = 63 (52-74) μm.

J2: L = 365 (346-400) μm; a = 24.0 (22.3-26.9); b = 2.5 (2.4-2.7); stylet = 16.7 (15.7-17.5) μm; hyaline region = 15.3 (12.4-17.8) μm; tail = 41.5 (33.0-48.0) μm.

Description: *Cyst*: lemon-shaped, vulval cone not protuberant, large egg masses persistent. In mature cysts, vulval aperture circumfenestrate, circular to rhomboidal, membranous lips persistent on mature cysts, underbridge and bullae absent. *J2*: stylet small with rounded knobs. Lip region rounded, slightly offset, with three annuli. Lateral field with four incisures. Tail elongate-conoid, tip narrowly rounded. The ITS-rRNA sequences clearly differentiate this species from all others.

10.11. References

Andersson, S. (1974) *Heterodera hordecalis* n. sp. (Nematoda: Heteroderidae) a cyst nematode of cereals and grasses in southern Sweden. *Nematologica* 20, 445–454.

Baldwin, J.G. and Mundo-Ocampo. M. (1991) Heteroderinae, cyst and non cyst-forming nematodes. In: Nickle, W.R. (ed), *Manual of Agricultural Nematology*. Marcel Dekker, New York, New York, pp. 275–362.

Bernard, E.C., Handoo, Z.A., Powers, T.O., Donald, P.A. and Heinz, R.D. (2010) *Vittatidera zeaphila* (Nematoda: Heteroderidae), a new genus and species of cyst nematode parasitic on corn (*Zea mays*). *Journal of Nematology* 42, 139–150.

Bridge, J., Luc, M. and Plowright, R.A. (1990) Nematode parasites of rice. In: Luc, M., Sikora, R.A. and Bridge, J. (eds), *Plant Parasitic Nematodes in Tropical and Subtropical Agriculture*. CAB International, Wallingford, UK, pp. 69–108.

Brown, E.B. (1969) Assessment of the damage caused to potatoes by potato cyst eelworm, *Heterodera rostochiensis* Woll. *Annals of Applied Biology* 63, 493–502.

Burrows, P.R. and Stone, A.R. (1985) *Heterodera glycines*. In: *CIH Descriptions of Plant-parasitic Nematodes*, Set 8, No. 118. Commonwealth Agricultural Bureaux, Farnham Royal, UK.

Canto Saenz, M. and de Scurrah, M.M. (1977) Races of the potato cyst nematode in the Andean region and a new system of classification. *Nematologica* 23, 340–349.

Coyne, D.L. and Plowright, R.A. (2000) *Heterodera sacchari*: field population dynamics and damage to upland rice in Cote d'Ivoire. *Nematology* 2, 541–550.

Elston, D.A., Phillips, M.S. and Trudgill, D.L. (1991) The relationship between initial population density of potato cyst nematode *Globodera pallida* and the yield of partially resistant potatoes. *Revue de Nématologie* 14, 231–219.

Endo, B.Y. (1978) Feeding plug formation in soybean roots infected with the soybean cyst nematode. *Phytopathology* 68, 1022–1031.

Esquibet, M., Mwangi, J.M., Kiewnick, S., Wang, X., Mimee, B., *et al.* (2024) Europe as a secondary distribution hub in the worldwide invasion of the potato cyst nematode *Globodera rostochiensis*. *Scientific Reports* 14(1), 13915. Doi: 10.1038/s41598-024-64617-0

Esser, R.P. (1992) Cactus cyst nematode, *Cactodera cacti*. Nematology Circular No. 197., Florida Department Agriculture and Consumer Service, Division of Plant Industry, Gainsville, Florida.

Evans, K. (1969) Changes in a *Heterodera rostochiensis* population through the growing season. *Annals of Applied Biology* 64, 31–41.

Evans, K. and Stone, A.R. (1977) A review of the distribution and biology of the potato cyst-nematodes *Globodera rostochiensis* and *G. pallida*. *International Journal of Pest Management* 23, 178–189.

Franco, J. (1977) *Studies on the Taxonomy and Biology of Potato Cyst Nematodes* Globodera *spp*. PhD. dissertation, University of London, London, UK.
Franco, J. and González, A. (2010) Pérdidas causadas por el nematodo Quiste de la papa (*Globodera* sp.) en Bolivia y Perú. *Revista Latinoamericana de la Papa* 16, 233–249.
Franco, J., Oros, R., Main, G. and Ortuno, N. (1998) Potato cyst nematodes (*Globodera* species) in South America. In: Marks, R.J. and Brodie, B.B. (eds), *Potato Cyst Nematodes, Biology, Distribution and Control*. CAB International, Wallingford, UK, pp. 239–269.
Franklin, M.T. (1951) *The Cyst-forming Species of* Heterodera. Commonwealth Agricultural Bureaux, Farnham Royal, UK.
Franklin, M.T. (1969) *Heterodera latipons* n. sp., a cereal cyst nematode from the Mediterranean region. *Nematologica* 15, 535–542.
Franklin, M.T. (1972) *Heterodera schachtii*. In: *CIH Descriptions of Plant-parasitic Nematodes*, Set 1, No. 1. Commonwealth Agricultural Bureaux, Farnham Royal, UK.
Golden, A.M., Epps, J.M., Riggs, R.D., Duclos, L.A., Fox, J.A., *et al.* (1970) Terminology and identity of infraspecific forms of the soybean cyst nematode (*Heterodera glycines*). *Plant Disease Reporter* 54, 544–546.
Handoo, Z.A. and Subbotin, S.A. (2018). Taxonomy, identification and principal species. In: Perry, R.N., Moens, M. and Jones, J.T. (eds), *Cyst Nematodes*. CAB International, Wallingford, UK, pp. 365–397.
Hashmi, S., Krusberg, L.R. and Sardanelli, S. (1993) Reproduction of *Heterodera zeae* and its suppression of corn plant growth as affected by temperature. *Journal of Nematology* 25, 55–58.
Hirschmann, H. and Riggs, R.D. (1969) *Heterodera betulae* n. sp. (Heteroderidae), a cyst-forming nematode from river birch. *Journal of Nematology* 1, 169–179.
Ichinohe, M. (1952) On the soybean nematode, *Heterodera glycines* n. sp., from Japan. *Magazine of Applied Zoology* 17, 1–4.
Ichinohe, M. (1988) Current research on the major nematode problems in Japan. *Journal of Nematology* 20, 184–190.
Jain, K.K., Sethi, C.L., Swarup, G. and Srivastava, A.N. (1982) *Heterodera sorghi* sp. n., a new cyst-forming nematode parasitizing. sorghum. *Revue de Nématologie* 5, 201–204.
Jones, F.G.W. (1950) Observations on the beet eelworm and other cyst forming species of *Heterodera*. *Annals of Applied Biology* 37, 407–440.
Kirjanova, E.S. and Krall, E.L. (1971) [*Parasitic Nematodes of Plants and Their Control*.] Vol. II. Nauka, Leningrad, USSR.
Kort, J., Ross, H., Rumpenhorst, H.J. and Stone, A.R. (1977) An international scheme for identifying and classifying pathotypes of potato cyst-nematodes *Globodera rostochiensis* and *G. pallida*. *Nematologica* 23, 33–339.
Koshy, P.K. (1967) A new species of *Heterodera* from India. *Indian Phytopathology* 20, 272–274.
Koshy, P.K., Swarup, G. and Sethi, C.L. (1970) *Heterodera zeae* n. sp. (Nematoda: Heteroderidae) a cyst forming species on *Zea mays*. *Nematologica* 16 (1971), 511–516.
Krusberg, L.R., Sardanelli, S. and Grybauskas, P. (1997) Damage potential of *Heterodera zeae* to *Zea mays* as affected by edaphic factors. *Fundamental and Applied Nematology* 20, 593–599.
Langeslag, M., Mugniéry, D. and Fayet, G. (1982) Développement embryonaire de *Globodera rostochiensis* et *G. pallida* en fonction de la temperature, en conditions contrôlées et naturelles. *Revue de Nématologie* 5, 103–109.
Lax, P., Rondan Dueñas, J.C., Andrade, A.J. and Doucet, M.E. (2021) *Punctodera achalensis* n. sp. (Nematoda: Punctoderidae), a new cyst nematode from Argentina. *European Journal of Plant Pathology* 161, 397–409. Doi: 10.1007/s10658-021-02331-7
Lilley, C.J., Atkinson, H.J. and Urwin, P.E. (2005) Molecular aspects of cyst nematodes. *Molecular Plant Pathology* 6, 577–588.
Lownsbery, B.F. and Lownsbery, J.W. (1954) *Heterodera tabacum* new species, a parasite of Solanaceaous plants in Connecticut. *Proceedings of the Helminthological Society of Washington* 21, 42–47.
Luc, M. and Brizuela, B.R. (1961) *Heterodera oryzae* n. sp. (Nematoda: Tylenchoidea) parasite du riz en Côte d'Ivoire. *Nematologica* 6, 272–279.
Luc, M. and Merny, G. (1963) *Heterodera sacchari* n. sp. (Nematoda: Tylenchoidea) parasite de la canne à sucre au Congo-Brazzaville. *Nematologica* 9, 31–37.
Luc, M. (1986) Cyst nematodes in equatorial and hot tropical regions. In: Lamberti, F. and Taylor, C.E. (eds), *Cyst Nematodes*. Plenum Press, New York and London, pp. 355–372.
Maas, P.W.Th., Schoemaker, A. and Stemerding, S. (1976) Een cysteaaltje bij suikerbieten in Brabant. *Gewasbescherming* 7, 10–11.

Miller, L.I. and Gray, B.J. (1968) Horsenettle cyst nematode, *Heterodera virginiae* n. sp., a parasite of solanaceous plants. *Nematologica* 14, 535–543.

Miller, L.I. and Gray, B.J. (1972) *Heterodera solanacearum* n. sp., a parasite of solanaceous plants. *Nematologica* 18, 404–413.

Mor, M., Cohn, E. and Spiegel, Y. (1992) Phenology, pathogenicity and pathotypes of cereal cyst nematodes, *Heterodera avenae* and *Heterodera latipons* (Nematoda: Heteroderidae) in Israel. *Nematologica* 38, 494–501.

Mulk, M.M. (1977) *Heterodera cacti. CIH Descriptions of Plant-parasitic Nematodes*, Set 7, No 96. Commonwealth Agricultural Bureaux, Farnham Royal, UK.

Müller, J. (1999) The economic importance of *Heterodera schachtii* in Europe. *Helminthologia* 36, 205–213.

Mulvey, R.H. (1959) Investigation on the clover cyst nematode, *Heterodera trifolii* (Nematoda: Heteroderidae). *Nematologica* 4, 147–159.

Mulvey, R.H. and Ebsary, B.A. (1980) *Dolichodera fluvialis* n. gen., n. sp. (Nematoda: Heteroderidae) from Québec, Canada. *Canadian Journal of Zoology* 58, 1697–1702.

Niblack, T.L., Arelli, P.R., Noel, G.R., Opperman, C.H., Orf, J.H., *et al.* (2002) A revised classification scheme for genetically diverse populations of *Heterodera glycines*. *Journal of Nematology* 34, 279–288.

Nicol, J. and Rivoal, R. (2008) Global knowledge and its application for the integrated control and management of nematodes on wheat. In: Ciancio, A. and Mukerji, K.G. (eds), *Integrated Management and Biocontrol of Vegetable and Grain Crops nematodes*. Springer, Dordrecht, Netherlands, pp. 251–294.

Ohshima, Y. (1974) *Heterodera elachista* n. sp., an upland rice cyst nematode from Japan. *Japanese Journal of Nematology* 4, 51–56.

Picard, D., Sempere, T. and Plantard, O. (2007) A northward colonisation of the Andes by the potato cyst nematode during geological times suggests multiple host-shifts from wild to cultivated potatoes. *Molecular Phylogenetics and Evolution* 42, 308–316.

Plantard, O., Picard, D., Valette, S., Scurrah, M., Grenier, E., *et al.* (2008) Origin and genetic diversity of Western European populations of the potato cyst nematode *Globodera pallida* inferred from mitochondrial sequences and microsatellite loci. *Molecular Ecology* 17, 2208–2218.

Riggs, R.D., Hamblen, M.L. and Rakes, L. (1981) Infra-species variation in reactions to hosts in *Heterodera glycines* populations. *Journal of Nematology* 13, 171–179.

Riggs, R.D. and Schmitt, D.P. (1988) Complete characterization of the race scheme for *Heterodera glycines*. *Journal of Nematology* 20, 392–395.

Riggs, R.D. (1992) Host range. In: Riggs, R.D. and Wrather, J.A. (eds), *Biology and Management of the Soybean Cyst Nematode*. American Phytopathological Society, St. Paul, Minnesota, pp. 107–114.

Rivoal, R. and Cook, R. (1993) Nematode pests of cereals. In: Evans, K., Trudgill, D.L. and Webster, J.M. (eds), *Plant Parasitic Nematodes in Temperate Agriculture*. CAB International, Wallingford, UK, pp. 259–303.

Scholz, U. (2001) *Biology, pathogenicity and control of the cereal cyst nematode* Heterodera latipons *Franklin on wheat and barley under semiarid conditions, and interactions with common root rot* Bipolaris sorokiniana (*Sacc.*) Shoemaker. PhD thesis, University of Bonn, Bonn, Germany.

Seinhorst, J.W. (1965) The relation between nematode density and damage to plants. *Nematologica* 11, 137–154.

Seinhorst, J.W. (1982) The relationship in field experiments between population density of *Globodera rostochiensis* before planting potatoes and yield of potato tubers. *Nematologica* 28, 277–284.

Siddiqi, M.R. (2000) *Tylenchida Parasites of Plants and Insects*, 2nd edn. CAB International, Wallingford, UK.

Singh, P.R., Karssen, G., Couvreur, M. and Bert. W. (2020) Morphological and molecular characterization of *Heterodera dunensis* n. sp. (Nematoda: Heteroderidae) from Gran Canaria, Canary Islands. *Journal of Nematology* 52, e2020–98

Smiley, R.W., Dababat, A.A., Iqbal, S., Jones, M.J.K., Tanha Maafi, Z. *et al.* (2017) Cereal cyst nematodes: a complex and destructive group of *Heterodera* species. *Plant Disease* 101, 1692–1770. DOI: 10.1094/PDIS-03-17-0355-FE

Stone, A.R. (1973a) *Heterodera rostochiensis. CIH Descriptions of Plant-parasitic Nematodes*, Set 2, No 16. Commonwealth Agricultural Bureaux, Farnham Royal, UK, 3 pp.

Stone, A.R. (1973b) *Heterodera pallida* n. sp. (Nematoda: Heteroderidae), a second species of potato cyst nematode. *Nematologica* 18(1972), 591–606.

Stone, A.R. (1975) Lip morphology of second-stage juveniles of some Heteroderidae (Nematoda: Tylenchoidea). *Nematologica* 21, 81–88.

Stone, A.R. (1979) Co-evolution of nematodes and plants. *Symbolae Botanicae Upsaliensis* 22, 46–61.

Stone, A.R. and Rowe, J.A. (1976) *Heterodera cruciferae*. In: *CIH Descriptions of Plant-parasitic Nematodes*, Set 6, No. 90. Commonwealth Agricultural Bureaux, Farnham Royal, UK.

Stone, A.R., Sosa Moss, C. and Mulvey, R.H. (1976) *Punctodera chalcoensis* n. sp. (Nematoda: Heteroderidae) a cyst nematode from Mexico parasitising *Zea mays*. *Nematologica* 22, 381–389.

Sturhan, D. (2002) Notes on the genus *Cactodera* Krall & Krall, 1978 and proposal of *Betulodera betulae* gen. nov., comb. nov. (Nematoda: Heteroderidae). *Nematology* 4, 875–882.

Sturhan, D., Wouts, W.M. and Subbotin, S.A. (2007) An unusual cyst nematode from New Zealand, *Paradolichodera tenuissima* gen. n., sp. n. (Tylenchida, Heteroderidae). *Nematology* 9, 561–571.

Subbotin, S.A., Mundo-Ocampo, M. and Baldwin, J.G. (2010) *Systematics of Cyst Nematodes (Nematoda: Heteroderinae). Nematology Monographs and Perspectives, 8A, B*. Brill, Leiden, Netherlands.

Subbotin, S.A., Toumi, F., Elekcioglu, I.H., Waeyenberge, L. and Tanha Maafi, Z. (2018) DNA barcoding, phylogeny and phylogeography of the cyst nematode species of the *Avenae* group from the genus *Heterodera* (Tylenchida: Heteroderidae). *Nematology* 20, 671–702.

Subbotin, S.A., Franco, J., Knoetze, R., Roubtsova, T.V., Bostock, R.M., *et al.* (2020) DNA barcoding, phylogeny and phylogeography of the cyst nematode species from the genus *Globodera* (Tylenchida: Heteroderidae). *Nematology* 22, 269–297.

Subbotin, S.A., Roubtsova, T.V., Bostock, R.M., Tanha Maafi, Z. and Chizhov, V.N. (2022) DNA barcoding, phylogeny and phylogeography of the cyst nematode species of the *Humuli* group from the genus *Heterodera* (Tylenchida: Heteroderidae). *Nematology* 24, 873–886.

Subbotin, S.A., Roubtsova, T.V., Bostock, R.M., Tanha Maafi, Z., Chizhov, V.N., *et al.* (2024) DNA barcoding, phylogeny and phylogeography of the cyst nematode species of the *Schachtii* group from the genus *Heterodera* (Tylenchida: Heteroderidae). *Nematology* 26, 71–97.

Tanha Maafi, Z., Sturhan, D., Handoo, Z., Mordehai, M., Moens M. *et. al*. (2007) Morphological and molecular studies of *Heterodera sacchari, H. golden* and *H. leuceilyma* (Nematoda: Heteroderidae) *Nematology* 9, 483–497.

Vovlas, N., Greco, N. and Di Vito, M. (1985) *Heterodera ciceri* sp. n. (Nematoda: Heteroderidae) on *Cicer arietinum* L from northern Syria. *Nematologia Mediterranea* 13, 239–252.

Williams, T.D. and Siddiqi, M.R. (1972) *Heterodera avenae*. In: *CIH Descriptions of Plant-parasitic Nematodes*, Set 1, No. 2. Commonwealth Agricultural Bureaux, Farnham Royal, UK.

Wouts, W.M. and Weischer, B. (1977) Eine Klassifizierung von fünfzehn in Westeuropa häufigen Arten der Heteroderinae auf Grund von Larvenmerkmalen. *Nematologica* 23, 289–310.

Wouts, W.M. (1985) Phylogenetic classification of the family Heteroderidae (Nematoda: Tylenchida). *Systematic Parasitology* 7, 295–328.

Wouts, W.M. and Baldwin, J.G. (1998) Taxonomy and identification. In: Sharma, S.B. (ed), *Cyst Nematodes*. Kluwer Academic Publishers, Dordrecht, Boston and London, pp. 83–122.

Wouts, W.M., Rumpenhorst, H.J. and Sturhan, D. (2001) *Heterodera betae* sp. n., the yellow beet cyst nematode (Nematoda: Heteroderidae). *Russian Journal of Nematology* 9, 33–42.

Wrather, J.A., Anderson, T.R., Arsyad, D.M., Tan, Y., Ploper, L.D., *et al.* (2001) Soybean disease loss estimates for the top ten soybean producing countries in 1998. *Canadian Journal of Plant Pathology* 23, 115–121.

11 Root-knot Nematodes

Zafar Handoo[1]*, David J. Hunt[2] and Alemayehu Habteweld[1]
[1]USDA-ARS Mycology and Nematology Genetic Diversity and Biology Laboratory, Building 010A, Room 118, BARC-West 10300 Baltimore Avenue, Beltsville, MD 20705, USA; [2]CABI, Silwood Park, Buckhurst Road, Ascot, Berkshire, UK

Abstract

The root-knot nematodes (RKN) are highly adapted plant parasites and have a very specialized mode of life marked by extreme sexual dimorphism, the females being massively swollen and sedentary whereas the males, when present, retain their mobility and are vermiform. The chapter provides an introduction to their general morphology, life cycle, plant-host relationships, classification, major species, and the various management strategies. There is a key to the genera of the Meloidogyninae, although only *Meloidogyne* is discussed. The importance of accurate identification and diagnostics for control/management strategies of RKN species is emphasised, morphological and morphometric criteria, in combination with biochemical and molecular diagnostics and other specialized techniques, being paramount. Ten major species of economic importance, including *M. chitwoodi*, *M. enterolobii*, *M. ethiopica*, *M. exigua*, *M. incognita* and *M. javanica*, are dealt with in more detail with information on morphology, molecular diagnostics, host range and distribution.

11.1. Introduction

Root-knot nematodes were first discovered by Berkeley (1855) on the root of cucumber growing in an English glasshouse (Figs 11.1, 11.2). Initially regarded more as a curiosity, their importance has gradually been recognized. Today, along with the cyst nematodes and pratylenchids, they are among the most economically important plant-parasitic nematodes in the world. Root-knot nematodes belong to the genus *Meloidogyne* Göldi, 1887, a genus originally created for a species found attacking coffee roots in Brazil (Göldi, 1887). Due mainly to poor descriptions and confusion with cyst nematodes, much uncertainty existed as to the status of root-knot nematodes until Chitwood (1949) established the genus on a firm footing, establishing, as first reviser, five species: *M. exigua* Göldi, 1887, the type species known for attacking coffee roots, together with *M. arenaria* (Neal, 1889) Chitwood, 1949, *M. hapla* Chitwood, 1949, *M. incognita* (Kofoid & White, 1919) Chitwood, 1949 and *M. javanica* (Treub, 1885) Chitwood, 1949. A full account of the early history of the genus is given by Hunt and Handoo (2009).

*E-mail: Zafar.Handoo@usda.gov

A revision of Hunt, D.J. and Handoo, Z. (2012) Root-knot nematodes. In: Manzanilla-López, R.H. and Marbán-Mendoza, N. (eds), *Practical Plant Nematology*. Mundi-Prensa and Biblioteca Básica de Agricultura, Montecillo, Mexico.

(eds R.H. Manzanilla-López, D.J. Hunt, N. Marbán-Mendoza and S. Infante Gil)
DOI: 10.1079/9781836990413.0011

Fig. 11.1. The Reverend Miles Joseph Berkeley FRS (1803-1889). Berkeley, a pioneering mycologist and plant pathologist, was the first to publish a paper on root-knot nematodes.

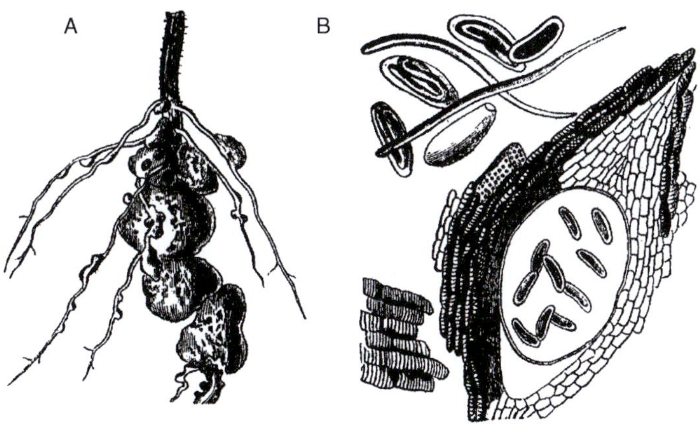

Fig. 11.2. The first illustration of root-knot nematodes on plant roots. A: Galled roots; B: section through gall showing nematodes and eggs. (After Berkeley, 1855.)

The root-knot nematodes are highly adapted and have a very specialized mode of life marked by extreme sexual dimorphism, with the females being massively swollen and sedentary, whereas the males retain their mobility and are vermiform. Although superficially similar to cyst nematodes, root-knot nematodes represent a separate evolutionary development, with similarities in morphology and biology being due to convergence. Whereas the cyst nematodes retain either all or most of the eggs inside the female body (one or two species, such as *Heterodera schachtii*, have a small egg-sac into which some eggs are deposited), root-knot nematodes deposit the majority of their eggs into an external gelatinous matrix, which is secreted by the rectal glands of the female. The egg masses are large and usually protrude through the root surface, and the jelly-like egg masses often coalesce with their neighbours. Although two other genera have been proposed, namely *Hypsoperine* Sledge & Golden, 1964 and *Spartonema* Siddiqi, 2000, these are usually treated as junior synonyms of *Meloidogyne* (see Hunt and Handoo, 2009).

The last decade or so has seen a great increase in the number of described species (Subbotin *et al*., 2021), although some of these are undoubtedly junior synonyms of common, highly variable species, such as *M. incognita* (see Hunt and Handoo, 2009). Because of their relatively conserved morphology, and the fact that the morphometrics and few morphological characters are often subject to seemingly endless variation, more objective diagnostic techniques, such as molecular characterization, must be applied to confirm or refute the status of the nominal species, and any proposal for a new species should include molecular data as an integral part of the description.

The most recent major monographs on the root-knot nematodes are *Systematics of Root-knot Nematodes* (Subbotin *et al*., 2021) and *Root-knot Nematodes* (Perry *et al*., 2009). These provide an excellent and up-to-date introduction to the genus and form compulsory reading for those who wish to know more about these fascinating nematodes. Other major works include those by Whitehead (1968), Franklin (1976, 1979), Sasser and Carter (1985), Barker *et al*. (1985), Jepson (1987), Karssen (2002), Karssen and Moens (2006), Hunt and Handoo (2009) and Karssen *et al*. (2013). The most recent species list, including full synonymies, is that of Subbotin *et al*. (2021).

11.2. General Morphology

There are many detailed accounts of the morphology of root-knot nematodes, the latest being that of Subbotin *et al*. (2021). Other excellent sources include those by Franklin (1957, 1971, 1976, 1979), Golden and Birchfield (1965), Whitehead (1968), Bird (1971), Esser *et al*. (1976), Taylor and Sasser (1978), Lamberti and Taylor (1979), Jepson (1983a, b, c; 1987), Eisenback (1985a, b; 1997), Hirschmann (1985a, b), Eisenback and Triantaphyllou (1991), Kleynhans (1991), Karssen (2002), Carta *et al*. (2006), Karssen and Moens (2006), Hunt and Handoo (2009) and Karssen *et al*. (2013). The following account focuses on the morphology of the infective second-stage juvenile (J2), adult male and adult female, as these are the stages most used in identification using classical morphology-/morphometric-based techniques. The general morphology, including the male, is shown in Figure 11.3 while the J2 is illustrated in Figure 11.4, and the adult female in Figures 11.5 and 11.6. For greater detail of the organ systems, the reader is referred to Eisenback and Hunt (2009).

11.2.1. Second-stage Juvenile (J2)

The J2 (Figs 11.3E-G, 11.4) is vermiform and is the stage that hatches from the egg and penetrates a host root in order to establish a feeding site to facilitate the further development of the nematode. The J2 is highly mobile and may move considerable distances in search of a host root. The vermiform body is quite slender, although the intestine may be replete with food reserves. The labial region (Fig. 11.3F) is relatively weakly cuticularized, much more so than in a J2 cyst nematode, and the stylet is also relatively weak, whereas in the cyst nematode, the J2 it is strongly cuticularized and has much larger basal knobs for muscle attachment. The pharynx (Fig. 11.4A) comprises the usual parts but is notable for the length of the pharyngeal gland lobe, the source of the complex secretions that enables the J2 to establish a feeding site within the host root. As previously mentioned, the intestine is replete with food reserves, at least in young individuals, although this dissipates over time as the nematode searches for a suitable host. The tail is elongated and tapers to a finely rounded tip in most species (Figs 11.3G, 11.4B). The terminal third or so of the tail comprises clear cuticle and is known as the hyaline region, the length of which as a proportion of the tail

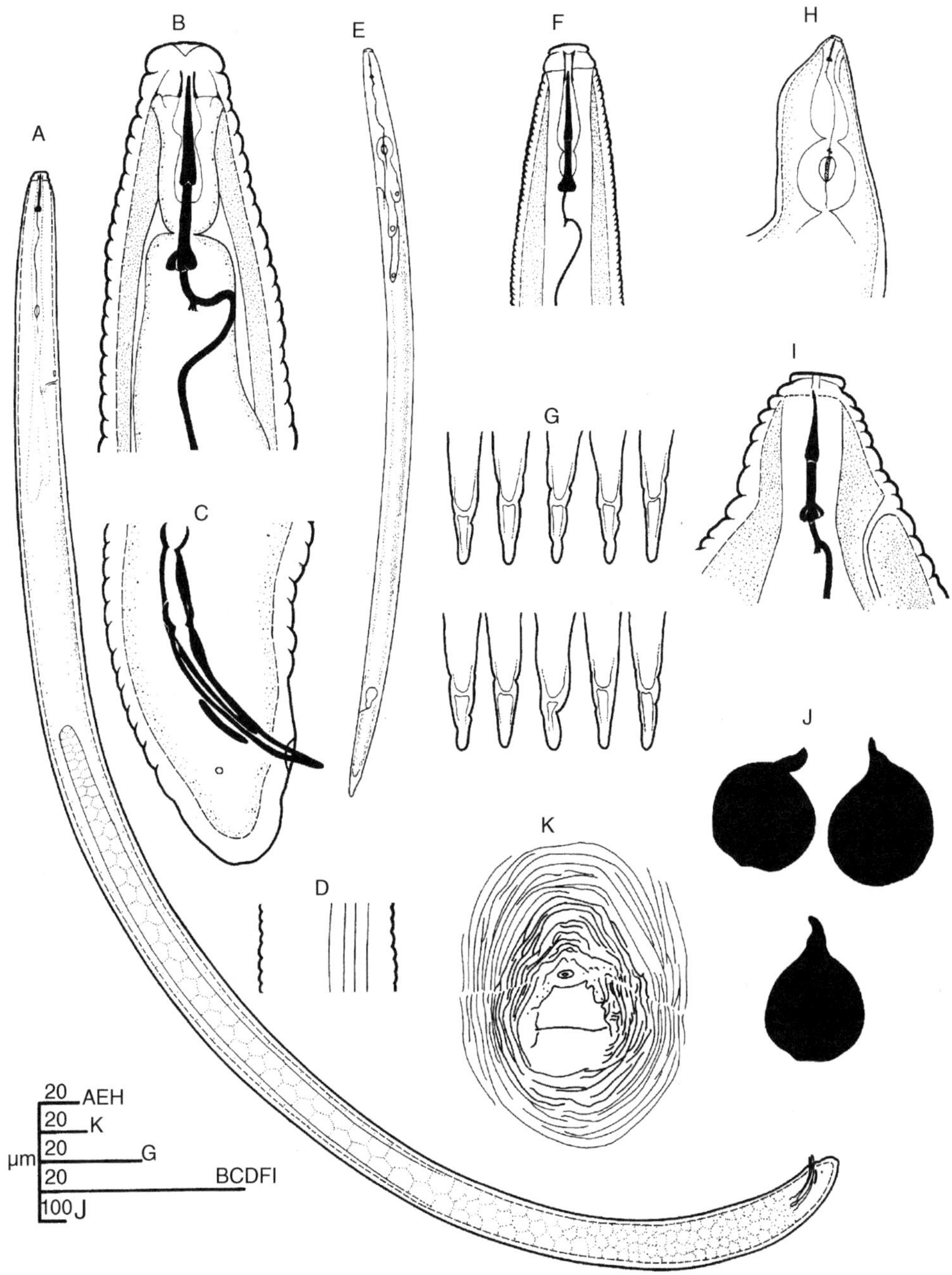

Fig. 11.3. General morphology of root-knot nematode. A: Male entire body; B: male anterior region; C: male tail region; D: male lateral field; E: entire second-stage juvenile (J2); F: J2 anterior region; G: J2 tail region; H, I: female anterior region, lateral view; J: entire females; K: perineal pattern. (After Jepson, 1985, courtesy CABI.)

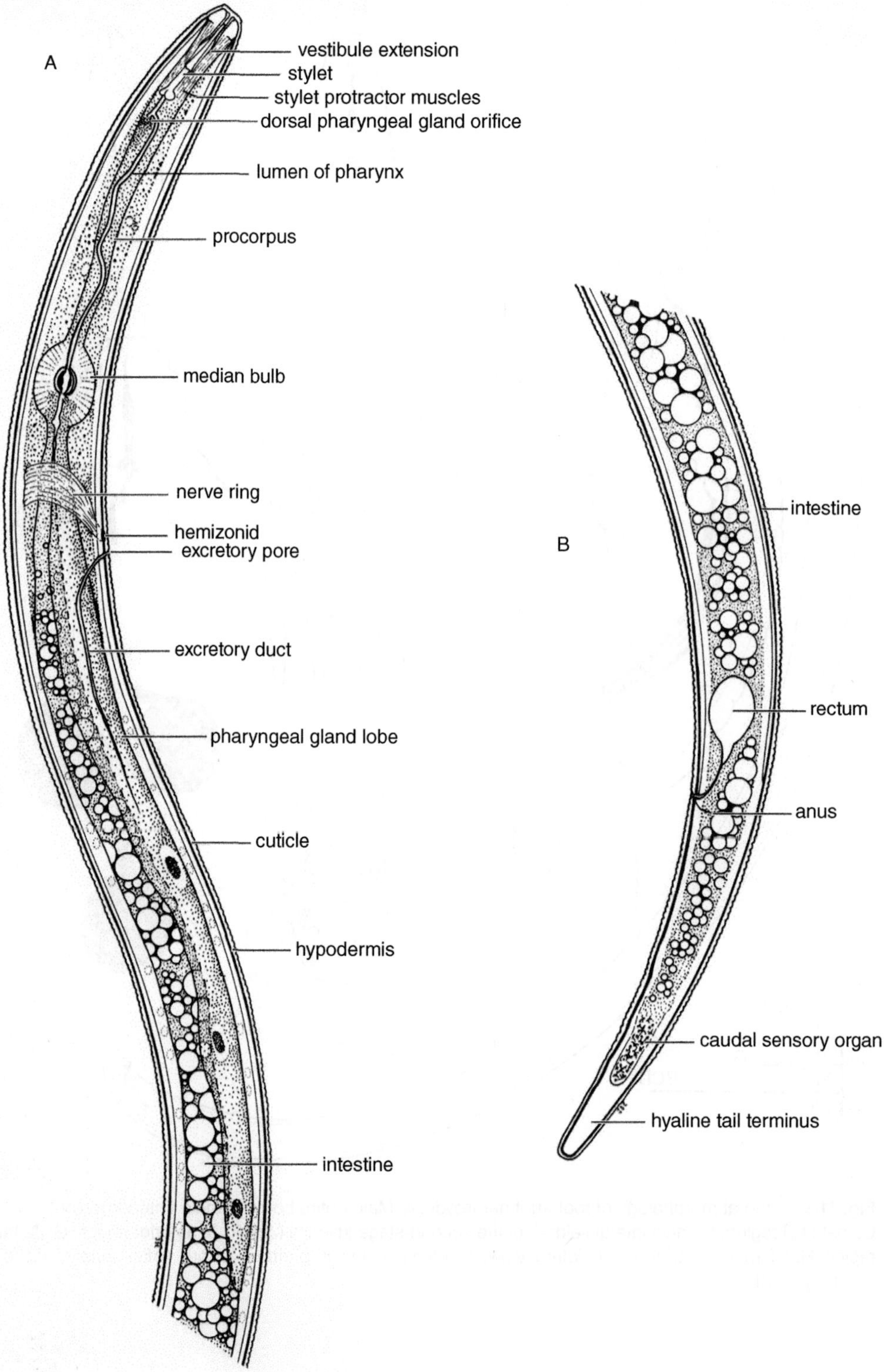

Fig. 11.4. General morphology of infective second-stage juvenile. (After Eisenback, 1985a.)

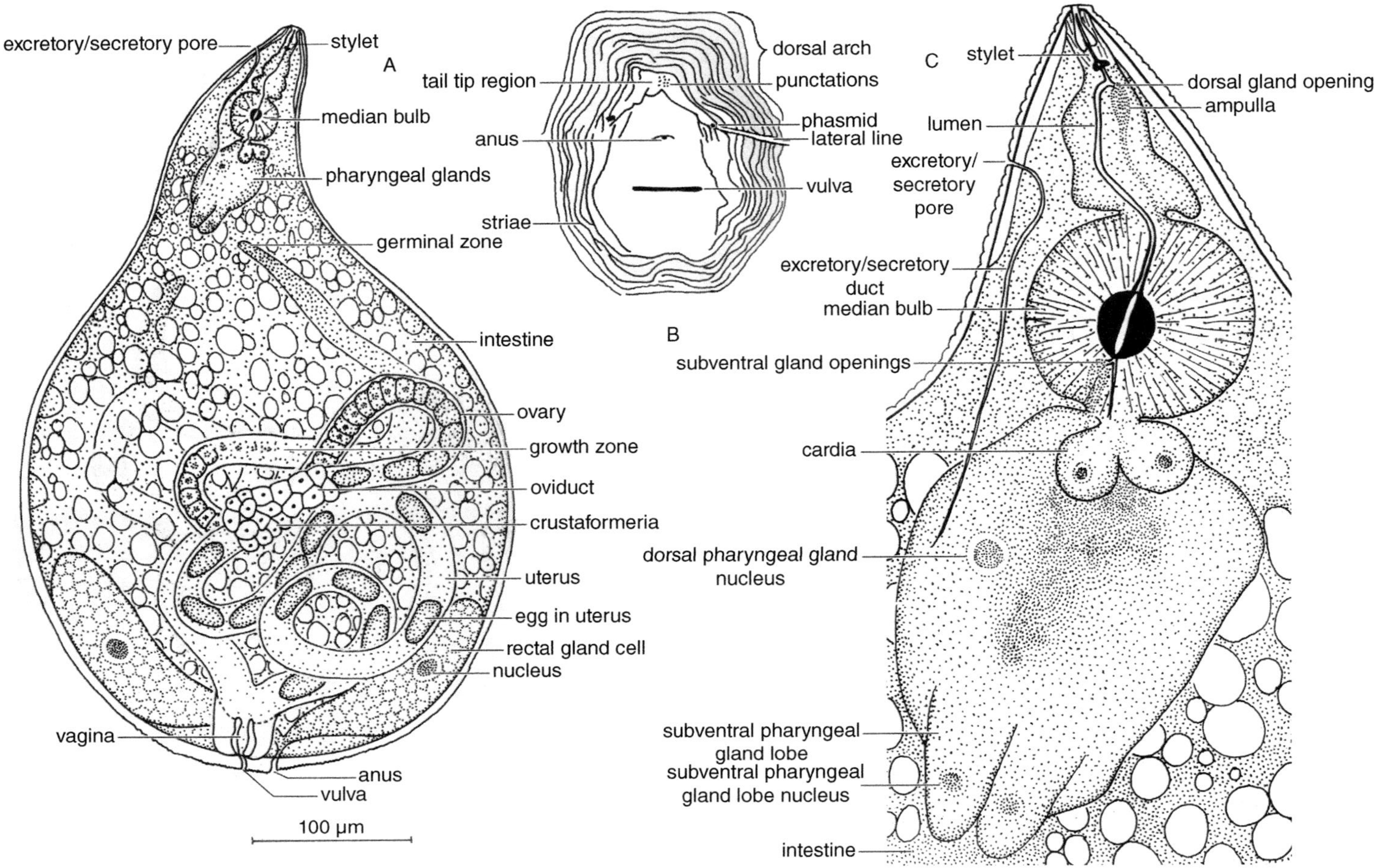

Fig. 11.5. General morphology of mature female root-knot nematode: A: Entire nematode, for clarity, only one of the two genital tracts have been shown; B: perineal pattern (stylized); C: anterior region. (Composite drawings after Triantaphyllou, 1960, and Eisenback, 1985a.)

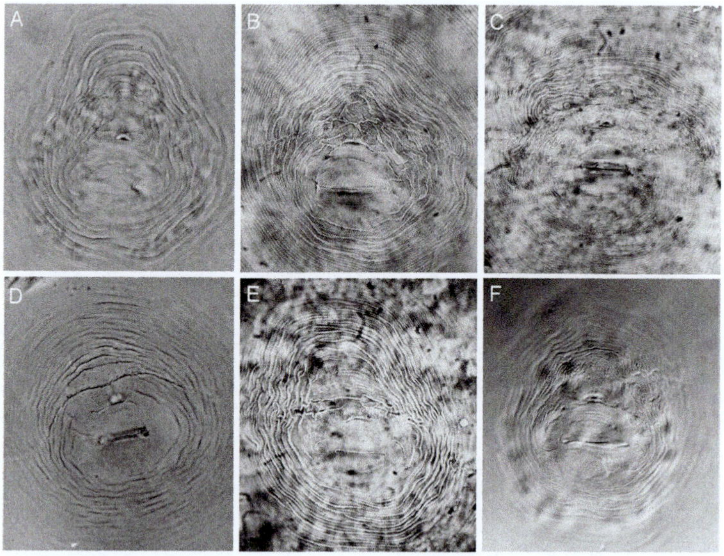

Fig. 11.6. Light micrographs of perineal patterns of selected *Meloidogyne* species. A: M. incognita; B: *M. javanica* (note broad lateral fields); C: *M. arenaria* (note low dorsal arch); D: *M. hapla* (note punctations around tail tip); E: *M. exigua*; F: *M. ethiopica* (paratype female). (All photographs courtesy D.J. Hooper, Rothamsted Research.)

itself is of some importance as a diagnostic aid (note that the hyaline region is longer and comprises about half the tail length in cyst nematode J2). The phasmids are located between the anus and the tail tip. A small, four-celled genital primordium, the precursor of the adult reproductive system, is visible at about 65% of the body length and is positioned towards the ventral surface and parallel with the body axis. There are usually three or four lines or incisures in the lateral field in the mid-body region, with these reducing anteriorly and posteriorly.

11.2.2. Male

The adult male is, like the J2, vermiform (Fig. 11.3A-D). The vermiform state does not occur until after the third moult; they then remain vermiform, although enclosed within the shed cuticles of previous stages until the fourth moult releases the fully formed adult. The mobile males emerge from the retained cuticles from earlier stages and enter the soil phase, in which they migrate in search of females. Many species lack males altogether, while in others, the male seems superfluous to reproduction, with the existence of males seemingly being an evolutionary throwback. In amphimictic species, however, males are common. Although possessing a well-developed stylet (Fig. 11.3B), the male does not feed and relies upon the reserves stored in its intestine from the period when it was a sedentary J2. The lateral field typically has four lines or incisures (Fig. 11.3D) and appears to encircle, or merge with, the tail tip. Areolations may be incomplete and/or irregular. The genital system typically consists of a single testis (monorchic), but in specimens where sex reversal has occurred, there may be two, with these representing the two female genital tracts. Sex reversal occurs under conditions of nutrient and temperature stress, with female nematodes altering their development to become males in an attempt to ensure that at least some genetic material 'escapes' from a deteriorating resource that may not be capable of satisfactorily supporting the development of a mature female. There are two spicules (Fig. 11.3C), and these are of typical tylenchid form. A bursa is absent. The actual tail (Fig. 11.3C) is very short, only a few microns long, as the cloacal aperture is located virtually at the end of the body, an adaptation to mating with the greatly swollen female, whose vulval aperture is also virtually terminal in position. The phasmids are situated near the cloacal aperture.

11.2.3. Female

The female stage is sedentary and, when mature, is usually pear-shaped or derived from that form (Figs 11.3J, 11.5A). The neck region (Figs 11.3H-I, 11.5C) protrudes and retains some mobility so that the nematode may feed from one or other of the several giant cells that were formed near her anterior region. The vulva opens in a more or less terminal position and is surrounded by the perineal pattern (Figs 11.3K, 11.5B, 11.6), which comprises of cuticular ridges (variations in the form of which have been used as diagnostic characters), the lateral fields, phasmids, vulval opening, anal opening and tail tip. The phasmids are located near the tail tip and occur within the perineal pattern, where they appear as minute dots. The expanded portion of the female body contains two genital tracts, both running anteriad because of the extreme posterior location of the vulva (i.e., they are didelphic, prodelphic). The tracts are very long and convoluted and intertwine around the intestine, with the enormous ovaries (comprising over half of the length of each genital tract) absorbing nutrients directly through the gut wall (Fig. 11.5A). The very large genital system can produce hundreds of eggs. The intestine itself lacks any connection to the rectum and thus ends in a blind sac. Six well-developed rectal gland cells (Fig. 11.5A) debouche via the anus, their secretions forming the gelatinous matrix into which the eggs are laid. The gelatinous matrix is a complex structure that may be as large or larger than the female herself. It consists of various proteins, mucopolysaccharides, antimicrobial compounds and enzymes. The eggs are just over twice as long as wide and have a resistant glycolipid layer.

11.3. Life Cycle

The life cycle of root-knot nematodes is typical of many other nematodes, in that there is an egg stage, four juvenile stages (one of which is found only in the egg) and the adult male and female. The male may be rare or absent in many species, as reproduction is often parthenogenetic, although some species are amphimictic. The life cycle of root-knot nematode is similar to that of cyst nematodes, the primary difference being that female root-knot nematodes do not turn into brownish-collared cysts formed from the tanned and hardened cuticle of the female but instead remain white with a soft cuticle. The female lays eggs into a large gelatinous matrix (egg-sac) that is usually deposited on the surface of galled roots, although sometimes they occur within the galls themselves (Fig. 11.7B). The first-stage juvenile (J1) moults in the egg and hatches as the J2. Hatching is temperature-driven and occurs without requiring stimulus from plant roots, although root diffusate sometimes stimulates hatching, and enzymes are thought to be involved in altering the eggshell structure. The J2 can survive in the soil in a quiescent state for an extended period and relies on the food reserves stored in the intestine. This motile stage, often called the pre-parasitic or infective J2, migrates through the soil, searching for roots of a suitable host plant. Once a suitable host is located, the J2 invades the root tip region (Fig. 11.7A) and moves intracellularly to the feeding site. After entering the roots, the J2 moves across the cortical tissue until it reaches its preferred site to initiate feeding. The feeding of the J2 on protoxylem and protophloem cells induces these cells to differentiate into specialized nurse cells – the giant cells. Giant cells are highly specialized cellular adaptations induced and maintained by the developing female in susceptible plants (Castillo *et al.*, 2001). The formation of giant cells is essential for successful parasitism (Fig. 11.7D). If a juvenile fails to induce the formation of giant cells in the root, it will either starve or migrate out. Once a giant cell is initiated, the nematode undergoes morphological changes, becomes sedentary and enlarges greatly to assume a sausage-shaped parasitic J2. Without further feeding, and under favourable conditions, the J2 stage moults to the third-stage juvenile (J3) after approximately 14 days, then to the fourth-stage juvenile (J4), both of which lack a buccal stylet, have a degenerate pharynx and do not feed. Finally, the moult to the adult stage takes place. The combined time for the J3 and J4 stages is much shorter than that for the J2 or the adult, typically lasting 4-6 days. After the last moult, the adult stylet is formed. Females continue feeding, and growth is extremely rapid; the nematodes become saccate and then commence egg-sac production and egg-laying (Fig. 11.7B, C). Large numbers of eggs (100-1000) are extruded into the gelatinous matrix. Males, when present, are vermiform and do not feed. They increase only in length

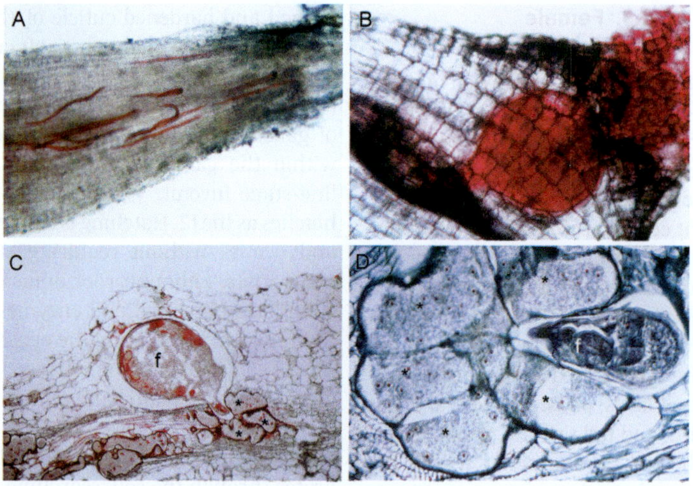

Fig. 11.7. Histopathology. A: Longitudinal section showing infective second-stage juveniles of *Meloidogyne incognita* accumulating near cotton root tip (R.S. Hussey); B: stained mature female *Meloidogyne* in root with egg sac and eggs protruding from galled tissue (R.S. Hussey); C: section showing *M. incognita* female (f) and giant cells (*) in tobacco root (H.A. Johnson); D: section showing female (f) *M. incognita* in tobacco root with a cluster of multinucleate giant cells (*) located around the anterior region of the nematode (N.T. Powell). (Individual figures reproduced with permission from Slide Set Nematology, 100 Slides. T.A. Niblack, Department of Plant Pathology, University of Missouri, Columbia, Missouri. Courtesy Society of Nematologists.)

and leave the root for mating. Males may be found in parthenogenetic species when conditions are unfavourable for female development, such as when population densities are very high and there is presumably a limited food supply (see Triantaphyllou, 1960). The length of the life cycle (egg to egg) is strongly influenced by temperature but can be completed within 3 weeks. Egg production, however, may continue for 2-3 months (Bird, 1959; Triantaphyllou and Hirschmann, 1960; Franklin, 1965; de Guiran and Ritter, 1979; Eisenback and Triantaphyllou, 1991). For a more complete account of the life cycle, the reader is referred to Moens *et al.* (2009) and Abad *et al.* (2009). In addition, further details can be found in Songul and Dickerson (1976), Taylor and Sasser (1978), Endo (1987), Veech (1991), Weischer and Brown (2000), Ferraz and Brown (2002), Karssen (2002), Karssen and Moens (2006), Khan (2008) and Karssen *et al.* (2013).

11.4. Plant-host Relationships

Root-knot nematodes are one of the most economically destructive groups of plant-parasitic nematodes worldwide, causing the formation of knots or galls on roots of susceptible host plants. Many histopathological studies have been conducted on *Meloidogyne* infestations in a wide range of host plants during the last 70 years. Infection is initiated by the J2, which penetrates the roots, mainly at the zone of elongation behind the root cap (Fig. 11.7A). Penetration involves mechanical action of the stylet (Linford, 1942) as well as enzymatic action from pharyngeal gland secretions (Bird *et al.*, 1975). Nematodes move intracellularly through the cortex to the region of cell differentiation, where they settle and begin feeding. The cells along the migration path become compressed but are not ruptured. Infrequently, feeding on cortical cells occurs. Preferred feeding sites are primary phloem or the adjacent parenchyma. The juvenile begins by feeding on a group of five or six cells. In response to this feeding, the host tissue undergoes pronounced morphological and physiological changes within a few hours after the injection of pharyngeal gland secretions. Some parenchyma cells develop into permanent nurse cells for the nematode and, as a result of secretions produced by the subventral pharyngeal gland cells of feeding J2, become hypertrophied, multinucleate, giant cells (Fig. 11.7C, D) (Dropkin, 1972; Hussey, 1987; Hussey and Grundler, 1998). The nematodes obtain nourishment from these nutritive cells and cannot continue their development to adulthood without them. Males, when formed, do not feed and leave the roots, but females continue feeding at the same giant cells for the remainder of their life span.

For a successful host-parasite relationship, giant cells are essential, and their active metabolism is maintained either through secretions of the dorsal pharyngeal gland (Bird, 1968) or the removal of solutes by the adult female (McClure, 1977). Photosynthates are mobilized to the giant cells in the roots and, as a result, plant growth and yield may be suppressed. Above-ground symptoms of infected plants include chlorosis of foliage and temporary wilting during periods of water stress, and nutrient and water absorption are greatly reduced by the damaged, galled root system. Cell wall dissolution and cell fusion may be involved in giant cell formation in some host-parasite relationships, depending upon the nematode and host species involved (Bird, 1974).

Root tissues around the nematode and the giant cells undergo hyperplasia and hypertrophy, resulting in the characteristic root gall, with the galls usually developing 1-2 days after juvenile penetration (Fig. 11.8). Gall size is commonly related to the host plant, the number of J2 and the nematode species. Most of the galls induced by *Meloidogyne* species are similar in

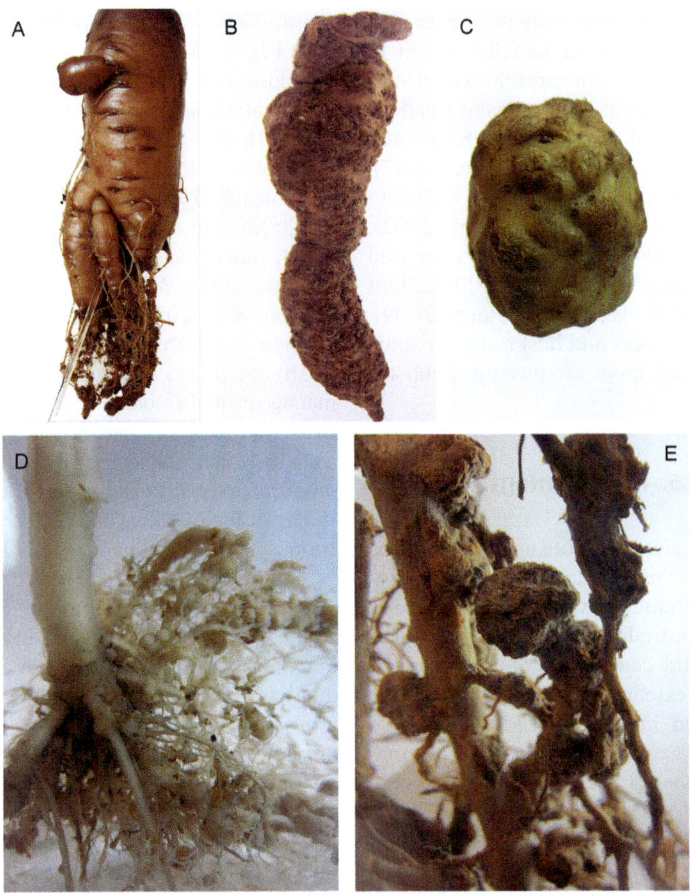

Fig. 11.8. Symptoms of root-knot nematode damage to roots. A: *Meloidogyne hapla* on carrot; B: *M. incognita* on sweet potato; C: *M. incognita* on potato; D: *M. javanica* on tomato; E: *Meloidogyne* sp. on sycamore. (Courtesy Joseph Mowery, USDA Nematode Collection, Beltsville, MD, USA.)

morphology. However, some species, such as *M. exigua*, *M. hapla* and *M. kikuyensis*, produce characteristic galls that may be useful in species identification.

Plant-growth regulators have been implicated in the development of giant cells and galls. Auxins (promoters of cell growth) have been identified in higher concentrations in root-knot infected tissue than in non-galled tissue. Cytokinins (promoters of cell division) may also increase in *Meloidogyne*-infected plants. Application of these plant-growth regulators to resistant plants reverses the resistant response and makes plants susceptible. Another plant-growth regulator, ethylene, has been associated with gall formation. In a few interactions, such as that occurring with parasitism of coffee trees by *M. coffeicola*, root knots are formed in young plants, but not in older ones (up to 5 years old). Accordingly, root knots are usually present in *Meloidogyne* infections, but exceptions do occur! For more details about host-parasite relationships, see Taylor and Sasser (1978), Sasser and Carter (1982), Hussey (1985), Endo, (1987), Eisenback and Triantaphyllou (1991), Di Vito *et al.* (1996, 2004), Ferraz and Brown (2002), Mitkowski and Abawi (2003), Gheysen and Jones (2006), Karssen and Moens (2006), Khan (2008), Manzanilla-López and Starr (2009). The current host races and host range for some major agricultural crops are given in Subbotin *et al.* (2021).

11.5. Management

Meloidogyne species constitute a major group of plant-pathogenic nematodes affecting crop production throughout the world, and, therefore, preventing the introduction and spread of these species is a vital component of management strategies. The extensive host range, worldwide distribution and involvement with fungi and bacteria in disease complexes place these nematodes high on the list of major plant pathogens affecting the world's food supply (Carter and Sasser, 1982). Collectively, various root-knot nematode species attack nearly every crop that is grown, thus reducing not only the yield but also the quality of underground crops, such as potatoes, carrots and peanuts. All methods for the control of plant pathogens, including parasitic nematodes, have been categorized under one or more principles in Table 1.1 of Moens *et al.* (2009), and all of the various methods for the control of nematodes fit within one of these principles. The basic objective in any pest management is to increase both quantity and quality of crop yield, and increasing the yield quality is sometimes the most important objective for *Meloidogyne*. Targeting species as quarantine organisms reduces the risk of spread through international trade. *Meloidogyne chitwoodi* is a serious pest of economically important crops, such as potatoes and carrots, and is a prohibited pest in many countries. In future, with many more new species being described, additional species are likely to be of regulatory concern. For more details about the management of nematodes, the following references may be consulted (Canto-Saenz, 1985; Fassuliotis, 1985; Imbriani, 1985; Jatala, 1985; Johnson, 1985a, b; Mendoza and Jatala, 1985; Raymundo, 1985; Sasser and Carter, 1985; Schmitt, 1985; Bird, 1986; Heald, 1986; Morgan-Jones and Rodriguez-Kabana, 1986; Pinochet, 1986; Thomason and Caswell, 1987; Bridge, 1996; Atkins *et al.*, 2003; Karssen and Moens, 2006; Coyne *et al.*, 2009; Starr and Mercer, 2009; Nyczepir and Thomas, 2009; Williamson and Roberts, 2009; Mahfouz *et al.*, 2010). *Plant Parasitic Nematodes in Subtropical and Tropical Agriculture* (Sikora *et al.*, 2018) also provides a good account of nematode management. In addition, another recent book, *Nematode Diseases of Crops and their Sustainable Management* (Khan and Quintanilla, 2023), includes several chapters detailing sustainable management strategies for root-knot and other nematodes on various crops (Shokoohia *et al.*, 2023).

The use of resistant plants to combat *Meloidogyne* species and the associated molecular studies on resistance are given in detail by Cook and Starr (2006). The nematode population reduction may permit growing of susceptible crops more often in the rotation scheme. Resistant cultivars inhibit root-knot nematode reproduction, and cultivars resistant to these species are known in both woody and non-woody plants as well as in perennial and annual plants of various families. Their resistance is monogenic or polygenic. As mentioned by Karssen and Moens (2006), they are sometimes resistant to a single *Meloidogyne* species (e.g., coffee and *M. exigua*), while other cultivars show resistance to several species (tomato and *M. arenaria*, *M. incognita* and *M. javanica*). The *Mi-1* gene, which confers resistance in tomato to root-knot nematodes, is inactivated at soil temperatures above 28°C. Resistance-'breaking' biotypes or *Mi*-virulent isolates of root-knot nematodes have been reported in many areas of the world. Heat-stable resistance genes have been found in tomatoes. Several cultural and physical control methods have been used with some degree of success, but these methods are often of local or regional relevance (see Gaur and Perry, 1991) and are only of use in regions where solar energy is available for long periods of time. The era of nematicides lasted from the 1950s to the 1980s, and their effectiveness is often cited as a reason why, for many years, other alternative management systems did not receive greater attention (Moens *et al.*, 2009). Starting in the 1970s, the use of fumigants was initially greatly restricted and then forbidden entirely, and suspension of some of the granular nematicides followed. There is now little prospect for new and effective nematicides being released in the future.

The principles and main components of effective control programmers and integrated pest control have been discussed in detail (Taylor and Sasser, 1978; Johnson and Fassuliotis, 1984; IFAS, 1989; Netscher and Sikora, 1990). We agree with the following main aspects that were considered important by Netscher and Sikora (1990):

1. Prevention of infestations by controlling nematode spread must be top priority.
2. Only root-knot nematode-free transplants should be used as planting material.
3. In view of the high multiplication rate of root-knot nematodes and difficulty in determining occurrence of low population densities, previously infected land should always be considered infested, even if the presence of *Meloidogyne* cannot be demonstrated by soil analysis.
4. Efforts should be made when planting vegetable crop rotations to select and develop pest management approaches that prevent the build-up of high nematode densities.
5. Integrated pest management should combine rotations with non-host crops, resistant, tolerant and susceptible cultivars as well as judicious use of nematicides, based on proper soil sampling estimations of damage threshold levels.
6. An integrated approach will control economically important nematodes, reduce pesticide costs and prevent unnecessary environmental contamination.
7. Proper selection of a combination of resistant, moderately resistant and tolerant vegetable crops can increase the number of vegetables in a short rotation cropping system.
8. Resistant cultivars should be used in rotation with susceptible cultivars and with other control techniques to prevent the development of resistance-breaking pathotypes.
9. All non-host crops and resistant cultivars should be challenged by local populations to determine true host status.
10. In regions where irrigated rice (or flooding) constitutes one of the components of the farming system, inundation can give good control.
11. Destruction of roots after harvest, soil drying and cultivation as a 'clean' fallow will significantly reduce population densities.
12. Time of planting may be effective in the cooler upland tropics and in the winter season in the subtropics.
13. Organic amendments or the use of non-host cover crops and green manures can reduce nematode densities.

11.6. Pathotypes

In his article on *The concept of race in phytonematology*, Dropkin (1988) refers the term 'pathotype' to a population whose members can reproduce on a host with genetic resistance to other parasite populations of the same species. The term does not imply partial isolation or uniform genetics, while 'race' designates a population with distinctive characters, either in morphology or in physiology, or both. A race is partially isolated from other intraspecific groups by geography or genetics. Races of phytonematodes are descriptively defined by their ability to reproduce on certain members of a set of differential hosts.

Meloidogyne spp. differ in their ability to attack certain host plants. Such pathotypes have been found to attack resistant peach rootstocks (Young and Sherman, 1977), and, because of the time required to develop and test new rootstock cultivars, they are a serious concern in rootstock breeding programmes for trees and wine crops. To breed nematode-resistant cultivars adapted to a wide geographic area, it is always necessary to identify commonly occurring pathotypes and locate sources of host resistance to each pathotype. The discovery of a new pathotype of root-knot nematode on grape rootstocks illustrates the necessity for identifying the frequency, incidence and geographic distribution of such pathotypes and the importance of searching for sources of resistance to them as part of a comprehensive programme for breeding grape rootstocks (Cain *et al.*, 1984). For more details on pathotypes and a pathotype system to describe intraspecific variation in pathogenicity of root-knot nematodes species, the reader is referred to Tzortzakakis and Gowen (1996), Van der Beek *et al.* (1999), Semblat *et al.* (2000), Mojtahedi *et al.* (2007), Van der Beek and Poleji (2008) and Verdejo-Lucas and Sorribas (2008).

11.7. Morphological and Morphometric Identification

The most important diagnostic features used for the identification of *Meloidogyne* spp. include the following characters. For the mature female, shape of the body, labial region morphology, stylet length, shape of stylet cone, shaft and basal knobs, nature of perineal pattern, including form of dorsal arch (Figs 11.6, 11.9), lateral

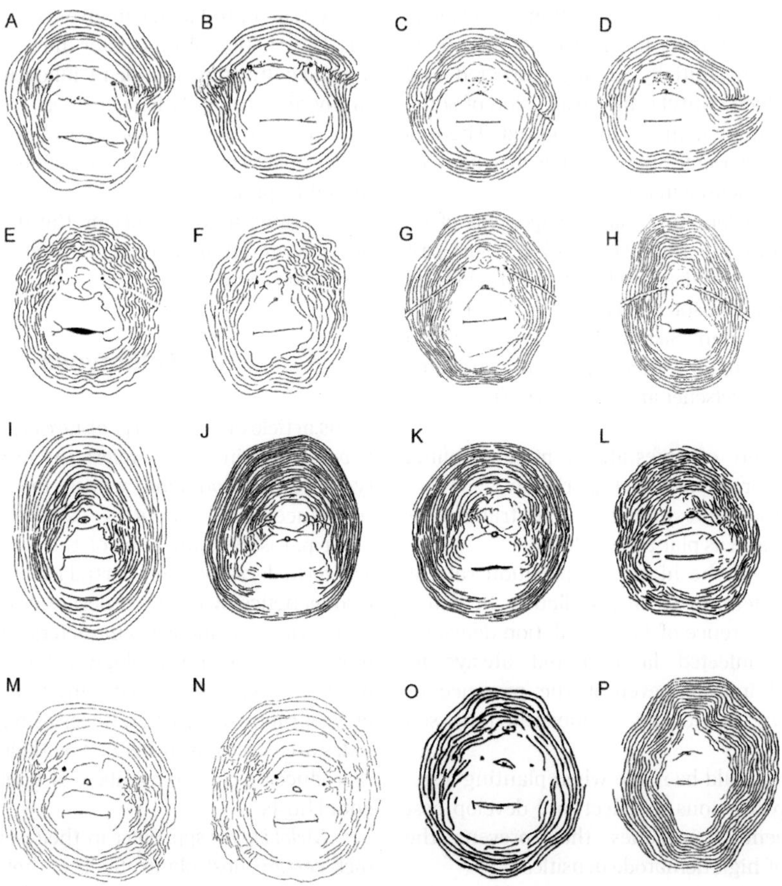

Fig. 11.9. Comparison of perineal patterns for 10 major species of *Meloidogyne*. A, B: *M. arenaria*; C, D: *M. hapla*; E, F: *M. incognita*; G, H: *M. javanica*; I: *M. chitwoodi*; J, K: *M. enterolobii*; L: *M. ethiopica*; M, N: *M. exigua*; O: *M. fallax*; P: *M. paranaensis*. Drawings not to scale. (A-H, after Orton Williams, 1972, 1973, 1974, 1975; I, after Jepson, 1985; J, K, after Rammah and Hirschmann, 1988; L, after Whitehead, 1968; M, N, courtesy Janet Machon; O, after Karssen, 1996; P, after Carneiro *et al.*, 1996.)

field, striae and tail terminus and excretory pore/stylet length ratios (EP/ST) are all important, if somewhat variable in nature. In the male, size, height and shape of the labial cap, number of annulations on the labial region, diameter of the labial region as compared to the first body annule, stylet length, form of stylet cone, shaft and basal knobs, distance of the dorsal gland orifice (DGO) from the stylet base and length and form of spicule are all important. For the infective juvenile (J2), body and stylet length, form of labial region and shape of stylet knobs, location of the hemizonid in relation to the excretory pore, distance of DGO from stylet base, number of lines in the lateral field and shape and length of the tail and hyaline terminus are important. For more details about these and other differentiating characters, see Whitehead (1968), Esser *et al.* (1976), Eisenback *et al.* (1981), Hirschmann (1985a, b), Kleynhans (1986), Jepson (1987), Eisenback and Triantaphyllou (1991), Karssen (2002), Karssen *et al.*, (2013) and Eisenback and Hunt (2009).

11.8. Biochemical and Molecular Diagnostics

As the taxonomy of the genus is now so complex, and as accurate identification is such an essential prerequisite for the selection of efficient management strategies, it is usually necessary (and certainly advisable) to perform biochemical or molecular characterization. Biochemical and DNA-based characterization involves the establishment of the isozyme phenotypes (Fig. 11.10) and nucleotide polymorphism, respectively, an

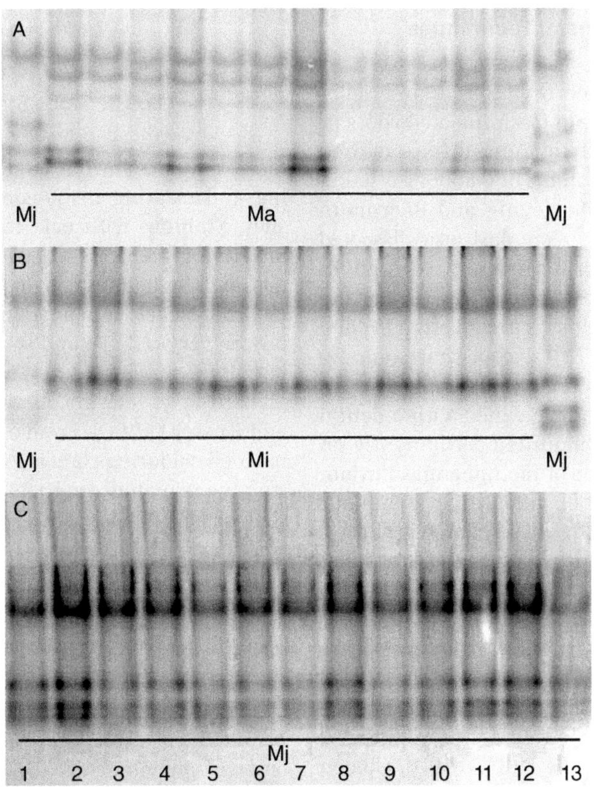

Fig. 11.10. Isozyme phenotypes. Single gels stained for both esterase (EST) and malate dehydrogenase (MDH). A: *Meloidogyne arenaria* (Ma) (lanes 2-13) (A2; N3); B: *M. incognita* (Mi) (I1; N1); C: *M. javanica* (Mj) (J3; N1) phenotypes. *Meloidogyne javanica* was used in each gel as the control. (Courtesy Janete Brito/CABI.)

excellent recent review being that of Subbotin *et al.* (2021). Isozyme patterns using esterase are relatively stable, although there are some exceptions that necessitate the use of an additional enzyme, usually malate dehydrogenase or glutamate dehydrogenase if an identification is to be confirmed. The use of molecular methodologies in taxonomy and phylogeny has been reviewed by Adams *et al.* (2009) and Subbotin *et al.* (2021). Details on the phylogeny of *Meloidogyne* spp., based on mitochondrial DNA sequences, are provided by Tigano *et al.* (2005). Multigene phylogeny of *Meloidogyne* spp. is discussed in Janssen *et al.* (2017), Álvarez-Ortega *et al.* (2019) and Subbotin *et al.* (2021).

11.9. Specialized Techniques

11.9.1. General Techniques

For morphological observation, J2 and males can be recovered from fresh infected roots or egg-sacs incubated in Petri dishes with a small amount of water. They may also be recovered from soil by sieving and Baermann funnel techniques. Females are dissected from infected roots after fixation overnight in 3% formaldehyde. Procedures for measuring and preparing specimens are given in Golden and Birchfield (1972), except that some females have the anterior and posterior ends cut with a sharp knife, cleaned with a dental root canal file and mounted permanently on a glass slide in a drop of lactophenol solution. Photomicrographs of female perineal patterns, J2 and males can be captured with a 35 mm or digital camera attached to a dissecting microscope.

For more details on killing, fixing, processing nematodes with glycerine, preparing temporary and permanent slide mounts and preserving nematode structures in a life-like manner, the reader is referred to Whitehead (1968), Hooper (1970, 1986, 1990), Golden (1990) and a book, *Techniques for Work with Plant and Soil Nematodes*, edited by Perry *et al.* (2021).

11.9.2. Perineal Pattern

The character most frequently used for *Meloidogyne* species identification is the morphology of the perineal pattern, which is located in the posterior body region of adult females. This area comprises the vulva-anus area (perineum), tail terminus, phasmids, lateral lines and surrounding cuticular striae. Preparation of perineal patterns (Fig. 11.11) for observation and identification has been covered by Taylor *et al.* (1955), Eisenback (1985a), Franklin (1965), Sasser and Carter (1982), Hartman and Sasser (1985), Hirschmann (1985a, b), Jepson (1987), Riggs (1990) and Charchar and Eisenback (2000). A more detailed account on root-knot perineal pattern development was given by Karssen (2002). Figure 11.9 summarizes the form of perineal pattern in the ten species of *Meloidogyne* that are discussed in greater detail in this chapter.

11.9.3. Root Staining

Many methods have been developed for staining and clearing nematode-infected root tissues. Staining with acid fuchsin-lactophenol or lacto-glycerol is the most widely used method. In addition, a method that utilizes chlorine bleach as a prestaining treatment has proven very reliable and is relatively simple to use (Byrd *et al.*, 1983). For more detail, see McBeth *et al.* (1941), Hooper (1986, 1990) and Hussey (1990). From a safety standpoint, red food colouring stain is a safer procedure for staining nematodes in roots and egg masses on root surfaces (Thies, 2002).

11.10. Root-knot or Cyst-forming Nematode?

Differentiating root-knot nematodes from cyst nematodes is usually an easy task, regardless of developmental stage. However, in the interest of completeness, these differences are summarized and compared in Table 11.1 and Figure 11.12.

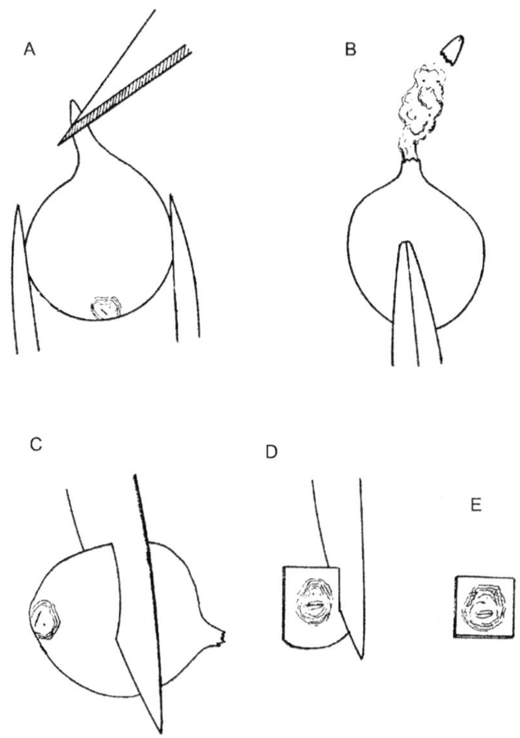

Fig. 11.11. How to cut perineal patterns. A, B: Excised female with neck region removed and body contents gently expelled; C: posterior body with perineal pattern removed; D: trimming surplus cuticle around perineal pattern; E: trimmed perineal pattern ready for mounting. (After Hartman and Sasser, 1985.)

Table 11.1. Comparison between root-knot and cyst-forming nematodes.

Stage	Character	Root-knot	Cyst-forming
Second-stage infective juvenile	Body form	Slender, anterior body tapering	Robust, anterior body more parallel
	Labial region	Weakly cuticularized, not offset	Strongly cuticularized, offset
	Stylet	Slender, <19 μm long	Robust, 20–30 μm long
	Hyaline region	Short, starting near tail tip	Long, well developed
Sedentary juvenile	Tail spike	Present	Absent
Male	Labial region	Weakly cuticularized, 2 annules	Strongly cuticularized, 4-5 annules
	Stylet	Slender	Robust
Mature female	Excretory pore	Anterior to median bulb valve plates	Posterior to median bulb valve plates
	Cuticle	Thin, white, not tanning to brownish colour on death	Thick, tanning to brownish colour on death
	Eggs	Deposited in external gelatinous egg-sac, few retained in body	Mostly retained within body, occasionally a few laid into small egg-sac
Biology	Host symptoms	Root galls almost always formed	No root galls formed
	Parasitic habit	Mature female usually endoparasitic	Mature female semi-endoparasitic
	Trophic system	Giant cells	Syncytia

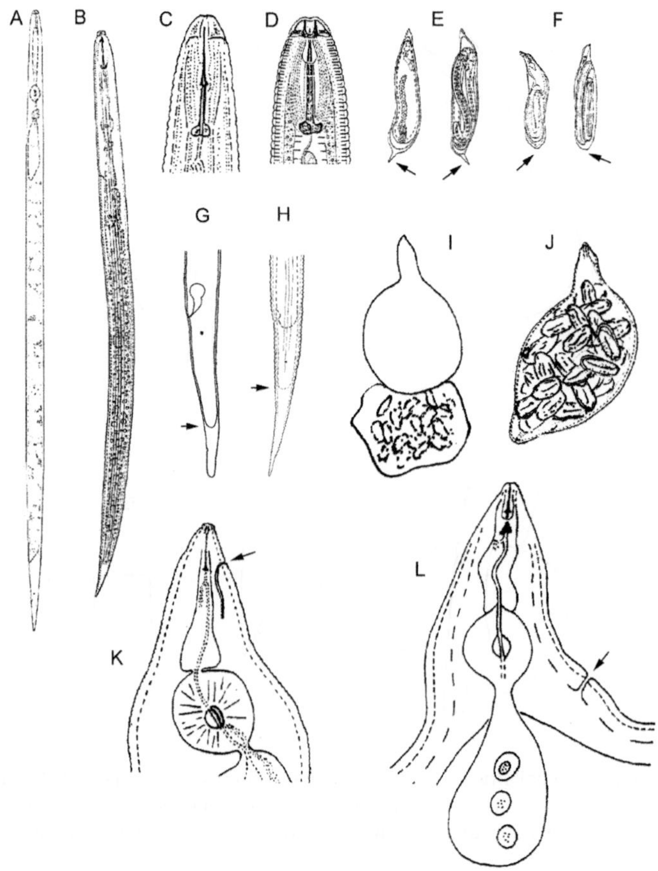

Fig. 11.12. Comparative morphology between root-knot and cyst-forming nematodes. A: *Meloidogyne* second-stage infective juvenile (J2); B: *Heterodera* infective J2 (note that the root-knot J2 is more slender); C: *Meloidogyne* J2 anterior region showing weak labial sclerotization and spear development; D: *Heterodera* J2 anterior region showing strong labial sclerotization and spear development; E: *Meloidogyne* immature stages showing retained tail spike (arrow); F: *Heterodera* immature stages without retained tail spike (arrow); G: *Meloidogyne* J2 tail region showing short hyaline region (arrow points to commencement); H: *Heterodera* J2 tail region showing much longer hyaline region (arrow points to commencement); I: mature female *Meloidogyne* showing eggs mostly laid in the external gelatinous sac; J: mature female *Heterodera* showing eggs retained in the body; K: *Meloidogyne* female anterior region. Note the position of the excretory pore (arrow) relative to the median bulb; L: *Heterodera* female anterior region (note the posterior position of the excretory pore (arrow) relative to the median bulb. (Drawings not to scale and adapted from various sources.)

11.11. Classification

(According to Manzanilla-López, Mejía-Madrid and Hunt, Chapter 3)

Order Rhabditida Chitwood, 1933
  Suborder Tylenchina Thorne, 1949
    Infraorder Tylenchomorpha De Ley & Blaxter, 2002
      Superfamily Tylenchoidea Örley, 1880
        Family Hoplolaimidae Filipjev, 1934
          Subfamily Meloidogyninae Skarbilovich, 1959

11.11.1. Subfamily Meloidogyninae Skarbilovich, 1959

Diagnosis (after Hunt and Handoo, 2009, itself based on Siddiqi, 2000): Hoplolaimidae. Root-gall inciting, female feeding producing multinucleate nurse cells. Marked sexual dimorphism. Cuticle striated. Lateral fields bearing four or five incisures. Labial region low, with one to four annules. In SEM, female labial disc dorsoventrally elongate, dumbbell-shaped, with oral opening a small round pore surrounded by six inner labial pits (sensilla). Framework moderately sclerotized, hexaradiate; lateral sectors equal to, or wider than, submedian sectors. Stylet moderately strong, male stylet longer and more robust than that of female. The orifice of dorsal pharyngeal gland located just posterior to stylet base. Median pharyngeal bulb oval or round, with large refractive thickenings. Pharyngeal glands elongate, extending over intestine mostly ventrally, but also laterally; subventral glands asymmetrical, extending past dorsal gland, with subventral nuclei (SVN) always posterior to dorsal nucleus (DN). Excretory pore in female located opposite or anterior to median bulb, in male, usually situated posterior to median bulb. No pre-adult vermiform female stage (except in *Meloinema*).

Mature female: Swollen, sedentary, rounded, oval to pear-shaped with a projecting neck. Cuticle moderately thick, striated, generally forming typical, finger-print-like perineal pattern terminally. Cyst stage absent. Vulva subterminal or terminal in position. Anus located near vulval lip; tail rudimentary or absent. Stylet less than 25 µm long in *Meloidogyne* (but 30-35 µm long in *Meloinema*). Median bulb oval or rounded, usually offset, with large refractive thickenings. Genital system didelphic-prodelphic; ovaries coiled. Most eggs are not retained in the body but laid externally. Large rectal glands present, gelatinous matrix present.

Male: Vermiform, migratory, generally non-feeding, over 1 mm long, posterior end twisted through 90-180°, developing by metamorphosis within a saccate juvenile. Labial region rather low and continuous; amphidial apertures large transverse slits; labial cap large, prominent; framework moderately sclerotized, lateral sectors wider than submedian sectors. Stylet strong, usually over 20 µm long, basal knobs prominent. Tail short or absent, lacking a bursa (except *Bursadera*). Spicules large (25-64 µm), distally pointed. Gubernaculum linear to trough-shaped, not protrusible. Cloacal lips non-tuboid, generally with hypoptygma.

Juveniles: J2 migratory and infective. J3 and J4 swollen, without stylet in type genus. In *Meloinema*, J3 and J4 vermiform. Labial region low, anteriorly flattened or rounded. Lateral sectors wider than submedian sectors, labial disc distinct in type genus. Stylet weak to moderately developed, less than 20 µm long in type genus, but strongly developed in *Meloinema*. Tail elongate-conoid, with minutely rounded tip and conspicuous terminal hyaline portion. Phasmids dot-like, located on tail, usually anterior to middle.

Remarks: Meloidogyninae differs from Heteroderinae in that its members incite root-galls and in having the lateral sectors of the labial framework wider than the submedian sectors, weaker sclerotization of the labial region and stylet, excretory pore of mature female located opposite or anterior to median bulb and J3 and J4 lacking a stylet in the type genus.

Type genus

Meloidogyne Göldi, 1887
= *Hypsoperine* Sledge & Golden, 1962
= *Spartonema* Siddiqi, 1986

Other genera

Meloinema Choi & Geraert, 1974
= *Nacobbodera* Golden & Jensen, 1974
= *Bursadera* Ivanova & Krall, 1985

Note: In this account, only *Meloidogyne* is discussed in detail. Further information on *Meloinema* and *Bursadera* can be obtained from other, more detailed taxonomic works, such as Siddiqi (2000).

11.11.2. Key to Genera of Meloidogyninae Skarbilovich, 1959

(Modified after Siddiqi, 2000)

1. Preadult female and female J3 and J4 saccate; mature female with anus at base of dorsal lip of vulva; male tail hemispherical .... *Meloidogyne*
Preadult female and female J3 and J4 vermiform; mature female with anus at some distance from base of dorsal lip of vulva; male tail conoid-rounded 2

2. Excretory pore located opposite anterior margin of median pharyngeal bulb or more anteriorly; male tail lacking bursa.. *Meloinema*
Excretory pore located opposite posterior margin of median pharyngeal bulb or more posteriorly; male tail with bursa .... *Bursadera*

11.11.3. Genus *Meloidogyne* Göldi, 1887

Diagnosis (after Hunt and Handoo, 2009, itself based on Siddiqi, 2000): Meloidogyninae. Root-gall inciting. Mature female: Round to pear-shaped with a short projecting neck, white, sedentary. No cyst stage. Vulva and anus located close together, terminal; perineum with a finger-print-like cuticular pattern, usually flattened, rarely elevated. Phasmids dot-like, slightly anterior to, and on either side of, anus. Cuticle striated. Stylet slender, generally 12-15 μm long, with small basal knobs. Excretory pore anterior to median bulb, often located just posterior to stylet base. Genital tracts paired, prodelphic, convoluted. Six large rectal glands secreting gelatinous material in which eggs are deposited[1]; eggs not retained in body.
Male: Vermiform, up to 2 mm long, tail end twisted, developing by metamorphosis within a swollen juvenile. Cuticle strongly annulated; lateral field with four incisures. Labial region not sharply offset, with distinct labial disc and few (1-3) annules; lateral sectors wider than submedian sectors, appearing as 'cheeks'. Stylet robust (18-25 μm), with large basal knobs. Pharyngeal glands lie mostly ventral to the intestine. Spicules slender, generally 25-33 μm long, gubernaculum 7-11 μm long. Testis single, but paired when sex reversal occurs. Tail rounded. Phasmids dot-like, located near the cloacal aperture which is subterminal. Bursa absent.
Juveniles: J1 with a blunt tail tip, moulting within egg; second and third moults occurring within J2 cuticle. J2 vermiform, migratory, infective, straight to arcuate habitus upon death. Labial region with coarse annules (1-4), a distinct labial disc, framework lightly sclerotized, lateral sectors wider than submedian sectors, stylet slender, under 20 μm, excretory pore posterior to hemizonid. Median bulb with large oval refractive thickenings. Tail with conspicuous hyaline region, tip narrow, irregular in outline. J3 sedentary, swollen, sausage-shaped with a short blunt tail. Stylet absent. J4 sedentary, swollen, with terminal anus. Stylet absent.

Type species: *Meloidogyne exigua*
Other species: About 100 species are currently regarded as valid, although many lack molecular or isozyme characterization and, being based on notoriously variable morphometric and morphological characters, their status as valid species may be doubtful. Recently, more species have been added to the list due to advanced molecular techniques for identification (Humphreys-Pereira *et al.*, 2014; Ali *et al.* 2015; Tao *et al.*, 2017; Maleita *et al.*, 2018; Yang *et al.* 2021; Singh *et al.*, 2023)

11.12. Major Species

The following account of ten representative species is based on the list and information provided by Hunt and Handoo (2009). The species have been selected on the basis of their common occurrence, economic importance, distribution and potential as emergent pests. A comparison of the perineal patterns is given in Figure 11.9. It must be emphasized that even the common species, which are often extremely variable, can be extremely difficult to identify using conventional techniques, including perineal patterns, and so the novice identifier should proceed with caution. Isozyme and/or molecular data are usually necessary to confirm identifications. It should also be stated that no proposal of a new species in this genus should be published without the support of appropriate molecular and isozyme data.

Meloidogyne arenaria **(Neal, 1889) Chitwood, 1949** (Figs 11.6C, 11.13)
Morphology and major morphometrics (after Hunt and Handoo, 2009)
Female: Pyriform or pear-shaped, posterior terminal protuberance absent. Stylet 13-17 μm long, cone curved dorsally, gradually tapering to a blunt tip; shaft broad, cylindrical, gradually widening posteriorly and bearing offset, rounded to tear drop-shaped basal knobs. Perineal pattern variable, rounded to ovoid, striae fine to coarse, dorsal arch low and flattened with striae smooth or

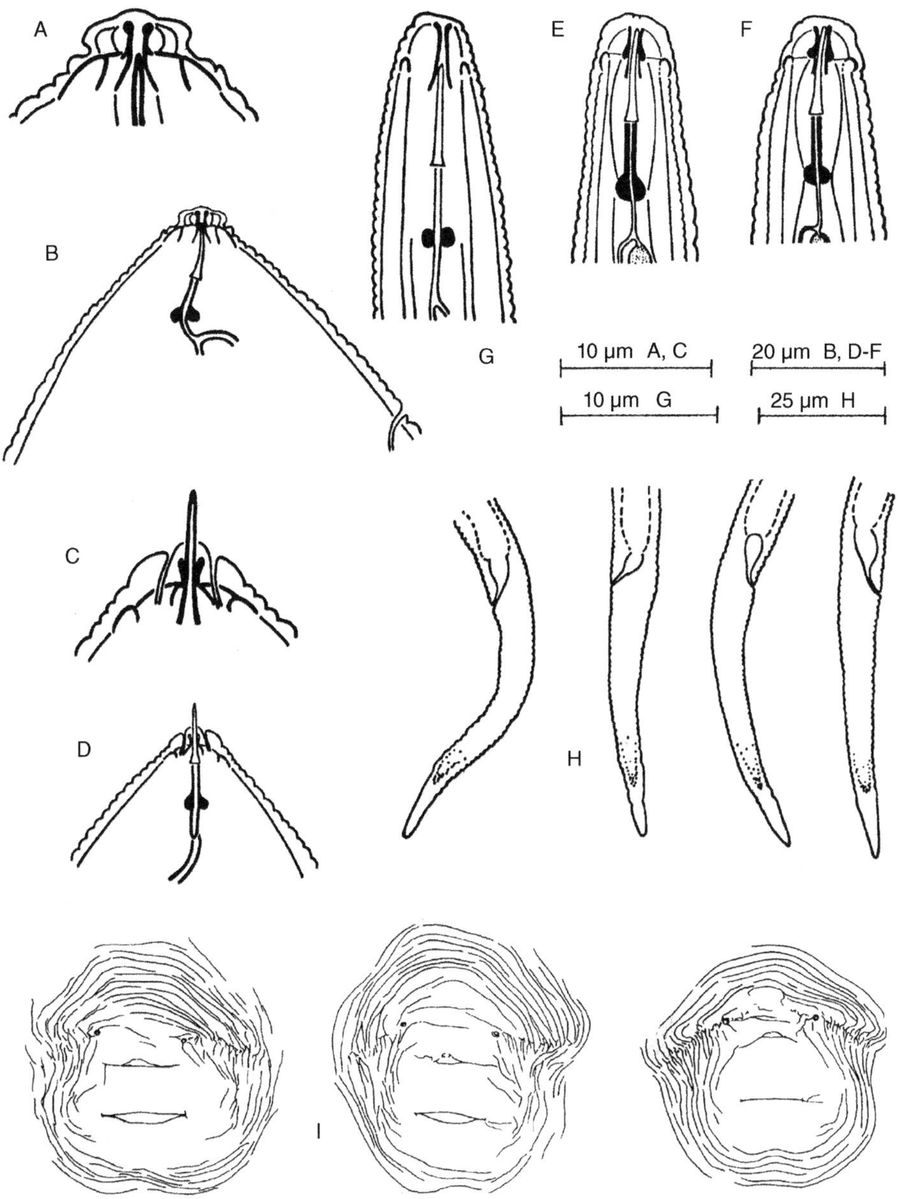

Fig. 11.13. *Meloidogyne arenaria*. A-D: Female anterior region; E, F: male anterior region; G: second-stage juvenile (J2) anterior region; H: J2 tail regions; I: perineal patterns. (A-H, after Whitehead, 1968, courtesy *Transactions of the Zoological Society of London*; I, after Orton Williams, 1975, courtesy CABI.)

slightly wavy, continuous or broken, slightly bent toward tail tip at lateral line and generally forming shoulders on lateral portion of the arch. Dorsal and ventral striae often meet at an angle at lateral lines, lateral fields distinct, slightly irregular.

Male: Labial region not offset, smooth, rarely with one or two incomplete annules, labial disc more or less rounded, slightly raised above the level of medial lips, lateral lips usually absent (remnants may occasionally be present). Stylet

20-28 μm long, basal knobs offset, either angular or more amalgamated, DGO = 4-8 μm.
Infective juvenile (J2): Body length of 392-605 μm, hemizonid located 1-3 annules anterior to excretory pore, tail 44-69 μm long with rounded to pointed tail tip and an indistinct hyaline region 6-13 μm long.
Hosts: Extremely polyphagous, attacking both monocotyledonous and dicotyledonous plants. Such a broad host range often makes this, and similar species, difficult to manage via crop rotation.
Distribution: This species is widely distributed throughout the world, being found in most of the warmer regions. It is also frequently encountered in cooler climates, where it is typically confined to glasshouses.
Isozymes: Populations of *M. arenaria* are variable in isozyme phenotype (Esbenshade and Triantaphyllou, 1985a, b). According to Hunt and Handoo (2009), there are three common phenotypes of esterase activity: A1, A2 and A3, although there are several other less common phenotypes. Phenotypes A1 and A2 include several cytological forms of *M. arenaria*, whereas phenotype A3 includes only the most typical cytological form with a somatic chromosome number of 51-56. Some populations of *M. arenaria* have atypical esterase phenotypes, namely, S1-M1, S2-M1 and M3-F1. The malate dehydrogenase phenotype is either N1 or N3, depending on the population (see Blok and Powers, 2009).
Molecular: Mitochondrial gene regions, such as *COII-16S* rDNA (Pagan *et al.* 2015) and *nad5*, can be good molecular markers for the identification of *M. arenaria* populations (Janssen *et al.*, 2016).
Remarks: One of the four common species reported worldwide (the other three being *M. incognita*, *M. javanica* and *M. hapla*) but part of a highly variable species complex. Typically inciting large, rather irregular, galls on the roots of infected plants.

***Meloidogyne chitwoodi* Golden, O'Bannon, Santo & Finley, 1980** (Fig. 11.14)

Morphology and major morphometrics (after Hunt and Handoo, 2009)
Female: Pyriform or pear-shaped, with slight posterior protuberance. Stylet relatively small, 11-12.5 μm long, basal knobs small, rounded to irregularly shaped, posteriorly sloping. Several vesicle-like structures usually present within median bulb and clustered around the lumen anterior to the valve plates (Golden *et al.*, 1980). Perineal pattern rounded to oval, dorsal arch low and rounded to high and angular; striae near perineal area broken, curved, twisted; lateral lines weak.
Male: Labial region not offset, labial disc elevated, lateral lips present. Stylet 18-19 μm long, basal knobs small, irregularly shaped, posteriorly sloping, offset, DGO = 2.2-3.4 μm.
Infective juvenile (J2): Body length of 336-417 μm, hemizonid located anterior or adjacent to the excretory pore, tail 39-47 μm long, with bluntly rounded tail tip and hyaline region 9-14 μm long.
Hosts: Good hosts include potato and tomato. The rather wide host range extends to plants from several families and includes both crop species and weeds (e.g., barley, carrots, sugar beet and peas; cereals such as maize, wheat; various members of the Poaceae family). Both monocotyledonous and dicotyledonous plants act as hosts.
Distribution: North America: Mexico, USA, where it is widely distributed (California, Colorado, Idaho, Nevada, New Mexico, Oregon, Texas, Utah, Virginia, Washington). South America: Argentina. Europe: Belgium, Germany, the Netherlands, Portugal. Africa: South Africa.
Isozymes: The esterase phenotype is S1 and the malate dehydrogenase phenotype is N1a (see Blok and Powers, 2009; Hunt and Handoo, 2009).
Molecular: Populations of *M. chitwoodi* can be identified using species-specific primers, as described by Zijlstra (2000) and Wishart *et al.* (2002). Anderson and Gleason (2023) also proposed real-time PCR to differentiate *M. chitwoodi* from related *Meloidogyne* species.
Remarks: *M. chitwoodi* is a species of major economic importance to potato. It typically incites relatively small galls, and the surface of infected potato tubers has numerous small, pimple-like, raised areas. It was added to the European list of quarantine organisms in 1998 in an effort to prevent further distribution within Europe.

***Meloidogyne enterolobii* Yang & Eisenback, 1983** (Fig. 11.15)

Morphology and major morphometrics (after Hunt and Handoo, 2009)
Female: Pyriform or pear-shaped, no posterior protuberance. Stylet 14-17 μm long, basal

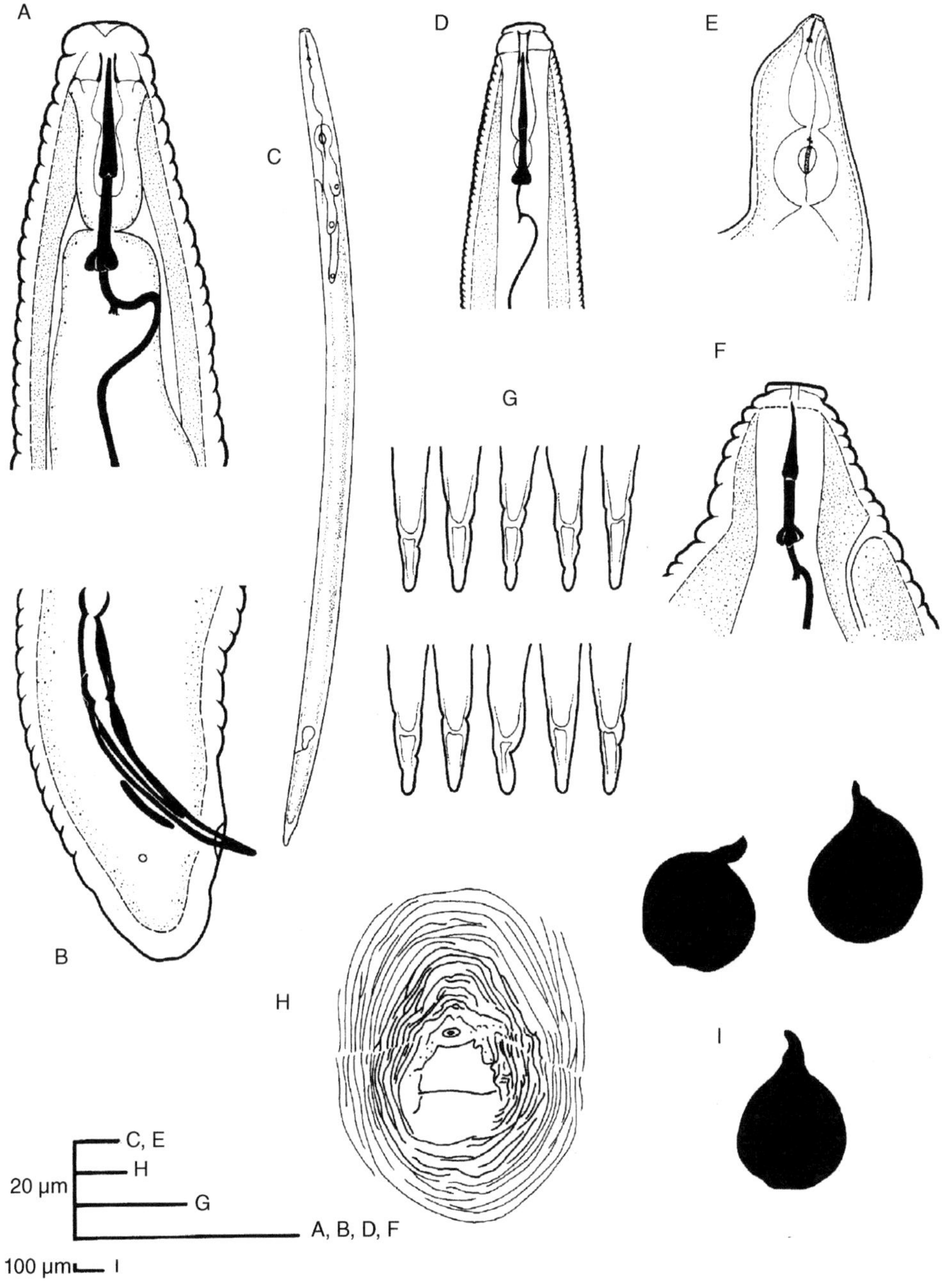

Fig. 11.14. *Meloidogyne chitwoodi*. A: Male anterior region; B: male tail region; C: entire second-stage juvenile (J2); D: J2 anterior region; E, F: female anterior region, lateral view; G: J2 tail regions; H: perineal pattern; I: entire females. (After Jepson, 1985, courtesy CABI.)

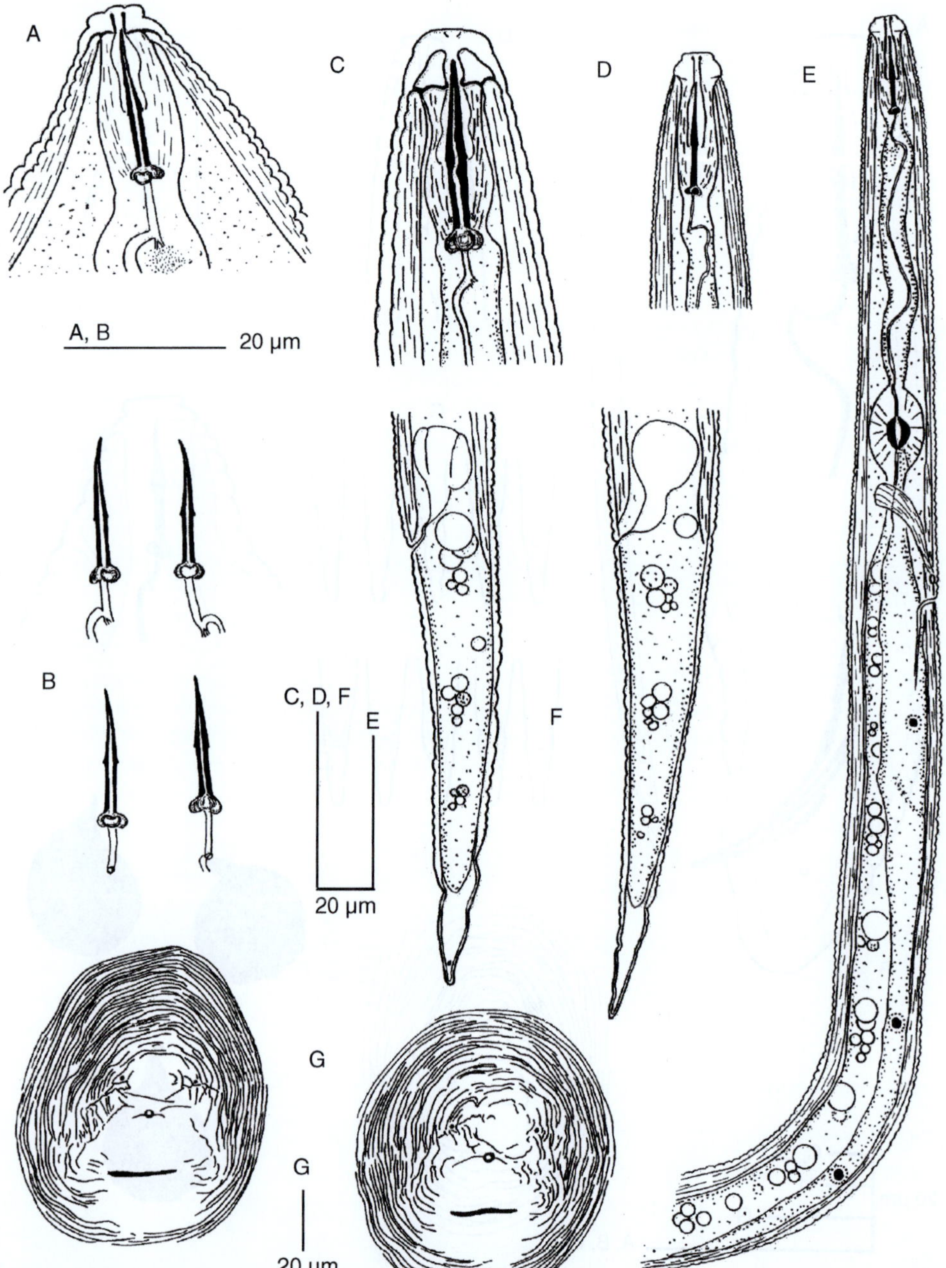

Fig. 11.15. *Meloidogyne enterolobii*. A: Female anterior region; B: female stylets; C: male anterior region; D: second-stage juvenile (J2) anterior region: E: J2 pharyngeal region; F: J2 tail regions; G: perineal patterns. (After Rammah and Hirschmann, 1988, courtesy *Journal of Nematology*.)

knobs kidney-shaped or reniform, indented, offset. Perineal pattern round to ovoid, dorsal arch rounded; striae fine, widely spaced, lateral field absent or with single line occurring at the junction of the dorsal and ventral arches.

Male: Labial region continuous, not offset, labial disc not elevated, lateral lips absent. Stylet 18-25 μm long, basal knobs rounded, sloping posteriorly, offset, DGO = 3-5 μm.

Infective juvenile (J2): Body length of 377-528 μm, hemizonid located two annules anterior to the excretory pore, tail 43-63 μm long, with bluntly pointed tail tip and hyaline region 5-15 μm long.

Hosts: There are many recorded hosts, including American black nightshade, ajuga, angel trumpet, basil, bean, beet, bell pepper, broccoli, celery, coffee, eggplant, glory bush, glory flower, guava, horse bean, parsley, potato, pumpkin, soybean, Spanish needle, sweet potato, tobacco, tomato, watermelon and wild poinsettia. Recent information on the host range for this species is given in Sikandar *et al.* (2023).

Distribution: The species is recorded from Africa (Burkina Faso, Ivory Coast, Malawi, Senegal, South Africa), North America (Florida, USA), South America (Brazil, Venezuela), the Caribbean (Cuba, Guadeloupe, Martinique, Puerto Rico, Trinidad and Tobago), Europe (France, Switzerland) and Asia (China). For detailed information on its distribution, see Philbrick *et al.* (2020) and Sikandar *et al.* (2023).

Isozymes: There are two major bands (VS1-S1 phenotype) of esterase activity and one strong malate dehydrogenase band (N1a) (see Blok and Powers, 2009; Hunt and Handoo, 2009).

Molecular: There are several ribosomal and mitochondrial gene markers that can be used for the identification of *M. enterolobii* populations. Ribosomal gene markers such as 18S, 28S, ITS and IGS can be used (Blok *et al.*, 1997; Brito *et al.* 2004; Kiewnick *et al.*, 2014), and mitochondrial gene markers such as *COI* and *COII* can also be used (Brito *et al.* 2004; Kiewnick *et al.*, 2014).

Remarks: This is a widely distributed emergent pest species with potential to cause great economic damage (see EPPO, 2008; 2011a; Sikandar *et al.*, 2023). It is virulent on tomato cultivars with the *Mi*-1 gene for resistance to other *Meloidogyne* species and on soybean Forrest and sweet potato varieties that are resistant to *M. incognita*. It is damaging to coffee in Cuba and reproduces on tomato with the *Mi* resistance gene. Galls are typically large and irregular. The root-knot nematode, formerly known as *M. mayaguensis*, is now widely accepted as a junior synonym of *M. enterolobii* (see Xu *et al.*, 2004; EPPO, 2008; Karssen *et al.*, 2012).

Meloidogyne ethiopica **Whitehead, 1968** (Figs 11.6F, 11.16)

Morphology and major morphometrics (after Hunt and Handoo, 2009)

Female: Elongate to pyriform or pear-shaped, terminal protuberance absent. Stylet 12-15 μm long, basal knobs rounded, tapering gradually into shaft. Perineal pattern oval to squarish, dorsal arch moderately high to high; striae coarse, widely separated, smooth to wavy, lateral field indistinct.

Male: Labial region not offset, labial disc distinct, lateral lips present. Stylet 23-27 μm long, basal knobs rounded to pear-shaped, offset, DGO = 3-5 μm.

Infective juvenile (J2): Body length of 326-510 μm, hemizonid located anterior to the excretory pore, tail 52-72 μm long with finely rounded to pointed terminus, hyaline region distinct, 12-15 μm long.

Hosts: Reported hosts include grapevine, kiwi, tomato, soybean and sugar cane, and it is known to attack at least 80 different hosts (see EPPO, 2011b, which also has symptomatology on various crops).

Distribution: The species was originally described from East Africa (Tanzania) but is also reported from Southern Africa and South America (Brazil, Chile). The nematode may have been introduced to Rio Grande do Sul, Brazil, in 1989, on kiwi fruit seedlings imported from Chile and then subsequently dispersed within Brazil on *Polymnia sonchifolia* (yacón or Peruvian ground apple). Somewhat curiously, it has also been reported from glasshouses in Slovenia, although it is not believed to occur there at present (see EPPO, 2011b).

Isozymes: The esterase phenotype E3 (Ki3) is species-specific, and this is the most useful character for differentiating *M. ethiopica* from other species of root-knot nematode. The esterase phenotype is E3 (*Rm* = 0.9, 1.05, 1.20), and the malate dehydrogenase is of the N1 type (see Blok and Powers, 2009; Hunt and Handoo, 2009).

Molecular: RAPD profiles were used by Carneiro *et al.* (2004) to identify conspecific populations

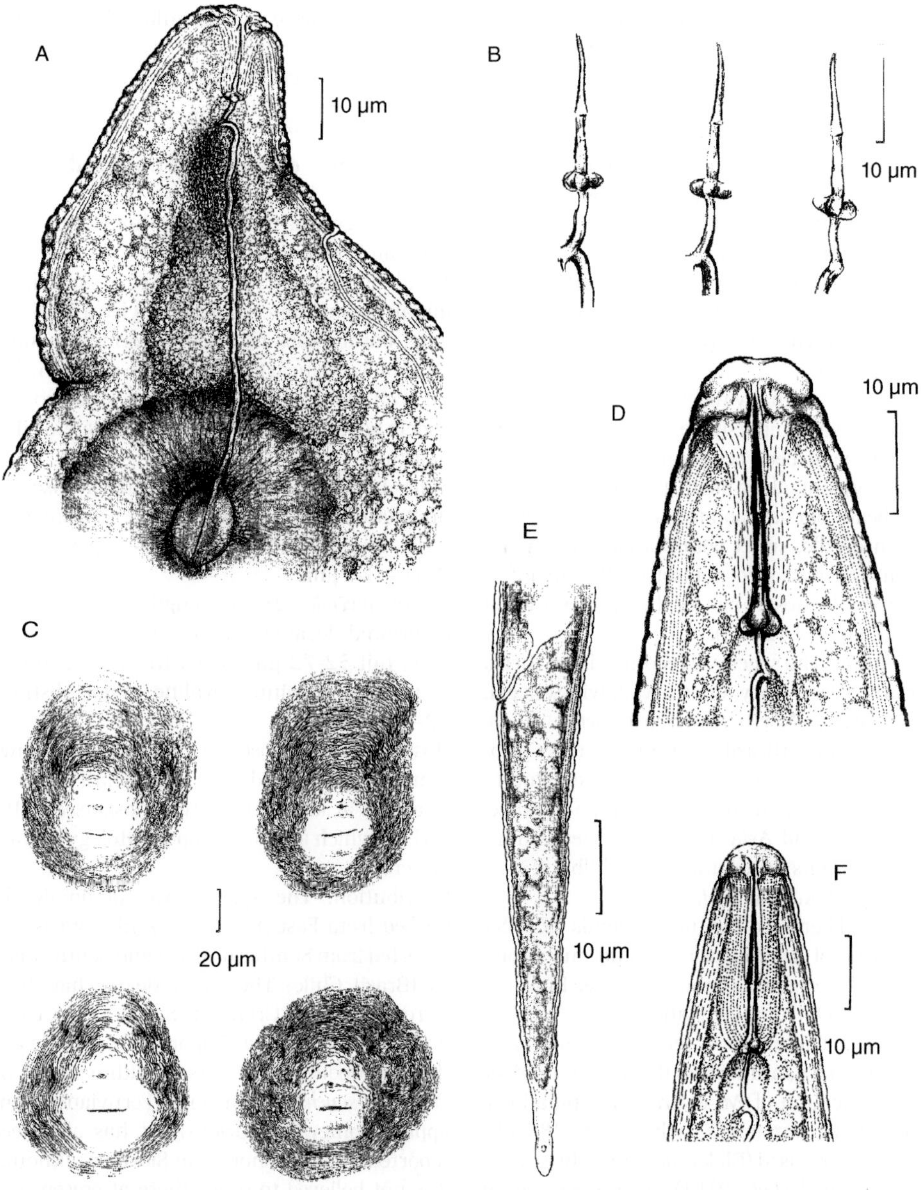

Fig. 11.16. *Meloidogyne ethiopica*. A: Female pharyngeal region, lateral view; B: female stylets; C: perineal patterns; D: male labial region, lateral view; E: second-stage juvenile (J2) tail, lateral view, F: J2 anterior region, lateral view. (Modified after Carneiro *et al*., 2004, courtesy *Nematology*.)

from Brazil, Chile and Kenya. In addition, ribosomal genes such as 18S, 28S and ITS, intergenic *COII-16S* mitochondrial gene can be used for the identification of *M. ethiopica* populations (Subbotin *et al*. 2021).

Remarks: This species has only recently been identified as an economically important and damaging species on grapevine and kiwi vines in Brazil and Chile (Carneiro *et al*., 2004, 2007). The species may have been in South America for at least 100 years and could be the species originally recorded as *Anguillula vialae* from vines in Chile by Lavergne (1901a, b). An intercept at the USDA dating to the mid-20th century also appears to be this species (Handoo, unpublished intercept report, USDA).

***Meloidogyne exigua* Göldi, 1887** (Figs 11.6E, 11.17)

Morphology and major morphometrics (after Hunt and Handoo, 2009)

Female: Pear-shaped, no posterior terminal protuberance. Stylet 12-14 μm long, basal knobs rounded, offset. Perineal pattern rounded/oval, low dorsal arch; striae smooth, widely spaced, coarse, broken and folded in lateral regions, lateral field absent.

Male: Labial region slightly offset, with distinct elevated labial disc and one annule, lateral lips present, trapezoidal or almost triangular. Stylet 18-20 μm long, basal knobs rounded, posteriorly sloping, offset, DGO = 0.0-3.0 μm.

Infective juvenile (J2): Body length of 290-370 μm, excretory pore located opposite posterior end of isthmus, tail 39-50 μm long with a narrowly rounded tip.

Hosts: Mostly attacks coffee, but also parasitizes common weeds growing in coffee plantations. Fairly wide host range, including tomato, rice, sugarcane, banana, citrus, onion and so on. Race 3 attacks rubber tree (but not coffee) in Brazil.

Distribution: Widely distributed in humid-coffee-producing areas of Central and South America. Occurs on coffee in Brazil, Guatemala, Peru, Suriname, Colombia, Venezuela, Costa Rica, El Salvador, Dominican Republic, Martinique. In Asia: India. Also found on other crops in Trinidad, French West Indies, China. In Europe: Greece, Italy.

Isozyme: The VF1 esterase phenotype of *M. exigua* is shared with *M. naasi*, although it can be differentiated from the latter by its N1 malate dehydrogenase phenotype.

Molecular: Ribosomal genes such as 18S and 28S and *COII-16S* mitochondrial sequence can be used for the identification of *M. exigua* populations (Tenente *et al*., 2004; Tigano *et al*., 2005; Humphreys-Pereira *et al*., 2014).

Remarks: Attacks coffee in Central and South America. Outside this region, it is a threat to all coffee-growing areas and is targeted in regulatory programmes. Typically incites severe root galling in heavily attacked coffee and other plants.

***Meloidogyne fallax* Karssen, 1996** (Fig. 11.18)

Morphology and major morphometrics (after Hunt and Handoo, 2009)

Female: Globular to pear-shaped, with slight posterior protuberance. Stylet 14-15 μm long, basal knobs large, rounded to transversally ovoid, offset. One or two large and several smaller vesicle-like structures located along lumen lining. Perineal pattern ovoid to oval shaped, or rectangular; dorsal arch low to moderately high; striae coarse, tail terminus and lateral field indistinct, resulting in a relatively large area without striae.

Male: Labial region slightly offset, labial disc rounded, elevated, fused with medial lips, lateral lips present. Stylet 19-21 μm long, basal knobs offset, large rounded, DGO = 3-6 μm.

Infective juvenile (J2): Body length of 381-435 μm, hemizonid situated at same level as excretory pore, tail 46-56 μm long with a broadly rounded tail tip and 12-16-μm long hyaline region.

Hosts: Attacks both monocotyledonous and dicotyledonous hosts, including potato, tomato, oyster plant and carrot.

Distribution: Australia; Europe (Belgium, France, Germany, the Netherlands); New Zealand.

Isozymes: Characterized by a unique malate dehydrogenase N1b phenotype and the lack of any major esterase band. All populations share this rare malate dehydrogenase phenotype and 'null' esterase phenotype. Prolonged esterase staining (for 60 min) revealed a very weak, three-banded pattern, which has been named F3.

Molecular: Ribosomal genes, such as ITS, 18S and 28S, and mitochondrial gene *COII* were used by many researchers for the identification of *M. chitwoodi* (Holterman *et al*., 2009; Nischwitz *et al*., 2013; Hodgetts *et al*., 2016).

Remarks: Morphologically closely related to *M. chitwoodi* and a pest of potato and cereals in Europe. Typically inciting small galls. Infected potato tubers have blister-like or raised swellings on surface. It was added to the European list of quarantine organisms in an effort to prevent further distribution within Europe.

***Meloidogyne hapla* Chitwood, 1949** (Figs 11.6D, 11.19)

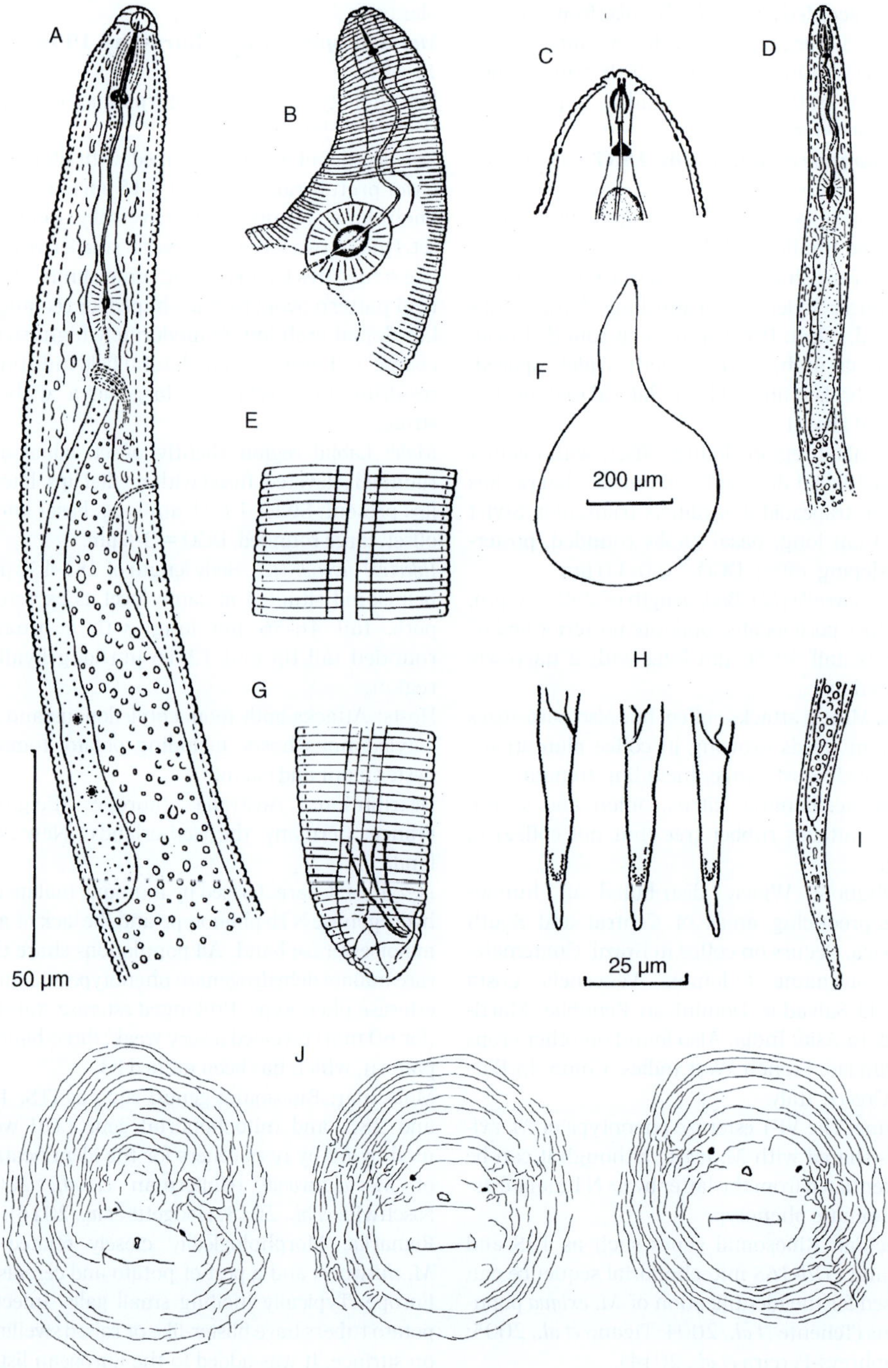

Fig. 11.17. *Meloidogyne exigua*. A: Male pharyngeal region; B: female pharyngeal region; C: female anterior region; D: second-stage juvenile (J2) pharyngeal region; E: male lateral field; F: entire female; G: male tail region; H, I: J2 tail regions; J: perineal patterns. (A, B, D-G, after Lordello and Zamith, 1958, courtesy of Proceedings of the Helminthological Society of Washington; C, H, after Whitehead, 1968, courtesy *Transactions of the Zoological Society of London*; J, courtesy Janet Machon.)

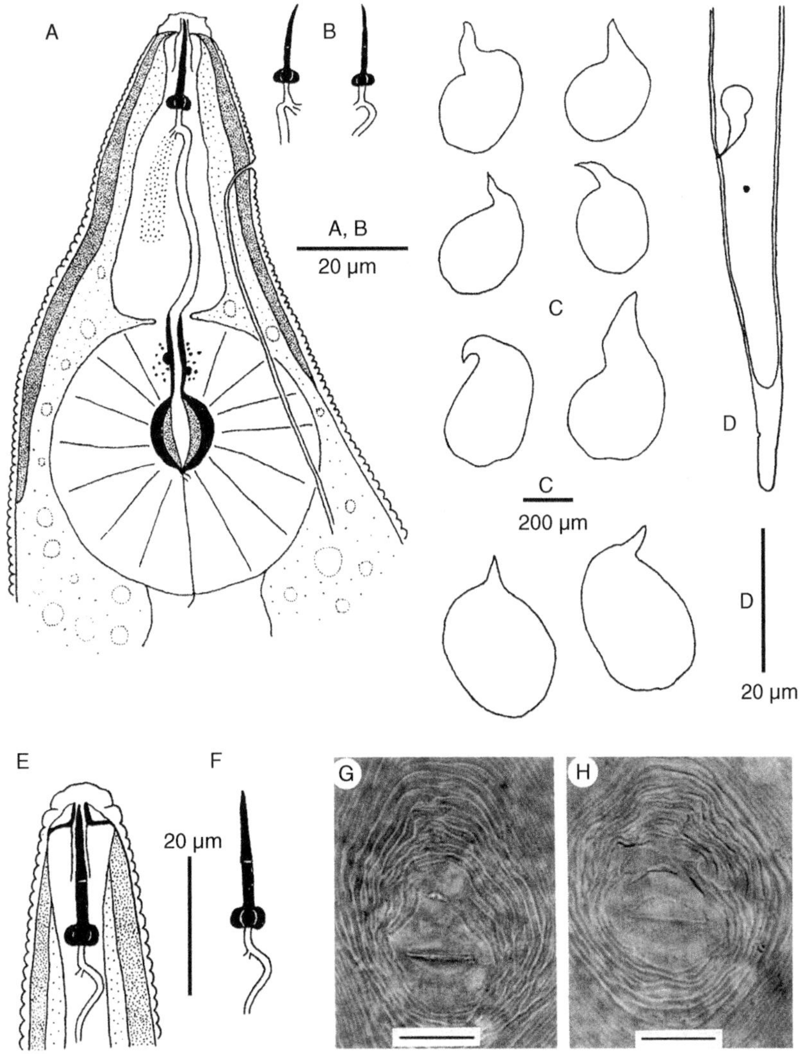

Fig. 11.18. *Meloidogyne fallax*. A: Female pharyngeal region, lateral view; B: female stylets; C: entire females; D: second-stage juvenile tail region, lateral view; E: male anterior region; F: male stylets; G, H: perineal patterns (scale bar = 25 μm). (Modified after Karssen, 1996, courtesy *Fundamental and Applied Nematology*.)

Morphology and major morphometrics (after Hunt and Handoo, 2009)

Female: Pear-shaped, terminal protuberance absent. Stylet 13-17 μm long, basal knobs small, rounded, offset. Perineal pattern rounded, low dorsal arch, characteristic punctations usually present near anus, fine striae, lateral field present.

Male: Labial region offset, labial disc not usually elevated, lateral lips present. Stylet 19-22 μm long, basal knobs small, rounded, offset, DGO = 4-5 μm.

Infective juvenile (J2): Body length of 360-500 μm, hemizonid located anterior to the excretory pore, tail 48-70 μm long with a finely rounded tail tip, hyaline region often irregular in form.

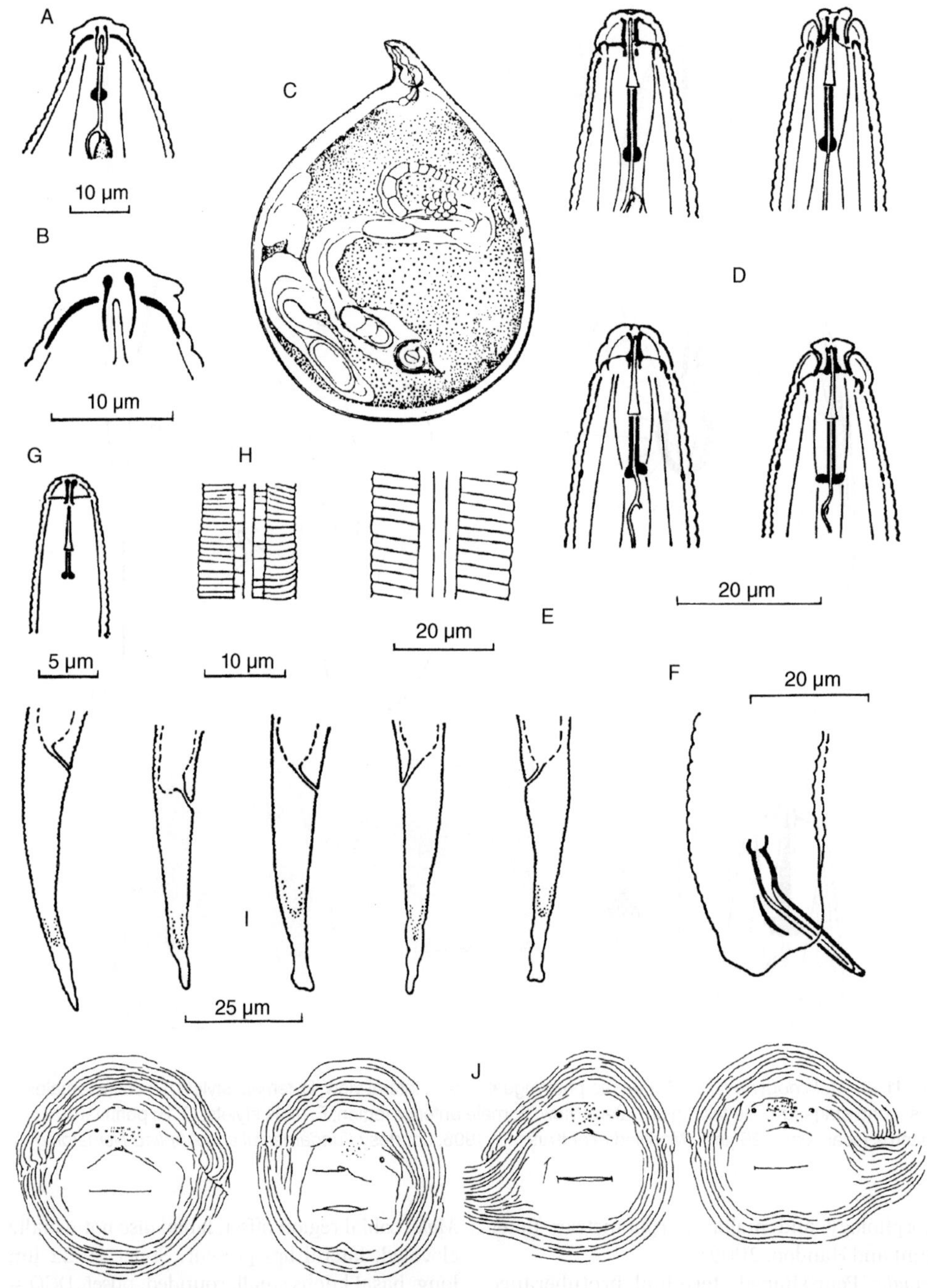

Fig. 11.19. *Meloidogyne hapla*. A, B: Female anterior region; C: entire female; D: male anterior region; E: male lateral field; F: male tail; G: second-stage juvenile (J2) anterior region; H: J2 lateral field at mid-body; I: J2 tail regions; J: perineal patterns. (A-I, after Whitehead, 1968, courtesy *Transactions of the Zoological Society of London*; J, after Orton Williams, 1974, courtesy CABI.)

Hosts: Mainly dicotyledonous plants and extremely polyphagous.
Distribution: Common in temperate areas and at higher, cooler, altitudes in the tropics.
Isozymes: Most populations show the H1 esterase phenotype (Esbenshade and Triantaphyllou, 1985b). One population from France does not have any major esterase activity, and another population from Minnesota, USA, has an A1 esterase phenotype identical to *M. arenaria*. The malate dehydrogenase H1 phenotype is unique for this species.
Molecular: Ribosomal genes, such as 18S, 28S, IGS and ITS, mitochondrial genes, such as *COI*, *COII* and *COII-16S* regions, can be used for the identification *M. hapla* populations (Subbotin *et al.*, 2021).
Remarks: One of the four most common species worldwide. Galls are usually relatively small and may have subsidiary roots emerging from the galled tissue.

Meloidogyne incognita **(Kofoid & White, 1919) Chitwood, 1949** (Figs 11.6A, 11.20)
Morphology and major morphometrics (after Hunt and Handoo, 2009)
Female: Pear-shaped, no posterior terminal protuberance. Stylet 15-16 µm long, basal knobs rounded, offset. Perineal pattern oval to rounded, typically with high, squared, dorsal arch; striae usually wavy, lateral field absent or weakly demarcated by forked striae.
Male: Labial region not offset, labial disc elevated, lateral lips usually absent. Stylet 23-26 µm long, basal knobs offset, rounded to transversely elongate, DGO = 2-4 µm.
Infective juvenile (J2): Body length of 350-450 µm, hemizonid located either anterior or adjacent, to excretory pore, tail 43-65 µm long with a rounded tip and 6-14-µm long hyaline region.
Hosts: Extremely polyphagous, attacking both monocotyledonous and dicotyledonous plants.
Distribution: Worldwide but restricted to protected cultivation in temperate regions.
Isozymes: The unique esterase I1 type and the malate dehydrogenase N1 type were described by Esbenshade and Triantaphyllou (1985a, b). The variability of esterase activity is small, with only one population not having the typical phenotype. The malate dehydrogenase phenotype N1 is like that of *M. javanica*, *M. exigua* and some populations of *M. arenaria*.
Molecular: Species-specific primers by Adam *et al.* (2007) and the mitochondrial gene nad5 (Janssen *et al.*, 2016) can serve as effective molecular markers for the identification of *M. incognita* populations.
Remarks: One of the four most common species worldwide. Typically inciting large, usually irregular, galls. Forms a highly variable and confusing complex.

Meloidogyne javanica **(Treub, 1885) Chitwood, 1949** (Figs 11.6B, 11.21)
Morphology and major morphometrics (after Hunt and Handoo, 2009)
Female: Pear-shaped, no posterior terminal protuberance. Stylet 14-18 µm long, basal knobs ovoid, offset. Perineal pattern rounded, low dorsal arch; striae smooth, tail whorl often distinct, lateral field distinct, clearly demarcated from striae by more or less parallel lines.
Male: Labial region not offset, labial disc not elevated, lateral lips absent. Stylet 19-24 µm long, basal knobs ovoid, offset, DGO = 3-5.5 µm.
Infective juvenile (J2): Body length of 400-560 µm, hemizonid located anterior, or adjacent, to the excretory pore, tail 47-60 µm with a finely rounded tip and 9-18-µm long hyaline region.
Hosts: Extremely polyphagous, attacking both monocotyledonous and dicotyledonous plants.
Distribution: Worldwide but restricted to protected cultivation in temperate regions.
Isozymes: The esterase phenotype is J3, J2 or J2a (Esbenshade and Triantaphyllou, 1985b; Tomaszewski *et al.*, 1994; Castro *et al.*, 2003), and the malate dehydrogenase is of the N1 type (Esbenshade and Triantaphyllou, 1985b). Most populations of *M. javanica* have a malate dehydrogenase phenotype N1, like that of *M. incognita*, *M. exigua* and some populations of *M. arenaria*. One population of *M. javanica* from Bangladesh and one from Korea have the N3 phenotype similar to that of some populations of *M. arenaria*.
Molecular: Several *M. javanica*-specific primers were used by several authors (Zijlstra *et al.*, 2000; Adam *et al.*, 2007; Janssen *et al.*, 2016).
Remarks: One of the four most common species worldwide. Typically inciting large, irregular, galls.

Meloidogyne paranaensis **Carneiro, Carneiro, Abrantes, Santos & Almeida, 1996** (Fig. 11.22)

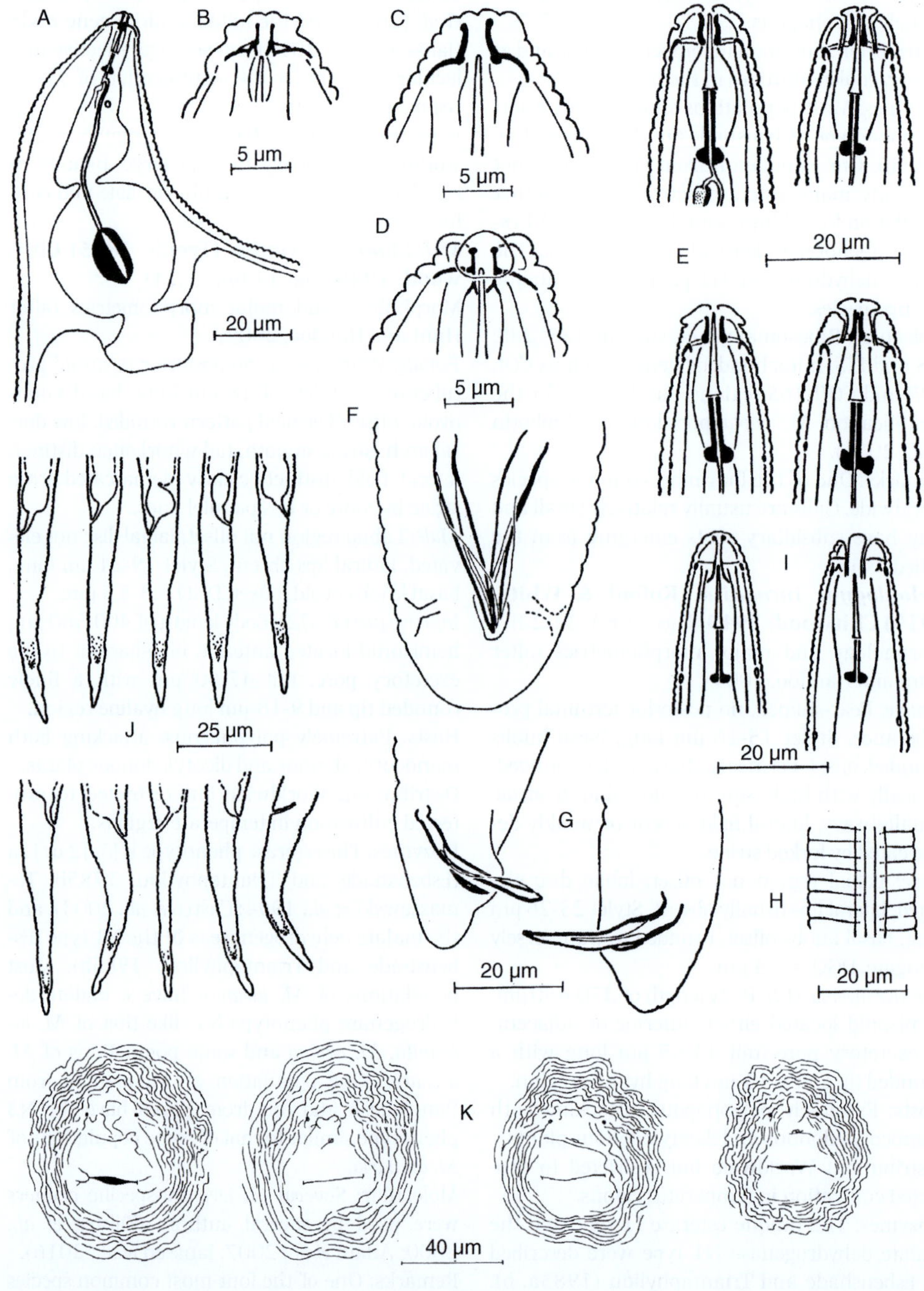

Fig. 11.20. *Meloidogyne incognita*. A: Female pharyngeal region; B-D: female anterior region; E: male anterior region; F-G: male tail region; H: male lateral field; I: second-stage juvenile (J2) anterior region; J: J2 tail regions; K: perineal patterns. (A-J, after Whitehead, 1968, courtesy *Transactions of the Zoological Society of London*; K, after Orton Williams, 1973, courtesy CABI.)

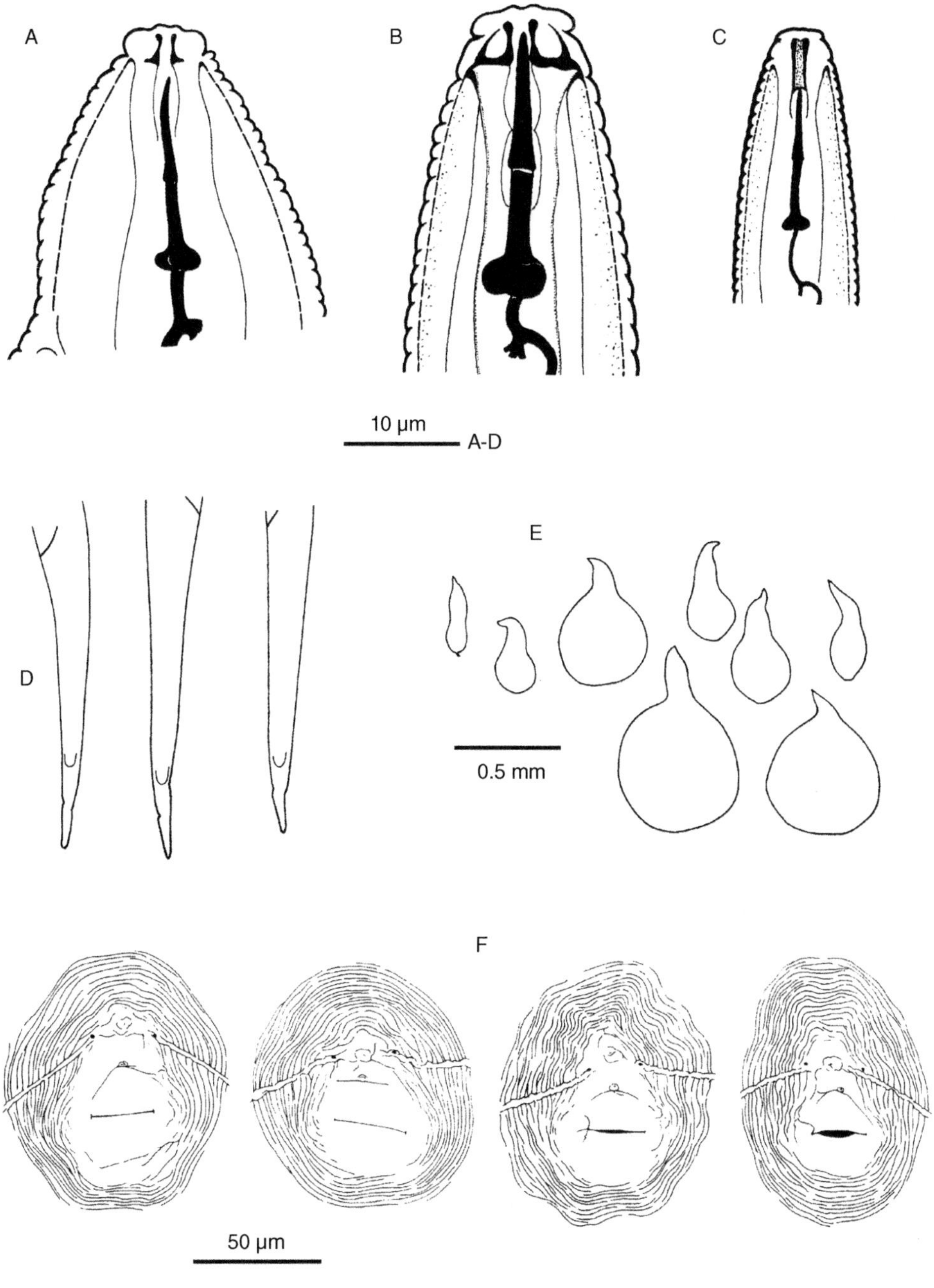

Fig. 11.21. *Meloidogyne javanica*. A: Female pharyngeal region, lateral view; B: male anterior region; C: second-stage juvenile (J2) anterior region; D: J2 tail regions; E: entire females; F: perineal pattern variation. (A-D after Jepson, 1987, and E, F after Orton Williams, (1972, courtesy CABI.)

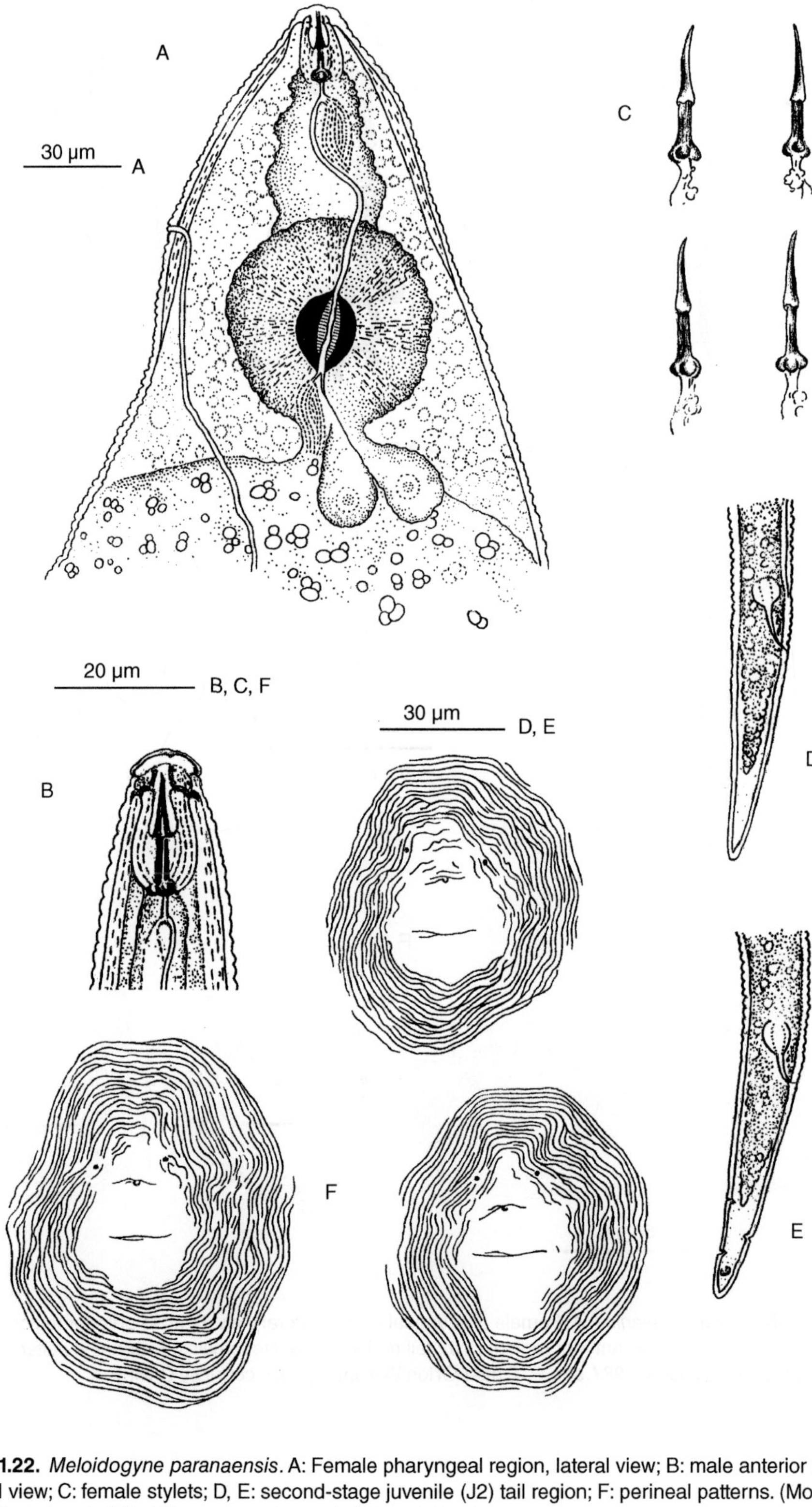

Fig. 11.22. *Meloidogyne paranaensis*. A: Female pharyngeal region, lateral view; B: male anterior region, lateral view; C: female stylets; D, E: second-stage juvenile (J2) tail region; F: perineal patterns. (Modified after Carneiro *et al.* (1996), courtesy *Journal of Nematology*.)

Morphology and major morphometrics (after Hunt and Handoo, 2009)

Female: Ovoid/pear-shaped, no posterior terminal protuberance. Stylet 15-17.5 μm long, basal knobs broad, offset. Perineal pattern rectangular to oval, high dorsal arch; striae fine/coarse, smooth/wavy, lateral field absent.

Male: Labial region not offset, labial disc elevated, lateral lips absent. Stylet 20-27 μm long, basal knobs round to transversally elongate, offset, DGO = 3.5-5.0 μm.

Infective juvenile (J2): Body length of 389-513 μm, hemizonid located anterior to excretory pore, tail 48-51 μm long with rounded tail tip and 9-10-μm long hyaline region.

Hosts: Attacks coffee and soybean. Also reproduces on tomato, tobacco, watermelon, mate (*Ilex paraguariensis*) and species of Solanaceae, Cucurbitaceae and Aquifoliaceae.

Distribution: Brazil, Guatemala.

Isozymes: The esterase phenotype has one fast-migrating band (F1) and the malate dehydrogenase phenotype is N1.

Molecular: *M. paranaensis*-specific primers were developed by Randig *et al.* (2002).

Remarks: Has the potential to adversely affect threatened and endangered plant species. Symptoms include a cracking and splitting of the tap root of coffee, but gall formation is lacking.

11.13. Acknowledgement

Alemayehu Habteweld was supported in part by an appointment to the Research Participation Program at the Mycology and Nematology Genetic Diversity and Biology Laboratory USDA, ARS, Northeast Area, Beltsville, MD, administered by the Oak Ridge Institute for Science and Education through an interagency agreement between the U.S. Department of Energy and USDA-ARS. Mention of trade names or commercial products in this publication is solely for the purpose of providing specific information and does not imply recommendation or endorsement by the U.S. Department of Agriculture. USDA is an equal opportunity provider and employer.

Note

[1] Rectal glands not developed and no egg-sac formed in *M. spartinae* and *M. kikuyensis*.

11.14. References

Abad, P., Castagnone-Sereno, P., Rosso, M.-N., de Almeida Engler, J. and Favery, B. (2009) Invasion, feeding and development. In: Perry, R.N., Moens, M. and Starr, J.L. (eds), *Root-knot Nematodes*. CAB International. Wallingford, UK, pp. 163–181.

Abd-Elgawad, M.M.M. and Kabeil, S.S.A. (2010) Management of root-knot nematode, *Meloidogyne incognita* on tomato in Egypt. *Journal of American Science* 6, 256–162.

Adam, M.A.M., Phillips, M.S. and Blok, V.C. (2007) Molecular diagnostic key for identification of single juveniles of seven common and economically important species of root-knot nematode (*Meloidogyne* spp.). *Plant Pathology* 56, 190–197. DOI: 10.1111/j.1365-3059.2006.01455.x

Adams, B.J., Dillman, A.R. and Finlinson, C. (2009) Molecular taxonomy and phylogeny. In: Perry, R.N., Moens, M. and Starr, J.L. (eds), *Root-knot Nematodes*. CAB International, Wallingford, UK, pp. 119–138.

Ali, N., Tavoillot, J., Mateille, T., Chapuis, E., Besnard, G., *et al.* (2015) A new root-knot nematode *Meloidogyne spartelensis* n. sp. (Nematoda: Meloidogynidae) in Northern Morocco. *European Journal of Plant Pathology* 143, 25–42. DOI: 10.1007/s10658-015-0662-3

Álvarez-Ortega, S., Brito, J.A. and Subbotin, S.A. (2019) Multigene phylogeny of root-knot nematodes and molecular characterization of *Meloidogyne nataliei* Golden, Rose & Bird, 1981 (Nematoda: Tylenchida). *Scientific Reports* 9, 11788. DOI: 10.1038/s41598-019-48195-0

Anderson, S.D. and Gleason, C.A. (2023) A molecular beacon real-time polymerase chain reaction assay for the identification of *M. chitwoodi*, *M. fallax*, and *M. minor. Front. Plant Science* 14, 1096239. DOI: 10.3389/fpls.2023.1096239.

Atkins, D.S., Hidalgo-Diaz, L., Kalisz, H., Mauchline, T.H., Hirsch, P.R., *et al.* (2003) Development of a new management strategy for the control of root-knot nematodes (*Meloidogyne* spp.) in organic vegetable production. *Pest Management Science* 59, 183–189.

Barker, K.R., Carter, C.C. and Sasser, J.N. (eds) (1985) *An Advanced Treatise on* Meloidogyne. *Volume II: Methodology*. Department of Plant Pathology and the United States Agency for International Development, Raleigh, North Carolina.

Berkeley, M.J. (1855) Vibrio forming cysts on the roots of cucumbers. *Gardener's Chronicle and Agricultural Gazette* 14, 220.

Bird, A.F. (1959) Development of the root-knot nematodes: *Meloidogyne javanica* (Treub) and *Meloidogyne hapla* Chitwood in tomato. *Nematologica* 4, 31–42.

Bird, A.F. (1968) Changes associated with parasitism in nematodes. *Journal of Nematology* 54, 475–489.

Bird, A.F. (1971) *The Structure of Nematodes*. Academic Press, New York, New York.

Bird, A.F. (1974) Plant response to root-knot nematodes. *Annual Review of Phytopathology* 16, 69–85.

Bird, A.F. (1986) The influence of the actinomycete, *Pasteuria penetrans*, on the host-parasite relationship of the plant-parasitic nematode, *Meloidogyne javanica*. *Parasitology* 93, 571–580.

Bird, A.F., Downton, W.S.J. and Hawker, J.S. (1975) Cellulase secretions by second stage larvae of the root-knot nematode (*Meloidogyne javanica*). *Marcellia* 38, 165–169.

Blok, V.C. and Powers, T.C. (2009) Biochemical and molecular identification. In: Perry, R.N., Moens, M. and Starr, J.L. (eds), *Root-knot Nematodes*. CAB International, Wallingford, UK, pp. 98–118.

Blok, V.C., Phillips, M. S. and Fargette, M. (1997) Comparison of sequence differences in the intergenic region of the ribosomal cistron of *Meloidogyne mayaguensis* and the major tropical root-knot nematodes. *Journal of Nematology* 29, 16–22.

Bridge, J. (1996) Nematode management in sustainable and subsistence agriculture. *Annual Review of Phytopathology* 34, 201–225.

Brito, J.A., Powers, T.O., Mullin, P.G., Inserra, R.N. and Dickson, D.W. (2004) Morphological and molecular characterization of *Meloidogyne mayaguensis* isolates from Florida. *Journal of Nematology* 36, 232–240.

Byrd Jr, D.W., Kirkpatrick, T. and Barker, K.R. (1983) An improved technique for clearing and staining plant tissues for detection of nematodes. *Journal of Nematology* 15, 142–143.

Cain, D.W., McKenry, M.V. and Tarailo, R.E. (1984) A new pathotype of root-knot nematode on grape rootstalks. *Journal of Nematology* 16, 207–208.

Canto-Saenz, M. (1985) The nature of resistance to *Meloidogyne incognita* (Kofoid & White, 1919) Chitwood, 1949. In: Barker, K.R., Carter, C.C. and Sasser, J.N. (eds), *An Advanced Treatise on* Meloidogyne. *Volume II: Methodology*. Department of Plant Pathology and the United States Agency for International Development, Raleigh, North Carolina, pp. 225–231.

Carneiro, R.M.D.G., Carneiro, R.G., Abrantes, I.M.O., Santos, M.S.N.A. and Almeida, M.R.A. (1996) *Meloidogyne paranaensis* n. sp. (Nemata: Meloidogynidae), a root-knot nematode parasitizing coffee in Brazil. *Journal of Nematology* 28, 177–189.

Carneiro, R.M.D.G., Randig, O., Almeida, M.R.A. and Gomes, A.C.M.M. (2004) Additional information on *Meloidogyne ethiopica* Whitehead, 1968 (Tylenchida: Meloidogynidae), a root-knot nematode parasitising kiwi fruit and grapevine from Brazil and Chile. *Nematology* 6, 109–123.

Carneiro, R.M.D.G., Almeida, M.R.A., Cofcewicz, E.T., Magunacelaya, J.C. and Aballay. E. (2007) *Meloidogyne ethiopica*, a major root-knot nematode parasitising *Vitis vinifera* and other crops in Chile. *Nematology* 9, 635–641.

Carta, L.K., Handoo, Z.A., Powers, T.O., Miller, S.A., Pérez-Zubiri, R., *et al.* (2006) Guidelines for isolation and identification of regulated nematodes of potato. *Revista Mexicana de Fitopatología* 23(2005), 211–222.

Carter, C.C. and Sasser, J.N. (1982) *Research on the Integrated Crop Protection System with Emphasis on the Root-knot Nematodes (*Meloidogyne *spp.) Affecting Economic Food Crops in Developing Nations*. International Meloidogyne Project (IMP) Contract No. AID/ta-c-1234. Department of Plant Pathology and the United States Agency for International Development, North Carolina State University, Raleigh, North Carolina.

Castillo, P., Di Vito, M., Vovlas, N. and Jimenez-Diaz, R.M. (2001) Host parasite relationships in root-knot disease of white mulberry. *Plant Disease* 85, 277–281.

Castro, J.M.C., Lima, R.D. and Carneiro, R.M.D.G. (2003) Variabilidae isoenzimática de populacões de *Meloidogyne* spp. provenientes de regiões brasileras produtoras se soja. *Nematologia Brasileira* 27, 1–13.

Charchar, J.M. and Eisenback, J.D. (2000) An improved technique to prepare perineal patterns of root-knot nematodes for SEM. *Nematologia Brasileira* 24, 245–247.

Chitwood, B.G. (1949) Root-knot nematodes - part 1. A revision of the genus *Meloidogyne* Göldi, 1887. *Proceedings of the Helminthological Society of Washington* 16, 90–104.

Cook, R. and Starr, J.L. (2006) Resistant cultivars. In: Perry, R.N. and Moens, M. (eds), *Plant Nematology*. CAB International, Wallingford, UK, pp. 370–391.

Coyne, D.L., Fourie, H.H. and Moens, M. (2009) Current and future management strategies in resource-poor farming. In: Perry, R.N., Moens, M. and Starr, J.L. (eds), *Root-knot Nematodes*. CAB International, Wallingford, UK, pp. 444–475.

Di Vito, M., Vovlas, N., Lamberti, F., Zaccheo, G. and Catalano, F. (1996) Pathogenicity of *Meloidogyne javanica* on Asian and African rice. *Nematologia Mediterranea* 25, 95–99.

Di Vito, M., Vovlas, N. and Castillo, P. (2004) Host-parasite relationship of *Meloidogyne incognita* on spinach. *Plant Pathology* 53, 508–514.

Dropkin, V.H. (1972) Pathology of *Meloidogyne*-galling, giant cell formation effects on host physiology. *EPPO Bulletin* 6, 23–32.

Dropkin, V.H. (1988) The concept of race in phytonematology. *Annual Review of Phytopathology* 26, 145–161.

Eisenback, J.D. (1985a) Detailed morphology and anatomy of second-stage juveniles, males, and females of the genus *Meloidogyne* (root-knot nematodes). In: Sasser, J.N. and Carter, C.C. (eds), *An Advanced Treatise on* Meloidogyne. *Volume I: Biology and Control*. Department of Plant Pathology and the United States Agency for International Development, Raleigh, North Carolina, pp. 47–77.

Eisenback, J.D. (1985b) Diagnostic characters useful in the identification of the four most common species of root-knot nematodes (*Meloidogyne* spp.). In: Sasser, J.N. and Carter, C.C. (eds), *An Advanced Treatise on* Meloidogyne. *Volume I: Biology and Control*. Department of Plant Pathology and the United States Agency for International Development, Raleigh, North Carolina, pp. 95–112.

Eisenback, J.D. (ed) (1997) *Root-knot Nematode Taxonomic Database*. CD-ROM. CAB International, Wallingford, UK.

Eisenback, J.D. and Hunt, D.J. (2009) General morphology. In: Perry, R.N., Moens, M. and Starr, J.L. (eds), *Root-knot Nematodes*. CAB International, Wallingford, UK, pp. 18–54.

Eisenback, J.D. and Triantaphyllou, H.H. (1991) Root-knot nematode: *Meloidogyne* spp. and races. In: Nickle, W.R. (ed), *Manual of Agricultural Nematology*. Marcel Dekker, New York, New York, pp. 191–274.

Eisenback, J.D., Hirschmann, H., Sasser, J.N. and Triantaphyllou, A.C. (1981) *A Guide to the Four most Common Species of Root-knot Nematodes (*Meloidogyne *Species) with a Pictorial Key*. Departments of Plant Pathology and Genetics, North Carolina State University and the United States Agency for International Development, Raleigh, North Carolina.

Endo, B.Y. (1987) Histopathology and ultrastructure of crops invaded by certain sedentary endoparasitic nematodes. In: Veech, J.A. and Dickson, D.W. (eds), *Vistas on Nematology*. Society of Nematologists, Hyattsville, Maryland, pp. 197–210.

EPPO. (2008) *Meloidogyne enterolobii*. http://www.eppo.org/QUARANTINE/Alert_List/nematodes/meloidogyne_enterolobii.htm (accessed March 2012).

EPPO (2011a) *Meloidogyne enterolobii*. *EPPO Bulletin* 41, 329–339. DOI: 10.1111/j.1365-2338.2011.02497.x.

EPPO (2011b) *Meloidogyne ethiopica*. http://www.eppo.org/QUARANTINE/Alert_List/nematodes/Meloidogyne_ethiopica.htm (accessed March 2012).

Esbenshade, P.R. and Triantaphyllou, A.C. (1985a) Use of enzyme phenotypes for identification of *Meloidogyne* species (Nematoda: Tylenchida). *Journal of Nematology* 17, 6–20.

Esbenshade, P.R. and Triantaphyllou, A.C. (1985b) Identification of major *Meloidogyne* species employing enzyme phenotypes as differentiating characters. In: Sasser, J.N. and Carter, C.C. (eds), *An Advanced Treatise on* Meloidogyne. *Volume I: Biology and Control*. Department of Plant Pathology and the United States Agency for International Development, Raleigh, North Carolina, pp. 135–140.

Esser, R.P., Perry, V.G. and Taylor, A.L. (1976) A diagnostic compendium of the genus *Meloidogyne* (Nematoda: Heteroderidae). *Proceedings of the Helminthological Society of Washington* 43, 138–150.

Fassuliotis, G. (1985) The role of nematologists in the development of resistant cultivars. In: Barker, K.R., Carter, C.C. and Sasser, J.N. (eds), *An Advanced Treatise on* Meloidogyne. *Volume II: Methodology*. Department of Plant Pathology and the United States Agency for International Development, Raleigh, North Carolina, pp. 233–240.

Ferraz, L.C.C.B. and Brown, D.J.F. (2002) *An Introduction to Nematodes: Plant Nematology: A Student's Textbook*. Pensoft, Sofia, Bulgaria.

Franklin, M.T. (1957) Review of the genus *Meloidogyne*. *Nematologica* 2(Supplement), 387–397.
Franklin, M.T. (1965) *Meloidogyne* - root-knot eelworm. In: Southey, J.F. (ed), *Plant Nematology*, 2nd edn. Technical Bulletin of the Ministry of Agriculture, Fisheries and Food, No. 7. Her Majesty's Stationery Office, London, UK, pp. 59–88.
Franklin, M.T. (1971) Taxonomy of Heteroderidae. In: Zuckerman, B.M., Mai, W.F. and Rohde, R.A. (eds), *Plant Parasitic Nematodes. Volume I. Morphology, Anatomy, Taxonomy and Ecology*. Academic Press, New York, New York, pp. 139–162.
Franklin, M.T. (1976) *Meloidogyne*. In: Southey, J.F. (ed), *Plant Nematology*. Her Majesty's Stationery Office, London, UK, pp. 98–124.
Franklin, M.T. (1979) Taxonomy of the genus *Meloidogyne*. In: Lamberti, F. and Taylor, C.E. (eds), *Root-knot Nematodes (Meloidogyne Species); Systematics, Biology and Control*. Academic Press, London and New York, pp. 37–54.
Gaur, H.S. and Perry, R.N. (1991) The use of soil solarization for control of plant-parasitic nematodes. *Nematological Abstracts* 60, 153–167.
Gheysen, G. and Jones, J.T. (2006) Molecular aspects of plant-nematode interactions. In: Perry, R.N. and Moens, M. (eds), *Plant Nematology*. CAB International, Wallingford, UK, pp. 234–254.
Golden, A.M. (1990) Preparation and mounting nematodes for microscopic observation. In: Zuckerman, B.M., Mai, W.F. and Krusberg, L.R. (eds), *Plant Nematology Laboratory Manual*. University of Massachusetts Agricultural Experiment Station, Amherst, Massachusetts, pp. 197–205.
Golden, A.M. and Birchfield, W. (1965) *Meloidogyne graminicola* (Heteroderidae) a new species of root-knot nematode from grass. *Proceedings of the Helminthological Society of Washington* 32, 228–231.
Golden, A.M. and Birchfield, W. (1972) *Heterodera graminophila* n. sp. (Nematoda: Heteroderidae) from grass with key to closely related species. *Journal of Nematology* 4, 147–154.
Golden, A.M., O'Bannon, J.H., Santo, G.S. and Finley, A.M. (1980) Description and SEM observations of *Meloidogyne chitwoodi* n. sp. (Meloidogynidae), a root-knot nematode on potato in the Pacific northwest. *Journal of Nematology* 12, 319–327.
Göldi, E.A. (1887) *Relatório sôbre a moléstia do cafeeiro na provincia do Rio de Janeiro*. Extrahido do VIII Vol. dos *Archivos do Museu Nacional*, Imprensa Nacional, Rio de Janeiro, pp. 1–121.
de Guiran, G. and Ritter, M. (1979) Comparison of the head patterns in Tylenchoidea (Nematoda). *Nematologica* 43, 283–294.
Hartman, K.M. and Sasser, J.N. (1985) Identification of *Meloidogyne* species on the basis of differential host test and perineal pattern morphology. In: Barker, K.R., Carter, C.C. and Sasser, J.N. (eds), *An Advanced Treatise on* Meloidogyne. *Volume II: Methodology*. Department of Plant Pathology and the United States Agency for International Development, Raleigh, North Carolina, pp. 69–77.
Heald, C.M. (1986) Classical nematode management practices. In: Veech, J.A. and Dickson, D.W. (eds), *Vistas on Nematology*. Society of Nematologists, Hyattsville, Maryland, pp. 100–104.
Hirschmann, H. (1985a) The classification of the family *Meloidogyne*. In: Sasser, J.N. and Carter, C.C. (eds), *An Advanced Treatise on* Meloidogyne. *Volume I: Biology and Control*. Department of Plant Pathology and the United States Agency for International Development, Raleigh, North Carolina, pp. 35–45.
Hirschmann, H. (1985b) The genus *Meloidogyne* and morphological characters differentiating its species. In: Sasser, J.N. and Carter, C.C. (eds), *An Advanced Treatise on* Meloidogyne. *Volume I: Biology and Control*. Department of Plant Pathology and the United States Agency for International Development, Raleigh, North Carolina, pp. 79–93.
Hodgetts, J., Ostojá-Starzewski, J.C., Prior, T., Lawson, R., Hall, J., *et al.* (2016) DNA barcoding for biosecurity: case studies from the UK plant protection program. *Genome* 59(11), 1033–1048. DOI: 10.1139/gen-2016-0010.
Holterman, M., Karssen, G., van den Elsen, S., van Megen, H., Bakker, J., *et al.* (2009) Small subunit rDNA-based phylogeny of the Tylenchida sheds light on relationships among some high-impact plant-parasitic nematodes and the evolution of plant feeding. *Phytopathology* 99, 227–235.
Hooper, D.J. (1970) Handling, fixing, staining and mounting nematodes. In: Southey, J.F. (ed), *Laboratory Methods for Work with Plant and Soil Nematodes*, 5th edn. Her Majesty's Stationery Office, London, UK, pp. 39–54.
Hooper, D.J. (1986) Preserving and staining nematodes in plant tissues. In: Southey, J.F. (ed), *Laboratory Methods for Work with Plant and Soil Nematodes*, 6th edn. Reference book 402. Her Majesty's Stationery Office, London, UK, pp. 81–85.
Hooper, D.J. (1990) Extraction and processing of plant and soil nematodes. In: Luc, M., Sikora, R.A. and Bridge, J. (eds), *Plant Parasitic Nematodes in Subtropical and Tropical Agriculture*. CAB International, Wallingford, UK, pp. 45–68.

Humphreys-Pereira, D.A., Flores-Chaves, L., Gómez, M., Salazar, L., Gómez-Alpízar, L., *et al.* (2014) *Meloidogyne lopezi* n. sp. (Nematoda: Meloidogynidae), a new root-knot nematode associated with coffee (*Coffea arabica* L.) in Costa Rica, its diagnosis and phylogenetic relationship with other coffee-parasitising *Meloidogyne* species. *Nematology* 16, 643–661.

Hunt, D.J. and Handoo, Z.A. (2009) Taxonomy, identification, and principal species. In: Perry, R.N., Moens, M. and Starr, J.L. (eds), *Root-knot Nematodes*. CAB International, Wallingford, UK, pp. 55–97.

Hussey, R.S. (1985) Host-parasite relationship and associated physiological changes. In: Sasser, J.N. and Carter, C.C. (eds), *An Advanced Treatise on* Meloidogyne. *Volume I: Biology and Control*. North Carolina State University, Raleigh, North Carolina, pp. 143–153.

Hussey, R.S. (1987) Secretions of esophageal glands of Tylenchida nematodes. In: Veech, J.A. and Dickson, D.W. (eds), *Vistas on Nematology*. Society of Nematologists, Hyattsville, Maryland, pp. 221–228.

Hussey, R.S. (1990) Staining nematodes in plant tissues. In: Zuckerman, B.M., Mai, W.F. and Krusberg, L.R. (eds), *Plant Nematology Laboratory Manual*. University of Massachusetts Agricultural Experiment Station, Amherst, Massachusetts, pp. 190–193.

Hussey, R.S. and Grundler, F.M.W. (1998) Nematode parasitism of plants. In: Perry, R.N and Wright, D.J. (eds), *The Physiology and Biochemistry of Free-living and Plant Parasitic Nematodes*. CAB International, Wallingford, UK, pp. 213–243.

IFAS (Institute of Food, Agriculture and Science) (1989) *Florida Nematode Control Guide*, Entomology Nematology Department Cooperative Extension Service, University of Florida, Gainesville, Florida, pp. 1–38.

Imbriani, J.L. (1985) The practical application of a nematode advisory service. In: Barker, K.R., Carter, C.C. and Sasser, J.N. (eds), *An Advanced Treatise on* Meloidogyne. *Volume II: Methodology*. Department of Plant Pathology and the United States Agency for International Development, Raleigh, North Carolina, pp. 309–323.

Janssen, T., Karssen, G., Topalović, O., Coyne, D. and Bert, W. (2017) Integrative taxonomy of root-knot nematodes reveals multiple independent origins of mitotic parthenogenesis. *PLoS One* 12(3), e0172190. DOI: 10.1371/journal.pone.0172190.

Janssen, T., Karssen, G., Verhaeven, M., Coyne, D. and Bert, W. (2016) Mitochondrial coding genome analysis of tropical root-knot nematodes (*Meloidogyne*) supports haplotype-based diagnostics and reveals evidence of recent reticulate evolution. *Scientific Reports* 6, 22591. DOI: 10.1038/srep22591.

Jatala, P. (1985) Biological control of nematodes. In: Barker, K.R., Carter, C.C. and Sasser, J.N. (eds), *An Advanced Treatise on* Meloidogyne. *Volume II: Methodology*. Department of Plant Pathology and the United States Agency for International Development, Raleigh, North Carolina, pp. 303–308.

Jepson, S.B. (1983a) Identification of *Meloidogyne*: a general assessment and a comparison of male morphology using light microscopy. *Revue de Nématologie* 6, 291–309.

Jepson, S.B. (1983b) The use of second-stage juvenile tails as an aid in the identification of *Meloidogyne* species. *Nematologica* 29, 11–28.

Jepson, S.B. (1983c) *Meloidogyne kralli* n. sp. (Nematode: Meloidogynidae) a root-knot nematode parasitizing sedge (*Carex acuta* L.). *Revue de Nématologie* 6, 239–245.

Jepson, S.B. (1985) *Meloidogyne chitwoodi. CIH Descriptions of Plant-parasitic Nematodes*, Set 8, No. 106. Commonwealth Agricultural Bureaux, Farnham Royal, UK.

Jepson, S.B. (1987) *Identification of Root-knot Nematodes (*Meloidogyne *Species)*. CAB International, Wallingford, UK.

Johnson, A.W. (1985a) The role of nematicides in nematode management. In: Barker, K.R., Carter, C.C. and Sasser, J.N. (eds), *An Advanced Treatise on* Meloidogyne. *Volume II: Methodology*. Department of Plant Pathology and the United States Agency for International Development, Raleigh, North Carolina, pp. 249–267.

Johnson, A.W. (1985b) Specific crop rotation effects combined with cultural practices and nematicides. In: Barker, K.R., Carter, C.C. and Sasser, J.N. (eds), *An Advanced Treatise on* Meloidogyne. *Volume II: Methodology*. Department of Plant Pathology and the United States Agency for International Development, Raleigh, North Carolina, pp. 283–301.

Johnson, A.W. and Fassuliotis, G. (1984) Nematode parasites of vegetable crops. In: Nickle, W.R. (ed), *Plant and Insect Nematodes*. Marcel Dekker, New York, New York, pp. 323–372.

Karssen, G. (1996) Description of *Meloidogyne fallax* n. sp. (Nematoda: Heteroderidae), a root-knot nematode from the Netherlands. *Fundamental and Applied Nematology* 19, 593–599.

Karssen, G. (2002) *The Plant-parasitic Nematode Genus* Meloidogyne *Göldi, 1892 (Tylenchida) in Europe*. Brill, Leiden, Netherlands.

Karssen, G. and Moens, M. (2006) Root-knot nematodes. In: Perry, R.N. and Moens, M. (eds), *Plant Nematology*. CAB International, Wallingford, UK, pp. 59–90.

Karssen, G., Liao, J., Kan, Z., van Heese, E.Y.J. and den Nijs, L.J.M.F. (2012) On the species status of the root-knot nematode *Meloidogyne mayaguensis* Rammah & Hirschmann, 1988. *ZooKeys* 181, 67–77.

Karssen, G., Wesemael, W.M.L. and Moens, M. (2013) Root-knot nematodes. In: Perry, R.N. and Moens, M. (eds), *Plant Nematology*, 2nd edn. CAB International, Wallingford, UK, pp. 73–108.

Khan, M.R. (2008) *Plant Nematodes, Methodology, Morphology, Systematics, Biology and Ecology*. Science Publishers, Enfield, New Hampshire.

Khan, M.R. and Quintanilla, M. (2023) *Nematode Diseases of Crops and Their Sustainable Management*. Elsevier, London, UK.

Kiewnick, S., Holterman, M.H.M., van den Elsen, S.J.J., van Megen, H.H.B., Frey, J.E., *et al.* (2014) Comparison of two short DNA barcoding loci (*COI* and *COII*) and two long ribosomal DNA genes (SSU and LSU rRNA) for specimen identification among quarantine root-knot nematodes (*Meloidogyne* spp.) and their close relatives. *European Journal of Plant Pathology* 140, 97–110. DOI: 10.1007/s10658-014-0446-1.

Kleynhans, K.P.N. (1986) Useful new characters for the identification of four *Meloidogyne* species. *Phytophylactica* 18, 93–94.

Kleynhans, K.P.N. (1991) *The Root-knot Nematodes of South Africa*. Technical Communication Department of Agricultural Development, No. 231, Pretoria, South Africa.

Lamberti, F. and Taylor, C.E. (eds) (1979) *Rootknot Nematodes (*Meloidogyne *Species); Systematics, Biology and Control*. Academic Press, London and New York.

Lavergne, G. (1901a) La Anguillula en Sud-América. *Revista Chilena de Historia Natural* 5, 85–91.

Lavergne, G. (1901b) L'anguillule du Chili (*Anguillula vialae*). *Revue de Viticulture* 16, 445–452.

Linford, M.B. (1942) The transient feeding of root-knot nematode larvae. *Phytopathology* 32, 580–589.

Lordello, L.G.E. and Zamith, A.P.L. (1958) On the morphology of the coffee root-knot nematode, *Meloidogyne exigua* Göldi, 1887. *Proceedings of the Helminthological Society of Washington* 24, 133–137.

Manzanilla-López, R.H. and Starr, J.L. (2009) Interactions with other pathogens. In: Perry, R.N., Moens, M. and Starr, J.L. (eds), *Root-knot Nematodes*. CAB International, Wallingford, UK, pp. 223–245.

Mendoza, H.A. and Jatala, P. (1985) Breeding potatoes for resistance to the root-knot nematode *Meloidogyne* species. In: Barker, K.R., Carter, C.C. and Sasser, J.N. (eds), *An Advanced Treatise on* Meloidogyne. *Volume II: Methodology*. Department of Plant Pathology and the United States Agency for International Development, Raleigh, North Carolina, pp. 217–224.

McBeth, C.W., Taylor, A.L. and Smith, A.L. (1941) Notes on staining nematodes in root tissue. *Proceedings of the Helminthological Society of Washington* 8, 26.

McClure, M.A. (1977) *Meloidogyne incognita*: a metabolic sink. *Journal of Nematology* 9, 88–90.

Mitkowski, N.A. and Abawi, G.S. (2003) Root-knot nematodes. *The Plant Health Instructor*. DOI: 10.1094/PHI-I-2003-0917-01.

Moens, M., Perry, R.N. and Starr, J.L. (2009) *Meloidogyne* species – a diverse group of novel and important plant parasites. In: Perry, R.N., Moens, M. and Starr, J.L. (eds), *Root-knot Nematodes*. CAB International, Wallingford, UK, pp. 1–17.

Morgan-Jones, G. and Rodríguez-Kábana, R. (1986) Fungi associated with cysts of potato cyst nematodes in Peru. *Nematropica* 16, 21–29.

Mojtahedi, H., Brown, C., Rigi, C. and Zhang, L. (2007) A new pathotype of *Meloidogyne chitwoodi* race 1 From Washington state. *Plant Disease* 91, 1051.

Netscher, C. and Sikora, R.A. (1990) Nematode parasites of vegetables. In: Luc, M., Sikora, R.A. and Bridge, J. (eds), *Plant Parasitic Nematodes in Subtropical and Tropical Agriculture*. CAB International, Wallingford, UK, pp. 237–284.

Nischwitz, C., Skantar, A., Handoo, Z.A., Hult, M.N., Schmitt, M.E., *et al.* (2013) Occurrence of *Meloidogyne fallax* in North America, and molecular characterization of *M. fallax* and *M. minor* from U.S. golf course greens. *Plant Disease* 97, 1424–1430.

Nyczepir, A.P. and Thomas, S.H. (2009) Current and future management strategies in intensive crop production systems. In: Perry, R.N., Moens, M. and Starr, J.L. (eds), *Root-knot Nematodes*. CAB International, Wallingford, UK, pp. 412–443.

Orton Williams, K.J. (1972) *Meloidogyne javanica*. *CIH Descriptions of Plant-parasitic Nematodes*, Set 1, No. 3. Commonwealth Agricultural Bureaux, Farnham Royal, UK.

Orton Williams, K.J. (1973) *Meloidogyne incognita*. *CIH Descriptions of Plant-parasitic Nematodes*, Set 2, No. 18. Commonwealth Agricultural Bureaux, Farnham Royal, UK.

Orton Williams, K.J. (1974) *Meloidogyne hapla. CIH Descriptions of Plant-parasitic Nematodes*, Set 3, No. 31. Commonwealth Agricultural Bureaux, Farnham Royal, UK.

Orton Williams, K.J. (1975) *Meloidogyne arenaria. CIH Descriptions of Plant-parasitic Nematodes*, Set 5, No. 62. Commonwealth Agricultural Bureaux, Farnham Royal, UK.

Pagan, C., Coyne, D., Carneiro, R., Kariuki, G., Luambuano, N., *et al.* (2015) Mitochondrial haplotype-based identification of ethanol-preserved root-knot nematodes from Africa. *Phytopathology* 105, 350–357.

Perry, R.N., Moens, M. and Starr, J.L. (eds) (2009) *Root-knot Nematodes*. CAB International, Wallingford, UK.

Perry, R.N., Hunt, D.J. and Subbotin, S.A. (eds) (2021) *Techniques for Work with Plant and Soil Nematodes*. CAB International, Wallingford, UK.

Philbrick, A.N., Adhikari, T.B., Louws, F.J. and Gorny, A.M. (2020) *Meloidogyne enterolobii*, a major threat to tomato production: current status and future prospects for its management. *Frontiers in Plant Science* 11, 606395. DOI: 10.3389/fpls.2020.606395.

Pinochet, J. (1986) A note on nematode control practices on bananas in Central America. *Nematropica* 16, 197–203.

Rammah, A. and Hirschmann, H. (1988) *Meloidogyne mayaguensis* n. sp. (Meloidogynidae), a root-knot nematode from Puerto Rico. *Journal of Nematology* 20, 58–69.

Randig, O., Bongiovanni, M., Carneiro, R.M.D.G. and Castagnone-Sereno, P. (2002) Genetic diversity of root-knot nematodes from Brazil and development of SCAR markers specific for the coffee-damaging species. *Genome* 45, 862–870.

Raymundo, S.A. (1985) Cropping systems research and root-knot nematode control. In: Barker, K.R., Carter, C.C. and Sasser, J.N. (eds), *An Advanced Treatise on* Meloidogyne. *Volume II: Methodology*. Department of Plant Pathology and the United States Agency for International Development, Raleigh, North Carolina, pp. 277–281.

Riggs, R.D. (1990) Making perineal patterns of root-knot nematodes and vulval cones of cyst nematodes. In: Zuckerman, B.M., Mai, W.F. and Krusberg, L.R. (eds), *Plant Nematology Laboratory Manual*. University of Massachusetts Agricultural Experiment Station, Amherst, Massachusetts, pp. 103–106.

Sasser, J.N. and Carter, C.C. (1982) Root-knot nematodes (*Meloidogyne* spp.): Identification, morphological and physiological variation, host-range, ecology, and control. In: Riggs, R.D. (ed), *Nematology in the Southern Region of the United States*. Southern Cooperative Series Bulletin 276, Arkansas Agricultural Experiment Station, Fayetteville, Arkansas, pp. 21–32.

Sasser, J.N. and Carter, C.C. (eds) (1985) *An advanced treatise on* Meloidogyne. *Volume I: Biology and Control*. Department of Plant Pathology and the United States Agency for International Development, Raleigh, North Carolina.

Schmitt, D.P. (1985) Preliminary and advanced evolution of nematicides. In: Barker, K.R., Carter, C.C. and Sasser, J.N. (eds), *An Advanced Treatise on* Meloidogyne. *Volume II: Methodology*. Department of Plant Pathology and the United States Agency for International Development, Raleigh, North Carolina, pp. 241–248.

Semblat, J.B., Bongiovanni, M., Wajnberg, E., Dalmasso, A., Abad, P., *et al.* (2000) Virulence and molecular diversity of parthenogenetic root-knot nematodes, *Meloidogyne* spp. *Heredity* 84, 81–89.

Shokoohia, E., Handoo, Z., Elshishka, M. and Salamad, A-M. (2023) Nematode problems in temperate fruits and their sustainable management. In: Khan, M. R. and Quintanilla, M. (eds), *Nematode Disease of Crops and Sustainable Management*. Elsevier, Cambridge, Massachusetts, pp. 375–426. DOI: 10.1016/B978-0-323-91226-6.00021-3.

Siddiqi, M.R. (2000) *Tylenchida Parasites of Plants and Insects*, 2nd edn. CAB International, Wallingford, UK.

Sikandar, A., Jia, L., Wu, H. and Yang, S. (2023) *Meloidogyne enterolobii* risk to agriculture, its present status and future prospective for management. *Frontiers in Plant Science* 13, 1093657. DOI: 10.3389/fpls.2022.1093657.

Sikora, R.A., Coyne, D., Hallmann, J. and Timper, P. (eds) (2018) *Plant Parasitic Nematodes in Subtropical and Tropical Agriculture*, 3rd edn. CAB International, Wallingford, UK.

Singh, P.R., Gitonga, D., Hajihassani, A., Verhage, A., van Aalst-Philipse, E., *et al.* (2023) A new root-knot nematode species, *Meloidogyne karsseni* n. sp. (Nematoda: Meloidogynidae), from Mexico and a taxonomic update on *M. paranaensis* from Guatemala. *Journal of Nematology* 55(1), 20230042

Songul, A.E. and Dickerson, O.J. (1976) Life cycle, pathogenicity, histopathology, and host range of race 5 of barley root-knot nematode. *Journal of Nematology* 8, 228–232.

Starr, J.L. and Mercer, C.F. (2009) Development of resistant varieties. In: Perry, R.N., Moens, M. and Starr, J.L. (eds), *Root-knot Nematodes*. CAB International, Wallingford, UK, pp. 326–337.

Subbotin, S.A., Palomares-Rius, J.E. and Castillo, P. (2021) *Systematics of Root-knot Nematodes (Nematoda: Meloidogynidae)*. In: *Nematology Monographs and Perspectives, 14* (Series eds: Hunt, D.J. and Perry, R.N.). Brill, Leiden, Netherlands.

Tao, Y., Xu, C., Yuan, C., Wang, H., Lin, B., Zhuo, K. and Liao, J. (2017) *Meloidogyne aberrans* sp. nov. (Nematoda: Meloidogynidae), a new root-knot nematode parasitizing kiwifruit in China. *PLoS One* 12(8), e0182627. DOI: 10.1371/JOURNAL.PONE.0182627.

Taylor, A.L. and Sasser, J.N. (1978) *Biology, Identification and Control of Root-knot Nematodes (*Meloidogyne *Species)*. Department of Plant Pathology and the United States Agency for International Development, Raleigh, North Carolina.

Taylor, A.L., Dropkin, V.H. and Martin, G.C. (1955) Perineal patterns of root-knot nematodes. *Phytopathology* 45, 26–34.

Tenente, G.C.M.V., De Ley, P., De Ley, I.T., Karssen, G. and Vanfleteren., J.R. (2004) Sequence analysis of the D2/D3 region of the large subunit rDNA from different *Meloidogyne* isolates. *Nematropica* 34, 1–12.

Thies, J.A. (2002) Red food coloring stain: Improved, safer procedure for staining nematodes in roots and egg masses on root surfaces. *Journal of Nematology* 34, 179–181.

Thomason, I.J. and Caswell, E.P. (1987) Principles of nematode control. In: Brown, R.H. and Kerry, B.R. (eds), *Principles and Practice of Nematode Control in Crops*. Academic Press, London, UK, pp. 87–130.

Tigano, M.S.T., Carneiro, R.M.D.G., Jeyaprakash, A., Dickson, D.W. and Adams, B.J. (2005) Phylogeny of *Meloidogyne* spp. based on mitochondrial DNA sequences. *Nematology* 7, 851–862.

Tomaszewski, E.K., Khalil, A.M.M., El-Deeb, A.A., Powers, T.O. and Starr, J.L. (1994) *Meloidogyne javanica* parasitic on peanut. *Journal of Nematology* 26, 436–441.

Triantaphyllou, A.C. (1960) Sex determination in *Meloidogyne incognita* Chitwood, 1949 and intersexuality in *M. javanica* (Treub, 1885) Chitwood, 1949. *Annals of the Institute of Phytopathology, Benaki* 3, 12–31.

Triantaphyllou, A.C. and Hirschmann, H. (1960) Post-infection development of *Meloidogyne incognita* Chitwood, 1949 (Nematoda: Heteroderidae). *Annual of the Institute of Phytopathology, Benaki* 3, 1–11.

Tzortzakakis, E.A. and Gowen, S.R. (1996) Occurrence of a resistance breaking pathotype of *Meloidogyne javanica* on tomatoes in Crete, Greece. *Fundamental and Applied Nematology* 19, 283–288.

Van der Beek, J.G. and Poleji, L.M. (2008) Evidence for pathotype mixture on *Solanum bulbocastanum* in *Meloidogyne chitwoodi* but not in *M. fallax*. *Nematology* 10, 575–584.

Van der Beek, J.G., Maas, P.W.Th., Janssen, G.J.W., Zijlstra, C. and Van Silfhout, C.H. (1999) A pathotype system to describe intraspecific variation in pathogenicity of *Meloidogyne chitwoodi*. *Journal of Nematology* 31, 386–392.

Veech, J.A. (1991) *Root-knot Nematode Parasitism of Cotton. III*. Florida Department of Agriculture and Consumer Services, Division of Plant Industry, *Nematology Circular,* No. 185, Gainesville, Florida.

Verdejo-Lucas, S. and Sorribas, F.J. (2008) Resistance response of the tomato rootstock SC 6301 to *Meloidogyne javanica* in plastic house. *European Journal of Plant Pathology* 121, 103–107.

Weischer, B. and Brown, D.J.F. (2000) *An Introduction to Nematodes: General Nematology: A Student's Textbook*. Pensoft, Sofia, Bulgaria.

Whitehead, A.G. (1968) Taxonomy of *Meloidogyne* (Nematoda: Heteroderidae) with descriptions of four new species. *Transactions of the Zoological Society of London* 31, 263–401.

Williamson, V.M. and Roberts, P.A. (2009) Mechanism and genetics of resistance. In: Perry, R.N., Moens, M. and Starr, J.L. (eds), *Root-knot Nematodes*. CAB International, Wallingford, UK, pp. 301–325.

Wishart, J., Phillips, M.S. and Blok, V.C. (2002) Ribosomal intergenic spacer: a polymerase chain reaction diagnostic for *Meloidogyne chitwoodi*, *M. fallax*, and *M. hapla*. *Phytopathology* 92, 884–892. DOI: 10.1094/PHYTO.2002.92.8.884.

Xu, J., Liu, P., Meng, Q. and Long, H. (2004) Characterisation of *Meloidogyne* species from China using isozyme phenotypes and amplified mitochondrial DNA restriction fragment length polymorphism. *European Journal of Plant Pathology* 110, 309–315.

Yang, Y., Hu, X., Liu, P., Chen, L., Peng, H., *et al.* (2021) A new root-knot nematode, *Meloidogyne vitis* sp. nov. (Nematoda: Meloidogynidae), parasitizing grape in Yunnan. *PLoS One* 16(2), e0245201. DOI: 10.1371/JOURNAL.PONE.0245201.

Young, M.J. and Sherman, W.B. (1977) Evaluation of peach rootstocks for root-knot nematode resistance. *Proceedings of the Florida State Horticultural Society* 90, 241–242.

Zijlstra, C., Donkers-Venne, D.T.H.M. and Fargette, M. (2000) Identification of *Meloidogyne incognita*, *M. javanica* and *M. arenaria* using sequence characterized amplified region (SCAR) based PCR assays. *Nematology* 2, 847–845.

12 Pratylenchidae – The Root-lesion Nematodes

Pablo Castillo[1]*, Antonio Archidona-Yuste[1], Jason Stanley[2], Renato N. Inserra[2] and Rosa H. Manzanilla-López[3]

[1]Instituto de Agricultura Sostenible (IAS), Consejo Superior de Investigaciones Científicas (CSIC), Córdoba, Spain; [2]Florida Department of Agriculture and Consumer Services, Division of Plant Industry, Gainesville, USA; [3]Centro de Desarrollo de Productos Bióticos, Instituto Politécnico Nacional, Yautepec, Mexico

Abstract

Members of the Pratylenchidae are migratory endoparasitic nematodes (except for the swollen sessile female of *Nacobbus*) parasitizing a wide range of plants, including important crops such as banana, cereals, citrus, coffee, corn, legumes, potato, rice, tomato and many fruit trees. Those of economic importance include the root-lesion or meadow nematodes (*Pratylenchus*), burrowing nematodes (*Radopholus*), rice root nematodes (*Hirschmanniella*), or the false root-knot nematode (*Nacobbus*). This chapter introduces the morphology, bionomics, economic importance, diagnostics and classification of the family and provides greater detail for the genera *Pratylenchus*, *Hirschmanniella*, *Radopholus*, and *Nacobbus*, including a dichotomous key for genus identification. Species of major economic importance, including ten species of *Pratylenchus* and two each of *Hirschmanniella*, *Radopholus* and *Nacobbus* are illustrated and receive more detailed coverage.

12.1. Introduction

Pratylenchids are obligate plant-parasitic nematodes included in the heterogeneous Pratylenchidae, ranking second only to root-knot nematodes as having the greatest economic impact on crops worldwide (Castillo and Vovlas, 2007; Geraert, 2013). Pratylenchids are migratory endoparasitic nematodes (except for the swollen sessile female of *Nacobbus*), parasitizing a wide range of plants, including important crops such as banana, cereals, citrus, coffee, corn, legumes, potato, rice, tomato and many fruit trees. Pratylenchids of economic importance include those commonly known as root-lesion or meadow nematodes (*Pratylenchus*), burrowing nematodes (*Radopholus*), rice root nematodes (*Hirschmanniella*) or the false root-knot nematode (*Nacobbus*). Nevertheless, there is controversy about the definition and delimitation of pratylenchid genera. Recent molecular and phylogenetic studies using sequences of the D2-D3 expansion regions of 28S rRNA (Subbotin *et al.*, 2006; Trinh *et al.*, 2009) and 18S rRNA (Holterman

*E-mail: ag1cascp@uco.es

A revision of Castillo, P., Stanley, J., Inserra, R.N. and Manzanilla-López, R.H. (2012) Pratylenchidae – the Root-Lesion nematodes. In: Manzanilla-López, R.H. and Marbán-Mendoza, N. (eds), *Practical Plant Nematology*. Mundi-Prensa and Biblioteca Básica de Agricultura, Montecillo, Mexico.

(eds R.H. Manzanilla-López, D.J. Hunt, N. Marbán-Mendoza and S. Infante Gil)
DOI: 10.1079/9781836990413.0012

et al., 2007; Bert *et al.*, 2008) have indicated that *Radopholus* and *Pratylenchoides* stand apart from other genera and do not belong to Pratylenchidae. These findings may therefore cast doubt on the inclusion of *Achlysiella*, *Hoplotylus* and *Zygradus* in the Pratylenchidae since they are morphologically related to *Radopholus* and *Pratylenchoides*. Similar studies also suggest that *Nacobbus* is related to Telotylenchinae and represents the earliest divergent lineage within Pratylenchidae (Ryss, 1988; Palomares-Rius *et al.*, 2010), and that *Pratylenchoides* (see Chapter 8, this volume) belongs in Merliniinae (Sturhan, 2012; Siddiqi and Sturhan, 2014).

The Pratylenchidae includes four major genera of agricultural importance: *Hirschmanniella*, *Nacobbus*, *Pratylenchus* and *Radopholus*, as well as other genera of minor economic interest such as *Achlysiella*, *Apratylenchoides*, *Apratylenchus*, *Hoplotylus* (= *Radopholoides*), *Zygotylenchus* and *Zygradus* (Maggenti *et al.*, 1987; Siddiqi, 2000; Sturhan, 2008; Trinh *et al.*, 2009).

12.1.2. Outline Systematic Scheme of the Pratylenchidae

(After Chapter 3; major genera in bold)

Family Pratylenchidae Thorne, 1949

- Subfamily Pratylenchinae Thorne, 1949
  - *Apratylenchus* Trinh, Waeyenberge, Nguyen, Baldwin, Karssen & Moens, 2009
  - ***Pratylenchus*** Filipjev, 1936
  - *Zygotylenchus* Siddiqi, 1963
- Subfamily Hirschmanniellinae Fotedar & Handoo, 1978
  - ***Hirschmanniella*** Luc & Goodey, 1964
- Subfamily Radopholinae Allen & Sher, 1967
  - *Achlysiella* Hunt, Bridge & Machon, 1989
  - *Apratylenchoides Sher*, 1973
  - *Hoplotylus* s'Jacob, 1960
  - = *Radopholoides* De Guiran, 1967
  - ***Radopholus*** Thorne, 1949
  - *Zygradus* Siddiqi, 1991
- Subfamily Nacobbinae Chitwood *in* Chitwood & Chitwood, 1950
  - ***Nacobbus*** Thorne & Allen, 1944

12.2. Morphology

Characterized by a vermiform body (except for mature females of *Nacobbus* and, to a lesser degree, those of *Achlysiella* and *Apratylenchus*). Labial region low, anteriorly flattened to broadly rounded, annulated; framework strongly sclerotized; labial disc inconspicuous; *en face* labial pattern highly variable, including within genera. Stylet strong, length not exceeding three labial region diameters (except in *Hirschmanniella*); conus about as long as shaft; knobs large, rounded, usually closely applied to shaft. Orifice of dorsal gland usually 2-3 μm posterior to stylet base. Lateral fields each with 4-6 incisures, very rarely areolated posterior to pharynx. Deirids absent. Phasmids pore-like, located on tail, far posterior to anal level, extending into caudal alae as pseudoribs, except in *Hirschmanniella*. Pharyngeal glands extending over intestine ventrally or dorsally; subventral glands asymmetrical, extending beyond dorsal gland; three gland nuclei usually lying in tandem. Vulva a transverse slit, positioned submedian to more posterior but not subterminal; lips not modified; lateral membranes absent. Didelphic, amphidelphic or pseudo-monoprodelphic. Ovaries outstretched; posterior ovary degenerate and forming a post-vulval uterine sac in pseudo-monodelphic forms. Spermatheca large, rounded, with small round or rod-like sperm when impregnated. Tail conoid, subcylindrical to elongate-conoid, ca two or more anal body diameter long, with round to pointed tip that may bear a mucro (*Hirschmanniella*), generally with inconspicuous hyaline terminal portion.

The main characters used for practical identification of genera within Pratylenchidae are (i) pharyngeal glands extending over intestine ventrally or dorsally; (ii) reproductive system with two (sometimes one) functional ovaries, and vulva position; (iii) presence or absence of deirids; (iv) tail similar or dissimilar between sexes; (v) sexual dimorphism in anterior region prominent or not; (vi) mature female swollen or vermiform and (vii) body annuli ornamented by transverse single or double rows of rounded tubercles (see Fig. 12.1, Table 12.1, dichotomous key).

12.3. Key to Genera of Pratylenchidae (Fig. 12.1)

1. Pharyngeal glands extending over intestine mostly ventrally and ventrolaterally, no marked sexual dimorphism in anterior region 2
Pharyngeal glands extending over intestine mostly dorsally and dorsolaterally, with marked sexual dimorphism in anterior region 3
2. Tails similar between sexes; phasmids near terminus .... *Hirschmanniella*
Tails dissimilar between sexes; phasmids not near terminus 4
3. Mature female spindle-shaped or batatiform, with numerous eggs within body, gall inciting .... *Nacobbus*
Mature female not spindle-shaped or batatiform, not with numerous eggs within body, not gall inciting 5
4. One functional ovary .... *Pratylenchus*
Two functional ovaries .... *Zygotylenchus*
5. Sexual dimorphism in anterior region prominent 6
Sexual dimorphism in anterior region not prominent 8
6. Posterior ovary normally developed, functional 7
Posterior ovary reduced or absent, non-functional 9
7. Mature female swollen, becoming sausage-shaped; juveniles moulting to adult without feeding .... *Achlysiella*
Mature female not swollen; juveniles developing to adult by feeding .... *Radopholus*
8. Posterior ovary normally developed, functional .... *Zygradus*
Posterior ovary reduced, non-functional .... *Apratylenchoides*
9. Body annuli without ornamentation .... *Hoplotylus*
Body annuli ornamented by transverse single or double rows of rounded tubercles arranged like kernels on an ear of corn. Mature female sedentary, swollen in posterior third but abruptly narrowing posterior to vulva .... *Apratylenchus*

12.4. Genus *Pratylenchus* Filipjev, 1936 (Fig. 12.1)

Diagnosis (modified after Siddiqi, 2000): Pratylenchinae. No marked sexual dimorphism in anterior region. Lateral fields with 4-6 lines, occasionally with oblique median markings. Deirids absent. Phasmids near mid-tail. Labial region low, flattened anteriorly or rarely rounded, continuous with body contour; sclerotization massive; labial disc inconspicuous (because of its fusion with the subdorsal and subventral labial sectors and also with the lateral labial sectors in some species), with six labial pits around a minute oral aperture; amphidial apertures pore-like, near labial disc, indistinct. Stylet 20 µm or less long, with round, anteriorly flat or indented basal knobs. Median bulb oval to round, very muscular. Pharyngeal glands usually less than two body diameters long, extending over intestine mostly ventrally. Vulva in posterior region (usually at 70-80%). **Pseudomonoprodelphic, with only anterior ovary functional.** Post-vulval uterine sac present, with or without rudiments of posterior ovary. Spermatheca large (in amphimictic species), rounded, usually axial. Female tail subcylindrical to conoid, usually ca 2-3 anal body diameters long; terminus smooth or annulated, devoid of a process or mucron. Bursa enclosing tail terminus. **Spicules with subterminal pore on dorsal side. Gubernaculum simple, trough-like, fixed.**

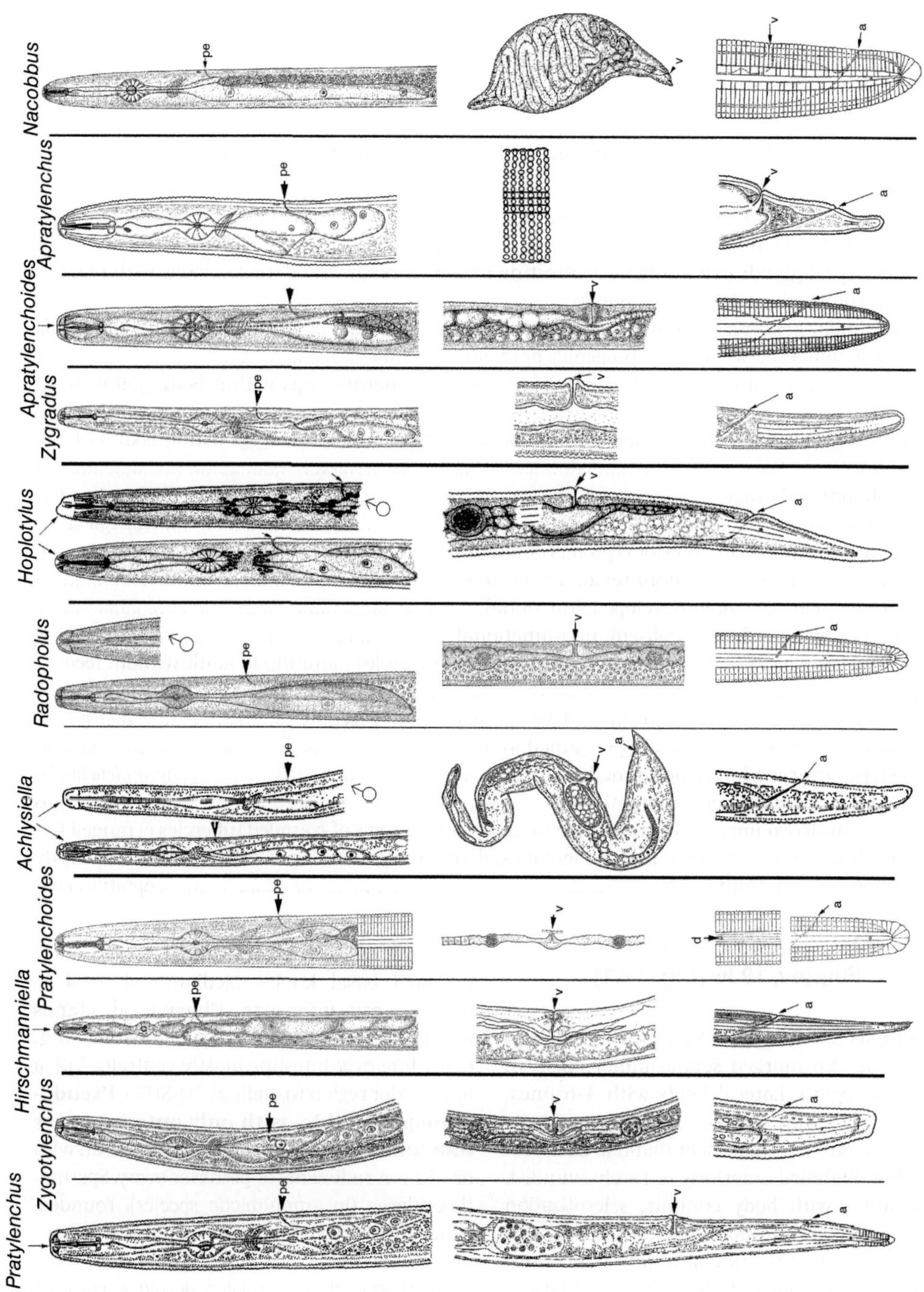

Fig. 12.1. Main diagnostic characters of Pratylenchidae genera.

Table 12.1. Diagnostic characters of Pratylenchidae genera.

Genus	Female body	Male secondary sexual dimorphism	Pharyngeal overlap	Female genital branch	Position of bursa
Pratylenchus	Vermiform	Absent	Ventral	Single	Terminal
Zygotylenchus	Vermiform	Absent	Ventral	Double	Terminal
Hirschmanniella	Vermiform	Absent	Ventral	Double	Subterminal
Achlysiella	Sausage-shaped	Present	Dorsal	Double	Terminal
Radopholus	Vermiform	Present	Dorsal	Double	Terminal
Hoplotylus	Vermiform	Present	Dorsal	Single	Subterminal
Zygradus	Vermiform	Present	Dorsal	Double	Terminal
Apratylenchoides	Vermiform	Present	Dorsal	Single	Terminal
Apratylenchus	Club-shaped	Present	Ventral	Single	Subterminal
Nacobbus	Saccate	Present	Dorsal	Single	Terminal

Adapted from Trinh *et al.* (2009).

Type species: *Pratylenchus pratensis* (de Man, 1880) Filipjev, 1936

Other species: There are more than 100 valid species and molecular studies have demonstrated the occurrence of cryptic species and species complexes. For synonyms and *species inquirendae*, as well as dichotomous and tabular keys for *Pratylenchus* spp., see Castillo and Vovlas (2007) and Geraert (2013). Species described since can be identified by the matrix code *sensu* Castillo and Vovlas (2007). Species have recently been reported from several crops, including *P. vandenbergae* on maize, soybean, sunflower and tomato plants in Kenya (Wanjau *et al.*, 2024), *P. dakotaensis* on soybean in North Dakota (Handoo *et al.*, 2021), *P. vovlasi* on raspberries in North Italy (Troccoli *et al.*, 2021), *P. smoliki* on corn and soybean in Central Great Plains, USA (Powers *et al.*, 2021), *P. capsici* on pepper in Israel (Qing *et al.*, 2019), *P. rwandae* on corn in Rwanda (Singh *et al.*, 2018), *P. haiduongensis* on carrot in Vietnam (Nguyen *et al.*, 2017), *P. parazeae* on sugarcane in China (Wang *et al.*, 2015), *P. quasiteroides* on cereals in Western Australia (Hodda *et al.*, 2014), *P. oleae* on olive in Spain and Tunisia, and pistachio in Iran (Palomares-Rius *et al.*, 2014; Naeini and Taheri, 2021), and *P. speijeri* parasitizing plantain in West Africa (De Luca *et al.*, 2012). These data confirm the hypothesis that application of integrative taxonomical approaches (morphological and molecular markers) can decipher the worldwide cryptic biodiversity of the genus. In this chapter we will concentrate on the bionomics of the major species.

12.4.1. Major Species of *Pratylenchus*

***Pratylenchus brachyurus* (Godfrey, 1929) Filipjev & Schuurmans Stekhoven, 1941 (Fig. 12.2)**

Common name: Godfrey's root-lesion nematode

Measurements (after Sher and Allen, 1953; Loof, 1960)

Female: L = 0.39-0.75 mm; a = 15-29; b = 5-10; c = 13-28; V = 82-89; stylet = 17-22 μm.

Male: L = 0.46-0.56 mm; a = 27-29; b = 6.0; c = 21; T = 51-53; stylet = 19 μm.

Description: *Female:* **Labial region offset from body, with angular anterior margin and two distinct annuli.** Face undivided (as seen at SEM), with labial sectors fused together and with a labial disc. Lateral fields usually bearing four lines in mid-body, occasionally dividing into five or six in vulval region, terminating as two lines near tail. Stylet stout, **with rounded basal knobs.** Excretory pore just posterior to hemizonid, usually in region of pharyngo-intestinal junction, which is often indistinct. Gonad outstretched, occasionally reflexed, rarely extending past pharyngeal lobe. **Vulva at 82-89% of body length; post-vulval uterine sac less than one body diameter** (10-30 μm) **long; spermatheca inconspicuous, non-functional.** Tail broadly conoid, smooth, with broadly rounded, truncate or spatulate tip, 13-24 ventral annuli; phasmids ca mid-tail. *Male:* Very rare. Similar in form to female. Spicules arcuate, gubernaculum simple.

Castillo and Vovlas (2007) matrix code: A1, B2, C4, D1, E4, F3, G3, H1, I4, J2-3, K1. It is close

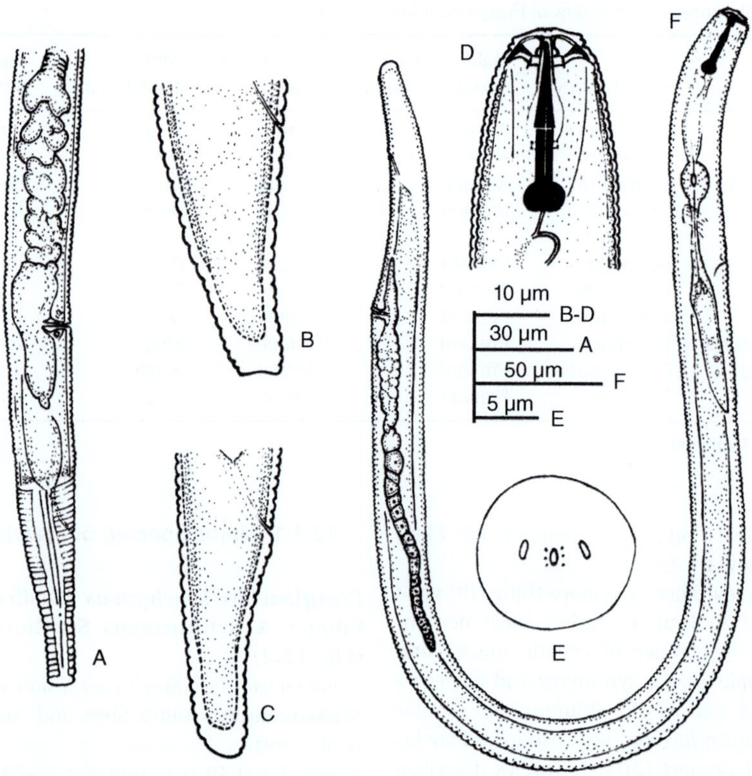

Fig. 12.2. *Pratylenchus brachyurus*. A: Female posterior region; B, C: female tails; D: female labial region; E: *En face* view; F: entire female. (After Corbett, 1976.)

to *P. japonicus* and *P. neobrachyurus*, from which it differs by body and stylet length, shape of labial region, shape of stylet knobs, vulva position and female tail shape.

Distribution and hosts: *Pratylenchus brachyurus* is widely distributed throughout the tropics (Africa, Asia and the Americas) and reported as damaging various crops and vegetables, that is, corn, cotton, cowpea, ginger, lychee, rubber, sugarcane, sweet potato, rice, plantain, soybean, tobacco, St. Augustine grass, peanut and pineapple. It has been recorded in Europe from Bulgaria and Russia on several hosts, and in Australia on cereals, peanuts and other crops (see Castillo and Vovlas (2007) for review).

Molecular and biochemical characterization: Characterization of *P. brachyurus* using isoelectric focusing electrophoresis (IEF) is useful for distinguishing it from *P. scribneri* (Payan and Dickson, 1990). Uehara (2003) separated *P. brachyurus* from *P. coffeae*, *P. loosi*, *P. penetrans*, *P. vulnus* and *P. zeae* by isoenzyme patterns, particularly with malate dehydrogenase (MDH) and phosphoglucose isomerase (PGI).

Populations of *P. brachyurus* have been characterized by rDNA sequences (D2-D3 expansion segments of 28S rRNA, internal transcribed spacers (ITS) and partial 18S rRNA gene sequences), and appear to be remarkably homogeneous at these phylogenetically important gene regions (Al-Banna *et al.*, 1997; Subbotin *et al.*, 2008). Machado *et al.* (2007) developed a single-step PCR using a species-specific reverse primer for the diagnosis of *P. brachyurus* based on ITS-1. Restriction fragment length polymorphisms (RFLP) of portions of the rDNA ITS separated *P. brachyurus*, *P. coffeae*, *P. goodeyi*, *P. loosi*, *P. neglectus*, *P. penetrans*, *P. thornei*, *P. vulnus* and *P. zeae* (Orui, 1996; Duncan *et al.*, 1999; Waeyenberge *et al.*, 2000), and Al-Banna *et al.* (2004) designed species-specific primers for *P. brachyurus*, *P. neglectus*, *P. scribneri*, *P. penetrans*, *P. thornei* and *P. vulnus* using the D3 expansion region of the 28S rDNA gene.

Pratylenchus coffeae **(Zimmermann, 1898) Filipjev & Schuurmans Stekhoven, 1941 (Fig. 12.3)**
Common name: Coffee root-lesion nematode
Measurements (after Sher and Allen, 1953; Loof, 1960)
Female: L = 0.45-0.70 mm; a = 25-35; b = 5-7; c = 17-22; V = 76-83; stylet = 15-18 μm.
Male: L = 0.45-0.70 mm; a = 26-40; b = 6-7; c = 17-24; T = 45-52; stylet = 15-17 μm.
Female (n = 69): L = 0.53 (0.37-0.70) mm; a = 23.7 (17.7-30.5); b = 6.8 (5.0-7.8); c = 19.0 (13.7-23.9); V = 80 (76-84); stylet = 14-17 μm.
Male (n = 10): L = 0.48 (0.41-0.56) mm; a = 27.4 (23.8-31.4); b = 6.5 (5.9-7.7); c = 19.1 (17.6-23.3); T = 48 (37-58); stylet = 14-15 μm.
Description: *Female*: Body rather slender. Lateral fields normally with four, sometimes with five or six lines. **Labial region slightly offset**

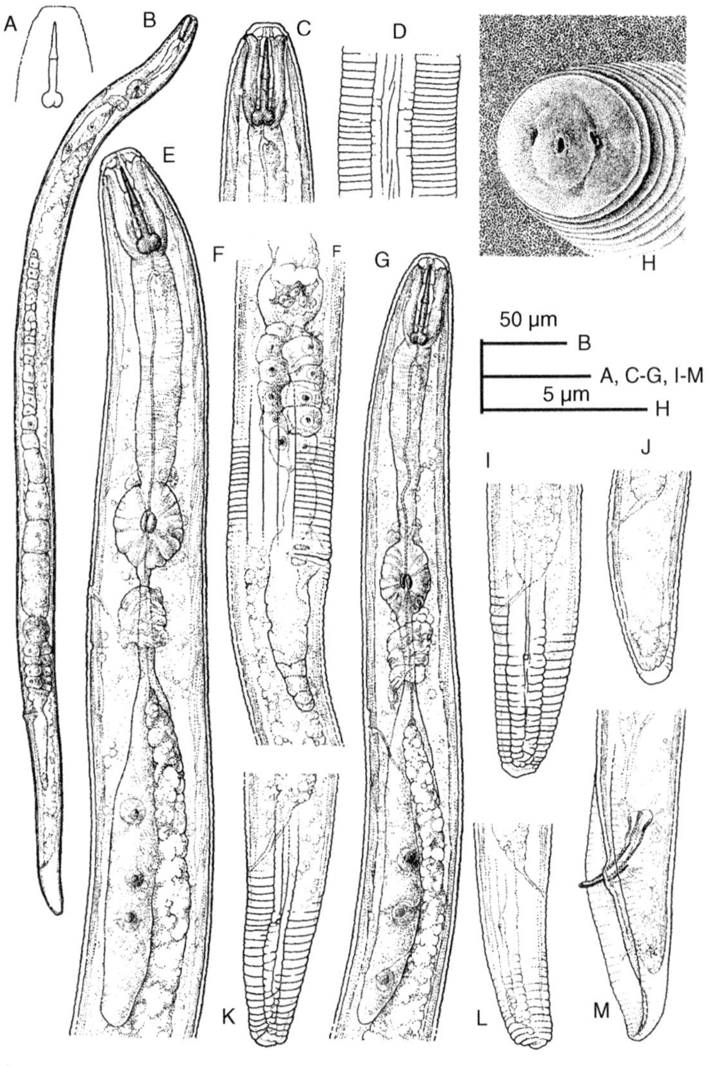

Fig. 12.3. *Pratylenchus coffeae*. A: Female stylet; B: female entire body; C: anterior female body; D: female lateral field. Note striae and crenate incisures; E: female pharyngeal region; F: vulval region; G: male pharyngeal region; H: female head pattern (SEM); I-L: female tail variations; M: male tail. Note faintly crenate bursa margins. (After Inserra *et al.*, 2001.)

from body, with two distinct annuli; occasionally three annuli on one side of labial region. **Face undivided (as seen at SEM) with labial sectors fused together and with labial disc.** Basal knobs of stylet round to oblong. **Post-vulval uterine sac 1.0-1.5 body diameters long, with a terminal rudimentary ovary that sometimes has distinct oocytes. Spermatheca large, broadly oval to nearly rounded**, usually with sperm. Tail 2.0-2.5 times anal body diameter in young females, 1.5-2.0 times anal body diameter in older females; **tail terminus indented**, sometimes appearing **smoothly rounded, truncate or irregularly crenate**. *Male*: Abundant. Spicules slender, with well-marked manubria and ventrally arcuate shaft, 16-20 µm long; gubernaculum 4-7 µm long; hypoptygma prominent; bursal margins faintly crenate.

Castillo and Vovlas (2007) matrix code: A1, B2, C2, D3, E3, F6, G2, H1, I2, J1-2-3, K1. Distinguished from the similar *P. loosi* by body annulation, vulva position, female tail shape and tail tip, which is truncate or hemispherical in *P. coffeae vs* bluntly or finely pointed in *P. loosi*.
Distribution and hosts: Recorded from coffee in many countries and citrus in India, Japan and USA but parasitizes many other crops such as sweet potato, mango, coconut, medicinal and aromatic plants, chrysanthemum, fruit crops, tea and banana. It is widely distributed in Asia and Central and South America. It has been reported from Africa on coffee, plantain and banana. It also occurs in Canada, Mexico (on coffee) and the USA (Florida and North Carolina) on aglaonema, citrus and *Ficus* spp. It has been recorded from several countries in Europe on fruit tree nurseries, carnation, vines and banana (see Castillo and Vovlas (2007) for review). Recently this species has been reported on banana in Tanzania (Luambano *et al.*, 2019), on corn and sesame in China (two unusual hosts for this species, Li *et al.*, 2020, 2022), and on yam (*Dioscorea* spp.) also in China (Abdulsalam *et al.*, 2021).
Molecular and biochemical characterization: Jaumot *et al.* (1997) determined the reliability and usefulness of total protein patterns for separating *P. coffeae* from *P. goodeyi*, *P. scribneri*, *P. thornei* and *P. vulnus*. Uehara *et al.* (1998) developed specific primer sets for the diagnosis of *P. coffeae* and *P. loosi* based on rDNA ITS regions. The genome of this species was studied by Burke *et al.* (2015), who identified 6712 protein-encoding genes.

Pratylenchus goodeyi **Sher & Allen, 1953 (Fig. 12.4)**

Common name: Banana root-lesion nematode
Measurements (after Sher and Allen, 1953)
Paratype females: L = 0.64-0.68 mm; a = 27-37; b = 5.5-6.1; c = 16-18; V = 73-75; stylet = 17 µm.
Male: L = 0.55-0.57 mm; a = 26; b = 5.4-5.8; c = 17-18; T = 54; stylet = 16 µm.
Description: *Female:* Body slender, heat-relaxed habitus almost straight but slightly ventrally arcuate posteriorly. **Lateral fields with four inconspicuous lines, two outer bands partially areolated. Labial region with four annuli. Face undivided (as seen at SEM), with labial sectors fused together and with labial disc. Stylet 16-18 µm long with pronounced, anteriorly flattened knobs**. Post-vulval uterine sac ca one vulval body diameter long. **Spermatheca large, sub-rectangular**, filled with sperm. **Tail conoid, ventrally concave, with dorsal contour sinuate just prior to tail tip**. Tail usually with 22-24 ventral annuli (range: 19-27). Phasmid usually conspicuous, 0-14 annuli from tail tip. *Male:* Common. Similar in general form to female. Single outstretched testis, spermatocytes in double row. Spicules slender, arcuate. Gubernaculum simple. Bursa envelops tail tip.

Castillo and Vovlas (2007) matrix code: A3, B2, C3, D4, E1, F1, G3, H3, I3, J1, K2. It is similar to *P. morettoi* and *P. typicus* from which it differs by stylet length and knob shape, vulva position, post-vulval uterine sac length, spermatheca shape and shape of tail and tail terminus.
Distribution and hosts: Recorded from numerous localities, especially in banana growing areas in Africa, Australia, Canary Islands, Spain, Madeira Island, Portugal, soil around boxwood (Greece) and in Russia on *Musa sapientum* L. *Pratylenchus goodeyi* has been found in several African countries associated with higher and cooler altitudes and is therefore considered an indigenous African nematode with an 'afro-montane' distribution (see Castillo and Vovlas (2007) for review). It has been recorded in Ethiopia on enset (*Ensete ventricosum*) (Kidane *et al.*, 2020), on banana in China (Zhang *et al.*, 2015), and in Kenya on banana and finger millet (Wanjau *et al.*, 2024).
Molecular and biochemical characterization: Andrés *et al.* (2000) found species-specific

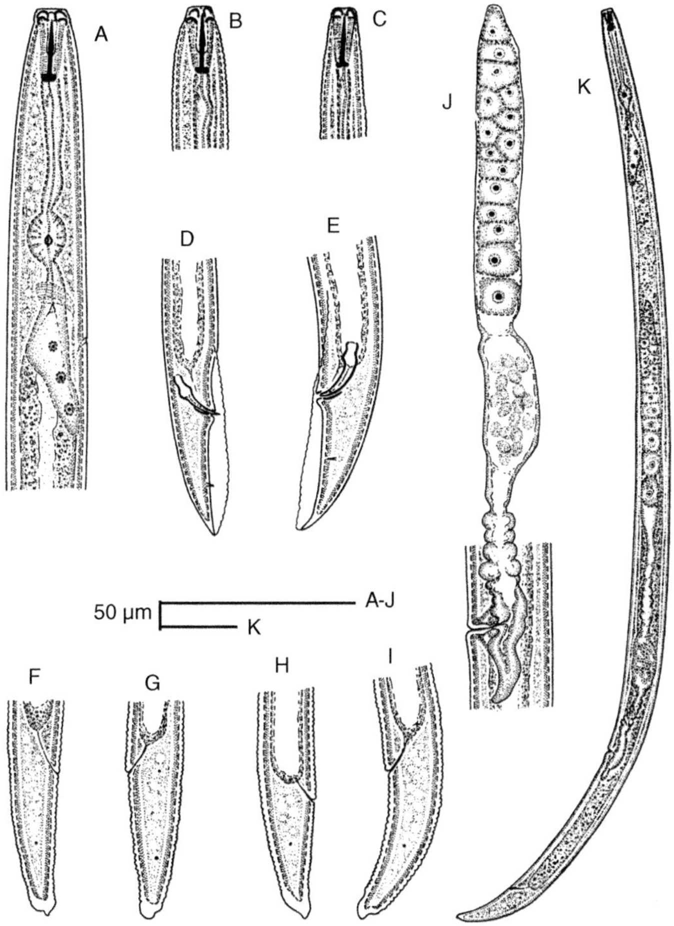

Fig. 12.4. *Pratylenchus goodeyi*. A: Female pharyngeal region; B: female labial region; C: male labial region; D, E: male tails; F-I: female tails; J: female reproductive system; K: entire female. (After Sher and Allen, 1953.)

enzyme markers that allowed the specific differentiation of *P. goodeyi* from *P. neglectus*, *P. penetrans*, *P. scribneri*, *P. thornei* and *P. vulnus*, from different geographic and host origins. In particular, analysis of phosphoglucomutase (PGM) was effective in differentiating these species. It can be also identified by ITS and partial 18S rRNA gene sequences (Meldal *et al.*, 2007).

Pratylenchus loosi **Loof, 1960 (Fig. 12.5)**

Common name: Tea root-lesion nematode

Measurements (after Loof, 1960)

Female (n = 34): L = 0.58 (0.48-0.64) mm; a = 32 (28-36); b = 6.4 (5.7-7.1); c = 20 (18-25); V = 83 (79-85); stylet = 14-18 µm.

Male (n = 15): L = 0.46 (0.38-0.58) mm; a = 35 (28-41); b = 5.9 (5.4-6.7); c = 20.8 (18.5-23.2); T = 42 (36-52); stylet = 12-16 µm.

Description: *Female:* Body very slender, almost straight after heat-relaxing, narrowing fairly suddenly just posterior to vulva to ca 0.8 times diameter anterior to vulva, then cylindrical to anus. **Lateral field broad with four (occasionally five or even six) lines anterior and three posterior to anus. Labial region rounded, with two annuli. Face undivided (as seen with SEM) with labial sectors fused together and with labial disc**. Posterior edges of labial framework extending one or two body annuli into body. Stylet knobs rounded, broadly fused to shaft. **Spermatheca rectangular, up to 20 µm long, rarely more rounded**; located 0.5-0.9 times vulva-anus distance anterior to vulva. **Post-vulval uterine sac 30-50% of vulva-anus distance. Tail** ca 2.5 anal body diameters long,

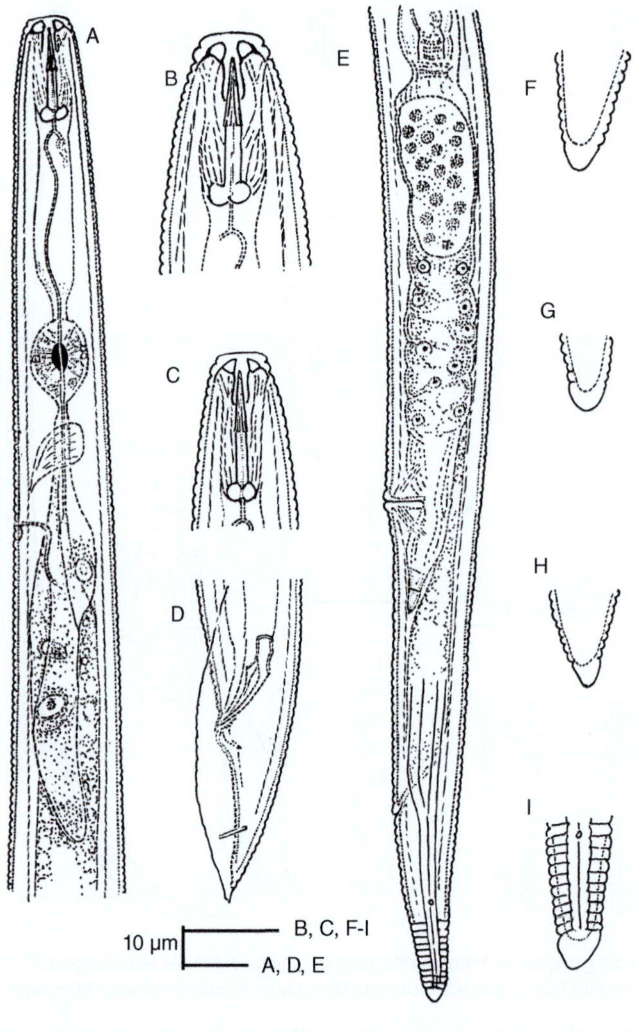

Fig. 12.5. *Pratylenchus loosi*. A: Female pharyngeal region; B: female labial region; C: male labial region; D: male tail; E: female posterior region; F-I: female tails. (After Seinhorst, 1977.)

tapering gradually to smooth conical tip, usually with narrowly rounded terminus. *Male:* Abundant. Spicules slender with well-marked manubria and ventrally arcuate shaft, 16-20 μm long gubernaculum 4-7 μm long, bursal margins faintly crenate.

Castillo and Vovlas (2007) matrix code: A1, B2, C3, D4, E3, F2, G3, H3, I2, J1, K1. It is similar to *P. coffeae*, *P. neobrachyurus*, *P. panamaensis* and *P. pseudocoffeae* from which it can be distinguished by labial region shape, body and stylet length, vulva position and female tail shape.

Distribution and hosts: It is an important pathogen of tea in Sri Lanka and other parts of the world and has been reported from citrus, apple, coffee, banana and fruit trees in several countries (see Castillo and Vovlas (2007) for review). Recently, it has been reported on tea in Iran (Mirghasemi *et al.*, 2019). The damage of *P. loosi* to tea crops has been estimated at 4-40% (Gnanapragasham and Mohotti, 2018).

Molecular and biochemical characterization: *Pratylenchus loosi* can be identified by D2-D3 and ITS rRNA gene sequences (Duncan *et al.*, 1999). Mirghasemi *et al.* (2019) developed a molecular characterization on an Iranian population based on ribosomal, mitochondrial and nuclear marker (D2-D3 expansion regions of the 28S rRNA gene, ITS region, COI and *hsp90*).

***Pratylenchus neglectus* (Rensch, 1924) Filipjev & Schuurmans Stekhoven, 1941 (Fig. 12.6)**

Common name: California root-lesion nematode

Measurements (after Rensch, 1924; Sher and Allen, 1953; Loof, 1964)

Females (n = 9): L = 0.31-0.58 mm; a = 16.5-32.2; b = 4.9-7.8; c = 20 (13.8-26.8); V = 81 (76-87); stylet = 15-19 μm.

Males (n = 2): L = 0.42-0.52 mm; a = 25.5-28.9; b = 6.2-6.3; c = 17.3-21.7; T = 42-56; stylet = 15 μm.

Females: L = 0.31-0.55 mm; a = 18-25; b = 4.0-6.3; c = 16-22; V = 80-88; stylet = 16-18 μm.

Male: L = 0.34 mm; a = 22; b = 4.8; c = 20; stylet = 14 μm.

Description: *Female*: Characterized by great variation in body length and diameter, tail shape and thickness and early maturity of adults, which is sometimes completed during final moult. Mature females extracted from roots are often markedly more robust than those from rhizosphere, which become more linear when relaxed. **Labial region bluntly rounded, with two annuli, second annulus wider than first**, apical one comprising the lips. **Face divided (SEM), with dorsal and ventral submedian labial sectors fused together and with labial disc forming a dumbbell-shaped large cap; lateral labial sectors delineated; amphidial apertures set at an oblique angle. Lateral fields with four lines, but median zone often marked by one or two longitudinal or several oblique striae. Stylet knobs**

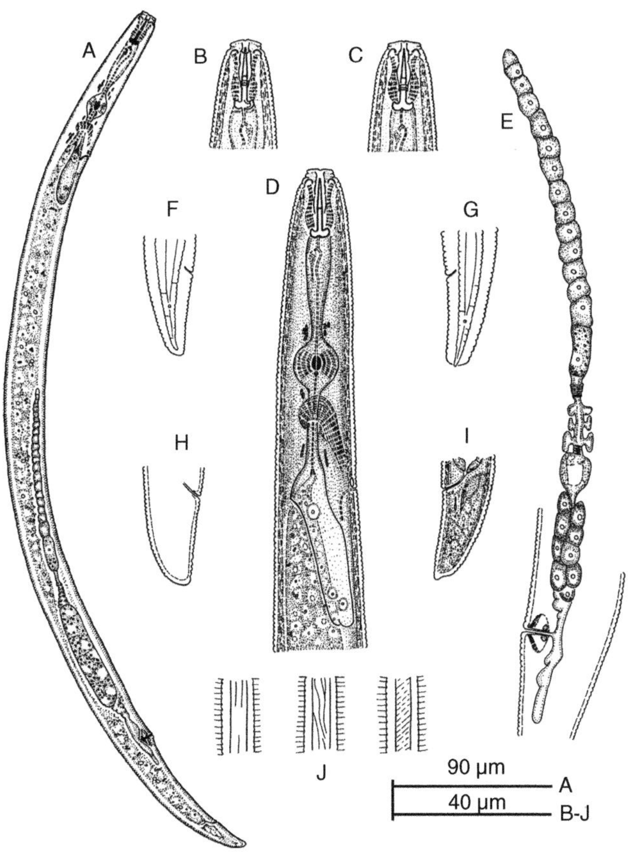

Fig. 12.6. *Pratylenchus neglectus*. A: Entire female; B, C: female labial region; D: female pharyngeal region; E: female reproductive system; F-I: female tails; J: variability of lateral field. (After Townshend and Anderson, 1976.)

typically indented on anterior surfaces. Pharyngeal gland overlapping intestine ventrally or ventrolaterally; subventral gland nuclei at end of lobes, not in tandem. Spermatheca oval, empty. **Post-vulval uterine sac less than or equal to body diameter**, 12-18 μm long, undifferentiated. Tail variable in shape, usually conoid with little curvature of ventral surface, usually with 15-20 annuli. **Tail terminus without annulation, usually rounded, but may be obliquely truncate or slightly digitate.** *Male:* Only three specimens known, of no diagnostic or reproductive significance. Similar to female. Phasmids slightly posterior to mid-tail, not appearing to extend into bursa.

Castillo and Vovlas (2007) matrix code: A1, B2, C3, D1, E2, F1, G3, H1, I1, J1, K1.

Distribution and hosts: Widely recorded on cereals, legumes and other crops in Europe, Africa and Asia, North, Central and South America, Western Australia and Tasmania (see Castillo and Vovlas (2007) for review). Also reported on wheat and corn in China (Li *et al.*, 2018; Xia *et al.*, 2021).

Molecular and biochemical characterization: Protein patterns and isozyme phenotypes of esterase, glucose-6-phosphate dehydrogenase (G6-PDH), isocitrate dehydrogenase (IDH), phosphoglucose isomerase (PGI) and PGM were used to discriminate *P. neglectus*, *P. penetrans* and *P. thornei* (Ibrahim *et al.*, 1995). It can also be identified by D2-D3, ITS and partial 18S rRNA gene sequences (De Luca *et al.*, 2004; Subbotin *et al.*, 2008; Holterman *et al.*, 2009). Fatemi *et al.* (2023) developed an RT-qPCR assay for identifying and quantifying *P. neglectus* within infected cereal roots.

***Pratylenchus penetrans* (Cobb, 1917) Filipjev & Schuurmans Stekhoven, 1941 (Fig. 12.7)**

Common name: Cobb's root-lesion nematode

Measurements (after Loof, 1960)

Females (n = 84): L = 0.35-0.81 mm; a = 19-32; b = 5.3-7.9; c = 15-24; V = 75-84; stylet = 15-17 μm.

Males (n = 34): L = 0.31-0.57 mm; a = 23-34; b = 5.4-7.3; c = 16-22; T = 36-58; stylet = 13-16 μm; spicules = 14-17 μm; gubernaculum = 3.9-4.2 μm.

Description: *Female:* Body moderately slender, almost straight when heat-relaxed. Lateral fields normally with four lines; outer bands may be partly areolated; central band sometimes with oblique striae near vulva, becoming areolated posterior to vulva, not extending to tail tip. **Labial region slightly offset from body, low, flat anteriorly with rounded outer margins, with three annuli. Face divided (as seen at SEM), with dorsal and ventral submedian labial sectors fused together and with labial disc, forming a distinct dumbbell-shaped large cap while small lateral labial sectors are delineated.** Labial region with strong, conspicuous skeleton. Basal knobs of stylet broadly rounded, sometimes cupped anteriorly. **Pharynx overlapping intestine ventrally for ca 1.5 body diameters.** Excretory pore more or less opposite pharyngo-intestinal junction, with hemizonid occupying ca two body annuli immediately anterior. **Post-vulval uterine sac short, undifferentiated, ca 1-1.5 vulval body diameters long. Spermatheca spherical**, or nearly so, and full of sperm. **Tail generally rounded, tip smooth**, with 15-27 annuli on ventral surface. *Male*: Common. Slightly smaller than female, but similar in form. Lateral fields with four lines ending on bursa, occasionally with oblique lines in central band near midbody. Spicules slender, with well-marked manubria and ventrally arcuate shafts, 14-17 μm long; gubernaculum simple, 3.9-4.2 μm long. Tail ca twice as long as cloacal body diameter; bursa irregularly crenate along margin, enveloping tail tip.

Castillo and Vovlas (2007) matrix code: A2, B2, C3, D2, E3, F4, G2, H1, I3, J1, K1. Distinguished from *P. fallax*, *P. pseudofallax* and *P. subpenetrans* by body and stylet length, number of labial annuli, labial framework, vulva position, spermatheca shape and tail terminus.

Distribution and hosts: Widely distributed and recorded from over 350 hosts, mainly in temperate areas in Europe, North America, Central and South America, Africa, Asia and Australia. It is a major pathogen of fruit and conifer nurseries in many areas and causes important losses in potato, tobacco, alfalfa, wheat, chrysanthemum, stone-, pome- and small-fruits, strawberries and roses. It has also been recorded in New Zealand and in Australia from agricultural soils and native vegetation (see Castillo and Vovlas, 2007 for review) and has been considered the causal agent of fern decline in Florida nurseries, although studies by Troccoli *et al.* (2021) have shown that such populations infesting fern were representatives of *P. bolivianus*.

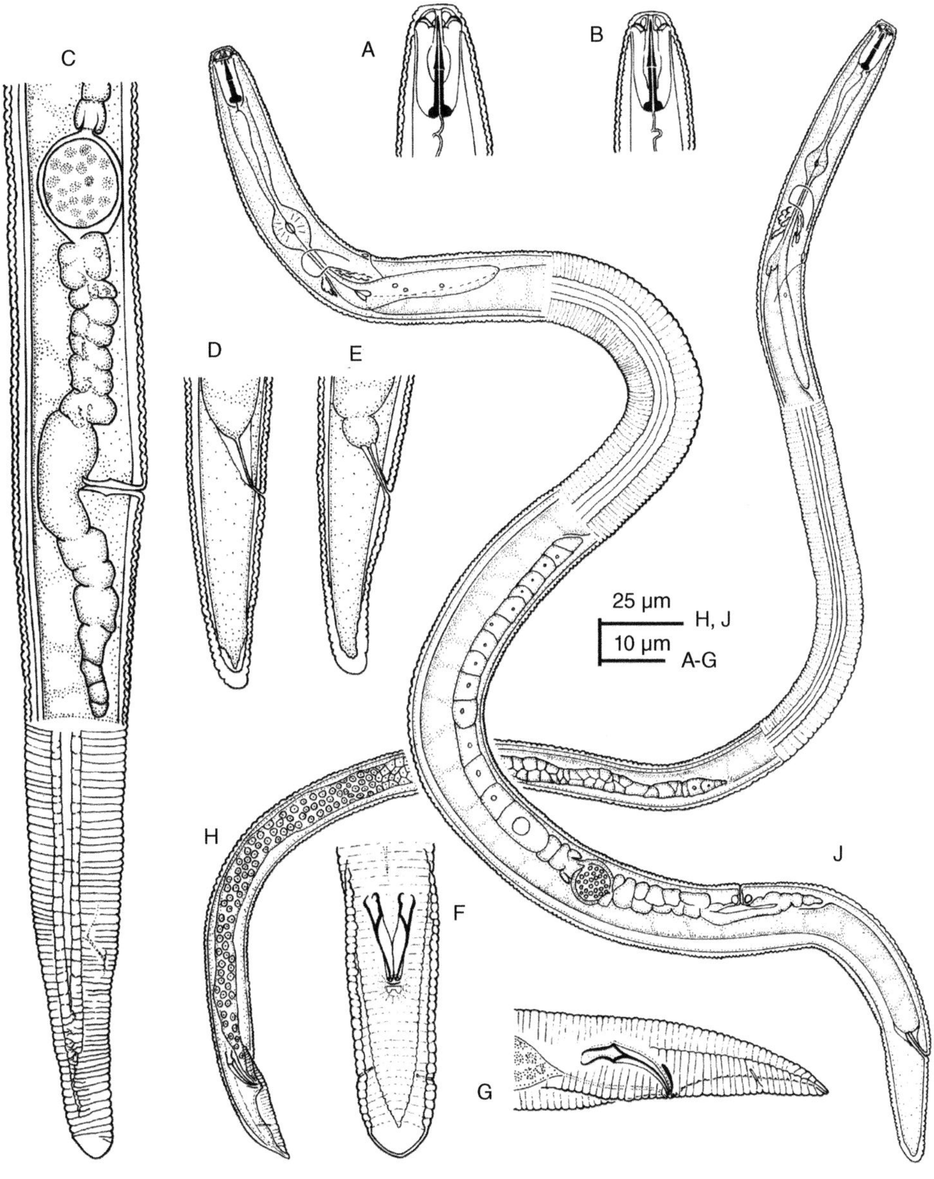

Fig. 12.7. *Pratylenchus penetrans*. A: Female labial region; B: male labial region; C: female posterior region; D, E: female tails; F, G: male tails; H: entire female; I: entire male. (After Corbett, 1973.)

Molecular and biochemical characterization: Identified by D2-D3, ITS and partial 18S rRNA gene sequences (De Luca *et al.*, 2004; Subbotin *et al.*, 2008; Holterman *et al.*, 2009). Waeyenberge *et al.* (2009) constructed an optimized sensitive species-specific duplex PCR for identifying this species. Mokrini *et al.* (2013) and Baidoo *et al.* (2017) developed a qPCR for the specific and quantitative detection of *P. penetrans*. Orlando *et al.* (2024) developed a TaqMan real-time PCR method focused on the D2-D3 expansion segments for identification and quantification of *P. penetrans*. The particular strength of these methods is their usefulness for quantifying *P. penetrans* in mixed populations of *Pratylenchus* spp., where visual identification of individual nematodes at species level is extremely difficult (Mokrini *et al.*, 2013; Fanelli *et al.*, 2018; Orlando *et al.*, 2024).

***Pratylenchus scribneri* Steiner *in* Sherbakoff & Stanley, 1943 (Fig. 12.8)**

Common name: Scribner's root-lesion nematode

Measurements (after Loof, 1985)

Females (n = 25): L = 0.50-0.66 mm; a = 23-33; b = 4.9-7.4; b' = 3.6-6.0; c = 14-18; c' = 2.3-3.3; V = 75-79; stylet = 14-15 µm.

Females (n = 25): L = 0.40-0.58 mm; a = 20-27; b = 4.8-7.6; b' = 3.7-5.9; c = 13-18; c' = 2.1-3.0; V = 74-79; stylet = 12-15 µm.

Females (n = 7): L = 0.42-0.52 mm; a = 22-30; b = 4.6-6.3; b' = 3.2-4.6; c = 14-22; c' = 2.1-2.7; V = 72-79; stylet = 15-16 µm.

Description: *Female:* Body stout, almost straight when heat-relaxed; young females more slender. Cuticular annulation fine, shallow, indistinct, ca 1 µm apart, tending to become more pronounced and further apart on tail. **Lateral fields with four longitudinal lines; occasionally a fifth line may be present in pre-vulval region, especially in younger specimens. Labial region slightly offset from body, composed of two annuli of ca same height**, anterior one distinctly narrower than second; angles rounded, anterior margin rather flat. Labial framework extending into body for one annulus. **Face divided (as seen at SEM), with dorsal and ventral submedian labial sectors fused together and with labial disc, forming a distinct dumbbell-shaped large cap, while lateral labial sectors are delineated. Stylet stout, with rounded knobs** varying little in shape. Orifice of dorsal pharyngeal gland 2-3 µm posterior to stylet base. Nucleus of dorsal gland just posterior to pharyngo-intestinal junction, nuclei of ventrosublateral glands close together near posterior end of gland lobe; in lateral view, all three nuclei lying in tandem. Pharyngeal overlap of medium length. Hemizonion and cephalids not seen. Anterior genital branch consisting of ovary with oocytes in single file except for a short zone near anterior end; short oviduct; **oblong, empty spermatheca**; and uterus a tricolumella (three rows of four cells each). **Post-vulval uterine sac generally 1.1-1.4 body diameter** long, occupying 25-40% of vulva-anus distance. **Tail tapering slightly, terminus mostly broadly rounded**, varying from somewhat narrower to almost truncate; terminus unstriated. *Male:* Spicules of normal shape, 17 µm long; gubernaculum slightly curved, 7 µm long. Phasmids 14 µm posterior to cloacal aperture.

Castillo and Vovlas (2007) matrix code: A1, B2, C2, D3, E2, F4, G3, H1, I2, J1-3, K1. Distinguished from *P. crassi*, *P. hexincisus* and *P. scribneri* by stylet length, lateral fields with four lines, often with oblique striae in the central band at mid-body; dorsal pharyngeal gland orifice position, tail shape, number of tail annuli.

Distribution and hosts: Widespread in the USA on corn, potato, alfalfa, roses, soybean, sorghum, white clover, cotton and pigeon pea. Records from other countries include Mexico, Guadeloupe, South America (Argentina, Venezuela, Brazil), Asia, Europe (Bulgaria, Italy, Slovenia) and Australia. It appears to be widespread in

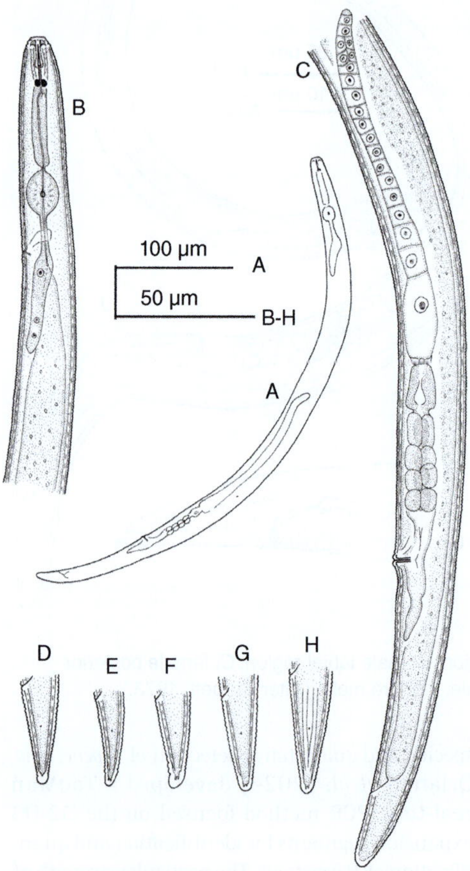

Fig. 12.8. *Pratylenchus scribneri.* A: Entire female; B: female pharyngeal region; C: female posterior region; D-H: female tails. (After Loof, 1985.)

Africa (Egypt, Nigeria, South Africa) (see Castillo and Vovlas (2007) for review) and has been reported on tomato and corn in China (Xia *et al.*, 2022a, b).

Molecular and biochemical characterization: Setterquist *et al.* (1996) constructed species-specific primers and probes derived from the major sperm protein gene region of *P. penetrans* and *P. scribneri* to be used in a species-specific PCR. It can also be identified by D2-D3, ITS and partial 18S rRNA gene sequences (Inserra *et al.*, 2007; Holterman *et al.*, 2009; Waeyenberge, Subbotin and Moens, unpublished). However, the results of these molecular analyses are controversial since they did not involve topotype specimens. Furthermore, putative *P. scribneri* from amaryllis in Florida was described as a new species, *P. hippeastri*. Until topotype material is collected and examined, the identity of *P. scribneri* remains unresolved. Arora *et al.* (2020) developed a qPCR for the specific and quantitative detection of *P. scribneri*, while Arora and Yan (2022) developed a qPCR assay for the detection and quantification of *P. scribneri* in potato roots.

Pratylenchus thornei **Sher & Allen, 1953 (Figs 12.9, 12.10)**

Common name: Cereal and legume root-lesion nematode

Measurements (after Sher and Allen, 1953; Loof, 1960)

Females: L = 0.45-0.77 mm; a = 26-36; b = 5.5-8.0; c = 18-22; V = 73-80; stylet = 17-19 μm.

Male: L = 0.48 mm; a = 32; b = 5.6; c = 20; T = 30; stylet = 16 μm.

Females (n = 37): L = 0.41-0.71 mm; a = 25.3-36.4; b = 5.4-8.3; c = 16.8-25.1; V = 74-79; stylet = 15-19 μm.

Male: L = 0.49 mm; a = 29; b = 6.2; c = 20.3; stylet = 16 μm.

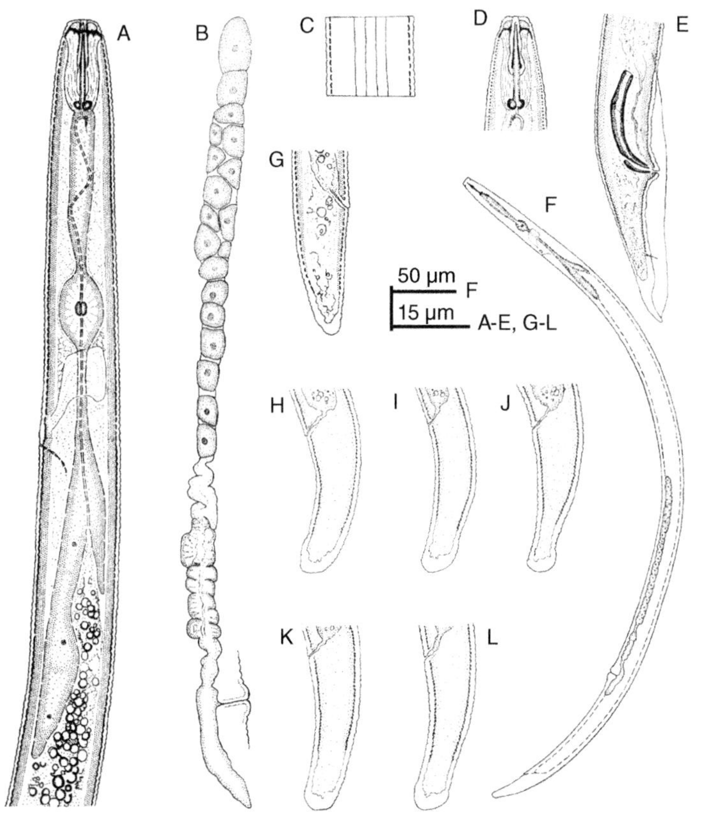

Fig. 12.9. *Pratylenchus thornei*. A: Female pharyngeal region; B: female reproductive system; C: lateral field at mid-body; D: male labial region; E: male tail; F: entire female; G-L: female tails. (After Fortuner, 1977.)

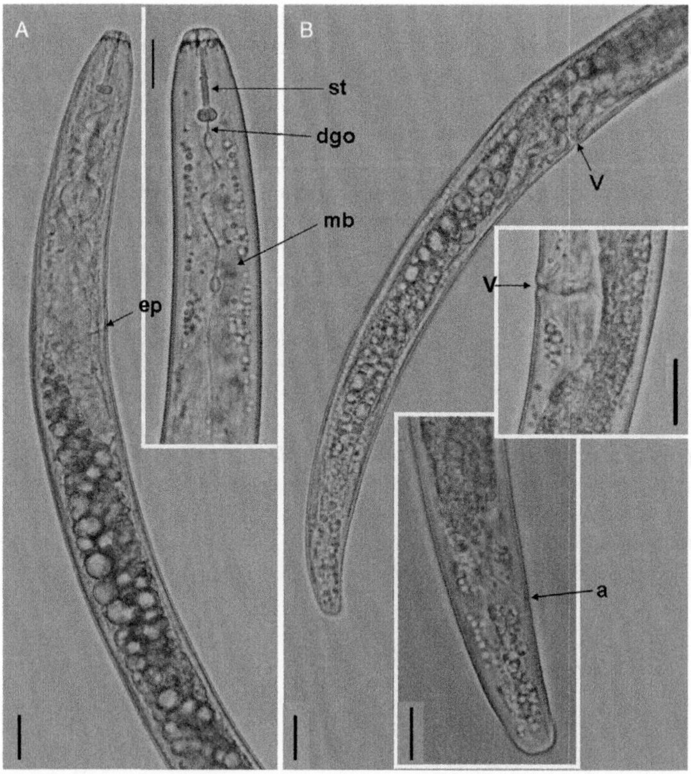

Fig. 12.10. Micrographs of *Pratylenchus thornei*. A: Female anterior region showing stylet (st), dorsal pharyngeal gland orifice (dgo), median pharyngeal bulb (mb) and excretory pore (ep); B: female posterior region showing vulva (v), tail and anus (a). (Scale bars = 10 µm.)

Males (n = 9; after carrot disc cultures, a population infecting chickpea in Jerez, Southern Spain, present study): L = 0.65 (0.62-0.69) mm; a = 34.2 (30.0-39.5); b = 7.5 (5.8-8.3); b' = 5.1 (4.6-5.9); c = 22.3 (18.1-26.0); c' = 2.1 (1.7-2.8); T = 45.0 (30.0-56.0); stylet = 16.0 (15.5-16.5) µm; spicules = 23.5 (21.5-26.0) µm; gubernaculum = 6.0 (5.0-7.5) µm.

Description: *Female:* **Body large and slender, assuming an open C-shape when killed by gentle heat**. Cuticle with transverse striae ca 1 µm apart, not conspicuous. **Lateral fields with four lines, outer ones straight or weakly crenate**. **Labial region slightly offset from body, consisting of four annuli.** Outer margin of sclerotized labial framework conspicuously extending ca two annuli into body and one annulus into labial region. **Face divided (as seen at SEM), with dorsal and ventral submedian labial sectors fused together and with labial disc, forming a distinct dumbbell-shaped large cap, while large lateral labial sectors are delineated**. **Stylet medium size (17-19 µm long), with broadly rounded to almost anteriorly flattened basal knobs**. Orifice of dorsal pharyngeal gland ca 3 µm posterior to stylet base. **Spermatheca difficult to see, not containing spermatozoa (males very rare)**; **post-vulval uterine sac slightly more than 1.5 vulval body diameter long**. Phasmids slightly posterior to mid-tail; all four incisures extending posterior to phasmids. **Tail dorsally convex-conoid, terminus bluntly rounded to truncate, unstriated**. *Male:* Very rare (only four specimens previously recorded). Similar to female. Testis outstretched, with spermatocytes in single row followed by a region of multiple rows. Phasmids slightly posterior to mid-tail, not extending into bursa. Spicules very long (21-26 µm), arcuate, shafted, resting upon trough-shaped gubernaculum (5-7.5 µm).

Castillo and Vovlas (2007) matrix code: A2, B2, C3, D1, E2, F2, G3, H1, I3, J1, K1. Distinguished from *P. mediterraneus*, *P. fallax*, *P. penetrans*, *P. pseudopratensis* and *P. sudanensis* by labial region shape, stylet length, the low proportion of males, spermatheca shape and tail shape.

Distribution and hosts: Recorded from various crops (cereals, legumes, tobacco, potato, corn, chrysanthemum, soybean) from several countries in Europe, Africa (Egypt, Nigeria, Morocco, South Africa and Sudan), North America (Canada and Mexico, on corn and wheat) and the USA on wheat and other hosts. It occurs in South America (Argentina, Chile, Venezuela) and several countries of Asia, and in Australia on cereals, clover and carrots (see Castillo and Vovlas (2007) for review).

Molecular and biochemical characterization: Carrasco Ballesteros *et al.* (2007) identified RAPD-PCR DNA fragments associated with *P. thornei* and designed specific primers for its SCAR-PCR diagnosis. It can also be identified by D2-D3, ITS and partial 18S rRNA gene sequences (De Luca *et al.*, 2004; Subbotin *et al.*, 2008). A real-time PCR protocol for quantitative identification of *P. thornei* was developed by Lin *et al.* (2020), and Orlando *et al.* (2024) developed a TaqMan real-time PCR method focused on the D2-D3 expansion segments for the identification and quantification of *P. thornei.*

Pratylenchus vulnus **Allen & Jensen, 1950 (Figs 12.11, 12.12)**

Common name: Walnut root-lesion nematode

Measurements (after Allen and Jensen, 1950; Loof, 1960)

Females: L = 0.46-0.91 mm; a = 26.6-39.5; b = 5.3-7.7; c = 14.2-27.7; V = 78-84; stylet = 16-18 μm.

Males: L = 0.46-0.73 mm; a = 28.3-39.2; b = 5.3-7.4; c = 17.5-29.4; T = 36-66; stylet = 15-18 μm; gubernaculum = 4-6 μm; spicules = 14-20 μm.

Females (n = 14): L = 0.47-0.72 mm; a = 27.8-37.6; b = 5.7-7.7; c = 18.4-24.7; V = 77-82; stylet = 13-16 μm.

Males (n = 9): L = 0.49-0.61 mm; a = 32.1-38.5; b = 6.3-7.2; c = 20.5-23.6; T = 34-53; stylet = 13-15 μm.

Description: *Female:* Body slender, nearly straight when heat-relaxed. **Labial region almost continuous with body contour, with three or four annuli**; heavily sclerotized skeleton extending into body for ca one annulus. **Face divided (as seen at SEM), with dorsal and ventral submedian labial sectors fused together and with labial disc, forming a distinct dumbbell-shaped large cap, while large lateral labial sectors are delineated. Lateral fields with four lines, outer ones smooth or weakly crenate, inner ones closer to each other than to outer ones**; occasionally oblique lines present in inner band. **Stylet with rounded basal knobs**, sometimes cupped anteriorly. Excretory pore ca opposite pharyngo-intestinal junction. **Pharynx overlapping intestine ventrally in a long lobe. Spermatheca functional, oblong** when distended with sperm. **Post-vulval uterine sac ca two vulval body diameters long**, with rudimentary ovary. **Tail tapering, with narrowly rounded to subacute smooth tip**, occasionally irregular, sometimes with one or two annulations. *Male:* Common; generally similar to female in appearance of labial region and pharynx. Lateral fields ending on bursa, which envelops tail. Spicules curved, cephalated; gubernaculum simple.

Castillo and Vovlas (2007) matrix code: A2, B2, C2, D3, E2, F6, G3, H3, I2, J1, K1. Distinguished from *P. pratensis* and *P. coffeae* by labial annuli number, absence of striae around female tail, lateral fields and the post-vulval uterine sac.

Distribution and hosts: Recorded from over 80 species and varieties of plants, many of them woody perennials (fruit crops, roses, olive, grapevine, sour orange, walnut). It is found in warmer climatic areas (southern Europe), although it also occurs in northern Europe as a pathogen of glasshouse crops. Recorded on several hosts from Canada and the USA, where it is a major pathogen of deciduous fruit and nut crops. It also occurs in Mexico on grapes and in Central and South America (Argentina, Brazil, Uruguay). It is recorded from Africa (Kenya and South Africa), several countries in Asia, and in Australia on grapevine and strawberry (see Castillo and Vovlas, 2007 for review). Also been reported in boxwood in Ohio, USA (Lopez-Nicora *et al.*, 2012), and on apple in Tunisia (Chihani-Hammas *et al.*, 2018).

Molecular and biochemical characterization: *Pratylenchus vulnus* can be identified by D2-D3, ITS and partial 18S rRNA gene sequences (De Luca *et al.*, 2004; Subbotin *et al.*, 2008). Hodson *et al.* (2021) developed a qPCR

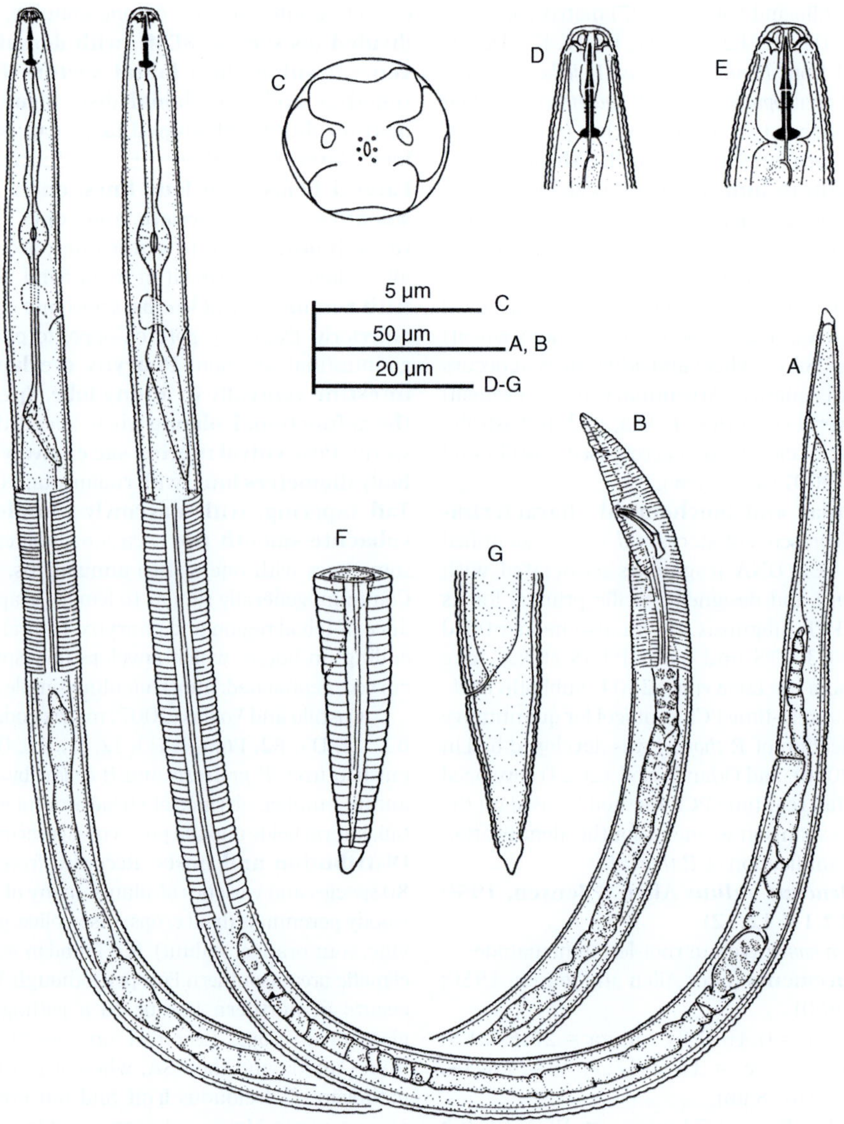

Fig. 12.11. *Pratylenchus vulnus*. A: Entire female; B: entire male; C: female *en face* view; D: male labial region; E: female labial region; F, G: female tails. (After Corbett, 1974.)

assay for detection and quantification of *P. vulnus* in soil.

Pratylenchus zeae **Graham, 1951 (Fig. 12.13)**

Common name: Corn root-lesion nematode

Measurements (after Graham, 1951)

Females: L = 0.54 (0.36-0.58) mm; a = 27 (25-30); b = 6.5 (5.4-8.0); c = 15.2 (17-21); V = 71 (68-76); stylet = 15.5 (15-17) µm.

Description: *Female:* Body slender, almost straight when heat-relaxed, marked by very faint annuli. **Labial region not offset from body, bearing three annuli. Face undivided (as seen at SEM), with labial sectors fused together and with labial disc**. Lateral fields with four lines extending along tail beyond phasmids; inner band showing slight irregularity in mid-body region, but no corresponding fifth line seen in transverse section. **Stylet 15-17 µm long, with broad, anteriorly flattened basal knobs. Spermatheca round, small, without**

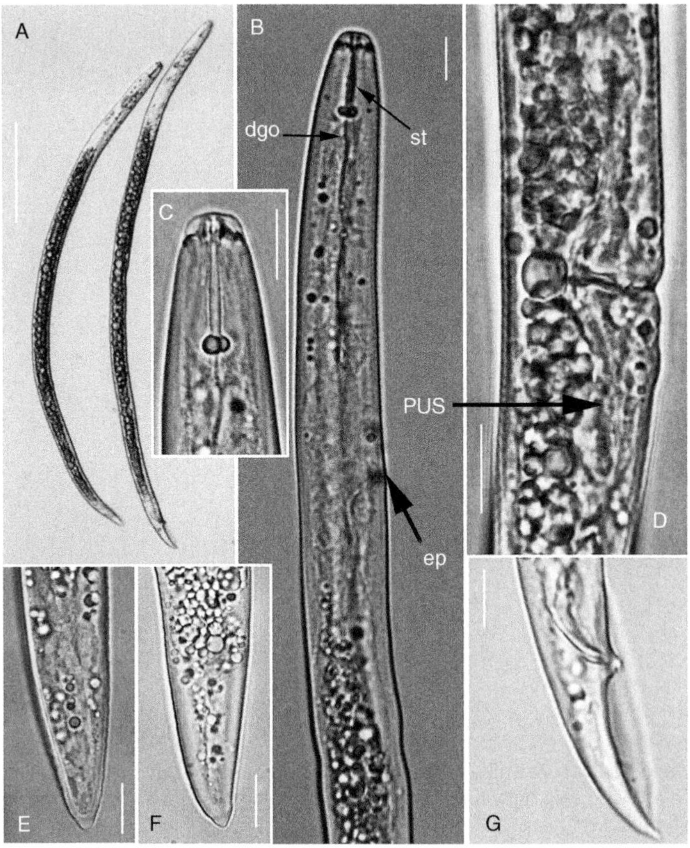

Fig. 12.12. Micrographs of *Pratylenchus vulnus*. A: Entire female and male; B: female anterior region showing stylet (st), dorsal pharyngeal gland outlet (dgo) and excretory pore (ep); C: Detail of labial region; D: female vulval region showing post-vulval uterine sac (PUS); E, F: female tails; G: male tail. (Scale bars: A = 100 μm; B-G = 10 μm.)

sperm (even in the only population recorded with males). **Vulva at 68-76% of body length. Post-vulval uterine sac short, 1-2 body diameters long**, with rudiments of ovary. Intestine with a short post-rectal extension dorsally. Phasmids slightly posterior to mid-tail. **Tail conoid, terminus variable, generally almost pointed, narrowly rounded to subacute, unstriated.** *Male:* Extremely rare, found only once in Ivory Coast (Merny, 1970). Similar to female except for sexual dimorphism. Spicules slender, ventrally arcuate, 14-15 μm long; gubernaculum 4-5 μm in length; bursa margins faintly crenate.

Castillo and Vovlas (2007) matrix code: A2, B2, C3, D1, E1, F5, G3, H3, I2, J1, K1. Distinguished from *P. bolivianus* and *P. curvicauda* by stylet length, vulva position, spermatheca shape and female tail shape.

Distribution and hosts: Parasitizes corn, blackberry, soybean, potato, cotton, coffee, sorghum and sugarcane; recorded from Canada and Mexico on alfalfa and from several states in the USA. It is widespread in several countries in Central and South America. In Europe, it has only been recorded on corn in Bulgaria, Russia and from several localities in Spain. It has been widely recorded on corn in Africa and from corn, sugarcane, tobacco and pulse crops in Asia, and in Australia on sugarcane, cereals, rice and white clover (see Castillo and Vovlas (2007) for review). It has also been reported in upland rice in India and Kenya (Pili *et al.*, 2016; Mondal *et al.*, 2022).

Molecular and biochemical characterization: *Pratylenchus zeae* can be identified by D2-D3, ITS and partial 18S rRNA gene sequences

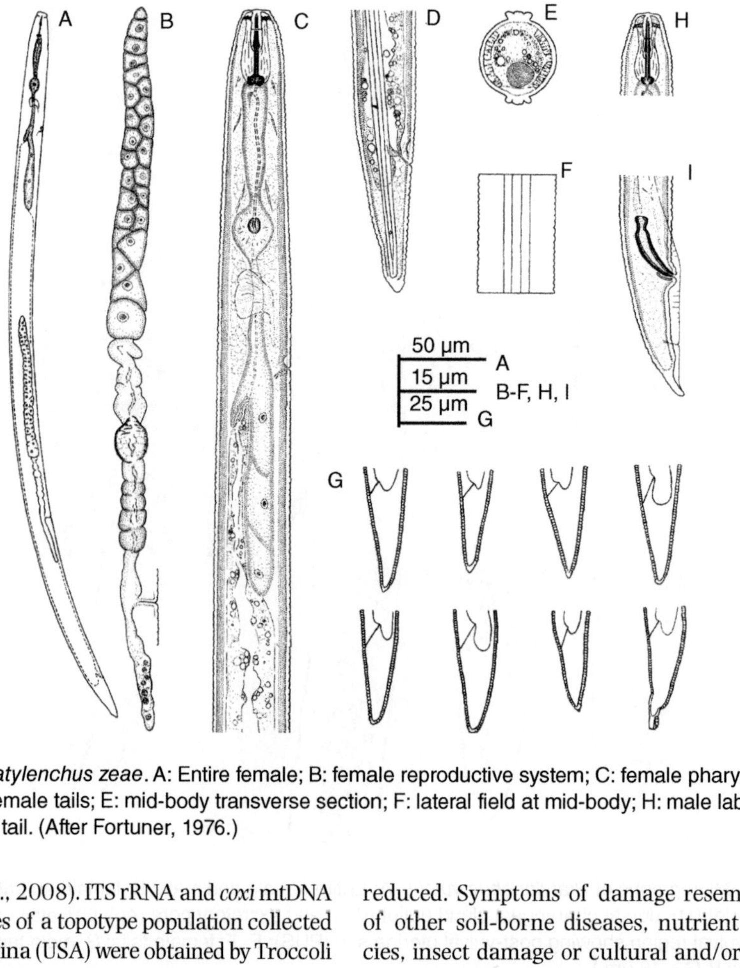

Fig. 12.13. *Pratylenchus zeae*. A: Entire female; B: female reproductive system; C: female pharyngeal region; D, G: female tails; E: mid-body transverse section; F: lateral field at mid-body; H: male labial region; I: male tail. (After Fortuner, 1976.)

(Subbotin *et al.*, 2008). ITS rRNA and *coxi* mtDNA gene sequences of a topotype population collected in South Carolina (USA) were obtained by Troccoli *et al.* (2016). Berry *et al.* (2008) developed a qPCR for the specific and quantitative detection of *P. zeae*, and Liu *et al.* (2017) developed a loop-mediated isothermal amplification based on the mitochondrial COI region to detect *P. zeae*.

12.4.2. Bionomics of *Pratylenchus*

Symptoms of attack/pathogenicity: The most obvious field symptoms of disease on herbaceous and woody plants are round to oval patches of stunted and chlorotic (yellowish) plants, which give the field a ragged appearance. Damage is often most severe in the centre of these areas, diminishing toward the edges where the appearance of plants becomes normal. There is a reduction in the number and size of leaves on heavily infected plants and yield can be substantially reduced. Symptoms of damage resemble those of other soil-borne diseases, nutrient deficiencies, insect damage or cultural and/or environmentally induced stress. The general response of nematode-colonized roots to parasitism by *Pratylenchus* spp. is necrosis or tissue death and a browning that typically involves the epidermis and cortical tissues (Fig. 12.14).

Pathogenicity studies show that lesion nematodes are very efficient parasites, as indicated by the fact that extremely high populations in soil do not kill their host plants. Seinhorst's equation has been accepted as a useful basis for describing the relationship between the initial population density and productivity or yield in several host plant-*Pratylenchus* associations, such as coffee-*P. brachyurus*, coffee-*P. coffeae*, faba bean-*P. neglectus*, faba bean-*P. thornei*, chickpea-*P. thornei* and *Rosa* spp.-*P. penetrans*. Nevertheless, damage thresholds are quite different among *Pratylenchus*-host plant combinations and range from 0.1 nematodes/cm^3 of soil in apple-*P. penetrans*

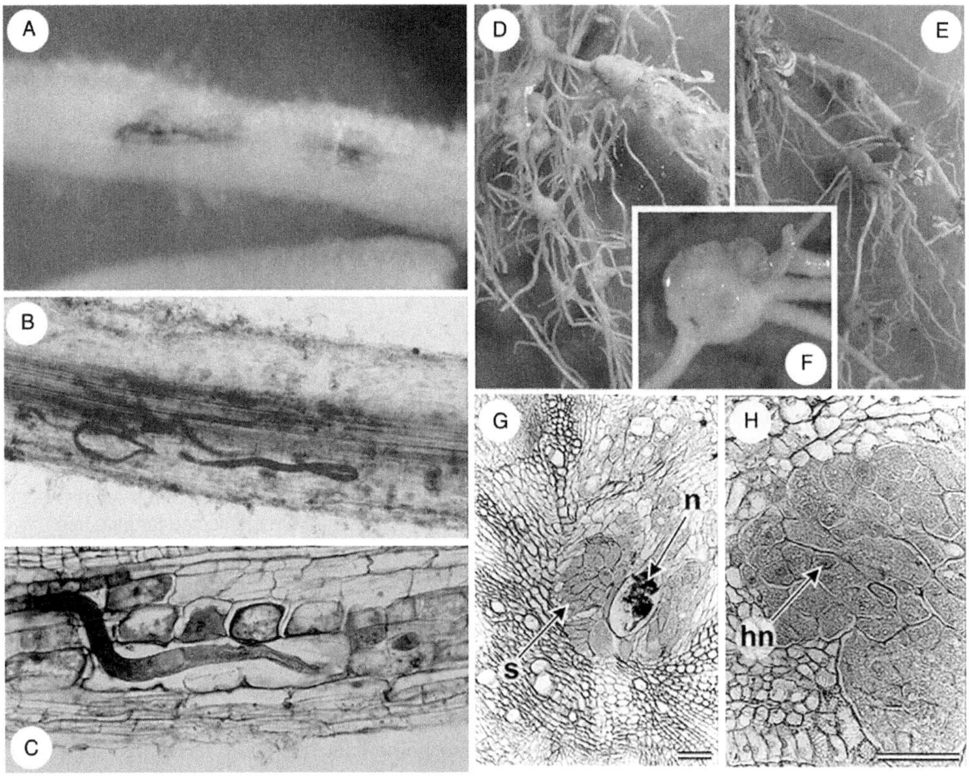

Fig. 12.14. A-C: *Pratylenchus thornei* on chickpea. A: Root necrosis; B: nematodes migrating within cortical parenchyma; C: histopathological changes induced by feeding and migration of nematode showing cavities, hypertrophied nuclei and granular cytoplasm of cells adjacent to nematode pathway. D-H: *Nacobbus aberrans*. D-F: tomato root systems ('platense') with several so-called 'false knots' induced by the nematode; G, H: anatomical changes induced in tomato roots showing multicellular syncytia (s) with mononucleate cells (n = nematode).

to 14 nematodes/cm^3 of soil in soybean-*P. scribneri* (Castillo and Vovlas, 2007).

Pratylenchus brachyurus is an economically important pathogen of peanut in most peanut-producing regions of the world. The nematode attacks fibrous roots, pegs and pods, causing necrotic lesions. Up to 30% soybean yield losses occur in North Carolina, and it has been reported damaging pineapple in several countries, resulting in decreased plant growth, delayed leaf emergence and reduction and yellowing of leaves (see Castillo and Vovlas, 2007 for review).

Pratylenchus coffeae is a major pathogen of coffee and is also pathogenic to many other crops such as banana, citrus, yam and potato in tropical and subtropical countries. In Florida, growth suppression of parasitized citrus trees has been observed in orchards and demonstrated in field trials (Figs 12.15, 12.16). Similar damage has been reported in India, where the nematode parasitizes banana, reducing bunch weight by 13-20% and number of fingers by 34% (see Castillo and Vovlas, 2007 for review). In Florida, infestations of *P. coffeae* were previously reported on St. Augustine grass and other turfgrasses, but subsequent morphological and molecular analyses of these populations indicated that they were actually *P. hippeastri*, the most common species on golf course turf grasses in the state.

Pratylenchus goodeyi is a primary pathogen of banana, causing stunting, delayed maturity and a reduction in plantation life. It predominates in the cooler, higher altitudes of Africa and the Canary Islands. Field studies have indicated that the host range of *P. goodeyi* is narrow, with 69 weed species as non-hosts and

Fig. 12.15. Influence of *Pratylenchus coffeae* on tree growth. A: Five-year-old tangelo (*Citrus tangelo*) grafted on rough lemon (*C. limon*) showing decline symptoms; B: five-year-old uninfected tangelo tree grafted on rough lemon. (Courtesy J.H. O'Bannon.)

only bean and corn as hosts (see Castillo and Vovlas, 2007 for review).

Pratylenchus loosi causes severe damage to tea plantations that becomes noticeable at the very early stages of infection by patches of unthrifty plants. Attack debilitates not only the available feeder roots (inducing slow decline) but also damages the storage roots, thus limiting carbohydrate reserves and therefore recovery from pruning (see Castillo and Vovlas, 2007 for review).

Pratylenchus neglectus and *P. thornei* have been reported to be pathogenic to cereals and legumes, resulting in chlorotic and stunted plants; the former is also pathogenic to alfalfa and wheatgrass. *Pratylenchus neglectus* causes economic damage to barley in North America (Pacific Northwest), where it reduces the yield of intolerant wheat cultivars by as much as 8-36%; it has also been reported in infested wheat fields in Australia. On susceptible wheat cultivars, *P. thornei* is reported to reduce yield by as much as 32% in Mexico and by 44-85% in Australia (see Castillo and Vovlas, 2007 for review).

Pratylenchus penetrans has been found closely associated with the decline of stone-fruits and grapevine. It is one of the most important root-lesion nematodes in western and central Europe and is considered a primary causal agent in replant problems in many parts of the world. *Pratylenchus penetrans* causes a potato disease similar to that caused by the golden cyst nematode and significant losses to tomato have also been reported. Nearly 400 plant species are parasitized by this nematode, including stone and pome fruit trees. Similarly, host-suitability studies showed many vegetables to be good hosts, whereas a few, such as pepper and asparagus, were poor or non-hosts (see Castillo and Vovlas, 2007 for review).

Pratylenchus scribneri parasitizes several hosts including corn, tomato, sweet onion, potato and snap bean. On potato, it causes a growth decline initiated by destruction of the root system, resulting in poor-quality tubers exhibiting scabby or sunken lesions. This disease is common in Florida but also occurs in the Netherlands. In Australia, *P. scribneri* is considered responsible for replant problems (see Castillo and Vovlas, 2007 for review).

Pratylenchus vulnus is regarded as economically important in stone-fruit crops in Mediterranean environments and can play an important role in replant problems. It is believed to be a threat to citrus orchards in Italy, where it has been associated with decline of sour orange seedlings and damage to young trees, has been implicated in olive decline in Italy, and has been frequently encountered in nurseries in Andalusia. In California, damage by *P. vulnus* to plum on peach and plum rootstocks is estimated to be responsible for about 16% reduction in marketable

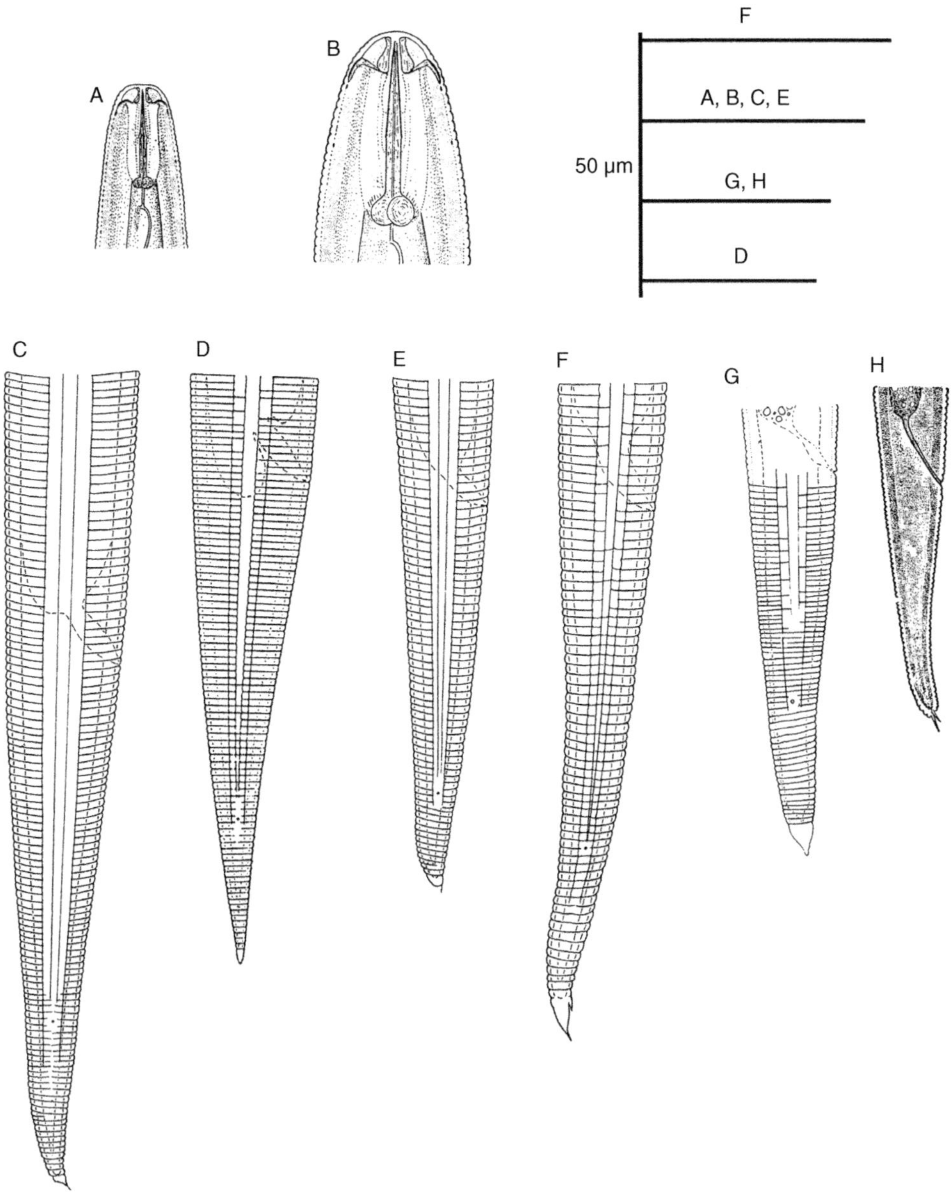

Fig. 12.16. *Hirschmanniella*. A, B: Two types of head shape. A: Flattened, stylet knobs low (*H. gracilis*); B: high, conoid, stylet knobs high (*H. spinicaudata*); C-H: female tails. C: *H. gracilis*, with axial mucro; D: *H. spinicaudata*, no mucro; E: *H. oryzae*, with ventral mucro; F: *H. caudacrena*, with axial mucro plus ventral notch; G: *H. miticausa*, with tip narrowed but no ventral mucro; H: *H. shamimi*, with two mucros. (Scale bars = 50 µm; from Loof, 1991, courtesy Marcel Dekker, Inc.)

fruit, and in Georgia, *P. vulnus* was reported associated with reduced peach tree vigour and rapid deterioration and reduction of feeder roots (see Castillo and Vovlas, 2007 for review).

Pratylenchus zeae can be a major pathogen of corn and tobacco and is also pathogenic to other crops such as sorghum, soybean and sugarcane. Host range studies showed that corn, crabgrass, millet, rye, sorghum, soybean and Sudan grass were very good or good hosts of *P. zeae*, while alfalfa, barley, bean, clover, lettuce, oat, carrot, chickpea, radish, tomato, vetch and watermelon were poor or non-hosts. Infection of corn results in stunted plants with yield losses near 10%, and yield losses ranging from 1243 to 2284 kg ha^{-1} were estimated when soil populations at harvest were 100-200 nematodes per 150 cm^3, respectively. It plays a key role in sugarcane yield decline, which is characterized by the diminishing ability of cane land to produce sugar per harvested hectare (see Castillo and Vovlas, 2007 for review).

Life cycle: The life cycle of *Pratylenchus* spp. is simple and all species attain several generations in a growing season. Eggs are laid in roots, corms or tubers. The first-stage juveniles moult inside the eggs, the second-stage juveniles (J2) hatch, moult through the third and fourth stages and become adults. Amphimictic reproduction occurs in species in which males are common (*P. coffeae*, *P. goodeyi*, *P. penetrans*, *P. vulnus*). Other species in which males are rare (*P. brachyurus*, *P. neglectus*, *P. thornei*, *P. zeae*) reproduce parthenogenetically. The duration of the *Pratylenchus* life cycle lasts from 4 to 8 weeks, but this may be influenced by environmental conditions such as temperature and moisture. Tropical species, such as *P. brachyurus* and *P. zeae*, can complete their life cycle in about 3-4 weeks at 30°C (Castillo and Vovlas, 2007). Species that prefer cooler climates, such as *P. penetrans*, require 6-7 weeks at 20°C (Castillo and Vovlas, 2007). Although hatching in *Pratylenchus* spp. is not influenced by plant root exudates, hatching in *P. brachyurus* and *P. zeae* was influenced by maize roots under glasshouse conditions (Castillo and Vovlas, 2007). In *P. goodeyi*, female fecundity was higher (30 eggs per female) in susceptible banana ('Nakyetengu') than in resistant ('Sukalindizi') banana (19 eggs per female) (Castillo and Vovlas, 2007).

Pratylenchus spp. migrate inter- and intracellularly in roots, corms or tubers and feed mainly on the cytoplasm of cortical cells. No developmental stage may be called exclusively the infective stage because apparently both adults and juvenile stages may enter and leave the roots; indeed, several studies have shown that all mobile life stages of *Pratylenchus* spp. are able to penetrate host roots (Castillo and Vovlas, 2007). *Pratylenchus* spp. move through tissues by breaking down successive cell walls, resulting in a loss of cell turgor pressure and a gradual increase in the size of the nucleus, followed by cell death along the nematode's route (Fig. 12.14A-C). Cell walls collapse and cause cavities and tunnels, which evolve as necrotic lesions that may extend to the entire cortex. Lesions on the root surface are dark brown, reddish-purple to black, elliptical in shape and arranged along the root axis (Fig. 12.14A-C). *Pratylenchus* spp. are cortical parasites, although *P. brachyurus* and *P. zeae* invade all parts of corn roots, including the stele (see Castillo and Vovlas, 2007 for review).

Pratylenchus nematodes move actively no more than 1-2 m from the root zone that they infest, although many agricultural operations favour their dispersal. Migration occurs in soil and between plants only when water is available or soil texture is adequate and soil temperatures lie between the lower and upper thresholds for activity (Castillo and Vovlas, 2007). However, this lack of long-distance movement does not mean *Pratylenchus* cannot spread rapidly from field to field by other means. Farm equipment and even muddy shoes contaminated with *Pratylenchus*-infested soil can rapidly disseminate the nematodes, which also may be dispersed by water during floods and irrigation. The movement of infected plants or propagative parts has been responsible for disseminating *Pratylenchus* long distances, even intercontinentally. In addition, since many *Pratylenchus* species are endemic to native vegetation in many locations, new planting sites may already be infested with the nematode prior to cultivation and this may be one of the main reasons why several species are now cosmopolitan (Castillo and Vovlas, 2007). In the absence of a host crop, *Pratylenchus* spp. survive on weeds or in the soil as eggs, juveniles or adults by means of several survival strategies (i.e., anhydrobiosis). It is well known that weeds play a significant role in the survival of many *Pratylenchus* species, and several have been reported as good hosts and help in the maintenance, multiplication and spread of the nematode within a field, thereby resulting in greater damage to susceptible crops (Castillo and Vovlas, 2007).

Interactions of *Pratylenchus* spp. with other pathogens: Damage caused by *Pratylenchus* species is not frequently obvious since interactions with other pathogens have been documented.

Some disease complexes and interactions between *Pratylenchus* spp. and pathogenic fungi and bacteria have demonstrated that infection by *Pratylenchus* spp. increases the incidence or severity of diseases on susceptible cultivars (i.e., interaction of *Pratylenchus* spp. with Verticillium wilt in potato causing potato early dieback), but the most important interactions between both pathogens are those in which *Pratylenchus* breaks down plant resistance (see Castillo and Vovlas, 2007 for review).

Management: *Pratylenchus* nematodes are difficult to control once introduced into the soil and much effort is spent on prophylactic measures. Most nematode management programmes focus on tactics involving the reduction of the initial nematode population density and/or the suppression of their reproduction during the growing season. Host plant resistance, cover crops, irrigation management, fallow periods, soil amendments and other cultural practices can all be used to minimize damage caused by *Pratylenchus*. Also, although *Pratylenchus* spp. are identified as polyphagous plant-parasitic nematodes with a wide host range, research has shown that the host range of certain *Pratylenchus* spp. can be a useful tool for the management of these species, allowing the implementation of crop rotation for management of soil infestations. Soil solarization has been shown to be effective in controlling *Pratylenchus*. In Australia, soil solarization reduced populations of *P. penetrans* in celery; in Syria, it reduced populations of *P. thornei* while increasing yields of chickpea and faba bean; and it was useful to disinfest nursery soils infested with *P. vulnus*. Hot water treatment may also be an option to control *Pratylenchus* in planting material, for example, infections of banana corms by *P. goodeyi* or infections of yam tubers by *P. coffeae*. Nevertheless, eradication of such infections may be very difficult.

Host resistance offers the easiest, safest and cheapest long-term approach to minimize the effects of *Pratylenchus* on crops. Resistant varieties have been identified in a number of important crops such as banana, potato and sweet potato, various cereals and forage crops, strawberry and some rootstocks of woody plants. Some resistance has been developed for *P. goodeyi* and *P. coffeae* on banana, and studies have shown that *Malus* germplasm varies in resistance or tolerance to apple replant disease caused by *P. penetrans*. Potential sources of resistance to *P. vulnus* have been detected in peach and some wild plum and apricot species and interspecific hybrids (see Castillo and Vovlas, 2007 for review).

Some plant growth-promoting rhizobacteria and bacterial endophytes have been reported to reduce populations of *P. zeae* in corn and *P. penetrans* in potato. Several studies have reported that *Glomus* arbuscular mycorrhizal fungi increased host tolerance in some host plant-*Pratylenchus* pathosystems, such as *P. penetrans* on bean, *P. vulnus* on rootstocks of plum, pear and peach, *P. goodeyi* on banana, *P. zeae* on corn, *P. coffeae* on coffee and banana and *P. penetrans* on carrot.

Soil treatments with chemical compounds to control *Pratylenchus* spp. are mainly used when cultural, physical and biological methods fail to reduce nematode populations sufficiently, but because of environmental and human health concerns there are increasing restrictions on their use. Additionally, due to the high cost and environmental concerns, chemical control is generally restricted to high-value crops such as coffee and banana. Pre-plant fumigation is the method most widely used to reduce population densities of *Pratylenchus* spp. before fruit trees and woody crops are planted in nursery stocks or infested fields. As the current agriculture and agri-food sector is expected to move toward a more sustainable system of production, management of *Pratylenchus* spp. should be based on integrated management programmes using combinations of available control strategies (see Castillo and Vovlas, 2007 for review).

Root-lesion nematodes are regulated in many countries. The tropical species such as *P. coffeae*, *P. loosi*, *P. scribneri* and *P. zeae* are more targeted by these regulations than temperate species. However, in 2022 the temperate *P. penetrans* was regulated in 17 countries (Hockland *et al.*, 2024). Some countries, like China and Egypt, regulate all root-lesion nematodes in the genus, while others provide a list of species that are regulated. South American countries like Argentina, Brazil and Chile regulate four or more *Pratylenchus* species (USDA-APHIS, 2022). In the USA, the State of California regulates *P. alleni*, *P. brachyurus*, *P. coffeae*, *P. flakkensis*, *P. penetrans*, *P. vulnus* and *P. zeae*.

12.5. Genus *Hirschmanniella* Luc & Goodey, 1964 (Figs 12.1, 12.16)

Hirschmanniella species are rather large (1-4 mm long) compared to other plant-parasitic nematodes in the Tylenchomorpha, which are usually <1 mm long. The type species, *H. gracilis*, is one of the oldest known plant-parasitic nematodes, being described by de Man (1880) as *Tylenchus gracilis*, and transferred to *Hirschmanniella* by Luc and Goodey (1964). The genus includes about 32 valid species, usually referred to as rice root nematodes. However, their ability to damage rice has been demonstrated for only seven species (*H. belli, H. gracilis, H. imamuri, H. mexicana, H. mucronata, H. oryzae* and *H. spinicaudata*). The other species have been detected inside rice or weed roots in rice fields, but their damage potential to rice has not been assessed (Peng *et al.*, 2018). *Hirschmanniella* spp. parasitize many aquatic plants, including those used as aquarium plants such as *Hydrophyllum* spp. and *Sagittaria subulata* (L.) Buch., which in Florida are infected by *H. caudacrena* and *H. oryzae* (Fig. 12.17). The majority of *Hirschmanniella* species are regulated by certain states and countries. These regulations are enacted to protect the rice industry from the spread of such damaging nematodes. Identification of *Hirschmanniella* spp. has been based on morphological analysis, but molecular approaches are very promising. Molecular phylogenetics of the genus were provided by Tandingan De Ley *et al.* (2007) and Van Den Berg *et al.* (2009) using D2-D3 and 18S rRNA sequences.

Diagnosis (after Loof, 1991; Siddiqi, 1973, 2000): **Root endoparasites characterized by a long body (1-4 mm). No sexual dimorphism in labial and pharyngeal regions.** Lateral field with four lines, often with partial areolation, occasionally with complete areolation. Labial region either low, flattened with rounded edges or higher and hemispherical, composed of 3-7 fine annuli, not offset from neck (Fig. 12.16). Stylet strong, 15-46 µm, with large basal knobs. Median bulb well developed. Excretory pore near pharyngo-intestinal junction, posterior to hemizonid. **Pharyngeal glands in tandem, forming a long, ventrally overlapping lobe. Tail in both sexes elongate, tapering to a terminus that often bears a mucro** (seldom two), and in some species has a ventral notch just anterior to terminus (Fig. 12.16). **Phasmids located in posterior third of tail.** *Female*: Vulva transverse, lips not modified. **Amphididelphic, outstretched.** Spermathecae round to oval, axial. Ovaries mostly with a single row of oocytes. *Male*: Testis outstretched. Spicules with apparently subterminal dorsal. Gubernaculum fixed. Bursa subterminal, simple, crenate, lacking phasmidial pseudoribs.

Type species: *H. spinicaudata* (Schuurmans Stekhoven, 1944) Luc & Goodey, 1964

Other species: About 32 other species are currently regarded as valid. See Loof (1991) and Siddiqi (2000) for *nomina nuda, species inquirendae* and synonyms. Recent new species include *H. dicksoni* in Florida (Habteweld *et al.*, 2019).

12.5.1. Major species of *Hirschmanniella*

***Hirschmanniella oryzae* (Van Breda de Haan, 1902) Luc & Goodey, 1964 (Figs 12.16A, 12.17)**

Common name: rice root nematode

Measurements (after Sher, 1968a)

Female topotypes (n = 17): L = 1.44 (1.14-1.63) mm; a = 60 (50-67); b = 10.7 (8.8-12.1); b' = 5.7 (4.5-7.2); c = 17 (15-19); V = 52 (50-55); stylet = 17 (16-19) µm; O = 17 (15-19).

Males (n = 10): L = 1.17 (1.01-1.40) mm; a = 59 (52-61); b = 10 (9.1-11.3); b' = 5 (4.6-5.7); c = 17 (16-18); stylet = 17 (16-18) µm; O = 17 (13-18); spicules = 23 (18-26) µm; gubernaculum = 8 (7-9) µm.

Description (after Siddiqi, 1973): *Female*: Body straight or slightly accurate ventrally; **transverse striae distinct, ca 1.6 µm apart near middle; lateral fields not areolated** except in neck region, though occasional incomplete areolation may occur in tail region. **Labial region low, continuous, flattened with rounded edges, with 3-5 (usually 4) annuli;** outer margins of head skeleton prominent, extending posteriorly about three body annuli. Anterior cephalids three annuli posterior to labial region, posterior ones indistinct; hemizonid 2-3 annuli long, 2-4 annuli anterior to excretory pore, usually a little anterior to pharyngo-intestinal junction. **Stylet knobs rounded, ca 3.2 µm across.** Spermathecae oval or sometimes round, with sperm. 'Thorneian cells' (Sher, 1968a), apparently of hypodermal origin, serve to connect intestinal epithelial cells with hypodermis and sometimes showing up to three nerve-like branches (Fig. 12.17G, H, J). **Intestine not overlapping**

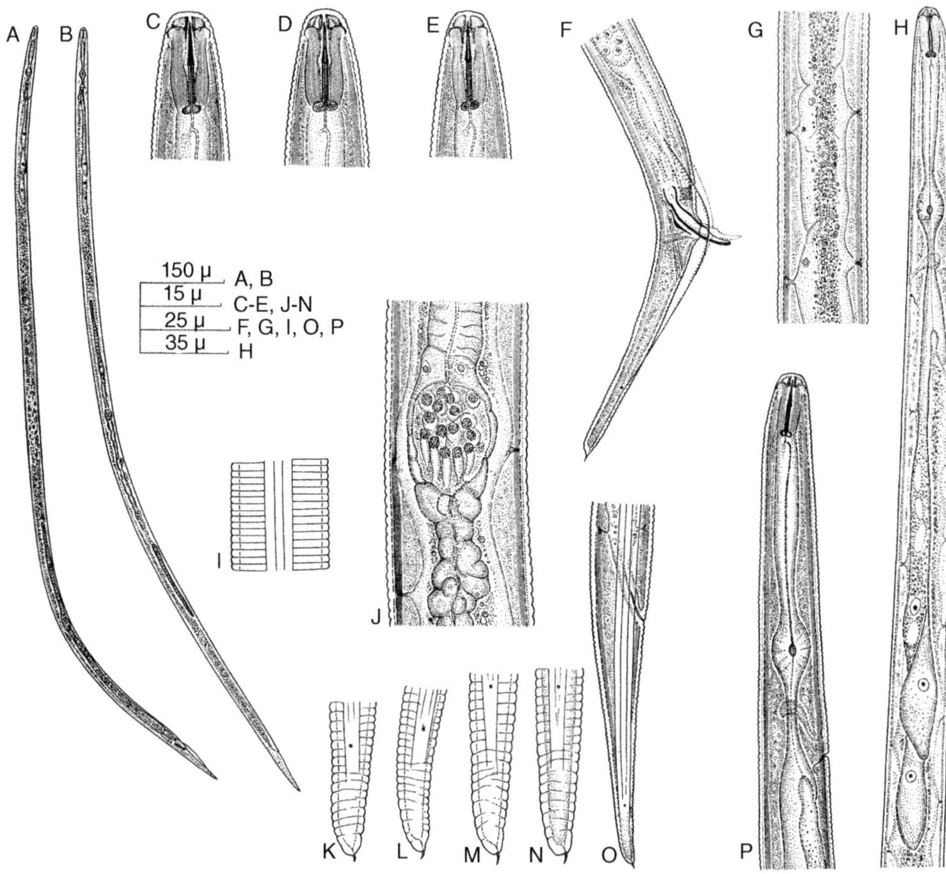

Fig. 12.17. *Hirschmanniella oryzae*. A, B: Adults; C-E: head ends; F, K-O: tail ends; G: mid-body portion showing 'Thornean cells' joining intestinal cells with hypodermis; H, P: pharyngeal regions; I: lateral field at mid-body; J: spermatheca with sperm. (From Siddiqi, 1973, courtesy CABI.)

rectum. Tail elongate-conoid, 4.3-5.5 time anal body diameters long, with rounded terminus bearing a sharp ventral mucro (Fig. 12.17K-O)**, with 49-54 annuli ventrally, striae extending almost to terminus**; without a hyaline terminal portion. Phasmids small, pore-like, 12-17 annuli from tail terminus. *Male:* Labial region, stylet and pharynx as in female. Bursa crenate, arising a little anterior to head of spicules and ending near phasmids. Spicules distinctly cephalated, slightly arcuate, ventrally flanged in distal third; gubernaculum simple, non-protruding. **Tail terminus rounded, with a ventral mucro or sharp projection.**

Hirschmanniella oryzae differs from other species primarily by having a rounded tail terminus with a sharply offset, needle-shaped, ventral mucro (Fig. 12.17K-O).

Distribution: Asia, Africa, the Americas (Brazil, Mexico, San Salvador, USA, Venezuela).

Molecular characterization: Partial sequences of 18S and 28S rRNA genes were provided by Chen *et al.* (2006). Katsuta *et al.* (2016) developed a qPCR assay for detection and quantification of this species.

Hirschmanniella spinicaudata **(Schuurmans-Stekhoven, 1944) Luc & Goodey, 1964 (Fig. 12.18)**

Common name: rice root nematode

Measurements (after Sher, 1968a)

Females (n = 18): L = 3.17 (2.88-3.49) mm; a = 60 (48-72); b = 16 (12-18); b' = 6.5 (5.2-9.5); c = 24 (21-26); c' = 3.9 (3.1-4.7); V = 53 (50-57); stylet = 47 (42-50) µm.

Males (n = 7): L = 2.72 (2.30-2.95) mm; a = 60 (52-71); b = 15 (14-17); b' = 7.3 (5.4-9.2);

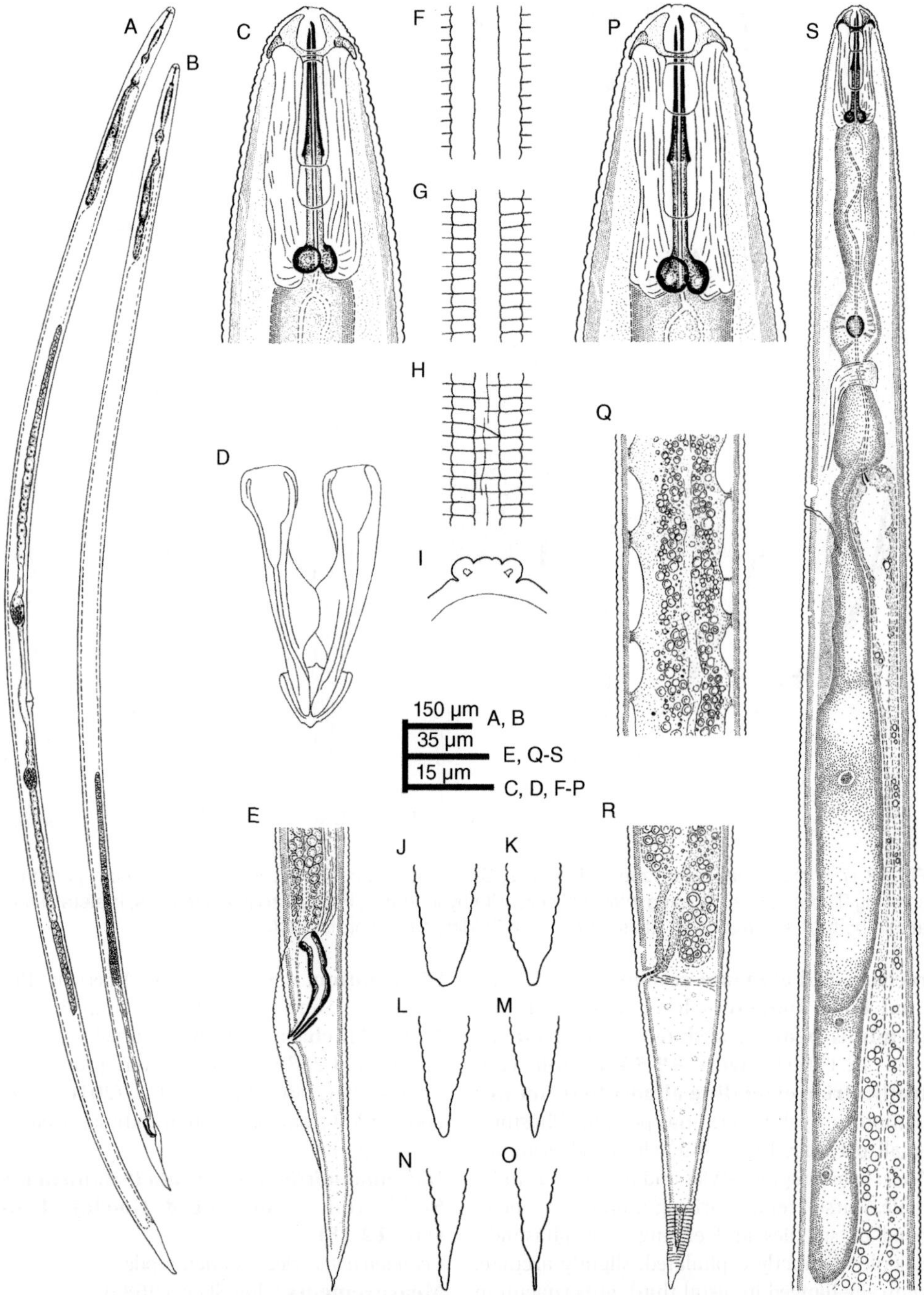

Fig. 12.18. *Hirschmanniella spinicaudata*. A: Female; B: male; C: male head end; D: spicules and gubernaculum (ventral view); E: male tail; F-H: female lateral field; I: cross section of female lateral field; J-O: female tail tips; Q: female mid-body portion showing 'Thornean cells'; R: female tail; S: female pharyngeal region. (From Luc and Fortuner, 1975, courtesy CABI.)

c = 23 (22-24); c' = 4.2 (3.2-5.1); stylet = 43 (42-45) µm; spicules = 51 (46-54) µm; gubernaculum = 18 (16-20) µm.

Description (after Luc and Fortuner, 1975): *Female:* Body elongate, straight or ventrally curved when heat-relaxed; cuticle transversely striated; striae ca 3 µm apart in middle of body; lateral fields with four incisures, **transverse striae crossing all three bands or only two outer ones**; less frequently, lateral fields plain; in some specimens a very faint irregular fifth longitudinal line appearing in central band (Fig. 12.18F-I). **Labial region high, hemispherical, continuous, with 4-6 (usually 5) annuli**; labial framework heavily sclerotized, extending posteriorly for three body annuli. **Stylet long (over 40 µm)**, knobs rounded, 5-7 µm high, 7-9 µm across. Intestine separated from hypodermal layer by 'Thorneian cells' (Fig. 12.18Q). **Intestine slightly overlapping rectum more or less distinctly** (Fig. 12.18R). **Tail elongate-conoid; extremity variable** (in same population) from **rounded to effilate, but never indented or with distinct mucro** (Fig. 12.18J-O); striae extending almost to terminus (Fig. 12.16D); phasmids pore-like, situated in posterior third of tail, 17-24 annuli from terminus. *Male:* Labial region, pharynx and intestine as in female; stylet slightly shorter. Spicules arcuate, cephalated, flanged on the inner edges; gubernaculum curved, simple (Fig. 12.16D). Bursa crenate, extending from level of distal end of spicules to posterior third of tail. Tail elongate-conoid; terminus as in female. Phasmids situated on posterior third of tail.

Hirschmanniella spinicaudata differs from other species by females having a long stylet (over 40 µm), intestine slightly overlapping rectum, tail terminus rounded to effilate but never indented of with a distinct mucro, and striae extending almost to terminus (see Luc and Goodey, 1962; Sher, 1968a; Luc and Fortuner, 1975).

Distribution and hosts: Mainly recorded parasitizing flooded rice (*Oryza sativa*) in North America, South America and Africa. In Nigeria, *H. spinicaudata* was recorded from soil around sugar cane (*Saccharum officinarum*) but not mentioned in roots of this crop (Sher, 1968a). In Ivory Coast, additional hosts include Cyperaceae growing in flooded rice fields, such as *Cyperus viridis*, *Cyperus* sp., *Fimbristylis dichotoma* and *Mariscus longibracteatus*. *Hirschmanniella spinicaudata* was also recorded in the rhizosphere of *Solanum lycopersicum* and *Hibiscus esculentus* in Senegal, but these two crops were growing in fields in which rice had been previously grown. Outside Africa, *H. spinicaudata* was recorded in Venezuela on rice (Sher, 1968a) and in California on *Typha* sp. (Sher, 1968a) and *Juncus* sp. *Hirschmanniella spinicaudata* was frequently found together with *H. oryzae* in flooded rice fields in Nigeria, Gambia, southern Senegal and Venezuela.

Molecular and biochemical characterization: No available data.

12.5.2. Bionomics of *Hirschmanniella*

Symptoms of attack/pathogenicity: Host plants of *Hirschmanniella* species are mainly monocots. *Hirschmanniella oryzae* parasitizes predominantly monocotyledons such as rice, corn (*Zea mays*) and sugar cane, but also dicotyledons (cotton). Several weeds are also hosts. *Hirschmanniella oryzae* is well distributed in many rice-growing areas. About 25 plant species, mostly monocotyledons, are reported as hosts in Indonesia and in Japan. It also occurs on cotton (*Gossypium* spp.) in Egypt, on sugarcane in Ghana, Nigeria (Sher, 1968a) and Indonesia, on corn in Nigeria, and also on aquatic and aquarium plants in Florida (Sher, 1968a; Gerber and Smart, 1986).

The damage and symptoms observed on rice infected by *H. spinicaudata* are similar to those caused by *H. oryzae*. In rice fields of north Cameroon, numerous patches of various sizes (from a few square meters to several hectares) where rice plants were yellowing and etiolated, but of the same size as healthy plants, were attributed to the presence of *H. spinicaudata*. Experiments with *Hirschmanniella* spp. have indicated that one and 10 *H. oryzae* g^{-1} of soil caused 27 and 34% yield loss respectively and the numbers of rice panicles and grain weight were reduced by 16 and 32% respectively, with populations of 1200 *Hirschmanniella*/plant. Above-ground symptoms of nematode damage in the field are not easily identifiable although occasionally yellowing of rice plants may be observed and the onset of flowering may be delayed. *Hirschmanniella* infection induces a reduction of growth, especially in early growth as well as a significant decrease of tillering.

Roots become yellowish brown, necrotic and eventually rot.

In the Ivory Coast, rice yields were increased by 20-53% after nematicide treatment for *H. spinicaudata*, although there was no significant correlation between yields and nematode populations. Symptoms of damage caused by *H. spinicaudata* are similar to those of *H. oryzae*. In micro-plots, *H. spinicaudata* had a depressive effect on tillering and reduced the number of rice panicles and the fresh weight (Merny, 1970).

Life cycle: *Hirschmanniella oryzae* is a migratory endoparasite of roots. The juveniles and adults always enter roots at some distance from the tip, moving freely in the air channels between the radial lamellae of the parenchyma and, in older roots, may be found anywhere from base to tip. Sometimes several have been found in the coleoptile, but none in the lower parts of the leaf sheaths (Van der Vecht and Bergman, 1952). A few days after entry, the female lays eggs which probably hatch in 4-5 days, the development to adult taking at least a month. The multiplication factor per generation is as high as 13 (Van der Vecht and Bergman, 1952), with up to two generations per year (Kuwahara and Tyatomi, 1970). The nematode increased 10- and 15-fold in 100 and 130 days respectively, and as many as 2510 individuals were recovered from 10 g of roots of rice 'Tjina' (Van der Vecht and Bergman, 1952).

Management: *Hirschmanniella oryzae*, *H. spinicaudata* and other species are disseminated with infected rice seedlings, irrigation and flood water and in soil adhering to machinery and workers' clothes and implements. *Hirschmanniella oryzae* survives for more than 12 months in the absence of hosts and in soils that are not continually wet. It persists on weeds of Cyperaceae and Gramineae. These nematodes should be managed by fallow, weed control, rotation with non-host plants (leguminous crops) and strict sanitation practices in the rice nurseries.

Rice-producing countries enact phytosanitary measures against *Hirschmanniella* species, these regulations mainly targeting *H. oryzae* and *H. spinicaudata*, which both are on the lists of regulated nematodes in Egypt and India (USDA-APHIS, 2022). Phytosanitary measures against *H. oryzae* are enacted by the State of California (USA).

12.6. Genus *Radopholus* Thorne, 1949 (Figs 12.19, 12.20)

Radopholus species are migratory root endoparasites native to Australasia. Their body is <1 mm (0.4-0.9 mm) long. Most of these species are of minor economic importance because they parasitize indigenous plants and their agricultural impact in the geographical areas where they have been found has not been demonstrated. *Radopholus similis* is the only species of major economic relevance. Much of the information presented in this section concerns this nematode.

Diagnosis (after Siddiqi, 2000): **Marked sexual dimorphism in anterior body region.** Males show a more offset labial region than females and poorly developed stylet and pharynx. Labial patterns of females, as observed at SEM, consist of submedian labial sectors fused together and with labial disc, and separated by lateral lips, which are wedge-like and extend to base of labial region. **Pharyngeal glands overlapping intestine dorsally,** their nuclei located posterior to pharyngeal-intestinal junction. Vulva at mid-body (50-70%). **Ovary didelphic, amphidelphic**, with two distinct rounded or oval spermathecae filled with rod-like or rarely round

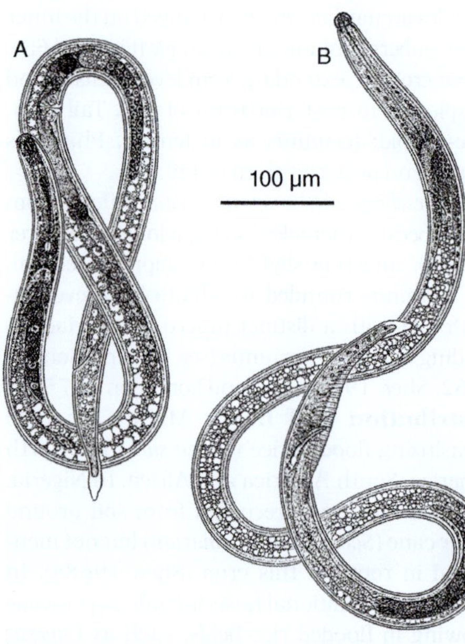

Fig. 12.19. *Radopholus similis*. A: Female; B: male. (From Orton Williams and Siddiqi, 1973, courtesy CABI.)

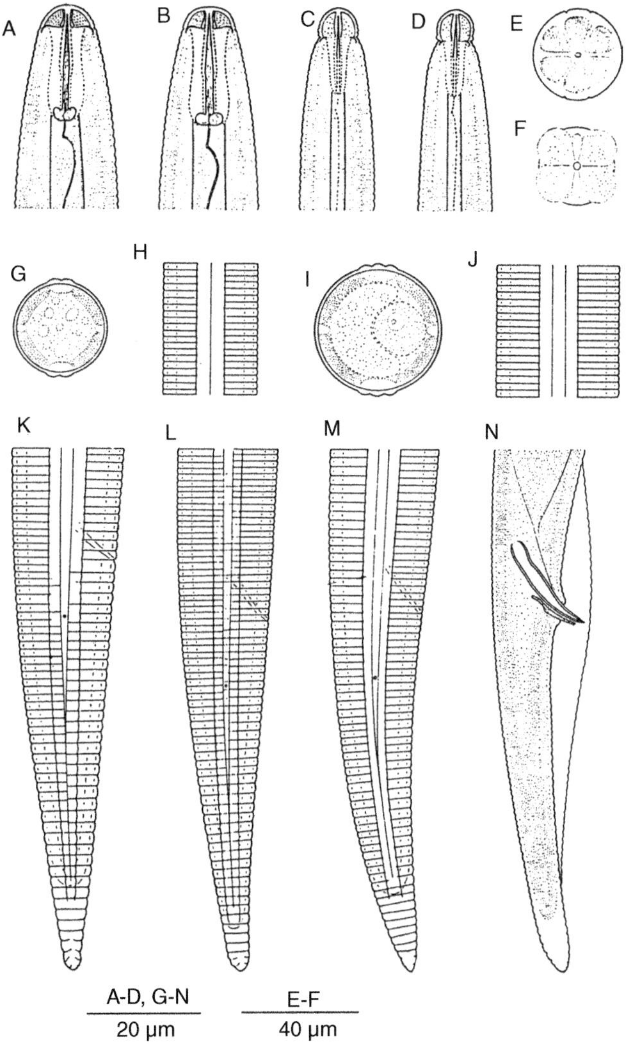

Fig. 12.20. *Radopholus similis*. A, B: Female anterior end; C, D: male anterior end; E: female face view; F: male face view; G: juvenile cross section near mid-body; H: juvenile lateral view near mid-body; I: female cross section near mid-body; J: female lateral view near mid-body; K-M: female posterior region; N: male posterior region. (From Orton Williams and Siddiqi, 1973, courtesy CABI.)

or kidney-shaped sperm. **Female tail conoid or cylindroid with narrow, bluntly pointed or hemispherical annulated or smooth terminus. Male tail more tapered and elongated than that of female and with sub-terminal bursa.** Spicules cephalated, slightly arcuate. Gubernaculum large, protrusible.

Type species: *Radopholus similis* (Cobb, 1893) Thorne, 1949

Other species: More than 50 species of *Radopholus* have been recognized although perhaps only about 21 are valid. See Ryss and Wouts (1997), Siddiqi (2000) and Trinh *et al.* (2004) for *species inquirendae* and synonyms. Ryss and Wouts (1997) also provided a key to species. *Radopholus daklakensis* was described parasitizing robusta coffee in Vietnam (Trinh *et al.*, 2012).

12.6.1. Major Species of *Radopholus*

Radopholus similis **(Cobb, 1893) Thorne, 1949 (Figs 12.19, 12.20)**
Common name: burrowing nematode
Radopholus similis is a migratory endoparasitic nematode primarily found in plant roots and widely known as a destructive pest of banana (*Musa* spp.), citrus (*Citrus* spp.) and black pepper (*Piper nigrum*). It has been found on more than 365 different plant species throughout the tropics and subtropics, with banana as the main host plant (Holdeman, 1986). This species is considered the most damaging nematode pest of citrus in Florida where extensive research and studies have been conducted on this nematode and its parasitism to citrus since the early 1950s. The results of these studies have also benefited the banana industry worldwide. *Radopholus similis* is a major target of phytosanitary measures in more than 50 countries, the aim being to prevent its introduction into citrus-growing areas.
Measurements (after Sher, 1968b)
Topotype females (n = 12): L = 690 (520-880) μm; a = 27 (22-30); b = 6.5 (4.7-7.4); b' = 4.5 (3.5-5.2); c = 10.6 (8-13); c' = 3.4 (2.9-4.0); V = 56 (55-61); stylet = 19 (17-20) μm; O = 18 (12-20).
Topotype males (n = 5): L = 630 (590-670) μm; a = 35 (31-44); b = 6.4 (6.1-6.6); b' = 4.8 (4.1-4.9); c = 9 (8-10); c' = 5.7 (5.1-6.7); stylet = 14 (12-17) μm; spicules = 20 (19-22) μm; gubernaculum = 9 (8-12) μm.
Description (after Orton Williams and Siddiqi, 1973): *Female:* Body straight to slightly arcuate ventrally; cuticle distinctly annulated. **Lateral field with two incisures or three bands, not areolated except towards extremities**, arising from near median pharyngeal bulb and ending near tail terminus; inner incisures coalescing near middle of tail. **Labial region hemispherical, sometimes offset, usually with 3-4 annuli**; sclerotization strong; dorsal and ventral arms of framework not wider than submedians; six equal lips. **Stylet ca 18 μm long, with well-developed round basal knobs** usually indented anteriorly; dorsal knob sometimes appearing larger than subventrals. Reproductive organs paired, opposed, outstretched. Spermathecae spherical, usually packed with small rod-shaped sperms. **Tail somewhat elongate-conoid with a narrow rounded or indented terminus.** *Male:* **Pharynx and stylet degenerate; median bulb and valvular apparatus indistinct, stylet with very small or without distinct knobs. Labial region elevated, 4-lobed, with lateral lips considerably reduced, not strongly sclerotized**, with 3-5 annuli posteriorly. Bursa coarsely crenate, enveloping ca two-thirds of tail. Spicules strongly cephalated, 18-22 μm long, with pointed distal ends. Gubernaculum rod-like, protrusible, with distinct sharp claw-like titillae at distal end.
Distribution and hosts: Widespread in tropical and subtropical regions and present in temperate countries under controlled environmental conditions. It is the major banana root-damaging organism in most banana and plantain-growing regions of the world (Gowen *et al.*, 2005). It has been reported on various hosts under field conditions and also on indoor ornamental plants. In Florida, USA, *R. similis* causes 'spreading decline of citrus'. The wide distribution of *R. similis* has been caused unintentionally by international trade involving nematode-infected plant material (especially banana setts) (Marin *et al.*, 1998).

The host range is wide and encompasses at least 365 species (Holdeman, 1986). Hosts of major economic importance include banana, black pepper, citrus, coffee (*Coffea arabica*), many palms (betel-nut palm (*Arecha catechu*), coconut (*Cocos nucifera*), oil palm (*Elaeis guineensis*) and parlour palm (*Chamadorea* spp.)) and tea (*Camellia sinensis*). Many ornamental plants in the Araceae, especially *Anthurium* spp., are seriously damaged. Aquatic Araceae, such as *Anubias* spp., are commonly parasitized by the nematode which invades and also damages the foliar tissues (Lehman *et al.*, 2000).
Molecular and biochemical characterization: According to Kaplan *et al.* (2000), the *R. similis* genome is highly conserved and there is little intraspecific variation in the DNA sequences of genomic regions used for taxonomic purposes among populations from distant geographical areas. However, recent studies by Múnera Uribe *et al.* (2010) on two Colombian populations parasitizing plantain showed a high genetic variability as confirmed by the high ITS1-5.8-ITS2 sequence divergence within these populations (5.4%), which was comparable to the divergence (4.2-6.9%) among the *Radopholus* species included in the analysis. These results, which are in conflict with those of Kaplan *et al.* (2000), could be due to the different rDNA regions analysed in the

two studies. Kaplan's study was based on ITS1, whereas Múnera Uribe's study was based on the IT1-5.8-ITS2 region, which contains more parsimony-informative substitutions than ITS1 (Múnera Uribe *et al.*, 2010).

These findings by Múnera Uribe *et al.* (2010) and those by other authors (Nguyen *et al.*, 2003) suggest that *R. similis* may be a species complex. Phylogenetic analyses conducted by Múnera Uribe *et al.* (2010) indicate that two newly described species, *R. arabocoffeae* (Trinh *et al.*, 2004) and *R. duriophilus* (Nguyen *et al.*, 2003), cluster in a clade closely related to *R. similis*. The genome of this species has been sequenced by Mathew and Opperman (2019) and Wram *et al.* (2019).

Radopholus musicola **Stanton, Mundo-Ocampo, Baldwin & Kaplan, 2001 (Fig. 12.21)**

Common name: burrowing nematode

Measurements (after Stanton *et al.*, 2001)

Paratype females (n = 18): L = 696 ± 52.3 μm (572-668); a = 28.9 ± 2.6 (24.0-33.3); b = 8.0 ± 0.7 (6.9-9.0); b' = 4.8 ± 0.3 (4.1-5.6); c = 9.8 ± 1.2(7.8-12.1); c' = 4.5 ± 0.5(3.8-5.8); V = 54.9 ± 1.5 (52.6-57.0); stylet = 17.8 ± 1.1 (15.0-21.0) μm; stylet knob width = 3.8 ± 0.6 (3.0-5.0) μm; DGO = 3.0 ± 1.0 (2.0-4.8); tail = 72.4 ± 11.3 μm (57-91); h = 7.8 ± 11.8 (5.5-13.5).

Paratype males (n = 18): L = 614 ± 27.6 μm (604-781); a = 35.5 ± 2.2 (29.8-39.2); b = 9.6 ± 1.0 (8.3-12.3); b' = 6.4 ± 0.8 (4.7-7.8); c = 7.5 ± 0.4 (7.0-8.8); c' = 7.5 ± 1.2 (5.6-10.5); T = 29.4 ±

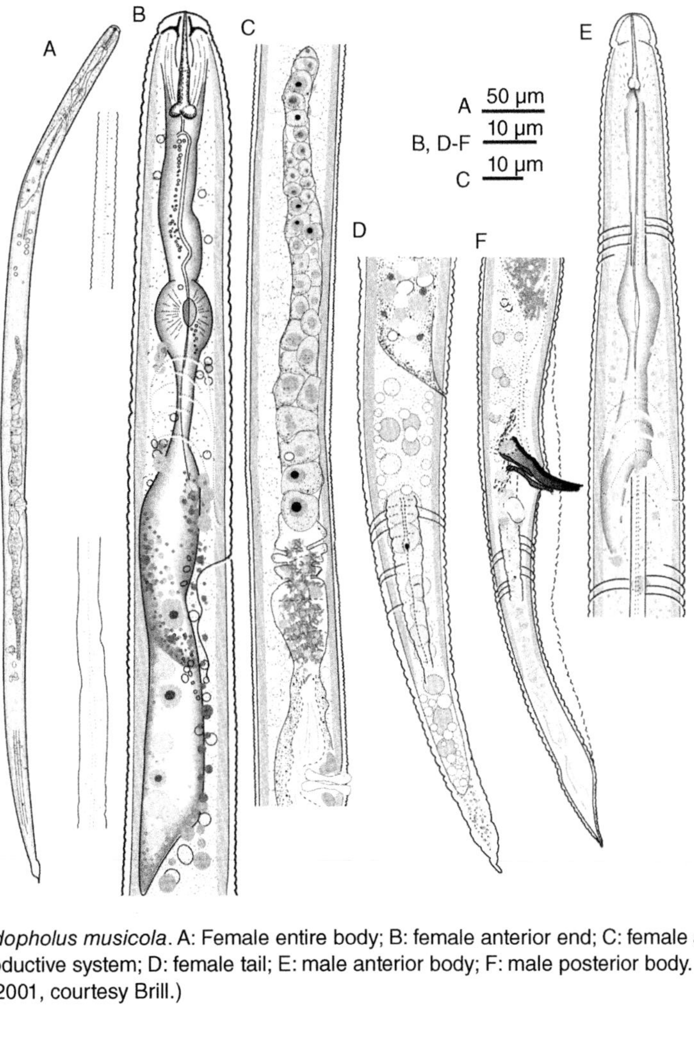

Fig. 12.21. *Radopholus musicola*. A: Female entire body; B: female anterior end; C: female anterior portion of reproductive system; D: female tail; E: male anterior body; F: male posterior body. (From Stanton *et al.*, 2001, courtesy Brill.)

3.9 (22.5-35.8); stylet = 10.5 ± 2.1 (9.0-12.0) μm; stylet knob width = 1.5 ± 0.4 (1.0-2.5) μm; tail = 82.1 ± 4.6 (74-93) μm; h = 7.5 ± 1.6 (5.0-9.5) μm; spicule = 18.9 ± 2.1 (15.0-22.0) μm; gubernaculum = 10.1 ± 0.7 (8.0-11.0) μm.

Description: *Female:* Body slightly curved ventrally. **Lateral field, as viewed by light microscopy (LM), with alae separated by two deep incisures extending nearly entire body length. Often with additional faint areolation and inner two incisures (best visible by SEM).** Anterior to level of excretory duct, and posterior to phasmid, inner two lines converging into a single line; faint areolation visible by SEM. **Labial region slightly offset, with 2-3 annuli**, some of which are incomplete (not extending around entire circumference of labial region). Stylet stout, cone typically slightly longer than shaft plus knobs. **Stylet knobs round (not pointed anteriorly or flattened), broad dorsal and subventral knobs about equal to one another in size and shape.** Pharyngeal corpus and isthmus about equal in length to gland lobe. Reproductive system didelphic, amphidelphic. Vulva slightly protruding and near middle of body; slit ventral, transverse. Nearer uterus, reproductive tract includes a non-offset spermatheca; **mature female spermatheca large, oval-shaped, filled with rod-shaped structures and a few vesicle-like spherical elements. Tail tapering, tip conoid, terminating as a sharp point or a minute spike.** Phasmid opening in anterior half of tail. Hyaline region of tail variable in length, less than anal body diameter. *Male:* Strongly sexually dimorphic. Labial region distinctly offset, wider than high, bulbous, with large subdorsal and subventral swellings and reduced lateral lips. Obscure amphid openings far anterior on lateral lips. Annulation on labial region weakly defined. **Stylet weakly developed, with little separation between cone and shaft; stylet knobs distinctly separated,** about twice diameter of shaft; stylet musculature weak. Pharynx weakly developed with corpus longer than isthmus and gland lobe. Reproductive system monorchic. Spicules well developed with distinct manubrium. Gubernaculum flattened with pronounced pair of titillae. Tail nearly cylindrical, becoming conical posterior to caudal alae; **tail terminus sharply pointed.**

Radopholus musicola morphologically resembles *R. similis* from which it differs by two deep outer folds and two faint, almost fused inner incisures of the lateral field (*R. similis* has lateral bands not folded and inner incisure distinct and separated), and also by the shorter male stylet (8.8-12 *vs* 13.0-13.5 μm) (Stanton *et al.*, 2001).

Distribution and hosts: Only reported from the type locality from roots of *Musa acuminata* 'Williams' (genome group AAA) at Berrimah Research Station, Darwin, Northern Territory, Australia.

Molecular and biochemical characterization: None available.

12.6.2. Bionomics of *Radopholus*

Symptoms of attack/pathogenicity: *Radopholus similis* is the only species in the genus that has been the subject of long-term field and laboratory research aimed at elucidating its biology, parasitism and pathogenic effects on different crops. It causes large losses to citrus in Florida and black pepper plantations in Indonesia. On banana, *R. similis* is a major pest in many countries, Wehunt and Edwards (1968) stating that, in Central America, uninfested banana plots yielded up to 17,000 lb/acre (19,000 kg/ha) per year more fruit than infested plots. The nematode has also been introduced into heated glasshouses in Europe by way of trade with ornamental plants, in particular *Anthurium* (EPPO, 2008). These ornamentals are severely damaged by the nematode (Fig. 12.22).

Roots parasitized by *R. similis* show dark lesions on their surface. Major anatomical alterations consist of large cavities and necrotic areas that expand from the cortical parenchyma into the vascular tissues, disrupting the flow of water and nutrients from the roots to the canopy and vice versa.

Root invasion, migration, feeding and tunnelling result in deterioration of the root system. Subsequent invasion of secondary pathogens destroys the infected root tissues with consequent debilitation of the infected plants and yield losses. In the case of parasitized banana trees, the root system is not able to anchor the plant to the ground causing toppling of the plants, especially when in bunch, and resulting in loss of saleable fruit. Other symptoms include lack of vigour, premature defoliation and reduction in bunch weight, size and number of leaves (Fig. 12.23A).

The syndrome induced by *R. similis* on citrus is called spreading decline as the disease spreads about 15 m/year. The symptoms consist of defoliation, reduction of leaf and fruit size and exposed branches (Fig. 12.23B), which are usually covered by thick foliage in healthy trees. Leaves and fruit are small and wilted especially in dry conditions due to the inability of the debilitated root system to deliver sufficient water to the canopy. Feeder roots in the deep soil layers below 30 cm are more damaged than those in the shallow layers above 30 cm. Spreading decline only occurs in the sandy soils of Florida where it causes crop losses ranging from 40 to 70%. In spite of these heavy losses, the economic impact of *R. similis* to the citrus industry worldwide is modest (0.1-0.2% of the total citrus production worldwide) due to the fact that the nematode damages citrus only in the coarse sandy soils of Florida (Tarjan and O'Bannon, 1984; Shokoohi and Duncan, 2018).

In glasshouse tests, similar symptoms were induced by *R. citri* on citrus seedlings (Machon and Bridge, 1996) and *R. musicola* on banana plantlets (Stanton *et al.*, 2001).

Fig. 12.22. A: Declining stand of anthurium (*Anthurium andreanum*) infested by *Radopholus similis*; B: healthy stand of anthurium. (Courtesy N. Vovlas.)

Fig. 12.23. A: Banana plantation with decline symptoms caused by *Radopholus similis*. Note leaning plants with compromised root system, prone to topple in windy conditions; B: Grapefruit (*Citrus paradisi*) orchard exhibiting spreading decline symptoms caused by *Radopholus similis*. Note defoliation exposing large branches of the infected trees. (Original.)

Life cycle: *Radopholus similis* and *R. musicola* share similar parasitic habits and biology. *Radopholus similis* is a migratory endoparasite and penetrates, develops and reproduces inside the root tissues where it remains until it has to leave by overcrowding and root decay. The life cycle is completed in 20-25 days at 24-32°C and in 18-20 days at 24-27°C (DuCharme and Price, 1966). All juvenile stages and females are infective. Mature males do not feed. Fertilization is usual, but hermaphroditism does occur (Kaplan and Opperman, 2000). Survival in soil and in absence of a host is short (less than six months), but may be long-lasting if residual host roots remain in the soil. At least two races of *R. similis* have been reported: the 'banana race' parasitizing banana but not citrus, and the 'citrus race' that infects both (DuCharme and Birchfield, 1956), the latter being elevated to species status (*R. citrophilus*) by Huettel *et al.* (1984), although this was later rejected (Kaplan and Opperman, 1997, 2000; Kaplan *et al.*, 2000).

Radopholus musicola reproduces by both cross fertilization and hermaphroditism as in *R. similis*. However, *R. musicola* does not parasitize citrus, but attains population levels in banana roots comparable to those of *R. similis* (Stanton *et al.*, 2001).

Association with other organisms: The incidence of Panama wilt caused by *Fusarium oxysporum* Schlechtend. f. sp. *cubense* doubled in 'Gros Michel' bananas when *R. similis* was added to the soil and wilt symptoms appeared faster on *R. similis*-infected bananas (Loos, 1959). Spreading decline of citrus involves an interaction of *R. similis* with *Fusarium*, *Sclerotium* and other soil-inhabiting organisms (DuCharme, 1968).

Management: Information on management practices is only available for *R. similis*. A major objective of recent management strategies of *R. similis* in banana plantations and citrus orchards is the adoption of pest exclusion programmes using certified nematode-free planting stock, vitroplants, sanitation practices and resistant rootstocks. These non-chemical approaches are effective in replanting situations when new land free from *R. similis* is available. A successful replanting programme was implemented in Florida when the citrus industry expanded into new land in south Florida that was free from *R. similis*. These nematode exclusion programmes are successful in banana operations in Martinique where bananas are grown in enclosed environments. However, chemical approaches (using applications of oxime carbamates and organophosphates) are still necessary in many banana-exporting countries in Central and South America where replanting of banana plantations is uncommon and where it remains necessary to suppress nematode populations in established banana plantations (Gowen *et al.*, 2005). This approach is also used in old and infested citrus orchards (Duncan, 2005).

In 2022, *R. similis* was included on the lists of regulated nematodes by 64 countries (Hockland *et al.*, 2024).

12.7. Genus *Nacobbus* Thorne & Allen, 1944

Nacobbus includes endoparasitic, gall-inciting nematode species endemic to North and South America from where it is reported predominantly from dry or semi-arid environments and at high altitudes. Although *Nacobbus*, the false root-knot nematode, is endemic to the American continent, it justifies international quarantine regulations due to its broad host range and economic importance. Its host range includes species of quarantine importance such as potato (*Solanum tuberosum*), sugarbeet (*Beta vulgaris*), tomato (*Solanum lycopersicum*), chili pepper (*Capsicum* spp.) and beans (*Phaseolus vulgaris*). It has not been reported from Central America. Occasional introductions to other countries have been reported, and records from India and China need confirmation (Manzanilla-López *et al.*, 2002).

Recent classifications (Siddiqi, 2000; Andrássy, 2007; Decraemer and Hunt, 2024) indicate *Nacobbus* as the only genus in the Nacobbinae Chitwood & Chitwood, 1950, which, with the Pratylenchinae, is included under the Pratylenchidae (Siddiqi, 2000).

Diagnosis (modified after Siddiqi, 2000): Nacobbinae. Strong sexual dimorphism. *Young female:* Vermiform, elongate-slender, ca 1 mm long. Cuticle distinctly annulated. Lateral fields with four incisures, irregularly areolated. Phasmid pore-like, anterior to middle of tail. Labial region broadly rounded, continuous, with 3-4 annuli. Median bulb rounded, with large refractive thickenings and prominent valve; pharyngeal glands overlapping intestine dorsally. Stylet strong, 21-25 µm long, conus about as long as posterior part, knobs rounded. Pharyngeal glands elongate, extending for more than two body diameters; subventral glands asymmetrical, extending past dorsal gland; gland nuclei lying in tandem posterior to pharyngo-intestinal junction. Vulva a transverse slit, located at 89-97% of body length, within 1-2.3 anal body diameters from anus. Ovary immature, short (with few tandem oocytes), followed by oviduct, which, at its junction with, uterus occasionally appears like a 'spermatheca'. Tail tapering to a broadly rounded terminus, 1-2 anal body diameters long. **Migratory juvenile stages, vermiform female and male in soil and roots.** *Mature female:* Body saccate, tapering anteriorly from median bulb and posteriorly from uterine region, often appearing spindle-shaped; early stage of mature female usually batatiform, with subterminal anus and vulva, **very long, tubular uterus and a long, serpentine ovary** reaching pharynx. Swollen body may contain eggs. **Sedentary endoparasites inciting root-galls.** *Male:* Vermiform, with well-developed labial sclerotization. Stylet 23-27 µm long, pharynx structurally similar to that of immature female. Pharyngeal glands elongate, mostly dorsal to intestine. Testis single, outstretched. Spicules cephalated, ventrally arcuate, 20-35 µm long. Gubernaculum simple, linear to trough-shaped, fixed. Cloacal lips not modified. Tail completely enveloped by a bursa. **Migratory in soil and roots.**

Type species: *Nacobbus dorsalis* Thorne & Allen, 1944 (= *species inquirenda*)

Other species: *N. aberrans* (Thorne, 1935) Thorne & Allen, 1944, *N. bolivianus* Lordello, Zamith & Boock, 1961, *N. celatus* Lax, González-Itting, Rondan Dueñas, Andrade, Gardenal, Franco & Doucet, 2021.

Note: Criteria to identify and separate *Nacobbus* species have been based mainly on adult stage morphology, especially of the vermiform immature female, and includes number of annuli between vulva and anus (VA) and vulva position expressed as percentage of the body length (V). Other features include host range and biological features. Although four species and one subspecies were proposed before Sher's revision of the genus (Sher, 1970), he recognized only *N. dorsalis* and *N. aberrans* as valid. Several characters have been used to separate the females of these two species, including VA, V, body shape of the swollen adults, and the retention or not of embryonated eggs within the obese females. Variation within *N. dorsalis*, the type species, remains largely unknown (it is usually regarded as a *species inquirenda*), but *N. aberrans sensu* Sher populations can differ greatly in morphology and host range (Manzanilla-López *et al.*, 2002). Neither morphometric partition of variation nor morphology has been successful in separating cryptic species, although subtle morphological differences in labial region, labial patterns and in stylet knob shape between populations from North and South America occur (Baldwin and Cap, 1992). Many authors, using molecular techniques, recognize the existence of a species complex rather than a single entity (Ibrahim *et al.*, 1997; Reid *et al.*, 2003; Anthoine and Mugniéry, 2005a; Lax *et al.*, 2007, 2021; Vovlas *et al.*, 2007; Ramirez-Suarez and Powers, 2011).

Nacobbus aberrans sensu Reid *et al.* (2003) is the major species, although this taxon may represent several cryptic species, while *N. bolivianus* (regarded as a synonym of *N. aberrans* by some authors) is apparently restricted to Bolivia where it is found mainly on potato in the high Andes. The difficulties in partitioning morphological and morphometric variation within *N. aberrans sensu* Sher have been attributed to sibling or cryptic species whose existence is more likely to be demonstrated by molecular methodologies rather than by relying upon classical methods alone (see Manzanilla-López, 2010 for a review). Application of integrative taxonomical approaches demonstrates that *N. aberrans* comprises a complex of almost cryptic species and resulted in the description of a new species, *N. celatus*, widely distributed on tomato in the lowlands of Argentina (Lax *et al.*, 2021).

12.7.1. Major Species of *Nacobbus*

***Nacobbus aberrans* (Thorne, 1935) Thorne & Allen, 1944 (Fig. 12.24)**

Common name: false root-knot nematode

Measurements (after Sher, 1970)

Mature female topotypes (n = 6): L = 1.0 (0.8-1.2) mm; stylet = 22 (20-24) µm.

Immature female topotypes (n = 12): L = 0.84 (0.71-0.93) mm; a = 27 (23-40); b = ?; b' = 3.8

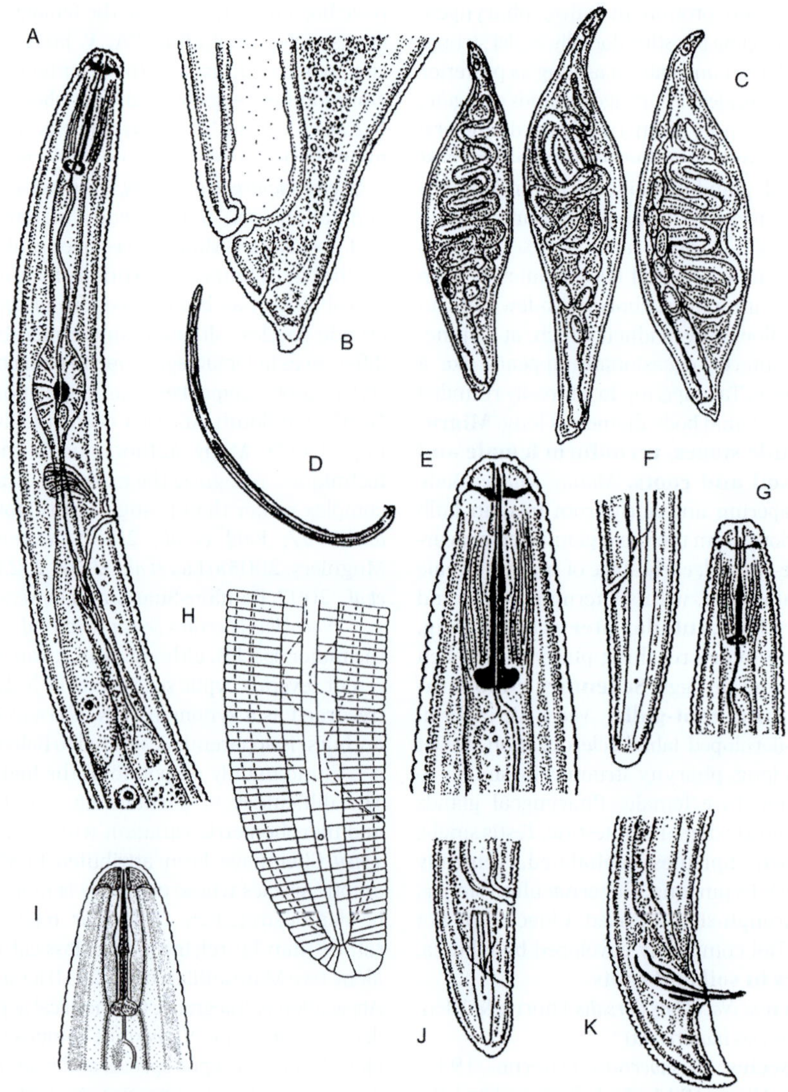

Fig. 12.24. *Nacobbus aberrans*. A: Male pharyngeal region; B: tail region of mature female; C: mature females; D: entire male; E, I: male cephalic region; F: tail region of second-stage juvenile; G: cephalic region of second-stage juvenile; H, J: immature female posterior region; K: male tail. (Line drawings are not to scale, courtesy CABI.)

(2.8-4.1); c' = 1.2 (0.9-1.5); V = 93 (92-94); VA = 20 (15-24); stylet = 23 (21-25) μm.

Male topotypes (n = 7): L = 0.86 (0.71-0.92) mm; a = 29 (24-31); b = 7.0 (6.4-7.2); b' = 3.6 (3.4-4.0); c = 38 (32-42); c' = 1.3 (1.2-1.4); stylet = 25 (23-27) μm; gubernaculum = 7 (6-8) μm; spicules = 27 (21-30) μm.

Description (after Siddiqi, 2000): *Young female:* Vermiform, elongate-slender, ca 1 mm long. Cuticle distinctly annulated. Lateral fields each with four incisures, irregularly areolated. Phasmid pore-like, anterior to middle of tail. Labial region broadly rounded, continuous, with 3-4 annuli. Median bulb rounded, with large refractive thickenings and prominent valve; pharyngeal glands overlapping intestine dorsally. Stylet strong. Pharyngeal glands elongate, extending for more than two body diameters; subventral glands asymmetrical, extending past dorsal gland; nuclei of all three glands lying in tandem posterior to pharyngo-intestinal junction. Vulva a transverse slit, located at 89-97% of body length, within 1-2.3 anal body diameters of anus; 15-24 annuli between VA. Ovary immature, short (with few tandem oocytes), followed by oviduct, which, at its junction with uterus, occasionally appears like a 'spermatheca'. Tail tapering to a broadly rounded terminus, 1-2 anal body diameters long. **Migratory in soil and roots.** *Mature female:* Body saccate; early stage of mature female usually 'batatiform', with subterminal anus and vulva. **Uterus tubular, ovary long, serpentine,** reaching pharynx. Unsegmented eggs in posterior portion of body. **Sedentary endoparasites inciting root-galls.** *Male:* Vermiform, with well-developed labial sclerotization. Stylet 23-27 µm long, pharynx structurally similar to that of immature female. Pharyngeal glands elongate, mostly dorsal to intestine. Testis single, outstretched. Spicules cephalated, ventrally arcuate, 20-35 µm long. Gubernaculum simple, linear to trough-shaped, fixed. Cloacal lips not modified. Male tail completely enveloped by bursa. **Migratory in soil and roots.** *Second-stage juvenile (J2).* Vermiform, with elongate body, pharynx like that of immature females, tail blunt, curved ventrally. **Migratory in soil and roots.** *Third (J3) and fourth (J4) stage juveniles.* Vermiform, with a C-shaped body. Less motile than J2. **Migratory in soil and roots.**

Molecular and biochemical characterization: Can be distinguished from *N. bolivianus* and *N. celatus* by molecular and biochemical techniques (see section 12.7.3).

Nacobbus bolivianus **Lordello, Zamith & Boock, 1961**

= *N. serendipiticus bolivianus* Franklin, 1959 (Lordello, Zamith & Boock, 1961)

Common name: potato rosary nematode

Measurements

Bolivian population (after Lordello *et al.*, 1961):

Immature female: L = 771 µm; stylet = 16 µm; a = 29.6; b = ?; c = 25.9; V = 92.4.

Adult female: L = 1.085-1.906 mm; stylet = 22 µm

Male: L = 788-902.7 µm; stylet = 21.4-22.9 µm; a = 18.8-25.4; b = 7.2-8.5; c = 34.3-46.8; spicule = 26.0-30.0 µm; gubernaculum = 7.6-9.2 µm.

Description: In view of the lack of information about the type of *N. bolivianus*, we present data from the description of *N. serendipiticus*, a junior synonym. *Nacobbus serendipiticus* mature females were described as irregularly spindle-shaped, without a posterior elongated region. Vulva-anus distance in young females 1.5 tail lengths long; phasmids adanal or immediately postanal. Stylet 19-22 µm. According to *N. bolivianus* (= *N. serendipiticus bolivianus*) description (Lordello *et al.*, 1961), it differed from *N. s. serendipiticus* Franklin, 1959 in having a narrower central ribbon in the lateral field of the first-stage juvenile and young female; and wider males (width = 35.2-45.9 µm).

Molecular and biochemical characterization: Can be distinguished from *N. aberrans* by molecular and biochemical techniques (see section 12.7.3).

12.7.2. Distribution and Hosts

Nacobbus has been found in association with numerous crops and native plants in temperate and subtropical areas of North America (USA, Mexico) and South America (Argentina, Bolivia, Chile, Ecuador, Peru). The wide distribution of *Nacobbus* in the American continent has probably resulted from its broad host range, which includes many weeds, and from the passive transport of the nematode with propagative plant material such as seed potatoes and other tuber-forming hosts (Manzanilla-López *et al.*, 2002).

12.7.3. Molecular and Biochemical Characterization

Although Sher's 'two species' classification scheme is a rather conservative approach for assessing *Nacobbus* variation, it remains in widespread use due in part to resistance against molecular diagnostic techniques. Molecular approaches (Ibrahim *et al.*, 1997; Reid *et al.*, 2003; Anthoine and Mugniéry, 2005a; Lax *et al.*, 2007, 2021; Vovlas *et al.*, 2007; Ramirez-Suarez and Powers, 2011), although consistently recognizing at least two groups within *N. aberrans sensu* Sher, have not yet found universal acceptance as to whether the differences found indicate separate taxa at the species rank or are more applicable at infraspecific categories, such as race.

Molecularly, within the scheme proposed by Reid *et al.* (2003), *N. dorsalis*, the type, was regarded as *species inquirenda* while *N. aberrans sensu* Sher was divided into two species, *viz.*, *N. aberrans sensu* Reid *et al.* (2003), representing most of the populations studied, and *N. bolivianus* for those from Bolivia, including from Cochabamba, the type locality. Subsequent studies (see Anthoine and Mugniéry, 2005a; Lax *et al.*, 2007, 2021; Vovlas *et al.*, 2007) have provided molecular evidence to support the existence of at least two major clades within South American populations of *Nacobbus*, these clades being congruent with the two groups that Reid *et al.* (2003) had earlier recognized at the species rank. Ramirez-Suarez and Powers (2011) identified over 50 diagnostic nucleotide characters for *Nacobbus* populations from western Nebraska, north and central regions of Mexico and the Argentinean lowlands, data that strongly support the inference that *N. aberrans* comprises at least four distinct lineages, as supported by the recent description of *N. celatus* by Lax *et al.* (2021). Although much progress has been made in understanding the complexity and variation within the *N. aberrans* species complex (Cabrera-Hidalgo *et al.*, 2019), and the growing consensus on the validity and use of molecular diagnostics, most populations and quarantine interceptions are still labelled simply as '*N. aberrans sensu* Sher', despite mounting evidence to the contrary.

Although no genome sequencing of *N. aberrans* has been published, the use of RNAseq and transcriptomics tools to describe the changes occurring in all expressed genes across the life cycle of *N. aberrans*, and a multigene phylogenetic analysis approach, confirmed the proximity of *N. aberrans* to both cyst and root-knot nematodes, and that genes considered to be cyst nematode or root-knot nematode 'specific' were also identified in *N. aberrans* (Eves-van den Akker *et al.*, 2014). Other transcriptomic studies have focused on transcript accumulation of defence genes in plants infected by *N. aberrans* (Cabrera-Hidalgo *et al.*, 2021).

12.7.4. Bionomics

Symptoms of attack/pathogenicity: Patches of poor growth are a common feature of affected crops (see Fig. 4.2A, B, Chapter 4). Above-ground symptoms include a reduction in apical growth (stunting), chlorotic leaves with rolled margins, wilting and, in tomato, the production of small fruits. The main underground symptoms on the various hosts are the presence on the roots of galls (see Fig. 4.9, Chapter 4 and Fig. 12.14D-F), which may bear lateral roots (Fig. 12.14F), and the cessation in growth of tertiary and fibrous roots. On potato roots, bead-like galls tend to be distributed along the root axis, which acquires the appearance of a rosary (see Fig. 4.10B, Chapter 4). Galls on tomato and chili pepper roots are more elongate. The wounds and necrosis caused by the juveniles on sugarbeet interfere with the function of the smaller branch roots of the taproots, often causing complete loss of stand. Any remaining plants are usually stunted (Manzanilla-López *et al.*, 2002).

Life cycle: Different *Nacobbus* populations can have shorter or longer embryogenesis and generation times, the duration of the life cycle being affected by temperature, diapause and host (Clark, 1967; Manzanilla-López *et al.*, 2002; Anthoine *et al.*, 2006). Life cycle begins with eggs laid in a gelatinous matrix secreted by the female. The first-stage juvenile (J1) develops inside the egg, the first moult occurring therein. After hatching, the second-stage juvenile (J2) penetrates a root, where it may remain or leave to re-enter the root at another site (Clark, 1967). The J2 moults either in the root or in the soil to the third-stage juvenile (J3), which is less active and tends to remain coiled in the root cortex. The J3 initially resembles the J2 but tends to adopt a C-shape or to coil up as it feeds and enlarges, a notable reduction occurring in the length of the pharyngeal glands. Gender can be distinguished shortly after the moult to the J3. Moulting of the fourth-stage juvenile (J4) to the adult may take place inside or outside the root, immature females being vermiform. It is likely that fertilization usually occurs after the female has become established in the root. Sedentary females produce a gelatinous matrix where eggs are laid. Although parthenogenesis has been rejected as a reproductive strategy for some South American populations of *Nacobbus* (Anthoine and Mugniéry, 2005b, 2006), other authors have suggested it as a possible mechanism (Canto-Sáenz, 1992; Reid *et al.*, 2003).

The life cycle of *N. aberrans* is similar on all of its non-tuber-forming hosts (Manzanilla-López *et al.*, 2002) but on potato is reported to be somewhat different (Costilla, 1985) from that on tomato and sugarbeet. Because tubers are harvested and stored for different periods of time, *Nacobbus* populations in the Andes (where temperatures can be below −10°C) developed an extended survival strategy. The nematode overwinters in the lenticels of stored crops or wild tubers in the ground. When conditions improve in the following spring, the life cycle resumes on either planted crops or sprouting wild tubers (see Fig. 4.10B, Chapter 4, this volume). In *Nacobbus*, a change to sedentary endoparasitism occurs when young and adult mature females induce the formation of large, specialized feeding cells (Fig. 12.14G-H, Chapter 12, this volume). These cells, grouped in a large syncytium, are derived by successive fusions of adjacent cells that are induced by the nematode feeding activity. These syncytia develop adjacent to the female body as a spindle of interconnected cells. Mitotic stimulation occurs in the surrounding cells outside the syncytium, giving rise to the gall (Fig. 12.14D-F; Chapter 12, this volume). Accentuated asymmetry of root structure, fragmentation of the stele, hyperplasia of vascular parenchyma and abnormal proliferation of lateral roots are common histological alterations induced by the nematode, resulting in reduced uptake of water and nutrients by the host (Manzanilla-López *et al.*, 2002; Vovlas *et al.*, 2007).

Physiological specialization of *N. aberrans s.l.* has been observed in some North and South American populations in response to clinal variation, host, cultivar and geographical region, supporting the possibility that biotypes and races occur (Inserra *et al.*, 1985; Manzanilla-López, 2010). Three major groups (previously known as races or pathotypes) are generally accepted: 1. *The sugarbeet group*, to include populations infecting sugarbeet but not potato in the lowlands of Argentina, Nebraska and the Pacific Northwest (USA), and probably Ecuador; 2. *The potato group*, which damages potato and also infects sugarbeet but not chili pepper. Populations of this type are present in the highland Andean regions of Argentina, Bolivia, Chile, Peru and Mexico; and 3. *The bean group*, to include populations that attack beans and chili pepper but are not able to infect potato or sugarbeet. Such populations are reported in Mexico (Manzanilla-López *et al.*, 2002). The potato group is one of the most important, not just economically, but also under quarantine regulations, as it can cause yield losses greater than 80% in Bolivia. The races/groups, as defined by a group of differential hosts, have been used until now as a 'guide' to identify populations that potentially represent a threat to potato and other crops because of their pathogenicity. However, these schemes have serious limitations (see Manzanilla-López *et al.*, 2002; Manzanilla-López, 2010; Lax *et al.*, 2011). Integration of morphological, physiological and molecular data and implementation of standard operation procedures are needed to support the geographical and physiological groups.

Interactions: Under field conditions, *N. aberrans* occurs together with other plant-parasitic nematodes and important soil-borne pathogens such as *Spongospora subterranea* (Wallr.), Lagerh and *Synchitrium endobioticum* (Schilb.) Perc. in potato, *Fusarium oxysporum* Schlechtendal, *F. roseum* Limk:Fries, and *Phytophthora capsici* Leo. in tomato, the interaction of the latter with *Nacobbus* being responsible for the failure of cultivars of chili (*Capsicum annuum* L.) resistant to *P. infestans* (Manzanilla-López *et al.*, 2002; see also Chapter 18, this volume).

Management: Control methods and management strategies largely depend on crop and availability of local 'technological packages' to deal with this pest. Some strategies include prevention of introductions and dispersal of infected tubers, clean seedlings (nursery stage control), crop rotation, planting of resistant/tolerant varieties, trap crops, burning of stubble between crops, weeding, cleaning of agricultural tools and machinery, various control methods ranging from chemical to biological control and epidemiology, farmer schools, knowledge dissemination and plant clinics (Manzanilla-López *et al.*, 2002; Cristóbal-Alejo *et al.*, 2006; Franco and Main, 2006; Lax *et al.*, 2022). Molecular tools are now also available to detect the nematode in soil and samples of tuber skin (Atkins *et al.*, 2005). However, accurate identification remains fundamental and remains a priority in planning control, management strategies and phytosanitary regulations within and between countries. In 2011, *Nacobbus aberrans* was included on the list of regulated nematodes by 52 countries, although this had been reduced to only ten in 2022 (Hockland *et al.*, 2024).

12.8. References

Abdulsalam, S., Peng, H., Yao, Y., Fan, L., Jiang, R., *et al.* (2021) Prevalence and molecular diversity of plant-parasitic nematodes of yam (*Dioscorea* spp.) in China, with focus on *Merlinius* spp. *Biology* 10, 1299. DOI: 10.3390/biology10121299.

Al-Banna, L., Williamson, V. and Gardner, S.L. (1997) Phylogenetic analysis of nematodes of the genus *Pratylenchus* using nuclear 26S rDNA. *Molecular Phylogenetics and Evolution* 7, 94–102.

Al-Banna, L., Ploeg, A.T., Williamson, V.M. and Kaloshian, I. (2004) Discrimination of six *Pratylenchus* species using PCR and species-specific primers. *Journal of Nematology* 36, 142–146.

Allen, M.W. and Jensen, H.J. (1950) *Pratylenchus vulnus*, new species (Nematoda: Pratylenchinae), a parasite of trees and vines in California. *Proceedings of the Helminthological Society of Washington* 18, 47–50.

Andrássy, I. (2007) *Free-living Nematodes of Hungary, II (Nematoda Errantia)*. In: *Pedozoologica Hungarica 4* (Series eds: Csuzdi, C. and Mahunka, S.). Hungarian Natural History Museum and Systematic Zoological Research Group of the Hungarian Academy of Sciences, Budapest, Hungary.

Andrés, M.F., Pinochet, J., Hernández Borrego, A. and Delibes, A. (2000) Detection and analysis of inter- and intraspecific diversity of *Pratylenchus* spp. using isozyme markers. *Plant Pathology* 49, 640–649.

Anthoine, G. and Mugniéry, D. (2005a) Variability of the ITS rDNA and identification of *Nacobbus aberrans* (Thorne, 1935) Thorne & Allen, 1944 (Nematoda: Pratylenchidae) by rDNA amplification. *Nematology* 7, 503–516.

Anthoine, G. and Mugniéry, D. (2005b) Obligatory amphimixis and variation in chromosome number within and among South American populations of *Nacobbus aberrans* (Thorne, 1935) Thorne & Allen, 1944 (Nematoda: Pratylenchidae). *Nematology* 7, 783–787.

Anthoine, G. and Mugniéry, D. (2006) Crossing experiments with South American populations of *Nacobbus aberrans* (Thorne, 1935) Thorne & Allen, 1944 (Nematoda: Pratylenchidae). *Nematropica* 36, 36–77.

Anthoine, G., Gauthier, J.P. and Mugniéry, D. (2006) Aspects of the biology of *Nacobbus aberrans* (Thorne, 1935) Thorne & Allen, 1944 (Nematoda: Pratylenchidae): 1 – Capacities of quiescence and diapause. *Bulletin OEPP* 36, 359–364.

Arora, D. and Yan, G. (2022) Early detection and temporal dynamics of *Pratylenchus scribneri* infection in potato roots determined using quantitative PCR and root staining. *Phytopathology* 112, 1776–1782. DOI: 10.1094/PHYTO-10-21-0412-R.

Arora, D., Yan, G. and Baidoo, R. (2020) Developing a realtime PCR assay for direct detection and quantification of *Pratylenchus scribneri* in field soil. *Nematology* 22, 1–12. DOI: 10.1163/15685411-00003336.

Atkins, S.D., Manzanilla-López, R.H., Franco, J., Peteira, B. and Kerry, B.R. (2005) A molecular diagnostic method for detecting *Nacobbus* in soil and in potato tubers. *Nematology* 7, 193–202.

Baidoo, R., Yan, G., Nagachandrabose, S. and Skantar, A.M. (2017) Developing a real-time PCR assay for direct identification and quantification of *Pratylenchus penetrans* in soil. *Plant Disease* 101, 1432–1441. DOI: 10.1094/PDIS-01-17-0117-RE.

Baldwin, J.G. and Cap, G.B. (1992) Systematics of *Nacobbus*, the false root-knot nematode. In: Gommers, F.J. and Maas, P.W.Th. (eds), *Nematology from Molecule to Ecosystem*. European Society of Nematologists, Dundee, UK, pp. 101–112.

Berry, S.D., Fargette, M., Spaull, V.W., Morand, S. and Cadet, P. (2008) Detection and quantification of root-knot nematode (*Meloidogyne javanica*), lesion nematode (*Pratylenchus zeae*) and dagger nematode (*Xiphinema elongatum*) parasites of sugarcane using real-time PCR. *Molecular and Cellular Probes* 22, 168–176. DOI: 10.1016/j.mcp.2008.01.003

Bert, W., Leliaert, F., Vierstraete, A.R., Vanfleteren, J.R. and Borgonie, G. (2008) Molecular phylogeny of the Tylenchina and evolution of the female gonoduct (Nematoda: Rhabditida). *Molecular Phylogenetics and Evolution* 48, 728–744.

Burke, M., Scholl, E.H., Bird, D.M., Schaff, J.E., Colman, S.D., *et al.* (2015) The plant parasite *Pratylenchus coffeae* carries a minimal nematode genome. *Nematology* 17, 621–637. DOI: 10.1163/15685411-00002901.

Cabrera-Hidalgo, A.J., Marban-Mendoza, N. and Valadez-Moctezuma, E. (2019) Phylogenetic relationships among Mexican populations of *Nacobbus aberrans* (Nematoda, Pratylenchidae) reveal the existence of cryptic (complex) species. *Nematropica* 49, 1–11.

Cabrera-Hidalgo, A.J., Valadez-Moctezuma, E., Bustamante-Ortiz, A.G., Camacho-Tapia, M. and Marban-Mendoza, E. (2021) Transcript accumulation of defense genes in tomato infected by the false-root-knot nematode *Nacobbus aberrans*. *Nematropica* 51, 17–26.

Canto-Sáenz, M. (1992) Life cycle and pathogenicity of *Nacobbus aberrans* (Thorne, 1935) Thorne & Allen, 1944. In: Gommers, F.J. and Maas, P.W.Th. (eds), *Nematology from Molecule to Ecosystem*. European Society of Nematologists, Dundee, UK, pp. 113–127.

Carrasco Ballesteros, S., Castillo, P., Adams, B.J. and Pérez Artés, E. (2007) Identification of *Pratylenchus thornei*, the cereal and legume root-lesion nematode, based on SCAR-PCR and satellite DNA. *European Journal of Plant Pathology* 118, 115–125.

Castillo, P. and Vovlas, N. (2007) Pratylenchus *(Nematoda: Pratylenchidae): Diagnosis, Biology, Pathogenicity and Management. Nematology Monographs and Perspectives 6* (Series eds: Hunt, D.J. and Perry, R.N.). Brill, Leiden, Netherlands.

Chen, D.Y., Ni, H.F., Yen, J.H., Chen, R.S. and Tsay, T.T. (2006) Distribution of rice root nematode *Hirschmanniella oryzae* and a new record *H. mucronata* (Nematoda: Pratylenchidae) in Taiwan. *Zhi Wu Bing Li Xue Hui Kan* 15, 197–210.

Chihani-Hammas, N., Hajji-Hedfi, L., Regaieg, H., Larayedh, A., Badiss, A., *et al.* (2018) First report of *Pratylenchus vulnus* associated with apple in Tunisia. *Journal of Nematology* 50, 579–586.

Chitwood, B.G. and Chitwood, M.B. (1950) *An Introduction to Nematology*. Monumental Printing Co., Baltimore, Maryland.

Clark, S.A. (1967) The development and life history of the false root-knot nematode, *Nacobbus serendipiticus*. *Nematologica* 13, 91–101.

Costilla, M.A. (1985) El falso nematode del nudo *Nacobbus aberrans* (Thorne, 1935) Thorne y Allen, 1944 y su relación con el cultivo de papa en el noreste argentino. *Revista Industrial y Agrícola de Tucumán* 62, 79–97.

Cristóbal-Alejo, J., Mora-Aguilera, G., Manzanilla-López, R.H., Marbán-Mendoza, N., Sánchez-G., P., *et al.* (2006) Epidemiology and integrated control of *Nacobbus aberrans* on tomato (*Lycopersicon esculentum* Mill.) in Mexico. *Nematology* 8, 727–737.

De Luca, F., Fanelli, E., Di Vito, M., Reyes, A. and De Giorgi, C. (2004) Comparison of the sequences of the D3 expansion of the 26S ribosomal genes reveals different degrees of heterogeneity in different populations and species of *Pratylenchus* from the Mediterranean region. *European Journal of Plant Pathology* 110, 949–957. DOI: 10.1007/s10658-004-0813-4

De Luca, F., Troccoli, A., Duncan, L.W., Subbotin, S.A., Waeyenberge, L., *et al.* (2012) *Pratylenchus speijeri* n. sp. (Nematoda: Pratylenchidae), a new root-lesion nematode pest of plantain in West Africa. *Nematology* 14, 987–1004. DOI: 10.1163/156854112X638424.

Decraemer, W. and Hunt, D.J. (2024) Structure and classification. In: Perry, R.N., Moens, M. and Jones, J.T. (eds), *Plant Nematology*, 3rd edn. CAB International, Wallingford, UK, pp. 3–40.

DuCharme, E.P. (1968) Burrowing nematode decline of citrus. A review. In: Smart, G.C. and Perry, V.G. (eds), *Tropical Nematology*, University of Florida Press, Gainesville, Florida, pp. 20–37.

DuCharme, E.P. and Birchfield, W. (1956) Physiological races of the burrowing nematode. *Phytopathology* 46, 615–616.

DuCharme, E.P. and Price, W.C. (1966) Dynamics of multiplication of *Radopholus similis*. *Nematologica* 12, 113–121.

Duncan, L.W. (2005) Nematode parasites of citrus. In: Luc, M., Sikora, R.A. and Bridge, J. (eds.) *Plant Parasitic Nematodes in Subtropical and Tropical Agriculture*, 2nd edn. CAB International, Wallingford, UK, pp. 437–465. DOI: 10.1079/9780851997278.0437

Duncan, L.W., Inserra, R.N., Thomas, W.K., Dunn, D., Mustika, L., *et al.* (1999) Molecular and morphological analysis of isolates of *Pratylenchus coffeae* and closely related species. *Nematropica* 29, 61–80.

EPPO. (2008) *Radopholus similis*. *EPPO Bulletin* 38, 374–378.

Eves-van den Akker, S., Liley, C.J., Danchin, E.G., Rancurel, C., Cock, P.J.A., *et al.* (2014) The transcriptome of *Nacobbus aberrans* reveals insights into the evolution of sedentary endoparasitism in plant-parasitism in plant-parasitic nematodes. *Genome Biology Evolution* 6, 2181–2194. DOI: 10.1093/gbe/evu/71.

Fanelli, E., Troccoli, A., Capriglia, F., Lucarelli, G., Vovlas, N., *et al.* (2018) Sequence variation in ribosomal DNA and in the nuclear *hsp90* gene of *Pratylenchus penetrans* (Nematoda: Pratylenchidae) populations and phylogenetic analysis. *European Journal of Plant Pathology* 152, 355–365. DOI: 10.1007/s10658-018-1480-1.

Fatemi, E., Melzer, S. and Jung, C. (2023) DNA-based assessment of root lesion nematode infections in cereal roots. *Scientific Reports* 13, 12602. DOI: 10.1038/s41598-023-39559-8.

Franco, J. and Main, G. (2006) Screening for resistance to *Nacobbus aberrans* and *Globodera* spp. in wild tomato species resistant to other pathogens. *Nematologia Mediterranea* 34, 165–169.

Geraert, E. (2013) *The Pratylenchidae of the World. Identification of the Family Pratylenchidae (Nematoda: Pratylenchidae)*. Academia Press, Ghent, Belgium.

Gerber, K. and Smart, G.C. (1986) Plant-parasitic nematodes associated with aquatic vascular plants. In: Veech, J.A. and Dickson, D.W. (eds), *Vistas on Nematology: A Commemoration of the Twenty-fifth Anniversary of the Society of Nematologists*. Society of Nematologists Inc., Hyattsville, Maryland, pp. 488–501.

Gnanapragasham, N.C. and Mohotti, K.M. (2018) Nematode parasites of tea. In: Sikora, R.A., Coyne, D., Hallmann, J. and Timper, P. (eds), *Plant Parasitic Nematodes in Subtropical and Tropical Agriculture*, 3rd edn. CAB International, Wallingford, UK, pp. 584–616.

Gowen, S.R., Quénéhervé, P. and Fogain, R. (2005) Nematodes parasites of bananas and plantains. In: Luc, M., Sikora, R.A. and Bridge, J. (eds), *Plant Parasitic Nematodes in Subtropical and Tropical Agriculture*, 2nd edn. CAB International, Wallingford, UK, pp. 611–643.

Graham, T.W. (1951) Nematode root rot of tobacco and other plants. In: *South Carolina Agricultural Experiment Station* (Bulletin 390). Clemson, South Carolina.

Habteweld, A.W., Akyazi, F., Joseph, S., Crow, W.T., Abebe, E., *et al.* (2019) Description of *Hirschmanniella dicksoni* n. sp. (Nematoda: Pratylenchidae) from rhizosphere soil of limpograss from Florida, USA. *Journal of Nematology* 51, e2019-83. DOI: 10.21307/jofnem-2019-083.

Handoo, Z.A., Yan, G., Kantor, M.R., Huang, D., Chowdhury, I.A., *et al.* (2021) Morphological and molecular characterization of *Pratylenchus dakotaensis* n. sp. (Nematoda: Pratylenchidae), a new root-lesion nematode species on soybean in North Dakota, USA. *Plants* 10, 168. DOI: 10.3390/plants10010168.

Hockland, S., Prior, T., Stanley, J.D., Inserra, R.N. and Khol, L.M. (2024) *International plant health - putting legislation into practice*. In: Perry, R.N., Moens, M. and Jones, J.T. (eds), *Plant Nematology*, 3rd edn. CAB International, Wallingford, UK, pp. 415–442.

Hodda, M., Collins, S.J., Vanstone, V.A., Hartley, D., Wanjura, W., *et al.* (2014) *Pratylenchus quasitereoides* n. sp. from cereals in Western Australia. *Zootaxa* 3866, 277–288. DOI: 10.11646/zootaxa.3866.2.6. PMID: 25283659.

Hodson, A.K., Cicchetto, A. and Fierro, F.A. (2021) Real time PCR assays to detect and quantify the nematodes *Pratylenchus vulnus* and *Mesocriconema xenoplax*. *Crop Protection* 145, 105617. DOI: 10.1016/j.cropro.2021.105617.

Holdeman, Q.L. (1986) The burrowing nematode *Radopholus similis sensu lato*. *Nematology Publication*. California Department of Food and Agriculture, Division of Plant Industry, Sacramento, California.

Holterman, M., van der Wurff, A., van den Elsen, S., van Megen, H., Bongers, T., *et al.* (2007) Phylum-wide analysis of SSU rDNA reveals deep phylogenetic relationships among nematodes and accelerated evolution toward crown clades. *Molecular Biology and Evolution* 23, 1792–1800.

Holterman, M., Karssen, G., van den Elsen, S., van Megen, H., Bakker, J., *et al.* (2009) Small subunit rDNA-based phylogeny of the Tylenchida sheds light on relationships among some high-impact plant-parasitic nematodes and the evolution of plant feeding. *Phytopathology* 99, 227–235.

Huettel, R.N., Dickson, D.W. and Kaplan, D.T. (1984) *Radopholus citrophilus* sp. n. (Nematoda), a sibling species of *Radopholus similis*. *Proceedings of the Helminthological Society of Washington* 51, 32–35.

Ibrahim, S.K., Baldwin, J.G., Roberts, P.A. and Hyman, B.C. (1997) Genetic variation in *Nacobbus aberrans*: an approach toward taxonomic resolution. *Journal of Nematology* 29, 241–249.

Ibrahim, S.K., Perry, R.N. and Webb, R.M. (1995) Use of isoenzyme and protein phenotypes to discriminate between six *Pratylenchus* species from Great Britain. *Annals of Applied Biology* 126, 317–327.

Inserra, R.N., Griffin, G.D. and Anderson, J.L. (1985) The false root-knot nematode *Nacobbus aberrans*. Research Bulletin, Utah Agricultural Experimental Station No. 510.

Inserra, R.N., Troccoli, A., Gozel, U., Bernard, E.C., Dunn, D., *et al.* (2007) *Pratylenchus hippeastri* n. sp. (Nematoda: Pratylenchidae) from amaryllis in Florida with notes on *P. scribneri* and *P. hexincisus*. *Nematology* 9, 25–42.

Jaumot, M., Pinochet, J. and Fernández, C. (1997) Protein analysis of root-lesion nematodes using SDS-PAGE. *Nematropica* 27, 33–39.

Kaplan, D.T. and Opperman, C.H. (1997) Genome similarity implies that citrus-parasitic burrowing nematodes do not represent a unique species. *Journal of Nematology* 29, 430–440.

Kaplan, D.T. and Opperman, C.H. (2000) Reproductive strategies and karyotypes of the burrowing nematode, *Radopholus similis*. *Journal of Nematology* 32, 126–133.

Kaplan, D.T., Thomas, W.K., Frisse, L.M., Sarah, J.L., Stanton, J.M., *et al.* (2000) Phylogenetic analysis of geographically diverse *Radopholus similis* via rDNA sequence reveals a monomorphic motif. *Journal of Nematology* 32, 134–142.

Katsuta, A., Toyota, K., Min, Y.Y. and Maung, T.T. (2016) Development of real-time PCR primers for the quantification of *Meloidogyne graminicola*, *Hirschmanniella oryzae* and *Heterodera cajani*, pests of the major crops in Myanmar. *Nematology* 18, 257–263.

Kidane, S.A., Meressa, B.H., Haukeland, S., Hvoslef-Eide, T., Magnusson, C., *et al.* (2020) Occurrence of plant-parasitic nematodes on enset (*Ensete ventricosum*) in Ethiopia with focus on *Pratylenchus goodeyi* as a key species of the crop. *Nematology* 23, 529–541. DOI: 10.1163/15685411-bja10058.

Kuwahara, M. and Iyatomi, K. (1970) Studies on the bionomics of rice-root nematode, *Hirschmanniella imamuri* Sher and *Hirschmanniella oryzae* (Soltwedel) Luc & Goodey, with special reference to its mode of life and population dynamics. *Japanese Journal of Applied Entomology and Zoology* 14, 117–121. DOI: 10.1303/jjaez.14.117 [in Japanese with English abstract].

Lax, P., Dueñas, J.C.R., Gardenal, C.N. and Doucet, M.E. (2007) Assessment of genetic variability in populations of *Nacobbus aberrans* (Thorne, 1935) Thorne & Allen, 1944 (Nematoda: Pratylenchidae) from Argentina. *Nematology* 9, 261–270.

Lax, P., Rondan Dueñas, J.C., Coronel, N.B., Bima, P. and Doucet, M.E. (2011) Host range study of Argentine *Nacobbus aberrans* sensu Sher populations and comments on the differential host test. *Crop Protection* 30, 1414–1420. DOI: 10.1016/j.cropro.2011.06.001.

Lax, P., Gonzalez Ittig, R., Rondan Dueñas, J.C., Andrade, A.J., Gardenal, C.N., *et al.* (2021) Decrypting species in the *Nacobbus aberrans* (Nematoda: Pratylenchidae) complex using integrative taxonomy. *Zoologica Scripta* 50, 667–688. DOI: 10.1111/zsc.12494.

Lax, P., Passone, M.A., Becerra, A.G., Sosa, A.L., Ciancio, A., *et al.* (2022) Sustainable strategies for management of the 'false root-knot nematode' *Nacobbus* spp. *Frontiers in Plant Science* 13, 1046315. DOI: 10.3389/fpls.2022.1046315.

Lehman, P.S., Vovlas, N., Inserra, R.N., Duncan, L.W. and Kaplan, D.T. (2000) Colonization of foliar tissues of an aquatic plant, *Anubias barteri* Schott, by *Radopholus similis*. *Nematropica* 30, 63–75.

Li, L., Wang, S., Du, J., Lu, Q.S., Wang, Z.Y., *et al.* (2018) First report of the root-lesion nematode (*Pratylenchus neglectus*) on wheat in Anhui Province, China. *Plant Disease* 102, 2663–2663. DOI: 10.1094/PDIS-05-18-0764-PDN.

Li, Y., Xia, Y.H., Liu, Y.K., Hao, P.H., Sun, B.J., *et al.* (2020) Discovery of root-lesion nematode, *Pratylenchus coffeae*, infesting sesame in China. *Plant Disease* 104, 1873.

Li, Y., Li, J., Hu, Y., Wang, K., Yuan, H., *et al.* (2022) Occurrence of root-lesion nematode *Pratylenchus coffeae* on corn in Xinjiang Uygur Autonomous Region, China. *Plant Disease* 106, 3219.

Lin, B., Tao, L., Wang, H., Liao, J. and Zhuo, E. (2020) Duplex real-time quantitative PCR for simultaneous detection and quantification of *Pratylenchus neglectus* and *P. thornei*. *European Journal of Plant Pathology* 157, 185–196. DOI: 10.1007/s10658-020-01999-7.

Liu, X., Wang, H., Lin, B., Tao, Y., Zhuo, K., *et al.* (2017) Loop-mediated isothermal amplification based on the mitochondrial COI region to detect *Pratylenchus zeae*. *European Journal of Plant Pathology* 148, 435–446.

Loof, P.A.A. (1960) Taxonomic studies on the genus *Pratylenchus* (Nematoda). *Tijdschrift voor Plantenziekten* 66, 29–90.

Loof, P.A.A. (1985) *Pratylenchus scribneri*. In: *CIH Descriptions of Plant-parasitic Nematodes*, Set 8, No. 110. Commonwealth Agricultural Bureaux, Farnham Royal, UK.

Loof, P.A.A. (1991) The family Pratylenchidae Thorne, 1949. In: Nickle, W.R. (ed), *Manual of Agricultural Nematology*. Marcel Dekker, New York, New York, pp. 363–421.

Loos, C.A. (1959) Symptom expression of Fusarium wilt of the Gros Michel banana in the presence of *Radopholus similis* (Cobb, 1893) Thorne, 1949 and *Meloidogyne incognita acrita* Chitwood, 1949. *Proceedings of the Helminthological Society of Washington* 26, 103–111.

Lopez-Nicora, H.D., Mekete, T., Taylor, N.J. and Niblack, T.L. (2012) First report of lesion nematode (*Pratylenchus vulnus*) on boxwood in Ohio. *Plant Disease* 96, 1385. DOI: 10.1094/PDIS-03-12-0272-PDN.

Lordello, L.G.E., Zamith, A.P.L. and Boock, O.J. (1961) Two nematodes found attacking potato in Cochabamba, Bolivia. *Anais da Academia Brasileira de Ciencias* 33, 209–215.

Luambano, N.D., Kashando, B.E., Masunga, M.M., Mwenisongole, A.E., Mziray, M.F., *et al.* (2019) Status of *Pratylenchus coffeae* in banana-growing areas of Tanzania. *Physiological and Molecular Plant Pathology* 105,102–109. DOI: 10.1016/j.pmpp.2018.08.002.

Luc, M. and Fortuner, R. (1975) *Hirschmanniella spinicaudata*. *CIH Descriptions of Plant-parasitic Nematodes*. Set 5, No. 68. Commonwealth Agricultural Bureaux, Farnham Royal, UK.

Luc, M. and Goodey, J.B. (1962) *Hirschmannia* n.g. differentiated from *Radopholus* Thorne, 1949 (Nematoda: Tylenchoidea). *Nematologica* 7, 197–202.

Luc, M. and Goodey, J.B. (1964) *Hirschmanniella* nom. nov. for *Hirschmannia*. *Nematologica* 9, 471.

Machado, A.C.Z., Ferraz, L.C.C.B. and De Oliveira, C.M.G. (2007) Development of a species-specific reverse primer for the molecular diagnostic of *Pratylenchus brachyurus*. *Nematropica* 37, 249–257.

Machon, J.E. and Bridge, J. (1996) *Radopholus citri* n. sp. (Tylenchida: Pratylenchidae) and its pathogenicity to citrus. *Fundamental and Applied Nematology* 19, 127–133.

Maggenti, A.R., Luc, M., Raski, D.J., Fortuner, R. and Geraert, E. (1987) A reappraisal of Tylenchina (Nemata). 2. Classification of the suborder Tylenchina (Nemata: Diplogasteria). *Revue de Nématologie* 10, 135–142.

Man, J.G. de (1880) Die einheimischen, frei in der reinen Erde und im süssen Wasser lebenden Nematoden. Vorläufiger Bericht and descriptiv-systematischer Theil. *Tijdschrift Nederlandsche dierkundige Vereeniging* 2, 78–196.

Manzanilla-López, R.H. (2010) Speciation within *Nacobbus*: consilience or controversy? *Nematology* 12, 321–334.

Manzanilla-López, R.H., Costilla, M.A., Doucet, M., Franco, J., Inserra, R.N., *et al.* (2002) The genus *Nacobbus* Thorne & Allen, 1944 (Nematoda: Pratylenchidae): systematics, distribution, biology and management. *Nematropica* 32, 149–227.

Marin, D.H., Sutton, T.B. and Barker, K.R. (1998) Dissemination of bananas in Latin America and the Caribbean and its relationship to the occurrence of *Radopholus similis*. *Plant Disease* 82, 964–974.

Mathew, R. and Opperman, C.H. (2019) The genome of the migratory nematode, *Radopholus similis*, reveals signatures of close association to the sedentary cyst nematodes. *PLoS ONE* 14, e0224391. DOI: 10.1371/journal.pone.0224391.

Meldal, B.H.M., Debenham, N.J., De Ley, P., De Ley, I.T., Vanfleteren, J.R., *et al.* (2007) An improved molecular phylogeny of the Nematoda with special emphasis on marine taxa. *Molecular Phylogenetics and Evolution* 42, 622–636.

Merny, G. (1970) Les nématodes phytoparasites des rizières inondées en Côte d'Ivoire 1. – les espèces observées. *Cahier ORSTOM, Série Biologie* 11, 3–43.

Mirghasemi, S.N., Fanelli, E., Troccoli, A., Jamali, S., Sohani, M.M., *et al.* (2019) Molecular variability of the root-lesion nematode, *Pratylenchus loosi* (Nematoda: Pratylenchidae), from tea in Iran. *European Journal of Plant Pathology* 155, 557–569.

Mokrini, F., Waeyenberge, L., Viaene, N., Abbad Andaloussi, F. and Moens, M. (2013) Quantitative detection of the root-lesion nematode, *Pratylenchus penetrans*, using qPCR. *European Journal of Plant Pathology* 137, 403–413. DOI: 10.1007/s10658-013-0252-1.

Mondal, S., Purohit, A., Chakraborti, D., Khan, M.R. and Mukherjee, A. (2022) First report of *Pratylenchus zeae* on upland rice from Jharkhand, India. *Plant Disease* 106, 1765.

Múnera Uribe, G.E., Bert, W., Vierstraete, A.R., de la Peña, E., Moens, M., *et al.* (2010) Burrowing nematodes from Colombia and their relationship with *Radopholus similis* populations, *R. arabocoffeae and R. duriophilus*. *Nematology* 12, 619–629.

Naeini, F.S. and Taheri, Z.M. (2021) First report of root-lesion nematode, *Pratylenchus oleae* from pistachio in Iran. *Journal of Nematology* 53, e2021-103. DOI: 10.21307/jofnem-2021-103.

Nguyen, C.N., Subbotin, S.A., Madani, M., Trinh, P.Q. and Moens, M. (2003) *Radopholus duriophilus* sp. n. (Nematoda: Pratylenchidae) from Western Highland of Vietnam. *Nematology* 5, 549–558.

Nguyen, T.D., Le, T.M.L., Nguyen, H.T., Nguyen, T.A.D., Liebanas, G., *et al.* (2017) Morphological and molecular characteristics of *Pratylenchus haiduongensis* sp. n., a new species of root-lesion nematodes associated with carrot in Vietnam. *Journal of Nematology* 49, 276–285.

Orlando, V., Roberts, D., Edwards, S., Prior, T., Neilson, R., *et al.* (2024) Development and validation of four TaqMan real-time PCR diagnostics for the identification and quantification of *Pratylenchus crenatus*, *Pratylenchus neglectus*, *Pratylenchus penetrans* and *Pratylenchus thornei*. *Nematology* 26, 433–445. DOI: 10.1163/15685411-bja10315.

Orton Williams, K.J. and Siddiqi, M.R. (1973) *Radopholus similis*. *CIH Descriptions of Plant-parasitic Nematodes*. Set 2, No. 27. Commonwealth Agricultural Bureaux, Farnham Royal, UK.

Orui, Y. (1996) Discrimination of the main *Pratylenchus* species (Nematoda: Pratylenchidae) in Japan by PCR-RFLP analysis. *Applied Entomology and Zoology* 31, 505–514.

Palomares-Rius, J.E., Castillo, P., Liébanas, G., Vovlas, N., Landa, B.B., *et al.* (2010) Description of *Pratylenchus hispaniensis* n. sp. from Spain and considerations on the phylogenetic relationship among selected genera in the family Pratylenchidae. *Nematology* 12, 429–451.

Palomares-Rius, J.E., Guesmi, I., Horrigue-Raouani, N., Cantalapiedra-Navarrete, C., Liébanas, G., *et al.* (2014) Morphological and molecular characterisation of *Pratylenchus oleae* n. sp. (Nematoda:

Pratylenchidae) parasitizing wild and cultivated olives in Spain and Tunisia. *European Journal of Plant Pathology* 140, 53–67.

Payan, L.A. and Dickson, D.W. (1990) Comparison of populations of *Pratylenchus brachyurus* based on isozyme phenotypes. *Journal of Nematology* 22, 538–545.

Peng, D., Gaur, H.S. and Bridge, J. (2018) Nematodes parasites of rice. In: Sikora, R.A., Coyne, D., Hallmann, J. and Timper, P. (eds), *Plant Parasitic Nematodes in Subtropical and Tropical Agriculture*, 3rd edn. CAB International, Wallingford, UK, pp. 120–162.

Pili, N.N., Kyndt, T., Gheysen, G., Janssen, T., Couvreur, M., *et al.* (2016) First report of *Pratylenchus zeae* on upland rice from Kwale County, Kenya. *Plant Disease* 100, 1022.

Powers, T., Todd, T., Harris, T., Higgins, R., MacGuidwin, A., *et al.* (2021) *Pratylenchus smoliki*, a new nematode species (Pratylenchidae: Tylenchomorpha) from the Great Plains region of North America. *Journal of Nematology* 53, e2021-100. DOI: 10.21307/jofnem-2021-100.

Qing, X., Bert, W., Gamliel, A., Bucki, P., Duvrinin, S., *et al.* (2019) Phylogeography and molecular species delimitation of *Pratylenchus capsici* n. sp., a new root-lesion nematode in Israel on pepper (*Capsicum annuum*). *Phytopathology* 109, 847–858.

Ramirez-Suarez, A. and Powers, T.O. (2011) Systematics and DNA character-based diagnosis of the false root-knot nematode *Nacobbus aberrans* (Nemata: Pratylenchidae). *Program of The Society of Nematologists. Fiftieth Annual Meeting*, Corvallis, Oregon, July 17–20, 2011, pp. 131–132. [Abstract].

Reid, A.P., Manzanilla-López, R.H. and Hunt, D.J. (2003) *Nacobbus aberrans* (Thorne, 1935) Thorne & Allen, 1944 (Nematoda: Pratylenchidae); a nascent species complex revealed by RFLP analysis and sequencing of the ITS-rDNA region. *Nematology* 5, 441–451.

Rensch, B. (1924) *Aphelenchus neglectus* sp. n., eine neue parasitäre Nematodenart. *Zoologischer Anzeiger* 59, 277–280.

Ryss, A.Y. (1988) *Root Parasitic Nematodes of the Family Pratylenchidae (Tylenchida) of the World Fauna*. Zoological Institute, Russian Academy of Sciences, Nauka, Leningrad, Russia.

Ryss, A.Y. and Wouts, W.M. (1997) The genus *Radopholus* (Nematoda: Pratylenchidae) from native vegetation in New Zealand, with description of two new species. *International Journal of Nematology* 7, 1–17.

Setterquist, R.A., Smith, G.K., Jones, R. and Fox, G.E. (1996) Diagnostic probes targeting the major sperm protein gene that may be useful in the molecular identification of nematodes. *Journal of Nematology* 28, 414–421.

Sher, S.A. (1968a) Revision of the genus *Hirschmanniella* Luc & Goodey, 1963 (Nematoda: Tylenchoidea). *Nematologica* 14, 243–275.

Sher, S.A. (1968b) Revision of the genus *Radopholus* Thorne, 1949 (Nematoda: Tylenchoidea). *Proceedings of the Helminthological Society of Washington* 35, 219–237.

Sher, S.A. (1970) Revision of the genus *Nacobbus* Thorne & Allen, 1944 (Nematoda: Tylenchoidea). *Journal of Nematology* 2, 228–235.

Sher, S.A. and Allen, M.W. (1953) Revision of the genus *Pratylenchus* (Nematoda: Tylenchidae). *University of California Publications in Zoology* 57, 441–470.

Shokoohi, E. and Duncan, L.W. (2018) Nematodes parasites of citrus. In: Sikora, R.A., Coyne, D., Hallmann, J. and Timper, P. (eds), *Plant Parasitic Nematodes in Subtropical and Tropical Agriculture*, 3rd edn. CAB International, Wallingford, UK, pp. 446–476.

Siddiqi, M.R. (1973) *Hirschmanniella oryzae*. *CIH Descriptions of Plant-parasitic Nematodes*. Set 2, No. 26. Commonwealth Agricultural Bureaux, Farnham Royal, UK.

Siddiqi, M.R. (2000) *Tylenchida Parasites of Plants and Insects*, 2nd edn. CABI Publishing, Wallingford, UK.

Siddiqi, M.R. and Sturhan, D. (2014) A remarkable new nematode genus *Telomerlinius* gen. n. (Tylenchida: Merliniidae), with descriptions of two new species. *International Journal of Nematology* 24, 40–48.

Singh, P.R., Nyiragatare, A., Janssen, T., Couvreur, M., Decraemer, W., *et al.* (2018) Morphological and molecular characterisation of *Pratylenchus rwandae* n. sp. (Tylenchida: Pratylenchidae) associated with maize in Rwanda. *Nematology* 20, 781–794. DOI: 10.1163/15685411-00003175.

Stanton, J., Mundo-Ocampo, M., Baldwin, J.G. and Kaplan, D.T. (2001) *Radopholus musicola* n. sp., a new pathogenic species from Australia (Nematoda: Pratylenchidae). *Nematology* 3, 689–698.

Sturhan, D. (2008) Review of the genera *Hoplotylus* and *Radopholoides* (Tylenchida, Pratylenchidae). *Journal of Nematode Morphology and Systematics* 11, 159–166.

Sturhan, D. (2012). Contribution to a revision of the family Merliniidae Ryss, 1998, with proposal of Pratylenchoidinae subfam. n., *Paramerlinius* gen. n., *Macrotylenchus* gen. n. and description of *M. hylophilus* sp. n. (Tylenchida). *Journal of Nematode Morphology and Systematics* 15, 127–147.

Subbotin, S.A., Sturhan, D., Chizhov, V.N., Vovlas, N. and Baldwin, J.G. (2006) Phylogenetic analysis of Tylenchida Thorne, 1949 as inferred from D2 and D3 expansion fragments of the 28S rRNA gene sequences. *Nematology* 8, 455–474.

Subbotin, S.A., Ragsdale, E.J., Mullens, T., Roberts, P.A., Mundo-Ocampo, M., *et al.* (2008) A phylogenetic framework for root-lesion nematodes of the genus Pratylenchus (Nematoda): Evidence from 18S and D2-D3 expansion segments of 28S ribosomal RNA genes and morphological characters. *Molecular Phylogenetics and Evolution* 48, 491–505.

Tandingan De Ley, I., Mundo-Ocampo, M., Yoder, M. and De Ley, P. (2007) Nematodes from vernal pools in the Santa Rosa Plateau Ecological Reserve, California I. *Hirschmanniella santarosae* sp. n. (Nematoda: Pratylenchidae), a cryptic sibling species of *H. pomponiensis* Abdel-Rahman & Maggenti, 1987. *Nematology* 9, 405–429.

Tarjan, A.C. and O'Bannon, J.H. (1984) Nematode parasites of citrus. In: Nickle, W.R. (ed), *Plant and Insect Nematodes*. Marcel Dekker, New York, New York, pp. 395–433.

Thorne, G. (1935) The sugar beet nematode and other indigenous nemic parasites of shadscale. *Journal of Agricultural Research* 51, 509–514.

Thorne, G. (1949) On the classification of the Tylenchida, new order (Nematoda: Phasmidia). *Proceedings of the Helminthological Society of Washington* 16, 37–73.

Thorne, G. and Allen, M.W. (1944) *Nacobbus dorsalis*, nov. gen. nov. spec. (Nematoda: Tylenchidae) producing galls on the roots of alfileria, *Erodium cicutarium* (L.) L'Her. *Proceedings of the Helminthological Society of Washington* 11, 27–31.

Trinh, P.Q., Nguyen, C.N., Waeyenberge, L., Subbotin, S.A., Karssen, G., *et al.* (2004) *Radopholus arabocoffeae* sp. n. (Nematoda: Pratylenchidae), a nematode pathogenic to *Coffea arabica* in Vietnam, and additional data on *R. duriophilus*. *Nematology* 6, 681–693.

Trinh, P.Q., Waeyenberge, L., Nguyen, C.N., Baldwin, J.G., Karssen, G., *et al.* (2009) *Apratylenchus vietnamensis* gen. n., sp. n. and *A. binhi* gen. n., sp. n., sedentary Pratylenchidae (Nematoda: Tylenchida) from coffee in Vietnam, with proposal of Apratylenchinae subfam. n. *Nematology* 11, 565–581.

Trinh, P.Q., Waeyenberge, L., Nguyen, C.N. and Moens, M. (2012) Morphological and molecular diversity of *Radopholus* on coffee in Vietnam and description of *R. daklakensis* sp. n. from Robusta coffee. *Nematology* 14, 65–83.

Troccoli, A., Subbotin, S.A., Chitambar, J.J., Janssen, T., Waeyenberge, L., *et al.* (2016) Characterization of amphimictic and parthenogenetic populations of *Pratylenchus bolivianus* Corbett, 1983 (Nematoda, Pratylenchidae) and their phylogenetic relationships with closely related species. *Nematology* 18, 651–678.

Troccoli, A., Fanelli, E., Castillo, P., Liébanas, G., Cotroneo, A., *et al.* (2021) *Pratylenchus vovlasi* sp. nov. (Nematoda: Pratylenchidae) on raspberries in North Italy with a morphometrical and molecular characterization. *Plants* 10, 1068. DOI: 10.3390/plants10061068.

Uehara, T. (2003) Molecular phylogenetic analysis and characterization of secretory enzyme genes of plant parasitic nematodes. *Journal of Genetic Plant Pathology* 69, 422.

Uehara, T., Mizukubo, T., Kushida, A. and Momota, Y. (1998) Identification of *Pratylenchus coffeae* and *P. loosi* using specific primers for amplification of ribosomal DNA. *Nematologica* 44, 357–368.

USDA-APHIS. (2022) Phytosanitary certificate issuance and tracking system. https://pcit.aphis.usda.gov/pcit/faces/signin.jsf

Van den Berg, E., Subbotin, S.A., Handoo, Z.A. and Tiedt, L.R. (2009) *Hirschmanniella kwazuna* sp. n. from South Africa with notes on a new record of *H. spinicaudata* (Schuurmans Stekhoven, 1944) Luc & Goodey, 1964 (Nematoda: Pratylenchidae) and on the molecular phylogeny of *Hirschmanniella* Luc & Goodey, 1964. *Nematology* 11, 523–540.

Van der Vecht, J. and Bergman, B.H.H. (1952) Studies on the nematode *Radopholus oryzae* (Van Breda de Haan) Thorne and its influence on the growth of the rice plant. *Pemberitaan Bali Besar Penjelidikan Pertanian Bogor* 131, 82.

Vovlas, N., Nico, A.I., De Luca, F., De Giorgi, C. and Castillo, P. (2007) Diagnosis and molecular variability of an Argentinean population of *Nacobbus aberrans* with some observations on histopathology in tomato. *Journal of Nematology* 39, 17–26.

Waeyenberge, L., Ryss, A., Moens, M., Pinochet, J. and Vrain, T.C. (2000) Molecular characterisation of 18 *Pratylenchus* species using rDNA restriction fragment length polymorphism. *Nematology* 2, 135–142. DOI: 10.1163/156854100509024.

Waeyenberge, L., Viaene, N. and Moens, M. (2009) Species-specific duplex PCR for the detection of *Pratylenchus penetrans*. *Nematology* 11, 847–857. DOI: 10.1163/156854109X428016.

Wang, H., Zhuo, K., Ye, W. and Liao, J. (2015) Morphological and molecular characterisation of *Pratylenchus parazeae* n. sp. (Nematoda: Pratylenchidae) parasitizing sugarcane in China. *European Journal of Plant Pathology* 143, 173–191. DOI: 10.1007/s10658-015-0674-z.

Wanjau, C., Knoetze, R., Girgan, C., Pili, N.N., Engelbrecht, G., *et al.* (2024) Characterization and pathogenicity of *Pratylenchus vandenbergae* stat. nov. (Tylenchina: Pratylenchidae), a highly pathogenic root-lesion nematode parasitizing crops in Kenya and South Africa. *Plant Pathology* 73, 706–723. DOI: 10.1111/ppa.13836.

Wehunt, E.J. and Edwards, D.I. (1968) *Radopholus similis* and other nematode species on banana. In: Smart, G.C. and Perry, V.G. (eds), *Tropical Nematology*. University of Florida Press, Gainesville, Florida, pp. 1–19.

Wram, C.L., Hesse, C.N., Wasala, S.K., Howe, D.K., Peetz, A.B., *et al.* (2019) Genome announcement: The draft genomes of two *Radopholus similis* populations from Costa Rica. *Journal of Nematology* 51, e2019-52. DOI: 10.21307/jofnem-2019-052.

Xia, Y.H., Liu, Y.K., Hao, P.H., Yuan, H.X., Wang, K., *et al.* (2021) Molecular and morphological characterization of the root-lesion nematode, *Pratylenchus neglectus*, on corn from Henan Province of China. *Helminthologia* 58, 385–393. DOI: 10.2478/helm-2021-0044.

Xia, Y.H., Li, J., Hao, P., Wang, K., Lei, B., *et al.* (2022a) Discovery of root-lesion nematode (*Pratylenchus scribneri*) on corn in Hainan Province of China. *Plant Disease* 106, 1999. DOI: 10.1094/PDIS-09-21-1960-PDN.

Xia, Y.H., Li, J., Sun, M.R., Lei, H.L., Li, Y., *et al.* (2022b) Identification and pathogenicity of *Pratylenchus scribneri* on tomato in Sichuan Province of People's Republic of China. *Journal of Helminthology* 96, e5. DOI: 10.1017/S0022149X21000754.

Zhang, F., Yan, S., Zhou, Y., Guo, G., Guo, S., *et al.* (2015) First report of *Pratylenchus goodeyi* on banana in Hainan Province, China. *Plant Disease* 99, 731. DOI: 10.1094/PDIS-08-14-0874-PDN.

13 Criconematoidea

Ignacio Cid del Prado Vera[1]*, Antonio Archidona-Yuste[2] and Sergei A. Subbotin[3,4]
[1]*Colegio de Postgraduados, Montecillo, Texcoco, Mexico;* [2]*Institute for Sustainable Agriculture (IAS), Spanish National Research Council (CSIC), Córdoba, Spain;* [3]*Plant Pest Diagnostics Center, California Department of Food and Agriculture, Sacramento, USA;* [4]*Center of Parasitology of A.N. Severtsov Institute of Ecology and Evolution of the Russian Academy of Sciences, Moscow, Russia*

Abstract

Criconematids are one of the major groups of plant-parasitic nematodes and occur from the tropics to the arctic. They include those commonly named as 'ring nematodes', 'sheath nematodes', 'pin nematodes' and 'the citrus nematode'. While a few species are adapted to a number of agricultural crops, the vast majority are confined to natural habitats with a relatively stable soil structure. As a group, they exhibit a relatively wide host range. Females and most juveniles are obligate root ectoparasites, though some females of Tylenchulidae may become secondarily semi-endoparasitic. The morphology, bionomics and classification of the superfamily Criconematoidea and families Criconematidae, Hemicycliophoridae and Tylenchulidae are discussed with keys to taxa.

13.1. Introduction

Criconematids are one of the major groups of plant-parasitic nematodes and occur from the tropics to the Arctic. They include those commonly named as 'ring nematodes', 'sheath nematodes', 'sheathoid nematodes', 'pin nematodes' and 'the citrus nematode'. While a few species are adapted to a number of agricultural crops, the vast majority are confined to natural habitats with a relatively stable soil structure. They exhibit a relatively wide host range as a group, although individual species can have an exceptionally high level of endemism. Females and most juveniles are obligate root ectoparasites, though some females of Tylenchulidae may become secondarily semi-endoparasitic. Several species, including the citrus nematode, *Tylenchulus semipenetrans* and the ring nematode, *Criconemoides xenoplax*, are considered to be of major agricultural importance in many countries.

There are more than 750 nominal species of criconematids, many of them with an ornamented cuticle, which has made their classification controversial, since numerous authors have separated new genera and subgenera according to these ornamentations. At a higher taxonomic rank, most authors include criconematids in Criconematoidea Taylor, 1936, as both morphological and molecular characters support the

*E-mail: iciddelpradovera@gmail.com

A revision of Cid del Prado Vera Talavera, M. (2012 Criconematoidea. In: Manzanilla-López, R.H. and Marbán-Mendoza, N. (eds), *Practical Plant Nematology*. Mundi-Prensa and Biblioteca Básica de Agricultura, Montecillo, Mexico.

 Practical Plant Nematology 2nd Edition
(eds R.H. Manzanilla-López, D.J. Hunt, N. Marbán-Mendoza and S. Infante Gil)

DOI: 10.1079/9781836990413.0013

monophyly of Criconematoidea (van Megen *et al.*, 2009; Archidona-Yuste *et al.*, 2024). Relationships within the superfamily, however, are not well resolved as none of the phylogenetic analyses of the D2-D3 of 28S rRNA gene dataset have allowed the resolution of the relationships between main lineages (Subbotin *et al.*, 2005). As a consequence, genera recognized by different taxonomists are still under discussion.

Currently, there are three classifications that are followed by most authors, although with some modifications, that is, Siddiqi (2000), Decraemer and Hunt (2024), who detailed to subfamily and genus the classification provided by De Ley and Blaxter (2004), and Andrássy (2007). In this chapter, we follow the system of De Ley and Blaxter (2004) and, with minor modification concerning the validity of *Mesocriconema*, that of Decraemer and Hunt (2024) to be consistent with the other chapters in this book. However, equivalence in genera and other taxonomic ranks between the various schemes is shown in Table 13.1 to broaden the understanding of how the schemes relate to one another.

We provide keys to families and genera within each family, all of them modified from Siddiqi (2000), as practical tools when trying to identify the genera found in a sample. Readers interested in a closer look at criconematids taxonomy and systematics may consult the chapter of Criconematina in Siddiqi (2000) or other publications (Raski and Luc, 1987; Raski, 1991; Wouts, 2006; Andrássy, 2007; Geraert, 2010; Chitambar and Subbotin, 2014). The monograph by Geraert (2010) illustrates and describes all species regarded as valid by the author, while Ghaderi *et al.* (2016) monograph the Tylenchulidae.

In this chapter, descriptions of nematode genera are listed within their subfamilies in alphabetical order to facilitate their location within the text. For outsource information on biology, distribution, symptoms and disease management, see Sikora *et al.* (2018) and Perry *et al.* (2024).

13.2. Superfamily Criconematoidea Taylor, 1936

Diagnosis: Tylenchina. Soil nematodes with **marked sexual dimorphism, particularly in anterior region**. Female body vermiform, sausage-shaped or obese, 0.20-2.00 mm long, but mostly less than 1 mm. Cuticle either with thin and fine annuli or thick and coarse annuli, smooth (Fig. 13.1A), crenate (Fig. 13.1B) or ornamented by scales (Fig. 13.1C), spines (Fig. 13.1D) or other cuticular outgrowths (Fig. 13.1E), some genera can have an extra cuticular body sheath (Fig. 13.1F). Lateral fields present or absent. Labial region apparently smooth or with one to three annuli and a hexaradiate framework, offset or continuous with body contour. Oral opening as a dorsoventral slit surrounded by six lip areas (two lateral and four submedian) that may be fused to form a labial disc, submedian lip areas with lobe-like outgrowths (submedian lobes) in several genera (Fig. 13.2). Amphidial apertures absent or very small. **Well-developed stylet with a relatively long cone**, rigid or flexible, basal knobs well developed, rounded, sloping forward (anchor shaped) or backwards. Deirids generally absent. Orifice of dorsal pharyngeal gland ca 3-6 μm from stylet base. **Pharyngeal procorpus broad, cylindroid, amalgamated with muscular metacorpus**, isthmus either broad and amalgamated with basal bulb (Fig. 13.1P) or slender and offset from a **small basal bulb** (Fig. 13.1Q). Pharyngo-intestinal valve (cardia) very small. Excretory pore at pharyngeal or post-pharyngeal level. Intestine syncytial, lacking a definite lumen. Rectum and anus inconspicuous. Reproductive system consisting of a single anterior branch (monodelphic and prodelphic). Vulva transversely oval (= open) (Fig. 13.1G) or slit-like (= closed) (Fig. 13.1H), located well posterior (V = 80-90), anterior vulval lip overlapping vulva in some genera (Fig. 13.1I). Vagina anteriorly directed. Spermatheca small, offset. Uterus with a muscular part and a glandular crustaformeria, post-uterine sac absent. Phasmids absent. Tail short or long and with rounded to pointed tail terminus (Fig. 13.1J-O). Male vermiform and thinner than female, degenerate. **Pharynx and stylet reduced or lacking**. Single anteriorly directed testis (monorchic), usually filled with minute round or amoeboid sperm. Bursa weakly developed or absent. Spicules elongate, straight, arcuate, U- or hook-shaped. Gubernaculum simple, fixed. Juvenile vermiform, generally similar to female in most characters.

Table 13.1. Classification of criconematids after Siddiqi (2000), Andrássy (2007) and Decraemer and Hunt (2024).

After Siddiqi, 2000	After Andrássy, 2007	After Decraemer & Hunt, 2024
Superfamily Criconematoidea Taylor, 1936	**Superfamily Criconematoidea Taylor, 1936**	**Superfamily Criconematoidea Taylor, 1936**
Family Criconematidae Taylor, 1936	**Family Criconematidae Taylor, 1936**	**Family Criconematidae Taylor, 1936**
Subfamily Criconematinae Taylor, 1936	**Subfamily Criconematinae Taylor, 1936**	**Subfamily Criconematinae Taylor, 1936**
Bakernema Wu, 1964	*Bakernema* Wu, 1964	*Bakernema* Wu, 1964
Criconema Hofmänner & Menzel, 1914	*Criconema* Hofmänner & Menzel, 1914	*Criconema* Hofmänner & Menzel, 1915
Lobocriconema De Grisse & Loof, 1965	*Lobocriconema* De Grisse & Loof, 1965	*Lobocriconema* De Grisse & Loof, 1966
Neolobocriconema Mehta & Raski, 1971	*Neolobocriconema* Mehta & Raski, 1971	*Neolobocriconema* Mehta & Raski, 1972
Blandicephalanema Mehta & Raski, 1971	*Blandicephalanema* Mehta & Raski, 1971	*Ogma* Southern, 1914
Crossonema Mehta & Raski, 1971	*Crossonema* Mehta & Raski, 1971	
Ogma Southern, 1914	*Croserinema* Khan, Chawla & Saha, 1976	
	Ogma Southern, 1914	
	Orphreyus Siddiqi, 2000	
	Pateracephalanema Mehta & Rasi, 1971	
Subfamily Macroposthoniinae Skarbilovich, 1959	**Subfamily Macroposthoniinae Skarbilovich, 1959**	**Subfamily Macroposthoniinae Skarbilovich, 1959**
Criconemoides Taylor, 1936	*Criconemoides* Taylor, 1936	*Criconemoides* Taylor, 1936
Macroposthonia de Man, 1880	*Mesocriconema* Andrássy, 1965	*Mesocriconema* Andrássy, 1965
Nothocriconemoides Maas, Loof & De Grisse, 1971	*Nothocriconemoides* Maas, Loof & De Grisse, 1971	*Discocriconemella* De Grisse & Loof, 1965
Discocriconemella De Grisse & Loof, 1965	*Discocriconemella* De Grisse & Loof, 1965	*Xenocriconemella* De Grisse & Loof, 1965
Xenocriconemella De Grisse & Loof, 1965	*Xenocriconemella* De Grisse & Loof, 1965	
Subfamily Hemicriconemoidinae Andrássy, 1979	**Subfamily Hemicriconemoidinae Andrássy, 1979**	**Subfamily Hemicriconemoidinae Andrássy, 1979**
Hemicriconemoides Chitwood & Birchfield, 1957	*Hemicriconemoides* Chitwood & Birchfield, 1957	*Hemicriconemoides* Chitwood & Birchfield, 1957
Superfamily Hemicycliophoroidea Skarbilovich, 1959		
Family Hemicycliophoridae Skarbilovich, 1959	**Family Hemicycliophoridae Skarbilovich, 1959**	**Family Hemicycliophoridae Skarbilovich, 1959**
Subfamily Hemicycliophorinae Skarbilovich, 1959	**Subfamily Hemicycliophorinae Skarbilovich, 1959**	**Subfamily Hemicycliophorinae Skarbilovich, 1959**
Aulosphora Siddiqi, 1980	*Aulosphora* Siddiqi, 1980	*Hemicycliophora* de Man, 1921
Hemicycliophora de Man, 1921	*Hemicycliophora* de Man, 1921	
Loofia Siddiqi, 1980	*Colbranium* Andrássy, 1979	
Colbranium Andrássy, 1979		*Colbranium* Andrássy, 1979

Family Caloosiidae Siddiqi, 1980		
Subfamily Caloosiinae Siddiqi, 1980	**Subfamily Caloosiinae Siddiqi, 1980**	**Subfamily Caloosiinae Siddiqi, 1980**
Caloosia Siddiqi & Goodey, 1964 *Hemicaloosia* Ray & Das, 1978	*Caloosia* Siddiqi & Goodey, 1964	*Caloosia* Siddiqi & Goodey, 1964 *Hemicaloosia* Ray & Das, 1978
Superfamily Tylenchuloidea Skarbilovich, 1947		
Family Tylenchulidae Skarbilovich, 1947	**Family Tylenchulidae Skarbilovich, 1947**	**Family Tylenchulidae Skarbilovich, 1947**
Subfamily Tylenchulinae Skarbilovich, 1947	**Subfamily Tylenchulinae Skarbilovich, 1947**	**Subfamily Tylenchulinae Skarbilovich, 1947**
Boomerangia Siddiqi, 1991 *Trophotylenchulus* Raski, 1957 *Tylenchulus* Cobb, 1913	*Trophotylenchulus* Raski, 1957 *Tylenchulus* Cobb, 1913	*Trophotylenchulus* Raski, 1957 *Tylenchulus* Cobb, 1913
Family Sphaeronematidae Raski & Sher, 1952	**Family Sphaeronematidae Raski & Sher, 1952**	
Subfamily Sphaeronematinae Raski & Sher, 1952	**Subfamily Sphaeronematinae Raski & Sher, 1952**	**Subfamily Sphaeronematinae Raski & Sher, 1952**
Goodeyella Siddiqi, 1986 *Sphaeronema* Raski & Sher, 1952 *Tumiota* Siddiqi, 1986	*Goodeyella* Siddiqi, 1986 *Sphaeronema* Raski & Sher, 1952 *Tumiota* Siddiqi, 1986	*Sphaeronema* Raski & Sher, 1952
Subfamily Meloidoderitinae Kirjanova & Poghossian, 1973		**Subfamily Meloidoderitinae Kirjanova & Poghossian, 1973**
Meloidoderita Poghossian, 1966	*Meloidoderita* Poghossian, 1966	*Meloidoderita* Poghossian, 1966
Family Paratylenchidae Thorne, 1947	**Family Paratylenchidae Thorne, 1947**	
Subfamily Paratylenchinae Thorne, 1949	**Subfamily Paratylenchinae Thorne, 1949**	**Subfamily Paratylenchinae Thorne, 1949**
Paratylenchus Micoletzky, 1922 *Cacopaurus* Thorne, 1943	*Paratylenchus* Micoletzky, 1922 *Gracilacus* Raski, 1962 *Cacopaurus* Thorne, 1943	*Paratylenchus* Micoletzky, 1922 *Cacopaurus* Thorne, 1943
Subfamily Tylenchocriconematinae Raski & Siddiqui, 1975	**Subfamily Tylenchocriconematinae Raski & Siddiqui, 1975**	
Tylenchocriconema Raski & Siddiqui, 1975	*Tylenchocriconema* Raski & Siddiqui, 1975	*Tylenchocriconema* Raski & Siddiqui, 1975

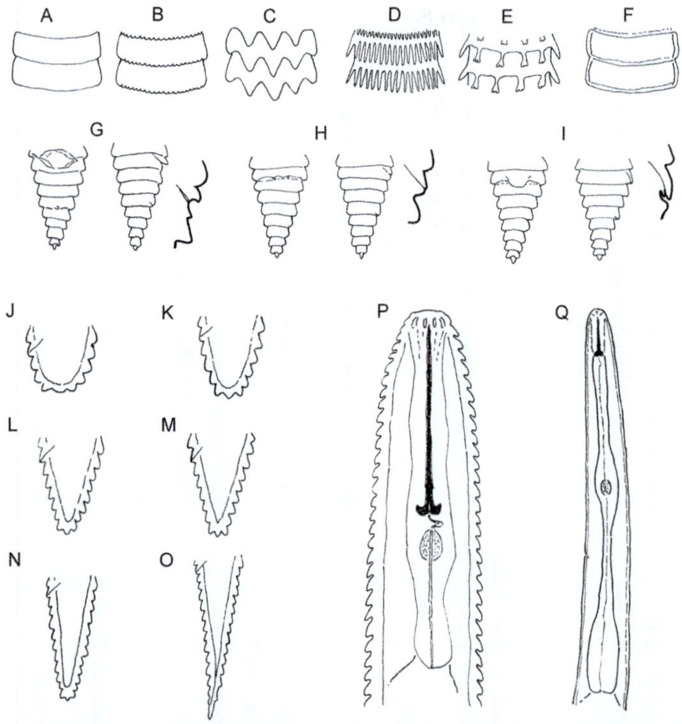

Fig. 13.1. A-F. Types of annule in Criconematoidea. A: Smooth; B: crenate; C: ornamented by scales; D: ornamented by spines; E: ornamented by other cuticular outgrowths; F: cuticular body sheath present. G-I. vulval opening in Criconematidae. G: Open; H: slit-like or closed; I: with overlapping anterior vulval lip. J-O: Types of tail in Criconematidae. J: Short rounded with rounded terminus; K: Rounded-conical with rounded terminus; L: conical-rounded with rounded terminus; M: conical-rounded with pointed terminus; N: elongated with rounded terminus; O: elongated with pointed terminus. P, Q: Pharyngeal region in Criconematoidea; P: isthmus broad and amalgamated with basal bulb; Q: isthmus slender and offset from a small basal bulb.

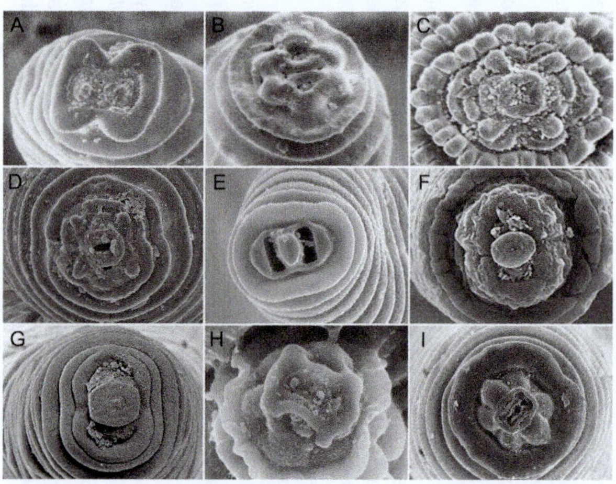

Fig. 13.2. Labial region. SEM. A: *Discocriconemella limitanea*; B: *Criconema mutabile*; C: *Criconemoides similicrenata*; D: *Criconemoides sphaerocephalus*; E: *Caloosia longicaudata*; F: *Hemicycliophora typica*; G: *Hemicycliophora onubensis*; H: *Ogma decalineatus*; I: *Criconema silvum*. (After Cid del Prado and Subbotin.)

13.2.1. Key to Families of Criconematoidea

1. Females and juveniles with thick cuticle, bearing coarse round or retrorse annuli, pharynx with isthmus broad and amalgamated with basal bulb.................... 2
   Females and juveniles with thin cuticle, bearing fine round annuli, pharynx with isthmus not amalgamated with basal bulb .... Tylenchulidae
2. Females and juveniles spindle- or sausage-shaped with retrorse annuli (except females of *Hemicriconemoides* with rounded annuli), annuli may have cuticular outgrowths in some genera. Males with short tail and generally lacking bursa, if bursa present, then terminal .... Criconematidae
   Females and juveniles elongate-vermiform, annuli without cuticular outgrowths. Males with elongate tail and generally subterminal bursa .... Hemicycliophoridae

13.3. Family Criconematidae Taylor, 1936

Diagnosis: Small nematodes, usually under 0.50 mm long. **Body stout, spindle- or sausage-shaped**, straight or slightly curved ventrally after death by gentle heat and fixation. **Cuticle thick with coarse, generally retrorse annuli**, smooth, crenate or ornamented by scales, spines or other appendages (*Hemicriconemoides* females have rounded annuli without appendages and a cuticular body sheath, but its juveniles have retrorse and scaled annuli). Lateral fields absent, observed at most as an irregular line because of junction where annuli anastomose. Labial region with one or two annuli, with or without submedian lobes. **Stylet strong, conus markedly longer than shaft, basal knobs prominent, anchor shaped**. Criconematoid pharynx, procorpus amalgamated with metacorpus, short isthmus, barely separated from small basal bulb. Vulva open or closed. Tail short, rounded or conoid with rounded to pointed terminus. Male vermiform, more slender than female. Cuticle with fine annuli without cuticular outgrowths. Reduced stylet and pharynx. Spicules elongate, straight to arcuate. Bursa terminal or absent, rarely subterminal. Juveniles generally similar to female.

13.3.1. Key to Subfamilies and Genera of Criconematidae

1. Female with cuticular body sheath of rounded annuli giving appearance of a double cuticle; juvenile without sheath but with scales or spines on posterior margins of annuli .... Hemicriconemoidinae, *Hemicriconemoides*
   Female without a cuticular body sheath 2
2. Female and juvenile annuli ornamented with scales, spines or other appendages (except females of *Criconema* and *Lobocriconema*) .... Criconematinae 3
   Female and juvenile with smooth or crenate annuli lacking scales, spines or other appendages .... Macroposthoniinae 7
3. Female body annuli with scales, spines or finger-like projections 4
   Female anterior body annuli without scales, spines or finger-like projections, although they may be present in tail region 5
4. Cuticular appendages membranous, transparent .... *Bakernema*
   Cuticular appendages not membranous .... *Ogma*
5. Labial region with submedian lobes 6
   Labial region without submedian lobes .... *Criconema*
6. Scales, spines or other outgrowths present in posterior region of female body, vulva closed .... *Neolobocriconema*
   Scales, spines, or other outgrowths not present in posterior region of female body, vulva mostly open .... *Lobocriconema*

7. Anterior labial annulus large disc-like, distinctly offset by a collar .... *Discocriconemella*
   Anterior labial annulus not disc-like, not distinctly offset 8
8. Stylet thin, very long and flexible, anterior labial annulus very small, surrounded by second annulus .... *Xenocriconemella*
   Stylet not so long and usually not flexible, anterior labial annulus not very small, not surrounded by second annulus.. *Criconemoides*

13.3.2. Subfamily Criconematinae Taylor, 1936

Criconematinae are characterized by the presence of spines or scales on the cuticle of the female and juvenile stages, except for *Criconema* and *Lobocriconema* in which only the juveniles have cuticular appendages.

Genus *Bakernema* Wu, 1964
Diagnosis: Criconematidae, Criconematinae. Female 0.50-0.55 mm long. **Slightly retrorse annuli with transparent spine-like appendages on head and body regions and leaf-like flaps on tail tip**, these appendages not being arranged in clear longitudinal rows over body and difficult to see in glycerin-mounted specimens. Labial region composed of two or three undifferentiated annuli, not set off and with elevated labial disc, pseudolips and submedian lobes weakly developed. Stylet 64-70 μm long in type species, rigid. Vulva closed with overhanging anterior lip. **Vagina sigmoid**. Tail conoid with rounded terminus. Male lateral field with four incisures. Bursa subterminal. Juvenile with annuli apparently smooth or crenate.
Type species: *Bakernema inaequale* (Taylor, 1936) Mehta & Raski, 1971.
Other species: One.
Distribution: Confined to North America.

Genus *Criconema* Hofmänner & Menzel, 1914
= *Nothocriconema* De Grisse & Loof, 1965
= *Nothocriconemella* Ebsary, 1981
= *Notholetus* Ebsary, 1981
= *Paracriconema* Ebsary, 1981
= *Cerchnotocriconema* Bernard, 1982
= *Nenocriconema* Darekar & Khan, 1982
Diagnosis: Criconematidae, Criconematinae. Female 0.20-0.80 mm long with 24-140 **round or retrorse annuli having smooth or crenate margins but without cuticular outgrowths in adult**. Labial region with one or two annuli, lip region elevated with six pseudolips, **submedian lobes poorly developed or absent**. Stylet 45-150 μm long, rigid or flexible. **Vulva closed.** Tail conoid with rounded to pointed terminus. Male lateral field with three or four incisures. Bursa terminal or absent. **J4 (fourth-stage juvenile) with 8-18 rows of scales.**
Type species: *Criconema guernei* (Certes, 1889) Menzel *in* Hofmänner & Menzel, 1914.
Other species: More than 90.
Distribution: Worldwide.
Major species:
Criconema mutabile (Taylor, 1936) Raski & Luc, 1985. L = 0.27-0.58 mm; V = 87-94; stylet = 48-60 μm. Female with 85-114 body annuli and very high rounded pseudolips but lacking submedian lobes. Juveniles with 17-20 longitudinal rows of scales. *Criconema mutabile* has been associated with bacterial canker of *Prunus* spp.

Genus *Lobocriconema* De Grisse & Loof, 1965
Diagnosis: Criconematidae, Criconematinae. Female 0.24-0.80 mm long with 24-73 **very coarse (8-18 μm wide) and retrorse annuli, posterior margins smooth or finely crenate**, without appendages. Two labial annuli, anterior one larger than posterior, **submedian lobes present**. Stylet 43-110 μm long, rigid. **Vulva generally open**. Tail conoid with a rounded terminus. Male lateral field with three or four incisures. Bursa terminal. **J4 with 8-18 rows of smooth scales.**
Type species: *Lobocriconema crassianulatum* (de Guiran, 1963) De Grisse & Loof, 1965.
Other species: More than 20.
Distribution: Africa, Asia, Europe, North America.
Bionomics: Mainly described from roots of woody plants.

Genus *Neolobocriconema* Mehta & Raski, 1971
= *Merocriconema* Raski & Pinochet, 1976
= *Crossonemoides* Eroshenko, 1981
= *Paralobocriconema* Minagawa, 1986
Diagnosis: Criconematidae, Criconematinae. Female 0.22-0.90 mm long with 36-52 **very coarse annuli (9-17 μm wide)**, slightly retrorse to rounded, crenate, finely serrated or irregularly

lobed and/or bearing small spines. Labial region with one or two annuli, high pseudolips, **submedian lobes present**. Stylet 57-140 μm long, rigid. Vulva closed. Tail conoid with a rounded terminus. Male rare. Lateral field with four incisures. Bursa terminal. Juveniles with more annuli than female and **J4 with 8-25 rows of spined scales**.
Type species: *Neolobocriconema laterale* (Khan & Siddiqi, 1963) Mehta & Raski, 1971.
Distribution: Africa, America, Asia.
Bionomics: Mainly described from roots of woody plants.

Genus *Ogma* Southern, 1914
= *Blandicephalanema* Mehta & Raski, 1971
= *Crossonema* Mehta & Raski, 1971
= *Pateracephalanema* Mehta & Raski, 1971
= *Variasquamata* Mehta & Raski, 1971
= *Croserinema* Khan, Chawla & Saha, 1976
= *Seriespinula* Mehta & Raski, 1976
= *Neocrossonema* Ebsary, 1981
= *Amphisbaenema* Orton Williams, 1982
= *Pseudocriconema* Minagawa, 1984
= *Syro* Orton Williams, 1985
= *Homogma* Siddiqi, 1986
= *Macrocriconema* Minagawa, 1986
= *Pellipecten* Siddiqi, 1986
= *Orphreyus* Siddiqi, 2000
(Fig. 13.3)
Diagnosis: Criconematidae, Criconematinae. Female 0.24-0.90 mm long, with 50-100 **annuli bearing outgrowths of different shapes arranged as a continuous fringe or in longitudinal rows** on most of body. Labial region with one or two annuli set off from rest of body. Submedian lobes present or absent. Stylet 48-125 μm, rigid or flexible. Vulva closed. Tail rounded or conical with a rounded to pointed terminus. Male lateral field with three or four incisures. Bursa terminal or lacking. **J4 with scales arranged in longitudinal rows**.
Type species: *Ogma murrayi* Southern, 1914.
Other species: More than 75.
Distribution: Worldwide. Species of *Ogma* have been reported from soils with moss, grass and woody plants.
Major species:
Ogma rhombosquamatum (Mehta & Raski, 1971) Andrassy, 1979. Female. L = 0.33-0.47 mm; V = 85-90; stylet = 93-116 μm. R = 74-85. Female with body annuli. Scales in 10 rows. Conoid tail with rounded terminus. *Ogma rhombosquamatum* was found in the rhizosphere of *Olea europaea* L., *Pinus* sp. and *Smilax* sp. in Italy. It formed colonies on olive roots and feeds in the cortical tissues. Feeding caused thickening of the cell walls and results in necrosis of epidermal and cortical tissue.

13.3.3. Subfamily Macroposthoniinae Skarbilovich, 1959

Macroposthoniinae are characterized by the absence of scales and spines on the cuticle of the female and juvenile stages.

Genus *Criconemoides* Taylor, 1936
= *Macroposthonia* de Man, 1880
= *Criconemella* De Grisse & Loof, 1965
= *Mesocriconema* Andrássy, 1965
= *Nothocriconemoides* Maas, Loof & De Grisse, 1971
= *Madinema* Khan, Chawla & Saha, 1976
= *Neobakernema* Ebsary, 1981
= *Seshadriella* Darekar & Khan, 1981
= *Pakcriconemoides* Shahina & Maqbool, 1993
(Fig. 13.4)
Diagnosis: Criconematidae, Macroposthoniinae. Female 0.20-1.00 mm long, sausage- or ring-like when relaxed with 40-200 fine or coarse annuli with smooth, rough or crenate margins, **but lacking scales, spines and other cuticular appendages**. Labial annuli two to three, smaller than and not differentiated or separated by a collar from body annuli. Some species with labial annulus appearing as a large, anteriorly flattened disc. Submedian lobes present or absent. Stylet 38-130 μm long, usually rigid. Vulva open or closed. Vagina sigmoid or straight. Tail rounded or conoid with a rounded to pointed terminus. Male lateral field with two to four incisures. Bursa terminal or subterminal. **Juvenile annuli smooth or crenate, never with scales or spines**.
Type species: *Criconemoides morgensis* (Hofmänner *in* Hofmänner & Menzel, 1914) Taylor, 1936.
Other species: More than 140.
Distribution: Worldwide.
Major species:
Criconemoides curvatus Raski, 1952. Female. L = 0.29-0.56 mm; V = 92-96; stylet = 47-68 μm; R= 74-106. Body stout, slightly curved ventrally

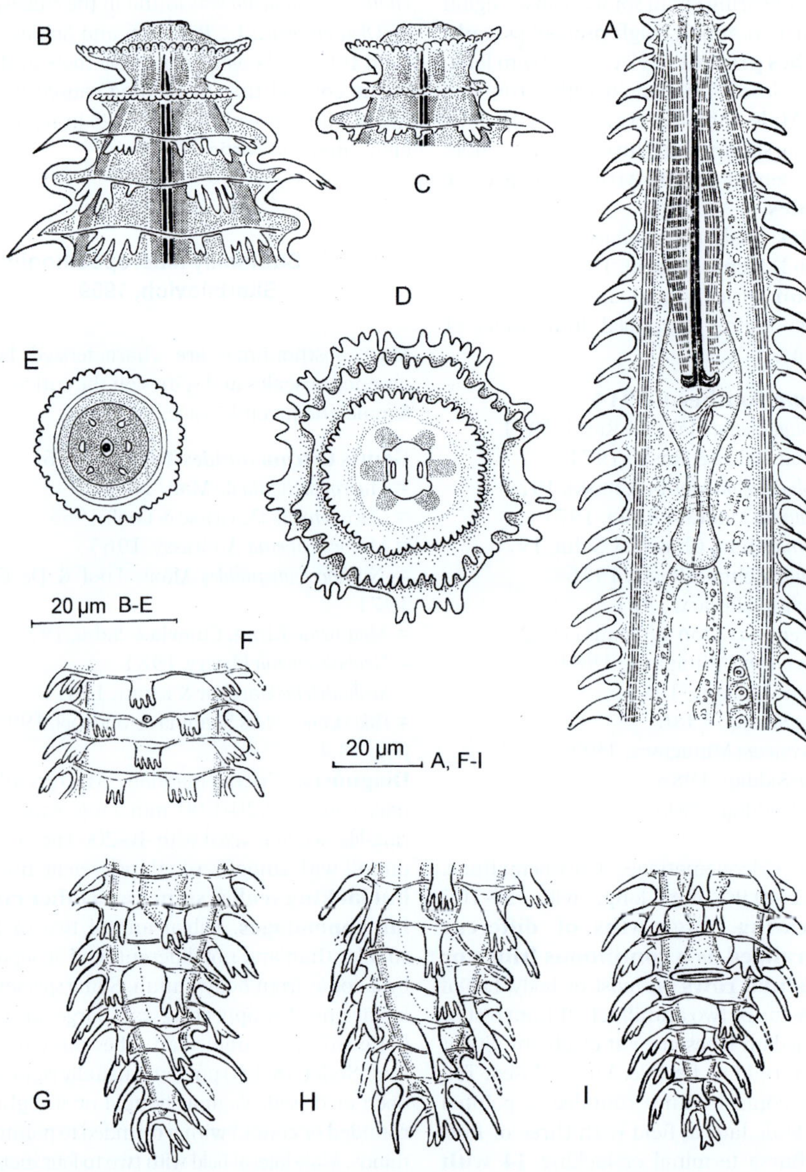

Fig. 13.3. *Ogma palmatum*. A: Pharyngeal region; B, C: labial region; D: *en face* view; E: optical section through second labial annule; F: ventral view of body showing excretory pore; G, H: posterior region, lateral; I: posterior region, ventral, showing vulva. (After Orton Williams (1974), courtesy CABI.)

and annuli with smooth posterior margins and generally without anastomoses. Vagina straight. Vulva open. Tail conoid-rounded, terminal annulus often only faintly indented. *C. curvatus* injured carnation and has been associated with decline of peach trees. Nematodes cause destruction of epidermis and necrosis of cortex and can reduce root weight of carnations by 33% in 6 months. In peach, they caused extensive lesions on roots.

Criconemoides onoensis Luc, 1959. Female. L = 0.39-0.67 mm; V = 89-94; stylet = 40-63 µm; R = 111-138. Body annuli with retrorse, smooth with occasional anastomoses. Vagina straight.

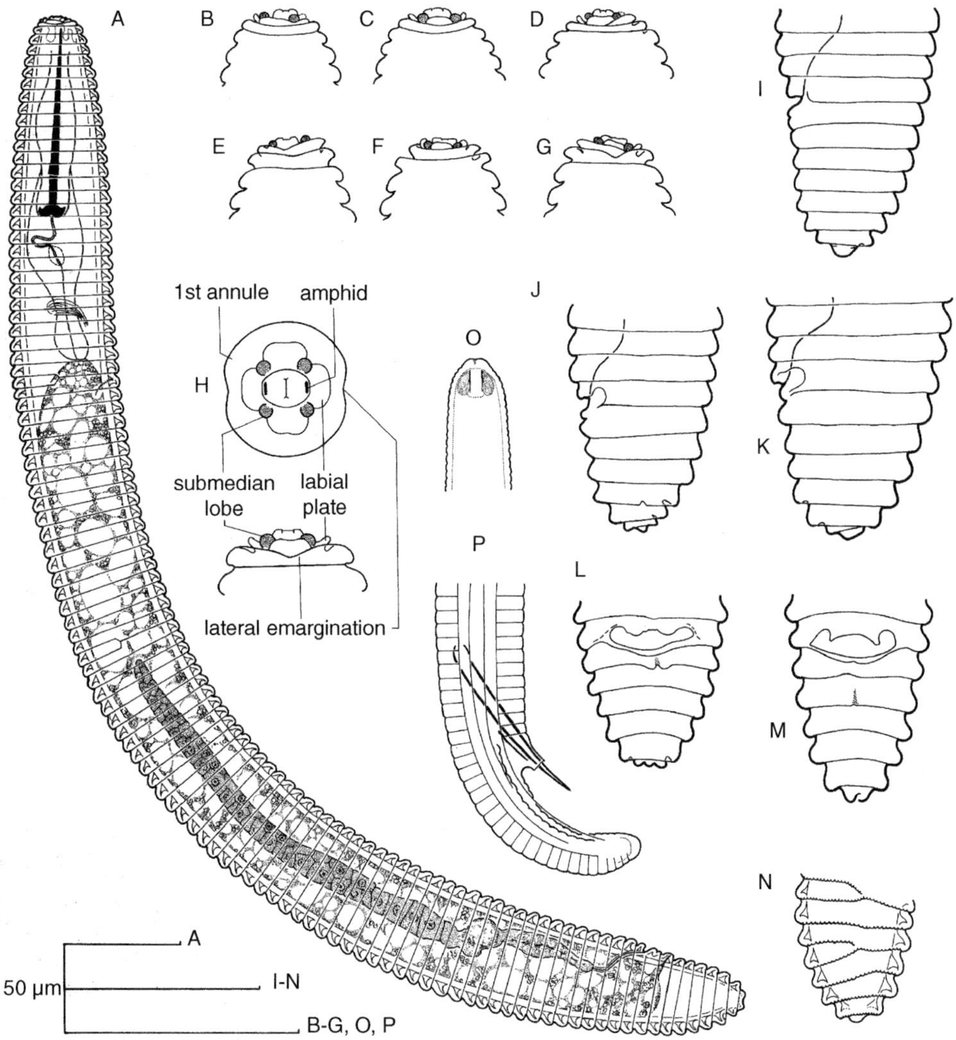

Fig. 13.4. *Criconemoides xenoplax*. A: Entire female; B-G: female head; H: *en face* and lateral view of lip area; I-K: female tail in lateral view; L, M: female tail in ventral view; N: juvenile tail in lateral view; O: male head; P: male tail. (After Orton Williams (1972), courtesy CABI.)

Vulva open. Tail terminus rounded and lobed. *C. onoensis* caused rice production losses in several countries through root decay, severe stunting and yellowing of plants.

Criconemoides ornatus Raski, 1958. Female. L = 0.33-0.52 mm; V = 92-96; stylet = 44-56 µm; R = 79-94. Female body with few or no anastomoses. Labial plates inconspicuous or very low. Vagina not sigmoid. Anterior vulva lip with two spines. *C. ornatus* has been reported causing damage to cabbage, carnation, grasses, lucerne, okra, peanut, sorghum and soybean.

Criconemoides sphaerocephalus Taylor, 1936. Female. L = 0.22-0.48 mm; V = 85-97; stylet = 43-77 µm. R = 55-82. Body with many anastomoses producing a zig-zag line laterally where they occur. Vulva open, with conspicuously large pre-vulval annulus on ventral side. Tail rounded or hemispherical. This species has been reported from different crops in Europe, America, Africa and Asia.

C. xenoplax Raski, 1952 (Fig. 13.4). L = 0.40-0.75 mm; V = 92-96; stylet = 54-87 µm; R = 77-114. Female body straight or slightly curved body with retrorse and smooth annuli. Few or no anastomoses. Labial region with distinct submedian lobes, separated by labial plates. Vulva open, sigmoid vagina and rounded tail terminus. *C. xenoplax* was reported from almonds, apple, apricot, carnation, cherry, grapefruit, guava, lettuce, peach, plum, grapevine and walnuts, producing interveinal chlorosis, leaf drop and stunted plants. Root systems parasitized by *C. xenoplax* show a lack of feeder roots. Damage caused by ring nematodes to stone fruits is two-fold: firstly, direct damage to the roots; and secondly, but more important, damage caused by interactions with other pathogens. In the south-eastern USA, *C. xenoplax* in association with *Pseudomonas syringae* predisposes peach trees to a disease complex called 'Peach Tree Short Life' (PTSL). Above-ground symptoms of PTSL become visible when trees are 3-6 years old and include chlorosis, wilting and sudden death of trees after bloom. Although tree loss due to PTSL varies from year to year, a total loss of US$6 million per year was estimated in South Carolina. In California, *C. xenoplax* in association with *P. syringae* causes the disease complex Bacterial Canker Complex (BCC) in peach, this being associated with delayed bloom and, in severe cases, collapse of trees. In addition, bacterial spot damage caused by *Xanthomonas campestris* pv. *pruni* is more severe on peach trees if the soil is infested with *C. xenoplax*. In grapevines, plant death does not occur, but reduction in vine growth is observed in sandy areas where *C. xenoplax* is present.

Management: In general, nematicide treatments, fumigant and non-fumigant, are considered effective against ring nematodes and have been recommended for control of *Criconemoides* spp. Nevertheless, they are only effective if fields are weeded, because many weeds are good hosts for these nematodes. For control of PTSL or BCC, nematicide treatments are recommended when *C. xenoplax* population densities exceed 40 nematodes per 100 cm^3 of soil. The rootstocks 'Lovell' and 'Guardian' have been recommended for their tolerance to *C. xenoplax* and longer survival in PTSL sites but resistance to *C. xenoplax* has not yet been detected in *Prunus* accessions. Cultural treatments such as pre-planting Bahia grass (*Paspalum notatum*) or wheat (*Triticum aestivum*) increased peach tree growth and survival from PTSL in soil infested with *C. xenoplax*. Soil solarization reduced populations of *C. xenoplax* in peach orchards and increased peach tree growth. Regarding biological control, one strain of *Pseudomonas* 'BG33R' isolated from suppressive PTSL soils showed some capacity for reducing *C. xenoplax* multiplication *in vivo* and egg hatch *in vitro*. In field experiments, *Pseudomonas* 'BG33R' maintained *C. xenoplax* populations below the economic threshold for nematicide treatment for up to 18 months (Kluepfel *et al.*, 2002).

Genus *Discocriconemella* De Grisse & Loof, 1965

= *Neocriconema* Diab & Jenkins, 1965
= *Acrozostron* Orton Williams, 1981

Diagnosis: Criconematidae, Macroposthoniinae. Female 0.18-0.49 mm long with 65-174 rounded or angular annuli, with smooth, rough or crenate margins **but lacking scales, spines and other cuticular appendages**, often with anastomoses. **Labial annulus appearing as a large, anteriorly flattened disc, separated by a collar from adjacent body annuli**, disc indented dorsally and ventrally or rarely circular, with submedian pseudolip areas projecting beyond its contour in some species, pseudolips indistinct, fused into this disc. Submedian lobes absent or present. Stylet 33-113 µm, rigid or flexible. **Vulva closed or open, anterior lip smooth or with bilobed** lip. Tail conoid with rounded to pointed terminus. Male lateral field narrow, with two, three or four incisures. Labial region tapering, smooth, **surmounted by a distinct disc or knob of thickened cuticle**. Bursa terminal or absent. Juvenile with annuli crenate, lacking scales or spines.

Type species: *Discocriconemella limitanea* (Luc, 1959) De Grisse & Loof, 1965.

Other species: More than 25.

Distribution: Africa, America, Asia, South Pacific.

Genus *Xenocriconemella* De Grisse & Loof, 1965

Diagnosis: Criconematidae, Macroposthoniinae. Female 0.2-0.4 mm long, with 100-130 retrorse annuli. Labial region with two annuli, second one larger than first. **Stylet 80-140 µm long, up to 40% of body length, thin and flexible. Vulva closed** with slightly projecting lips. Tail short,

conoid with a rounded terminus. Male lateral field with three to four incisures. Bursa subterminal. Juvenile annuli with crenate margins.
Type species: *Xenocriconemella macrodora* (Taylor, 1936) De Grisse & Loof, 1965.
Other species: More than six (Archidona-Yuste *et al.*, 2024; Cantalapiedra-Navarrete *et al.*, 2024; Peraza-Padilla *et al.*, 2024).
Distribution: Worldwide, especially in North America and Europe, species have close association with woodland forest.

13.3.4. Subfamily Hemicriconemoidinae Andrássy, 1979

The Hemicriconemoidinae are characterized by a double cuticle, formed by a sheath of round annuli that surrounds the body.

Genus *Hemicriconemoides* Chitwood & Birchfield, 1957
= *Iota* Cobb, 1913
(Fig. 13.5A-F)
Diagnosis: Criconematidae, Hemicriconemoidinae. **Female elongate-cylindrical enclosed in a cuticular sheath**, 0.29-0.75 mm long, with 50-164 **rounded or rarely retrorse annuli lacking any appendages**. Labial region with two, rarely three annuli, continuous or offset, variable in shape. Pseudolips inconspicuous, submedian lobes usually absent. **Stylet elongate** (41-100 µm), **with anteriorly cupped knobs appearing anchor shaped**, rigid or flexible. Vulva open or closed, occasionally with cuticular flaps. Tail short, subcylindroid to conoid with rounded to pointed terminus. Male with body sheath absent. Lateral field usually with four incisures. Tail conoid to subcylindroid and bursa, when present, subterminal or terminal. **Juvenile with single cuticle, with smooth, dentate or spined scales**, arranged on body in longitudinal, usually alternating rows, or irregularly.
Type species: *Hemicriconemoides wessoni* Chitwood & Birchfield, 1957.
Other species: More than 50.
Distribution: Sheathoid nematodes, *Hemicriconemoides* spp. are widely distributed in warmer areas, particularly in Africa, the Americas, Australia, South and Southeast Asia and southern Europe.
Major species: The sheathoid nematodes have been reported associated with many crops. A few of them have been implicated with the decline of some crops, but most are not considered aggressive parasites.

Hemicriconemoides cocophillus (Loos, 1949) Chitwood & Birchfield, 1957. Female. L = 0.41-0.54 mm; R = 99-120; V = 91-94; stylet = 47-58 µm. Truncate lip region. Tail convex-conoid to attenuated. *H. cocophillus* was described from soil around roots of grass and coconut roots in Sri Lanka, associated with sugarcane decline in Martinique and since then reported from mainly African and Asian countries as well as the Americas.

Hemicriconemoides kanayaensis Nakasono & Ichinohe, 1961. Female. L = 0.41-0.67 mm; R = 125-164; V = 87-91; stylet = 66-78 µm. Conoid and continuous lip region. Tail elongated conoid, with a smoothly rounded tip. *H. kanayaensis* is an important pest of tea bushes in Japan as the continuous feeding of the nematodes results in extensive damage to the root cortex.

Hemicriconemoides strictathecatus Esser, 1960 (Fig. 13.5A-E). Female. L = 0.49-0.59 mm; R = 127-152; V = 90-93; stylet = 73-83 µm. Rounded and slightly offset lip region. Tail rounded or convex-conoid. This species attacks citrus, banana, mango, pineapple and may cause significant yield loss to these crops. *Hemicriconemoides mangiferae* is usually considered as a junior synonym of *H. strictathecatus* (see Van den Berg *et al.*, 2015, 2018).

Hemicriconemoides wessoni Chitwood & Birchfield, 1957. Female. L = 0.37-0.62 mm; R = 76-96; V= 89-96; stylet = 50-61 µm. Tail conoid, tapering abruptly behind anus. In Florida, *H. wessoni* has economic relevance for the sod grass industry because it suppresses the growth and vigour of highly maintained grasses at population levels ranging from 300 to1000 specimens/100 cm^3 of soil. Disease symptoms induced by these parasites consist of stunting, premature wilting, leaf yellowing, root malformation, necrosis of cortical root tissues and related signs characteristic of nutrient deficiencies
Management: Pre-plant fumigation with 1-3 dichloropropene in countries where its use is still allowed, improves effectively the performance of the replants of various tropical fruit trees in infested areas.

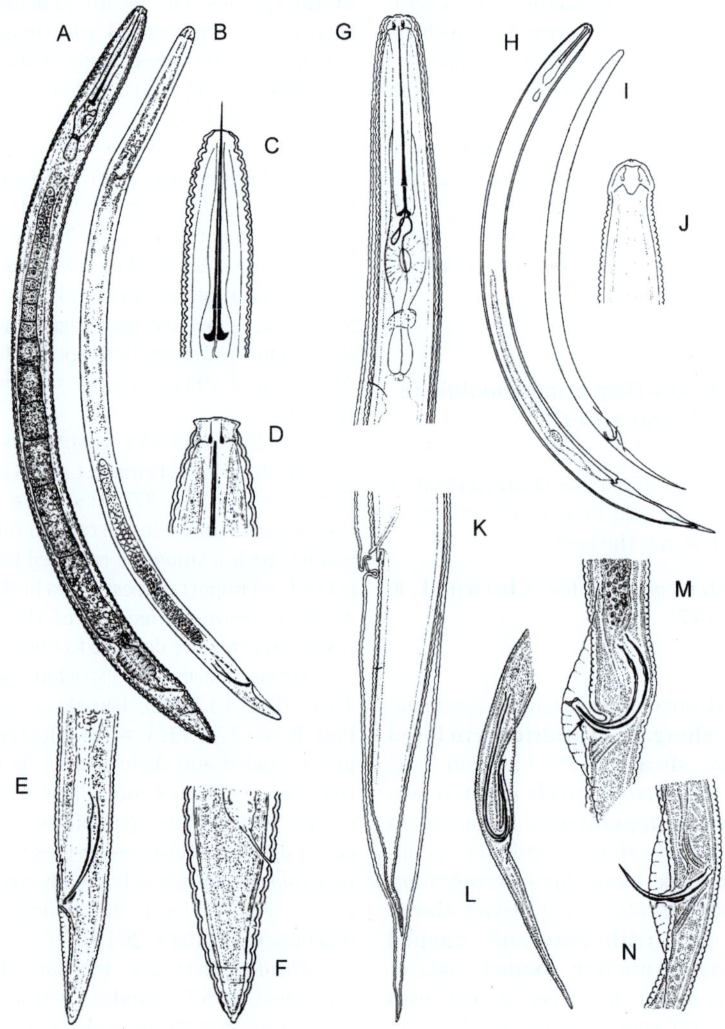

Fig. 13.5. A-F. *Hemicriconemoides mangiferae*. A: Entire female; B: entire male; C, D: female labial region; E: male tail; F: female tail. G-L. *Hemicycliophora chathami*; G: female pharynx; H: entire female; I: entire male; J: male labial region; K: female posterior region; L: male tail. *Hemicycliophora penetrans*. M: Male tail. *Hemicycliophora thienemanni*. N: Male tail. Line drawings are not to scale. (A-E, after Siddiqi (1977), courtesy CABI.)

13.4. Family Hemicycliophoridae Skarbilovich, 1959

Diagnosis: Criconematoidea. Female 0.60-2.00 mm long, vermiform, straight to arcuate upon relaxation, with **thick cuticle** and usually more than 200 **rounded annuli**. Juveniles and females usually have an **extra cuticle as a protective sheath** (except in *Caloosia*). Lateral fields present in forms with sheath. Labial region with one to three annuli, with a labial disc but lacking submedian lobes. **Stylet elongated (over 50 μm long in female), with rounded to backwardly sloping knobs,** not anchor shaped. Pharynx criconematoid, with a **short broad isthmus amalgamated with basal bulb**. Monodelphic, prodelphic. Vulva transversely oval with usually projecting lips, posterior (V

= 80-90). Vagina sigmoid, straight or curved. Tail usually elongate-conoid to filiform but may be cylindroid or rarely hemispherical. Male with reduced pharynx and without stylet. Elongated tail and large adanal bursa. Spicules, straight, arcuate, semicircular, U- or hook-shaped. Cloacal lips elongated to form a penial tube in some species. Juvenile similar to female in most cases.

13.4.1. Key to Subfamilies and Genera of Hemicycliophoridae

1. Female and juveniles with a membranous body sheath or without sheath, labial annuli separated, vulva flush with body contour and overhung by its anterior lip, male spicules straight .... Caloosiinae - 2
Female and juveniles with a thick body sheath, labial annuli not separated, vulva rarely flush with body contour, not overhung by its anterior lip, male spicules arcuate, semicircular, U- or hook-shaped .... Hemicycliophorinae – *Hemicycliophora*
2. Female and juveniles with body sheath and lateral fields .... *Hemicaloosia*
Female and juveniles without body sheath and lateral fields .... *Caloosia*

13.4.2. Subfamily Caloosiinae Siddiqi, 1980

Caloosiinae are characterized by females and juveniles that have an elongate, vermiform body, with or without a thin body sheath, annuli lacking spines or scales and straight spicules in the male.

Genus *Caloosia* Siddiqi & Goodey, 1964
Diagnosis: Hemicycliophoridae, Caloosiinae. Female 0.4-1.2 mm long. **Body sheath and lateral fields in female and juveniles absent**. Stylet 74-92 µm long. Labial annuli separated. Vulva transversely oval, less than half body width long, partially overhung by its dorsal lip, thus appearing flush with body contour. Tail elongate filiform. Vagina sigmoid. Male labial region annulated, not marked off from body. **Spicules straight 42-45** µm long. **Cloacal lips not elongated to form a penial tube**. Large bursa covering more than one-third of tail.
Type species: *Caloosia longicaudata* (Loos, 1948) Siddiqi & Goodey, 1964.
Other species: Seven.
Bionomics: *Caloosia heterocephalus*, *Caloosia paxi* and *Caloosia triannulata* feed on rice roots and can disturb the apical growth. *Caloosia nudata* causes some galls at citrus root tips when the nematodes are present in large numbers in the rhizosphere.
Distribution: *Caloosia* occurs widely in Asia, Africa and Australia.

Genus *Hemicaloosia* Ray & Das, 1978
Diagnosis: Hemicycliophoridae, Caloosiinae. Female 0.5-1.4 mm long. A membranous body sheath and lateral fields in female and juveniles present. **Annuli usually separated by grooves, cuticle not folded between annuli**. Lateral field on body sheath marked by two incisures. Labial annuli separated. Stylet 60-64 µm long. Vulva transversely oval, less than half body width long, partially overhung by its dorsal lip, thus appearing flush with body contour. Vagina sigmoid. Tail filiform. Male labial region smooth, marked off from body by a discontinuity. **Spicules straight, cloacal lips not elongated to form a penial tube**. Large bursa covering more than one-third of tail.
Type species: *Hemicaloosia americanae* Ray & Das, 1978.
Other species: Eight.
Distribution: *Hemicaloosia* occurs in Asia, South and North America and Australia.

13.4.3. Subfamily Hemicycliophorinae Skarbilovich, 1959

Hemicycliophorinae are characterized by females and juveniles that have an elongate, vermiform body. Extra cuticular sheath always present, loosely or closely fitting. Spicules arcuate, semicircular, U- or hook-shaped.

Genus *Hemicycliophora* de Man, 1921
= *Procriconema* Micoletzky, 1925
= *Colbranium* Andrassy, 1979
= *Aulosphora* Siddiqi, 1980
= *Loofia* Siddiqi, 1980
(Fig. 13.5G-N)
Diagnosis: Hemicycliophoridae, Hemicycliophorinae. Extra cuticular sheath always present,

loosely or closely fitting, never membranous. Body distinctly, slightly or not recessed immediately posterior to vulva. Vulva a transverse slit over 0.5 body diameter long, lips either modified and projecting, or not modified and rounded. Tail elongate-conoid, filiform or cylindroid, rarely hemispherical. Vulva lips rounded, pointed or elongated. **Male spicules arcuate, semicircular or hook-shaped**. Cloacal lips elongated to form a penial tube. Bursa covering less than one-third of tail.

Type species: *Hemicycliophora typica* de Man, 1921

Other species: More than 135.

Distribution: Sheath nematodes, *Hemicycliophora* spp. are widely, although not uniformly, distributed throughout the world.

Major species: *Hemicycliophora* species are obligate ectoparasites of plants and inhabit moist soils and aquatic environments. Many species have been found in cultivated soils associated with agricultural crops, fruit and nut trees and ornamental plants. They are sedentary during their feeding process while remaining attached to plant roots, and can induce the formation of root galls or stunted and terminal thickenings. Several *Hemicycliophora* species, namely, *Hemicycliophora arenaria*, *H. conida*, *Hemicycliophora parvana*, *Hemicycliophora poranga*, *Hemicycliophora similis*, *H. typica* and *Hemicycliophora ripa*, are known to damage agricultural crops in different countries.

H. arenaria Raski, 1958. Female. L = 0.6-0.9 mm; V = 89-92. R = 141-206 annuli. Tail terminus broadly rounded, almost hemispherical, annulated. The citrus sheath nematode, *H. arenaria*, was first reported in 1957 parasitizing lemon seedling roots in a nursery in the Coachella Valley, California, USA. The species has also been reported from Australia, Argentina and Brazil. Plants infected by *H. arenaria* are easily recognized by swellings or galls produced on lateral and terminal roots

H. parvana Tarjan, 1952. Female. L = 0.8-1.2 mm; V =79-90, R = 234-293. Tail elongated conical, terminus rounded. This species was recovered in field soil from Florida, USA and found parasitizing on celery, citrus, cabbage, grapefruit, queen palm, sweet corn, common bean and other plants.

H. poranga Monteiro & Lordello, 1978. Female. L = 0.7-1.2 mm; V = 79-86, R =266-345. Tail elongate-conoid, tapering gradually to a finely rounded terminus, distal portion symmetrical or offset in lateral view. This species was first reported from Brazil infesting the rhizosphere of cabbage. Presently, it was also reported from USA, Argentina, Venezuela, Italy and Spain and associated with several agricultural crops.

Management: *Hemicycliophora* spp. tend to be polyphagous, so management by crop rotation is difficult. Applications of organic material to sandy soil inhibited population increase. Nematicides are generally effective in the sandy soils in which these nematodes typically occur. *H. arenaria* can be eradicated from root systems with hot water dips (10 min at 46°C). Pre-plant fumigation of soil is very effective in reducing numbers of *Hemicycliophora* and several citrus rootstocks are resistant to the nematode.

13.5. Family Tylenchulidae Skarbilovich, 1947

Diagnosis: Criconematoidea. Female vermiform curved ventrally (Paratylenchinae), swollen to obese (Tylenchulinae), or spherical or subspherical (Sphaeronematinae and Meloidoderitinae), 0.10-0.70 mm long. **Cuticle thin, finely annulated in non-obese forms**. Lateral fields present. Labial region rounded, with or without perioral elevation. Stylet well developed with rounded knobs, conus abnormally elongated, often more than twice as long as shaft, in Paratylenchinae. Pharynx with broad procorpus, well-developed muscular metacorpus, **slender isthmus not amalgamated with a small offset basal bulb containing pharyngeal glands**. Excretory pore in pharyngeal region or well posterior to pharynx in Tylenchulinae. In Tylenchulinae, the excretory cell (renette) is enormously developed, producing a gelatinous matrix. Vulva a large transverse slit. Post-uterine sac absent. Ovary coiled in obese females. Tail cylindroid, subcylindroid or short or lacking in obese forms. Male vermiform. Labial region symmetrical or asymmetrical. Stylet reduced or absent. Pharynx reduced, often completely so. Testis outstretched, sometimes reduced. Spicules small and arcuate. Bursa present or absent, if present subterminal. Juvenile slender, straight to arcuate when relaxed. Deirids often present. Stylet well developed, but occasionally later-stage juveniles may have a reduced pharynx and no stylet. Tail elongate-conoid with terminus minutely rounded. Note: Ghaderi *et al.* (2016) provide a monograph of the family.

13.5.1. Key to Subfamilies and Genera of Tylenchulidae

1. Females spherical to elongate-obese, stylet short (usually under 15 μm and with conus not abnormally elongated), procorpus slender 2
   Females usually vermiform, but in some genera may become slightly swollen, stylet long (usually over 20 μm and with elongated conus), procorpus broad........ Paratylenchinae, 5
2. Females spherical or subspherical, lacking a post-vulval region, excretory pore in pharyngeal region........ 3
   Females swollen, but not spherical or subspherical, with a distinct post-vulval region, excretory pore much behind pharyngeal region,........ Tylenchulinae 4
3. Adult female with neck, uterine walls not forming a protective cystoid body for the eggs .... Sphaeronematinae, *Sphaeronema*
   Adult female without a neck, uterine walls or body forming a protective cystoid body for the eggs .... Meloidoderitinae, *Meloidoderita*
4. Excretory pore in female well posterior, at 68-85% of body length, surrounded by papilla-like outgrowths .... *Tylenchulus*
   Excretory pore in female not so posterior, at less than 61% of body length, not surrounded by outgrowths .... *Trophotylenchulus*
5. Female elongate-slender (L ≥ 0.5 mm, a = 40-70), male with asymmetrical head and distinct bursa peloderan .... *Tylenchocriconema*
   Female not elongate-slender (L < 0.5 mm, a < 35), male with symmetrical head and bursa leptoderan indistinct or absent........ *Paratylenchus*

13.5.2. Subfamily Tylenchulinae Skarbilovich, 1947

Tylenchulinae are characterized by having swollen females that are not spherical or subspherical and which have a distinct post-vulval region. The excretory pore is located some distance posterior to the pharyngeal region and produces a gelatinous matrix into which the eggs are laid.

Genus *Tylenchulus* Cobb, 1913
(Fig. 13.6)
Diagnosis: Tylenchulidae, Tylenchulinae. Pre-adult females migratory but adult females sedentary, males non-feeding. **Mature female elongate-obese, enlarging posterior to metacorpus, and ventrally arcuate**, maximum width up to 100 μm, **post-vulval part elongate and tapering**. Cuticle thick. Stylet 11-15 μm long, knobs prominent. **Excretory system well developed, producing a gelatinous matrix**, renette cell occupying about 30% of body cavity in type species, excretory pore at 68-85% of body length from anterior end, **papilla-like outgrowths around excretory pore present**. Criconematoid pharynx. Isthmus elongate-slender, basal bulb offset from intestine. Vulva a distinct transverse slit, lips bulging. Post-uterine sac absent. Spermatheca present. Ovary coiled, or with one to two flexures. Anus obscure or absent. Tail tapering, tip rounded or with a peg. Immature female vermiform, slender, straight to arcuate upon relaxation, usually under 0.50 mm long. Cuticle distinctly striated. Stylet and pharynx as in mature female. Excretory pore at 76-84% of body length. Vulval lips bulging. Anus obscure. Male slender, straight or slightly arcuate ventrally, 0.33-0.46 mm long. Stylet and pharynx reduced but not completely. Excretory pore at 53-60% of body length. Testis outstretched. Spicules slender and arcuate. Gubernaculum simple, fixed. **Bursa absent**. Tail elongate-conoid, tip rounded or with a peg. Juvenile slender, straight to slightly arcuate when relaxed. Lateral fields narrow, each with two incisures. Excretory pore at 50-60% of body length. Labial region rounded, labial disc indistinct, framework moderately sclerotized. Stylet 11-15 μm long, knobs rounded. Metacorpus oval or fusiform, with elongated refractive thickenings. Basal bulb offset from isthmus and intestine. Rectum and anus obscure. Tail elongate-conoid, tip rounded or deeply furcate.

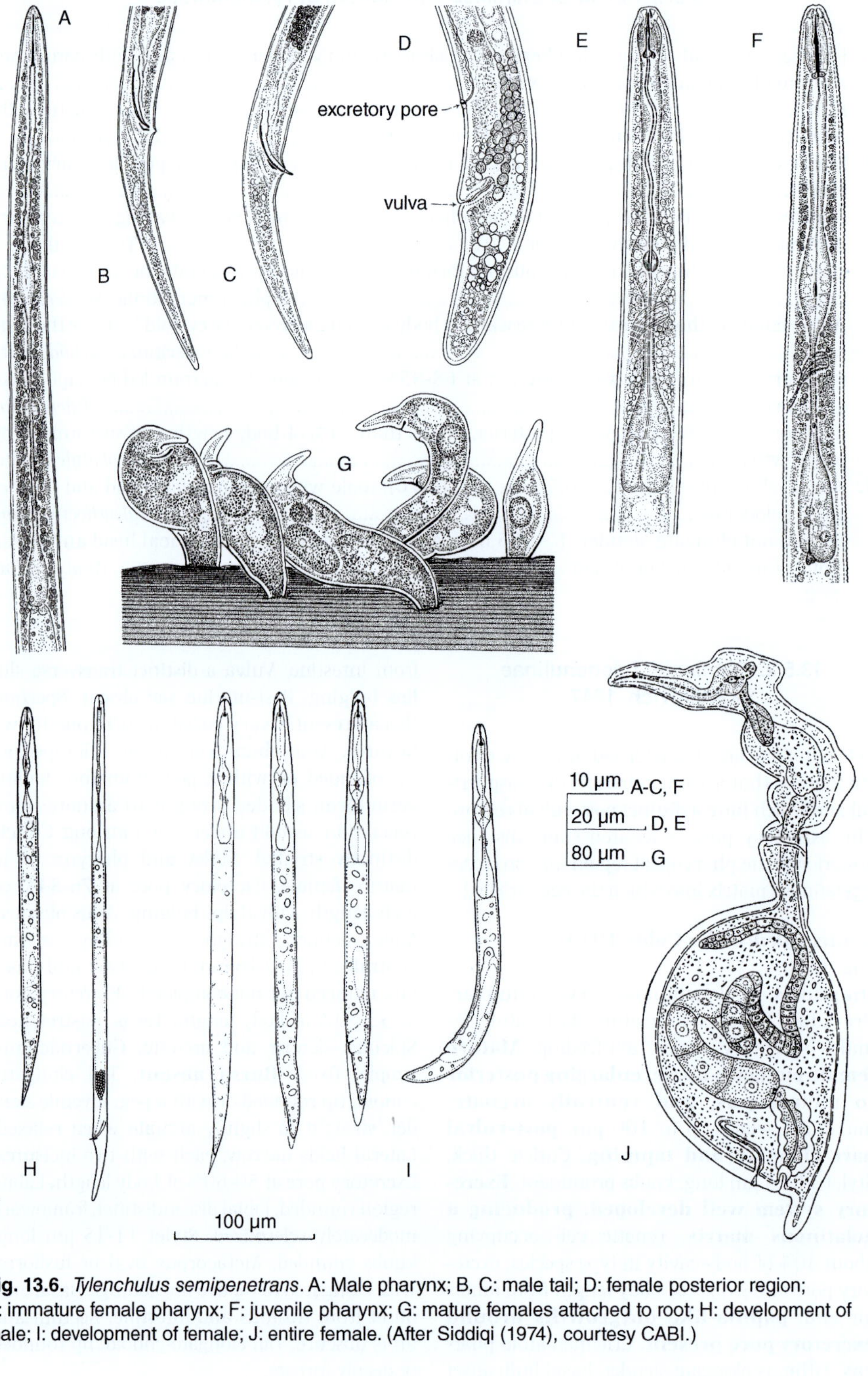

Fig. 13.6. *Tylenchulus semipenetrans*. A: Male pharynx; B, C: male tail; D: female posterior region; E: immature female pharynx; F: juvenile pharynx; G: mature females attached to root; H: development of male; I: development of female; J: entire female. (After Siddiqi (1974), courtesy CABI.)

Type species: *Tylenchulus semipenetrans* Cobb, 1913
Other species: About five.
Distribution: Worldwide. *T. semipenetrans* is distributed throughout the world where citrus is cultivated.
Major species: The citrus nematode, *T. semipenetrans*. Adult female L = 0.25-0.47 mm; stylet = 11-15 µm. Lip region continuous with body contour, head skeleton weak. Mature female a sedentary semi-endoparasite with anterior third of body embedded in root tissue, posterior body swollen, irregular in shape, vulva and excretory pore with prominent lips, excretory cell well developed, producing a gelatinous matrix.

T. semipenetrans is one of the most important and widely distributed plant parasites of citrus-growing areas throughout the world. The nematode is a sessile, obligate, highly specialized parasite of citrus and causes a disease called 'slow decline' because of the slow development of symptoms in the host and the slow rate of nematode population growth in newly replanted orchards. The citrus nematode does not cause galling on roots, although soil particles adhere to the gelatinous material excreted by females, thereby giving the roots a dirty appearance. Nematode feeding on roots causes cortical tissues to slough off and the eventual death of the portion of roots affected. However, aboveground symptoms may not be noticeable until sometime later and take the form of poor growth and vigour of trees, yellowing and shedding of leaves, undersized fruits and a marked crop loss. Affected trees usually do not die, but have unthrifty appearance and yield fewer than uninfested trees. Most studies estimate yield losses due to *T. semipenetrans* to be in the range of 10-30% depending on the level of infection, susceptibility of the rootstock, climatic conditions, soil characteristics, presence of other pathogens or natural enemies and orchard management practices. Several races or biotypes of *T. semipenetrans* are recognized based on their ability to multiply on *Poncirus trifoliata*, *Vitis* spp. and *Olea europaea*. The 'citrus' race is widespread in the USA and Italy, the 'Poncirus' races occur in California and the 'Mediterranean' race in Mediterranean countries and South Africa.

The complete life cycle from egg to egg requires from 4 to 8 weeks, depending on soil temperature. The eggs hatch in 12-14 days at 24°C, the first moult occurring within the egg. The second-stage juvenile, which hatches from the egg, is the infective stage. In moist soil at 15°C, juveniles can survive without a host for up to 2.5 years. Citrus nematodes have been observed to survive, after their host trees were removed, for up to 9 years on surviving roots deep in the ground. J2 (second-stage juvenile), J3 (third-stage juvenile) and early J4 (fourth-stage juvenile) of *T. semipenetrans* are ectoparasites and do not swell. For further development and long-term survival of the species, host roots in the soil are required. The juveniles undergo four moults to the male or female stages. The females bury the anterior region into the root, up to several cortical cells deep, and establish a feeding site, which includes six to ten cortical cells grouped around the nematode head and modified to form the so-called 'nurse cells'. Once a feeding site is established, the nematode becomes sessile, starts swelling and transforms into a mature female with its posterior portion exterior to the root and exposed to the soil. The mature female is therefore a sedentary semi-endoparasite. The adult female excretes a gelatinous matrix from the excretory pore and into which the eggs are deposited. The eggs are often deposited in batches of 10-20. The nematode is usually spread by citrus nurseries since the seedlings and rootstocks are carried to virgin soil with infested soil.

Tylenchulus furcus Van den Berg & Spaull, 1982 was described in South Africa from sugarcane, *Saccharum* hybrid plants and poorly growing grass on a golf course (Van den Berg and Spaull, 1982). *Tylenchulus palustris* Inserra, Vovlas, O'Bannon & Esser, 1988 was described from native plants in Florida, USA and also found infecting peach in Alabama, Arkansas, Georgia and Virginia and some non-cultivated plants. *Tylenchulus musicola* Tanha Maafi, Amani, Stanley, Inserra, Van den Berg & Subbotin, 2012 was described from banana plantations in southeast Iran (Tanha Maafi *et al.*, 2012).
Management: Management of the citrus nematode remains difficult as no one strategy provides adequate control of the nematode once it is established in the field. Therefore, the nematode should be controlled mainly by preventing its spread to infested soil. Bare-root dips in hot water (at 45°C for 25 min) or in suitable concentrations of nematicides can control the spread of disease. Grafting on resistant *Poncirus trifoliata*

hybrids rootstock has been the main control method for new plantations, although there are some biotypes able to break the resistance provided by these rootstocks and therefore a resistance management programme becomes necessary. There are also some grapevine rootstocks, obtained from *Vitis champini* or *Vitis riparia*, with resistance to *T. semipenetrans*. Pre-plant fumigation with available fumigant nematicides of old orchard-sites with a history of citrus nematode is fundamental to prevent a rapid infection of young trees. Post-planting nematicides have been also widely used in citriculture against the citrus nematode, carbamates (oxamyl and carbofuran) and organophosphates (fenamiphos and etoprophos) have controlled *T. semipenetrans* in orchards. They are applied either in bands down the tree rows or through the irrigation system to reduce nematode population levels when nematodes invade new roots.

Genus *Trophotylenchulus* Raski, 1957
= *Trophonema* Raski, 1957
= *Ivotylenchulus* Hashim, 1984
Diagnosis: Tylenchulidae, Tylenchulinae. Mature female small 0.38-0.48 mm, **elongate-saccate, coiled, curving more than 360°, post-vulval body region remaining slender**. Cuticle thin, finely annulated, lateral fields obscure. Labial region continuous, anteriorly tapering, usually **surmounted by a distinct circumoral elevation or disc**. Excretory pore located near nerve ring or posterior to pharyngeal base, at 33-61% of body length from anterior end, **papilla-like outgrowths around excretory pore absent. Excretory cell (renette) enormously developed, located laterally, producing gelatinous matrix**. Stylet usually 11-15 µm long, knobs prominent and rounded. Orifice of dorsal gland 5-7 µm posterior to stylet base. Procorpus elongate-slender. Metacorpus large, with prominent refractive thickenings. Isthmus slender. Basal bulb pyriform, enclosing pharyngeal glands. Ovary extending anteriorly, with flexures. Post-uterine sac absent. Tails elongate-conoid with rounded terminus. Male stylet and pharynx reduced. Spicules arcuate with ventrally bent tip, 14-20 µm long. Cloacal lips elevated. Bursa absent. Juvenile slender, vermiform, straight to arcuate.
Type species: *Trophotylenchulus floridensis* Raski, 1957
Other species: 13.
Distribution: Worldwide.
Raski (1957) proposed *Trophotylenchulus*, which was separated from *Tylenchulus* by the more anterior position of the excretory pore and the presence of a prominent circumoral elevation, lacking in *Tylenchulus*. The species of both genera have a large excretory and secretory cell that exudes a gelatinous matrix through the excretory pore.

T. floridensis has been found parasitizing roots of several woody plants and *Trophotylenchulus obscurus* (Colbran, 1961) Cohn & Kaplan, 1983 has been cited damaging cortical cells of coffee roots in Africa. *Trophotylenchulus piperis* Mohandas, Ramana & Raski, 1985 reduced the root growth, causing yellowing and die-back, of black pepper in India. *Trophotylenchulus saltensis* Hashim, 1984 has been associated with *Olea europaea* in Jordan. Only the anterior part of the female is within the root tissue, feeding inducing a syncytium with hypertrophied nuclei and dense cytoplasm. The mature females are usually encapsulated in a round, capsule-like structure apparently produced by host reaction. As the females of *T. floridensis* and *T. piperis* feed and mature, they develop a protective covering that later turns brown. In *T. piperis*, it is initiated as a transparent jelly, probably produced by the excretory system of the nematode. The protective matrix or capsule starts developing 30-40 days after inoculation and eggs are laid inside it, the number of eggs per capsule ranging from 25 to 35 for *T. piperis* and the life cycle from egg to egg taking up to 55 days at 24-32°C.

13.5.3. Subfamily Sphaeronematinae Raski & Sher, 1952

Sphaeronematinae are characterized by the spherical or subspherical female with a neck, absence of a post-vulval region and with the excretory pore located in the pharyngeal region. The uterus does not form a cystoid body for the eggs.

Genus *Sphaeronema* Raski & Sher, 1952
= *Goodeyella* Siddiqi, 1986
(Fig. 13.7)
Diagnosis: Tylenchulidae, Sphaeronematinae. **Mature female subspherical to spherical**, with a slender neck, 0.10-0.65 mm long. Cuticle thick, smooth or marked by a well-defined reticulate pattern. Stylet 8-21 µm long, basal

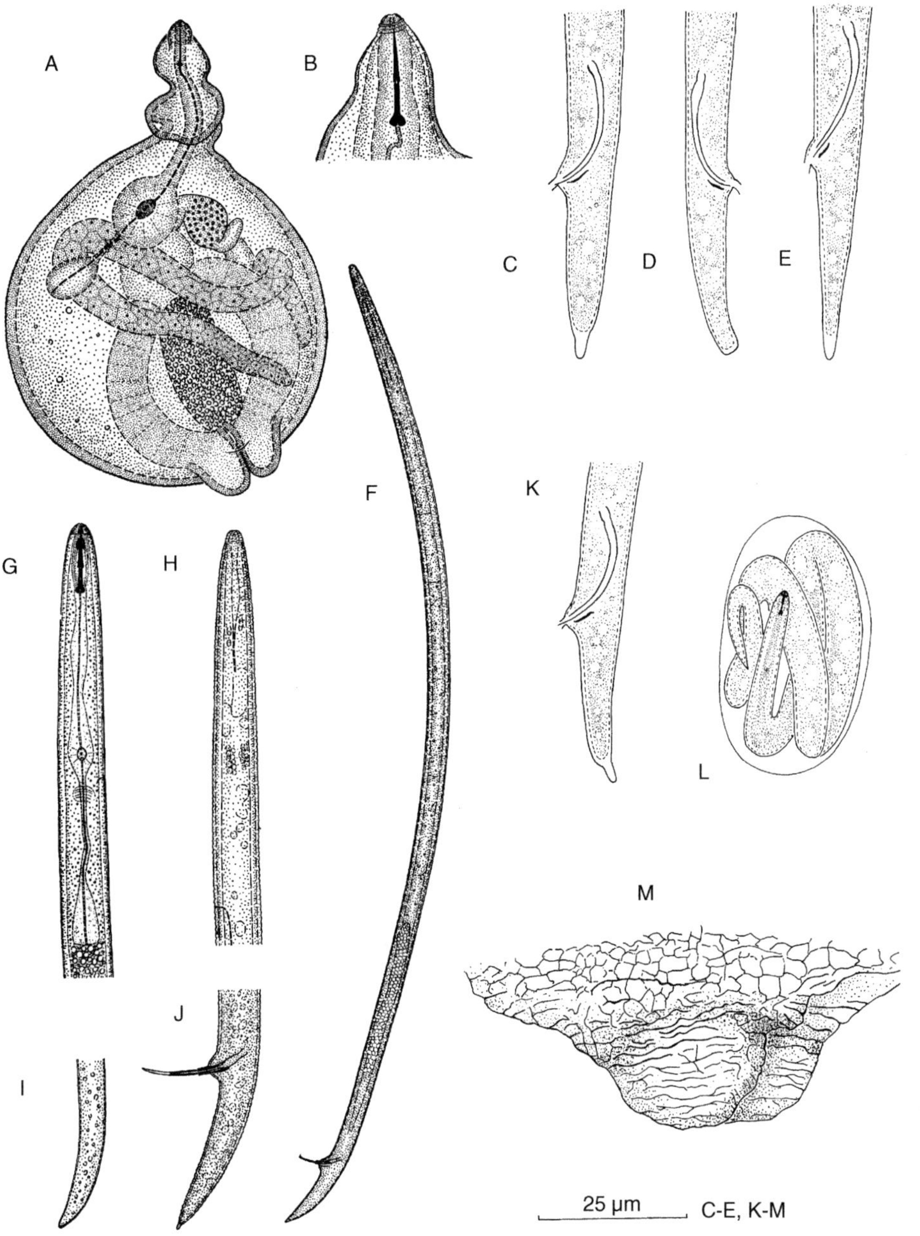

Fig. 13.7. *Sphaeronema californicum*. A: Entire female; B: female head; C-E: male tail; F: entire male; G: pharyngeal region of immature female; H: pharyngeal region of male; I: juvenile tail; J, K: male tail; L: egg containing juvenile; M: female vulval region. (After Raski and Sher (1952), courtesy *Proceedings of the Helminthological Society of Washington*.)

knobs well developed, rounded or slightly sloping posteriorly. Muscular and rounded metacorpus, with refractive thickenings, isthmus short or slender set off from basal bulb. Pharyngo-intestinal junction and intestine obscure. Excretory pore opposite or near metacorpus. Vulva terminal, flush with body surface or on a cone-like protuberance of body. Uterus thick-walled, containing one or more eggs at a time, but not filling entire body cavity or turning into a cystoid body, eggs laid in a gelatinous matrix. Anus and tail absent. Male vermiform, 0.30-0.47 mm long. Pharynx reduced. Labial region smoothly rounded, continuous with body contour or slightly offset. Stylet short or reduced. Spicules slender, slightly arcuate, 19-21 µm long. Gubernaculum simple, 3-4 µm long. Cloacal lips forming a short penial tube. Bursa absent. Tail elongate-conoid, pointed. Juvenile body as long as that of male. Cuticle finely and indistinctly annulated. Labial region continuous, rounded, sclerotization moderate. Stylet 13-23 µm, conus forming 53-57% of stylet length, knobs rounded. Metacorpus muscular, merging anteriorly with procorpus. Isthmus slender, basal bulb pyriform, offset from intestine. Excretory pore in isthmus region. Tail arcuate, tapering into a bluntly rounded tip, anus obscure. Female an obligate ectoparasite of plant roots.

Type species: *Sphaeronema californicum* Raski & Sher, 1952.

Other species: Seven.

Distribution: Worldwide.

Females of *S. californicum* were found inside the roots of *Umbellularia californica* and the nematode occurred in colonies. The feeding site is probably initiated by the early J4 stage. Once a feeding site has been established, the J4 gradually increases in size and transforms into a somewhat swollen pre-adult female in about 7 days. The posterior part of the body remains exposed outside the root. In the last moult, the tail is lost and the adult sessile female swells symmetrically and becomes an almost spherical body with minute, terminally located, vulval cone lips. The eggs are laid in a gelatinous matrix.

13.5.6. Subfamily Meloidoderitinae Kirjanova & Poghossian, 1973

Meloidoderitinae are characterized by the spherical or subspherical female without a neck, lacking a post-vulval region and the excretory pore located in the pharyngeal region. The uterus forms a cystoid body in which the eggs are retained.

Genus *Meloidoderita* Poghossian, 1966
= *Tumiota* Siddiqi, 1986
(Fig. 13.8)

Diagnosis: Tylenchulidae, Meloidoderitinae. **Mature female fully swollen, without a neck or tail**, pear-shaped or oval, 0.20-0.50 mm long. Longitudinal axis of body from head to vulva, anus shifted to dorsal side. Cuticle thick, may have spine-like outgrowths. Stylet ca 15-19 µm long, knobs prominent. Orifice of dorsal gland 3-7 µm from stylet base. Excretory pore opposite metacorpus, latter very muscular with large refractive thickenings. **Vulva terminal, on cone-like elevation of body**. Uterus spheroidal, with very thick walls, filling one-third to a half of body cavity in young, and most of body in old, females. In older female, uterus becoming patchily sclerotized and transforming into a **cystoid body**, which serves as a protective sac for the retained eggs. Some eggs retained, others laid in a gelatinous matrix. In *Meloidoderita whittoni* (Sledge & Christie, 1962) Troccoli *et al.* 2017, females also deposit eggs in the uterine sac, but this uterine sac has a thin wall and does not become a cystoid body. It remains permanently inside the female body, which does not disintegrate but has a thick, tanned and sclerotized cuticle, like that of *Globodera* cysts. Male vermiform, under 0.50 mm long. Labial region continuous. Stylet absent. Pharynx reduced, testis atrophied in mature males. Tail elongate, subcylindroid or conoid. Bursa absent. Spicules and gubernaculum typical of the family. Juvenile straight to arcuate upon relaxation, 0.31-0.44 mm long. Cuticle finely annulated, lateral field with three or four incisures. Labial region conoid-rounded to truncate, continuous, labial disc indistinct, framework lightly sclerotized. Stylet well developed, 12-14 µm long in type species, conus often a little longer than shaft, knobs large, rounded. Metacorpus oval, with prominent refractive thickenings. Isthmus long, slender. Pharyngeal glands enclosed in terminal bulb, offset from or with base slightly extending over intestine. Excretory pore opposite isthmus. Rectum and anus indistinct. Tail tip spicate, or minutely rounded.

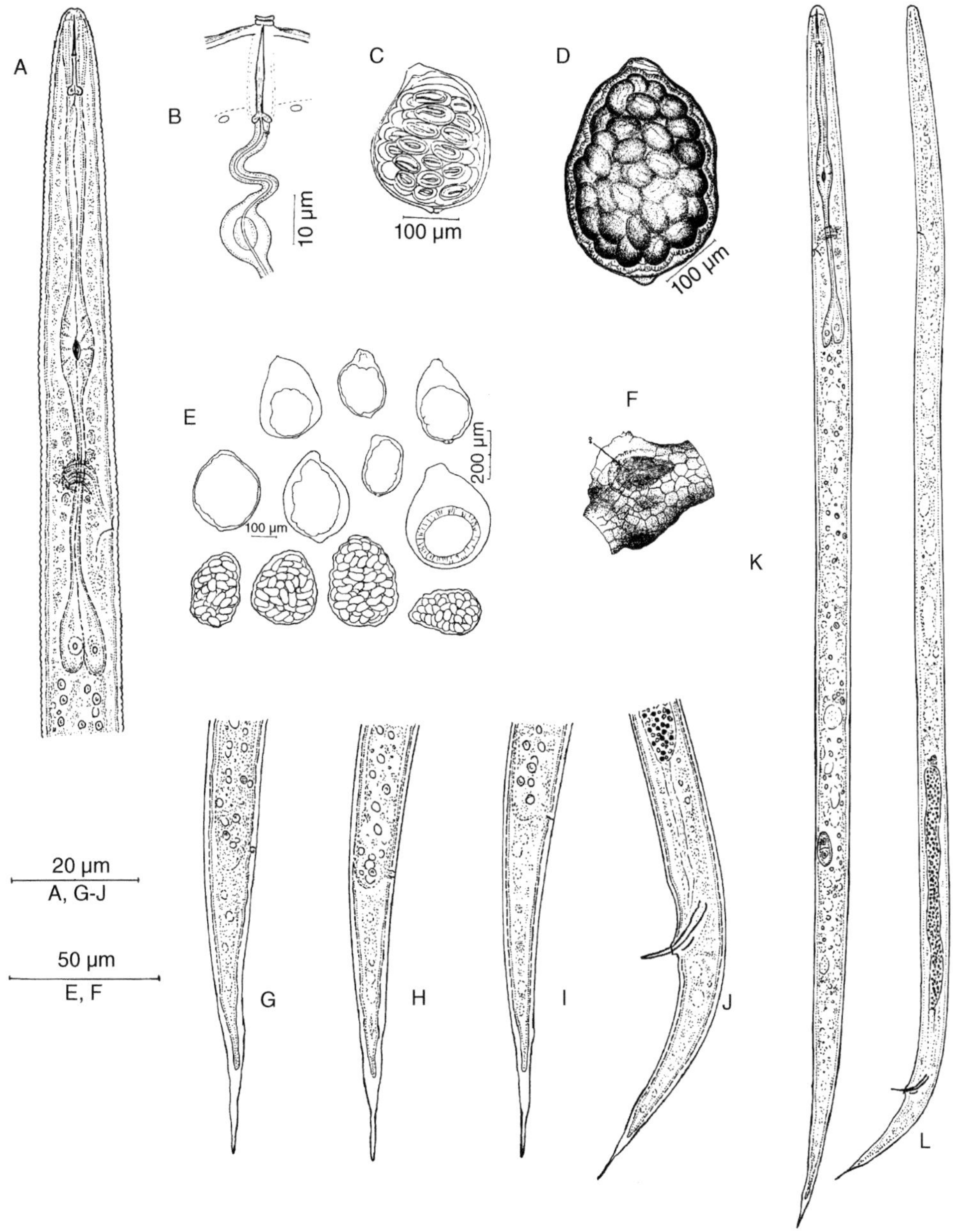

Fig. 13.8. *Meloidoderita kirjanovae*. A: Second-stage juvenile (J2) pharyngeal region; B: female anterior region; C, D: mature females; E: development of cystoid body via uterine enlargement; F: female (arrow) in root; G-I: J2 tails; J: male tail; K: J2, entire; L: male, entire. (After Poghossian (1966), Kirjanova and Poghossian (1973) and Siddiqi (1985), courtesy CABI.)

Type species: *Meloidoderita kirjanovae* Poghossian, 1966

Other species: Four.

Distribution: *Meloidoderita* spp. have been reported from USA, the Azores (Portugal), Europe, Israel, Iran and South Africa.

M. kirjanovae are sedentary semi-endoparasites of *Mentha*. *Meloidoderita polygoni* Poghossian, 1966, parasitizes several species of *Polygonum*, while *Meloidoderita safrica* Van den Berg & Spaull, 1982 parasitizes sugarcane in South Africa.

13.5.5. Subfamily Paratylenchinae Thorne, 1949

Paratylenchinae are characterized by having vermiform or slightly swollen females and a relative long stylet, usually over 20 µm.

Genus *Paratylenchus* Micoletzky, 1922
= *Cacopaurus* Thorne, 1943
= *Gracilacus* Raski, 1962
= *Paratylenchoides* Raski, 1973
= *Gracilpaurus* Ganguly & Khan, 1990
(Figs 13.9, 13.10)

Diagnosis: Tylenchulidae, Paratylenchinae. **Female vermiform or swollen, obese ventrally curved**, often C-shaped on death and usually less than 0.50 mm long. Annuli fine or moderately coarse, round, smooth, not ornamented with tubercles. Lateral field with two, three or often four incisures. Labial region continuous, round, conoid or truncate, smooth or finely annulated, with or without four submedian lobes, framework with light to heavy sclerotization. **Stylet 12-120 µm long**, knobs spheroidal. Isthmus slender, basal bulb round to pyriform. **Excretory pore at level of nerve ring or more anterior, can be as far posterior as pharynx base**. Ovary outstretched. Post-uterine sac absent. Lateral vulval membranes present or absent. Male stylet reduced or absent. Pharynx reduced. Bursa absent. Spicules and gubernaculum usually 15-23 µm and 3-5 µm long, respectively. Juvenile similar to female.

Type species: *Paratylenchus bukowinensis* Micoletzky, 1922

Other species: More than 140.

Distribution: The nematodes of the genus *Paratylenchus*, commonly known as pin nematodes, are obligate ectoparasites of a wide variety of plants, including herbs, shrubs and trees, that are distributed worldwide and can damage roots at high population density.

Major species: *Paratylenchus epacris* (Allen & Jensen, 1950) Goodey, 1963. Small obese females. L = 0.30 mm; V = 85-87. Long and flexible stylet ca 110 µm long. Head conoid to rounded. Lateral field with four incisures. Lateral vulval membranes present. Tail rounded. *P. epacris* was implicated in a disease of California black walnut (*Juglans hindsii*).

Paratylenchus hamatus Thorne & Allen, 1950. Female vermiform. L = 0.30-0.47 mm; V = 78-84; stylet = 27-34 µm. Head rounded and continuous with body contour. Lateral field with four incisures. Tail conoid-rounded. *P. hamatus* has been reported as a pest that might be the cause of tree decline in some fig orchards in California, USA. This species parasitizes peach, plum and prune trees, reducing vigour and yield. High populations of *P. hamatus* on rose reduced both production and quality. There are also cases in which *P. hamatus* has considerably reduced the yield of vegetable crops.

Paratylenchus neoamblycephalus Geraert, 1965. Female vermiform. L = 0.34-0.48 mm; V = 80-84; stylet = 27-36 µm. Head rounded, continuous with body contour. Lateral field with four incisures. Tail slender conoid. *P. neoamblycephalus* occurs on apple, apricot, peach and on herbaceous plants in Australia, Europe and the Americas. Nematodes were observed feeding ectoparasitically. They were associated with small lesions and dead lateral roots. Inoculated plum seedlings had smaller, darker and reduced root systems and poor growth.

Paratylenchus pestis (Thorne, 1943) Goodey, 1963. Female cylindroid obese. L = 0.20-0.30 mm; V = 92-95. Cuticle ornamented with minute refractive tubercles. Stylet long and curved, ca 97 µm long. Lateral field with four incisures. *P. pestis* parasitizes roots of *Juglans regia* in California, USA. It occurs in southern France on *Rosa indica major* L. and on *Citrus aurantium* L. and in Spain on *Syringa vulgaris* L. Males, which do not feed, and juveniles, occur in the soil, while females are permanently attached to their feeding sites.

Management: Nematicide treatments, fumigant and non-fumigant are considered effective against pin nematodes.

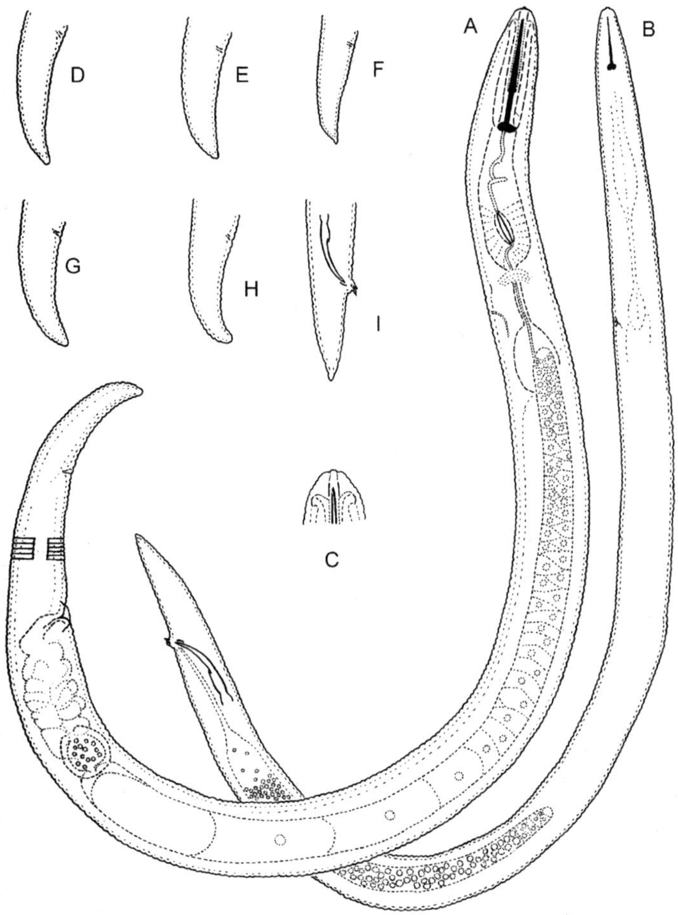

Fig. 13.9. *Paratylenchus bukowensis*. A: Entire female; B: entire male; C: female labial region; D-H: female tail; I: male tail. (After Brzeski (1976), courtesy CABI.)

Genus *Tylenchocriconema* Raski & Siddiqui, 1975

Diagnosis: Tylenchulidae, Paratylenchinae. Medium-sized nematodes with **slender body** (L = 0.42-0.83 mm, a = 40-70), both sexes vermiform. Marked sexual dimorphism in anterior region. Female labial region very low, truncate with an I-shaped oral aperture flanked by two lateral liplets and four small submedian lobes or pseudolips at corners of oral plate, amphidial apertures in the form of dorso-ventral slits at base of lateral liplets, framework weak. Stylet ca 20-30 µm long, with conus longer than shaft, knobs rounded. Orifice of dorsal gland 5-6 µm posterior to stylet base. Excretory pore near basal bulb. Deirids absent. Vulva slightly overhung by anterior lip flap and with small lateral membranes, well marked by a sudden narrowing of body contour posterior to it, V = 78-81. **Vulva overhung by anterior lip flap**. Vagina, cuticularized, directed inward and anteriad. Post-uterine sac absent. Spermatheca thin-walled, oval, offset, with minute, round sperm. Female tail straight, elongate (over five anal body widths long), subcylindroid to conoid-pointed tip. Male 0.42-0.65 mm long, shorter and more slender than female with an asymmetrical, **ventrally inclined, labial region**, reduced pharynx and no stylet. Lateral field with four incisures. Testis outstretched. Cloacal lips forming a penial tube. Tail elongate-conoid, arcuate ventrally, **bursa peloderan,** with crenate margins. Spicules arcuate. Gubernaculum small, fixed. Juvenile similar to female in body shape, labial region,

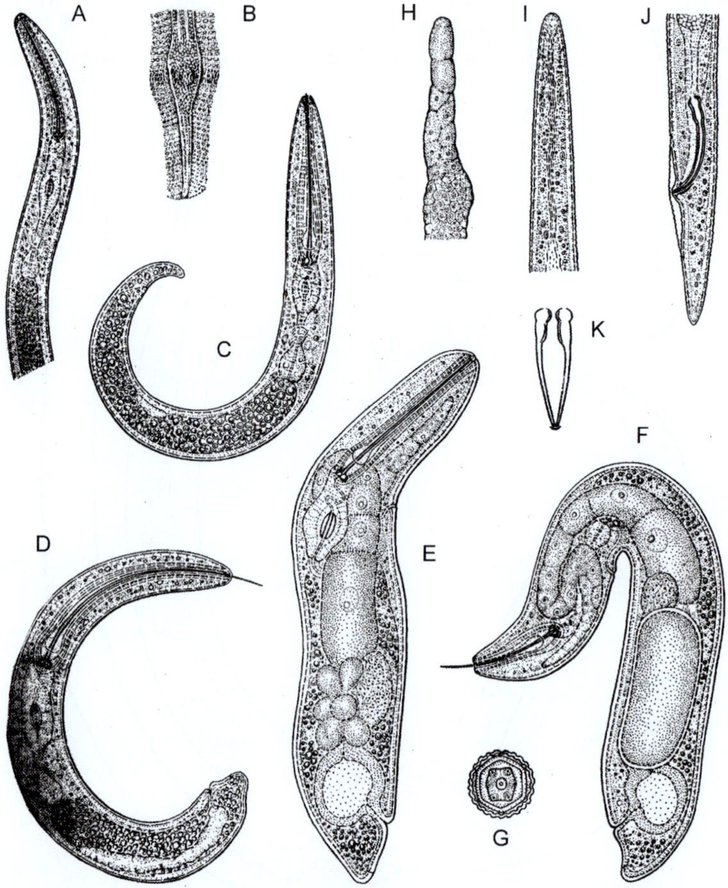

Fig. 13.10. *Paratylenchus pestis*. A: Anterior part of young juvenile; B: lateral field of female in vulval region; C, D: young females; E, F: adult females; G: female head in face view; H: anterior part of testis; I: anterior part of male; J: posterior part of male; K: spicules and gubernaculum. (After Thorne (1943), courtesy *Proceedings of the Helminthological Society of Washington*, and Franklin (1974), courtesy CABI.)

stylet and pharynx, tail evenly tapering to an acute tip.

Type and only species: *Tylenchocriconema alleni* Raski & Siddiqui, 1975.

Bionomics: The type species occurred in large numbers in the crowns of the bromeliad, *Tillandsia flabellata* Bak., feeding on leaf surfaces and in the crown tissues.

13.6. References

Andrássy, I. (1979) Revision of the subfamily Criconematinae Taylor, 1936 (Nematoda). *Opuscula Zoologica Instituti Zoosystematici Universitatis Budapestinensis* 16, 11–57.

Andrássy, I. (2007) *Free-living nematodes of Hungary (Nematoda Errantia), II.* In: *Pedozoologica Hungarica, 4* (Series eds: Csuzdi, C. and Mahunka, S.). Hungary, Hungarian Natural History Museum and Systematic Zoology Research Group of the Hungarian Academy of Sciences, Budapest, Hungary.

Archidona-Yuste, A., Clavero-Camacho, I., Ruiz-Cuenca, A.N., Cantalapiedra-Navarrete, C., Liebenas, G., *et al.* (2024) The more we search, the more we find: discovering and expanding the biodiversity in the ring nematode genus *Xenocriconemella* De Grisse & Loof, 1965 (Nematoda: Criconematidae). *Zoological Letters* 10, 8. DOI: 10.1186/s40851-024-00230-3.

Brzeski, M.W. (1976) *Paratylenchus bukowinensis. CIH Descriptions of Plant-parasitic Nematodes.* Set 6, No. 79. Commonwealth Agricultural Bureaux, Farnham Royal, UK.

Cantalapiedra-Navarrete, C., Clavero-Camacho, I., Criado-Navarro, I., Salazar-García R., García-Velázquez, A., *et al.* (2024) Another new ring nematode, *Xenocriconemella andreae* sp. nov. (Nematoda, Criconematidae), from the Iberian Peninsula. *Zoosystematics and Evolution* 100(3), 1175–1190. DOI: 10.3897/zse.100.129009.

Chitambar, J.J. and Subbotin, S.A. (2014) *Systematics of the Sheath Nematodes of the Superfamily Hemicycliophoroidea. Nematology Monographs and Perspectives, 10* (Series eds: Hunt, D.J. and Perry, R.N.). Brill, Leiden, Netherlands.

Decraemer, W. and Hunt, D.J. (2024) Structure and classification. In: Perry, R.N., Moens, M. and Jones, J.T. (eds), *Plant Nematology*, 3rd edn. CAB International, Wallingford, UK, pp. 3–40.

De Ley, P. and Blaxter, M. (2004) A new system for Nematoda: Combining morphological characters with molecular trees and translating clades into ranks and taxa. In: Cook, R. and Hunt, D.J. (eds), *Nematology Monographs and Perspectives, 2. Proceedings of the Fourth International Congress of Nematology. 8–13 June 2002. Tenerife, Spain.* Brill, Leiden, Netherlands, pp. 633–653.

Franklin, M.T. (1974) *Cacopaurus pestis. CIH Descriptions of Plant-parasitic Nematodes.* Set 3, No. 44. Commonwealth Agricultural Bureaux, Farnham Royal, UK.

Geraert, E. (2010) *The Criconematidae of the World. Identification of the Family Criconematidae (Nematoda).* Academia Press, Ghent, Belgium.

Ghaderi, R., Geraert, E. and Karegar, A. (2016) *The Tylenchulidae of the World.* Academia Press, Ghent, Belgium.

Kirjanova, E.S. & Poghossian, E.E. (1973) [Redescription of *Meloidoderita kirjanovae* Poghossian, 1966 (Nematoda: Meloidoderitidae fam. n.).] *Parazitologiya* 7, 280–285.

Kluepfel, D.A., Nyczepir, A.P., Lawrence, J.E., Wechter, W.P. and Leverentz, B. (2002) Biological control of the phytoparasitic nematode *Mesocriconema xenoplax* on peach trees. *Journal of Nematology* 34, 120–123.

Orton Williams, K.J. (1972) *Macroposthonia xenoplax. CIH Descriptions of Plant-parasitic Nematodes.* Set 1, No. 12. Commonwealth Agricultural Bureaux, Farnham Royal, UK.

Orton Williams, K.J. (1974) *Criconema palmatum. CIH Descriptions of Plant-parasitic Nematodes.* Set 4, No. 57. Commonwealth Agricultural Bureaux, Farnham Royal, UK.

Peraza-Padilla, W., Aráuz-Badilla, J., Cantalapiedra-Navarrete, C., Palomares-Rius, JE., Archidona-Yuste, A., *et al.* (2024) A new ring nematode, *Xenocriconemella costaricense* sp. nov., (Nematoda: Criconematidae) from Costa Rica. *Journal of Helminthology* 98, e39. DOI: 10.1017/S0022149X24000294.

Perry, R.N., Moens, M. and Jones, J.T. (eds) (2024) *Plant Nematology*, 3rd edn. CAB International, Wallingford, UK.

Poghossian, E.E. (1966) [A new nematode genus and species of the family Heteroderidae from the Armenian SSR.] *Doklady Akad. Nauk armyan, SSR* 42, 117–123.

Raski, D.J. (1991) Tylenchulidae in agricultural soils. In: Nickle, W.R. (ed), *Manual of Agricultural Helminthology.* Marcel Dekker, New York, New York, pp. 761–794.

Raski, D.J. and Luc, M. (1987) A reappraisal of Tylenchina (Nemata). 10. The superfamily Criconematoidea Taylor, 1956. *Revue de Nématologie* 10, 409–444.

Raski, D.J. and Sher, S.A. (1952) *Sphaeronema californicum*, nov. gen., nov. spec., (Criconematidae: Sphaeronematinae, nov. subfam.) an endoparasite of the roots of certain plants. *Proceedings of the Helminthological Society of Washington* 19, 77–80.

Siddiqi, M.R. (1974) *Tylenchulus semipenetrans. CIH Descriptions of Plant-parasitic Nematodes.* Set 3, No. 34. Commonwealth Agricultural Bureaux, Farnham Royal, UK.

Siddiqi, M.R. (1977) *Hemicriconemoides mangiferae. CIH Descriptions of Plant-parasitic Nematodes.* Set 7, No. 99. Commonwealth Agricultural Bureaux, Farnham Royal, UK.

Siddiqi, M.R. (1985) *Meloidoderita kirjanovae. CIH Descriptions of Plant-parasitic Nematodes.* Set 8, No. 113. CAB International, Wallingford, UK.

Siddiqi, M.R. (2000) *Tylenchida Parasites of Plants and Insects,* 2nd edn. CAB International, Wallingford, UK.

Sikora, R.A., Coyne, D., Helder, J. and Timper, P. (eds) (2018) *Plant Parasitic Nematodes in Subtropical and Tropical Agriculture*, 3rd edn. CAB International, Wallingford, UK.

Subbotin, S., Vovlas, N., Crozzoli, R., Sturhan, D., Lamberti, F., *et al.* (2005) Phylogeny of Criconematina Siddiqi, 1980 (Nematoda: Tylenchida) based on morphology and D2-D3 expansion segments of the 28S-rRNA gene sequences with application of a secondary structure model. *Nematology* 7, 927–944.

Tanha Maafi, Z., Amani, M., Stanley, J.D., Inserra, R.N., Van den Berg, E., *et al.* (2012) Description of *Tylenchulus musicola* sp. n. (Nematoda: Tylenchulidae) from banana in Iran with molecular phylogeny and characterization of species of *Tylenchulus* Cobb, 1913. *Nematology* 14, 353–369.

Thorne, G. (1943) *Cacopaurus pestis*, nov. gen., nov. spec. (Nematoda: Criconematinae), a destructive parasite of the walnut, *Juglans regia* Linn. *Proceedings of the Helminthological Society of Washington* 10, 78–83.

Troccoli, A., Subbotin, S.A., Stanley, J.D. and Inserra, R.N. (2017) *Meloidoderita whittoni* (Sledge & Christie, 1962) comb. n. (Tylenchida: Sphaeronematidae) and its parasitic habits on sweet gum (*Liquidambar styraciflua* L.). *Nematology* 19, 709–722.

Van den Berg, E. and Spaull, V.W. (1982) Two new species of Tylenchoidea on sugar cane in South Africa. *Phytophylactica* 14, 131–144.

Van den Berg, E., Tiedt, L.R., Inserra, R., Stanley, J.D., Vovlas, N., *et al.* (2015) Characterisation of a topotype and other populations of *Hemicriconemoides strictathecatus* Esser, 1960 (Nematoda: Criconematidae) from Florida with description of *H. phoenicis* sp. n. from the USA *Nematology* 17, 265–300.

van Megen, H., van den Elsen, S., Holterman, M., Karssen, G., Mooyman, P., *et al.* (2009) A phylogenetic tree of nematodes based on about 1200 full-length small subunit ribosomal DNA sequences. *Nematology* 11, 927–950.

Wouts, W.M. (2006) *Criconematina (Nematoda: Tylenchida).* In: *Fauna of New Zealand* 55. Landcare Research, Lincoln, New Zealand.

14 Anguinidae – Stem and Gall Forming Nematodes

Sergei A. Subbotin[1,2]* and Ian T. Riley[3]

[1]Plant Pest Diagnostics Center, California Department of Food and Agriculture, Sacramento, CA, USA; [2]Centre of Parasitology of A. N. Severtsov Institute of Ecology and Evolution of the Russian Academy of Sciences, Moscow, Russia; [3]School of Agriculture, Food and Wine, Adelaide University, South Australia, Australia

Abstract

The chapter deals with the Anguinidae, a family containing stem and gall forming nematodes. The taxonomy of this group is particularly complex and several recent advances have been made through the application of molecular methodologies, including the proposal of new genera. This chapter deals with their systematics, morphology, life cycle, karyology and phylogenetic relationships and has been updated to include the latest research. The diagnostics of Anguinidae are covered as well as nematode-bacteria and nematode-fungal associations. The morphology, bionomics, economic importance and management of the major genera are dealt with in greater detail and selected representative species, particularly those of economic importance, highlighted. Six species of *Anguina*, two of *Afrina*, three each of *Ditylenchoides* and *Ditylenchus*, and representative species of *Cotylenchus*, *Halenchus*, *Litylenchus*, *Mesoanguina*, *Orrina* and *Pterotylenchus* and *Subanguina* are described and illustrated. Information on the control/management of major species is provided.

14.1. Introduction

The Anguinidae Nicoll, 1935, a family under the Superfamily Anguinoidea Nicoll, 1935, includes mycophagous and plant-parasitic nematodes. The latter are obligate specialized parasites of higher plants, mosses and seaweeds on which they often induce swellings and galls. Several species: *Anguina agrostis*, *Anguina funesta*, *Anguina pacificae*, *Anguina paludicola*, *Anguina tritici*, *Afrina wevelli*, *Ditylenchoides africanus*, *Ditylenchoides destructor*, *Ditylenchus angustus*, *Ditylenchus dipsaci* and *Ditylenchus gigas* are considered of economic importance as agricultural and quarantine pests in various countries. The name for the family Anguinidae was first proposed by Nicoll (1935) to replace Anguillulinidae Baylis & Daubney, 1926, and it included genera that are now in other families of the order Tylenchida (Fortuner and Maggenti, 1987). Practically all nematologists recognize the family Anguinidae, discrepancies only concern its composition of subfamilies and genera. The validities and positions of several genera and subfamilies have been the subject of intensive discussions and speculation (Paramonov, 1962, 1970;

*E-mail: sergei.a.subbotin@gmail.com

A revision of Subbotin, S.A. and Riley, I.T. (2012) Stem and gall forming nematodes. In: Manzanilla-López, R.H. and Marbán-Mendoza, N. (eds), *Practical Plant Nematology*. Mundi-Prensa and Biblioteca Básica de Agricultura, Montecillo, Mexico.

(eds R.H. Manzanilla-López, D.J. Hunt, N. Marbán-Mendoza and S. Infante Gil)
DOI: 10.1079/9781836990413.0014

Siddiqi, 1971, 1980, 1986, 2000; Brzeski, 1981, 1991; Chizhov and Subbotin, 1985, 1990; Fortuner and Maggenti, 1987; Subbotin and Ryss, 2024) and remain unresolved. Comprehensive morphological and molecular analyses are still needed to understand relationships within these nematodes and replace the present artificial classification by a natural one reflecting evolutionary trends in this group. According to Siddiqi's (2000) classification, the family Anguinidae together with the family Sychnotylenchidae forms the infraorder Anguinata, superfamily Anguinoidea, Anguinidae contained two subfamilies Anguininae and Halenchinae, the latter includes exclusively marine, parasitic nematodes forming galls on sea algae. However, molecular phylogeny does not confirm this subfamily division. In addition, molecular phylogeny does not justify the Sychnotylenchidae as a separate lineage quite different from all other representatives of the superfamily Anguinoidea *sensu* Siddiqi, 2000.

14.2. Systematics

14.2.1. Family Anguinidae Nicoll, 1935

Diagnosis (modified after Siddiqi, 2000). Adults about 0.4-4.8 mm long, slender or obese. Labial region low and smooth. Muscular valvate median bulb present or absent. Basal pharyngeal bulb small or large, offset from intestine or dorsal gland may become enlarged and extend over intestine as a lobe. Vulva generally at less than 85% of body length. Post-vulval uterine sac present, rarely absent. Ovary outstretched or flexed. Tail similar between sexes, filiform, conical, subconical or subcylindroid. Bursa variable from adanal to peloderan, enclosing tail tip. Fungal feeders or parasites of plants, sometimes inducing swellings and galls. Some species are associated with insects.

Type genus: *Anguina* Scopoli, 1777

Other genera:

Afrina Brzeski, 1981
Cotylenchus Álvarez-Ortega & Subbotin, 2024
Diptylenchus Khan, Chawla & Seshadri, 1969
Ditylenchoides Subbotin & Ryss, 2024
Ditylenchus Filipjev, 1936
Ficotylus Davies, 2009
Halenchus N.A. Cobb *in* M.V. Cobb, 1933
Heteroanguina Chizhov, 1980
Indoditylenchus Sinha, Choudhury & Baqri, 1985
Litylenchus Zhao, Davies, Alexander & Riley, 2011
Mesoanguina Chizhov & Subbotin, 1985
Nothanguina Whitehead, 1959
Nothotylenchus Thorne, 1941
Orrina Brzeski, 1981
Palaeoanguina Poinar, 2011
Paraditylenchus Subbotin & Ryss, 2024
Pseudohalenchus Tarjan, 1958
Pterotylenchus Siddiqi & Lenné, 1984
Safianema Siddiqi, 1980
Subanguina Paramonov, 1967
Zeatylenchus Zhao, Davies, Alexander & Riley, 2013

Important morphological and biological characters useful for distinguishing the plant-parasitic genera are given in Table 14.1.

14.3. Morphology

Nematodes have a vermiform body and, in plant-parasitic species especially as a result of a strong development of the sexual system, the body becomes swollen so much that adult females are practically motionless (Fig. 14.1). The intraspecific variations of body length in plant-parasitic species may be significant; for example, in *Anguina agropyri*, the measurements of adult individuals vary from 1.5 mm in the first generation to 4.8 mm in the second generation, body length depending on host plant and numbers of specimens in a gall. With heat fixation, males of almost all species straighten and only individuals more than 3 mm in length are sometimes bent ventrally. Female body shape varies depending on the species and generation to which the female belongs, being almost straight or C-shaped. Cuticle is finely annulated. Lateral field incisures vary from four to six in different species. They are usually constant in young females of the same species, but may disappear completely or reach up to 20 lines in the stretched cuticle of the stout, gravid females of the plant parasites. Labial framework always present but rather weak. Labial region with three to five annuli. Stylet short and thin. Procorpus cylindrical. Median bulb present, with a

Table 14.1. Some comparative morphological and biological characters for the plant-parasitic genera of Anguinoidea.

Genus	Symptoms on plants	Infective stage	Dorsal pharyngeal gland	Median bulb	Incisures in lateral field	Ovary	Oocyte arrangement	Crustaformeria structure	Tail
Afrina	Obligate plant parasites inducing galls	J2	Slightly overlapping intestine	Muscular, valvate	4 or more	Without or with one or two flexures	One row	4 rows, 14 cells per a row	Conoid
Anguina	Obligate plant parasites inducing galls	J2	Slightly overlapping intestine	Muscular, valvate	4 or more	With flexures	Two or more rows	6-12 rows, 20-60 cell per a row	Conoid
Cotylenchus	Obligate plant parasites inducing chlorotic spots on leaves	No data	Not overlapping intestine	Muscular, valvate	6	Without flexures	One row	4 rows, 4 cells per a row	Conoid
Ditylenchoides	Fungal feeders or/and plant parasites infecting underground plant organs	Juveniles and adults	Not or slightly overlapping intestine	Muscular, valvate or non-muscular, non-valvate	4-6	Without flexures	One row	4 rows, 4 cells per a row	Conical to subcylindroid
Ditylenchus	Obligate plant parasites inducing swellings and deformations	Juveniles and adults	Not or slightly overlapping intestine	Muscular valvate	4-6	Without flexures	One row	4 rows, 4 or more cells per a row	Elongate-conoid
Halenchus	Obligate plant parasites inducing galls on sea weeds	No data	Forming a long lobe overlapping intestine	Non-muscular, non-valvate	4	Without flexures	One row	4 rows, 4 cells per a row	Elongate-conoid
Heteroanguina	Obligate plant parasites inducing galls	J4	Slightly overlapping intestine	Muscular valvate	4 or more	With or without flexures	One row	4-6 rows, 12-18 cells per a row	Conoid
Litylenchus	Obligate plant parasites inducing chlorotic spots on leaves	No data	Slightly overlapping intestine	Non-muscular, valvate	4	With or without flexures	One row	4 rows, 4 cells per a row	Conoid

Continued

Table 14.1. Continued.

Genus	Symptoms on plants	Infective stage	Dorsal pharyngeal gland	Median bulb	Incisures in lateral field	Ovary	Oocyte arrangement	Crustaformeria structure	Tail
Mesoanguina	Obligate plant parasites inducing galls	J3	Slightly overlapping intestine	Muscular, valvate	4 or more	With or without flexures	Two or more rows	4-5 rows, 8-12 cells per a row	Conoid
Nothanguina	Obligate plant parasites inducing galls	J2	Slightly overlapping intestine	Non-muscular, non-valvate	No data	With flexures	Two or three rows	4 rows, more than 4 cells per a row	Conoid
Orrina	Obligate plant parasites inducing galls	J4	Forming a long lobe overlapping intestine	Non-muscular, non-valvate	4	With flexures	One row	4 rows, 4 cells per a row	Conoid
Paraditylenchus	Obligate plant parasites inducing swellings and deformations	No data	Not or slightly overlapping intestine	Muscular, valvate	4	With flexures	One row	4 rows, 4 cells per a row	Conical
Pterotylenchus	Obligate plant parasites inducing galls	No data	Forming a long lobe overlapping intestine	Non-muscular, non-valvate	4	Without flexures	One row	4 rows, 8-9 cells per a row	Elongate-conoid
Subanguina	Obligate plant parasites inducing galls	J2	Not forming a long lobe overlapping intestine with nucleus lying anterior to pharyngo-intestinal junction	Muscular, valvate	4 and more	With flexures	One or two row	4-5 rows, 1 2-20 cell per a row	Conoid
Zeatylenchus	Obligate plant parasites inducing chlorotic spots on leaves	Juveniles and adults	Not or slightly overlapping intestine	Muscular, or non-muscular, valvate	3	Without flexures	One row	4 rows, 4 cells per a row	Conical

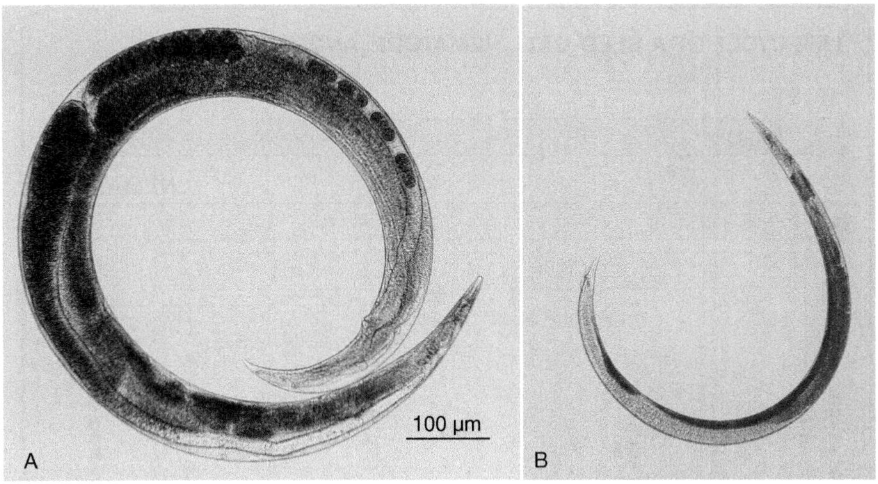

Fig. 14.1. *Anguina pacificae*. A: Female; B: male. (Courtesy of M. McClure.)

valve in most genera but without a valve in *Nothanguina*, *Nothotylenchus* and *Orrina*. Pharyngeal glands may extend over intestine with nucleus lying anterior to pharyngo-intestinal junction or form a long lobe overlapping intestine with nucleus lying posterior to this junction (Table 14.1). Ovary outstretched or with one or two flexures, with one or several rows of oocytes in maturation zone. Crustaformeria of *Ditylenchoides*, *Ditylenchus*, *Diptylenchus* and other genera forming a quadricolumella of four rows of four cells each, or, for example, in *Anguina* forming a long tube with a large number of cells in multiple irregular rows. Post-vulval uterine sac present in all genera, except for *Diptylenchus*. Bursa adanal to subterminal, never enclosing tail tip. Tail elongate-conoid to subcylindrical or filiform.

14.4. Life Cycle

Temperature has a great influence on egg production, hatching and life cycle (Fig. 14.2). *Ditylenchoides myceliophagus* thrives and breeds optimally at a temperature of between 18 and 20°C. The life cycle of *D. myceliophagus* is completed in 40 days at 13°C, 26 days at 18°C and 11 days at 23°C. The life cycle of *D. dipsaci* ranged from 17 to 23 days at 13-22°C. Development of one generation of *D. destructor* takes 18 days at 27-28°C or 20-26 days, at 20-24°C. Optimal temperature for development of *D. africanus* is 28°C, its life cycle lasting 6-7 days. The life cycle of *A. pacificae* is completed in about 32 days at a mean temperature of 20°C (McClure *et al.*, 2008). If the plant-parasitic *Ditylenchus* species complete multiple cycles within a season in plant tissues, the gall forming anguinins may likewise progress through one (most *Anguina* species), or two or more life cycles (*A. agropyri*, *Subanguina*, *Heteroanguina* and *Mesoanguina*) per season. Anguinids having several life cycles per season, either with one generation per gall, the progeny then leaving to induce a further cohort of galls (*Subanguina radicicola* and *Heteroanguina*), or with two generations per gall (*Mesoanguina* and *A. agropyri*).

The plant-parasitic anguinins are largely characterized by life cycles that have particular and robust survival and infective stages, these being consistent and specific for each species. The survival stages are able to enter a cryptobiotic (anabiotic) state and survive long periods of desiccation or freezing. For example, *Ditylenchoides myceliophagus*, *Ditylenchus dipsaci* and *A. tritici* may remain viable more than 25 years in a cryptobiotic state. In a desiccated state, they are heat and chemical resistant. Therefore, many species are parasites of above-ground plant parts and are found in semi-arid environments with hot, dry summers. The survival stages normally

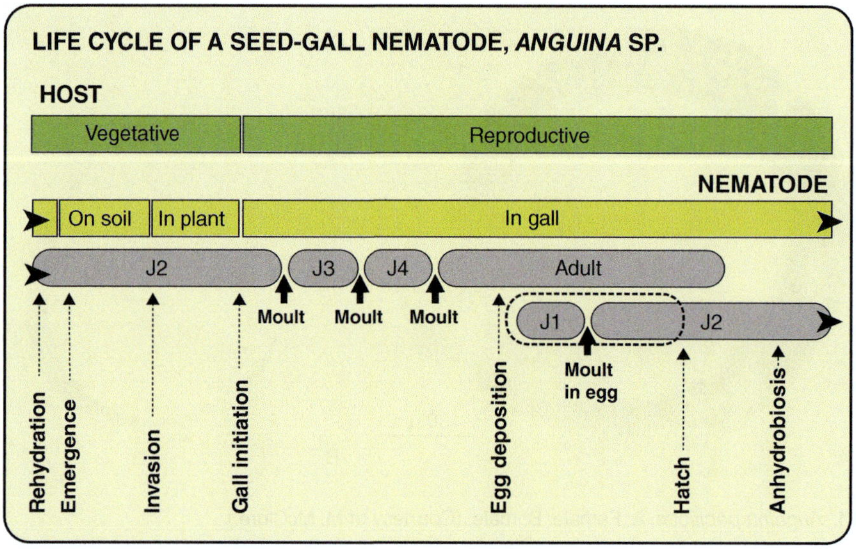

Fig. 14.2. Life cycle of seed-gall *Anguina* species such as *Anguina funesta* and *Anguina tritici*. (After Riley and Barbetti, 2008.)

exhibit only facultative quiescence and are revived by the onset of seasonal rains to become the infective stage. Survival and infective stages may be different in some species. In *Anguina australis*, which forms leaf galls on a rapidly maturing annual grass, the adults are the survival stage and after the galls rehydrate in the autumn, eggs and then infective second-stage juveniles (J2) are produced. This is similar to the life history of *Anguina danthoniae*. In *Anguina microleanea* both eggs and J2, and in *S. radicicola*, eggs, J2 and adults, can survive unfavourable conditions, but in these species only the J2 are able to infect plants. The infective stage is always a specific stage for these nematodes, for example, it is the J2 for *Anguina*, *Nothanguina* and *S. radicicola*, the third-stage juvenile (J3) for *Mesoanguina* and the fourth-stage juvenile (J4) for some plant-parasitic *Ditylenchus*, *Heteroanguina* and *Orrina*.

14.5. Symptomatology

Plant-parasitic representatives of this family cause necrosis, swelling, deformation, distortion or galls on leaves, stems, inflorescence or roots (Figs 14.3-14.7). Anguinins forming galls induce intensive plant cell hypertrophy and proliferation. Mature galls contain a cavity lined with parenchymatous nutritive tissue comprising intercellular spaces and actively dividing hypertrophied cells (Fig. 14.8). These cells contained granular cytoplasm, hypertrophied nuclei and brightly stained large nucleoli. Galls caused by some nematodes are pigmented.

14.6. Karyology

Basic haploid chromosome numbers for *Anguina* and *Ditylenchus* are n = 9 and n = 12, respectively. Other species may be polyploid or aneuploid forms derived from these basic numbers (Triantaphyllou and Hirschmann, 1980; Sturhan and Brzeski, 1991). Haploid chromosome numbers for some species are as follows: *A. paludicola*, 18; *Heteroanguina caricis*, 18-19; *A. tritici*, 19; *A. funesta*, 22; *D. dipsaci*, 12; *Ditylenchus weischeri*, 26; *D. gigas*, 24-30; *Mesoanguina plantaginis*, 27.

14.7. Anguinid Fossil in Amber

Amber is fossilized tree resin that it has been found to preserve some nematodes including anguinid representative. Due to the inability to

Fig. 14.3. Galls caused by nematodes of the subfamily Anguininae. A: Galls on a stem induced by *Anguina woodi* in *Ehrharta villosa* var. *villosa*; B: leaf galls induced by *Heteroanguina graminophila* on *Calamagrostis* sp.; C: leaf galls induced by *Anguina graminis* on *Festuca rubra*; D: leaf galls induced by *Mesoanguina millefolii* on *Achillea millefolium*; E: bulbous galls induced by *Anguina agropyri* at the base of the stems of *Agropyron repens*; F: root galls induced by *Subanguina radicicola* on *Poa annua*. (B-F: Courtesy of V.N. Chizhov.)

Fig. 14.4. Symptoms of infection of *Ditylenchus dipsaci*. A: Onion; B: sugar beet; C: tulip; D: alfalfa; E: red clover; F: strawberry. (Courtesy of EPPO (http://photos.eppo.org) (A-C), G. D. Jesperson (D), M. McClure (E), V.N. Chizhov (F).)

Fig. 14.5. Symptoms of infection caused by *Ditylenchus gigas* on *Vicia faba* L. A: Apical stems and leaves necrotized and deformed by severe attacks; B: symptomatic range of necrotic areas on stems; C: longitudinal sections of healthy (first on left) and damaged (two on right) stems showing internal necrosis; D, E: deformed and undersized pods; F: deformed (middle and bottom) and uninfected (top) seeds. (After Vovlas *et al.*, 2011.)

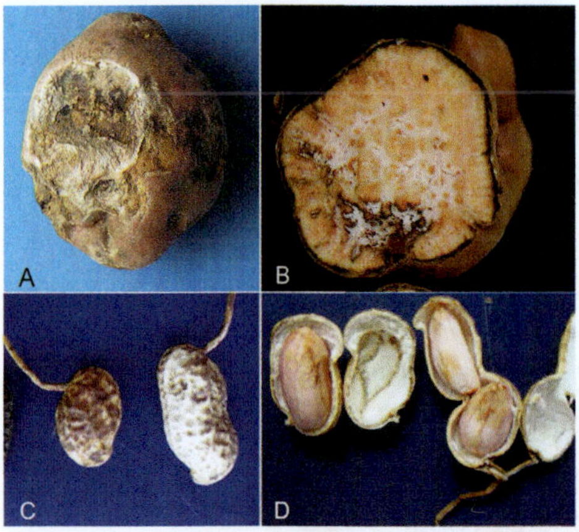

Fig. 14.6. Symptoms of infection caused by *Ditylenchoides destructor* on potato (A), sweet potato (B). (Courtesy of V.N. Chizhov.) Symptoms of infection caused by *Ditylenchoides arachis* on peanut (C, D). (After Zhang *et al.*, 2014.)

reveal morphological diagnostic characters, it is still debatable to make correct assignment of fossil nematodes to a certain taxonomic taxon. Obese immature female with curved body adjacent to a monocot seed described as *Palaeoanguina dominicana* from Dominican amber by Poinar (2011) (Fig. 14.9). This fossil is estimated to be 15-20 million years old.

14.8. Phylogenetic Relationships

The phylogenetic relationships between genera and species of Anguinidae have been studied by using morphological, biological and biochemical and molecular datasets. Relationships among some gall forming nematodes have been established on the basis of allozymes (Riley *et al.*, 1988) and the rRNA gene sequences (Powers *et al.*, 2001; Subbotin *et al.*, 2004; Bert *et al.*, 2008; Giblin-Davis *et al.*, 2010; Zhao *et al.*, 2011; Cid del Prado Vera *et al.*, 2018). Phylogenetic analysis using the ITS-rRNA gene sequence datasets supported, for example, monophyly of *Anguina*, and a narrow concept of *Subanguina* with but a single species, *S. radicicola*, as proposed

Fig. 14.7. Symptoms of infection caused by *Mesoanguina chilensis* on *Nothofagus obliqua* (A), *Halenchus fucicola* on *Ascophyllum nodosum* (B), *Litylenchus crenatae* on *Fagus grandifolia* (C), *Cotylenchus cleo* on *Polystichum munitum* (D), *Anguina danthoniae* on *Danthonia californica* (E), *Orrina phyllobia* on *Solanum elaeagnifolium* (F). (Courtesy of W.M. Ciesia (A), D. Fenwick, S.A. Subbotin (C, D), M.A. Voeltz (E), J.A. Skinner (F).)

by Paramonov (1970), and not the broad concept proposed by Brzeski (1981). However, phylogenetic analysis revealed paraphylies for *Mesoanguina* and *Heteroanguina*. Molecular analysis of seed-gall nematodes supported the concept of a narrow trophic specialization in this group (Southey *et al.*, 1990) and showed that *A. agrostis* occurred only with *Agrostis* spp. and *Poa* spp., whereas other seed-gall nematodes parasitizing grasses belonged to other, still undescribed, species.

14.9. Nematode-bacteria Associations

A number of gall forming nematodes are found in association with species of coryneform bacteria of the genus *Rathayibacter*, previously placed in *Clavibacter* or *Corynebacterium*. *Rathayibacter* causes gumming diseases, usually characterized by yellow bacterial slime on seed-heads, stems and leaves of the plant host. Presently, *Rathayibacter* contains 10 valid species and several undescribed species, with several species associated with nematodes. The well-known plant-parasitic species (*Rathayibacter rathayi*, *Rathayibacter tritici*, *Rathayibacter iranicus* and *Rathayibacter toxicus*) affect wheat and several grasses in the family Poaceae. In nature, this nematode-bacteria association appears to be specific and coevolved. For example, *R. tritici* is found exclusively with *A. tritici* and *R. toxicus* with *A. funesta* and *A. paludicola*. The bacteria adhere to the cuticle of infective-stage juveniles

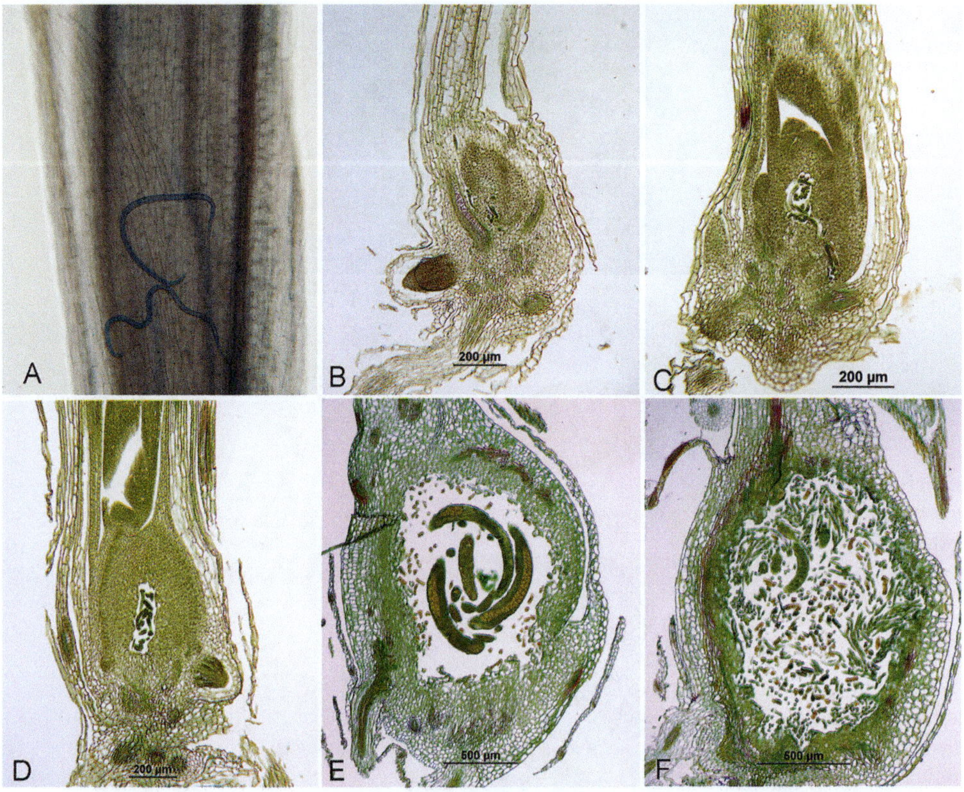

Fig. 14.8. Histopathology of *Anguina pacificae* on *Poa annua*. A: Infective J2 passing between leaves of the shoot, 2 mm above crown, after 2 days; B-F. longitudinal sections through crown of plant; B: fourth day, J2 have initiated a cavity just below the shoot meristem; C, D: hyperplasia of shoot parenchyma cells, after 6 and 15 days, results in visible swelling at base of stem. Cavity volume increases; E: adults in cavity at onset of egg laying, after 29 days. Cavity contains loose parenchyma cells and a few eggs; F: eggs begin to hatch, after 50 days, and a layer of cells, filled with cytoplasm and 3- to 6-cells thick, surrounds the cavity. (After McClure *et al.* (2008) with modification.)

Fig. 14.9. *Palaeoanguina dominicana* in Dominican amber. (After Poinar, 2011.)

in compatible interactions of bacterial adhesins and nematode receptors (Murray *et al.*, 2017). However, cross-inoculation experiments and *in vitro* adhesions test show that, although some degree of specificity occurs, associations not found in nature can occur: *A. australis* and *A. tritici* have been shown to be potential vectors of some strains of *R. toxicus*. *Rathayibacter* spp. seem to be fully dependent on a nematode vector and direct inoculation attempts have only rarely been successful.

Although *Anguina* species can have an economic impact as agricultural pests in their own right, when they carry *Rathayibacter* the impact is increased. *A. tritici* reduces the yield of wheat and barley by replacing seed with galls. The admixture of galls also reduces grain quality, adding cleaning costs and limiting market access. However, in seed-heads concomitantly infected with *R. tritici*, seed set in florets not infested by the nematode will be prevented if the bacterial gummosis affects the entire inflorescence. A similar situation occurs for commercial seed production of the forage grass, *Dactylis glomerata*, when colonized by *R. rathayi* carried by its associated *Anguina* sp. However, in the case of *R. toxicus*, its vectors are of no direct economic impact, although when the bacterium colonizes the host grass, *Lolium rigidum*, it produces a neurotoxigenic tunicaminyluracil antibiotic that can result in fatal poisoning of animals that graze on diseased plant, a syndrome now referred to as annual ryegrass toxicity, and which affects grazing animals in Australia. Putative tunicaminyluracil biosynthetic gene clusters were also identified in genome sequences of *R. iranicus*, *Rathayibacter agropyri* and an undescribed South African *Rathayibacter* species (Tancos *et al.*, 2020).

In addition to anguinid-*Rathayibacter* associations, other bacteria from *Curtobacterium*, *Microbacterium*, *Agreia* and *Leifsonia* belonging to the *Microbacteriaceae* have been isolated from nematode galls (Evtushenko *et al.*, 1994, 2001). The stem nematode, *D. dipsaci*, is known to increase the transmission of the actinobacterium, *Clavibacter michiganensis* subsp. *insidiosum*, responsible for a bacterial wilt in *Medicago sativa*, although it is not an obligate vector.

14.10. Nematode-fungal Associations

Just as *Rathayibacter* spp. adhere to the cuticle of *Anguina*, so do the spores of the plant pathogenic fungus, *Dilophospora alopecuri*. Unlike the bacterium, this fungus appears to be a true plant pathogen able to directly infect a host, albeit with limited aggressiveness. However, when spores are carried into the apex of the plant, the infection is much more effective and the impact can be severe, completely inhibiting any seed set in the affected tiller and preventing the nematode from completing its life cycle. Although populations of *Anguina* vary in their receptiveness to adhesion by spores of *D. alopecuri*, there appears to be no species level specificity. *Dilophospora* is a monotypic genus and, although genetically variable, this is not strongly associated with vector specificity.

The fungal disease was important in wheat in Europe when *A. tritici* was common. Nowadays it is of limited importance, although it still occurs in some parts of the current geographic range of the vector, thereby exacerbating its impact. Because the fungus is more aggressive than *R. toxicus* in its colonization of the host, it also acts as a more effective regulator of vector populations. It has been observed as a strong contributing factor to the decline of *A. funesta* populations in Australia and has been mass produced and applied as a biocontrol agent in areas where natural decline had not occurred.

14.11. Control

The gall forming nematodes have both a direct and an indirect economic impact and so approaches to control vary depending on the species and the context. Galls of *A. tritici* are readily cleaned from cereal grain by normal seed grading practices in mechanized agriculture. This, in conjunction with crop rotation, has provided highly effective control and has made the nematode essentially extinct over much of its former range. In small scale, traditional, farming systems in west and south Asia, the nematode persists as this control method cannot be as efficiently applied. Some work in India has focused on selecting resistant cultivars and on alternative seed-cleaning methods.

For grass seed production, for example *Agrostis* and *Dactylis*, the impact of *Anguina* is mostly through reduction in seed germination, quality and market access (quarantine restrictions). In some grass-*Anguina* combinations, the infested diaspores are of the same size and only of slightly lower mass than those with normal seed, so cleaning is unlikely to be fully effective. Therefore, to eliminate the nematode, complete removal of the host for a period and resowing with uninfested seed would be needed. Alternatively, early mowing or crash grazing can be effective because the nematodes do not reach their survival stage until well after initial tiller elongation. This is why grass seed production systems favour build-up of the nematode, as the nematode is protected through to maturity. In contrast, in grazed pasture or hay production systems, development of the majority of nematodes is prevented.

For *A. funesta*, the host is a weed and so the nematode is able to reach maturity protected within the crop (Riley and Barbetti, 2008). In this case, control has focused on eliminating the host from the crop with herbicides, but herbicide resistant populations of *L. rigidum* constrain this approach. Where the cropping system includes pasture leys, heavy grazing, mowing and/or herbicides in the pasture phase are used to control the nematode. As mentioned above, *D. alopecuri* is exploited for its biocontrol potential. Also, resistant *L. rigidum* cultivars are available but these are only adopted where the value of the grass as a pasture component is greater than its impact as a crop weed, and effective deployment depends on highly effective initial control of the naturalized host ecotypes.

Other gall nematodes have had a minor, localized, inconsistent or transient impact, so control options have not been given as much attention as the *Anguina* spp. discussed above.

Most *Ditylenchoides* species are fungal feeders and considered to be beneficial members of soil ecosystems; however, there are several problematic plant pathogens in the genus *Ditylenchus* that need to be managed in cropping systems. As these species can infest seeds and bulbs, primary control is based on selecting uninfested propagating material, especially, but not only, for planting in uninfested fields. Various treatments such as hot water dips, fumigation and other chemical treatments can reduce infestation levels and provide worthwhile control, but are unlikely to eliminate the risk of introducing these pests to uninfested fields. Where quarantine is important, a combination of sourcing propagating material from uninfested fields/regions, testing and treatments is necessary to provide an acceptable level of risk. In some host species, seeds are not infested, but harvested fodder (as in the case of *D. dipsaci* in *Avena sativa*) can be a means of spread, as can infested crop residues in other hosts and contaminated farm machinery. Irrigation water can also be an effective means of spread, as is the case with *D. angustus* in deepwater and lowland rice production.

In infested fields, crop rotation would be the first choice for control, but this relies on control of all potential hosts, including volunteers and weeds, in the non-host phase. For example, *Gallium aparine* (bedstraw) is a weed that can maintain *D. dipsaci* populations at problem densities between host crops. Where populations have been lowered by crop rotation, reinfestation from adjacent fields should be considered, as anguinid nematodes can be located on the soil surface and in above-ground crop residues, facilitating their dispersal by surface water flows and wind.

Given that field applied fumigants and nematicides are limited by efficacy, cost and environmental/regulatory concerns, breeding/selecting resistant crop cultivars is the other important approach to control plant-parasitic *Ditylenchus* spp. Resistance to *D. dipsaci* has been identified and deployed in commercial cultivars for crops such as alfalfa (lucerne) and oats, but is complicated by variation in pathogenicity of the populations and races of the nematode. For the other plant-parasitic species, progress has been more variable. Resistance found to *D. africanus* appears to fail in heavy infestations, indicating that tolerance and resistance may be needed for workable control. Therefore, incorporation of all or several (depending on the crop and pest species) of the above approaches into an integrated control programme is likely to provide the most effective control.

14.12. Genera and Major Species

14.12.1. Genus *Anguina* Scopoli, 1777

= *Anguillulina* Gervais & Van Beneden, 1859
= *Paranguina* Kirjanova, 1955
= *Cynipanguina* Maggenti, Hart & Paxman, 1974

Diagnosis: (modified after Chizhov and Subbotin, 1990). Female after heat relaxation crescentic or spirally coiled. Ovary reflexed two or three times. Oocytes in zone of maturation usually in two or more rows (in second-generation female of *A. agropyri* in a single row). Crustaformeria gland irregular, consisting of 180-600 cells in 6-12 rows (20-60 cells in each row). Up to 18 synchronous eggs present in crustaformeria gland. Infective juveniles represented by J2. Inducing usually pigmented galls with an internal cavity on above-ground parts or sometimes on the rhizomes (underground stems) of monocotyledons of the family Poaceae (grasses), subfamilies Pooideae and Bambusoideae.

Type species: *A. tritici* (Steinbuch, 1799) Filipjev, 1936.

Number of other species: more than 20.

Adults of *Anguina* species are morphologically almost indistinguishable, although they develop on different hosts in widely separated geographical areas. A key to species identification is given by Chizhov and Subbotin (1990).

***Anguina tritici* (Steinbuch, 1799) Chitwood, 1935 – the wheat seed-gall nematode (Figs 14.10, 14.11)**

The wheat seed-gall nematode was the first plant-parasitic nematode to be described in the literature. In 1743, the English theologian J.T. Needham examining seed galls of wheat observed a soft white fibrous substance within the galls which, after adding water, started moving in a disorderly fashion. Needham named these unfamiliar animals as 'worms, eels or serpents'. J.G. Steinbuch was the first to name this species,

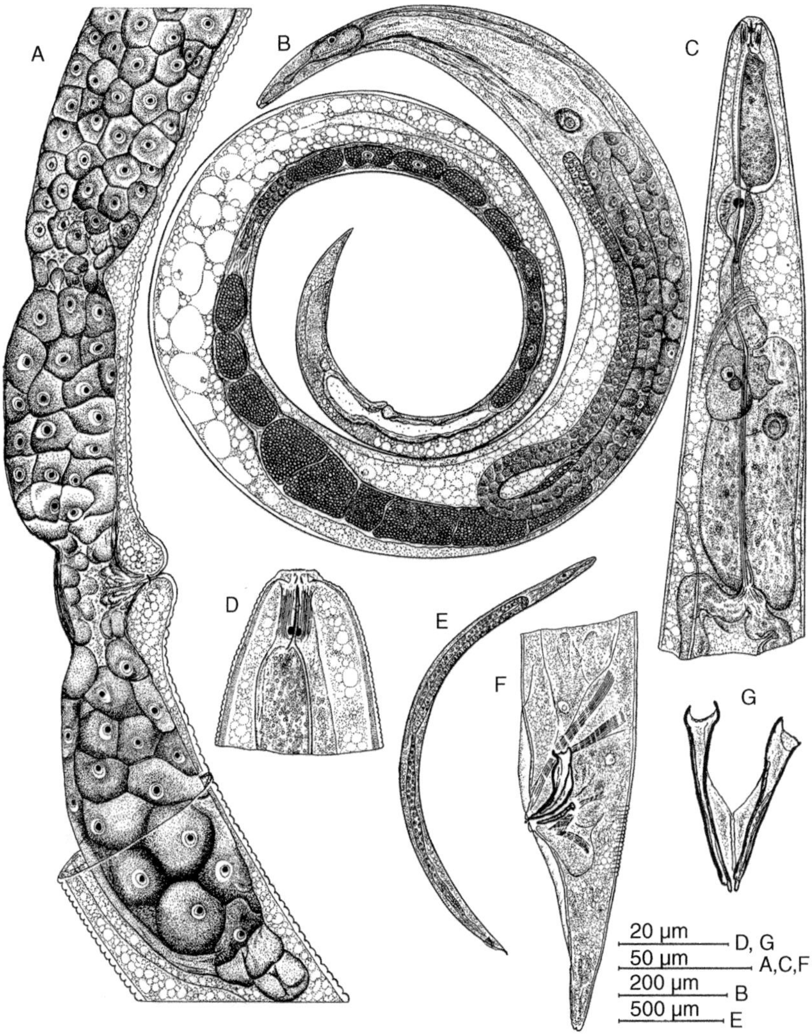

Fig. 14.10. *Anguina tritici*. A: Female; B: male; C: female pharyngeal region; D: male tail; E: spicules and gubernaculum. (After Goodey, 1932.)

Fig. 14.11. *Anguina tritici*. A: Seed galls with J2; B: infected wheat seed head; C: seed galls and healthy seeds of wheat. (Courtesy of M. McClure (A, B) and V.N. Chizhov (C).)

although an earlier description of this nematode was made by Roffredi in 1775. *A. tritici* has been recorded in major wheat-growing areas of all five continents, where it causes a disease in wheat and rye called earcockle or seed gall. Application of effective mechanical seed-cleaning methods have practically eradicated this nematode from many countries where it was once common.

Measurements: After Goodey (1932), Kirjanova and Krall (1971) and Southey (1972).

Female: L = 2.6-5.2 mm; a = 13-30; b = 9.8-25.0; c = 24-63; V = 70-95; stylet = 8-11 µm.

Male: L = 1.9-2.5 mm; a = 21-30; b = 6.3-13.0; c = 17-28; stylet = 8-11 µm; spicules = 35-40 µm; gubernaculum = 10 µm.

J2 infective: L = 0.75-0.95 mm; a = 47-54; b = 4.0-6.3; c = 23-28; stylet = 10 µm.

Description: Female body obese, spirally coiled ventrally when relaxed by heat. Cuticular annules very fine, usually visible only in pharyngeal region; lateral fields of adults with four or more lines, visible only on young specimens. Procorpus swollen but constricted at junction with median bulb. Isthmus sometimes posteriorly swollen, then offset from glandular basal bulb by a deep constriction. Anterior branch of genital tract greatly developed; ovary with two or more flexures and very many oocytes in multiple rows around a rachis. Simple post-vulval sac present. Tail conoid, tapering to an obtuse or rounded tip, not mucronate. Male body sometimes curved dorsally when killed by heat. Bursa extending from just anterior to spicules to just anterior to tail tip. This nematode is characterized by the large size of the adults and form of the procorpus and median bulb of the pharynx compared with related species.

Type host: Wheat, *Triticum aestivum*.

Other hosts: *Secale cereale*, *Triticum dicoccum*, *Triticum durum*, *Triticum monococcum*, *Triticum spelta* and *Triticum ventricosum*. Oats and barley are poor hosts (Krall, 1991).

Symptoms and pathogenicity: Symptoms in infected plants are illustrated in Figure 14.11. The infective juveniles enter young seedlings and move upwards within the plant to the meristem or concentrate in the leaf axis of shoots where they remain until the ears develop. Infected plants become stunted and bear shorter and deformed stems and leaves. Severely infested plants do not form ears or form only stunted ears on stunted stems. Infested ears are generally smaller, shorter and thicker than healthy ones, and have abnormally spreading glumes. Some or all of the grains are replaced by galls. The galls are light to dark brown to almost black in colour, are hard to the touch and are filled with a mass consisting of infective juveniles. They may remain in anhydrobiosis for decades (Fig. 14.12A). Each gall initially contains 40 or more adults of each sex which produce 30,000 or more eggs and/or juveniles. Yield loses up to 70% have been reported in wheat grain.

Distribution: Formerly widespread in wheat-growing areas of Asia, Australia, Europe, North Africa and North America. Earcockle disease has almost been eliminated in most countries. Outbreaks of the disease have been reported more recently in India, North Africa and west Asia.

Fig. 14.12. *Anguina agrostis*. A: Inflorescence of fowl bluegrass, *Poa palustris* with normal spikelets (left) and inflorescence with nematode seed galls (right); B: nematode seed galls with second-stage juveniles of *A. agrostis*, C: seed galls induced by *A. agrostis* on *Agrostis* sp. (Courtesy of S.A. Subbotin (A, B) and V.N. Chizhov (C).)

Disease complex: Wheat seed-gall nematode infection is associated with a bacterial yellow gumming disease caused by *R. tritici*, which is especially important in India, where it has been known for many years as tundu. The characteristic symptom is the presence of a bright yellow slime, consisting of massed bacteria, on the abortive ears and unsheathing leaves while the ears are still in boot. The fungal pathogen, *D. alopecuri*, is also associated with *A. tritici* and causes diliophosphorosis of wheat, which is characterized by the spike being enveloped in a sticky black mass.

Control: The disease may be effectively controlled by seed cleaning and crop rotation.

***Anguina agrostis* (Steinbuch, 1799) Filipjev, 1936 – the bentgrass seed-gall nematode**

The bentgrass seed-gall nematode was described by J.G. Steinbuch in 1799, from specimens of a grass collected in a wood in the vicinity of Erlangen, Bavaria, Germany. He thought that grass belonged to *Agrostis silvatica* Huds., but after dissection of the enlarged structures of the inflorescence he discovered parasitic nematodes and came to the conclusion that the grass was actually *Agrostis capillaris*. Another species, now known as *Anguina phalaridis*, and from *Phleum phleoides*, was also described together with the bentgrass seed-gall nematode. There are numerous reports of galls found in various species and genera of grasses, but without satisfactory diagnostic descriptions of the associated nematodes. The grasses *Apera*, *Arctagrostis*, *Calamagrostis*, *Dactylis*, *Eragrostis*, *Festuca*, *Hordeum*, *Koeleria*, *Lolium*, *Phalaris*, *Phleum*, *Poa*, *Puccinellia*, *Sporobolus* and *Trisetum* have been included in the list of host plants for *A. agrostis* (Kirjanova and Krall, 1971). Southey (1973), after reviewing literature on the host specificity of seed-gall nematode populations and their morphological and morphometric differences from the type host, suggested that a thorough revision of species causing galls in the flowers of grasses was required. He suggested that *A. agrostis sensu stricto* appeared to be restricted to populations causing characteristic elongate galls and abnormally elongated floral structures in *Agrostis* grasses. Analysis of the ITS-rRNA gene sequences supported Southey's suggestion and the concept of a narrow specialization of seed-gall nematodes and confirmed that *A. agrostis* infected *A. capillaris* and *Poa palustris*. Other grasses known previously as hosts for *Anguina agrostis* are actually hosts for several other, still undescribed, species of

Anguina (Subbotin *et al.*, 2004). *A. agrostis* is considered to be a serious or potentially important nematode pest of bent grass, especially in the Pacific Northwest of the USA and in New Zealand (Southey, 1973).

Measurements: After Goodey (1932), Southey (1973) and Chizhov (1980).

Female: L = 1.39-2.60 mm; a = 13.8-25.4; b = 8-28.7; c = 25.2-44.0; V = 87-92; stylet = 8-12 μm.

Male: L = 1.05-1.68 mm; a = 23.8-38.0; b = 6.5-9; c = 20.5-28.4; stylet = 10-12 μm; spicules = 25-40 μm; gubernaculum = 10-14 μm.

J2 infective: L = 0.55-0.82 mm; a = 47.2-65.0; b = 3.2-4.5; c = 11.7-20.0; stylet = 10 μm.

Description: Female body obese, coiled or crescentic when relaxed by heat with ventral surface inwards; tapered towards both ends. Procorpus cylindrical, somewhat enlarged in the middle part with a constriction before joining median bulb. Ovary reflexed two or three times. Postvulval uterine branch reaching halfway from vulva to anus, filled with sperm. Tail short, conoid, with acute terminus. Male smaller and more slender than female, body more or less arcuate ventrally when heat relaxed, that is, dorsal surface outermost. Bursa subterminal, not reaching tail tip.

Type host: *A. capillaris.*

Other hosts: *P. palustris.* Under experimental conditions, *A. agrostis* infected several *Agrostis* species.

Life cycle: Once in the plant crown, J2 remain between the leaf sheaths or rhizomes until spring, when floral primordia form. Upon invading the floret ovaries, J2 rapidly develop to adults and reproduce. Juveniles moult once in the egg stage, hatch quickly as J2 and become resistant before the galls dry. The life cycle requires 3-4 weeks and is often completed before the inflorescence opens.

Symptoms and pathogenicity: Forms characteristic cigar-shaped galls, at first green, later dark purple-brown, in florets of bent grass and bluegrass (Fig. 14.12). Galled florets have the floral parts (glumes, lemma and palea) greatly elongated to three to five or more times normal length, the lemma projecting beyond the glumes as a sharp point (Hooper and Southey, 1978). Fresh galls of *A. agrostis* associated with *Rathayibacter* from *A. capillaris* have been found to be poisonous to some animals.

***Anguina agropyri* Kirjanova, 1955 – the couch grass gall nematode**

The couch grass nematode was described from *Elytrigia repens.* This nematode is distributed in several European countries and may damage some cereals.

Measurements: After Kirjanova and Krall (1971), Chizhov and Berezina (1986) and Krall (1991).

Female (first generation): L = 2.23-4.78 mm; a = 16.6-25.7; b = 14.5-24.4; c = 21.2-47.7; V = 84-90; stylet = 10-12 μm.

Male (first generation): L = 1.91-3.11 mm; a = 22.7-29.5; b = 10.6-21.3; c = 18.6-34.0; stylet = 10-12 μm; spicules = 33-40 μm; gubernaculum = 13-15 μm.

J2 infective (first generation): L = 0.81-1.13 mm; a = 51.5-75.5; b = 3.8-5.2; c = 9.7-11.8; stylet = 10-12 μm.

Female (second generation): L = 1.94-3.23 mm; a = 26.0-36.5; b = 9.3-14.6; c = 16.9-23.2; V = 77-84; stylet = 10-12 μm.

Male (second generation): L = 1.53-2.27 mm; a = 28.6-44.0; b = 7.2-11.4; c = 14.0-21.0; stylet = 10-12 μm; spicules = 31-37 μm; gubernaculum = 13-15 μm.

J2 infective (second generation): L = 0.83-1.15 mm; a = 52.3-67.4; b = 4.0-5.2; c = 10.0-12.6; stylet = 10-12 μm.

Description: Female of first generation considerably more obese than vermiform adults of second generation. Cuticle with fine striations intercepting folds of the lateral fields. Procorpus swollen. Pharyngeal glands overlapping intestine only slightly. Ovaries reflexed. Post-vulval uterine branch long, extending up to halfway or more from vulva to anus. Tail conoid, tapering to a finely rounded tip. Male considerably more slender than females. Spicules elongate, almost straight. Bursa narrow, seeming to extend to tail tip.

Type host: Couch grass, *E. repens.*

Other hosts: *S. cereale, Triticum vulgare, Hordeum sativum* and *Elymus arenarius.*

Symptoms and pathogenicity: *A. agropyri* induces bulbous galls at the base of the stems of couch grass, and sometimes also on rhizomes (Fig. 14.3E). After nematode infection of cereals, the lower parts of the stems of wheat and rye become stout and swollen. In rye, small bulbous thickenings at the base of the stem may also be present. In the phase of development when the grains emerge in the tubule, the lower internodes of infested plants remain abnormally short and thickened. As a result of the severe reduction of the lower internodes, the leaves of infested cereals become bunched and rosette-like. Depending on the degree of infestation, the plants either do not form ears or give an extremely low yield, the number and weight of grains being much smaller than from uninfested plants. As a result of their stoutness, the infested shoots are easily broken. The grain yield can be estimated up to 46% of the average harvest in infested areas (Krall, 1991).

***Anguina funesta* Price, Fisher & Kerr, 1979 – the ryegrass seed-gall nematode**

A. funesta is a seed-gall nematode found in *L. rigidum* and associated with annual ryegrass toxicity (ARGT) in Australia and South Africa. Although described as a new species by Price *et al.* (1979), Stynes and Bird (1980) considered that it was not morphologically distinguishable from *A. agrostis*. A growing body of biological, ecological, isozyme and molecular data eventually confirmed the validity of *A. funesta* as a distinct species. *A. funesta* is a vector for *R. toxicus* and a key component of the multi-partite interactions that cause ARGT. The earliest known outbreaks of ARGT were in South Australia in the mid-1950s and it has remained an economic problem for over five decades. The nematode and associated organisms spread to Western Australia and South Africa and generally follow a pattern of build-up, impact, then decline. The decline seems to be associated with a subsequent build-up of *D. alopecuri* and changes in land management practices that had supported high host densities protected through to seed set. As an economic problem, it now largely persists in the northern cropping areas of Western Australia.

Measurements: After Price *et al.* (1979), Stynes and Bird (1980) and Bird and Stynes (1981).

Female: L = 1.65-2.44 mm; a = 16.8-20.1; b = 9.3-34.0; c = 18.1-41.2; V = 87-94; stylet = 7-10 μm.

Male: L = 0.78-1.52 mm; a = 20.3-30.9; b = 6.3-9.5; c = 16.1-24.9; stylet = 7-10 μm; spicules = 16-28 μm; gubernaculum = 9-14 μm.

J2: L = 0.76-0.91 mm; a = 48-53; b = 4.2-5.3; c = 12-14; stylet = 9-11 μm.

Description: Female obese, coiled with ventral surface inwards; head narrower than body and offset. Wide procorpus opening to a muscular median bulb, ovate to spheroid, with three crescentic thickenings. Prodelphic ovary with one or two flexures. Post-vulval sac similar length to uterus, sometimes muscular or with small straight process. Male smaller than female, cylindrical, slender, straight to slightly curved dorsally when heat relaxed, with a single reflexed testis.

Type host: Annual ryegrass, *L. rigidum*.

Other hosts: Inoculation experiments demonstrated that the host range includes various *Lolium*, *Festuca* and *Vuplia* species, but only *L. rigidum* and *V. myuros* are known as hosts from natural infestations.

Symptoms and pathogenicity: The J2, having over-summered anhydrobiotically in last season's galls, invade host seedlings following the onset of autumn rains. The juveniles congregate at the vegetative meristem and induce galls after floral initiation. The galls are smaller and lighter than the seed they replace and there are no morphological changes to the inflorescence as seen with *A. agrostis* and *A. paludicola*. Infested florets can only be detected by examining diaspores with transmitted light under a microscope or on a light table. Although *A. funesta* will reduce seed set in the host, galled florets are mostly removed with normal seed-cleaning procedures so direct economic impact comes through quarantine restrictions on trade (indirect impact is discussed below).

Distribution: *A. funesta* is found in winter rainfall areas of southern Australia, especially in pasture-ley farming systems with *L. rigidum* as the main grass weed in the cereal cropping phase. It is most common in South Australia and Western Australia, but also occurs in Victoria. Recently, this species was reported in Oregon, USA.

Disease complex: *A. funesta* is vector for *R. toxicus* and *D. alopecuri*. *R. toxicus* colonizes the galls and gums the seed head of the host, causing distortions. On senescence of the host, the bacterium produces toxin, paralleled by bacteriophage multiplication. The glycolipid toxins inhibit glycosylation of proteins, which compromises membrane integrity leading to neurological symptoms and finally death of the grazing animal. The toxins are cumulative and, as symptoms are not usually observed in livestock before a lethal dose is consumed, mortality rates can be high.

Control: Control has focused on the host. The nematode invades tillers produced early in the growing season, so effective control is achieved by preventing these tillers from maturing by crash grazing, mowing or herbicide in the pasture phase. In cereal crops, herbicides often only reduce the host density so that the nematode population is concentrated in fewer, larger plants and thus nematode control is only partially achieved. Nematode resistant ryegrass cultivars have been released but are not widely adopted. *D. alopecuri* appears to contribute to natural decline in ARGT outbreaks in some areas and has been mass produced for application in other areas in order to accelerate this decline.

Anguina pacificae **Cid del Prado & Maggenti, 1984 – the Pacific shoot gall nematode (Figs 14.1, 14.8)**

The Pacific gall nematode was found on grasses on *Poa annua* in 1978 from several golf courses along the Coast of Central California, USA and was initially identified as *S. radicicola*. Subsequently, it was characterized morphologically and described as a new species by Cid del Prado Vera and Maggenti (1984). *A. pacificae* is distributed along a narrow strip on the Pacific coast of Northern California where it forms galls on the shoots of *P. annua* and causes significant damage to golf course greens.

Measurements: After Cid del Prado Vera and Maggenti (1984).

Female: L = 1.44-2.58 mm; a = 19.4-33.2; b = 8.1-13.1; c = 19.9-34.4; stylet = 8.9-12.4 µm; V = 82-89 µm.

Male: L = 1.22-1.84 mm; a = 21.1-33.0; b = 5.8-10.4; c = 19.8-29.8; stylet = 9.3-12.7 µm; spicules = 26.2-40.7 µm; gubernaculum = 10.6-15.5 µm.

J2 infective: L = 0.63-0.97 mm; a = 37.0-56.4; b = 3.5-7.3; c = 8.8-11.4; stylet = 10.0-12.7 µm.

Description: Body cylindrical, assuming C-shape after death. Labial region without visible annuli by light microscope, not offset. Lateral field with four main distinct lines and many longitudinal fine lines between them. Ovary with one or two flexures, sometimes straight and extending to pharynx. Post-vulval uterine sac containing degenerate cells. Vulval lips distinct. Tail conical, sharply pointed tip. Male body cylindrical, arcuate with ventral surface outermost after killing. Bursa beginning anterior to spicules and ending before tail tip. Tail with acute terminus.

A. pacificae differs from *A. agropyri* by the pharyngeal glands being well separated from the median bulb in the females. They also differ in the position of the excretory pore; in *A. agropyri*, it is situated a little anterior to the pharyngo-intestinal valve, and in *A. pacificae* the excretory pore is situated at the level of posterior end of the isthmus. *A. pacificae* has a sharply pointed tail tip, while *A. agropyri* has a conical tail tip.

Type host: *P. annua*.

Other hosts: *Poa trivialis* and several *Agrostis* cultivars proved to be poor hosts for *A. pacificae* (McClure *et al.*, 2008).

Symptoms and pathogenicity: The nematode induces galls on stems. Initial symptoms on turf consist of small yellow patches, 25-75 cm in diameter, which enlarge and may coalesce as the nematodes spread. Young, infected plants may die and, when the infestation is severe, a rough, uneven putting surface results. *A. pacificae* is considered as a major threat to turf in coastal areas of California (McClure *et al.*, 2008).

Anguina paludicola **Bertozzi & Davies, 2009**

A. paludicola is a seed-gall nematode described by Bertozzi and Davies (2009) from two hosts from areas subject to occasional seasonal flooding in eastern Australia. In *Polypogon monspeliensis* (type host, non-native grass), it occurs in the southeast of South Australia where pastures can be subject to extended, annual flooding. In this host, shoot meristem galls are also formed in inundated plants. In *Lachnagrostis filiformis* (an Australian native grass), it occurs along the inland river systems of northern New South Wales in semi-arid pastures subject to transient, infrequent flooding. Morphological differences occur between the populations but are variable and not considered as justification to separate at the species level. In both hosts and areas, the nematode acts as a vector for *R. toxicus*, which colonizes the nematode galls and grass seed-heads, producing a toxin responsible for poisoning and death of livestock.

Measurements: After Bertozzi and Davies (2009).

Female: L = 1.51-4.40 mm; a = 18.8-33.2; b′ = 8.5-22.1; c = 21.8-54.0; V = 81-94; stylet = 9-13 µm.

Male: L = 1.50-2.49 mm; a = 18.3-37.3; b′ = 8.9-14.4; c = 18.0-35.3; stylet = 8-12 µm; spicules = 27-37 µm; gubernaculum = 7-13 µm.

Second-stage juveniles: L = 603-771 mm; a = 49.0-60.8; b′ = 3.3-5.1; c = 9.9-15.2; stylet = 9-11 µm.

Description: Female obese, coiled with ventral surface inwards; head narrower than body and offset. Procorpus wide, metacorpus oval with refractive thickenings slightly anterior to middle, isthmus shorter and narrower than procorpus. Ovary usually reflexed twice. Post-vulval uterine sac similar in length to uterus, frequently with a reflexed process. Tail conical with irregular and variable terminus, appearing lobed in some specimens (diagnostic feature of the species). Male smaller than female, cylindrical, slender, C-shaped when heat relaxed with dorsal surface inwards, with a single reflexed testis.
Type host: *P. monspeliensis*.
Other hosts: *L. filiformis*.
Symptoms and pathogenicity: In populations of *P. monspeliensis*, *A. paludicola* juveniles invade the host with the onset of autumn rains and rapidly forms galls in the shoot meristems, a complete life cycle being completed before the field is flooded in winter. The new cohort of juveniles invades tillers that develop as flooding recedes to form seed galls and a small proportion of panicle branch galls. When these galls mature, they contain a second cohort of J2 that over-summer in an anhydrobiotic state. In *L. filiformis*, *A. paludicola* only induces seed galls and has one cycle per year. In both hosts, the seed galls formed are larger than normal seed and the floral bracts are enlarged five times or more (McKay *et al.*, 1993).
Distribution: This species is found in Australia but only in situations where flooding allows its host grass to become the dominant species in the sward. It is found in South Australia in the introduced weedy grass, *P. monspeliensis*, and in New South Wales in the native grass, *L. filiformis*.
Disease complex: *A. paludicola* acts as a vector for both *R. toxicus* and *D. alopecuri*. As with the association of *A. funesta* and *R. toxicus*, toxins are produced and grazing animals poisoned (the syndrome in this case is called flood plain staggers or FPS). The bacterium adheres to invasive juveniles and is carried into the host, colonizing nematode galls and whole inflorescences, and in the latter situation causing gummosis and distortion. In the early 1990s, serious stock losses were recorded, which prompted research to identify the cause and develop controls. It now appears that conditions favouring the high nematode population densities and bacterial infection are relatively uncommon, and major stock losses have not been recorded since. *D. alopecuri* is likewise carried into the plant with *A. paludicola*, producing leaf lesions and distorted floral development.
Control: Bertozzi and Davies (2009) investigated control options, mostly focused on host control and some consideration of the biological control potential of *D. alopecuri*. However, given the infrequent occurrence of major outbreaks of toxicity, there has been no impetus to apply or further develop control measures. Also, in the southeast of South Australia, drainage schemes have reduced the winter flooding of pasture and thereby the conditions that favour high population densities of *A. paludicola*. In northern New South Wales, extended periods of drought have dominated climatic patterns for a decade or more.

14.12.2. Genus *Afrina* Brzeski, 1981

Brzeski (1981) proposed this genus to accommodate nematodes having a crustaformeria of four rows with 14 cells, a lobed bursa and mucronate tail in both sexes. He considered two species to belong to this genus: *Anguina hyparrheniae* and *Anguina tumefaciens*. Later, Van den Berg (1985) described *A. wevelli* and transferred *Anguina spermophaga* (Steiner, 1937) to *Afrina*.
Diagnosis: (after Brzeski, 1981, with modifications). Mature female swollen. Median bulb with refractive thickenings. Isthmus separated by a constriction from terminal bulb. Crustaformeria formed by four rows of cells, 14 cells in a row. Testes often with two flexures. Tail of both sexes conical, tail tip mucronate. Bursa notched at posterior end, almost reaching tip but does enclose a mucro. Forms seed and leaf galls on grasses of subfamily Panicoideae and Chloridoideae. Distributed in Africa, Australia, Asia and North America.
Type species: *A. hyparrheniae* (Corbett, 1966) Brzeski, 1981.
Number of other species: 5.
***Afrina sporoboliae* Barrantes-Infante, Schroeder, Subbotin & Murray, 2018.**
This seed-gall nematode species was found in Idaho, USA from *Sporobulus cryptandrus*.
Measurements: After Barrantes-Infante *et al.* (2018).
Female: L = 1.70-2.57 mm; a = 18.0-27.9; b = 7.0-13.0; V = 90.6-95.3; stylet = 7.5-10 μm.

Male: L = 1.32-1.79 mm; a = 30.7-47.7; b = 7.5-11.2; stylet = 7.5-10 µm; spicules = 40.5-51 µm.
Description: Female coiled, very stout. Lips not annulated, set off. Lateral field consisting of two marginal lines. Basal bulb large, forming a long bode overlapping. Tail tapering gradually to a minutely pointed terminus. Bursa reaching tail tip.
Type host: *Sporobulus cryptandrus*
Distribution: USA (Idaho).
Symptoms: Induced seed galls. Mature seed galls varied in colour from purplish to light brown and usually 1.5-2.2 mm in a length. Infected plants showed rolling or twisting leaves. Rathayibacter species was found in this seed galls and was associated with nematodes.

***Afrina wevelli* Van den Berg, 1985 – the weeping lovegrass seed-gall nematode (Fig. 14.13)**

A. wevelli was described from South Africa in 1985, from seed galls on weeping lovegrass (*Eragrostis curvula*), a very important forage grass in semi-arid regions. This nematode became an important nematode pest of this grass in South Africa and significantly affected the seed export industry as seed must be free of nematodes.

Measurements: After Van den Berg (1985).
Female: L = 1.61-1.95 mm; a = 19.9-26.9; b = 8.1-8.7; c = 27.0-42.2; V = 90-94; stylet = 8.1-10.3 µm.
Male: L = 1.01-1.42 mm; a = 30.0-41.9; b = 5.9-8.1; c = 20.1-25.7; stylet = 8.0-11.8 µm; spicules = 34.2-40.8 µm; gubernaculum = 15.8-19.9 µm.
Second-stage juveniles: L = 675-781 mm; a = 36.7-43.9; b = 4.7-5.3; c = 13.1-15.8; stylet = 8.0-9.6 µm.
Description: Female body posture ranging from an open circle to tightly coiled. Labial region flattened. Basal bulb large. Tail long, tapering to narrow rounded tip with slight irregularities just anterior to tip of most specimens. Bursa extending from about one and a half times cloacal body width anterior to cloaca to tail tip. Tail ending in an irregular projection.
Type host: *E. curvula*.
Distribution: South Africa.
Symptoms: Nematodes induced seed galls, which are slightly larger than the normal seeds. Galls are shrivelled and mostly a very dark brown to almost black in colour.

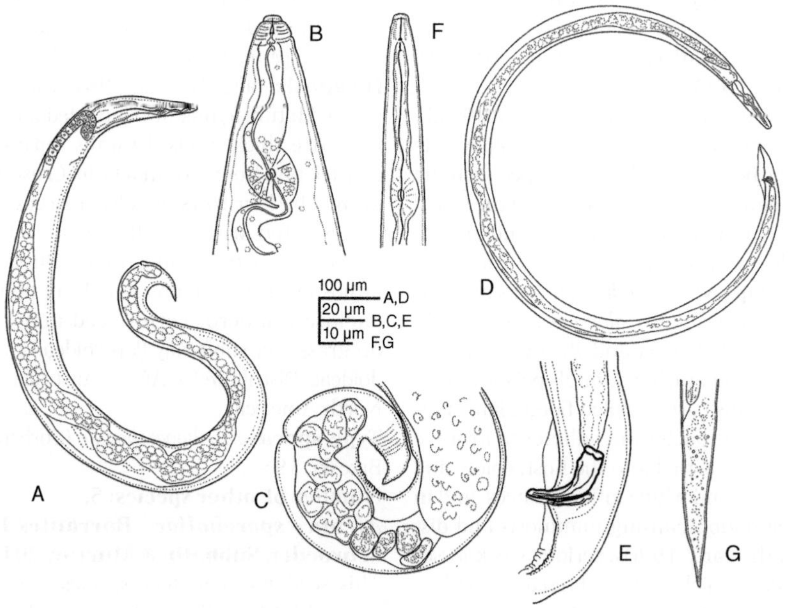

Fig. 14.13. *Afrina sporoboliae*. A: Female entire body; B: female anterior region; C: female posterior region; D: male entire body; E: spicules; F: second-stage juvenile anterior region; G: second-stage juvenile tail region. (After Barrantes-Infante *et al*., 2018.)

14.12.3. Genus *Cotylenchus* Álvarez-Ortega & Subbotin, 2024

This genus was erected by Álvarez-Ortega and Subbotin (2024) for nematodes parasitizing Western sword fern growing in rain forests in Washington State, USA.

Diagnosis: (after Álvarez-Ortega and Subbotin, 2024). Body short, slender, straight to slightly arcuate upon relaxation. Lateral field with six incisures. Lip region low, continuous with body contour. Stylet robust, 12.0-13.5 µm, knobs distinctly rounded with anterior surface slightly sloped posteriorly. Corpus nearly cylindrical; median bulb oval, distinct, with well-developed valve; basal pharyngeal bulb pyriform to elongate, abutting intestine. Post-vulval uterine sac present. Spermatheca oblong, with or without rounded sperms. Ovary outstretched, oocytes in single row. Tail elongate-conical, terminus pointed. Bursa well developed, extending less than half length of the tail.

Type species and only species: *Cotylenchus cleo* gen. n. sp. n.

Cotylenchus cleo **Álvarez-Ortega & Subbotin, 2024 (Fig. 14.14)**

Measurements: After Álvarez-Ortega and Subbotin (2024).

Female: L = 1.03-1.16 mm, a = 51.9-61.4; b = 7.3-8.5; c = 12.2-14.2; c′ = 6.1-7.3; V = 73.9-78.6; stylet = 12.2-13.6 µm; tail = 79.7-94.8 µm.

Male: L = 1.07-1.21 mm, a = 55.3-64.3; b = 7.2-9.3; c = 12.1-14.3; c′ = 5.0-6.4; stylet = 12.3-13.1 µm, spicules = 20.2-23.2 µm.

Description: Body cylindrical, long, slender, slightly ventrally arcuate or straight when relaxed. Lateral field with six incisures. Stylet robust, knobs large. Median bulb oval, muscular, with distinct valve. Basal pharyngeal bulb pyriform to elongated and abutting intestine. Crustaformeria consisting of four rows of four cells, forming a quadricolumella. Tail long conical, tapering to a pointed terminus. Bursa with crenate margin, well developed, starting anterior to cloacal opening and extends less than half of the tail.

Type and only host: Western sword fern, *Polystichum munitum*.

Symptoms and pathogenicity: Necrotic and chlorotic symptoms characterized by yellowing, light to brown interveinal areas were observed in diseased leaves of fern (Fig. 14.7D) Leaf symptoms in fern caused by the new species were similar to those induced by other foliar plant-parasitic nematodes, *Aphelenchoides fragariae* and *Aphelenchoides ritzemabosi*.

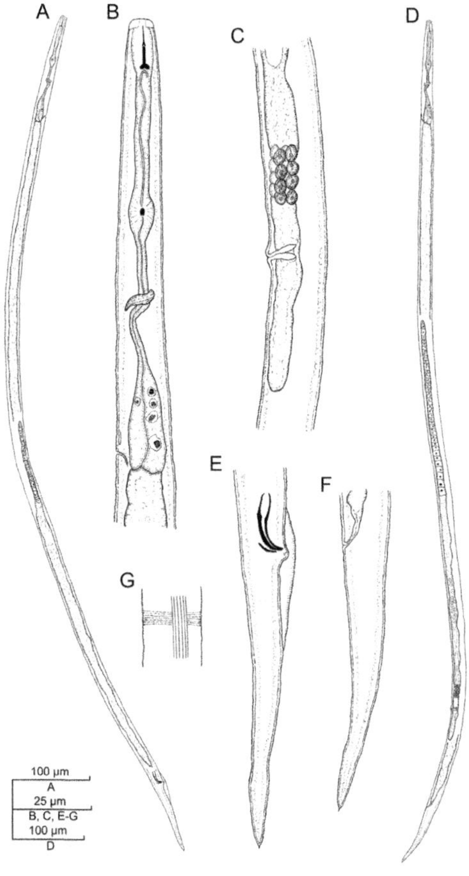

Fig. 14.14. *Cotylenchus cleo*. A: Male entire body; B: female anterior region; C: vulval region and post-vulval uterine sac; D: female entire body; E: male posterior body region; F: female posterior body region; G: lateral field. (After Álvarez-Ortega and Subbotin, 2024.)

Distribution: USA (Washington State).

14.12.4. Genus *Ditylenchoides* Subbotin & Ryss 2024

Mycophagous species including those associated with insects and previously considered in the genus *Ditylenchus* were transferred by Subbotin and Ryss (2024) to new genus *Ditylenchoides*. Adults of the genus *Ditylenchoides* differ from those of *Ditylenchus* in body length, number of incisures in lateral field, tail terminus shape and types of plant symptoms induced.

Diagnosis: (after Subbotin and Ryss, 2024). Female body 0.4-1.9 mm, almost straight or slightly curved ventrally when relaxed. Lip region continuous, high, cephalic framework moderately sclerotized. Median bulb muscular, with indistinct or strongly refractive valve. Basal bulb offset from intestine and compact or irregular, not overlapping intestine or overlapping up to 0.5 body diameter. Secretory-excretory pore in the middle part of basal bulb, posterior to isthmus. Lateral field mostly with six incisures, rarely four or seven incisures. Ovary straight, without distinct flexures, its posterior part with one row of oocytes. Spermatheca elongate, with small sperm cells with nuclei occupying most of cells. Crustaformeria in the form of quadricolumella, four row of four cells each. Vulva is a transverse slit, vagina perpendicular to ventral body wall. Post-vulval uterine sac 0.5-3.1 times longer than the vulval body diameter. Female tail conical or subcylindrical, terminus mostly rounded or rarely pointed. Male bursa leptoderan, adanal to subterminal, not reaching tail tip. Mostly fungal feeders, parasites of underground plant parts. Some species are associated with insects. Infective stages are juveniles and adults. Distributed in temperate and tropical regions.

Type species: *D. destructor* (Thorne, 1945) Subbotin & Ryss 2024

= *Ditylenchus destructor* Thorne, 1945.

Number of other species: More than 17.

***Ditylenchoides destructor* (Thorne, 1945) Subbotin & Ryss 2024 – the potato rot nematode (Fig. 14.15)**

Before *D. destructor* was described in 1945 it was regarded as a race of *D. dipsaci*. Potato rot nematode is of an economic importance and causes great losses in potato, sweet potato, flower bulbs and tubes. It is on the list of quarantine organisms of many countries and organizations (Esser, 1985).

Measurements: After Hooper (1973).

Female: L = 0.69-1.89 mm, a = 18-49, b = 4-12, c = 9-30, V = 73-90, stylet = 10-15 μm.

Male: L = 0.63-1.35 mm, a = 24-50, b = 4-11, c = 11-21, stylet = 10-12 μm, spicules = 24-27 μm, gubernaculum = 9-12 μm.

Description: Female body with slight ventral curvature when killed by heat. Labial region low and flattened, slightly set off from body, hexagonal to squarish in outline, medial lip large, subdivided into submedial lobes by a small bulging area. Lateral fields with six incisures throughout most of body length, gradually reducing to two towards extremities. Glandular pharyngeal bulb overlapping intestine on dorsal side for half to one body width. Vulva distinct, anterior ovary outstretched sometimes reaching pharyngeal region, oocytes in a double row in anterior part but a single row nearer uterus. Post-vulval sac extending about three-quarters of distance to anus. Tail conoid with slight ventral curvature with a narrowly rounded terminus.

Type host: Potato, *Solanum tuberosum*.

Other hosts: Economically important crops are *Ipomoea batatas*, *Iris* spp., *Tulipa* spp., *Dahlia* spp., *Gladiolus* spp., *Rheum rhabarbarum*, *Trifolium* spp., *Zea mays* and *Daucus carota*. The known host range comprises more than 100 species of plants from a wide variety of families. *D. destructor* is also capable of reproducing on the mycelium of many fungi.

Symptoms and pathogenicity: The nematode attacks the subterranean parts of plants (tubers, stolons, bulbs, rhizomes and roots), but may occasionally invade above-ground parts, mainly the base of stem (Fig. 14.6A, B). On heavily infected tubers, there are typically slightly sunken areas with cracked and wrinkled skin which is detached in places from the underlying flesh. The flesh has a dry and mealy appearance, varying in colour from light to dark brown or black.

Distribution: The species is recorded from many countries from Europe, North and South America and parts of Asia and Australia (presently considered as eradicated), mainly from temperate regions.

***Ditylenchoides africanus* (Wendt, Swart, Vrain & Webster, 1995) Subbotin & Ryss 2024 – the peanut pod nematode**

The peanut pod nematode was first reported as *D. destructor* from infested peanut fields in South Africa in 1987. Populations of this nematode were tested on South African potato cultivars but no damage was done to the potato tubers and potato was a poor host. Consequently, the populations parasitizing peanut were considered a distinct race of *D. destructor* with a limited host range (De Waele *et al.*, 1989, 1991).

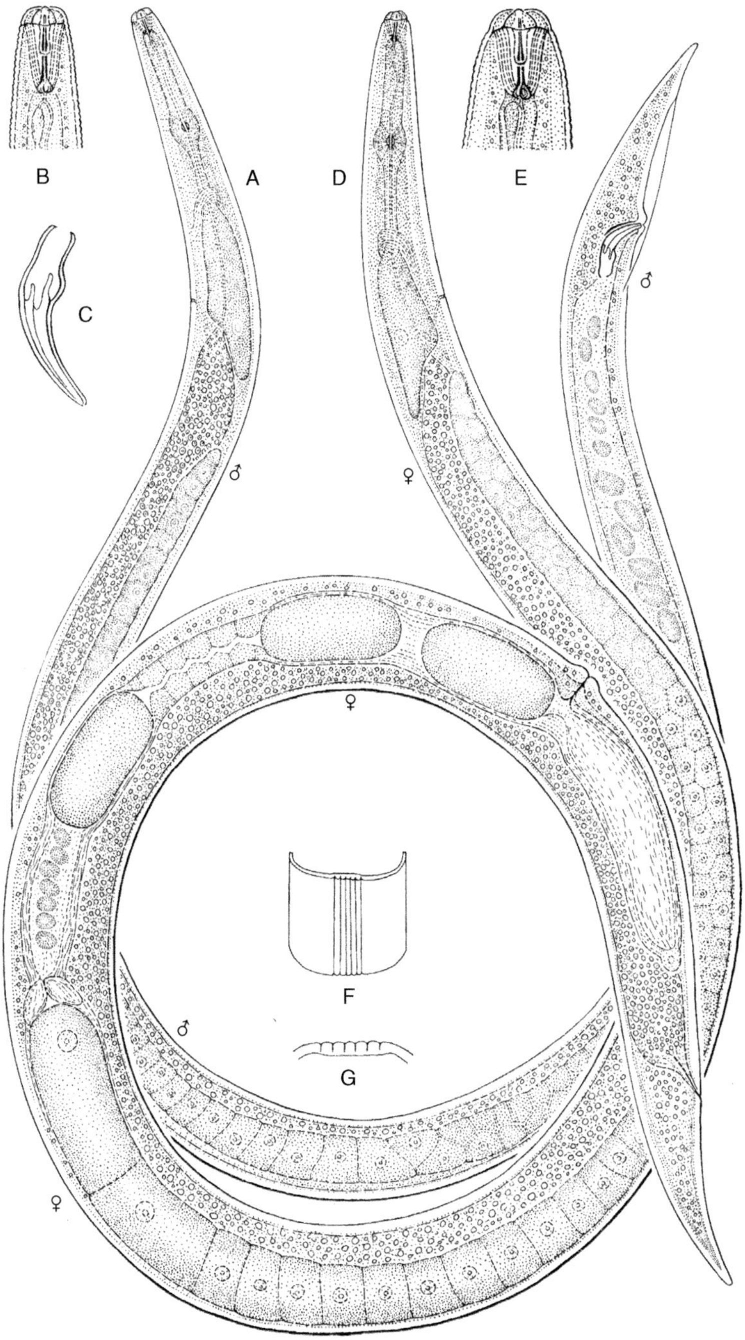

Fig. 14.15. *Ditylenchoides destructor*. A: Male; B: head of male; C: spicule; D: female; E: head of adult female; F: section of cuticle at mid-body showing six incisures in lateral field; G: cross section of lateral field. (After Thorne, 1945.)

After analysis of rDNA, it was established as a separate species (Wendt *et al.*, 1995). *D. africanus* has been found in all the major peanut production areas of South Africa and can destroy 40-60% of the seeds in heavily infested fields.

Measurements: After Wendt *et al.* (1995).

Female: L = 0.7-1.14 mm; a = 24.2-40.4; b = 7.1-11.8; c = 8.8-16.9; stylet = 8-10 µm; V = 77-81.

Male: L = 0.8-1.0 mm; a = 31-42.4; b = 7.4-10.1; c = 13-15.4; stylet = 8-9.5 µm; spicules = 17-21 µm; gubernaculum = 6-8 µm.

Description: Female head flattened, hexagonal in outline, medial lips large each indented to form two submedial lobes. Stylet delicate, knobs distinct, separated, sloping backwards (Fig. 14.16A). Lateral field with six lines, becoming subdivided to a maximum of 15 lines (Fig. 14.16E). Post-vulval uterine sac 50-143 µm. Tail elongate-conoid, tapering in posterior one-third to a finely rounded terminus. Bursa covering 48-66% of tail length.

Type host: Peanut, *Arachis hypogaea*.

Other hosts: Survives on a variety of crops without causing damage.

Symptoms and pathogenicity: Soon after the peg enters the soil and pod formation is initiated, the nematodes enter the plant tissues at the base of the pod near the point of connection with the peg. The first symptom is the appearance of dark brown tissues with a corky appearance at the pod base. The pod may break off during harvesting. In the advanced stage of infection, longitudinal stripes of black or brown tissues appear on the side of the pods and expand into large dark areas. Infected seeds are shrunken. *D. africanus* can survive in stored seeds, which may be symptomless (De Waele *et al.*, 1997).

Distribution: South Africa and other southern African countries.

***Ditylenchoides myceliophagus* (Goodey, 1958) Subbotin & Ryss 2024 – the mushroom spawn nematode (Fig. 14.17)**

D. myceliophagus has been reported from most mushroom growing countries throughout the world, predominantly in temperate areas. It is considered the most voracious and damaging nematode for mushroom cultures.

Measurements: After Goodey (1958), Hesling (1974) and Brzeski (1991).

Considerable variation in measurements of juveniles and adults occurs and is dependent on their age, host, nutrition and environment.

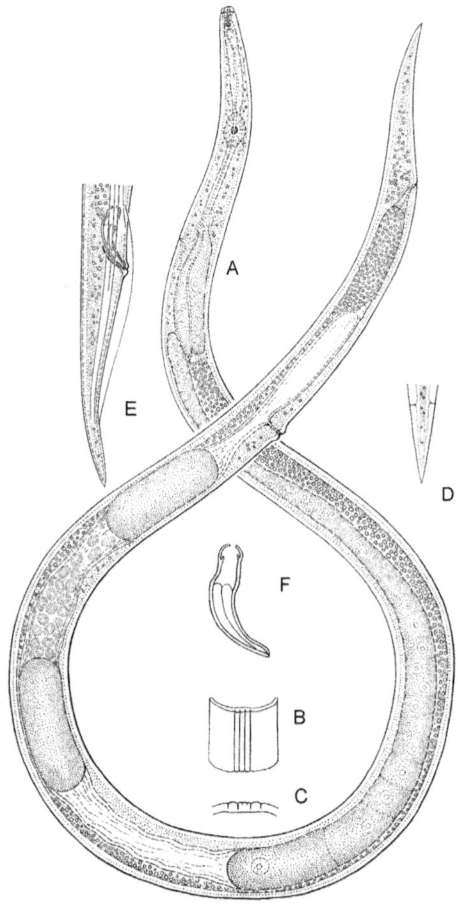

Fig. 14.16. *Ditylenchus dipsaci*. A: Female; B: section of cuticle at mid-body showing four incisures; C: cross section of lateral field; D: tail; E: posterior portion of male; F: spicules. (After Thorne, 1945.)

Female: L = 0.54-1.4 mm, a = 23-50, b = 5.4-11.7, c = 9.8-20.5, c′ = 3.1-6.7, V = 73-90, stylet = 6.5-8.5 µm.

Male: L = 0.50-0.88 mm, a = 21.4-46, b = 4.5-8.7, c = 8.3-15.3, c′ = 3.9-6.2, stylet = 7-9 µm.

Description: Relaxed specimens straight or ventrally arcuate. Labial region flattened and cap-like, not offset, without visible striations, hexagonal in outline, medial lips large, each indented to form two submedial lobes. Lateral field with six equally spaced incisures, in some species becoming subdivided to a maximum of 12 lines. Stylet thin and

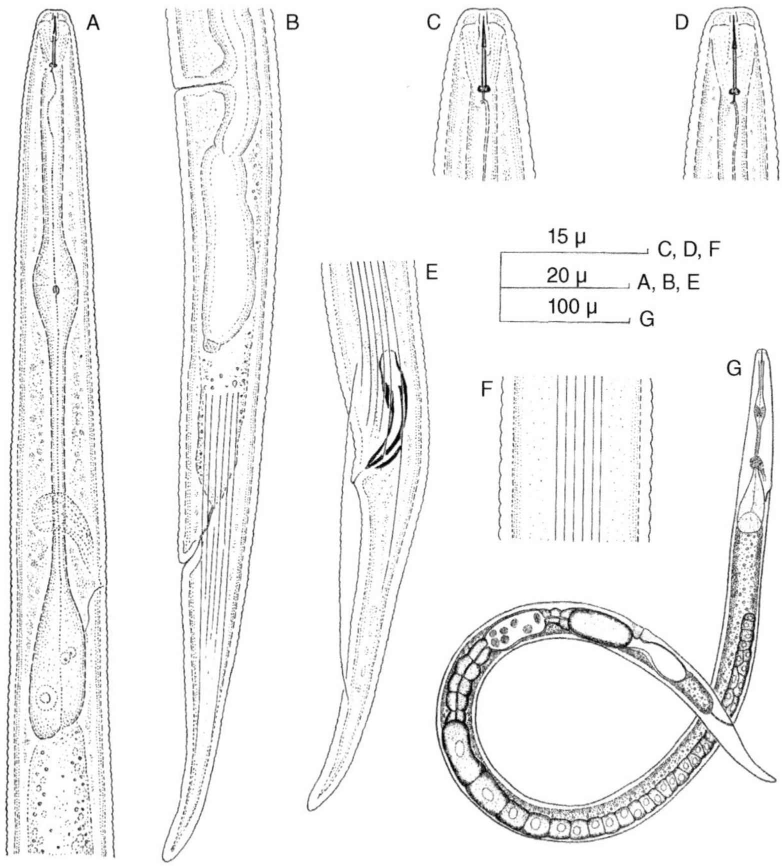

Fig. 14.17. *Ditylenchoides myceliophagus*. A: Pharyngeal region of female; B: posterior portion of female; C: head end of female; D: head end of male; E: tail end of male; F: lateral field: G. female. (A-F: Courtesy of M.R. Siddiqi; G: after Goodey, 1958 and Hooper 1972.)

delicate, cone about one-third of stylet length, knobs rounded. Median bulb fusiform and muscular; basal, glandular bulb broadly spatulate; basal portion slightly overlapping intestine, usually dorsally, but sometimes ventrally; extent of overlap variable. Post-vulval sac present, 28-50 µm long, extending to about halfway between vulva and anus. Bursa extending posteriad from just anterior to base of retracted spicules to about three-quarters of tail length.

Type host: Cultivated mushroom *Agaricus bisporus*.

Other hosts: *D. myceliophagus* is a mycophagous nematode and there appears to be no accepted evidence that it can feed on higher plants, although successful culturing on callus tissue has been reported. The nematode has been cultured on a wide range of other fungi, including saprophytic, plant-parasitic, predaceous and animal pathogenic forms (Sturhan and Brzeski, 1991).

Symptoms and pathogenicity: *D. myceliophagus* pierces a fungal cell with its stylet and sucks out the contents. Mushroom or other fungus mycelium may appear unaffected until a critical population level is reached. The growth then stops and thereafter the mycelium degenerates rapidly. Most of the mycelium disappears and the compost in the affected area becomes sunken, soggy and foul-smelling (Hesling, 1974).

14.12.5. Genus *Ditylenchus* Filipjev, 1936

Since Filipjev proposed *Ditylenchus*, this genus has remained one of the most problematic in systematics. The systematic position of the genus within the Tylenchida has changed several times and it has been treated differently by recent authors. Fortuner (1982), Fortuner and Maggenti (1987) and Sturhan and Brzeski (1991) considered *Nothotylenchus*, *Diptenchus Safianema* and *Orrina* as synonyms of *Ditylenchus*, whereas Siddiqi (2000) kept these genera valid. Subbotin and Ryss (2024) proposed a narrower definition of *Ditylenchus* than that used in previous classifications of Anguinoidea. This genus included only plant-parasitic species.

Diagnosis: (after Subbotin and Ryss, 2024). Female body 0.9-2.2 mm, almost straight or slightly curved ventrally, mature adults slender. Lip region continuous, high, cephalic framework moderately sclerotized. Median bulb muscular, with strongly refractive valve. Basal bulb with 0.5 or less overlapping intestine, dorsal and subventral gland nuclei anterior to pharyngeal-intestinal junction. Secretory-excretory pore at posterior half of isthmus to anterior conical or middle part of basal bulb. Lateral field with four or rarely six incisures. Ovary straight, without distinct flexures, its posterior part with one row of oocytes. Spermatheca elongate, with small sperm cells and nuclei occupying most of cells. Crustaformeria in a form of quadricolumella, four row of four cells each. Vulva is a transverse slit, vagina perpendicular to ventral body wall. Post-vulval uterine sac more than three times longer than vulval body diameter. Female tail conical with pointed to mucronate terminus. Male bursa leptoderan, adanal to subterminal, not reaching tail tip. Testis outstretched. Parasites of higher plants inducing swellings and deformations of plant tissues. Infective stages are juveniles and adults. Distributed in temperate regions.

Type species: *D. dipsaci* (Kuhn, 1957) Filipjev, 1936.

Number of other species: More than 10.

***Ditylenchus dipsaci* (Kuhn, 1857) Filipjev, 1936 – the stem and bulb nematode (Fig. 14.16)**

D. dipsaci is among the phytonematodes of greatest economic significance and is on the list of quarantine organisms of many countries around the world. This nematode is distributed worldwide, especially in temperate regions. It is known to attack over 450 different plant species and cause serious problems on onions, garlic, cereals, legumes, strawberries, sugar beet, carrot and many other agricultural and vegetable crops, as well as many horticultural plants, especially phlox, flower bulbs, such as narcissus, hyacinth and tulip. It occurs in several biological races, some of which have a limited host range (Hooper, 1972). According to present morphological, karyological and molecular analyses, *D. dipsaci* has to be considered a species complex. Several species were described: *D. weischeri* from parasitize *Cirsium* spp. and *D. gigas* from *Vicia faba*, *Ditylenchus oncogenus* from *Sonchus bulbosus* and *Ditylenchus paraoncogenus* from *Gundelia tournefortii*. *D. gigas*, formerly considered as a giant race of the stem and bulb nematode, was designated and described (Vovlas *et al.*, 2011). Several highly specialized races and populations parasitizing *Plantago maritima*, *Taraxacum officinale*, *Crepis praemorsa*, *Pilosella* spp. and other plants are still grouped under the nominal name *D. dipsaci sensu lato*, although they actually represent separate species.

Measurements: After Sturhan and Brzeski (1991).

Female: L = 1.1-1.7 mm; a = 30.2-64; b = 6.0-8.8; c = 13.5-19.5; V = 79-86; stylet = 9-12 μm.

Male: L = 1.0-1.5 mm; a = 35.4-53.5; b = 5.5-7.8; c = 12-17.3; stylet = 9-12 μm; spicules = 20-28 μm; gubernaculum = 8-11 μm.

Description: Female body almost straight when killed by heat. Labial region low, unstriated, slightly flattened, barely set off from the body, squarish in outline, medial lips large, each subdivided into two submedial lobes by a small, medial bulging area (Fig. 14.16D). Lateral field with four lines (Fig. 14.16H). Basal bulb offset or overlapping intestine. Vulva distinct, anterior ovary outstretched with oocytes usually in a single, occasionally double, row that sometimes reaches the pharyngeal region. Post-vulval part of uterine sac about half of vulva-anus distance long or slightly more. Tail of both sexes conical, always pointed.

Type host: Fuller's teasel, *Dipsacus fullonum*.

Other hosts: More than 500 plant species from over 40 angiosperm families are known as hosts for this nematode.

Symptoms and pathogenicity: *D. dipsaci* lives mostly as an endoparasite in aerial parts of plants (stems, leaves and flowers) but also attacks bulbs, tubers and rhizomes. Common symptoms of infestation are swelling, distortion, discoloration and stunting of above-ground plant parts and necrosis of bulbs and tubers (Fig. 14.4). The economic thresholds of *D. dipsaci* are mostly very low and even population densities of only 20 nematodes per kg of soil may lead to significant crop losses.

Distribution: Cosmopolitan.

***Ditylenchus angustus* (Butler, 1913) Filipjev, 1936 – the rice stem nematode (Fig. 14.18)**

The rice stem nematode was first described by E.J. Butler causing 'ufra' disease of rice in the territory of modern Bangladesh. *D. angustus* is known to occur in rice-growing regions in Asia and Africa, where it is considered as an important pest.

Measurements: After Seshadri and Dasgupta (1975).

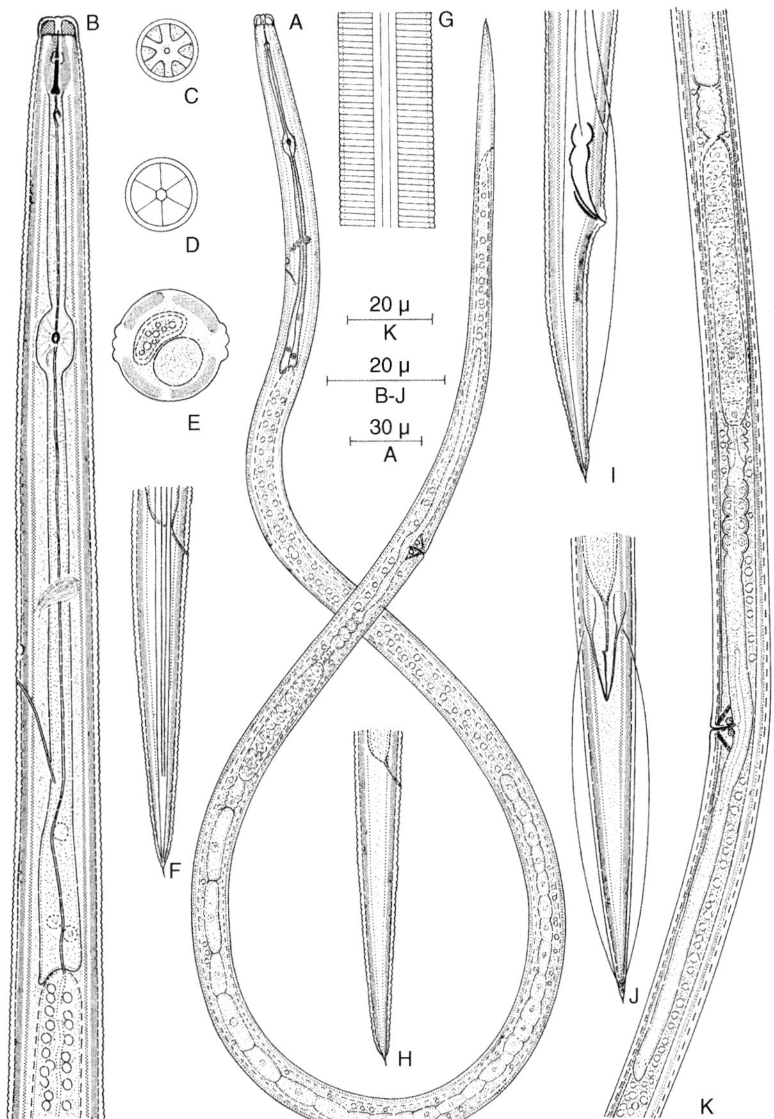

Fig. 14.18. *Ditylenchus angustus*. A: Female, entire; B: female pharyngeal region; C: *en face* view; D: labial framework; E: female, cross section at mid-body; F: female tail; G: lateral field at mid-body; H: juvenile tail; I: male tail, lateral view; J: male tail, ventral view; K: female vulval region. (After Seshadri and Dasgupta, 1975.)

Female: L = 0.7-1.2 mm; a = 47-62; b = 6-9; c = 15-24; stylet = 9 = 11 µm; V = 70-80.
Male: L = 0.6-1.2 mm; a = 36-62; b = 7-8; c = 17-26; stylet = 9-10 µm; spicules = 16-21 µm; gubernaculum = 6-9 µm.

Description: Female body slender, almost straight to slightly arcuate ventrally when relaxed. Lateral field with four incisures, outer incisures more distinct than inner ones. Median pharyngeal bulb oval, with a distinct valvular apparatus anterior to centre. Vulva a transverse slit, vaginal tube somewhat oblique, reaching more than halfway across body. Spermatheca very long, vagina somewhat oblique, more than half of vulval body-width long. Post-vulval uterine sac 2-2.5 vulval body width or 50-67% of vulva-anus distance long. Tail conoid, tapering to a sharply pointed terminus resembling a mucro. Male morphology similar to female. Bursa narrow extending almost to tail tip. Spicules curved ventrally. Gubernaculum short and simple.

Type host: Rice, *Oryza sativa*.

Other hosts: *Oryza alta*, *Oryza cubensis*, *Oryza eichingeri*, *Oryza globerrima*, *Oryza latifolia*, *Oryza meyriana*, *Oryza minuta*, *Oryza nivara*, *Oryza officinalis*, *Oryza perennis*, *Oryza rufipogon* and *Oryza spontanea*. The wild rice *Leersia hexandra* has been recorded as a host in Madagascar and Myanmar, and the graminaceous hosts *Echinochloa colona* and *Sacciolepis interrupta* were reported from rice fields in Vietnam. Propagation of the rice stem nematode is also possible on various fungi (Sturhan and Brzeski, 1991).

Symptoms and pathogenicity: *D. angustus* occurs in the leaves, inflorescences, young seeds and rolled stems of growing plants. The nematodes feed ectoparasitically causing malformation of host tissues. Infected plants are stunted and the leaves often wilt and become chlorotic. Necrotic brown strains may develop on leaves and leaf sheaths. The leaf sheath and the limb above the last internode become distorted and crimped. The peduncle coils and pedicels bear distorted sterile spikelets. Dark brown patches of infected plants can be observed within fields, normally after panicle initiation. Yield loss caused by this disease has been reported up to 10-15% in India, 20-90% in Thailand, 50-100% in Vietnam and 40-60% or occasionally 100% in Bangladesh deepwater, irrigated and lowland rice. *D. angustus* can survive for several months in plant material and seeds (Ibrahim and Perry, 1993).

Distribution: Found in several Asian countries: Myanmar, Bangladesh, India, Malaysia, Thailand and Vietnam and in Madagascar.

***Ditylenchus gigas* Vovlas, Troccoli, Palomares-Rius, De Luca, Liébanas, Landa, Subbotin & Castillo, 2011 – the giant stem nematode (Fig. 14.19)**

The first description of a disease on broad beans caused by the stem nematode, whose adults were considerably larger than *D. dipsaci* adults, was published more than 100 years ago and was named as a giant race of *D. dipsaci*. The giant race causes a severe disease of broad bean in several European and African countries and is generally more damaging than any of the other *D. dipsaci* races as it causes more severe symptoms to field broad beans and produces more infested seed. Several studies have shown that this race differs from the normal race in its biology. Compared with the normal race, the giant race has a limited host range. The normal race of *D. dipsaci sensu stricto* is assumed to have diploid chromosome number 2n = 24, whereas the giant race parasitizing broad beans is tetraploid with 2n = 48-60. Crossing experiments with diploid races of *D. dipsaci sensu stricto* and the giant race showed that F1 hybrids were formed, but they were infertile. Several studies including PCR-RFLP, protein isoelectric focusing, RAPD and ITS-rRNA gene sequence analysis also confirmed that the giant race is different from normal races. These results led to conclusion that the giant race should be considered as a separate species — *D. gigas*.

Measurements: After Vovlas *et al.* (2011).
Female: L = 1.5-1.9 mm; a = 43.0-56.4; b = 7.3-9.3; c = 16.8-27.6; V = 80-83; stylet = 11.5-13.0 µm.
Male: L = 1.3-1.7 mm; a = 34.3-63.0; b = 6.7-10.7; c = 15.7-20.0; stylet = 11.0-12.5 µm; spicules = 23.5-28 µm; gubernaculum = 8.0-11.0 µm.

Description: Labial region flattened, separated from rest of body by a slight constriction, appearing smooth in anterior two-thirds and with a basal annulus in posterior third. Stylet delicate, knobs distinctly sloping backwards. Median bulb oval. Lateral field with four incisures, inner two sometimes faint and indistinct. Post-vulval uterine sac well developed, occupying two vulval body widths. Tail elongate-conoid, tapering posteriorly to a finely rounded terminus. Bursa leptoderan, extending anterior to cloaca

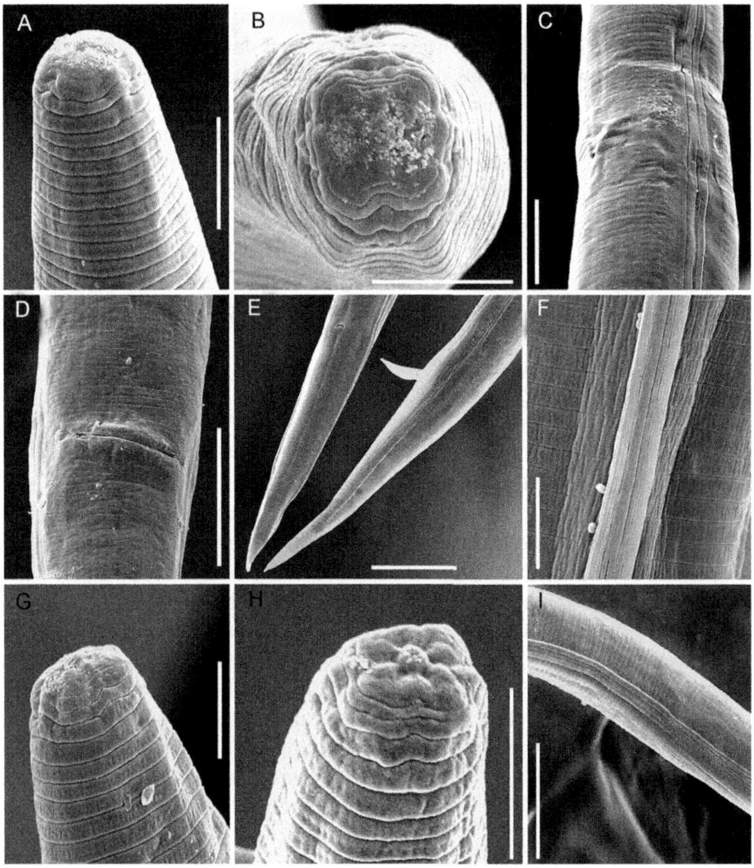

Fig. 14.19. Scanning electron microscope photographs of *Ditylenchus gigas*. A-F: Female; G-I: male. A: Anterior end; B: *en face* view; C: excretory pore; D: vulva in ventral view; E: female (left) and male (right) tail (anus opening arrowed); F, I: lateral field, showing four incisures; G, H: anterior end in different views. (Scale bars: A, B, F-H = 5 μm; C = 10 μm; D, E, I = 20 μm.) (After Vovlas *et al.*, 2011.)

for about 1.5 anal body widths and covering 72-76% of tail length. Spicules arcuate ventrad, slightly cephalated. Gubernaculum simple, slightly thickened in the central part.

Type host: Broad beans, *V. faba*.

Other hosts: *Lamium* spp., *Ranunculus arvensis*, *Convolvulus arvensis* and *Avena sterilis*.

Symptoms and pathogenicity: Stems infested by this nematode are often stunted, twisted and blistered and darker brown or black in colour. Infestations in the pod-bearing region are more common and leaf petioles and pods may be distorted (Fig. 14.5). Extensive, brown necrotic zones were observed on the surface of stem towards the end of the crop season. Secondary rotting at the stem base sometimes causes the whole plant to collapse. The lowest pods on plants tend to become infested. Ripe infested pods contain many dried, resistant, J4 attached to the seeds, especially in the slit in the hilum. Several thousand live nematodes can be recovered from individual seeds, even several years after harvest (Hooper, 1971; Vovlas *et al.*, 2011).

Distribution: European and North African countries, mainly bordering the Mediterranean Sea.

14.12.6. Genus *Halenchus* N.A. Cobb *in* M.V. Cobb, 1933

Diagnosis: (after Siddiqi, 2000). About 0.5-1.5 mm long, slender, vermiform. Cephalic region continuous or slightly offset, smoothly rounded. Stylet well developed, 14-25 μm long. Corpus cylindrical, non-muscular, with a basal non-valvate fusiform swelling. Dorsal pharyngeal gland large, extends over intestine dorsally or subdorsally. Post-vulval uterine sac about one body-width long in type species. Quadricolumella present. Tail elongate-conoid. Tip finely rounded and hooked ventrally or laterally. Bursa adanal, extending to about middle of tail. Obligate endoparasites of marine algae (seaweeds), inciting and inhabiting galls.

Type species: *Halenchus fucicola* (de Man & Barton *in* de Man, 1892) N.A. Cobb *in* M.V. Cobb, 1933.

Other species: 2.

***Halenchus fucicola* (de Man & Barton *in* de Man, 1892) N.A. Cobb *in* M.V. Cobb, 1933 (Fig. 14.20)**

The type species was found by Barton (1892) in galls of *Ascophyllum nodosum*, a brown alga, on the east and west coast of Scotland.

Measurements: After Coles (1958).

Female: L = 0.98-1.4 mm; a = 32.5-49.5; c = 10.2-15; V = 59-68; stylet = 15-18 μm.

Male: L = 0.8-1.0 mm; a= 34.6-51.1; c = 10.9-13.1; stylet = 15-16 μm; spicules = 20-24 μm.

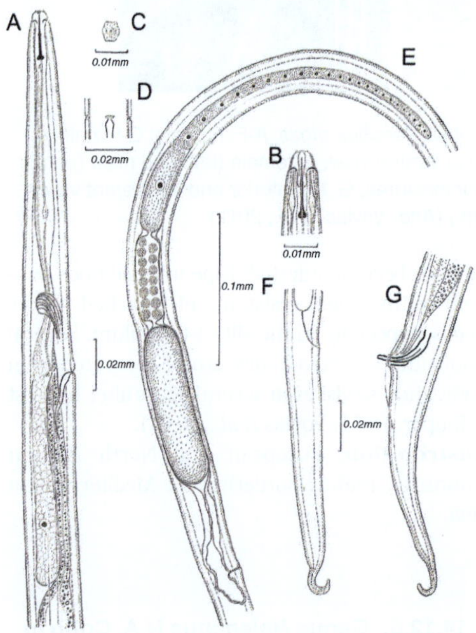

Fig. 14.20. *Halenchus fucicola*. A. Anterior region, B: head dorsal view; C: *en face* view; D: ventral view of excretory pore and deirids; E: gonad region; F: female tail; G: male tail. (After Goodey, 1963.)

Description: Body slender. Lip region set off. Stylet well developed. Excretory pore and duct well defined and heavily cuticularized. Dorsal pharyngeal glands extend over intestine. Ovary straight. Tip finely rounded and hooked ventrally or laterally. Bursa extending to about midway between cloaca and tail tip.

Type host: *A. nodosum.*

Symptoms: Nematodes induce falls on a brown alga (Fig. 14.7B).

Distribution: East and west coast of Scotland.

14.12.7. Genus *Litylenchus* Zhao, Davies, Alexander & Riley, 2011

Diagnosis: (after Zhao *et al.*, 2011). Slender to semi-obese, cylindrical nematodes, barely curved. Stylet short, 9-12 μm, robust, with rounded knobs. Pharynx with non-muscular fusiform median bulb. Female with monoprodelphic gonad with quadricolumella. Bursa arising 1-2 cloacal body diameter anterior to cloacal aperture, extending nearly to tail tip. Tail medium, conoid, tip shape variable. Induced chlorotic and necrotic spots on leaves.

Type species: *Litylenchus coprosma* Zhao, Davies, Alexander & Riley, 2011.

Other species: 1.

***Litylenchus crenatae* Kanzaki, Ichihara, Aikawa, Ekino & Masuya, 2018 – the beech leaf nematode (Fig. 14.21)**

Litylenchus crenatae subsp. *crenatae* was isolated from a leaf of *Fagus crenata* from Japan. A disease of beech trees (*Fagus* spp.) called beech leaf disease (BLD) induced by *L. crenatae* subsp. *mccannii* has increasingly been observed in forest areas in Eastern USA and Canada and is raising serious concerns. It has been suggested that *L. crenatae* may not be the sole cause of BLD, but rather that the disease is a complex, comprising one or more microbial components.

Measurements: After Kanzaki *et al.* (2019).

Female (mature): L = 0.76-0.87 mm, a = 30-41; b = 6.1-7.6; c = 21.8-28.1; c′ = 2.5-3.3; V = 79-83; stylet = 9.9-11.3 μm.

Male (mature): L = 0.77-0.84 mm; a = 37-44; b = 5.9-7.3; c = 21.4-30.3; c′ = 2.4-3.3; stylet = 9.9-11.3 μm, spicules = 16.7-20.2 μm.

Description: Body cylindrical, vermiform to semi-obese, but not clearly obese, smoothly

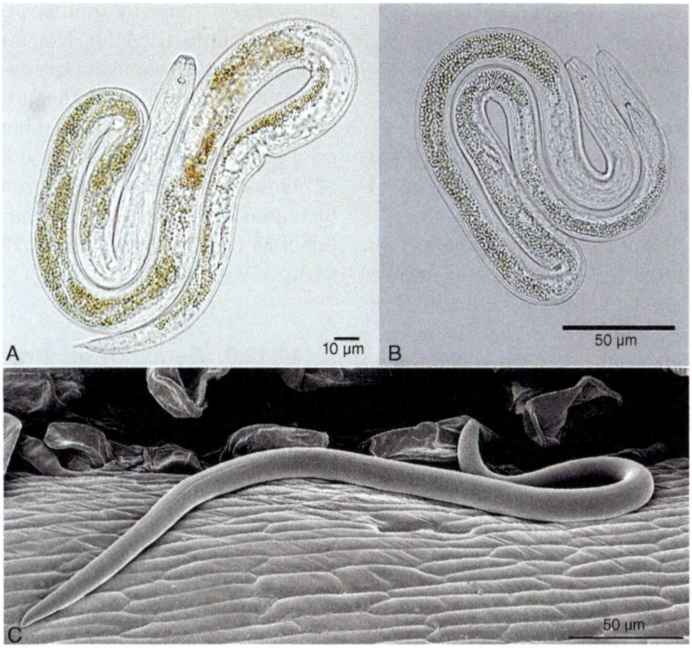

Fig. 14.21. *Litylenchus crenatae* A: Mature Female; B: male; C: young female. (After Handoo *et al.*, 2020.)

ventrally arcuate when killed by heat. Stylet prominent with rounded basal knobs. Pharynx consists of cylindrical procorpus, distinctive and weakly muscular median bulb with small metacorpal valve at mid-bulb length. Six or more incisures in lateral field. Post-uterine sac well developed, with thin wall. Tail short, broad. Distal end of tail abruptly narrowing to form a very short, conoid and bluntly pointed tip.

Type and other hosts: American beech (*Fagus grandifolia*), European (*Fagus sylvatica*), oriental (*Fagus orientale*) and Chinese beech (*Fagus engleriana*).

Symptoms and pathogenicity: Early symptoms of BLD include dark-green striped bands between lateral veins of leaves and reduced leaf size (Fig. 14.7C). Banded areas usually become leathery-like and leaf curling is also observed. As symptoms progress, aborted buds, reduced leaf production and premature leaf drop. Affected buds are aborted and no new leaves are produced. Limited mortality has been documented on large trees after 6 years, but swifter mortality for young sapling (Ewing *et al.*, 2018).

Distribution: *L. crenatae crenatae* reported in Japan and *L. crenatae mccannii* was detected in the Great Lakes Region (Canada and USA) and along the Atlantic Coast of USA (Fitza *et al.*, 2024).

14.12.8. Genus *Mesoanguina* Chizhov & Subbotin, 1985

Diagnosis: (modified after Chizhov and Subbotin, 1985). During each growing season, usually two generations of nematodes develop in each gall. Adults of the first generation large, after heat relaxation crescentic or spirally coiled. Adults of second generation smaller, not curved remarkably. Ovary in female of first generation with two or three flexures. Oocytes in several rows. Crustaformeria consisting of 32-60 cells in four to five rows, 8-12 eggs in each, and may contain up to 16 synchronous eggs. Ovary in female of second generation outstretched or with only one flexure. Oocytes in zone of multiplication generally in single file. Crustaformeria shorter than in female of first generation and may contain up to four synchronous eggs. Infective juvenile of third stage. Induce unpigmented galls with well-expressed internal cavity on above-ground parts of dicotyledons, mostly on species of the family Asteraceae (Compositae)

(Fig. 14.8), but also on Boraginaceae and Plantaginaceae.

Type species: *Mesoanguina millefolii* (Low, 1874) Chizhov & Subbotin, 1985 (Fig. 14.3D).

Number of other species: 6.

***Mesoanguina picridis* (Kirjanova, 1944) Chizhov & Subbotin, 1985 – the knawel cyst nematode**

Mesoanguina picridis is the only plant pathogen, along with the fungus *Puccinia acroptili*, known to be effective as a biological control agent against Russian knapweed, *Acroptilon repens*. *A. repens* is a persistent herbaceous perennial. The weed is toxic to horses and difficult to control with herbicides. The nematode is endemic to Asia where its use for biocontrol was pioneered (Ivanova, 1966) *M. picridis* is not restricted to Russian knapweed and some members of the Cynareae tribe of the Asteraceae family are also hosts.

Measurements: Dimensions of *M. picridis* from different hosts varied considerably. After Kirjanova and Ivanova (1968) and Watson (1986a).

Female: L = 1.5-2.1 mm; a = 22-28; b = 6.0-12.5; c = 14.0-25.0; V = 81-84; stylet = 12 μm.

Male: L = 1.3-1.8 mm; a = 36-40; b = 6.7-7.0; c = 14.0-18.7; stylet = 12 μm; spicules = 40-43 μm; gubernaculum = 10-15 μm.

Description: Head region with three annuli. Lateral field variable within a single specimen, in middle-body region smooth or with six or more up to 20 incisures. Procorpus slightly swollen posteriad with a constriction before spherical median bulb. Ovary with two flexures. Post-vulval uterine branch long, with reduced ovary consisting of one to two cells. Tail conoid with pointed terminus.

Type host: Russian knapweed *A. repens*.

Other hosts: It has a wide range of host plants within the tribe Cynareae of the Asteracea, including *Centaurea squarrosa, Chartolepis biebersteinii, Cousinia* spp. and *Centaurea leucophylta*.

Symptoms: The nematode induces galls on the stems, leaves and root collars causing reduction in plant growth and seed production.

Distribution: Tajikistan, Turkmenistan, Iran, Russia, Armenia and Turkey. Introduced into Canada.

Biological control potential: Ivanova (1966) reported that over 90% of stems of *A. repens* inoculated with crushed gall material (100 g/m^2) were galled, nearly 20% of them died, and there was severe damage to another 30%. *M. picridis* has been little used for biocontrol in North America, but is promising in sites that are irrigated or naturally moist in the spring. The nematode can be used as a classical biocontrol agent or applied as a bioherbicide and large numbers of nematodes can be obtained from field plots or increased 140 to 200-fold in 3 months on shoot tips of *M. repens* in tissue culture (Watson, 1986b; Ou and Watson, 1992).

14.12.9. Genus *Orrina* Brzeski, 1981

Diagnosis: (after Siddiqi, 2000). Body about 1 mm or less long, slender (a = 20-30), straight to ventrally arcuate when relaxed. Lateral fields narrow, with four incisures. Labial region flat, cap-like, slightly offset by narrowing of contour. Corpus cylindroid, muscular median bulb absent. Pharyngeal glands extending over intestine. Ovary outstretched, oocytes in one or two rows. Bursa subterminal. J4 is infective stage.

Type and only species: *Orrina phyllobia* (Thorne, 1934) Brzeski, 1981.

***Orrina phyllobia* (Thorne, 1934) Brzeski, 1981 – the nightshade gall nematode (Fig. 14.22)**

The nightshade gall nematode is of economic interest because of its potential as a biological control agent of the cosmopolitan silverleaf nightshade, which is an economically important perennial weed species and, in particular, troublesome in cotton production and difficult to control (Sturhan and Brzeski, 1991).

Measurements: After Thorne (1961) and Brzeski (1991).

Female: L = 0.59-0.84 mm, a = 20-32; b = 7.4-10.5; c = 11.4-17.6; c′ = 2.9-4.5; V = 78-84; stylet = 9-11 μm.

Male: L = 0.67-0.85 mm; a = 27-34; b = 7.0-9.7; b′ = 4.7-7.5; c = 16.3-19.2; c′ = 2.5-3.1; stylet = 9-10 μm.

Description: Female cuticle finely striated. Lateral field with four incisures. Labial region set off. Stylet short, slightly longer than labial region width. Ovary a single line of oocytes except for a short section of multiplication. Post-vulval uterine branch extending two-thirds distance from vulva to anus. Tail of both sexes conical, tip rounded. Spicules slightly arcuate, somewhat

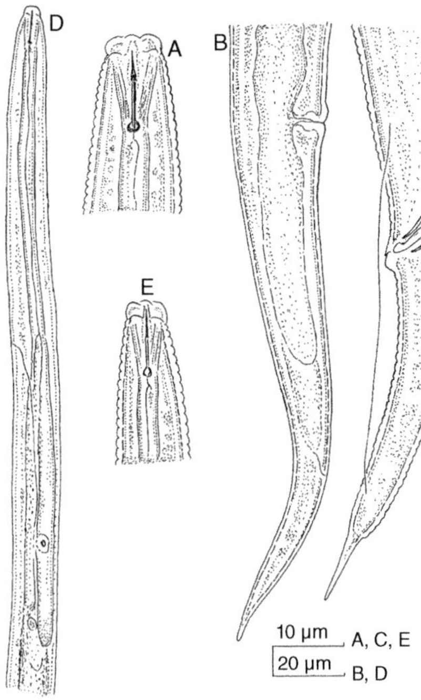

Fig. 14.22. *Orrina phyllobia.* A: Anterior end of female; B: tail of female; C: tail of male; D, E: anterior end of J2. (After Siddiqi, 2000.)

cephalated. Gubernaculum thin, trough-like. Bursa not extending to tail terminus.

Type host: Silverleaf nightshade, *Solanum elaeagnifolium*.

Symptoms and pathogenicity: Visual indications of infection are usually evident after 10-14 days. Galls arise on the leaf surfaces, increase in size, and often engulf the entire leaf surface (Fig. 14.7F). Hundreds of thousands of nematodes can be found in a single gall. Heavily infected plants are severely stunted or even killed.

Distribution: Known from the southwest USA (Arizona and Texas), Mexico and southern India (Sturhan and Brzeski, 1991).

Biological control potential: Its extreme host specificity, the capacity to survive desiccation and the detrimental effects it causes on its host make *O. phyllobia* a promising agent for biological weed control (Robinson *et al.*, 1978). Under field conditions in the USA, and in heavily infested soil, 50% of the infected plants were killed each year while growth and reproduction of the surviving plants were reduced by 50% (Krall, 1991).

14.12.10. Genus *Pterotylenchus* Siddiqi & Lenné, 1984

Diagnosis: (after Siddiqi and Lenné, 1984, and Krall, 1991). Female straight to slightly arcuate. Lateral field with four incisures. Corpus nearly cylindrical, not differentiated into pre- and post-corpus; posterior region slightly swollen, lacking musculature and valve plates. Basal bulb elongate-saccate, with dorsal pharyngeal gland extending over intestine for about one body width. Vulva a long transverse slit flanked and partly covered by large prominent cuticular flaps. Post-vulval uterine sac present. Crustaformeria comprising 32-36 cells in four rows of eight to nine cells each. Spermatheca elongate, empty. Ovary outstretched, with oocytes in one or two rows. Tail elongate-conoid with a pointed tip.

Type and only species: *Pterotylenchus cecidogenus* Siddiqi & Lenné, 1984.

***Pterotylenchus cecidogenus* Siddiqi & Lenné, 1984 (Fig. 14.23)**

The nematode was first detected in 1981 from galls on the stems of a tropical pasture legume, *Desmodium ovalifolium*. The stem galls are formed on nodes and stem divisions and are not easily detectable.

Measurements: After Siddiqi and Lenné (1984).
Female: L = 0.59-0.80 mm; a = 22-35; b = 4.4-5.8; c = 9.6-12.5; V = 80-84; stylet = 8-11 μm.
J2: L = 0.21-0.31 mm; a = 18-26; c = 6.1-7.5; stylet = 6-7.5 μm.
Male: not known.

Description: Female body straight or slightly ventrally arcuate, often angular at vulva. Vulva about half body-width long; partially covered by large flaps composed of the cuticle of about six enlarged annuli. Juvenile similar to female in most details including dorsal glandular lobe and tail shape.

Type host: *D. ovalifolium*.

Symptoms and pathogenicity: The first symptoms associated with *P. cecidogenus* in *Desmodium* are shoot leaves with chlorosis, which later wilt and finally causes necrosis and defoliation. Galls 5-20 mm in diameter induced on stems of *D. ovalifolium* and *D. barbatum*. Young galls are small and light brown; however, they form a suberose area that cracks and dries out, turning the gall a dark, almost black colour. Nematodes

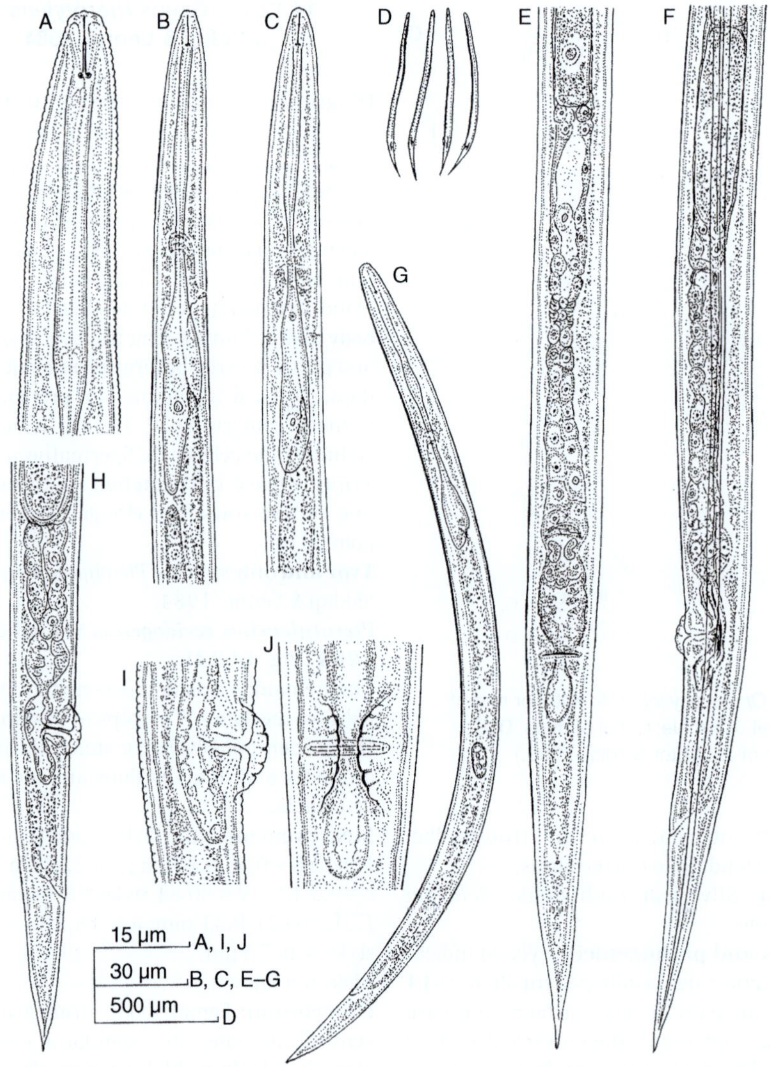

Fig. 14.23. *Pterotylenchus cecidogenus*. A-C: Head of female; D: female; E, F, H: tail of female; G: J2; I, J: vulval region showing lateral vulval flaps. (After Siddiqi and Lenné, 1984.)

more rapidly colonize the wounded stems resulting from grazing or animal trampling.

Distribution: Colombia.

14.12.11. Genus *Subanguina* Paramonov, 1967

Paramonov (1967, 1970) proposed *Subanguina*, with *S. radicicola* as type species and stated that it retained the ancestral type of ontogeny. *S. radicicola* is the only species of the Anguinidae that forms galls on roots of many grass species. Based on similarity in form of the crustaformeria structure, Brzeski (1991) transferred 17 species to this genus. However, molecular phylogenetic analysis using the ITS-rRNA sequence data (Subbotin *et al.*, 2004) supports the narrow concept of *Subanguina* by Paramonov (1970) and Chizhov and Subbotin (1990) and not the broad concept proposed by Brzeski (1991).

Diagnosis: (after Chizhov and Subbotin, 1990). Adult specimens almost straight after heat relaxation. Ovary outstretched or with one to two flexures. Oocytes in the zone of maturation in a single row. Crustaformeria consisting of 48-80 cells in four to five rows, 12-20 cells in each, and may contain up to four synchronous eggs. Infective juveniles belong to second stage. Induce hook-like galls with well-developed internal cavity on roots of grasses. Only one generation developing in gall.

Type and only species: *S. radicicola* (Greeff, 1872) Paramonov, 1967.

Subanguina radicicola **(Greeff, 1872) Paramonov, 1967 – the grass root gall nematode (Fig. 14.24)**

S. radicicola causes characteristic root galls on graminaceous hosts. It damages golf courses sown with *Poa* species, and barley, rye and other cereals. It also occurs on grasses in various regions of the world.

Measurements: After Goodey (1932) and Chizhov and Marjenko (1984).

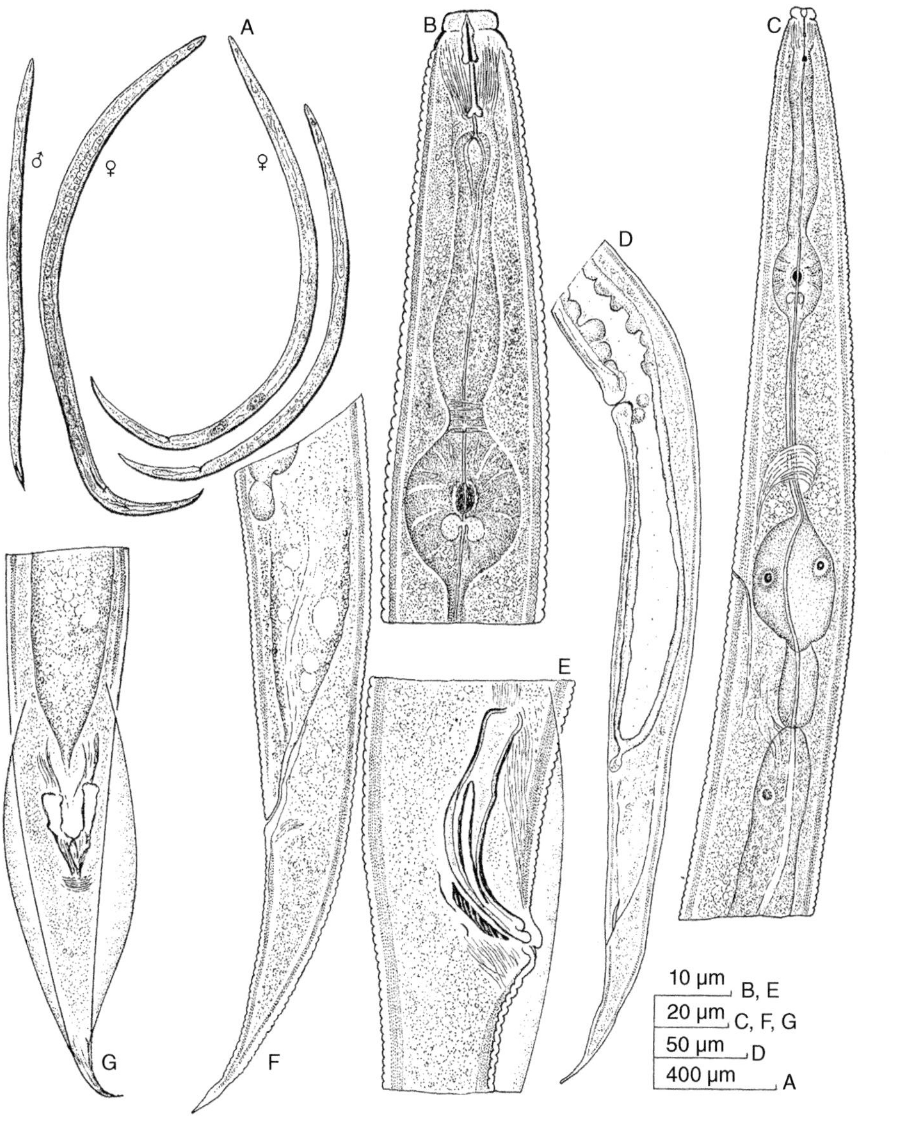

Fig. 14.24. *Subanguina radicicola*. A: Adults; B, C: female pharyngeal region; D, F: female tail end; E: spicular region; G: male tail end. (After Siddiqi, 2000.)

Female: L = 1.11-3.15 mm; a = 26.2-37.6; b = 6.8-13.8; c = 13.5-21.5; V = 75-81; stylet = 12-16 µm.
Male: L = 0.86-2.04 mm; a = 26.0-36.9; b = 6-10.1; c = 13.6-20.8; stylet = 12-16 µm; spicules = 25-31 µm; gubernaculum = 6-11 µm.
J2 infective: L = 390-530 mm; a = 24.7-40.0; b = 3.4-4.1; c = 6.4-7.7; stylet = 10-11 µm.

Description: Female labial region offset. Cuticle finely annulated with 10-12 incisures in lateral field. Procorpus cylindrical, median bulb ovoid. Isthmus short and narrow. Pharyngeal glands as a wide bulb slightly overlapping the intestine dorsally. Ovary simple or once or twice reflexed. In maturation zone, oocytes arranged in one row. Post-vulval uterine branch long, more than half vulva-anus distance. Tail tip sharply pointed. Male testis outstretched, sometimes with one flexure. Bursa not extending to sharply pointed tail tip.

Type host: Meadow grass, *P. annua*.

Other hosts: This species parasitizes roots of grasses belonging to 20 genera and more than 25 species, including rye, wheat, oats and barley.

Symptoms and pathogenicity: The J2 infest root tips of grasses and induce hypertrophy of parenchymatous tissue with subsequent formation of a cavity between the parenchyma and the root central cylinder. Extensive galling can occur throughout the plant roots, galls being slender and twisted in appearance and up to several millimetres long (Fig. 14.3F). Chlorotic patches may be observed on newly established *P. annua* golf courses. Only one generation develops in each gall but several subsequent generations may be observed per growing season.

Races: Five races are presently recognized: barley race; *Elymus* race; Saskatchewan race parasitizing Smith's couch grass; *Poa* race; Rhode Island race parasitizing *Ammophila*; and *Phalaris* race. The *Poa* race is the most widespread in Europe and New Zealand. A New Zealand population heavily infesting *P. annua* did not infect barley or wheat, whereas a Dutch population caused only a few galls on barley roots. The barley race and a Scandinavian population have been reported as pathogenic to cereals in northern Europe. The *Elymus* race attacks *E. arenarius* and may be transferred to barley, rye, wheat and oat (Krall, 1991).

Distribution: Europe, North America and New Zealand.

14.13. References

Álvarez-Ortega, S. and Subbotin, S.A. (2024) *Cotylenchus cleo* gen. n., sp. n., a new plant-parasitic nematode (Tylenchida: Anguinidae) parasitising on leaves of western sword fern plants from rainforests in Washington State, USA. *Nematology* 26, 579–592.

Barrantes-Infante, B.L., Schroeder, B.K., Subbotin, S.A. and Murray, T.D. (2018) *Afrina sporoboliae* sp. n. (Nematoda: Anguinidae) associated with *Sporobolus cryptandrus* (Torr.) A. Gray from Idaho, USA. Phylogenetic relationships and population structure. *Phytopathology* 108, 768–779.

Bert, W., Leliaert, F., Vierstraete, A.R., Vanfleteren, J.R. and Borgonie, G. (2008) Molecular phylogeny of the Tylenchina and evolution of the female gonoduct (Nematoda: Rhabditida). *Molecular Phylogenetics and Evolution* 48, 728–744.

Bertozzi, T. and Davies, K.A. (2009) *Anguina paludicola* sp. n. (Tylenchida: Anguinidae): The nematode associated with *Rathayibacter toxicus* infection in *Polypogon monspeliensis* and *Lachnagrostis filiformis* in Australia. *Zootaxa* 2060, 33–46.

Bird, A.F. and Stynes, B.A. (1981) The life cycle of *Anguina agrostis*: post-embryonic growth of the second stage larva. *International Journal of Parasitology* 11, 243–250.

Brzeski, M.W. (1981) The genera of Anguinidae (Nematoda, Tylenchida). *Revue de Nématologie* 4, 23–34.

Brzeski, M.W. (1991) Review of the genus *Ditylenchus* Filipjev, 1936 (Nematoda: Anguinidae). *Revue de Nématologie* 14, 9–59.

Chizhov, V.N. (1980) [On the taxonomic status of several species of the genus *Anguina* Scopoli, 1777.] *Byulleten Vsesoyznogo Instituta Gel'mintologii K.I. Skryabina* 26, 83–92.

Chizhov, V.N. and Berezina, N.V. (1986) [*Poa annua* L. – a new host of *Paranguina agropyri* (Nematoda: Tylenchida).] *Byulleten Vsesoyznogo Instituta Gel'mintologii K.I. Skryabina* 45, 68–73.

Chizhov, V.N. and Marjenko, A.Y. (1984) [Morphology and biology of *Subanguina radicicola*]. *Zoologichesky Zhurnal* 63, 767–769.

Chizhov, V.N. and Subbotin, S.A. (1985) [Revision of the nematode from the subfamily Anguininae (Nematoda, Tylenchida) on the basis of their biological characteristics.] *Zoologichesky Zhurnal* 64, 1476–1486.

Chizhov, V.N. and Subbotin, S.A. (1990) [Phytoparasitic nematodes of subfamily Anguininae (Nematoda, Tylenchida). Morphology, trophic specialization, taxonomic.] *Zoologichesky Zhurnal* 69, 15–26.

Cid del Prado Vera, I. and Maggenti, A.R. (1984) A new gall-forming species of *Anguina* Scopoli, 1777 (Nemata: Anguinidae) on bluegrass, *Poa annua* L., from the coast of California. *Journal of Nematology* 16, 386–392.

Cid del Prado Vera, I., Chizhov, V.N. and Subbotin, S.A. (2018) Molecular characterisation of gall-forming nematodes, *Mesoanguina amsinckiae* and *Anguina danthoniae* (Anguinidae: Tylenchida) from California, USA. *Russian Journal of Nematology* 26, 109–113.

Coles, J.W. (1958) Nematodes parasitic on sea weeds of the genera *Ascophyllum* and *Fucus*. *Journal of the Marine Biology Association* 37, 145–155.

Corbett, D.C.M. (1966) Central African Nematodes. III. *Anguina hyparrheniae* n. sp. associated with 'witches' broom of *Hyparrhenia* spp. *Nematology* 12, 280–286.

De Waele, D., Jones, B.L., Bolton, C. and Van den Berg, E. (1989) *Ditylenchus destructor* in hulls and seeds of peanut. *Journal of Nematology* 21, 10–15.

De Waele, D., Venter, C. and Mc Donald, A.H. (1997) The peanut pod nematode, *Ditylenchus africanus*. Nematology Circular 218. Division of Plant Industry, Florida Department of Agriculture and Consumer Services, Gainesville, FL.

De Waele, D., Wilken, R. and Lindique, J.M. (1991) Response of potato cultivars to *Ditylenchus destructor* isolated from groundnut. *Revue de Nématologie* 14, 123–127.

Esser, R.P. (1985) Characterization of potato rot nematode, *Ditylenchus destructor* Thorne, 1945 (Tylenchidae) for regulatory purposes. *Nematology Circular* 124, 3.

Evtushenko, L.I., Dorofeeva, L.V., Dobrovolskaya, T.G. and Subbotin, S.A. (1994) Coryneform bacteria from plant galls induced by nematodes of the subfamily Anguininae. *Russian Journal of Nematology* 2, 99–104.

Evtushenko, L.I., Dorofeeva, L.V., Dobrovolskaya, T.G., Streshinskaya, G.M., Subbotin, S.A., *et al.* (2001) *Agreia bicolorata* gen. nov., sp. nov., to accommodate actinobacteria isolated from narrow reed grass infected by the nematode *Heteroanguina graminophila*. *International Journal of Systematic and Evolutionary Microbiology* 51, 2073–2079.

Ewing, C.J., Hausman, C.E., Pogacnik, J., Slot, J. and Bonello, P. (2018) Beech leaf disease: an emerging forest epidemic. *Forest Pathology* 49(2), e12488. DOI: 10.1111/efp.12488.

Fitza, K.N.E., Allison, J., Slippers, B., Chingandu, N. and Reed, S.E. (2024) Diversity and potential sources of introduction of the beech leaf nematode (*Litylenchus crenatae mccannii*) to Ontario, Canada. *Canadian Journal of Plant Pathology* 46(4), 356–366. DOI: 10.1080/07060661.2024.2312150.

Fortuner, R. (1982) On the genus *Ditylenchus* Filipjev, 1936 (Nematoda: Tylenchida). *Revue de Nématologie* 5, 17–38.

Fortuner, R. and Maggenti, A.R. (1987) A reappraisal of Tylenchina (Nemata). 4. The family Anguinidae Nicoll, 1935 (1926). *Revue de Nématologie* 10, 163–176.

Giblin-Davis, R.M., Erteld, C., Kanzaki, N., Ye W., Zeng, Y., *et al.* (2010) *Ditylenchus halictus* n. sp. (Nematoda: Anguinidae), an associate of the sweat bee, *Halictus sexcinctus* (Halictidae), from Germany. *Nematology* 12, 891–904.

Goodey, J.B. (1958) *Ditylenchus myceliophagus* n. sp. (Nematoda: Tylenchidae). *Nematologica* 3, 91–96.

Goodey, J.B. (1963) *Soil and freshwater nematodes*, 2nd edn. Methuen, London, UK.

Goodey, T. (1932) The genus *Anguillulina* Gerv. & v. Ben., 1859, vel *Tylenchus* Bastian, 1865. *Journal of Helminthology* 10, 75–180.

Handoo, Z., Kantor, M. and Carta, L. (2020) Taxonomy and identification of principal foliar nematode species (Aphelenchoides and Litylenchus). *Plants* 9, 1490. DOI: 10.3390/plants9111490.

Hesling, J.J. (1974) *Ditylenchus myceliophagus*. *CIH Descriptions of Plant-parasitic Nematodes*, Set 3, No. 36. Commonwealth Agricultural Bureaux, Farnham Royal, UK.

Hooper, D.J. (1971) Stem eelworm (*Ditylenchus dipsaci*), a seed and soil-borne pathogen of field beans (*Vicia faba*). *Plant Pathology* 20, 25–27.

Hooper, D.J. (1972) *Ditylenchus dipsaci*. *CIH Descriptions of Plant-parasitic Nematodes*, Set 1, No. 14. Commonwealth Agricultural Bureaux, Farnham Royal, UK.

Hooper, D.J. (1973) *Ditylenchus destructor*. *CIH Descriptions of plant-parasitic nematode*, Set 2, No. 21. Commonwealth Agricultural Bureaux, Farnham Royal, UK.

Hooper, D.J. and Southey, J.F. (1978) *Ditylenchus*, *Anguina* and related genera. In: Southey, J.F. (ed), *Plant Nematology*. HMSO, London, UK, pp. 78–97.

Ibrahim, S.K. and Perry, R.N. (1993) Desiccation survival of the rice stem nematode *Ditylenchus angustus*. *Fundamental and Applied Nematology* 16, 31–38.

Ivanova, T.S. (1966) [Biological control of mountain bluet (*Acroptilon picridis* C.A.M.).] *Izvestiya Academil Nauk Tadzhikoi SSR* 2, 51–63.

Kanzaki, N., Ichihara, Y., Aikawa, T., Ekino, T. and Masuya, H. (2019) *Litylenchus crenatae* n. sp. (Tylenchomorpha: Anguinidae), a leaf gall nematode parasitising *Fagus crenata* Blume. *Nematology* 21, 5–22. DOI: 10.1163/15685411-00003190.

Kirjanova, E.S. and Ivanova, T.S. (1968) [New species of the genus *Paranguina* Kirjanova, 1955 (Nematoda: Tylenchidae) from Tadzhikistan.] The Gorge of Kondor, Danin, Dushambe, pp. 200–217.

Kirjanova, E.S. and Krall, E.L. (1971) [*Plant Parasitic Nematodes and Their Control.*] Vol. 2. Nauka, Leningrad, USSR.

Krall, E.L. (1991) Wheat and grass nematodes: *Anguina, Subanguina*, and related genera. In: Nickle, W.R. (ed), *Manual of Agricultural Nematology*. Marcel Dekker, New York, New York, pp. 721–760.

Lax, P., Gonzalez Ittig, R., Rondan Dueñas, J.C., Andrade, A.J., Gardenal, C.N., *et al.* (2021) Decrypting species in the *Nacobbus aberrans* (Nematoda: Pratylenchidae) complex using integrative taxonomy. *Zoologica Scripta* 50, 667–688. DOI: 10.1111/zsc.12494.

McClure, M., Schmitt, M.E. and McCullough, M.D. (2008) Distribution, biology and pathology of *Anguina pacificae*. *Journal of Nematology* 40, 226–239.

McKay, A.C. (1993) Livestock deaths associated with *Clavibacter toxicus*/*Anguina* sp. infection in seedheads of *Agrostis avenacea* and *Polypogon monspeliensis*. *Plant Disease* 77, 635–641.

Murray, T.D., Schroeder, B.K., Schneider, W.L., Luster, D.G., Sechler, A., *et al.* (2017) *Rathayibacter toxicus*, other *Rathayibacter* species inducing bacterial head blight of grasses, and potential for livestock poisonings. *Phytopathology* 107, 804–815.

Nicoll, W. (1935) Vermes. *Zoological Record* 72, 105.

Ou, X. and Watson, A.K. (1992) *In vitro* culture of *Subanguina picridis* in Russian knapweed callus, excised roots and shoot tissues. *Journal of Nematology* 24, 199–204.

Paramonov, A.A. (1962) [*Fundamentals of Plant Helminthology.*] Vol. I. Nauka, Moscow, USSR.

Paramonov, A.A. (1967) A critical review of the suborder Tylenchina (Filipjev, 1934) (Nematoda: Secernentea). *Trudy Gel'mintologicheskoi laboratorii Akademii Nauk SSSR* 18, 78–101.

Paramonov, A.A. (1970) [*Fundamentals of Plant Helminthology. Taxonomy of Nematodes of the Superfamily Tylenchoidea*.] Vol. 3. Nauka, Moscow, USSR.

Poinar Jr, G.O. (2011) *The Evolutionary History of Nematodes. Nematology Monographs and Perspectives,* 9 (Series eds: Hunt, D.J. and Perry, R.N.). Brill, Leiden, Netherlands.

Powers, T.O., Szalanski, A.L., Mullin, P.G., Harris, T.S., Bertozzi, T., *et al.* (2001) Identification of seed gall nematodes of agronomic and regulatory concern with PCR-RFLP of ITS1. *Journal of Nematology* 33, 191–194.

Price, P.C., Fisher, J.M. and Kerr, A. (1979) On *Anguina funesta* n. sp. and its association with *Corynebacterium* sp., in infecting *Lolium rigidum*. *Nematologica* 25, 76–85.

Riley, I.T. and Barbetti, M.J. (2008) Australian anguinids: their agricultural impact and control. *Australasian Plant Pathology* 37, 289–297.

Riley, I.T., Reardon, T.B. and McKay, A.C. (1988) Electrophoretic resolution of species boundaries in seed-gall nematodes, *Anguina* spp. (Nematoda: Anguinidae), from some graminaceous hosts in Australia and New Zealand. *Nematologica* 34, 401–411.

Robinson, A.F., Orr, C.C. and Abernathy, J.R. (1978) Distribution of *Nothanguina phyllobia* and its potential as a biological control agent for silver-leaf nightshade. *Journal of Nematology* 10, 363–366.

Seshadri, A.R. and Dasgupta, D.R. (1975) *Ditylenchus angustus. CIH Descriptions of Plant-parasitic Nematodes*, Set 5, No. 64. Commonwealth Agricultural Bureaux, Farnham Royal, UK.

Siddiqi, M.R. (1971) Structure of the oesophagus in the classification of the superfamily Tylenchoidea (Nematoda). *Indian Journal of Nematology* 1, 25–43.

Siddiqi, M.R. (1980) The origin and phylogeny of the nematode orders Tylenchida Thorne, 1949 and Aphelenchida n. ord. *Helminthological Abstracts, Series B* 49, 143–170.

Siddiqi, M.R. (1986) *Tylenchida Parasites of Plants and Insects*. Commonwealth Agricultural Bureaux, Farnham Royal, UK.

Siddiqi, M.R. (2000) *Tylenchida Parasites of Plants and Insects*, 2nd edn. CAB International, Wallingford, UK.

Siddiqi, M.R. and Lenné, J.M. (1984) *Pterotylenchus cecidogenus* n. gen., n. sp., a new stem-gall nematode parasitizing *Desmodium ovalifolium* in Colombia. *Journal of Nematology* 16, 62–65.

Southey, J.F. (1972) *Anguina tritici. CIH Descriptions of Plant-parasitic Nematodes*, Set 1, No. 13. Commonwealth Agricultural Bureaux, Farnham Royal, UK.

Southey, J.F. (1973) *Anguina agrostis. CIH Descriptions of Plant-parasitic Nematodes*, Set 2, No. 20. Commonwealth Agricultural Bureaux, Farnham Royal, UK.

Southey, J.F., Topham, P.B. and Brown, D.J.F. (1990) Taxonomy of some species of *Anguina* Scopoli, 1777 (*sensu* Brzeski, 1981) forming galls on Gramineae: value of diagnostic characters and present status of nominal species. *Revue de Nématologie* 13, 127–142.

Sturhan, D. and Brzeski, M.W. (1991) Stem and bulb nematodes, *Ditylenchus* spp. In: Nickle, W.R. (ed), *Manual of Agricultural Nematology*. Marcel Dekker, New York, New York, pp. 423–464.

Stynes, B.A. and Bird, A.F. (1980) *Anguina agrostis,* the vector of annual rye grass toxicity in Australia. *Nematologica* 26, 475–490.

Subbotin, S.A. and Ryss, A.Y. (2024) Revision of the genus *Ditylenchus* Filipjev, 1936: *Ditylenchoides* gen. n. and *Paraditylenchus* gen. n. (Nematoda: Anguinoidea). *Russian Journal of Nematology* 32, 91–102.

Subbotin, S.A., Krall, E.L., Riley, I.T., Chizhov, V.N., Staelens, A., *et al.* (2004) Evolution of the gall-forming plant parasitic nematodes (Tylenchida: Anguinidae) and their relationships with hosts as inferred from Internal Transcribed Spacer sequences of nuclear ribosomal DNA. *Molecular Phylogenetics and Evolution* 30, 226–235.

Tancos, M.A., Sechler, A.J., Davis 2nd, E.W., Chang, J.H., Schroeder, B.K., *et al.* (2020) The identification and conservation of tunicaminyluracil-related biosynthetic gene clusters in several *Rathayibacter* species collected from Australia, Africa, Eurasia, and North America. *Frontiers in Microbiology* 10, 2914. DOI: 10.3389/fmicb.2019.02914.

Thorne, G. (1945) *Ditylenchus destructor* n. sp., the potato rot nematode, and *Ditylenchus dipsaci* (Kühn, 1857) Filipjev, 1936, the teasel nematode. *Proceedings of the Helminthological Society of Washington* 12, 27–33.

Thorne, G. (1961) *Principles of Nematology*. McGraw-Hill, New York, New York.

Triantaphyllou, A.C. and Hirschmann, H. (1980) Cytogenetics and morphology in relation to evolution and speciation of plant-parasitic nematodes. *Annual Review of Phytopathology* 18, 333–359.

Van den Berg, E. (1985) Notes on the genus *Afrina* Brzeski, 1981 (Anguinidae: Nematoda) with description of new and known species. *Phytophylactica* 17, 69–79.

Vovlas, N., Troccoli, A., Palomares-Rius, J.E., De Luca, F., Liébanas, G., *et al.* (2011) *Ditylenchus gigas* n. sp. parasitizing broad bean: a new stem nematode singled out from the "*Ditylenchus dipsaci* species complex" based on polyphasic identification and molecular phylogeny. *Plant Pathology* 60, 762–775.

Watson, A.K. (1986a) Morphological and biological parameters of the knapweed nematode, *Subanguina picridis. Journal of Nematology* 18, 154–158.

Watson, A.K. (1986b) Biology of *Subanguina picridis*, a potential biological control agent of Russian knapweed. *Journal of Nematology* 18, 149–154.

Wendt, K.R., Swart, A., Vrain, T. and Webster, J.M. (1995) *Ditylenchus africanus* sp. n. from South Africa; a morphological and molecular characterization. *Fundamental and Applied Nematology* 18, 241–250.

Zhang, S.L., Liu, G.K., Janssen, T., Zhang, S.S., Xiao, S., *et al.* (2014). A new stem nematode associated with peanut pod rot in China: morphological and molecular characterization of *Ditylenchus arachis* n. sp. (Nematoda: Anguinidae). *Plant Pathology* 63, 1193–1206. DOI: 10.1111/ppa.12183.

Zhao, Z.Q., Davies, K., Alexander, B. and Riley, I.T. (2011) *Litylenchus coprosma* gen. n., sp. n. (Tylenchida: Anguinata), from leaves of *Coprosma repens* (Rubiaceae) in New Zealand. *Nematology* 13, 29–44.

15 Longidoridae and Trichodoridae

Wilfrida Decraemer* and Wim Bert
Ghent University, Department of Biology, Ghent, Belgium

Abstract
The nematodes of the families Longidoridae and Trichodoridae can be important parasites in their own right, although some species are also effective virus vectors and pose substantial risks to susceptible crops. The genera of the Longidoridae include *Xiphinema*, *Longidorus* and *Paralongidorus*. These three genera are diagnosed and illustrated, a number of species being discussed in more detail with information on their morphology, morphometrics, distribution, host range and relevance to quarantine as virus vectors, where appropriate. Within the Trichodoridae, three genera are dealt with in detail: *Nanidorus*, *Trichodorus* and *Paratrichodorus*. Once again, the major species have information on their morphology, morphometrics, distribution, host range and relevance as virus vectors, including the identification of genera and major species of quarantine and economic importance.

15.1. Introduction

Longidoridae Thorne, 1935 (Meyl, 1961) and Trichodoridae Thorne, 1935 (Siddiqi, 1961) are polyphagous root ectoparasites that cause direct damage through feeding. The Longidoridae are characterized by their considerable length (1.5-13 mm), and feed using a very thin stylet. This stylet, which can exceed 300 μm in length, consists of a fine hollow cylinder (odontostyle), which has a dorsal opening, and a spear extension (odontophore) formed by the sclerotization of the pharynx's most anterior part. In contrast, the Trichodoridae are notably smaller (350-1800 μm) and are equipped with a much shorter (typically around 40 μm), ventrally curved spear (onchiostyle). The so-called 'spear' in Trichodoridae is a stylettiform tooth (onchium) closed at the tip, allowing the nematode to pierce plant cells. This onchium is inserted on a support (onchiophore) formed by a thickening of the dorsal wall of the pharyngostom.

Significantly, the Longidoridae and Trichodoridae are distinguished within the Nematoda in comprising species that act as vectors for plant viruses, specifically Nepoviruses and Tobraviruses, respectively. The first documented instance of a nematode-plant virus association was presented by Hewitt *et al.* (1958) who demonstrated that *Xiphinema index* Thorne & Allen, 1950

*E-mail: wilfrida.decraemer@ugent.be
A revision of Decraemer, W. and Chaves, E. (2012) Longidoridae and Trichodoridae. In: Manzanilla-López, R.H. and Marbán-Mendoza, N. (eds), *Practical Plant Nematology*. Mundi-Prensa and Biblioteca Básica de Agricultura, Montecillo, Mexico.

 Practical Plant Nematology 2nd Edition
(eds R.H. Manzanilla-López, D.J. Hunt, N. Marbán-Mendoza and S. Infante Gil)
DOI: 10.1079/9781836990413.0015

(Longidoridae) served as a vector for the grapevine fanleaf virus (GFLV). Subsequently, Sol *et al.* (1960) reported that the tobacco rattle virus (TRV) was transmitted by *Trichodorus pachydermus* Seinhorst, 1954, now *Paratrichodorus pachydermus* (Trichodoridae). Despite the few vector species identified within these families, their role as virus vectors grants them the status of major pests. Both nematode families belong to unrelated subclass taxa: Longidoridae to Dorylaimia and Trichodoridae to Enoplia. This distinction also extends to the virus groups they vector, supporting the hypothesis that the virus-vector relationship in nematodes has evolved independently at least twice.

15.1.2. Simplified Systematic Scheme

Order Dorylaimida Inglis, 1983
  Family Longidoridae Thorne, 1935
    Subfamily Longidorinae Thorne, 1935
      Longidorini Coomans, 1996
      Longidoroides Khan, Chawla & Saha, 1978
      Longidorus Micoletzky, 1922
      Paralongidorus Siddiqi, Hooper & Khan, 1963[1]
      Xiphidorini Coomans, 1996
      Australodorus Coomans, Olmos, Casella & Chaves, 2004
      Paraxiphidorus Coomans & Chaves, 1995
      Xiphidorus Monteiro, 1976
    Subfamily Xiphinematinae Dalmasso, 1969
      Xiphinema Cobb, 1913
Order Triplonchida Cobb, 1920
  Family Trichodoridae Thorne, 1935
      Allotrichodorus Rodriguez-M., Sher & Siddiqi, 1978
      Ecuadorus Siddiqi, 2002
      Monotrichodorus Andrássy, 1976
      Nanidorus Siddiqi, 1974
      Paratrichodorus Siddiqi, 1974
        = *Atlantadorus* Siddiqi, 1974
      Trichodorus Cobb, 1913

15.2. Longidoridae Thorne, 1935

The Longidoridae, the dagger nematodes, and the largest of the virus vector nematode families, has nearly 500 described species, and is the only plant-parasitic family within the Dorylaimoidea (Dorylaimida). In general, the morphology of longidorid nematodes is quite conserved (similar), a fact that often makes it difficult to determine species boundaries and for the number of valid species to be agreed upon. Jairajpuri and Ahmad (1992) and Hunt (1993) recognized three subfamilies: Xiphinematinae Dalmasso, 1969, Longidorinae Thorne, 1935 and Xiphidorinae Khan, Chawla & Saha, 1978. Based on a phylogenetic analysis, Coomans (1996) proposed two subfamilies within the Longidoridae, the Longidorinae and Xiphinematinae. The Xiphinematinae contain only one genus, *Xiphinema* Cobb, 1913, in which two species groups are distinguished: the *Xiphinema americanum* group (65 species, *sensu* Lamberti *et al.*, 2000) characterized by symbiotic bacteria in the ovaries and the remainder (233 species) without such symbionts. The subfamily Longidorinae is far more differentiated, containing two tribes, the Longidorini with the genera *Longidorus* Micoletzky, 1922 (188 species), *Paralongidorus* Siddiqi, Hooper & Khan, 1963 (64 species of *Paralongidorus*, *sensu stricto*) and *Longidoroides* Khan, Chawla & Saha, 1978 (14 species). The latter genus was considered a synonym of *Paralongidorus* by Siddiqi *et al.* (1993) but is still accepted as valid by others until more conclusive morphological and/or molecular sequence evidence can be provided (Coomans, 1996; Decraemer and Geraert, 2006). The tribe Xiphidorini harbours the genera *Xiphidorus* Monteiro, 1976 (eight species); *Paraxiphidorus* Coomans & Chaves, 1996 (three species) and

Australodorus Coomans, Olmos, Casella & Chaves, 2004 (one species). Except for the genera of the Xiphidorini, which are recorded only from Central and South America, the other genera have a more worldwide distribution (Decraemer and Robbins, 2007). Although following the systematic scheme of this volume (see Chapter 3), herein we also adopt the tribe classification used by Coomans (1996) whereby the Longidorinae is subdivided into the Longidorini Coomans, 1996, and the Xiphidorini Coomans, 1996.

Diagnosis: Elongated and thin nematodes, between 1.5 and 13 mm long. Body C shaped or spiral, occasionally straight upon fixation. Tail usually similar in both sexes: short, conoid or hemispherical, with or without a terminal mucron, elongated conoid or long and filiform. Smooth body cuticle. Body pores present along body laterally (may be absent in the vulval region) and ventrally (absent in *Xiphidorus*); dorsally, pores restricted to the anterior pharyngeal region. Lip region continuous or separated from body contour by an expansion, constriction or depression. Lips largely or completely fused, with two circles of sensilla: six inner labial sensilla and 10 sensilla (six outer labial + four cephalic), all papilliform or at level with head surface and visible as pores, flush with head surface or surrounded by a cuticular rim. Amphids post-labial with stirrup-, goblet- to pouch-shaped fovea and aperture in the shape of a slit or a pore. Guiding ring of odontostyle simple or double, varying in position from close to anterior end (Longidorini) to far posteriorly near spear base (Xiphidorini, *Xiphinema*). Long and slender odontostyle, 50-220 μm long, with a thin lumen and dorsal aperture, and simple or complex (= forked) at connection level with its support, the odontophore. Odontophore simple, rod-shape or with knob-like thickened end (= flanges). Pharynx (= oesophagus) dorylaimid, with a tubular anterior part, not muscular or weakly muscular, separated from a bottle-shaped posterior bulb containing three glands: one dorsal and two ventrosublateral, and with a thickened lumen wall, formed by six triangular platelets in transverse section. Intestine with differentiated prerectum. Female reproductive system amphidelphic-didelphic with median vulva, monodelphic (one uterus) when the anterior branch of the reproductive system has been reduced and lost and pseudomonodelphic when the reduction of the anterior genital branch is partial – in these cases, the vulva is displaced anteriad. Males with two testes (diorchic), posterior testis reflexed, both testes described as opposite and joining in a common *vas deferens*. Spicules, dorylaimoid, arcuate, well developed; gubernaculum consisting of two lateral guide pieces or crura. Precloacal supplements (PS) separated from adanal pair (rarely 2-3 pairs) and formed by a medioventral series of 1-20 papillae, rarely staggered, either forming a continuous series or showing a gap or hiatus between the adanal pair and the series.

Type genus: *Longidorus* Micoletzky, 1922.

15.2.1. Key to Subfamilies

1. Amphids with pouch-like, funnel- or stirrup-shaped fovea, aperture a pore or minute to a large slit. Odontostyle forked or non-forked; odontophore without well-developed flanges. Guide ring simple. Dorsal pharyngeal gland nucleus located posteriorly, at a considerable distance from its aperture. Female reproductive system amphidelphic. Tail short, conoid, similar in both sexes Longidorinae (Fig. 15.1B-G)

Amphidial fovea stirrup or funnel-like, aperture large slit-like. Odontostyle forked; odontophore flanged. Guide ring double. Dorsal pharyngeal gland nucleus located anteriorly, at about level of its outlet. Female reproductive system monodelphic, pseudomonodelphic or amphidelphic. Tail short, conoid, long or filiform; usually similar in both sexes Xiphinematinae (Fig. 15.1A)

15.2.2. Key to Tribes of Longidorinae

1. Guide ring simple, located in the posterior part of odontostyle, forked odontostyle, odontophore with weak to moderate basal thickenings.. Xiphidorini (Fig. 15.1B-D)

Guide ring simple, located in the anterior part of odontostyle, non-forked odontostyle; odontophore without basal thickenings, at most a slight thickening .... Longidorini (Fig. 15.1E-G)

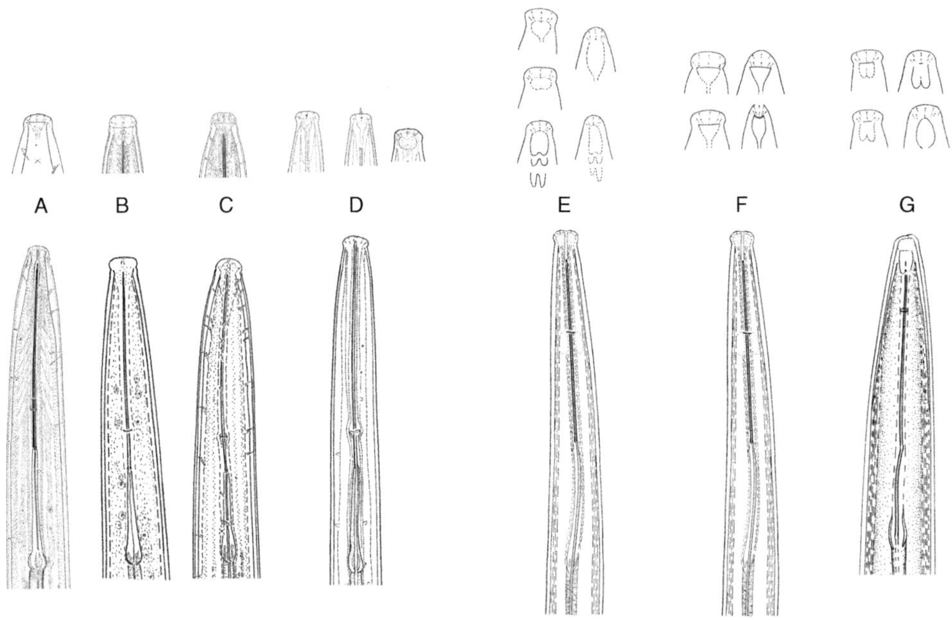

Fig. 15.1. Generic comparison of head and odontostyle region of Longidoridae showing variation of amphidial fovea and opening. A: *Xiphinema*; B: *Australodorus*; C: *Paraxiphidorus*; D: *Xiphidorus*; E: *Longidorus*; F: *Paralongidorus*; G: *Longidoroides*. Not to scale.

15.2.3. Genera of Longidoridae

In addition to the features used in the keys to subfamilies and tribes, the presence or absence at guide ring level of compensation sacs (difficult to observe by light microscopy and thus not included in key) differentiates the Longidorini from the Xiphidorini and Xiphinematinae. For identification to genus level, use Table 15.1 together with illustrations in Figure 15.1.

Confusable genera: In general, the genera within the Longidoridae are well defined. However, exceptions do occur. For example, the guide ring can be more posterior in position in some *Longidorus* species (such as *Longidorus martini* Merny, 1966) and more anteriorly located in some *Paralongidorus* species (see *Paralongidorus bullatus* Sharma & Siddiqi, 1980, *P. esci*) compared to the genus diagnosis. Furthermore, the guide ring in some species of the *X. americanum* group can be slightly more anterior compared to other *Xiphinema* species and appear simple due to a minute guiding sheath. In some *Xiphidorus* specimens, however, the guide ring can be interpreted as double because of the refraction of the thick guide ring.

Other variations in the genus diagnoses concern the arrangement of ventromedian PS with respect to the adanal pair in the male. In *Longidorus* and *Paralongidorus* Siddiqi, Hooper & Khan, 1963, PS are more or less evenly spread. Exceptionally, the posteriormost PS may be at a larger distance from the adanal pair presenting a transitional step towards a hiatus (*L. martini*, *P. iberis* Escuer & Arias, 1997), or the hiatus typical for *Xiphinema* can be absent or small in some *Xiphinema* species (e.g., *X. americanum*, *Xiphinema macrostylus*). When we consider the fact that identification is preferably based on a combination of diagnostic features, genus identification does not cause a problem. Apart from the occurrence of variation, misidentification can be due to non-precise observations, for example, of the amphidial opening as a pore or a transverse slit (*Longidorus iberis*, *Longidorus spiralis* (Khan, Saha & Seshadri, 1972) Khan, Chawla & Saha, 1978 (see Decraemer and Coomans, 2007; Clavero-Camacho *et al.*, 2022).

Table 15.1. Overview of diagnostic features for genus identification within the Longidoridae.

Taxonomic features	*Longidorus* Fig. 15.1E	*Longidoroides* Fig. 15.1G	*Paralongidorus* Fig. 15.1F	*Xiphidorus* Fig. 15.1D	*Paraxiphidorus* Fig. 15.1C	*Australodorus* Fig. 15.1B	*Xiphinema* Fig. 15.1A
Amphidial opening	**Pore**	Transverse slitsmall	Slit large to medium-sized	Minute slit or pore	Large slit	Large slit	Slit large to medium-sized
Amphidial fovea	**Pouch***	**Pouch***	Stirrup-shaped, funnel or goblet-shaped	Wide pouch or fovea narrow pocket-like or wide to narrow cup-shaped	Wide stirrup-shaped	Large pouch with mid-rib	Wide stirrup-, funnel-, goblet-, heart-shaped or elongate stirrup-shaped
Guiding ring	Single	Single	Single	Single	Single	Single	**Double**
Position guiding ring (% of odontostyle from tip)	Up to 40% from anterior end	Up to 60% from anterior end	Up to 60% from anterior end	Around posterior third odontostyle	Around posterior third odontostyle	Around posterior third odontostyle	Around posterior third odontostyle
Odontostyle base	**Non-furcate**	**Non-furcate**	**Non-furcate**	furcate	furcate	furcate	furcate
Odontophore base	Not flanged	Not flanged, at most slightly widened	Not flanged, at most slightly widened	Flanges weak to moderately developed	Flanges weak to moderately developed	Flanges weak to moderately developed	Flanges well developed to moderate
Dorsal pharyngeal gland nucleus: position	More posteriorly in bulb, posterior to opening of dorsal gland	Far posterior to opening of dorsal gland	Far posterior to opening of dorsal gland	Far posterior to opening of dorsal gland	Far posterior to opening of dorsal gland	Far posterior to opening of dorsal gland	**Close to dorsal gland opening** and far anteriorly in bulb
Size dorsal pharyngeal gland nucleus (DN) *vs* ventrosublaterals (SN)	DN < SN	DN < SN	DN < SN	DN < SN	DN < SN	DN < SN	**DN > SN**
Arrangement precloacal supplements (SP) in male	Hiatus absent between SP and adanal pair of pre-cloacal papillae	Hiatus absent	Hiatus absent	Hiatus present between SP and adanal pair of precloacal papillae	Hiatus present between SP and adanal pair of precloacal papillae	Hiatus present between SP and adanal pair of precloacal papillae	Hiatus present between SP and adanal pair of precloacal papillae
Female reproductive system	Didelphic	Didelphic	Didelphic	Didelphic	Didelphic	Didelphic	**Didelphic, mono or pseudo-monodelphic**

*lobed (symmetrical/asymmetrical) or non-lobed.

15.2.4. Species Identification and Major Species

At the species level, identification is partially based on measurements such as body length, tail length, odontostyle length, spear length, the distance of the guide ring from the anterior, lip region diameter and spicule length in males, ratios (a, b, c, c′), vulva position and the number of ventromedian PS in males, and partially on morphological features. Depending on the genus, the contribution of morphological features to identification can be very limited; this is exemplified in the *X. americanum* group. As major species, we refer primarily to the economically important virus vector species.

The genus *Longidorus* Micoletzky, 1922

For *Longidorus*, species identification is based upon the following morphological features: (i) shape of pouch-like amphidial fovea; (ii) shape of lip region (rounded, flattened, offset by constriction, depression or continuous); (iii) tail shape and (iv) males present or absent. Of more restricted use is the number of ventromedian PS in males. Currently, seven vector species are known for *Longidorus* (*Longidorus apulus* Lamberti & Bleve-Zacheo, 1977, *Longidorus arthensis* Brown, Grunder, Hooper, Klinger & Kunz, 1994, *Longidorus attenuatus* Hooper, 1962, *Longidorus diadecturus* Eveleigh & Allen, 1982, *Longidorus elongatus* de Man, 1878, *Longidorus fasciatus* Roca & Lamberti, 1981, *Longidorus macrosoma* Hooper, 1961, *L. martini* Merny, 1966). Males are only common in *L. apulus*, *L. macrosoma* and *L. martini*. A polytomous identification and references of *Longidorus* species were given by Chen *et al.* (1997).

***Longidorus elongatus* de Man, 1878 (Fig. 15.2A-C)**

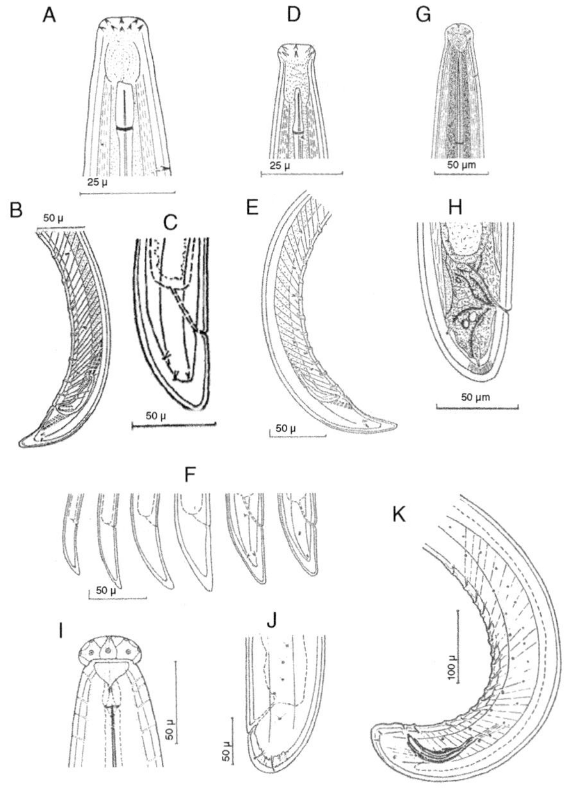

Fig. 15.2. A-C. *Longidorus elongatus*. A: Head region in surface view; B: male posterior body region; C: female tail. D-F. *Longidorus attenuatus*. D: Head region in surface view; E: male posterior body region; F: tails of J1, J2, J3, J4, female (after Hooper, 1961). G, H. *Longidorus diadecturus*. G: Head region in surface view; H: female tail region (after Eveleigh & Allen, 1982). I-K. *Paralongidorus maximus*. I: Head region in surface view; J: female tail region; K: male posterior body region (after Sturhan, 1963).

Body slender, more or less open C-shape upon heat killing. Lip region flattened, continuous with body contour or showing a minute depression. Amphidial fovea a large elongated pouch, hardly bilobed in lateral view. Odontostyle 94 µm (average, British population), odontophore 50 µm (average, British population). Female reproductive system didelphic-amphidelphic; vulva at 49% (average, British population). Tail dorsally convex conoid, ventrally straight or slightly concave, rounded terminus; two to three caudal pores on each side. Male very rare. Spicules 58 µm. Next to adanal pair, a ventral series of 7-10 PS. Measurements after Hooper (1961).

L. elongatus is a widespread plant-parasite in temperate regions worldwide in various types of soil. It has a wide host range including many crops and weeds. It transmits raspberry ringspot virus (RRSV), tomato black ring virus Scottish strain and spoon leaf virus. The species shows substantial variation in morphometrics between populations and is thought by many to represent a species complex meaning that many records require validation as they could refer to other, closely related, species.

***Longidorus attenuatus* Hooper, 1962 (Fig. 15.2D-F)**

Body an open C-shape upon heat relaxation, 6.1 mm long (average paratype females). Lip region anteriorly flattened and clearly expanded. Amphidial fovea a bilobed pouch extending halfway to guide ring; amphid opening a pore. Odontostyle 78 µm (average paratype females), 82 µm (average paratype males), odontophore average length 41 and 39 µm (in female and male paratypes respectively). Female reproductive system didelphic- amphidelphic, unipartite uteri, vulva at 49% (average paratypes). Tail dorsally convex conoid, 1.5-1.7 anal body diameter long, a pair of caudal pores on each side. Male rare. Spicule 45 µm long; a series of 10 mid-ventral PS, regularly spaced without hiatus with adanal pair.

L. attenuatus is widespread in Europe and the former USSR and occurs regularly in well-drained sandy soils. It causes damage to many herbaceous plants, including carrots, potatoes, strawberry, barley, wheat and sugar beet. It acts as a vector of tomato black ring virus English/German strain to sugar beet.

***Longidorus diadecturus* Eveleigh & Allen, 1982 (Fig. 15.2G-I)**

Body ventrally arcuate, 3.71 mm long. Lip region rounded, expanded, set off by depression. Amphidial fovea an obscure undivided pouch, amphidial opening a lateral pore just posterior to level of second crown of labial and cephalic sensilla. Odontostyle 115 µm, odontophore 60 µm. Female reproductive system didelphic-amphidelphic; vulva 46% from anterior end. Tail dorsally convex conoid with bluntly rounded end; two caudal pores on each side. Males unknown. Measurements represent average of paratype females.

L. diadecturus is widespread in central USA and extends northwards into Canada. It is a vector of peach rosette mosaic virus (PRMV). The other virus vector species have a more restricted distribution.

The genus *Paralongidorus* Siddiqi, Hooper & Khan, 1963 (Fig. 15.2I-K)

For *Paralongidorus*, species differentiation is based on the following morphological features: (i) shape of amphidial fovea; (ii) shape and diameter of lip region; (iii) tail shape in female and (iv) width of amphidial aperture.

Paralongidorus maximus (Bütschli, 1874) Siddiqi, 1964 is the only vector species of the genus. It can be easily recognized by the large body (9.84 mm average females), the rounded and enlarged lip region, sharply offset at level of the wide slit-like amphid opening, the long odontostyle (171.5 µm, average females) and odontophore (53.5 µm average females), bluntly rounded short tail about 0.5 anal body diameter long. Males very rare, with stout spicules (103 µm average) and 14-15 ventromedian PS. Morphometric data for a German population.

P. maximus shows a sporadic distribution in Europe where it can be locally widespread, for example, in Germany, but it has also been recorded from South Africa. The species prefers sandy and loamy-sand soils. It is broadly polyphagous, attacking a wide range of woody and herbaceous plants and is a vector of RRSV (grapevine strain). A polytomous identification key and references of species were provided by Escuer and Arias (1997).

The genus *Xiphinema* Cobb, 1913

Xiphinema can be divided into two main species groups, the *americanum* group and the remainder of the species. It is possible that the latter group will one day be recognized as a separate genus, the type species of *Xiphinema, sensu lato*, falling into the first group.

***Xiphinema americanum* group**

Among the nine *Xiphinema* virus vector species, the members of the *X. americanum* group can be easily differentiated from the other species, mainly because of the presence of symbiotic bacteria obscuring the individual oocytes in the ovaries, the short non-differentiated uteri in several species restricted to an ovejector, the long reflexed ovaries often reaching the vaginal region, the shorter body length (<3 mm), the rounded lip region continuous or marked by depression or constriction, odontostyle rarely longer than 150 μm and short conoid tail with a more or less acute end. Males are extremely rare, and species reproduce parthenogenetically. Three or four juvenile stages can be present. Within the *X. americanum* group, species identification is notoriously difficult. Polytomous identification keys (Lamberti *et al.*, 2000, 2004; OEPP/EPPO, 2017) have been based largely on measurements that usually show interspecific overlaps: (odontostyle length, V, ratios c and c′, body length, tail length) and a few morphological characters such as the shape of the tail tip (pointed or rounded) and lip region (continuous with body or marked by depression or constriction). In the most recent polytomous key in OEPP/EPPO (2017), the presence *vs* absence of verrumicrobial bacteria in the ovaries and their location is added as character. Attention must be paid to the different use of characters and coding in the polytomous identification keys and the reference to the key used in species descriptions.

There is no general agreement on the definition of the *X. americanum* group morphologically or molecularly and, therefore, also on the number of species included in the species group. Lamberti and Ciancio (1993) distinguished, based on a hierarchical cluster analysis with five morphometric variables (body length, ratios a, b, c of de Man (1880), odontostyle length and odontophore length), five species subgroups, among them a Xiphinema *pachtaicum* (Tulaganov, 1938) group including Xiphinema *pachydermum*. The latter group differs from the typical X. *am*ericanum species group by a female reproductive system with few or no bacteria visible in ovary by LM, a well-developed sphincter muscle and longer uteri and males being common. Species numbers vary from 34 in Coomans *et al.* (2002), considering the species of the *X. pachydermum* group with males and mostly without symbiotic bacteria as a separate intermediate species group, to 51 in Lamberti *et al.* (2000) all species included.

Because of the difficulties with morphology-based identification, current quarantine regulations refer to *X. americanum sensu lato* as a 'pathogen agent' until more precise identification tools such as molecular based analyses become widely available. At present, six vector species have been indicated: *X. americanum* (*sensu stricto*), *Xiphinema bricolense* Ebsary, Vrain & Graham, 1989, *Xiphinema californicum* Lamberti & Bleve-Zacheo, 1979, *Xiphinema rivesi* Dalmasso, 1969, *Xiphinema intermedium* Lamberti & Bleve-Zacheo, 1979 and *Xiphinema tarjanense* Lamberti & Bleve-Zacheo, 1979. For the latter two species, vectoring ability has only been shown under laboratory conditions. *X. rivesi* is the only vector species of the *X. americanum* group that is widespread in America but also occurs in Europe.

***Xiphinema americanum*, *sensu stricto* (Fig. 15.3A-E)**

Body an open spiral, 1.68 mm long (average topotypes). Lip region rounded, marked by depression. Amphidial fovea stirrup shaped, opening a transverse slit half as wide as lip region. Odontostyle 71.5 μm (average of topotypes), odontophore 46 μm (average of topotypes). Female reproductive system didelphic-amphidelphic. Male rare, spicule 38 μm long (neotype); five mid-ventral PS (neotype). Tail dorsally convex conoid, 1.7 anal body diameter long (average of female topotypes), tail tip rounded; two pairs of caudal pores (topotypes) (based on Siddiqi, 1973).

It is widespread in the Eastern USA and was also recorded from Arkansas, California and South Africa (Taylor and Brown, 1997). It acts as a vector for cherry rasp leaf (CRLV), tobacco ringspot (TRSV) and tomato ringspot viruses (ToRSV).

***Xiphinema bricolense* Ebsary, Vrain & Graham, 1989 (Fig. 15.3F-H)**

Body an open spiral. Lip region marked by a depression. Amphidial fovea stirrup shaped, opening two-thirds of lip diameter. Odontostyle 88 μm (average of types), odontophore 51 μm (average of types). Female reproductive system didelphic-amphidelphic. Male rare, spicules 39 μm long (paratype); 11 mid-ventral PS. Tail dorsally convex conoid, 1.5 anal body diameter

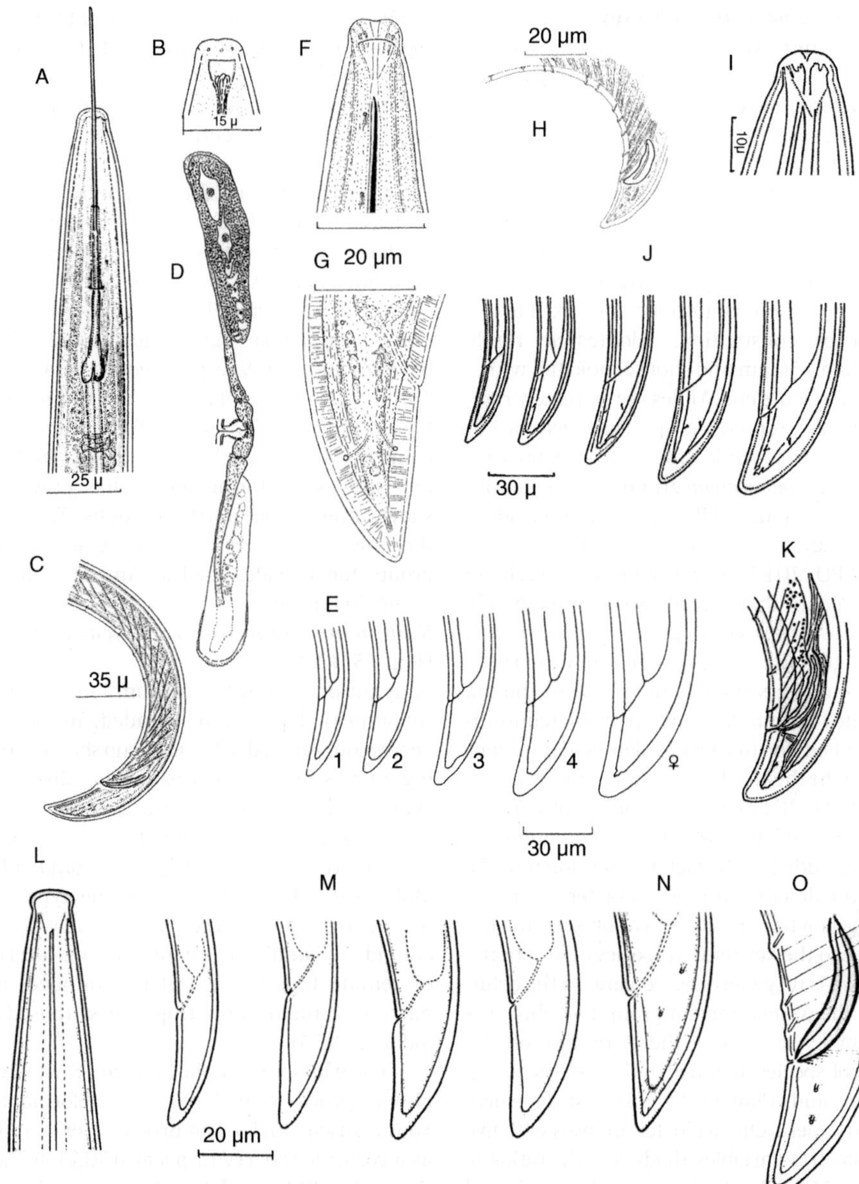

Fig. 15.3. A-E. *Xiphinema americanum*. A: Odontostyle region neotype male; B: head region in surface view; C: male posterior body region; D: female reproductive system (after Siddiqi, 1973 except D, after Coomans & Claeys, 2001); E: tails of J1, J2, J3, J4 and female (after Coomans *et al.*, 2002). F-H. *Xiphinema bricolense*. F: Odontostyle region; G: female tail region; H: male posterior body region (after Ebsary *et al.*, 1989). I-K. *Xiphinema rivesi*. I: Head region; J: tails of J1, J2, J3, J4, female; K: male tail region (after Dalmasso, 1969). L-O. *Xiphinema californicum*. L: Head region in surface view; M: tails of J1, J2, J3, J4; N: female tail; O: male tail region (after Lamberti and Blève-Zacheo, 1979).

long (average of types); two pairs of caudal pores near mid-tail region.

Reported from Canada (British Colombia) and Pacific coast states of North America. It is a vector of ToRSV.

***Xiphinema californicum* Lamberti & Bleve-Zacheo, 1979 (Fig. 15.3L-O)**

Body characterized by a 2 mm (average of types) long body, slender, ventrally arcuate, more pronounced in the posterior region. Lip region well expanded, set off by clear constriction. Amphidial fovea stirrup shaped and opening a wide transverse slit. Odontostyle 90 µm long (average of types), odontophore 48 µm (average of types) weakly sclerotized. Female reproductive system didelphic, uterus short, not differentiated. Male extremely rare; spicules 35 µm long (paratype) and with seven mid-ventral PS. Tail elongate-conoid, dorsally convex, in male slightly concave ventrally. Two pairs of caudal pores.

It was described from California and Mexico and has also been recorded from Eastern USA, Chile, Peru, Brazil. It is a vector of CRLV, TRSV and ToRSV.

***Xiphinema rivesi* Dalmasso, 1969 (Fig. 15.3I-K)**

Body C shaped, 1.96 mm long (average of types). Lip region rounded, marked by depression or appearing continuous. Amphidial fovea stirrup shaped, amphidial opening a wide transverse slit. Odontostyle 96 µm (average of types), odontophore 51 µm (average of types). Reproductive system didelphic, short, non-differentiated uteri, sphincter obscure, vulva at about mid-body (52% average). Tail conoid, dorsally convex and with rounded end, c′ 1.38 (average of types). Male rare, spicules 44.5 µm (Chilean population); six mid-ventral PS.

X. rivesi was originally described from France but has also been recorded from the USA, Canada, South America, West Indies and several countries in Europe. It is the most widespread species of the *X. americanum* group in North America. It is a vector of CRLV, PMRV and TRSV.

The *Xiphinema* 'non-*americanum*' group

Species identification within the *Xiphinema* non-*americanum* group is based in the first place on the type of reproductive system and type of uterus differentiation, subdividing the species into eight species groups. Group 1: absence of anterior genital branch; Group 2: absence of anterior ovary; Group 3: anterior genital branch present but reduced; Group 4: two equally developed genital branches with the presence of a Z-organ; Group 5: as in Group 4 but with a pseudo-Z-organ with or without uterine spines; Group 6: similar to Group 4 but without Z-organ or other uterine sclerotized structures; Group 7: as in Group 4 but uterus not differentiated and tail elongate to conical; and Group 8: as in Group 7 but tail short, rounded. Additional to these characters, species identification is also based on ratios c′ and c, the presence or absence of a terminal canal in the cuticle at the tail tip, V, body length, length of total spear, outline of lip region, habitus, presence of males and tail shape in first- and fourth-stage juveniles (Coomans *et al.*, 2002).

Three virus vector species have been identified: *Xiphinema diversicaudatum* (Micoletzky, 1927), *X. index* and *Xiphinema italiae* Meyl, 1953. Polytomous identification keys and references to species can be found in Loof and Luc (1993), Santos *et al.* (1997) and Tenente *et al.* (2002).

***Xiphinema. diversicaudatum* (Micoletzky, 1927) Thorne, 1939 (Fig. 15.4F)**

Body ventrally curved, 4.9 mm long (average). Lip region low rounded, continuous with body contour. Amphidial fovea stirrup shaped, opening a wide (about as long as diameter lip region) transverse slit. Odontostyle 143 µm long (average), odontophore 85 µm (average) with prominent flanges. Female reproductive system didelphic-amphidelphic, vulva at 43% (average) from anterior end, uterus differentiation with a characteristic pseudo-Z-organ composed of 10-20 rounded flower-like globules in each uterus. Tail short (about one anal body diameter long) dorsally convex-conoid, ventrally somewhat flattened with terminal peg (digitate) variable in length; two to four pairs of lateral body pores on tail. Male common, spicules 76 µm (average) long, guide piece 18 µm long; three well developed and one rudimentary ventromedian PS. Measurements are averages of the British fauna.

Xiphinema diversicaudatum is widespread in the temperate region of Europe under cool and humid conditions and is the most widespread longidorid in Europe. The species has been introduced into North America and New Zealand. It has a wide host range and is found in association with arable crops, grasslands, deciduous woodland

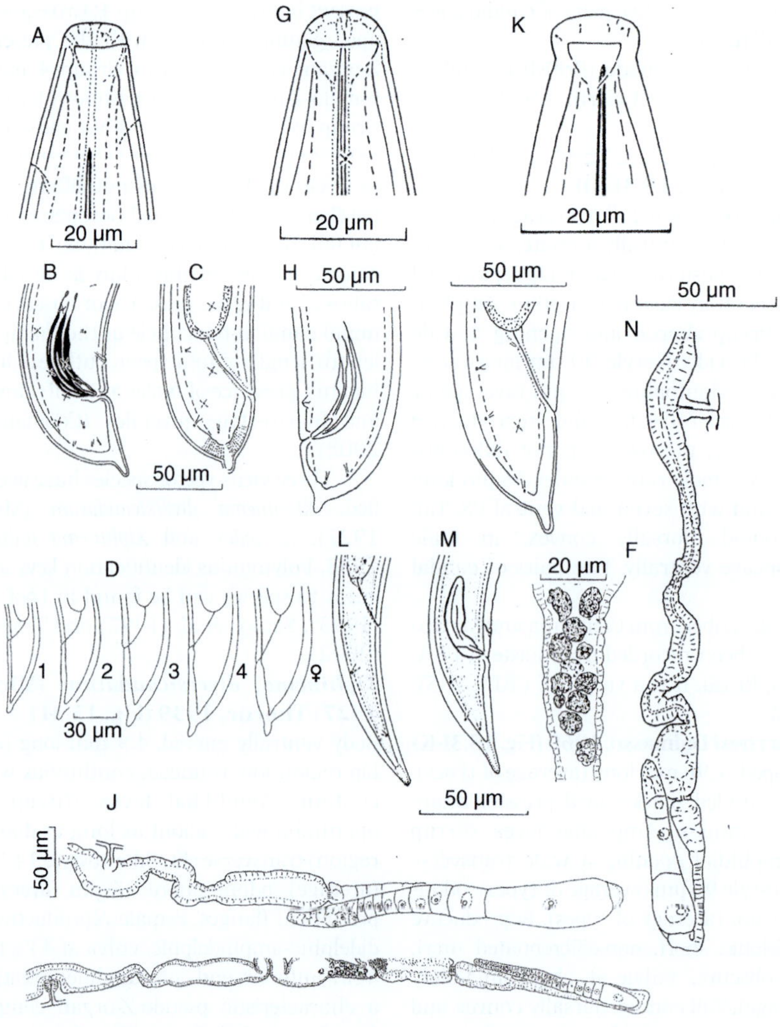

Fig. 15.4. A-F. *Xiphinema diversicaudatum*. A: Head region in surface view; B: male tail region; C: female tail region; D: tails of J1, J2, J3, J4, female; E: female reproductive system anterior branch (after Coomans *et al*., 2002); F: detail of pseudo-Z-organ in uterus. G-J. *Xiphinema index*. G: Head region in surface view; H: male tail region; I: female tail region; J: female reproductive system, anterior branch. K-N. *Xiphinema italiae*. K: Head region in surface view; L: female tail region; M: male tail region; N: female reproductive system, anterior branch (after Coomans *et al*., 2002).

and woody plants such as apple, grapevine, hops, peach and roses. It is a vector of *Arabis* mosaic and strawberry latent ringspot virus.

Xiphinema index Thorne & Allen, 1950 (Fig. 15.4G-J)

This species is smaller (about 3 mm) than *X. diversicaudatum* and forms an open spiral upon fixation. Lip region and amphidial fovea largely similar to *X. diversicaudatum*, odontostyle and odontophore slightly smaller, 126 and 70 µm long, respectively. Female reproductive system didelphic-amphidelphic, vulva at similar position (38-40% of the total body length from anterior end) but uterus not differentiated and without sclerotized structures. Tail 1-1.3 anal body diameter long, convex-conoid with a digitate ventral or terminal peg, 9-13 μm long and inner protoplasmic core not extending into peg; peg may be absent. Male extremely rare, spicules 63 μm long; apart from adanal pair of PS, a ventromedian series of seven pairs of supplements outside the range of the retracted spicules.

X. index occurs worldwide, and its distribution coincides with the distribution of grapevine, its major host. It has been spread in soil accompanying propagation material from its geographic centre of origin, the Middle East where it has been found in Iranian natural forests where wild grapevines may be common (Nguyen *et al.*, 2019). Other hosts are fruit trees such as citrus, *Prunus* spp. and fig. It is a vector of GFLV.

Xiphinema italiae **Meyl, 1953 (Fig. 15.4K-N)**
This species is comparable with *X. index* in body length (3.0 mm), but the body is only curved ventrad in the posterior region. Lip region rounded, set off by gentle constriction. Odontostyle 91 μm (average) long, odontophore 59 μm (average). Female reproductive system didelphic-amphidelphic with vulva near mid-body (46%). Uteri without Z- or pseudo-Z-organ but showing a differentiated short, widened part near the prominent sphincter between oviduct and uterus. Tail somewhat variable in length/shape but mostly elongate-conoid with slight dorsal and ventral constriction near terminal end, in some populations tail end bluntly rounded or nearly subdigitate; two or three pairs of caudal pores. Male rare, spicule 46 μm long, lateral guide pieces 14 μm and two ventromedian supplements next to the adanal pair. Measurements of topotypes, except for male (Hunt, 1993).

Xiphinema italiae has grapevines as the major host, although it also attacks a wide range of fruit trees such as apricot, peach, plum and other woody plants. In France, it was mostly limited to grapevine but in Italy and Sicily, it has been recorded from citrus, olive, mulberry, grapevine and conifers. In nature, it is associated with grapevines. It is widely distributed in Mediterranean countries but is also found in South Africa and Nigeria. It reproduces parthenogenetically and is a vector of GFLV in Israel.

The tribe Xiphidorini

Among the Xiphidorini genera, *Xiphidorus* has the widest distribution in Central and South America. Currently, the species have no major economic impact and virus transmission has not been detected. Dichotomous identification keys for *Xiphinema* and *Xiphidorus* recorded in Brazil are provided by Oliveira and Neilson (2006).

15.2.5. Molecular Identification

Accurate identification of many Longidoridae species, particularly those of the *X. americanum* group, requires an integrative taxonomic approach that includes molecular techniques for the identification and inference of relationships. Sequences of nuclear ribosomal RNA genes, D2- D3 of 28S rRNA genes and ITS1, as well as mitochondrial COI gene, have proved successful for species identification, except for certain species of the *X. americanum* group (Gutiérrez-Gutiérrez *et al.*, 2010; Kumari *et al.*, 2010; Palomares-Rius *et al.*, 2017). Furthermore, (partial) 18S rRNA has shown a potential for differentiating between species in Longidoridae. However, phylogenetic relationships at the genus level could not be resolved based on *cox1* and D2-D3 expansion regions of 28S rRNA (Palomares-Rius *et al.*, 2017).

Phylogenetic studies of the Longidoridae confirmed the monophyly of *Xiphidorus* (Oliveira *et al.*, 2004). Molecular phylogenetic analyses of the *X. americanum* species group based on rRNA show two main clades: Clade I grouping the majority of typical *X. americanum* species with characteristic female reproductive systems and Clade II including the *X. pachtaicum* species group and several recently described species from Spain (Archidona-Yuste *et al.*, 2016). The two main phylogenetic lineages are primarily correlated with their bacterial endosymbionts, that is, Clade I with *Candidatus Xiphinemabacter* and Clade II with *Candidatus Xiphinematincola* (family Burkholderiaceae) (Palomares-Rius *et al.*, 2016). So far, the relationship between *Paralongidorus* species, including *Longidoroides*, remains under discussion depending on the molecular marker used. Molecular data (18S rRNA, D2-D3 of 28S rRNA) are only available for a single *Longidoroides* species *Longidoroides. bikanerensis* (Lal & Mathur, 1987) that sees them clustered with three typical *Paralongidorus* species with slit-like apertures and stirrup-shaped foveae (*Paralongidorus sacchari*, *Paralongidorus sali* Siddiqi, Hooper &

Khan, 1963, *Paralongidorus koreanensis* Mwamula, Decraemer, Kim, Ko & Lee, 2019) but separately from the other *Paralongidorus* species (Cai *et al.*, 2020 Mwamula *et al.*, 2020; Clavero-Camacho *et al.*, 2021).

Recent molecular studies based on D2-D3 of 28S rRNA and 18S rRNA have confirmed the synonymization of *Siddiqia* Khan, Chawla & Saha, 1978 with *Paralongidorus sensu stricto* species (Cai *et al.*, 2018; Clavero-Camacho *et al.*, 2021).

15.3. Trichodoridae Thorne, 1935 – the Stubby Root Nematodes

The Trichodoridae belong to the Diphtherophoroidea within the Triplonchida (Enoplia) and currently include 127 species classified within six genera: *Trichodorus* Cobb, 1913 (72 spp.), *Paratrichodorus* Siddiqi, 1974 (36 spp.), *Nanidorus* Siddiqi, 1974 (seven spp.), *Monotrichodorus* Andrássy, 1976 (four spp., one subsp.), *Allotrichodorus* Rodriguez-M., Sher & Siddiqi, 1978 (six spp.) and *Ecuadorus* Siddiqi, 2002 (two spp.). The first three genera are referred to as the didelphic genera and occur worldwide. The other three genera are monodelphic and are only recorded from Central America and northern South America. Siddiqi (1974) recognized three subgenera within *Paratrichodorus*: *Paratrichodorus*, *Atlantadorus* and *Nanidorus*. In 1980, he raised them to genus level, but other authors did not accept this due to a lack of convincing morphological support (Decraemer, 1995). Recently, based on DNA sequences, Duarte *et al.* (2010) approved *Nanidorus* as a valid genus. Morphologically, two species groups can be recognized within *Paratrichodorus*: a *Paratrichodorus allius* group with small sperm and vulva a transverse slit and a *Paratrichodorus hispanus* group with large sperm and vulva a pore; the latter group corresponds largely with *Atlantadorus*. At present, *Atlantadorus* must still be considered as a synonym of *Paratrichodorus* until additional support can be provided through molecular sequence data (Decraemer *et al.*, 2024).

Diagnosis: Body rather short (average about 700 μm), plump, cigar shaped or J shaped (depending on genus and sex), tail short rounded, maximum one anal body diameter long in male and minute/subterminal in female. Lip region rounded, continuous with body contour; lips amalgamated. Anterior sensilla papilliform, arranged in two circlets, the anterior circle with six inner labial papillae, the outer with 10 papillae of which four protruding 'double' papillae are formed by the two subventral and two subdorsal outer labial papillae coming together with the four cephalic papillae. Amphidial fovea medium sized, cup shaped and amphid opening a transverse slit located just posterior to the lip region. Feeding apparatus or onchiostyle ventrally curved, composed of a free stylettiform dorsal tooth (onchium) on a support (onchiophore) formed by the thickened dorsal wall of the pharyngostom lumen; guide ring simple, formed by folding of the pharyngeal wall at the level of its fusion with the onchium. Pharynx with narrow mid-part and swollen posterior glandular bulb with five gland nuclei (a single large dorsal nucleus, a nearly equally large posterior ventrosublateral pair and a smaller anterior ventrosublateral pair); pharyngo-intestinal junction variable (pharyngeal bulb offset from intestine or with a ventral overlap of the pharyngeal glands over the intestine or a dorsal overlap by the intestine over the pharynx or a combination of both types of overlap). Intestine without prerectum. Male monorchic, with a single outstretched anterior testis, sperm variable in shape and size. Spicules straight to ventrally curved, variable in shape and ornamentation; spicule protractor muscles transformed into a capsule of suspensor muscles surrounding the spicular pouch. One to four (usually two or three) mid-ventral PS, one subventral pair (rarely two) of postcloacal papillae. Male with or without weakly developed caudal alae. Female reproductive system didelphic-amphidelphic or monodelphic-prodelphic, ovaries (ovary) reflexed; oviduct consisting of two finely granular cells; sperm located in spermatheca or distributed along the uterus, rarely located at flexure of ovaries (ovarial sac).

Type genus: *Trichodorus* Cobb, 1913

15.3.1. Trichodoridae Genera

The main diagnostic characters at genus level are for the male: (i) presence or absence of caudal alae; (ii) degree of development of copulatory muscles and related body habitus (J shaped or straight), and (iii) degree of development of capsule of spicule suspensor muscles; and for the female: (i) reproductive system (didelphic or monodelphic); (ii) length of vagina with respect to corresponding body diameter (cbd); (iii) size of vaginal sclerotized pieces in optical section, and (iv) presence or absence of advulvar (sub)lateral or ventral body pores. Polytomous identification keys were provided by Decraemer and Baujard (1998).

15.3.2. Generic Pictorial Key

Males (Fig. 15.5) are illustrated by the pharyngeal region and posterior body region with copulatory apparatus; females (Fig. 15.6) by the reproductive system. In addition, the pharynx is usually offset in *Trichodorus*, *Monotrichodorus*, *Allotrichodorus*, *Ecuadorus* and *Nanidorus*, while in *Paratrichodorus*, the pharyngo-intestinal junction usually shows overlaps that may be pronounced, especially the ventral overlap of the pharyngeal glands over the intestine.

Males (Fig. 15.5: males unknown for *Ecuadorus* Siddiqi, 2002).

Without caudal alae

1. *Trichodorus* Cobb, 1913: three PS, usually none or one PS opposite retracted spicule; spicules ventrally curved, shape variable, blade often with ornamentation of striae, bristles or velum;

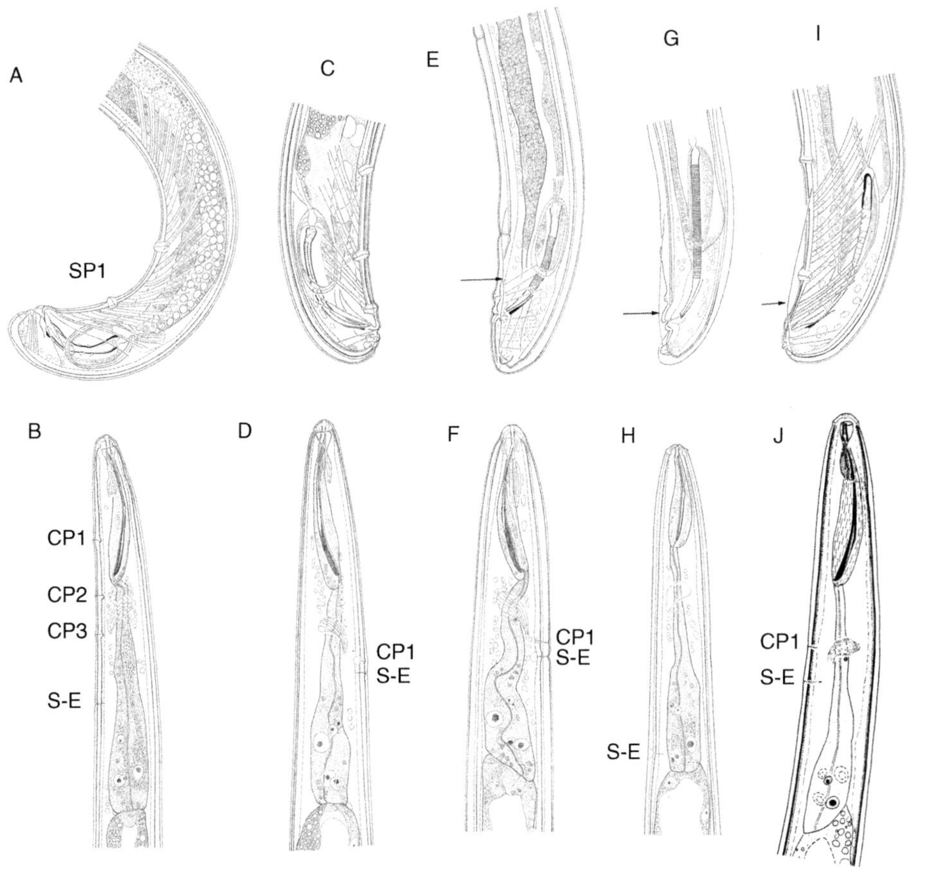

Fig. 15.5. Male posterior and anterior body region within the Trichodoridae. A, B: *Trichodorus*; C, D: *Monotrichodorus*; E, F: *Paratrichodorus*; G, H: *Nanidorus*; I, J: *Allotrichodorus*. (Not to scale, after Decraemer.) Arrows point to bursa. CP = cervical papilla; S-E = secretory-excretory pore.

spicule suspensor muscles and copulatory muscles well developed (Fig. 15.5A, B).
2. *Monotrichodorus* Andrássy, 1976: three PS, usually two PS within spicule region; spicules ventrally curved, blade striated; spicule suspensor muscles and copulatory muscles of medium development (Fig. 15.5C, D).

With caudal alae

1. *Paratrichodorus* Siddiqi, 1974: two to three (exceptionally four) PS, usually two PS opposite retracted spicule; spicules more or less straight, blade smooth or with striae; suspensor muscles and copulatory muscles weakly developed (Fig. 15.5E, F).
2. *Allotrichodorus* Rodriguez-M., Sher & Siddiqi, 1978: two to three PS, usually two PS opposite retracted spicule; spicules more or less straight, blade smooth or with striae, rarely with capitular extension; spicule suspensor muscles and copulatory muscles medium developed (Fig. 15.5G, H).
3. *Nanidorus* Siddiqi, 1974: a single PS; males rare; spicules long; spicule suspensor muscles and copulatory muscles weakly developed; no caudal pores or cervical papillae (CP) (Fig. 15.5I, J).

Females (Fig. 15.6).
Reproductive system didelphic

1. *Trichodorus*: sclerotized pieces well developed, vagina about 50% of cbd; vulva at mid-body; sperm large, spermatheca present (Fig. 15.6A).
2. *Paratrichodorus*: sclerotized pieces very small; vagina at most one-third of cbd; vulva at mid-body and either a pore or a transverse slit; spermatheca present (Fig. 15.6B) or absent (Fig. 15.6C); sperm size and shape variable (Fig. 15.6B, C).
3. *Nanidorus*: sclerotized pieces very small; vagina at most one-third of cbd; vulva posterior to mid-body at 60-64%; usually non-functional striated sperm along uteri (Fig. 15.6D).

Reproductive system monodelphic

1. *Ecuadorus* Siddiqi, 2002, males unknown: sclerotized pieces very small; vagina at most one-third of cbd; vulva posterior to mid-body at 60-66%; no sperm observed (Fig. 15.6E).
2. *Monotrichodorus* Andrássy, 1976: sclerotized pieces well developed; vagina long, anteriorly orientated; vulva at 79-84%; lateral advulvar body pore present; spermatheca present; sperm large (Fig. 15.6F).
3. *Allotrichodorus* Rodriguez-M., Sher & Siddiqi, 1978: sclerotized pieces well developed, vagina long, anteriorly orientated; vulva at 81-89%; without advulvar body pores; spermatheca present; sperm large (Fig. 15.6G).

Confusable genera: Males of *Monotrichodorus* could be confused with those of *Trichodorus* because both lack caudal alae and the spicules are curved and ornamented. However, in *Monotrichodorus*, the capsule of the suspensor muscles is less developed and the copulatory muscles extend less far anteriorly, resulting in a weaker ventral curvature of the posterior body region. *Monotrichodorus* males have, at most, a single ventromedian CP.

Females of the didelphic *Nanidorus* and *Paratrichodorus* can easily be confused since both possess a short vagina and small vaginal sclerotized pieces in lateral optical section. *Nanidorus* species used to belong to *Paratrichodorus* but were later differentiated from it by the lack of body pores, a slightly more posterior position of the vulva (> 60%) and uteri with thread-like and apparently non-functional sperm (one exception), males usually rare.

Females of the monodelphic *Allotrichodorus* and *Monotrichodorus* can only be distinguished from one another by the presence of an advulvar lateral body pore in *Monotrichodorus* that is absent in *Allotrichodorus*.

Although *Trichodorus* and *Paratrichodorus* are usually well characterized, their differentiation may be rather confusing in some *Trichodorus* species with males possessing caudal alae, as in *Trichodorus cylindricus* Hooper, 1962 and *Trichodorus paracedarus* Xu & Decraemer, 1995, or a slightly swollen subventral body wall suggesting rudimentary caudal alae (*Trichodorus pseudobursatus* Decraemer, Radivojević, de la Peña, 2008, *T. sparsus* Szczygiel, 1968) and females with a short vagina (*Trichodorus elefjohnsoni* Bernard, 1982) or sperm dispersed along the uteri (*Trichodorus nanjingensis* Liu & Cheng, 1990) or with small vaginal sclerotized pieces.

15.3.3. Species Identification and Major Species

Identification to species level is often hampered by the impact of fixation on morphological fea-

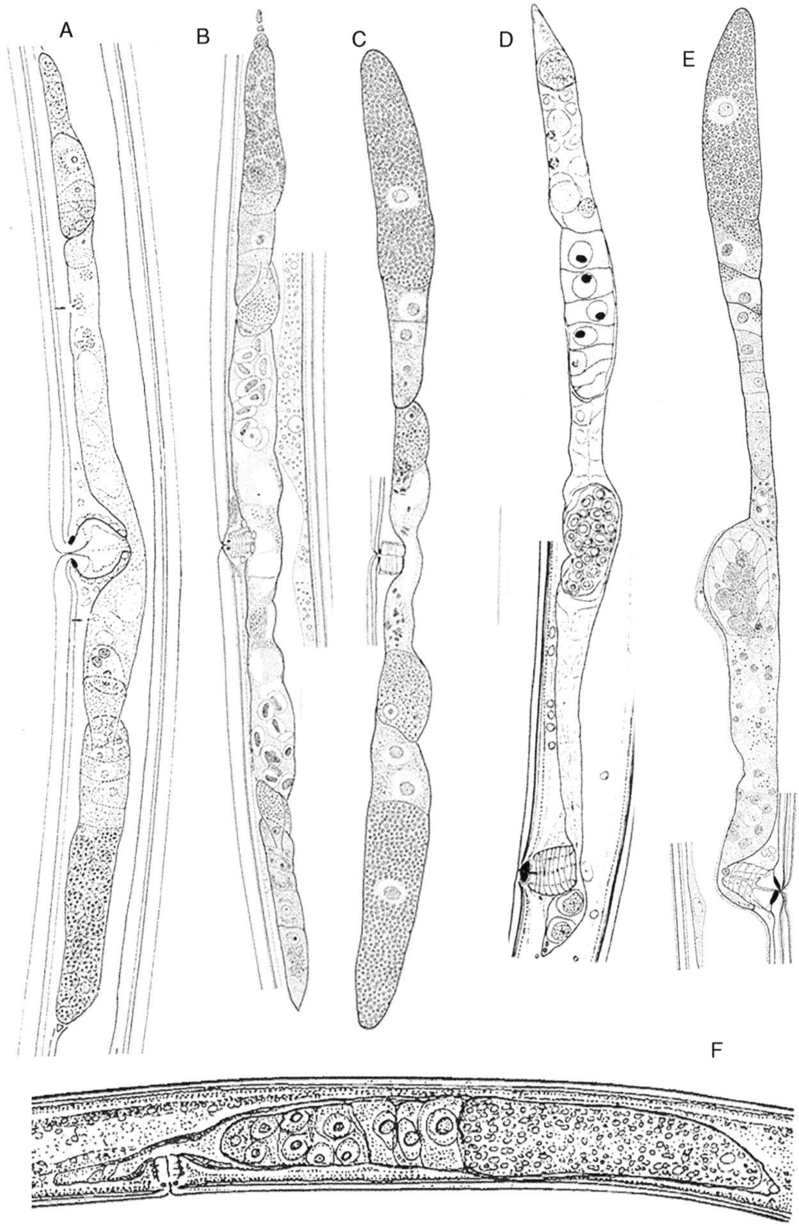

Fig. 15.6. Female reproductive system within the Trichodoridae. A: *Trichodorus*; B: *Paratrichodorus*; C: *Nanidorus*; D: *Allotrichodorus*; E: *Monotrichodorus* (after Decraemer); F: *Ecuadorus* (after Siddiqi). (Not to scale.)

tures and measurements, the necessity for specimens to be in lateral view, overlapping morphometric data and the co-occurrence of cryptic species. Multivariate morphometric analysis can thus represent a valuable complementary tool in species delimitation (Decraemer *et al.*, 2024). The main morphological features at the species level are, in the female, the shape and size of the vaginal sclerotized pieces in optical section; and, in the male, the shape and ornamentation of the spicules. Other diagnostic features include morphometrics such as body length, onchiostyle length, male spicule length, number and position of PS, number and position of ventromedian CP, size of female sclerotized vaginal pieces, length of vagina and number and position of female pre- and post-vulval body pores. In addition, for *Paratrichodorus*, presence/absence of male, sperm structure and size and position of sperm in the female are also useful indicators. As major species, we consider the virus vector species due to their economic impact. Despite no virus vector association being proven for the monodelphic genera, we have included the most common species in Tables 15.2-15.4 as monodelphic genera are considered endemic for Central and South America, whereas the didelphic genera are considered to have been introduced. A dichotomous identification key is provided by Decraemer (1995) and a tabular identification key by Decraemer and Baujard (1998). An illustrated key to Trichodoridae vector species based on male tail region and copulatory apparatus and female vagina region (Figs 15.7, 15.8 together with Tables 15.2-15.4); and common and monodelphic species (Fig. 15.7 and Table 15.5) are provided.

Molecular identification

Molecular data of known species including from specimens from the type locality are important for reliable species identification. Within the Trichodoridae, sequence data from D2-D3 of 28S rRNA are only available from about one-third of *Trichodorus* species and sequence data of 18S rRNA from one-fifth of the species; very few ITS rRNA data or COI sequences are available. For *Paratrichodorus*, about 40% of valid species sequence data of both the small and large subunit of ribosomal RNA gene are known, while mitochondrial COI data are only available for three species. For *Nanidorus*, only three species possess sequence data information of 18S rRNA and/or D2-D3 of 28S rRNA. Molecular data of the monodelphic genera are only present for *Monotrichodorus vangundyi* Rodriguez-M., Sher & Siddiqi, 1978. Other molecular identification methods include polymerase chain reaction-restriction fragment length polymorphism (PCR-RFLP), end-point single or duplex PCR with specific primers and quantitative real-time PCR (Huang *et al.*, 2019). More recently, one-step multiplex PCR assays have been developed for the rapid detection of major trichodorid pest species in the USA such as *Trichodorus obtusus* Cobb, 1913, *Nanidorus minor* (Colbran, 1956), *P. allius* (Jensen, 1963) and *Paratrichodorus porosus* (Allen, 1957).

15.4. Bionomics of Virus Vector Families

15.4.1. Feeding

Trichodoridae possess a rather short tooth-like spear (onchiostyle) with a closed tip and feed on epidermal cells near the root tip and on root hairs using a feeding tube, produced anteriorly in the buccal cavity by secretions from the dorsal and anterior ventrosublateral pharyngeal glands. Trichodorids feed by a continuous thrusting of the onchiostyle throughout all phases of a complete feeding cycle, representing a unique behaviour for plant parasites. They often damage the host's meristematic tissue, resulting in a cessation of root elongation and the development of terminal swelling, thereby forming the characteristic 'stubby root.' When the main root is damaged, lateral roots can take over or more lateral roots are developed. Trichodorid nematodes often puncture a large number of cells and interrupt feeding before damage to the plant cell and provide a route for viral transmission.

Longidoridae, however, possess a hollow spear or odontostyle with a dorsal opening at the tip. This much longer spear allows them to feed progressively deeper on a series of cells, ingesting the cell content during short periods, which alternate with periods of secretion release from the dorsal pharyngeal gland. Feeding time is much longer compared with trichodorids and can last from 1 to 3 h. *Longidorus*/*Paralongidorus* feed preferentially at, or just behind, the root tip on procambium and primary meristem, while most *Xiphinema* species feed on the cortical cells further away from the root tip. Attacked roots show a lack of lateral roots and display tissue necrosis at the feeding sites. Moreover, in the case of *Xiphinema* (except for the *X. americanum* group), parasitism often induces the formation of elongated or curled terminal root galls. Root galls caused by *X. index* contain enlarged multinucleate cells, reminiscent of giant cells formed

Table 15.2. Illustrated key for *Trichodorus* vectors with main features in bold (Fig. 15.7).

Species	Habitus male; spic (μm)	L male and female (μm)	Stylet male and female (μm)	Nr SP, position SP1	Number CP and location	Spicule shape	Shape and length of vagina	Shape and size vaginal sclerotized pieces	Sperm shape and location	Vulva
Trichodorus cylindricus (Fig. 15.7A, B)	**Straight; caudal alae;** 28–49	500–800; 420–770	37–48; 35–52	3 (SP1 opposite spicule)	3 anterior EP; **CP1 in onchiostyle region**	**Blade posteriorly enlarged; striated, bristles**	**Barrel/ pear-shaped**; >1/3rd cbd	Well developed, rounded triangular	Sausage-shaped nucleus; in spermatheca	Transverse slit
Trichodorus primitivus (Fig. 15.7C, D)	J-shaped; 32–54	500–1000; 500–950	47–59; 34–57	3 (SP1 opposite head spicule)	3; **CP1 and CP2 in onchiostyle region**	Enlarged head; **blade very narrow, with bristles**	Rhomboid to elongate; >1/3rd cbd	**Rods, well separated, parallel with vagina lumen**	Sausage-shaped nucleus; in spermatheca	Transverse slit
Trichodorus similis (Fig. 15.7E, F)	J-shaped; 30–44	550–1060; 600–1000	35–50; 36–52	3; SP1 at head spicule	3; **CP1 in onchiostyle region**	**Knob-like head offset**, calomus narrower with few obscure bristles blade stout ventrally curved,	**Rhomboid;** >1/3rd cbd	Medium-sized, **rounded triangular, pointing to vulva**	Sausage-shaped nucleus; in spermatheca	Transverse slit
Trichodorus viruliferus (Fig. 15.7G, H)	J-shaped; 22–37	615–890; 650–920	32–53; 36–56	3; **SP1 just anterior to head spicule**	3; **CP1 and CP2 in onchiostyle region**	Widened head, **blade with constriction and bristles**	Rhomboid; > 1/3rd cbd	**Oval, oblique, well separated**	Sausage-shaped nucleus; in spermatheca	Transverse slit

Table 15.3. Pictorial key *Paratrichodorus* vectors with main features in bold (Fig. 15.8).

Species	Habitus male and spicule length (μm)	L male and female (μm)	Stylet male and female (μm)	Nr SP and position SP1	Number CP and location	Spicule shape	Shape and length vagina	Shape and size vaginal sclerotized pieces	Sperm shape and location	Vulva
Paratrichodorus anemones; (Fig. 15.8A, B)	Straight; caudal alae; 46–53	710–900; 650–970	47–52; 42–53	3; **SP1, SP2 opposite distal half spicule**	1 anterior to EP	**Slender, calomus with irregular outline,** striated	Barrel with **mid-way constriction**	Small narrow triangular, oblique, close	Sausage-shaped nucleus; in spermatheca	Pore
Paratrichodorus divergens; (Fig. 15.8C, D)	Straight; caudal alae; 45–56	590–855 600–1070	43–55 44–55.5	3; 2 opposite distal third spicule	1 anterior to EP	**Slender, calomus with irregular outline,** striated	Rounded rectangular with slight mid-way constriction	**Large drop-like or oval triangular oblique, tips diverging outwards**	Sausage-shaped nucleus; **no spermathecae**	Pore
Paratrichodorus hispanus; (Fig. 15.8E, F)	Straight; caudal alae; 44–64	690–1105; 660–1100	49–64; 43–64	3; two opposite distal half spicule	1 anterior EP	**Straight; kinking calamus**; striated blade	Trapezoid, without indentation	**Small, round triangular, oblique and close**	Sausage-shaped nucleus; **in spermatheca**	Pore
Paratrichodorus allius (male rare) (Fig. 15.8G, H)	Straight; caudal alae; 31	590–680; 400–1045	42; 35–53	2 in spicule region	0	straight	Broad rounded; oval	**Small, well spaced**	**Small globular; in uteri** and at tip flexure ovaries	Longitudinal slit
Paratrichodorus porosus; male rare (Fig. 15.8I, J)	Straight; caudal alae; 36–39	530–770; 420–867	43–48; 39–58	**2 in spicule region, well separated**	1 anterior to EP	Straight; blade striated	Short, wide rounded rectangular to trapezoid	Small oval well separated; **ventral body pores**	Sausage shaped; rarely present	Pore
Paratrichodorus pachydermus (Fig. 15. 8K, L)	Straight; caudal alae; 37–54	610–990; 600–1020	45–60; 44–58	3; two in spicule region, well separated	1 anterior to EP; not opposite onchiostyle	Straight, blade striated	Barrel-shaped	**Small rounded triangular to** oval	Sausage-shaped nucleus; in uterus	Pore
Paratrichodorus teres; male rare (Fig. 15.8M, N)	Straight; caudal alae; 36–58 μm	535–1030; 615–1325	41–64; 41–61	3; two in spicule region	**0**	Straight, slightly cephalated, blade striated	Rhomboid, widened mid-way	Minute oval, well separated	Small rounded, **fibrillar; in uterus**	Longitudinal slit
Paratrichodorus tunisiensis (Fig. 15.8O, P)	Straight; caudal alae; 45–50	715–900; 705–1100	47–52; 48–51	3, two in spicule region, spread	**1 anterior to EP**, not in onchiostyle region	Slightly cephalated; blade striated	Short (<1/3rd cbd), oval	Minute oval, well separated	**Small oval, dispersed in uteri**	**Longitudinal slit**

Table 15.4. Pictorial key to *Nanidorus* vectors with main features in bold (Fig. 15.7 A-D).

Species	Habitus male and spicule length (μm)	L male and female (μm)	Stylet male and female (μm)	Nr SP and position SP1	Number CP and location	Spicule shape	Shape and length of vagina	Shape and size vaginal sclerotized pieces	Sperm shape and location	Vulva
Nanidorus minor (Fig. 15.7A, B)	Straight; caudal alae; **48–73**	540–670; 440–1530	28–39; 20–47; inner onchium in adult	1, near cloacal opening	none	Long, straight, except curved distal end	Quadrangular/ rectangular; at most 1/3rd cbd	**Rod-shaped, parallel with longitudinal body axis**	**Non-functional, striae; in uterus**	Transverse slit
Nanidorus nanus (Fig. 15.7C, D)	Straight; caudal alae; 42–50	430–550;430–600	25–26; 21–26; inner onchium in adult	1, near cloacal opening	none	Long, straight, striated	Barrel-shaped; at most 1/3rd cbd	**Rounded triangular, close**	Non-functional, striae; in uterus	Transverse slit

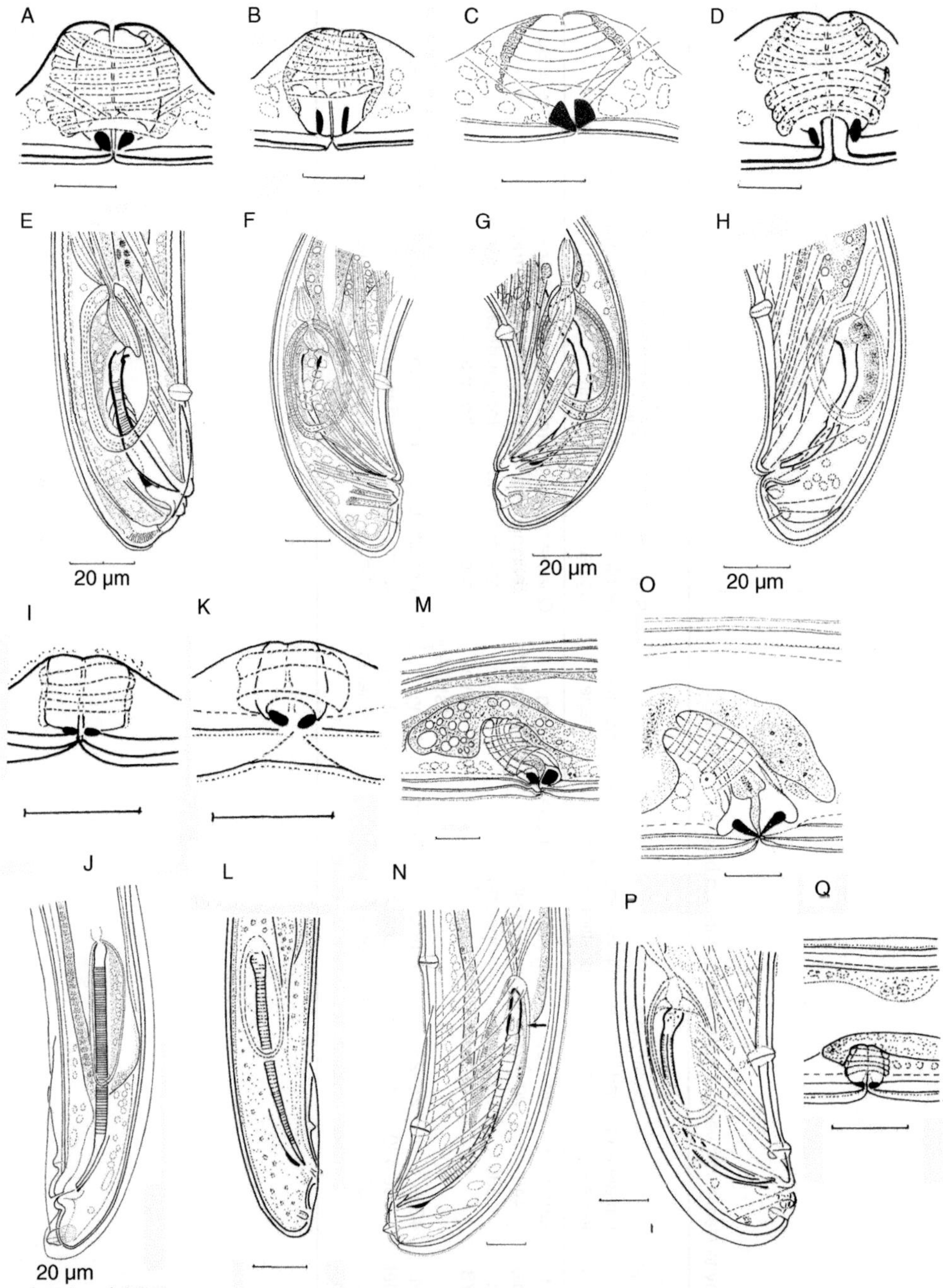

Fig. 15.7. Vaginal region (optical section, lateral) and posterior male body region. A, E: *Trichodorus cylindricus*; B, F: *Trichodorus primitivus*; C, G: *Trichodorus similis*; D, H: *Trichodorus viruliferus*; I, J: *Nanidorus minor*; K, L; *Nanidorus nanus*: M, N: *Allotrichodorus campanullatus*; O, P: *Monotrichodorus monohystera*; Q: *Ecuadorus* (after Siddiqi).

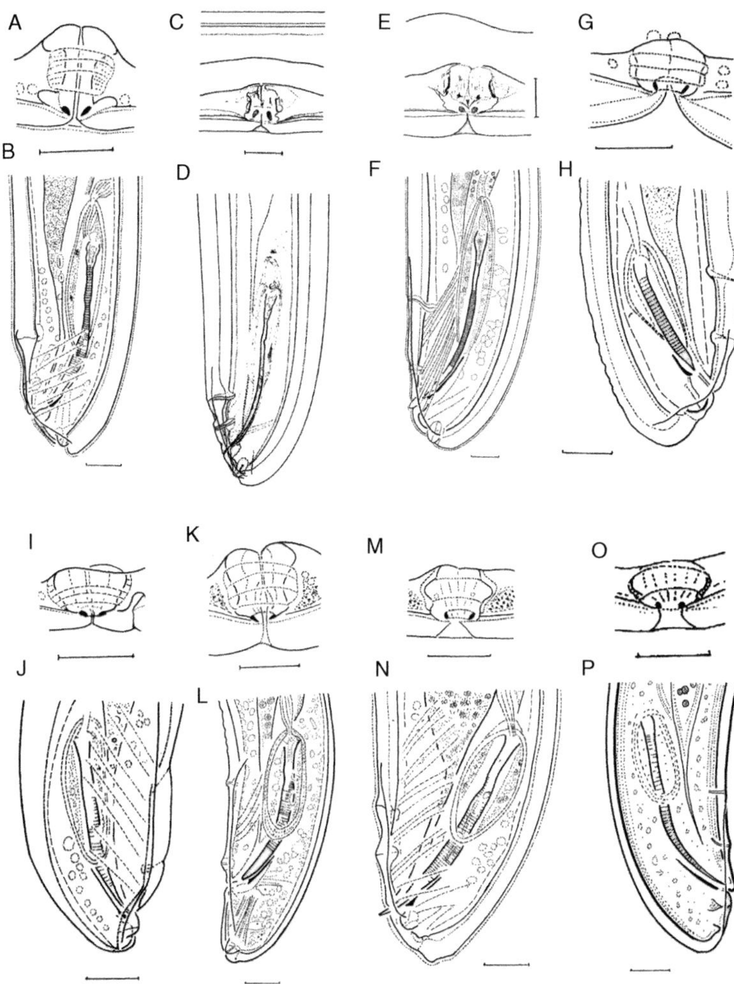

Fig. 15.8. Vaginal region (optical section, lateral) and posterior male body region in *Paratrichodorus*. A, B: *Paratrichodorus anemones*; C, D: *Paratrichodorus divergens*; E, F*: Paratrichodorus hispanus*: G, H: *Paratrichodorus allius*; I, J: *Paratrichodorus porosus*; K, L: *Paratrichodorus teres*; M, N: *Paratrichodorus pachydermus*; O, P: *Paratrichodorus tunisiensis* (after Decraemer).

by root-knot nematodes. Within *Longidorus* and *Paralongidorus*, root galling also occurs, but frequently the root continues to grow beyond the gall and no multinucleate cells are observed. In general, *Longidorus* and related genera are more often found in association with herbaceous plants, whereas *Xiphinema* is better adapted to feed on the roots of woody plants.

Virus vector species in both families acquire virus particles during feeding on an infected plant cell. In trichodorid vectors, acquired virus particles adhere to the pharyngeal wall in the onchiostyle region, the narrow pharyngeal region and the pharyngeal bulb anterior to the outlets of the posterior ventrosublateral glands. In *Longidorus*, virus particles adhere as a single layer on the inner surface of the odontostyle (e.g., *L. macrosoma* and CRLV) but may also be found between the odontostyle and guiding sheath (*L. elongatus* and ToRSV) (Taylor and Brown, 1997). In *Xiphinema* vectors, nepoviruses are associated with the pharyngeal lumen at the level of the odontophore, the narrow part of the pharynx and the pharyngeal bulb. Pharyngeal gland secretion during feeding induces the release of virus particles in the vector through

Table 15.5. Pictorial key to common monodelphic species with key features in bold (Fig. 15.7E-J).

Species	Habitus male and spicule length (μm)	L male and female (μm)	Stylet male and female (μm)	Nr SP and position SP1	Number CP and location	Spicule shape	Shape and length vagina	Shape and size vaginal sclerotized pieces	Sperm shape and location	Vulva
Allotrichodorus campanulatus (Fig. 15.7E, F)	Straight; caudal alae; XX μm	530–780; 530–750	45–64.5; 45–65.5	**3 in spicule region**	1 anterior EP	Slender, with **capitular extension**, striae, bristles	>1/2, anteriorly directed; vulva a transverse slit; V = 83–86.5	Triangular to trapezoid	Rounded;	Transverse slit
Ecuadorus westindicus male unknown (Fig. 15.7G, H)		380–500; 430–600	30–37; 21–26; **inner onchium in adult**				Barrel-shaped; at 1/3rd cbd; rounded; vulva a transverse slit; **V = 60–65**	**Rounded minute triangular, close**	Non-small cuneate, in uterus	Transverse slit
Monohystera monohystera (Fig. 15.7I, J)	J-shaped, no alae	490–920; 480–940	41–62; 40–63	3 SP of **which 1 in spicule region**	1 anterior EP	**Slender, curved, striae**	½ **mbd, anteriorly directed; vulva a transverse slit; V = 75–85.5**	Rod-like, tips close, oblique	Rounded, in spermatheca	Transverse slit

changes in pH or via enzymatic effects, and transport and consequent injection of the virus particles into the plant cell occur together with the secretions of pharyngeal glands.

Viruses do not replicate in the vector species, are not passed transovarially through nematode eggs and are shed when juvenile stages of the nematode pass through a moult, as at this time the feeding apparatus and inner lining of the pharynx are shed or partially resorbed.

15.4.2. Habitat Preference

In general, Longidoridae and Trichodoridae appear to be more abundant in coarse-structured soils with a high percentage of sand, although this does not preclude their occurrence in heavier soils (e.g., *X. diversicaudatum* or *Trichodorus primitivus* (de Man, 1880)). Trichodoridae appear to be highly susceptible to desiccation, whereas longidorids can tolerate both very dry and waterlogged soils. For example, *Xiphinema bergeri* Luc, 1973, a parasite of rice roots in West Africa, lives for a third of the year in flooded soils yet survives the rest of the time in hot sunny conditions, while *Paralongidorus australis* Stirling & McCulloch, 1984 is restricted to very wet soils.

Both families may show differences in soil depth between species that are not always related to the host plant. Trichodoridae are often found at 30-40 cm depth in the dry summers in temperate regions or below 40 cm in tropical/subtropical regions. However, in temperate and wet climatic conditions such as those encountered in Scotland, trichodorids occur near the surface, between 10 and 20 cm depth. Longidorid populations are often found to be greater at depths below 20 cm. Both trichodorids and longidorids may occur at great depths (e.g., deeper than 2 m for *X. index*). Trichodoridae also appear sensitive to mechanical disturbance and, as a result, are usually found below the normal depth of soil cultivation. In both families, it is common to find a mixture of several species occurring together in the rhizosphere of a host plant.

15.4.3. Reproduction

Trichodoridae and Longidoridae are K-strategists. They display a somewhat low reproduction rate, resulting in a slow increase in population density and rather long adult life expectancy. Within the Longidoridae, Longidorini (*Longidorus*, *Paralongidorus*) and Xiphidorini (*Xiphidorus*, *Paraxiphidorus* and *Australodorus*), most species reproduce through amphimixis. In *Xiphinema*, 40% of the species possess an amphimictic reproduction, but 60% reproduce by thelytokous parthenogenesis. Of the latter, 25% occasionally have males and reproduce amphimictically, while in 35% of the species, males are unknown. The *X. americanum* group possesses two genera of maternally inheritable verrumicrobial bacteria with similar ecological niches (Palomares-Rius *et al.*, 2021). They are potentially responsible for the induction of parthenogenesis in this group, a phenomenon which has also been observed with the presence of *Wolbachia* in various groups of arthropods (Coomans *et al.*, 2002). Parthenogenesis as a model of reproduction allows a rapid build-up of populations when environmental conditions are unfavourable. Longidoridae species have, in general, a longer life cycle compared to Trichodoridae, although substantial variation based on laboratory experiments or due to different climatic conditions can be found in the literature. *L. elongatus* (de Man, 1876) was estimated to have one generation per year in the UK, while in Israel at 30°C, *Longidorus africanus* Merny, 1966 completed its life cycle in less than 3 months. The life cycle of *X. index* has likewise been reported to last between 2 and 14 months depending on the soil temperature and related growth of the host plant.

Within *Longidorus* and *Xiphinema*, several examples are available of species that can survive for long periods in moist soils without a host plant (e.g., *L. elongatus* survived more than 2 years). *X. pachtaicum* (Tulaganov, 1938) can overcome dry periods by going into a quiescent phase by coiling up. In *Longidorus* and *Xiphinema*, some species have only three juvenile stages instead of the typical four stages found in nematodes. Only in the *X. americanum* species group do the species with three juvenile stages appear clearly related, representing an early evolutionary separation from the species in the 'old world' with four juvenile stages (Taylor and Brown, 1997). According to Coomans *et al.* (2002), progenesis as a phenomenon is not uncommon in parthenogenetic organisms.

In Trichodoridae, reproduction (i) is amphimictic in *Trichodorus*, *Monotrichodorus*, *Allotrichodorus* and *Paratrichodorus* (*partim*); (ii) occurs by thelytokous parthenogenesis in the absence of a male and without sperm in the female reproductive system *Nanidorus* (*partim*, e.g., *N. minor*), *Ecuadorus*, *Paratrichodorus* (*partim*, e.g., *Paratrichodorus teres* (Hooper, 1962)), and (iii) is assumed automixis with male rare or absent and sperm present in the female reproductive system (ovarial sac or ovary) as in *Nanidorus* (e.g., *Nanidorus nanus* (Allen, 1957)) and *Paratrichodorus* (*partim*, e.g., *P. allius*). In *Paratrichodorus*, species of the *P. hispanus* Roca & Aria, 1986 species group with large sperm cells are gonochoristic, while species of the *P. allius* species group with small sperm cells are either gonochoristic or assumed hermaphrodites/parthenogenetic (Decraemer *et al.*, 2024).

Little information is available on the embryogenesis of Trichodoridae. In general, four juvenile stages are accepted but first-stage juveniles have rarely been observed. Also, moulting specimens are rarely found in fixed samples, for example, *Trichodorus marylandi* Decraemer, Kantor, Handoo & Subbotin, 2022.

15.5. Biogeography of Virus Vector Families

Within the Longidoridae and Trichodoridae, a parallel is observed in their biogeography, with genera exhibiting a worldwide distribution *vs* those with distribution limited to Central and South America. The latter taxa appear to have evolved more recently, that is, after South America separated from Africa as a result of continental drift. For the Longidoridae, such genera belong to the Xiphidorini (*Australodorus*, *Paraxiphidorus* and *Xiphidorus*) and for the Trichodoridae, the monodelphic genera *Monotrichodorus*, *Allotrichodorus* and *Ecuadorus*. All these genera have only a few described species.

Within the Longidoridae, *Xiphinema* and to a lesser extent *Longidorus* have the most extensive worldwide distribution and are most numerous in terms of species description. However, most species from both genera have a restricted distribution, for example, within the *X. americanum* group, 50% of the species are either rare or taxonomically questionable. *Xiphinema* species appear more numerous in tropical areas than in temperate ones; the highest diversity of the *Xiphinema non-americanum* species group has been observed in South Africa (>50 spp.), while the highest diversity of the *X. americanum* group was recorded from Europe (28 spp.), followed by Asia (24 spp.) and North America (16 spp.) (OEPP/EPPO, 2017). The largest percentage of *Longidorus* species (56%) have been recorded from Europe, followed by 50% from Asia. The largest percentage of *Paralongidorus* species (including *Longidoroides*) are reported from India (40%) and South Africa (33%). Within the Xiphidorini, *Xiphidorus* is the most widespread. However, the records are biased by the interest in the taxa and surveys conducted, including or excluding natural environments, as illustrated by the increase in new taxa described from Iran.

In the Trichodoridae, the didelphic genera are also widespread, occurring in all six zoogeographic regions, whereas the most species-rich genus, *Trichodorus*, is cosmopolitan, being predominant in more temperate regions, with no known records from Australia and Antarctica. The genus is assumed to have originated in former Laurasia, with Europe as the main centre of speciation. *Paratrichodorus* is also a cosmopolitan genus with some species characteristic of temperate regions (e.g., *P. pachydermus*, *Paratrichodorus anemones* ((Loof, 1965)) and others for tropical/subtropical regions (e.g., *P. porosus*, *Paratrichodorus lobatus* (Colbran, 1965)). Some species have a wide distribution as in *P. pachydermus* (northern hemisphere), while others have a more regional distribution as in *P. anemones* (Western Europe, Portugal). The occurrence in South America is considered the result of anthropogenic interference. *Nanidorus* is also a cosmopolitan genus, but with only a small number of species (seven) described. Nevertheless, it includes the most widespread trichodorid pest species, *N. minor* (typical for tropical and subtropical regions), while other species are more regionally distributed (e.g., *Paratrichodorus caribbensis* Marais, Decraemer & Quénéhervé, 1996 is restricted to Martinique).

The monodelphic genera are only known from the Neotropical region, with *Monotrichodorus* being most widely recorded from Central and South America (north and northwest of the Amazon Basin) (Decraemer *et al.*, 2019) while *Allotrichodorus* has only been recorded from Bahia State, Brazil, and *Ecuadorus* from Brazil (Bahia state), Martinique and Ecuador. Their association with native plants supports the theory that monodelphic genera are endemic to Central America and the tropical South.

15.6. Management

15.6.1. Direct Damage

Trichodoridae and Longidoridae cause direct damage to a wide variety of plants when their population densities are high. In the field, direct feeding results in stunted plants with discoloured leaves appearing in patches of irregular shape and size, reflecting the distribution of the nematodes. Although both families are polyphagous, the host range may vary significantly between the families, genera and even among species within a genus. *N. minor* (Trichodoridae), for instance, has been associated with more than 100 plant species, including several crops of economic importance, while other species, like *X. index*, have a narrow host range and mainly parasitize grapevine. The economic impact of crop losses due to direct damage by Longidoridae and Trichodoridae is a global issue, particularly when population densities are high. Reports of yield losses are available for only a few species. In Europe, losses as high as 50-60% were estimated in sugar beet crops showing symptoms of the disease 'Docking disorder' caused by *L. attenuatus* Hooper, 1961 and *P. anemones* (Taylor and Brown, 1997). Some species, like *N. minor*, are considered more economically significant due to their widespread distribution and broad host ranges. The extent of root system damage varies not only with the host species (as not all hosts are equally susceptible) but also appears to be influenced by seasonal factors (correlated with precipitation) and soil type, with damage more severe in lighter soils. *X. index* is a notable pathogen of vineyards (Fig. 15.9; Andret-Link *et al.*, 2004). Soil fumigation, once a

Fig. 15.9. Damage (galling) caused by direct feeding of *Xiphinema index* on root system of *Vitis* rootstock (courtesy Pablo Castillo).

quick and effective method to prevent virus infection, is no longer an option due to environmental concerns; incorporating virus resistance or tolerance into crops appears to represent a promising new approach (Halbrendt, 2021). *X. americanum* has a broad host range, causing severe stunt in soybean in the USA (Delaware). In this study, reduction in the dagger nematode population was achieved through crop rotation with severe stunt virus-tolerant soybean, corn and wheat, or via a 2-year fallow period (Evans *et al.*, 2007). It is noteworthy that integrated management programs for *X. americanum sensu lato* are currently under investigation (Halbrendt, 2021; Zasada and Forge, 2022).

15.6.2. Damage as Virus Vectors

The major pest status of Longidoridae and Trichodoridae is as virus vectors, despite the remarkably low number of vector species recorded from both families. However, the minimum density at which a virus vector species is considered to cause economic damage (damage-level threshold) is equivalent to a single specimen. Although the distribution of virus vector species differs between continents and zoogeographic regions, all four *Trichodorus* vector species are common in Europe and naturally transmit TRV and/or pea early browning virus. Apparently, some or all of them have been introduced to other regions such as North America, southern South America, the Middle East and Oceania. Within *Paratrichodorus*, apart from *P. allius*, all vector species are common in Western Europe and the Mediterranean region. In the USA, four trichodorid vector species, *P. allius*, *P. porosus* and *N. minor* occur in several states and show a broad host range, including potato, vegetables, common grain and turf grasses (Huang *et al.*, 2019). The need for a rapid, more efficient diagnostic and low-cost method resulted in the development of a one-step multiplex PCR assay (Huang *et al.*, 2019).

15.7. Quarantine Species

Of all Trichodoridae, *N. minor* is the most widely spread trichodorid virus vector species in tropical and subtropical regions. It is a vector of TRV, which is a quarantine pathogen in many countries (e.g., USA, Brazil and Turkey) and whose spread is controlled by seed-potato regulation in many countries of the European and Mediterranean Plant Protection Organization (EPPO). However, the nematode vector is not listed as a pest within the Plant Health Directive. CSL pest risk analysis for *N. minor* considers this species of low risk for the UK. The virus it vectors is already present on potatoes in the UK, along with many other established vectors. Its pathways are restricted to growing media around plant material, and for this reason, this nematode is of low risk for most field-grown crops in the UK (UK Plant Health Information Portal). By contrast, the OEPP/EPPO A1 and A2 lists of pests recommended for regulation as quarantine pests (PM 1/2 (19) PM ½ (19) September 2010 do include *X. americanum, sensu stricto* (A1/150), *X. bricolense* (A1/260), *X. californicum* (A1/261) and *X. rivesi* (A2/262) (OEPP/EPPO 2017; PM7/95 (2). *L. diadecturus* is still listed on the EU Directive 2000/29/EC but due to insufficient analysis used to prove its vector capability, it should not be considered a quarantine listed pest (Jeger *et al.*, 2018).

Note

[i] *Siddiqia* Khan, Chawla & Saha, 1978 is not included as a valid genus (see Peña-Santiago, 2021) as *Paralongidorus* is paraphyletic with species arranged in two subclades clustering within *Longidorus* (Mwamula *et al.*, 2020).

15.8. References

Andret-Link, P., Laporte, C., Valat, L., Ritzenthaler, G., Vigne, E., *et al.* (2004) Grapevine fanleaf virus: still a major threat to the grapevine industry. *Journal of Plant Pathology* 86, 183–195.

Archidona-Yuste, A., Navas-Cortés, J.A., Cantalapiedra-Navarrete, C., Palomares-Rius, J.E., *et al.* (2016) Cryptic diversity and species delimitation in the *Xiphinema americanum*-group complex (Nematoda:

Longidoridae) as inferred from morphometrics and molecular markers. *Zoological Journal of the Linnean Society* 176, 231–265.

Cai, R., Archidona-Yuste, A., Cantalapiedra-Navarrete, C., Palomares-Rius, J.E. and Castillo, P. (2020) New evidence of cryptic speciation in the family Longidoridae (Nematoda: Dorylaimida). *Journal of Zoological Systematics and Evolutionary Research* 58, 869–899.

Chen, Q.W., Hooper, D.J., Loof, P.A.A. and Xu, J. (1997) A revised polytomous key for the identification of species of the genus *Longidorus* Micoletzky, 1922 (Nematoda: Dorylaimoidea). *Fundamental and Applied Nematology* 20, 15–28.

Clavero-Camacho, I., Liébanas, G., Escuer, M., Cantalapiedra, Navarrete, C., *et al.* (2021) Studies on *Longidorus iberis* (Escuer & Arias, 1997) n. comb. (Nematoda: Longidoridae) from Spain. *Nematology* 23, 871–885.

Coomans, A. (1996) Phylogeny of the longidoridae. *Russian Journal of Nematology* 4, 51–60.

Coomans, A., Huys, R., Heyns, J. and Luc, M. (2002) Character analysis, phylogeny, and biogeography of the genus *Xiphinema* Cobb, 1873 (Nematoda, Longidoridae). *Annales du Musée Royal de l'Afrique Centrale (Zoologie), Tervuren, Belgique* 287, 1–240.

Decraemer, W. (1995) *The Family Trichodoridae: Stubby Root and Virus Vector Nematodes*. In: *Developments in Plant Pathology*, 6. Kluwer Academic Publishers, Dordrecht, Boston, London.

Decraemer, W. and Baujard, P. (1998) A polytomous key for the identification of the family Trichodoridae (Thorne, 1935) Clark, 1961 (Nematoda: Triplonchida). *Fundamental and Applied Nematology* 21, 37–62.

Decraemer, W. and Coomans, A. (2007) Revision of some species of the genus *Paralongidorus sensu* Siddiqi *et al.* (1993) with a discussion on the relationships within the family Longidoridae (Nematoda: Longidoridae). *Nematology* 9, 643–662.

Decraemer, W. and Geraert, G. (2006) Ectoparasitic nematodes. In: Perry, R.N. and Moens, M. (eds) *Plant Nematology*. CAB International, Wallingford, UK., pp. 153–184. DOI: 10.1079/9781780641515.0179

Decraemer, W. and Robbins, R.T. (2007) The who, what, and where of Longidoridae and Trichodoridae. *Journal of Nematology* 39, 295–297.

Decraemer, W., Cantalapiedra-Navarrete, C., Archidona-Yuste, A., Varela-Benavides, I., Gutiérrez-Gutiérrez, C., *et al.* (2019) Integrative taxonomy unravels cryptic diversity in the *Paratrichodorus hispanus*-group complex and resolves two new species of the genus and the molecular phylogeny of the family (Nematoda: Trichodoridae). *Zoological Journal of the Linnean Society* 185, 656–692.

Decraemer, W., Achidona-Yuste, A., Clavero-Camacho, I., Vovlas, A., Cantalapiedra-Navarrete, C., *et al.* (2024) Unravelling cryptic diversity in the *Paratrichodorus allius*-group species complex to solve eight new species of the genus and new insights on the molecular phylogeny (Nematoda: Trichodoridae). *Zoological Journal of the Linnean Society,* zlad94. DOI: 10.1093/zoolinnean/zlad194.

Duarte, I.M., Almeida, de, M.T., Brown, D.J.F., Marques, I., Neilson, R., *et al.* (2010) Phylogenetic relationships, based on SSU rDNA sequences, among the didelphic genera of the family Trichodoridae from Portugal. *Nematology* 11, 171–180.

Escuer, M. and Arias, M. (1997) *Paralongidorus iberis* sp. n. and *P. monegrensis* sp. n. from Spain with a polytomous key to the species of the genus *Paralongidorus* Siddiqi, Hooper & Khan, 1963 (Nematoda: Longidoridae). *Fundamental and Applied Nematology* 20, 135–148.

Evans, T.A., Miller, L.C., Vasilas, B.L., Taylor, R.W. and Mulrooney, R.P. (2007) Management of *Xiphinema americanum* and soybean severe stunt in soybean using crop rotation. *Plant Disease* 91, 216–219.

Gutiérrez-Gutiérrez, C., Palomares-Rius, J., Cantalapiedra-Navarrete, C., Landa, B.B., Esmenjaud, D., *et al.* (2010) Molecular analyses and comparative morphology to resolve a complex of cryptic *Xiphinema* species. *Zoologica Scripta* 39, 483–498.

Gutiérrez-Gutiérrez, C., Palomares-Rius, J., Jiménez-Diaz, R.M. and Castillo, P. (2010) Host suitability of *Vitis* rootstocks to root-knot nematodes (*Meloidogyne* spp.) and the dagger nematode *Xiphinema index*, and plant damage caused by infections. *Plant Pathology.*60, 575–585.

Halbrendt, J.M. (2021) A threat to stone fruit and grape production: tomato ringspot virus (ToRSV) transmission by *X. americanum* s.l. (sensu lato). In: Sikora R.A., Desaeger, J. and Molendijk, L.P.G. (eds), *Integrated Nematode Management: State-of-the-art and Visions for the Future*. CAB International, Wallingford, UK, pp. 207–214.

Hewitt, W.B., Raski, D.J. and Gowen, A.C. (1958) Nematode vector of soil borne-virus of grapevines. *Phytopathology* 48, 586–595.

Huang, D., Yan, G., Gudmestad, N., Ye, W., Whitworth, J., *et al.* (2019) Developing a one-step multiplex PCR Assay for rapid detection of four stubby-root nematode species, *Paratrichodorus allius*, *P. minor*, *P. porosus*, and *Trichodorus obtusus*. *Plant Disease* 103, 404–410.

Hunt, D.J. (1993) *Aphelenchida, Longidoridae and Trichodoridae: their Systematics and Bionomics*. CAB International, Wallingford, UK.

Jairajpuri, M.S. and Ahmad, W. (1992) *Dorylaimida. Free-living, Predaceous and Plant-Parasitic Nematodes*. Brill, Leiden, Netherlands.

Jeger, M., Bragard, C., Caffier, D., Candresse, T., Chatzivassiliou, E., *et al.* (2018) Pest categorisation of *Xiphinema americanum* sensu lato. *EFSA Journal* 16, e05298.

Kumari, S., Decraemer, W., De Luca, F. and Tiefenbrunner, W. (2010) Cytochrome c oxidase subunit 1 analysis of *Xiphinema diversicaudatum*, *X. pachtaicum*, *X. simile* and *X. vuittenezi* (Nematoda, Dorylaimida). *European Journal of Plant Pathology* 127, 493–499.

Lamberti, F. and Ciancio, A. (1993) Diversity of *Xiphinema americanum*-group species and hierarchical cluster analysis of morphometrics. *Journal of Nematology* 25, 332–343.

Lamberti, F., Molinari, S., Moens, M. and Brown, D.J.F. (2000) The *Xiphinema americanum* group. I. Putative species, their geographical occurrence and distribution, and regional polytomous identification keys for the group. *Russian Journal of Nematology* 8, 65–84.

Lamberti, F., Hockland, S. Agostinelli, A., Moens, M. and Brown, D.J.F. (2004) The *Xiphinema americanum* group. III. Keys to species identification. *Nematologia Mediterranea* 32, 53–56.

Loof, P.A.A. and Luc, M. (1993) Revised polytomous key for the identification of species of the genus *Xiphinema* Cobb, 1913 (Nematoda: Longidoridae) with exclusion of the *X. americanum*-group. *Systematic Parasitology* 16, 35–66.

Mwamula, O.K., Decraemer, W., Kim, Y.H., Ko, H.R. and Lee, D.W. (2020) Description of a new needle nematode, *Paralongidorus koreanensis* n. sp., and two known *Xiphinema* spp. Cobb, 7 1913, from turfgrass in Korea. *European Journal of Plant Pathology* 156, 1–20.

Nguyen, V.C., Villate, L., Gutiérrez-Gutiérrez, C., Castillo, P., Van Ghelder, C., *et al.* (2019) Phylogeography of the soil-borne vector nematode *Xiphinema index* highly suggests Eastern origin and dissemination with domesticated grapevine. *Scientific Reports* 9, 7313. DOI: 10.1038/s41598-019-43812-4.

OEPP/EPPO. (2017) Diagnostics. PM 7/95 (2): *Xiphinema americanum sensu lato*. *Bulletin OEPP/EPPO Bulletin* 47, 198–210.

Oliveira, C.M.G. and Neilson, R. (2006) Taxonomy of longidorid nematodes and dichotomous keys for the identification of *Xiphinema* and *Xiphidorus* species recorded in Brazil. *Arquivos do Institutos Biologico Sao Paulo* 73, 131–141.

Oliveira, C.M.G., Hübschen, J., Brown, D.J.F., Ferraz, L., Wright, F., *et al.* (2004) Phylogenetic relationships among *Xiphinema* and *Xiphidorus* nematode species from Brazil inferred from 18S rDNA sequences. *Journal of Nematology* 36, 153–159.

Palomares-Rius, J., Archidone-Yuste, A., Cantalapiedra-Navarrete, C., Prieto, P. and Castillo, P. (2016) Molecular diversity of bacterial endosymbionts associated with dagger nematodes of the genus *Xiphinema* (Nematoda: Longidoridae) reveals a high degree of phylogenetic congruence with their host. *Molecular ecology* 25, 6225–6247.

Palomares-Rius, J.E., Cantalapiedra-Navarrete, C., Archidona-Yuste, A., Subbotin, S.A. and Castillo, P. (2017) The utility of mtDNA and rDNA for barcoding and phylogeny of plant-parasitic nematodes from Longidoridae (Nematoda, Enoplea). *Scientific Reports* 7, 10905.

Palomares-Rius, J.E., Gutiérrez-Gutiérrez, C., Mota, M., Bert, W., Claeys, M., *et al.* (2021) '*Candidatus Xiphinematincola pachtaicus*' gen. nov., sp. nov., an endosymbiotic bacterium associated with nematode species of the genus *Xiphinema* (Nematoda, Longidoridae). *International Journal of Systematic and Evolutionary Microbiology* 71(7), 004888.

Peña-Santiago, R. (2021) Morphology and Bionomics of Dorylaims (Nematoda, Dorylaimida). In: *Nematology Monographs and Perspectives, 13* (Series eds: Hunt, D.J. and Perry, R.N.). Brill, Leiden, Netherlands.

Santos, M.S.N., de, Abrantes, I.M. de O., Brown, D.J.F. and Lemos, R.M. (eds) (1997) *An Introduction to Virus Vector Nematodes and their Associated Viruses*. Gráfica de Coimbra Lida, Coimbra, Portugal.

Siddiqi, M.R. (1973) *Xiphinema americanum. CIH Descriptions of Plant-parasitic Nematodes,* Set 2, No 29. CAB, Farnham Royal, UK.

Siddiqi, M.R. (1974) Systematics of the genus *Trichodorus* Cobb, 1913 (Nematoda: Dorylaimida), with descriptions of three new species. *Nematologica* 19, 259–278.

Siddiqi, M.R., Baujard, P. and Mounport, D. (1993) Description of *Paratylenchus pernoxius* sp. n. and *Paralongidorus duncani* sp. n. from Senegal, and the synonymization of *Longidoroides* with *Paralongidorus*. *Afro-Asian Journal of Nematology* 3, 81–89.

Sol, H.H., van Heuven, J.C. and Seinhorst, J.W. (1960) Transmission of rattle virus and *Atropa belladonna* mosaic virus by nematodes. *Tijdschrift over Plantenziekten* 66, 228–231.

Taylor, C.A. and Brown, D.J.F. (1997) *Nematode Vectors of Plant Viruses*. CAB International, Wallingford, UK.

Tenente, R., Vilardi, C., Abrantes, I., de Oliveira, M., Tenente, G. and Munhoz Vilardi, C. (2002) Nematóide - o gênero *Xiphinema*. *RAPP Revisão anual de Patologia de Plantas* 10, 111–153.

Zasada, I. and Forge, T. (2022) Ectoparasitic nematodes: emerging challenges to wine grape production in the Pacific Northwest of North America. In: Sikora, R.A., Desaeger, J. and Molendijk, L. (eds) Integrated nematode management: state-of-the-art and visions for the future. CAB International, Wallingford, UK, 192–198. DOI: 10.1079/9781789247541.0027.

16 Other Soil Nematodes

Oleksandr Holovachov[1]*, Ingrid Varela-Benavides[2], Walter Peraza-Padilla[3] and Alejandro Esquivel[3]

[1]Department of Zoology, Swedish Museum of Natural History, Stockholm, Sweden; [2]Doctorado en Ciencias Naturales para el Desarrollo, Instituto Tecnológico de Costa Rica, Cartago, Costa Rica; [3]Laboratorio de Nematologia, Escuela de Ciencias Agrarias, Universidad Nacional, Heredia, Costa Rica

Abstract

Although not usually considered as plant parasites (although a few are, or suspected of being so), the other soil nematodes are dealt with in summary form to aid the student as such nematodes are commonly encountered in soil extractions and, often being abundant, need to be eliminated as potential plant parasites. The chapter presents an introduction to the morphology, classification and ecological importance of the most common groups of soil nematodes belonging to the Orders Araeolaimida, Chromadorida, Desmoscolecida, Dorylaimida, Enoplida, Isolaimiida, Monhysterida, Mononchida, Plectida, Rhabditida and Triplonchida, representatives of which can often be found during studies of soil-dwelling plant-parasitic nematodes. The groups are diagnosed and well-illustrated using light micrographs.

16.1. Introduction

This chapter is intended to give a brief introduction to the most common families of terrestrial nematodes that can be encountered during studies of plant-parasitic species. Currently, the phylum Nematoda includes 17 orders, of which four are exclusively parasites of animals (Dioctophymatida, Mermithida, Muspiceida and Trichinellida). Of these four, only the Mermithidae (Mermithida) have free-living life stages found in freshwater sediments and in soil. In all other nematode orders, members of at least one family can be found in soil on a regular basis and will be discussed in this chapter.

16.2. Order Enoplida Filipjev, 1929

The Enoplida includes about 230 genera and over 2000 known species. They are bacterial and algal feeders, carnivores and scavengers. A detailed account of the morphology and systematics of the order was provided by Smol *et al.* (2014). Most species are marine, but several independently evolved freshwater and terrestrial lineages exist. In addition to the three most common families described herein, there are less common and some extremely rare taxa, including Andrassyidae (two genera), Campydoridae (two genera), Rhabdolaimidae (five genera) and Thalassogeneridae (a single genus).

*E-mail: oleksandr.holovachov@nrm.se

A revision of Holovachov, O. and Esquivel, A. (2012) Other soil nematodes. In: Manzanilla-López, R.H. and Marbán-Mendoza, N. (eds), *Practical Plant Nematology*. Mundi-Prensa and Biblioteca Básica de Agricultura, Montecillo, Mexico.

DOI: 10.1079/9781836990413.0016

16.2.1. Family Alaimidae Micoletzky, 1922

This family includes 13 genera and over 140 valid species (Fig. 16.1) that are commonly found in soils throughout the world (Andrássy, 2002). Small to large nematodes, between 0.5 and 4 mm in size. Cuticle appears smooth. Inner and outer labial and cephalic sensilla papilliform. Pore-like or transverse crescent-shaped amphids. Stoma very small (Fig. 16.1F, G). Pharynx strongly muscular, gradually widening posteriorly (Fig. 16.1E). Female reproductive system with either two or a single anterior or posterior gonad, ovary(-ies) antidromously reflexed, vulva opening at midbody. Male copulatory receptors papilliform, few in number. Spicules present, gubernaculum usually absent (Fig. 16.1H). Tail conoid to filiform. Caudal glands and spinneret absent.

16.2.2. Family Ironidae de Man, 1876

The terrestrial genus *Ironus* (Fig. 16.2) with about 30 species (Andrássy, 2011) and eight marine genera are united under the Ironidae. Possibly predatory in nature. Medium size nematodes, 1-4 mm. Cuticle finely annulated. Inner and outer labial sensilla papilliform, cephalic sensilla setiform. Slit-like or oval amphid. Stoma a long tube, with three teeth, one dorsal and two subventral in its anterior part (Fig. 16.2B-D). Pharynx strongly muscular, cylindrical along most of its length. Includes both amphimictic and parthenogenetic species. Female reproductive system with either two or a single posterior gonad, ovary(-ies) antidromously reflexed, vulva opening at midbody (Fig. 16.2E). Single setiform male copulatory receptor. Spicules and gubernaculum present (Fig. 16.2F). Tail elongate conoid to filiform (Fig. 16.2H), caudal glands and spinneret absent in Ironus.

16.2.3. Family Trischistomatidae Andrássy, 2007

A small family with two genera and about 40 species (Fig. 16.3), commonly found in moist soil

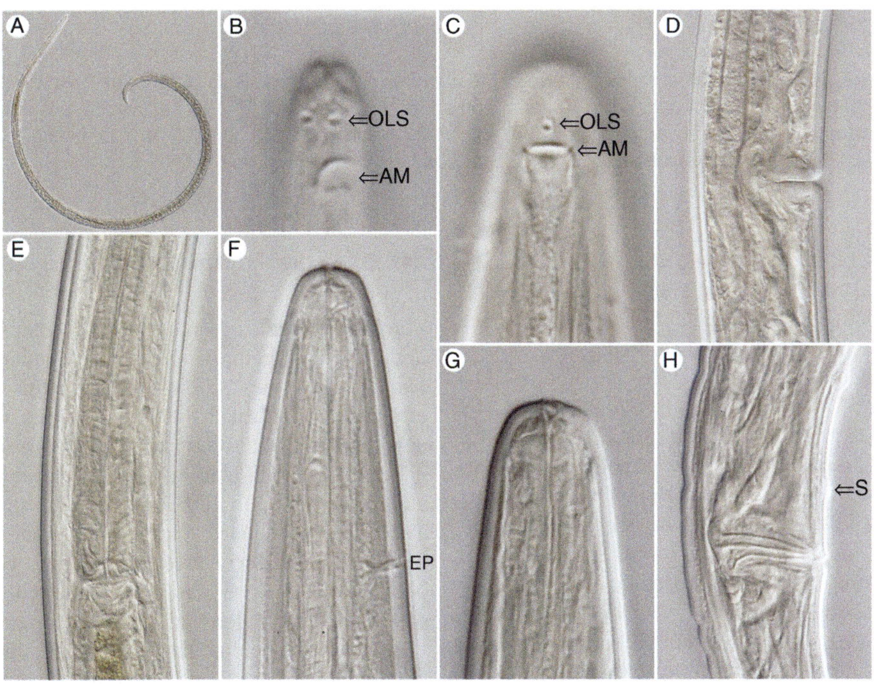

Fig. 16.1. *Amphidelus* (A, C-H) and *Laxamphidelus* (B) – representatives of the Alaimidae. A: Entire female; B, C: anterior end (surface view) showing outer labial sensilla (OLS) and amphid (AM); D: vulval region; E: pharynx base; F: anterior end showing stoma, anterior part of pharynx and excretory pore (EP); G: anterior end showing stoma; H: male cloacal region showing spicules, gubernaculum and precloacal copulatory receptors (S).

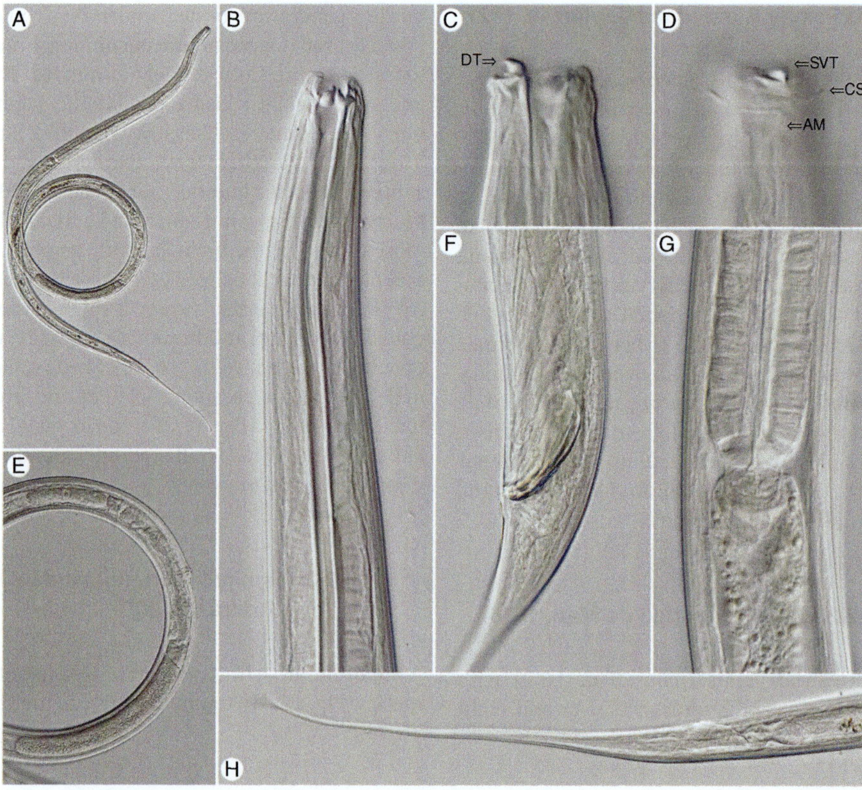

Fig. 16.2. *Ironus* – terrestrial representative of the Ironidae. A: Entire female; B: anterior end showing stoma and anterior part of pharynx; C: anterior end showing dorsal tooth (DT) and stoma; D: anterior end (surface view) showing subventral tooth (SVT), cephalic sensilla (CS) and amphid (AM); E: female gonads; F: male cloacal region showing spicules; G: base of pharynx; H: female tail.

(Cid del Prado-Vera *et al.*, 2016). Similar to Tripylidae in general morphology (except for the reproductive system) and hence used to be grouped together. Medium size nematodes, 1-2 mm in size, body often bent to the dorsal side (Fig. 16.3A, B). Cuticle finely annulated. Inner labial sensilla papilliform, outer labial and cephalic sensilla setiform. Slit-like amphid. Stoma narrow, with one or more teeth in its posterior part (Fig. 16.3D). Pharynx strongly muscular, cylindrical along most of its length. Includes both amphimictic and parthenogenetic species. Female reproductive system with a single anterior gonad, ovary antidromously reflexed, vulva opening along posterior body end (Fig. 16.3C). Male copulatory receptors papilliform. Spicules and gubernaculum present (Fig. 16.3E). Tail conoid, caudal glands and spinneret present.

16.3. Order Triplonchida Cobb, 1919

The Triplonchida is represented by freshwater and terrestrial taxa, including some important plant parasites described elsewhere in this book (see Chapter 15, this volume). It includes about 40 genera and over 400 valid species, with Tobrilidae and Trichodoridae the most diverse and species-rich families. In addition to plant parasites, Triplonchida includes microbivores, algal and fungal feeders, carnivores and scavengers. In addition to the three most common families described below, Triplonchida includes four other taxa, of which Onchulidae is the least common, restricted to tropical areas. The other three, Prismatolaimidae, Bastianiidae and Odontolaimidae, are common in soils but not abundant (Fig. 16.4). They are small, slender nematodes, where

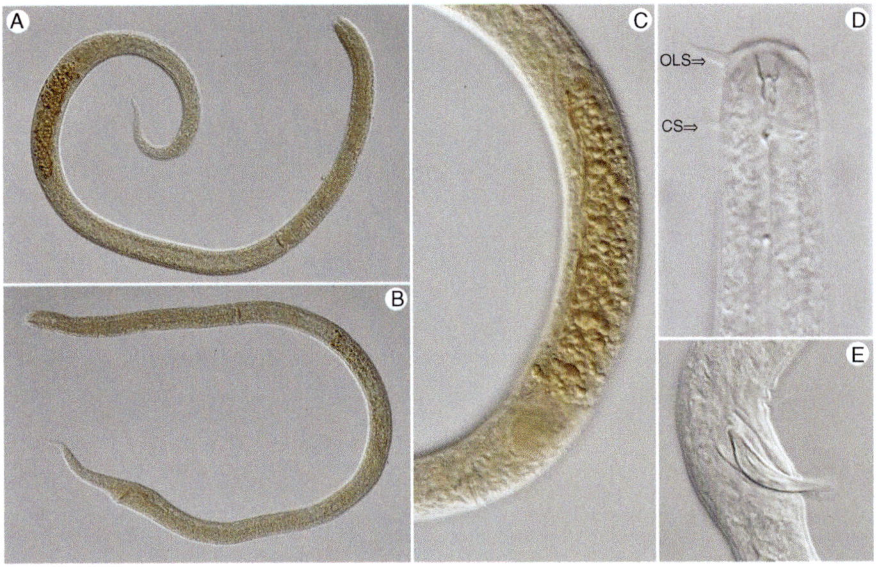

Fig. 16.3. *Trischistoma* – representative of the Trischistomatidae. A: Entire female; B: entire male; C: female gonad; D: anterior end showing outer labial (OLS) and cephalic sensilla (CS) and stoma; E: male cloacal region showing spicules.

the major differences between families (and genera) lie in the structure of the stoma (Fig. 16.4E, G, L) and amphid (Fig. 16.4F, H, M). These have been reviewed in detail in Zullini (2006) and in Holovachov and Shoshin (2014).

16.3.1. Family Tobrilidae de Coninck, 1965

Mostly freshwater inhabitants (Fig. 16.5), some species of this diverse family (about 100 species grouped in 14 genera) are found in wet soils (Holovachov and Shoshin, 2014). Medium to large size nematodes, 1-7 mm in size. Cuticle finely annulated. Inner labial sensilla papilliform. Outer labial and cephalic sensilla setiform. Slit-like amphid. Stoma funnel-shaped and strongly refractive, with at least two subventral (exceptionally three) teeth located at its base, in many species in special 'pockets' (Fig. 16.5C, D). Pharynx strongly muscular, cylindrical along most of its length (Fig. 16.5A). Most described species are amphimictic, but parthenogenetic species definitely occur in this family. Female reproductive system with two gonads, ovaries antidromously reflexed, vulva opening at midbody (Fig. 16.5B). Male copulatory receptors single papilliform or complex 'echinulate' (Fig. 16.5E, F), few in number. Spicules and gubernaculum present. Tail elongate conoid, caudal glands and spinneret present (Fig. 16.5G).

16.3.2. Family Tripylidae de Man, 1876

This family includes three genera and over 50 valid species of predatory nematodes, inhabiting a variety of terrestrial and freshwater habitats all over the world (Fig. 16.6), except for Antarctica, with *Tripyla* (Asghari *et al.*, 2017) being the most common representative in soils. Medium size nematodes, 1-2 mm in size. Cuticle finely annulated. Inner labial sensilla papilliform. Outer labial and cephalic sensilla either papilliform or setiform, depending on the species. Slit-like or oval amphid (Fig. 16.6D). Stoma narrow, with dorsal tooth and two subventral denticles in its posterior part (Fig. 16.6C). Pharynx strongly muscular, cylindrical along most of its length (Fig. 16.6G). *Tripyla* is amphimictic, while *Tripylella* and *Tobrilia* are parthenogenetic. Female reproductive system with two gonads, ovaries antidromously reflexed, vulva opening at midbody. Male copulatory receptors papilliform, extending into pharyngeal region (Fig. 16.6C). Spicules and gubernaculum

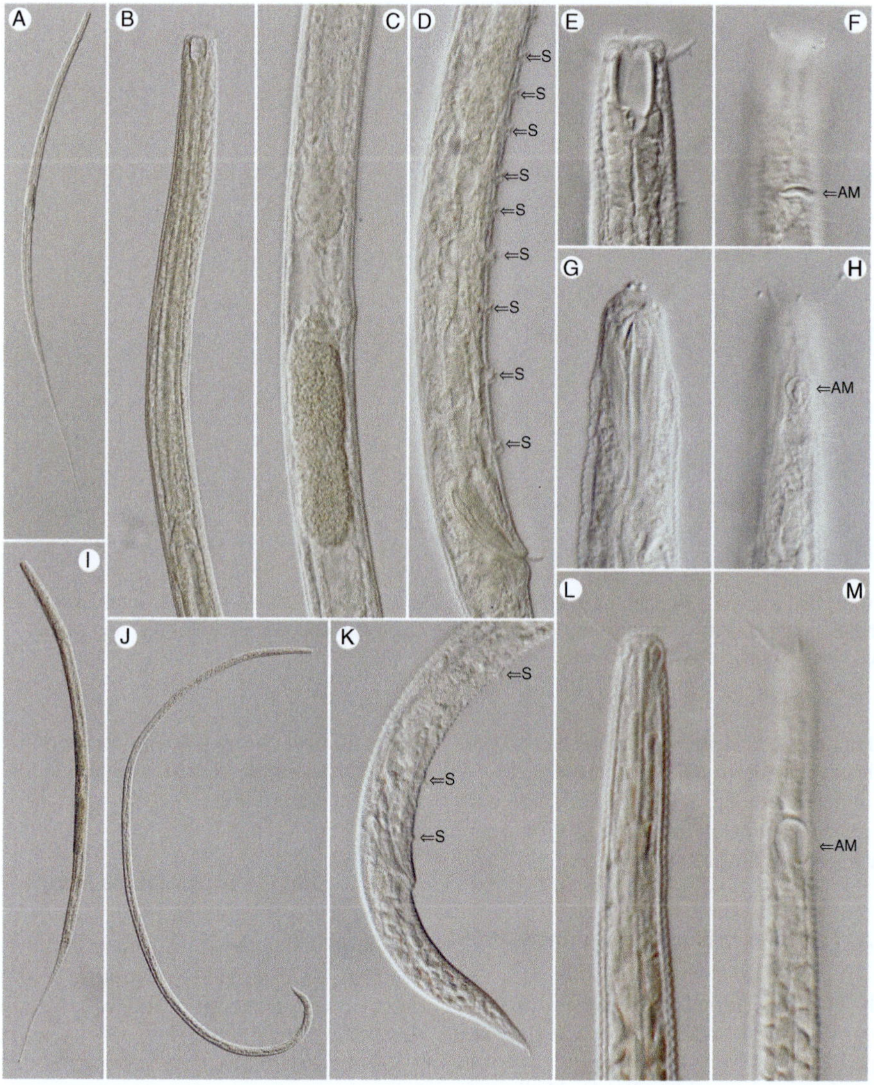

Fig. 16.4. *Prismatolaimus* (A-F), *Odontolaimus* (G-I) and *Bastiania* (J-M) – representatives of the Prismatolaimidae, Odontolaimidae and Bastianiidae, respectively. A, I: Entire females; B: pharyngeal region; C: female gonads; D, K: male cloacal region showing copulatory receptors (S) and spicules; E, G, L: anterior end showing stoma; F, H, M: anterior end (surface view) showing anterior sensilla and amphid (AM); J: entire male.

present (Fig. 16.6E). Tail elongate conoid, caudal glands and spinneret present.

16.3.3. Family Diphtherophoridae Micoletzky, 1922

This family includes three genera and about 50 species (Ghaderi *et al.*, 2017), found in soils on all continents except for Antarctica. Most known species are less than 1 mm in length. Cuticle tightly adpressed to the body in *Tylolaimophorus* or loosely fitted in *Diphtherophora* (Fig. 16.7) and *Longibulbophora*. Inner labial, outer labial and cephalic sensilla papilliform. Oval amphids (Fig. 16.7B). Stoma in the shape of a complex stylet, the base of which has swellings for muscle attachments (Fig. 16.7A). Pharynx strongly muscular, gradually expanding posteriad (Fig. 16.7E). Amphimictic and parthenogenetic

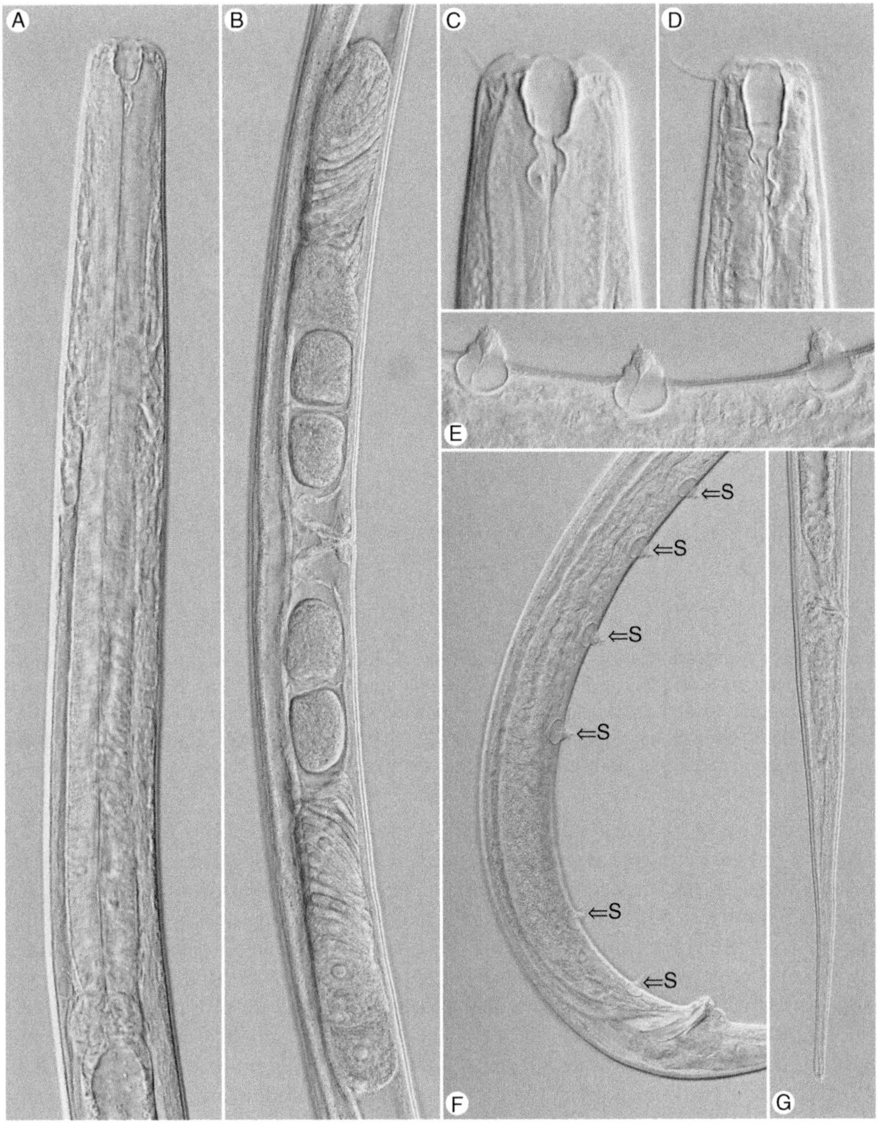

Fig. 16.5. *Eutobrilus* (A-C, E, G), *Neotobrilus* (D) and *Epitobrilus* (F) – representatives of the Tobrilidae. A: Pharyngeal region; B: female gonads; C, D: anterior end showing anterior sensilla and stoma; E: copulatory receptors; F: male posterior end showing copulatory receptors (S) and spicules; G: female tail.

species. Female reproductive system with two gonads, ovaries antidromously reflexed, vulva opening at midbody (Fig. 16.7F). Male copulatory receptors papilliform. Spicules and gubernaculum present (Fig. 16.7G). Tail short, rounded or conoid, caudal glands and spinneret absent (Fig. 16.7H).

16.4. Order Dorylaimida de Man, 1876

Dorylaims include one of the most diverse orders of the phylum Nematoda. They are common inhabitants of all soil types and freshwater sediments worldwide. They can also be found in other types of substrates above the soil surface

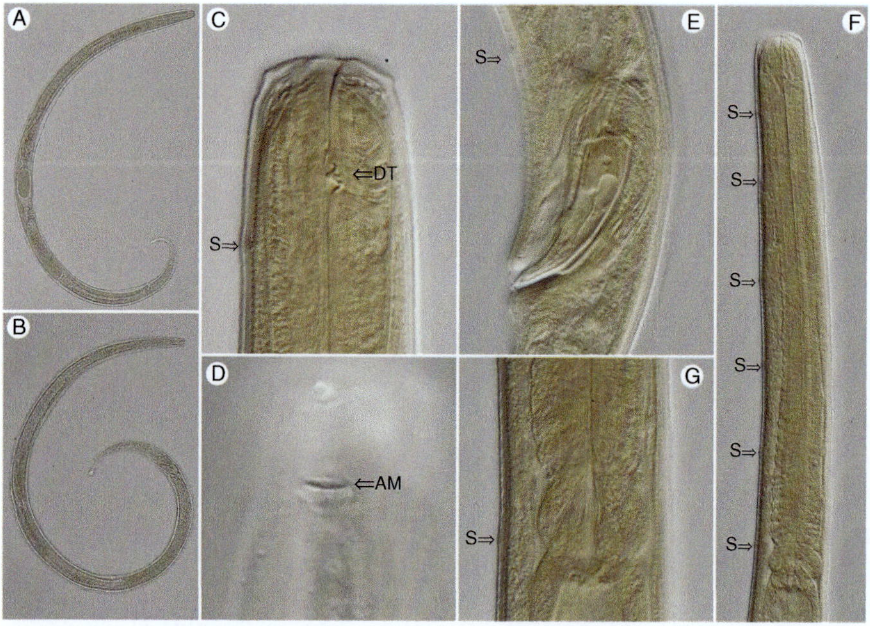

Fig. 16.6. *Tripyla* – representative of the family Tripylidae. A: Entire female; B: entire male; C: anterior end showing stoma, dorsal tooth (DT) and anteriormost male copulatory receptor (S); D: anterior end (surface view) showing amphid (AM); E: male cloacal region showing spicules and copulatory receptors (S); F: male pharyngeal region showing copulatory receptor (S); G: base of pharynx showing pharyngo-intestinal junction (cardia) and associated gland cells, and copulatory receptor (S).

such as litter, mosses, lichens and epiphytes in tropical rain forest. To date, approximately 300 valid genera and over 3000 species have been described (Peña-Santiago, 2021a). The majority of dorylaims are free-living omnivorous nematodes, however, there are also many predatory and plant-parasitic species. Some species belonging to *Xiphinema* and *Longidorus* are very important in agriculture as parasites of crops and as virus vectors (see Chapter 15). Many dorylaims are very sensitive to environmental alterations. In general, they are more abundant and diverse in undisturbed areas rather than in agricultural fields, and are therefore used as bioindicators of soil health. For comprehensive information about Dorylaimida see Jairajpuri and Ahmad (1992), Jiménez-Guirado *et al.* (2007), Peña-Santiago (2006, 2014a, 2021a, b) and Vinciguerra (2006).

16.4.1. General Morphology

Dorylaims are nematodes of variable size, ranging from less than 0.5 to 12 mm in length (Peña-Santiago, 2021b). Lip region of Dorylaimida highly variable and important for identification, in the majority of species it is wider and marked off by depression or set off by distinct constriction. Liplets or disc-like structure can surround the oral opening in some dorylaims, being modifications of inner portion of the lips (Fig. 16.8A). Amphids are located behind the lateral lips (Fig. 16.8F), one on each side. In dorylaims the shape as well as size of amphid is variable, but most commonly is cup-, pouch- or goblet-shaped. Amphidial aperture a narrow or wide transverse slit or small pore. Oral aperture variable in shape: pore-like, hexagonal or a slit; followed by cylindrical stoma connecting to the pharyngeal lining. The guiding sheath corresponds to the intermediate region of stoma. Under the light microscope the guiding ring appears as a refringent structure and varies in morphology. Odontostyle located inside the guiding sheath and is a distinctive characteristic of Dorylaimina. It is a hollow (tube-like) protrusible structure with pointed anterior end and a sloping dorsal aperture. The odontostyle is attached by its base to the wall of the guiding

Fig. 16.7. *Diphtherophora* – representative of the Diphtherophoridae. A: Anterior end showing stoma; B: anterior end (surface view) showing amphid (AM); C: entire female; D: entire male; E: pharyngeal region; F: female gonads; G: male caudal region showing spicules and copulatory receptor (S); H: female tail.

sheath, connected with the posterior portion of the stoma. In the suborder Nygolaimina the feeding apparatus consists of a solid mural tooth. Differences in length, shape, width and thickness of odontostyle and mural tooth are useful characters for identification of species and genera. The odontophore is the posteriormost portion of the stoma. It is a tubular structure attached to the base of the odontostyle anteriorly and extending into the pharyngeal lining posteriorly. Odontophore simple, rod-like in most dorylaims (Fig. 16.8G); however, it varies in length, thickness (knobs at the base, Fig. 16.8H) and shape (angular or arcuate), which is of taxonomic importance. Pharynx bipartite, bottle-shaped and muscular, anterior portion slender and usually muscular, gradually extending into a well-muscularized expanded posterior portion. Variations in the basic morphology of the pharynx are used as an important taxonomic character to define major taxa. Dorylaimoid type of pharynx consists of two parts that are muscular and similar in size; the transition between anterior and posterior parts is usually gradual (Fig. 16.8B). Belondiroid type is characterized by the presence of a spiral muscle sheath surrounding the posterior pharyngeal expansion (Fig. 16.8C). In the leptonchoid type the pharyngeal expansion is particularly short and more or less pear-shaped, forming a true basal bulb (Fig. 16.8D).

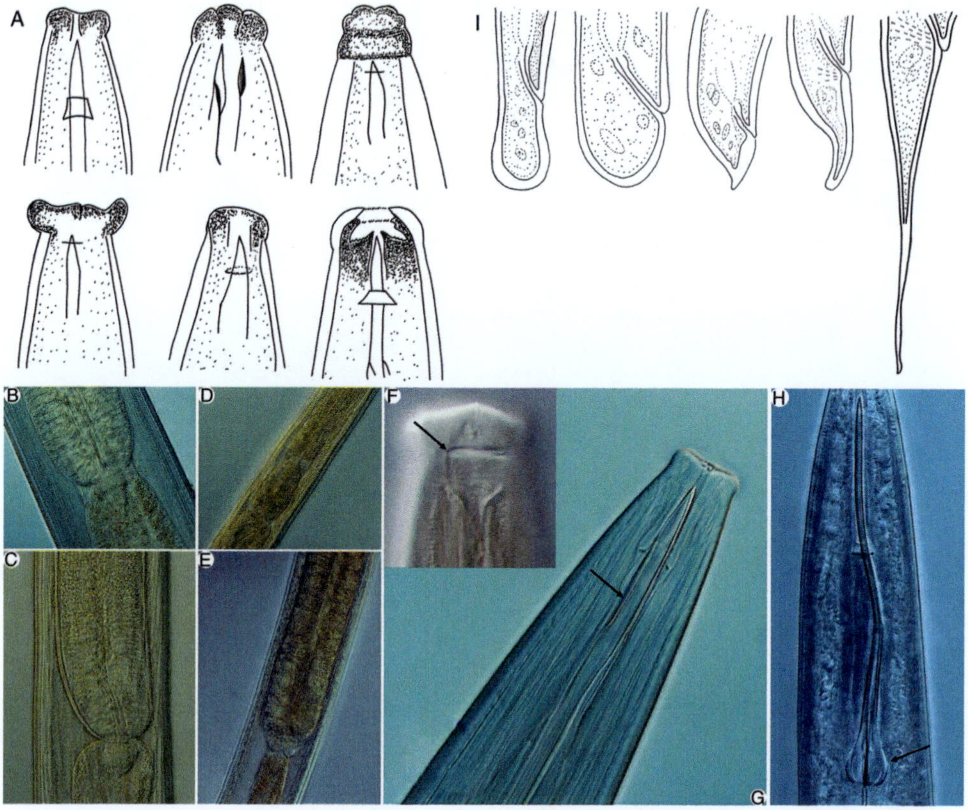

Fig. 16.8. Morphology of Dorylaimida. A: Shapes of lip region, feeding apparatus and guiding ring types (single and double); B-E: variation in pharynx morphology; B: dorylaimoid type; C: belondiroid type; D: leptonchoid type with true basal bulb (bracket); E: nygolaimoid type with three large cardiac cells at pharyngo-intestinal junction; F: cup-like amphidial fovea with a well-marked slit-like opening (arrow); G: rod-like odontophore (arrow); H: flange-like odontophore (arrow); I: variation in tail shape.

The nygolaimoid type, on the other hand, is characterized by the presence of three large cardiac cells at the pharyngo-intestinal junction (Fig. 16.8E). The intestine is divided into anterior and posterior (prerectum) sections. The last portion of digestive tract is rectum: a short tube that connects with the exterior via the ventral anus. Ejaculatory duct in males opens in a common channel with the alimentary tract to form cloaca. Female reproductive system is usually didelphic, amphidelphic. Monodelphic condition is present in some genera with either only anterior (mono-prodelphic) or only posterior (mono-opisthodelphic) branch developed (Fig. 16.9A). Each genital branch consists of a reflexed ovary and a gonoduct. Vagina a short tube linking uterus with exterior and composed of three regions: (i) *pars proximalis vaginae* is a translucent part, often longer than wide, connected to the uterus; (ii) *pars refringens vaginae* (not always present), an intermediate region distinguished by the presence of two refracting pieces variable in size and shape, whose morphology has a taxonomic value and (iii) *pars distalis vaginae*, a short tube opening to the outside (Fig. 16.9B, C). Vulva most commonly a transverse slit (Fig. 16.9D) but can also be a longitudinal slit or pore-like. In amphidelphic females, the vulva is generally located at midbody; its position in either opisthodelphic or prodelphic females varies considerably. Male reproductive system diorchic, with opposed testes connected

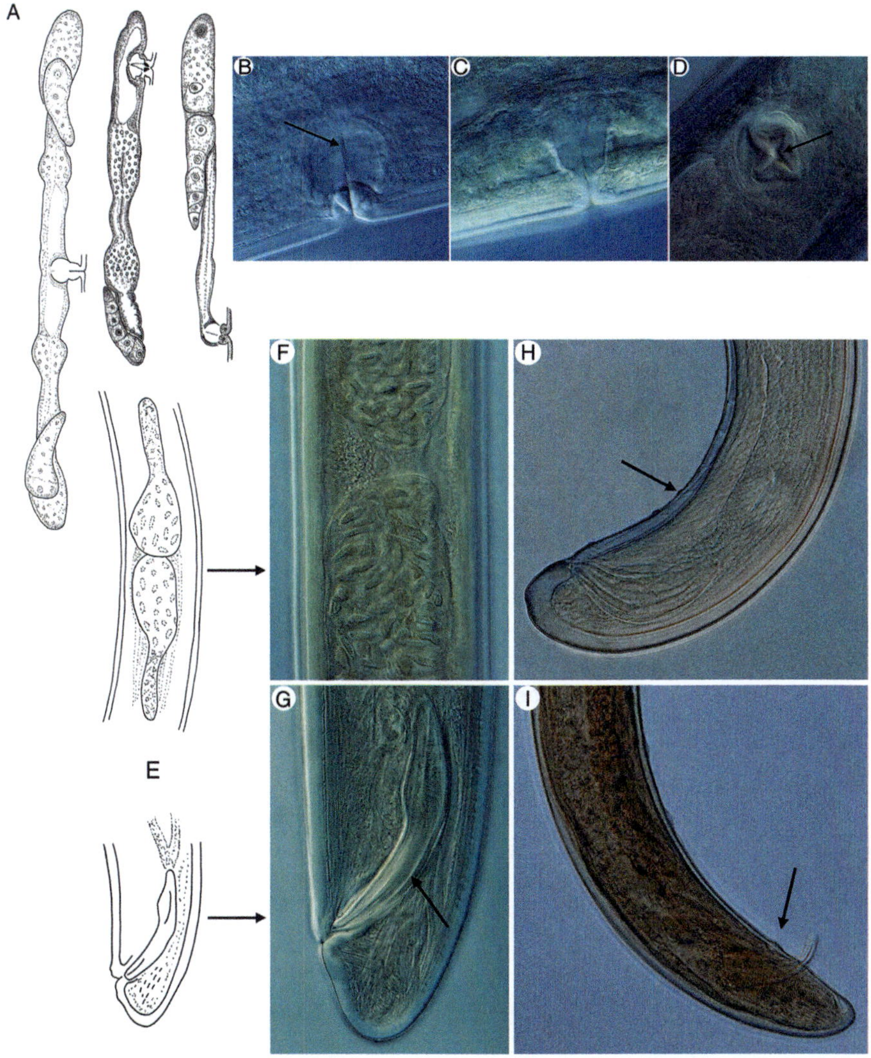

Fig. 16.9. Morphology of Dorylaimida, female and male reproductive system. A: Didelphic, opisthodelphic and prodelphic types of female reproductive system (left to right); B: vagina with pars refringens vaginae (arrow); C: vagina lacking pars refringens vaginae; D: transverse vulva in ventral view (arrow); E: diorchic male reproductive system; F: maturation zone of testis; G: spicule; H, I: arrangement of ventromedian copulatory receptors in male (arrows).

with a common genital tract that joins the rectum to open via the cloacal aperture (Fig. 16.9E, F). Copulatory apparatus consisting of a pair of tubular spicules lodged in pouches of the dorsal cloacal wall (Fig. 16.9G). Lateral guiding pieces (crura) and gubernaculum present. Copulatory receptors forming a series of small sensilla, one ad-cloacal pair (close to the cloacal opening) and a series of 1-100 precloacal ventromedian sensilla. Their number and arrangement (contiguous or separated in groups) have great taxonomic value (Fig. 16.9H-I). Tail very variable in size and shape, ranging from long to filiform, short, rounded, conical, convex conical, cylindrical (Fig. 16.8I). Most commonly, females and males have a similar tail shape, but sexual dimorphism is present in certain genera when the female tail is more or less long and the male tail short and rounded.

16.4.2. Family Belondiridae Thorne, 1939

Nematodes of variable size, usually more than 1 mm long. Lip region narrow, continuous or set off from body contour. Odontostyle short, rarely longer than lip region width, linear or fusiform, with aperture of variable size (Fig. 16.10A). Anterior slender portion of pharynx more or less muscular, posterior pharyngeal expansion surrounded by a more or less developed spiral muscular sheath. Female reproductive system didelphic or monodelphic. Tail similar or dissimilar in sexes: long and filiform to short and rounded. This family includes 36 genera and 282 valid species (Peña-Santiago, 2021a). Type genus: *Belondira* (Fig. 16.10).

16.4.3. Family Tylencholaimidae Filipjev, 1934

Nematodes variable in size, but usually less than 1 mm long. Cuticle with loose inner layer and radial refractive elements. Lip region set off from adjacent body, usually cap-like, but variable in shape (Fig. 16.11A). Odontostyle short, odontophore rod-like, often bearing basal knobs, or rarely flanged. Anterior region of pharynx more or less muscular, pharyngeal expansion occupying more than one-third of total neck length. Female genital system didelphic or monodelphic. Vulva transverse, longitudinal or pore-like. Spicules dorylaimoid, copulatory receptors present, usually spaced (Fig. 16.11E), rarely contiguous. Tail similar in sexes, variable in shape and size. This family includes 21 genera and 125 valid species (Peña-Santiago, 2021a). Type genus: *Tylencholaimus* (Fig. 16.11).

16.4.4. Family Dorylaimidae de Man, 1876

Small to large nematodes. Cuticle smooth or with fine transverse striae; longitudinal ridges present in a few genera (Fig. 16.12B). Lip region set off or continuous with body contour. Odontostyle straight or slightly sinuate, with wide lumen; aperture occupying about one-third of odontostyle length. Odontophore rod-like, without basal knobs or flanges. Expanded part of pharynx comprising about half of total neck length. Female reproductive system didelphic; vulva transverse or longitudinal; *pars refringens vaginae* present and sclerotized. Spicules dorylaimoid; copulatory receptors variable in number, spaced or contiguous (Fig. 16.12E) or grouped. Tail attenuated or filiform in female; short and rounded in male (Fig. 16.12D); in few cases long and filiform

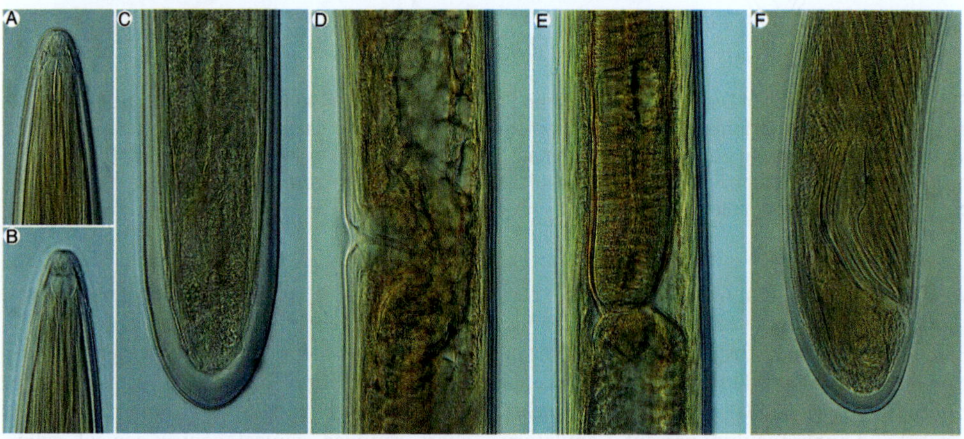

Fig. 16.10. *Belondira* – representative of the Belondiridae. A: Female anterior end; B: amphid; C: female tail; D: vulva; E: basal part of pharynx; F: male tail and spicules.

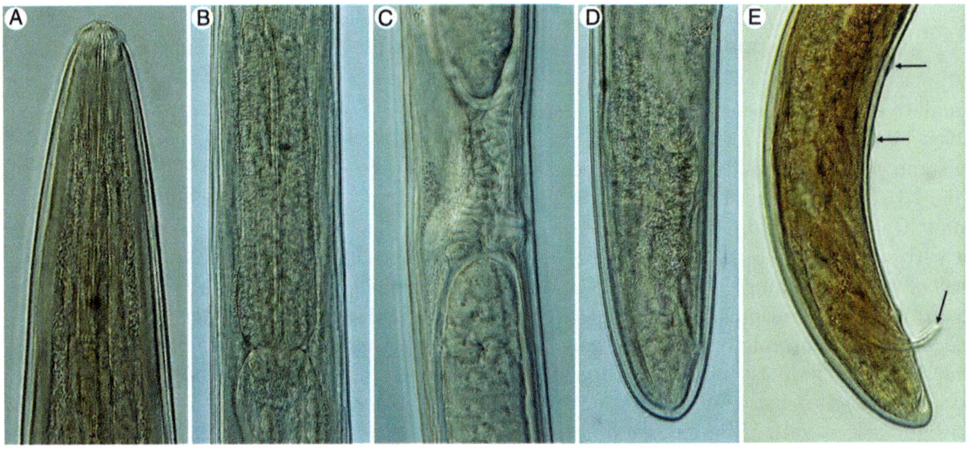

Fig. 16.11. *Tylencholaimus* – representative of the Tylencholaimidae. A: Female anterior end; B: basal part of pharynx; C: vulval region; D: female tail; E: male caudal region, with protruded spicules and spaced copulatory receptors (arrows).

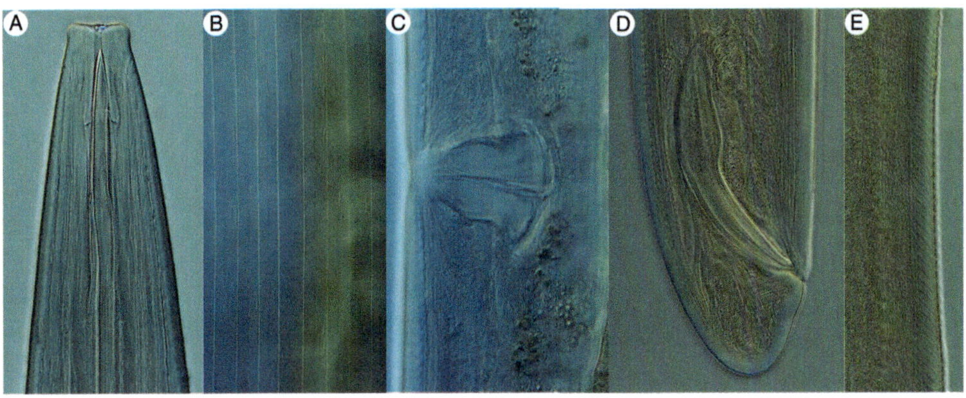

Fig. 16.12. *Dorylaimus* – representative of the Dorylaimidae. A: Female anterior end; B: cuticle showing longitudinal ridges; C: vulval region; D: male tail and spicules; E: numerous ventromedian copulatory receptors.

in both sexes. This family includes 40 genera and 550 valid species (Peña-Santiago, 2021a). Type genus: *Dorylaimus* (Fig. 16.12).

16.4.5. Family Actinolaimidae Thorne, 1939

Large nematodes. Lip region continuous or set off from adjacent body. An anterior cuticularized ring present around oral opening, often appearing corrugated. Cheilostom wide, armed with four large onchia, with or without smaller denticles on cheilostom walls (Fig. 16.13A). Odontostyle with distinct aperture. Expanded part of pharynx about half of total neck length. Female reproductive system didelphic; vulva transverse, longitudinal or pore-like; *pars refringens vaginae* with sclerotized pieces. Spicules dorylaimoid, copulatory receptors arranged in two or three fascicles. Tail variable, elongate to filiform in both sexes or elongate to filiform in females (Fig. 16.13C) and short and rounded in males (Fig. 16.13D). This family includes 20 genera and 171 valid species (Peña-Santiago, 2021a). *Paractinolaimus* (Fig. 16.13) is a common terrestrial genus of this family.

16.4.6. Family Qudsianematidae Jairajpuri, 1965

Small to large nematodes. Lip region generally offset by a depression or constriction, discoid in some genera. Amphids stirrup-shaped (Fig. 16.14B). Odontostyle straight, its aperture less than 50% of total length. Odontophore rod-like or with sclerotized flanges. Posterior expanded part of pharynx comprising about half of neck length. Cardia hemispheroid to elongate conoid. Female reproductive system generally didelphic, sometimes mono-opisthodelphic; *pars refringens vaginae* with (Fig. 16.14C) or without sclerotization. Spicules dorylaimoid; copulatory receptors variable in number, spaced or continuous. Tail similar in both sexes (Fig. 16.14D, E). This family includes 41 genera and 400 valid species (Peña-Santiago, 2021a). Type genus: *Eudorylaimus* (Fig. 16.14).

16.4.7. Family Aporcelaimidae Heyns, 1965

Nematodes of medium, large to very large size. Cuticle thick, smooth or with fine transverse striation (Fig. 16.15D), often with criss-cross lines or punctations and with numerous body pores. Lip region clearly set off by a constriction from adjacent body (Fig. 16.15B). Amphids often with sclerotized median support, fovea often duplex. Odontostyle robust and relatively short (Fig. 16.15A), generally axial, with wide aperture that is longer than half of odontostyle

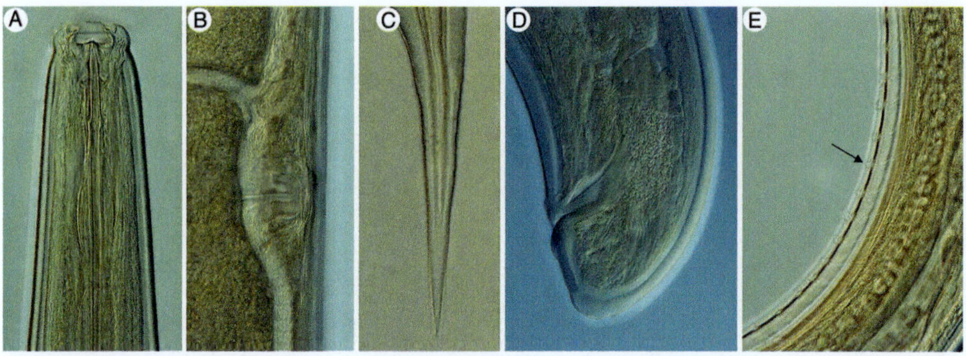

Fig. 16.13. *Paractinolaimus* – representative of the Actinolaimidae, inhabiting wet soils. A: Female anterior end; B: vulval region; C: female tail; D: male tail and spicules; E: male copulatory receptors (arrow).

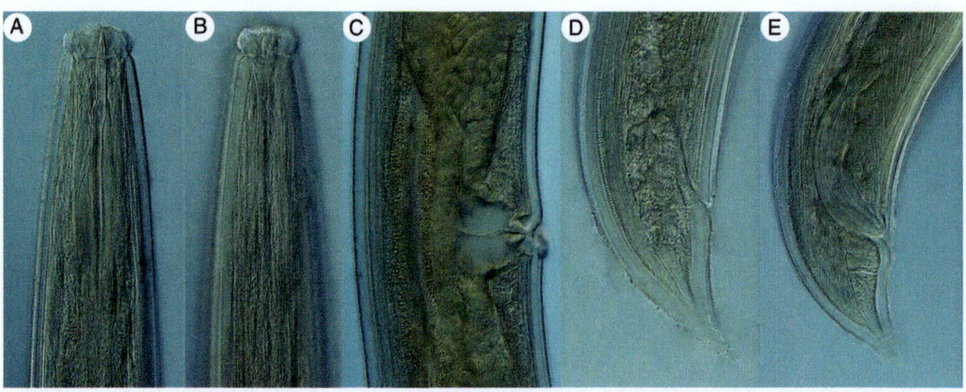

Fig. 16.14. *Eudorylaimus* – representative of the Qudsianematidae. A: Female anterior end; B: lips and amphid; C: vulval region; D: female tail; E: male tail and spicules.

length; rarely as a mural tooth. Female reproductive system amphidelphic; vulva transverse, longitudinal or pore-like. Spicules dorylaimoid, copulatory receptors spaced. Tail similar in both sexes, generally short, rounded or conoid, exceptionally filiform. This family includes 15 genera and 199 valid species (Peña-Santiago, 2021a). Type genus: *Aporcelaimus* (Fig. 16.15).

16.4.8. Family Nordiidae Jairajpuri & Siddiqi, 1964

Nematodes of variable size. Lip region continuous with body contour or offset by a depression or constriction; lips distinct or fused. Amphids with stirrup-shaped fovea and slit-like aperture (Fig. 16.16B). Odontostyle slender, attenuated, generally long, with narrow lumen and aperture. Pharynx muscular, expanded part comprising about half of total neck length. Female reproductive system usually didelphic, sometimes monodelphic; vulva transverse; *pars refringens vaginae* well developed (Fig. 16.16C). Tail similar in both sexes, short and rounded or elongate to filiform. This family includes 30 genera and 241 valid species (Peña-Santiago, 2021a). *Oriverutus* (Fig. 16.16) is a common tropical genus of this family.

16.4.9. Family Nygolaimidae Thorne, 1935

Nematodes of medium to large size, rarely less than 1 mm long. Lip region offset or continuous

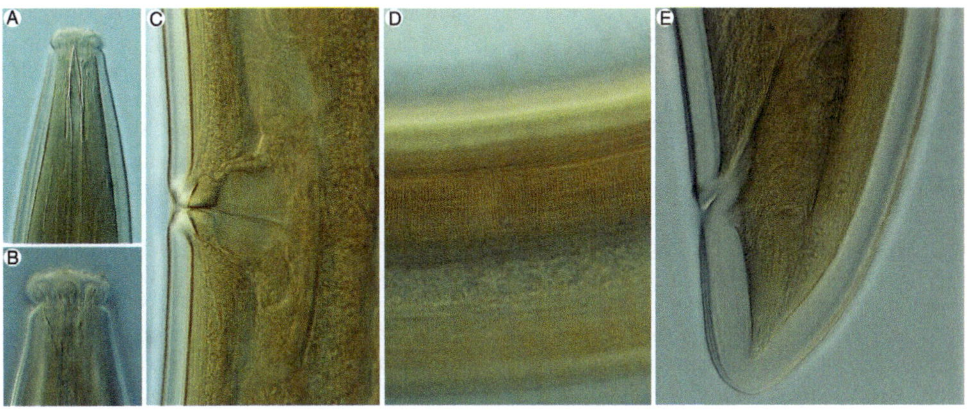

Fig. 16.15. *Aporcelaimus* – representative of the Aporcelaimidae. A: Female anterior end; B: lips and amphid; C: vulval region; D: cuticle with fine transverse striations; E: female tail.

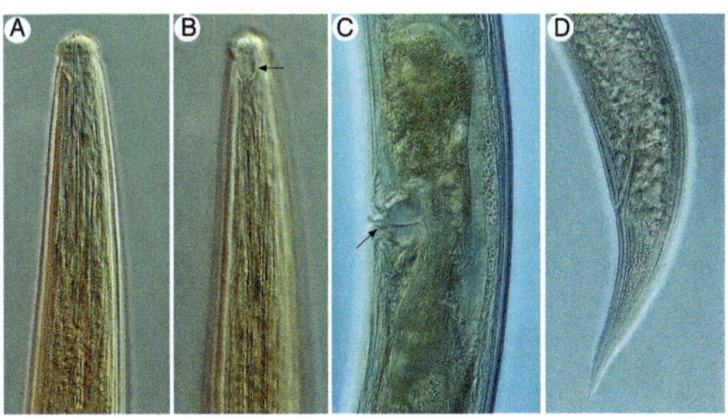

Fig. 16.16. *Oriverutus* – common tropical representative of the Nordiidae. A: Female anterior end; B: amphid (arrow) and labial papillae; C: vulval region with *pars refringens* (arrow); D: female tail.

with body contour, lips usually amalgamated. Cheilostom simple, without sclerotization. Mural tooth variable in shape and size (Fig. 16.17A): deltoid, solididentoid, acicular, dorylaimoid or linear. Expanded part of pharynx comprising half or more of total neck length. Three large cardiac cells present at pharyngo-intestinal junction. Female reproductive system didelphic. *Pars refringens vaginae* usually absent (Fig. 16.17B). Vulva transverse, exceptionally longitudinal. Spicules dorylaimoid, gubernaculum present or absent. Copulatory receptors weakly developed, few in number (Fig. 16.17D). Tail similar in both sexes (Fig. 16.17C, D) but variable in shape and size. This family includes 12 genera and 111 valid species (Peña-Santiago, 2021a). Type genus: *Nygolaimus* (Fig. 16.17).

16.5. Order Mononchida Chitwood, 1937

Mononchs are nematodes of wide geographical distribution and found on all continents in both aquatic and terrestrial environments, but more frequent in soils than in freshwater sediments. Some genera and species are restricted to tropical and subtropical areas, while others are more common in temperate regions. They are very active predators, feeding on a variety of micro- and mesofauna, including free-living and plant-parasitic nematodes, protozoa, rotifers, tardigrades, oligochaetes, and may also feed on their own species. Although their population densities are much lower than other soil nematodes, the Mononchida have been considered potential biological control agents against plant-parasitic nematodes. Some diagnostic features of the Mononchida include relatively large and robust body; barrel-shaped stoma with strongly sclerotized walls, armed with teeth or denticles; cylindrical muscular pharynx and no prerectum. Further information on this group can be found in Jiménez-Guirado *et al.* (2007), Ahmad and Jairajpuri (2010) and Peña-Santiago (2014a).

16.5.1. General Morphology

The Mononchida includes a group of large-sized nematodes, usually over 1 mm in length. Mononchs have a large, cylindrical and robust body, truncated at the anterior end and tapering towards the posterior extremity. A distinctive character of Mononchida is the presence of cuticular pores along the body, which are sometimes difficult to detect. They show species-specific arrangement. In some mononchs, the lip region is almost continuous, while in others it is separated from the adjacent body by a simple depression or set off by a constriction. Amphids located posterior to the lateral lips, amphidial aperture slit-like (Fig. 16.18G). Stoma and its structures (tooth, teeth or denticles) are very important diagnostic characters; all genera are distinguished on the basis of shape, size and position of buccal armament. Stoma barrel-shaped, somewhat flattened at base in Anatonchidae

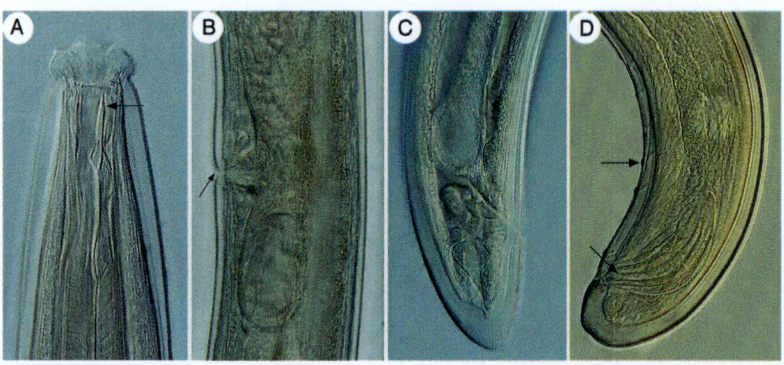

Fig. 16.17. *Nygolaimus* – representative of the Nygolaimidae. A: Female anterior end with prominent mural tooth; B: vulval region; C: female tail; D: male tail, spicules and copulatory receptors (arrows).

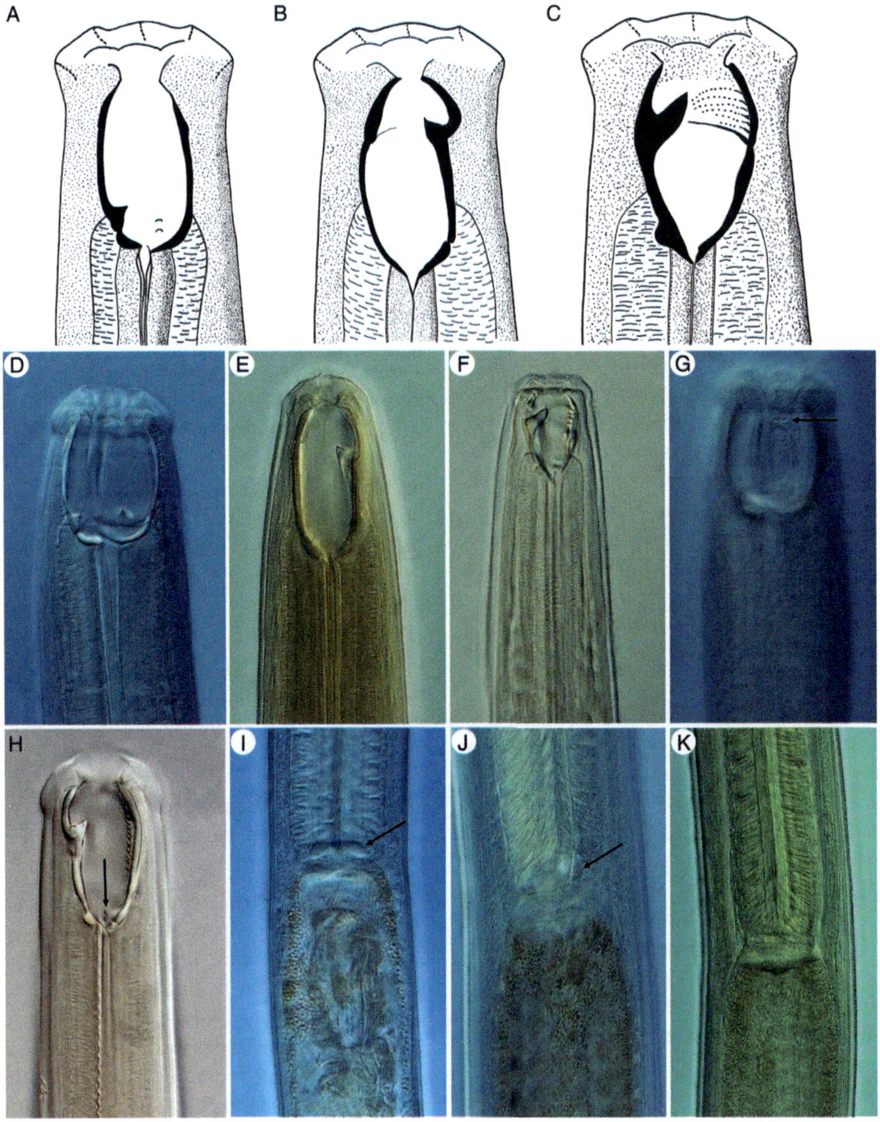

Fig. 16.18. Morphology of Mononchida. A-H: General morphology of buccal cavity and associated structures (tooth, teeth or denticles) in Mononchida. A, D: stoma of Iotonchidae and Anatonchidae; B, E: stoma of Mononchidae; C, F: stoma of Mylonchulidae; G: amphidial slit-like aperture (arrow); H: small peg-like 'foramina' at base of stoma (arrow); I, J: tuberculate type pharyngo-intestinal junction in Anatonchidae and Iotonchidae (arrows); K: non-tuberculate type junction in Mononchidae.

(Fig. 16.18A, D) and funnel-shaped in Mononchidae and Mylonchulidae (Fig. 16.18B, C, E, F). Oral aperture and vestibulum (= cheilostom) hexaradial; both characterized by the presence of folds or liplets. An important feature of Mononchida is the presence of a large, sharply pointed dorsal tooth (except for *Nullonchus*) of variable size, shape and location. Usually, the tooth apex is directed anteriad, however in some genera of Anatonchidae the tooth or teeth can be posteriorly directed. Other teeth of similar size or smaller than the dorsal tooth, as well as denticles, are often present on the ventro-sublateral plates, either scattered or organized in transverse or longitudinal rows. Base of stoma composed of three oblique basal plates surrounded by anterior part of pharynx. The 'foramina' (Fig. 16.18H) are small peg-like structures located on basal plates. A series of muscles is associated with the buccal cavity, and play an important role in capturing and ingesting the prey. Pharynx strongly muscular, cylindrical in shape without any constrictions or swellings. Pharyngeal lumen appearing strongly cuticularized and thick in cross section. Structure of pharyngo-intestinal junction of high taxonomic value and the main distinguishing feature between Anatonchoidea and Mononchoidea. In the Anatonchoidea, the pharyngo-intestinal junction is characterized by the presence of three prominent nodules or tubercles at the end of the pharyngeal lumen (Fig. 16.18I). These structures are formed by the widening of the pharynx lumen and cardia. The non-tuberculate junction is characteristic of Mononchoidea, the tubercles being absent or underdeveloped (Fig. 16.18K). The intestine is a simple long and straight tube, consisting of a single layer of polygonal (mostly hexagonal) cells. In preserved specimens it is common to find entire bodies of nematodes as well as cutitcularized parts of digested animals (rotifer mastax, oligochaete spines, nematode spears) inside the intestine. Female reproductive system is usually didelphic amphidelphic, each genital branch comprising an ovary, oviduct and uterus. Uterus long, sac-like, often with one or two developed intrauterine eggs (Fig. 16.19C). Oviduct and uterus connected by a sphincter; both structures may also function as sperm reservoirs. Vagina a thick-walled muscular tube showing clear cuticularized pieces at the vulva-vagina junction or just behind the vulval orifice in some species (Fig. 16.19D). Male reproductive system diorchic, with outstretched testes. The seminal vesicle stores developing sperm and there is also an ejaculatory duct and four pairs of uninucleate ejaculatory glands; vas deferens composed of tubular and glandular regions and ejaculatory duct, joining the rectum to form cloaca. Cloacal aperture pore-like. Spicules paired, ventrally arcuate and sclerotized (Fig. 16.19F); of variable morphology and size. Ventromedian copulatory receptors located anterior to cloaca at level of spicules and anteriorly, variable in size and number (Fig. 16.19F). Tail morphology highly variable. Marked sexual tail dimorphism not observed, tail shape similar in both sexes except that the male tail is shorter and more curved ventrad.

16.5.2. Family Iotonchidae Jairajpuri, 1969

Medium to large-sized nematodes, often over 6 mm long. Stoma broad, flattened at base, often very roomy and heavily sclerotized (Fig. 16.20A). Dorsal tooth present, small to moderate in size, predominantly directed forward but in some cases backward; located either in anterior or in posterior part of stoma. Subventral teeth, minute denticles or longitudinal ridges of various types and arranged in longitudinal rows (*Nullonchus* is an exception, lacking teeth or denticles). Pharyngeal lumen ending in three short vesicular extensions (tubercula). This family includes 11 valid genera and 124 valid species according to Ahmad and Jairajpuri (2010), with *Iotonchus* (Fig. 16.20) being a typical predatory nematode in tropical and subtropical habitats.

16.5.3. Family Mononchidae Chitwood, 1937

Medium to large-sized nematodes, generally well over 1 mm. Stoma spacious, more or less oval, tapering at base, with large dorsal tooth, predominantly located in anterior part of stoma, and directed forward (Fig. 16.21A). Subventral walls either smooth or provided with large teeth or small denticles arranged in longitudinal ridges or scattered irregularly. Pharyngeal lumen not forming vesicular extensions (tubercula). This family includes eight genera and 89 valid species according to Ahmad and Jairajpuri (2010). Type genus: *Mononchus* (Fig. 16.21).

16.5.4. Family Mylonchulidae Jairajpuri, 1969

Medium sized nematodes. Stoma strongly tapering at base, more funnel-shaped than in other families. Dorsal tooth large, claw-like, located in anterior part of stoma, sharply pointed directed forward

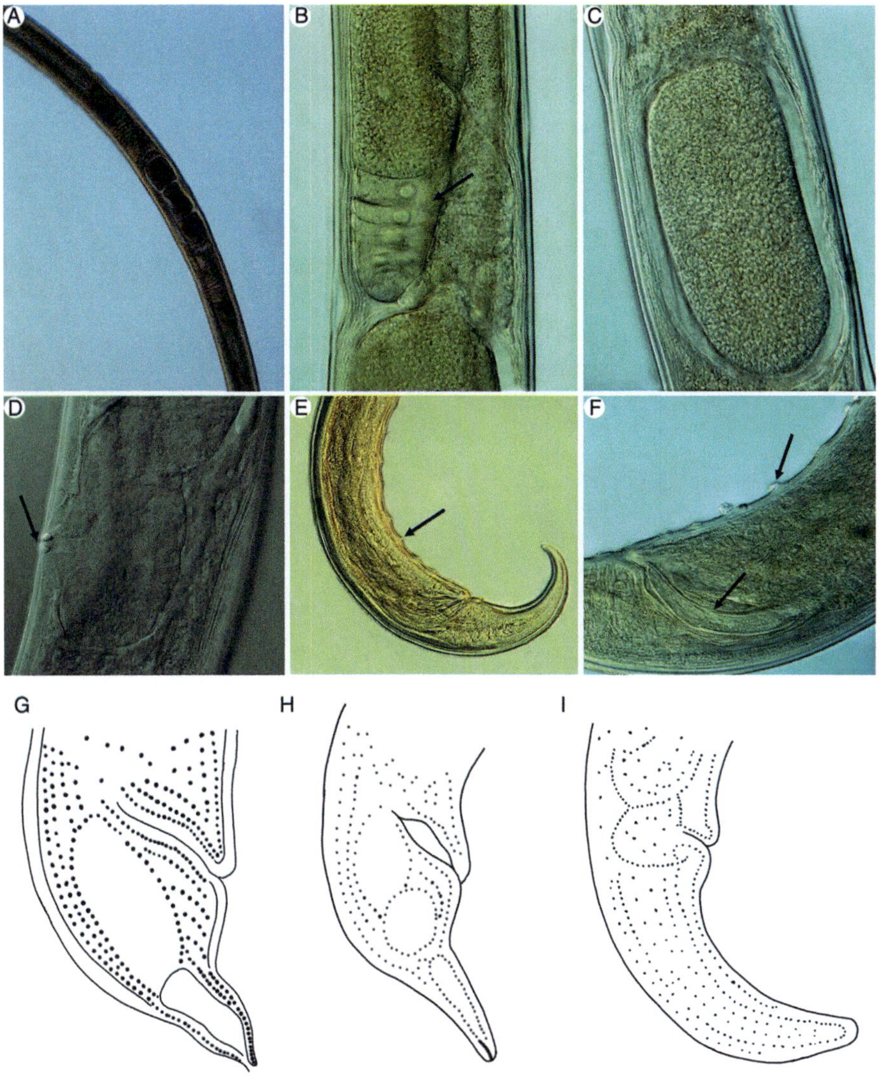

Fig. 16.19. Morphology of Mononchida, female and male reproductive system and tail. A: Didelphic female gonads; B: germinal zone of the ovary (arrow); C: single intrauterine egg; D: *Pars refringens vaginae* at vulva-vagina junction (arrow); E, F: arrangement of ventromedian copulatory receptors and sclerotized spicules (arrows); G-I: some variations in tail shape of Mononchida.

(Fig. 16.22A). Ventrosublateral denticles rasp-like, generally numerous, arranged in transverse rows (Fig. 16.22B). Two small subventral teeth behind the rasp-fields may be present. Pharyngo-intestinal junction without tubercula. This family includes 14 genera and 99 valid species according to Ahmad and Jairajpuri (2010). Type genus: *Mylonchulus* (Fig. 16.22), the largest genus of the family.

16.5.5. Family Anatonchidae Jairajpuri, 1969

Medium to large-sized nematodes. Stoma broad, flattened at base, strongly sclerotized. Three teeth, two subventral and one dorsal, nearly equal in size and usually retrorse. Pharyngeal lumen ending in three short vesicular extensions

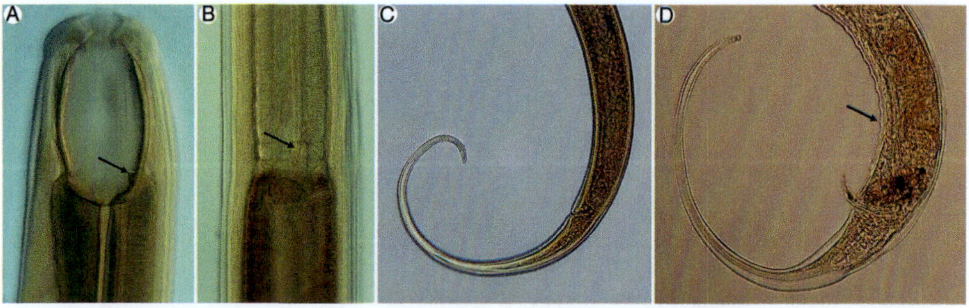

Fig. 16.20. *Iotonchus* – representative of the Anatonchidae, common in tropical soils. A: Female stoma with tooth at base (arrow); B: pharyngo-intestinal junction with tubercles (arrow); C: female tail; D: male tail, spicules and copulatory receptors (arrow).

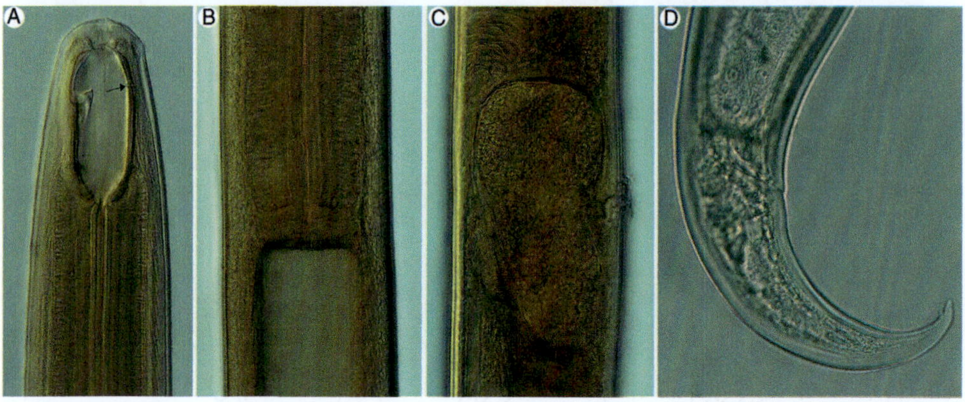

Fig. 16.21. *Mononchus* – representative of the Mononchidae. A: Female stoma with dorsal tooth and opposite transverse rib (arrow); B: posterior end of pharynx without tubercles; C: vulva and intrauterine egg; D: female tail.

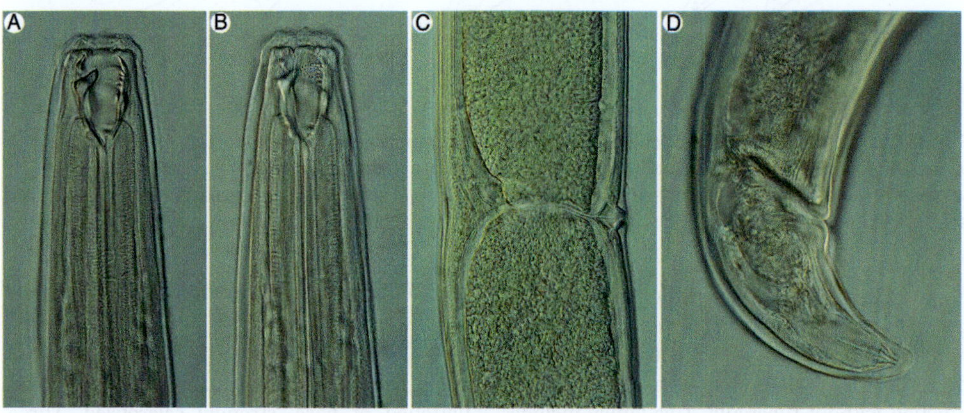

Fig. 16.22. *Mylonchulus* – representative of the Mylonchulidae. A: Female stoma with dorsal tooth; B: rasp-like denticles (arrow); C: vulval region with intrauterine eggs; D: female tail with spinneret.

(tubercula). This family includes nine valid genera and 71 valid species according to Ahmad and Jairajpuri (2010), *Miconchus* having the highest number of species.

16.6. Order Chromadorida Chitwood, 1933

Order Chromadorida includes over 100 valid genera and over 900 known species, most of which inhabit marine habitats, with the highest known diversity in the tropics (Tchesunov, 2014a). Terrestrial fauna is represented by several independently evolved lineages, which remain largely restricted to freshwater bodies and moist soil (Chromadoridae, Achromadoridae, Cyatholaimidae and Ethmolaimidae) or occupy transitional habitats between the sea and the land, such as coastal dunes (Selachinematidae). All of these nematodes can be identified by having an ornamented cuticle with distinct refractive dots, and by having a stoma with distinct teeth or jaws. Of the five families, Chromadoridae is the most diverse and Achromadoridae is the most common in soil, while the remaining three are much less common. Out of nearly 20 genera and close to 150 species of Cyatholaimidae (Fig. 16.23A, B), only seven species from *Cyatholaimus* occur in terrestrial habitats but can be locally very common – an undescribed species of *Cyatholaimus* is abundant in freshwater lakes and inflows of salt ponds throughout the Atacama Desert in Chile. *Ethmolaimus* (Fig. 16.23C, D) is a single typical terrestrial inhabitant, common in brackish and freshwater habitats, among ten genera and about 70 species of the family Ethmolaimidae. Selachinematidae, with eight genera and some 50 species, is represented in soils by two genera, *Choanolaimus* and *Synonchium* (Fig. 16.23E, F), which are rare but locally abundant in coastal sand dunes and found alongside typical soil nematodes, sometimes observed feeding on plant-parasitic nematodes. Morphologically, all three families are rather similar to *Achromadora*, especially in amphid shape (multispiral) and cuticle structure with transverse rows of evenly distributed refractive

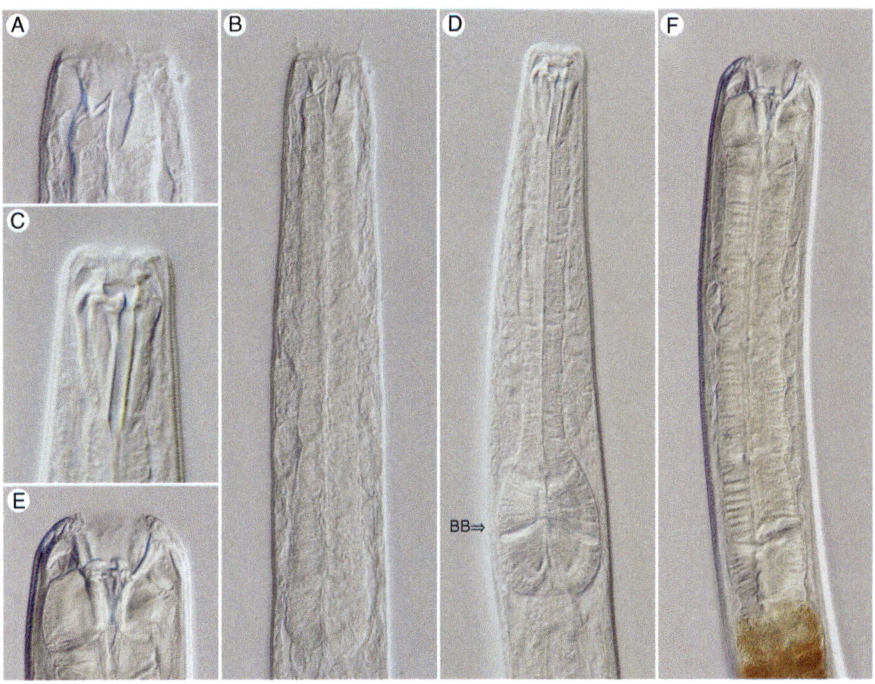

Fig. 16.23. *Cyatholaimus* (A, B), *Ethmolaimus* (C, D) and *Synonchium* (E, F) – terrestrial representatives of Cyatholaimidae, Ethmolaimidae and Selachinematidae, respectively. A, C, E: Anterior end showing stoma; B, D, F: pharyngeal region showing basal bulb (BB) in *Ethmolaimus*.

dots. *Cyatholaimus* is the most similar, differing only by the absence of basal pharyngeal bulb (Fig. 16.23B). *Ethmolaimus* – stoma with three equal teeth located at its anterior end (Fig. 16.23C); *Synonchium* – stoma with three movable mandibles (Fig. 16.23E); *Choanolaimus* – funnel-shaped stoma with three sets of small teeth directed posteriad.

16.6.1. Family Chromadoridae Filipjev, 1917

The most diverse taxon of the order, it includes over 40 genera and 450 species (Fig. 16.24). It also has the largest number of terrestrial representatives, with six genera and about 20 species found in freshwater habitats and moist soil. All of these species appear to be algal feeders. Most species are rather small, 1 mm or less, with few reaching over 2 mm in size. Cuticle annulated, with transverse rows of refractive dots (Fig. 16.24H), which are often larger along lateral body; many marine species with complex refractive structures along lateral body sides, often differing in various areas of the body. Inner and outer labial sensilla papilliform, cephalic sensilla setiform. Slit-like amphids, hard to observe (Fig. 16.24F). Some species have dark photosensitive eyespots along the anterior part of the pharynx. Stoma funnel-shaped, with large dorsal tooth in its

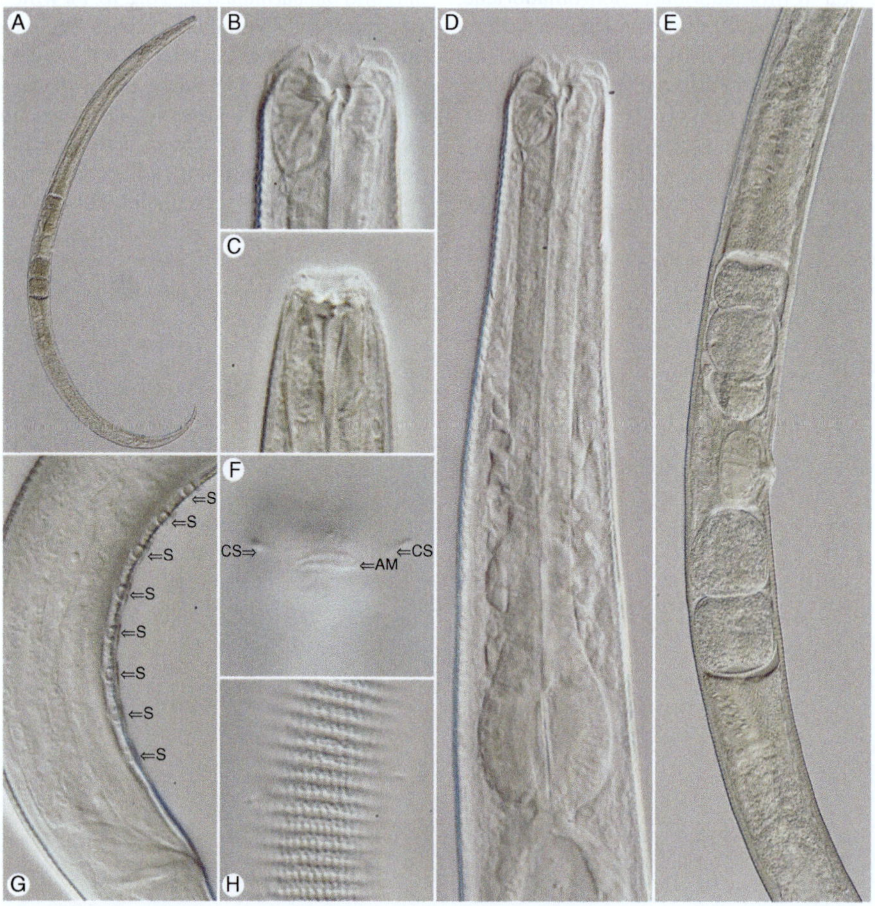

Fig. 16.24. *Chromadorita* (A, B, D-H) and *Punctodora* (C) – representatives of the Chromadoridae. A: Entire female; B, C: anterior end showing stoma; D: pharyngeal region; E: female gonads; F: anterior end (surface view) showing cephalic sensilla (CS) and amphid (AM); G: male cloacal region showing copulatory receptors (S); H: surface view of cuticle.

anterior part and small subventral denticles opposite (Fig. 16.24B, C). Pharynx strongly muscular, cylindrical along most of its length, often with well-developed basal bulb, but without a grinder (Fig. 16.24D). Amphimictic species, males are common and necessary for species identification. Female reproductive system with two gonads, ovaries antidromously reflexed, vulva opening at midbody (Fig. 16.24E). Male copulatory receptors cup-shaped (Fig. 16.24G) or absent. Spicules and gubernaculum present. Tail elongate conoid, caudal glands and spinneret present.

16.6.2. Family Achromadoridae Gerlach & Riemann, 1973

This family includes the single genus *Achromadora* with just over 20 valid species (Fig. 16.25), inhabiting freshwater, moist soil and mosses, but never in marine habitats and probably feeding on algae. They are relatively small nematodes, rarely reaching over 1 mm in size. Cuticle with transverse rows of evenly distributed refractive dots (Fig. 16.25C). Inner labial sensilla papilliform, outer labial and cephalic sensilla setiform. Multispiral amphids (Fig. 16.25B). Stoma funnel-shaped, with large dorsal tooth in its anterior part and small subventral denticles at its base (Fig. 16.25A). Pharynx strongly muscular, cylindrical along most of its length, with well-developed basal bulb, but without a grinder (Fig. 16.25D). Most species parthenogenetic, males rare or unknown. Female reproductive system with two gonads, ovaries antidromously reflexed, vulva opening at midbody. Male copulatory receptors either setiform or cup-shaped (Fig. 16.25G). Spicules and gubernaculum present. Tail elongate conoid, caudal glands and spinneret present.

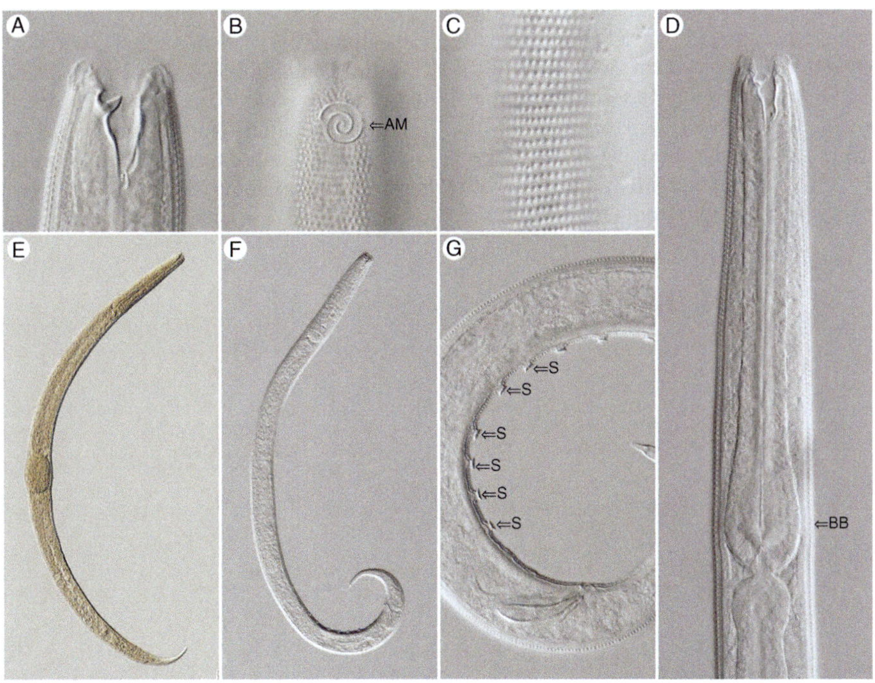

Fig. 16.25. *Achromadora* – representative of the Achromadoridae. A: Anterior end showing stoma; B: anterior end (surface view) showing amphid (AM); C: surface view of cuticle; D: pharyngeal region showing basal bulb (BB); E: entire female; F: entire male; G: male cloacal region showing copulatory receptors (S) and spicules.

16.7. Order Desmodorida De Coninck, 1965

The Desmodorida is rarely mentioned in soil nematology – nearly all of its 100 genera and approximately 1000 species are marine (Tchesunov, 2014b). The exceptions include *Prodesmodora*, which is the only typical freshwater and soil-inhabiting member of the family Desmodoridae, *Microlaimus* (Microlaimidae), some species of which are common in coastal salt marshes, and *Haliplectus* (Haliplectidae), found in coastal sand dunes. Morphologically, all three genera are very similar and can easily be confused, despite their dissimilar ecology. Terrestrial species are less than 2 mm in length, usually much smaller. Cuticle annulated and may appear yellowish or brownish. Inner and outer labial sensilla papilliform, cephalic sensilla setiform in *Prodesmodora* and *Microlaimus* and papilliform in *Haliplectus*. Amphids circular in all three genera. Stoma funnel-shaped, with distinct dorsal tooth in its anterior part and small subventral denticles at its base in *Prodesmodora* and *Microlaimus*, and simple tubular in *Haliplectus*. Pharynx strongly muscular, cylindrical along most of its length, with well-developed basal bulb in *Prodesmodora* and *Microlaimus* (Fig. 16.26A, F), while pharynx in *Haliplectus* is uniquely guitar-shaped (Fig. 16.26C). Female reproductive system with two gonads, ovaries antidromously reflexed in *Prodesmodora* (Fig. 16.26B) and *Haliplectus* (Fig. 16.26D), and outstretched in *Microlaimus* (Fig. 16.26E), vulva opening at midbody. Male copulatory receptors absent in *Prodesmodora* or small papilliform in *Microlaimus* and *Haliplectus*. Spicules and gubernaculum present. Tail elongate conoid, caudal glands and spinneret present.

16.8. Order Isolaimiida Cobb, 1920

One of the smallest orders of terrestrial nematodes, it includes two families: Aulolaimidae (three genera) and Isolaimiidae (one genus, *Isolaimium*), and just over 30 species. They are rarely found in soil, but some species of *Aulolaimus* can be locally abundant. *Isolaimium* are found very rarely and in few numbers, but when found are very noticeable. Members of *Aulolaimus* are easily identified by their fusiform body, cuticle with longitudinal ridges (Fig. 16.27C, F), long cylindrical stoma that is often longer than the muscular pharynx (Fig. 16.27A), refractive structures supporting vulva (Fig. 16.27B), and tail without caudal glands and spinneret (Abolafia and Peña-Santiago, 2018). *Isolaimium* is distinct from other nematodes by its large body, up to 7 mm, cuticle marked with refractive dots arranged in longitudinal and transverse rows, long cylindrical stoma equal in length to half the length of the muscular pharynx, and short

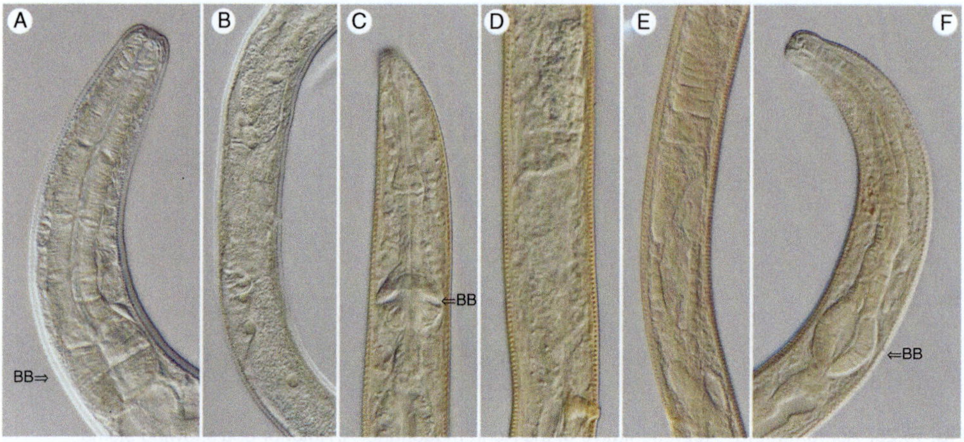

Fig. 16.26. *Prodesmodora* (A, B), *Haliplectus* (C, D) and *Microlaimus* (E, F) – representatives of Desmodoridae, Haliplectidae and Microlaimidae, respectively. A, C, F: Pharyngeal region showing basal bulb (BB); B, D, E: female gonads.

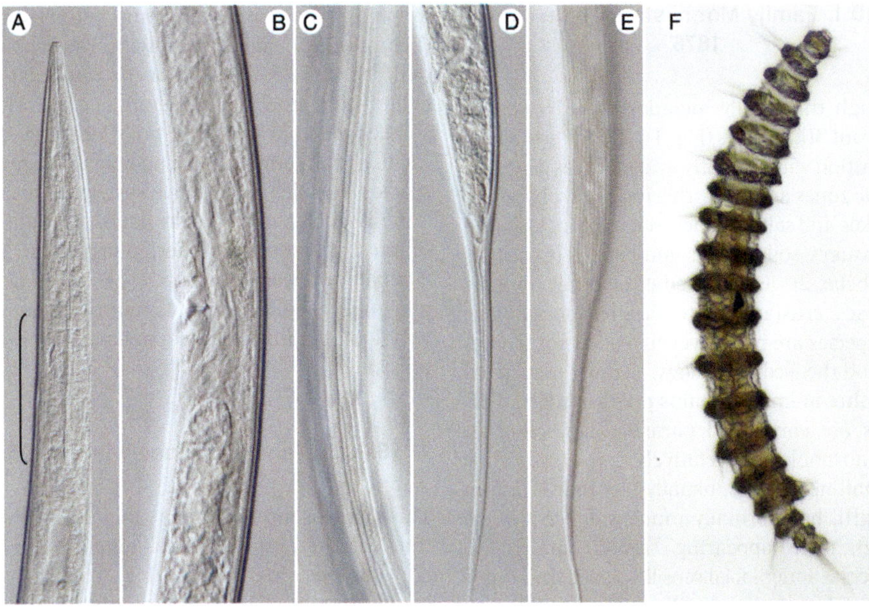

Fig. 16.27. *Aulolaimus* (A-E) and *Desmoscolex* (F) – representatives of the Aulolaimidae and Desmoscolecidae. A: Pharyngeal region; B: female gonads; C: surface view of longitudinal ridges of cuticle along middle of body; D: female tail; E: surface view of longitudinal ridges of cuticle along tail; F: entire male.

rounded tail without caudal glands and spinneret (Holovachov, 2014a).

16.9. Order Desmoscolecida Schuurmans Stekhoven, 1950

The Desmoscolecida is nearly exclusively marine and has around 30 genera and 300 species (Decraemer and Rho, 2014). Nevertheless, some species can be found in soil, specifically in salt marshes and in tropical rainforests. The appearance of these species is unmistakable once one realizes they are actually nematodes, since they are often confused with annelids, showing obvious 'segmentation' and distinct setae (Fig. 16.27F). Terrestrial species are small, with fusiform bodies ornamented by desmen – distinct refractive rings composed of mineral particles (concretion rings), separated by interzones of clear cuticle. Their anterior end is offset, with four cephalic setae and 'vesicular' amphids. Long somatic setae cover the entire body in a species-specific pattern. The internal morphology is hard to observe; the stoma is minute, and the pharynx is glandular. Female reproductive system with two gonads, ovaries antidromously reflexed, vulva opening at midbody. Males without copulatory receptors. Spicules and gubernaculum present. Tail conoid, caudal glands and spinneret present.

16.10. Order Monhysterida De Coninck & Schuurmans Stekhoven, 1933

The Monhysterida includes about 110 valid genera and over 1300 known species (Coomans and Abebe, 2006; Fonseca and Bezerra, 2014b). They are typical aquatic nematodes, with the highest diversity in oceans and seas, but also common and often ubiquitous in freshwater habitats, wet soil, litter and mosses. Round amphids, funnel-shaped stoma and single anterior gonad in females are good diagnostic characters to separate terrestrial Monhysterida from other nematodes, but differentiating between Monhysteridae and Xyalidae is challenging – the only difference between the two is the position of the gonad relative to the intestine: on the right side in Monhysteridae and on the left in Xyalidae.

16.10.1. Family Monhysteridae de Man, 1876

Although this family includes only 18 genera and about 300 species (Fig. 16.28), its ecological distribution varies greatly and includes deep sea, hypoxic zones and hydrothermal vents, hypersaline lakes and salt marshes, sea ice, brackish and fresh waters, soil, mosses and lichens (Coomans and Abebe, 2006) and bodies of other animals (molluscs, crustaceans, turtles, baleen whales). Most species are bacterivorous, a few are predatory, and the feeding biology of those associated with other animals remains poorly studied. Few species are considered parasitic. Nevertheless, their morphology is relatively conserved. They are small nematodes, usually less than 1.5 mm in length, but normally much smaller. Cuticle striated, often appearing smooth under the microscope. Inner labial sensilla papilliform, outer labial and cephalic sensilla setiform. Circular amphids (Fig. 16.28D). Stoma funnel-shaped (Fig. 16.28C), sometimes with thicker walls and/or small denticles. Pharynx muscular and cylindrical, gradually widening posteriorly, in smaller genera with a basal bulb. Monhysteridae includes both amphimictic and parthenogenetic species. Female reproductive system with single anterior gonad, ovary outstretched, on right side of intestine, post-vulval uterine sac absent. Male copulatory receptors usually absent. Spicules and gubernaculum present. Tail short conoid to very long filiform, caudal glands and spinneret present.

16.10.2. Family Xyalidae Wieser, 1956

This family is much more taxonomically diverse than the Monhysteridae, including over 50 genera and over 550 species (Fig. 16.29), but is mostly restricted in its distribution to marine

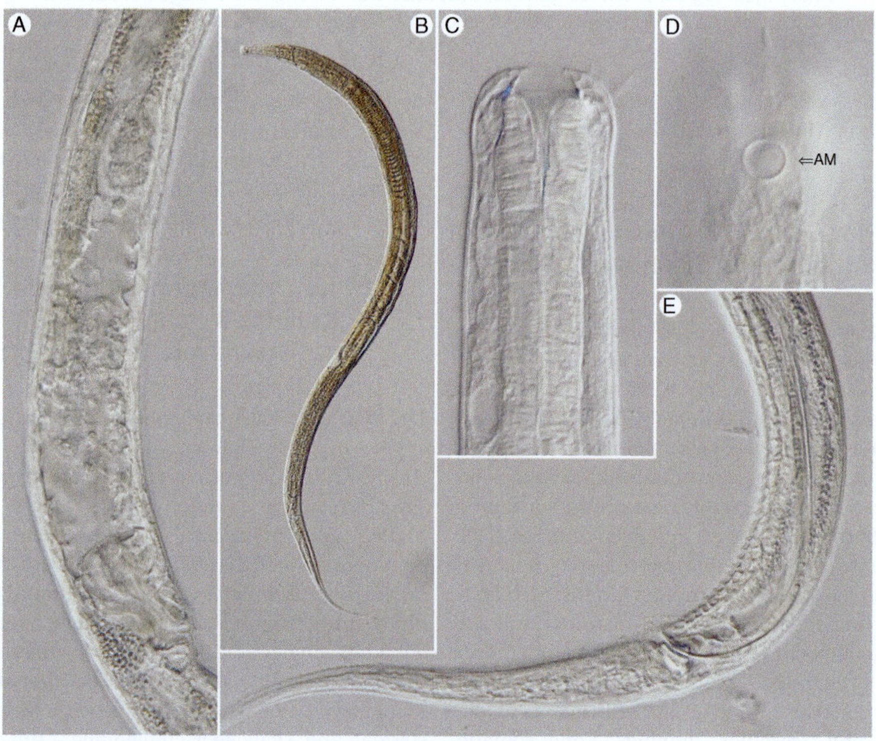

Fig. 16.28. *Monhystera* – representative of the Monhysteridae. A: Female gonad; B: entire female; C: anterior end showing stoma; D: anterior end (surface view) showing anterior sensilla and amphid (AM); E: male caudal region showing spicules.

and brackish habitats. Only some species of *Daptonema* and *Theristus* are commonly found in freshwater and in soil, and thus must be differentiated from Monhysteridae that are also common in the same habitats. Terrestrial species are bacterivorous, while the feeding biology of many marine genera is not clearly understood. Most known species are less than 2 mm in length, usually much smaller. Cuticle striated or annulated and may have distinct longitudinal ridges or spines. Inner labial sensilla papilliform, outer labial and cephalic sensilla setiform. Circular amphids (Fig. 16.29D). Stoma funnel- or barrel-shaped (Fig. 16.29C), sometimes with thicker walls or denticles. Pharynx muscular and cylindrical, gradually widening posteriorly. Amphimictic. Female reproductive system with single anterior gonad, ovary outstretched, on left side of intestine; post-vulval uterine sac can be present (Fig. 16.29A). Male copulatory receptors usually absent. Spicules L-shaped in many species (Fig. 16.29E), gubernaculum present. Tail short conoid to very long filiform, caudal glands and spinneret present.

16.11. Order Araeolaimida De Coninck & Schuurmans Stekhoven, 1933

The Araeolaimida comprises nearly exclusively marine species and accounts for around 60 genera and over 580 species (Muthumbi and Vanreusel, 2006; Fonseca and Bezerra, 2014a). The genera *Cylindrolaimus* (22 species) and *Edalonema* (single species) are exceptions. *Cylindrolaimus* is common in both mesic and xeric soils (Fig. 16.30). Its species are small, usually under 1 mm in length. Cuticle is finely striated, without any ornamentation. Inner and outer labial sensilla papilliform, cephalic sensilla setiform. Circular amphids. Stoma tubular but rather short (Fig. 16.30B). Pharynx muscular, gradually widening posteriorly, without basal bulb (Fig. 16.30B). Species either amphimictic or asexual (parthenogenetic). Female reproductive system with either two or single anterior gonad, ovary(-ies) outstretched; post-vulval uterine sac or non-functional ovary can be present. Male with single precloacal papilliform copulatory receptor (Fig. 16.30D). Spicules and gubernaculum present

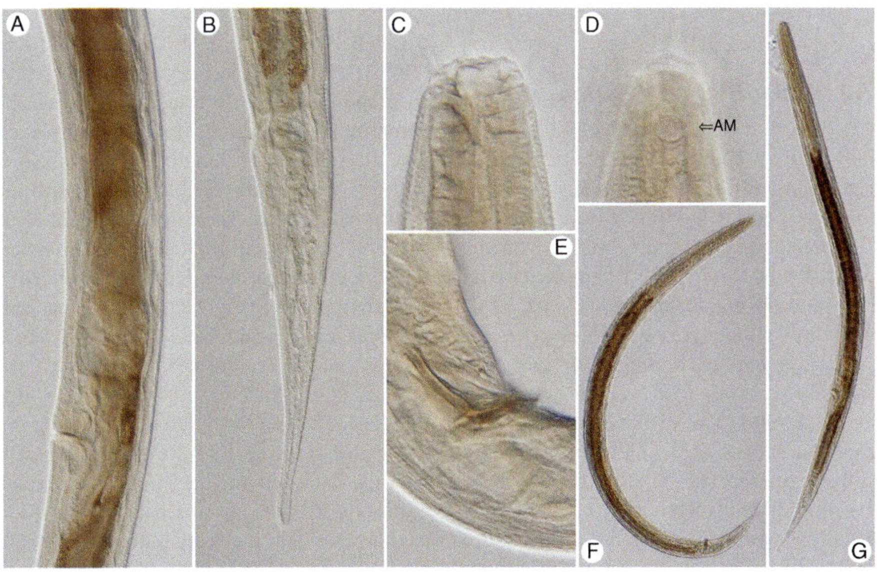

Fig. 16.29. *Daptonema* – terrestrial representative of the Xyalidae. A: Female gonad; B: female tail; C: anterior end showing stoma; D: anterior end (surface view) showing anterior sensilla and amphid (AM); E: male cloacal region showing spicules; F: entire male; G: entire female.

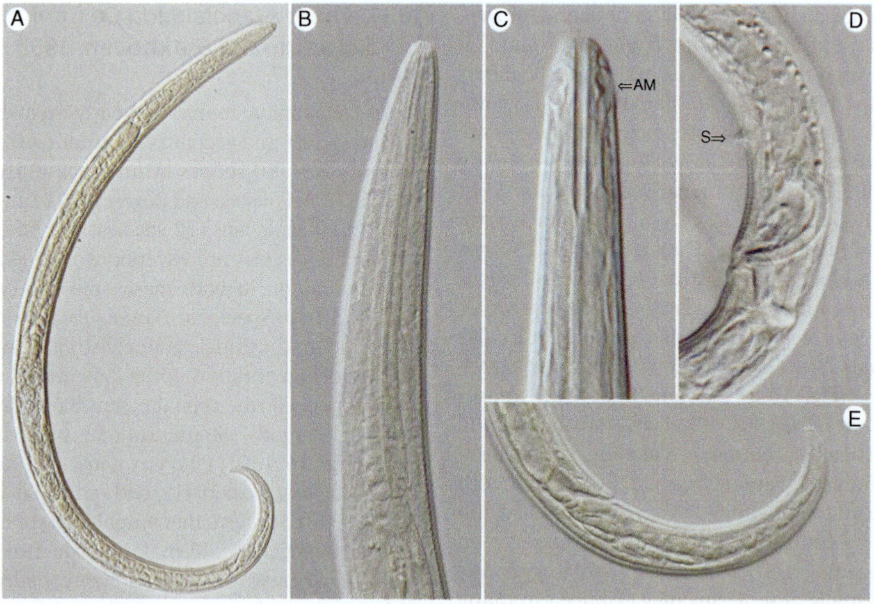

Fig. 16.30. *Cylindrolaimus* – terrestrial representative of the Diplopeltidae. A: Entire female; B: pharyngeal region; C: anterior end showing stoma and amphid (AM); D: male cloacal region showing spicules and precloacal copulatory receptors (S); E: female tail.

(Fig. 16.30D). Tail elongate conoid, caudal glands and spinneret present.

16.12. Order Plectida Gadea, 1973

Plectids are a common and widely distributed group of nematodes, occurring in marine, freshwater and terrestrial habitats throughout the world. Some species are specialists of extreme polar ecosystems, others live in epiphytes and frequently desiccating mosses and lichens, while yet others are symbionts or parasites of protozoans and invertebrates. Many terrestrial representatives of this group feed on bacteria or unicellular eukaryotes. The group includes about 60 genera and over 500 valid species, being more diverse in marine habitats (Holovachov and Boström, 2010; Holovachov, 2014b). Morphologically they are diverse and it is difficult to provide a set of unique characters uniting them all. Five families are found in soil, of which the Ohridiidae are rare symbionts of earthworms.

16.12.1. Family Plectidae Örley, 1880

Ten genera and almost 400 species are known (Fig. 16.31). Most species are thought to be cosmopolitan, with several *Plectus* species being common in Antarctica. Most known species are less than 1 mm in length. Cuticle annulated, without ornamentation but with lateral alae – lateral bands of cuticle extending from anterior third of pharyngeal region to posterior part of tail and encompassing deirids (papilliform sensory structures) just posterior to the nerve ring, also present in species from the Rhabditida. Most species have a typical hexaradial anterior end with six equal lips and setiform cephalic sensilla (Fig. 16.31B), not too different from other nematodes, but some genera are characterized by a laterally flattened and dorso-ventrally expanded anterior end with specialized structures and projections evolved to sweep and capture bacteria, like *Tylocephalus* (Fig. 16.31C) and *Wilsonema* (Fig. 16.31D). Amphids unispiral or transverse slit. Stoma funnel-shaped. Pharynx muscular,

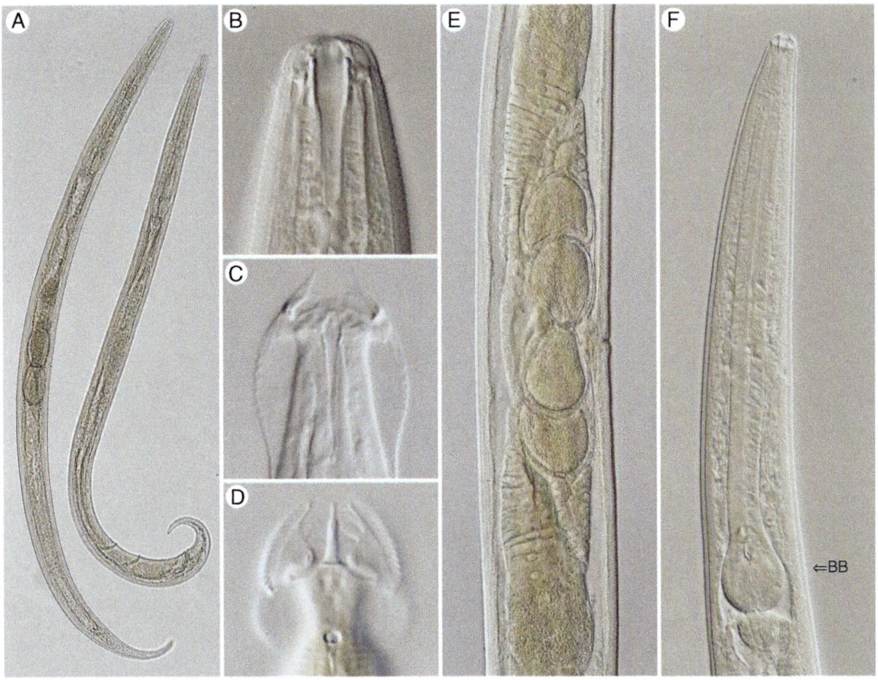

Fig. 16.31. *Plectus* (A, B, E, F), *Tylocephalus* (C) and *Wilsonema* (D) – representatives of the Plectidae. A: Entire female and male; B-D: anterior end; E: female gonads; F: pharyngeal region showing basal bulb (BB).

with distinct subdivision into corpus, isthmus and basal bulb; basal bulb with a grinder (Fig. 16.31F). Plectidae includes both amphimictic and parthenogenetic species. Female reproductive system with two gonads, ovaries antidromously reflexed, vulval opening at midbody (Fig. 16.31E). Male copulatory receptors tubular or absent. Spicules and gubernaculum present. Tail elongate conoid, caudal glands and spinneret present.

16.12.2. Family Metateratocephalidae Eroshenko, 1973

Very small group with just two genera and nine species (Fig. 16.32), which can be locally abundant in freshwater habitats, litter and mosses (Holovachov, 2014b). Most known species are less than 1 mm in length. They are unique among plectids in having the cuticle with transverse rows of evenly distributed refractive dots (Fig. 16.32F). Labial region very typical, with leaf-like lips supported by U-shaped refractive elements (Fig. 16.32A-D). Inner labial sensilla papilliform, outer labial and cephalic sensilla setiform. Amphids with pore-like opening (Fig. 16.32E). Stoma barrel-shaped, with refractive walls (Fig. 16.32A, C). Pharynx muscular, with distinct subdivision into corpus, isthmus and basal bulb; basal bulb with a grinder (Fig. 16.32G). All species are parthenogenetic, males are extremely rare or unknown. Female reproductive system with two gonads, ovaries antidromously reflexed, vulval opening at midbody. Male with single precloacal papilliform copulatory receptor. Spicules and gubernaculum present. Tail elongate conoid, caudal glands and spinneret absent.

16.12.3. Family Chronogastridae Gagarin, 1975

There are six genera with about 80 species (Fig. 16.33) known from tropical regions and inhabiting freshwater habitats, with the exception of *Cynura* and *Caribplectus*, which are marine

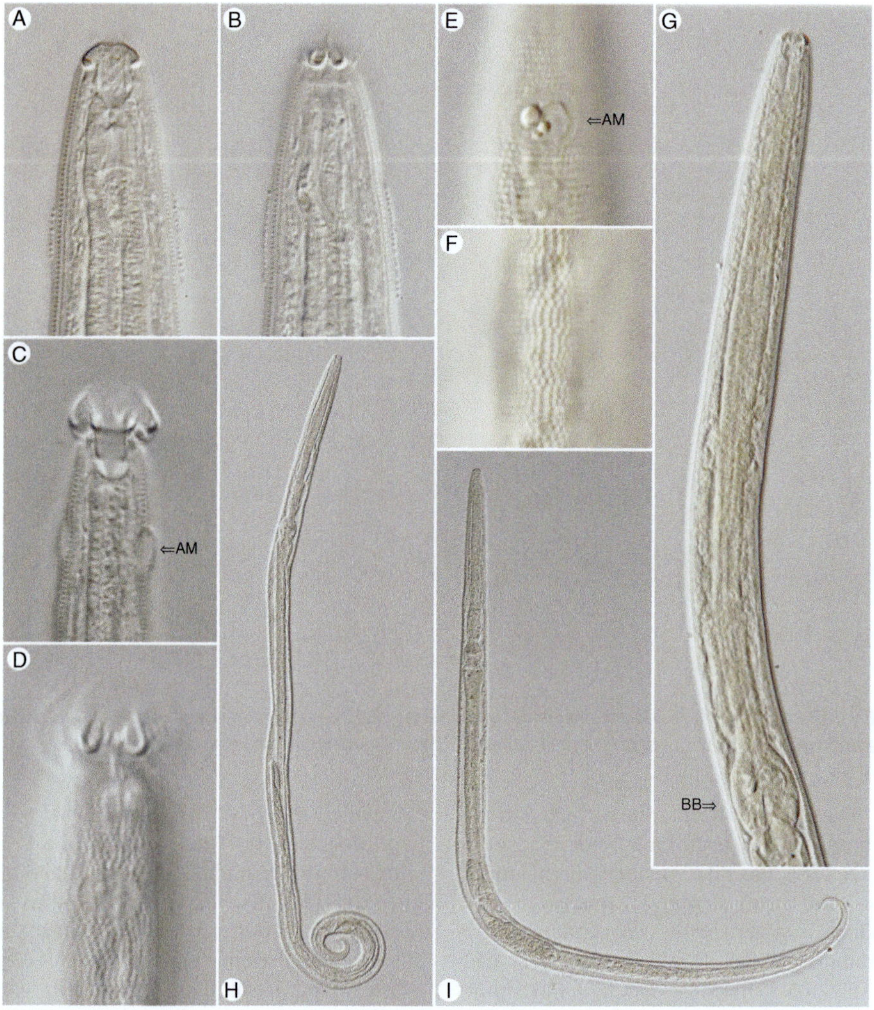

Fig. 16.32. *Euteratocephalus* (A, B, E-I) and *Metateratocephalus* (C, D) – representatives of the Metateratocephalidae. A, C: Anterior end showing stoma and amphid (AM); B, D: anterior end (surface view) showing refractive elements of labial region; E: amphid (AM); F: surface view of cuticle ornamentation along middle of body; G: pharyngeal region showing basal bulb (BB); H: entire male; I: entire female.

(Holovachov, 2014b). Marine species are larger, reaching 3 mm, while terrestrial species are smaller, 1 mm on average. Cuticle annulated, without lateral alae but with longitudinal lines or ridges in some species. Inner and outer labial sensilla papilliform, cephalic sensilla setiform. Unispiral, slit-like (Fig. 16.33B) or horseshoe-shaped amphids. Stoma funnel-shaped (Fig. 16.33C). Pharynx muscular, with distinct subdivision into corpus, isthmus and basal bulb; basal bulb with zip-like cuticularized valves (Fig. 16.33E). This family includes both amphimictic and parthenogenetic species. Marine species have a female reproductive system with two gonads, while terrestrial ones have a single anterior gonad (Fig. 16.33D), ovary(-ies) antidromously reflexed, post-vulval uterine sac present in monodelphic species, vulva opening at midbody. Male copulatory receptors tubular or entirely absent. Spicules and gubernaculum

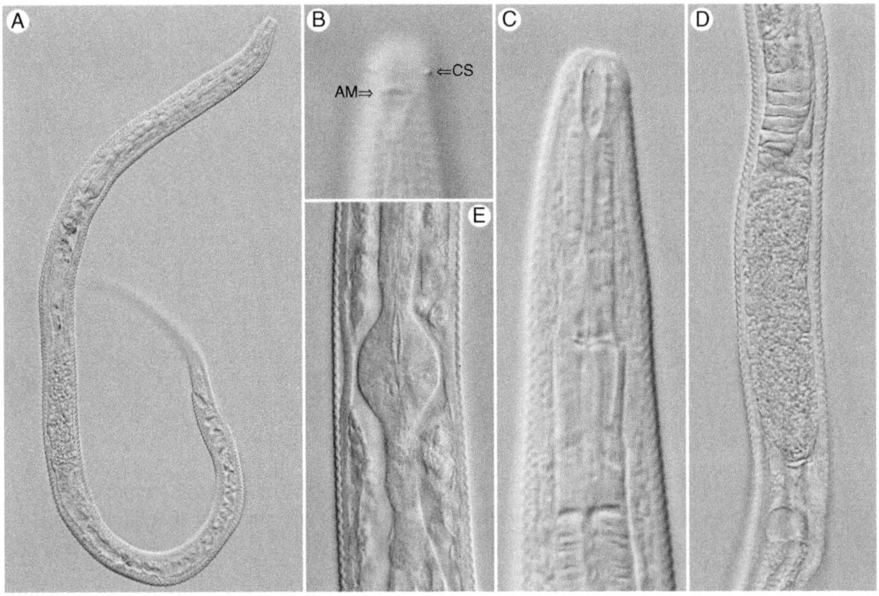

Fig. 16.33. *Chronogaster* – representative of the Chronogastridae. A: Entire female; B: anterior end showing cephalic sensilla (CS) and amphid (AM); C: anterior end showing stoma and anterior part of the pharynx; D: female gonad; E: base of the pharynx showing basal bulb.

present. Tail elongate conoid to filiform, caudal glands and spinneret present in some genera and absent in others, tail tip may be armed with spines or hooks.

16.12.4. Family Aphanolaimidae Chitwood, 1936

There are only four genera in this family and almost 100 species (Fig. 16.34), most of which inhabit freshwater and moist soil although a few species live in brackish river estuaries and lagoons (Holovachov, 2014b). Small to medium sized nematodes, 0.5-2 mm in size. Cuticle annulated, with lateral alae, with punctation or fine longitudinal striation in some species. Cuticle along anterior end not annulated, forming a cephalic capsule. Inner and outer labial sensilla papilliform, cephalic sensilla setiform. Unispiral amphids (Fig. 16.34D, E). Stoma very small in some species and large barrel-shaped with six digitate projections in its anterior part in others. Pharynx muscular and cylindrical. Most known species are amphimictic. Female reproductive system with either two or single anterior gonad, ovary(-ies) antidromously reflexed, post-vulval uterine sac present in monodelphic species, vulval opening at midbody. Typical for this group is the presence of two offset, sac-like spermathecae located on right and left sides of each gonoduct, with four spermathecae in total. Male copulatory receptors alveolar (may be absent) and tubular (Fig. 16.34C). Spicules and gubernaculum present. Tail elongate conoid, caudal glands and spinneret present.

16.13. Order Rhabditida Chitwood, 1933

In the current classification the Rhabditida includes all taxa that were formerly assigned in the Class Secernentea and includes a large number of morphologically and ecologically divergent groups of nematodes: free-living terrestrial species, plant, invertebrate and vertebrate parasites. Free-living rhabditids are distributed in every terrestrial habitat on Earth, including Antarctic deserts. They are also found in many freshwater

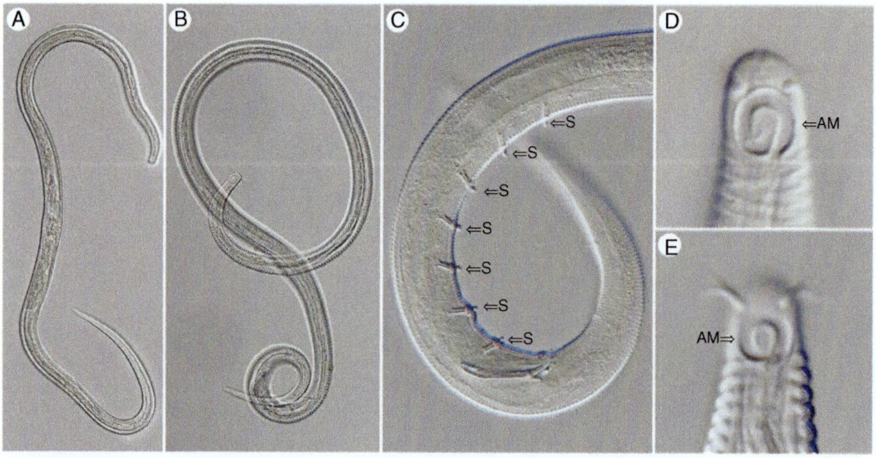

Fig. 16.34. *Aphanolaimus* – representative of the Aphanolaimidae. A: Entire female; B: entire male; C: male caudal region showing precloacal copulatory receptors (S) and spicules; D, E: anterior end (surface view) showing anterior sensilla and amphid (AM).

habitats and several species have re-colonized marine coastal habitats. Most terrestrial species are bacterial feeders or predators, often feeding on other nematodes and constituting the majority of the nematode fauna in saprobic habitats. Some species utilize plant organisms and fungi as a major food source and many of them are agriculturally important plant parasites, which are described elsewhere in this book. Animal parasitic species are endoparasites of terrestrial invertebrates, terrestrial and aquatic vertebrates, and humans, and are a subject of parasitology and veterinary studies; they are beyond the scope of this publication. The rate of species discovery in Rhabditida is higher than in any other nematode lineage and any statistics provided in this chapter will be outdated before it is published.

In addition to the taxa described below, species from these three families can sometimes be found in soil or associated with plants: (i) Brevibuccidae (four genera and about 20 species, Fig. 16.35A, B), some of which are found in soil and decomposing organic matter, while others are pathogenic to spiders; morphologically similar to Panagrolaimidae, differing mostly by a vulva located close to posterior body end; (ii) Chambersiellidae (five genera and about 20 species, Fig. 16.35D-G), most of which are found on aboveground parts of plants and in epiphytes (lichens and bromeliads in particular), are unique among rhabditids in having setiform outer labial and cephalic sensilla and (iii) Myolaimidae (two genera and 18 species, Fig. 16.35H-K), easily identifiable by having a loose cuticular sheath and no spicules in males; found in decomposing plant matter.

16.13.1. Family Panagrolaimidae Thorne, 1937

There are currently 13 genera and about 70 species in this family (Fig. 16.36), inhabiting freshwaters, mesic and xeric soils, extremely arid environments, decomposing organic matter, weakened plants, and feeding on bacteria. Some species are associated with various species of insects. Most species are rather small, 1 mm or less. Cuticle annulated, with lateral alae and deirids. Inner and outer labial, and cephalic sensilla papilliform. Small oval amphids. Labial region with six lips, rarely lips are leaf-shaped with refractive margins and 12 setiform projections. Stoma cylindrical: cheilostom wide; gymnostom wide and relatively long, often longer than stegostom and with strongly cuticularized rhabdia; stegostom funnel-shaped, with weakly developed rhabdia (Figs 16.36, 16.37C, D). Pharynx muscular, with distinct subdivision into corpus,

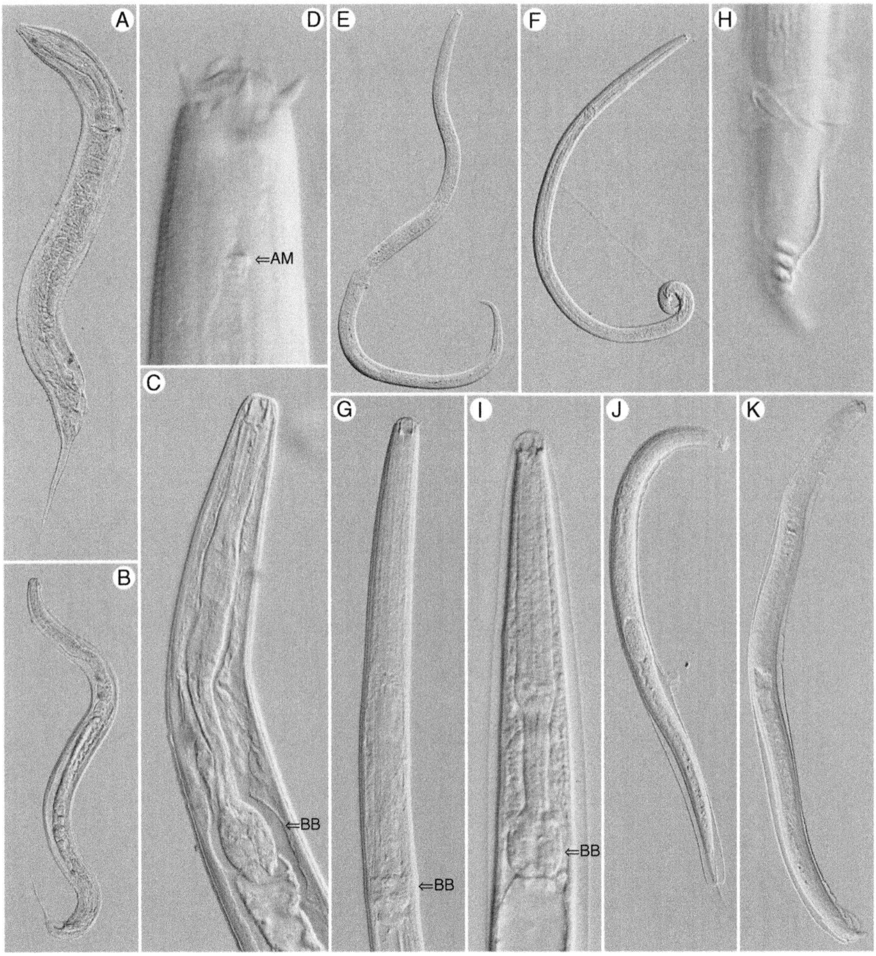

Fig. 16.35. *Brevibucca* (A-C), *Diastolaimus* (D-G) and *Myolaimus* (H-K) – representatives of the Brevibuccidae, Chambersiellidae and Myolaimidae. A, E, J: Entire female; B, F, K: entire male; C, G, I: pharyngeal region showing basal bulb (BB); D: anterior end (surface view) showing structures of labial region and amphid (AM); H: male caudal region (surface view) showing caudal sensilla.

isthmus and basal bulb; basal bulb with a grinder (Figs 16.36, 16.37E). Most described species are amphimictic, but some unnamed species of *Panagrolaimus* are triploids and reproduce asexually. Female reproductive system with single anterior gonad, ovary antidromously reflexed, spermatheca axial in most of species, post-vulval uterine sac can be present (Fig. 16.37F). Male copulatory receptors papilliform, arranged in subventral and subdorsal rows (Fig. 16.36). Spicules and gubernaculum present (Fig. 16.37H). Tail variable in size, conoid to filiform, caudal glands and spinneret absent.

16.13.2. Family Cephalobidae Filipjev, 1934

Within this taxon we include all genera and species formerly placed in the Alirhabditidae, Bicirronematidae, Daubayliidae, Elaphonematidae and Osstellidae, since all phylogenetic analyses place them within Cephalobidae (Holovachov *et al.*, 2009). Almost 40 genera and over 270 species (Fig. 16.38) are known, mostly from arid environments in the rhizosphere of desert plants, but feeding on the associated microbiome. Most species are rather small, 1 mm or less.

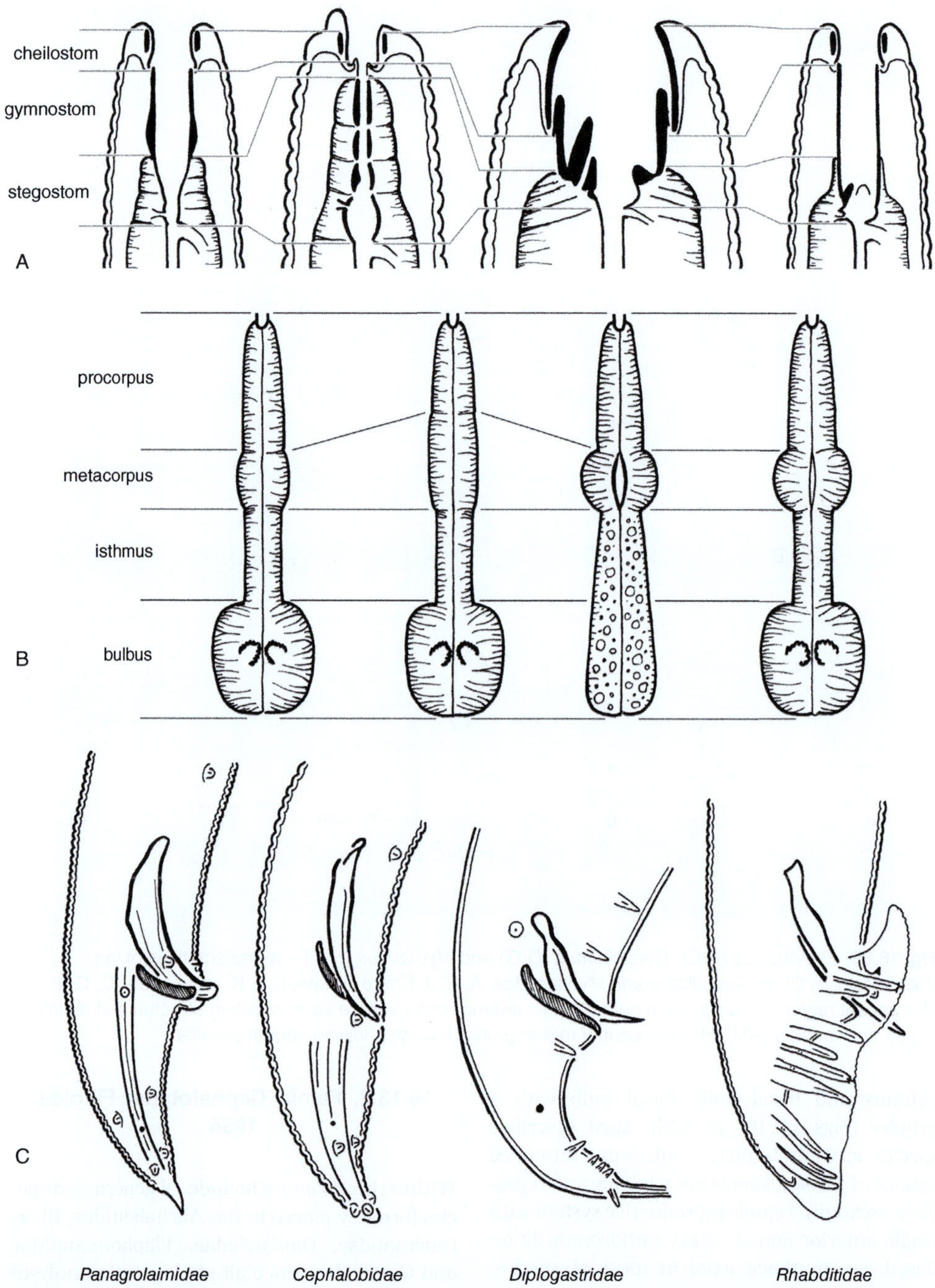

Fig. 16.36. Morphology of Rhabditida exemplified by the four most common families: Panagrolaimidae, Cephalobidae, Diplogastridae and Rhabditidae. A: Stoma and its sections; B: pharynx and its sections; C: male caudal region.

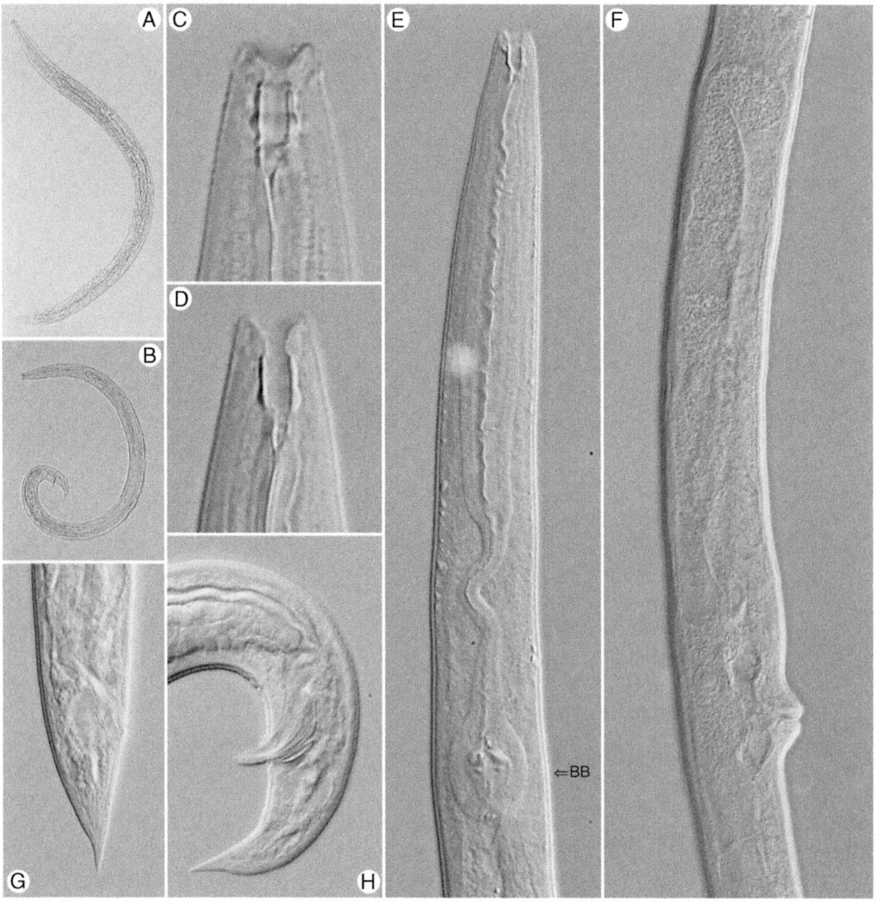

Fig. 16.37. *Panagrolaimus* – representative of the Panagrolaimidae. A: Entire female; B: entire male; C, D: anterior end showing stoma; E: pharyngeal region showing basal bulb (BB); F: female gonad; G: female tail; H: male caudal region showing spicules.

Cuticle annulated (Fig. 16.38J), with lateral alae (Fig. 16.38L) and deirids, in many species with dots, longitudinal lines or ridges (Fig. 16.38K). Inner and outer labial, and cephalic sensilla papilliform. Small oval amphids. Labial region either with six equal moderately developed lips, or with complex labial and cephalic projections called probolae, used to capture bacterial food (Fig. 16.38A-D). Stoma cylindrical: cheilostom with strongly refractive rhabdia; gymnostom narrow and very short; stegostom tubular, distinctly divided into four refractive subsections (Fig. 16.36). Pharynx muscular, with distinct subdivision into corpus, isthmus and basal bulb; basal bulb with a grinder (Fig. 16.36), basal bulb and grinder reduced in a few species. Majority of described species are amphimictic, but there are a number of parthenogenetic species. Female reproductive system with single anterior gonad, ovary antidromously reflexed, spermatheca offset from the ovary-uterus junction, post-vulval uterine sac present (Fig. 16.38N). Male copulatory receptors papilliform, arranged in subventral and subdorsal rows (Fig. 16.36). Spicules and gubernaculum present (Fig. 16.38M). Tail variable in size, rounded, cylindrical, conoid to filiform (Fig. 16.38G-I), often with variously shaped mucro on its tip, caudal glands and spinneret absent.

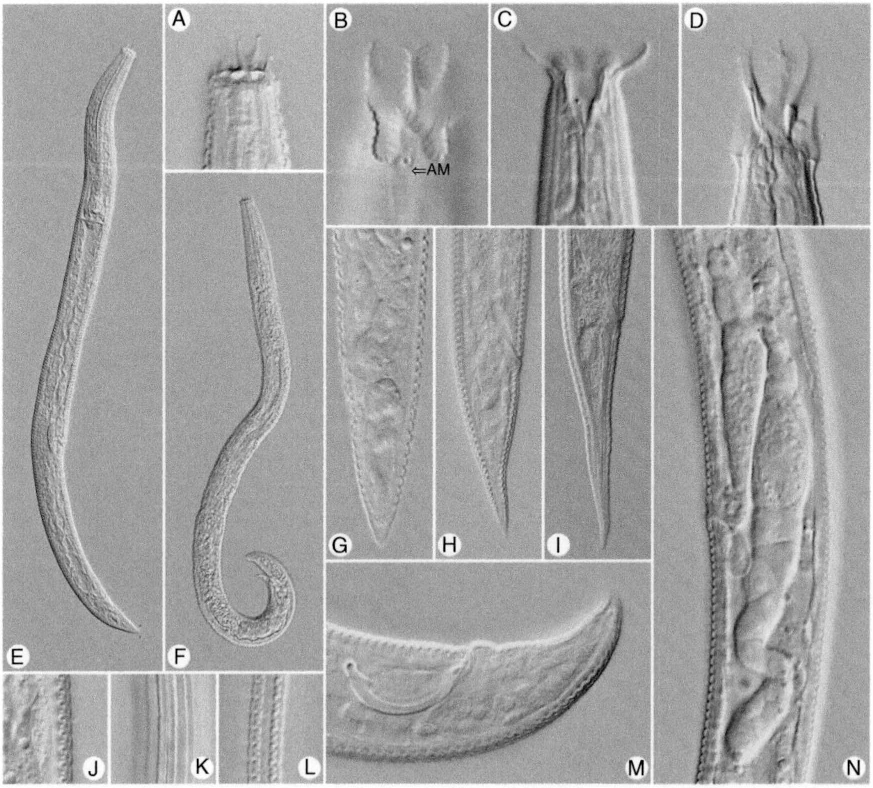

Fig. 16.38. *Nothacrobeles* (A, E-G), *Bicirronema* (C), *Acrobeles* (B, D, H-J, L-M), *Stegelleta* (K) and *Cervidellus* (N) – representatives of the Cephalobidae. A, B: Anterior end (surface view) showing structures of labial region; C, D: anterior end showing stoma; E: entire female; F: entire male; G-I: female tail; J: cuticle; K, L: lateral alae; M: male caudal region showing spicules; N: female gonad.

16.13.3. Family Teratocephalidae Andrássy, 1958

This small family, consisting of the single genus *Teratocephalus* with about 20 species (Fig. 16.39), is phylogenetically important – molecular phylogenies place it at the base of the entire Rhabditida. They are bacterivorous nematodes found in soil and mosses throughout the world, including the Antarctic. Very small nematodes. Cuticle annulated, with lateral alae and sometimes with longitudinal lines or ridges. Inner and outer labial sensilla papilliform, cephalic sensilla setiform. Small oval amphids. Labial region with six leaf-shaped lips with refractive margins (Fig. 16.39B). Stoma narrow and weakly differentiated. Pharynx muscular, with distinct subdivision into corpus, isthmus and basal bulb; basal bulb with a grinder (Fig. 16.39D). There are both amphimictic and parthenogenetic species in this genus. Female reproductive system with single anterior gonad, ovary antidromously reflexed, post-vulval uterine sac present. Male copulatory receptors small papilliform, arranged in subventral rows. Spicules present, gubernaculum absent (Fig. 16.39E). Tail elongate conoid to filiform, caudal glands and spinneret absent.

16.13.4. Family Diplogastridae Micoletzky, 1922

Older classifications divide this group into many families: Cephalobiidae, Cylindrocorporidae, Diplogastridae, Diplogasteroididae, Mehdine-

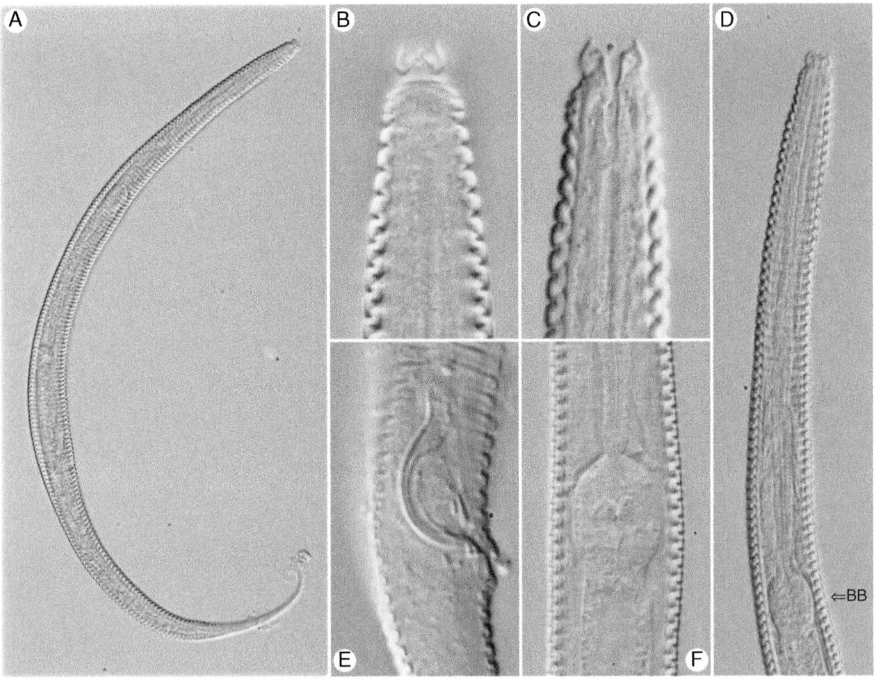

Fig. 16.39. *Teratocephalus* – representative of the Teratocephalidae. A: Entire female; B: anterior end (surface view) showing structures of labial region; C: anterior end showing stoma; D: pharyngeal region showing basal bulb (BB); E: male cloacal region showing spicules; F: base of pharynx showing basal bulb.

matidae, Neodiplogastridae, Odontopharyngidae, Pseudodiplogasteroididae and Tylopharyngidae. In reality, they represent a coherent unit consisting of about 30 genera and over 300 species (Sudhaus and Fürst von Lieven, 2003), including free-living bacterivorous, fungivorous and predatory species, insect associates and parasites. Small to medium sized nematodes, 0.5-2 mm in size. Cuticle annulated, with lateral alae and deirids, sometimes with longitudinal lines. Inner and outer labial sensilla papilliform, cephalic sensilla either papilliform or setiform. Small oval amphids. Stoma variable in shape: cheilostom and gymnostom refractive and wide, stegostom with one, two or three large and movable teeth (Figs 16.36, 16.40A-D); in some species stoma is a narrow simple cylinder; in *Tylopharynx* it is reminiscent of a stylet. Nevertheless, this morphological variability of stoma is continuous, and differences between genera are minute. Pharynx muscular only in its anterior part, with distinct subdivision into corpus with distinct median (metacorpal bulb), isthmus and glandular basal bulb, without grinder (Figs 16.36, 16.40D). Most described species are amphimictic. Female reproductive system with either two or single anterior gonad, ovary(-ies) homodromously reflexed. In monodelphic species the vulval opening is usually close to midbody and with a post-vulval uterine sac. Male copulatory receptors papilliform or setiform, arranged in subventral and subdorsal rows, bursa present in some species (Fig. 16.36). Spicules and gubernaculum present (Fig. 16.40F). Tail variable, elongate conoid to filiform, caudal glands and spinneret absent.

16.13.5. Family Rhabditidae Örley, 1880

Similarly to the previous taxon, older classifications recognize at least five families (Agfidae, Diploscapteridae, Mesorhabditidae, Peloderidae and Rhabditidae) while phylogenetic studies fail

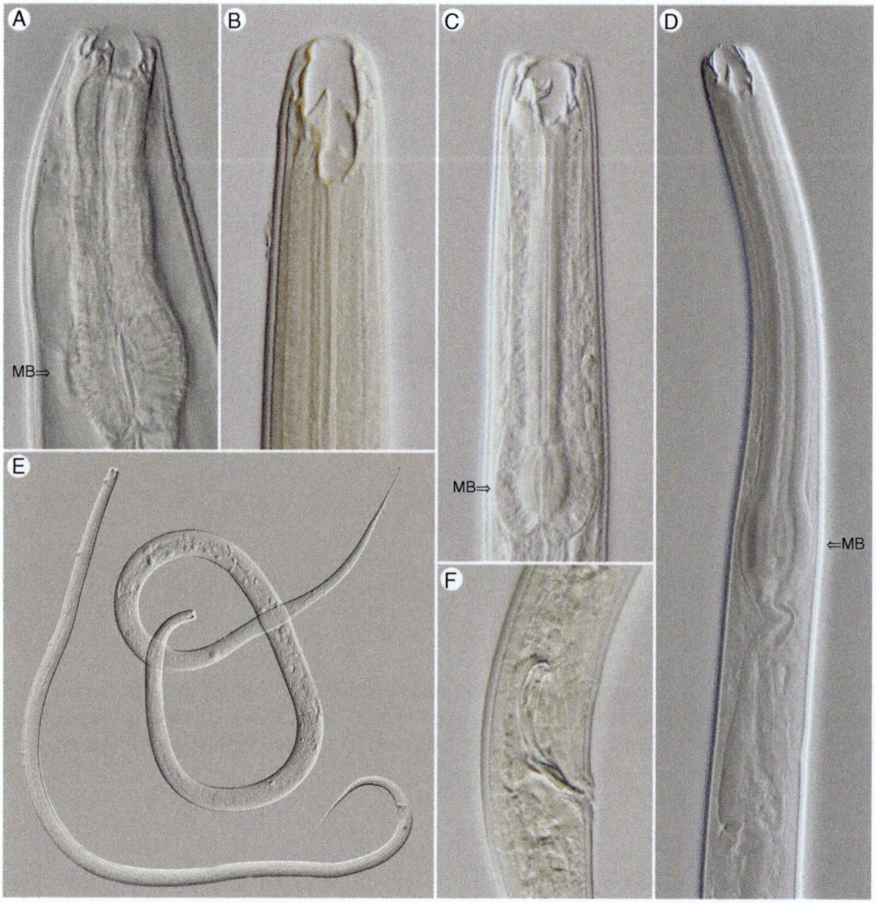

Fig. 16.40. *Pristionchus* (A), *Odontopharynx* (B, F), *Mononchoides* (C) and *Diplogaster* (D, E) – representatives of the Diplogastridae. A-C: Anterior end showing stoma and median bulb (MB); D: pharyngeal region showing median bulb (MB); E: entire female and male; F: male cloacal region showing spicules.

to provide a stable alternative (Sudhaus, 2011, 2023). For the purpose of this review, we will treat nearly 50 genera and around 300 species together. Rhabditids are common in soils, especially in saprobic habitats, in compost, dung, decomposing plant and animal remains and so on. Members of this family are opportunistic bacterial feeders that quickly colonize habitats rich in bacteria. Many are associated with invertebrates, are necromenic and parasitic. Small to medium sized nematodes, 0.5-2 mm in size. Cuticle striated, with lateral alae and deirids, sometimes with refractive dots. Inner and outer labial and cephalic sensilla papilliform. Small oval amphids. Anterior end with six equal lips, exceptionally with bi-radially symmetrical projections as in *Diploscapter*. Stoma cylindrical, usually with glottoid apparatus at its base (Figs 16.36, 16.41A). Pharynx muscular, with distinct subdivision into corpus (often with distinct median bulb), isthmus and basal bulb; basal bulb with grinder (Figs 16.36, 16.41D). Most described species are amphimictic, but there are also hermaphroditic and parthenogenetic species. Female reproductive system with either two (Fig. 16.41B) or single (Fig. 16.41C) anterior gonad, ovary(-ies) homodromously reflexed. In monodelphic species the vulval opening is usually close to the posterior body end and a post-vulval uterine sac is absent, but there are

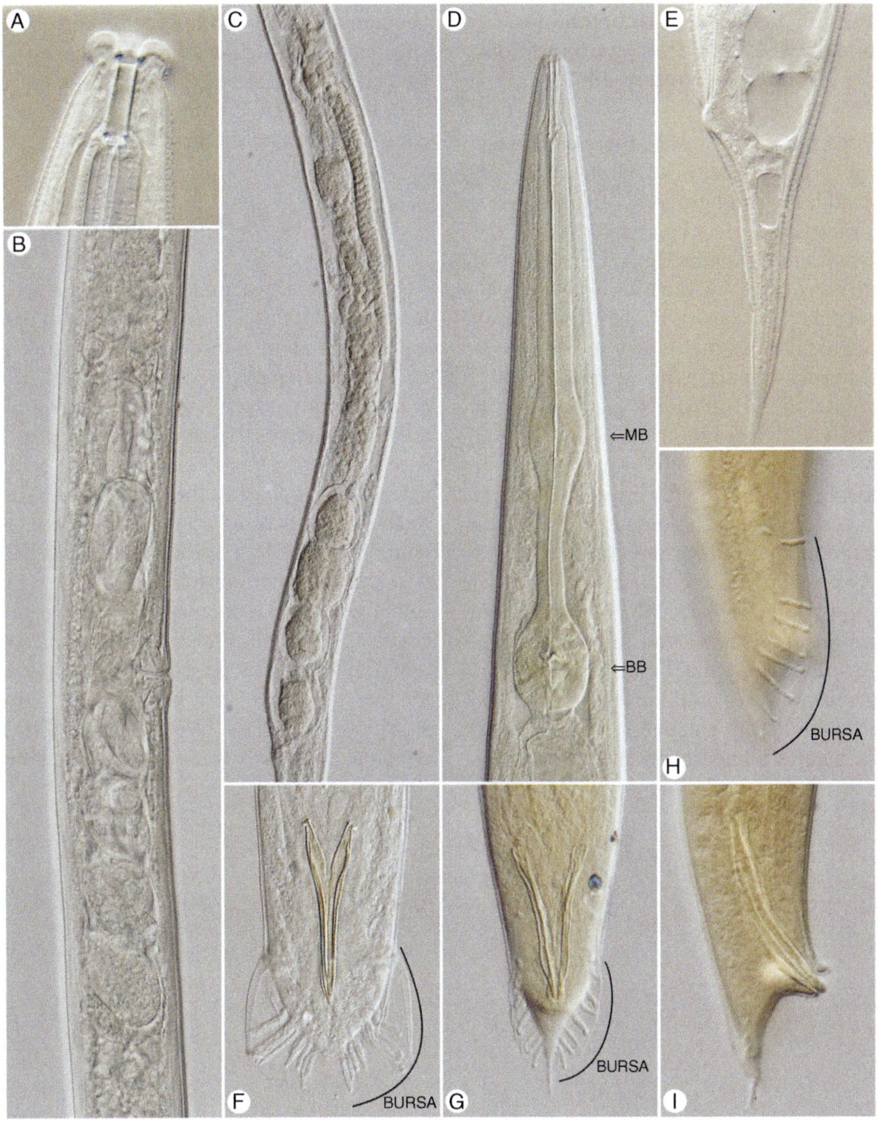

Fig. 16.41. *Cruznema* (A, E), *Cephaloboides* (B), *Distolabrellus* (C), *Metarhabditis* (D), *Mesorhabditis* (F) and an unidentified rhabditid (G-I) – representatives of the Rhabditidae. A: Anterior end showing stoma; B, C: female gonads; D: pharyngeal region showing median (MB) and basal (BB) bulbs; E: female tail; F, G: male caudal region showing spicules and bursa, ventral view; H: male caudal region showing bursa, lateral view; I: male caudal region showing spicules, lateral view.

exceptions. Male copulatory receptors papilliform or finger-like, arranged in subventral and subdorsal rows, usually enclosed in the thin cuticular membrane forming a bursa (Figs 16.36, 16.41F-H). Spicules and gubernaculum present (Fig. 16.41I). Female tail variable in shape, from rounded to elongate conoid to filiform, caudal glands and spinneret absent. Male tail shorter than female tail, conoid, with or without filiform tip.

16.13.6. Families Bunonematidae Micoletzky, 1922 and Pterygorhabditidae Goodey & Goodey, 1963

These two small families (Bunonematidae with seven genera and about 50 species and Pterygorhabditidae with single genus and six species) are unmistakable in their appearance (Fig. 16.42): the cuticle on the right side of the body is much more complex than on the left side, the internal body symmetry is right-left laterally bisymmetrical (bilateral, except for the gonads) while external symmetry (lip region and cuticle) is bisymmetric but shifted 90° (dorso-ventral). They are commonly found in soil, mosses and decomposing organic matter, some species are phoretically associated with insects. Very small nematodes. Cuticle with complex refractive ornamentation, more elaborate on the right body side. Anterior end bisymmetric, with complex projections. Stoma cylindrical. Pharynx muscular, with distinct subdivision into corpus with median bulb, isthmus and basal bulb; basal bulb with grinder. Species amphimictic or parthenogenetic. Female reproductive system with two gonads, ovaries homodromously reflexed. Male copulatory receptors finger-like, arranged in subventral and subdorsal rows, enclosed in the thin cuticular membrane forming a bursa. Spicules and gubernaculum present. Tail conoid, caudal glands and spinneret absent.

16.14. Conclusion

Nematodes are an abundant and diverse group of terrestrial organisms found in all soil types and aquatic sediments. Substrates such as litter, mosses, bark and epiphytes also contain interesting nematode diversity. A small amount of soil can contain hundreds of individuals and dozens of species, most of them free-living. This book discusses and emphasizes the importance of nematodes as plant parasites, however there is a higher abundance and diversity of free-living nematodes that play an important role in nutrient cycling in soil. They are key organisms in ecological processes as well as in the sustainability of ecosystems. This chapter represents a good opportunity for the readers to get a better insight into the enormous diversity of free-living terrestrial nematodes not directly associated with plants.

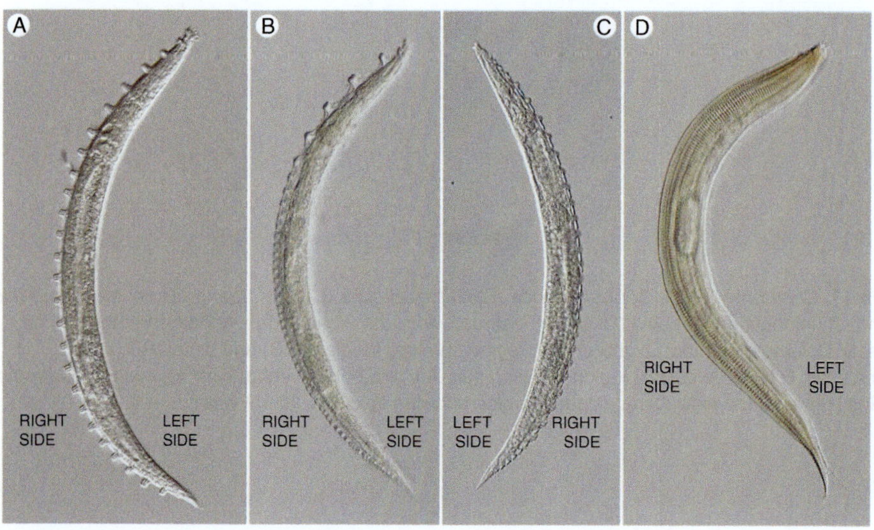

Fig. 16.42. Females of *Bunonema* (A), *Rhodolaimus* (B), *Aspidonema* (C) and *Pterygorhabditis* (D) – representatives of the Bunonematidae and Pterygorhabditidae.

16.15. References and Further Reading

Abolafia, J. and Peña-Santiago, R. (2018) Morphology, taxonomy and phylogeny of the enigmatic genus *Aulolaimus* de Man, 1880 (Nematoda, Aulolaimidae). *Zoologischer Anzeiger* 277, 101–115. DOI: 10.1016/j.jcz.2018.09.001.

Ahmad, W. and Jairajpuri, M.S. (2010) *Mononchida. The Predaceous Nematodes*. Brill, Leiden, Netherlands.

Ahmed, M. and Holovachov, O. (2021) Twenty years after De Ley and Blaxter – how far did we progress in understanding the phylogeny of the phylum Nematoda? *Animals* 11, 3479. DOI: 10.3390/ani11123479.

Andrássy, I. (2002) The genus *Cristamphidelus* Siddiqi and Vinciguerra, 1991 and a general survey of the family Alaimidae (Nematoda). *Journal of Nematode Morphology and Systematics* 4, 51–82.

Andrássy, I. (2005) *Free-living Nematodes of Hungary, Volume I (Nematoda Errantia).* In: *Pedozoologica Hungarica, 3* (Series eds: Csuzdi, C. and Mahunka, S.). Hungarian Natural History Museum and Systematic Zoology Research Group of the Hungarian Academy of Sciences, Budapest, Hungary.

Andrássy, I. (2007) *Free-living Nematodes of Hungary, Volume II (Nematoda Errantia).* In: *Pedozoologica Hungarica, 4* (Series eds: Csuzdi, C. and Mahunka, S.). Hungarian Natural History Museum and Systematic Zoology Research Group of the Hungarian Academy of Sciences, Budapest, Hungary.

Andrássy, I. (2009) *Free-living Nematodes of Hungary, Volume III (Nematoda Errantia).* In: *Pedozoologica Hungarica, 5* (Series eds: Csuzdi, C. and Mahunka, S.). Hungarian Natural History Museum and Systematic Zoology Research Group of the Hungarian Academy of Sciences, Budapest, Hungary.

Andrássy, I. (2011) A study on *Ironus* Bastian, 1865 (Nematoda: Enoplida) with descriptions of two new and two known species. *International Journal of Nematology* 21, 27–41.

Asghari, R., Eskandari, A., Tanha Maafi, Z., Álvarez-Ortega, S. and Zhao, Z. (2017) Morphological and molecular characterisation of new and known species of *Tripyla* Bastian, 1865 (Triplonchida: Tripylidae) from northern Iran, with phylogenetic relationships, compendium and identification key. *Nematology* 19, 21–56. DOI: 10.1163/15685411-00003030.

Cid del Prado-Vera, I., Ferris, H. and Nadler, S.A. (2016) Five new species of the family Trischistomatidae (Nematoda: Enoplida) from North and Central America, with keys to the species of *Trischistoma* and *Tripylina*. *Zootaxa* 4109, 173–197. DOI: 10.11646/zootaxa.4109.2.4.

Coomans, A. and Abebe, E. (2006) Order Monhysterida. In: Eyualem, A., Andrássy, I. and Traunspurger, W. (eds), *Freshwater Nematodes: Ecology and Taxonomy.* CAB International, Wallingford, UK, pp. 574–603.

Decraemer, W. and Rho, H. (2014) Order Desmoscolecida. In: Schmidt-Rhaesa, A. (ed), *Handbook of Zoology. Gastrotricha, Cycloneuralia, Gnathifera. Volume 2: Nematoda*. de Gruyter, Berlin, Germany, pp. 351–372. DOI: 10.1515/9783110274257.351.

Decraemer, W. and Smol, N. (2006) Orders Chromadorida, Desmodorida and Desmoscolecida. In: Eyualem, A., Andrássy, I. and Traunspurger, W. (eds), *Freshwater Nematodes: Ecology and Taxonomy.* CAB International, Wallingford, UK, pp. 497–573.

De Ley, P. and Blaxter, M. (2004) A new system for Nematoda: combining morphological characters with molecular trees, and translating clades into ranks and taxa. In: *Proceedings of the Fourth International Congress of Nematology, Tenerife, Spain, 8–13 June 2002, Nematology Monographs and Perspectives, Volume 2* (Series eds: Cook, R. and Hunt, D.J.). Brill, Leiden, Netherlands, pp. 633–653.

Eyualem, A., Andrássy, I. and Traunspurger, W. (eds) (2006) *Freshwater Nematodes: Ecology and Taxonomy*. CAB International, Wallingford, UK.

Fonseca, G. and Bezerra, T.N. (2014a) Order Araeolaimida De Coninck and Schuurmans Stekhoven, 1933. In: Schmidt-Rhaesa, A. (ed), *Handbook of Zoology. Gastrotricha, Cycloneuralia, Gnathifera. Volume 2: Nematoda*. de Gruyter, Berlin, Germany, pp. 467–486. DOI: 10.1515/9783110274257.467.

Fonseca, G. and Bezerra, T.N. (2014b) Order Monhysterida Filipjev, 1929. In: Schmidt-Rhaesa, A. (ed), *Handbook of Zoology. Gastrotricha, Cycloneuralia, Gnathifera. Volume 2: Nematoda*. de Gruyter, Berlin, Germany, pp. 435–466. DOI: 10.1515/9783110274257.435.

Ghaderi, R., Kashi, L., Karani, H.M. and Karegar, A. (2017) A new and four known species of *Diphtherophora* (Nematoda: Diphtherophoridae) from Iran, with a diagnostic compendium of its species. *Zootaxa* 4365, 311–330. DOI: 10.11646/zootaxa.4365.3.3.

Hodda, M. (2022) Phylum Nematoda: A classification, catalogue and index of valid genera, with a census of valid species. *Zootaxa* 5114, 1–289. DOI: 10.11646/zootaxa.5114.1.1.

Holovachov, O. (2014a) Isolaimida Cobb, 1920. In: Schmidt-Rhaesa, A. (ed), *Handbook of Zoology. Gastrotricha, Cycloneuralia, Gnathifera. Volume 2: Nematoda*. de Gruyter, Berlin, Germany, pp. 313–316. DOI: 10.1515/9783110274257.313.

Holovachov, O. (2014b) Order Plectida Gadea, 1973. In: Schmidt-Rhaesa, A. (ed), *Handbook of Zoology. Nemathelminthes and Gnathifera. Volume 2: Nematoda*. de Gruyter, Berlin, Germany, pp. 487–535. DOI: 10.1515/9783110274257.487.

Holovachov, O. and Boström, S. (2010) *Identification of Plectida (Nematoda).* EUMAINE, Gent and Nematology, UC Riverside, California.

Holovachov, O. and De Ley, P. (2006) Order Plectida. In: Eyualem, A., Andrássy, I. and Traunspurger, W. (eds), *Freshwater Nematodes: Ecology and Taxonomy.* CAB International, Wallingford, UK, pp. 611–647.

Holovachov, O. and Shoshin, A. (2014) Order Triplonchida Cobb, 1919. In: Schmidt-Rhaesa A. (ed), *Handbook of Zoology. Gastrotricha, Cycloneuralia, Gnathifera. Volume 2: Nematoda*. de Gruyter, Berlin, Germany, pp. 251–276. DOI: 10.1515/9783110274257.251.

Holovachov, O., Tandingan De Ley, I. and De Ley, P. (2009) *Identification of Cephaloboidea (Nematoda).* EUMAINE, Gent and Nematology, UC Riverside, California.

Jairajpuri, M.S. and Ahmad, W. (1992) *Dorylaimida. Free-living, Predaceous and Plant Parasitic Nematodes.* Brill, Leiden, Netherlands.

Jiménez-Guirado, D., Peralta, P.M. and Peña-Santiago, R. (2007) Nematoda, Mononchida. Dorylaimida I. In: Ramos, M.A. *et al.* (eds), *Fauna Iberica*, Vol. 30. Museo Nacional de Ciencias Naturales, CSIC, Madrid, Spain.

Muthumbi, A. and Vanreusel, A. (2006) Order Araeolaimida. In: Eyualem, A., Andrássy, I. and Traunspurger, W. (eds), *Freshwater Nematodes: Ecology and Taxonomy.* CAB International, Wallingford, UK, pp. 604–610.

Peña-Santiago, R. (2006) Dorylaimida part I: superfamilies Belondiroidea, Nygolaimoidea and Tylencholaimoidea. In: Eyualem, A., Andrássy, I. and Traunspurger, W. (eds), *Freshwater Nematodes: Ecology and Taxonomy.* CAB International, Wallingford, UK, pp. 326–391.

Peña-Santiago, R. (2014a) Order Dorylaimida Pearse, 1942. In: Schmidt-Rhaesa, A. (ed), *Handbook of Zoology. Gastrotricha, Cycloneuralia, Gnathifera. Volume 2: Nematoda*. de Gruyter, Berlin, Germany, pp. 277–298. DOI: 10.1515/9783110274257.277.

Peña-Santiago, R. (2014b) Order Mononchida Jairajpuri, 1969. In: Schmidt-Rhaesa, A. (ed), *Handbook of Zoology. Gastrotricha, Cycloneuralia, Gnathifera. Volume 2: Nematoda*. de Gruyter, Berlin, Germany, pp. 299–312. DOI: 10.1515/9783110274257.299.

Peña-Santiago, R. (2021a) *Dorylaimida Mundi (Nematoda): Checklist of Genera and Species, with their Records*. Editorial Universidad de Jaén, Jaén, Spain.

Peña-Santiago, R. (2021b) *Morphology and Bionomics of Dorylaims (Nematoda, Dorylaimida)*. In: *Nematology Monographs and Perspectives, 13* (Series eds: Hunt, D.J. and Perry, R.N.). Brill, Leiden, Netherlands. DOI: 10.1163/9789004447851.

Smol, N., Muthumbi, A. and Sharma, J. (2014) Order Enoplida. In: Schmidt-Rhaesa, A. (ed), *Handbook of Zoology. Gastrotricha, Cycloneuralia, Gnathifera. Volume 2: Nematoda*. de Gruyter, Berlin, Germany, pp. 193–250. DOI: 10.1515/9783110274257.193.

Sudhaus, W. (2011) Phylogenetic systematisation and catalogue of paraphyletic 'Rhabditidae' (Secernentea, Nematoda). *Journal of the Nematode Morphology and Systematics* 14, 113–178.

Sudhaus, W. (2023) An update of the catalogue of paraphyletic 'Rhabditidae' (Nematoda) after eleven years. *Soil Organisms* 95, 95–116. DOI: 10.25674/so95iss1id312.

Sudhaus, W. and Fürst von Lieven, A. (2003) A phylogenetic classification and catalogue of the Diplogastridae (Secernentea, Nematoda). *Journal of Nematode Morphology and Systematics* 6, 43–90.

Tchesunov, A. (2014a) Order Chromadorida Chitwood, 1933. In: Schmidt-Rhaesa, A. (ed), *Handbook of Zoology. Gastrotricha, Cycloneuralia, Gnathifera. Volume 2: Nematoda*. de Gruyter, Berlin, Germany, pp. 373–398. DOI: 10.1515/9783110274257.373.

Tchesunov, A. (2014b) Order Desmodorida De Coninck, 1965. In: Schmidt-Rhaesa, A. (ed), *Handbook of Zoology. Gastrotricha, Cycloneuralia, Gnathifera. Volume 2: Nematoda*. de Gruyter, Berlin, Germany, pp. 399–434. DOI: 10.1515/9783110274257.399.

Vinciguerra, M.T. (2006) Dorylaimida Part II: Superfamily Dorylaimoidea. In: Eyualem, A., Andrássy, I. and Traunspurger, W. (eds), *Freshwater Nematodes: Ecology and Taxonomy.* CAB International, Wallingford, UK, pp. 392–467.

Zullini, A. (2006) Order Triplonchida. In: Eyualem A., Andrássy, I. and Traunspurger, W. (eds), *Freshwater Nematodes: Ecology and Taxonomy.* CAB International, Wallingford, UK, pp. 293–325.

Zullini, A. (2021) Nematodi d'acqua dolce. Manuale di identificazione al genere e metodi di raccolta. *Biologia Ambientale* 35 (2 suppl.), 1–250. DOI: 10.30463/ao211.009.

Zullini, A. and Peneva, V. (2006) Order Mononchida. In: Eyualem, A., Andrássy, I. and Traunspurger, W. (eds), *Freshwater Nematodes: Ecology and Taxonomy.* CAB International, Wallingford, UK, pp. 468–498.

17 Nematodes as Bioindicators

Sara Sánchez-Moreno[1]* and Howard Ferris[2]

[1]*Department of Biodiversity and Evolutionary Biology, National Museum of Natural Sciences (CSIC), Madrid, Spain;* [2]*Department of Entomology and Nematology, University of California, Davis, CA, USA*

Abstract

Nematodes are among the simplest and abundant of the multicellular animals. They are found in all soil and aquatic systems in any environment that provides a source of organic carbon, under all climatic conditions, and in habitats that vary from pristine to those disturbed by human activities, natural disasters and pollution. Nematodes vary in their responses to disturbance; some species being very sensitive to pollutants, other chemical stressors or mechanical disruption of their environment, while others are extremely tolerant. As such, and being diverse in feeding habit and biology, they are ideal organisms for bioindicator studies of environmental health. The chapter addresses topics such as to why nematodes are good bioindicators as well as reviewing the various indices of ecosystem structure and diversity, indices of ecosystem function, biomass and metabolic footprints. Research opportunities and current and future trends in this area are briefly discussed.

17.1. Introduction – Why Nematodes Are Good Bioindicators

Bioindicators, or biological indicators, are organisms, considered as individual species or groups of species, whose function, abundance, diversity or physiological status reflects ecosystem or environmental condition.

Nematodes are among the simplest, and are considered the most abundant, of the multicellular animals. They are aquatic organisms that occur in open water or in water films. They are found in all soil and aquatic systems in any environment that provides a source of organic carbon, under all climatic conditions, and in habitats that vary from pristine to those disturbed by human activities, natural disasters and pollution. Nematodes vary in their responses to disturbance; some species are very sensitive to pollutants and other chemical stressors while others are extremely tolerant; some species are quite tolerant of mechanical disruption of their environment while others are very sensitive (Bongers and Ferris, 1999).

Nematodes play major roles in component processes of most of the ecosystem services that provide food, fibre, clean water and air, and which regulate pest and disease organisms. In soil,

*E-mail: sarasm@inia.csic.es

A revision of Ferris, H. (2012) Nematodes as bioindicators. In: Manzanilla-López, R.H. and Marbán-Mendoza, N. (eds), *Practical Plant Nematology*. Mundi-Prensa and Biblioteca Básica de Agricultura Montecillo, Mexico.

 Practical Plant Nematology 2nd Edition
(eds R.H. Manzanilla-López, D.J. Hunt, N. Marbán-Mendoza and S. Infante Gil)
DOI: 10.1079/9781836990413.0017

nematodes live in capillary water; their permeable cuticle provides direct contact with their microenvironment. Their populations respond rapidly to disturbance and enrichment, and individuals do not rapidly migrate from stressful conditions. In soil and aquatic food webs, nematodes are involved in the transformation of organic matter into mineral and organic nutrients which are available to plants and enhance plant growth and crop productivity (Ferris *et al.*, 2004a, b; Lazarova *et al.*, 2021). Analysis of nematode faunal composition provides information on succession and changes in decomposition pathways in the food web, nutrient status, environmental quality and the effects of contaminants. Because of their abundance, diversity of function and differing sensitivities to environmental change, they serve as useful indicators of ecosystem disturbance and recovery from perturbation (Porazinska *et al.*, 1999; Ferris *et al.*, 2001; Neher, 2001; Yeates *et al.*, 2009; Du Preez *et al.*, 2022).

In natural ecosystems, nematodes influence spatial and temporal diversity of plant-associated communities both above- and below-ground (De Deyn and Van der Putten, 2005). Plant identity, vegetation diversity and the functional composition of plant assemblages are relevant drivers of nematode community composition (Wardle *et al.*, 2003), and plant diversity might mitigate the negative effects of environmental perturbation on nematodes (Cesarz *et al.*, 2017). Interestingly, however, plant community diversity is not necessarily indicative of the diversity of the below-ground nematode assemblage. While plant species certainly determine the presence and abundance of host-specific plant-feeding nematode species, the structure of the complete soil nematode assemblage apparently cannot be inferred from the diversity or characteristics of the plant community (Sánchez-Moreno *et al.*, 2008).

The transparent nature of nematodes allows easy microscopic observation of diagnostic feeding and ingestion structures. There is a clear relationship between structure and function; the feeding behaviour is inferred from the structure of the mouth cavity and pharynx. Different nematode taxa exhibit specificities of food sources (Fig. 17.1). For example, many families within the order Tylenchida feed exclusively on the roots of higher plants while others feed either exclusively on fungi or are able to feed on both fungal hyphae and roots of higher plants. In the order Rhabditida, nematodes in the families Rhabditidae, Cephalobidae and Plectidae (Fig. 17.2) feed

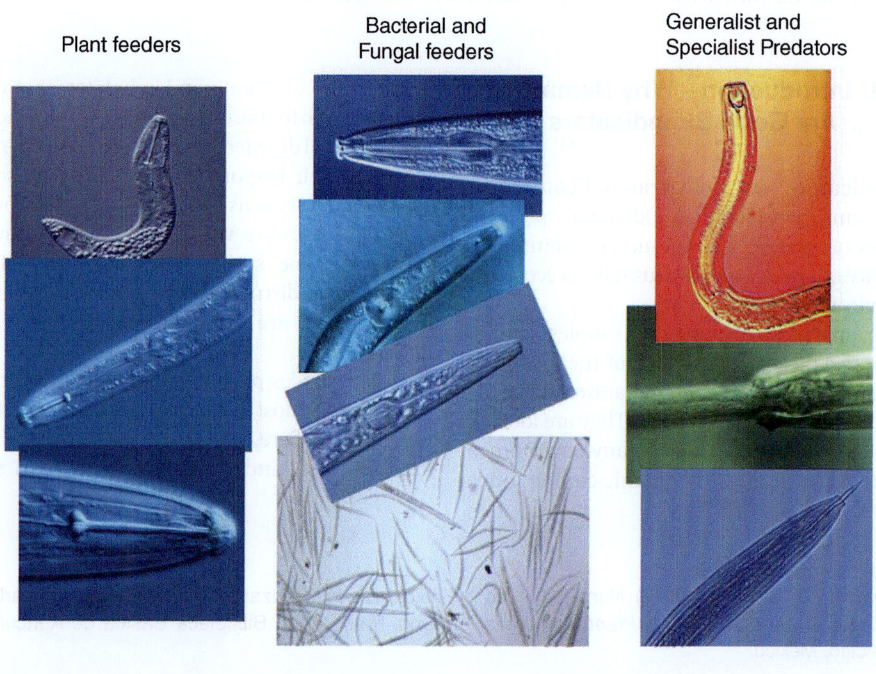

Fig. 17.1. Nematodes of different feeding habits and life-history characteristics form the basis for environmental indicators.

on bacteria but not on higher plants or fungi. In the order Mononchida, representatives of the families Mononchidae and Anatonchidae are specialist predators of other nematodes and do not feed on higher plants or fungi. Many representatives of the order Dorylaimida are designated omnivorous and can be considered generalist predators of various prey species (Ferris and Bongers, 2009). Changes in the food webs in which they participate are mirrored by shifts among nematode feeding groups.

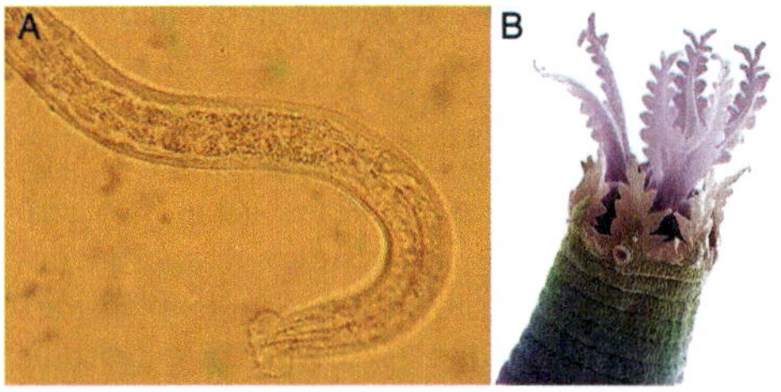

Fig. 17.2. A: Anterior of *Wilsonema* sp. (Plectidae) showing flanges on either side of the head that may scrape bacteria from surfaces of soil particles and stir the soil solution so that more bacteria are in suspension; B: The complex probolae of *Acrobeles* sp. probably have a similar agitation function. (Both are Ba2 functional guild.)

Nematodes vary in life span; some species have a generation time of days, others months or even a year. The matrix of feeding habit, generation time and sensitivity to environmental disturbance allows the designation of functional guilds of different nematode taxa that are similar in response characteristics (Bongers and Bongers, 1998) (Fig. 17.3).

Extraction of all taxonomic and functional groups of nematodes from soil and aquatic environments can be standardized through efficient, routine procedures. Extracted nematodes can be preserved and stored for future analyses. Intensive repeated sampling of monitoring sites is possible because the required sample size is quite small and therefore relatively non-disruptive. The large numbers of species and individuals recovered, and the diversity of trophic groups and functional guilds they represent, give each sample a high intrinsic information value. Routine analysis of nematode fauna affords the possibility of rapid assessment of effects of environmental stress, including physical soil disturbance, nutrient enrichment, soil erosion and heavy metal

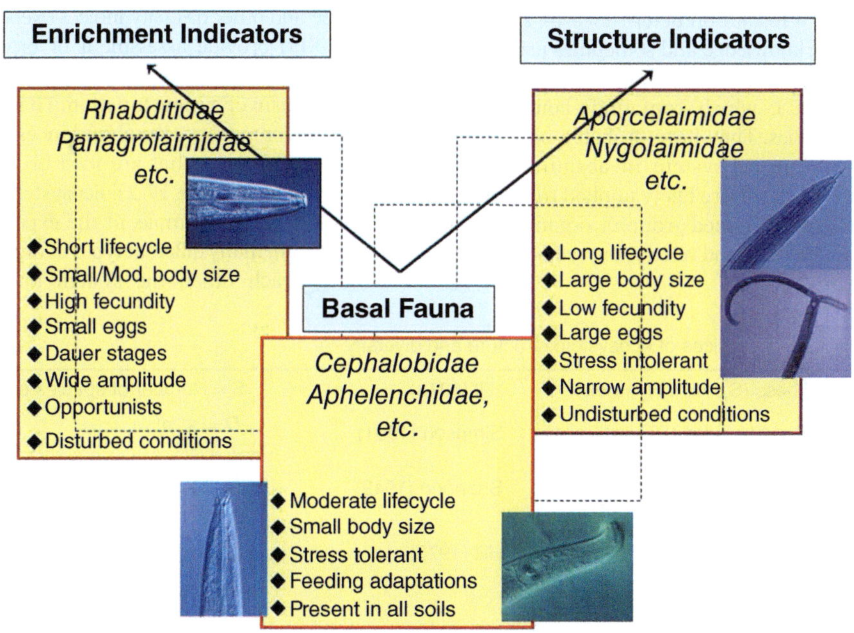

Fig. 17.3. Nematodes separate into functional guilds based on feeding habits, life history characteristics and environmental sensitivities.

pollution (Baxter *et al.*, 2013; Šalamún *et al.*, 2015; Bongiorno *et al.*, 2019; Betancur-Corredor *et al.*, 2022). In recent years, nematode communities have been successfully utilized to assess the effects of emerging environmental threats such as microplastic pollution, antimicrobial soil contamination and climate change, on the environment (Lin *et al.*, 2020; Zheng *et al.*, 2021; Zhou *et al.*, 2022). Nematode community analysis provides decision criteria for management, conservation and remediation (Bongers and Ferris, 1999).

Indices of environmental condition based on nematode assemblages may be categorized as those that attempt to quantify ecosystem structure and diversity of organisms and those that attempt to describe function of the ecosystem.

17.2. Indices of Ecosystem Structure and Diversity

Many indices have been developed and applied to assess the diversity of ecosystems (Neher and Darby, 2009). In the purest sense, they are based on the relative abundance of all organisms at the species level, as indicated by terms such as *species richness*. In practice, indices are usually applied at a resolution determined by available taxonomic knowledge and sometimes using data of differing taxonomic resolution (species, genera, families, etc.), which confounds comparisons among studies. They have probably never been applied to an ecosystem in absolute terms; rather they are more likely applied to numbers of species of a defined group of organisms, for example, aboveground vertebrates, soil nematodes and so on.

Indices of taxonomic diversity are applied to describe the *structure* of the nematode assemblage. They include indices of species richness, diversity and evenness.

17.2.1. Calculation of Indices of Structure and Diversity

The indices are quite interrelated, and all depend on the proportional representation of each taxon in the system, as apparent from the examples in Table 17.1. In very diverse systems, Simpson's diversity index (D), which is sensitive to the abundance of common species, is close to zero while in homogeneous systems it approaches the value 1. For that reason, Simpson's index is often expressed as 1-D so that systems with greater diversity have a higher index value. Shannon's index is more sensitive to the occurrence of rare species than Simpson's index. Hill's N_1 index is the exponential of Shannon's index and Hill's N_2 index is the reciprocal of Simpson's index. Pielou's evenness index can be derived from a combination of Shannon's index and species richness (Shannon, 1949; Simpson, 1951; Pielou, 1966; Hill, 1973).

These and other diversity indices (Neher and Darby, 2009) provide assessment of organism heterogeneity in a system but no direct indication of organism or system function. The indices do not assess or measure the abundances or proportion of organisms that are food or feeders, prey or predators (Table 17.2). Ecosystem function is inferred, for example, in the expectation that if there are many different types of organisms feeding on each other, the system has some

Table 17.1. Some indices of ecosystem structure and diversity.

Species richness (S); Hill's N_0 index	Hill (1973)	S = Number of species present
Simpson's diversity index (D)	Simpson (1951)	$D = \sum_{i=1,S} p_i^2$
Shannon's diversity index (H')	Shannon (1949)	$H' = -\sum_{i=1,S} p_i (\log p_i)$
Hill's N_1 index	Hill (1973)	$N1 = e^{H'} = \exp\left(-\sum_{i=1,S} p_i (\log p_i)\right)$
Hill's N_2 index	Hill (1973)	$N2 = D^{-1} = \left(\sum_{1,S} p_i^2\right)^{-1}$
Pielou's evenness index (J')	Pielou (1966)	$J' = H' / \log(S)$

In each index, p_i is the number of individuals of a taxon as a proportion of the total number of individuals.

Table 17.2. Diversity index calculations for two nematode assemblages (Samples A and B) of the same number of taxa and the same number of individuals.

			Sample A Diversity Calculations			Sample B Diversity Calculations		
Nematode Taxon	Sample A	Sample B	p_i	p_i^2	$\log(p_i)$	p_i	p_i^2	$\log(p_i)$
Hoplolaimidae	**5**	**15**	0.045	0.002	–3.109	0.134	0.018	–2.010
Pratylenchidae	**5**	**15**	0.045	0.002	–3.109	0.134	0.018	–2.010
Aphelenchidae	**15**	**5**	0.134	0.018	–2.010	0.045	0.002	–3.109
Cephalobidae	**15**	**5**	0.134	0.018	–2.010	0.045	0.002	–3.109
Plectidae	**2**	**15**	0.018	0.000	–4.025	0.134	0.018	–2.010
Rhabditidae	**5**	**50**	0.045	0.002	–3.109	0.446	0.199	–0.806
Dorylaimidae	**50**	**5**	0.446	0.199	–0.806	0.045	0.002	–3.109
Discolaimidae	**15**	**2**	0.134	0.018	–2.010	0.018	0.000	–4.025
Total	**112**	**112**						
Hill N_0	**8**	**8**						
Simpson	**0.26**	**0.26**						
Shannon	**1.66**	**1.66**						
Hill N_1	**1.30**	**1.30**						
Hill N_2	**0.60**	**0.60**						
Pielou J′	**0.80**	**0.80**						

p_i is the number of individuals of a taxon as a proportion of the total number of individuals.
From Ferris and Bongers (2009).

internal regulation and therefore overall stability. While that conclusion may bear weight in Sample A, with high proportions of omnivores (Dorylaimidae) and predators (Discolaimidae), it would be less valid for Sample B (Table 17.2). In Sample B, there are greater proportions of opportunists, including bacterial-feeding Rhabditidae and plant feeders (Hoplolaimidae and Pratylenchidae), which may be responding to enhanced host vigour in the more enriched environment. Both samples have identical indices of diversity but their functional attributes are quite different.

17.3. Indices of Ecosystem Function

Knowledge or inference of the feeding habits, and consequently the possible ecosystem function, of nematode taxa led to the development of indices which reflect the functional diversity and inferred stability and resilience of ecosystem function. Several series of studies on ecological succession in cow dung and on nematode assemblages in a variety of ecosystems provided the basis for categorizing nematode taxa into five groups along an r-K continuum (Ferris and Bongers, 2009). That arrangement evolved into the Maturity Index (MI) for terrestrial, freshwater and marine nematodes (Bongers, 1990; Bongers *et al.*, 1991). The cataloguing of nematode feeding habits (Yeates *et al.*, 1993) extended the functional components of the system.

The MI is based on relative abundance of nematodes categorized, by life course characteristics, into a 1-5 colonizer-persister (*c-p*) series (Table 17.3) ranging from extreme *r*- to extreme *K*-strategists. 'Colonizer' nematodes (*r-strategists*) at the lower end of the c-p scale are considered enrichment opportunists and therefore indicate resource availability; 'persister' nematodes (*K-strategists*) at the high end of the scale indicate system stability, food web complexity and connectance. Each nematode taxon, usually at family level, is classified into one of the five c-p classes. Genera and species within a taxon have the same c-p value as their family, or genus in the case of some marine taxa. A family of indices with different attributes has emerged based on the MI, including MI2-5, PPI, ΣMI and others (Table 17.4) (Ferris and Bongers, 2009).

17.3.1. Calculation and Use of the Maturity Index Family

All the indices are based on the weighted proportion of nematodes in the fauna that meet the index criteria. A generic formula for calculation of indices in the MI family is $XI = \sum_{i=1,f} v_i n_i / \sum_{i=1,f} n_i$, where XI is the index of interest, v_i is the colonizer-persister (c-p) value assigned to taxon i and n_i is the number of nematodes in each of the f taxa *that meet the criteria of the index.*

Sometimes the MI has been expressed as $MI = \sum_{i=1,f} v_i p_i$, which has led to calculation errors. One common error is that proportions of all taxa present in a sample are calculated in a spreadsheet, as for the calculation of ΣMI, and then the same proportions, excluding those that are not relevant, are used to calculate the other indices in the family. To obtain the correct index values, it is necessary to recalculate the proportions to be weighted with respect to the total number of nematodes in the sample which meet the specific criteria of each index.

All the indices of ecosystem function of the MI family show differences between the two samples of identical species richness and abundance (Table 17.5) that were not apparent in the diversity and evenness indices calculated for the same data (Table 17.2).

The MI is calculated as the sum of products of the weighted proportions of nematodes in each c-p category and their c-p values. Low c-p value nematodes are responsive to, and reflective of, enrichment while high c-p taxa are affected by disturbance. Consequently, after environmental disturbance, the number of nematodes in lower c-p groups often becomes proportionally greater. That is, the environment appears more enriched. Similarly, a recovery in abundance of higher c-p taxa effectively reduces the proportional representation of lower c-p groups, which suggests an apparently less enriched environment. However, the actual abundance of enrichment indicators may not have changed. The notion that enrichment should be independent of diversity of the nematode assemblage and the proportion of indicators of disturbance (Fig. 17.3) led to development of separate trajectories of enrichment and structure to assess the relative magnitudes of disparate ecosystem services (Ferris *et al.*, 2001).

Table 17.3. Characteristics of nematodes in the colonizer-persister series (Bongers, 1990).

c-p1	Short generation time, proportionally large gonads producing many small eggs. Rapid population growth in response to resource enrichment. Mainly bacterial feeders with high metabolic activity. Tolerant of environmental disturbance. Form dauer juveniles when food resources decrease.
c-p2	Short generation time and reproduction rates lower than c-p1. Slower to respond to environmental enrichment than c-p1. Include bacterial feeders, fungal feeders and a few predators. Occur in all environments; very tolerant of disturbances. Do not form dauer juveniles.
c-p3	Longer generation time than c-p2 nematodes. Greater sensitivity to disturbances. Include bacterial feeders, fungal feeders and some predators.
c-p4	Small dorylaimids and large non-dorylaimids. Low ratio of gonad to body volume. Long generation time, permeable cuticle, high pollutant sensitivity. Predators actively seek prey. Larger carnivores, smaller omnivores and some bacterial feeders.
c-p5	Large dorylaimids with long life span, low reproduction rates, low metabolic activity and slow movement. Gonads small relative to body volume, a few large eggs. Permeable cuticle, very sensitive to pollutants and other disturbances. Larger omnivores and predators.

Table 17.4. The Maturity Index Family (see detailed descriptions in Ferris and Bongers, 2009).

MI	The Maturity Index is based on non-plant-feeding taxa, dauer juveniles excluded. Low values indicate a disturbed and/or enriched environment. High values indicate a stable environment.
PPI	The Plant-Parasite Index responds as the inverse of the MI but based only on plant-feeding nematodes. High PPI suggests vigorous host plants which reflect system enrichment. Low PPI indicates poor host growth supporting nematodes assigned to lower c-p categories.
PPI/MI	The PPI/MI ratio is lower under nutrient-poor conditions than under nutrient-rich conditions. It is a sensitive indicator of enrichment in agroecosystems.
MI2-5	Identical to MI but excludes c-p1 enrichment opportunists. It avoids the situation of enrichment opportunists responding to organic material dead and decomposing as a result of pollutants.
ΣMI	The MI for all nematodes in the system, including plant feeders, based on the notion that the complete assemblage provides integral information with regard to disturbance and environmental condition. The tolerance to pollution of many plant-feeders that are c-p3 offsets the expected decrease of MI is offset by their inclusion in ΣMI.
ΣMI2-5	The MI for all nematodes in the c-p2-5 range. Recognizes that higher c-p value plant-feeding species also provide information on environmental stress. The index bears some of the conflict issues of ΣMI.

17.3.2. Evolution of Indices of Ecosystem Function

Building upon the Bongers (1990) model, Ferris *et al.* (2001) considered general opportunist c-p2 nematodes to be representative of organisms that persist in most soil food webs, almost always present and capable of surviving the most adverse conditions. Two axes are conceived as emerging from this basal state of the nematode fauna: one defines an enrichment index (EI), indicated by the weighted abundance of the proportion

Table 17.5. Calculation of the MI, MI2-5, ΣMI, ΣMI2-5, PPI and PPI/MI for two nematode assemblages (Samples A and B) of the same number of taxa and the same number of individuals (cf. Table 17.2).

Nematode Taxon	Sample A	Sample B	c-p	Feeding	A MI c-p wtd propn	B MI c-p wtd propn	A MI2-5 c-p wtd propn	B MI2-5 c-p wtd Propn	A ΣMI c-p wtd propn	B ΣMI c-p wtd propn	A ΣMI2-5 c-p wtd propn	B ΣMI2-5 c-p wtd propn	A PPI c-p wtd propn	B PPI c-p wtd propn
Hoplolaimidae	**5**	**15**	3	H					0.1	0.4	0.1	0.8	1.5	1.5
Pratylenchidae	**5**	**15**	3	H					0.1	0.4	0.1	0.8	1.5	1.5
Aphelenchidae	**15**	**5**	2	F	0.3	0.1	0.3	0.3	0.3	0.1	0.3	0.2		
Cephalobidae	**15**	**2**	2	B	0.3	0.1	0.3	0.1	0.3	0.0	0.3	0.1		
Plectidae	**2**	**15**	2	B	0.0	0.4	0.0	1.0	0.0	0.3	0.0	0.5		
Rhabditidae	**2**	**50**	1	B	0.0	0.6			0.0	0.5				
Dorylaimidae	**50**	**5**	4	O	2.0	0.3	2.1	0.7	1.8	0.2	1.9	0.3		
Discolaimidae	**15**	**2**	5	P	0.8	0.1	0.8	0.3	0.7	0.1	0.7	0.2		
Relevant total	**109**	**109**			**99**	**79**	**97**	**29**	**109**	**109**	**107**	**59**	**10**	**30**
MI	**3.4**	**1.6**												
MI2-5	**3.5**	**2.6**												
ΣMI	**3.4**	**2.0**												
ΣMI2-5	**3.4**	**2.8**												
PPI	**3.0**	**3.0**												
PPI/MI	**0.9**	**1.9**												

H = plant feeders, F = fungal feeders, B = bacterial feeders, O = omnivores, P = predators. In the columns where each index is calculated, the total number of nematodes meeting the criteria of the index is first determined (relevant total) and then the proportion of that total in taxa meeting the criteria is weighted by the c-p values of those taxa. The index value is the sum of the weighted proportions.
From Ferris and Bongers (2009).

of nematodes that are c-p1 bacterivores and c-p2 fungivores, and the other a structure index (SI), derived from the proportional contribution of the weighted c-p3-5 nematodes to the c-p2-5 grouping (Fig. 17.4).

Also included in the indices of ecosystem function is a basal index (BI), the relative proportion of the basal (c-p2) component to that of all nematodes present (Berkelmans *et al.*, 2003). Further resolution to the enrichment component is provided by the channel index (CI), which indicates the relative flow of resources into the food web through decomposition channels mediated by fungi and bacteria (Ruess and Ferris, 2004).

The EI and SI, both based on the indicator importance of functional guilds of nematodes, are descriptors of food web condition. Functional guilds are defined as a matrix of nematode feeding habits with the biological, ecological and life history characteristics embodied in the c-p classification. Thus, the Ba1 functional guild is comprised of c-p1 bacterivores such as those in the family Rhabditidae (Fig. 17.5A). The Ba2 guild includes general opportunist c-p2 bacterivores such as those in the Plectidae and Cephalobidae (Fig. 17.2), while the Fu2 guild includes c-p2 fungivores such as those in the Aphelenchidae (Fig. 17.5B). The Ba3 guild includes c-p3 bacterivores such as those in the Teratocephalidae or Prismatolaimidae, and the Ca3 guild includes c-p3 carnivores (predator nematodes such as those in the Tripylidae (Fig. 17.5C, D)). The Ca4 guild includes c-p4 specialist predators of other nematodes such as those in the Mononchidae (Fig. 17.6A, B), while the Ca5 guild includes c-p5 generalist predators such as those in the Discolaimidae (Fig. 17.6C, D).

17.3.3. Calculation of Functional Indices

Nematodes of all feeding habits classified as c-p2 are considered basal (b) to both enrichment and structure trajectories. Bacterial-feeding c-p1 and fungivores in c-p2 are indicators of enrichment (e), while nematodes of all feeding habits in c-p3-5 are indicators of structure (s). Functional guild indicators are weighted according to growth and metabolic rates (resource utilization) on the enrichment axis and to estimates of the degree of connectance, as determined by numbers of nematodes in higher c-p classes, in food webs of increasing complexity (Ferris *et al.*, 2001).

Functional guilds are designated based on feeding habits and c-p classification and weighted according to their indicator characteristics.

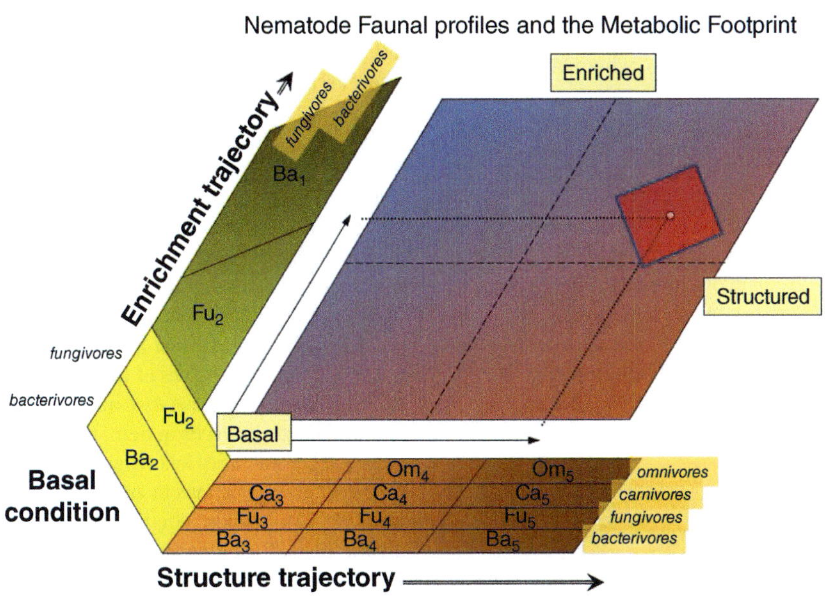

Fig. 17.4. A graph of the Enrichment Index and Structure Index provides an assessment of ecosystem condition and level of activity based on the weighted proportions of functional guilds.

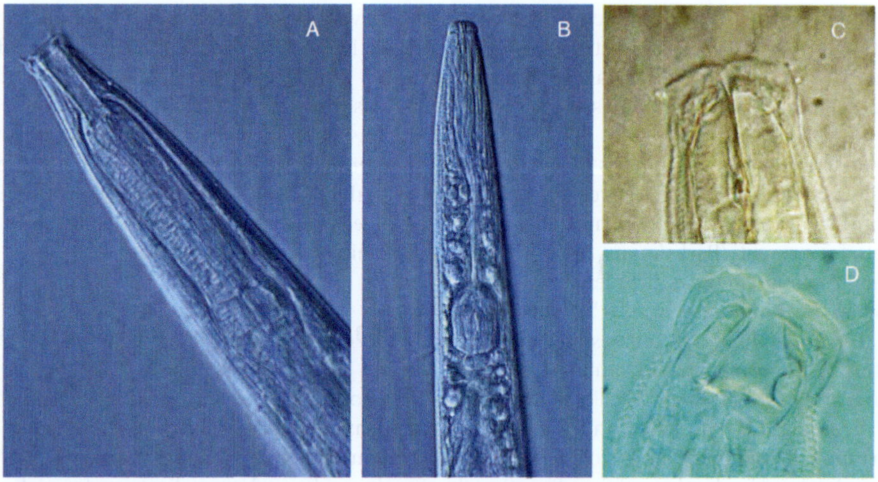

Fig. 17.5. A: Open cylindrical stoma of a nematode of the family Rhabditidae, Ba1 functional guild; enrichment opportunist c-p1 nematodes prevalent in moist, bacteria-enriched conditions; B: Anterior portion of a fungal-feeding nematode in the Aphelenchoididae (Fu 2 functional guild). The delicate stylet with small basal knob is inserted into fungal hyphae; cell contents of the hyphae are removed by pumping action of muscles in enlarged metacorpus; C: Generalist predator nematode of the family Tripylidae (Ca3 functional guild) with stoma relaxed and dorsal tooth embedded in pocket of stoma wall; and D: With stoma distended (micrographs by Ignacio Cid del Prado.)

Enrichment (EI), Structure (SI), Basal (BI) and Channel (CI) indices are calculated from weighted faunal components (b, e, s) of the nematode assemblage:

$b = (Ba_2 + Fu_2) * W_2$, where $W_2 = 0.8$;
$e = (Ba_1 * W_1) + (Fu_2 * W_2)$, where $W_1 = 3.2$ and $W_2 = 0.8$;
$s = (Ba_n * W_n + Ca_n * W_n + Fu_n * W_n + Om_n * W_n)$

where $n = 3\text{-}5$; $W_3 = 1.8$; $W_4 = 3.2$; $W_5 = 5.0$ and Ba_n, Fu_n, Ca_n are respectively functional guild designations of bacterial-feeding, fungal-feeding and carnivore (predator) nematodes.

The Enrichment (EI), Structure (SI), Basal (BI) and Channel (CI) indices are calculated from the weighted faunal components (Ferris *et al.*, 2001; Berkelmans *et al.*, 2003; Ferris and Bongers, 2009):

$EI = 100 * e/(e + b)$
$SI = 100 * s/(s + b)$
$BI = 100 * b/(e + s + b)$
$CI = 100\, Fu2 * W_2/(Ba1 * W_1 + Fu2 * W_2)$.

The EI, SI and BI provide higher resolution to the enrichment, disturbance and contamination assessments of the ecosystem (Table 17.6). Clearly, Sample A represents an environment with an abundance of omnivore and predator nematodes, suggesting greater connectance in the soil food web and the probable top-down regulation of opportunistic species (Sánchez-Moreno and Ferris, 2007). Sample B represents a disturbed and enriched condition in which the disturbance has had detrimental effects on higher trophic levels. Faunal analyses based on these indices provide insights into food web enrichment and structure and allow derivation of testable hypotheses based on the relative enrichment and structure of the system (Table 17.7).

The CI provides a means to partition flow of resources through fungal and bacterial decomposition channels. Indices of fungal and bacterial activity, based on the relative abundance of fungal- and bacterial-feeding nematodes, have been proposed several times following the calculation of their relative proportions in grasslands, woodlands and cultivated fields. The indices have included a ratio of F/B (Sohlenius and Boström, 1984), which has been defined most recently as NCR = B/(B + F) where NCR is the Nematode Channel Ratio, and B and F represent the abundance of

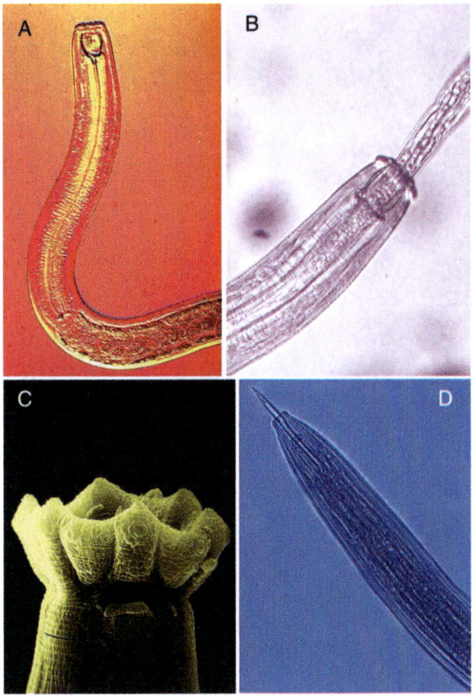

Fig. 17.6. A: Classic photograph of a mononchid nematode (Ca4 functional guild); B: Mononchid nematode ingesting a prey nematode; C: Scanning electron micrograph of the lip region of a nematode in the Discolaimidae; D: The piercing spear of a large dorylaimid nematode (Both Ca5 functional guild.)

bacterial- and fungal-feeding nematodes, respectively (Yeates, 2003). The CI differs in including weighting parameters for the size and metabolic rates of the nematode indicators.

17.4. Biomass and Metabolic Footprints

While diversity and functional indices are useful descriptive tools for assessment of food web and ecosystem condition, they do not provide information on the magnitude of ecosystem functions. For example, different assemblages with an abundance of nematodes or with a few nematodes may have the same diversity indices, the same MI or the same SI and EI (Tables 17.5, 17.6). Documentation of metabolic activity and biomass levels of different indicator guilds of nematodes would convey more information on the importance of the food web or ecosystem attributes suggested by the indices.

The evolution of indices of food web structure and function, and the accumulation of information on nematode biology and behaviour, confer greater value on biomass and metabolism in ecological studies. When other constraints are not limiting, carbon (C) and energy are the resources that determine food web size and activity. Besides utilization of C in body and egg production, nematodes have size-dependent metabolic costs (Ferris *et al.*, 1995). Ferris (2010) built on the concept of functional indices to determine the metabolic footprints of nematodes associated with each functional attribute of the food web. The metabolic footprint has a production component and a respiration component. The production component is the lifetime amount of C partitioned into growth and egg production and the respiration component assesses C utilization in metabolic activity.

The metabolic footprint is an estimator of nematode contribution to various ecosystem services and functions (Table 17.8). As an example, the Enrichment and Structure Footprints may be centred on nematode faunal profiles (Fig. 17.7) to add magnitude to the ecosystem service potential of the nematode assemblage. In the example, the aggregate footprint is also provided as a measure of total nematode presence in the soil food web.

17.4.1. Calculation and Interpretation of Biomass and Metabolic Footprints

Production component of the metabolic footprint

Nematode biomass can be calculated using the method of Andrássy (1956): $W = (L * D^2)/(1.6 * 10^6)$, where W is the fresh weight (μg) per individual, L is the nematode length (μm) and D is the greatest body diameter (μm). Body length and width measurements have been among the standard morphometric descriptors for nematode taxa for over 50 years. Among the standard parameters are L, the body length, and a, the ratio of length to maximum body diameter. Thus, the formula for nematode volume can be restated as $V = (L^3/a^2)/1.7$. Andrássy (1956) determined the average specific gravity of nematodes as 1.084,

Table 17.6. Calculation of the BI, EI and SI for two nematode assemblages (samples A and B) of the same number of taxa and the same number of individuals (cf. Tables 17.2 and 17.5).

Nematode Taxon	Sample A	Sample B	c-p	feeding	b wt	e wt	s wt	A: b	A: E	A: s	B: b	B: e	B: s
Hoplolaimidae	**5**	**15**	3	H				0.00	0.00	0.00	0.00	0.00	0.00
Pratylenchidae	**5**	**15**	3	H				0.00	0.00	0.00	0.00	0.00	0.00
Aphelenchidae	**15**	**5**	2	F	0.80	0.80		12.00	12.00	0.00	4.00	4.00	0.00
Cephalobidae	**15**	**2**	2	B	0.80			12.00	0.00	0.00	1.60	0.00	0.00
Plectidae	**2**	**15**	2	B	0.80			1.60	0.00	0.00	12.00	0.00	0.00
Rhabditidae	**2**	**50**	1	B		3.20		0.00	6.40	0.00	0.00	160.00	0.00
Dorylaimidae	**50**	**5**	4	O			3.20	0.00	0.00	160.00	0.00	0.00	16.00
Discolaimidae	**15**	**2**	5	P			5.00	0.00	0.00	75.00	0.00	0.00	10.00
Total	**109**	**109**											
Index components								25.60	18.40	235.00	17.60	164.00	26.00
BI	**9.18**	**8.48**											
EI	**41.82**	**90.31**											
SI	**90.18**	**59.63**											

H = plant feeders, F = fungal feeders, B = bacterial feeders, O = omnivores, P = predators. In the columns where each index is calculated, the total number of nematodes in each taxon meeting the criteria of the index is weighted in terms of the basal, enrichment and structure characteristics of that taxon. The index values are calculated from the index components, that is the sum of the b, e and s values (see text).

From Ferris and Bongers (2009).

Table 17.7. Inferred condition of the soil food web and its environment based on weighted nematode faunal analysis. Quadrats refer to faunal ordination in the faunal profile (Fig. 17.3).

General diagnosis	Quadrat A	Quadrat B	Quadrat C	Quadrat D
Disturbance	High	Low to Mod.	Undisturbed	Stressed
Enrichment	N-enriched	N-enriched	Moderate	Depleted
Decomposition channels	Bacterial	Balanced	Fungal	Fungal
C-to-N ratio	Low	Low	Mod. to High	High
Food web condition	Disturbed	Maturing	Structured	Degraded

From Ferris *et al.*, (2001).

Table 17.8. Characteristics of metabolic footprints based on biomass and metabolic activity of various functional components of the nematode assemblage (Ferris, 2010).

Footprint designation	Description
Enrichment	The metabolic footprint of those nematodes most rapidly responsive to resource enrichment.
Structure	The metabolic footprint of higher trophic levels which may have a regulatory function in the food web and which are indicative of the abundance of organisms of similar functions in non-nematode taxa.
Functional	The total area of enrichment and structure functional footprints.
Herbivore/Bacterivore/ Fungivore	The nematode indicators of C and energy entering the soil food web respectively through channels mediated by herbivores, bacterivores and fungivores.
Aggregate	The metabolic footprint of the complete nematode assemblage, regardless of trophic role or ecosystem function.

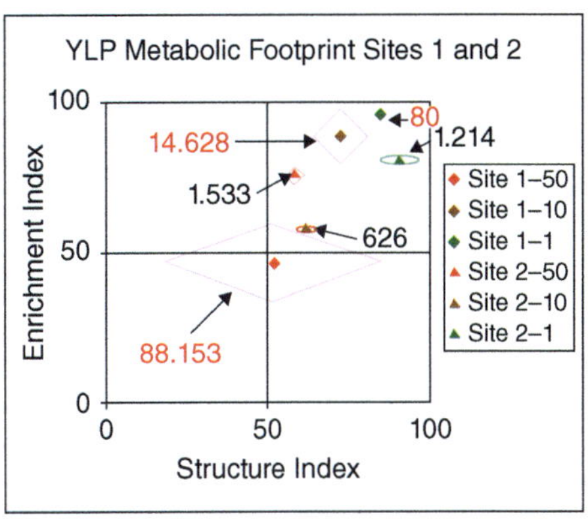

Fig. 17.7. Enrichment and structure footprints at 50m, 10m and 1m from water sources at two locations in Yolo County, California, USA. Numbers indicate the functional footprint of the nematode assemblage in each sample. Adapted from a study by Culman *et al.*, 2010.

based on the specific gravity of liquids in which they neither rose nor sank. From the product of specific gravity and volume, he determined the weight (W) of a nematode in terms of L and D. The formula can be rewritten to reflect available parameters as $W = (L^3/a^2)/(1.6 * 10^6)$ µg.

Respiration component of the metabolic footprint

Nematode respiration rate per individual decreases with body size according to the relationship $R = cW^b$, where R is the respiration rate, W is the fresh weight of the individual and c and b are regression parameters, such that b is close to 0.75 for nematodes and other invertebrates (Klekowski *et al.*, 1974). In studies so far, the c values for each nematode species of the relationship $R = cW^b$ increase to maxima at soil temperatures between 20 and 30°C and decline at higher temperatures (Ferris *et al.*, 1995). If all species present in the same environment are similarly adapted to ambient conditions, we might assume that the c values of the relationships between respiration rates and body weight change similarly with temperature (e.g., Ferris, 2010). Consequently, the species and temperature-specific coefficient c is omitted from the respiration-body weight relationship and the aggregate respiration rate is calculated as $\Sigma R = N_t W^{0.75}$, where N_t is the number of individuals in each of the t taxa of interest.

Since we may be more interested in resource availability and C flow through the food web than CO_2 evolution, the weight of lifetime C mineralized by each taxon and, by summation, by each functional guild or the complete nematode assemblage, is derived from the molecular weights of C and O_2 as 12/44 or 0.273 of the mass of CO_2 evolved.

The metabolic footprint calculation

The expanded equation for the metabolic footprint of nematodes (F), as an index of C utilization of component taxa, is the sum of the production and respiration components, F = P + R, and expanded as:

$$F = \Sigma(N_t(0.1(W_t/m_t) + 0.273(W_t^{0.75}))$$

for each of the t taxa involved in the summation. Then, from the formula of Andrássy (1956) and the L and a values of each nematode species, the W_t parameter can be replaced by $(L^3/a^2)/(1.6{*}10^6)$.

17.5. Data Gaps, Research Opportunities, Current and Future Developments

The framework for structural and functional indices and for metabolic footprints of nematode assemblages is based on the c-p classes of Bongers (1990) and Bongers and Bongers (1998), which are essentially structural guild assignments at the family level. While those assignments are defensible with regard to life history strategies and response characteristics (Bongers and Ferris, 1999; Ferris and Bongers, 2009), there may be greater plasticity with regard to feeding habits within families, genera and even species.

We infer feeding habits from stomal and pharyngeal structure. However, even among the well-studied stomatostyle-bearing Tylenchida, the feeding habits of some groups are uncertain (higher plant roots, fungi, algae or all?). With cavalier abandon, we describe many of the Dorylaimida as omnivores and sometimes as generalist predators of other animals, including nematodes. But some Dorylaimida have intestinal contents that appear to have a green coloration. Is that from feeding on algae and, if so, exclusively or opportunistically? And, if they are opportunistic feeders on algae, what then about higher plants? Indeed, some Dorylaimida may behave as opportunistic colonizers in certain conditions (Austin *et al.*, 2009; Martínez *et al.*, 2018), apparently contradicting their high c-p value and omnivore-predator habits. Some apparently specialist predator Mononchida can be cultured, at least in juvenile stages, on bacteria (Yeates *et al.*, 2009). Some diplogastrids feed on both bacteria and protozoans. So, what is the appropriate level of taxonomic resolution for assigning nematodes to functional guilds? Resolution will improve with further observation and experimentation on nematode biology and feeding habits. With the increase in our understanding of nematode auto-ecology, the description of new species, and the continuous revision of the taxonomic position of nematode species, genera and families in the phylum phylogenetic tree, assignment of nematode families to trophic and c-p groups have occasionally changed and future adjustments may be necessary (Du Preez *et al.*, 2022). Although so far there is not an updated list of current nematode functional group assignations, the Nematode Ecophysiological Parameter database of Nemaplex (http://nemaplex.ucdavis.edu) lists the generally accepted functional assignments of thousands of nematode taxa and is an internationally recognized source of nematode physiological parameters. Currently, most nematode ecologists calculate nematode-based indices using the online tool NINJA (Sieriebriennikov *et al.*, 2014; www.wur.nl/en/product/ninja-nematode-indicator-joint-analysis.htm). NINJA uses the Nemaplex database on nematode ecophysiological parameters to describe nematode community functional characteristics and to calculate nematode-based indices.

When we superimpose biomass calculations on nematode assemblage assessments to determine metabolic footprints, we risk an additional error of unknown magnitude. Firstly, validation studies suggest that biomass determinations using the Andrássy (1956) formula may be high or low by 5-31% depending on the morphotype of the nematode (Freckman, 1982). Secondly, definitive measurements for nematode species are based on adult females. Where species measurements are taken from taxonomic descriptions, nematode biomass must be overestimated unless every individual in the assemblage is an adult female. The alternatives are to measure every nematode in the sample, which usually exceeds time constraints, or to make the heroic assumption that the inherent bias is equally applied across all comparable samples. While that may be true for a single study with samples taken at one point in time, it cannot be true across multiple studies. Although more accurate inferences of nematode biomass based on measurement of individuals have been proposed (Zhao *et al.*, 2019; Ma *et al.*, 2024), and new technologies (Mazurkiewicz *et al.*, 2016) will probably automate measurements in the future, nematode body mass measurements still have a degree of uncertainty. Thirdly, there is the issue of size variability among species or genera. How much variability is there among the species of nematodes identified at only the genus or family level? In a database of 6325 nematode taxa in which each genus is represented by more than one species, the average coefficient of variation (standard deviation as a percentage of the mean calculated mass) across females of the species in a genus was 24% and as high as 99% of the mean for some genera. Among the 6486 taxa in which there were data for more than one species in each family, the average coefficient of variation across females of the species in a family was 30% and as high as 100% of the mean for some families. This database is used by NINJA (Sieriebriennikov *et al.*, 2014) to calculate NBIs. Clearly, there is room for greater efficiency and precision in determination of nematode biomass by functional category.

Finally, the full diversity of nematode species is unknown; perhaps we have just scratched the surface for some taxa. The approximate number of nematode species described so far, around 20,000, surely represents a small fraction of the actual number of nematode species present on Earth. Different estimates calculate a total nematode species richness between 1 million (Scheffers *et al.*, 2012) and 41 million (Larsen *et al.*, 2017). Characterization of new species requires extensive measurement and description of morphological, anatomical and molecular features (see Chapter 2, this volume). As more species are added to genera, and more genera to families, taxonomic keys as identification aids become more complex. Additionally, few taxonomists and systematists have been trained in recent years, so there is a dearth of expertise. However, there is enormous potential for the application of molecular analyses of gene sequences in nematode identification, taxonomy and systematics.

'DNA fingerprinting' methods evolved for description of microbial communities and their functional attributes (Machulla *et al.*, 2005; Mou *et al.*, 2008), and their application to assignment of nematodes to structural and functional guilds is just emerging (Powers *et al.*, 2009; Holterman *et al.*, 2011). At the present time, molecular identification of soil nematode communities is used to assess nematode diversity in many types of ecosystems and under varying levels of environmental perturbations (Kerfahi *et al.*, 2017; Marais *et al.*, 2020). Although classical morphological and new molecular nematode identification approaches have sometimes provided dissimilar results (Wu *et al.*, 2009), there have been numerous efforts to integrate morphological and molecular techniques for comprehensive assessment of soil nematode diversity (Geisen *et al.*, 2018). Clearly, it is both important and aesthetically satisfying to recognize visually, and to be able to name, the organisms that are expressing genes of interest, and one could argue that molecular and morphometric approaches must go hand-in-hand lest we define all biological activity in terms of gene frequencies. However, diversity within the Nematoda is orders of magnitude greater than the number of taxonomists available for identification of taxa. The development of high-throughput DNA sequencing methodology allows metagenetic analysis and the rapid definition of molecular operational taxonomic units, first proposed by Floyd *et al.* (2002), as an alternative to traditional taxonomy (Creer *et al.*, 2010).

17.6. Conclusion

The value of nematodes as bioindicators of environmental condition, and of the status of ecosystem functions and services, has been advanced by integration of nematode feeding behaviour classification with guilds based on life history traits. Undoubtedly, refinement and fine-tuning of the system is warranted and will occur as further information is developed on feeding habits, life history traits and appropriate assignment of taxa to functional guilds. There are other examples of the use of the community structure of various organism groups for environmental monitoring. The advantage of those based on nematode functional guilds derives from the abundance and ubiquity of nematodes, their diversity of function, relationships between form and function, differences in sensitivity to environmental disturbance, and the ease with which they can be separated from substrate and categorized into taxonomic groups or functional guilds.

17.7. References

Andrássy, I. (1956) Die rauminhalst und gewichtsbestimmung der fadenwurmer, (Nematoden). *Acta Zoologica Academi Sciences* 2, 1–15.

Austin, E., Semmens, K., Parsons, C. and Amy, T. (2009) Granite rock outcrops: an extreme environment for soil nematodes? *Journal of Nematology* 41, 84–91.

Baxter, C., Rowan, J.S., McKenzie, B.M. and Neilson, R. (2013) Understanding soil erosion impacts in temperate agroecosystems: bridging the gap between geomorphology and soil ecology using nematodes as a model organism. *Biogeosciences* 10, 7133–7145.

Berkelmans, R., Ferris, H., Tenuta, M. and Van Bruggen, A.H.C. (2003) Effects of long-term crop management on nematode trophic levels other than plant feeders disappear after one year of disruptive soil management. *Applied Soil Ecology* 23, 223–235.

Betancur-Corredor, B., Lang, B. and Russell, D.J. (2022) Reducing tillage intensity benefits the soil micro- and mesofauna in a global meta-analysis. *European Journal of Soil Science* 73, e13321. DOI: 10.1111/ejss.13321.

Bongers, T. (1990) The maturity index: an ecological measure of environmental disturbance based on nematode species composition. *Oecologia* 83, 14–19.

Bongers, T. and Bongers, M. (1998) Functional diversity of nematodes. *Applied Soil Ecology* 10, 239–251.

Bongers, T. and Ferris, H. (1999) Nematode community structure as a biomonitor in environmental monitoring. *Trends in Ecology and Evolution* 14, 224–228.

Bongers, T., Alkemade, R. and Yeates, G.W. (1991) Interpretation of disturbance-induced maturity decrease in marine nematode assemblages by means of the Maturity Index. *Marine Ecology Progress Series* 76, 135–142.

Bongiorno, G., Bodenhausen, N., Bünemann, E.K., Brussaard, L., Geisen, S., *et al.* (2019) Reduced tillage, but not organic matter input, increased nematode diversity and food web stability in European long-term field experiments. *Molecular Ecology* 28, 4987–5005.

Cesarz, S., Ciobanu, M., Wright, A.J., Ebeling, A., Vogel, A., *et al.* (2017) Plant species richness sustains higher trophic levels of soil nematode communities after consecutive environmental perturbations. *Oecologia* 184, 715–728.

Creer, S., Fonseca, V.G., Porazinska, D.L., Giblin-Davis, R.M., Sung, W., *et al.* (2010) Ultrasequencing of the meiofaunal biosphere: practice, pitfalls and promises. *Molecular Ecology* 19(Suppl. 1), 4–20.

Culman, S.W., Young-Mathews, A., Hollander, A.D., Ferris, H., Sánchez-Moreno, S., *et al.* (2010) Biodiversity is associated with indicators of soil ecosystem functions over a landscape gradient of agricultural intensification. *Landscape Ecology* 25, 1333–1348.

De Deyn, G.B. and Van der Putten, W.H. (2005) Linking above- and below-ground biodiversity. *Trends in Ecology and Evolution* 20, 625–633.

Du Preez, G., Daneel, M., De Goede, R., Du Toit, M.J., Ferris, H., *et al.* (2022) Nematode-based indices in soil ecology: application, utility, and future directions. *Soil Biology and Biochemistry* 169, 08640.

Ferris, H. (2010) Form and function: metabolic footprints of nematodes in the soil food web. *European Journal of Soil Biology* 46, 97–104.

Ferris, H. and Bongers, T. (2009) Indices developed specifically for analysis of nematode assemblages. In: Wilson, M.J. and Kakouli-Duarte, T. (eds), *Nematodes as Environmental Bioindicators*. CAB International, Wallingford, UK, pp. 124–145.

Ferris, H., Bongers, T. and De Goede, R.G.M. (2001) A framework for soil food web diagnostics: extension of the nematode faunal analysis concept. *Applied Soil Ecology* 18, 13–29.

Ferris, H., Lau, S. and Venette, R. (1995) Population energetics of bacterial-feeding nematodes: respiration and metabolic rates based on carbon dioxide production. *Soil Biology and Biochemistry* 27, 319–330.

Ferris, H., Bongers, T. and De Goede, R.G.M. (2004a) Nematode faunal analyses to assess food web enrichment and connectance. In: Cook, R. and Hunt, D.J. (eds), *Proceedings of the Fourth International Congress of Nematology. Nematology Monographs and Perspectives, 2*. Brill, Leiden, Netherlands, pp. 503–510.

Ferris, H., Venette, R.C. and Scow, K.M. (2004b) Soil management to enhance bacterivore and fungivore nematode populations and their nitrogen mineralization function. *Applied Soil Ecology* 24, 19–35.

Floyd, R., Abebe, E., Papert, A. and Blaxter, M. (2002) Molecular barcodes for soil nematode identification. *Molecular Ecology* 11, 839–850.

Freckman, D.W. (1982) Parameters of the nematode contribution to ecosystems. In: Freckman, D.W. (ed), *Nematodes in Soil Ecosystems*. University of Texas Press, Austin, Texas, pp. 81–97.

Geisen, S., Snoek, L.B., Ten Hooven, F.C., Duyts, H., Kostenko, O., *et al.* (2018) Integrating quantitative morphological and qualitative molecular methods to analyse soil nematode community responses to plant range expansion. *Methods in Ecology and Evolution* 9, 1366–1378.

Hill, M.O. (1973) Diversity and evenness: a unifying notation and its consequences. *Ecology* 54, 427–432.

Holterman, M.H.M., Korthals, G.W., Doruszuk, A., van Megen, H.B., Bakker, J., *et al.* (2011) A strategy in searching for stress tolerance-correlated characteristics in nematodes while accounting for phylogenetic interdependence. *Nematology* 13, 261–275.

Kerfahi, D., Park, J., Tripathi, B.M., Singh, D., Porazinska, D.L., *et al.* (2017) Molecular methods reveal controls on nematode community structure and unexpectedly high nematode diversity, in Svalbard high Arctic tundra. *Polar Biology* 40, 765–776.

Klekowski, R.Z., Wasilewska, L. and Paplinska, E. (1974) Oxygen consumption in the developmental stages of *Panagrolaimus rigidus*. *Nematologica* 20, 61–68.

Larsen, B.B., Miller, E.C., Rhodes, M.K. and Wiens, J.J. (2017) Inordinate fondness multiplied and redistributed: the number of species on earth and the new pie of life. *Quarterly Review of Biology* 92, 229–265.

Lazarova, S., Coyne, D., Rodriguez, M.G., Peteira, B. and Ciancio, A. (2021) Functional diversity of soil nematodes in relation to the impact of agriculture – a review. *Diversity-Basel* 13, 64. DOI: 103390/d13020064.

Lin, D., Yang, G., Dou, P., Qian, S., Zhao, L., *et al.* (2020) Microplastics negatively affect soil fauna but stimulate microbial activity: insights from a field-based microplastic addition experiment. *Proceedings of the Royal Society B-Biological Sciences* 287, 20201268. DOI: 10.1098/rspb.2020.1268.

Ma, Q., Zhu, Y., Chen, Y., Wu, W., Qing, X., *et al.* (2024) Simplified estimates of soil nematode body mass using maximum diameter: insights from large-scale grasslands across China. *Soil Biology and Biochemistry* 191, 109349. DOI: 10.1016/soilbio.2024.109349.

Machulla, G., Bruns, M.A. and Scow, K.M. (2005) Microbial properties of mine spoil materials in the initial stages of soil development. *Soil Science Society of America Journal* 69, 1069–1077.

Marais, E., Maggs-Kolling, G., Sherman, C., Doniger, T., Rentao, L., *et al.* (2020) Profiling soil free-living nematodes in the Namib Desert, Namibia. *Journal of Arid Land* 12, 130–143.

Martinez, J.G., Torres, M.A., Dos Santos, G. and Moens, T. (2018) Influence of heavy metals on nematode community structure in deteriorated soil by gold mining activities in Sibutad, southern Philippines. *Ecological Indicators* 91, 712–721.

Mazurkiewicz, M., Gorska, B., Jankowska, E. and Wlodarska-Kowalczuk, M. (2016) Assessment of nematode biomass in marine sediments: a semi-automated image analysis method. *Limnology and Oceanography-Methods* 14, 816–827.

Mou, X., Sun, S., Edwards, R.A., Hodson, R.E. and Moran, M.A. (2008) Bacterial carbon processing by generalist species in the coastal ocean. *Nature* 451, 708–711.

Neher, D.A. (2001) Role of nematodes in soil health and their use as indicators. *Journal of Nematology* 33, 161–168.

Neher, D.A. and Darby, B.J. (2009) General community indices that can be used for analysis of nematode assemblages. In: Wilson, M.J. and Kakouli-Duarte, T. (eds), *Nematodes as Environmental Bioindicators*. CAB International, Wallingford, UK, pp. 107–123.

Pielou, E.C. (1966) The measurement of diversity in different types of biological collections. *Journal Theoretical Biology* 13, 131–144.

Porazinska, D.L., Duncan, L.W., McSorley, R. and Graham, J.H. (1999) Nematode communities as indicators of status and processes of a soil ecosystem influenced by agricultural management practices. *Applied Soil Ecology* 13, 69–86.

Powers, T.O., Neher, D.A., Mullin, P., Esquivel, A. Giblin-Davis, R.M., *et al.* (2009) Tropical nematode diversity: vertical stratification of nematode communities in a Costa Rican humid lowland rainforest. *Molecular Ecology* 18, 985–996.

Ruess, L. and Ferris, H. (2004) Decomposition pathways and successional changes. In: Cook, R. and Hunt, D.J. (eds), *Proceedings of the Fourth International Congress of Nematology, Tenerife, Spain, 8–13 June 2002. Nematology Monographs and Perspectives, 2*. Brill, Leiden, Netherlands, pp. 547–556.

Šalamún, P., Brázová, T., Miklisová, D. and Hanzelová, V. (2015) Influence of selected heavy metals (As, Cd, Cr, Cu) on nematode communities in experimental soil microcosm. *Helminthologia* 52, 341–347.

Sánchez-Moreno, S. and Ferris, H. (2007) Suppressive service of the soil food web: effects of environmental management. *Agriculture, Ecosystem and Environment* 119, 75–87.

Sánchez-Moreno, S., Smukler, S., Ferris, H., O'Geen, A.T. and Jackson, L.E. (2008) Nematode diversity, food web condition, and chemical and physical properties in different soil habitats of an organic farm. *Biology and Fertility of Soils* 44, 727–744.

Scheffers, B.R., Joppa, L.N., Pimm, S.L. and Laurance, W.F. (2012) What we know and don't know about Earth's missing biodiversity. *Trends in Ecology and Evolution* 27, 501–510.

Shannon, C.E. (1949) *The Mathematical Theory of Communication*. University of Illinois Press, Urbana, Illinois.

Sieriebriennikov, B., Ferris, H. and De Goede, R.G.M. (2014) NINJA: an automated calculation system for nematode-based biological monitoring. *European Journal of Soil Biology* 61, 90–93.

Simpson, E.H. (1951) The interpretation of interaction in contingency tables. *Journal of the Royal Statistical Society, Series B* 13, 238–241.

Sohlenius, B. and Boström, S. (1984) Colonization, population development and metabolic activity of nematodes in buried barley straw. *Pedobiologia* 27, 67–78.

Wardle, D.A., Yeates, G.W., Williamson, W. and Bonner, K.I. (2003) The response of a three trophic level soil food web to the identity and diversity of plant species and functional groups. *OIKOS* 102, 45–56.

Wu, T., Ayres, E., Li, G., Bardgett, R.D., Wall, D.H., *et al.* (2009) Molecular profiling of soil animal diversity in natural ecosystems: incongruence of molecular and morphological results. *Soil Biology and Biochemistry* 41, 849–857.

Yeates, G.W. (2003) Nematodes as soil indicators: functional and biodiversity aspects. *Biology and Fertility of Soils* 37, 199–210.

Yeates, G.W., Bongers, T., De Goede, R.G.M., Freckman, D.W. and Georgieva, S.S. (1993) Feeding habits in soil nematode families and genera – an outline for soil ecologists. *Journal of Nematology* 25, 315–331.

Yeates, G.W., Ferris, H., Moens, T. and Van der Putten, W.H. (2009) The role of nematodes in ecosystems. In: Wilson, M.J. and Kakouli-Duarte, T. (eds), *Nematodes as Environmental Bioindicators*. CAB International, Wallingford, UK, pp. 1–44.

Zhao, J., Xiao, J., Zhang, W., Fu, Z., Zhang, M., *et al.* (2019) A method for estimating nematode body lengths for use in the calculation of biomass via a simplified formula. *Soil Biology and Biochemistry* 134, 36–41.

Zheng, F., Bi, Q.F., Giles, M., Neilson, R., Chen, Q.L., *et al.* (2021) Fates of antibiotic resistance genes in the gut microbiome from different soil fauna under long-term fertilization. *Environmental Science and Technology* 55, 423–432.

Zhou, J., Wu, J., Huang, J., Sheng, X., Dou, X., *et al.* (2022) A synthesis of soil nematode responses to global change factors. *Soil Biology and Biochemistry* 165, 108538. DOI: 10.1016/soilbio.2021.108538.

18 Plant-nematode Interactions

Cynthia Gleason[1]* and Edgar Villar-Luna[2]
[1]*Department of Plant Pathology, Washington State University, Pullman, Washington, USA;* [2]*Instituto Politécnico Nacional, CIIDIR Michoacán, Jiquilpan, Michoacán, Mexico*

Abstract
Throughout their life cycle, plants are exposed to many potential interactions with plant-parasitic micro-organisms, including nematodes, which can lead to plant disease. Most plants, however, remain healthy, and disease is the exception rather than the rule. Many plant-pathogen interactions fail due to the pathogen's inability to locate a suitable host plant or generate the necessary array of pathogenicity molecules crucial for parasitic success. Through co-evolution, plants have evolved a fully innate immune system, contrasting with animals, which possess both adaptive and innate immune mechanisms. This chapter provides an introduction to plant-nematode interactions at the molecular level and how disease is induced. Plant structural, metabolic and chemical (effectors) defence mechanisms against nematode infestation, as well as disease resistance genes and resistance mechanisms are also discussed.

18.1. Introduction

Throughout their life cycle, plants are exposed to many potential interactions with plant-parasitic microorganisms, including nematodes, which can lead to plant disease. However, most plants remain healthy, and disease is the exception rather than the rule. Many plant-pathogen interactions fail due to the pathogen's inability to locate a suitable host plant or generate the necessary array of pathogenicity molecules crucial for parasitic success. Through co-evolution, plants have evolved a fully innate immune system, in contrast to animals, which possess both adaptive and innate immune mechanisms (Zipfel and Felix, 2005).

For the plant-nematode interaction to be successful, a sustained exchange of physical and chemical signals between both organisms is important. This communication might start even before physical contact between the plant and the parasite occurs. The communication or dialogue may or may not continue once the nematode has established contact with the cells of the host. Should this dialogue endure, the initial recognition between them can lead to nematode feeding and reproduction, signifying a compatible interaction. If the process is interrupted,

*E-mail: cynthia.gleason@wsu.edu

A revision of Zavaleta-Mejía, E. and Kaloshian, I. (2012) Plant-nematode interactions. In: Manzanilla-López, R.H. and Marbán-Mendoza, N. (eds), *Practical Plant Nematology*. Mundi-Prensa and Biblioteca Básica de Agricultura, Montecillo, Mexico.

 Practical Plant Nematology 2nd Edition
(eds R.H. Manzanilla-López, D.J. Hunt, N. Marbán-Mendoza and S. Infante Gil)
DOI: 10.1079/9781836990413.0018

presumably due to plant immune responses, the interaction ends, and this is considered an incompatible interaction. A successful parasite is one that is capable of penetrating, colonizing and reproducing in the host tissue. This interaction is possible due to a co-evolution process with the plant, through which the nematodes have accumulated the necessary genetic information (i.e., pathogenicity or parasitism genes) that confers the capacity to carry out infection, to derive nutrients from the host tissue and to have the ability to overcome, escape or neutralize the barriers or physical and chemical defences of the host. In the case of those plant-parasite interactions in which a high degree of specificity exists, and there is a strong dominant resistance response, there is likely a gene-for-gene interaction (Flor hypothesis). The gene-for-gene interaction implies that for every resistance (*R*) gene in the host, there is an avirulence gene (*Avr*) in the parasite (Flor, 1971). The *Avr* genes encode secreted molecules called effectors, which, in the absence of the cognate *R*-gene, can facilitate parasitism, but in the presence of the *R*-gene, trigger a resistance response.

The ever-evolving interaction between plants and pathogens is sometimes referred to as an 'arms race'. This description implies that the plant and pathogen are each trying to outdo the other in terms of success. Therefore, the use in this chapter of the term 'armoury' refers to the different plant defence mechanisms used against the attack of pathogens. The term 'weapons' will describe the nematode effectors that are used to overcome those host defence mechanisms. In this chapter, we will present aspects related to plant-parasitic nematodes (PPN), including how PPN induce disease and the alterations that occur in host cells. There will be an overview of the recent advances in nematology, particularly about the 'weapons' that nematodes use to attack the host plant, followed by an overview of plant resistance responses ('armoury'). The information presented corresponds mainly, but not exclusively, to that generated by studying major PPN species, which are the root-knot nematodes of the genus *Meloidogyne* and the cyst nematodes (*Globodera* and *Heterodera*). The topics presented herein have been dealt with extensively and in greater depth in the books *Root-knot Nematodes* (Perry *et al.*, 2009), *Cell Biology of Plant Nematode Parasitism* (Berg and Taylor, 2009) and *Genomics and Molecular Genetics of Plant-nematode Interactions* (Jones *et al.*, 2011). As our understanding progresses of how the interaction mechanisms at the genetic, molecular, metabolic and physiological levels take place during the plant-nematode interaction, we will be in a better position to find and develop better strategies to manage PPN.

18.2. How Plant-parasitic Nematodes Infect and Affect Plants

18.2.1. Host Penetration and Recognition

To establish the parasitic relationship with its host plants, nematodes use their stylet, a cuticularized, hollow, protrusible stomatostyle (see Chapter 2, this volume). Endoparasitic PPN (e.g., root-knot and cyst nematodes) mechanically penetrate the host roots. After penetration, PPN secrete via the stylet a mixture of enzymes that are synthesized in the cells of the pharyngeal gland. These enzymes degrade the cell wall polymers, thus facilitating their migration within the plant root tissues and thereby breaching the plant's first line of defence. In addition, they release secretions into host cells, near the vasculature, that promote the formation of specialized feeding sites necessary for them to complete their life cycle.

Several genes that encode for proteins produced in the pharyngeal glands of cyst and root-knot nematodes exhibit characteristics of pathogenicity factors (i.e., effectors), which are required for successful plant infection, feeding site formation and nematode reproduction. The effectors may be helping the nematodes overcome or suppress several lines of plant defence. The first line of defence is the physical barriers established by the plant, such as the cell wall. The subsequent lines of defence are inducible and include the recognition of pathogen-associated molecular patterns (PAMP), which are recognized by plant receptors (Zipfel and Robatzek, 2010). This recognition sets off a series of defensive reactions, collectively known as PAMP-triggered immunity (PTI). If the plant has resistance (R) protein(s), it may recognize specific nematode effectors, and these effectors are sometimes referred to as avirulence (Avr) proteins. The nematode Avr proteins are recognized directly or indirectly by the associated R-protein and trigger a strong defence response, typically characterized by cell death, also known as a hypersensitive response (HR). This *R*-gene-mediated level of plant immunity is otherwise known as effector-triggered immunity (ETI) (Cui *et al.*, 2015). To date, there are only a few cloned nematode Avr proteins that elicit *R*-gene-dependent plant defences. The other *Avr* genes in nematodes are yet to be identified. The relationship between nematodes and plants is marked by an ongoing struggle, where the outcome hinges on the coordination and effectiveness of the defence mechanisms employed by the plant and the corresponding responses from the nematode. Compatibility or incompatibility is determined by the extent to which the plant can trigger its defence strategies, as well as the intensity with which these defence genes are activated. In incompatible interactions, these changes occur more swiftly and with greater intensity compared to compatible ones.

18.2.2. Physical and Chemical Weapons (Effector Molecules) Used by the Nematode to Establish Infection in Plants

Pathogens possess several mechanisms (weapons) to carry out penetration, infection and induction of disease. Some pathogens mainly use mechanical force (i.e., mechanical weapons), and others also possess chemical weapons such as toxins, enzymes and growth regulators. PPN can use morphological structures, e.g., the nematode stylet, to facilitate infection. They can also synthesize metabolites with fundamental physiological roles in the interaction with its host. In the case of endoparasitic nematodes, mechanical force and enzymatic secretions appear to be involved in the root penetration and movement of the nematode within the roots. There are multiple enzymes that modify the cell walls, some participate in the invasion, while others may be important in the development of the feeding sites. Specifically, cyst nematodes form feeding sites called syncytia, and root-knot nematodes form feeding sites called giant cells. Throughout the parasitism process, changes occur in both cyst and root-knot nematodes in the morphology, activity and content of the pharyngeal glands (Davis *et al.*, 2004). Transcriptional activity within the subventral glands is predominant during the migratory stages of these nematodes, while in subsequent sedentary stages, the activity in the dorsal gland becomes dominant (Hussey and Grundler, 1998). Many of the proteins identified in the pharyngeal glands are referred to as effectors. These effector proteins are packaged and secreted out of the nematode, typically via the stylet, during parasitism. They function to suppress plant defence mechanisms and/or promote host structural changes and metabolite accumulation that facilitate nematode development and reproduction (Sato *et al.*, 2019).

Proteins that modify cell walls

Several of the PPN pharyngeal gland genes encode for enzymes that degrade or modify plant cell walls (Khan *et al.*, 2023). In fact, the first cellulase gene, endo-1,4-β-glucanase, from an animal was identified from *Globodera. rostochiensis* and *Heterodera glycines* (Smant *et al.*, 1998). A number of endoglucanases that can degrade cellulose and/or hemicellulose have been found in cyst and root-knot nematodes, facilitating their penetration and migration through the plant roots (Davis *et al.*, 2011; Mitsumasu *et al.*, 2015; Holbein *et al.*, 2016). Many of the nematode-secreted glucanases belong to the glycosyl hydrolase (GH) family (Rai *et al.*, 2015), which were acquired from bacteria by horizontal gene transfer (Jones *et al.*, 2005). In addition to sedentary endoparasitic nematodes, GH genes have also been identified in migratory endoparasites such as *Radopholus similis*, *Ditylenchoides africanus* and *Pratylenchus* species (Uehara *et al.*, 2001; Haegeman *et al.*, 2008; Kyndt *et al.*, 2008). Xylanase enzymes, which can degrade xylans, have also been identified in a small number of PPN, mainly from *Meloidogyne* species and *R. similis* (Mitreva-Dautova *et al.*, 2006; Abad *et al.*, 2008; Opperman *et al.*, 2008; Haegeman *et al.*, 2009). Pectate lyase, which catalyses the cleavage of the α-1,4-linkage of unesterified polygalacturonate (pectate) and plays in important role in degrading pectin, has also been identified in PPN. The genes encoding pectate lyases have been cloned from cyst and root-knot nematodes as well as from the migratory PPN *Bursaphelenchus xylophilus* (Kikuchi *et al.*, 2007; Peng *et al.*, 2016).

Expansins are the non-enzymatic proteins that are secreted by nematodes for cell wall modification. The expansins function to weaken the cell walls by breaking the non-covalent bonds between the fibres of the cell wall, thus allowing some fibres to slide over others (Cosgrove, 2000). Several nematode-encoded expansins were identified in root-knot nematodes and in cyst nematodes (Qin *et al.*, 2004). Expansin-like proteins from PPN may function as effectors. Expansins were found in the *Pratylenchus* transcriptome, where they were predicted to serve as effectors involved in parasitism (Vieira and Nemchinov, 2020). In addition, the cyst nematode *G. rostochiensis* expansin-like protein (*Gr*EXPB2) not only modifies the host cell walls but also suppress defence responses (Ali *et al.*, 2015). In the case of the cyst nematode, transient expression of *H. avenae* HaEXPB2 in *Nicotiana benthamiana* leaves induced cell death, suggesting that the cell modifications triggered by HaEXPB2 may also be eliciting plant immune responses (Liu *et al.*, 2016a).

Protection from host responses

After getting past the host's physical defences (i.e., the cell wall), the nematodes will trigger host-induced defences, which is the second layer of defence called PTI. For PTI, the plant defence response is based on the recognition of nematode PAMP by plant receptors on the plasma membrane. One such example of nematode PAMP is the ascarosides, which are nematode pheromones involved in social cueing. Studies of root-knot and cyst nematodes found that ascaroside 18 was the predominant PPN ascaroside, and the detection of this ascaroside by the plant triggers PTI (Manosalva *et al.*, 2015; Manohar *et al.*, 2020). The plant receptor NILR1 is involved in the recognition of ascaroside 18 (Huang *et al.*, 2023). Thus, nematodes produce PAMP, and the recognition of PAMP by the plant triggers basal defence responses.

Hallmarks of PTI include, but are not limited to, a calcium influx in the plant cell, reactive oxygen species (ROS) burst, protein phosphorylation and transcriptional reprogramming (Bigeard *et al.*, 2015). To combat these plant defence responses, the nematodes secrete effectors. The details of nematode effectors involved in plant defence suppression are constantly evolving, but manipulating ROS production and/or downstream signalling seems to be a major focus of many PPN effectors. Although cyst nematodes stimulate ROS production during syncytia formation, typically, ROS is associated with cell death and defence. It is generally agreed that too much and uncontrolled ROS would be harmful to nematodes, because they are biotrophs. Therefore, it is not surprising that most identified effectors can reduce or fine-tune ROS production. This includes the *Heterodera* effector 10A06, which targets the host plant spermidine synthase and ultimately helps elevate the expression of antioxidant genes (Hewezi *et al.*, 2010). The *G. rostochiensis* effector 29D09 suppresses elicitor-induced ROS production when transiently expressed in *N. benthamiana* leaves (Chen *et al.*, 2024). In root-knot nematodes, the MjTTL5 effector activates host ROS scavenging activity (Lin *et al.*, 2016). Another root-knot nematode effector called MiCTL1a interacts with catalases, which are enzymes involved in the reduction of ROS (Zhao *et al.*, 2021). Nematode effectors fine-tune host ROS levels in order to stimulate specific downstream responses and promote parasitism (Chopra *et al.*, 2021).

Host defence to PPN also includes the induction of pathogenesis-related (PR) plant proteins, which play important roles in plant resistance (Dos Santos and Franco, 2023). Various PR proteins are induced in response to nematode infections and are also targets of nematode effectors. For example, the *Heterodera* effector 30C02 was shown to bind to PR protein β-1,3-endoglucanase, and the *M. graminicola* effector MO237 was shown to bind to several host PR proteins (Chen *et al.*, 2018).

Usually, the effectors are secreted from the pharyngeal glands, because the pharynx is connected to the stylet. However, effectors could be secreted from any secretory organ, including the amphid and hypodermis. In cells of the nematode hypodermis, several genes are expressed that code for proteins on the cuticle surface, and as a result, these nematode proteins are in direct contact with the plant. Such proteins could be involved in enhancing plant defence response. In *G. rostochiensis*, peroxidase genes are expressed in the nematode hypodermis (Robertson *et al.*, 2000; Jones *et al.*, 2004), and protein products of gene expression accumulate over the nematode body surface, presumably to detoxify ROS generated by the host defence response. Another example is the root-knot nematode MiMIF effectors, which were shown to be secreted by the nematode hypodermis and interfere with host annexin-mediated immune responses (Zhao *et al.*, 2019).

Identifying the host target is useful in understanding the effector's role *in planta*. However, in many cases, the nematode effectors are shown to suppress host immunity, but their plant targets and mechanism of action are not fully elucidated. For example, the root-knot nematode Mg01965 is secreted into the apoplast and suppresses the plant basal immunity, but the precise mechanism of function is unknown (Zhuo *et al.*, 2019).

There are only a handful of nematode effectors that have been characterized as avirulence proteins (Avr, recognized by host R-proteins). One example comes from the venom allergen-like protein (VAP) family. The VAP are found in a variety of PPN, but one cyst nematode VAP called VAP1 can act as an Avr in the presence of the immune receptor Cf-2 (Gao *et al.*, 2001; Vanholme *et al.*, 2006; Lin *et al.*, 2011; Yan *et al.*, 2012; Duarte *et al.*, 2014; Lozano-Torres *et al.*, 2014). In a susceptible plant, the cyst nematode VAP1 plays a role in suppressing basal immune responses by binding to an apoplastic papain-like cysteine protease (PLCP) called $Rcr3^{pim}$ (Lozano-Torres *et al.*, 2014). However, when the resistance gene *Cf-2* is also present in the host, the VAP1-induced perturbations of $Rcr3^{pim}$ are detected by the R-protein and an HR is triggered (Lozano-Torres *et al.*, 2012, 2014).

Nematode effectors that facilitate feeding site formation

Nematodes secrete a cocktail of effector proteins, and the majority of characterized nematode effectors play important roles in plant immune suppression, likely owing to the defined assays available for measuring plant defence responses (Bali and Gleason, 2024). It is likely that nematode effectors manipulate multiple host responses at once. Some effectors have been characterized to affect host metabolism. For example, nematodes express parasitism genes that code for chorismate mutase, a fundamental enzyme that converts chorismate to prephenate (Romero *et al.*, 1995; Lambert *et al.*, 1999; Huang *et al.*, 2005), and isochorismate mutase (ICM) (Bauters *et al.*, 2020). Chorismate-derived compounds include auxins and salicylic acid (SA), and secondary metabolites derived from prephenate include flavonoids that can act as auxin transport inhibitors (Gheysen and Mitchum, 2009). The activity of chorismate mutase represents a key mechanism that determines the cell balance of compounds such as phenylalanine, tyrosine and tryptophan aromatic amino acids (Romero *et al.*, 1995). It has been suggested that chorismate mutases secreted by the nematode could reduce the cell cytoplasmic pool of chorismate, thereby increasing chorismate exports from plastids to cytoplasm, which in turn would result in an effective reduction in the synthesis of metabolites (auxins and SA) that depend upon the chorismate produced in the plastid (Djamei *et al.*, 2011). It is considered that a reduction in cell SA or phenylpropanoids, caused in response to the chorismate mutase injected by the nematode into the cytoplasm of the host cell, could result in a reduction of defence regulation mechanisms against the invading nematode (Bauters *et al.*, 2021). ICM catalyses the hydrolysis of isochorismate to 2,3-dihydro-2,3-dihydroxybenzoate (DDHB) to limit the flow of isochorismate into SA biosynthesis, and thus, the nematode ICM may interfere with SA biosynthesis (Bauters *et al.*, 2021).

Annexin genes that are expressed in the pharyngeal gland (Fioretti *et al.*, 2001; Gao *et al.*, 2003) represent a large family that codes for calcium-dependent phospholipid-binding proteins, proteins which have a broad range of functions. Annexin-like effectors have been identified from PPN, including *H. glycines* (Gao *et al.*, 2003), *H. avenae* (Chen *et al.*, 2015) and *H. schachtii* (Patel *et al.*, 2010). The data suggest that these effectors mimic plant annexins and manipulate the plant responses to promote parasitism. In addition to mimicking plant annexins, there is also at least one effector from *M. incognita*, called MiMIF, that is secreted from the nematode hypodermis and can directly bind to plant annexins and promote parasitism by interfering with plant immune responses (Zhao *et al.*, 2020).

The nematode effectors that mimic annexins are an example of host mimicry, a strategy employed by PPN to exploit the plant's endogenous functions to promote parasitism. Another such example would be the cyst nematode CLE-like peptides. These nematode peptides share sequence similarity to plant CLAVATA3/ESR (CLE) ligand peptides. The CLE-like effectors can be processed by the plant protein system and secreted like a plant CLE to support cyst nematode parasitism

(Mitchum *et al.*, 2012). There are several examples of nematode-secreted peptide mimics (e.g., CLE-like, IDA) and proteins (e.g., chorismate mutase) that facilitate PPN parasitism (Goverse and Mitchum, 2022).

Some parasitism genes encode proteins that have a direct effect on the host cytoskeleton. The root-knot nematode effector PFN3 was shown to be secreted and to disrupt actin filaments (Caillaud *et al.*, 2008a; Leelarasamee *et al.*, 2018). Cyst nematodes have SPRY domain-containing proteins, called SPRYSECs, which have been shown in *Globodera* to suppress PTI and ETI, and in one case, the effector GpSPRY-414-2 targets microtubule-associated proteins of the cytoskeleton (Mei *et al.*, 2018).

Some nematode effectors are utilizing the host proteosome, as suggested by the fact that effector genes that code for ubiquitin are expressed in the pharyngeal gland cells of *H. glycines* (Gao *et al.*, 2003). Ubiquitins are small proteins whose function is to target other proteins for destruction, a process known as proteolysis. Ubiquitin molecules anchor themselves to the protein to be destroyed and then move towards the proteosome, where proteolysis is carried out. This system not only functions to degrade abnormal proteins but also involved in other processes, such as cell cycle control, protein traffic, organelle biogenesis or transcription control. Nematode ubiquitins could also be involved in the polyubiquitination of target proteins in the host cells, so that, through degradation, the nematode can modulate defence and cell signalling and, in this way, either influence the cell cycle of the host cell or provide itself with an appropriate substrate from which to obtain nutrients. One example of a nematode ubiquitin effector is from *G. pallida*. The effector GpRHA1B is a ubiquitin ligase that targets resistance genes for degradation, as well as suppressing plant basal immune responses (Kud *et al.*, 2019). Recently, a ubiquitin carboxyl extension protein (UBCEP12) effector was identified from the potato cyst nematode *G. rostochiensis*. This effector is processed in the plant into free ubiquitin and a peptide that suppresses basal immunity (Chen *et al.*, 2013; Chronis *et al.*, 2013).

Other genes of the nematode code for proteins of the family 14-3-3. In *Meloidogyne*, the genes that code for two isoforms of the 14-3-3 protein were cloned, and the analysis of the *in situ* hybridization indicated that one isoform (14-3-3a) was expressed in the genital primordium, and the other (14-3-3b) was expressed inside the dorsal pharyngeal gland (Jaubert *et al.*, 2004). Interestingly, one of the key aspects of this effector is the lack of the predicted signal peptide sequence for secretion, a bioinformatic hallmark of effectors. It is likely that nematodes utilize alternative secretion pathways for some effectors. Different members of the 14-3-3 family are associated with parasites and have diverse functions in primary metabolism, such as in the regulation of responses to cell stress and defence, in organelle traffic and in the cell cycle, where 14-3-3 often acts as an interacting essential protein or a chaperone (del Mar Siles-Lucas and Gottstein, 2003). Chaperone proteins bind to their client proteins to help their folding, assembling and intracellular transport to other parts of the cell where the proteins perform their function. Some nematode effectors are chaperones, and some effectors directly target host enzymes. Some of the cyst nematode effectors target 'interacting plant kinases' (IPK) to utilize their phosphorylation activities on host proteins. The activities modulate host responses that ultimately facilitate infections (Hewezi *et al.*, 2015; Yang *et al.*, 2022).

Effectors that reprogramme host genetic expression

Some effectors contain a nuclear localization signal (NLS) (Gao *et al.*, 2003; Huang *et al.*, 2003) and are trafficked to the host cell nucleus, thus suggesting a potential regulatory control of the specialized feeding cells induced by the nematode in the host (Jaouannet *et al.*, 2012; Jaouannet and Rosso, 2013; Harris *et al.*, 2023; Molloy *et al.*, 2023). For example, in the subventral pharyngeal gland cells of *M. incognita*, there is a gene encoding the peptide 16D10. This peptide, featuring a CLE domain, is characteristic of the SCARECROW (SCL) family of transcription factors. These factors are predominantly active in plant roots, where they play a crucial role in regulating growth pathways and cell differentiation (Huang *et al.*, 2006). Work on 16D10 provides the first evidence that a product of a nematode parasitism gene could regulate genetic activity by binding to the plant's intracellular transcription factor, thereby modulating the growth of root cells during the initial development of the feeding site. Another example of a nuclear-localized effector is GLAND4, which binds to and acts as a suppressor of lipid transfer protein (*LTP*) gene expression (Barnes *et al.*, 2018). Some nuclear-targeted effectors have been shown to be involved in mRNA processing and mediating alternative gene splicing in the nucleus (Mejias *et al.*, 2021).

Host cell reprogramming also occurs through nematode effectors that alter the epigenomes of the host plant (Hewezi, 2020). The *H. schachtii* effector Hs32E03 modulates the acetylation of histones in plants, which is important in controlling host gene expression to support parasitism (Vijayapalani *et al.*, 2018).

18.3. Structural, Metabolic and Physiological Changes in the Host Cells (Giant Cells and Syncytia)

Cells destined to form the specialized feeding sites exhibit changes in structure, metabolism and physiology (Hewezi and Baum, 2013). They function as transfer cells with cell wall and membrane finger-like projections along the xylem wall, thereby facilitating water transport towards the feeding cell. The high metabolic activity and constant extraction of nutrients by the nematode transform these host plant cells into metabolic sinks. For root-knot nematodes, the nematodes choose 6-10 host cells to convert into 'giant cells'. These cells are manipulated by the nematode effectors and will enlarge in size. The nuclei of these cells undergo successive rounds of mitosis and endo-reduplication cycles, but with absent or incomplete cytokinesis, which leads to enlarged, multinucleate giant cells (de Almeida Engler *et al.*, 2015). The giant cells also exhibit dense cytoplasm filled with numerous organelles, small vacuoles and cell wall ingrowths (Rodiuc *et al.*, 2014). The cyst nematode feeding sites are called syncytia. They are formed when the nematode chooses one plant cell as the initial feeding site. Neighbouring plant cells are fused to this cell through the dissolution of the cell walls. The syncytium can contain hundreds of plant cells (Grundler *et al.*, 1998). Like giant cells, the syncytium exhibits increased cytoplasmic density, a proliferation of organelles and cell wall modifications, such as thickening and ingrowths (de Almeida Engler *et al.*, 2004).

During nematode infections, the feeding sites (giant cells and syncytia) undergo dramatic transcriptional changes (Gheysen and Fenoll, 2002). Many of the over-expressed genes are involved in modifications of the cell wall or biosynthesis of secondary metabolites that are crucial for the establishment of compatible parasitic interaction. For example, genes involved in the strengthening of the cell wall (extensin, lignin biosynthesis and peroxidase genes) are over-expressed in the plant during infections. These genes may be involved in the extensive modifications of the cell that are fundamental in the formation of the nematode's specialized feeding sites (Gheysen and Mitchum, 2009). In the same way, the accumulation of flavonoids, isoflavonoids and other phenolic compounds are a typical response of the plant, although in compatible interaction, flavonoid inhibition of auxin transport could favour the accumulation of auxin and subsequently the development of the feeding site (Gheysen and Mitchum, 2009). The nematodes must suppress plant defences during a compatible interaction, and they use their effectors to cause a down-regulation of defence-regulated responses. One example is the suppression of ROS production by the *M. javanica* effector called MjTTL5, which is involved in ROS scavenging. In cyst nematodes, there is an interesting phenomenon describing the neofunctionalization of the nematode's own housekeeping genes into effectors; the cyst nematode housekeeping glutathione synthetase (GS) family of enzymes were repurposed by the nematode to act as secreted effectors that modulate redox homeostasis in the plant (Lilley *et al.*, 2018).

18.3.1. Cytoskeleton

Cells that will become the specialized feeding site undergo changes in the cytoskeleton. Both giant cells and syncytia exhibit rearrangements and fragmentation of the plant cytoskeleton. In the giant cells, the cortical actin filaments become denser while the cytoplasmic actin appears fragmented. These changes in actin filament morphology are coupled with changes in actin and actin-regulating genes. Actin genes are highly expressed in these sites, where a considerable increase in cell volume occurs, requiring an internal extensive transport (de Almeida Engler *et al.*, 2004). Formin *AtFH6* was expressed two days after inoculation with root-knot nematodes, and the protein accumulated uniformly throughout the plasma membrane (Favery *et al.*, 2004). Therefore, it was suggested that the distribution of protein AtFH6 could be involved with the isotropic growth of the giant cells. The fragmented actin could be a result of the increased expression of actin depolymerizing genes (ADF) in giant cells. ADF/cofilins sever actin filaments and increase the rate of actin depolymerization. Recently, specific ADF genes were shown to be up-regulated in *M. incognita*-infected cucumber (*Cucumis sativus* L.) roots (Liu *et al.*, 2016b). Interestingly, the root-knot nematodes secrete an effector called profilin 3 (MiPFN3), which affects the polymerization of soluble actin (Leelarasamee *et al.*, 2018). The data suggest that root-knot nematodes are modulating the actin network to facilitate their giant cell formation.

The reorganization of the microtubule network is also critical for giant cell and syncytia formation. The cytoskeleton protein 65-3 (MAP-65-3) is essential for cytokinesis and is required for giant cell ontogenesis (Caillaud *et al.*, 2008b), and the gene that codes for it was over-expressed in giant cells. Moreover, work on the *Arabidopsis* γ-tubulin complex showed that the complex helps control microtubular network remodelling during the development of root-knot and cyst nematode feeding cells (Banora *et al.*, 2011; Różańska *et al.*, 2018). The nematode may help the microtubule reorganization with effectors. For instance, the effector called Minc00344 can interact directly with HUB-10, a kinesin light chain-related protein that is involved in maintaining the stability of cortical microtubules (Godinho Mendes *et al.*, 2021).

18.3.2. Cell Wall

The nematodes secrete effectors that include cell wall-degrading enzymes that facilitate their penetration and migration through the root tissues. In addition, a wide array of host gene-encoding enzymes responsible for modifying plant cell walls display distinctive regulation within giant cells and syncytia. Giant cells are formed by hypertrophy of an individual parenchymal cell, in contrast to a syncytium, which is formed by progressive dissolving of the cell wall and the fusion of adjacent cells. Although giant cells and syncytia differ in the degree of degradation that occurs in the cell wall, both types of feeding structure require loosening of the cell wall for expansion and subsequent strengthening and thickening through callose and lignin deposition in order to support the increase in turgor pressure inside the feeding structures (Bohlmann and Sobczak, 2014; Bozbuga *et al.*, 2018). The extensive structural modifications of the cell wall, e.g., loosening, dissolution, growth and thickening, are mediated by a variety of cell wall-modifying enzymes (CWME), which are either produced through nematode-induced regulation of host genes or injected into the plant as nematode effectors (e.g., pectate lyase from cyst nematodes).

Host proteins associated with cell wall loosening, including expansins, cellulases, pectate lyases and polygalacturonases, play crucial roles in the formation of feeding sites by root-knot and cyst nematodes. Expansins play an important role in root-knot and cyst nematode infections by weakening the bonds between cellulose and hemicellulose (Jammes *et al.*, 2005; Wieczorek *et al.*, 2006; Bohlmann and Sobczak, 2014). Plant expansins were found to be differentially expressed in root-knot and cyst nematode-infected roots (Jammes *et al.*, 2005; Gal *et al.*, 2006; Wieczorek *et al.*, 2006).

Cellulases are enzymes that catalyse the hydrolysis of 1,4-β-glucosidic linkages between glucose residues in cellulose; these enzymes function in a broad range of developmental processes that involve changes to the architecture of the cell wall. During feeding site formation for both root-knot and cyst nematodes, there is a specific up-regulation of host cellulase genes (Wang *et al.*, 2007; Rai *et al.*, 2015). For example, the genes encoding β-1,4-endoglucanases exhibit significant over-expression in root tissue infected by either root-knot or cyst nematodes compared to non-infected roots (Goellner *et al.*, 2001; Karczmarek *et al.*, 2008; Wieczorek, 2015). A knockdown of two endo-1,4-β-glucanases showed reduced female development, further implicating the role of plant cellulases in feeding site formation and nematode parasitic success (Wieczorek, 2015).

Pectin is a component of the cell wall that needs to be modified or degraded during feeding site formation. Over-expression of the pectin acetylesterase (*PAE*) gene was observed in *Arabidopsis* at the beginning of the formation of the giant cells, syncytium and in cells around the nematode (Vercauteren *et al.*, 2002). Pectin degradation can occur through the action of host polygalacturonases and pectate lyases; both host enzymes have been observed to be up-regulated in plants during nematode infections (Szakasits *et al.*, 2009; Wieczorek *et al.*, 2014).

The expression of endo-glucanases and PAE in both root-knot nematode and cyst nematode feeding cells suggests that similar mechanisms are at play in loosening the plant cell walls. However, the varying expression of specific gene members within CWME families could lead to differences in the extent of cell wall loosening observed in each feeding site type (giant cell *vs* syncytium). The development of the feeding sites is likely a complex process requiring coordinated regulation by various types of plant CWME and the distinct expression patterns of individual gene family members.

One of the modifications of the cyst and root-knot nematode feeding sites is the formation of pronounced ingrowths at the interface with xylem vessels. These ingrowths likely help expand the surface area of the plasma membrane and may help facilitate solute transfer in and out of the feeding cells (Wieczorek, 2015). In *Arabidopsis*, it was found that two genes involved in cell wall polysaccharide production (UDP-glucose dehydrogenase genes *UGD2* and *UGD3*) are needed to produce syncytial cell wall ingrowths (Siddique *et al.*, 2012).

18.3.3. Transport Pathways of Nutrients and Water

As sedentary endoparasitic plant nematodes develop, they continuously take nutrients from metabolically active feeding sites. Because of this, the feeding sites represent an important place that demands sugars and amino acids, which are imported either symplastically from the phloem through plasmodesmata (Hofmann *et al.*, 2010; Xu *et al.*, 2021) or apoplastically via trans-membrane transport proteins (Hammes *et al.*, 2006). In early experiments, galls induced by root-knot nematodes were found to be connected by plasmodesmata to the phloem (Dorhout *et al.*, 1993). However, it should be noted that whether giant cells are connected to the phloem by plasmodesmata may be dependent on the type of plant and the age of the infection. For example, work on *Arabidopsis* suggested that the giant cells are **symplastically** isolated from their neighbouring cells, and sucrose import into the giant cells is mediated by the sucrose transporter gene *AtSUC4* (Hoth *et al.*, 2008; Hofmann *et al.*, 2009). However, in infected tobacco, the giant cells exhibited plasmodesmata in the walls between the giant cells and neighbouring cells (Hofmann *et al.*, 2010). For syncytia, there is more consistency in the plasmodesmata phenotype between plants. For both *H. schachtii* in *Arabidopsis* and *Globodera tabacum* and *H. schachtii* in tobacco, there are few plasmodesmata in the walls of the syncytia at early stages, but the number increases at later stages (Hofmann *et al.*, 2010).

Supply of the large volumes of water required by the feeding sites is possible thanks to high over-expression of genes that code for water channel proteins (aquaporins). For example, the gene *TobRB7* in tobacco was induced in giant cells 4-35 days after inoculation (Opperman *et al.*, 1994), and there was also a high over-expression of the gene in tomato giant cells (He *et al.*, 2005). Similarly, in the syncytium induced by *H. glycines* on soybean, the expression of the gene *GmPIP2.2*, which codes for an aquaporin, was increased sevenfold (Klink *et al.*, 2005).

18.3.4. Changes in Metabolic Activity

Root-knot and cyst nematode feeding sites have been shown to be hyper-metabolically active and serve as nutrient sinks for the nematodes (Bird and Loveys, 1975). Genes and pathways related to primary metabolism are noticeably increased in expression in giant cells and syncytia. In the syncytium, as well as in giant cells, sugar levels and metabolism rates increase considerably via the glycolytic and pentose phosphate pathways. Through glycolysis, glucose is oxidized to pyruvic acid to generate energy in the form of adenosine triphosphate (ATP) and nicotinamide adenine dinucleotide (NADH) and intermediary compounds of three and six carbons for other metabolic pathways.

In *Arabidopsis*, the gene phosphoglycerate mutase/bisphosphoglycerate mutase (*PGM/bPGM*) was over-expressed early in the development of the giant cells and syncytium, and its expression increased through the development of the feeding site (Mazarei *et al.*, 2003). PGM catalyses key steps in glycolysis and gluconeogenesis. In soybean, over-expression of genes that code for enzymes that participate in each step of the glycolytic and pentose phosphate pathways was also detected (Ithal *et al.*, 2007). This pathway has several functions in plant metabolism, for example, in maintaining the pool of reduced molecules in the form of nicotinamine adenine dinucleotide phosphate (NADPH) to protect against stress. In the pentose phosphate pathway, intermediary carbohydrates are produced, such as ribose-5-phosphate and deoxyribose, for the synthesis of nucleic acids, and erythrose-4-phosphate, a metabolite precursor of the shikimate pathway through which phenolic compounds are produced, such as aromatic amino acids for the synthesis of proteins and indol acetic acid (IAA), flavonoids, alkaloids, lignin and other secondary metabolites. The pentose phosphate pathway is essential for maintaining the feeding site cells, and Favery *et al.* (1998) reported that in *Arabidopsis*, the gene *RPE*, which codes for ribulose-3-phosphate-epimerase (RPE), an enzyme that catalyses the reversible conversion of ribulose-5-phosphate and xylulose-5-phosphate, was over-expressed in giant cells and during the late stages of syncytium development, thus suggesting an increase in flow through the pentose phosphate pathway.

An investigation into glutathione (GSH) and homoglutathione (hGSH) metabolism revealed their critical importance in facilitating nematode infection in *Medicago truncatula* (Baldacci-Cresp *et al.*, 2012). Diminishing the levels of hGSH hindered the formation of nematode egg masses and altered the nematode sex ratio in the next generation. Furthermore, analyses of gene expression and metabolites indicated significant alterations in starch and gamma-aminobutyrate metabolism, as well as changes in malate and glucose levels within (h)GSH-depleted galls. These findings underscore the vital role of adapted GSH metabolism in gall development and function, while highlighting how depletion of the GSH content disrupts nematode reproduction and modifies the sex ratio.

Metabolic analyses of syncytia formed by cyst nematodes and giant cells induced by root-knot nematodes reveal common features, including elevated sucrose and other sugars, as well as increased levels of various amino acids and phosphorylated metabolites (Hofmann and Grundler, 2007; Hofmann *et al.*, 2007). The alterations in giant cell primary metabolism mirror those observed in syncytia, suggesting functional similarities between both feeding sites despite their distinct origins (Kyndt *et al.*, 2013). Results of recent experiments on sugar-processing enzyme families invertase (INV) and sucrose synthase (SUS) in *Arabidopsis* further support this notion of commonalities in feeding sites, since disrupting the function of these enzymes enhances susceptibility to both cyst and root-knot nematodes (Cabello *et al.*, 2013). SWEET sucrose transporters were shown to be induced in root-knot nematode giant cells and in soybeans infected with *Rotylenchulus reniformis* (Redding *et al.*, 2018; Zhou *et al.*, 2023), indicating that these genes are involved in broad host-nematode interaction.

In addition to sugars, amino acids are also essential in the nematode diet, leading to changes in the expression of amino acid transporters during nematode infections. The amino acid permease (AAP) family of amino acid transporters is induced in syncytia (Gheysen and Fenoll, 2002; Grunewald *et al.*, 2009; Elashry *et al.*, 2013). In the *Arabidopsis*/root-knot nematode interaction, AAP3 and AAP6 play significant roles, as knock-out plants exhibit significantly reduced susceptibility compared to wild-type plants (Marella *et al.*, 2013).

18.3.5. Growth Hormones

Evidence suggests that various plant hormones, including auxin, cytokinin and ethylene, play roles in the interaction between plants and nematodes. Many cellular changes triggered by nematodes, such as endo-reduplication, akinetic mitosis, cell elongation, cell wall degradation and lateral root development, can be partially replicated through the application of the auxin 3-IAA. Auxins regulate the expression of genes involved in the formation of giant cells and syncytia, including those related to the cell cycle, expansin synthesis and cell wall-degrading enzymes. Increase in auxin content at feeding sites may stem from the secretion of auxin-like compounds by nematodes or from enhanced plant auxin biosynthesis and transport. The localized auxin increase at feeding sites can be attributed, at least in part, to the heightened expression of AUX1 and LAX3 auxin influx proteins during early infection stages (Grunewald *et al.*, 2009; Kyndt *et al.*, 2016). Additionally, the PIN-FORMED (PIN) family of auxin efflux carriers is also involved in feeding site formation. For example, PIN3 facilitates local auxin transport from the feeding sites towards surrounding cells to bring about syncytium expansion and gall initiation and development (Kyndt *et al.*, 2016).

There is also evidence for the role of cytokinins in nematode infections. Plants mutated in their cytokinin signalling are less susceptible to root-knot and cyst nematodes, indicating that cytokinins are needed for formation of the syncytia and giant cells (Lohar *et al.*, 2004; Siddique *et al.*, 2015; Shanks *et al.*, 2016; Dowd *et al.*, 2017). Interestingly, De Meutter *et al.* (2003) demonstrated that *Meloidogyne* spp. secrete biologically active cytokinins (De Meutter *et al.*, 2003), indicating that some of the cytokinin signalling that occurs during giant cell formation is due to nematode-derived cytokinin. Results have not been very clear regarding the roles of other hormones such as SA and jasmonic acid (JA) during plant-nematode interactions. These hormones play an important role in defence signalling after pathogen or insect attack (Cooper *et al.*, 2005; Bhattarai *et al.*, 2008; Wubben *et al.*, 2008; Gleason *et al.*, 2016), and there is evidence that they may help and/or hinder nematode infections, depending on the nematode species and host plant (Gheysen and Mitchem, 2019).

18.4. Plant Resistance to Nematodes

18.4.1. Disease Resistance Genes

Plant resistance is the genetic ability of plants to reduce or stop the growth, development or reproduction of a potential pathogen. This response involves the inability of the PPN to establish functional feeding sites and to subsequently develop into reproductive females on the host plant (Williamson and Kumar, 2006). Various studies have reported genes conferring resistance to different species of PPN, leading to promising germplasm resources, with wild crop relatives being particularly notable (Fuller *et al.*, 2008; Molinari, 2011). During the second half of the last century, numerous sources for nematode resistance have been identified in wild relatives of cultivated plants, and using these sources, nematode-resistant crop varieties have been developed (Williamson and Roberts, 2009). However, a subset of these resistance sources has not been utilized in breeding programmes due to genetic incompatibility with cultivated relatives or linkage drag of undesirable agronomic characteristics. The way the resistance is inherited also plays a role in the introgression of these sources into crops. Although the best-known root-knot and cyst nematode resistances are inherited in a dominant manner, inheritance of resistance to these nematode species does not conform to a single pattern. Nematode resistance could also be inherited as a recessive trait as well as additive traits and quantitative trait loci (QTL). The ease with which single dominant genes can be bred into crops has skewed the search for this type of resistance.

To date, various *R*-genes have been mapped and demonstrated to confer resistance to root-knot or cyst nematodes. Five *R*-genes encode proteins belonging to the largest class of R-proteins with central nucleotide-binding (NB) and C-terminal leucine-rich repeat (LRR) domains. These include *Mi-1.2*, *Mi-9*, *Hero*, *Gpa2* and *Gro1-4*, all of which are from solanaceous plant species (Milligan *et al.*, 1998; van der Vossen *et al.*, 2000; Ernst *et al.*, 2002; Paal *et al.*, 2004; Jablonska *et al.*, 2007). None of these proteins have localization signals, suggesting that they are localized in the cell cytoplasm. Except for Gro1-4, these R-proteins have N-terminal coiled-coil (CC) domains, while Gro1-4 has an N-terminal toll-interleukin-1 receptor (TIR) domain. $Hs1^{pro\text{-}1}$, the first cloned nematode *R*-gene, encodes for a protein with an extracellular leucine-rich region with a transmembrane domain, which bears little similarity to other R-proteins (Cai *et al.*, 1997; Hauge *et al.*, 2001; Lightfoot and Meksem, 2002). The Rhg1 (resistance to *Heterodera glycines* 1) locus in the soybean accession PI 88788 is the primary source of resistance to soybean cyst nematodes (SCN). This locus is a ≈30-kb region that contains four genes, which are repeated up to 10 times, creating a ≈300-kb resistance locus. Among the genes involved in resistance, three key ones are identified: one encodes a putative amino acid transporter (*Glyma.18G022400*); another encodes a soluble N-ethylmaleimide sensitive fusion (NSF)-associated protein (α-SNAP), *Glyma.18G022500*; and the third encodes a protein with minimal annotation but a putative wound-inducible domain (*Glyma.18G022700*) (Cook *et al.*, 2012). Peking-type resistance to SCN requires the presence of two key components: a low-copy version of the rhg1 resistance allele (rhg1-a) and a resistant allele at the Rhg4 locus. Both of these are necessary for effective resistance in Peking soybean (Shaibu *et al.*, 2020).

In tomato, *Mi-1*, *Mi-2*, *Mi-3*, *Mi-4*, *Mi-5*, *Mi-6*, *Mi-7*, *Mi-8*, *Mi-HT* and *Mi-9* genes have been identified (El-Sappah *et al.*, 2019). The homologous *Mi-1.2* gene confers resistance to root-knot nematodes, *M. arenaria*, *M. incognita*, *M. javanica*, *M. luci* and *M. ethiopica* (Dropkin, 1969; Roberts and Thomason, 1986; Veremis *et al.*, 1999), while *Mi-9* confers resistance to *M. arenaria*, *M. incognita* and *M. javanica* (Dropkin, 1969; Roberts and Thomason, 1986; Veremis *et al.*, 1999). The resistance mediated by *Mi-1.2* in tomato is inactive at soil temperatures above 28°C, whereas *Mi-9*-mediated resistance in the tomato wild relative species *Solanum arcanum* is functional at soil temperatures as high as 32°C (Dropkin, 1969; Ammati *et al.*, 1986; Veremis *et al.*, 1999). *Mi-HT*, conferring resistance to *M. incognita* at high temperatures (32°C) in line ZN17, was a single dominant gene and located on the short arm of chromosome 6 (Wang *et al.*, 2013). Molecular data indicated that *Mi-HT* might be an orthologue of *Mi-1* and *Mi-9* or a new gene (Wang *et al.*, 2013). The tomato gene *Hero A* confers resistance to *G. rostochiensis* populations and partial resistance to *G. pallida* pathotype $Pa_{2/3}$, while the potato *Gro1-4* confers resistance to *G. rostochiensis* pathotype Ro1. The sugar beet gene $Hs1^{pro\text{-}1}$ confers resistance to *H. schachtii*. The soybean gene *Rhg1* requires an unlinked modifier gene to confer resistance to *H. glycines* race 3 and race 14 across many soybean genotypes, while *Rhg4* confers resistance to *H. glycines* type 0 across a few soybean genotypes, indicating the requirement for additional *R*-genes or modifying genes for resistance (Matson and Williams, 1965; Concibido *et al.*, 2004).

R-genes are often found in multigene clusters within the genome, arising from tandem gene duplication events (Michelmore and Meyers, 1998; Ellis *et al.*, 2000; Dodds *et al.*, 2001; Martin *et al.*, 2003; Seah *et al.*, 2004; McDowell and Simon, 2006; Radwan *et al.*, 2008; Lawrence *et al.*, 2010). Gene duplication combined with diversifying selection is thought to give rise to novel recognition specificities in response to ever-changing pathogen/pest populations. Therefore, it is not surprising that four of the eight cloned nematode *R*-genes are also members of multigene clusters. The potato *Gpa2* gene lies within a cluster of four structurally related genes, of which two are described as functional *R*-genes. Interestingly, these two genes mediate recognition to phylogenetically distinct types of organisms. While *Gpa2* confers resistance to *G. pallida*, *Rx* confers resistance to the potato virus X (van der Vossen *et al.*, 2000). The *Gro-1* locus contains about 14 closely related genes, but only *Gro1-4* is known to confer disease resistance (Paal *et al.*, 2004).

Tomato *Hero A* is the only member with a described role in defence of a cluster of 14 NB-LRR genes localized within a 118 kb region (Ernst *et al.*, 2002). The *Mi* gene family consists of seven paralogs arranged into two clusters of three and four members, respectively (Vos *et al.*, 1998; Seah *et al.*, 2004; Seah *et al.*, 2007). This genomic organization is thought to have occurred via a genetic duplication event (Seah *et al.*, 2004). Two of the seven paralogs are pseudogenes and one encodes for a truncated protein indicating that they are unable to encode functional R-proteins. Although the rest of the genes seem to be intact and are transcribed, only one of these genes, *Mi-1.2*, confers broad-spectrum resistance, including resistance to three species of root-knot nematodes (Milligan *et al.*, 1998), potato aphids (Rossi *et al.*, 1998), whiteflies (Nombela *et al.*, 2003) and psyllids (Casteel *et al.*, 2006). The functional roles of the remaining three genes are unknown.

Davies and Elling (2015) reported a non-exhaustive list of resistance genes to root-knot nematodes and cyst nematodes in various host plants. Not all nematode *R*-genes are clustered. Several additional root-knot nematode resistance genes, *Mi-3*, *Mi-2* and *Mi-8*, from accession of the tomato wild relative species *Solanum peruvianum*, assort independently from *Mi-1.2*, indicating that they are located in distinct genomic region(s) (Veremis and Roberts, 1996). In fact, *Mi-3* is localized on chromosome 12 of tomato and assorts independently from *Mi-2* (Yaghoobi *et al.*, 1995). On the other hand, in chilli pepper, *Me4*, *Mech1* and *Mech2* genes are specific to certain *Meloidogyne* species or populations, whereas *Me1*, *Me3* and *Me7* are effective against a wide range of *Meloidogyne* species, including *M. arenaria*, *M. javanica* and *M. incognita*, the most common species in Mediterranean and tropical areas (Djian-Caporalino *et al.*, 2007). In various cases, the resistance conferred by these genes is overcome by *M. enterolobii* (= *M. mayaguensis*). *Meloidogyne enterolobii* is an emerging species and is considered highly dangerous by the European and Mediterranean Plant Protection Organization (EPPO) due to its increasing geographic distribution, wide host range and its remarkable aggressiveness in overcoming resistance to common root-knot nematode species (*M. incognita*, *M. arenaria* and *M. javanica*) (Brito *et al.*, 2007; Castagnone-Sereno, 2012; EPPO, 2014). Thus, this species successfully reproduces in crop genotypes, carrying resistance genes to major *Meloidogyne* species, such as cotton, sweet potato, tomatoes (*Mi-1* gene), potatoes (*Mh* gene), soybeans (*Mir1* gene), bell peppers (*N* gene), sweet peppers (*Tabasco* gene) and cowpea (*Rk* gene) (Castagnone-Sereno, 2012; Philbrick *et al.*, 2020). To date, certain crop genotypes have been reported to exhibit resistance to *M. enterolobii*, such as *Capsicum* spp. (Carrillo-Fasio *et al.*, 2020; Marques *et al.*, 2020), *Ipomoea batatas* (Rutter *et al.*, 2021; Schwarz *et al.*, 2021), *Psidium* spp. (Freitas *et al.*, 2014), *Vigna unguiculata* (Dareus *et al.*, 2021), *Solanum stramonifolium* (Pinheiro *et al.*, 2020) and *Glycine soja* (Schwarz and Gorny, 2024).

18.4.2. Interaction of R-protein with Nematode Avr Effector

As part of their immune system, plants possess a reservoir of resistance genes (*R*-genes), whose products function as sentinel proteins to protect them from pathogens. Once the plant detects the presence of an effector produced by a potential pathogen, the corresponding *R*-gene triggers a series of defensive responses (e.g., HR) at the infection site (Williamson and Kumar, 2006). HR is the typical manifestation of a gene-for-gene relationship, where, for each resistance gene in the host plant (*R*-gene), there is a corresponding gene in the pathogen that conditions pathogenicity (*Avr* gene) (Flor, 1971).

There are few examples of nematode Avr effectors that trigger HR via plant *R*-gene. In one case, Sacco *et al.* (2009) showed that the *G. pallida* effector called RBP-1 is recognized by the potato *R*-gene *Gpa2*. It is known that the N-terminal CC domain of the Gpa2 R-protein binds to Ran GTpase-activating protein 2 (RanGAP2), and Sacco *et al.* (2009) tested whether *Gp*-RBP-1 could be involved in this interaction (Rehman *et al.*, 2009; Sacco *et al.*, 2009). They used a commonly used plant system, *Nicotiana benthamiana*, for protein expression. Transiently expressing the *Gp*-RBP-1 protein in leaves of transgenic *N. benthamiana* expressing *Gpa2* resulted in activation of resistance responses and HR. This indicates that *Gp*-RBP-1 is the Avr effector recognized by the *Gpa2*-mediated resistance. Similarly, eliminating *RanGAP2* expression in *N. benthamiana* abolished the HR formed by the expression of Avr *Gp*-RBP-1 in the presence of the R-protein *Gpa2*, confirming an essential role for RanGAP2 in Gpa2 function (Sacco *et al.*, 2009). Interestingly, RanGAP2 contributes to defences mediated by two different resistance genes: the viral resistance gene *Rx1* in addition to *Gpa2*. A recent study found that both Gp-RBP-1 and PVX-CP can physically interact with RanGAP2. In fact, reducing the levels of RanGAP homologues (RanGAP1 and RanGAP2) significantly diminishes infection by cyst nematodes and PVX. One hypothesis is that the docking of the effectors on the RanGAPs brings them in close proximity to the immune receptor to trigger immune responses (Sukarta *et al.*, 2021).

18.4.3. Resistance Mechanisms

A common mechanism that affects the life cycle of PPN is the HR (Albuquerque *et al.*, 2010; Sato *et al.*, 2019). Hypersensitive reaction is a rapid and localized cell death that occurs at the site of pathogen penetration (Balint-Kurti, 2019); HR is a common phenomenon in higher plants and is induced by various types of pathogens (e.g., fungi, oomycetes, bacteria, viruses and nematodes). This defence mechanism is generally effective against biotrophic and hemibiotrophic pathogens but is less effective against those with a necrotrophic lifestyle (organisms that require dead host tissue to complete their life cycle, where HR can even be beneficial to them) (Balint-Kurti, 2019).

Davies and Elling (2015) explain that mechanisms of resistance to nematodes can be divided into two categories: those associated with HR and a HR-independent resistance response. According to them, HR is rapid and effective against root-knot nematodes; in contrast, resistance to cyst nematodes is characterized by a delayed response that occurs once a feeding site has been established. Cyst and root-knot nematodes are able to penetrate and migrate towards the vascular bundle of roots of resistant hosts as efficiently as in susceptible genotypes (Melillo *et al.*, 2006; Das *et al.*, 2008; Dhandaydham *et al.*, 2008; Cai *et al.*, 1997). In resistant plants, recognition of the nematode is thought to occur during initiation of the feeding site after they pierce host cells and inject pharyngeal gland secretions. In most cases, resistance to cyst and root-knot nematodes is characterized by the lack of successful formation of the specialized feeding site by the nematode because of a localized HR at this site. In terms of effects of the resistance on the nematode, responses vary from complete inhibition of nematode development to a significant reduction in the numbers of adult females and cysts. During the incompatible interactions, HR can be seen as early as 24 h after nematode inoculation. Early HR, near the head of the infective feeding-stage nematode, is a hallmark of *Mi-1.2*-mediated resistance in tomato (Riggs and Winstead, 1959; Dropkin, 1969; Paulson and Webster, 1972). In addition, necrosis has been observed in epidermal and cortical cells of root-knot nematode-resistant pepper roots; thus, Pegard *et al.* (2005) found that HR blocked migration, development and reproduction of *M. incognita, M. arenaria* and *M. javanica*, and HR was associated with chlorogenic acid accumulation. Hypersensitive reaction has also been evidenced in other incompatible interactions, such as *Solanum lycopersicum-M. incognita* (Melillo *et al.*, 2006), *S. sparsipilum-M. fallax* (Kouassi *et al.*, 2004), *Coffea arabica-M. incognita* (Albuquerque *et al.*, 2010), *C. canephora-M. exigua* (Silva *et al.*, 2013), *Avena sativa-M. incognita* (Marini *et al.*, 2016), *Oryza sativa-M. graminicola* (Phan *et al.*, 2018) and so on.

In contrast to the rapid HR response mediated by *Mi-1.2*, resistance can be manifest later after successful establishment of a feeding site. Infection of tomato plants containing *Hero A* with *G. rostochiensis* or *G. pallida*, potato plants containing *Gpa2* with *G. pallida* or sugar beet plants containing $Hs1^{pro1}$ with *H. schachtii* results in the establishment of a syncytium and allows the nematodes to become sedentary, a status that is indistinguishable from compatible interaction (Cai *et al.*, 1997; Holtmann *et al.*, 2000; Sobczak *et al.*, 2005; Koropacka, 2010). However, in *Hero A-*, *Gpa2- or* $Hs1^{pro1}$-containing plants, cells surrounding the syncytia eventually become necrotic, followed by the deterioration of the syncytia, preventing successful completion of the nematode life cycle. During these delayed resistance responses, the intruding juvenile nematodes predominantly develop into males, an epigenetic phenomenon observed under unfavourable conditions (Trudgill, 1967).

Resistance to root-knot nematodes can also occur without an obvious HR or delayed necrosis. In roots of resistant *Medicago truncatula* accession DZA045, infective juveniles of the root-knot nematode either die or develop as males (Dhandaydham *et al.*, 2008). In resistant cowpea containing the *R*-gene *Rk* and *M. incognita*, normal nematode feeding occurred, and nematodes developed up to later juvenile stages but never reached maturity (Das *et al.*, 2008). This diversity in mechanisms of resistance indicates the complex nature of plant interactions with nematodes.

18.5. Plant Defences against Nematodes

Plants defend themselves from attack by plant pathogens via pre-existing physical and chemical barriers (pre-formed or pre-infective defence), and those that are formed from the moment that interaction occurs between the nematode and the plant (induced defences) (Holbein *et al.*, 2016; Desmedt *et al.*, 2020). Among the first barriers plants have against pathogen entry are the cell wall and the cuticle. Regarding pathogens that penetrate the plant cell wall to access the nutrients of the plant protoplast, small wounds are created at the penetration sites through enzymatic or physical alteration of the plant cell wall (Hématy *et al.*, 2009; Underwood, 2012). Therefore, the integrity of the cell wall is a mechanism by which plants can detect pathogen attacks (Hématy *et al.*, 2009).

PPN have developed a protrusible stylet in their oral cavity for successful entry into host cells and subsequent intra- or intercellular migration. Nematodes use their stylet to pierce plant cell walls, thereby delivering glandular secretions into host cells for feeding. Reciprocal adaptations in plants have given rise to defence strategies, and as already mentioned, the first line of defence consists of the PTI and ETI systems, and downstream signalling (Goverse and Smant, 2014; Cui *et al.*, 2015; Holbein *et al.*, 2016). Basal defence in plants involves a broad and robust repertoire of rapid chemical responses, as mitogen-activated protein kinase (MAPK) cascades, calcium flux, release of ROS, phytohormone signalling, antimicrobial secondary metabolites, hydrolytic enzymes, protease inhibitors and local fortifications of plant cell walls by lignification and callose deposits (Goverse and Smant, 2014; Yuan *et al.*, 2021).

Plant hormone signalling pathways and the crosstalk between these pathways are responsible for regulating plant growth and development, as well as their defence responses (Berry and Argueso, 2022). Following attack by biotrophic or necrotrophic pathogens, plants respond by increasing the levels of SA or JA, and their extensive crosstalk (Berry and Argueso, 2022). SA regulates many genes encoding PR proteins, including PR1 and most acidic PR proteins, whereas the JA signalling pathway affects the expression of genes encoding the proteins defensin, thionin, PR3 and PR8 (Przybylska and Obrępalska-Stęplowska, 2020; Khan and Khan, 2021). Considering that PPN are biotrophic, it is common to assume that plant defences are mainly mediated by SA-sensitive pathways (Rutter *et al.*, 2022); however, recent studies have suggested that JA may play a dominant role in pathogen interactions with both monocotyledonous and dicotyledonous plants, including PPN (Nahar *et al.*, 2011; Przybylska and Obrępalska-Stęplowska 2020).

Expression analysis of defence-related genes in nematode-infected roots suggests that basal defence is activated during migration and around expanding feeding structures in both susceptible and resistant host plants, whereas it appears to be suppressed within feeding structures in susceptible host plants (Goverse and Smant, 2014). In this sense, Hamamouch *et al.* (2011) reported the disruption of five genes encoding PR proteins (PR-1 to PR-5) in *Arabidopsis thaliana* at 5, 9 and 14 days after infection with *M. incognita*; and demonstrated that up-regulation of PR-1 in transgenic *A. thaliana* plants reduced susceptibility to *M. incognita*. The function associated with PR-1 is cell wall reinforcement, PR-2 is a β-1,3 glucanase, PR-3 and PR-4 are chitinases and PR-5 is a thaumatin-like protein that alters cell permeability of fungal pathogens, although there may be other functions that are still unknown (Vidhyasekaran, 2008). In tomato plants, this type of alterations induced by *Meloidogyne* has also been documented (Bar-Or *et al.*, 2005; Molinari *et al.*, 2014). For instance, Portillo *et al.* (2013) found that in tomato at 1, 7 and 14 days after inoculation (dai) with *M. javanica*, the expression of the TPX1 gene (peroxidase) was reduced in the susceptible genotype 'Moneymaker', and the opposite was recorded in the resistant genotype ('Motelle'); furthermore, in transgenic tomato lines over-expressing TPX1, the reproduction of the nematode was severely affected. TPX1 (PR-9) codes for a peroxidase associated with the lignification of cell walls. This behaviour strengthens the assertion that the resistance and susceptibility of plants to a potential pathogen depends on the speed and magnitude with which the defence mechanisms are activated. Also, in the *Coffea arabica-M. incognita* interaction, Albuquerque *et al.* (2017) compared the transcription patterns of defence-related genes in resistant and susceptible genotypes after nematode challenge. They found that in susceptible plants, most genes were suppressed starting at 5 dai, when nematode feeding sites are induced. In resistant plants, gene expression was mainly activated at 6 dai, when a HR is detectable. Resistance was associated with the over-expression of genes encoding Miraculin (defence), RGLG1 (protein degradation), SENA (senescence-associated cell death), NLR (immune signalling), CaWRK11 (transcription factor) and OBP (cell wall modification).

Also, during nematode infection, there is up-regulation of other genes involved in wounding and defence response (Gheysen and Fenoll, 2002), including those that code for phenylalanine ammonia lyase (PAL) (Zhang *et al.*, 2023; Gupta *et al.*, 2024) and 3-hydroxy-3-methyl-glutaryl-CoA reductase (HMG-CoAr), key enzymes in the two main metabolic pathways (phenylpropanoid and isoprenoid pathways, respectively) through which metabolites are synthesized to build up the physical and chemical barriers with which plants defend themselves from nematode attack. The phenylpropanoid pathway branches into different routes where the synthesis of lignin and suberin involved in the strengthening and repairing of the cell wall, phenolic compounds, coumarins and isoflavonoid phytoalexins with antimicrobial activity and signalling compounds such as SA occurs (Shadle *et al.*, 2003; Gómez-Vásquez *et al.*, 2004; Zinoveva *et al.*, 2004; Pegard *et al.*, 2005). Tomczak *et al.* (2009) have questioned the efficacy of pre-existing physical barriers to impede penetration by PPN because they are provided with a stylet and have the capacity to secrete enzymes that degrade cell walls, thereby facilitating root penetration and subsequent migration of endoparasitic nematodes. In the tomato-*G. rostochiensis* interaction, Uehara *et al.* (2010) found that the transcript levels of PAL and Myb transcription factor-related genes were up-regulated at 3 dai in the incompatible interaction. A similar behaviour occurred with the transcription of genes encoding pyruvate decarboxylase (PDC) and alcohol dehydrogenase (ADH). In contrast, all the genes analysed were down-regulated in the compatible interaction. In a similar way, when they compared the expression levels of several PR protein genes in tomato roots between the incompatible and compatible interactions, they found that SA-dependent PR genes were induced in the incompatible interaction.

Other defence mechanisms have been reviewed by Desmedt *et al.* (2020), Gupta *et al.* (2023) and Gupta *et al.* (2024), where it is shown how the defence of PPN is multifaceted. These mechanisms involve phenolic compounds, terpenoids, saponins, benzoxazinoids, organosulfur compounds, alkaloids and glucosinolates. Various phenolic compounds, such as chlorogenic acid, trans-cinnamic acid, 2-hydroxy-naphthoic acid, caffeic acid and ferulic acid, are toxic to nematodes (Mahajan *et al.*, 1992; Wuyts *et al.*, 2006), although toxicity mechanisms are still unknown. As we know, resistance to nematodes is associated with a HR and with increases in the levels of phenolic compounds with antimicrobial properties; for example, the CM334 chilli pepper genotype exhibits thermostable resistance to *M. incognita*, *M. arenaria* and *M. javanica*, and such resistance is expressed as a HR (Pegard *et al.*, 2005). Pegard *et al.* (2005) suggest that CM334 could exhibit two types of defence, the first represented by physical barriers or the exudation of some nematostatic compound (which could explain the reduced percentage of root penetration recorded), while the second type of defence prevents the formation of galls and egg masses by blocking the migration, development and reproduction of the nematode; it was in the latter that the HR and the accumulation of chlorogenic acid were involved. In addition, in CM334 plants infected with *Nacobbus aberrans*, there was less PAL activity (Godínez-Vidal *et al.*, 2008; López-Martínez *et al.*, 2011), this being associated with lower amounts of soluble phenolic acids, particularly chlorogenic acid, and in *in vitro* assays, the phenolic acids caused the mortality of second-stage juveniles (J2) of *N. aberrans* (López-Martínez *et al.*, 2011).

Isoprenoid compounds synthesized by plants through the isoprenoid pathway, such as ristin, gossypol and tomatine phytoalexins, can be toxic to the nematode. Others, such as azadirachtin, saponins and sesquiterpene lactones, can have antifeeding effects, affect growth, moulting (azaradichtin, phytoecdysone and sesquiterpene lactoses) or block nematode ability to recognize the host (i.e., citral and menthol) (Zinoveva *et al.*, 2004; Desmedt *et al.*, 2020).

The ability of plants to produce phytoalexins in response to nematode invasion is directly and positively correlated with plant resistance to the parasite (Desmedt *et al.*, 2020). For instance, Veech and McClure (1977) observed that the expression of incompatibility in cotton (*Gossypium hirsutum*) to *M. incognita* was correlated with post-infection increments in the concentration of terpenoid aldehyde compounds, which accumulated at the infection site, and that their crude extract was toxic *in vitro* to *M. incognita* (Veech, 1978). In soybean, it was found that in the incompatible interaction of soybean 'Centenal' – *M. incognita* – there was a significant accumulation of glyceolin in the stele of incompatible roots (Kaplan and Keen, 1977), while in the compatible interaction with cultivars such as 'Pickett', glyceolin accumulation was slight.

Both soybean cultivars were compatible with *M. javanica*, and there was giant cell development and no significant glyceolin accumulation. Glyceolin at low concentrations had a nemastatic effect *in vitro* on *M. incognita* but not on *M. javanica*. Such results suggest that glyceolin accumulation could be responsible for incompatibility expression and that the plant's failure to accumulate the phytoalexin can be related with the compatible interaction (Kaplan and Keen, 1977). Also, in the interaction *Pratylenchus scribneri* – lima beans (*Phaseolus lunatus*), the start of production of the phytoalexin coumestrol coincided with the time at which the endoparasitic nematode *P. scribneri* was immobilized, and this production was maintained for 72 h (Rich *et al.*, 1977). It was also reported that in the compatible interaction chilli pepper CM-334 – *N. aberrans*, the concentrations of the capsidiol phytoalexin were similar to those in control plants (Godínez-Vidal *et al.*, 2010). The deleterious effect of capsidiol on *N. aberrans* was already demonstrated (Godínez-Vidal *et al.*, 2010).

Another active defence mechanism that is triggered during the infection by pathogens is the synthesis of PRs, such as β-1,3-glucanases, chitinases, peroxidases and PR1, among others. Information about β-1,3-glucanase induction in plant-nematode interactions is scarce and contradictory, and is seemingly dependent on the nematode species with which the interaction is established. For instance, in the interaction *G. pallida* – potato, in both compatible and incompatible interactions, the levels of β-1,3-glucanases were increased (Rahimi *et al.*, 1996), whereas Oka *et al.* (1997) reported that in the compatible interaction tomato – *M. javanica*, there was no increment in the accumulation of these enzymes. In a similar way, in the compatible interaction chilli pepper CM334 – *N. aberrans*, the levels of the gene *GLU* transcripts and corresponding β-1,3-glucanase activity were like those observed in non-inoculated plants, whereas such levels were increased significantly in the incompatible interaction CM334 – *P. capsici* (Fernández-Herrera *et al.*, 2012). The β-1,3-glucanases have been associated with plant defence against fungi and oomycetes, because they degrade polymers of the mycelial cell wall. Although glucanases do not have a direct effect on nematodes, their activity on the cell wall of host cells can generate endogen elicitors that warn neighbouring cells, a feature that could result in an indirect negative effect against the nematode. In roots of chilli pepper CM334 inoculated with *N. aberrans*, the expression levels of genes encoding peroxidases (*POX*) and PR1 protein (*PR1*) also increased in comparison to the control. However, transcript accumulation was significantly lower in comparison to the CM334 – *P. capsici* incompatible interaction. A slight accumulation of the peroxidase transcripts was associated with the small activity in the interaction CM334-nematode, in comparison to that found in the incompatible interaction, and was statistically equal to that of the non-inoculated group. It is not known what the level of involvement, if any, of the PR1 is in the defence against *N. aberrans*.

Usually, the production of PR proteins is considered an important defence mechanism, and they are linked to plant-pathogen resistance (Linthorst, 1991; Van Loon *et al.*, 2006). However, in the plant-nematode interaction literature, information is controversial, with some researchers indicating that there is no direct correlation between PR accumulation and degree of resistance (Rahimi *et al.*, 1996), while others report that such a relationship exists. Information about the protective effect of PR proteins against PPNs is also scarce.

During the infection by the nematode, some genes are up-regulated or down-regulated, depending upon whether the host is resistant or susceptible. For instance, Godínez-Vidal *et al.* (2010) found in the compatible interaction chilli pepper CM334 – *N. aberrans*, modifications in transcript accumulation of isoforms of the gene *hmg* responsible to code for HMGCoA-R, a key enzyme in the pathway of the terpenoid compounds, such as sterols and capsidiol. The expression of the gene *hmg1*, associated with the synthesis of phytosterols (fundamental in nematode development), was increased, while production of *hmg2*, associated with the synthesis of the phytoalexin capsidiol, was reduced. In addition to a lower expression of the gene *hmg2* in the interaction chilli pepper CM334 – *N. aberrans*, a reduction in other responses associated with defence, such as lower activity of the PAL (Godínez-Vidal *et al.*, 2008) and lower accumulation of total soluble phenols (López-Martínez *et al.*, 2011) in comparison to non-inoculated plants, was also found.

Genes involved in the strengthening of the cell wall, such as the extensin gene, genes for the biosynthesis of lignin and peroxidase, were up-regulated in a plant infected by a nematode. In a similar way, extensive modifications to the cell wall are also fundamental in the development of the feeding site. Accumulation of flavonoids, isoflavonoids and other phenolic compounds is a typical plant defence response. However, flavonoids inhibit auxin transport, and local production could trigger the accumulation of auxins, a growth regulator considered important to the development of the giant cells or syncytium. There is also evidence that specific plant defence pathways can be suppressed locally within specific sites. For example, various genes that code for key enzymes involved in the biosynthesis of JA, an important defence molecule, were found to be inhibited in the soybean syncytium induced by *H. glycines* (Ithal *et al.*, 2007).

18.6. Conclusions and Perspectives

Plant-nematode interactions are intricate and highly specialized, posing significant challenges due to the substantial agricultural losses they can cause. Understanding these interactions at the molecular level is crucial for developing new management strategies. As growers increasingly shift away from chemical controls, genetic resistance offers a more sustainable and environmentally friendly solution. While some *R*-genes have been cloned, providing insights into resistance signalling, the identification of *Avr* genes in PPN remains limited. This gap hinders our understanding of how nematodes trigger resistance responses or evolve to circumvent them. Additionally, the scarcity of resistance genes suggests the need for alternative strategies. One promising approach is to investigate how nematodes exploit plant susceptibility. By understanding how nematodes use effectors to bypass plant defences and manipulate plant responses to establish feeding sites, we can engineer plants with enhanced resistance. To achieve this goal, there are currently various techniques used to study the molecular basis of the plant-nematode interaction, such as high-throughput transcriptomics and genomics technologies and computational resources as well as gene cloning and silencing techniques. These approaches contribute to the continuous enrichment of our knowledge of this complex and sophisticated interaction. Also, approaches related to epigenetic studies (e.g., RRNA) and genome editing (i.e., CRISPR/Cas9) can also be essentially useful in our efforts to develop novel resistance to nematodes.

18.7. References

Abad, P., Gouzy, J., Aury, J.M., Castagnone-Sereno, P., Danchin, E.G., *et al.* (2008) Genome sequence of the metazoan plant-parasitic nematode *Meloidogyne incognita*. *Nature Biotechnology* 26, 909–915.

Albuquerque, E.V.S., Carencro, R.M.D.G., Costa, P.M., Gomes, A.C.M. M., Santos, M., *et al.* (2010) Resistance to *Meloidogyne incognita* expresses a hypersensitive-like response in *Coffea arabica*. *European Journal of Plant Pathology* 127, 365–373.

Albuquerque, E.V., Petitot, A.S., da Silva, J.P., Grossi-de-Sa, M.F. and Fernandez, D. (2017) Early responses of coffee immunity-related genes to root-knot nematode infection. *Physiological and Molecular Plant Pathology* 100, 142–150.

Ali, S., Magne, M., Chen, S., Côté, O., Stare, B.G., *et al.* (2015) Analysis of putative apoplastic effectors from the nematode, *Globodera rostochiensis*, and identification of an expansin-like protein that can induce and suppress host defenses. *PLoS One* 10, e0115042.

Ammati, M., Thomason, I.J. and McKinney, H.E. (1986) Retention of resistance to *Meloidogyne incognita* in *Lycopersicon* genotypes at high soil temperature. *Journal of Nematology* 18, 491–495.

Baldacci-Cresp, F., Chang, C., Maucourt, M., Deborde, C., Hopkins, J., *et al.* (2012) (Homo)glutathione deficiency impairs root-knot nematode development in *Medicago truncatula*. *PLoS Pathogens* 8, e1002471.

Bali, S. and Gleason, C. (2024) Unveiling the diversity: Plant parasitic nematode effectors and their plant interaction partners. *Molecular Plant-Microbe Interactions* 37, 179–189.

Balint-Kurti, P. (2019) The plant hypersensitive response: Concepts, control and consequences. *Molecular Plant Pathology* 20, 1163–1178.

Banora, M.Y., Rodiuc, N., Baldacci-Cresp, F., Smertenko, A., Bleve-Zacheo, T., *et al.* (2011) Feeding cells induced by phytoparasitic nematodes require γ-tubulin ring complex for microtubule reorganization. *PLoS Pathogens* 7, e1002343.

Barnes, S.N., Wram, C.L., Mitchum, M.G. and Baum, T.J. (2018) The plant-parasitic cyst nematode effector GLAND4 is a DNA-binding protein. *Molecular Plant Pathology* 19, 2263–2276.

Bar-Or, C., Kapulnik, Y. and Koltai H. (2005) A broad characterization of the transcriptional profile of the compatible tomato response to the plant parasitic root knot nematode *Meloidogyne javanica*. *European Journal of Plant Pathology* 111, 181–192.

Bauters, L., Kyndt, T., De Meyer, T., Morreel, K., Boerjan, W., *et al.* (2020) Chorismate mutase and isochorismatase, two potential effectors of the migratory nematode *Hirschmanniella oryzae*, increase host susceptibility by manipulating secondary metabolite content of rice. *Molecular Plant Pathology* 21, 1634–1646.

Bauters, L., Stojilković, B. and Gheysen, G. (2021) Pathogens pulling the strings: Effectors manipulating salicylic acid and phenylpropanoid biosynthesis in plants. *Molecular Plant Pathology* 22, 1436–1448.

Berg, R.H. and Taylor, C.G. (2009) *Cell Biology of Plant Nematode Parasitism*. Plant Cell Monograph (CELLMONO, Volume 15). Springer-Verlag Berlin, Heidelberg, Germany.

Berry, H.M. and Argueso, C.T. (2022) More than growth: Phytohormone-regulated transcription factors controlling plant immunity, plant development and plant architecture. *Current Opinion in Plant Biology* 70, 102309.

Bhattarai, K.K., Xie, Q.G., Mantelin, S., Bishnoi, U., Girke, T., *et al.* (2008) Tomato susceptibility to root-knot nematodes requires an intact jasmonic acid signaling pathway. *Molecular plant-microbe interactions* 21, 1205–1214.

Bigeard, J., Colcombet, J. and Hirt, H. (2015) Signaling mechanisms in pattern-triggered immunity (PTI). *Molecular Plant* 8, 521–539.

Bird, A.F. and Loveys, B.R. (1975) The incorporation of photosynthates by *Meloidogyne javanica*. *Journal of Nematology* 7, 111–113.

Bohlmann, H. and Sobczak, M. (2014) The plant cell wall in the feeding sites of cyst nematodes. *Frontiers in Plant Science* 5, 89. DOI: 10.3389/fpls.2014.00089.

Bozbuga, R., Lilley, C.J., Knox, J.P. and Urwin, P.E. (2018) Host-specific signatures of the cell wall changes induced by the plant parasitic nematode, *Meloidogyne incognita. Scientific Reports* 8, 17302.

Brito, J.A., Stanley, J.D., Kaur, R., Cetintas, R., Di Vito, M., *et al.* (2007) Effects of the *Mi-1*, N and Tabasco genes on infection and reproduction of *Meloidogyne mayaguensis* on tomato and pepper genotypes. *Journal of Nematology* 39, 327.

Cabello, S., Lorenz, C., Crespo, S., Cabrera, J., Ludwig, R., *et al.* (2013) Altered sucrose synthase and invertase expression affects the local and systemic sugar metabolism of nematode-infected *Arabidopsis thaliana* plants. *Journal of Experimental Botany* 65, 201–212.

Cai, D., Kleine, M., Kifle, S., Harloff, H.J., Sandal, N.N., *et al.* (1997) Positional cloning of a gene for nematode resistance in sugar beet. *Science* 275, 832–834.

Caillaud, M.C., Abad, P. and Favery, B. (2008a) Cytoskeleton reorganization, a key process in root-knot nematode-induced giant cell ontogenesis. *Plant Signaling & Behavior* 3, 816–818.

Caillaud, M.C., Lecomte, P., Jammes, F., Quentin, M., Pagnotta, S., *et al.* (2008b) MAP65-3 microtubule-associated protein is essential for nematode-induced giant cell ontogenesis in *Arabidopsis*. *Plant Cell* 20, 423–437.

Carrillo-Fasio, J.A., Martínez-Gallardo, J.A., Ayala-Tafoya, F., López-Orona, C.A., Allende-Molar, R., *et al.* (2020) Screening for resistance to *Meloidogyne enterolobii* in *Capsicum annuum* landraces from Mexico. *Plant Disease* 104, 817–822.

Castagnone-Sereno, P. (2012) *Meloidogyne enterolobii* (= *M. mayaguensis*): Profile of an emerging, highly pathogenic, root-knot nematode species. *Nematology* 14, 133–138.

Casteel, C.L., Walling, L.L. and Paine, T.D. (2006) Behavior and biology of the tomato psyllid, *Bactericerca cockerelli*, in response to the *Mi-1.2* gene. *Entomologia Experimentalis et Applicata* 121, 67–72.

Chen, C., Liu, S., Liu, Q., Niu, J., Liu, P., *et al.* (2015) An ANNEXIN-like protein from the cereal cyst nematode *Heterodera avenae* suppresses plant defense. *PLoS One* 10, e0122256.

Chen, J., Hu, L., Sun, L., Lin, B., Huang, K., *et al.* (2018) A novel *Meloidogyne graminicola* effector, MgMO237, interacts with multiple host defence-related proteins to manipulate plant basal immunity and promote parasitism. *Molecular Plant Pathology* 19, 1942–1955.

Chen, S., Chronis, D. and Wang, X. (2013) The novel GrCEP12 peptide from the plant-parasitic nematode *Globodera rostochiensis* suppresses flg22-mediated PTI. *Plant Signaling & Behavior* 8(9), e25359. DOI: 10.4161/psb.25359.

Chen, S., Tran, T.T.T., Yeh, A.Y., Yang, H., Chen, J., *et al.* (2024) The *Globodera rostochiensis* Gr29D09 effector with a role in defense suppression targets the potato hexokinase 1 protein. *Molecular Plant-microbe Interactions* 37, 25–35.

Chopra, D., Hasan, M.S., Matera, C., Chitambo, O., Mendy, B., *et al.* (2021) Plant parasitic cyst nematodes redirect host indole metabolism via NADPH oxidase-mediated ROS to promote infection. *New Phytologist* 232, 318–331.

Chronis, D., Chen, S., Lu, S., Hewezi, T., Carpenter, S.C.D., *et al.* (2013) A ubiquitin carboxyl extension protein secreted from a plant-parasitic nematode *Globodera rostochiensis* is cleaved in planta to promote plant parasitism. *The Plant Journal* 74, 185–196.

Concibido, V.C., Diers, B.W. and Arelli, P.R. (2004) A decade of QTL mapping for cyst nematode resistance in soybean. *Crop Science* 44, 1121–1131.

Cooper, W.R., Jia, L. and Goggin, L. (2005) Effects of jasmonate-induced defenses on root-knot nematode infection of resistant and susceptible tomato cultivars. *Journal of Chemical Ecology* 31, 1953–1967.

Cosgrove, D.J. (2000) Loosening of plant cell walls by expansins. *Nature* 407, 321–326.

Cui, H., Tsuda, K. and Parker, J.E. (2015) Effector-triggered immunity: From pathogen perception to robust defense. *Annual Review of Plant Biology* 66, 487–511.

Cook, D.E., Lee, T.G., Guo, X., Melito, S., Wang, K., *et al.* (2012) Copy number variation of multiple genes at Rhg1 mediates nematode resistance in soybean. *Science* 338(6111), 1206–1209.

Dareus, R., Porto, A.C.M., Bogale, M., DiGennaro, P., Chase, C.A., *et al.* (2021) Resistance to *Meloidogyne enterolobii* and *Meloidogyne incognita* in cultivated and wild cowpea. *HortScience* 56, 460–468.

Das, S., DeMason, D.A., Ehlers, J.D., Close, T.J. and Roberts, P.A. (2008) Histological characterization of root-knot nematode resistance in cowpea and its relation to reactive oxygen species modulation. *Journal of Experimental Botany* 59, 1305–1313.

Davies, L.J. and Elling, A.A. (2015) Resistance genes against plant-parasitic nematodes: A durable control strategy? *Nematology* 17, 249–263.

Davis, E., Haegeman, A. and Kikuchi, T. (2011) Degradation of plant cell wall by nematodes. In: Jones, J., Gheysen, G. and Fenoll, C. (eds), *Genomics and Molecular Genetics of Plant-Nematode Interactions*. Springer, Dordrecht, Netherlands, pp. 255–272.

Davis, E.L., Hussey, R.S. and Baum, T.J. (2004) Getting to the roots of parasitism by nematodes. *Trends in Parasitology* 20, 134–141.

de Almeida Engler, J., Van Poucke, K., Karimi, M., De Groodt, R., Gheysen, G., *et al.* (2004) Dynamic cytoskeleton rearrangements in giant cells and syncytia of nematode-infected roots. *The Plant Journal* 38, 12–26.

de Almeida Engler, J., Vieira, P., Rodiuc, N., Grossi-de-Sá, M.F. and Engler, G. (2015) The plant cell cycle machinery: Usurped and modulated by plant-parasitic nematodes. *Advances in Botanical Research* 73, 91–118. DOI: 10.1016/bs.abr.2014.12.003.

De Meutter, J., Tytgat, T., Witters, E., Gheysen, G., Van Onckelen, H., *et al.* (2003) Identification of cytokinins produced by the plant parasitic nematodes *Heterodera schachtii* and *Meloidogyne incognita*. *Molecular Plant Pathology* 4, 271–277.

del Mar Siles-Lucas, M.A. and Gottstein, B. (2003) The 14-3-3 protein: A key molecule in parasites as in other organisms. *Trends in Parasitology* 19, 575–581.

Desmedt, W., Mangelinckx, S., Kyndt, T. and Vanholme, B. (2020) A phytochemical perspective on plant defense against nematodes. *Frontiers in Plant Science* 11, 602079.

Dhandaydham, M., Charles, L., Zhu, H., Starr, J.L., Huguet, T., *et al.* (2008) Characterization of root-knot nematode resistance in *Medicago truncatula*. *Journal of Nematology* 40, 46–54.

Djamei, A., Schipper, K., Rabe, F., Ghosh, A., Vincon, V., *et al.* (2011) Metabolic priming by a secreted fungal effector. *Nature* 478, 395–398.

Djian-Caporalino, C., Fazari, A., Arguel, M.J., Vernié, T., VandeCasteele, C., *et al.* (2007) Root-knot nematode (*Meloidogyne* spp.) *Me* resistance genes in pepper (*Capsicum annuum* L.) are clustered on the P9 chromosome. *Theoretical and Applied Genetics* 114, 473–486.

Dodds, P.N., Lawrence, G.J. and Ellis, J.G. (2001) Contrasting modes of evolution acting on the complex *N* locus for rust resistance in flax. *The Plant Journal* 27, 439–453.

Dorhout, R., Gommers, F.J. and Kollöffel, C. (1993) Phloem transport of carboxyfluorescein through tomato roots infected with *Meloidogyne incognita*. *Physiological and Molecular Plant Pathology* 43, 1–10.

Dos Santos, C. and Franco, O.L. (2023) Pathogenesis-related proteins (PRs) with enzyme activity activating plant defense responses. *Plants* 12(11), 2226. DOI: 30.3390/plants12112226.

Dowd, C.D., Chronis, D., Radakovic, Z.S., Siddique, S., Schmülling, T., *et al.* (2017) Divergent expression of cytokinin biosynthesis, signaling and catabolism genes underlying differences in feeding sites induced by cyst and root-knot nematodes. *The Plant Journal* 92, 211–228.

Dropkin, V.H. (1969) Cellular responses of plants to nematode infections. *Annual Review of Phytopathology* 7, 101–122.
Duarte, A., Curtis, R., Maleita, C., Tiago, I. and Abrantes, I. (2014) Characterization of the venom allergen-like protein (vap-1) and the fatty acid and retinol binding protein (far-1) genes in *Meloidogyne hispanica*. *European Journal of Plant Pathology* 139, 825–836.
Elashry, A., Okumoto, S., Siddique, S., Koch, W., Kreil, D.P., *et al.* (2013) The AAP gene family for amino acid permeases contributes to development of the cyst nematode *Heterodera schachtii* in roots of *Arabidopsis*. *Plant Physiology and Biochemistry* 70, 379–386.
El-Sappah, A.H., M.M., I., El-Awady, H.H., Yan, S., Qi, S., *et al.* (2019) Tomato natural resistance genes in controlling the root-knot nematode. *Genes* 10(11), 925. DOI: 10.3390/genes10110925.
Ellis, J., Dodds, P. and Pryor, T. (2000) Structure, function and evolution of plant disease resistance genes. *Current Opinion in Plant Biology* 3, 278–284.
EPPO (2014) *Meloidogyne enterolobii*. *European and Mediterranean Plant Protection Organization Bulletin* 44, 159–163.
Ernst, K., Kumar, A., Kriseleit, D., Kloos, D.U., Phillips, M.S., *et al.* (2002) The broad-spectrum potato cyst nematode resistance gene (Hero) from tomato is the only member of a large gene family of NBS-LRR genes with an unusual amino acid repeat in the LRR region. *The Plant Journal* 31, 127–136.
Favery, B., Lecomte, P., Gil, N., Bechtold, N., Bouchez, D., *et al.* (1998) RPE, a plant gene involved in early developmental steps of nematode feeding cells. *The EMBO Journal* 17, 6799–6811.
Favery, B., Chelysheva, L.A., Lebris, M., Jammes, F., Marmagne, A., *et al.* (2004) *Arabidopsis* formin AtFH6 is a plasma membrane-associated protein upregulated in giant cells induced by parasitic nematodes. *Plant Cell* 16, 2529–2540.
Fernández-Herrera, E., Rojas-Martínez, R.I., Guevara-Olvera, L., Rivas-Dávila, M.E., Valadez-Moctezuma, E., *et al.* (2012) Defense in chilli CM-334 inoculated with *Phytophthora capsici* and infected by *Nacobbus aberrans*. *Nematropica* 42, 96–107.
Flor, H.H. (1971) Current status of the gene-for-gene concept. *Annual Review of Phytopathology* 9, 275–296.
Fioretti, L., Warry, A., Porter, A., Haydock, P. and Curtis, R. (2001) Isolation and localisation of an annexin gene (gp-nex) from the potato cyst nematode, *Globodera pallida*. *Nematology* 3, 45–54.
Freitas, V.M., Correa, V.R., Motta, F.C., Sousa, M.G., Gomes, A.C.M.M., *et al.* (2014) Resistant accessions of wild *Psidium* spp. to *Meloidogyne enterolobii* and histological characterization of resistance. *Plant Pathology* 63, 738–746.
Fuller, V.L., Lilley, C.J. and Urwin, P.E. (2008) Nematode resistance. *New Phytologist* 180, 27–44.
Gal, T.Z., Aussenberg, E.R., Burdman, S., Kapulnik, Y. and Koltai, H. (2006) Expression of a plant expansin is involved in the establishment of root knot nematode parasitism in tomato. *Planta* 224, 155–162.
Gao, B., Allen, R., Maier, T., Davis, E.L., Baum, T.J., *et al.* (2001) Molecular characterisation and expression of two venom allergen-like protein genes in *Heterodera glycines*. *International Journal of Parasitology* 31, 1617–1625.
Gao, B., Allen, R., Maier, T., Davis, E.L., Baum, T.J., *et al.* (2003) The parasitome of the phytonematode *Heterodera glycines*. *Molecular Plant-Microbe Interactions* 16, 720–726.
Gheysen, G. and Fenoll, C. (2002) Gene expression in nematode feeding sites. *Annual Review of Phytopathology* 40, 191–219.
Gheysen, G. and Mitchum, M.G. (2009) Molecular insights in the susceptible plant response to nematode infection. In: Berg, R.H. and Taylor, C.G. (eds), *Cell Biology of Plant Nematode Parasitism*. Springer Berlin, Heidelberg, Germany, pp. 45–81.
Gheysen, G. and Mitchum, M.G. (2019) Phytoparasitic nematode control of plant hormone pathways. *Plant Physiology* 179, 1212–1226.
Gleason, C., Leelarasamee, N., Meldau, D. and Feussner, I. (2016) OPDA Has key role in regulating plant susceptibility to the root-knot nematode *Meloidogyne hapla* in *Arabidopsis*. *Frontiers in Plant Science* 7, 1565.
Godínez-Vidal, D., Rocha-Sosa, M., Sepúlveda García, E.B., Lara-Reyna, J., Rojas-Martínez, R., *et al.* (2008) Phenylalanine ammonia lyase activity in chilli cm-334 infected by *Phytophthora capsici* and *Nacobbus aberrans*. *European Journal of Plant Pathology* 120, 299–303.
Godínez-Vidal, D., Soto-Hernández, M., Rocha Sosa, M., Lozoya-Gloria, E., Rojas-Martínez, R.I., *et al.* (2010) Contenido de capsidiol en raíces de chile CM-334 infectadas por *Nacobbus aberrans* y su efecto en juveniles del segundo estadio. *Nematropica* 40, 227–237.
Godinho Mendes, R.A., Basso, M.F., Fernandes de Araújo, J., Paes de Melo, B., Lima, R.N., *et al.* (2021) Minc00344 and Mj-NULG1a effectors interact with GmHub10 protein to promote the soybean parasitism by *Meloidogyne incognita* and *M. javanica*. *Experimental Parasitology* 229, 108153.

Goellner, M., Wang, X. and Davis, E.L. (2001) Endo-beta-1,4-glucanase expression in compatible plant-nematode interactions. *Plant Cell* 13, 2241–2255.

Gómez-Vásquez, R., Day, R., Buschmann, H., Randles, S., Beeching, J.R., *et al.* (2004) Phenylpropanoids, phenylalanine ammonia lyase and peroxidases in elicitor-challenged cassava (*Manihot esculenta*) suspension cells and leaves. *Annals of Botany* 94, 87–97.

Goverse, A. and Mitchum, M.G. (2022) At the molecular plant-nematode interface: New players and emerging paradigms. *Current Opinion in Plant Biology* 67, 102225.

Goverse, A. and Smant, G. (2014) The activation and suppression of plant innate immunity by parasitic nematodes. *Annual Review of Phytopathology* 52, 243–265.

Grundler, F.M.W., Sobczak, M. and Golinowski, W. (1998) Formation of wall openings in root cells of *Arabidopsis thaliana* following infection by the plant-parasitic nematode *Heterodera schachtii*. *European Journal of Plant Pathology* 104, 545–551.

Grunewald, W., Cannoot, B., Friml, J. and Gheysen, G. (2009) Parasitic nematodes modulate PIN-mediated auxin transport to facilitate infection. *PLoS Pathog* 5, e1000266.

Gupta, R., Mfarrej, M.F.B., Elnour, R.O., Hashem, M. and Ahmad, F. (2023) Defence response of host plants for cyst nematode: A review on parasitism and defence. *Journal of King Saud University-Science* 35, 102829.

Gupta, R., Mfarrej, M.F.B., Xhemali, B., Khan, A., Nadeem, H., *et al.* (2024) Metabolic responses of plants to *Meloidogyne* species parasitism: A review on molecular events and functions. *Journal of King Saud University-Science* 36, 103083.

Haegeman, A., Jacob, J., Vanholme, B., Kyndt, T. and Gheysen, G. (2008) A family of GHF5 endo-1,4-beta-glucanases in the migratory plant-parasitic nematode *Radopholus similis*. *Plant Pathology* 57, 581–590.

Haegeman, A., Vanholme, B. and Gheysen, G. (2009) Characterization of a putative endoxylanase in the migratory plant-parasitic nematode *Radopholus similis*. *Molecular Plant Pathology* 10, 389–401.

Hamamouch, N., Li, C., Seo, P.J., Park, C.M. and Davis, E.L. (2011) Expression of *Arabidopsis* pathogenesis-related genes during nematode infection. *Molecular Plant Pathology* 12, 355–364.

Hammes, U., Schachtman, D., Berg, R., Nielsen, E., Koch, W., *et al.* (2006) Nematode-induced changes of transporter gene expression in *Arabidopsis* roots. *Molecular Plant-Microbe Interactions* 18, 1247–1257.

Harris, W., Kim, S., Völz, R. and Lee, Y.H. (2023) Nuclear effectors of plant pathogens: Distinct strategies to be one step ahead. *Molecular Plant Pathology* 24, 637–650.

Hauge, M.B., Wang, L.M., Parsons, D.J. and Parnell, D.L. (2001) Nucleic acid molecules and other molecules associated with soybean cyst nematode resistance. U.S.P. No. 20030005491.

He, B., Magill, C. and Starr, J.L. (2005) Laser capture microdissection and Real-Time PCR for measuring mRNA in giant cells induced by *Meloidogyne javanica*. *Journal of Nematology* 37, 308–312.

Hématy, K., Cherk, C. and Somerville, S. (2009) Host-pathogen warfare at the plant cell wall. *Current Opinion in Plant Biology* 12, 406–413.

Hewezi, T. (2020) Epigenetic mechanisms in nematode-plant interactions. *Annual Review of Phytopathology* 58, 119–138.

Hewezi, T. and Baum, T.J. (2013) Manipulation of plant cells by cyst and root-knot nematode effectors. *Molecular Plant-Microbe Interactions* 26, 9–16.

Hewezi, T., Howe, P.J., Maier, T.R., Hussey, R.S., Mitchum, M.G., *et al.* (2010) *Arabidopsis* spermidine synthase is targeted by an effector protein of the cyst nematode *Heterodera schachtii*. *Plant Physiology* 152, 968–984.

Hewezi, T., Juvale, P.S., Piya, S., Maier, T.R., Rambani, A., *et al.* (2015) The cyst nematode effector protein 10A07 targets and recruits host posttranslational machinery to mediate its nuclear trafficking and to promote parasitism in *Arabidopsis*. *Plant Cell* 27, 891–907.

Hofmann, J. and Grundler, F. (2007) How do nematodes get their sweets? Solute supply to sedentary plant-parasitic nematodes. *Nematology* 9, 451–458.

Hofmann, J., Szakasits, D., Blöchl, A., Sobczak, M., Daxböck-Horvath, S., *et al.* (2007) Starch serves as carbohydrate storage in nematode-induced syncytia. *Plant Physiology* 146, 228–235.

Hofmann, J., Kolev, P., Kolev, N., Daxböck-Horvath, S. and Grundler, F.M.W. (2009) The *Arabidopsis thaliana* sucrose transporter gene AtSUC4 is expressed in *Meloidogyne incognita*-induced root galls. *Journal of Phytopathology* 157, 256–261.

Hofmann, J., Youssef-Banora, M., de Almeida-Engler, J. and Grundler, F.M. (2010) The role of callose deposition along plasmodesmata in nematode feeding sites. *Molecular Plant-Microbe Interactions* 23, 549–557.

Holbein, J., Grundler, F.M. and Siddique, S. (2016) Plant basal resistance to nematodes: An update. *Journal of Experimental Botany* 67, 2049–2061.

Holtmann, B., Kleine, M. and Grundler, F.M.W. (2000) Ultrastructure and anatomy of nematode-induced syncytia in roots of susceptible and resistant sugar beet. *Protoplasma* 211, 39–50.

Hoth, S., Stadler, R., Sauer, N. and Hammes, U.Z. (2008) Differential vascularization of nematode-induced feeding sites. *Proceedings National Academy of Sciences of the United States of America* 105, 12617–12622.

Huang, G., Gao, B., Maier, T., Allen, R., Davis, E.L., *et al.* (2003) A profile of putative parasitism genes expressed in the esophageal gland cells of the root-knot nematode *Meloidogyne incognita*. *Molecular Plant-Microbe Interactions* 16, 376–381.

Huang, G., Dong, R., Allen, R., Davis, E.L., Baum, T.J., *et al.* (2005) Two chorismate mutase genes from the root-knot nematode *Meloidogyne incognita*. *Molecular Plant Pathology* 6, 23–30.

Huang, G., Allen, R., Davis, E.L., Baum, T.J. and Hussey, R.S. (2006) Engineering broad root-knot resistance in transgenic plants by RNAi silencing of a conserved and essential root-knot nematode parasitism gene. *Proceedings National Academy of Sciences of the United States of America* 103, 14302–14306.

Huang, L., Yuan, Y., Lewis, C., Kud, J., Kuhl, J.C., *et al.* (2023) NILR1 perceives a nematode ascaroside triggering immune signaling and resistance. *Current Biology* 33, 3992–3997.e3993.

Hussey, R.S. and Grundler, F.M. (1998) *Nematode Parasitism of Plants*. CAB International, Wallingford, UK.

Ithal, N., Recknor, J., Nettleton, D., Maier, T., Baum, T.J., *et al.* (2007) Developmental transcript profiling of cyst nematode feeding cells in soybean roots. *Molecular Plant-Microbe Interactions* 20, 510–525.

Jablonska, B., Ammiraju, J.S., Bhattarai, K.K., Mantelin, S., Martinez de Ilarduya, O., *et al.* (2007) The *Mi-9* gene from *Solanum arcanum* conferring heat-stable resistance to root-knot nematodes is a homolog of *Mi-1*. *Plant Physiology* 143, 1044–1054.

Jammes, F., Lecomte, P., de Almeida-Engler, J., Bitton, F., Martin-Magniette, M.L., *et al.* (2005) Genome-wide expression profiling of the host response to root-knot nematode infection in *Arabidopsis*. *The Plant Journal* 44, 447–458.

Jaouannet, M., Perfus-Barbeoch, L., Deleury, E., Magliano, M., Engler, G., *et al.* (2012) A root-knot nematode-secreted protein is injected into giant cells and targeted to the nuclei. *New Phytologist* 194, 924–931.

Jaouannet, M. and Rosso, M.N. (2013) Effectors of root sedentary nematodes target diverse plant cell compartments to manipulate plant functions and promote infection. *Plant Signaling and Behaviour* 8(9), e25507. DOI: 10.4161/psb.25507.

Jaubert, S., Laffaire, J.B., Ledger, T.N., Escoubas, P., Amri, E.Z., *et al.* (2004) Comparative analysis of two 14-3-3 homologues and their expression pattern in the root-knot nematode *Meloidogyne incognita*. *International Journal for Parasitology* 34, 873–880.

Jones, J., Gheysen, L. and Fenoll, C. (eds) (2011) *Genomics and Molecular Genetics of Plant-Nematode Interactions*. Springer, Dordrecht, Netherlands.

Jones, J.T., Reavy, B., Smant, G. and Prior, A.E. (2004) Glutathione peroxidases of the potato cyst nematode *Globodera Rostochiensis*. *Gene* 324, 47–54.

Jones, J.T., Furlanetto, C. and Kikuchi, T. (2005) Horizontal gene transfer from bacteria and fungi as a driving force in the evolution of plant parasitism in nematodes. *Nematology* 7, 641–646.

Kaplan, D.T. and Keen, N.T. (1977) *De novo* synthesis of an isoflavonoid compound by soybeans in response to *Meloidogyne incognita* infection. *Journal of Nematology* 9, 274.

Karczmarek, A., Fudali, S., Lichocka, M., Sobczak, M., Kurek, W., *et al.* (2008) Expression of two functionally distinct plant endo-β-1, 4-glucanases is essential for the compatible interaction between potato cyst nematode and its hosts. *Molecular Plant-microbe Interactions* 21, 791–798.

Khan, A., Chen, S., Fatima, S., Ahamad, L. and Siddiqui, M.A. (2023) Biotechnological tools to elucidate the mechanism of plant and nematode interactions. *Plants* 12, 2387.

Khan, M. and Khan, A.U. (2021) Plant parasitic nematodes effectors and their crosstalk with defense response of host plants: A battle underground. *Rhizosphere* 17, 100288.

Kikuchi, T., Aikawa, T., Kosaka, H., Pritchard, L., Ogura, N., *et al.* (2007) Expressed sequence tag (EST) analysis of the pine wood nematode *Bursaphelenchus xylophilus* and *B. mucronatus*. *Molecular and Biochemical Parasitology* 155, 9–17.

Klink, V.P., Alkharouf, N., MacDonald, M. and Matthews, B. (2005) Laser capture microdissection (LCM) and expression analyses of *Glycine max* (soybean) syncytium containing root regions formed by the plant pathogen *Heterodera glycines* (soybean cyst nematode). *Plant Molecular Biology* 59, 965–979.

Koropacka, K.B. (2010) Molecular contest between potato and the potato cyst nematode *Globodera pallida*: Modulation of *Gpa2*-mediated resistance. Thesis, Wageningen University, Netherlands. https://edepot.wur.nl/16380.

Kouassi, A., Kerlan, M., Sobczak, M., Dantec, J., Rouaux, C., *et al.* (2004) Resistance to the root-knot nematode *Meloidogyne fallax* in *Solanum sparsipilum*: Analysis of the mechanisms. *Nematology* 6, 389–400.

Kud, J., Wang, W., Gross, R., Fan, Y., Huang, L., *et al.* (2019) The potato cyst nematode effector RHA1B is a ubiquitin ligase and uses two distinct mechanisms to suppress plant immune signaling. *PLoS Pathog* 15, e1007720.

Kyndt, T., Haegeman, A. and Gheysen, G. (2008) Evolution of GHF5 endoglucanase gene structure in plant-parasitic nematodes: No evidence for an early domain shuffling event. *BMC Evolutionary Biology* 8, 305.

Kyndt, T., Vieira, P., Gheysen, G. and de Almeida-Engler, J. (2013) Nematode feeding sites: Unique organs in plant roots. *Planta* 238, 807–818.

Kyndt, T., Goverse, A., Haegeman, A., Warmerdam, S., Wanjau, C., *et al.* (2016) Redirection of auxin flow in *Arabidopsis thaliana* roots after infection by root-knot nematodes. *Journal of Experimental Botany* 67, 4559–4570.

Lambert, K.N., Allen, K.D. and Sussex, I.M. (1999) Cloning and characterization of an esophageal-gland-specific chorismate mutase from the phytoparasitic nematode *Meloidogyne javanica*. *Molecular plant-microbe interactions* 12, 328–336.

Lawrence, G.J., Anderson, P.A., Dodds, P.N. and Ellis, J.G. (2010) Relationships between rust resistance genes at the *M* locus in flax. *Molecular Plant Pathology* 11, 19–32.

Leelarasamee, N., Zhang, L. and Gleason, C. (2018) The root-knot nematode effector MiPFN3 disrupts plant actin filaments and promotes parasitism. *PLoS Pathog* 14, e1006947.

Lightfoot, A.D. and Meksem, K. (2002) Isolated polynucleotides and polypeptides relating to loci underlying resistance to soybean cyst nematode and soybean sudden death syndrome and methods employing same. U.S.P. No. 20020144310.

Lilley, C.J., Maqbool, A., Wu, D., Yusup, H.B., Jones, L.M., *et al.* (2018) Effector gene birth in plant parasitic nematodes: Neofunctionalization of a housekeeping glutathione synthetase gene. *PLoS Genetics* 14(4), e1007310.

Lin, B., Zhuo, K., Chen, S., Hu, L., Sun, L., *et al.* (2016) A novel nematode effector suppresses plant immunity by activating host reactive oxygen species-scavenging system. *New Phytologist* 209, 1159–1173.

Lin, S., Jian, H., Zhao, H., Yang, D. and Liu, Q. (2011) Cloning and characterization of a venom allergen-like protein gene cluster from the pinewood nematode *Bursaphelenchus xylophilus*. *Experimental Parasitology* 127, 440–447.

Linthorst, H.J.M. (1991) Pathogenesis-related proteins of plants. *Critical Reviews in Plant Sciences* 10, 123–150.

Liu, B., Liu, X., Liu, Y., Xue, S., Cai, Y., *et al.* (2016a) The infection of cucumber (*Cucumis sativus* L.) roots by *Meloidogyne incognita* alters the expression of actin-depolymerizing factor (ADF) genes, particularly in association with giant cell formation. *Frontiers in Plant Science* 7, 1393.

Liu, J., Peng, H., Cui, J., Huang, W., Kong, L., *et al.* (2016b) Molecular characterization of a novel effector expansin-like protein from *Heterodera avenae* that induces cell death in *Nicotiana benthamiana*. *Scientific Reports* 6, 35677.

Lohar, D.P., Schaff, J.E., Laskey, J.G., Kieber, J.J., Bilyeu, K.D., *et al.* (2004) Cytokinins play opposite roles in lateral root formation, and nematode and rhizobial symbioses. *The Plant Journal* 38, 203–214.

López-Martínez, N., Colinas-León, Ma.T., Peña-Valdivia, B.C., Salinas-Moreno, Y., Fuentes-Montiel, P., *et al.* (2011) Alterations in peroxidase activity and phenylpropanoid metabolism induced by *Nacobbus aberrans* Thorne & Allen, 1944 in chilli (*Capsicum annuum* L.) CM334 resistant to *Phytophthora capsici* Leo. *Plant and Soil* 338, 399–409.

Lozano-Torres, J.L., Wilbers, R.H., Gawronski, P., Boshoven, J.C., Finkers-Tomczak, A., *et al.* (2012) Dual disease resistance mediated by the immune receptor Cf-2 in tomato requires a common virulence target of a fungus and a nematode. *Proceedings of the National Academy of Sciences of the United States of America* 109, 10119–10124.

Lozano-Torres, J.L., Wilbers, R.H., Warmerdam, S., Finkers-Tomczak, A., Diaz-Granados, A., *et al.* (2014) Apoplastic venom allergen-like proteins of cyst nematodes modulate the activation of basal plant innate immunity by cell surface receptors. *PLoS Pathogens* 10, e1004569.

Mahajan, R., Kaur, D.J. and Bajaj, K.L. (1992) Nematicidal activity of phenolic compounds against *Meloidogyne incognita*. *Nematologia Mediterranea* 20, 217–219.

Manohar, M., Tenjo-Castano, F., Chen, S., Zhang, Y.K., Kumari, A., *et al.* (2020) Plant metabolism of nematode pheromones mediates plant-nematode interactions. *Nature Communications* 11, 208.
Manosalva, P., Manohar, M., von Reuss, S.H., Chen, S., Koch, A., *et al.* (2015) Conserved nematode signalling molecules elicit plant defenses and pathogen resistance. *Nature Communications* 6, 7795.
Marella, H., Nielsen, E., Schachtman, D. and Taylor, C. (2013) The amino acid permeases AAP3 and AAP6 are involved in root-knot nematode parasitism of *Arabidopsis*. *Molecular Plant-Microbe Interactions* 26, 44–54.
Marini, P.M., Garbuglio, D.D., Dorigo, O.F. and Machado, A.C. (2016) Histological characterization of resistance to *Meloidogyne incognita* in *Avena sativa*. *Tropical Plant Pathology* 41, 203–209.
Marques, M.L.S., Oliveira, M.F., Pereira, P.S. and Rocha, M.R. (2020) Penetration and development of *Meloidogyne enterolobii* in resistant and susceptible *Capsicum* spp. *European Journal of Horticultural Science* 85, 86–91.
Martin, G.B., Bogdanove, A.J. and Sessa, G. (2003) Understanding the functions of plant disease resistance proteins. *Annual Review of Plant Biology* 54, 23–61.
Matson, A.L. and Williams L.F. (1965) Evidence of a fourth gene for resistance to the soybean cyst nematode. *Crop Science* 5, 477.
Mazarei, M., Lennon, K.A., Puthoff, D.P., Rodermel, S.R. and Baum, T.J. (2003) Expression of an *Arabidopsis* phosphoglycerate mutase homologue is localized to apical meristems, regulated by hormones, and induced by sedentary plant-parasitic nematodes. *Plant Molecular Biology* 53, 513–530.
McDowell, J.M. and Simon, S.A. (2006) Recent insights into *R* gene evolution. *Molecular Plant Pathology* 7, 437–448.
Mei, Y., Wright, K.M., Haegeman, A., Bauters, L., Diaz-Granados, A., *et al.* (2018) The *Globodera pallida* SPRYSEC effector *Gp*SPRY-414-2 that suppresses plant defenses targets a regulatory component of the dynamic microtubule network. *Frontiers in Plant Science* 9. DOI: 10.3389/fpls.2018.01019.
Mejias, J., Bazin, J., Truong, N.M., Chen, Y., Marteu, N., *et al.* (2021) The root-knot nematode effector MiEFF18 interacts with the plant core spliceosomal protein SmD1 required for giant cell formation. *New Phytologist* 229, 3408–3423.
Melillo, M.T., Leonetti, P., Bongiovanni, M., Castagnone-Sereno, P. and Bleve-Zacheo, T. (2006) Modulation of reactive oxygen species activities and H_2O_2 accumulation during compatible and incompatible tomato-root-knot nematode interactions. *New Phytologist* 170, 501–512.
Michelmore, R.W. and Meyers, B.C. (1998) Clusters of resistance genes in plants evolve by divergent selection and birth-and-death process. *Genome Research* 8, 1113–1130.
Milligan, S.B., Bodeau, J., Yaghoobi, J., Kaloshian, I., Zabel, P., *et al.* (1998) The root-knot nematode resistance gene *Mi* from tomato is a member of leucine zipper, nucleotide binding, leucine-rich repeat family of plant genes. *Plant Cell* 10, 1307–1319.
Mitchum, M.G., Wang, X., Wang, J. and Davis, E.L. (2012) Role of nematode peptides and other small molecules in plant parasitism. *Annual Review of Phytopathology* 50, 175–195.
Mitreva-Dautova, M., Roze, E., Overmars, H., de Graaff, L., Schots, A., *et al.* (2006) A symbiont-independent endo-1,4-beta-xylanase from the plant-parasitic nematode *Meloidogyne incognita*. *Molecular Plant Microbe Interactions* 19, 521–529.
Mitsumasu, K., Seto, Y. and Yoshida, S. (2015) Apoplastic interactions between plants and plant root intruders. *Frontiers in Plant Science* 6. DOI: 10.3389/fpls.2015.00617.
Molinari, S. (2011) Natural genetic and induced plant resistance, as a control strategy to plant-parasitic nematodes alternative to pesticides. *Plant Cell Reports* 30, 311–323.
Molinari, S., Fanelli, E. and Leonetti, P. (2014) Expression of tomato salicylic acid (SA)-responsive pathogenesis-related genes in Mi-1-mediated and SA-induced resistance to root-knot nematodes. *Molecular Plant Pathology* 15, 255–264.
Molloy, B., Baum, T. and Eves-van den Akker, S. (2023) Unlocking the development- and physiology-altering 'effector toolbox' of plant-parasitic nematodes. *Trends in Parasitology* 39, 732–738.
Nahar, K., Kyndt, T., De Vleesschauwer, D., Höfte, M. and Gheysen, G. (2011) The jasmonate pathway is a key player in systemically induced defense against root knot nematodes in rice. *Plant physiology* 157, 305–316.
Nombela, G., Williamson, V.M. and Muñiz, M. (2003) The root-knot nematode resistance gene *Mi-1.2* of tomato is responsible for resistance against the whitefly *Bemisia tabaci*. *Molecular Plant-Microbe Interactions* 16, 645–649.
Opperman, C.H., Bird, D.M., Williamson, V.M., Rokhsar, D.S., Burke, M., *et al.* (2008) Sequence and genetic map of *Meloidogyne hapla*: A compact nematode genome for plant parasitism. *Proceedings of the National Academy of Sciences of the United States of America* 105, 14802–14807.

Opperman, C.H., Taylor, C.G. and Conkling, M.A. (1994) Root-knot nematode-directed expression of a plant root-specific gene. *Science* 263, 221–223.

Oka, Y., Chet, I. and Spiegel, Y. (1997) Are pathogenesis-related proteins induced by *Meloidogyne javanica* or *Heterodera avenae* invasion? *Journal of Nematology* 29, 501–508.

Paal, J., Henselewski, H., Muth, J., Meksem, K., Menéndez, C.M., *et al.* (2004) Molecular cloning of the potato *Gro1-4* gene conferring resistance to pathotype Ro1 of the root cyst nematode *Globodera rostochiensis*, based on a candidate gene approach. *The Plant Journal* 38, 285–297.

Patel, N., Hamamouch, N., Li, C., Hewezi, T., Hussey, R.S., *et al.* (2010) A nematode effector protein similar to annexins in host plants. *Journal of Experimental Botany* 61, 235–248.

Paulson, R.E. and Webster, J.M. (1972) Ultrastructure of the hypersensitive reaction in roots of tomato, *Lycopersicon esculentum* L., to infection by the root-knot nematode, *Meloidogyne incognita. Physiological Plant Pathology* 2, 227–234.

Pegard, A., Brizzard, G., Fazari, A., Soucaze, O., Abad, P., *et al.* (2005) Histological characterization of resistance to different root-knot nematode species related to phenolics accumulation in *Capsicum annuum. Phytopathology* 95, 159–165.

Peng, H., Cui, J., Long, H., Huang, W., Kong, L., *et al.* (2016) Novel pectate lyase genes of *Heterodera glycines* play key roles in the early stage of parasitism. *PLoS One* 11, e0149959.

Perry, R.N., Moens, M. and Starr, J.L. (eds) (2009) *Root-knot Nematodes*. CAB International, Wallingford, UK.

Phan, N.T., De Waele, D., Lorieux, M., Xiong, L. and Bellafiore, S. (2018) A hypersensitivity-like response to *Meloidogyne graminicola* in rice (*Oryza sativa*). *Phytopathology* 108, 521–528.

Philbrick, A.N., Adhikari, T.B., Louws, F.J. and Gorny, A.M. (2020) *Meloidogyne enterolobii*, a major threat to tomato production: Current status and future prospects for its management. *Frontiers in Plant Science* 11, 606395.

Pinheiro, J.B., da Silva, G.O., Macedo, A.G., Biscaia, D. and de Castro e Melo, R.A. (2020) Evaluation of sources of resistance to *Meloidogyne enterolobii* in *Solanum stramonifolium* and *S. scuticum* as potential rootstocks for cultivated Solanaceae. *Nematropica* 50, 144–150.

Portillo, M., Cabrera, J., Lindsey, K., Topping, J., Andrés, M.F., *et al.* (2013) Distinct and conserved transcriptomic changes during nematode-induced giant cell development in tomato compared with *Arabidopsis*: A functional role for gene repression. *New Phytologist* 197, 1276–1290.

Przybylska, A. and Obrępalska-Stęplowska, A. (2020) Plant defense responses in monocotyledonous and dicotyledonous host plants during root-knot nematode infection. *Plant and Soil* 451, 239–260.

Qin, L., Kudla, U., Roze, E.H.A., Goverse, A., Popeijus, H., *et al.* (2004) A nematode expansin acting on plants. *Nature* 427, 30–30.

Radwan, O., Gandhi, S., Heesacker, A., Whitaker, B., Taylor, C., *et al.* (2008) Genetic diversity and genomic distribution of homologs encoding NBS-LRR disease resistance proteins in sunflower. *Molecular Genetics & Genomics* 280, 111–125.

Rahimi, S., Perry, R.N. and Wright, D.J. (1996) Identification of pathogenesis-related proteins induced in leaves of potato plants infected with potato cyst nematodes, *Globodera* species. *Physiological and Molecular Plant Pathology* 49, 49–59.

Rai, K.M., Balasubramanian, V.K., Welker, C.M., Pang, M., Hii, M.M., *et al.* (2015) Genome wide comprehensive analysis and web resource development on cell wall degrading enzymes from phyto-parasitic nematodes. *BMC Plant Biology* 15, 187.

Redding, N.W., Agudelo, P., and Wells, C.E. (2018) Multiple nodulation genes are up-regulated during establishment of reniform nematode feeding sites in soybean. *Phytopathology* 108, 275–291.

Rehman, S., Postma, W., Tytgat, T., Prins, P., Qin, L., *et al.* (2009) A secreted SPRY domain-containing protein (SPRYSEC) from the plant-parasitic nematode *Globodera rostochiensis* interacts with a CC-NB-LRR protein from a susceptible tomato. *Molecular Plant-Microbe Interactions* 22, 330–340.

Rich, J.R., Keen, N.T. and Thomason, I.J. (1977) Association of coumestans with the hypersensitivity of lima bean roots to *Pratylenchus scribneri*. *Physiological Plant Pathology* 1, 105–116.

Riggs, R.D. and Winstead, N.N. (1959) Studies on resistance in tomato to root-knot nematodes and on occurrence of pathogenic biotypes. *Phytopathology* 49, 716–724.

Roberts, P.A. and Thomason, I.J. (1986) Variability in reproduction of isolates of *Meloidogyne* incognita and *M. javanica* on resistant tomato genotypes. *Plant Disease* 70, 547–551.

Robertson, L., Robertson, W.M., Sobczak, M., Helder, J., Tetaud, E., *et al.* (2000) Cloning, expression and functional characterisation of a peroxiredoxin from the potato cyst nematode *Globodera rostochiensis*. *Molecular and Biochemical Parasitology* 111, 41–49.

Rodiuc, N., Vieira, P., Banora, M.Y. and de Almeida Engler, J. (2014) On the track of transfer cell formation by specialized plant-parasitic nematodes. *Frontiers in Plant Science* 5, 160.

Romero, R., Roberts, M. and Phillipson, J. (1995) Chorismate mutase in microorganisms and plants. *Phytochemistry* 40, 1015–1025.

Rossi, M., Goggin, F.L., Milligan, S.B., Kaloshian, I., Ullman, D.E., *et al.* (1998) The nematode resistance gene *Mi* of tomato confers resistance against the potato aphid. *Proceedings of the National Academy of Sciences of the United States of America* 95, 9750–9754.

Różańska, E., Czarnocka, W., Baranowski, Ł., Mielecki, J., de Almeida Engler, J., *et al.* (2018) Expression of both *Arabidopsis* γ-tubulin genes is essential for development of a functional syncytium induced by *Heterodera schachtii*. *Plant Cell Reports* 37, 1279–1292.

Rutter, W.B., Wadl, P.A., Mueller, J.D. and Agudelo, P. (2021) Identification of sweet potato germplasm resistant to pathotypically distinct isolates of *Meloidogyne enterolobii* from the Carolinas. *Plant Disease* 105, 3147–3153.

Rutter, W.B., Franco, J. and Gleason, C. (2022) Rooting out the mechanisms of root-knot nematode-plant interactions. *Annual Review of Phytopathology* 60, 43–76.

Sacco, M.A., Koropacka, K., Grenier, E., Jaubert, M.J., Blanchard, A., *et al.* (2009) The cyst nematode SPRYSEC protein RBP-1 elicits Gpa2- and RanGAP2-dependent plant cell death. *PLoS Pathogens* 5, e1000564.

Sato, K., Kadota, Y. and Shirasu, K. (2019) Plant immune responses to parasitic nematodes. *Frontiers in Plant Science* 10, 1165.

Schwarz, T., and Gorny, A. (2024) Evaluation of soybean genotypes (*Glycine max* and *G. soja*) for resistance to the root-knot nematode, *Meloidogyne enterolobii. Plant Disease* 108, 694–699.

Schwarz, T.R., Li, C., Yencho, G.C., Pecota, K.V., Heim, C.R., *et al.* (2021) Screening sweetpotato genotypes for resistance to a North Carolina isolate of *Meloidogyne enterolobii*. *Plant Disease* 105, 1101–1107.

Seah, S., Yaghoobi, J., Rossi, M., Gleason, C.A. and Williamson, V.M. (2004) The nematode-resistance gene, *Mi-1*, is associated with an inverted chromosomal segment in susceptible compared to resistant tomato. *Theoretical and Applied Genetics* 108, 1635–1642.

Seah, S., Telleen, A.C. and Williamson, V.M. (2007) Introgressed and endogenous *Mi-1* gene clusters in tomato differ by complex rearrangements in flanking sequences and show sequence exchange and diversifying selection among homologues. *Theoretical and Applied Genetics* 114, 1289–1302.

Shadle, G.L., Wesley, S.V., Korth, K.L., Chen, F., Lamb, C., *et al.* (2003) Phenylpropanoid compounds and disease resistance in transgenic tobacco with altered expression of L-phenylalanine ammonia-lyase. *Phytochemistry* 64, 153–161.

Shaibu, A.S., Li, B., Zhang, S. and Sun, J. (2020) Soybean cyst nematode-resistance: Gene identification and breeding strategies. *The Crop Journal* 8, 892–904. DOI: 10.1016/j.cj.2020.03.001.

Shanks, C.M., Rice, J.H., Zubo, Y., Schaller, G.E., Hewezi, T., *et al.* (2016) The role of cytokinin during infection of *Arabidopsis thaliana* by the cyst nematode *Heterodera schachtii*. *Molecular Plant-Microbe Interactions* 29, 57–68.

Siddique, S., Radakovic, Z.S., De La Torre, C.M., Chronis, D., Novák, O., *et al.* (2015) A parasitic nematode releases cytokinin that controls cell division and orchestrates feeding site formation in host plants. *Proceedings of the National Academy of Sciences of the United States of America* 112, 12669–12674.

Siddique, S., Sobczak, M., Tenhaken, R., Grundler, F.M. and Bohlmann, H. (2012) Cell wall ingrowths in nematode induced syncytia require UGD2 and UGD3. *PLoS One* 7, e41515.

Silva, R.V., Oliveira, R.D., Ferreira, P.S., Ferreira, A.O. and Rodrigues, F.A. (2013) Defense responses to *Meloidogyne exigua* in resistant coffee cultivar and non-host plant. *Tropical Plant Pathology* 38, 114–121.

Smant, G., Stokkermans, J.P., Yan, Y., de Boer, J.M., Baum, T.J., *et al.* (1998) Endogenous cellulases in animals: Isolation of beta-1, 4-endoglucanase genes from two species of plant-parasitic cyst nematodes. *Proceedings of the National Academy of Sciences of the United States of America* 95, 4906–4911.

Sobczak, M., Avrova, A., Jupowicz, J., Phillips, M.S., Ernst, K., *et al.* (2005) Characterization of susceptibility and resistance responses to potato cyst nematode (*Globodera* spp.) infection of tomato lines in the absence and presence of the broad-spectrum nematode resistance *Hero* gene. *Molecular Plant-Microbe Interactions* 18, 158–168.

Sukarta, O.C.A., Diaz-Granados, A., Slootweg, E.J., Overmars, H., van Schaik, C., *et al.* (2021) Two evolutionary distinct effectors from a nematode and virus target RanGAP1 and 2 via the WPP domain to promote disease. *bioRxiv*, p. 2021.06.24.449730. DOI: 10.1101/2021.06.24.449730

Szakasits, D., Heinen, P., Wieczorek, K., Hofmann, J., Wagner, F., *et al.* (2009) The transcriptome of syncytia induced by the cyst nematode *Heterodera schachtii* in *Arabidopsis* roots. *The Plant Journal* 57, 771–784.

Tomczak, A., Koropacka,K., Smant, G., Govers, A. and Bakker, E. (2009) Resistant plant responses. In: Berg, R.H. and Taylor, C.G. (eds) *Cell biology of plant nematode parasitism*. Springer Verlag, Heidelberg, Germany, pp. 83–113.

Trudgill, D.L. (1967) The effect of environment on sex determination in *Heterodera rostochiensis. Nematologica* 13, 263–272.

Uehara, T., Kushida, A. and Momata, Y. (2001) PcCR-based clonning of two beta-1,4-endoglucanases from the root-leison nematode *Pratylenchus penetrans*. *Nematology* 3, 335–341.

Uehara, T., Sugiyama, S., Matsuura, H., Arie, T. and Masuta, C. (2010) Resistant and susceptible responses in tomato to cyst nematode are differentially regulated by salicylic acid. *Plant and Cell Physiology* 51, 1524–1536.

Underwood, W. (2012) The plant cell wall: A dynamic barrier against pathogen invasion. *Frontiers in Plant Science* 3, 85.

van der Vossen, E.A., Van Der Voort, J.N.R., Kanyuka, K., Bendahmane, A. and Sandbrink, H. (2000) Homologues of a single resistance gene cluster in potato confer resistance to distinct pathogens: A virus and a nematode. *The Plant Journal* 23, 567–576.

Vanholme, B., Mitreva, M., Van Criekinge, W., Logghe, M., Bird, D., *et al.* (2006) Detection of putative secreted proteins in the plant-parasitic nematode *Heterodera schachtii*. *Parasitology Research* 98, 414–424.

Van Loon, L.C., Rep, M. and Pieterse, C.M.J. (2006) Significance of inducible defense-related proteins in infected plants. *Annual Review of Phytopathology* 44, 135–162.

Veech, J.A. and McClure, M.A. (1977) Terpenoid aldehydes in cotton roots susceptible and resistant to the root-knot nematode, *Meloidogyne incognita. Journal of Nematology* 9, 225–229.

Veech, J.A. (1978) An apparent relationship between methoxy-substituted terpenoid aldehydes and the resistance of cotton to *Meloidogyne incognita. Nematologica* 24, 81–87.

Vercauteren, I., de Almeida Engler, J., De Groodt, R. and Gheysen, G. (2002) An *Arabidopsis thaliana* pectin acetylesterase gene is upregulated in nematode feeding sites induced by root-knot and cyst nematodes. *Molecular Plant-Microbe Interactions* 15, 404–407.

Veremis, J.C. and Roberts, P.A. (1996) Relationships between *Meloidogyne incognita* resistance genes in *Lycopersicon peruvianum* differentiated by heat sensitivity and nematode virulence. *Theoretical and Applied Genetics* 93, 950–959.

Veremis, J.C., van Heusden, A.W. and Roberts, P.A. (1999) Mapping a novel heat-stable resistance to *Meloidogyne* in *Lycopersicon peruvianum*. *Theoretical and Applied Genetics* 98, 274–280.

Vidhyasekaran, P. (2008) Fungal pathogenesis in plants and crops: Molecular biology and host defense mechanisms. CRC Press. Boca Raton, Florida.

Vieira, P. and Nemchinov, L.G. (2020) An expansin-like candidate effector protein from *Pratylenchus penetrans* modulates immune responses in *Nicotiana benthamiana*. *Phytopathology* 110, 684–693.

Vijayapalani, P., Hewezi, T., Pontvianne, F. and Baum, T.J. (2018) An effector from the cyst nematode *Heterodera schachtii* derepresses host rRNA genes by altering histone acetylation. *The Plant Cell* 30, 2795–2812.

Vos, P., Simons, G., Jesse, T., Wijbrandi, J., Heinen, L., *et al.* (1998) The tomato *Mi-1* gene confers resistance to both root-knot nematodes and potato aphids. *Nature Biotechnology* 16, 1365–1369.

Wang, X., Replogle, A., Davis, E.L. and Mitchum, M.G. (2007) The tobacco Cel7 gene promoter is auxin-responsive and locally induced in nematode feeding sites of heterologous plants. *Molecular Plant Pathology* 8, 423–436.

Wang, Y., Yang, W., Zhang, W., Han, Q., Feng, M., *et al.* (2013) Mapping of a heat-stable gene for resistance to southern root-knot nematode in *Solanum lycopersicum*. *Plant Molecular Biology Reporter* 31, 352–362.

Wieczorek, K. (2015) Cell wall alterations in nematode-infected roots. In: Escobar, C. and Fenoll, C. (eds), *Plant Nematode Interactions: A View on Compatible Interrelationships. Advances in Botanical Research Vol. 73.* Academic Press, London, UK, pp. 61–90. DOI: 10.1016/bs.abr.2014.12.002

Wieczorek, K., Golecki, B., Gerdes, L., Heinen, P., Szakasits, D., *et al.* (2006) Expansins are involved in the formation of nematode-induced syncytia in roots of *Arabidopsis thaliana*. *The Plant Journal* 48, 98–112.

Wieczorek, K., Elashry, A., Quentin, M., Grundler, F., Favery, B., *et al.* (2014) A distinct role of pectate lyases in the formation of feeding structures induced by cyst and root-knot nematodes. *Molecular Plant-Microbe Interactions* 27, 901–912.

Williamson, V.M. and Kumar, A. (2006) Nematode resistance in plants: The battle underground. *Trends in Genetics* 22, 396–403.

Williamson, V.M. and Roberts, P.A. (2009) Mechanisms and genetics of resistance. In: Perry, R.N, Moens, M. and Starr, J.L. (eds), *Root-knot Nematodes*. CAB International. Wallingford, UK, pp. 301–325.

Wubben, M.J., Jin, J. and Baum, T.J. (2008) Cyst nematode parasitism of *Arabidopsis thaliana* is inhibited by salicylic acid (SA) and elicits uncoupled SA-independent pathogenesis-related gene expression in roots. *Molecular Plant-microbe Interactions* 21, 424–432.

Wuyts, N., Swennen, R. and De Waele, D. (2006) Effects of plants phenylpropanoid products and selected terpenoids and alkalids on the behaviour of the plant-parasitic nematodes *Radopholus similis*, *Pratylenchus penetrans* and *Meloidogyne incognita*. *Nematology* 8, 89–101.

Xu, L.H., Xiao, L.Y., Xiao, Y.N., Peng, D.L., Xiao, X.Q., *et al.* (2021) Plasmodesmata play pivotal role in sucrose supply to *Meloidogyne graminicola*-caused giant cells in rice. *Molecular Plant Pathology* 22, 539–550.

Yan, X., Cheng, X.Y., Wang, Y.S., Luo, J., Mao, Z.C., *et al.* (2012) Comparative transcriptomics of two pathogenic pinewood nematodes yields insights into parasitic adaptation to life on pine hosts. *Gene* 505, 81–90.

Yaghoobi, J., Kaloshian, I., Wen, Y. and Williamson, V.M. (1995) Mapping a new nematode resistance locus in *Lycopersicon peruvianum*. *Theoretical and Applied Genetics* 91, 457–464.

Yang, N., Yu, Q., Li, W., Chen, D., Jian, J., *et al.* (2022) Functional characterization of the soybean cyst nematode effector SCN-27D09 using the model plant pathogenic fungus *Magnaporthe* oryzae-mediated delivery system. *Phytopathology Research* 4, 33.

Yuan, M., Ngou, B.P.M., Ding, P. and Xin, X.F. (2021) PTI-ETI crosstalk: An integrative view of plant immunity. *Current Opinion in Plant Biology* 62, 102030.

Zhang, F., Wang, J., Li, X., Zhang, J., Liu, Y., *et al.* (2023) Genome-wide identification and expression analyses of phenylalanine ammonia-lyase gene family members from tomato (*Solanum lycopersicum*) reveal their role in root-knot nematode infection. *Frontiers in Plant Science* 14, 1204990.

Zhao, J., Li, L., Liu, Q., Liu, P., Li, S., *et al.* (2019) A MIF-like effector suppresses plant immunity and facilitates nematode parasitism by interacting with plant annexins. *Journal of Experimental Botany* 70, 5943–5958.

Zhao, J., Mao, Z., Sun, Q., Liu, Q., Jian, H., *et al.* (2020) MiMIF-2 effector of *Meloidogyne incognita* exhibited enzyme activities and potential roles in plant salicylic acid synthesis. *International Journal of Molecular Sciences* 21(10), 3507. DOI:10.3390/ijms21103507.

Zhao, J., Sun, Q., Quentin, M., Ling, J., Abad, P., *et al.* (2021) A *Meloidogyne incognita* C-type lectin effector targets plant catalases to promote parasitism. *New Phytologist* 232, 2124–2137.

Zhuo, K., Naalden, D., Nowak, S., Xuan Huy, N., Bauters, L., *et al.* (2019) A *Meloidogyne graminicola* C-type lectin, Mg01965, is secreted into the host apoplast to suppress plant defence and promote parasitism. *Molecular Plant Pathology* 20, 346–355.

Zhou, Y., Zhao, D., Duan, Y., Chen, L., Fan, H., *et al.* (2023) AtSWEET1 negatively regulates plant susceptibility to root-knot nematode disease. *Frontiers in Plant Science* 14, 1010348.

Zipfel, C. and Felix, G. (2005) Plants and animals: A different taste for microbes? *Current Opinion in Plant Biology* 8, 353–360.

Zipfel, C. and Robatzek, S. (2010) Pathogen-associated molecular pattern-triggered immunity: Veni, vidi . . .? *Plant Physiology* 154, 551–554.

Zinoveva, S.V., Vasyukova, N.I. and Ozeretskovskaya, O.L. (2004) Biochemical aspects of plant interactions with phytoparasitic nematodes: A review. *Applied Biochemistry and Microbiology* 40, 111–119.

19 Chemical and Non-chemical Tactics to Control and Manage Plant-parasitic Nematodes

Nahum Marbán-Mendoza[1], Anselmo J. Cabrera-Hidalgo[2] and Rosa H. Manzanilla-López[3]*

[1]*Departamento de Parasitología Agrícola, Universidad Autónoma Chapingo, Chapingo, Estado de México, Mexico;* [2]*TecNM-Tlatlauquitepec (ITSTL), Tlatlauquitepec, Puebla, Mexico;* [3]*Centro de Desarrollo de Productos Bióticos, Instituto Politécnico Nacional, Yautepec, Morelos, Mexico*

Abstract

This chapter is an introduction to the basic concepts of integrated pest-management (IPM) and their application to reduce and control plant-parasitic nematode populations in an agroecological, and environmentally sustainable way in order to achieve integrated-nematode management (INM). The general INM strategies and control tactics, which can be used alone or in combination to limit crop damage caused by nematodes, are reviewed. They include physical methods, chemical methods ('old' and 'next generation' nematicides), non-chemical or cultural methods, biological control agents, biological-based products and high-input agriculture technologies, as well as low-input technologies for small-medium-growers. The inclusion of sophisticated technologies, e.g., molecular and biotechnological approaches, are discussed with respect to soil health and integrated soil biology management within the framework of an integrated-nematode management programme.

19.1. Introduction

Integrated pest management (IPM) refers to the management of pests (including nematodes) via a holistic approach, which includes higher organization levels (i.e., communities and ecosystems) to support the sustainable production of food, fibre and climate change mitigation. Nematodes are considered pest organisms and, within the context of either an IPM or an integrated nematode management (INM) approach, nematode control aims 'to prevent significant loss of yield and quality in vulnerable crop plants, and in the long term, to keep nematode populations small' (Whitehead, 1998). Nematode control has evolved considerably, from an initial dependency on 'old-generation nematicides' (Desaeger *et al.*, 2020; Oka, 2020) to the present INM approach (see Chapter 20, this volume), which combines and applies different strategies and tactics to limit crop damage caused by plant-parasitic nematodes (PPN) in a sustainable and environmentally friendly way,

*E-mail: rosa.manzanilla@gmail.com

A revision of Nahúm Marbán-Mendoza, N. and Rosa H. Manzanilla-López (2012) Chemical and non-chemical strategies to control nematodes. In: Manzanilla-López, R.H. and Marbán-Mendoza, N. (eds), *Practical Plant Nematology*. Mundi-Prensa and Biblioteca Básica de Agricultura Montecillo, Mexico.

 Practical Plant Nematology 2nd Edition (eds R.H. Manzanilla-López, D.J. Hunt, N. Marbán-Mendoza and S. Infante Gil)
DOI: 10.1079/9781836990413.0019

where the use of dangerous pesticides and nematicides has been reduced or been replaced by 'next-generation' nematicides with lesser or minimal toxicity to other organisms (including humans) and the environment.

The yield loss due to nematode attack is proportional to nematode population density, whereas nematode reproduction is inverse to population density (Whitehead, 1998). To control nematode population density, four general strategies can be implemented: (i) exclusion or avoidance; (ii) reduction of nematode inoculum; (iii) suppression of nematode reproduction rate and (iv) restriction of crop damage. Each strategy (or plan) includes a set of different tactics or methods (Barker, 1997; Whitehead, 1998).

The diagnosis and accurate identification of the nematode species responsible for crop damage or disease are essential to devise any control plan, since strategies and tactical effectiveness can vary according to nematode species. Depending on the target nematode species, crop damage severity, nematode population density and action threshold (see McSorley and Duncan, 1995; Whitehead, 1998), different control strategies and chemical or non-chemical tactics can be selected and combined into an INM approach. The exclusion strategy can be applied, for example, through local or international quarantine regulations for specific nematode species (see Chapter 23, this volume), elimination of the initial infestation focus and the use of nematode-free certified plant germplasm (seeds, plantlets, tubers). Chemical tactics to reduce nematode initial population density mainly rely on the application of different nematicides active substance, whereas non-chemical tactics include physical methods, crop resistance, biological control and cultural practices, such as crop rotation, inter- and intra-cropping, cover and trap crops, organic amendments, fallow, tillage, sowing/harvest dates, disinfection of tools and equipment and staff hygiene practices. Among the available physical control methods to treat infested soils are solarization, flooding, tillage and disposal of roots or other nematode infested plant material. Resistant and tolerant cultivars can be used to suppress/reduce nematode reproduction and restrict crop damage (see Chapters 1 and 20, this volume); biological control includes a range of biological control agents (parasites and predators) and beneficial microbes (see Chapter 21, this volume). However, individual tactic efficiency varies according to nematode species, host crop susceptibility, environment and soil health conditions, among others. The importance of carrying out analysis of edaphic factors (e.g., soil type and structure, texture, pH, salinity, moisture, temperature, organic carbon and nitrogen content, heavy metal content) is worth mentioning as this information will be needed to plan INM tactics, since some edaphic factors affect not only nematode populations (Aseffa *et al.*, 2024) but also microbial control agents, crop nutrient availability (K, N, P), crop nematode tolerance and nematicide performance (e.g., pH, temperature and soil texture conditions), among others (Desaeger *et al.*, 2020; Oka, 2020).

INM is being improved constantly due to the increasing availability of knowledge on the biology and ecology of PPN and free-living nematodes, which has revealed the importance of free-living nematodes in ecosystem services and their use as soil health indicators (Neher, 2010; Ferris *et al.*, 2012; see Chapter 17, this volume). INM currently requires a multidisciplinary and interdisciplinary effort to achieve holistic management of nematode populations under different agroecosystem and pathosystem conditions and needs economically viable and sustainable soil health practices to be available for farmers and agroindustry (see Chapter 20, this volume). Within this framework, we will introduce in this chapter some of the commonest chemical and non-chemical tactics used to control and manage PPN.

19.2. Control Strategies Development

The availability of efficient techniques for counting populations and determining crop damage with precision allowed entomologists to rationalize the control of pests and to develop action thresholds, which in turn helped to maintain pest populations below crop and environment damage threshold levels and more successfully manipulate the pest system. By contrast, plant nematology took longer to develop such techniques, except for a few cases, such as the cyst nematodes (Norris *et al.*, 2003), despite the progress in nematode-control alternatives strongly encouraged by the banning and phasing out of methyl bromide (MB) due to its incorporation in the Montreal Protocol of products that destroy the ozone layer. As a result, modified formulations of fumigant nematicides and transition technologies were developed (McSorley, 2009; Krueger and McSorley, 2011) in addition to, more recently, the development and commercial production of next-generation nematicides (Desaeger *et al.*, 2020; Oka, 2020; Back *et al.*, 2024). In practice, the rationale behind the use of MB and MB alternatives is to exclude or prevent the entrance of any pest, and, as long as their use is legal, they will continue as the equivalent of an insurance policy to all types of farmers growing crops of high economic value. In this context, PPN population levels are of little importance. However, there is an increasing consciousness in consumers, who demand higher food security and environmental quality, where chemical and non-chemical alternatives (e.g., biological, organic amendments, crop rotations, biofumigation, resistance, etc.), either in single or in combination (as part of an INM), are viable solutions to PPN, particularly in the large community of small farmers worldwide. The adoption of this approach has been slow yet sustained over more than four decades (see Whitehead, 1998; Sikora *et al.*, 2018; 2021). Recently, the European Commission adopted a proposal for a new regulation on the Sustainable Use of Plant Protection Products to reduce by 50% the use and risk of chemical pesticides by 2030 (https://food.europa.eu/plants/pesticide/sustainable-use-pesticides/evaluation-and-impact-assessment_en).

19.2.1. Chemical Control: Fumigant Nematicides

Chemical control of PPN has had two important phases: the use of fumigant and non-fumigant compounds (reviewed by Desaeger *et al.*, 2020; Oka, 2020; Back *et al.*, 2024). The importance of nematode damage to crops was shown by the nematicidal activity of chemical soil fumigants (Desaeger *et al.*, 2020; Oka, 2020). Fumigant nematicides owe their name to the conditions in which farming practices are normally made, that is to say, they volatilize when applied to soils due to their physicochemical characteristics (i.e., high vapour pressure). In general terms, the response of soil microorganisms to fumigants falls within the limits of the function: concentration *vs* time of exposure to the compound. Disadvantages of soil fumigants include high application doses, difficult application and being restricted for pre-planting use (Oka, 2020).

Fumigants (Table 19.1) are generally formulated as liquids and are divided into two major groups: halogenated hydrocarbons and methyl isothiocyanate (MITC) generators. Nematodes exposed to halogenated hydrocarbon nematicides initially show a hyperactive phase, which is diminished until the nematode is immobilized by being either narcotized or killed. Nevertheless, the majority of these products cause a rapid death, thus reducing nematode populations dramatically. They could be applied in large volumes to soils (300-2000 kg ha^{-1}) to reduce nematode populations, but their use has declined for a number of reasons, the main factors being toxic effects on man and the environment. As for MIT generators' mode of action, for example, active substance (a.s.) in Metam® and Dazomet®, they decompose when in contact with water to produce MIT, which penetrates the nematode cuticle, reacting with amino acids, oxidases and nucleophilic sites of proteins (Ebone *et al.*, 2019).

19.2.2. Non-fumigant Nematicides

These products were developed in the mid-1950s, and their main advantages were low toxicity, easy application, no requirement for sophisticated application equipment, nematode control at lower dosages (2-5 kg a.s. ha^{-1}), lower residual characteristics, systemic action, broad spectrum of activity (nematicides, insecticides and acaricides) and pre- and post-planting application due to low phytotoxicity (Oka, 2020). Non-fumigant nematicides of the organophosphate and carbamate groups are acetylcholinesterase inhibitors (Table 19.1) resulting, at recommended application doses, in a nematistat rather than a nematicidal effect, as treated nematodes can potentially recover and infect plants (Oka, 2020). Frequently, non-fumigant nematicides offer an advantage over fumigants when used in grafted plants, since they can be applied before and at transplant, or throughout the duration of the crop cycle when a pathogen is detected at damaging levels. Few non-fumigant nematicides (commonly formulated as granules and emulsifiable solutions) are available in the market without restrictions

Table 19.1. Chemical nematicides available in the international market (fumigants and non-fumigants).

Chemical structure*	Common name*	Mode or action mechanism**
Avermectins	Abamectin	Glutamate-gated chloride channel allosteric modulators
Alkyl halides	Methyl bromide	Soil fumigant; penetrates into the nematodes via mouth and cuticle
	Methyl iodide	Respiratory action, highly toxic
Benzimidazole	Benomyl	Inhibition of mitosis and cell division (Beta-tubulin assembly in mitosis)
Carbamates	Carbofuran Carbosulfan • **Oxime carbamate:** Aldicarb, Aldoxycarb (aldicarb-sulfone), Benfuracarb, Oxamyl, thioaldicarb • **Phenyl carbamate:** Cloethocarb	Acetylcholinesterase inhibitors
Chloropicrin	Chloropicrin	Presumed multi-site inhibitor
Fluoroalkene	Fluensulfone	Unknown or uncertain
	Trifluenfuronate	A β-oxidation inhibitor based on mitochondrial fatty acid
Halogenated hydrocarbon	1,2-dibromo-3-chloropropane (DBCP) Ethylene dibromide 1,3-dichloropropene	Presumed multi-site inhibitor
Methyl-isothiocyanate generators	Dazomet Metam-potassium	Presumed multi-site inhibitor. Releases methyl isothiocyanate
	Allyl isothiocianate	Soil fumigant, binds to oxygen-carrying molecules and prevents tissues from using oxygen
Organophosphates	• **Organophosphate:** Phosphamidon (trialkyl phosphate) • **Organothiophosphate:** ➢ Aliphatic organothiophosphate: Cadusafos, Dimethoate, Ethoprophos, Isamidofos, Phorate, Terbufos ➢ Aryl organothiophosphate: Thionazin, Triazophos ➢ Phenyl organothiophosphate: Dichlofenthion, Fensulfothion, Heterophos*** • **Phosphoramidate:** Fenamiphos, Fosthiazate, Imicyafos	Acetylcholinesterase inhibitors
Phenylpyrazole	Acetoprole	Gamma-Aminobutyric Acid (GABA)–gated chloride channel antagonist
Pyridinecarboxamide	Cyclobutrifluram	Inhibitor of the mitochondrial electron transport chain complex II
Pyridinylethylbenzamide	Fluopyram	Succinate dehydrogenase inhibitor
Sulfonamide	Fluazaindolizine	Unknown or uncertain. Highly selective nematicides
Thiocarbonate	Sodium tetrathiocarbonate	Non-systemic, fumigant action
Unclassified	Furfural	Fumigant action
	Tioxazafen	Broad spectrum

*Chemical structure classification based on the Insecticide Resistance Action Committee (IRAC) (http://www.bcpcpesticidecompendium.org/class_nematicides.html).

**PPDB: Pesticide Properties DataBase (https://sitem.herts.ac.uk/aeru/ppdb/en/index.htm); Chemical methods of nematode management (http://nemaplex.ucdavis.edu/Mangmnt/Chemical.htm).

***Not approved in Europe.

from the Environmental Protection Agency (EPA) or Insecticide Resistance Action Committee (IRAC). Restriction is mainly due to high toxicity to mammals, persistence in soils and contamination of the water table due to leaching.

19.3. Simple Chemical Formulations

19.3.1. Methyl Bromide

MB is a halogenated hydrocarbon used for agricultural purposes that goes back to the 1940s. A colourless gas, volatile at temperatures above 4°C, MB is the perfect biocide, because it rapidly kills any form of life. The phasing out of MB has had a significant impact on soil fumigation, and the alternative replacements for quarantine/statutory uses have been much discussed, although some developing countries can still extend their MB time limit under critical usage provision (https://www.epa.gov/ods-phaseout/methyl-bromide). During the phasing out process, new formulations were developed, including combining a percentage of MB with a percentage of another compound to achieve synergistic efficacy and the use of high-gauge plastics, to prevent emissions from reaching the atmosphere, along with carbon catalysts (https://www.epa.gov/ods-phaseout/methyl-bromide#alternatives).

19.3.2. Chloropicrin

Chloropicrin (CPi) is a liquid fumigant (boiling point 112°C) that is injected into the soil under plastic sheets. It is usually applied on its own at 15-25 g m^{-2} or at 10-15 g m^{-2} when used in a mixture with MB. CPi is an excellent fungicide with a small nematicidal activity, although not as efficient as 1,3-dichloropropene (1,3-D). In the majority of cases, CPi is mixed with 1,3-D, resulting in better herbicidal and nematicidal efficiency (Minuto *et al.*, 2006; Gilreath and Santos, 2008) and has been considered for use in land where weeds and nematode populations are relatively low.

19.3.3. 1,3-Dichloropropene

1,3-D is a good alternative as a fumigant to control PPN in production systems where the main pest problem is nematodes. CPi mixtures are the principal fumigation system presently adopted as an alternative to MB in most countries. It is commercialized as a fumigant alone (Telone II® and Cordon® at 96% and 97%, respectively) or in combination with CPi (i.e., Telone C-17® at 17%; Telone C-35® and Inline® at 35%). Its use has been restricted in California (USA) due to security thresholds and reduced efficacy in fine-texture soils, and in Florida (USA), it is restricted or banned depending on the county.

19.3.4. Generators of Methyl Isothiocyanate

Generators of MITC, such as metam-sodium (Vapam®), metam-potassium (K-PAM® HL™) and Dazomet (Basamid®), are very effective against PPN. However, despite their broad spectrum of efficacy, they are limited in their application as fumigants since efficacy can vary because of soil preparation, application method and other factors. It has often been reported to have a poor distribution in soil and plant parts, such as roots and tubers (Triky-Dotan *et al.*, 2009). A waiting period of several weeks after application is necessary, since its slow dissipation, particularly at low temperatures, can cause phytotoxic effects.

Metam-sodium: Metam-sodium, alone or in combination with other alternatives (CPi and 1,3-D), is as effective as MB; hence, it is adopted by growers as a key alternative in many pre-sowing soil applications against nematodes and other pests in strawberries, tomatoes, peppers, cucurbits, miscellaneous horticultural crops and ornamentals (Porter *et al.*, 2006).

Dazomet: Dazomet is a granulated generator of MIT. To achieve a good distribution and efficacy of the product, it is necessary to distribute it uniformly in the soil by mechanical means and for adequate humidity to be present in the soil. Under these conditions, it has been shown that dosages of 35-50 g m^{-2} can give a better control than, or at least equivalent to, that of metam-sodium (López-Aranda *et al.*, 2000). As Dazomet is a granular formulation, it is easier to apply than most fumigants and is therefore an adequate substitute for small-scale applications, where use of heavy machinery is difficult.

19.3.5. Methyl Iodide

Methyl iodide (MI) or iodomethane is a chemical liquid (boiling point 42°C) for pre-sowing application. It has been considered as being the most similar to MB and tested on several crops; with drip irrigation it can be efficient at controlling a broad range of phytopathogenic fungi, nematodes and weeds (Schneider *et al.*, 2008; Schneider and Hanson, 2009; Yakabe and MacDonald, 2010). It can be used against nematodes and other pests in crops such as tomatoes, strawberry, bell pepper, fruit trees and vines (Zasada *et al.*, 2010).

19.3.6. Dimethyl Disulfide

Dimethyl disulfide (DMSD) is highly effective against several types of nematodes, including root-knot nematodes (RKN) (*Meloidogyne* spp.), and is more effective when applied with low-density polyethylene (LDPE) and virtually impermeable film (VIF) mulches (Santos *et al.*, 2008; López-Aranda *et al.*, 2009). DMSD is a volatile compound produced naturally by allinase-mediated enzymatic degradation and has a complex mode of action, that is, interfering with mitochondrial function and inhibiting cytochrome oxidase. It has been evaluated in the USA and the EU for the control of nematodes, such as *Meloidogyne incognita*, *M. javanica* and *Heterodera schachtii*, by soil injection and drip irrigation. It has been approved for use in 29 EU Member States (https://sitem.herts.ac.uk/aeru/ppdb/en/Reports/2886.htm).

19.4. Combination of Chemical Alternatives

Research has shown that fumigant mixtures and continued applications of these products, whether integrated with other non-chemical products or not, can give better results in pest control and increase crop yield (Gullino *et al.*, 2003; Guo *et al.*, 2005; Minuto *et al.*, 2006; Porter *et al.*, 2006). Fumigant combinations can help reduce the susceptibility of each fumigant to accelerated degradation in soils where this phenomenon has been reported (Ashworth *et al.*, 2015). For example, in a soil where accelerated degradation and loss of metam-sodium activity has been observed, the use of a mixture of metam-sodium with formalin resulted in effective control of *Verticillium* and other diseases (Triky-Dotan *et al.*, 2009).

19.4.1. 1,3-D/CPi

The mixture of 1,3-D with CPi (1,3-D/CPi) has been widely accepted commercially to control soil nematodes and fungal diseases (Ajwa *et al.*, 2003). Formulations of 1,3-D/CPi (i.e., TC35 and TC17, which contain ca 35% and 17% of CPi, respectively) are registered in different countries to be used against nematodes and other soil-borne pests associated with diverse crops. The use of 1,3-D and CPi has expanded rapidly in some crops, including strawberries, melons and ornamentals (carnations).

19.4.2. MB/CPi

One of the most effective measures in MB phasing out has been the adoption of MB/CPi formulations with lower concentrations of MB (e.g., MB:CPi 50:50 or less). Its use can be achieved by application machinery that allows co-injection of MB and CPi, or through the use of mixed formulations. These formulations have been shown to be as efficient as MB at higher concentrations (i.e., 98:2; 67:33) to control soil pathogens, especially when used with low-permeability films (Porter *et al.*, 1999).

19.4.3. 1,3-D and MITC

Combinations of 1,3-D with metam-sodium or MITC have demonstrated improved control of soil pests and weeds (Ajwa *et al.*, 2003). However, application of 1,3-D and metam-sodium has been shown to have limitations, such as a delay in the waiting period before planting, increased degradation in sandy soils and compatibility problems with some fumigants.

19.5. Use of Plastics and Fumigants

LDPE and VIF mulches are used in combination with all kind of fumigants in several countries. Their combination with partial or completely impermeable sheeting (Luo *et al.*, 2013) has been an efficient strategy to reduce application dosages of chemical products. Reduction in the dosage of 1,3-D/CPi under VIF can control *Pratylenchus penetrans* and *Meloidogyne hapla* at a level similar to that shown by MB/CPi (Medina-Mínguez *et al.*, 2012).

19.6. Methods of Application of Chemical Alternatives

Fumigants must be applied to soil that has been well prepared and with adequate moisture. The majority of commercial fumigants, with the exception of MB and Dazomet, are applied in a liquid state at a defined soil depth by equipment loaded on tractors. The objective of chemical vaporizations reaching all niches where pests live in the soil is achieved via a liquid phase diffusing in the soil.

19.6.1. Injection of Soil Fumigants

Fumigant chisel injection is the equivalent of 'spraying the subsoil' with a liquid formulation. The level of pest control depends on the fumigant distribution in the soil, dosage and injection depth. The latter varies according to the target and its location. For the majority of crops, the injection is usually limited to the top 30 cm of soil. By contrast, the fumigation for replanting orchards requires a deeper application at a depth of 50-60 cm. One of the main limitations in applying soil fumigants is the potential for vapours to escape to the atmosphere, leading to subsequent reduction of the effective concentration of the fumigant. Vapour escape can be reduced by covering/sealing devices that are added to the injection equipment, like the power roller or rotovator. Gas retention in soil fumigation is also improved by covering the soil with a tarp or, to a lesser extent, through using a water application to seal the soil surface (Wang *et al.*, 2005).

19.6.2. Drip Irrigation

Application of soil fumigants through drip irrigation systems ensures uniformity in the application of a fumigant. This method, however, requires a special emulsifiable formulation and an adequate irrigation system to achieve an effective and uniform application. As a result, emulsifiable formulations of 1,3-D, CPi and 1,3-D/CPi have been developed with the aim of allowing application through irrigation systems (Apt and Caswell, 1988). Drip irrigation in subsoil and plastic mulching are used to improve distribution and reduce fumigant volatility to a minimum (Sullivan *et al.*, 2004). Applying nematicides (volatile or non-volatile) by drip irrigation is quite different from injecting volatile fumigants, as success of drip irrigation in field applications depends on water movement, that is, the wetted surface and the wetting depth, to ensure proper compound penetration to the root zone surface (Apt and Caswell, 1988).

19.6.3. Precision Agriculture

Precision agriculture (PA) refers to geographical information system (GIS), global positioning system (GPS) and remote sensing (RS) technologies applied to manage soil-driven, yield-limiting problems, including biotic crop-yield limiting factors such as PPN (Melakeberhan, 2002). These technologies help to diagnose, quantify the cause of plant stress and reduce nematicidal application under a site-specific management approach. PA technologies are mainly applied for PPN control on high-value crops, for example, soybean nematode (*Heterodera glycines*), potato cyst nematode (*Globodera rostochiensis*, *G. pallida*) and reniform nematode (*Rotylenchulus reniformis*) on cotton (Nutter *et al.*, 2002; Evans and Barker, 2004; Lawrence *et al.*, 2004). To be successful, PA requires accurate knowledge about the presence, distribution (spatial and temporal), mapping (including proper geostatistical tools) and intensity of any yield-reducing factor(s) within the field. Collecting and producing such information is labour-intensive and costly, but thanks to new technologies and lower costs, PA is becoming more accessible (Melakeberhan, 2002; Nutter *et al.*,

2002; Webster, 2010; Plumblee and Mueller, 2021). Data gathered from RS platforms (i.e., ground, aircraft and satellite), combined with GIS, GPS and proper data analysis tools, can be cost-effective in helping farmers manage diseases in high-value crops in a timely manner (Nutter *et al.*, 2002). It is foreseen that PA will soon take advantage of unmanned aerial vehicles and satellite imagery to measure crop health through Normalized Difference Vegetative Index and determine soil texture and variability across a field (Plumblee and Mueller, 2021).

19.7. Next-generation Nematicides

The 'new' or 'next-generation' nematicides were developed in response to demand for less toxic, more selective (i.e., only targeting nematodes), less expensive, easy to apply and safer products (reviewed by Desaeger *et al.*, 2020; Oka, 2020; Back *et al.*, 2024; see also Chapter 20, this volume). An important issue for nematicide registration is their soil behaviour, which refers to leaching potential, soil persistence, effect on beneficial organisms and a.s. degradation and metabolism pathways (Desaeger *et al.*, 2020). The three new nematicides, fluensulfone, fluopyram and fluazindolizine (Table 19.1), have a trifluoro (3-F) group in their molecular structure (Desaeger *et al.*, 2020). Fluensulfone a.s. 5-choro-2-(3,4,4-trifluorobut-3-enylsulfonyl(7-1,3-thiazole) affects different nematode functions, for example, locomotion, pharyngeal pumping, egg-laying, hatching and development (Oka, 2020). Fluopyram a.s. *N*-[2-[3-chloro-5-(trifluoromethyl)-2-pyridinyl]ethyl]-2-(trifluoromthyl)benzamide) inhibits complex II (succinate-ubiquinone reductase) involved in mitochondrial respiration. The mode of action of fluazaindolizine a.s. 8-chloro-*N*-[(2-chloro-5-methoxyphenyl)sulfonyl]-6-(trifluoromethyl) imidazo[1,2-a]pyridine-2-carboxamide is unknown (Desaeger *et al.*, 2020; Oka, 2020).

Based on the mode of action, the IRAC Nematode Working Group or Nematodes Team has recently released a nematicide classification that includes information about the chemical structure group for both old- and next-generation nematicides, bacterial and fungal agents, botanical or animal-derived agents (including synthetic), extracts and unrefined oils (https://irac-online.org/mode-of-action/nematicides-classification-online/ and https://irac-online.org/documents/nematicides-poster/).

19.8. Enhanced Microbial Biodegradation of Nematicides

Repeated application of the same or related nematicide group to the same field can result in decreased nematicidal activity, caused by soil microorganisms using nematicides as an energy substrate, an effect known as enhanced microbial biodegradation (EMB), which should not be confused with nematicide resistance (Oka, 2020). Rotation of nematicides from different chemical classes, as well as employing other control methods, is encouraged (IRAC, 2024).

19.9. Non-chemical Tactics

19.9.1. Resistant Cultivars

Plant resistance is an attractive approach for controlling nematode populations. This may be naturally occurring and transferred to crop cultivars from wild relatives or breeding lines through either conventional breeding or molecular engineering methods. Most resistance reports so far deal with *Heterodera*, *Meloidogyne* or *Globodera* species, which reflects their importance as pests, but no resistance has been reported for other important PPN, such as *Belonolaimus* or *Hoplolaimus* (Starr *et al.*, 2002).

Multiple resistance to soil-borne pathogen diseases is becoming increasingly common and has been made possible thanks to disease resistance breeding programmes. Single, dominant, vertical genes that fitted perfectly with hybrid reproduction programmes are nowadays common worldwide, and attributing most of the resistant characteristics to crops (see Chapter 20, this volume). It is important to point out that the development of resistant varieties, if there is gene availability, requires substantial investment in research and development (Thies *et al.*, 2003). The main limitations of using resistant varieties include the appearance of new races, the high population levels of pathogens and environmental conditions that can limit the resistance level

(Devran and Sogut, 2010; Takken and Rep, 2010). Resistance to RKN can be inactivated at higher temperatures or through interaction with other pathogens (Manzanilla-López and Starr, 2009).

In tomatoes, the *Mi* resistance gene can be an efficient and economical alternative in glasshouses infested with RKN (*Meloidogyne* spp.), but it has to be used in the context of an integrated management system in order to preserve its durability and prevent the selection of virulent populations due to *Meloidogyne* isolate variability and environmental conditions (Thies *et al.*, 2003). This gene was identified in the wild tomato species, *Solanum peruvianum*, and was introduced into cultivated tomato using embryo culture of an interspecific cross of *S. lycopersicum* and *S. peruvianum*. The progeny of a single F1 plant is currently the sole source of nematode resistance in fresh-market and processing tomato cultivars (Bosland and Barchenger, 2024).

Resistance genes from natural resources have been cloned from numerous plant species and transferred to other plant species, for instance, *Hs1pro-1* from sugar beet (*Beta vulgaris*) against *H. schachtii*, *Gpa-2* from potato against *Globodera pallida* and *Hero A* from tomato against *G. rostochiensis* (reviewed by Fuller *et al.*, 2008). The over-expression of different protease inhibitors (PI), such as cowpea trypsin inhibitor (CpTI), PIN2, cystatins and serine proteases, has been used for producing nematode-resistant plants (Lilley *et al.*, 1999). Transgenic *Musa* approaches have explored cystatins, to provide partial resistance to plants from a wide range of nematodes, and synthetic peptides, to disrupt the chemosensation of nematodes in a non-lethal manner, thus suppressing the number of nematodes that can locate and invade the roots (Atkinson *et al.*, 2010). Another potential strategy is the targeted suppression of important nematode effectors in plants using RNA interference (RNAi) approach. An overview of various transgenic strategies aimed at nematode resistance is provided by Ali *et al.* (2017).

An increased and reliable production of agricultural crops, within the present area of arable land and without further expansion of the agricultural frontier, is a priority to ensure crop security. In this scenario, new molecular and omics approaches (see Chapter 18, this volume) may be called upon to improve yield by mitigating biotic constraints, such as pest and disease damage and competition from weeds.

19.9.2. Grafting

Grafted plants, combined with other non-chemical INM tactics or even alternative fumigants, have been adopted in some regions, particularly for vegetables such as tomatoes, aubergines, peppers and cucurbits. The cost of grafted plants can be higher than that for ungrafted plants, but their advantages are responsible for their present extended use. Grafting is the commonest method used in glasshouse production systems, soil or substrates (Davis *et al.*, 2008; Maršić and Jakše, 2010). It provides an excellent protection against damage caused by soil pathogens to horticultural and fruit crops in soils infested with RKN (Fig. 19.1) (Cortada-González, 2009) and plant pathogenic fungi (e.g., *Fusarium* spp., *Verticillium dahliae*, *Phytophthora* spp.) and is as effective as the resistant cultivars (López-Aranda *et al.*, 2002). Natural resistance to RKN has been reported, for example, in wild lines of watermelon germplasm derived from *Citrullus lanatus* var. *citroides*, which can be used as resistant grafting stocks to manage PPN of watermelon (Thies *et al.*, 2015). The use of *C. metuliferus* as a rootstock to graft RKN-susceptible melons lowered root galling and nematode numbers at harvest (Sigüenza *et al.*, 2005).

Resistance in grafting stocks can be lost under conditions of high population pressure from the pathogen, either when new races of the pathogens evolve or under environmental conditions such as high temperature and salinity. Although grafting of RKN-susceptible tomato cultivars on resistant rootstocks has been effective to control RKN in naturally infested fields (Rivard *et al.*, 2010), rootstock resistance may not be uniformly stable due to the sensitivity of the *Mi* gene to temperature. The viability of grafted plants can also be limited by the tolerance of grafting stocks to pests and diseases. Sweet pepper rootstocks resistant to *P. capsici* and *M. incognita* in soils treated with MB alternatives have led to the selection of virulent populations of *M. incognita*, but not *P. capsici*. Grafted plants in soil treated with 1,3-D + CPi or metam-sodium resulted in levels of disease control similar to MB. However, metam-sodium treatment showed

Fig. 19.1. Grafting. A: The scion (fruit-producing shoot) plant and the disease-resistant rootstock cultivar are cut with a razor at the same angle; B: a silicone band is placed at the grafting of scion and rootstock to tighten the union, increase contact of vascular tissues and promote cicatrization. (Images courtesy of L. Cortada-González.)

a higher incidence of *M. incognita*, and reduced plant growth and marketable yield. Grafted plants in biofumigated and solarized soils were not affected either by *P. capsici* or by *M. incognita*, and marketable yield was higher than that in non-grafted plants (Ros *et al.*, 2005).

Pepper (*Capsicum annuum* L.) cultivars possessing the *N*-gene, which controls resistance to *M. incognita*, *M. arenaria* and *M. javanica*, have been effective as rootstocks to control RKN (Thies and Fery, 2000; Oka *et al.*, 2004). Grafting a commercial bell pepper variety scion on nematode-resistant rootstock has shown potential to reduce the damage caused by *M. incognita* and contributes to RKN management as a component in sustainable crop production systems (Kokalis-Burelle *et al.*, 2009).

19.9.3. Vapour

Soil vaporization is used in conventional agriculture, especially in glasshouses, to produce horticultural crops and ornamentals in order to reduce the incidence of soil pathogens (i.e., fungi, bacteria and nematodes) and weed seeds. The usually recommended treatment is to keep the soil at a temperature of 70°C for at least for 30 min to control plant diseases and weeds, although some treatments can be applied at 60-80°C for about an hour. Soil temperature and duration of the treatment determine the complete elimination (= sterilization) or partial elimination (= Pasteurization) of soil microflora. When it is done correctly, the vapour is efficient against all soil pests, having a broad activity spectrum and not requiring a waiting period before replanting (Triolo *et al.*, 2004). However, although the high temperatures achieved can eradicate pathogenic agents (Sosnowski *et al.*, 2009), they can also negatively impact on key components of the soil ecosystem (Roux-Michollet *et al.*, 2008, 2010). The use of Pasteurization with vapour has continuously increased in intensive culture systems and in protected and high-value crops, such as vegetables and ornamentals. This is largely due to the availability of more efficient equipment equipped with negative vapour pressure or a vaporization hood (for seed beds) (O'Neill and Green, 2010) and the use of substantially less water and fuel in vaporization methods.

19.9.4. Solarization

Solarization is a simple, non-chemical, non-hazardous method that avoids the use of any toxic materials, does not contaminate the site and is suited to organic farming or other low-input agricultural systems (Dai *et al.*, 2016). Solarization is produced when the heat of solar radiation is trapped under films or sheets of clear plastic in order to raise the temperature of damp soil to a lethal level for nematodes and other pests, such as weeds, insects and acari. Solarization can achieve pathogen control close to that achieved by MB (Di Vito *et al.*, 2000; Gamliel *et al.*, 2000;

Minuto *et al.*, 2000). The use of double-layer plastics and VIF has improved the efficacy of solarization. Solarization can also be used in nursery soils to disinfect soil containers in glasshouses (Stapleton *et al.*, 2001). However, several important pests are not controlled in a consistent manner by solarization alone, including RKN (*Meloidogyne* spp.), yellow nut-grass (*Cyperus* spp.) and fungi *Monosporascus* and *Macrophomina* spp. (Gilreath *et al.*, 2004). Solarization has significant limitations to be considered as a MB universal replacement, for example, soil solarization is most effective close to the soil surface under climatic weather conditions of high air temperature and requires long day lengths for adequate soil heating. It may require land to be taken out of production for 3-6 weeks during the summer months for treatment (possibly interfering with planting schedules) and also requires disposal of used plastic film.

Several studies have examined the long-term effects of the large-scale use of soil solarization and organic amendments on weed populations, nematodes, yield and soil fertility in peppers (*Capsicum annuum*) and cucumbers (*Cucumis sativus*) (Ozores-Hampton *et al.*, 2004, 2005). Many beneficial organisms are also negatively affected, especially bacterivores and fungivores, although others, like *Trichoderma*, *Pseudomonas* and *Bacillus*, can survive the high temperatures generated by solarization (Suryawanshi and Shinde, 2015). The practical value of soil solarization should be assessed by several factors, including extent and predictability of pesticide efficacy, effect on crop growth, yield and economic cost/benefit. In addition, the extensive use of plastics in agriculture leads to the generation of large amounts of waste, such as LDPE, the plastic used most in agriculture, greenhouse farming and mulching. The presence of microplastics as pollutants has detrimental effects on the soil, surface and underground water resources, thus threatening human health and food security (see Sa'adu and Farsang, 2023).

19.10. Combination of Chemical and Non-chemical Tactics

The combination of chemical and non-chemical methods or control tactics has been acknowledged as part of an efficient strategy to overcome problems derived from the narrow spectrum of activity of the individual components. Soil solarization and grafting of vegetable crops has been shown to be a valuable non-chemical alternative. Similarly, the effect of grafted plants can be improved through the combination of grafting with biofumigation, green manures and/or chemicals as MIT, 1,3-D or non-fumigant nematicides (Spotti, 2004). Even in warmer climates, solarization is more efficient when it is combined with other methods.

19.10.1. Solarization Combined with Chemicals

The combination of soil solarization with smaller dosages of fumigants improves consistency and allows its use under a broad range of conditions, even shortening the required mulching period, and improving control efficacy (López-Aranda *et al.*, 2002; Montealegre *et al.*, 2005). In particular, liquid formulations of MIT generators (metam-sodium, metam-potassium) and 1,3-D or 1,3-D/CPi are easy to deliver through drip irrigation and offer opportunities for improved control of fungi, nematodes and weeds. Large-scale field experiments conducted with commercial growers showed that root gall severity induced by RKN was lower when soil solarization was combined with 1,3-dichloropropane + CPi (16.2 + 3.4 g m^{-2}) and a single layer of film impermeable to gas (Chellemi *et al.*, 1997) than with solarization alone (Fig. 19.2).

19.10.2. Plant Grafting Combined with Chemical and Non-chemical Methods

As previously mentioned, plant grafting on resistant stocks is expanding, especially in Mediterranean countries, East Europe, the Far East and Central and South America. Grafting is used to control pathogens, such as RKN and those causing vascular wilt and root rot, in several crops. However, under certain climatic and operational circumstances, for example, in zones where inoculum pressure is severe or when suitable grafting stocks are not available, grafting is not viable unless it is combined with chemical products. In Italy, grafted plants are widely used

Fig. 19.2. A: Mechanical application of plastics for chemical fumigation (N. Marbán-Mendoza); B: manual application of plastics for biofumigation plus solarization; C: *Brassica rapa*; D: *Brassica rapa* incorporated into soil in preparation for biofumigation plus solarization; E-F: glasshouse biofumigation with fresh cattle manure. (B-D images courtesy of César Ornat, Xavier Rubio and Arnau Alsedà; E, F images courtesy of C. Ornat and Albert Lacunza, Universitat Politècnica de Catalunya, Spain.)

in soil that has been previously fumigated with alternative fumigants (1,3-D or CPi). Results have shown that, in terms of commercial yield, galling index or disease severity was equivalent to MB (Spotti, 2004).

19.11. Phytochemical-based Tactics

Phytochemical-based tactics have been proposed as an alternative for nematode control (Chitwood, 2002). Phytochemical research on plant extracts has been conducted to identify potential active substances (e.g., organic compounds, secondary metabolites) that can not only control PPN (see also Mwamula *et al.*, 2002; Ntalli *et al.*, 2010) but also be safer to use and have minimal side effects on non-target organisms. However, a.s. of botanicals may have a larger variation in the qualitative and quantitative composition than synthesized chemicals (European Commission, 2014). As with other novel strategies, the development and registration of new plant-based products or botanical derivatives is increasingly being required to provide accurate identification and characterization of extracts of botanical a.s., mode of action, standardization of good agricultural practices, extraction and purification methods, product formulation, extract application rate and frequency, cost/benefit, field tests and addressing effectiveness inconsistencies to control nematode populations, among others.

19.11.1. Biostimulants

Commercially available products include biostimulants (organic materials), which are applied in small quantities to enhance plant growth and development by helping to deliver nutrients to the roots by increasing root permeability and by influencing metabolic processes, such as respiration, photosynthesis, nucleic acid synthesis and ion uptake, and enabling antioxidant activity in times of stress caused by pests, nematodes and drought or disease, as well as promoting soil-beneficial microbes and reducing PPN. Biostimulant effects can be produced by seaweed extracts, humic acid, triazole, fungicides, amino acids, potassium silicate, salicylic acid, plant hormones and mustard and garlic extracts. Biostimulants and humates can also improve chlorophyll content, thereby helping the plant to produce greener leaves and ameliorating problems such as chlorosis. The formulation of this type of products usually contains natural sugars, sugar plant extracts, humectants, surfactants, beneficial fungi (*Trichoderma* spp.), microbes and bacteria, vitamins, amino acids, humic substances, langbeinite, enzymes, sea kelp extracts and additives; biostimulants can be also formulated with microorganisms. At least 40 plant biostimulants/strengtheners, which have an effect on the control of PPN, have been listed as commercially available in Italy (D'Addabbo *et al.*, 2019).

19.11.2. Botanical/Animal Derivative Products to Suppress Nematodes

Botanical/animal derivatives include glucosinolate derivatives (mustard and other brassicas), molasses and complex carbohydrates (various plants), polysulphides (garlic), essential oils (reviewed by Ntalli and Caboni, 2012), chitin and so on. Several of these products have been used in amenity turf and vegetable crops to control nematodes. The a.s. of BioMass-Sugar® (registered as an organic soil amendment) is a hydrolysate of sugarcane (furfural). A commercial formulation of furfural (2-furfuraldehyde) improved turf health but did not reduce PPN densities (Crow and Luc, 2014). Eagle Green Care™ a.s. (Eco Spray, Norfolk, UK) is derived from garlic plant extracts, the product is registered in the UK as a turf grass nematicide, which has been trialled for controlling *Meloidogyne fallax* and other PPN (Fleming, 2016).

19.12. Effective Technologies for Small Growers

Throughout the world, a large number of growers cannot afford fumigants or non-fumigant nematicides. Instead, their nematological problems should be confronted in a different way. Some alternatives listed below are based on contributions made by researchers who, in collaboration with farmers, helped to mitigate in an effective manner their nematode problems. There are reviews available on the subject (see Barker and Koening, 1998; Halbrendt and Lamondia, 2004; Sikora and Roberts, 2018; Sikora *et al.*, 2021) with diverse examples for solving problems in different crops through crop rotations, fallow, farming practices, sowing date and so on (see Chapter 20, this volume). We will refer to the most common ones below.

19.12.1. Biofumigation and Biodisinfestation using Cultures that Release Volatile Toxic Compounds

The term 'biofumigation' refers to the process where volatile toxic gases are released during the degradation process of organic amendments, roots and other plant tissues. These gases control soil-borne diseases, nematodes and weeds. They are used at a commercial level in some developed and developing countries, generally in small areas to control soil-borne pathogens (Njoroge *et al.*, 2009; Dutta *et al.*, 2019). Incorporation in the soil of residues from some species of *Brassicacea* or *Asteraceae* results in the release of a series of volatile compounds that have herbicidal, fungicidal, insecticidal and/or nematicidal properties (Villar *et al.*, 1990). In almost all cases, application of such amendments results in an enormous increase in soil microorganism populations, while populations of most plant pathogenic, and probably some non-pathogenic, microorganisms are substantially reduced.

Biodisinfestation is 'the process generated after the incorporation of organic amendments followed by a plastic cover to control soil-borne diseases' (Gandariasbeitia *et al.*, 2021), thus helping to increase soil temperature and trap the volatile compounds. Biodisinfestation combined with solarization (i.e., biosolarization) considerably reduces the time needed to achieve pest control by solarization alone (Guerrero *et al.*, 2006). Biofumigation and biosolarization are among the main non-chemical alternatives used in pepper and tomato production in some countries (Díez Rojo *et al.*, 2011; Gandariasbeitia *et al.*, 2021).

Members of the *Brassicacea* contain high levels of glucosinolates that breakdown to form biocidal volatile compounds, mainly isothiocyanates and nitriles, which are involved in the biofumigation process. A biofumigation effect is also found in sorghum (*Poaceae*) and sudangrass (sorghum × drummondi) via enzymatic hydrolysis of a cyanogenic precursor and marigold α-terthienyl (Dutta *et al.*, 2019). Drying of plants such as *Rapistrum rugosum*, *Cleome hassleriana*, *Brassica juncea*, *Iberis amara* and *Lepidium sativum* allows the production of biocide pellets that can then be used in organic treatments (Lazzeri *et al.*, 2003).

19.12.2. Substrates

Substrates are commonly used worldwide as plant growth media to evade pathogenic agents that may exist in natural soils. They include artificial substrates and natural materials such as rock wool, clay granules, solid foams (i.e., polyurethane), glass wool, peat, coconut fibre, volcanic gravel, pine bark, industrial waste from vine, coffee and sugar cane, as well as many other materials. It is advisable to look for locally available substrate sources so as to make the system more efficient and cheaper. Using healthy commercial substrates is critical for maintaining low PPN infestation levels, as this 'exclusion approach' in plant nurseries contributes to prevent accidental pest introduction (Landi *et al.*, 2024). The use of coconut fibre substrate pots in soilless farming resulted in a greater reduction in the number of PPN in *Acer palmatum* ('Japanese maple') and *Cupressus* × *leylandii* ('Leyland cypress') than cultivation in natural soil, and only a few Telotylenchidae individuals were found. Hugo and Malan (2010) found that the risk of substrate contamination with PPN was low only when capped boreholes were the source of irrigation water. According to the same authors, the important PPN found in irrigation water belonged mainly to *Meloidogyne*, *Xiphinema*, *Tylenchulus*, *Trichodorus*, *Criconemoides* and *Pratylenchus*.

19.12.3. Crop Rotation and Cover Crops

Crop rotation, including cover crops and fallow, has been used for a long time as an important part of non-chemical practices employed to manage soil pathogens. Non-host crops planted in a crop rotation help to reduce nematode population, as they may starve without a suitable host. Fallow soil also helps to keep nematode populations at lower levels because of the absence of a host crop, but it may lead to soil erosion and other problems. Removing weeds after a fallow period can improve nematode-control results, as weeds can be alternate hosts to PPN.

It is recommended that the chosen rotation crops and cover crop species are first confirmed as not being good hosts for the target nematode or other PPN species. Examples of cover crops include: ricinus (*Ricinus communis*), oats (*Avena sativa*), sorghum (*Sorghum bicolor*), crotalaria (*Crotalaria spectabilis*, *C. longirostrata*), sunn hemp (*C. juncea*), velvet bean (*Mucuna pruriens*), marigold (*Tagetes* spp.), jointvetch (*Aeschynomene americana*), crimson clover (*Trifolium incarnatum*), vetch (*Vicia* spp.) and lupin (*Lupinus angustifolius*); some of these species are also suppressive to RKN and other PPN (Villar and Zavaleta-Mejía, 1990; Barker and Koening, 1998; Hartwig and Ammon, 2002).

Crop rotation and cover crops (Fig. 19.3) are more efficient when considered as components in an IPM of soil pests. Abdul-Baki *et al.* (2004) found that alternating tomato cultivars with the nematode-resistant gene *Mi* with *Leguminosae* cover crops, which are non-hosts to the RKN (*Meloidogyne* spp.), can be used as a MB substitute. Legumes used as cover crops supply nitrogen to subsequent crops and can also be

Fig. 19.3. A: *Canavalia* as cover crop (A) and mulching (B). (Images courtesy R. Quiroga Madrigal.)

used as high-quality forage in silage production (Hartwig and Ammon, 2002).

19.12.4. Trap Crops

Some weeds and other non-cash crops can be used to trap nematodes. Plant crops are planted to act as a host for PPN, where they can feed and develop for only part of their life cycle since crops are removed and destroyed before sedentary females can start egg-laying. This practice helps reduce damage to the following crop (Sasanelli *et al.*, 2021). The practice has been used for *Meloidogyne* spp. (Westerdane, 2020) and *Nacobbus aberrans*. Some *Gramineae* species (*Hordeum vulgare*, *Bromus unioloides*) can be planted as part of a crop rotation to control *Nacobbus*. The juvenile stages (J2, J3, J4) can penetrate the trap crop roots but are not able to reach the adult stage (Cespedes *et al.*, 1998). *Solanum sisymbrifolium*, a resistant trap crop for potato cyst nematode (*Globodera pallida*, *G. rostochiensis*), can reduce nematode populations by 90% in the next potato crop season (Mhatre *et al.*, 2021).

19.12.5. Organic Amendments, Green Manures and Compost

Organic amendments can provide numerous benefits, for example, to improve soil tilth, retention of water and nutrients and act as substrates to support beneficial organisms that allow natural competition, antagonistic and mutualistic relationships (Neher, 2010). The basic mechanisms through which organic amendments reduce pathogens are frequently of a chemical nature (Lazarovits *et al.*, 2005). A high concentration of volatile fatty acids, including formic acid, acetic, propionic and so on, has been found in many organic materials stored anaerobically, including cattle manure, liquid swine manure, fish emulsion, some composts and so on (Conn *et al.*, 2005). Organic amendments, such as green compost and animal manures, and subproducts from agriculture, silviculture and the food industry, have been used in many countries to manage certain soil pests (e.g., fungi, nematodes) in various crops (Ozores-Hampton *et al.*, 2005; Zinati, 2005; Sasanelli *et al.*, 2021). As a result, this tactic represents a long-term possibility to reduce the use of soil pesticides. Present limitations on the use of organic amendments include: lack of large-scale manufacturers, inconsistency in production parameters due to lack of consistent quality standards, the requirement for large quantities to be added to the soil, high transport costs and regulatory norms (lax or stringent) for their use.

19.12.6. Pest-suppressive Crops: *Tagetes* spp.

Tagetes spp. (calendula or marigold) suppress populations of endomigratory nematodes, such as *Pratylenchus penetrans*, and endoparasitic nematodes, like *Meloidogyne* spp. (Castro *et al.*, 1990;

Zavaleta-Mejía *et al.*, 1993; Hooks *et al.*, 2010). Suppression of nematodes by *Tagetes* is considered to be caused by the presence of thiophenes, which are abundant molecules that contain heterocyclic sulphur and, when activated, produce oxygen radicals. *Tagetes* roots release this biocidal compound, which is activated in the soil, impacting microbial populations in the rhizosphere. However, *Tagetes* has no effect on fungal diseases and weeds and is therefore a selective disinfestant. It has been shown that nematode populations decreased below economic thresholds with two consecutive plantings of *Tagetes*, with yields also increasing significantly (Reynolds *et al.*, 2000). *Tagetes* also suppresses PPN by enhancing nematode antagonistic microorganisms or acting as a poor host or a trap crop (Hooks *et al.*, 2010). However, *Tagetes* host status for local PPN populations (i..e., *Meloidogyne* spp.) should be assessed prior to planting.

19.12.7. Biocontrol Agents

Biocontrol agents (BA), including antagonists such as bacteria and fungi, have been shown to be efficient in the biological control of PPN and other soil pests (see Chapter 21, this volume). Entomopathogenic nematodes and their symbiotic bacteria have also received considerable attention not only against insects (Tarasco *et al.*, 2023), but also against other nematodes (Kenney and Eleftherianos, 2016). Some examples of commercially available BA are given below. The bacterium *Pasteuria penetrans* is a parasite that is effective in controlling RKN (*Meloidogyne* spp.) and other PPN (Mateille *et al.*, 2010; Mohan *et al.*, 2012). *Pasteuria usgae* (Econem™) is registered against *Belonolaimus longicaudatus* in golf courses and *P. nishizawe* Clariva™ pn (Syngenta) as a seed treatment against soybean cyst nematode.

Some plant growth-promoting bacterial species of *Bacillus*, *Agrobacterium*, *Azotobacter* and *Pseudomonas* produce toxins that can kill nematodes (Walia *et al.*, 2000). *Bacillus cereus*, *B. firmus*, *B. thuringiensis*, *B. licheniformis* and *B. nematocida* have been found to control PPN efficiently (Furmanczyk and Malusà, 2023). Nortica™ (Bayer) is a biological product that suppresses nematode populations (e.g., cyst, lance, lesion, sheath, spiral, sting and stunt nematodes) through the activity of *Bacillus firmus*, which creates a protective layer around the plant roots, thereby inhibiting the ability of the pest to feed. Nortica™ is temperature-activated and is effective when roots are also active. Poncho® Votivo® Precise seed treatment (clothianidin + *Bacillus firmus* I-1582) (BASF Agricultural Solutions) and Votivo® Prime are biological nematicides that provide tolerance to nematodes in soybean, cotton, maize and rice.

Pochonia chlamydosporia (= *Verticillium chlamydosporium*) is an endophytic, nematophagous fungus that parasitizes several PPN, including species of *Heterodera*, *Globodera*, *Meloidogyne*, *Nacobbus*, *Tylenchulus* and *Rotylenchulus*. The fungus penetrates the eggshell by producing extracellular enzymes and appressoria (Manzanilla-López *et al.*, 2013). *Pochonia*-based products are available in the biopesticide market under different commercial brands (e.g., Klamic®, Rhizotec®). *Mirothecium verrucaria* (Ditera®) is registered against *Meloidogyne* spp., *Heterodera* spp., *Globodera* spp., *Pratylenchus* spp., *Tylenchulus semipenetrans*, *Xiphinema* spp. *Radopholus* spp. and other nematodes that affect important economic crops. Some strains of *Purpureocillium lilacinum* (= *Paecilomyces lilacinus*) are commercially available against *Meloidogyne* spp. and other nematode genera, including *Ditylenchus*, *Helicotylenchus*, *Heterodera*, *Globodera*, *Pratylenchus*, *Radopholus*, *Rotylenchulus* and *Tylenchulus*.

Suppressive soils are good places to find potential BA, and several strains of fungi and antagonistic bacteria have been isolated from them. For a BA to succeed, it is necessary to know not only its mode of action but also the environmental conditions required for their establishment and expression of antagonistic activities (Stirling, 2014). Finally, according to present European Union legislation, BA have to be registered in a similar way to chemical pesticides. The register is based on efficacy and on risk assessment related to toxicity to man, animals and the environment.

19.12.8. Hybrid Systems and Organic Agriculture

There is a wide range of cover crops and non-chemical alternatives for managing PPN problems that can be used in sustainable and subsistence farming system practices or by organic growers when nematicides are not an option (Bridge, 1996; Krueger and McSorley, 2008; McSorley, 2009). Hybrid systems involve decreasing the level of conventional practices while increasing the level of organic practices in order to lessen the impact of crop losses during the transition to organic crop production period. According to Krif *et al.* (2020) in Morocco, organic growers practice crop rotation, application of soil amendments and biological nematicides to control PPN, especially *Meloidogyne* spp. Crop rotation is practised to lower the nematode population density below economic threshold and to increase the efficacy of bionematicides such as Sesamin EC (derived from the extracts of specific cultivars of hybrid sesame plant) and NemGuard DE (microgranule formulation) containing 45% garlic extract.

19.13. Soil Health and Integrated Soil Biology Management

The soil is a complex, multifaceted environment. This complexity has prompted the evolution and adaptation of a highly diverse biotic community that uses the soil as its permanent or temporary habitat or refuge. Many thousands of species of animals and microorganisms live in soils, ranging in size from the almost invisible microbiota (e.g., bacteria, fungi and protozoans) to the more conspicuous macrofauna and megafauna (e.g., earthworms, termites, millipedes, moles and rats). The activities of this wide range of soil biota contribute to many critical ecosystem services (soil formation, organic matter decomposition, nitrogen fixation and plant nutrient uptake, suppression or induction of plant diseases and pests, restoration of soil physical, chemical and biological properties and processes, etc.). Most tactics used in INM or IPM can be detrimental to the soil biota. Hence the need for a management approach that is more sustainable, that is, where control practices detrimental to the soil biological community are not included (Neher, 2010; Stirling, 2014; Sprunger and Martin, 2023).

Many soil health assessment methods are being developed, although they often lack assessment of soil-borne diseases. To address better management strategies for soil-borne disease (including nematodes) and overall soil and plant health, the integrated soil health management (ISHM) approach is increasingly being called upon. The ISHM offers a framework for developing and implementing biointensive soil health management strategies for a particular agroecosystem (Muramoto *et al.*, 2022).

Soil fumigation, intensive tillage, solarization and bare fallows may markedly reduce populations of PPN, but they can also deplete soil carbon levels, kill beneficial organisms, reduce the biological buffering capacity of the soil and have a negative impact on other soil health parameters. Thus, the most appropriate response is to build an active and diverse biological community capable of providing a full array of ecosystem services. This can be done by increasing carbon inputs, minimizing carbon losses and reducing management impacts on soil organisms, an approach that is termed integrated soil biology management (ISBM). To illustrate this, Stirling (2014) proposed a series of compatible comparisons between nematode management tactics and ISBM (Fig. 19.4).

19.14. Future Perspectives

It is acknowledged that sustainable nematode-control methods not are always effective enough, especially for cash crops and intensive agriculture (Oka, 2020) and that non-chemical methods require a better understanding of nematode biology and ecology than traditional chemical approaches (McSorley, 2009; Neher, 2010). INM components also require rethinking and restructuring (see Chapter 20, this volume).

Many researchers believe that the main nematode problems can be solved, and more realistic and effective sustainable production systems can be developed with the help of modern tools and technologies such as PA, biotechnology, molecular biology and artificial intelligence. It may even be possible to produce tolerant or resistant plant varieties against most nematode pathogenic species that are grown in production systems under protected or open field conditions. Nematode BA can be better assessed and improved for virulence, antagonism and so on.

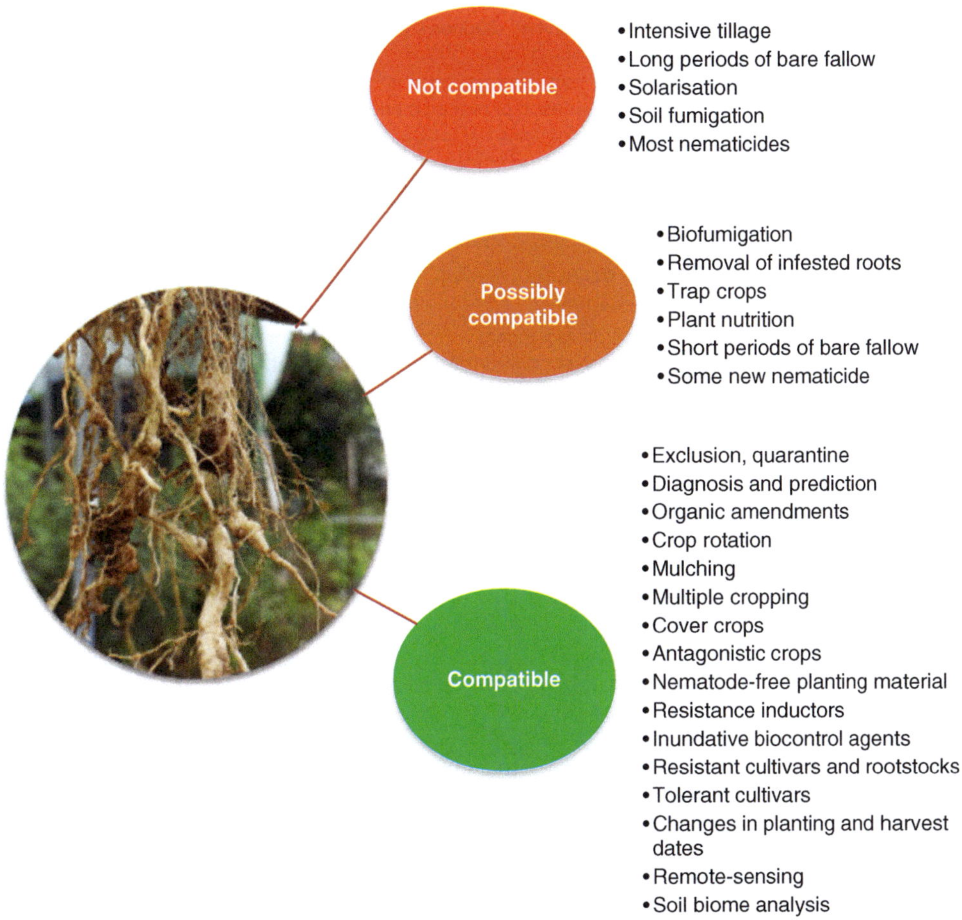

Fig. 19.4. Compatibility between **nematode-antagonistic crop** management tactics according to integrated soil health management concept. (Image courtesy Anselmo J. Cabrera-Hidalgo.)

To speed up research and knowledge dissemination programmes in both INM and soil health, it is necessary to integrate multi- and interdisciplinary teams that include social scientists, technical assistants, extensionists, growers and so on. A work system plan (Norris *et al.*, 2023) can be prepared (Fig. 19.5) in order to achieve IPM and/or Integrated Soil Health Management (ISHM) outputs similar to those defined by Muramoto *et al.* (2022) for plant health and 'one health'. Finally, although IPM/INM goals and purposes can occasionally seem to be unreachable in the short term, given enough time, significant success can be achieved, as has been shown in many production systems in a number of countries (Peshin *et al.*, 2009).

19.15. Health and Safety

Many of the chemicals and procedures used in nematode control are potentially hazardous. All procedures should undergo a risk analysis, and recommended procedures, including the use of personal protective equipment, should be rigorously implemented and followed. Hazard data sheets for the chemicals involved should be made available to staff to help minimize risk. Individual countries have own legal requirements in these matters, and it is the responsibility of the user to ensure that the appropriate legislation and procedures are followed.

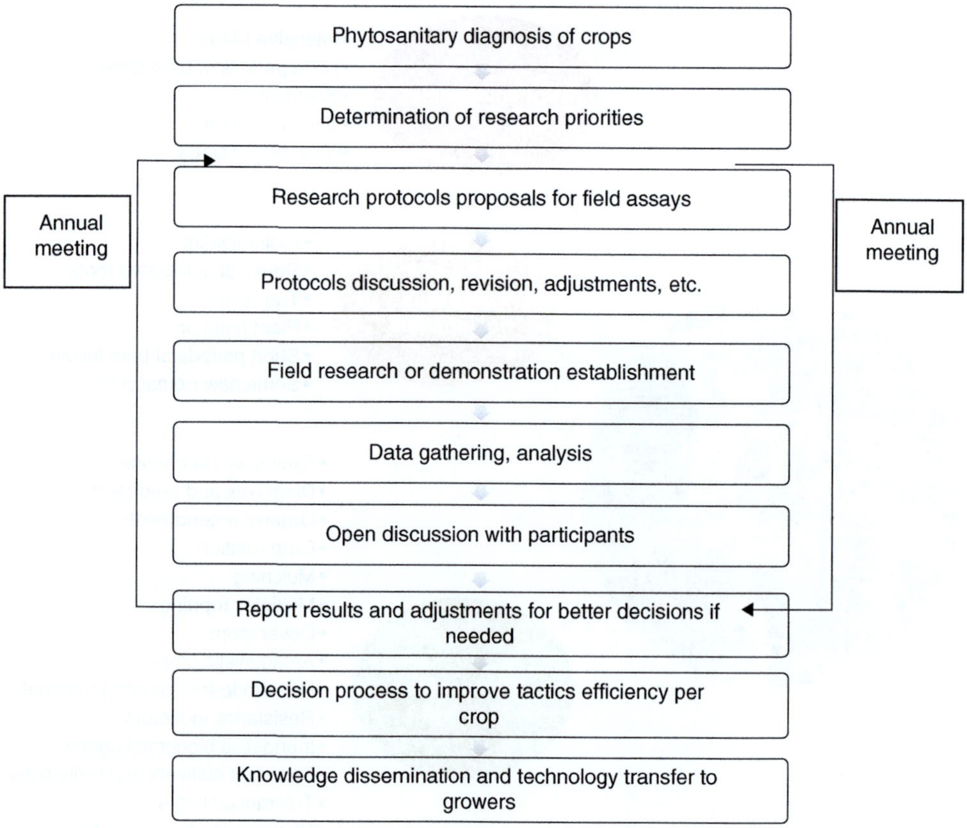

Fig. 19.5. Flow chart diagram to develop a work system for integrated pest management, integrated nematode management or integrated soil health management under a multidisciplinary approach.

19.16. References

Abdul-Baki, A., Bryan, H., Klassen, W., Carrera, L., Li, Y.C., *et al.* (2004) Low production cost alternative systems are the avenue for future sustainability of vegetable growers in the U.S. *Acta Horticulturae* 638, 197–200.

Ajwa, H.A., Klose, S., Nelson, S.D., Minuto, A., Gullino, M.L., *et al.* (2003) Alternatives to methyl bromide in strawberry production in the United States of America and the Mediterranean region. *Phytopathologia Mediterranea* 42, 220–244.

Ali, M.A., Azeem, F., Abbas, A., Joyia, F.A., Li, H., *et al.* (2017) Transgenic strategies for enhancement of nematode resistance in plants. *Frontiers in Plant Science* 8, 750. DOI: 10.3389/fpls.2017.00750.

Apt, W.J. and Caswell, E.P. (1998) Application of nematicides via drip irrigation. *Annals of Applied Nematology* 2, 1–10.

Aseffa, A.W., Meressa, B.H., Tefera, T., Green, K.K., Anderson, P., *et al.* (2024) Plant-parasitic nematodes associated with maize, and their relationship with soil properties in southern and southwestern Ethiopia. *Nematology* 26, 719–729.

Ashworth, D.J., Yates, S.R., Van Wesenbeeck, I.J. and Stanghellini, M. (2015) Effect of co-formulation of 1,3-dichloropropene and chloropicrin on evaporative emissions from soil. *Journal of Agriculture and Food Chemistry* 63, 415–421. DOI: 10.1021/jf506010b.

Atkinson, H.J., Arinaitwe, G., Kiggundu, A. and Tripathi (2010) Designing nematode resistant crops from Africa: progress and constraint. In: Bruce, T., Foyer, C., Halford, N., Keys, A., *et al.* (eds), *Agriculture: Africa's "Engine for Growth" – Plant Science & Biotechnology Hold the Key. Aspects of Applied Biology 96*. Association of Applied Biologists, Warwick, UK, pp. 119–122.

Back, M.A., Hare, M.C. and Thoden, T. (2024) Chemical control of nematodes. In: Perry, R.N., Moens, M. and Jones, J.T. (eds), *Plant Nematology*, 3rd edn. CAB International, Wallingford, UK, pp. 514–538.

Barker, K.R. (1997) Opportunities for integrated management of plant-parasitic nematodes in the Near East. https://www.fao.org/4/V9978E/v9978e0c.htm.

Barker, K.R. and Koening, S.R. (1998) Developing sustainable systems for nematode management. *Annual review of Phytopathology* 36, 165–205.

Bosland, P.W. and Barchenger, D.W. (2024) Resistance: plant-parasitic nematodes. In: Bosland, P.W. and Barchenger, D.W. (eds), *Breeding Disease-Resistant Horticultural Crops*, Academic Press, Cambridge, Massachusetts, pp. 263–282.

Bridge, J. (1996) Nematode management in sustainable and subsistence agriculture. *Annual Review of Phytopathology* 34, 201–225.

Castro, A.A.E., Zavaleta-Mejía, E., Cid del Prado, V.I. and Zamudio, G.V. (1990) Rotación e incorporación de *Tagetes erecta* L. para el manejo de *Meloidogyne incognita* (Kofoid y White) Chitwood en el cultivo de tomate (*Lycopersicon esculentum* Mill.) en Tecamachalco, Puebla. *Revista Mexicana de Fitopatología* 8, 173–180.

Cespedes, L., Franco, J. and Montalvo, R. (1998) Comportamiento de diferentes especies vegetales a la invasión y desarrollo de *Nacobbus aberrans* (Thorne, 1935), Thorne & Allen, 1944. *Nematropica* 28, 165–171.

Chellemi, D.O., McSorley, R., Rich, J.R. and Olson, S.M. (1997) Field validation of soil solarisation for fall production of tomato. *Proceedings of the Florida State Horticultural Society* 110, 330–332.

Chitwood, D.J. (2002) Phytochemical based strategies for nematode control. *Annual Review of Phytopathology* 40, 221–249. DOI: 10.1146/annurev.phyto.40.032602.130045.

Conn, K.L, Tenuta, M. and Lazarovits, G. (2005) Liquid swine manure can kill *Verticillium dahliae* microsclerotia in soil by volatile fatty acid, nitrous acid, and ammonia toxicity. *Phytopathology* 95, 28–35.

Cortada-González, L. (2009) *Tomato Rootstock for the Control of Meloidogyne spp. Characterization and Evaluation of the Resistance Response Conferred by the* Mi-1 *Gene in Tomato Rootstocks.* Ph.D. Dissertation, Universitat Politècnica de Catalunya, Department d' Enginyeria Agroalimentaria i Biotecnologia, Barcelona, Spain.

Crow, W.T. and Luc, J.E. (2014) Field efficacy of furfural as a nematicide on turf. *Journal of Nematology* 46, 8–11.

D'Addabbo, T., Laquale, S., Perniola, M. and Candido, V. (2019) Biostimulants for plant growth promotion and sustainable management of phytoparasitic nematodes in vegetable crops. *Agronomy* 9, 616.

Dai, Y., Senge, M., Yoshiyama, K., Zhang, P. and Zhang, F. (2016) Influencing factors, effects and development prospect of soil solarization. *Reviews in Agricultural Science* 4, 21–35.

Davis, R.A., Perkins-Veazie, P., Sakata, Y., Lopez-Galarza, S., Maroto, J.V., *et al.* (2008) Cucurbit grafting. *Critical Reviews in Plant Sciences* 27, 50–74.

Desaeger, J., Wram, C. and Zasada, I. (2020) New reduced-risk agricultural nematicides – rationale and review. *Journal of Nematology* 52, e2020-91. DOI: 10.21307/jofnem-2020-091.

Devran, Z. and Sogut, M.A. (2010) Occurrence of virulent root-knot nematode populations on tomatoes bearing the Mi gene in protected vegetable-growing areas of Turkey. *Phytoparasitica* 38, 245–251.

Di Vito, M., Zaccheo, F., Catalano, F. and Campanelli, R. (2000) Effect of soil solarisation and low dosages of fumigants on control of the root knot nematodes *Meloidogyne incognita*. *Acta Horticulturae* 532, 171–173.

Díez Rojo, M., López Pérez, J., Urbano Terrón, P. and Bello Pérez, A. (2011) *Biodesinfección de Suelos y Manejo Agronómico*. Ministerio Medio Ambiente, Medio Rural y Marino, Madrid, Spain.

Dutta, T.K, Khan, M.R. and Phani, V. (2019) Plant-parasitic nematode management via biofumigation using brassica and non-brassica plants: current status and future prospects. *Current Plant Biology* 17, 17–32. DOI: 10.1016/j.cpb.2019.02.001.

Ebone, L.A., Kovaleski, M. and Deuner, C.C. (2019) Nematicides: history, mode, and mechanism action. *Plant Science Today* 6, 91–97. DOI: 10.14719/pst.2019.6.2.468.

European Commission. (2014) Guidance document on botanical active substances used in plant protection products. SANCO/11470/2012. Health and consumer protection directorate-general. Safety of the food chain. Chemical, contaminants, pesticides, Brussels, Belgium. https://food.ec.europa.eu/system/files/2016-10/pesticides_ppp_app-proc_guide_doss_botanicals-rev-8.pdf.

Evans, K. and Barker, A.D.P. (2004) Economies in nematode management from precision agriculture-limitations and possibilities. In: Cook, R. and Hunt, D.J. (eds), *Nematology Monographs and Perspectives, 2. Proceedings of the Fourth International Congress of Nematology, 8–13 June 2002. Tenerife, Spain.* Brill, Leiden, Netherlands, pp. 23–32.

Ferris, H., Griffiths, B.S., Porazinska, D.L., Powers, T.O., Wang, K-H., *et al.* (2012) Reflections on plant and soil nematode ecology: past, present and future. *Journal of Nematology* 44, 115–126.

Fleming, C. (2016) Root-knot nematodes – a notifiable nuisance. 19 December 2016. https://www.pitchcare.com/blogs/news/root-knot-nematodes.

Fuller, V.L., Lilley, C.J. and Urwin, P.E. (2008) Nematode resistance. *New Phytologist* 180, 27–44.

Furmanczyk, E.M. and Malusà, E. (2023) Control of nematodes in organic horticulture exploiting the multifunctional capacity of microorganisms. *Horticulturae* 9, 920.

Gamliel, A., Grinstein A., Zilberg, A.V. and Katan, J. (2000) Control of soil borne diseases by combining soil solarisation and fumigants. *Acta Horticulturae* 532, 157–164.

Gandariasbeitia, M., López-Pérez, J.A., Juaristi, B., Abaunza, L. and Larregla, S. (2021) Biodisinfestation with agricultural by-products developed long-term suppressive soils against *Meloidogyne incognita* in lettuce crops. *Frontiers in Sustainable Food Systems* 5. DOI: 10.3389/fsufs.2021.663248.

Gilreath, J.P., Jones, J.P., Santos, B.M. and Overman, A.J. (2004) Soil fumigant evaluations for soil borne pest and *Cyperus rotundus* control in fresh market tomato. *Crop Protection* 23, 889–893.

Gilreath, J.P. and Santos, B.M. (2008) Managing weeds and nematodes with combinations of methyl bromide alternatives in tomato. *Crop Protection* 27, 648–652.

Guerrero, M.M., Ros, C., Martínez, M.A., Martínez, C., Bello, A., *et al.* (2006) Biofumigation vs. biofumigation plus solarization to control *Meloidogyne incognita* in sweet pepper. Integrated control in protected crops, Mediterranean Climate. *IOBC/wprs Bulletin* 29, 313–318.

Gullino, M.L., Camponogara, A., Gasparrini, A., Rizzo, V., Cini, C., *et al.* (2003) Replacing methyl bromide for soil disinfestations: the Italian experience and its implications for other countries. *Plant Disease* 87, 1012–1019.

Guo, M., Yates, S., Papierink, S. and Zheng, W. (2005) Incompatibility of metam sodium with halogenated fumigants. *Pesticide Management Science* 61, 467–476.

Halbrendt, J.M. and LaMondia, J.A. (2004) Crop rotation and other cultural practices. In: Chen, Z.X., Chen, S.Y. and Dickson, D.W. (eds), *Nematology: Advances and Perspectives. Vol. II: Nematode Management and Utilization*. Tsinghua University Press and CAB International, Wallingford, UK, pp. 63–1234.

Hartwig, N.L. and Ammon, H.U. (2002) Cover crops and living mulches. *Weed Science* 50, 688–699.

Hooks, C.R.R., Wang, K.H., Ploeg, A. and McSorley, R. (2010) Using marigold (*Tagetes* spp.) a cover crop to protect crops from plant-parasitic nematodes. *Applied Soil Ecology* 46, 307–320.

Hugo, H.J. and Malan, A.P. (2010) Occurrence and control of plant-parasitic nematodes in irrigation water: a review. *South African Journal of Enology and Viticulture* 31, 169–180.

Insecticide Resistance Action Committee (IRAC). (202) Nematodes team. https://irac-online.org/teams/nematodes/.

Kenney, E. and Eleftherianos, I. (2016) Entomopathogenic and plant pathogenic nematodes as opposing forces in agriculture. *International Journal for Parasitology* 46, 13–19. DOI: 10.1016/j.ijpara.2015.09.005.

Kokalis-Burelle, N., Bausher, M.G. and Rosskopf, E.N. (2009) Greenhouse evaluation of *Capsicum* rootstocks for management of *Meloidogyne incognita* on grafted bell pepper. *Nematropica* 39, 121–132.

Krif, G., Mokrini, F., Elaissami, A., Laasli, S-E., Imren, M., *et al.* (2020) Diversity and management strategies of plant parasitic nematodes in Morocco organic farming and their relationships with soil physic-chemical properties. *Organic Agriculture* 10, 447. DOI: 10.3390/agriculture10100447.

Krueger, R. and McSorley, R. (2008) Nematode management in organic agriculture: ENY-58/NG047,1/2008. University of Florida, IFAS Extension. https://journals.flvc.org/edis/article/view/119072 (accessed 20 August 2024).

Landi, S., Carletti, B., Binazzi, F., Cacini, S., Nesi, B., *et al.* (2024) Impact of pot farming on plant-parasitic nematode control. *Soil Systems* 8, 60.

Lawrence, G.W., Kelley, A.T., King, R.L., Vickery, J., Lee, H.K., *et al.* (2004) Remote sensing and precision nematicide applications for *Rotylenchulus reniformis* management in cotton. In: Cook, R. and Hunt, D.J. (eds), *Nematology Monographs and Perspectives, 2. Proceedings of the Fourth International Congress of Nematology, 8–13 June 2002. Tenerife, Spain*. Brill, Leiden, Netherlands, pp. 13–21.

Lazarovits, G., Conn, K.L., Abbasi, P.A. and Tenuta, M. (2005) Understanding the mode of action of organic soil amendments provides the way for improved management of soilborne plant pathogens. *Acta Horticulturae* 689, 215–224.

Lazzeri, L., Baruzzi, G., Malaguti, L. and Antoniacci, L. (2003) Replacing methyl bromide in annual strawberry production with glucosinolate-containing green manure crops. *Pest Management Science* 59, 983–990.

Lilley, C.J., Devlin, F., Urwin, P.E., and Atkinson, H.J. (1999) Parasitic nematodes, proteinases and transgenic plants. *Parasitology Today* 15, 414–417.

López-Aranda, J.M., Lacasa, A., Guirao, P. and Hernandez, F. (2002) Non-chemical alternatives to methyl bromide in greenhouse-grown sweet pepper in Spain. In: Batchelor, T.A. and Bolivar, J.M. (eds), *Proceedings of the International Conference on Alternatives to Methyl Bromide; 5–8 March 2002; Sevilla, Spain*. Office for Official Publications of the European Communities, Luxembourg, pp. 281–282.

López-Aranda, J.M., Medina, J.J., Miranda, L. and Dominguez, F. (2000) Three years of short-term alternatives to MB on Huelva strawberries. In: *Annual International Research Conference on Methyl Bromide Alternatives and Emissions Reduction 2000*. Orlando, Florida, pp. 1–10.

López-Aranda, J.M., Miranda, L., Medina, J.J., Soria, C., Santos, B.D.L., *et al.* (2009) Methyl bromide alternatives for high tunnel strawberry production in southern Spain. *HortTechnology* 19, 187–192.

Luo, L, Yates, S.R., Ashworth, D.J., Xuan, R. and Becker, J.O. (2013) Effect of films on 1,3-dichloropropene and chloropicrin emission, soil concentration, and root-knot nematode control in a raised bed. *Journal of Agricultural Food and Chemistry* 61, 2400–2406. DOI: 10.1021/jf3034368.

Manzanilla-López, R.H., Esteves, I., Finetti-Sialer, M.M., Hirsch, P.R., Ward, E., *et al.* (2013) *Pochonia chlamydosporia*: advances and challenges to improve its performance as a biological control agent of sedentary endoparasitic nematodes. *Journal Nematology* 45, 1–7.

Manzanilla-López, R.H. and Starr, J.L. (2009) Interactions with other pathogens. In: Perry, R.N., Moens, M. and Starr, J.L. (eds), *Root-knot Nematodes*. CAB International, Wallingford, UK, pp. 223–245.

Maršić, N.K. and Jakše, M. (2010) Growth and yield of grafted cucumber (*Cucumis sativus* L.) on different soilless substrates. *Journal of Food, Agriculture and Environment* 8, 654–658.

Mateille, T., Roch Dabiré, K., Fould, S. and Thiam Diop, M. (2010) Host-parasite soil communities and environmental constraints: modelling of soil functions involved in interactions between plant-parasitic nematodes and *Pasteuria penetrans*. *Soil Biology and Biochemistry* 42, 1193–1199.

McSorley, R. (2009) Alternative practices for managing plant-parasitic nematodes. *American Journal of Alternative Agriculture* 13, 98–104.

McSorley, R. and Duncan, L.W. (1995) Economic thresholds and nematode management. *Advances in Plant Pathology* 11, 147–171. DOI: 10.1016/S0736-4539(06)80010-3

Medina-Mínguez, J.J., Miranda, E.L., Domínguez, P., Soria, C., Pérez-Jiménez, R.M., *et al.* (2012) Comparison of different chemicals and non-chemical alternatives to methyl bromide for strawberry in Huelva (Spain). *Journal of Berry Research* 2, 113–121.

Melakeberhan, H. (2002) Embracing the emerging precision technologies for the site-specific management of yield-limiting factors. *Journal of Nematology* 34, 185–188.

Mhatre, P.H., Divya, K.L., Venkatasalam, E.P., Bairwa, A., Sudha, R., *et al.* (2021) Evaluation of trap crop, *Solanum sisymbrifolium* and antagonistic crops against potato cyst nematodes, *Globodera* spp. *South African Journal of Botany* 138, 242–248. DOI: 10.1016/j.sajb.2021.01.001.

Minuto, A., Gilardi, G., Gullino, M.L. and Garibaldi, A. (2000) Combination of soil solarisation and dazomet against soil borne pathogens of glasshouse-grown basil, tomato and lettuce. *Acta Horticulturae* 532, 165–170. DOI: 10.17660/ActaHortic.2000.532.20.

Minuto, A., Gullino, M.L., Lamberti, F., D'Addabbo, T., Tescari, E., *et al.* (2006) Application of an emulsifiable mixture of 1,3-dichloropropene and chloropicrin against root knot nematodes and soilborne fungi for greenhouse tomatoes in Italy. *Crop Protection* 25, 1244–1252.

Mohan, S., Mauchline, T.H., Rowe, J., Hirsch, P.R. and Davies, K.G. (2012) *Pasteuria* endosporess from *Heterodera cajani* (Nematoda: *Heteroderidae*). *FEMS Microbiology Ecology* 79, 675–684. DOI: 10.1111/j.1574-6941.2011.01249.x.

Montealegre, J.R., Herrera, R., Velásquez, J.C., Silva, P., Besoain, X., *et al.* (2005) Biocontrol of root and crown rot in tomatoes under greenhouse conditions using *Trichoderma harzianum* and *Paenibacillus lentimorbus*: additional effect of solarization. *Electronic Journal of Biotechnology* 8, 249–257.

Muramoto, J., Parr, D.M., Perez, J. and Wong, D.G. (2022) Integrated soil health management for plant health and one health: lessons from histories of soil-borne disease management in California strawberries and arthropod pest management. *Frontiers in Sustainable Food Systems* 6, 839648.

Mwamula, A.O., Kabir, M.F. and Lee, D.A. (2002) Review of the potency of plant extracts and compounds from key families as an alternative to synthetic nematicides: history, efficacy, and current developments. *Plant Pathology Journal* 38, 53–77.

Neher, D.A. (2010) Ecology of plant and free-living nematodes in natural and agricultural soil. *Annual Review of Phytopathology* 48, 371–394.

Njoroge, S.M.C., Kabir, Z., Martin, F.N., Koike, S.T. and Subbarao, K.V. (2009) Comparison of crop rotation for Verticillium wilt management and effect on *Pythium* species in conventional and organic strawberry production. *Plant Disease* 93, 519–527.

Norris, R.F., Caswell-Chen, E.P. and Kogan, M. (2003) *Concepts in Integrated Pest Management*. Prentice Hall, Delhi, India.

Ntalli, G. and Caboni, P. (2012) Botanical nematicides: a review. *Journal of Agricultural and Food Chemistry* 60, 9929–9940. DOI: 10.1021/jf303107j.

Ntalli, G., Serrari, T., Giannakou, K. and Menkissoglu-Spiroudi, U. (2010) Phytochemistry and nematicidal activity of the essential oils from 8 Greek Lamiaceae aromatic plants and 13 terpene components. *Journal of Agriculture and Food Chemistry* 58, 7856–7863.

Nutter, F.W., Jr., Tylka, G.I., Guan, J., Moreira, A.J.D., Marett, C.C., *et al.* (2002) Use of remote sending to detect soybean cyst nematode-induced plant stress. *Journal of Nematology* 34, 222–231.

Oka, J. (2020) From an old-generation to next-generation nematicides. *Agronomy* 10, 1387. DOI: 10.3390/agronomy10091387.

Oka, Y., Offenbach, R. and Pivonia, S. (2004) Pepper rootstock graft compatibility and response to *Meloidogyne javanica* and *M. incognita*. *Journal of Nematology* 36, 137–141.

O'Neill, T.M. and Green, K.R. (2010) Evaluation of some pre-plant soil treatment and chemical disinfectants form control of Fusarium wilt diseases in protected cut flowers. *Acta Horticulturae* 883, 215–222. DOI: 10.17660/ActaHortic.2010.883.26.

Ozores-Hampton, M., McSorley, R., Stansly, P.A., Roe, N.E. and Chellemi, D.O. (2004) Long term large scale soil solarization as a low-input production system for Florida vegetables. *Acta Horticulturae* 638, 177–188.

Ozores-Hampton, M., Stansly, P.A., McSorley, R. and Obreza, T.A. (2005) Effects of long-term organic amendments and soil solarization on pepper and watermelon growth, yield, and soil fertility. *HortScience* 40, 80–84.

Peshin, R., Bandral, R.S., Zhang, W., Wilson, L. and Dhawan, A.K. (2009) Integrated pest management: a global overview of history, programs and adoption. In: Peshin, R. and Dhawan, A.K. (eds), *Integrated Pest Management: Innovation-Development Process*. Springer, Dordrecht, Netherlands, pp. 1–49. DOI: 10.1007/978-1-4020-8992-3_1.

Plumblee, M.T. and Mueller, J.D. (2021) Implementing precision agriculture concepts and technologies into crop production and site-specific management of nematodes. In: Sikora, R.A., Desaeger, J. and Molendij, K. (eds), *Integrated Nematode Management: State-of-the-art and Visions for the Future*. CAB International, Wallingford, UK, pp. 421–427.

Porter, I.J., Brett, R.W. and Wiseman, B.M. (1999) Alternatives to methyl bromide: chemical fumigants or integrated pest management systems? *Australasian Plant Pathology* 28, 65–71.

Porter, I.J., Trinder, L. and Partington, D. (2006) Special report validating the yield performance of alternatives to methyl bromide for preplant fumigation. TEAP/MBTOC Special Report, UNEP, Nairobi, Kenya, May 2006.

Reynolds, L.B., Potter, J.W. and Ball-Coelho, B.R. (2000) Crop rotation with *Tagetes* sp. is an alternative to chemical fumigation for control of root-lesion nematodes. *Agronomy Journal* 92, 957–966.

Rivard, C.L., O'Connell, S., Peet, M.M. and Louws, F.J. (2010) Grafting tomato with interspecific rootstock to manage diseases caused by *Sclerotium rolfsii* and Southern Root-Knot Nematode. *Plant Disease* 94, 1015–1021.

Ros, C., Guerrero, M.M., Martínez, M.A., Barceló, N., Martínez, M.C., *et al.* (2005) Resistant sweet pepper rootstocks integrated into the management of soilborne pathogens in greenhouse. *Acta Horticulturae* 698, 305–310.

Roux-Michollet, D., Czarnes, S., Adam, B., Berry, D., Commeaux, C., *et al.* (2008) Effects of steam disinfestation on community structure, abundance and activity of heterotrophic, denitrifying and nitrifying bacteria in an organic farming soil. *Soil Biology and Biochemistry* 40, 1836–1845.

Roux-Michollet, D., Dudal, Y., Jocteur-Monrozier, L. and Czarnes, S. (2010) Steam treatment of surface soil: How does it affect water-soluble organic matter, C mineralization, and bacterial community composition? *Biology and Fertility of Soils* 46, 607–616.

Sa'adu, I. and Farsang, A. (2023) Plastic contamination in agricultural soils: A review. *Environmental Sciences Europe* 35, 13.

Santos, B.M., Mora-Bolanos, J.E. and Solorzano-Arrova, J.S. (2008) Impact of solarization and soil fumigants on hot pepper production in high tunnels. *Asian Journal of Plant Sciences* 7, 113–115.

Sasanelli, N., Konrat, A., Migunova, V., Toderas, I., Iurcu-Straistaru, E., *et al.* (2021) Review on control methods against plant-parasitic nematodes applied in Southern member states (C zone) of the European Union. *Agriculture* 11, 602. DOI: 10.3390/agriculture11070602.

Schneider, S.M., Ajwa, H.A., Trout, T.J. and Gao, S. (2008) Nematode control from shank- and drip applied fumigant alternatives to methyl bromide. *HortScience* 43, 1826–1832.

Schneider, S.M. and Hanson, B.D. (2009) Effects of fumigant alternatives to methyl bromide on pest control in open field nursery production of perennial fruit and nut plants. *HortTechnology* 19, 526–532.

Sigüenza, C., Schochow, M., Turini, T. and Ploeg, A. (2005) Use of *Cucumis metuliferus* as a rootstock for melon to manage *Meloidogyne incognita*. *Journal of Nematology* 37, 276–280.

Sikora, R.A., Coyne, D., Hallmann, J. and Timper, P. (eds) (2018) *Plant Parasitic Nematodes in Subtropical and Tropical Agriculture*, 3rd edn. CAB International, Wallingford, UK.

Sikora, R.A., Desaeger, J. and Molendijk, L. (eds) (2021) *Integrated Nematode Management: State-of-the-art and Visions for the Future*. CAB International, Wallingford. https://www.cabi.org/bookshop/book/9781789247541/.

Sikora, R.A. and Roberts, P.A. (2018) Management practices: an overview of integrated nematode management technologies. In: Sikora, R.A., Coyne, D., Hallmann, J. and Timper, P. (eds), *Plant Parasitic Nematodes in Subtropical and Tropical Agriculture*, 3rd edn. CAB International, Wallingford, UK, pp. 795–838.

Spotti, C. (2004) The use of fumigants and grafted plants as alternatives to Methyl Bromide for the production of tomatoes and vegetables in Italy. In: Batchelor, T.A. and Alfarroba, F. (eds), *Proceedings of International Conference on Alternatives to Methyl Bromide*. European Commission, Brussels, Belgium, pp. 57–59.

Sosnowski, M.R., Fletcher, J.D., Daly, A.M., Rodoni, B.C. and Viljanen-Rollinson, S.L.H. (2009) Techniques for the treatment, removal and disposal of host material during programmes for plant pathogen eradication. *Plant Pathology* 58, 621–635.

Sprunger, C.D. and Martin, T.K. (2023) Chapter three – An integrated approach to assessing soil biology health. In: Sparks, D.L. (ed.) *Advances in Agronomy* 182, 131–168.

Stapleton, J.J., Ruiz, T.S., McKenry, M.V. and Ferguson, L. (2001) An additional time/temperature solarisation treatment approved in California to ensure against nematode pest infestation of containerized nursery stock. In: *Proceedings 2001 Annual International Research Conference on Methyl Bromide Alternatives and Emissions Reductions*. San Diego, California, pp. 12-1, 12-2.

Starr, J.L., Cook, R. and Bridge, J. (2002) Resistance to plant-parasitic nematodes: History, current use and future potential. In: Starr, J.L., Cook, R. and Bridge, J. (eds), *Plant Resistance to Parasitic Nematodes*. CAB International, Wallingford, UK, pp. 1–22.

Stirling, G.R. (2014) Integrated soil biology management: the pathway to enhanced natural suppression of plant-parasitic nematodes. In: Stirling, G.R. (ed), *Biological Control of Plant-parasitic Nematodes: Soil Ecosystem Management in Sustainable Agriculture*, 2nd edn. CAB International, Wallingford, UK, pp. 304–341.

Sullivan, D.A., Holdsworth, M.T. and Hlinka, D.J. (2004) Control of off-gassing rates of Methyl Isothiocyanate from the application of metam-sodium by chemigation and shank injection. *Atmospheric Environment* 38, 2457–2470.

Suryawanshi, K.K. and Shinde, V.B. (2015) Nematode management strategies for organic farming and precision farming. *Indian Farmer* 2, 783–788.

Takken, F. and Rep, M. (2010) The arms race between tomato and *Fusarium oxysporum*. *Molecular Plant Pathology* 11, 309–314.

Tarasco, E., Fanelli, E., Salvemini, C., El-Khoury, Y., Troccoli, A., *et al.* (2023) Entomopathogenic nematodes and their symbiotic bacteria: From genes to field uses. *Frontiers in Insects Science. Sect. Insect Health and Pathology* 3. DOI: 10.3389/finsc.2023.1195254

Thies, J.A., Ariss, J.J., Hassell, R.L., Buckner, S. and Levi, A. (2015) Accessions of *Citrullus lanatus* var. *citroides* are valuable rootstocks for grafted watermelon in fields infested with root-knot nematodes. *HortScience* 50, 4–8.

Thies, J.A. and Fery, R.L. (2000) Characterization of resistance conferred by the N gene to *Meloidogyne arenaria* races 1 and 2, *M. hapla*, and *M. javanica* in two sets of isogenic lines of *Capsicum annuum* L. *Journal of the American Society for Horticultural Science* 25, 71–75.

Thies, J.A., Fery, R.L., Mueller, J.D., Miller, G. and Varne, J. (2003) Response of bell pepper cultivars near-isogenic for the N gene to *Meloidogyne incognita* in field trials. *HortScience* 38, 1394–1396.

Triky-Dotan, S., Austerweil, M., Steiner, B., Peretz-Alon, Y., Katan, J. and Gamliel, A. (2009) Accelerated degradation of metham sodium in soil and consequences or root-disease management. *Phytopathology* 99, 362–368.

Triolo, E., Materazzi, A. and Luvisi, A. (2004) Exothermic reactions and steam for the management of soil-borne pathogens: five years of research. *Advances in Horticultural Science* 18, 89–94.
Villar, E.M.J. and Zavaleta-Mejía, E. (1990) Efecto de *Crotalaria longirostrata* Hook y Arnott sobre nematodos agalladores (*Meloidogyne* spp.). *Revista Mexicana de Fitopatología* 8, 166–172.
Villar, E.M.J., Zavaleta-Mejía, E. and García, E.R. (1990) Efecto de la incorporación de residuos de cruciferas sobre fitopatógenos del suelo. II Efecto de la incorporación de col y brócoli sobre la pudrición blanca de la cebolla (*Sclerotium cepivorum* Berk) bajo condiciones de invernadero. *Revista Mexicana de Fitopatología* 8, 160–165.
Walia, R.K., Sharma, S.B. and Vats, R. (2000) Bacterial antagonists of phytonematodes. In: Upadhyay, R.K., Mukerji, K.G. and Chamola, B.P. (eds), *Biocontrol Potential and Its Exploitation in Sustainable Agriculture: Crop Diseases, Weeds, and Nematodes*. Springer, Boston, Massachusetts, pp. 173–186.
Wang, D., Juzwik, J., Fraedrich, S.W., Spokas, K., Zhang, Y., *et al.* (2005) Atmospheric emissions of methyl isothiocyanate and chloropicrin following soil fumigation and surface containment treatment in bare-root forest nurseries. *Canadian Journal of Forest Research* 35, 1202–1212.
Webster, R. (2010) Weeds, worms and geostatistics. In: Oliver, M.A. (ed), *Geostatistical Applications for Precision Agriculture*. Springer, Dordrecht, Netherlands, pp. 221–241.
Westerdane, B.B. (2020) Evaluation of trap cropping for management of root-knot nematode on annual crops. *Acta Horticulturae* 1270, 141–146. DOI: 10.17660/ActaHortic.2020.1270.15.
Whitehead, A.G. (1998) *Plant Nematode Control*. CAB International, Wallingford, UK.
Yakabe, L.E. and Macdonald, J.D. (2010) Soil treatments for the potential elimination of *Phytophthora ramorum* in ornamental nursery beds. *Plant Disease* 94, 320–324.
Zasada, I.A., Halbrendt, J.M., Kokalis, B.N., Lamondia, J., McKenry, M.V., *et al.* (2010) Managing nematodes without methyl bromide. *Annual Review of Phytopathology* 48, 311–328.
Zavaleta-Mejía, E., Castro, A.A.E. and Zamudio, G.V. (1993) Efecto del cultivo e incorporación de *Tagetes erecta* L. sobre la población e infección de *Meloidogyne incognita* (Kofoid y White) en chile (*Capsicum annuum* L.). *Nematropica* 23, 49–56.
Zinati, G.M. (2005) Compost in the 20th century: a tool to control plant diseases in nursery and vegetable crops. *HortTechnology* 15, 61–66.

20 Integrated Nematode Management – Rethinking and Restructuring the Different Components

Johan Desaeger[1]*, Richard Sikora[2], Koon-Hui Wang[3] and Tristan Watson[4]
[1]University of Florida, Wimauma, Florida, USA; [2]University of Bonn, Bonn, Germany; [3]University of Hawaii, Manoa, Hawaii, USA; [4]Louisiana State University, Baton Rouge, Louisiana, USA

Abstract
This chapter highlights the different components of integrated nematode management and the factors that impact the development of effective integrated nematode management (INM) programmes across crop types, and the wide spectrum of production systems that vary from small-scale to large scale. Development of INM programmes requires a solid ecological understanding of the target nematode pests and the crop production needs and limitations. INM tools although greatly affected by environment, are also influenced by the socio-political laws that control agriculture within a country. In addition, many tools cannot be used due to either a lack of farmer knowledge or financial resources. Specific examples of several currently used crop-nematode INM systems, with low to high numbers of management tools, are provided.

20.1. Introduction

In this chapter, we want to highlight the different components of integrated nematode management (INM) and the factors that impact the development of effective INM programmes across crop types and the wide spectrum of production systems that vary from small scale to large scale. Development of INM programmes requires a solid ecological understanding of the target nematode pests and the crop production needs and limitations (Desaeger *et al.*, 2021a). We want to stress that the INM tools shown in Figure 20.1 are greatly affected not only by environment but also by the socio-political laws controlling agriculture in a country. In addition, many tools cannot be used due to a lack of either farmer knowledge or financial resources (Cortada, 2021). At the end of this chapter, we will give specific examples of several currently used crop-nematode INM systems with low to high numbers of management tools.

20.1.1. Economic Importance of INM

Plant-parasitic nematodes are part of what has been called the 'big giveaway' by farmers who unwillingly feed a large proportion of the food they produce to insects, nematodes and diseases,

*E-mail: jad@ufl.edu

(eds R.H. Manzanilla-López, D.J. Hunt, N. Marbán-Mendoza and S. Infante Gil)
DOI: 10.1079/9781836990413.0020

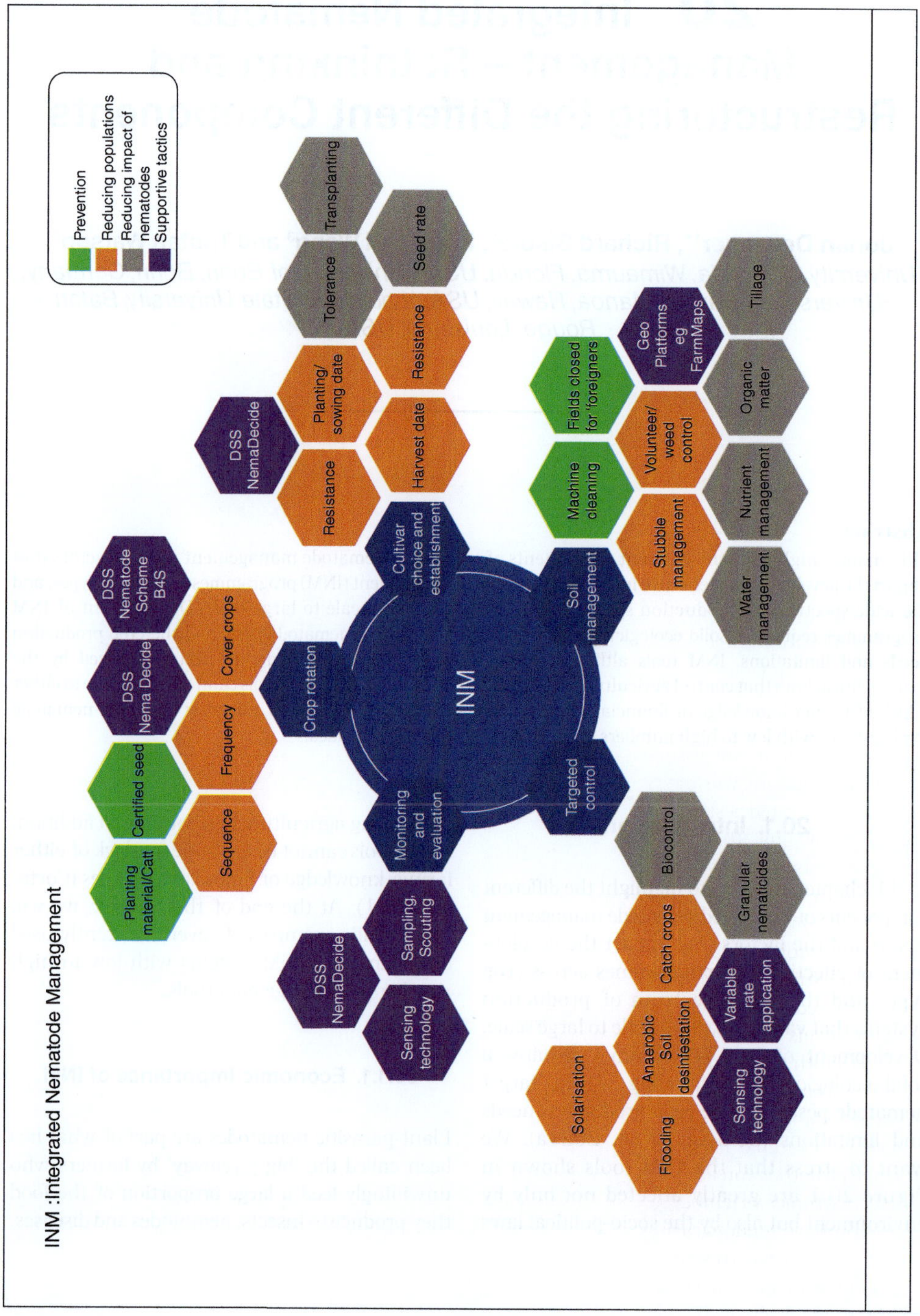

Fig. 20.1. The five pillars and management tools available for integrated nematode management (from Sikora *et al.*, 2021); DSS = decision support systems; B4S = Best for Soil (EU database).

among others, without compensation for their efforts (Savary *et al.*, 2019; Sikora *et al.*, 2020). Plant-parasitic nematodes have been calculated to reduce up to 10% of the world's agricultural output, causing economic losses valued at over US$125 billion each year (Chitwood, 2003). The amount of damage varies greatly depending on the (i) nematode-crop interaction involved, (ii) cropping systems followed, (iii) predominant environment factors and (iv) nematode management tools available. Damage due to plant-parasitic nematodes is usually more severe in subtropical and tropical climates, and more importantly, often mistaken for fertility, diseases or even abiotic factors. Nematode management is often neglected because of this lack of knowledge, and when it is practised, INM programmes can vary from non-existent to highly developed.

20.1.2. Integrated Pest Management

This highly recognized, and often used umbrella term, harbours more specific pest management programmes that deal with nematodes, weeds, insects and pathogens among others. The term often gives the impression that integrated management is widely used and that it is an effective management tool. Surprisingly, the opposite is currently occurring worldwide. In fact, there are very few integrated pest management (IPM) programmes, if any, that simultaneously target multiple pests and those that do seldom have components that target nematodes. The standard IPM definition used by the Food and Agriculture Organization is 'A management system that utilizes all suitable techniques in a compatible manner to reduce pest populations and maintain them at levels below those causing economic injury' (Smith and Reynolds, 1966).

The phrase – suitable techniques – is important because INM means different things to different people. For example, INM varies greatly depending on (i) whether you are rich or poor, (ii) where you farm (tropics vs temperate), (iii) what nematodes you have (endoparasites vs ectoparasites) and (iv) your access to resources (large commercial *vs* small family farms). These factors determine whether INM is a realistic option or a figment of a scientist's imagination. INM is also not a goal in itself but should be integrated with other tools to manage a plethora of microscopic nemesis of plants in the soil or true IPM because most crop producers are not challenged by nematode pests alone.

20.1.3. Rating INM Systems

There are many tools available that can be used in union, to develop and achieve effective INM programmes. The uniqueness of an INM programme is not necessarily related to the number of tools used. INM can be achieved with one tool: a fumigant (root-knot on vegetables), a nematicide (burrowing nematodes in banana), a resistant cultivar (soybean cyst nematode) or with up to seven tools (beet cyst nematode in sugar beet). Examples of these combinations will be given later in section 20.8. Ideally, INM systems should also be based on an ecological understanding of the target nematode pests and aim to improve the soil's natural capacity to reduce damage from plant-parasitic nematodes. However, in many cases, INM is not even attempted, due to the convenience of therapeutic nematicide use, or the inability of a farmer to use INM tools due to cost, availability or the highly specialized nature of the crop that is produced. Therefore, management efficacy ranges from non-existent to highly technical. The goal of INM should be a system that results in (i) a reduction in nematode damage, (ii) improved yield, (iii) financial profit and (iv) improved environmental health. INM can be made up of both independent and interdependent tools that were allocated to specific pillars of use (Sikora *et al.*, 2021a; Fig. 20.1).

The ability to combine these tools for nematode management is challenging and requires knowledge not only of the tools selected for use but also of crop production, the nematode targeted and of course the ability of the farmer to use these tools effectively. This chapter will attempt to show how best these tools can be used and why INM is needed to improve crop health worldwide.

20.2. Socio-political and Environmental Impact

The design of an INM programme is often strongly influenced by political and environmental constraints: hunger, degraded ecosystems, poor infrastructure, lack of access to finances, climate change, wars and political unrest. These factors limit the ability to produce food effectively and, in many cases, increase the intensity of nematode damage by limiting INM resources and management options. In addition, in many countries, there is increasing political pressure to regulate and reduce pesticide use, which is still an important tool in many INM programmes. These new government laws that are a reaction to negative human perception of agriculture as seen in the European Union (EU) will affect how we should restructure INM (Sikora *et al.*, 2023). Restructuring INM will require new information generated through good science that will allow transformation from commercial near monoculture cropping to multiple crop rotations that include non-hosts, break and trap crops, biocontrol and manipulation of the soil's antagonistic potential.

20.2.1. Organic Agriculture, Genetics and the Environment

Organic agriculture has a powerful lobby in the EU and in other parts of the world, and strongly influences both the present and future forms of agriculture. A sizeable segment of the population is now consuming organic food, and they are lobbying for a large reduction in the use of pesticides. This is being justified not only by the desire to have pesticide-free food but also by environmental side effects. The local populations in the EU, for example, are aware of the reduction in bee, butterfly, bird and flowering plant biodiversity. This form of community pressure will lead to changes in agricultural production systems and in how we develop INM. The future development of INM systems, therefore, must take sociological and government policy into consideration. A similar perception exists worldwide with regard to genetic manipulation. This perception has led to the successful banning of genetic manipulation in many countries where the technology is needed.

To reduce the impact of agriculture on the environment, laws have been proposed (The Green Deal and Farm to Fork) to set a goal of at least 25% organic farming by 2030 and a reduction of pesticide use by 50% (Farm to Fork Strategy, 2024). Although this bill recently failed to pass this year within the EU parliament, pressure to pass these types of laws will be promoted again in the future. Should these changes become law, agriculture and INM approaches will have to adapt. Some of the proposed changes in the EU include the mandatory monitoring of soil health indicators such as bacteria, fungus and worm populations as well as the quality and amount of biomass. These indicators could be of value to certain INM approaches. In other places like Brazil, the number of biological nematicides has expanded rapidly in recent years, while in many other countries, including the USA, there has been an increasing effort from large retailers to reduce the use of chemicals and grow more organic produce (Desaeger *et al.*, 2021b). All these trends emphasize the need for more and versatile INM programmes.

20.2.2. Re-Structuring INM and Organic Agriculture

The increased transformation of commercial agricultural production into organic will require new scientific knowledge and knowhow including the following: (i) nematodes presence, (ii) rotation crop host status, (iii) nematode population dynamics, (iv) effects of organic amendments and (v) shifts and manipulation of antagonistic potential. Wang and Uchida (2014) outlined five sustainable approaches for managing soil-borne pathogens that could be foundation principles for organic INM. These principles include (i) enhancing high biological diversity through polyculture instead of conventional monoculture systems, (ii) increasing ecosystem community stability by promoting natural enemies of multiple pests and pathogens, (iii) stimulating inherent plant defences, (iv) improving plant health by maintaining nutrient cycling and energy flow and (v) targeting vulnerable stages of pests or pathogens through the understanding of their ecology. This type of research is data-intensive and time-consuming. For some nematologists, it might look like we are moving backwards, but for others, we are moving forward to protect the environment. Interestingly, organic farming is almost a common practice on small family farms worldwide where pesticides are often too expensive. We know little about the nematode problems existing in these subsistence farming systems, and information is lacking on how to design INM for consistent performance.

20.2.3. Climate Change and Climate Variability

Environmental impacts such as heat waves and severe droughts will continue to affect crop yields and how INM is designed. While climate change is a long-term process (multi-decadal), climate variability refers to short-term or inter-annual or seasonal changes in weather. Such changes will have an immediate impact on agriculture and nematodes; therefore, it will be vital to develop management approaches to react positively to inter-annual changes that cannot be readily predicted (Sikora *et al.*, 2021b). There are many ways in which a changing climate can affect nematodes and the damage they cause; shorter life cycles, geographical shifts, impact on resistance genes, increased microbial diversity, to name a few. The outcome of climate impact will be hard to predict; however, the greatest impact will be seen in climate hotspots such as the semi-arid and arid regions of the world. In these regions, it will be especially important to develop INM practices that increase resilience to climate variability such as developing crop cultivars with combined drought and heat resistance, improving tolerance and resistance to nematodes, adjusting planting dates and optimizing soil moisture management.

20.3. Designing Crop Rotation Practices Against Nematodes

Crop rotation has been used since ancient times to combat soil degradation or what has often been called 'soil fatigue' or 'soil exhaustion'. Such soils are characterized by poor nutrient levels, low levels of organic matter, poor structure and often high populations of plant-parasitic nematodes. When crop rotation is abandoned, as is often the case in highly specialized near monoculture production systems, nematode problems often become worse. In addition to their potential to help manage nematodes and other soil pests, crop rotation and cover crops offer multiple benefits to growers such as: increasing soil organic matter, soil fertility and moisture holding capacity, reducing soil erosion, nutrient run-off and soil compaction. In addition, many cover crops can be used for cattle grazing or the biomass collected for feed. Because of its multiple benefits, crop rotation and cover crops are also more easily adopted by growers and should be an integral part of any INM system. In the words of Howard G. Buffett, a billionaire philanthropist and Illinois farmer "*Cover crops to me are just the next natural step in trying to have a broader system, and I think the single biggest issue we have as farmers in this country is we don't farm with a system in mind.*"

20.3.1. Different Types of Crop Rotation

Many examples exist of crop rotation programmes that are primarily used for nematode management, and their efficacy depends on the type of nematode and the selection of crops that are available. This approach to nematode management can be designed very differently based on whether the targeted nematode has a narrow versus a wide host range. Specialized plant-parasitic nematodes such as potato cyst nematodes (PCN), *Globodera rostochiensis* and *Globodera pallida*, and soybean cyst nematode, *Heterodera glycines*, are very host specific, typically only infecting crops within the same plant family. Because the eggs inside the cysts can survive for many years in the soil, long crop rotations with non-host crops like cereal crops are usually required. For instance, seven years of bare fallow have been required to manage PCN, *G. rostochiensis*, below economic threshold. Excellent control of PCN was also obtained using 'litchi tomato' (*Solanum sisymbriifolium*) as a rotation crop, which acts as a trap crop of *G. pallida* by stimulating juveniles in cysts to hatch but not allowing them to reproduce. Establishing a litchi tomato, however, can be a challenge due to its high temperature requirement and slow growth, as well as limited seed supply (Dandurand and Knudsen, 2016; Augustin, 2021). Ease of establishment and growth rate can make or break a cover crop, and the most successful cover crops are almost always fast-growing species that produce a high amount of biomass in a relatively short time.

Root-knot (*Meloidogyne* spp.) and reniform (*Rotylenchulus reniformis*) nematodes are sedentary endoparasites that have a much wider host range than cyst nematodes. Designing effective crop rotations for INM is, therefore, much more challenging. One successful example is the corn-cotton-peanut rotation used in the south-eastern USA (Davis and Kemerait, 2021; Lawrence, 2021). One of the most popular cover crops to help manage root-knot and reniform nematodes around the world is *Crotalaria* spp., fast-growing leguminous crops, that are known poor or non-hosts to root-knot nematodes, as well as to reniform nematodes (McSorley and Dickson, 1995). While some varieties of 'sunn hemp' (*Crotalaria juncea*) produce harmful alkaloids (most highly concentrated in seeds), others are non-toxic and make excellent forage.

Lesion nematodes, *Pratylenchus* spp., have a wide host range and, therefore, are also more difficult to manage with crop rotations. Some examples of rotations to manage lesion nematodes include rotations with the poor hosts, oat and sorghum, to manage *Pratylenchus thornei* in the grain growing region of eastern Australia (Owen, 2021), and rotation with 'Saia' oat (*Avena strigosa*) to reduce *Pratylenchus penetrans* in the north-eastern USA (LaMondia, 2006).

20.3.2. Cover Crops With Nematicidal Properties

In addition to being poor or non-hosts, cover crops may impact soil nematodes by directly killing them via the production of nematicidal or allelopathic compounds, either during their growth phase or following incorporation of its biomass. This type of cover crop may offer a better crop rotation strategy when dealing with nematodes with wide host ranges, that is, root-knot and lesion nematodes. A list of the well-known cover crops with allelopathic effects against plant-parasitic nematodes was reviewed by Wang and Uchida (2014). Allelopathic compounds are found in Brassica crops such as mustard, radish and rapeseed that produce glucosinolates and convert it to isothiocyanates upon hydrolysis (Brown and Morra, 1997); sunn hemp (*C. juncea*) that produce monocrotaline (Warnke *et al.*, 2008); sorghum (*Sorghum bicolor*) that produce dhurrin which is converted into hydrogen cyanide upon hydrolysis (Abawi and Widmer, 2000); marigold (including *Tagetes erecta, Tagetes minuta, Tagetes patula, Tagetes polynema*) produce α-terthienyl (Hooks *et al.*, 2010), velvet bean (*Mucuna pruriens*) that produces L-dopa or 3,4-dihydroxy phenylalanine (Nogueira *et al.*, 1996) and horsebean (*Canavalia ensiformis*) that produce concanavalin A (Rocha *et al.*, 2017). Not all nematode allelopathic cover crops suppress the targeted nematodes the same way. Some of them leach out their allelopathic compounds after its residues are soil incorporated such as sunn hemp (Wang *et al.*, 2002, 2003), some of them such as marigold release their nematicidal compounds through living root systems (Hooks *et al.*, 2010), whereas sorghum and brown mustard need to be tissue macerated/chopped and hydrolysed into biofumigants and soil incorporated to suppress the nematodes in the soil (Paudel *et al.*, 2021; Waisen *et al.*, 2020). For example, marigold (*T. erecta* 'Crackerjack') as a nematode-allelopathic rotational crop reduced the abundance of *P. penetrans* on potato (Kimpinski *et al.*, 2000) and carrot (Kimpinski and Sanderson, 2004).

There are still many questions as to how and under which conditions these nematode allelopathic cover crops suppress nematodes. The allelopathic compound in sunn hemp is more prominent in the leaves or other shoot tissues than the roots, and younger sunn hemp tissue (<3 months old) is more toxic than that from older plants (Wang and Zasada, 2011). Marahatta *et al.* (2012a) showed that marigold as a rotation crop was more effective against *Meloidogyne* spp. if planted when these nematodes are in their actively moving stage (i.e., planted right after termination of a susceptible crop rather than planted in fields that had been dry fallow), whereas *C. juncea* as a green manure suppressed reniform nematodes more effectively following a susceptible host than when incorporated into soil that has been dry fallowed for months (Marahatta *et al.*, 2012b).

20.3.3. Limitations of Cover Crops

The main limitation of crop rotation and cover crops is that agricultural fields will have many different types of plant-parasitic nematodes and that by reducing populations of certain plant-parasitic nematode species, other species may fill the void. *Crotalaria* spp., for instance, will often increase populations of *Pratylenchus* spp., as well as spiral nematodes (Desaeger and Rao, 2000, 2002). The key is to identify the most damaging species and focus on reducing this species. Knowing the nematode host status and selecting the right cover crop species are critical, if not, a cover crop could make the nematode problem worse. In many cases, the nematode host status of cover crops is species-specific and will require accurate identification of the target nematode (Bui and Desaeger, 2021). Conversely, nematode host status can also vary significantly among various cultivars of the same cover crop species (e.g., cowpea and oats) (Gallaher and McSorley, 1993).

Cover crops are mostly used in annual crops but can also be integrated into perennial crops (banana, fruit and nut trees, etc.). They can be grown as intercrops and groundcovers and be either permanent (annual reseeding legumes, perennial sods), or temporary (green manure crops). Competition for water and nutrients is naturally a concern in such systems, so having ample rainfall or irrigation is important. Most of the research on intercrops and groundcovers in perennial crops has focused on issues such as soil conservation, biodiversity enhancement and weed control, and little has been done investigating the effect on nematodes. As for annual crops, it is critical to select the right cover crop species based on the target nematode pest(s). Lamondia (2021) clearly demonstrated the dilemma of multiple key plant-parasitic nematodes coexisting in an apple orchard. Millet (*Pennisetum glaucum*), black-eyed Susan (*Rudbeckia hirta*) and sudangrass 'Trudan 8' (*Sorghum sudanense*) as an annual green manure effectively controlled *P. penetrans*, but increased population densities of *Xiphinema americanum*, whereas rapeseed 'Dwarf Essex' (*Brassica napus*) reduced *X. americanum* densities but increased *P. penetrans* (Lamondia, 2021). In addition, Lamondia (2021) showed differences within *Brassica juncea* where 'Pacific Gold' but not 'Caliente' could reduce population densities of *X. americanum*. Selecting the right crop is especially important when a permanent groundcover is planted, which will have a long-term impact on the crop, and could aggravate nematode problems if the wrong species is selected.

In many places, there has been a push towards planting cover crop mixtures rather than single species cover crops. Cover crop mixtures can provide more benefits in terms of soil nutrient management. For instance, mixtures that contain legumes and cereals combine the nitrogen-fixing capability of legumes with the nitrogen-scavenging ability of cereal crops. Such mixtures may also provide benefits for nematode management, as they can be expected to have more diverse soil nematode populations and are less likely to have any of the more damaging nematode species build up to high levels (Desaeger and Rao, 2001).

A new area of research has focused on the effect that cover crops have on the soil microbiome. Recent studies have shown that cover crop species, even cultivars, have their own microbial signature (at least in part) depending on the C/N ratio (Cazzaniga *et al.*, 2023; Aiyer *et al.*, 2024). These 'cover crop microbiomes' are regulated by specific root secretions during plant growth as well as residue decomposition after cover crop termination. This type of agricultural soil microbiome research could be of great value to INM in the future and may help us better understand and manage the nematode-suppressive potential of agricultural soils.

20.4. Improving Plant Health Through Soil Management

Free-living nematodes play an important role in soil nutrient cycling and mineralizing soil nutrients from soil organic matter for plant uptake. It is often thought that increasing free-living nematodes, especially ones in the higher hierarchy of the soil food web, such as omnivorous and predatory nematodes, may help to keep the abundance of plant-parasitic nematodes in check. This has been called the 'Linford Hypothesis' after M.B. Linford who was the first to report amendment-induced nematode suppression (Linford *et al.*, 1938; Stirling, 1991; McSorley, 2011). However, top-down regulation of omnivorous or predatory nematodes is often diluted by changes in other soil microbes or fauna as the succession of soil food web structures evolved over time (Paudel *et al.*, 2021).

20.4.1. Soil Conservation for More Stable Communities

Frequent soil tillage disturbs nematode community structures, especially those of nematodes in the higher hierarchy of the soil food web, which include most predatory nematodes (McSorley *et al.*, 2007). Conservation tillage such as the use of strip tillage and roller-crimpers is growing in popularity, especially for field crops. It is a practice that can improve the community structure and increase the abundance of omnivorous or predatory nematodes, but it takes time for such a build-up to influence PPN (Dupont *et al.*, 2009). To speed up this process, Wang *et al.* (2011) added organic matter with low C:N ratio and compost tea, in conjunction with conservation tillage. By growing sunn hemp and terminating the cover crop through strip tillage, leaving rows of sunn hemp intercropped with cucumber or zucchini, and periodic clipping of the sunn hemp as surface organic mulch, combined with weekly drenching of vermicompost tea, they increased the structure of soil food web within one cropping cycle (Wang *et al.*, 2014).

20.4.2. Soil Amendments and Nematodes

It is widely accepted that nematode damage tends to be more severe in sandy soils. It is also often said that by increasing organic matter content, nematode damage can be reduced. This can be attributed to stimulation of nematode parasites and antagonists and increasing non-parasitic or free-living nematodes. Amendments can include manures (solids, slurries, extracts), composts, crop and food processing residues, green manures, seed meals and nematicidal plants and extracts. Regular inputs of animal manures or organic wastes, combined with crop rotations, reduced tillage, residue retention, cover cropping enhanced the performance of naturally occurring nematode biocontrol agents against lesion and root-knot nematodes (Stirling, 2011). The nematicidal activity of an organic amendment has been related to its C/N ratio, with lower C/N ratios generally considered to be more nematicidal (Rodriguez-Kabana, 1986).

Several reviews have been written on the use of organic amendments for the management of plant-parasitic nematodes (Rodriguez-Kabana, 1986; Oka, 2010; McSorley, 2011; Thoden *et al.*, 2011) with often contradictory results. McSorley (2011) concluded that there is a lack of evidence supporting the Linford hypothesis and that amendments mostly cause a positive crop growth response, but not so much actual nematode suppression. This may be partly because nematode suppression is not always easy to measure and amendments not only impact just nematodes but also plant growth. Indices such as the nematode structure index and functional metabolic footprint (see Chapter 17, this volume) have been used as indicators of the suppressiveness of a soil towards certain nematodes (Ferris *et al.*, 2001; Ferris, 2010), but they probably oversimplify the picture. Soil is an extremely complicated and competitive environment. The chance of successfully introducing suppressiveness for long-term nematode management in arable soil is negligible. However, raising the density of soil antagonists (see Chapter 21, this volume) together with the activation of metabolically inactive biocontrol agents could be a promising way forward in biological nematode control (Mazzola, 2007; Helder and Heuer, 2021). Recently, studies have underlined the difference between the resident and the active microbial community in agroecological settings (Cazzaniga *et al.*, 2023), and this implies that soils might have an unrealized potential for suppressiveness vis-à-vis plant-parasitic nematodes.

20.4.3. Soil Microbiome Is the New Frontier

Until recently, the detection or monitoring of the abundance and/or the activity of relevant antagonistic bacteria and fungi would have been costly and technically demanding. With the boom in soil microbiome research and access to new DNA sequencing technologies over the last two decades, we can now map and monitor changes in soil biota more efficient and better identify biota that are related to disease suppressiveness (Sikora *et al.*, 2023). Certain biota that have shown to be especially rich in nematode antagonists are the fungal families *Orbiliaceae* and *Ophiocordycipitaceae*. With more affordable DNA and RNA extractions and sequencing technology, determining the soil antagonistic potential present in any field may become more attainable. This could further help to more precisely monitor the effectiveness of soil management such as crop rotation and cover crops on the soil's antagonistic potential against plant-parasitic nematodes. Potentially, this could become a next-generation biocontrol tool that will aid farmers in more durable nematode management (Sikora *et al.*, 2023).

20.5. Nematicides and Biologicals for Targeted Control

Chemical nematicides can be credited for demonstrating the large amount of crop loss plant-parasitic nematodes can cause, starting in the 1920s, and for the boom in nematode research in the decades afterwards (Taylor, 2003). Several comprehensive reviews on nematicides have been written in the past (Taylor, 2003; Desaeger *et al.*, 2020). Nematicides remain the preferred nematode management strategy for many growers, especially for high-value cash crops such as fruits and vegetables. The demand for nematicides continues to grow, probably because nematodes are getting increasingly problematic (warmer climate, more intensive agriculture) but also because farmers are becoming more aware of nematodes.

The first nematicides were all broad-spectrum products (see Chapter 19, this volume), either fumigants (soil sterilants), organophosphates or carbamates (neural toxins). Only since 2010 have less toxic and more selective nematicides such as fluensulfone, fluopyram, fluazaindolizine and cyclobutrifluram become widely available (Table 20.1). This influx of new nematicides is the result of a loss of old nematicides (e.g., methyl bromide, carbamate and organophosphate nematicides) and a growing demand for safer and more selective products. The new nematicides differ from previous products in that (i) they are more selective, often only targeting nematodes and (ii) they are less toxic and safer to use. They are also less likely to have a negative impact on non-target beneficial soil organisms and more amenable to INM programmes (Desaeger *et al.*, 2020). Their use will

Table 20.1. Chemicals currently used as nematicides in different parts of the world.

Common name[a]	First use (Country)	Product type/Chemistry	Mode-of-action[b]	Signal words[c]
Chloropicrin	1920/1936	Fumigant	Multi-site	Danger
Methyl bromide	1932/1961	Fumigant	Multi-site	Danger*
1,3-D	1954	Fumigant	Multi-site	Danger
Metam sodium	1954	MIT generator	Multi-site	Danger
Ethoprop	1963 (USA)	Organophosphate	AChE	Danger
Aldicarb	1965 (USA)	Carbamate	AChE	Danger*
Dazomet	1967	MIT generator	Multi-site	Danger
Carbofuran	1969	Carbamate	AChE	Danger*
Fenamiphos	1968 (DE)	Organophosphate	AChE	Danger*
Oxamyl	1972 (USA)	Carbamate	AChE	Danger
Terbufos	1974 (USA)	Organophosphate	AChE	Danger*
Cadusafos	1990? (USA)	Organophosphate	AChE	Danger*
Imicyafos	2010 (JPN)	Organophosphate	AChE	Danger*
Fosthiazate	1992 (JPN)	Organophosphate	AChE	Danger*
Ivermectin / Abamectin	1981 (JPN)	Lactone	GluCl	Danger
Spirotetramat	2008 (USA)	Tetramic acid	LBI	Caution
DMDS	2010 (USA)	Fumigant	Multi-site	Danger*
Allyl ITC	2013 (USA)	Fumigant	Multi-site	Danger
Fluensulfone	2014 (USA)	Thizaole	Unknown	Caution
Fluopyram	2010 (USA)	Benzamide	SDHI	Caution
Fluazaindolizine	2024 (USA)	Carboxamide	Unknown	Caution
Cyclobutrifluram	2024 (ARG)	Benzamide	SDHI	Caution

[a]1,3-D = 1,3-dichloropropene; MIT = methyl isothiocyanate generator; Allyl ITC = allyl isothiocyanate; DMDS = dimethyl disulfide

[b]AChE = Acetylcholinesterase inhibitors; inhibition is reversible for carbamates, and irreversible for organophosphates; GluCl = Glutamate-gated chloride channel allosteric modulators; SDHI = succinate dehydrogenase inhibitors; LBI = Lipid Biosynthesis Inhibitor.

[c]*Limited registration. Sources: Newhall, 1955; Chitwood, 2003; http://nemaplex.ucdavis.edu/Mangmnt/Chemical.htm; https://sitem.herts.ac.uk/aeru/iupac/Reports/19.htm.

Adapted from Desaeger *et al.* (2020).

likely increase in the next decades, and they will gradually replace older nematicides.

20.5.1. Risks With New Nematicides

A concern with especially succinate dehydrogenase inhibitor nematicides, being more selective, having long soil persistence and similar mode-of-action towards fungi and nematodes is the potential selection of nematicide-resistant (less-sensitive) populations (Desaeger *et al.*, 2020). This was confirmed recently when (for the first time in the history of nematicides), resistant root-knot and sting nematode (*Belonolaimus longicaudatus*) populations against fluopyram were reported from Florida and Spain (Kammerer *et al.*, 2023; Braun *et al.*, 2024). Nematode resistance was never a major concern for nematicides in the past, probably due to the broad-spectrum nature and relatively limited use of most of the old products. The recent reports on resistant nematode populations are a reminder to use any nematicide judiciously, to not exceed labelled rates and use them in combination with other methods and as part of an INM programme.

In the past, metam-based fumigants and many organophosphate and carbamate nematicides have also been reported to be prone to accelerated degradation in the soil caused by microbial adaptation (Smelt *et al.*, 1996). This simply means that microbe populations in these fields have started metabolizing the nematicides at a faster rate, reducing the nematicide's efficacy. This is something that will have to be monitored for the new nematicides as well. Recently, a new Insecticide Resistance Action Committee (IRAC) Nematode Working Group was established by companies that have a stake in nematicides to investigate the resistance risk of new nematicides and to develop a mode-of-action classification scheme similar to insecticides and acaricides (IRAC, 2024). Another concern is related to the fact that the new nematicides are fluorinated compounds. Fluorination is a common practice in agrochemical discovery and brings favourable properties such as selectivity, quick action and longer residual activity (Jeschke, 2004). However, potential environmental and health concerns of polyfluoroalkyl substances have brought on increased scrutiny on fluorinated compounds, including many agrochemicals (Zhang and Liang, 2023).

Despite these concerns, new nematicides can be a valuable addition to INM programmes as they provide safer and more selective chemical options than have been available in the past. Especially for high-value cash crops and when other management tools are not an option, they can be a good fit. While they offer safer and needed solutions for growers that can afford it, they are likely too expensive and not within reach of the many small farmers around the world. Also, many nematicides are formulated into liquid form and are easy to apply through chemigation methods, especially drip irrigation, which is often not available in subsistence agriculture.

20.5.2. Biologicals Gain Momentum

Many natural nematicidal toxins can be derived from either plant (e.g., thyme oil, mustard oil, neem oil and other essential oils) or bacteria (e.g., *Burkholderia*) and fungi (e.g., *Myrothecium*) In addition, many nematode antagonistic microorganisms have been identified as potential nematode biocontrol agents. This includes fungi such as *Purpureocillium lilacinum*, *Pochonia chlamydosporia*, several *Trichoderma* species, and bacteria such as *Pasteuria* spp., and many *Bacillus* and *Pseudomonas* spp. Extensive reviews have been published on natural-derived compounds and biological control organisms (Stirling, 1991; Sikora, 1992; Chitwood, 2002; Ntalli and Caboni, 2012; Poveda *et al.*, 2020; Topalovic *et al.*, 2020).

Sesame seed, thyme, garlic and other essential oils are a popular source for nematicidal products. Mushroom compost waste from cultivating mushrooms in the *Pleurotaceae* (e.g., *Pleurotus*, *Hohenbuehelia* and *Resupinatus*) were found to exude a toxin, trans-2-decenediotic acid, from their fungal hyphae and paralyses and colonizes nematodes *in vitro* (Kwok *et al.*, 1992) and in pot trials (Okorie *et al.*, 2011). Soap bark tree, *Quillaja saponaria*, has been formulated into a bionematicide to control various plant-parasitic nematodes in vineyards, orchards, field crops, turf and ornamental plants (D'Addabbo *et al.*, 2020). Chitinous materials (shrimp and crab shell meal) are popular soil amendments in organic production (Sikora and Roberts, 2018). Chitinous materials enrich soil chitinolytic microorganisms, which in turn digest the chitin on fungal cell wall and nematode eggshells.

Some nematode antagonistic microorganisms have been formulated into commercial products (see Chapter 21, this volume). For example, *Myrothecium verrucaria* and *Burkholderia rinojensis* are fermented killed mycelium toxic to plant-parasitic nematodes (EPA, 1998). Conidia of a nematode egg parasitic fungus, *Purpureocillium lilacinum* (= *Paecilomyces lilacinus*) have been formulated into biological nematicides and are sold in many countries (Kiewnick and Sikora, 2006). *Bacillus firmus* has been formulated commercially and was shown to produce toxins that can damage the peripheral egg pellicle of gall-forming nematodes, thus preventing the hatching of the nematode (d'Errico *et al.*, 2019). Plant growth-promoting rhizobacteria like *Streptomyces saraciticus* SS31 produce amylase, protease, cellulase and chitinase, as well as indole-acetic acid and siderophores, and suppress a wide range of soil pathogens (Chen, 2008; Saeed *et al.*, 2021). Wu *et al.* (2021) made a concoction of SS31 with several agricultural wastes, including soybean meal, crab and shrimp shells, neem seed oilcake, sugar and dolomite, and formed a bioorganic soil amendment (LTM-m) providing a good example of INM strategy against *Rhizoctonia solani* and *Meloidogyne incognita* concurrently.

One of the main challenges is how to integrate nematode biological control agents into industry-scale crop production. Most biological control agents are used in conventional agriculture and a better understanding of how to combine them with other inputs is needed. Some of the best examples of biological and chemical integration can be found in seed treatments of crops such as soybean, cotton and corn. *B. firmus* is now commonly used as a nematicide seed treatment together with fungicide and insecticide seed treatments. The challenge is how to boost integration in other crops, especially ones where nematodes can be extremely challenging, like in intensive vegetable and fruit production. There are many questions that need to be addressed, such as (i) are biological products compatible with chemical nematicides, (ii) should they be used before or after the chemical, (iii) can they be mixed with a chemical or (iv) are they to be used as a rotational partner, just to name a few.

20.6. Host Resistance

Within the context of nematode management, host resistance refers to the ability of a plant host to prevent or limit nematode reproduction. When available, host resistance is often the preferred strategy to limit nematode-associated yield losses because it is cost-effective and environmentally benign. In some cases, host resistance can be conferred through a single dominant resistance gene (R gene) in the plant host that interacts with a corresponding avirulence gene in the nematode, also known as a 'gene-for-gene' interaction (see Chapter 18, this volume). This interaction initiates a cascade of defence responses that often leads to degradation of the nematode feeding site resulting in complete resistance (i.e., qualitative resistance). An example of this is the *Mi-1* resistance gene in tomato, which provides a high level of resistance to *M. incognita*, *Meloidogyne javanica* and *Meloidogyne arenaria* due to a hypersensitive response that prevents feeding site development (Milligan *et al.*, 1998). In other cases, host resistance is a result of complex traits inherited in a polygenic manner, with multiple genes involved in providing incomplete or partial resistance (i.e., quantitative resistance). For example, in cotton, partial resistance to *M. incognita* is conferred through two independent resistance genes located on chromosomes 11 and 14 (Gutiérrez *et al.*, 2010). One gene is responsible for reducing gall formation and the other gene reduces egg production (Wubben *et al.*, 2020).

20.6.1. The Old and the New

Conventional breeding for nematode resistance often involves identifying natural host resistance in non-elite varieties and wild relatives and introgressing the resistance traits into high-yielding commercially acceptable varieties. This process has been used to introduce the *Mi-1* resistance gene from *Solanum peruvianum* (wild relative) into cultivated tomato (*S. lycopersicum*) to confer root-knot nematode resistance (Milligan *et al.*, 1998). Although natural host resistance to nematodes has been identified in several species of crop wild relatives, the number of reports of natural resistance to nematodes far exceeds the number of reports of successful deployment of new nematode-resistant varieties. This is often because introgression through conventional breeding is highly time-consuming, can add unwanted deleterious genes and can be difficult on crops with complex genomes such as sugarcane.

New biotechnology tools now allow for the introduction of nematode resistance genes directly into elite cultivars, difficult-to-breed polyploid crops, sterile plants and parental lines used in breeding programmes in a single well-characterized region of the genome. Transgenic expression of proteinase inhibitors in potato has successfully been used to reduce damage by *G. pallida* (PCN) and has shown potential for engineering host resistance in other crops as well (Fuller *et al.*, 2008). The use of *in planta* RNA interference (RNAi) has also been explored as a means of engineering plants to silence key nematode genes associated with parasitism (Lilley *et al.*, 2007). More recently, CRISPR/Cas9-mediated genome editing has been deployed to confer new modes of resistance to *H. glycines* in soybean through a loss-of-function mutation in a key plant gene involved in successful nematode parasitism (Usovsky *et al.*, 2023). Although genetic engineering of nematode host resistance is still a relatively new technology, commercialization of transgenic soybean varieties with resistance to *H. glycines* is currently underway (Kahn *et al.*, 2021).

Preserving the durability of host resistance is a key goal of INM programmes. When host resistance is conferred through a single dominant R gene, there is selective pressure on the nematode population that can lead to the development of resistance-breaking genotypes. A classic example of this is in soybean, where overuse of a single source of resistance (PI88788) led to the selection of resistance-breaking populations of soybean cyst nematode (*H. glycines*) in the USA (Niblack *et al.*, 2008). The addition of multiple resistance genes into a single cultivar (i.e., gene stacking) may reduce the likelihood of the development of resistance-breaking genotypes. Combining host resistance with additional management practices, such as crop rotation and nematicide use, may prevent the development of resistance-breaking nematode populations by presenting the nematode population with multiple barriers to overcome. Current research is aimed at developing soybean varieties with new sources of host resistance and developing INM programmes that can preserve the durability of host resistance.

20.6.2. A Case for Tolerance

The difference between resistance and tolerance is that resistant genotypes will greatly reduce nematode reproduction, while tolerant genotypes will not reduce the nematode population, but the crop will suffer less or no yield loss. With nematode-resistant cultivars not available for most crops, identifying tolerant cultivars is the next best option. The problem with this strategy is that nematode populations will remain high and can still cause damage to more susceptible crops. Still, for many field crops, tolerant genotypes are an important tool to help with nematode management. Several grain cultivars exist that have good tolerance for lesion nematodes, which is a common problem in many grain-growing regions. In the grain production region of eastern Australia, *P. thornei* is managed mostly using tolerant cultivars (Owen, 2021). In Brazil, *M. incognita* and *R. reniformis* (reniform nematode) are the main nematode problems in cotton, and while *M. incognita*-resistant cotton cultivars are available, for reniform nematodes only tolerant cultivars can be used (Machado, 2021). Tolerant cultivars are also an important component of INM of *Heterodera schachtii* (beet cyst nematode BCN), one of the main pests of sugar beet in Europe (Daub, 2021). In this case, resistant cultivars for BCN are available, but they are unattractive to growers due to their low yield potential.

20.7. Monitoring Programmes and Decision Support Systems

One of the reasons why INM is lacking in many places is that INM is much more knowledge-driven and data-intensive than more simple management options like soil fumigation or planting a resistant cultivar. Developing INM programmes requires information on nematode host status of the crop(s), crop sensitivity, nematode density and distribution, and effectiveness of individual management measures (Molendijk and Sikora, 2021). Most of these programmes were developed in countries with modern agricultural structures and a sizeable scientific community that are dedicated to nematode management. Also, government policies and incentives to farmers to embrace more sustainable practices are almost always an essential driver in developing INM programmes.

20.7.1. Show Me the Data

One of the first steps for successful INM is to set up nematode databases. In many countries, databases for quarantine nematodes have been developed, for example, that of the European Plant Protection Organization (https://gd.eppo.int/), the Widely Prevalent Nematode list in North America (https://www.prevalentnematodes.org/) as well as other databases maintained by other plant protection organizations around the world. These programmes are aimed at preventing the introduction and spread of invasive nematode pests. With increasing globalization and trade, there is a greater risk of spreading harmful nematodes with infested plants and soil. This is especially the case for endoparasitic nematodes, which can be easily spread inside infected planting material. PCNs have been spread across the world with infected tubers. Many other quarantine nematodes, like *Meloidogyne enterolobii*, are now commonly intercepted by plant inspection services, especially in the global trade of ornamental and nursery plants.

To support the adoption of INM, decision support systems (DSS) have been developed in countries that have a well-developed agricultural research and extension service. This is a requirement for the development and implementation of DSS, and something that is not present in most countries. DSS programmes are most likely to start with high-value horticultural and industrial crops, and in regions where information is highly developed and accessible to farmers. A good example is Best4Soil, a European network DSS on prevention and reduction of nematodes and soilborne diseases. The programme provides an INM system by developing 'clever crop rotations' (www.best4soil.eu/). Other websites are available from universities, extension services and government agencies across the world that offer information on diagnostics, sampling and management on a wide array of nematodes and crops (Molendijk and Sikora, 2021).

To collect the enormous amount of data that is required, more and better tools are needed. We need to improve our ability to accurately measure nematode populations and predict the risk of damage. Measuring nematode soil populations is time-consuming and not always very reliable due to variability in the field and limitations of sampling and extraction methods. Predictive sampling before planting can give deceptively low counts as nematodes may be dormant, present as eggs (which, except for cyst nematodes, are not extracted with current methods) or hiding deeper in the soil. Molecular identification and quantification tools like qPCR may give more accurate information but are limited by the small amount of soil that is analysed. Other than improving our ability to measure nematode populations, monitoring and predicting nematode development are needed. The use of soil moisture and temperature probes can provide real-time monitoring of temperature and moisture levels in the soil that influences nematode population development over time. These data can be used to estimate population increase, and time the use of nematicide and biocontrol treatments, as well as for timely destruction of trap crops before egg laying begins.

20.7.2. Remote Sensing and Precision Agriculture

Remote sensing could be a game-changer in upgrading INM if it works for the so-called 'unseen enemy or nematodes' in the soil ecosystem (Kuska *et al.*, 2021). The use of infrared (IR) cameras was used in the 1970s to spot poorly growing plants. Since then, major improvements in cameras and sensor as well as computer analysis have improved remote sensing and precision application technology in pest control. Remote sensing can vary from simply walking a field with a handheld device, to using drones, to analysing satellite images. Many different vegetation indices (e.g., Normalized Differential Vegetation Index) can be used to characterize a field and reveal where the crop is thriving and where it is under stress. In addition to multispectral and hyperspectral sensors that record the electromagnetic spectrum, thermal sensors can be used (Mahlein, 2018).

Remote sensing is already used for the detection and management of foliar pathogens and insects. Conversely, the use of remote sensing for soilborne nematodes and pathogens has been tested but is not being used extensively. The technology can be used to identify pockets of nematode damage in fields, but the infestations must be verified with ground truth sampling. In some cases, the use of soil electrical conductivity to determine soil textures, coupled with Global Positioning Systems and Geographic Information Systems technology, has been used for site-specific nematicide applications (Mueller *et al.*, 2010; Plumblee and Muller 2021). Field populations of the Columbia lance nematode (*Hoplolaimus columbus*) in cotton and soybean in the south-eastern USA are highly correlated with the distribution of coarse-textured sands in a field, making this nematode a prime candidate for site-specific management (Mueller, 2021).

More important is the fact that at the time of detection, the crop is already damaged and in-crop treatment is usually not possible or economical. However, since nematode mobility is almost non-existent, detection of sites in fields with damage year after year should lead to precision application of nematicides, biocontrol agents and/or resistant cultivars. This is still not happening on a large scale. Where resistant cultivars are used, e.g., in soybean, remote sensing will obviously not be used. The same applies to horticultural crops where fumigation and nematicides are used repeatedly at planting. New multispectral and thermal cameras and high-powered mathematical analysis could lead to the determination of population densities in the standing crop and thereby the prediction and selection of INM tools in the following crops (Hillnhütter *et al.*, 2010; Sikora and Roberts, 2018).

Limiting the use of remote sensing and precision application of management tools for INM is the cost of the detection and application technology. Satellite images are to some extent available for free, but easy image analysis is still not available to most farmers. Of course, the major question with remote sensing for nematodes is still how to separate nematode damage from other crop stressors. The hope is that with better sensors and more accuracy to measure nematodes, we will be able to more precisely discover specific patterns related to nematode damage in the future.

20.8. Case Studies

In this final section, we give some examples of nematode management for specific crop-nematode combinations, including low and high-value crops (Table 20.2). While INM can be very common in some situations (BCN in EU), for many other nematodes and crops, INM is not really practised and only one or two, mostly chemical, strategies are the main practice.

20.8.1. The Sugar-beet Cyst Nematode, *Heterodera schachtii* in Europe: Five INM Tools

When INM is rated by the number of tools used to manage a nematode problem, the integrated control programme used for sugar beet cyst nematode, *H. schachtii* in the EU, is an excellent example. INM of this cyst nematode is now used on a practical level by most growers with nematode infestations (Daub, 2021). It is a programme that is constantly improving through research, and is an INM programme, that gives the growers of this high-value crop a great deal of flexibility in developing their rotations to take advantage of the markets for sugar.

The Problem: The sugar beet cyst nematode *H. schachtii* is a major factor limiting the yield of the beet crop in Central Europe. It has been estimated that 10-20% of the EU growing area is infested and in some areas with a long history of sugar beet up to 50% of the fields are infested. In the 1990s, estimated losses were Euro 90 million annually (Müller, 1999).

Current Practice: INM of *H. schachtii* has been based on using a combination of the following five management tools: (i) soil sampling for estimation of the damage threshold and need for control input, (ii) 2 years of a non-host, winter wheat and barley, each reducing the egg and juvenile population 30%, (iii) the use of a nematode resistant white mustard or oilseed radish as a catch crop which reduces nematode populations 20-60% and (iv) the planting of tolerant

Table 20.2. Nematode problems in different crops and INM tools employed for their management.

Nematode	Host	Sanitation/ Roguing	Rotation/Cover crops	Resistance/ Tolerance	Seed treatment	Nematicide	Timed planting	Flooding/BSD[1]
Heterodera glycines	Soybean		X	X	X			
H. schachtii	Sugar beet		X	X	X	X		
H. avenae	Wheat		X	X			X	
Rotylenchulus reniformis	Cotton		X	X	X	X		
Globodera rostochiensis	Potato	X	X	X		X	X	
G. pallida	Potato		X	X		X	X	
Meloidogyne spp.	Tomato		X	X		X	X	X
M. incognita	Cotton		X	X	X	X		
M. enterolobii	Vegetables		X			X		
M. hapla	Vegetables		X			X		
M. graminicola	Rice		X					X
M. chitwoodi	Potato		X			X		
M. arenaria	Groundnut		X	X		X		
Meloidogyne spp.	Coffee			X		X		
Pratylenchus spp.	Cereals		X	X				
Ditylenchus dipsaci	Beet		X	X		X	X	
D. dipsaci	Onion	X	X			X		X
Xiphinema spp.	Grape vine			X		X		
Trichodorids	Potato	X	X			X		
Belonolaimus longicaudatus	Strawberry		X			X		
Aphelenchoides[2]	Strawberry	X	X					
Bursaphelenchus xylophilus	Pine trees	X						
B. cocophilus	Palm trees	X						
Radopholus similis	Banana					X		

[1]*BSD = biological soil disinfestation (anaerobic, biosolarization).*
[2]*A. ritzemabosi, A. besseyi, A. fragariae.*

sugar beet cultivars. The development of INM programmes now use tolerant cultivars over resistant ones, which improves yield significantly. The development of these cultivars, by plant breeders, has been responsible for the removal of nematicides as a standard tool from management programmes. However, growers often ignore INM programmes and plant tolerant cultivars in ever shorter rotations, which could lead to tolerance-breaking populations of *H. schachtii* (Daub, 2021).

20.8.2. Soybean Cyst Nematode, *H. glycines* in the US Midwest: Two or Three INM Tools

Soybean is one of the most important grain crops, with cultivation now occupying more than 6% of the world's arable land. Soybean is important for both human food and animal feed and the USA is a major producer of this crop.

The Problem: The soybean cyst nematode (*H. glycines*) is the most yield-reducing pathogen of soybean in the USA (Bandara *et al.*, 2020) and is also a major pathogen in Brazilian and Chinese soybean production (Peng *et al.*, 2021). Although aboveground damage can be indistinct in many infestations, considerable yield loss still occurs. Soybean cyst nematode is widely distributed throughout soybean production regions in the USA and Canada (Bent, 2022). As an obligate sexually reproducing nematode, genetic diversity within a population is often high. Soybean cyst nematode has a relatively narrow host range; however, the cysts it produces can remain viable in soil for up to 10 years (Davis and Tylka, 2000).

Current Practice: The primary management tactic involves selecting a suitable resistant soybean variety for a particular field based on knowledge of the virulence of the resident nematode population on established soybean indicator lines (i.e., HG type) (Niblack *et al.*, 2002). The HG type bioassay is conducted for a population of soybean cyst nematode inhabiting a particular field and is useful for making cultivar recommendations to minimize yield loss. The most widely utilized source of resistance is PI 88788, which has subsequently been bred into thousands of soybean varieties and planted for decades over millions of acres (Bent, 2022). Another important source of host resistance is in the cultivar 'Peking', which is associated with multiple loci, but overuse is also selecting for resistance-breaking nematode populations (Meinhardt *et al.*, 2021). Additional management tactics need to be implemented alongside host resistance to minimize the development of virulence in the nematode population. In addition to proper rotation of host resistance sources, crop rotations to a non-host (i.e., corn or cotton) can help reduce the number of juveniles in soil between soybean plantings and reduce the selective pressure on nematode resistance traits. Although the use of nematicides has shown only moderate benefit with respect to yield increase and suppression of soybean cyst nematode, nematicides may still help minimize the development of virulence on new sources of host resistance.

20.8.3. Root-knot Nematodes, *Meloidogyne* spp. in Tomato: One to Three INM Tools

Root-knot nematodes are common and major pathogens in all open-field tomato-growing regions of the world. Root-knot nematode management in tomatoes is very different depending on what type of tomatoes are grown and where they are grown.

The Problem: Tomatoes are very susceptible to root-knot nematodes which can cause early senescence and in severe cases complete yield loss if not managed. Tomatoes are grown for fresh market (highest value) and for processing (lower value). Many species of root-knot nematode can damage tomato, and one species that is especially concerning is *M. enterolobii* (guava root-knot), which is increasingly found in tomato regions across the globe and will infect and damage current root-knot-resistant tomato cultivars. Co-infection of root-knot with Fusarium wilt is common in many tomato fields, and often the damage is solely attributed to Fusarium as most growers do not bother to dig up roots and check for nematode damage (Desaeger, 2021).

Current Practice: Root-knot nematodes on tomatoes are managed in different ways, sometimes consisting of various INM tools, but often only a single tool. Nematode-resistant cultivars (*Mi* gene) have been available for many decades and are standard practice, especially for processing tomatoes, as is the case in the USA in California. However, this is the only resistance available in tomato, and it only protects against three species of *Meloidogyne*, and not against *M. enterolobii*. Also, for higher-value fresh market tomatoes, fewer commercially important cultivars are available, often because the focus is on other needed resistance such as disease and virus. Also, introducing *Mi* often does come with collateral damage in that it weakens the stem base, negatively affecting the quality of the fruit. Soil fumigants and nematicides therefore often are the main and only practice, as is the case with most large tomato growers in Florida and other places where nematodes, soilborne diseases and weeds like nutsedge (*Cyperus esculentus*) are major problems. In southern Europe where since recently fumigation is no longer allowed, tomato growers are using a combination of drip-applied nematicides (see earlier) and bio-solarization, which is applying high amounts of plant and animal-based soil amendments and covering them with clear plastic. In some regions, cover crops are also used as an INM tool, like in Florida, where many tomato growers will plant sunn hemp or sorghum-sudangrass to help reduce root-knot nematodes.

20.8.4. Potato Cyst Nematode in Upland Potato in the Philippines: Five INM Tools

The Problem: PCN *G. rostochiensis* and *G. pallida* (PCN) are economically important problems in potato on a worldwide basis. In most of the major temperate potato-producing countries having PCN, one crop is effectively produced per year using INM with land quarantine, resistant cultivars, nematicides and rotation (Augustin, 2021). PCN is, however, prevalent in small farms in tropical and subtropical lowland and upland areas, e.g., the Philippines, India and South America. Unique to these climatic regions, potatoes can be grown two times in a 12-month cycle. Small farmers do not have financial access to nematicides and often not even resistant cultivars and this limits their ability to manage PCN.

Current Practice: Sikora (1984) reported that this unique cropping pattern can be used to take advantage of specific weak links in the PCN life cycle: (i) a single lifecycle per crop, (ii) a 4-month diapause, (iii) quick penetration of next potato crop and (iv) death due to early trap cropping. Using a combination of non-host crops, susceptible and/or resistant potato cultivars, early maturing potato cultivars and INM-suitable rotation patterns, it would be possible to reduce the population and keep it below the threshold density. INM is attained by planting a short-duration cultivar after a standard susceptible cultivar. The juveniles in the eggs are dormant; therefore, by planting the short-duration cultivar directly thereafter, early root penetration is delayed. At harvest, the delay in juvenile penetration and development results in the production of, or only a few, eggs that are subsequently trapped and killed by harvesting the early maturing crop. The planting of a resistant cultivar thereafter would further reduce the population density. We often do not, or cannot, take advantage of weak links in nematode biology for INM.

20.8.5. Root-knot and Reniform Nematodes in Organic Vegetables in Hawaii: Two INM Tools

The Problem: Root-knot nematodes (*Meloidogyne* spp.) and reniform nematodes (*R. reniformis*) are the main PPN affecting a wide range of crops in Hawaii, including zucchini (*Cucurbita pepo*) and sweet potato (*Ipomoea batatas*). Nematode damage on sweet potato can include galls on lateral roots, cracking on storage roots, and slender storage roots despite healthy vegetative growth. Many other pests and diseases, such as sweet potato weevil, can cause significant damage as well, which is why an integrated approach is needed.

Current Practice/Recommendation: The current recommendation is a combination of allelopathic cover crops with post-plant organic nematicide chemigation. The allelopathic cover crop of choice is sunn hemp, which is (i) a poor host to both target nematodes; (ii) produces monocrotaline, which is nematostatic and

(iii) adds significant organic matter to the soil, thus enhancing indigenous nematode antagonists. It is recommended to grow sunn hemp as a cover crop and terminate within 3 months after planting in a strip-till system and plant the cash crops in the strip-tilled strip to receive the allelopathic benefit of sunn hemp. For short-term crops such as zucchini and cucumber, this practice, combined with monthly chemigation with azadirachtin, reduced root-knot damage and improved zucchini growth. For a longer-term crop, like 5 months of 'Okinawan' sweet potato, this practice reduced reniform nematodes but did not increase sweet potato tuber yield. Having a long growing season of 5 months, nematodes need to be managed longer, and other pests like sweet potato weevils need to be managed as well. For sweet potato in this system, a sunn hemp cover crop, combined with monthly applications of *P. lilacinum*, in rotation with azarirachtin starting at 2 months after planting (for nematode control) and *Beauveria bassiana* (for weevil control), provided the best yield protection.

20.8.6. The Banana Burrowing Nematode, *Radopholus similis*: Zero to One INM Tool

The Problem: Bananas are lowland tropical crops where rainfall exceeds 1250 mm/year. The most widespread cultivated are dessert bananas of the Cavendish. *Radopholus similis* is responsible for what is called 'toppling disease' and is present in all major banana-growing regions of the world. Furthermore, yield responses following nematicide treatments have reached 275% over untreated controls (Sikora *et al.*, 2018).

Current Practice: The number of INM tools used to manage the burrowing nematode, which is a major pest of banana worldwide, is based on one tool – nematicides. All major banana growers and commercial companies who grow commercial bananas use nematicides, to different degrees, applied 2-3 times in a growing cycle. They are used to reduce nematode root damage caused by this lesion nematode and to reduce root rot caused by interrelationships with fungal pathogens. Conversely, small growers and subsistence farmers do not have any good alternatives for management (Sikora *et al.*, 2018). Nematologist have not been successful in developing alternative biological control or plant resistance tools for the management of *R. similis*.

20.9. Conclusion

Developing and implementing INM programmes is complex and more easily said than done. We have shown some examples where an INM programme has become standard practice, but there are many more where INM remains a pipe dream. We have tried to explain that INM is knowledge-intensive, requiring easy access to information, databases and the ability to properly monitor nematodes in the field. INM research has mainly been carried out in arable farming systems in the temperate climate zones. These countries have a strong foundation of nematology facilities and personnel already present, and official (government) programmes that support INM. In most (sub)tropical regions, INM programmes have proven to be more difficult to implement. In addition, the favourable climate allows nematodes (and other pests) to reproduce all year-round, making them more difficult to manage. Also, there is often a lack of nematode expertise and funding to conduct INM research and extension. These are certainly extenuating circumstances, but other things are also play. We tried to explain that no matter how good a farmer can predict nematode damage, there are aspects they cannot control and that can greatly impact INM. Nematode management decisions are not made in isolation but are influenced by perceptions outside of the farmer's control, like those of the agricultural industry, politicians and consumers. Navigating all these factors is complicated and farmers have to consider many factors other than 'simply' managing nematodes.

20.10. Health and Safety

Many of the chemicals and procedures used in nematode control are potentially hazardous. All procedures should be the subject of risk analysis, and recommended procedures, including the use of personal protective equipment, should be rigorously implemented and followed. Hazard data sheets for the chemicals involved should also be available to staff to minimize risk. Individual countries have own legal requirements in these matters and it is the responsibility of the user to ensure that the appropriate legislation and procedures are followed.

20.11. References

Abawi, G.S. and Widmer, T.L. (2000) Impact of soil health management practices on soilborne pathogens, nematodes and root diseases of vegetable crops. *Applied Soil Ecology* 15, 37–47.

Aiyer, H.S., McKenzie-Gopsill, A., Mills, A. and Foster, A.J. (2024) Select cover crop residue and soil microbiomes contribute to suppression of Fusarium root and crown rot in barley and soybean. *Microorganisms* 12(2), 404. DOI: 10.3390/microorganisms12020404.

Augustin, B. (2021) The need for new approaches for management of potato cyst nematodes: the view from the Rhineland-Palatinate. In: Sikora, R.A., Desaeger, J. and Molendijk, L.P.G. (eds), *Integrated Nematode Management: State-of-the-art and Visions for the Future*. CAB International, Wallingford, UK, pp. 319–326. https://www.cabi.org/bookshop/book/9781789247558/

Bandara, A.Y., Weerasooriya, D.K., Bradley, C.A., Allen, T.W. and Esker, P.D. (2020) Dissecting the economic impact of soybean diseases in the United States over two decades. *PLoS One* 15, e0231141.

Bent, A.F. (2022) Exploring soybean resistance to soybean cyst nematode. *Annual Review of Phytopathology* 60, 379–409.

Best4soil. A network of practioners, for sharing knowledge on prevention and reduction of soil borne diseases. EU database. Available at: https://www.best4soil.eu/ (accessed September 23 2025).

Braun, A.C., Nauen, R. and Rist, M. (2024) Plant pathogenic nematodes behave badly: uncovering the first case of field-relevant nematicide resistance. *35th Symposium of the European Society of Nematologists, Cordoba, Spain, 15–19 April, 2024*. Abstract 147. https://esn24cordoba.com/wp-content/uploads/2024/04/Abstracts-Book.pdf.

Brown, P.D. and Morra, M.J. (1997) Control of soil-borne plant pests using glucosinolate-containing plants. *Advances in Agronomy* 61, 167–231. DOI: 10.1016/S0065-2113(08)60664-1.

Bui, H.X., and Desaeger, J. (2021) Host suitability of summer cover crops to *Meloidogyne arenaria*, *M. enterolobii*, *M. incognita* and *M. javanica*. *Nematology* 24, 1–9. DOI: 10.1163/15685411-bja10122.

Cazzaniga, S.J., van den Elsen, S., Lombaers, C., Kroonen, M., Visser, J., *et al.* (2023) On the legacy of cover crop-specific microbial footprints. *Soil Biology and Biochemistry* 184, 109080. DOI: 10.1016/j.soilbio.2023.109080.

Chen, Y.C. (2008) Application of Streptomyces spp. for controlling disease caused by plant-pathogenic fungi and plant-parasitic nematodes. M.S. Thesis, National Chung Hsing University, Taichung, Taiwan.

Chitwood, D.J. (2002) Phytochemical based strategies for nematode control. *Annual Review of Phytopathology* 40, 221–249.

Chitwood, D.J. (2003) Research on plant-parasitic nematode biology conducted by the United States department of agriculture – Agricultural research service. *Pest Management Science* 59, 748–753.

Cortada, L. (2021) Technologies for INM in smallholder farming systems: no one-size-fits-all. In: Sikora, R.A., Desaeger, J. and Molendijk, L.P.G. (eds), *Integrated Nematode Management: State-of-the-art and Visions for the Future*. CAB International, Wallingford, UK, pp. 457–462. https://www.cabi.org/bookshop/book/9781789247558/.

D'Addabbo, T., Argentieri, M.P., Żuchowski, J., Biazzi, E., Tava, A., *et al.* (2020) Activity of saponins from *Medicago* species against phytoparasitic nematodes. *Plants* (Basel) 9(4), 443. DOI: 10.3390/plants9040443.

Dandurand, L.M. and Knudsen, G.R. (2016) Effect of the trap crop *Solanum sisymbriifolium* and two biocontrol fungi on reproduction of the potato cyst nematode, *Globodera pallida*. *Annals of Applied Biology*, 169(2), 180–189. DOI: 10.1111/aab.12295.

Daub, M. (2021) The beet cyst nematode (*Heterodera schachtii* Schmidt): An ancient threat to sugar beet crops in Central Europe has become an invisible actor. In: Sikora, R.A., Desaeger, J. and Molendijk, L.P.G. (eds), *Integrated Nematode Management: State-of-the-art and Visions for the Future*. CAB International, Wallingford, UK, pp. 394–400. https://www.cabi.org/bookshop/book/9781789247558/.

Davis, E.L. and Tylka, G.L. (2000) Soybean cyst nematode disease. *The Plant Health Instructor*. DOI: 10.1094/PHI-I-2000-0725-01.

Davis, R.F. and Kemerait, R.C. (2021) Integrated management of *Meloidogyne incognita*, the most economically damaging pathogen of cotton in the south-eastern United States. In: Sikora, R.A., Desaeger, J. and Molendjik, L.P.G. (eds), *Integrated Nematode Management: State-of-the-art and Visions for the Future*. CAB International, Wallingford, UK, pp. 87–93. DOI: 10.1079/9781789247541.0013.

d'Errico, G., Marra, R., Crescenzi, A., Davino, S.W., Fanigliulo, A., *et al.* (2019) Integrated management strategies of *Meloidogyne incognita* and *Pseudopyrenochaeta lycopersici* on tomato using a *Bacillus firmus*-based product and two synthetic nematicides in two consecutive crop cycles in greenhouse. *Crop Protection* 122, 159–164. DOI: 10.1016/j.cropro.2019.05.004.

Desaeger, J. (2021) A root-knot paradise made in plastic – the case of Florida vegetables. In: Sikora, R.A., Desaeger, J. and Molendijk, L.P.G. (eds), *Integrated Nematode Management: State-of-the-art and Vision for the Future*. CAB International, Wallingford, UK, pp. 247–254. https://www.cabi.org/bookshop/book/9781789247541/.

Desaeger, J. and Rao, M.R. (2000) Parasitic nematode populations in natural fallows and improved cover crops and their effects on subsequent crops in Kenya. *Field Crops Research* 65, 41–56.

Desaeger, J. and Rao, M.R. (2001) The potential of mixed cover crops of *Sesbania*, *Tephrosia* and *Crotalaria* for minimizing nematode problems on subsequent crops. *Field Crops Research* 70, 111–125.

Desaeger, J. and Rao, M.R. (2002) Significance of lesion and spiral nematodes (*Pratylenchus* spp.) in crotalaria-maize rotation in western Kenya. *Nematropica* 33, 27–39.

Desaeger, J., Sikora, R.A. and Molendijk, L.P.G. (2021a) Outlook: a vision of the future of integrated nematode management. In: Sikora, R.A., Desaeger, J. and Molendijk, L.P.G. (eds), *Integrated Nematode Management: State-of-the-art and Visions for the Future*. CAB International, Wallingford, UK, pp. 475–483. https://www.cabi.org/bookshop/book/9781789247558/.

Desaeger, J., Williams, K. and Rosskopf, E. (2021b) Organic management strategies for nematode control in Florida plasticulture (Chapter 12). In: Chaudhary, K.K. and Meghvansi, M.K. (eds), *Sustainable Management of Nematodes in Agriculture, Vol. 1: Organic Management*. Sustainability in Plant and Crop Protection 18. Springer, Dordrecht, Netherlands, pp. 293–325.

Desaeger, J., Wram, C. and Zasada, I. (2020) New reduced-risk agricultural nematicides - rationale and review. *Journal of Nematology* 52, 1–16. DOI: 10.21307/jofnem-2020-091.

Dupont, S.T., Ferris, H. and Van Horn, M. (2009) Effects of cover crop quality and quantity on nematode-based soil food webs and nutrient cycling. *Applied Soil Ecology* 41, 157–167.

EPA (1998) Available at: https://www3.epa.gov/pesticides/chem_search/reg_actions/registration/related_PC-119204_1-Aug-98.pdf.

Farm to Fork Strategy. (2024) EU Commission. Available at: https://food.ec.europa.eu/horizontal-topics/farm-fork-strategy_en (accessed August 29 2024).

Ferris, H. (2010) Form and function: metabolic footprints of nematodes in the soil food web. *European Journal of Soil Biology* 46, 97–104. DOI: 10.1016/j.ejsobi.2010.01.003.

Ferris, H., Bongers, T. and de Goede, R. (2001) A framework for soil food web diagnostics: extension of the nematode faunal analysis concept. *Applied Soil Ecology* 18, 13–29.

Fuller, V.L., Lilley, C.J. and Urwin, P.E. (2008) Nematode resistance. *New Phytologist* 180, 27–44.

Gallaher, R.N. and McSorley, R. (1993) Population densities of *Meloidogyne incognita* and other nematodes following seven cultivars of cowpea. *Nematropica* 23, 21–26.

Gutiérrez, O.A., Jenkins, J.N., McCarty, J.C., Wubben, M.J., Hayes, R.W., *et al.* (2010) SSR markers closely associated with genes for resistance to root-knot nematode on chromosomes 11 and 14 of Upland cotton. *Theoretical and Applied Genetics* 121, 1323–1337.

Helder, H. and Heuer, H. (2021) Let's be inclusive - the time of looking at individual plant parasitic nematodes is over, and new technologies allow for it. In: Sikora, R.A., Desaeger, J. and Molendijk, L.P.G. (eds), *Integrated Nematode Management: State-of-the-art and Visions for the Future*. CAB International, Wallingford, UK, pp. 403–407. https://www.cabi.org/bookshop/book/9781789247558/.

Hillnhütter, C., Schweizer, A., Kühnhold, V. and Sikora, R.A. (2010) Remote sensing for the detection of soil-borne plant parasitic nematodes and fungal pathogens. In: Oerke, E.C., Gerhards, R., Menz, G. and Sikora, R.A. (eds), *Precision Crop Protection – The Challenge and use of Heterogeneity*. Springer, Dordrecht, Netherlands, pp. 151–165. DOI: 10.1007/978-90-481-9277-9_10.

Hooks, C.R.R., Wang, K.H., Ploeg, A. and McSorley, R. (2010) Using Marigold (*Tagetes* spp.) as a cover crop to protect crops from plant-parasitic nematodes. *Applied Soil Ecology* 46, 307–320. DOI: 10.1016/j.apsoil.2010.09.005.

Insecticide Resistance Action Committee (IRAC). (2024). Nematodes Team. Available at: https://irac-online.org/teams/nematodes/ (Accessed May 20 2025).

Jeschke, P. (2004) The unique role of fluorine in the design of active ingredients for modern crop protection. *ChemBioChem* 5, 570–589.

Kahn, T.W., Duck, N.B., McCarville, M.T., Schouten, L.C., Schweri, K., *et al.* (2021) A *Bacillus thuringiensis* Cry protein controls soybean cyst nematode in transgenic soybean plants. *Nature Communications* 12, 3380.

Kammerer, C.L., Harmon, P.F. and Crow, W.T. (2023) Reduced sensitivity to fluopyram in *Meloidogyne graminis* following long-term exposure in golf turf. *Journal of Nematology* 55(1), 20230048. DOI: 10.2478/jofnem-2023-0048.

Kiewnick, S. and Sikora, R.A. (2006) Biological control of the root-knot nematode, *Meloidogyne incognita* by *Paecilomyces lilacinus* strain 251. *Biological Control* 38, 179–187.
Kimpinski, J. and Sanderson, K. (2004) Effects of crop rotations on carrot yield and on the nematodes *Pratylenchus penetrans* and *Meloidogyne hapla*. *Phytoprotection* 85, 13–17. DOI: 10.7202/008901ar.
Kimpinski, J., Arsenault, W.J., Gallant, C.E. and Sanderson, J.B. (2000) The effect of marigolds (*Tagetes* spp.) and other cover crops on *Pratylenchus penetrans* and on following potato crops. *Journal of Nematology* 32(4S), 531–536.
Kuska, M.T., Daub, M. and Mahlein, A.K. (2021) Emerging technologies for integrated nematode management: remote sensing or proximal sensing as a potential tool to detect and identify nematode infestation. In: Sikora, R.A., Desaeger, J. and Molendijk, L.P.G. (eds), *Integrated Nematode Management: State-of-the-art and Visions for the Future*. CAB International, Wallingford, UK, pp. 414–420. DOI: 10.1079/9781789247541.0058.
Kwok, O.C.H., Plattner, R., Weisleder, D. and Wicklow, D.C. (1992) A nematicidal toxin from *Pleurotus ostreatus* NRRL 3526. *Journal of Chemical Ecology* 18, 127–136. DOI: 10.1007/BF00993748.
LaMondia, J.A. (2006) Management of lesion nematodes and potato early dying with rotation crops. *Journal of Nematology* 38, 442–448.
LaMondia, J.M. (2021) Management of lesion and dagger nematodes with rotation crops. *Nematropica* 51, 1. https://journals.flvc.org/nematropica/article/view/128983.
Lawrence, K.S. (2021) Reniform nematodes (*Rotylenchulus reniformis*) and its interactions with cotton (*Gossypium hirsutum*). In: Sikora, R.A., Desaeger, J. and Molendijk, L.P.G. (eds), *Integrated Nematode Management: State-of-the-art and Visions for the Future*. CAB International, Wallingford, UK, pp. 94–99. https://www.cabidigitallibrary.org/doi/abs/10.1079/9781789247541.0014.
Lilley, C.J., Bakhetia, M., Charlton, W.L. and Urwin, P.E. (2007) Recent progress in the development of RNA interference for plant parasitic nematodes. *Molecular Plant Pathology* 8, 701–711.
Linford, M.B., Yap, F. and Oliveira, J.M. (1938) Reduction of soil populations of the root-knot nematode during decomposition of organic matter. *Soil Science* 45, 127–142.
Machado, A.C.Z. (2021) Problems and solutions to integrated nematode management of root-knot, reniform and lesion nematodes in cotton in Brazil. In: Sikora, R.A., Desaeger, J. and Molendijk, L.P.G. (eds), *Integrated Nematode Management: State-of-the-art and Visions for the Future*. CAB International, Wallingford, UK, pp. 73–79. https://www.cabi.org/bookshop/book/9781789247558/.
Mahlein, A.K., Kuska, M.T., Behmann, J., Polder, G. and Walter, A. (2018) Hyperspectral sensors and imaging technologies in phytopathology: State of the art. *Annual Review of Phytopathology* 56, 535–558.
Marahatta, S.P., Wang, K.H., Sipes, S.P. and Hooks, C.R.R. (2012a) Effects of *Tagetes patula* on active and inactive stages of root-knot nematodes. *Journal of Nematology* 44, 26–30.
Marahatta, S.P., Wang, K.H., Sipes, S.P. and Hooks, C.R.R. (2012b) Effect of *Crotalaria juncea* on the anhydrobiotic state of *Rotylenchulus reniformis*. *Nematropica* 42, 34–40.
Mazzola, M. (2007) Manipulation of rhizosphere bacterial communities to induce suppressive soils. *Journal of Nematology* 39, 213–220.
McSorley, R. (2011) Overview of organic amendments for management of plant-parasitic nematodes, with case studies from Florida. *Journal of Nematology* 43, 69–81.
McSorley, R. and Dickson, D.W. (1995) Effect of tropical rotation crops on *Meloidogyne incognita* and other plant-parasitic nematodes. *Journal of Nematology* 27, 535–544.
McSorley, R., Wang, K.H. and Church, G. (2007) Suppression of root-knot nematodes in natural and agricultural soils. *Applied Soil Ecology* 39, 291–298.
Meinhardt, C., Howland, A., Ellersieck, M., Scaboo, A., Diers, B., *et al.* (2021) Resistance gene pyramiding and rotation to combat widespread soybean cyst nematode virulence. *Plant Disease* 105, 3238–3243.
Milligan, S.B., Bodeau, J., Yaghoobi, J., Kaloshian, I., Zabel, P., *et al.* (1998) The root knot nematode resistance gene Mi from tomato is a member of the leucine zipper, nucleotide binding, leucine-rich repeat family of plant genes. *The Plant Cell* 10, 1307–1319.
Molendijk, L.P.G. and Sikora, R.A. (2021) Decision support systems in integrated nematode management: the need for a holistic approach. In: Sikora, R.A., Desaeger, J. and Molendijk, L.P.G. (eds), *Integrated Nematode Management: State-of-the-art and Visions for the Future*. CAB International, Wallingford, UK, pp. 428–438. https://www.cabi.org/bookshop/book/9781789247558/.
Mueller, J. (2021) *Hoplolaimus columbus*: a prime candidate for site-specific management in cotton and soybean production. In: Sikora, R.A., Desaeger, J. and Molendijk, L.P.G. (eds), *Integrated Nematode Management: State-of-the-art and Visions for the Future*. CAB International, Wallingford, UK, pp. 80–86. https://www.cabi.org/bookshop/book/9781789247558/.

Mueller, J.D., Khalilian, A., Monfort, W.S., Davis, R.F., Kirkpatrick, T.L., *et al.* (2010) Site-specific detection and management of nematodes. In: Oerke, E.C., Gerhards, R., Menz, G. and Sikora, R.A. (eds), *Precision Crop Protection – the Challenge and Use of Heterogeneity*. Springer, Dordrecht, Netherlands, pp. 385–402.

Müller, J. (1999) The economic importance of *Heterodera schachtii* in Europe. *Helminthologia* 36, 205–213.

Newhall, A.G. (1955) Disinfestation of soil by heat, flooding, and fumigation. *Botany Review* 29, 189–250.

Niblack, T.L, Arelli, P.R., Noel, G.R., Opperman, C.H., Orf, J.H., *et al.* (2002) A revised classification scheme for genetically diverse populations of *Heterodera glycines*. *Journal of Nematology* 34, 279.

Niblack, T.L., Colgrove, A.L., Colgrove, K. and Bond, J.P. (2008) Shift in virulence of soybean cyst nematode is associated with use of resistance from PI 88788. *Plant Health Progress* 9, 29.

Nogueira, M.A., De Oliveira, J.S., Ferraz, S. and Dos Santos, M.A. (1996) Nematicidal constituents in *Mucuna aterrima* and its activity on *Meloidogyne incognita* race 3. *Nematologia Mediterranea* 24, 249–252.

Ntalli, N.G. and Caboni, P. (2012) Botanical nematicides: a review. *Journal of Agricultural and Food Chemistry* 60, 9929–9940.

Oka, Y. (2010) Mechanisms of nematode suppression by organic soil amendments – a review. *Applied Soil Ecology* 44, 101–115.

Okorie, C.C., Ononuju, C.C. and Okwujiako, I.A. (2011) Management of *Meloidogyne incognita* with *Pleurotus ostreatus* and *P. tuberregium* in soybean. *International Journal of Agriculture and Biology* 13, 401–405.

Owen, K. (2021) A triumph of tolerance: managing the threat to wheat production by the root lesion nematode *Pratylenchus thornei* in the subtropical grain region of eastern Australia. In: Sikora, R.A., Desaeger, J. and Molendijk, L.P.G. (eds), *Integrated Nematode Management: State-of-the-art and Visions for the Future*. CAB International, Wallingford, UK, pp. 13–19. https://www.cabi.org/bookshop/book/9781789247558/.

Paudel, R., Waisen, P. and Wang, K.H. (2021) Exploiting the innate potential of sorghum/sorghum-sudangrass cover crops to improve soil microbial profile that can lead to suppression of plant-parasitic nematodes. *Microorganisms* 9, 1831. Doi: 10.3390/microorganisms9091831.

Peng, D., Jiang, R., Peng, H. and Liu, S. (2021) Soybean cyst nematodes: a destructive threat to soybean production in China. *Phytopathology Research* 3, 19.

Plumblee, M.T. and Mueller, J. (2021) Implementing precision agriculture concepts and technologies into crops production and site-specific management of nematodes. In: Sikora, R.A., Desaeger, J. and Molendijk, L.P.G. (eds), *Integrated Nematode Management: State-of-the-art and Visions for the Future*. CAB International, Wallingford, UK, pp. 421–427. https://www.cabi.org/bookshop/book/9781789247558/.

Poveda, J., Abril-Urias, P. and Escobar, C. (2020) Biological control of plant-parasitic nematodes by filamentous fungi inducers of resistance: *Trichoderma*, mycorrhizal and endophytic fungi. *Frontiers in Microbiology* 11, 992. DOI: 10.3389/fmicb.2020.00992.

Rocha, T.L., Soll, C.B., Boughton, B.A., Silva, T.S., Oldach, K., *et al.* (2017) Prospection and identification of nematotoxic compounds from *Canavalia ensiformis* seeds effective in the control of the root knot nematode *Meloidogyne incognita*. *Biotechnology Research and Innovation* 1, 87–100. DOI: 10.1016/j.biori.2017.10.003.

Rodriguez-Kabana, R. (1986) Organic and inorganic nitrogen amendments to soil as nematode suppressants. *Journal of Nematology* 18, 129–135.

Saeed, Q., Xiukang, W., Haider, F.U., Kučerik, J., Mumtaz, M.Z., *et al.* (2021) Rhizosphere bacteria in plant growth promotion, biocontrol, and bioremediation of contaminated sites: a comprehensive review of effects and mechanisms. *International Journal of Molecular Sciences* 22(19), 10529. DOI: 10.3390/ijms221910529.

Savary, S., Willocquet, L., Pethybridge, S.J., Esker, P., McRoberts, N., *et al.* (2019) The global burden of pathogens and pests on major food crops. *Nature Ecology & Evolution* 3, 430–439. https://doi.org/10.1038/s41559-018-0793-y

Sikora, R.A. (1984) Importance of diapause, resistant and early maturing potato cultivars, chicken manure and non-host crops for *Globodera rostochiensis* integrated control in the upland tropics. *Communications in Agriculture and Applied Biological Sciences* 49, 613–620.

Sikora, R.A. (1992) Management of the antagonistic potential in agricultural ecosystems for the control of plant parasitic nematodes. *Annual Review of Phytopathology* 30, 245–270.

Sikora, R.A. and Roberts, P.A. (2018) Management practices: an overview of integrated nematode management technologies. In: Sikora, R.A., Coyne, D., Hallmann, J. and Timper, P. (eds), *Plant Parasitic Nematodes in Subtropical and Tropical Agriculture*, 3rd edn. CAB International, Wallingford, UK, pp. 795–838.

Sikora, R.A., Coyne, D. and Quénéhervé, P. (2018) Nematode parasites of bananas and plantains. In: Sikora, R.A., Coyne, D., Hallmann, J. and Timper, P. (eds), *Plant Parasitic Nematodes in Subtropical and Tropical Agriculture*, 3rd edn. CAB International, Wallingford, UK, pp. 617–657.

Sikora, R.A., Terry, E.R., Vlek, P. and Chitja, J. (2020) *Transforming Agriculture in Southern Africa: Constraints, Technologies, Policies and Processes*. Taylor & Francis, Abingdon. UK.

Sikora, R.A., Desaeger, J. and Molendijk, L. (2021a) The unpredictability of adapting integrated nematode management to climate variability. In: Sikora, R.A, Desaeger, J. and Molendijk, L.P.G. (eds), *Integrated Nematode Management: State-of-the-Art and Visions for the Future*. CAB International, Wallingford, UK, pp. 463–472. https://www.cabi.org/bookshop/book/9781789247558/.

Sikora, R.A., Desaeger, J. and Molendijk, J.P.G. (eds) (2021b) *Integrated Nematode Management: State-of-the-art and Visions for the Future*. CAB International, Wallingford, UK. https://www..org/bookshop/book/9781789247558/.

Sikora, R.A., Helder, J., Molendijk, L., Desaeger, J., Eves-van den Akker, S., *et al.* (2023) Integrated nematode management in a world in transition: constraints, policy, processes and technologies for the future. *Annual Review of Phytopathology* 61, 209–230. DOI: 10.1146/annurev-phyto-021622-113058.

Smelt, J.H., Van de Peppel-Groen, A.E., Van der Pas, L.J.T. and Dijksterhuis, A. (1996) Development and duration of accelerated degradation of nematicides in different soils. *Soil Biology and Biochemistry* 28, 1757–1765. DOI: 10.1016/S0038-0717(96)00281-7.

Smith, R.F. and Reynolds, H.T. (1966) FAO Symposium on integrated pest control. *Proceedings FAO Symposium on Integrated Pest Control* 1, 11–17.

Stirling, G.R. (1991) *Biological Control of Plant Parasitic Nematodes: Progress, Problems and Prospects*. CAB International, Wallingford, UK.

Stirling, G.R. (2011) Biological control of plant-parasitic nematodes: an ecological perspective, a review of progress and opportunities for further research. In: Davies, K. and Spiegel, Y. (eds), *Biological Control of Plant-parasitic Nematodes*. Springer, Dordrecht, Netherlands, pp. 1–38.

Taylor, A.L. (2003) Nematocides and nematicides – a history. *Nematropica* 33, 225–232.

Thoden, T.C., Korthals, G.W. and Termorshuizen, A.J. (2011) Organic amendments and their influences on plant-parasitic and free-living nematodes: a promising method for nematode management? *Nematology* 13, 133–153.

Topalovic, O., Hussain, M. and Heuer, H. (2020) Plants and associated soil microbiota cooperatively suppress plant-parasitic nematodes. *Frontiers in Microbiology* 11, 313. DOI: 10.3389/fmicb.2020.00313.

Usovsky, M., Gamage, V.A., Meinhardt, C.G., Dietz, N., Triller, M., *et al.* (2023) Loss-of-function of an α-SNAP gene confers resistance to soybean cyst nematode. *Nature Communications* 14, 7629.

Waisen, P., Cheng, Z., Sipes, B.S., DeFrank, J., Marahatta, S.P., *et al.* (2020) Effects of biofumigant crop termination methods on suppression of plant-parasitic nematodes. *Applied Soil Ecology* 154, 103595. DOI: 10.1016/j.apsoil.2020.103595.

Wang, K.H. and Uchida J. (2014) Plant disease prevention and management in sustainable agricultural systems. In: Nandwani, D. (ed.), *Sustainable Development and Biodiversity Volume 2: Sustainable Horticultural Systems*. Springer, Cham, Switzerland, pp. 353–384.

Wang, K.H. and Zasada, I. (2011) Allelopathic effects of *Crotalaria* spp. against *M. incognita* as affected by crop age, biomass, heat and species. *Society of Nematologists Conference, Corvallis, OR. Journal of Nematology* 43, 292.

Wang, K.H., McSorley, R. and Gallaher, R.N. (2003) Effect of *Crotalaria juncea* amendment on nematode communities in soil with different agricultural histories. *Journal of Nematology* 35, 294–301.

Wang, K.H., Hooks, C.R.R. and Marahatta, S.P. (2011) Can using a strip-tilled cover cropping system followed by surface mulch practice enhance organisms higher up in the soil food web hierarchy? *Applied Soil Ecology* 49, 107–117.

Wang, K.H., Sipes, B.S. and Schmitt, D.P. (2002) *Crotalaria juncea* as a cover crop for nematode management: a review. *Nematropica* 32, 35–57.

Wang, K.H., Radovich, T., Pant, A. and Cheng, Z. (2014) Integration of cover crops and vermicompost tea for soil and plant health management in a short-term vegetable cropping system. *Applied Soil Ecology* 82, 26–37.

Warnke, S., Chen, S., Wyse, D., Johnson, G. and Porter, P. (2008) Effect of rotation crops on hatch, viability and development of *Heterodera glycines*. *Nematology* 10, 869–882. Doi: 10.1163/156854108786161391.

Wu, P.H., Tsay, T.T. and Chen, P. (2021) Evaluation of *Streptomyces saraciticas* as soil amendments for controlling soil-borne plant pathogens. *Plant Pathology Journal* 37, 596–606.

Wubben, M.J., Gaudin, A.G., McCarty, J.C. and Jenkins, J.N. (2020) Analysis of cotton chromosome 11 and 14 root-knot nematode resistance quantitative trait loci effects on root-knot nematode postinfection development, egg mass formation, and fecundity. *Phytopathology* 110, 927–932.

Zhang, W. and Liang, Y. (2023) The wide presence of fluorinated compounds in common chemical products and the environment: a review. *Environmental Science and Pollution Research* 30, 108393–108410. DOI: 10.1007/s11356-023-30033-6.

21 Biological Control of Plant-parasitic Nematodes

Senyu Chen[1]*, Donald W. Dickson[2] and Deepak Haarith[2]

[1]Department of Plant Pathology, St. Paul, MN, USA; [2]Entomology and Nematology Department, University of Florida, Gainesville, FL USA

Abstract

Biological control of nematodes may be defined as the action of natural antagonists maintaining nematode population densities at lower levels than would otherwise occur. Development of harmful densities of plant-parasitic nematodes in soil is regulated by numerous factors, including host suitability and natural antagonists. Because of the complexity of the soil environments, there are limitations to capitalize on these antagonists. Biological control of nematodes is often visualised as the mass introduction of a biological control agent that results in an immediate reduction of nematode population densities and an immediate response with increased crop yields and an economic return on investment. In other words, this involves inundating nematode-infested soil with one or more antagonists that act as a biopesticide. Another approach is to enhance the activity of indigenous antagonists by manipulating the environment through cultural practices. Both strategies face challenges and are unlikely to be practical in agricultural production systems unless used as part of an integrated management programme.

21.1. Introduction

During the past century, agricultural productivity has increased dramatically because of genetically improved plant species and cultivars, modern farming equipment and extensive chemical fertilizers and pesticides. The globalization of economies is driving crop production towards higher level of specialization. Some producers grow only one crop, whereas others grow just a few. These extensive and often highly intensive cropping systems tend to increase pest and pathogen pressures, with plant-parasitic nematodes becoming increasingly important. Industrialization in agriculture, particularly in developing countries, exacerbates the spread of soilborne diseases. This is especially true with plant-parasitic nematodes. They are continuously being moved nationally and internationally via movement of farm machinery and planting materials, including seed. To maintain high productivity, effective pest and pathogen management programmes must be in place for producers to remain competitive.

Over the past 30 years, many of the most effective chemical nematicides have been banned, restricted or phased out because of environmental,

*E-mail: chenx099@umn.edu

A revision of Chen, S. and Dickson, D.W. (2012) Biological control of plant-parasitic nematodes. In: Manzanilla-López, R.H. and Marbán-Mendoza, N. (eds), *Practical Plant Nematology*. Mundi-Prensa and Biblioteca Básica de Agricultura Montecillo, Mexico.

(eds R.H. Manzanilla-López, D.J. Hunt, N. Marbán-Mendoza and S. Infante Gil)
DOI: 10.1079/9781836990413.0021

animal and human health concerns (see Chapters 19 and 20, this volume). Consequently, plant-parasitic nematodes have become one of the most difficult crop pathogens to manage. Using genetic resistance, where available, is among the most cost-effective methods for managing nematodes. Unfortunately, using resistant cultivars often results in the selection of more virulent populations, which eventually results in significant yield suppression. Furthermore, there is little effective resistance against ectoparasitic nematodes, and in many important agricultural crops, no resistance sources have been found for any nematode species. Crop rotation schemes, while highly advantageous, are often greatly influenced by agroeconomics; consequently, they are challenging to employ in many cases to manage nematodes. There is scientific consensus on the need to use a combination of methods to sustainably manage disease-causing plant-parasitic nematodes. Biological control agents offer great promise, albeit requiring further scientific investigations.

The concept of biological control varies in different disciplines and circumstances. A broad definition may include host genetic resistance; however, generally, this is not included in plant nematology. Hence, biological control of nematodes may be defined as the action of natural nematode antagonists maintaining population densities at lower levels than would occur in their absence (Stirling, 2011). Development of harmful densities of plant-parasitic nematodes in soil is regulated by numerous factors, including host suitability and natural antagonists (Fig. 21.1). Because of the complexity of the soil environment, there are several limitations on our ability to capitalize on these antagonists. Traditionally, biological control of nematodes is often understood as the mass introduction of a biological control agent that results in an immediate reduction of nematode population densities and an immediate response with increased crop yields and an economic return on investments. In other words, this involves inundating nematode-infested soil with one or more antagonists that would act as a biopesticide. Another approach to biological control is enhancing the activity of indigenous antagonists by manipulating the environments through cultural practices. However, both strategies face many challenges. Furthermore, biological control of nematodes is unlikely to be practical in agricultural production systems, unless used as part of an integrated management programme.

21.2. Biological Antagonists of Nematodes

Nematodes are possibly the most abundant multicellular organisms on Earth, especially among the plethora of complex soil inhabitants. These nematodes interact with other soil organisms in several ways, including as both prey and predators of each other. The organisms that are known to adversely affect nematodes are collectively referred to as natural antagonists. These include fungi, bacteria, viruses, rickettsiae, plants, protozoans, turbellarians, tardigrades, enchytraeids, mites, insects and even other nematodes. Among these, fungal and bacterial antagonists have been investigated most extensively because of their accessibility and demonstrated effectiveness in suppressing soilborne nematodes. Invertebrate predators are unlikely to be effective biocontrol agents because they are difficult to mass produce. There is little evidence supporting their roles in the induction of nematode soil suppressiveness. Our focus herein is on fungal and bacterial antagonists. A summary of all soil organisms reported affecting nematode population densities and their potential for commercialization is listed in Table 21.1.

21.3. Fungal Antagonists of Nematodes

Fungi are abundant in soil, and some of them are antagonists of nematodes, including non-nematophagous and nematophagous fungi. Non-nematophagous fungi may not rely entirely on nematodes as a food source; nevertheless, they are reported to adversely affect nematodes, whereas nematophagous fungi use nematodes for nutrition either obligatorily or saprophytically. Based on their mode of action, nematode fungal antagonists are grouped as following: (i) predacious fungi, (ii) endoparasites of vermiform nematodes, (iii) parasites of sedentary females and eggs, (iv) fungi producing antibiotic substances and (v) vesicular-arbuscular mycorrhizal (VAM) fungi. Some fungal antagonists may be placed in more than one category.

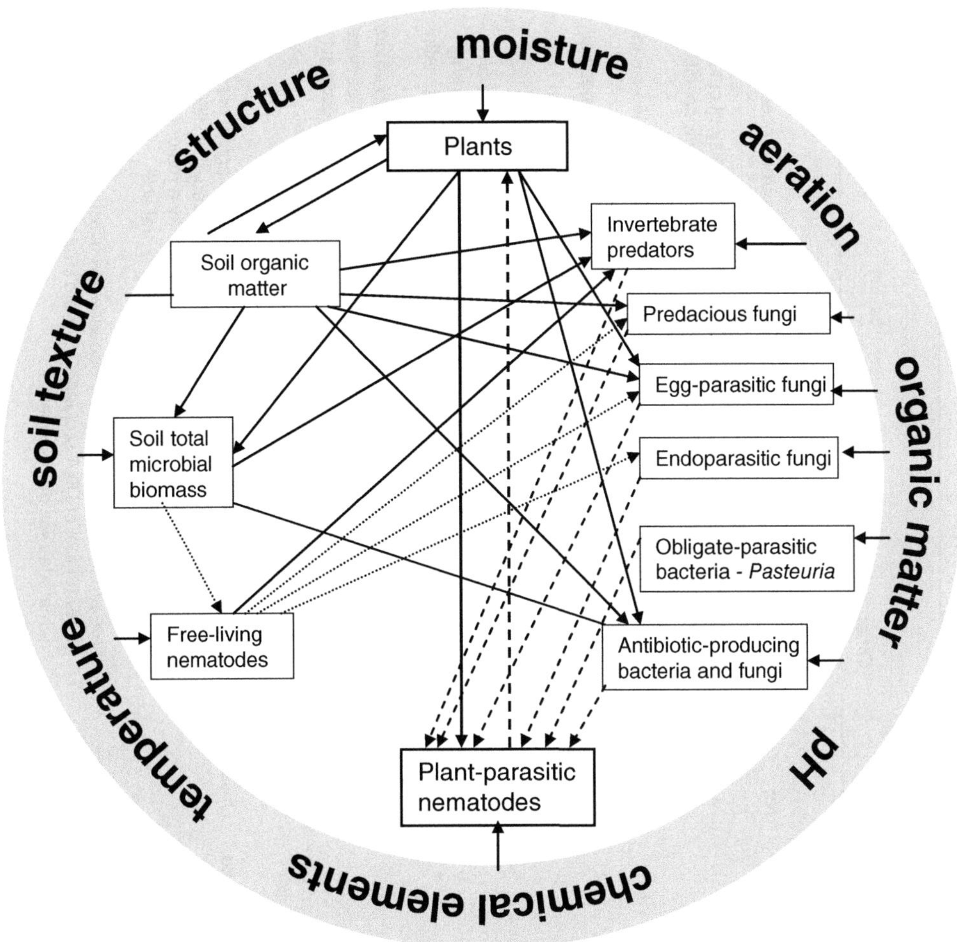

Fig. 21.1. Major interactions of plant-parasitic nematodes with soil biotic and abiotic factors. The circle illustrates major components of soil physical and chemical environments that may affect nematodes and their antagonists. The solid arrows indicate that there may be significant effect, either positive or negative. The dashed arrows indicate an antagonistic effect. The dotted arrows indicate a positive effect.

21.3.1. Predacious Fungi

These fungi are common in soils having evolved special devices for capturing nematodes. These include adhesive hyphae, branches, nets, knobs, stephanocysts, non-constricting rings and constricting rings (Fig. 21.2) (Barron, 1977; Chen and Dickson, 2012).

Producing physical adhesive traps for consuming nematodes has been a commonly used method by some ascomycete fungi, especially in the classes *Orbiliomycetes* and *Leotiomycetes*. These adhesive traps formed by the members of *Dactylella* spp., *Arthrobotrys* spp. and *Dactylaria* spp. could be as simple as adhesive hyphae or as complex as a mesh-like network of sticky hyphae (Figs 21.2, 21.3). One class of such contraptions is the adhesive knobs, which are morphologically distinct cells that are considered highly specialized trapping devices (Figs 21.2A-C, 21.3B). They are covered by a thin layer of adhesive material and are either sessile or borne at the apex of a short non-adhesive stalk. When a nematode touches a knob, a flattened mass of adhesive material is produced at the point of attachment that forms a thick pad. This increases the area of

Table 21.1. Nematode antagonists and their potential as biocontrol agents (Chen and Dickson, 2012).

Group of organisms	Mode of action	Importance in natural soil	Potential as biocontrol agent	Other comments
Invertebrate predators				
Protozoa				
Amoebae	Engulf nematodes	Poor	No	Poor biocontrol potential because they move slowly, do not survive desiccation well and are non-specific predators.
Ciliate protozoans	Ingest moribund nematodes	Poor	No	Poor biocontrol potential because they have limited ability of catching nematodes, do not survive desiccation well and are non-specific predators.
Turbellarids	Ingestion	Poor	No	No apparent biocontrol potential in field due to their distribution in the upper layer of soil.
Tardigrades	Ingestion	Poor	No	They are in moist land, not common in mineral soil and not specific predators.
Oligochaeta				
Enchytraeids	Ingestion	Poor	Unlikely	*Fredericia* sp. feeds on *Heterodera trifolii* females.
Earthworms	Ingestion	Unclear	Unlikely	Ingestion of nematodes by earthworm is common. The studies on influence of earthworms on nematode population are increasing. Some studies showed significant reduction of nematode population by earthworm.
Mites	Ingestion	Unclear	Unlikely	Many species have been reported. Most work was done in laboratory.
Insects	Catch and ingest	Unclear	Unlikely	Various insects including Collembola, Diptera, ants and beetles have been reported to feed on nematodes. In natural soil, insects may have difficulty to follow nematodes in the small soil pores.
Crustaceans	Catch and ingest	Unknown	Unlikely	Copepod preys on a variety of nematodes in Petri dish.
Nematodes	Ingest; kill and take body fluid using stylet	Not fully understood. Probably poor but depends on soil environments.	Unlikely	It seems unlikely to use predatory nematodes as biological control agents because many nematodes may be needed, it is difficult to mass culture and individual species do not play a major role in controlling plant-parasitic nematodes in natural soil.
Microbial antagonists				
Virus	Infection	Unknown	Unlikely	There are few reports probably because it is difficult to detect. With powerful tool of DNA sequencing, more viral infection of nematodes will be discovered.

Rickettsiae	Little known; infection or symbiosis	Unknown	Unknown	Few reports of microscopic observations.
Bacteria				
Pasteuria	Obligate parasitism	Important	High	This group of bacteria is obligate parasite and highly host specific. Culture on artificial media has been reported, and commercial biocontrol products are available.
Other nematode-parasitic bacteria	Infection	Unknown	Unknown	*Paenibacillus*, for example.
Bacteria producing antibiotic substances	Antibiotic	Important	High	*Streptomyces avermitilis* produces metabolites, avermectins highly toxic to nematodes; many rhizobacteria may produce nematicidal compounds.
Fungi				
Predacious fungi	Catch with various trapping devices, kill and colonize	Important, depending on species and environments	Possible, but there is no successful agent	Predacious fungi can undergo saprophytic activity, and their dependency on nematodes for formation of trapping device varies among diverse types and species. The trapping activities are highly influenced by environments.
Endoparasites of vermiform nematodes	Infection with various means, depending on species	Important	High	More specific than predacious fungi. Except for obligate parasites, most of them can be cultured on artificial media. Their population density depends on their host nematode population density.
Parasites or colonizers of sedentary females and eggs	Infection or saprophytic colonization	Important	High	Most of them are facultative parasites and easy to culture.
Vesicular-arbuscular mycorrhizal fungi	Competition, infection	Important	Unlikely	Obligate symbiosis with plants. Difficult to mass produce.
Fungi producing antibiotic substances	Antibiotic	Important	High	These fungi are generally easy to culture and mass produce. The studies of this group have increased in the past decades.

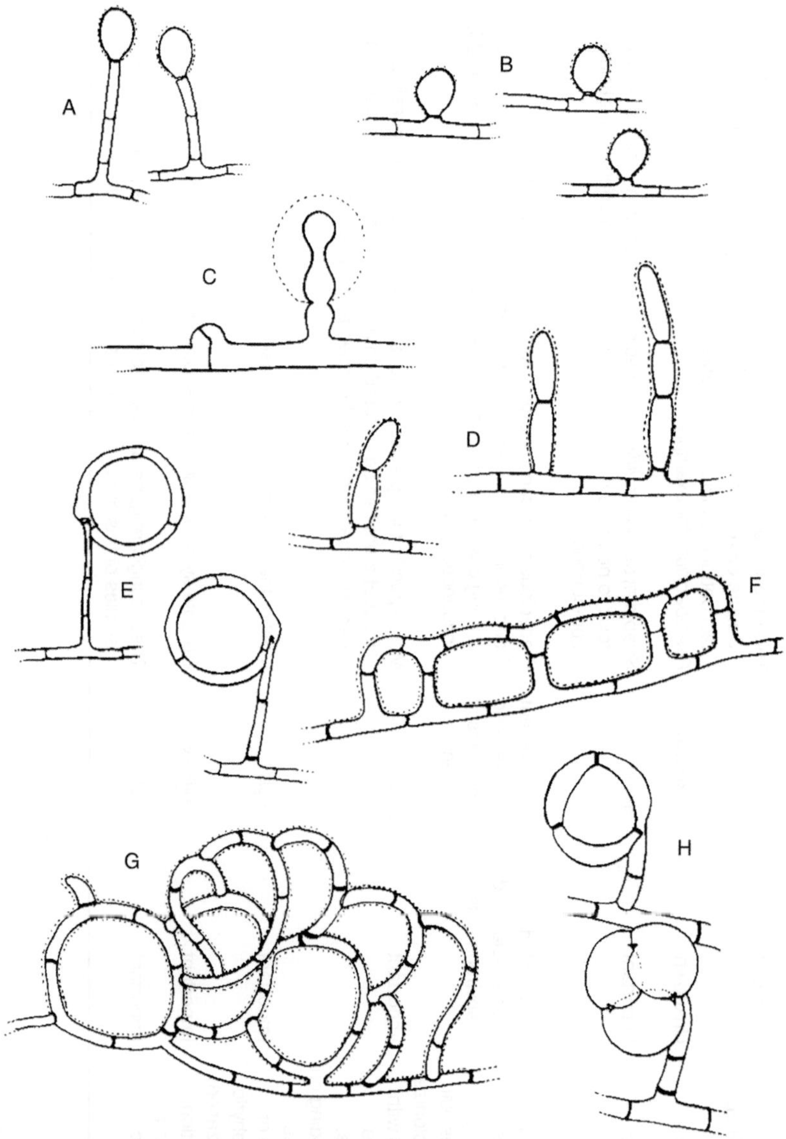

Fig. 21.2. Organs of capture in predacious nematophagous fungi. A: Stalked adhesive knobs; B: sessile adhesive knobs; C: hour-glass adhesive knob of *Nematoctonus*; D: adhesive branches; E: non-constricting rings; F: two-dimensional scalariform adhesive net; G: three-dimensional adhesive net; H: constricting rings. (After Barron, 1977.)

attachment several fold, thus ensuring that the nematode is firmly held (Gray, 1988). The more advanced trapping devices are non-constricting rings (Fig. 21.2E). When a nematode enters the non-constricting ring and its body diameter is only slightly larger than the inner diameter of the ring, the ring will firmly hold the nematode. A more sophisticated trapping device for predacious fungi is the constricting ring (Fig. 21.2H), which detects a nematode passing through and uses an inflation mechanism to constrict instantly upon the nematode body (Fig. 21.3D). Once a nematode is trapped by a constricting ring, the fungus quickly penetrates their cuticle and other

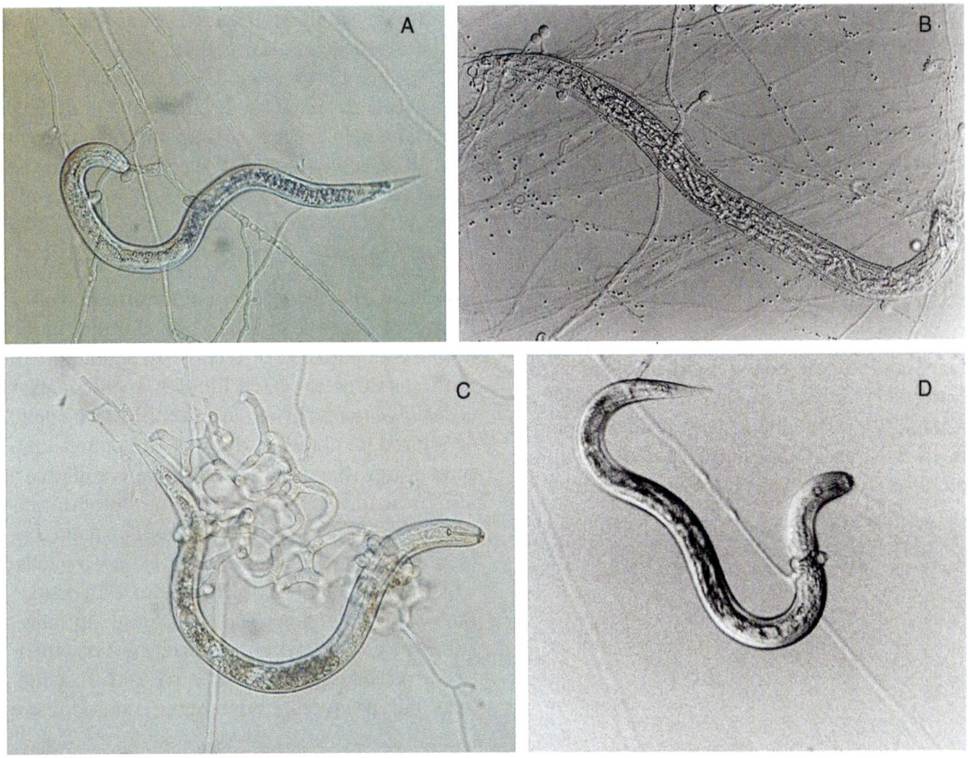

Fig. 21.3. The soybean cyst nematode second-stage juveniles (J2) caught by predacious fungi. A: J2 caught by adhesive branches; B: J2 caught by adhesive knobs; C: J2 caught by three-dimensional nets; D: J2 caught by a constricting ring.

natural body openings to successfully devour their internal contents as a food source.

21.3.2. Endoparasites of Vermiform Nematodes

Endoparasitic fungi have no special trapping devices. Most are obligate parasites with some having limited saprotrophic capability. There is no extensive hyphal development. Some are facultative parasites and can survive as saprobes in the absence of nematodes as a food source.

Obligate endoparasites of vermiform nematodes

The obligate endoparasitic fungi are found among lower and higher classes of fungi. They can be placed into four groups based on their mechanism of parasitism: (i) encysting species, (ii) species forming adhesive conidia, (iii) species with conidia that may be ingested and (iv) species with gun cells.

Encysting species that produce zoospores are found in *Chytridiomycota* and *Oomycota* (not true fungi), both predominantly aquatic taxa. The zoospores can swarm for a short time, and if they reach their host nematode, they encyst on the nematode cuticle. *Catenaria* spp., which belongs to Chytridiomycetes, are the most common endoparasitic fungi encountered parasitizing vermiform nematodes.

The endoparasitic fungi forming adhesive conidia are found among the *Zygomycota*, *Basidiomycota* and *Deuteromycota*. A species of *Meristacrum* (*Zygomycota*) forms adhesive conidia that are forcibly discharged at their maturity. If the primary conidia fail to contact a nematode, they may germinate and produce a secondary conidium. Endoparasitic basidiomycetes, for example, *Nematoctonus* spp., also infect nematodes using adhesive conidia, but the conidia formed by *Nematoctonus* spp. are similar to the adhesive knobs formed on hyphae by the predacious species. Most endoparasitic fungi forming adhesive conidia are found in *Deuteromycota*, namely, *Pochonia* spp. and *Hirsutella* spp. (Fig. 21.4). The parasitic conidia produced by these fungi either have adhesive mucilage or a mucous sheath on parts of the surface of the conidia. Most of these fungi can be extensively cultured on artificial media and produce infective conidia, even though their saprotrophy is not fully understood. Thus, they may be considered facultative endoparasites.

Some species of endoparasites exclusively in the genus *Harposporium* (*Deuteromycota*) have developed morphologically adapted conidia, which, when ingested by a non-plant-parasitic nematode, become lodged in either its buccal cavity or pharynx. The conidia are crescent, helicoid or irregular shaped, and they are well adapted for their parasitic behaviour. These fungi do not infect the stylet-bearing plant-parasitic nematodes because the small stylet lumen opening prevents conidia from passing through.

Species of *Haptoglossa* (*Oomycota*) have evolved a special infective cell called a 'gun' cell (Fig. 21.5A). Once a nematode contacts the gun cell, it triggers the cell to fire a 'missile' (projectile) through the host cuticle, thereby serving as a hypodermic conduit through which the parasite's cytoplasm is injected into the nematode host (Fig. 21.5B, C).

Facultative parasites of vermiform nematodes

Although the saprotrophic capabilities of many fungal endoparasites of vermiform nematodes in nature are unclear, a few fungi, namely, *Catenaria anguillulae*, can grow saprotrophically on dead or injured nematodes. Many of the Deuteromycetes fungi that produce adhesive conidia can also grow in various substrates and produce infective conidia. These fungi can be considered as facultative parasites. For example, *Esteya vermicola*, which was isolated from the pinewood nematode *Bursaphelenchus xylophilus*, produced infective, adhesive conidia when cultured on less nutritious media or grown from a parasitized nematode, but produced non-infective, non-adhesive

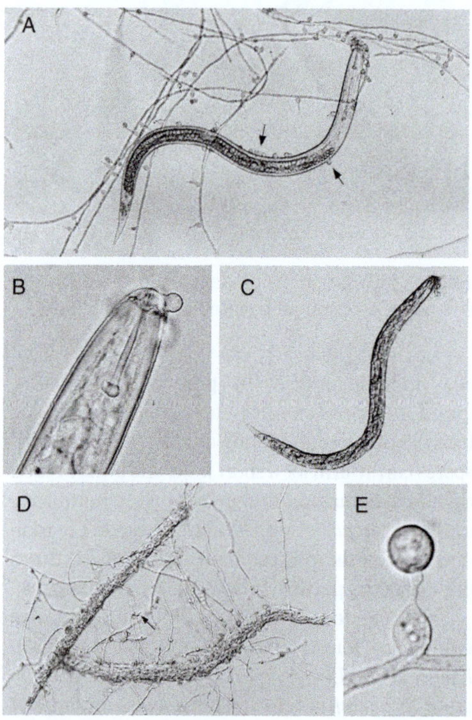

Fig. 21.4. Parasitism of second-stage juveniles (J2) of *Heterodera glycines* by the endoparasitic fungus *Hirsutella minnesotensis*. A: J2 with conidia attached to the cuticle on cornmeal agar, arrows; B: J2 with attached conidium that is penetrating the nematode cuticle; C: J2 filled with mycelium; D: the fungus growing from the nematode body and sporulate; a conidiogenous cell with an attached conidium, arrow; E: a conidiogenous cell with a conidium. (After Chen *et al.*, 2000.)

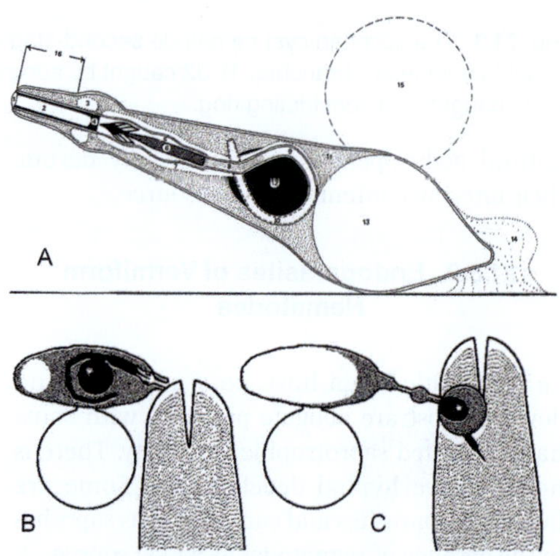

Fig. 21.5. A: Diagrammatic representation of the gun cell of *Haptoglossa mirabilis*. 1. Muzzle; 2. bore with extra wall layer; 3. apical vacuolar system; 4. plug; 5. projectile chamber; 6. projectile; 7. lateral vacuolar system; 8-10. lower flexuous tube; 11. protoplasm; 12. nucleus; 13. basal vacuole; 14. adhesive pad; 15. empty zoospore cyst; 16. apical cone (after Barron, 1987); B: a triggered infection cell; C: fired infection cell and infected nematode. (After Beakes and Glockling, 1998.)

conidia when cultured on enriched media (Liou *et al.*, 1999). Like predacious fungi, *E. vermicola* has evolved both a saprophytic and a parasitic growth habit. It is noteworthy that this fungus has a bacterial endosymbiont. These endosymbionts could alter the fungal biology and ecology (Wang *et al.*, 2017). Some predacious fungi, such as those with adhesive knobs, act similarly to the endoparasitic fungi that form adhesive conidia. Consequently, the predacious fungi with adhesive knobs can be considered facultative endoparasites of vermiform nematodes. Many fungi, such as *Pochonia chlamydosporia*, that colonize cysts or egg masses also can attack second-stage juveniles (J2) contained within cysts or egg masses. These fungi are also facultative parasites of vermiform-stage juveniles.

21.3.3. Parasites of Sedentary Females and Eggs

Sedentary nematode life stages, for example, females, cysts, eggs and eggs with masses, are subjected to parasitism by fungi (Fig. 21.6). Active mobile vermiform nematodes have an increased probability of contacting predacious or endoparasitic fungi, whereas sedentary life stages have less probability of contact with parasitic fungi, unless those fungi have developed mechanism to reach the nematodes. Fungi attacking sedentary life stages of nematodes are biologically and ecologically diverse. Although a few of them are obligate parasites, most fungi within this group can live in soil as saprobes and are facultative nematode parasites.

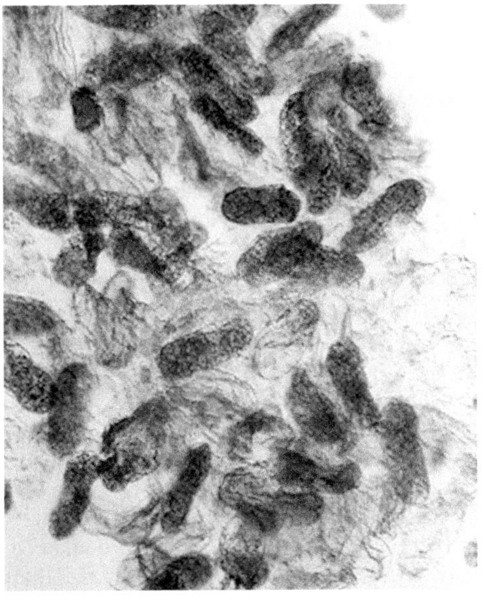

Fig. 21.6. Nematode eggs colonized by a fungus.

Obligate parasites of sedentary females and eggs

These obligate parasitic fungi cannot develop mycelium in soil; therefore, they must have a mobile or disseminative life stage that provides them a mechanism to target their prey. *Catenaria auxiliaris* and *Nematophthora gynophila* are two examples of species that have evolved such mechanisms. They produce thick-walled resting spores that remain in soil from one season to the next. The zoospores released from the resting spores are mobile. They often contact nematode sedentary life stages, penetrate the body wall and assimilate the body contents. *C. auxiliaris* and *N. gynophila* are known to attack nematode eggs contained within the cyst life stage of nematode species within the family Heteroderidae.

Facultative parasites of eggs and sedentary females

Numerous fungi have been isolated from different life stages of nematodes, for example, eggs, sedentary females, egg masses and cysts of Heteroderidae. The facultative parasitic fungi produce an expanding mycelial growth that enables them to reach the sedentary stage of their nematode target. These fungi belong to taxonomically divergent groups, but they are encountered more frequently in the Deuteromycetes. The fungal species most isolated are limited to a few genera, which indicate there is a certain degree of specialization towards the host nematode. These fungi include *Acremonium*, *Alternaria*, *Catenaria*, *Cylindrocarpon*, *Exophiala*, *Fusarium*, *Gliocladium*, *Nematophthora*, *Purpureocillium*, *Penicillium*, *Phoma*, *Pochonia* and *Verticillium* (Haarith *et al.*, 2020a). Besides these genera, *Neocosmospora vasinfecta* and *Stagonospora heteroderae* were frequently isolated from *Heterodera glycines* in warm climatic conditions (Chen *et al.*, 1994).

21.3.4. Fungi Producing Antibiotic Substances

Fungal metabolic enzymes and secondary metabolites are excellent sources of bioactive compounds. Scientist continues to provide studies that support the diverse impacts these metabolites have on nematodes, including inhibiting egg hatch, paralysing and killing juveniles (Haarith *et al.*, 2020b). Fungi that form close associations with nematodes, such as those in the genera *Purpureocillium*, *Pochonia*, *Fusarium*, *Aspergillus*, *Trichoderma*, *Myrothecium* and *Penicillium*, have been demonstrated to produce antibiotically active compounds. Numerous secondary metabolites have been identified from species of *Trichoderma*, of which some such as trichodermin, viridin, gliotoxin and peptaibols have nematicidal activities (Yang *et al.*, 2010). Greenhouse studies have demonstrated that nematode density was correlated negatively with activities of chitinase, collagenase and proteinase of some soil microorganisms, including fungi (Rodríguez-Kábana *et al.*, 1989; Galper *et al.*, 1991). These enzymatic active compounds are often linked to the infection of nematodes by nematophagous fungi (Huang *et al.*, 2004).

21.3.5. Vesicular-arbuscular Mycorrhizal Fungi

The VAM fungi invade deeply into roots. Both antagonistic and beneficial effects of VAM on nematode populations have been reported. VAM fungi may compete for nutrition and root space, modify root exudates, alter plant physiology, colonize nematode feeding sites, reduce the number of giant cells, release nematoxins or antibiotics and induce plant resistance (Ingham, 1988; Abarca *et al.*, 2024). Indirectly, the effect may be mediated through improved plant growth, offsetting yield suppression normally caused by nematode parasitism.

21.4. Bacterial Antagonists of Nematodes

Bacteria are ubiquitous organisms in ecosystems, and they interact with nematodes in modes of action that range from neutralism, mutualism, protocooperation, commensalism, competition and amensalism to parasitism (Sayre, 1988; Chen and Dickson, 2004). Although these bacterial nematode antagonists make up only a small portion of total soil bacteria, a number of genera have been reported to be antagonistic to nematodes (Table 21.2). Most nematode-antagonistic bacteria are non-parasitic rhizobacteria; however, a few, mainly *Pasteuria* spp., have evolved sophisticated mechanisms of parasitism. There are numerous publications available that provide detailed information on biocontrol with antagonistic bacteria (Stirling, 1991; Chen and Dickson, 1998, 2004; Siddiqui and Mahmood, 1999; Li *et al.*, 2015; Migunova and Sasanelli, 2021; Aioub *et al.*, 2022; Pires *et al.*, 2022).

21.4.1. Endoparasites – *Pasteuria*

Pasteuria spp. are gram-positive, endospore-forming bacteria (Dickson *et al.*, 2009) with a worldwide occurrence. They are most extensively studied for their biocontrol potential against a number of plant-parasitic nematodes, especially the root-knot nematodes (RKN), *Meloidogyne* spp. (Chen and Dickson, 1998). Studies have shown *Pasteuria* spp. to be effective in the development of nematode soil suppressiveness. The *Pasteuria* endospore, which is almost exclusively found in soil, lies passively waiting for an active nematode to wander into its domain. Apparently, the endospore recognizes the nematode as a host, and attaches tightly, cupping onto the cuticle of the nematode (Fig. 21.7). *Pasteuria* endospores have a mass of fibrous outgrowths extruding from the central body. These presumably function in an adhesive process whereby the endospores bind tightly to a host nematode (Persidis *et al.*, 1991; Davies *et al.*, 1994; Brito *et al.*, 2003; Preston *et al.*, 2003; Schmidt *et al.*, 2004). Research data are variable, but generally, it is considered that when a nematode has 1 to 15 endospores attached, they remain infective, but greater numbers reduce their infectivity (Fig. 21.8) (Kariuki *et al.*, 2006). After the nematode initiates feeding within the plant root, the endospore germinates and penetrates the cuticle by means of a single germ tube (Fig. 21.9). The germ tube extends through the nematode body wall into the nematode pseudocoelomic cavity. The life cycle of *Pasteuria penetrans* in *Meloidogyne* spp. is illustrated in Figure 21.10 (Sayre and Starr, 1988). After entering the nematode body, *Pasteuria* begins to form microcolonies that resemble clusters of grapes. The bacterium then

Table 21.2. Taxonomic groups of nematode-antagonistic bacteria.

Phylum	Class	Order	Family	Genus
Actinomycetota	*Actinomycetia*	*Micrococcales*	*Microbacteriaceae*	*Agromyces*
		Streptomycetales	*Streptomycetaceae*	*Streptomyces*
Bacillota	*Bacilli*	*Bacillales*	*Bacillaceae*	*Bacillus*
				Cytobacillus
				Lysinibacillus
				Priestia
			Paenibacillaceae	*Paenibacillus*
			Pasteuriaceae	*Pasteuria*
	Clostridia	*Eubacteriales*	*Lachnospiraceae*	*Clostridium*
Bacteroidota	*Sphingobacteriia*	*Sphingobacteriales*	*Sphingobacteriaceae*	*Sphingobacterium*
Pseudomonadota	*Alphaproteobacteria*	*Hyphomicrobiales*	*Brucellaceae*	*Brucella*
			Rhizobiaceae	*Agrobacterium*
				Ensifer
				Rhizobium
		Rhodospirillales	*Azospirillaceae*	*Azospirillum*
	Betaproteobacteria	*Burkholderiales*	*Alcaligenaceae*	*Alcaligenes*
			Burkholderiaceae	*Burkholderia*
		Neisseriales	*Neisseriaceae*	*Chromobacterium*
	Gammaproteobacteria	*Enterobacterales*	*Enterobacteriaceae*	*Enterobacter*
			Erwiniaceae	*Pantoea*
			Morganellaceae	*Photorhabdus*
				Xenorhabdus
			Yersiniaceae	*Serratia*
		Pseudomonadales	*Moraxellaceae*	*Acinetobacter*
			Pseudomonadaceae	*Azotobacter*
				Pseudomonas
		Xanthomonadales	*Xanthomonadaceae*	*Stenotrophomonas*
Thermodesulfobacteriota	*Desulfovibrionia*	*Desulfovibrionales*	*Desulfovibrionaceae*	*Desulfovibrio*

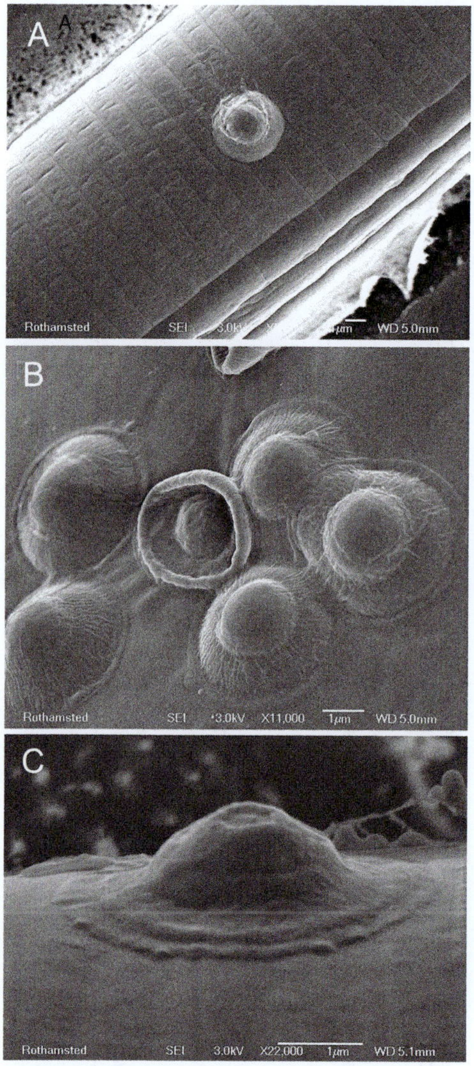

Fig. 21.7. A-C: Scanning electron micrographs of *Pasteuria penetrans* endospores attached to body of *Meloidogyne* second-stage juveniles. (Copyright Rothamsted Research.)

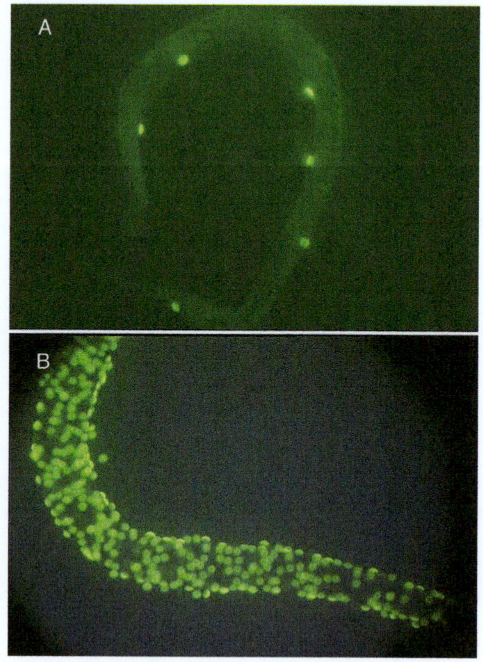

Fig. 21.8. Fluorescing endospores of *Pasteuria penetrans* attached to the second-stage juveniles of a root-knot nematode. A: Nematode with 1-15 endospores attached to nematode body can infect its host plant and the bacterium can develop and sporulate within the nematode body; B: nematodes encumbered with greater than 15 endospores lose mobility and infectivity.

enters its vegetative endoparasitic developmental stages, followed by the sporulation phase and, finally, endospore maturation. Female nematodes that are infected experience suppressed fecundity, generally producing no or occasionally a few eggs. Their bodies become filled with endospores. More details of the chronological fragmentation of the vegetative microcolony inside the nematode body have been reported (Phani and Rao, 2018).

Several species of *Pasteuria* have been recognized as valid, namely, *Pasteuria ramosa*; *P. penetrans* (hosts include RKN, *Meloidogyne* spp.); *Pasteuria thornei* (hosts include lesion nematodes, *Pratylenchus* spp.); *Pasteuria nishizawae* (hosts include cyst nematodes, *Heterodera* and *Globodera* spp.); '*Candidatus* Pasteuria usgae' (host includes sting nematode, *Belonolaimus* sp.); '*Candidatus* Pasteuria aldrichii' (host includes the bacterivorous nematode, *Bursilla*); and *Pasteuria hartismeri* (host includes *Meloidogyne ardenensis*).

21.4.2. Rhizobacteria

Most nematode-antagonistic bacteria that have been reported are rhizobacteria because they interact with nematodes and plant roots. These bacteria are generally non-parasites or occasionally

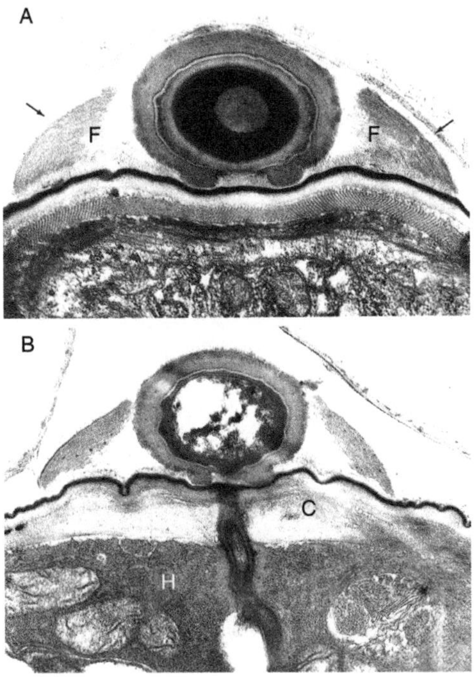

Fig. 21.9. Cross section of endospore of *Pasteuria penetrans* attached to the cuticle of *Meloidogyne incognita*. Note the endospore central core, basal ring, parasporal fibres (F) and the tight adhesion to the nematode body wall, and penetrating germ tube through cuticle (C) and hypodermis (H). (After Sayre and Wergin, 1977.)

opportunistic parasites of nematodes. The most common mode of antagonistic action against nematodes is antibiosis, which involves releasing nematicidal metabolites, including lytic enzymes, antibiotics, toxins and volatile organic compounds. Induction of plant resistance is another important mechanism of nematode biocontrol by rhizobacteria. Additionally, some of rhizobacteria are plant growth-promoting bacteria (PGPB) that provide indirect benefits in managing plant-parasitic nematodes by promoting plant growth and alleviating damage (Aioub *et al.*, 2022).

Over the past decade, there has been an increase in studies on *Bacillus* and related species for the management of plant-parasitic nematodes. Studies have also included research on biocontrol of plant diseases and have increasingly focused on rhizobacteria such as *Bacillus*, *Pseudomonas* and *Streptomyces* species (Bhattacharyya & Jha, 2012). Phylogenetically, *Bacillus* and *Pasteuria* are closely related groups within the order *Bacillales* (Table 21.2). Actinomycetes, particularly the genus *Streptomyces*, are another group of bacteria known for producing antibiotics and exhibiting high potential for biocontrol of plant-parasitic nematodes.

21.5. Potential Biocontrol Agents

The main goal of biological control research is to develop commercial products that can be used either as inoculum or as a biopesticide. Numerous studies have explored potential commercialization of biological control agents with the endeavour increasing yearly. Several species of nematode-antagonistic fungi and bacteria or their metabolites have been patented. Some of these have been widely tested in the field, with reports indicating they have high potential as commercial products. Unfortunately, only a few commercial products have been brought to the agricultural market (Chen and Dickson, 2012; Li *et al.*, 2015; Migunova and Sasanelli, 2021). The following are examples of fungal and bacterial biocontrol agents that have been tested to some extent in the greenhouse and field studies and are thought to have potential as commercial agents.

21.5.1. *Purpureocillium* lilacinum

Purpureocillium lilacinum (= *Paecilomyces lilacinus*) has been isolated worldwide from eggs, egg masses, females and cysts of numerous species of plant-parasitic nematodes. The fungus first colonizes the gelatinous matrices of sedentary endoparasitic nematodes, for example, *Meloidogyne* and *Heterodera* spp., and then infects their eggs and/or the female body. Enzymes such as serine protease and chitinase are involved in the penetration of the eggshells (Khan *et al.*, 2004; Dong *et al.*, 2007). Although the fungus primarily attacks nematode eggs and sedentary stages of females, it can also colonize and immobilize the vermiform stages and J2 within the eggshell. Pathogenicity among different isolates has been reported to be variable. Some are virulent, whereas others are less so or non-virulent. Several secondary metabolites, including the production of

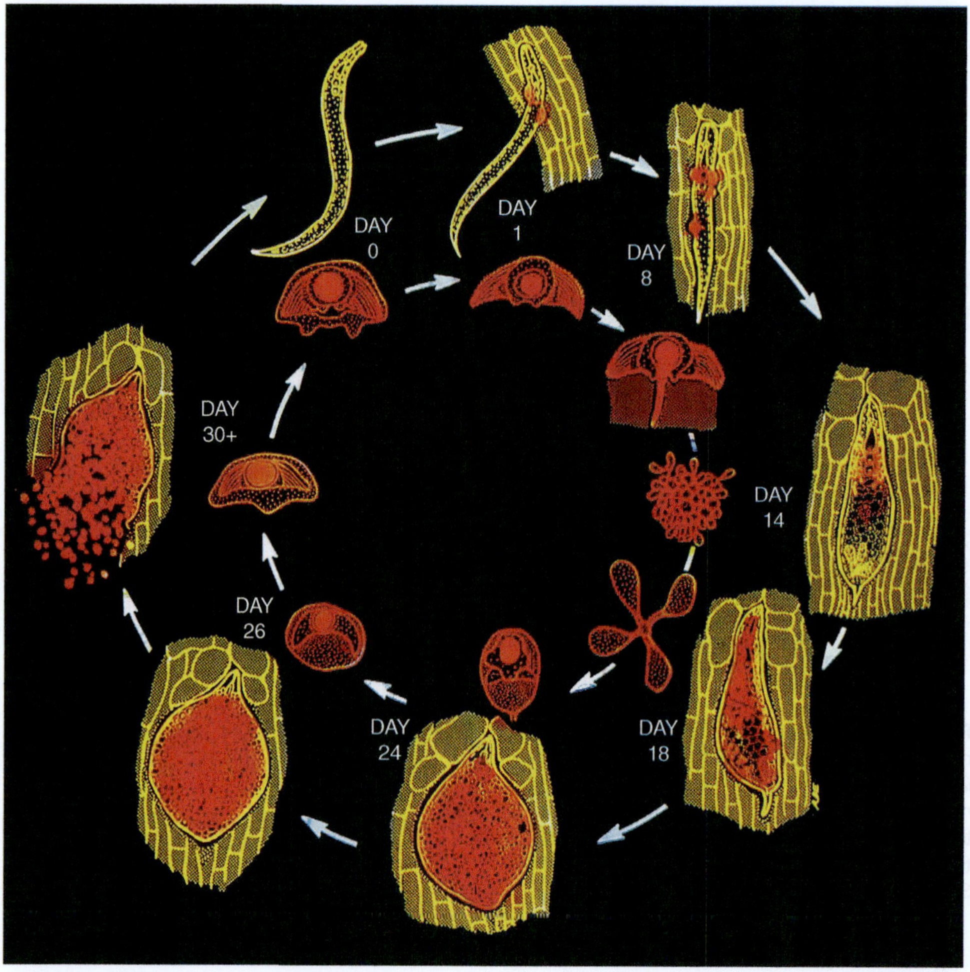

Fig. 21.10. Diagrammatic sketch of the life cycle of *Pasteuria penetrans* in *Meloidogyne* spp. (After Sayre and Starr, 1988.)

mycotoxins and acetylcholinesterase inhibitors, have been identified (Chen and Hu, 2022). They are potentially nematicidal, but their role in suppressing plant-parasitic nematodes has not been fully investigated. The organism is a good rhizosphere competitor, root colonizer and plant-growth promoter (Khan and Tanaka, 2023).

P. lilacinum has been widely tested for its efficacy as a biological control agent against plant-parasitic nematodes. Several commercial biocontrol products have been developed from the fungus and are now available in several countries. Biocontrol efficacy of the fungus varies in different trials, which may be attributed to the variation in virulence among isolates (Stirling and West, 1991), nematode species and experimental conditions.

21.5.2. Pochonia chlamydosporia

P. chlamydosporia (= *Verticillium chlamydosporium*, *Diheterospora chlamydosporia*) is another important facultative nematode-parasitic fungus found worldwide. It is well studied for its potential in biological control of plant-parasitic nematodes, notably RKN and cyst nematodes (Manzanilla-López and Lopez-Llorca, 2017). The fungus primarily parasitizes nematode eggs but also can parasitize J2 and other nematode stages. *P. chlamydosporia* enters nematode cysts through natural body openings or penetrates the cyst wall directly. The fungus forms a branched mycelial network and penetrates eggs by simple branches of hyphae or by the formation of appressoria (Lopez-Llorca and Claugher, 1990; Morton *et al.*, 2004). Several enzymes, including serine proteinases, chitinases and collagenases, have been identified involving infection of nematodes by *P. chlamydosporia* (Aranda-Martinez *et al.*, 2018; Yi *et al.*, 2021). Virulence to nematodes varies among different isolates of the fungus and has nematode-host preference at the infraspecific level (Mauchline *et al.*, 2004). *P. chlamydosporia* is an endophytic fungus and can induce plant resistance, but its capability of endophytism and resistance induction depends on interactions of genotypes of plant species and fungal isolates (Ghahremani *et al.*, 2019; Yi *et al.*, 2021; Rosso *et al.*, 2023). The fungus produces nematicidal secondary metabolites such as aurovertin (Niu *et al.*, 2010), phomalactone (Khambay *et al.*, 2000), ketamine (Ferreira *et al.*, 2020) and glycosylated resorcylic acid lactones (Li *et al.*, 2024), all of which can aid in killing nematodes. In addition, the fungus promotes host plant growth and potentially increases plant tolerance to the infection of nematodes (Zavala-Gonzalez *et al.*, 2016).

P. chlamydosporia has been extensively tested for its potential as a biocontrol agent in greenhouse and field plots throughout the world. Several commercial products from the fungus are on the market. The effectiveness of *P. chlamydosporia* can be affected by soil type, organic matter content and the presence of other soil microorganisms, which highlights the importance of understanding local soil ecology if one is to use the product successfully (Kerry and Hirsch, 2011; Manzanilla-López *et al.*, 2017).

21.5.3. *Trichoderma*

Trichoderma spp. are ubiquitous hyphomycete soil fungi that are widely studied for their biocontrol capabilities. They exhibit a broad spectrum of activities against various pathogens and pests such as fungi, bacteria, viruses, nematodes and insects, making them highly valuable as biocontrol agents in agricultural and horticultural settings (Guzmán-Guzmán *et al.*, 2023). The potential of these fungi as biocontrol agents against plant-parasitic nematodes has been well demonstrated (Sharon *et al.*, 2011). The modes of action against plant-parasitic nematodes are similar to that against microbial pathogens, including direct parasitism, antibiosis, competition and induction of plant defence systems (TariqJaveed *et al.*, 2021; Guzmán-Guzmán et *al.*, 2023). Numerous secondary metabolites have been identified from the fungi, and some of them, such as trichodermin, viridin, gliotoxin and peptaibols, have nematicidal activities (Yang *et al.*, 2010).

Several commercially available products have been developed from *Trichoderma* spp. These fungi primarily act as saprophytes, are fast-growing and are easy to mass produce for commercial use. Although numerous *Trichoderma* species are found in soil, most biocontrol studies of plant pathogens have focused on a few species, namely, *Trichoderma harzianum*, *Trichoderma atroviride*, *Trichoderma virens*, *Trichoderma asperellum* and *Trichoderma longibrachiatum* (Guzmán-Guzmán *et al.*, 2023). However, the efficacy of *Trichoderma* products for managing plant-parasitic organisms in field production has not been fully studied.

21.5.4. *Pasteuria* spp.

Field soil naturally infested with endospores of *Pasteuria* spp. often provides natural nematode soil suppressiveness. Certainly, there are limitations, but one large advantage is the stability of endospores. They will endure many environmental conditions, ranging from hot, dry soils to varying soil types. It is thought that the endospores are capable of being stored for long periods without a diminish in their viability. However, there exists a large and very real limitation to their use, that is, the high degree of specificity exhibited by various species of *Pasteuria* that have been described. For example, *P. penetrans* taken from the peanut RKN, *Meloidogyne arenaria*, tends to be highly specific to the infection of this single RKN species, and this specificity even applies down to the host race 1 level. *M. arenaria* race 2 is not a host for the *Pasteuria* that parasitizes race 1 (Oostendorp *et al.*, 1990). This implies that many strains of *Pasteuria* would have to be produced, each with specificity to a certain RKN species. Another limitation is that RKN often occurs as mixtures of species within many agricultural fields. While one species might be brought under control, another species might increase in numbers (Cetintas and Dickson, 2004). *In vitro* mass production of *Pasteuria* endospores is challenging, limiting the use of this organism for biological control of plant-parasitic nematodes. Nevertheless, a commercialized *P. nishizawae* product is being marketed for control of the soybean cyst nematode. However, there is scarce public data on the field control efficacy of the product.

21.5.5. *Bacillus* spp.

Bacillus spp., along with *Pseudomonas* spp., are well-known plant growth-promoting rhizobacteria (Bhat *et al.*, 2022). *Bacillus* spp. are probably the most widely studied bacteria for biocontrol of plant-parasitic nematodes and have shown promise. *Bacillus*, a genus of gram-positive, rod-shaped bacteria, is well known for its ability to produce a range of secondary metabolites and enzymes with nematicidal properties. A few *Bacillus* spp. have been shown to have nematode-antagonistic activities (Engelbrecht *et al.*, 2018). They produce antibiotics and other toxic compounds that are detrimental to nematodes. For instance, *Bacillus thuringiensis* produces crystal proteins that are toxic to various nematodes, disrupting their digestive processes (Yu *et al.*, 2015). Certain *Bacillus* spp. and strains are opportunistic parasites that can parasitize nematode eggs, juveniles and adults (Tian *et al.*, 2007). Enzymes, for example, chitinases, proteases and lipases, produced by *Bacillus* can degrade nematode eggshells and cuticles, thus directly reducing nematode population densities (Tian *et al.*, 2007). *Bacillus* spp. can trigger systemic resistance in plants, enhancing their defensive capabilities against nematodes. This involves the activation of plant defence pathways, leading to the production of defence-related proteins and compounds that suppress nematode infection and reproduction. *Bacillus* offers a multifaceted approach to managing plant-parasitic nematodes through mechanisms such as antibiosis, parasitism, induction of systemic resistance and enzyme production. Their efficacy has been validated in both greenhouse and field studies (Migunova and Sasanelli, 2021; Antil *et al.*, 2023), making them valuable potential biocontrol agents aimed at reducing nematode-induced crop damage.

21.5.6. *Pseudomonas* spp.

Pseudomonas is one of the bacterial genera most frequently found in agricultural soils and the rhizosphere. *Pseudomonas* spp., as representative rhizobacteria, have been extensively studied for their ability to promote plant growth, control plant microbial pathogens, induce resistance to abiotic stress and remediate contaminated soil (Saeed *et al.*, 2021; Bhat *et al.*, 2022). Various species of *Pseudomonas* have been frequently reported to be antagonistic to plant-parasitic nematodes (Chen and Dickson, 2004; Li *et al.*, 2015; Aioub *et al.*, 2022), but their efficacy depends on the bacterial species and strains. Among these, *Pseudomonas fluorescens* has been the most extensively studied. *Pseudomonas* can adversely affect nematodes through several mechanisms: (i) producing siderophores and antibiotics like pyrrolnitrin and pyoverdine; (ii) the nematicidal activity of metabolites such as 2,4-diacetylphloroglucinol and hydrogen cyanide; (iii) altering root exudate chemicals to affect egg hatching, juvenile attraction and penetration; (iv) binding the lipopolysaccharide layer of *P. fluorescens* cell membranes to root surface lectins, impacting host recognition by nematodes; (v) causing nematode egg mortality and premature hatching by producing chitinase enzymes; (vi) inducing systemic resistance through the production of peroxidase and chitinase enzymes and (vii) inhibiting nematode feeding site formation via 1-aminocyclopropane-1-carboxylic acid deaminase activity, which affects nematode-induced ethylene production in roots (Nagachandrabose, 2020). Commercial products developed from *Pseudomonas*, specifically for managing plant-parasitic nematodes, are scarce, while numerous products are available on the market for managing microbial diseases of plants and as biofertilizers. Due to the multifunctional nature of many *Pseudomonas* species and strains, more research is warranted to test these products for managing plant-parasitic nematodes, especially for the nematode-microbe disease complex.

21.5.7. *Streptomyces* – Avermectins

Streptomyces are actinomycete bacteria, which are well known for their antibiotic production. *Streptomyces avermitilis* is an excellent example in this group of antibiotic-producing microorganisms for biocontrol of nematodes. This bacterium was first isolated from soil in Japan in 1975. Soon after its discovery, avermectins, a family of macrocyclic lactones, were isolated from its culture, and their nematicidal activities were demonstrated (Egerton *et al.*, 1979). Avermectins are highly toxic to nematodes as well as invertebrate animals, including insects and mites (Lasota and Dybas, 1991). Numerous commercial products have been developed from avermectins, and most of them are used for the control of nematodes and insects in livestock and household animals. Ivermectin, an avermectin derivative, was subsequently commercialized, entering the veterinary, agricultural and aquaculture markets in 1981. Other nematicidal compounds, for example, actinomycines, nonane, funichromin, milbermectin and spectinabilin, have been identified in various species of *Streptomyces* (Seong *et al.*, 2021; Mogollon-Ortiz *et al.*, 2024). Parasitism of nematodes by *Streptomyces* has also been demonstrated (Mogollon-Ortiz *et al.*, 2024). Some *Streptomyces* spp. are endophytic, promoting plant growth and inducing host resistance to pathogens (Sharma *et al.*, 2020; Bhat *et al.*, 2022). The role of *Streptomyces* in sustainable pest control and agriculture has been documented (Diab *et al.*, 2024).

21.6. Nematode-suppressive Soils

Study of nematode-suppressive soils is an important step in searching for and identifying effective biocontrol agents and developing nematode management plans. A nematode-suppressive soil may be defined as one in which a nematode population that was once at an economically damaging level becomes suppressed and remains at a level that causes little or no economic damage to a nematode-susceptible plant cultivar. Generally, the reported suppressive soils involve specific suppression, and in some situations, one or more species of organisms have been identified as being at least partially responsible for the suppression. The suppression may come about in a subtle manner, or it may come about in a more dramatic manner, depending on the organisms involved and environmental conditions. A classic example of nematode-suppressive soil is the suppression of *Heterodera avenae* by *P. chlamydosporia*, along with several other fungal species (Kerry, 1978). Suppression of RKN by *Pasteuria* has been demonstrated (Dickson *et al.*, 1994; Weibelzahl-Fulton *et al.*, 1996). Soils suppressive to the soybean cyst nematode has been reported from the USA, China and Japan (Chen and Dickson, 2012; Yang *et al.*, 2023). Soybean field soils from Minnesota are documented to have both fungal and bacterial antagonists that suppressed soybean cyst nematode (Chen, 2007; Hu *et al.*, 2017). Another example is the natural suppression of the sugar beet cyst nematode, *Heterodera schachtii*, in California (Borneman and Becker, 2007).

21.7. Enhancement of Biocontrol

It is important to understand why some soils become suppressive to plant-parasitic nematodes while others do not. Is it practical to develop nematode suppressiveness in agricultural soils? Answering these questions requires much more knowledge about soil ecology and the biological mechanisms inherent in the development of nematode-suppressive soils. Nevertheless, scientists have been exploring various methods for decades to enhance the natural biological management of plant-parasitic nematodes, with both successes and failures reported.

Enhancing biological control or biological suppressiveness includes two aspects: general suppression and specific suppression. Enhancement of general suppression may be achieved by increasing the total biomass of soil microorganisms, which can alter the soil food web to suppress plant-parasitic nematode population densities. In contrast, enhancement of specific suppression involves encouraging the activities of certain biological antagonists. This type of enhancement requires much more knowledge about the types and species of antagonists in the soil and the factors affecting their antagonistic activities.

21.7.1. Organic Amendments

Most studies on the enhancement of biocontrol of nematodes involve soil organic amendments

(McSorley, 2011; Rosskopf *et al.*, 2020). Unfortunately, most such studies have been primarily greenhouse pot experiments. Generally, organic amendments enhance plant growth and crop yield by improving soil nutrients and controlling plant pests and pathogens through the toxic compounds released during decomposition (Akhtar and Malik, 2000; McSorley, 2011). Organic amendments significantly influence soil microbial communities, which can result in the reduction of nematode population densities. Biocontrol of nematodes by the microorganisms belonging to categories of facultative parasites or those producing antibiotics may be more likely enhanced by organic amendments than the obligate endoparasites.

Linford *et al.* (1938) first demonstrated that adding organic matter to soil significantly altered indigenous nematode populations. They suggested that the organic supplement increased microbial population density, which subsequently boosted the density of free-living nematodes and the activity of nematophagous fungi in the soil. This increased activity of nematophagous species would stabilize and eventually reduce soil nematode population densities, including those of plant-parasitic nematodes. However, this hypothesis is not universally true and depends on the species of nematophagous fungi and soil environments (Cooke, 1963).

Several later studies, mostly from greenhouse experiments, have associated the reduction of plant-parasitic nematodes with increased activities of microbial antagonists. For example, sunn hemp (*Crotalaria juncea*) used as green manure in a greenhouse study suppressed reniform nematodes (*Rotylenchulus reniformis*) on pineapple and increased predacious fungi and nematode egg parasitism (Wang *et al.*, 2003). Green manure increased egg-parasitic activities of fungi on *H. schachtii* in soil from a marigold-monoculture field and reduced activities in soil from a sugar beet-wheat rotation field (Pyrowolakis *et al.*, 1999). Straw fertilizer favoured more fungal parasitism of *H. schachtii* eggs than manure (Sosnowska and Banaszak, 2000). Green manure from rapeseed and oil radish enhanced the control of the RKN *Meloidogyne chitwoodi* in potato by the bacterium *Bacillus megaterium* (Al-Rehiayani *et al.*, 1999). Organic matter may increase predacious nematode population densities, contributing to the reduction of plant-parasitic nematodes (Sidhu and Kanwar, 2020). In field experiments, applying residue from chinaberry (*Melia toosendan*) reduced *Meloidogyne incognita* population density in balloonflower (*Platycodon grandiflorum*), likely due to the increased population density of *Trichoderma* spp., which parasitized nematode eggs (Zhang and Zhang, 2009). Sugarcane trash amendments induced suppressiveness to plant-parasitic nematodes in field trials, likely due to enhanced activities of several predacious fungi (Stirling *et al.*, 2005). A two -year field experiment demonstrated that organic fertilizer increased the richness of nematophagous fungi, including *Mortierella* and *Chaetomium*, which were associated with a reduction in the cereal nematode population density (Qiu *et al.*, 2020).

Chitin may be a useful organic soil amendment for controlling plant-parasitic nematodes. It is hypothesized that adding chitin to soil enhances chitinolytic activities by bacteria and fungi, which can destroy nematode eggs due to the presence of chitin in the nematode eggshell. The addition of chitin to nematode-infested soil suppressed root knot on cotton. There were increases in population densities of several bacteria in soil and root rhizosphere of cotton (Hallmann *et al.*, 1999).

21.7.2. Plants and Cropping Sequences

The decomposition of plant materials provides food and energy for the development of soil microorganisms. The exudates from plant roots have large effects on the composition and abundance of the soil microbial communities. These plant exudates may be correlated with plant species and their developmental growth stages. Although plants may play a key role in soil biological interactions, there is limited knowledge on how plant species and cropping sequences alter the biological composition of natural control agents that could be detrimental to plant-parasitic nematodes. It is difficult to predict the results of using certain cropping sequences for the enhancement of natural biocontrol organisms. For example, tomato root rhizosphere soil supported a higher population density of *P. chlamydosporia* than non-amended bulk soil (Bourne and Kerry, 2000). Similarly, the population densities of predacious fungi, *Arthrobotrys dactyloides*,

Arthrobotrys superba and *Monacrosporium ellipsosporum*, were greater in soil taken from tomato root rhizosphere than in root-free soil (Persson and Jansson, 1999).

Selection of plant species and cultivars that may favour different types and species of nematophagous fungi or bacteria may be one viable way to enhance soil with biological control agents. For example, rhizobacteria isolated from nematode-antagonistic plant species have a greater antagonistic effect on *H. glycines* and *M. incognita* than bacteria isolated from the nematode-susceptible soybean host crop (Kloepper *et al.*, 1992).

Brachyphoris oviparasitica, which is parasitic on nematode eggs, was a factor in natural biological control in some peach orchards in California. The fungus parasitized most of the eggs in smaller egg masses produced by *M. incognita* on peach trees, whereas fewer egg were parasitized in larger egg masses produced on tomato or grape (Stirling *et al.*, 1979).

Colonization of roots by *P. chlamydosporia* differed among plant species and cultivars (Bourne *et al.*, 1996). There was greater soil suppressiveness of *H. schachtii* in nematode tolerant sugar beet cultivar plantings than in plantings of susceptible sugar beet genotypes grown in monoculture (Eberlein *et al.*, 2020). With the gatherings of more knowledge, we may be able to select cultivars to maintain high population density of the nematophagous fungi in the roots and root rhizosphere.

Crops grown in monoculture or often repeatedly planted in the same fields supports the establishment of nematode pathogens specific for that crop, which, in turn, supports an abundance of nematode parasites. For example, peanut is often parasitized by *M. arenaria*, the peanut RKN, which serves as a good host of the bacterial parasite, *P. penetrans*. The dynamics of a specific parasite, such as *Pasteuria*, depends on the nematode-host plant relationship. Continuous planting of peanut leads to enhanced soil suppressiveness of *M. arenaria* due to increased soil densities of *Pasteuria* (Cetintas and Dickson, 2004; Timper, 2009).

A similar mechanism was reported for maintaining the endoparasitic fungus *Hirsutella rhossiliensis*, in which continuous plantings of soybean maintained a higher percentage of *H. glycines* J2 parasitized by the fungus (Chen and Reese, 1999). Over time, the soil becomes suppressive to the nematode (Chen, 2007). A later study suggested that other fungi that were antagonists of *H. glycines* eggs were also associated with the nematode suppression (Hu *et al.*, 2017, 2019).

Monoculture of sugar beet resulted in a higher percentage of the eggs of *H. schachtii* being parasitized by fungi (Steudel *et al.*, 1990). In most crop production systems, monoculture may not be the best choice for integrated disease and pest management, but in some cropping systems, choices may exist such as the interplanting of susceptible plants when a resistant cultivar or non-host crops are used and planting a tolerant or partially resistant cultivar to maintain nematode antagonist population densities. Planting nematode-susceptible cover crops during the period of non-crop production season is also a choice, but the effect of nematode population densities on the following crop should be evaluated before using this tactic.

21.7.3. Tillage

Although numerous studies have been reported on tillage effects on plant-parasitic nematodes, few of them were about the tillage effect on nematode antagonists. Bernard *et al.* (1997) investigated the incidence of fungi invading females, cysts and eggs of *H. glycines* in various cropping-tillage plots. Although the percentage of eggs parasitized by total fungi did not differ among the cropping-tillage treatments, *P. lilacinum* parasitized more eggs in the disc-cultivated treatment than in any of the no-till treatments, and *P. chlamydosporia* occurred more frequently where the mouldboard plough was used than in any other treatment. In a greenhouse study, soil disturbance resulted in a reduction in the proportion of J2 of *H. glycines* parasitized by the endoparasitic fungi *Hirsutella* and increased the nematode egg population density (Bao *et al.*, 2011). In a field study, however, the reduction of J2 parasitized by *Hirsutella* in conventional tillage as compared with no-till was observed only at one of the 11 sampling times and one of the two sites, indicating that the tillage effect on J2 parasitized by *Hirsutella* is minimal under field conditions (Chen and Liu, 2007). The obligate parasitic bacterium *P. nishizawae* does not appear to be affected by tillage (Noel *et al.*, 2010). More studies are needed to investigate the potential of using tillage practices to enhance the natural biological control of plant-parasitic nematodes.

21.7.4. Introducing Effective Antagonists

The type and species of antagonists in soil are key to the successful enhancement of biocontrol of nematodes. If there is no effective natural antagonist in the field soil, it may be difficult to enhance biocontrol with any cultural practices. Therefore, it may be necessary to introduce one or more species of effective biocontrol agents into the soil with a feasible inoculum application rate leaving the organisms to increase over time.

The successful transfer has been demonstrated for the bacterium *Pasteuria* spp. from suppressive sites to new field sites and established by growing susceptible hosts for the nematode (Kariuki and Dickson, 2007; Noel *et al.*, 2010). Because *Pasteuria* spores are highly resistant to environmental stresses such as dry soils and high or low temperatures, it is an excellent candidate for seed treatment. Even if the seed treatment cannot suppress nematode population density in the first year, it has been shown that the bacteria will build up its population density over three years, thereby increasing the soil's suppressiveness to specific plant-parasitic nematodes (Kariuki and Dickson, 2007). Many other organisms may also be suitable for introduction to field soils. Nematode-density-dependent endoparasites such as *Pasteuria* and *Hirsutella* may have high potential as inoculum agents. But facultative parasites, for example, the highly pathogenic strains of *Pochonia*, should be included.

21.8. Concluding Remarks

The importance of biological control of plant-parasitic nematodes in agriculture and horticulture is increasing. Scientists have made tremendous progress in exploring the possibility of using biological antagonists to control plant-parasitic nematodes. There is a wide range of natural enemies of nematodes, some of which have shown potential as biological control agents. However, there has been limited success in developing commercial biopesticides for this purpose. Much more research, education and extension are needed for future advances in this field.

Both challenges and opportunities exist for the biological control of plant-parasitic nematodes. A major challenge is the high production cost of commercial biocontrol agents. Extensive research is needed to explore technologies for the efficient and economical mass production of these agents. The soil ecosystem is complex, and much more knowledge is needed about the ecology of nematode-suppressive soils and the interaction of biological control agents with soil microbial communities and environments. Continuous efforts should be made to discover new, highly effective biocontrol agents. Advances in molecular biology and biotechnology provide excellent opportunities for studying the ecology of nematode-suppressive soils, the genetics of biocontrol agents and for genetically engineering novel, highly effective biocontrol agents. While biopesticides are a key focus of nematode biocontrol, enhancing the efficacy of natural antagonists may be practical and equally important. Biological control of plant-parasitic nematodes should be part of an integrated pest and disease management programme.

21.9. References

Abarca, C., Bidondo, L.F., Bompadre, J. and Velázquez, M.S. (2024) Arbuscular mycorrhizal fungi in tomato tolerance to pathogens and nematodes: a comprehensive review. *Scientia Horticulturae* 329, 112969. DOI: 10.1016/j.scienta.2024.112969.

Aioub, A.A.A., Elesawy, A.E. and Ammar, E.E. (2022) Plant growth promoting rhizobacteria (PGPR) and their role in plant-parasitic nematodes control: a fresh look at an old issue. *Journal of Plant Diseases and Protection* 129, 1305–1321.

Akhtar, M. and Malik, A. (2000) Roles of organic soil amendments and soil organisms in the biological control of plant-parasitic nematodes: a review. *Bioresource Technology* 74, 35–47.

Al-Rehiayani, S., Hafez, S.L., Thornton, M. and Sundararaj, P. (1999) Effects of *Pratylenchus neglectus*, *Bacillus megaterium*, and oil radish or rapeseed green manure on reproductive potential of *Meloidogyne chitwoodi* on potato. *Nematropica* 29, 37–49.

Antil, S., Kumar, R., Pathak, D.V. and Kumari, A. (2023) Recent advances in utilizing bacteria as biocontrol agents against plant parasitic nematodes emphasizing *Meloidogyne* spp. *Biological Control* 183, 105244. DOI: 10.1016/j.biocontrol.2023.105244.

Aranda-Martinez, A., Grifoll-Romero, L., Aragunde, H., Sancho-Vaello, E., Biarnés, X., *et al.* (2018) Expression and specificity of a chitin deacetylase from the nematophagous fungus *Pochonia chlamydosporia* potentially involved in pathogenicity. *Scientific Reports* 8, 2170. DOI: 10.1038/s41598-018-19902-0.

Bao, Y., Neher, D.A. and Chen, S.Y. (2011) Effect of biocides and soil disturbance on nematode community and extracellular enzyme activity in soybean cyst nematode suppressive soil. *Nematology* 13, 687–699.

Barron, G.L. (1977) *The Nematode-destroying Fungi*. Lancaster Press, Lancaster, Pennsylvania.

Barron, G.L. (1987) The gun cell of *Haptoglossa mirabilis*. *Mycologia* 79, 877–833.

Beakes, G.W. and Glockling, S.L. (1998) Injection tube differentiation in gun cells of a *Haptoglossa* species which infects nematodes. *Fungal Genetics and Biology* 24, 45–68.

Bernard, E.C., Self, L.H. and Tyler, D.D. (1997) Fungal parasitism of soybean cyst nematode, *Heterodera glycines* (Nematoda: Heteroderidae), in differing cropping-tillage regimes. *Applied Soil Ecology* 5, 57–70.

Bhat, B.A., Tariq, L., Nissar, S., Islam, S.T., Ul Islam, S., *et al.* (2022) The role of plant-associated rhizobacteria in plant growth, biocontrol and abiotic stress management. *Journal of Applied Microbiology* 133, 2717–2741.

Bhattacharyya, P.N. and Jha, D.K. (2012) Plant growth-promoting rhizobacteria (PGPR): emergence in agriculture. *World Journal of Microbiology & Biotechnology* 28, 1327–1350.

Borneman, J. and Becker, J.O. (2007) Identifying microorganisms involved in specific pathogen suppression in soil. *Annual Review of Phytopathology* 45, 153–172.

Bourne, J.M. and Kerry, B.R. (2000) Observations on the survival and competitive ability of the nematophagous fungus *Verticillium chlamydosporium* in soil. *International Journal of Nematology* 10, 9–18.

Bourne, J.M., Kerry, B.R., Leij, F.d. and De-Leij, F.A.A.M. (1996) The importance of the host plant on the interaction between root-knot nematodes (*Meloidogyne* spp.) and the nematophagous fungus, *Verticillium chlamydosporium* Goddard. *Biocontrol Science and Technology* 6, 539–548.

Brito, J.A., Preston, J.F., Dickson, D.W., Giblin-Davis, R.M., Williams, D.S., *et al.* (2003) Temporal formation and immunolocalization of an endospore surface epitope during *Pasteuria penetrans* sporogenesis. *Journal of Nematology* 35, 278–288.

Cetintas, R. and Dickson, D.W. (2004) Persistence and suppressiveness of *Pasteuria penetrans* to *Meloidogyne arenaria* race 1. *Journal of Nematology* 36, 540–549.

Chen, S. and Dickson, D.W. (2012) Biological control of plant-parasitic nematodes. In: Manzanilla-López, R.H. and Marbán-Mendoza, N. (eds), *Practical Plant Nematology*. Colegio de Postgraduados and Mundi-Prensa, Biblioteca Básica de Agricultura, Guadalajara, Jalisco, Mexico, pp. 761–811.

Chen, S. and Liu, S. (2007) Effects of tillage and crop sequence on parasitism of *Heterodera glycines* juveniles by *Hirsutella* spp. and on juvenile population density. *Nematropica* 37, 93–106.

Chen, S.Y. (2007) Suppression of *Heterodera glycines* in soils from fields with long-term soybean monoculture. *Biocontrol Science and Technology* 17, 125–134.

Chen, S.Y., Dickson, D.W., Kimbrough, J.W., McSorley, R. and Mitchell, D.J. (1994) Fungi associated with females and cysts of *Heterodera glycines* in a Florida soybean field. *Journal of Nematology* 26, 296–303.

Chen, S.Y., Liu, X.Z. and Chen, F.J. (2000) *Hirsutella minnesotensis* sp. nov. – a new pathogen of the soybean cyst nematode. *Mycologia* 92, 819–824.

Chen, S.Y. and Reese, C.D. (1999) Parasitism of the nematode *Heterodera glycines* by the fungus *Hirsutella rhossiliensis* as influenced by crop sequence. *Journal of Nematology* 31, 437–444.

Chen, W. and Hu, Q.B. (2022) Secondary metabolites of *Purpureocillium lilacinum*. *Molecules* 27, 18. DOI: 10.3390/molecules27010018.

Chen, Z.X. and Dickson, D.W. (1998) Review of *Pasteuria penetrans*: biology, ecology, and biological control potential. *Journal of Nematology* 30, 313–340.

Chen, Z.X. and Dickson, D.W. (2004) Biological control of nematodes with bacterial antagonists. In: Chen, Z.X., Chen, S.Y. and Dickson, D.W. (eds), *Nematology, Advances and Perspectives. Vol. II: Nematode Management and Utilization*. Tsinghua University Press and CAB International, Cambridge, Massachusetts, pp. 979–1039.

Cooke, R.C. (1963) Ecological characteristics of nematode-trapping Hyphomycetes. *Annals of Applied Biology* 52, 431–437.

Davies, K.G., Redden, M. and Pearson, T.K. (1994) Endospore heterogeneity in *Pasteuria penetrans* related to adhesion to plant-parasitic nematodes. *Letters in Applied Microbiology* 19, 370–373.

Diab, M.K., Mead, H.M., Khedr, M.M.A., Abu-Elsaoud, A.M. and El-Shatoury, S.A. (2024) Actinomycetes are a natural resource for sustainable pest control and safeguarding agriculture. *Archives of Microbiology* 206, 38762847. DOI:10.1007/s00203-024-03975-9.

Dickson, D.W., Oostendorp, M., Giblin-Davis, R.M. and Mitchell, D.J. (1994) Control of plant-parasitic nematodes by biological antagonists. In: Rosen, D., Bennett, F.D. and Capinera, J.L. (eds), *Pest Management in the Subtropics. Biological Control - a Florida Perspective*. Intercept, Andover, UK, pp. 575–601.

Dickson, D.W., Preston III, J.F., Giblin-Davis, R.M., Noel, G.R., Ebert, D., *et al.* (2009) Family V. Pasteuriaceae Laurent 1890, 756[AL]. *Pas.teu.ri.a'ce.ae*. N.L. fem. n. *Pasteuria* type genus of the family; suff. *-aceae* ending denoting family; N.L. fem. pl. n. *Pasteuriaceae* the family of *Pasteuria*. In: De Vos, P., Garrity, G.M., Jones, D., Krieg, N.R., Ludwig, W., *et al.* (eds), *Bergey's Manual of Systematic Bacteriology*, 2nd edn. Vol. Three, The Firmicutes. Springer, New York, New York, pp. 328–347.

Dong, L.Q., Yang, J.K., and Zhang, K.Q. (2007) Cloning and phylogenetic analysis of the chitinase gene from the facultative pathogen *Paecilomyces lilacinus*. *Journal of Applied Microbiology* 103, 2476–2488.

Eberlein, C., Heuer, H., and Westphal, A. (2020) Biological suppression of populations of *Heterodera schachtii* adapted to different host genotypes of sugar beet. *Frontiers in Plant Science* 11, 812. DOI: 10.3389/fpls.2020.00812.

Egerton, J.R., Ostlind, D.A., Blair, L.S., Eary, C.H., Suhayda, D., *et al.* (1979) Avermectins, new family of potent anthelmintic agents – efficacy of the B1A component. *Antimicrobial Agents and Chemotherapy* 15, 372–378.

Engelbrecht, G., Horak, I., van Rensburg, P.J.J. and Claassens, S. (2018) *Bacillus*-based bionematicides: development, modes of action and commercialisation. *Biocontrol Science and Technology* 28, 629–653.

Ferreira, S.R., Machado, A.R.T., Furtado, L.F., Gomes, J.H.D., de Almeida, R.M., *et al.* (2020) Ketamine can be produced by *Pochonia chlamydosporia*: an old molecule and a new anthelmintic? *Parasites & Vectors* 13(1), 527. DOI: 10.1186/s13071-020-04402-w.

Galper, S., Cohn, E., Spiegel, Y. and Chet, I. (1991) A collagenolytic fungus, *Cunninghamella elegans*, for biological control of plant-parasitic nematodes. *Journal of Nematology* 23, 269–274.

Ghahremani, Z., Escudero, N., Saus, E., Gabaldón, T. and Sorribas, F.J. (2019) *Pochonia chlamydosporia* induces plant-dependent systemic resistance to *Meloidogyne incognita*. *Frontiers in Plant Science* 10, 945. DOI: 10.3389/fpls.2019.00945.

Gray, N.F. (1988) Fungi attacking vermiform nematodes. In: Poinar Jr, G.O. and Jansson, H.B. (eds), *Diseases of Nematodes*. Vol. II. CRC Press, Boca Raton, Florida, pp. 3–38.

Guzmán-Guzmán, P., Kumar, A., de los Santos-Villalobos, S., Parra-Cota, F.I., Orozco-Mosqueda, M.D., *et al.* (2023) *Trichoderma* species: our best fungal allies in the biocontrol of plant diseases – a review. *Plants* 12, 432. DOI: 10.3390/plants12030432.

Haarith, D., Bushley, K.E. and Chen, S.Y. (2020a) Fungal communities associated with *Heterodera glycines* and their potential in biological control: a current update. *Journal of Nematology* 52, 1–17. DOI: 10.21307/jofnem-2020-022.

Haarith, D., Kim, D.G., Strom, N.B., Chen, S.Y. and Bushley, K.E. (2020b) *In vitro* screening of a culturable soybean cyst nematode cyst mycobiome for potential biological control agents and biopesticides. *Phytopathology* 110, 1388–1397.

Hallmann, J., Rodriguez-Kabana, R. and Kloepper, J.W. (1999) Chitin-mediated changes in bacterial communities of the soil, rhizosphere and within roots of cotton in relation to nematode control. *Soil Biology & Biochemistry* 31, 551–560.

Hu, W., Kidane, E., Neher, D.A. and Chen, S. (2019) Field and greenhouse evaluations of soil suppressiveness to *Heterodera glycines* in the Midwest corn-soybean production systems. *Journal of Nematology* 51, e2019–32. DOI: 10.21307/jofnem-2019-032.

Hu, W.M., Samac, D.A., Liu, X.Z. and Chen, S.Y. (2017) Microbial communities in the cysts of soybean cyst nematode affected by tillage and biocide in a suppressive soil. *Applied Soil Ecology* 119, 396–406.

Huang, X.W., Zhao, N.H. and Zhang, K.Q. (2004) Extracellular enzymes serving as virulence factors in nematophagous fungi involved in infection of the host. *Research in Microbiology* 155, 811–816.

Ingham, R.E. (1988) Interactions between nematodes and vesicular-arbuscular mycorrhizae. *Agriculture Ecosystems & Environment* 24, 169–182.

Kariuki, G.M., Brito, J.A. and Dickson, D.W. (2006) Effects of *Pastueria penetrans* endospore rate of attachment on root penetration and fecundity of *Meloidogyne arenaria* Race 1. *Nematropica* 36, 261–267.

Kariuki, G.M. and Dickson, D.W. (2007) Transfer and development of *Pasteuria penetrans*. *Journal of Nematology* 39, 55–61.

Kerry, B.R. (1978) Natural control of the cereal cyst-nematode by parasitic fungi. *ARC Research Review* 4, 17–21.

Kerry, B.R. and Hirsch, P.R. (2011) Ecology of *Pochonia chlamydosporia* in the rhizosphere at the population, whole organism and molecular scales. In: Davies, K. and Spiegel, Y. (eds), *Biological Control of*

Plant-parasitic Nematodes: Building Coherence between Microbial Ecology and Molecular Mechanisms. Springer, Dordrecht, Netherlands, pp. 171–182.

Khambay, B.P.S., Bourne, J.M., Cameron, S., Kerry, B.R. and Zaki, M.J. (2000) A nematicidal metabolite from *Verticillium chlamydosporium*. *Pest Management Science* 56, 1098–1099.

Khan, A., Williams, K.L. and Nevalainen, H.K.M. (2004) Effects of *Paecilomyces lilacinus* protease and chitinase on the eggshell structures and hatching of *Meloidogyne javanica* juveniles. *Biological Control* 31, 346–352.

Khan, M. and Tanaka, K. (2023) *Purpureocillium lilacinum* for plant growth promotion and biocontrol against root-knot nematodes infecting eggplant. *PLoS One* 18, e0283550. DOI: 10.1371/journal.pone.0283550.

Kloepper, J.W., Rodriguez-Kabana, R., McInroy, J.A. and Young, R.W. (1992) Rhizosphere bacteria antagonistic to soybean cyst (*Heterodera glycines*) and root-knot (*Meloidogyne incognita*) nematodes: Identification by fatty acid analysis and frequency of biological control activity. *Plant and Soil* 139, 75–84.

Lasota, J.A. and Dybas, R.A. (1991) Avermectins, a novel class of compounds – implications for use in arthropod pest control. *Annual Review of Entomology* 36, 91–117.

Li, J., Zou, C.G., Xu, J.P., Ji, X.L., Niu, X.M., *et al.* (2015) Molecular mechanisms of nematode-nematophagous microbe interactions: Basis for biological control of plant-parasitic nematodes. *Annual Review of Phytopathology* 53, 67–95.

Li, Z.Y., Luo, N., Zhang, W.W., Khan, R.A.A., Ling, J., *et al.* (2024) Nematicidal glycosylated resorcylic acid lactones from the fungus *Pochonia chlamydosporia* PC-170 and their key biosynthetic genes. *Frontiers in Microbiology* 15, 1385255. DOI: 10.3389/fmicb.2024.1385255.

Linford, M.B., Yap, F. and Oliveira, J.M. (1938) Reduction of soil populations of the root-knot nematode during decomposition of organic matter. *Soil Science* 45, 127–141.

Liou, J.Y., Shih, J.Y. and Tzean, S.S. (1999) *Esteya*, a new nematophagous genus from Taiwan, attacking the pinewood nematode (*Bursaphelenchus xylophilus*). *Mycological Research* 103, 242–248.

Lopez-Llorca, L.V. and Claugher, D. (1990) Appressoria of the nematophagous fungus *Verticillium suchlasporium*. *Micron and Microscopica Acta* 21, 125–130.

Manzanilla-López, R.H., Esteves, I. and Devonshire, J. (2017) Biology and management of *Pochonia chlamydosporia* and plant-parasitic nematodes. In: Manzanilla-López, R.H. and Lopez-Llorca, L.V. (eds), *Perspectives in Sustainable Nematode Management through Pochonia chlamydosporia Applications for Root and Rhizosphere Health*. Springer, Cham, Switzerland, pp. 47–76.

Manzanilla-López, R.H. and Lopez-Llorca, L.V. (eds) (2017) *Perspectives in Sustainable Nematode Management through Pochonia chlamydosporia Applications for Root and Rhizosphere Health*. Springer, Cham, Switzerland.

Mauchline, T.H., Kerry, B.R. and Hirsch, P.R. (2004) The biocontrol fungus *Pochonia chlamydosporia* shows nematode host preference at the infraspecific level. *Mycological Research* 108, 161–169.

McSorley, R. (2011) Overview of organic amendments for management of plant-parasitic nematodes, with case studies from Florida. *Journal of Nematology* 43, 69–81.

Migunova, V.D. and Sasanelli, N. (2021) Bacteria as biocontrol tool against phytoparasitic nematodes. *Plants* 10, 389. DOI: 10.3390/plants10020389.

Mogollon-Ortiz, A.M., Monteiro, T.S.A., de Freitas, L.G. and de Queiroz, M.V. (2024) Potential of different species of actinobacteria in the management of *Meloidogyne javanica*. *Archives of Microbiology* 206(4), 160. DOI: 10.1007/s00203-024-03874-z.

Morton, O.C., Hirsch, P.R. and Kerry, B.R. (2004) Infection of plant parasitic nematodes by nematophagous fungi – a review of the application of molecular biology to understand infection processes and to improve biological control. *Nematology* 6, 161–170.

Nagachandrabose, S. (2020) Management of potato cyst nematodes using liquid bioformulations of *Pseudomonas fluorescens*, *Purpureocillium lilacinum* and *Trichoderma viride*. *Potato Research* 63, 479–496.

Niu, X.M., Wang, Y.L., Chu, Y.S., Xue, H.X., Li, N., *et al.* (2010) Nematodetoxic aurovertin-type metabolites from a root-knot nematode parasitic fungus *Pochonia chlamydosporia*. *Journal of Agricultural and Food Chemistry* 58, 828–834.

Noel, G.R., Atibalentja, N. and Bauer, S.J. (2010) Suppression of *Heterodera glycines* in a soybean field artificially infested with *Pasteuria nishizawae*. *Nematropica* 40, 41–52.

Oostendorp, M., Dickson, D.W. and Mitchell, D.J. (1990) Host range and ecology of isolates of *Pasteuria* spp. from the southeastern United States. *Journal of Nematology* 22, 525–531.

Persidis, A., Lay, J.G., Manousis, T., Bishop, A.H. and Ellar, D.J. (1991) Characterisation of potential adhesins of the bacterium *Pasteuria penetrans*, and of putative receptors on the cuticle of *Meloidogyne incognita*, a nematode host. *Journal of Cell Science* 100, 613–622.

Persson, C. and Jansson, H.B. (1999) Rhizosphere colonization and control of *Meloidogyne* spp. by nematode-trapping fungi. *Journal of Nematology* 31, 164–171.

Phani, V. and Rao, U. (2018) Revisiting the life-cycle of *Pasteuria penetrans* infecting *Meloidogyne incognita* under soil-less medium, and effect of streptomycin sulfate on its development. *Journal of Nematology* 50, 91–98.

Pires, D., Vicente, C.S.L., Menéndez, E., Faria, J.M.S., Rusinque, L., *et al.* (2022) The fight against plant-parasitic nematodes: current status of bacterial and fungal biocontrol agents. *Pathogens* 11, 1178. DOI: 10.3390/pathogens11101178.

Preston, J.F., Dickson, D.W., Maruniak, J.E., Nong, G., Brito, J.A., *et al.* (2003) *Pasteuria* spp.: Systematics and phylogeny of these bacterial parasites of phytopathogenic nematodes. *Journal of Nematology* 35, 198–207.

Pyrowolakis, A., Schuster, R.P. and Sikora, R.A. (1999) Effect of cropping pattern and green manure on the antagonistic potential and the diversity of egg pathogenic fungi in fields with *Heterodera schachtii* infection. *Nematology* 1, 165–171.

Qiu, W., Su, H.Q., Yan, L.Y., Ji, K.Y., Liu, Q., *et al.* (2020) Organic fertilization assembles fungal communities of wheat rhizosphere soil and suppresses the population growth of *Heterodera avenae* in the field. *Frontiers in Plant Science* 11, 1225. DOI: 10.3389/fpls.2020.01225.

Rodríguez-Kábana, R., Boube, D. and Young, R.W. (1989) Chitinous materials from blue crab for control of root-knot nematode. I. Effect of urea and enzymatic studies. *Nematropica* 19, 53–74.

Rosskopf, E., Di Gioia, F., Hong, J.C., Pisani, C. and Kokalis-Burelle, N. (2020) Organic amendments for pathogen and nematode control. *Annual Review of Phytopathology* 58, 277–311.

Rosso, L.C., Romano, F., Pentimone, I., Ciancio, A. and Colagiero, M. (2023) Effect of *Pochonia chlamydosporia* endophytism and *Meloidogyne incognita* parasitism on WRKYs and defense gene expression in tomato roots. *Plants* 12(6), 1284. DOI: 10.3390/plants12061284.

Saeed, Q., Xiukang, W., Haider, F.U., Kucerik, J., Mumtaz, M.Z., *et al.* (2021) Rhizosphere bacteria in plant growth promotion, biocontrol, and bioremediation of contaminated sites: a comprehensive review of effects and mechanisms. *International Journal of Molecular Sciences* 22, 10529. DOI: 10.3390/ijms221910529.

Sayre, R.M. (1988) Bacterial diseases of nematodes and their role in controlling nematode populations. *Agriculture, Ecosystems and Environment* 24, 263–279.

Sayre, R.M. and Starr, M.P. (1988) Bacterial diseases and antagonisms of nematodes. In: Poinar Jr, G.O. and Jansson, H.B. (eds), *Diseases of Nematodes*. Vol. 1. CRC Press, Boca Raton, Florida, pp. 69–101.

Sayre, R.M. and Wergin, W.P. (1977) Bacterial parasite of a plant nematode: morphology and ultrastructure. *Journal of Bacteriology* 129, 1091–1101.

Schmidt, L.M., Preston, J.F., Nong, G., Dickson, D.W. and Aldrich, H.C. (2004) Detection of *Pasteuria penetrans* infection in *Meloidogyne arenaria* race 1 in planta by polymerase chain reaction. *FEMS Microbiology Ecology* 48, 457–464.

Seong, J., Shin, J., Kim, K. and Cho, B.K. (2021) Microbial production of nematicidal agents for controlling plant-parasitic nematodes. *Process Biochemistry* 108, 69–79.

Sharma, N., Khanna, K., Manhas, R.K., Bhardwaj, R., Ohri, P., *et al.* (2020) Insights into the role of *Streptomyces hydrogenans* as the plant growth promoter, photosynthetic pigment enhancer and biocontrol agent against *Meloidogyne incognita* in *Solanum lycopersicum* seedlings. *Plants* 9(9), 1109. DOI: 10.3390/plants9091109.

Sharon, E., Chet, I. and Spiegel, Y. (2011) *Trichoderma* as a biological control agent. In: Davies, K. and Spiegel, Y. (eds), *Biological Control of Plant-parasitic Nematodes: Building Coherence between Microbial Ecology and Molecular Mechanisms*. Springer, Dordrecht, Netherlands, pp. 183–201. DOI: 10.1007/978-1-4020-9648-8_8.

Siddiqui, Z.A. and Mahmood, I. (1999) Role of bacteria in the management of plant parasitic nematodes: a review. *Bioresource Technology* 69, 167–179.

Sidhu, H.S. and Kanwar, R.S. (2020) Integration of some organic amendments and the predatory nematode, *Fictor composticola*, for the management of *Meloidogyne incognita* in cucumber. *Egyptian Journal of Biological Pest Control* 30, 146. DOI: 10.1186/s41938-020-00349-3.

Sosnowska, D. and Banaszak, H. (2000) The effect of organic fertilizers on fungi parasitization of beet cyst nematode (*Heterodera schachtii* Schmidt) eggs in sugar beet cultivated in a three years rotation. *Journal of Plant Protection Research* 40, 73–79.

Steudel, W., Muller, J. and Schlang, J. (1990) Investigations on fungal egg parasites of the sugar beet nematode (*Heterodera schachtii* Schmidt) in two sugar beet-cereal-rotations. *Journal of Phytopathology* 129, 316–326.

Stirling, G.R. (1991) *Biological Control of Plant Parasitic Nematodes: Progress, Problems and Prospects*. CAB International, Wallingford, UK.

Stirling, G.R. (2011) Biological control of plant-parasitic nematodes: an ecological perspectives, a review of progress and opportunities for future research. In: Davies, K. and Spiegel, Y. (eds), *Biological Control of Plant-parasitic Nematodes: Building Coherence between Microbial ecology and Molecular Mechanisms*. Springer Science+Business Media, Dordrecht, Netherlands, pp. 1–38. DOI: 10.1007/978-1-4020-9648-8_1.

Stirling, G.R., McKenry, M.V. and Mankau, R. (1979) Biological control of root-knot nematodes (*Meloidogyne* spp.) on peaches. *Phytopathology* 69, 806–809.

Stirling, G.R. and West, L.M. (1991) Fungal parasites of root-knot nematode eggs from tropical and sub-tropical regions of Australia. *Australasian Plant Pathology* 20, 149–154.

Stirling, G.R., Wilson, E.J., Stirling, A.M., Pankhurst, C.E., Moody, P.W., *et al.* (2005) Amendments of sugarcane trash induce suppressiveness to plant parasitic nematodes in a sugarcane soil. *Australasian Plant Pathology* 34, 203–211.

TariqJaveed, M., Farooq, T., Al-Hazmi, A.S., Hussain, M.D. and Rehman, A.U. (2021) Role of *Trichoderma* as a biocontrol agent (BCA) of phytoparasitic nematodes and plant growth inducer. *Journal of Invertebrate Pathology* 183, 107626. DOI: 10.1016/j.jip.2021.107626.

Tian, B., Yang, J. and Zhang, K.Q. (2007) Bacteria used in the biological control of plant-parasitic nematodes: populations, mechanisms of action, and future prospects. *FEMS Microbiology Ecology* 61, 197–213.

Timper, P. (2009) Population dynamics of *Meloidogyne arenaria* and *Pasteuria penetrans* in a long-term crop rotation study. *Journal of Nematology* 41, 291–299.

Wang, K.H., Sipes, B.S. and Schmitt, D.P. (2003) Enhancement of *Rotylenchulus reniformis* suppressiveness by *Crotalaria juncea* amendment in pineapple soils. *Agriculture Ecosystems & Environment* 94, 197–203.

Wang, R.Z., Dong, L.M., Chen, Y.Q., Qu, L.J., Wang, Q.H., *et al.* (2017) *Esteya vermicola*, a nematophagous fungus attacking the pine wood nematode, harbors a bacterial endosymbiont affiliated with *Gammaproteobacteria*. *Microbes and Environments* 32, 201–209.

Weibelzahl-Fulton, E., Dickson, D.W. and Whitty, E.B. (1996) Suppression of *Meloidogyne incognita* and *M. javanica* by *Pasteuria penetrans* in field soil. *Journal of Nematology* 28, 43–49.

Yang, Y.Y., Wu, J.N., Perry, R.N. and Toyota, K. (2023) Evaluation of soil suppressiveness of various Japanese soils against the soybean cyst nematode *Heterodera glycines* and its relation with the soil chemical and biological properties. *Agronomy* 13, 2826. DOI: 10.3390/agronomy13112826.

Yang, Z.S., Li, G.H., Zhao, P.J., Zheng, X., Luo, S.L., *et al.* (2010) Nematicidal activity of *Trichoderma* spp. and isolation of an active compound. *World Journal of Microbiology & Biotechnology* 26, 2297–2302.

Yi, X., Guo, Y.C., Khan, R.A.A. and Fan, Z.Y. (2021) Understanding the pathogenicity of *Pochonia chlamydosporia* to root knot nematode through omics approaches and action mechanism. *Biological Control* 162, 104726. DOI: 10.1016/j.biocontrol.2021.104726.

Yu, Z.Q., Xiong, J., Zhou, Q.N., Luo, H.Y., Hu, S.B., *et al.* (2015) The diverse nematicidal properties and biocontrol efficacy of *Bacillus thuringiensis* Cry6A against the root-knot nematode *Meloidogyne hapla*. *Journal of Invertebrate Pathology* 125, 73–80.

Zavala-Gonzalez, E.A., Rodríguez-Cazorla, E., Escudero, N., Aranda-Martinez, A., Martínez-Laborda, A., *et al.* (2016) *Arabidopsis thaliana* root colonization by the nematophagous fungus *Pochonia chlamydosporia* is modulated by jasmonate signaling and leads to accelerated flowering and improved yield. *New Phytologist* 213, 351–364. DOI: 10.1111/nph.14106.

Zhang, S.X., and Zhang, X. (2009) Effects of two composted plant pesticide residues, Incorporated with *Trichoderma viride*, on root-knot nematode in Balloonflower. *Agricultural Sciences in China* 8, 447–454.

22 Nematology Knowledge Transfer: Creating Awareness and Understanding

Danny Coyne[1,2*], James Kisaakye[1] and Solveig Haukeland[3,4]
[1]International Institute of Tropical Agriculture (IITA), Kasarani, Nairobi, Kenya; [2]Nematology Research Unit, Faculty of Sciences, Ghent University, Ghent, Belgium; [3]International Centre for Insect Physiology and Ecology (ICIPE), Kasarani, Nairobi, Kenya; [4]Norwegian Institute of Bioeconomy Research, Oslo, Norway

Abstract

Typical above-ground symptoms of nematode infection are mostly non-specific and are often attributed to other problems and/or crop malaise. Farmer awareness of nematode problems therefore needs to be raised by utilising knowledge transfer techniques so that symptoms of nematode damage can be recognised in the field. The chapter addresses this need by discussing various approaches such as demonstration and/or hands on workshops of nematode control methods suitable for Farmer Field Schools (FFS) and farmer participatory research, the planning, implementing and long-term benefits of Integrated season-long Pest Management-FFS, nematology capacity building initiatives, nematology networks, societies and platforms. Creating successful knowledge transfer to farmers requires a critical evaluation of the whole communication chain, including via governmental policies.

22.1. Introduction

A rising global population, changing diets and climate change create ever-present challenges to our agricultural systems. Pests and diseases are among the key factors impacting crop production (Savary *et al.*, 2012), and managing them within economic thresholds presents a continuous and ever-changing battle, as agriculture modifies and adapts to ever-shifting demands. Our ability to manage such threats depends upon numerous factors, critical among these being the creation of farmer awareness on what the threats are, how to recognize them and what is required to manage the problems.

Difficulties in recognizing existing and emerging nematode pest threats prevent timely intervention strategies. A critical under-representation of nematologists, compared with other disciplines, and a universally poor recognition of nematode problems exacerbate the situation, especially in the tropics and global south (Coyne *et al.*, 2018). Consequently, creating awareness on pest and disease issues and providing accurate, updated information on current and emerging threats and their management is of immense value to farmers. Unfortunately, national extension services in the global south are very often under-equipped with either resources or knowledge and are poorly supported by the government. Even in developed nations, the extension services often struggle to maintain standards and reach the farmers that need their support. Although somewhat dated, it remains

*E-mail: d.coyne@cgiar.org

(eds R.H. Manzanilla-López, D.J. Hunt, N. Marbán-Mendoza and S. Infante Gil)
DOI: 10.1079/9781836990413.0022

prescient to quote Wilson (1982) that 'Nematodes are always working havoc, taking their toll on crops, sometimes causing complete destruction. We blind ourselves by accepting them as a matter of course. The problem is here; we live with it; but there is no solution. There must be a general awakening all along the line to the magnitude of this situation'. So, in terms of nematology, which is already under-represented, creating successful knowledge transfer to farmers requires a critical evaluation of the whole communication chain, including via governmental policies.

22.2. Stepping Back

In order to understand the situation regarding the capacity of extension services in relation to nematology, and consequently how this impacts farmers' understanding of nematology issues, we need to step back and determine the basis behind this. To generate and improve farmer awareness of nematology at the grassroots level, national advisory services need to be consciously informed and need themselves to be aware of nematode pest issues. A substantial component of future-proofed nematology must be a more holistic integration of nematodes into a broader and more inclusive narrative around crop improvement strategies, education curricula, training programmes and policies (Coyne *et al.*, 2018). This requires adequate educational exposure and agricultural policy support to create more interest, understanding and uptake (Cortada *et al.*, 2019). In other words, the lack of attraction to nematology effectively results from the limited early-career exposure of the discipline, in particular at the undergraduate level. Respondents to a survey for the study by Cortada *et al.* (2019) specifically highlighted the need for greater inclusion of nematology at the undergraduate level. Consequently, this requires a transformational shift in our approach to nematology knowledge transfer, a shift that undoubtedly takes a lengthy period of time to transcend and impact across the system as a whole. Cortada *et al.* (2019), for example, were able to demonstrate that over a period of two decades, despite nematology being a little-known discipline in sub-Saharan Africa, remarkable progress and increased awareness in nematology resulted from a range of initiatives and training programmes. Fast forward to 2024, and in order to address this very specific aspect, the Nematology Education in Sub-Saharan Africa (NEMEDUSSA) project (www.nemedussa.ugent.be), led by Ghent University, was designed to inject nematology into the hearts and minds of students across a range of selected universities in sub-Saharan Africa. This developed as a result of the unprecedented experience of Ghent University in implementing the International MSc in Agro- and Environmental Nematology (IMaNema) programme (www.imanema.ugent.be) (sponsored by the Flemish Interuniversity Council – university development cooperation (VLIR-UOS)). IMaNema (and its previous phases) has delivered over 300 MSc nematology trained students to all parts of the world since 1992. With this exceptional experience to learn from, Ghent University, through NEMEDUSSA, embarked on the ambitious task of infusing nematology into university curricula, to expose the discipline to a greater audience, in particular at the undergraduate level. Meanwhile, a range of nematology initiatives (see below) have created their own mark in sub-Saharan Africa, with the aim of raising local disciplinary capacity. The integration of nematode research in broader scientific programmes, through direct training or participatory approaches, can lead to improvements in pest management and provide substantial benefits to farmers (Pretty *et al.*, 2011; Ciancio, 2015). Through tailored, demand-driven approaches ('bottom-up'), extension services can also be improved to benefit rural farmers (Davis, 2008). Conversely, however, this approach can be flawed, in that many farmers, and indeed crop extensionists and protectionists, are ignorant of nematode pest problems and thus, by default, ignore agricultural problems due to nematology (e.g., Wairegi *et al.*, 2010; Abang *et al.*, 2014). Additionally, professional nematology (or phytopathology) networks and societies provide for a critical mass of nematology information, knowhow and expertise. The massive reach of the internet often provides immediate access to such platforms with minimal effort, enabling access to a wealth of material to support nematology. Nevertheless, there is no substitute for physical and personal interaction between an extension agent and farmer!

Success in addressing future nematode pest problems requires a more holistic integration into broader and more inclusive crop improvement strategies (see Chapters 19 and 20, this volume), which is undoubtedly dependent on creating public awareness and understanding of nematodes across the agricultural landscape. Nematology, as a discipline, is shrouded in mystery and obscurity. To overturn the ignorance and oversight of nematode pest problems, huge efforts are necessary to create visibility and understanding of nematode issues. To achieve this, greater efforts are necessary to introduce the topic at the undergraduate level or through agricultural training colleges and institutes, if not earlier. Through a combination of greater exposure, and a more facilitatory environment, nematology, as a discipline, can be made more attractive and attitudes fundamentally transformed for future generations of extensionists, researchers or professors towards better nematode pest management in farmers' fields (Sikora *et al.*, 2018; Cortada *et al.*, 2019).

22.3. Recognizing Symptoms of Nematode Damage in the Field

Typical above-ground symptoms of nematode infection are mostly unspecific and generally attributed to other problems and/or crop malaise (see Chapter 4, this volume). Symptoms may include stunting, wilting, chlorosis, premature senescence and low yields of poor-quality fruits. Poor soil fertility, lack of fertilizer, low water availability (drought) or infection by other pathogens are regularly accredited as the causal reason for nematode infections. Damage to roots may provide more specific symptomatic damage, such as root galling and deformation by root-knot nematodes (*Meloidogyne* spp.). However, farmers may not see this damage, as the roots more often remain in the ground and are not visible. For root and tuber crops though, this is different, and farmers will visually observe damage of infected crops. This damage may also be quite generic, such as cracked, deformed, rotted or reduced size of crop, and thus may also be difficult for farmers to determine nematodes as the causal agent. Symptoms that can be used to determine nematode problems above and below the ground have been documented in practical pest identification manuals or on websites, for example, and can be used to much greater effect than can be reproduced here (e.g., see Coyne *et al.*, 2007). It is important to note, however, that the nematode infection of a crop is usually not in isolation and that crops are mostly subjected and succumb to combined pest and disease attacks, creating symptom and disease complexes. Infection by nematodes can also predispose plants to attack by other pathogens that would otherwise not occur. Well-documented nematode-pathogen complexes include *Fusarium* or *Verticillium* wilt diseases and root-knot nematodes. In combination, damage to crops by the complex may be much greater than the additive effects of *either in isolation*. The following are some key elements that can be used to assess for nematode damage:

- Is the crop/variety susceptible to nematodes?
- Is the crop being irrigated by water flowing through soil channels?
- Is the soil sandy and lacking in organic matter?
- Has the farmer been growing the same or related crop in this field more than once in four seasons?
- Is the crop stunted? Are the stunted areas in patches within the field?
- Are plant leaves showing signs of mineral deficiencies, e.g., chlorosis (yellowing) despite having received adequate fertilizer?
- Are plant leaves wilting, even though the soil is moist?
- Are plant leaves deformed and have interveinal chlorosis patches?
- If a tree crop, is the yield slowly declining?
- Are roots and/or tubers galled, cracked, deformed, knobbly, dying prematurely or overly rotted?

22.4. Mechanisms for Knowledge Transfer

Various mechanisms have been successfully employed in the transfer of agricultural knowledge to farmers. Here, we highlight the farmer field school (FFS) system in particular, as a farmer-oriented participatory approach, and examine some others, especially in light of nematological information transfer. We further assess and evaluate the elements along the research-to-development continuum and refer to initiatives and case studies that demonstrate improved approaches or mechanisms to deliver nematology services and information to beneficiaries for improving nematode pest management.

22.4.1. Farmer Field Schools

FFS are an informal, group-based learning method that serve as a knowledge transfer bridge between science and farm practice. FFS originated in the late 1980s in Southeast Asia as a response to the pressing challenges of pest management (Van de Fliert *et al.*, 1995; Winarto, 1995). With the rise of chemical pesticide use leading to environmental degradation and health issues, a new approach was needed. The Food and Agriculture Organization (FAO) spearheaded the initiative, focusing on experiential learning and participatory methods (Fig. 22.1). In most cases, FFS are convened by local extension workers, who have undergone season-long 'training of trainers' (ToT) sessions. However, technical experts play a key role in facilitating and transfer of new knowledge and scientific innovations to the trainees. FFS provide a platform for farmers to learn about nematodes through observation, experimentation and collaboration. The hands-on, participatory nature of FFS makes them particularly effective for addressing complex issues, such as nematode management.

22.4.2. Nematode Management Methods Suitable for FFS

To develop and sustain an operational and economically viable management programme for nematode pests, the integration of a range of management options, which are context-specific and tuned to local conditions, is necessary. It also needs to be continuously monitored so that modifications based on new advances and technologies can be incorporated. The key constraint towards the effective control and management of nematodes, especially for resource-constrained farmers, is that there is no single effective management option applicable across cropping systems. Thus, management options need to be adapted and developed to suit the target cropping system. Below, we briefly highlight various options recommended for the control and management of nematode pests, which can be employed at the start, during or between cropping cycles and demonstrated with FFS. More detailed information is available in other chapters in this book (also see Page, 2012).

Plant host resistance: The use of resistant crop cultivars is among the most cost-efficient and effective strategies for managing nematode pests and improving crop yields (see Chapters 18-20, this volume). Significant efforts have been made to incorporate resistance into commercially produced crop cultivars, with a particular focus on

Fig. 22.1. Farmers discussing their records of input costs.

breeding for resistance against root-knot nematodes (Pradhan *et al.*, 2023). Cortada *et al.* (2009) demonstrated the impact of using root-knot nematode-resistant tomato cultivars compared to susceptible cultivars. Through the use of resistant cultivars, farmers can achieve better crop performance while reducing reliance on synthetic chemicals. However, the adoption of resistant cultivars is dependent upon their availability and farmer willingness to adopt them. Consequently, adoption is also heavily dependent on farmer knowledge of nematode pests and the reason for adopting the use of such cultivars. Moreover, discussions facilitated by FFS instructors encourage farmers to share their experiences with resistant cultivars, fostering a collaborative environment where knowledge is exchanged. This peer-to-peer learning helps build confidence in adopting new, resistant cultivars, thus leading to broader acceptance and implementation across communities. Continuous breeding efforts are essential to keep pace with evolving nematode pests and the changing environmental conditions, ensuring that resistance remains effective.

Healthy planting material: Using resistant cultivars where possible, the selection and use of healthy planting material is a key requirement to reducing pest and disease levels in general. Nematodes are key limiting factors for many vegetatively propagated crops, often as a consequence of the planting material being infected. Tuber crops, such as potato and yam, among others, can be heavily impacted by nematodes. By ensuring that the planting material is healthy and free of nematode infection, the crop will yield much better, even if planted into a nematode-infected field. Similarly, bananas, bulb crops (onion, flowers, garlic, etc.), tree seedlings and vegetable seedlings can all be heavily impacted if the planting material is infected. Selecting certified seed material, disinfecting planting material or propagating seedlings in sterile media are key considerations to reduce nematode pest problems and improve yields. Simple protocols and techniques for ensuring healthy planting material are available, which are highly suitable to demonstration via FFS.

Biologically based control: The use of natural enemies to suppress nematode densities has been a commonly studied subject and has become increasingly attractive over time (see Chapter 21, this volume). Biological control agents (BCA) offer an eco-friendly alternative to chemical nematicides. This method leverages natural predators, parasites or pathogens to manage nematode pests. Various microbial organisms that parasitize or are antagonistic against nematodes, have shown promise in suppressing nematode densities. For example, *Pasteuria penetrans*, a bacterium that specifically targets nematodes, can drastically reduce their populations (Devi, 2023). Similarly, nematophagous fungi, such as *Arthrobotrys* spp., trap and feed on nematodes, offering a natural method of control (Wu *et al.*, 2023). Other microorganisms may not primarily parasitize or feed on nematodes but suppress nematodes through a combination of mechanisms, such as the production of nematotoxic metabolites that kill or inactivate nematodes, competition for resources and space, production of chitinolytic enzymes that destroy nematode eggs or, alternatively, they may stimulate resistance to nematode infection by the host plant. *Bacillus subtilis*, non-pathogenic *Fusarium oxysporum*, *Pochonia chlamydosporia*, *Purpureocillium lilacinum*, *Trichoderma asperellum* and arbuscular mycorrhizal fungi are examples of antagonistic BCA that have shown promise in suppressing nematode densities and improving crop production systems in the field (see Viaene *et al.*, 2024). However, the efficacy of BCA can very much be influenced by environmental conditions, mode of delivery and the specific nematode species present, highlighting the need for clear understanding of the circumstances at hand. Workshops within the FFS framework can include demonstrations on how to identify effectively and apply BCA, with participants engaging in field observations to recognize signs of nematode damage and assess the impact of BCA interventions. Such practical hands-on experience enables farmers to better understand the conditions that favour BCA and how to integrate them into their pest management strategies. The collaborative atmosphere of FFS fosters innovation, as farmers are encouraged to experiment with different BCAs and share their outcomes and experiences. By learning from each other's successes and failures, participants can refine their approaches and develop tailored solutions that suit their unique agricultural contexts, ultimately nurturing a sense of empowerment and innovation.

Synthetic chemical nematicides: Chemical pesticides have long been the cornerstone of nematode management, providing rapid and effective control for severe infestations (see Chapters 19 and 20, this volume). While they can be highly effective, concerns about environmental impact and human health have led to many of the most dangerous compounds being removed from the global market. This has substantially reduced the availability and range of products available to farmers, resulting in over-reliance on a limited number of options. This creates the risk of the rapid development of nematode resistance. Currently, however, interest by the industry, driven by the demand from the agricultural sector, is leading to a surge in the emergence of new (safer) chemistries. Farmers need to consider several factors when employing synthetic chemical nematicides, including the timing of application, proper dosage and environmental safety. Integrated pest management (IPM) approaches, which combine chemical applications with cultural practices and resistant cultivars, can enhance effectiveness while minimizing negative impacts. It is essential for farmers to receive training on the responsible use of chemical pesticides, to prevent their misuse and overuse, lessen over-reliance and thereby mitigate resistance development. FFS provide a platform for accurate transfer of such knowledge. Through FFSs, farmers learn to evaluate the economic thresholds of nematode population densities, helping them to determine when chemical interventions are necessary. Through field demonstrations, farmers can gain insights into the use of nematicides, application techniques and timing, all while minimizing risks to beneficial organisms and the environment. Crucially, FFS encourage farmers to think critically about using synthetic chemicals. By discussing the potential consequences of relying too heavily on chemical solutions, farmers become more mindful of their choices. This knowledge equips them to make informed decisions that balance their immediate needs with long-term sustainability, fostering a healthier farming ecosystem.

Cultural methods: Cultural practices constitute a foundational approach that aims to modify agricultural practices so as to create an environment less conducive to nematode multiplication. Crop rotation and intercropping with resistant or non-host crops, cover cropping, application of organic amendments to improve soil health, removal of (infected) crop residue are all examples of common farming practices that can promote nematode management. Through FFS, farmers can be trained on the different cropping practices that play a key role in suppressing nematodes. For example, FFS participants can engage in practical exercises to design intercropping or crop rotation systems, thereby promoting an understanding of effective crop combinations or cropping cycles that enhance yield.

Botanicals and plant extracts: The application of botanicals and plant extracts as alternatives to synthetic nematicides is an area of growing interest in nematology. This section refers to the use of plants that can be grown or gathered locally and used for preparations that are applied to crops for protection against nematodes (or other pests and diseases). An expanding number of marketed products based on botanicals have been developed, and these are becoming increasingly available for farmers to purchase. The well-known botanical neem, for example, contains the active compound azadirachtin in leaves and seeds, which inhibits nematode development and reproduction (Aminisarteshnizi, 2021). Where neem trees are present, material can be gathered fresh for preparing extracts for soil drenches or foliar sprays. Alternatively, products based on neem extracts can be purchased. There are numerous plant species that are known to possess or are locally viewed as possessing pesticidal properties. Some, such as neem, have been verified and are well established as being effective, while the information for many needs to be substantiated. The use of locally harvested botanicals can provide for a cost-effective and environmentally friendly option to protect crops against nematodes or other pests and diseases. FFS can be an effective mechanism for exchanging information on the identity, preparation, application method, efficacy and performance of locally available botanicals.

22.4.3. FFS Impact Potential – Getting the Message Across!

Although the initial FFS may be costly in terms of donor-funding, the cost per farmer reduces as inspired graduates of successful FFS replicate the discovery-learning process. The long-term benefits of FFSs have been clearly highlighted, in terms of reducing the use of pesticide applications and raising crop yields, with a sevenfold recovery of the overall ToT and FFS training costs within a single season (Van den Berg, 2004). In addition to the wide dissemination of technical skills, FFS also provide many long-term social benefits, such as the creation of marketing groups, community tree planting groups and soil erosion control activities. Political benefits are realized when female and male FFS graduates are sufficiently empowered to take on leadership roles and begin negotiating with policy-makers.

22.4.4. National Systems

The capacity to provide extension services by various national programmes around the world varies enormously. This includes the provision of information, knowledge and management of nematology. Especially in the lower-income countries, extension services are particularly under-resourced and, for nematology, often poor or non-existent. The same scenario applies to the national research institutes and the services they provide to national agriculture programmes. The aim of this section is not to lament the limited resources for such services, but to acknowledge their existence and that they present a readily available human resource network that can, under the right conditions, be used to great effect. With additional resources, they can be mobilized to engage with farmers, organize farmer meetings and workshops, establish meeting points and conduct farmer demonstrations and so on. They often have a very good knowledge of the local farmer network and serve as a useful intermediary for contact with farmers, who usually show immense respect. They, however, likely lack a confident understanding of nematode pest issues. It is not the intention to provide an extensive overview or critique of national extension services here, as there are numerous articles, book chapters and books that address this issue, but by leveraging such resources and changing the format of engagement with national agricultural services, impact can be made. One example is through the FFS system, as outlined above. Another is the Plant Clinics, under the CABI Plantwise programme (https://www.plantwise.org), which have been introduced to 34 countries in Africa, Asia and the Americas. As with human health clinics, people (farmers) attend and discuss their ailments with the doctor, in this case, trained plant doctors, who provide assistance on diagnosis and a prognosis of how to deal with the issue. By working closely with national agricultural advisory services, the capacity of the extension staff to deliver quality plant health advice through the clinics and complementary extension approaches is strengthened. Farmers find practical plant health advice at the clinics, which are strategically placed and positioned. For instance, clinics are held at regular farmer meeting points, such as the local weekly market, establishing a regularity of where farmers can seek plant health advice from extension staff and plant doctors. Farmers, therefore, come to the extension agent, as opposed to waiting for them to visit their farms, which may be erratic or uncommon – a common weakness of systems that are under-resourced and financially stretched. Since its inception in 2011, Plantwise has had a positive impact on the lives of over 50 million smallholder farmers. All major funding contracts for Plantwise ended in 2020; however, in most countries, elements such as plant clinics and plant doctor training continue to run, as are adapted data management systems in a few countries and the use of social media apps to support advice delivery on plant health. Although focused on plant health in its entirety, plant nematology is an integral part of plant health. Plant clinics highlight how national programmes can engage in information delivery by shifting how this is conducted, including for nematology. There are promising signs of the institutionalization and scaling up of Plantwise-supported approaches in a number of countries, at both the public and private sector levels.

22.4.5. Nematology Capacity Building Initiatives

To elevate nematology awareness and attract future generations of nematologists, greater numbers of active nematologists are necessary to provide training and stimulate interest in the younger generation. With a focus on sub-Saharan Africa, a number of initiatives or projects have been implemented over the last couple of decades, aimed at augmenting nematology expertise and awareness. While support for such initiatives is much welcomed, a downside to such time-defined project-based undertakings is that once the project timeline has elapsed, so does the funding, with most projects being of short-to-medium duration. The NEMEDUSSA project, for instance, aimed at creating fundamental changes in university curricula to include and embody nematology as a standalone discipline. While the project was exceptionally successful in achieving its objectives, such fundamental changes in established systems require long-term timeframes and funding support. Conversely, the IMaNema programme and its various earlier phases have been training nematology MSc students from across the globe since 1992, with financial support from the Belgian Government. Consequently, it has been recognized as one of the most critical driving forces for training African nematology students (Cortada *et al.*, 2019). Over a five-year period, the Nematology Initiative for East and Southern Africa (NIESA), funded by the Gatsby Charitable Trust, was implemented to advance nematology research and to build plant nematology capacity in the region (Talwana *et al.*, 2016). While again being highly successful in achieving its objectives, the project closed when the investment was beginning to make progress and pay dividends. Numerous other projects have understandably been implemented that encompass or include nematology and contribute to the overall presence, capacity and creation of awareness in nematology, as well as reducing losses to nematode pests. The projects highlighted above specifically underscore knowledge transfer through awareness and capacity creation and are emphasized as such, while many more make their mark in advancing nematology awareness.

Within the International Agricultural Research Centres network (CGIAR; https://www.cgiar.org/), IITA and CIMMYT and other international research institutes, such as *icipe* and CABI, nematologists are important drivers in coordinating and stimulating research on food crops in Africa and Asia. Capacity building is central to these international centres in addition to and through the delivery of improved agricultural innovations. These international centres are well placed to provide the technical and infrastructural support and channel financial support over the long term for capacity development across agriculture. They are also well placed to provide the momentum to strengthen links between technical groups, as was suggested for nematologists working in the tropics by Sikora *et al.* (2018). They suggested this for both moral and financial reasons, as well as further suggesting that it may be necessary to develop 'virtual hubs' or 'centres of excellence' for nematology in the tropics and subtropics. This was partly based on the success of *Nem*Africa, the inter-institutional collaborative nematology research resource developed between IITA and *icipe* in East Africa (see Box 22.1). As an integral part of providing soil health and nematology research, numerous students are provided the opportunity to become immersed in nematology, within an active, research-based lab alongside a dynamic team at *Nem*Africa. This facilitatory environment creates a harmonious team spirit, which is very important for attracting staff and students and for generating output. An enjoyable atmosphere is a productive one! Furthermore, *Nem*Africa serves as a satellite outpost to the IMaNema MSc programme, to teach the Kenya Track option, which provides students with the practical experience of working in a tropical region. As a result of on-farm visits, students experience the reality of the situation towards understanding nematode pest problems in relation to the various other challenges that farmers are faced with (Fig. 22.2). This experience provides students with a holistic real-life scenario to better prepare them for postgraduate life. The Kenya Track also delivers the Basic Crash Course in Nematology (BCCN; https://imanema.ugent.be/basic-crash-course-nematology-kenya-bccn/), a one-week crash course in basic and practical nematology. It is designed for technical staff, who would benefit from a practical grounding in and knowledge of nematology for their work, such as technicians, extensionists, farm managers, lecturers and so on. The BCCN is very popular and highly praised by the participants and provides an effective mechanism for creating awareness and understanding of nematodes in people who can benefit directly from it.

While it is preferable to have hands-on practical training, it is not always possible as it can be costly, logistically difficult and the number of trainees limited. A number of online platforms provide

Fig. 22.2. Kenya Track students take to the field (A) and lab (B) for hands-on training with *Nem*Africa.

free and easy access to nematology tutorials, training, services and materials (see below), which help overcome the obstacles of physical training courses. But are they effective? In order to provide nematology training to county extension employees in Arkansas Delta district, USA, a set of online nematology modules was administered, and its efficacy as a training tool assessed, based on participants' changes in knowledge and evaluative feedback (Gentry *et al.*, 2017). They showed that levels of nematology knowledge and comfort with nematology topics increased and that participants found using the modules to be an effective method of learning.

Box 22.1. *Nem*Africa case study

*Nem*Africa, an inter-institute collaboration between the International Institute of Tropical Agriculture (IITA) and the International Centre for Insect Physiology and Ecology (*icipe*), was established in 2015 as a pioneering nematology lab for the East Africa region, based at *icipe* Nairobi. *Nem*Africa was established in recognition of the limited capacity for nematology in the region and is jointly led by two international nematologists. By joining the competitive forces of two international institutes, the infrastructural and financial resources to create a fully functional, well-equipped laboratory was harnessed and a trusted hub was created. This has provided convening and participation power, allowing interchange of knowledge with diverse stakeholders, creating greater visibility and critical mass than under a single institute. The dedicated *Nem*Africa team comprises a diverse group of professionals and aspiring researchers, including numerous PhD and MSc students. The *Nem*Africa lab bridges the gap between scientific knowledge and practical solutions for the challenges posed by nematodes in tropical ecosystems. It is devoted to unravelling the complexities of nematodes to achieve substantial agricultural and ecological impacts. Through interlinkages within *icipe* and IITA, collaborative partnerships and community outreach, *Nem*Africa is driving positive change in global agriculture and environmental sustainability. Moreover, with strong links to local universities, *Nem*Africa is providing training and capacity development in nematology for enthusiastic students, interns and attachees. With substantial and international links, students and visiting scientists bring a fresh perspective to the lab's dynamic environment. The constant flow of students and researchers through the lab continuously broadens and strengthens the lab's partnerships and network. To date, 78 (58% female) students (Fig. 22.3) are graduates of *Nem*Africa, having gained crucial practical experience and research insight in nematology. *Nem*Africa is also an integral member of the International Nematology MSc (IMaNema) coordinated by Ghent University, with MSc students received annually. To date, 55 students have attended the module on tropical nematology, and also provided nematology training to 73 local partners. Through diverse partnerships and collaborations with universities, national programmes and the private sector, *Nem*Africa has an extensive network across the region and around the world. *Nem*Africa serves to demonstrate how, in the span of one decade, a combination of strong institutional support, solid partnerships, an accommodating environment, enthusiastic student participation and a harmonious team spirit provides a highly effective platform to attract funding, conduct research, create capacity and transfer knowledge on nematology in a regional context. *Nem*Africa has proved to be a dynamic example of a triangular model of knowledge transfer – south–south, north–south, south–north. *Nem*Africa has also proved to be a dynamic tool in advancing the discipline of nematology as a leading nematology training hub in Africa.

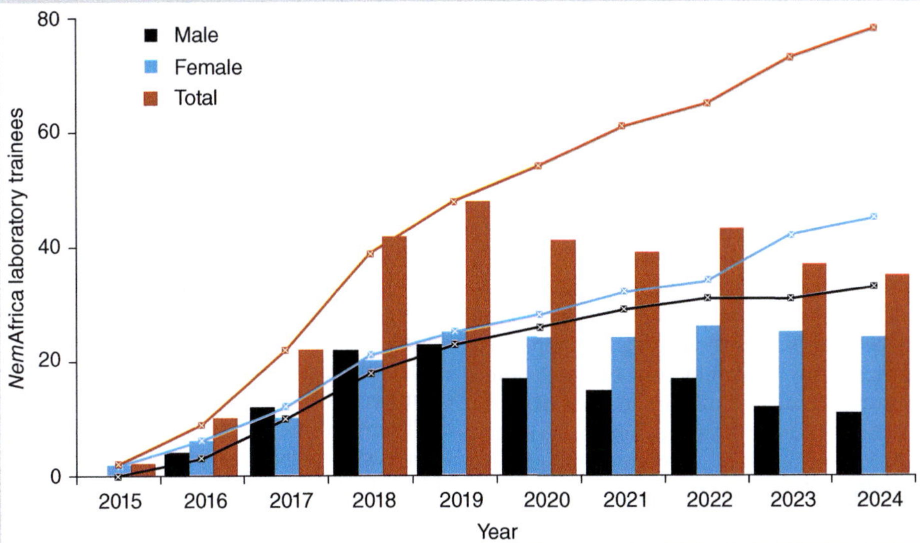

Fig. 22.3. Student training at *Nem*Africa over 10 years since establishment. Bars represent numbers present during each year; lines indicate cumulative numbers.

22.4.6. Nematology Networks, Societies and Platforms

There are currently 16 nematology societies registered within the International Federation of Nematology Societies (IFNS). Additionally, numerous phytopathology or crop protection and pest management societies promote nematology, such as the American Phytopathological Society (APS), Association of Applied Biologists, etc. The International Association for the Plant Protection Sciences (https://www.plantprotection.org/) promotes the development of a multidisciplinary integrated approach to research on plant protection and brings together plant protection specialists, including nematologists, from around the world every four years through the International Plant Protection Congress (IPPC) to discuss the science and practice of plant protection. A similar initiative was carried out by IFNS, which has so far sponsored more than 150 nematology graduates and early-career nematologists to attend the International Congress of Nematology conferences held in South Africa (2014) and France (2022). These societies and their symposia are excellent meeting points for students, researchers, as well as industry and extension services. However, do they contribute significantly to getting the message across? In the study by Cortada *et al.* (2019), networking featured prominently as a means of promoting nematology both within and outside Africa and towards generating greater collaboration globally. Indeed, Ciancio (2015) concluded that one feasible role that can be played by the community of nematologists is linked to their capacity to set up and develop cooperative research work at the international scale. Hence, associations, professional societies and other platforms, such as WhatsApp groups, provide important foundations for cooperation between individuals and teams. A significant amount of collaboration exists between scientists working in temperate countries who are conducting research and supporting capacity building in the tropics. A successful example of such collaboration is the *Nem*Africa nematology laboratory based in Kenya (Box 22.1), which has significantly increased the visibility and capacity of nematology in the East African region. Capacity building projects, such as the aforementioned NEMEDUSSA, can have long-lasting impacts, with project outputs that remain available for download from the project website. In conjunction with IMaNema, the Nematology Digital Learning Platform (NDLP; https://imanema.ugent.be/ndlp/) was developed to facilitate global and free access to training content relevant for nematology. Through this e-learning portal, lecturers, researchers, agricultural workers and nematology professionals globally have access to online tutorials, webinars and training materials, such as protocols, manuals, virtual microscopy and demonstration videos. It also acts as a point of contact and networking for nematologists. Similarly, the Nemaplex portal (http://nemaplex.ucdavis.edu/Uppermnus/topmnu.htm) provides access to a wonderful array of training materials, pictures, tools and learning resources. Non-specialized platforms, such as The African Agricultural Improvement Scientist Database (source: http://hub.africabiosciences.org/scientists/searchdb.php), provide a centralized repository for researcher contacts. Other digital platforms are also available, to varying degrees of content, providing quality nematology information, tools and materials (Table 22.1).

In addition to introducing nematology into university curricula, NEMEDUSSA also developed a pan-African nematology network (PANEMA) towards promoting exchange between nematologists in Africa. The focus has been towards sharing good educational practices, creating networks between higher education institutes to apply for educational projects and additional learning opportunities. Such networks and their continuing existence and activity are critical to the creation, dissemination and uptake of nematology knowledge, as well as fostering the interaction of nematologists to support this process.

Networks bringing nematology awareness and technologies to farmers are usually placed within broader initiatives than nematology alone, such as the CABI Plantwise programme (https://www.plantwise.org) or extension services that often are connected with research institutes or universities for access to nematology expertise. There are few platforms that are specialized in or designed specifically around nematology, although popular WhatsApp groups can arise from nematology training events, where the communication among participants and trainers continues. In initiatives and networks related to soil health, such as the Africa soil health

Table 22.1. Non-exhaustive list of some key nematology resources.

Nematology societies	Afro-Asian Society of Nematologists; Australian Association of Nematologists; Brazilian Nematological Society (Sociedade Brasileira de Nematologia); Chinese Society of Plant Nematologists; Egyptian Society of Agricultural Nematology; European Society of Nematologists; International Federation of Nematology Societies; Italian Society of Nematologists (Societa Italiana di Nematologia); Japanese Nematological Society; Nematological Society of India; Nematological Society of Southern Africa (Nematologiese Vereniging van Suidelike Afrika); Organization of Nematologists of Tropical America (Organización de Nematólogos de America Tropical); Pakistan Society of Nematologists; Russian Society of Nematologists; Society of Nematologists; Nigerian Society of Nematology; Young Nematologist Network
Online resource sites	EPPO: https://www.eppo.int/ABOUT_EPPO/panel_composition/pm_diag_nemato https://www.wur.nl/en/research-results/chair-groups/plant-sciences/laboratory-of-nematology/about-us.htm https://www.youngnematologists.net/ https://www.vliruos.be/en/scholarships https://www.nifa.usda.gov/topics/nematology Basic Crash Course in Nematology (BCCN) Clemson University
Digital platforms	NEMEDUSSA: https://nemedussa.ugent.be/ Nematology Digital Learning Platforn (NDLP): https://imanema.ugent.be/ndlp/ UCDavis: http://nemaplex.ucdavis.edu/Uppermnus/topmnu.htm CABI Plantwise: https://www.cabi.org/plantwiseplus/impact/plantwise/ Plant and insect parasitic nematodes: https://nematode.unl.edu/index.html Nematicide stewardship programme: https://nspstewardship.co.uk/
Technology platforms	Syngenta: https://www.syngenta.com/en/company/media/syngenta-news/year/2022/syngenta-launches-worlds-first-commercial-digital-tool-detect#:~:text=The%20introduction%20of%20the%20digital,preserving%20soil%20biodiversity%20and%20health

(https://www.cabi.org/projects/africa-soil-health/), nematodes are rarely mentioned or included, and efforts to join such initiatives should be strenuously encouraged to enable onward progress. Digital networks and tools directed at farmers are useful and reach a wide audience; for nematology, however, they have their limits due to the often unspecific symptoms of nematode damage, yet are extremely valuable in creating awareness. Increasing efforts to promote nematology within new and existing soil health- and agricultural-related digital platforms and tools will greatly assist in getting the message across at all levels.

22.5. Going Forward

Access to timely, appropriate and actionable extension advice is a fundamental requirement *and* challenge for farmers in reducing their crop losses. National extension and research services can be adapted and reformatted to delivering improved services to this end, while various online services are increasingly facilitating distance or virtual learning in nematology. However, central to realizing improved knowledge transfer of nematode problems and their management is exposure at an early stage of education. That is, earlier than is currently standard, which will require a major shift in the educational curricula across academic institutions. In order to generate interest in embracing nematology as a discipline, students need to be exposed to it – ideally, at an early stage and within a harmonious and facilitatory environment. Connectivity and interaction are also important. The internet and social media have facilitated communication, but physical interaction and physically attending training courses remain crucial to creating quality relationships and gaining quality knowledge and knowhow.

Reduction of the losses due to nematode pests requires that farmers know what nematodes are and how to control them. Public awareness, and an understanding of nematodes across the agricultural landscape, is a necessary component of any pest and disease management strategy. Nematology has typically been overshadowed by other disciplines but is becoming increasingly more visible. Improved and more dynamic knowledge transfer mechanisms have supported this progress towards reducing nematode pest damage. While the outlook is positive, it ultimately rests upon nematologists to attract the relevant attention and support to address the seriousness of the issue (Coyne *et al.*, 2018).

22.6. References

Abang, A.F., Kouamé, C.M., Abang, M., Hanna, R. and Fotso, A.K. (2014) Assessing vegetable farmer knowledge of diseases and insect pests of vegetable and management practices under tropical conditions. *International Journal of Vegetable Science* 20, 240–253. DOI: 10.1080/19315260.2013.800625.

Aminisarteshnizi, M. (2021) Efficacy of neem (*Azadirachta indica*) formulations on suppression of root-knot nematodes (*Meloidogyne javanica*) and growth of eggplant (*Solanum melongena*). *Research on Crops* 22, 158–162.

Ciancio, A. (2015) Nematology, crop protection and food security. *Nematology* 17, 1127–1140. DOI: 10.1163/15685411-00002942.

Cortada, L., Dehennin, I., Bert, W. and Coyne, D.L. (2019) Integration of nematology as a training and research discipline in sub-Saharan Africa: progress and prospects. *Nematology* 22, 1–21. DOI: 10.1163/15685411-00003291.

Cortada, L., Sorribas, F.J., Ornat, C., Andrés, M.F. and Verdejo-Lucas, S. (2009) Response of tomato rootstocks carrying the Mi-resistance gene to populations of *Meloidogyne arenaria*, *M. incognita* and *M. javanica*. *European Journal of Plant Pathology* 124, 337–343. DOI: 10.1007/s10658-008-9413-z.

Coyne, D.L., Cortada, L., Dalzell, J.J., Claudius-Cole, A.O., Haukeland, S., *et al.* (2018) Plant-parasitic nematodes and food security in Sub-Saharan Africa. *Annual Review of Phytopathology* 56, 381–403. DOI: 10.1146/annurev-phyto-080417-045833.

Coyne, D.L., Nicol, J.M. and Claudius-Cole, B. (2007) *Practical Plant Nematology: A Field and Laboratory Guide*. SP-IPM Secretariat, International Institute of Tropical Agriculture (IITA), Cotonou, Benin. https://cgspace.cgiar.org/items/5237082b-ccd4-4726-9c95-91c2c3c7e262.

Davis, K.E. (2008) Extension in Sub-Saharan Africa: overview and assessment of past and current models, and future prospects. *Journal of International Agricultural and Extension Education* 15, 15–28.

Devi, G. (2023) *Pasteuria* spp. as biocontrol agent against plant-parasitic nematodes: an overview. *International Journal of Plant & Soil Science* 35, 134–144. DOI: 10.9734/IJPSS/2023/v35i203793.

Gentry, M., Edgar, L.D., Graham, D.L. and Kirkpatrick, T. (2017) Assessment of an online nematology training for county extension agents. *Journal of Extension, Research in Brief* 55(1), rb7. https://archives.joe.org/joe/2017february/rb7.php.

Page, S.L.J. (2012) Farmer schools and knowledge transfer. In: Manzanilla-López, R.H. and Marbán-Mendoza, N. (eds), *Practical Plant Nematology*. Mundi-Prensa and Biblioteca Básica de Agricultura, Montecillo, Mexico, pp. 813–828.

Pradhan, P., Naresh, P., Barik, S., Acharya, G.C., Bastia, R., *et al.* (2023) Breeding for root-knot nematode resistance in fruiting *Solanaceous* vegetable crops: a review. *Euphytica* 219, 1–17. DOI: 10.1007/s10681-023-03204-2.

Pretty, J., Toulmin, C. and Williams, S. (2011) Sustainable intensification in African agriculture. *International Journal of Agricultural Sustainability* 9, 5–24.

Savary, S., Ficke, A., Aubertot, J.N. and Hollier, C. (2012) Crop losses due to diseases and their implications for global food production losses and food security. *Food Security* 4, 519–537. DOI: 10.1007/s12571-012-0200-5.

Sikora, R., Coyne, D., Hallman, J. and Timper, P. (2018) Reflections and challenges: nematology in subtropical and tropical agriculture. In: Sikora, R.A., Coyne, D., Hallmann, J. and Timper, P. (eds), *Plant Parasitic Nematodes in Subtropical and Tropical Agriculture*, 3rd edn. CAB International, Wallingford, UK. pp. 1–19.

Talwana, H., Sibanda, Z., Wanhoji, W., Kimenju, W., Luambuano-Nyoni, N., *et al.* (2016) Agricultural nematology in East and Southern Africa: problems, management strategies and stakeholder linkages. *Pest Management Science* 72, 226–245. DOI: 10.1002/ps.4104.

Van De Fliert, E., Pontius, J. and Röling, N. (1995) Searching for strategies to replicate a successful extension approach: training of IPM trainers in Indonesia. *European Journal of Agricultural Education and Extension* 1, 41–63. DOI: 10.1080/13892249485300331.

Van den Berg, H. (2004) *IPM farmer field schools: A synthesis of 25 impact evaluations*. Wageningen University, Prepared for the Global IPM Facility. http://www.fao.org/docrep/006/ad487e/ad487e01.htm#TopOfPage.

Viaene, N., Coyne, D.L. and Davies, K. (2024) Biological and cultural control. In: Perry, R.N., Moens, M. and Jones, J.T. (eds), *Plant Nematology*, 3rd edn. CAB International, Wallingford, UK, pp. 443–471.

Wairegi, L.W.I., van Asten, P.J.A., Tenywa, M.M. and Bekunda, M.A. (2010) Abiotic constraints override biotic constraints in East African highland banana systems. *Field Crops Research* 117, 146–153. DOI: 10.1016/j.fcr.2010.02.010.

Wilson, C.T. (1982) Development of an expanded program of teaching and research in nematology in the Southern region. In: Riggs, R.D. (ed), *Nematology in the Southern United States. Southern Cooperative Series Bulletin* 276, pp. 1–7. Arkansas Agricultural Experiment Station, University of Arkansas, Fayetteville, Arkansas.

Winarto, Y.T. (1995) State intervention and farmer creativity: integrated pest management among rice farmers in Subang, West Java. *Agriculture and Human Values* 12, 47–57. DOI: 10.1007/BF02218566.

Wu, Y., Yang, Z., Jiang, Z., Nizamani, M.M., Zhang, H., *et al.* (2023) Isolation, identification, and evaluation of the predatory activity of Chinese *Arthrobotrys* species towards economically important plant-parasitic nematodes. *Journal of Fungi* 9, 1–15. DOI: 10.3390/jof9121125.

23 International Phytosanitary Measures to Prevent the Spread of Plant-parasitic Nematodes

Renata Tenente[1]*, Vilmar Gonzaga[2], Juvenil Cares[3], Thomas Prior[4], Sue Hockland[5] and Juliano Prendi[6]

[1]SHIN, Brasília, DF, Brazil; [2]Embrapa Dairy Cattle, Juiz de Fora, MG, Brazil; [3]University of Brasília, Institute of Biological Sciences, Department of Plant Pathology, Brasília, DF, Brazil; [4]Department of Plant Protection, Sand Hutton, York, UK; [5]Independent Plant Nematology consultant, UK; [6]University of Brasília, Brazil

Abstract

The introduction and spread of nematode pests can have serious economic consequences for governments, farmers and consumers. Once pest species are established, their eradication is very difficult and controlling them is often a significant percentage of the cost of food production. It is the aim of international phytosanitary measures to prevent such introductions. This requires legislation and implementation of phytosanitary measures at country level (the example of Brazil is used herein), nematology techniques commonly used in the plant quarantine stations and nematology laboratories, phytosanitary control programmes to reduce the risk of infestation in regional or national germplasm stocks and the development of expert skills and systems to support the quarantine laboratories.

23.1. Introduction

The introduction and spread of pests can have serious economic consequences for governments, farmers and consumers. Once pest species are established, their eradication is very difficult, and controlling them often constitutes a significant percentage of the cost of food production. The International Plant Protection Convention (IPPC) was created in 1951 in an attempt to increase international cooperation regarding the management of international plant pest threats (www.ippc.int). The organization has since evolved to deal with an ever-increasing world trade in planting material and plant products and has subsequently been recognized by the WTO-SPS (World Trade Organization – Sanitary and Phytosanitary Measures) Agreement of 1995 as the standard-setting body for developing international phytosanitary standards, guidelines and recommendations. The IPPC is now an international agreement on plant health with 185 signatories and aims to protect cultivated and wild plants by preventing the introduction

*E-mail: renatatenente2@gmail.com

A revision of Tenente, R., Gonzaga, V. and Hockland, S. (2012) International phytosanitary measures to prevent the spread of plant-parasitic nematodes. In: Manzanilla-López, R.H. and Marbán-Mendoza, N. (eds), *Practical Plant Nematology*. Mundi-Prensa and Biblioteca Básica de Agricultura, Montecillo, Mexico.

(eds R.H. Manzanilla-López, D.J. Hunt, N. Marbán-Mendoza and S. Infante Gil)

DOI: 10.1079/9781836990413.0023

and spread of pests without restricting international trade or travel by people. The IPPC is governed by the Commission on Phytosanitary Measures (CPM), which meets annually to promote cooperation to help implement the objectives of the IPPC. In particular, the CPM reviews the state of plant protection around the world. The IPPC encourages countries to analyse risks to their national plant resources and to use science-based measures to safeguard their cultivated and wild plants. It recommends methods to control the spread of pests into new areas by developing and establishing international standards, providing guidance for resolving disputes and the sharing of phytosanitary information, and cooperates with international organizations on matters covered by the Convention.

The IPPC Secretariat is responsible for the coordination of these core activities under the IPPC work programme and is resourced by the Food and Agriculture Organization (FAO) of the United Nations. The focus of this effort is the development of International Standards for Phytosanitary Measures (the ISPM), which include standards for pest surveillance, survey and monitoring, import regulations and pest risk analysis, compliance procedures and phytosanitary inspection methods, pest management, post-entry quarantine, exotic pest emergency response, control and eradication and export certification. In line with the IPPC principles, national governments around the world develop their own plant health and quarantine regulations, these typically being enforced by an integral National Plant Protection Organization (NPPO). Regional Plant Protection Organizations (RPPO) are inter-governmental organizations functioning as a coordinating body for both the NPPO and for implementing ISPM at a regional level. More detailed information on the history and development of international plant health is given by Ebbels (2003) and for plant-parasitic nematodes in particular by Hockland *et al.* (2024).

23.2. Plant-parasitic Nematodes Listed in Legislation

Many countries do not have the same native fauna and therefore differ in the plant-parasitic nematodes listed in their legislation. Most, however, tend to target those species that have endoparasitic habits as they can be easily transported via roots, tubers and above-ground parts of plants. The prohibition of soil in trade, apart for some derogations, should minimize the spread of most nematodes. There is, however, a core group of nematode species listed in legislations across the world as being economically important, including *Anguina tritici*, *Aphelenchoides besseyi*, *A. fragariae*, *Bursaphelenchus xylophilus*, *B. cocophilus*, *Ditylenchoides destructor* (= *Ditylenchus destructor*), *Ditylenchus dipsaci*, *Globodera pallida*, *G. rostochiensis*, *Heterodera glycines*, *H. schachtii*, *Meloidogyne* species, *Nacobbus aberrans*, *Radopholus similis*, *Xiphinema americanum sensu lato* and *X. index* (Hockland *et al.*, 2024). Increased trade of plants around the world has resulted in the interception of nematodes that are newly considered to pose an economic threat, with these being classified as regulated non-quarantine pests (RNQP), such as *Helicotylenchus dihystera*, a species commonly intercepted on ornamental plants. All pests requiring repeated statutory action should be subjected to a pest risk analysis (PRA) of one sort or another, a procedure of increasing importance in plant health (FAO, 1996, 2006). In line with IPPC principles, the production of a PRA is a critical part of the decision-making process as to whether action should be taken on a particular interception. Such procedures demand the skills of specialized personnel, not only in the recognition of a wide range of species and the ability to distinguish them from native fauna but also with the knowledge and experience to enable sound judgements to be made. This is particularly important as relatively little is known about the biology or potential economic importance of a significant number of nematode species intercepted in trade. More information on the development and implementation of PRA procedures is set out by Hockland *et al.* (2024) and on the IPPC website.

23.3. Implementing Phytosanitary Measures at Country Level

An example of RPPO is the 'Comité de Sanidad Vegetal del Cono Sur' (COSAVE), which was established in 1980 (EPPO, 1990). State members include Argentina, Bolivia, Brazil, Chile, Paraguay and Uruguay. In the following sections of this chapter, we will describe the Brazilian system as an example of how phytosanitary measures have been developed and implemented.

Plant protection activities began in Brazil in 1909, and in 1934, the Plant Sanitary Defence Regulation was established in order to prevent the introduction of exotic pests and diseases, and to limit their spread into the country. The Plant Quarantine Station (PQS) of Embrapa (Brazilian Agricultural Research Corporation) was created later with the purpose of supporting the Brazilian plant quarantine programme (Rocha *et al.*, 1984).

Plant germplasm introduction has been one of the most effective actions for agricultural development in the tropics. Coffee and soybean in Brazil and pastures in Latin America are good examples of successful introductions. However, the movement of germplasm involves the risk of introducing pests that may be carried by plants or plant materials. Therefore, to avoid the introduction of quarantine nematodes associated with seeds, propagating plant material or plants for planting into Brazil, new strategies have been implemented to ensure and maintain the phytosanitary status of the country with regard to quarantine nematodes (Tenente and Manso, 1987; Tenente *et al.*, 1996a, b, 1999a, 2002; Tenente and Gonzaga, 1999).

In Brazil, germplasm quarantine is different from commercial germplasm quarantine and is done by Embrapa Genetic Resources and Biotechnology (Cenargen), a research unit of Embrapa (Oliveira *et al.*, 2003). The Embrapa-Cenargen PQS works with exotic pests through detection, identification and eradication of plant-parasitic nematodes present in germplasm imported into Brazil (Marques and Marinho, 2007).

The development of international standards in plant health has resulted in the international and national accreditation of growers to ensure that standards are universally maintained. The Ministry of Agriculture and Livestock (MAPA), through the Normative Instruction No. 29 (MAPA, 2016), provides all the information needed for the accreditation of PQS working with living plant pests in Brazil (https://www.gov.br/agricultura). A new PQS was opened in November 2021 at Embrapa-Cenargen, Brasília (DF). The new PQS has the experts and capacity to detect and identify quarantine pest species in acarology, bacteriology, entomology, molecular biology, mycology, nematology, virology and weeds.

23.3.1. Plant Quarantine Station

Embrapa's new PQS facility building is provided with the equipment and enhanced security, following MAPA's regulation. The PQS is classified under Level 1, which means that it can support all quarantine-related activities to detect and identify quarantine pests. The PQS integrates with laboratories and professionals in all quarantine-related areas, thus providing a complete team for quarantine analysis diagnostics. The facilities also include indoor growth chambers and quarantine greenhouses, which are very important to give the right infrastructure support to conduct the analysis of imported plant materials. Furthermore, by belonging to the Level 1 category for analysis accreditation, quality standards need to be constantly improved, thus minimizing the risk of introducing new pests to the country.

The lists of regulated pests in Brazil (including plant-parasitic nematodes) were updated in 2022 by MAPA, and must be strictly considered during quarantine analysis. The lists are presented in separate acts, which include: (i) Absent Quarantine Pests (PQA) list, i.e., pests not present in Brazil but with potential for establishment, which is published in the Ordinance SDA/No. 617 (MAPA, 2022); and (ii) Present Quarantine Pests (PQP) list, i.e., pests already established but not widely spread or restricted to certain areas and are under official control; the list is published in the Normative Instruction No. 38 (MAPA, 2018). Such lists alert exporting countries to the necessary procedures for compliance with Brazilian regulations, so that only certified plant material is traded, i.e., plant material that is accompanied by a phytosanitary certificate with an assurance that it has been tested in the exporting country as being 'substantially free' from pests that are listed in Brazil. The PQA lists of pest and nematodes for Brazil can be accessed at https://www.gov.br/agricultura/pt-br/assuntos/sanidade-animal-e-vegetal/sanidade-vegetal/quarentena.

The PQA list for plant-parasitic nematodes includes: *Anguina agrostis*, *Anguina pacificae*,

Anguina tritici; *Aphelenchoides blastophthorus*; *Belonolaimus longicaudatus*; *Bursaphelenchus mucronatus*, *B. xylophilus*; *Criconema mutabile*; *Ditylenchoides africanus* (= *Ditylenchus africanus*), *Ditylenchoides destructor* (= *Ditylenchus destructor*); *Ditylenchus angustus*, *D. dipsaci* (all races, except the garlic race), *D. emus*, *D. equalis*, *D. fotedari*; *Globodera pallida*, *G. rostochiensis*; *Heterodera avenae*, *H. cajani*, *H. ciceri*, *H. goettingiana*, *H. mediterranea*, *H. oryzae*, *H. oryzicola*, *H. sacchari*, *H. schachtii*, *H. trifolii*, *H. zeae*; *Longidorus attenuatus*, *L. elongatus*; *Meloidogyne chitwoodi*, *M. fallax*; *Nacobbus aberrans*, *N. dorsalis*; *Paratylenchus hamatus*; *Pratylenchus fallax*, *P. goodeyi*, *P. neglectus*, *P. pratensis*, *P. scribneri*, *P. thornei*; *Punctodera chalcoensis*, *P. punctata* (= *Heterodera punctata*); *Rotylenchulus macrodoratus*, *R. parvus*; *Subanguina radicicola*; *Trichodorus viruliferus*; *Tylenchorhynchus claytoni*; *Xiphinema diversicaudatum*, *X. italiae*, *X. rivesi*, *X. vuittenezi* and *Zygotylenchus guevarai*.

23.4. Techniques Used in the PQS Nematology Laboratory of Embrapa-Cenargen

One specific point in plant pest quarantine actions refers to following standard techniques that laboratories use for sample analysis. In the PQS Nematology Laboratory, attention is given to the type of material (e.g., plant, soil) to be examined, and then the best procedure to apply the nematode analysis is chosen, depending on the objective of the study: taxonomy, biology, population density and so on.

According to the quarantine objective, the analysis of germplasm (propagating material) has to be supported by at least two different extraction techniques, but, generally, more than one extraction method is used in Embrapa PQS for nematodes, seeds, plant vegetative materials or any other associated substrate imported into Brazil for research purposes (Tenente *et al.*, 1994, 1996a, b). The PQS techniques have been standardized for all laboratories that analyse imported materials to guarantee the quality and security of the analysis. To optimize the analysis precision, the nematology techniques were modified and added to PQS procedures by the Nematology Laboratory (Nematology Unit) (Tenente *et al.*, 1994, 1996a, b, 1999b, see also http://pragawall.cenargen.embrapa.br/aiqweb/nemhtml/nembanco01_i.asp).

The techniques used for nematode detection and identification in the Nematology Laboratory of Embrapa Quarantine Station are listed and briefly described below:

Baermann funnel. Used to isolate vermiform nematodes from many materials, e.g., plant tissue, seeds, soil, etc. (Southey, 1986; Zuckerman *et al.*, 1990).

Centrifugation-flotation technique. Modified and used for nematode extraction from soil or other imported substrates (Jenkins, 1964).

Cyst flotation. This technique uses a flat-bottomed glass flask, plus sucrose solution (400 g/l + 12.5 ppm of Separan) to float the cysts (Carvalho *et al.*, 1953).

Direct examination. Used for propagative materials, and observation are made under a stereomicroscope.

Fenwick elutriator. Used for cyst extraction from dry soil samples treated with successive washings under a sieve placed in the Fenwick elutriator (also known as the Fenwick can). Cyst collection is made with a 100 mesh/inch (= 150 μm mesh aperture) sieve placed at the base of the equipment. Following extraction, the sample can be observed under the stereomicroscope (Zuckerman *et al.*, 1990). Using this method, it is also possible to recover cysts associated with imported seeds, such as cysts of *H. glycines* on the surface of soybean seed.

Germination paper. The technique consists of preparing layers of germination paper, where seeds are placed and wetted with distilled water. The paper containing the seeds is rolled and incubated at 25°C at high humidity (near 100%) for 8 days. Subsequently, nematode extraction is done using the Baermann funnel. This technique allows nematodes to rehydrate after dormancy. This protocol is used for leguminous seeds to avoid fermentation when shredded and placed in a Baermann funnel (Tenente *et al.*, 1994).

Seed handling peel. It is used for samples with a small quantity of seeds (about 50 seeds/sample), especially for *Panicum* and rice seeds (Silveira and Curi, 1983).

Tray technique. Designed originally by Whitehead and Hemming (1965), this technique includes the use of a tray (53 × 21 × 5 cm) and one layer of plastic net (2 mm mesh aperture) to support a sheet of Kimwipes® tissue (see also 23.4.1).

Sieving technique. It can be used with dry (to remove impurities) or wet (to extract mobile nematodes) soil samples, and sieving is used in combination with other techniques, such as a Whitehead tray or Baermann funnel, for better recovery of nematodes. The technique involves sieving the sample through sieves of different apertures, from larger to smaller mesh apertures (Southey, 1986; Zuckerman *et al.*, 1990), to collect the nematodes in 400- or 500-mesh aperture sieves (see Chapter 4, this volume).

Trituration in blender. This is not recommended for all type of seeds, such as leguminous seeds (which can ferment after blending), and where the recovery of nematodes has been shown to be more difficult. It is not used for chemically treated seeds (Tenente *et al.*, 1994).

Extraction of large nematodes in soil samples. Standard methods for extracting nematodes from soil are often inefficient at extracting large nematodes (i.e., over 3 mm long), such as species of *Xiphinema, Longidorus* and other dorylaimids (see Chapter 4, this volume). To increase the efficiency in extracting these nematodes from soil samples, Flegg (1967) made some adaptations to Cobb's (1918) decantation and sieving method. Further modifications to Flegg extraction method have been made by Whitehead and Hooper (1970), Evans (1979) and Machado *et al.* (2019). To avoid physical damage to nematodes due to direct sieving, the soil suspension should be poured through sieves, with the mesh slightly submerged.

Extraction of large nematodes (Furlanetto *et al.*, 2012)

For improving efficiency in the extraction of *Tubixaba tuxaua* (Aporcelaimidae) and other larger nematodes, the authors recommend 100-200 g aliquots of undisturbed soil blocks.

The procedure is as follows:

Using a 5-l bucket or beaker, place the soil aliquot in the bucket and add water until the volume reaches 3 l (soil must be handled gently).

Shake the suspension vigorously by hand in a circular pattern, and after total suspension of the soil particles is achieved, immediately pour the supernatant into a set of 10-70-mesh aperture sieves, and collect the supernatant in a beaker (do not let it settle, as the nematodes remain suspended for only a few seconds).

After collecting the suspension, it is possible, with the aid of a fine pin or needle, to collect the individuals directly from the suspension for subsequent morphological and morphometric analyses.

23.4.1. Modifications to Nematode Extraction and Diagnostic Techniques

Modifications have been made to some nematode extraction techniques (see section 23.4) and have been adopted by PQS to enable inspection when the amount of available germplasm material is reduced or to maximize analysis results. Such modifications include:

Addition of 2% H_2O_2 to the water used to treat plant extraction material; this improves the movement of nematodes after rehydration of seeds or plant tissue due to the increased oxygen in the water solution. With this modification, the fermentation of plant material is reduced.

Plant growth in a germination chamber under quarantine greenhouse conditions. The germination chamber is made of plastic and foam to maintain approximately 100% humidity before nematode extraction in the laboratory.

Smaller trays to extract nematodes from seeds. Smaller trays measuring ca 30 × 20 × 6 cm are used in this technique (see section 23.4 Tray technique). A mesh screen and sheet of paper towel are placed in this order to line the bottom of the tray; shredded seeds are then placed on top of the paper towel, soaked with distilled water, to extract the nematodes.

Fenwick flotation-elutriation technique adapted to extract cyst nematodes associated with seeds. Seeds are placed inside a volumetric flask containing enough water to cover them and agitated for 10 min. The liquid is poured and sieved simultaneously through a set of combined 100-, 200- and 300-mesh sieves to collect the nematodes.

Pre-soaking seeds prior to blender trituration. Used for extraction of vermiform nematodes in anhydrobiosis by soaking the seeds for 16 h before trituration; pre-soaking promotes nematode rehydration and initiates nematode activity.

Treatment of seeds before planting for quarantine analysis. Seeds that have been treated with chemicals have to be rinsed with water before planting in the quarantine greenhouses in order to reduce interference of chemicals in the nematode analysis.

23.5. Species Identification

International standards related to the diagnosis and precise identification of plant-parasitic nematodes have been developed, but as a wide range of species needs to be identified, local facilities and expertise are valuable assets. To identify intercepted nematodes, a combination of publications available on morphological characters and biochemical methods is used, including isoenzyme and Polimerase Chain Reaction (PCR) molecular methods, e.g., Random Amplified Polymorphic DNA (PCR-RAPD), Restrictive Fragment Length Polymorphism (PCR-RFLP), Amplified Fragment Length Polymorphism (PCR-ALFP) (Andrade *et al.*, 2009), which are especially used for recovered juveniles. DNA samples stored in the laboratory have also been evaluated using the PCR technique, with the use of primers as Ribo1/Ribo 2, Universal (Vrain *et al.*, 1992) and DitNF1/rDNA2, specific for *D. dipsaci sensu stricto* (Subbotin *et al.*, 2005). These molecular techniques are used to identify genera and species of nematodes, based on the list of absent and present nematodes from MAPA.

Due to the increasing demand for quality, it was necessary to implement a standard procedure in Embrapa, related to the accreditation and traceability of results. Hence, from 2005 onwards, some of the laboratories of Embrapa-Cenargen, which included the Quarantine Laboratory, implemented a quality system that was sufficient to meet trade requirements, the purpose being to obtain the certification NBR ISO/IEC 17.025 and laboratory technical competence. Both certifications are granted by Brazilian Agencies: National Institute of Metrology, Quality and Technology (INMETRO) (https://www.gov.br/inmetro/pt-br).

The Nematology Laboratory of Plant Quarantine of Embrapa, located in Brasília, DF, has been officially acknowledged by INMETRO for its excellence in the identification and diagnosis of nematodes in national and international forums, for the provision of specialist technical and research services and the development of Embrapa.

23.6. Phytosanitary Control Programmes in Brazil

When interceptions or outbreaks occur, decisions have to be made according to the PRA as to whether destruction, eradication or suppression treatments are applied. Another important step is the eradication treatment, ensuring that the essential plant germplasm material is sent to plant breeders located in different states of Brazil, clear of any pest organism that could spread or disrupt their studies. Whenever contaminated accessions of plant germplasm are found, they are subjected to thermal treatment (dry or wet heat) or tissue culture, when necessary, for nematode elimination. For eradication, the treatment used is thermotherapy with wet heat: 40°C/12-30 min followed by 55-60°C/8-10 min. When it is not possible to treat the germplasm material, it is incinerated, and a new search begins to find non-contaminated material from another country (Tenente *et al.*, 1994, 1996a, 1999b).

During the time that PQS of Embrapa has provided support to MAPA for plant germplasm analysis, between 20,000 and 30,000 accessions/year were analysed in the Nematology Laboratory. The nematode detection rate was 1-7.3% of the yearly accessions, and these came from different countries.

23.7. Reducing Risk of Infestation in Regional or National Stocks in Brazil

Certification and marketing schemes are administrative systems for phytosanitary and quality control of commercially produced plant material. They can be applied to both the production of true seed and vegetative propagation. While some schemes are more concerned with genetic purity and related aspects, such as germination potential, most include pest-free aspects, such as *Ditylenchus dipsaci*. The phytosanitary analysis of plant germplasm, developed by the Quarantine Station of Embrapa, has actively reduced the risk of introducing new plant-parasitic nematode species or races into Brazil, and the treatments for nematode eradication from the imported germplasm have, to date, not caused serious damage to the seed germination and vigour of these plant materials.

23.8. Development of Expert Skills and Systems to Support the PQS Nematology Laboratory of Embrapa-Cenargen

Computer (IT) support is vital to organize accurate databases and make them available to people interested in a specific subject. In this context, via a computer system, a database of germplasm nematode analysis of plant materials imported into Brazil, and based at Embrapa Genetic Resources and Biotechnology (Brasília, Brazil), was developed. Relevant information and references related to the nematology unit can be found at http://pragawall.cenargen.embrapa.br/aiqweb/nemhtml/nempublic01_1.asp.

The study for the potential establishment of plant-parasitic nematode species in Brazil includes some of the nematodes placed in the quarantine list of MAPA and is of great help in nematological analysis. In this context, to obtain updated data, the current nematologist in PQS of Embrapa must be consulted.

Some databases are not publicly available, but they give essential support in quarantine analysis, such as the Database of Nematode Analysis (Tenente *et al.*, 1981), which is only available for Embrapa staff and is only possible to consult through the person in charge of nematode analysis at Embrapa Quarantine Facility.

Another database, 'The Expert System' (ES), was created to help users to identify certain nematode species, although the system is no longer in use.

23.8.1. Development of a System to Capture and Store Nematode Morphological Images

In March 2006, the development of a system to store nematode images as part of the Pest Image Database Collection was started, the purpose being to improve the reference nematode collection (quarantine pests for Brazil). The system stored and organized all the images of nematodes detected in the Quarantine Laboratory at the time. It also helped to detect species of *Ditylenchus* from collected and exchanged plant materials, and ca 17 *Ditylenchus* species were included in the database (Tenente, 1985; Tenente and Manso, 1987; Tenente, 1996; Tenente *et al.*, 2007; Gonzaga *et al.*, 2008).

23.9. Conclusions

Many of the plant-parasitic nematodes associated with genetic plant material can cause damage to important crops in agroecosystems or to plants in natural environments, leading to adverse economic and environmental impact. International collaboration is essential to develop IPPC standards in plant health and improve the skills of scientists in implementing legislation. In Brazil, the increasing trade of plant products is of concern to the NPPO, which in turn is under the auspices of the Secretary of Agriculture and Livestock Defence of the Ministry of Agriculture, Livestock and Food Supply. Another concern of the NPPO is with the Brazilian agribusiness, which has increased tremendously over the last two decades (https://www.gov.br/agricultura/pt-br).

Plant quarantine plays an important phytosanitary role against exotic pests and assists in plant protection of a region, an area or a country. In Brazil, germplasm plant quarantine, for research purposes, is managed by Embrapa Genetic Resources and Biotechnology, a research unit of the Brazilian Agricultural Research Corporation (Embrapa), which received permission from the Ministry of Agriculture, Livestock and Food Supply on 3 May 1977 and reaffirmed its position on 15 February 2002 (Oliveira *et al.*, 2003). Since its creation, more than 700,000 accessions of germplasm have been exchanged and analysed. Important quarantine pests have been intercepted and identified, and, to date, pests have not escaped or been introduced in the country through the exchange of plant germplasm, thus emphasizing the importance of plant quarantine to the country.

23.10. References

Andrade, E.P., Cordeiro, M.C., Gonzaga, V. and Cares, J.E. (2009) Molecular characterization of populations of *Pratylenchus* spp. through PCR-RFLP. *Nematologia Brasileira* 33, 292.

Carvalho, J.C., Lordello, L.G.E. and Boock, O.J. (1953) Considerações acerca do nematóide dourado da batatinha. *O Biológico* 19, 196–200.

Cobb, N.A. (1918) Estimating the nema population of the soil. *Agricultural Technical Circular, Bureau of Plant Industry, No. 1*. USDA, Washington, D.C.

Ebbels, D.L. (2003) *Principles of Plant Health and Quarantine.* CAB International, Wallingford, UK.

EPPO. (1990) Global database – formal establishment of COSAVE. *EPPO Reporting Service* no. 05 – 199 Num. article: 1990/04. https://gd.eppo.int/reporting/article-5604 (accessed 27 September 2024).

Evans, K. (1979) Nematode problems in the Woburn key-arable experiment, and changes in *Longidorus leptocephalus* population density associated with time, depth, cropping and soil type. *Rothamsted Experimental Station Report for 1978*. Part 2, 27–45.

FAO (1996) *Guidelines for Pest Risk Analysis*. Secretariat of the International Plant Protection Convention of the Food and Agriculture Organization (FAO) of the United Nations. ISPM Publication No 2, FAO, Rome, Italy.

FAO (2006) Glossary of phytosanitary terms. *Secretariat of the International Plant Protection Convention of the Food and Agriculture Organization (FAO) of the United Nations.* ISPM No 5. Secretariat of the International Plant Protection Convention, Rome, Italy.

Flegg, J.J.M. (1967) Extraction of *Xiphinema* and *Longidorus* species from soil by a modification of Cobb's decanting and sieving technique. *Annals of Applied Biology* 60, 429–437.

Furlanetto, C., Enrique, C.J., Ferreira da Cruz, M.I. and Guimãraes, R. (2012) Development of winter cover crops in soil infested with the nematode *Tubixaba tuxaua* in Western Paraná, Brazil. *Nematropica* 42, 314–319.

Gonzaga, V., Tenente, R.C.V., Rissoli, V.R.V., Conceição, F.N.B. da and Cares, J.E. (2008) *Rastreabilidade em Banco de dados sobre Nematóides Interceptados pela Estação de Quarentena Vegetal da Embrapa em Culturas Envolvidas na Produção de Agroenergia*. Embrapa Recursos Genéticos e Biotecnologia, Comunicado Técnico 185, Brasília, Brazil.

Hockland, S., Prior, T., Stanley, J.D., Inserra, R.N. and Kohl, L.M. (2024) International plant health – putting legislation into practice. In: Perry, R.N., Moens, M. and Jones, J.T. (eds), *Plant Nematology*, 3rd edn. CAB International, Wallingford, UK, pp. 327–345.

Jenkins, W.R. (1964) A rapid centrifugal-flotation technique for separating nematodes from soil. *Plant Disease Report* 48, 692.

Machado, A.C.Z., Amaro, P.M. and Silva, S.A. (2019) Two novel potential pathogens for soybean. *PLoS One* 4(8): 1–13.

MAPA (2016) Instrução Normativa n° 29, MAPA, Agosto 24, 2016 (DOU 25/08/2016). https://www.gov.br/agricultura/pt-br/assuntos/sanidade-animal-e-vegetal/sanidade-vegetal/arquivos-quarentena/IN29_2016_atualizada662019.pdf/view.

MAPA (2018) Instrução Normativa n° 38, MAPA, Outubro 01, 2018 (DOU 04/10/2018). https://www.gov.br/agricultura/pt-br/assuntos/insumos-agropecuarios/insumos-agricolas/sementes-e-mudas/publicacoes-sementes-e-mudas/INN38de1deoutubrode2018PQP.pdf.

MAPA (2022) Portaria n° 617, MAPA, Julho 12, 2022 (DOU 12/07/2022. https://www.gov.br/agricultura/pt-br/assuntos/sanidade-animal-e-vegetal/sanidade-vegetal/analise-de-riscos-de-pragas-arquivos/copy_of_PRAGASQUARENTENRIASAUSENTES_SITEMAPA_25.7.2022.pdf.

Marques, A.S.A. and Marinho, V.L.A. (2007) Movimentação de Germoplasma Vegetal no Brasil – Intercâmbio e Quarentena. In: Nass, LL. (ed), *Recursos Genéticos Vegetais*. Embrapa Recursos Genéticos e Biotecnologia, Brasília, Brazil, pp. 145–168.

Oliveira, M.R.V., Martins, O.M., Marinho, V.L.A., Mendes, M.A.S., Fonseca, J.N.L., *et al.* (2003) *O mandato da quarentena vegetal da Embrapa Recursos Genéticos e Biotecnologia*. Serie Documentos, Embrapa Recursos Genéticos e Biotecnologia, Brasília, DF, Brazil.

Rocha, H.M., Fonseca, J.N.L., Tenente, R.C.V., Urben, A.F., Costa Manso, E.S.B.G., *et al.* (1984) *Quarentena de Pós-entrada de Germoplasma Para Pesquisa. Anais do III Encontro Nacional de Fitossanitaristas. Florianópolis, Santa Catarina, Brazil,* 1984, pp. 159–168.

Silveira, S.G.P. and Curi, S.M. (1983) Incidência do nematóide *Aphelenchoides besseyi* Christie em sementes de arroz no Est. de São Paulo em 1981. Publication No 7, *Sociedade Brasileira de Nematologia*, Piracicaba, Brazil, pp. 91–98.

Southey, J.F. (ed) (1986) *Laboratory Methods for Work with Plant and Soil Nematodes*, 6th edn. Her Majesty's Stationery Office, London, UK.

Subbotin, S.A., Madani, M., Krall, E., Sturhan, D. and Moens, M. (2005) Molecular diagnostics, taxonomy, and phylogeny of the stem nematode *Ditylenchus dipsaci* species complex based on the sequences of the internal transcribed spacer r-DNA. *Phytopathology* 95, 1308–1315.

Tenente, R.C.V. (1985) Inspeção e quarentena de germoplasma vegetal importado. *Nematologia Brasileira* 9, 53–61.

Tenente, R.C.V. (1996) Nematode problems of bulb, with special reference to *Ditylenchus dipsaci*. *Nematropica* 26, 91–99.

Tenente, R.C.V. and Gonzaga, V. (1999) Táticas quarentenárias como medida preventiva de controle de doenças de plantas. *Revisão Anual de Patologia de Plantas* 7, 259–270.

Tenente, R.C.V. and Manso, E.S.C. (1987) Nematóides em sementes. In: *Patologia de Sementes*. Fundação Cargill, São Paulo, Brazil, pp. 107–145.

Tenente, R.C.V., Bettiol, E.M. and Carvalho, E.R. (1981) *Bibliografia Brasileira de nematóides*. Embrapa, Brasília, Brazil.

Tenente, R.C.V., Costa Manso, E.S.B.G., Mendes, M.A.S. and Marques, A.S.A. (1994) Seed health testing for nematode detection and treatment of plant germplasm in Brazil. *Seed Science Technology* 22, 415–420.

Tenente, R.C.V., Manso, E.S.C. and Figueira Filho, E. (1996a) Inspeção e detecção de fitonematóides em introduções de germoplasma no Brasil no período de 1992–1994. *Nematologia Brasileira* 19, 25–28.

Tenente, R.C.V., Manso, E.S.C., Mendes, M.A.S., Marques, A.S.A. and Figueira Filho, E. (1996b) Quarantine detection of nematodes and procedures for their eradication from vegetative materials imported by Brazil. *Nematropica* 26, 187–191.

Tenente, R.C.V., Gonzaga, V., Fonseca, J.N.L. and Bustamante, P.G. (1999a) Intercâmbio de germoplasma de cucurbitaceas. *Horticultura Brasileira* 17, 6–9.

Tenente, R.C.V., Gonzaga, V., Pinheiro, F.P., Rodrigues, V. and Tarchetti, P. (1999b) Eradication technique of plant parasitic nematodes from infested maize, oat and rice seeds. *Nematropica* 29, 17–24.

Tenente, R.C.V., Gonzaga, V., Melo, L.P. and Tenente, M.S.M. (2002) *Bibliografia Brasileira de Nematóides*, 2. Embrapa, Brasília, Brazil.

Tenente, R.C.V., Rissoli, V.R.V., Ferreira, D.N.M., Cares, J.E., Gonzaga, V., *et al.* (2007) *Banco de Imagens de Praga. Identificação de Ácaros e Nematóides de Expressão Econômica e Quarentenária para a Agricultura Brasileira*. Embrapa Recursos Genéticos e Biotecnologia, 2007, Comunicado Técnico, 155, Brasília, DF, Brazil.

Vrain, T.C., Wakarchuk, D., Lévesque, A.C. and Hamilton, R. (1992) Intraspecific rDNA Restriction Fragment Length Polymorphism in the *Xiphinema americanum* group. *Fundamental and Applied Nematology* 15, 563–573.

Whitehead, A.G. and Hemming, J.R. (1965) A comparison of some quantitative methods of extracting small vermiform nematodes from soil. *Annals of Applied Biology* 55, 25–38.

Whitehead, A.G. and Hooper, D.J. (1970) Needle nematodes (*Longidorus* spp.) and stubby-root nematodes (*Trichodorus* spp.) harmful to sugar beet and other field crops in England. *Annals of Applied Biology* 65, 339–350.

Zuckerman, B.M., Mai, W.F. and Harrison, M.B. (1990) *Plant Nematology: A Laboratory Manual*. University of Massachusetts Agricultural Experimental Station, Amherst, Massachusetts.

24 Statistics in Nematology

Stephen J. Powers*
Stats Powers Ltd, South Petherton, Somerset, UK

Abstract
The importance of statistical practice in nematological experimentation is explored. The appropriate design of experiments is an essential, and all too often overlooked, area when attempting to elucidate and quantify nematode-induced problems to plants. The chapter introduces topics such as design of nematology experiments, data collection and spreadsheet design, data analysis and statistical 'dos and don'ts'. Rather than being a comprehensive text on the subject of statistics, the purpose of the chapter is to provide an overview of important guidelines together with particularly relevant aspects that may be encountered during the course of nematological studies. Throughout this chapter, the aim is to provide information that encourages the correct application of statistics, and to foster an awareness of the importance of statistical science at all stages of an experiment, although the information presented should in no way detract from the potential benefit gained by involving the input of a trained statistician.

24.1. Introduction

In this chapter, the importance of statistical practice in nematological experimentation will be explored and discussed. Rather than being a comprehensive text on the subject of statistics, for which particular books for the reader's further investigation are referenced, important guidelines are given here, particularly relevant aspects that may be encountered in nematological studies. Throughout this chapter, the aim is to give information that encourages the correct application of statistics and to foster an awareness of the importance of statistical science at all stages of an experiment. The information here should in no way detract from the potential benefit of the involvement and input of a trained statistician. For an introductory text to statistics as applied to the general field of agricultural biology, the author recommends Mead *et al.* (2003).

As in any field of research where comparative experiments are designed to answer specific questions, experiments in nematology require careful planning at the design stage. Application of statistics can be vital in the assessment of adequate replication for an experiment, while planning for the economical use of resources. The application at this stage can also use available

*E-mail: powerssj@gmail.com
A revision of Powers, S.J. (2012) Statistics in nematology. In: Manzanilla-López, R.H. and Marbán-Mendoza, N. (eds), *Practical Plant Nematology*. Mundi-Prensa and Biblioteca Básica de Agricultura, Montecillo, Mexico.

(eds R.H. Manzanilla-López, D.J. Hunt, N. Marbán-Mendoza and S. Infante Gil)
DOI: 10.1079/9781836990413.0024

information derived from an initial 'pilot' study. The design of the experiment should be statistically sound, involving the 'three *statistical* Rs': Replication, Randomization and Blocking – that one of them does not begin with 'R' is an *aide-memoire* to these most important principles of experimental design. These concepts will be discussed with reference to examples in nematology. It is vital that an experiment is designed so that the resulting data acquired from it can be analysed to suggest answers to the questions posed by the researcher. At this point, it is noted that the results of an experiment can only ever provide *evidence*, not proof, of a hypothesis posed.

The measurements to be taken from an experiment should also be decided at the design stage. It is important to include measurement of all potential variables that may be affected by treatments imposed in a comparative experiment, although obviously within the usual constraints of time and money. It may also be useful, particularly in field trials, to record any environmental characteristics that may interact with the treatment effect. The efficient and accurate recording of data is also vital to subsequent statistical analysis. For any experiment, thought must be given to designing a spreadsheet to record the data, a spreadsheet that is set up in the same order as the design layout of the experiment. This is often an overlooked issue but is a vital one to avoid mistakes that can occur during extensive manipulation of data, the setting up of indicator variables (design and treatment factors) and so on.

The type of data (i.e., continuous or discrete) and the statistical distribution that may be assumed for the data collected from an experiment allow a decision to be made about the type of analysis that is appropriate. This analysis could comprise a simple statistical test or, if appropriate, more complex linear modelling of the data, invoking analysis of variance (ANOVA) or regression including, as appropriate, generalized linear modelling (GLM).

The results of any analysis must then be interpreted to propose a biological meaning. A statistical test is notorious for providing a 'p-value', this being, in a common situation, *the probability that the difference between two treatments arose just by chance given the evidence (data) and a number of assumptions made in applying the test.* However, it is the size of the difference between the two treatment means that is also important, and the variability of that difference, known as the standard error of the difference (SED). There is statistical significance, measured by the p-value, and there is biological significance, measured by whether the size of the difference is meaningful.

This chapter is therefore split into four major sections. The first considers aspects of statistical design of experiments relevant to nematology. The second concerns collection of data and spreadsheet design. The third explores the possible analyses that can be applied to data arising. The fourth section takes the form of a list of 'statistical dos and don'ts', or else may be seen as answers to frequently asked statistical questions.

24.2. Design

Useful books on the design of experiments are Mead (1988) and Cochran and Cox (1957).

24.2.1. Treatments

The first statistical (and biological) question asked before embarking on a design for a comparative experiment: What *treatments* are to be used? In statistical terms, 'treatments' is the name given to the 'things' to be compared. For example, an experiment may be required to compare treatments being types of plant such as wheat, tomato or potato. Alternatively, treatments may be species and cultivars of a particular type of plant, such as 'Maris Piper', 'Desiree' and 'Sante' potato cultivars, all to be subjected to infection by potato cyst nematode (PCN) in order to compare performance. In the context of wanting to compare fungal biocontrol agents of nematodes, the treatments might be species of fungus such as *Paecilomyces lilacinus* and *Pochonia chlamydosporia* for a given species of nematode (e.g., *Meloidogyne incognita*). When there is more than one set of treatments to compare, for example, a number of tomato cultivars each at a number of different temperatures (perhaps with a number of nematode species), then the experiment is termed *factorial* and comprises a set of *treatment combinations*. The individual sets of treatments are termed *treatment factors*. The benefit of using a full set of all combinations of treatment factors (with or without added control treatments) is in the ability this gives to assess the effect due to the separate factors independently, i.e., the *main effects, and* the interaction between them, i.e., synergistic effect of the treatment combinations.

24.2.2. The Experimental Unit

The second statistical question: What is the *experimental unit*? By definition, the experimental unit is the *independent* structure to which a single treatment is applied in an experiment. In field trials this will be a plot, but in glasshouse or controlled environment studies with nematodes in soil or other media, the experimental unit will likely be a pot or a soil bag. In laboratory experiments, the experimental unit may be a Petri dish, with a number of treatments being applied to different dishes, or perhaps a single well in a 96-well plate, to which a given dose of some substance is applied to nematodes in a media. Experimental units *must* be independent from one another: two separate, independent experimental units (plots, pots, Petri dishes, wells, etc.) applied with the same treatment are termed (biological) replicates. Multiple observations taken from the same experimental unit at the same time are termed *technical replicates* or *pseudo-replicates*. Variation in technical replicates cannot be used to assess the differences between treatments, as they are not from independent visualizations of the effect of a given treatment. It is noted that multiple observations from the same experimental unit over time form a time series of repeated measures. Here again, multiple experimental units need to be observed over time for a given treatment, in order to have replication.

24.2.3. Replication

The third statistical question: How many replicates do I need? To answer this question, prior to designing the full experiment, it can be very important to do a 'pilot' study, especially if little or nothing is known about the variance that may be expected to be seen across a number of experimental units applied with the same treatment. Obviously, whether it is practical for a pilot study to be done will depend on resources. However, a pilot study need not use all the treatments proposed for the experiment, but must have replication, so that at least some idea of expected variation can be acquired. Here we note that variance is a value calculated as the sum of squared deviations from the overall mean of the set of data and divided by (n −1), this being the so-called *degrees of freedom*. This name comes from the fact that the deviations must sum to zero, since they are deviations from the mean. So, in theory, (n − 1) deviations can be chosen freely, whereas one of them must make all of them collectively sum to zero. This imposes *one* restriction on the calculation of the variance given n observations, so that there are (n − 1) degrees of freedom.

In other situations, the variance observed in previous, similar experiments where a certain number of replicates were used may help determine the number of replicates for new experiments. Without some prior knowledge of the variance that might be observed, the estimation of the number of replicates required cannot be done with any great confidence, beyond stating the usual necessity for three replicates as a minimum in experiments with three or more treatments, and four-to-five replicates as a minimum in experiments with only two treatments. These minimum requirements are based on the theory of Student's t-distribution, this being the theoretical statistical distribution of data sampled from the bell-shaped Normal distribution. The t-distribution is also bell-shaped, but with fatter tails depicting the lack of knowledge about the variance. The distribution is defined in terms of the mean and, most importantly, in terms of the variance of the data, thus invoking degrees of freedom. There is, therefore, a theoretical t-distribution for any degrees of freedom, 1, 2, 3, etc. However, after about 60 degrees of freedom the t-distribution rapidly approaches the Normal distribution. So, assuming that data to be analysed from the experiment *will be* from a Normal distribution, the degrees of freedom for the underlying (residual) variance should ideally be sufficient, i.e. six or more, to give a reliable statistical test (in this case the t-test), as will be explained below.

The theory of the t-distribution can be used to estimate the replication required in an experiment given three pieces of information: (1) the difference in treatment means to be detected, d, i.e. the size of the 'signal'; (2) the measure of overall residual variation, i.e. the 'noise' expected, s^2, i.e. the proposed standard deviation and (3) the associated probability at which significance is desired, α, usually 0.05. Clearly, it is s^2 which may be the unknown quantity here, for which information from previous experiments may be helpful.

Next, a suitable critical value from a t-distribution, for α = 0.05 (5%), is required – see, for example, Appendix C of Gomez and Gomez (1984). These critical values decrease less rapidly after about 6 degrees of freedom (2.45) than they do for values up to 6 degrees of freedom. So, the value 2.45 as the critical t-value may be taken for initiation of sample size calculation, this being the t-value on 6 degrees of freedom at α = 0.05 (5%) level of significance for a t-test. This would automatically imply a minimum replication of four, because, with two treatments, eight observations would be made, two means and variances calculated, and hence six degrees of freedom would remain for residual.

For simplicity here, it is assumed that the variances for the treatments are equal, and that equal replication is required for the treatments. Unequal replication for individual treatments can always be considered in the light of an initial calculation assuming equal replication. Starting with the difference between the sample means, d, divided by its *standard error* (SE),

$$\sqrt{\frac{2\times s^2}{n}},$$

where s^2 is the proposed variance and the n is the replication, the t-statistic (see, e.g., Snedecor and Cochran, 1989) is given by:

$$t = \frac{d}{\sqrt{\frac{2\times s^2}{n}}}$$

on 2(n − 1) degrees of freedom (i.e., for two treatments). Rearranging this for n, we obtain,

$$n = \frac{2\times t^2 \times s^2}{d^2}$$

which gives us a formula for the sample size required for significance of a given difference between means, d, a variance, s^2 and the critical value of the t-distribution, t (here we are using 2.45).

Now, having plugged all the required values into the formula, and obtained a first estimate of n, a total of 2 × n observations would be required and the degrees of freedom for a t-test in this case would be 2 × (n − 1). So, a *new* α = 0.05 (5%) critical t-value on *these* 2 × (n − 1) degrees of freedom can then be used in the formula, and so a re-calculation of our estimate for n can be made. It is clearly possible to iterate a few times in this way until convergence to a final value of n, the proposed replication to use. This straightforward, 'back-of-an-envelope' method may be fine for calculating replication in the situation of two treatments, and indeed for giving a reasonable idea of replication when there are more than two, but to be more scientific about the calculation, the issue of *power* needs to be included. In simple terms, the power is the probability that the difference, d, *will* actually be detected as significant at a specified level, such as α = 0.05 (5%). The method above assumes a power of only 50%, so generally, if a higher power is required then the method is clearly underestimating the replication. Usually, requiring a power of 80%, or 90%, is sensible, and it is recommended that a statistical package, such as GenStat® (VSN International Ltd., Hemel Hempstead, UK), Minitab® (State College, Pennsylvania, USA) or R (R Core Team; www.R-project.org) is used to enable specification of a specific power and hence appropriate estimation of replication. The assistance of a trained statistician can also be very useful here.

Obviously, large experiments may be limited by time and money to a number of replicates that is economical. In any case, with more and more treatments the degrees of freedom for the residual increase rapidly. This is demonstrated in the following example.

Suppose a simple experiment using second-stage juvenile nematodes (J2) of three different species in Petri dishes to test the effect of five different media on the potency of a biocontrol agent. Fifty J2 are used in each dish and the numbers not moving (dead) after 24 h are to be recorded. If three replicates are proposed for this three-by-five factorial experiment, there will be 45 observations altogether across the 15 treatment combinations. It is noted that in this experiment the Petri dish is the experimental unit, i.e. the level at which biological replication is seen. If they were assessed, individual J2 nematodes would be technical replicates. This is because nematodes within dishes are not independent, whereas separate dishes are. The variance for the data will be on 44 degrees of freedom. Moreover, there will be 14 degrees of freedom for the variance due to treatments. This variance is derived as the sum of squared deviations of the 15 treatment means from the (overall) grand mean divided by 14, but note that this variance can be partitioned into three portions: the variance due to the main effect of nematode species, that due to the main effect of media, and that due to the interaction between these two factors. In an analysis of variance (ANOVA) therefore, there are 44 − 14 = 30 degrees of freedom for the underlying residual variance (s^2). So, this many degrees of freedom should be ample with which to test the significance of differences between treatment combinations. However, this is not to say that use of four, five or more replicates might be better if the underlying residual variance, s^2, is expected to be great – i.e., if each individual treatment or treatment combination is likely to have an inconsistent effect giving rise to three observations that vary wildly, then clearly three replicates are not enough to pin down the average effect of that treatment or treatment combination. It should be emphasized that cutting corners with replication can end up in the 'signal' being swamped by the underlying 'noise' in the experiment.

For experiments done involving nematodes in glasshouse, controlled environment and laboratory conditions, usually with more than two treatments, the minimum of three replicates is almost always imposed, and in some cases five replicates are used due to the known variation observed in certain areas of the nematological research – such as in dose-response studies using potential biocontrol chemicals.

A further issue of design that arises in the example above, and relevant to many experiments with nematodes, is how many nematodes should be used in an experimental unit, i.e., Petri dish, single Y-tube assay, well, etc.? The number may usually be related to the numbers available at the time of the experiment and to the maximum that can quickly and easily be counted. In the example above, 50 is a *Binomial total* given that observations will fall into one of two groups (dead or alive). This was a reasonable number per dish, with 45 dishes to process in a limited time period. Also, an observed proportion of 25 nematodes not moving out of 50 is far more useful than an observed proportion of 1 out of 2. What should also be considered though is the number of nematodes that would naturally be in a given unit area (or volume), and, if sensible, extrapolate this to the area of the experimental unit. The biological issue here is that too many nematodes could lead to adverse effects due to competition between them. The statistical issue here is in the variation across the replicate observed proportions – too many nematodes in each experimental unit could easily contribute to incorrectly lowering the underlying variation between biological replicates, such that subsequent analysis highlights differences as statistically significant that are not biologically meaningful.

24.2.4. Randomization

The benefit of randomizing treatments to experimental units is in the avoidance of bias. In simple terms, if each experimental unit has the same chance as any other to receive a particular treatment, then bias has been avoided. Randomization disallows situations such as 'all the best-looking plants receiving the "most important" nematode species' and so on. Any one treatment should not have an unfair advantage over any other. Similarly, randomization ensures that all the pots of one treatment will not occur at one end of the bench in the glasshouse, and that all the Petri dishes of one treatment will not occur at one side of an incubator. There should be randomization in any experiment, regardless of whether it is deemed 'necessary' or not. The only occasion where randomization is not done would be where all experimental units *are* the same before applying the treatment – for example, in the wells of some 96-well plates loaded with exactly the same substance, and where randomization would make it virtually impossible not to make a mistake in administering the treatments.

A random allocation of treatments to experimental units does not, of course, prevent a treatment being allocated to 'better' units, giving an overestimate of that treatment's effect. Clearly, that could happen in an experiment, but over many experiments randomization ensures that comparison between treatments is fair, and allows a valid statistical analysis of the resulting data.

Most statistical packages (e.g., Genstat) allow experiments to be designed, and often lead the user through a series of menus to input the required information before a randomization is output. This is always better than a 'by-eye' randomization, which risks being biased – for example, the larger plants, or the aliquots with the most active nematodes, may always get picked first because they stand out. The randomization that the statistical package supplies should then be used without alteration; re-randomizing until a 'good-looking' design is achieved introduces bias. Also, the same design should not be used twice, e.g., for a repeat experiment, however tempting it may be to do so. This is because bias could then be introduced, even if the experimental units are a new set.

The simplest design for a comparative experiment with two or more treatments, or with two or more treatment factors, is called a *completely randomized design*. So, returning to the example above, with J2 of three nematode species with each of five media in Petri dishes, one version of the design of the experiment might be as in Figure 1. To be exact, this is called a *two-way completely randomized design*, the 'two-way' referring to there being two treatment factors. Designs are usually formed in rectangles, as this is the most compact way to arrange a set of any experimental units, but non-rectangular shapes will arise, if necessary, due to restrictions of space. However, all experimental units in a completely randomized design should be in the same 'environment'.

In this design, alphabetic and numeric labels are used to arbitrarily label the treatment combinations, and the Petri dishes (in statistical terms, the 'plots') are numbered 1 to 45, in row order. Here, assigning labels to the names of the species and the names of the media rather than using the actual names gives a much tidier way of physically labelling the Petri dishes. This system can save time in larger experiments, for example using pots with plastic labels, or 96-well plates where, in any case, there may be little room for labelling. Ideally, the actual names should be randomized to the labels. As well as being helpful for the accurate recording of data from the experiment into a spreadsheet (see main section below), numbering the Petri dishes is also worthwhile when a selection of them is to be removed for inspection, so that dishes can be returned to exactly the same spot in the design. This is essential, for if time has been taken to construct a design for experimental units, then they must be retained in that design over the full course of the experiment. Ideally, measurements should be taken in a strict order, row by row, or column by column, rather than from all the experimental units pertaining to the first treatment, then from all those pertaining to the second treatment and so on.

In situations where there is a suspected trend across the area allocated for a design, whether that area is in a glasshouse, a laboratory bench, the shelves of an incubator, or those of a refrigerator, the completely randomized design is inappropriate. However, such trends can be accounted for in designing the experiment, by blocking.

24.2.5. Blocking and More Complex Designs

In nematology, a typical situation where blocking is required would be in a field experiment where there are known underlying trends in the soil structure or other characteristics across the field proposed for the experiment, but where it is possible to prescribe that certain areas can be marked out as having similar soil structure and characteristics within each of them. The experimental units in this case are plots in the areas, and a design can be derived, known as a *randomized block design*, if each area can contain as many units (plots) as there are treatments, or treatment combinations in a two-way or multi-way treatment structure. The treatment combinations are then randomized within each of the blocks. In statistical

1. E2	2. A1	3. A3	4. A3	5. B3	6. B3	7. E1	8. C2	9. D3
10. C1	11. A2	12. E3	13. C1	14. D3	15. B2	16. B1	17. B1	18. C3
19. D2	20. D1	21. E2	22. D2	23. B2	24. B2	25. D1	26. C2	27. A2
28. E2	29. A1	30. E1	31. C1	32. A1	33. B3	34. C3	35. C3	36. D2
37. D1	38. E1	39. C2	40. A3	41. E3	42. D3	43. B1	44. E3	45. A2

Fig. 24.1. A completely randomized designed experiment for comparing J2 of three nematode species (labelled 1-3) with each of five different media (labelled A-E) in Petri dishes, generated using the Genstat statistical package, selecting a two-way design (no blocking), choosing three replicates, one treatment factor with three levels and one treatment factor with five levels. The Petri dishes are numbered 1-45.

terms such a design is called *balanced*, since all comparisons of treatments can be made with the same precision, as they all occur together in every block. Importantly, under the assumption of additive effects for the blocks, in the ANOVA for data coming from a randomized block design, the variance due to the blocks can be removed from the residual variance, which is used in the comparison of treatments. The averages for the blocks can also be compared to see where the variance is coming from, i.e., which blocks are behaving differently from others. This allows an assessment of how good the blocking scheme was, and whether refinements are required for future experiments.

Blocks can also be formed for experiments in the glasshouse and controlled environment cabinet or room, where gradients are thought to exist. In these last two cases, the opening and closing of the door is a usual source of variance, suggesting that blocks should be formed, blocking away from the door. So, suppose in the example above, where J2 of three nematode species are to be subjected to a biocontrol agent in five different media with three replicates, the Petri dishes are to be placed for 24 h in a fixed-temperature incubator possessing seven shelves. Furthermore, suppose it transpires that 45 Petri dishes will not all fit on one shelf, but that 15, a complete set (i.e., replicate) of treatment combinations, will. It then makes sense to have three blocks, one on each of three shelves, with a complete set of 15 treatment combinations on each shelf. Clearly, to save space, it may have been possible to have two blocks on one shelf and the third on a second shelf, if 30 Petri dishes could be accommodated on one shelf. This too is fine, although the two blocks on the same shelf need to be separate. In other words, it is not correct to have a completely randomized design on the first shelf and a single block on the second.

A randomized block design for the experiment is given in Figure 2, as generated using the Genstat statistical package. By statistical convention, blocks are numbered using capital Roman numerals (I-III in this case). The Petri dishes within the blocks are numbered by a three-digit code, the first digit being the block, and the remaining digits being the dish number within block. Although blocks are preferably

I

101. D1	102. A1	103. B1	104. D2	105. E2
106. C2	107. B3	108. C1	109. E3	110. D3
111. A3	112. B2	113. C3	114. E1	115. A2

II

201. E3	202. B2	203. A1	204. A3	205. C1
206. A2	207. E2	208. B1	209. E1	210. D2
211. C3	212. C2	213. B3	214. D3	215. D1

III

301. D2	302. B2	303. A1	304. A2	305. D1
306. B3	307. E1	308. B1	309. E3	310. D3
311. C2	312. E2	313. C3	314. A3	315. C1

Fig. 24.2. A randomized block designed experiment for comparing J2 of three nematode species (1-3) with each of five different media (labelled A-E), generated using the Genstat statistical package, selecting a two-way design (in randomized blocks), choosing three blocks, one treatment factor with three levels and one treatment factor with five levels. The blocks are numbered I-III and the Petri dishes are numbered by a three-digit code, the first digit being the block, the second and third digits being the Petri dish within each block, in row order within blocks.

formed as rectangles for compactness, different shape blocks will be used so that the area within a given block is homogeneous – all experimental units within a given block must be subject to the same 'environment'. It should also be mentioned that there is still an issue with the experiment here, in that the same incubator is used for all three blocks. In other words, the results seen for this experiment are particular to the incubator used. This caveat is usually one which has to be swallowed, as clearly the constraints of time and money may not allow a number of incubators to be used in order to have blocks as incubators, as a first level of blocking, and perhaps blocks within each incubator as a second level of blocking.

It is important that data are recorded from the design block-by-block, taking complete blocks in turn. This is even more vital if the recording of data from the experiment is time-consuming, such that the last block is visited a long time after the first. A benefit of the design is that this variance due to a 'time-lag' can also be extracted, this variance being included in the overall block-to-block variance. Moreover, for large designs, if necessary, different researchers can be assigned different blocks, and all such variation can again be extracted at the analysis stage – the variation due to blocks therefore including the variation due to researchers.

A drawback to the randomized block design is that the area to be allocated to a block does need to be large enough to allow one experimental unit of each treatment in the design. However, there are so-called *balanced incomplete block designs*, where a different set of treatments are left out of each block across the design, but where the number of blocks used ensures balance in treatment comparisons overall. Finally, *unbalanced block designs*, where not all treatments can be compared with equal precision, can be used. However, these more complex block designs are generally only used in specific situations where a simpler design will not suffice.

In some experiments with more than one treatment factor, it may be necessary to use experimental units of different sizes for the treatment factors. Experiments in which one treatment factor is applied to larger experimental units (called *main plots*) and another treatment factor is applied to smaller experimental units (called *sub-plots*) within the main plots are called *split-plot experiments*. However, they should only be used if there are very good practical reasons. For example, suppose a field trial where the degree of nematode infection is to be compared for six different potato varieties treated with each of three nematicides. The machinery for planting the potatoes does not allow for very small plots, so a solution is to use a split-plot design and to apply the different nematicide treatments within main plots of each potato variety. This is done for each of a number of blocks, with randomization, to ensure adequate replication of all treatment combinations.

A version of the design layout for the example described above is given in Figure 3. It can be seen that three contiguous sub-plots comprise a main plot. By statistical convention, blocks are numbered using capital Roman numerals (I-III in this case). The sub-plots are numbered by a three-digit code, the first digit being the block, the second digit being the main plot and the third digit being the sub-plot.

The problematic issue with split-plot experiments is that random variation is likely to be different for plots of different sizes. For example, in Figure 3 the underlying variation for measurements made on the main plots will probably be larger than that for measurements made on the split-plots. This implies that any test for differences between main plot treatment means will need to use a different estimate of random variation than a test for differences between the means of the split-plot treatment. This is accounted for in an ANOVA of a split-plot experiment.

24.3. Data Collection and Spreadsheet Design

24.3.1. Accuracy in the Recording of Data

After deciding what measurement (count, length, weight, etc.) to take from the experimental units, be they Petri dishes, pots, plots, wells or whatever, the degree of accuracy to record data needs to be decided. Clearly, this is not an issue for counts, but for many measurements, such as weights and lengths, accuracy to two decimal places is usually sufficient, but can be more than sufficient. For example, if recording in centimetres (cm), an accuracy of one decimal place effectively gives precision to millimetres (mm)

I					
111. E2	121. C1	131. D1	141. A1	151. B2	161. F1
112. E3	122. C2	132. D3	142. A3	152. B1	162. F3
113.E1	123. C3	133. D2	143. A2	153. B3	163. F2
II					
211. F3	221. E3	231. A3	241. C1	251. B1	261. D2
212. F2	222. E1	232. A2	242. C3	252. B2	262. D3
213. F1	223. E2	233. A1	243. C2	253. B3	263. D1
III					
311. B3	321. F3	331. A1	341. C1	351. E3	361. D1
312. B2	322. F1	332. A2	342. C3	352. E1	362. D3
313. B1	323. F2	333. A3	343. C2	353. E2	363. D2

Fig. 24.3. A split-plot designed experiment in three blocks for comparing six potato varieties (A-F) with three different nematicides (1-3), generated using the Genstat statistical package, selecting a split-plot design, three blocks, the main plot treatment factor with six levels and the split-plot treatment factor with three levels. The blocks are numbered I-III, the main plots are indicated by the solid lines surrounding each set of three sub-plots and sub-plots are indicated by dotted lines. The sub-plots are numbered by a three-digit code, the first digit being the block, the second digit being the main plot and the third digit being the sub-plot within each main plot, in column order.

rather than cm, so using two decimal places actually gives precision to tenths of millimetres (hundredths of cm), which may seem far too precise for the particular experimental units being measured. It may be desirable to measure the length of nematodes to a high degree of accuracy in micrometres (μm) and plant roots in millimetres (mm) rather than micrometres. However, the plant roots would effectively be measured in micrometres too if the measurement is in millimetres to three decimal places.

With this in mind, the instrument being used to give the measurements should be considered too. For example, if a ruler is used to measure root lengths, then one is restricted to using the scale it provides (e.g., mm). However, more sophisticated instruments such as weighing balances and automatic data-recorders have the capability to store data, for subsequent download to a spreadsheet, to many decimal places. The documentation of these instruments should state the accuracy associated with their recording; for example, the accuracy may be ±1%. This information allows one to decide the degree of accuracy (decimal places) to keep for data taken from the instrument. Although an instrument may present the recorded data to eight or nine decimal places, this is spurious accuracy if the documentation says that the instrument records to ±1%. In this particular case, all values should have at most two decimal places. Also, fewer than this (one or no decimal places) would be used depending on what is being measured, e.g., the length of plant roots rather than the length of nematodes, as considered above.

A rule of thumb for the accuracy (decimal places) to record data is to use the SE of the overall (grand) mean of data (e.g., from a pilot study) divided by four. An alternative is not to use any more than one decimal place when the variability of data is 3% to 5% of the mean.

24.3.2. Accuracy and Precision

Note that accuracy should not be confused with precision. Accuracy concerns how far recorded measurements are from the expected, true measurements. Measurements are 'good' if they are accurate or free from bias. An instrument is accurate if it does not produce any systematic error, such as would occur, for example, in the case of a weighing balance with some unknown foreign body attached to its plate; the extra weight is then always included in the weight of anything being weighed. Precision concerns how repeatable a measurement is. The variance between technical replicates from the same experimental unit informs about the precision of the measuring instrument as well as other factors associated with the preparation of the technical replicates from the experimental unit.

24.3.3. Technical Replicates and the Travesty of Pooling

As a further note, the variability due to technical replicates should be less than the variation due to biological replicates, as a set of technical replicates are all from the same biological replicate whereas biological replicates should be more different from one another. Also, replicate samples from a pool of material cannot be seen as biological replicates, but are also technical replicates. Although separate pools of material (from separate sets of experimental units) will be biological replicates, the problem with pooling is that the variation due to separate entities (single experimental units, such as a single plant, or a single nematode cyst and so on) cannot be extracted. Statistical analysis can only be effectively undertaken when there are independent (biological) replicates. In ANOVA where technical replicates have been taken from biological replicates, the analysis must ensure that the residual variance is based on the variance and degrees of freedom due to biological replicates only, and does not include the variance and degrees of freedom due to the technical replicates.

24.3.4. Construction of Spreadsheets

There are three parts to describing an experiment in a spreadsheet before the data in it can be analysed using a statistical package such as Genstat or Minitab. First, the spreadsheet needs columns that indicate the design of the experiment; second, it needs columns that indicate the treatments applied and third, it needs columns that hold the recorded data. The distinction between these three types of columns is important. A spreadsheet should be made column by column. This is because the experimental design, treatments and recorded data can be stored far more efficiently in columns rather than rows, and this aids the use of the statistical package to analyse the data. Hence, all the information about a particular experimental unit should usually have just one row in the spreadsheet, the values along that row denoting, first, items about the design, second, the items about the treatment applied to it, and third, the observations recorded. If each experimental unit has multiple observations (i.e., technical or pseudo-replicates) then multiple rows should be used per experimental unit, with an extra column in the design part of the spreadsheet to indicate the number of each pseudo-replicate within the experimental unit.

The columns describing the design of the experiment indicate how sets of experimental units are grouped together in the actual layout in the field, glasshouse, controlled environment or laboratory. The information about the treatments applied is then recorded separately in more columns. Incredible quantities of time are saved through setting up and sticking to a spreadsheet for the designed experiment: if data are entered once and do not need subsequent manipulation, for example, from rows into columns, then analysis is facilitated and no mistakes are made due to multiple cutting and pasting, transposition and so on.

For example, suppose a simple fixed-temperature glasshouse experiment to consider the number of nematodes of a certain species invading roots of wheat plants in pots, for which four concentrations of a chemical biocontrol treatment (0, 50, 100, 250 units) are to be trialled with and without a water stress treatment (i.e., watering or no watering) for the plants. There are eight possible treatment combinations here (four chemical concentrations by two water condition levels). There are three replicates of each treatment combination. This gives a total of 24 possible observations (one per pot) of each of a number of desired measurements. Each pot has one plant and 100 J2 nematodes are added to it. After a period of time, the plants are sacrificed and the roots processed to count the number of nematodes that invaded, post inoculation. Each complete set of eight treatment combinations (pots) will be arranged on a glasshouse bench, three benches forming three statistical blocks, blocking away from the access door of the glasshouse.

So, for this randomized block design, there should be at least six columns in the spreadsheet: one to indicate all the pots (coded '11', '12', ..., '38') according to the blocks, i.e. the *experimental units*; one to indicate the *blocks* (i.e., bench 1, 2 or 3); one for the *pots* within each bench (1, 2...8); one for the *chemical treatment* (with levels 0, 50, 100 or 250) and one for the *drought treatment* (with two levels, watered or stressed). There will also be one column for the measurement (count of nematodes invaded) and probably other columns such as root length, root fresh weight, etc, as appropriate, that are taken from each pot. The spreadsheet is given in Figure 4, as constructed in Excel (© Microsoft Inc., Washington, USA). It has 24 rows.

The first three columns (Unit, Bench and Pot) relate to the design, are in a systematic nested order, and are in a fixed pattern according to the layout of the design. So, the pots would be numbered '11' through to '38', in strict order, row by row or column by column, on the benches, so that the position of each is correct according to the design. The combinations of levels of the two treatment factors are randomized to the pots within each bench, so the Chemical and Drought columns are in random order as assigned in the construction of the design. The information in this form, along with at least one other column for the measurements made on the individual pots, i.e. Count (of nematodes), can now be transferred to a statistical package for analysis.

The spreadsheet can be presented in an unrandomized order by sorting all the data in it on the treatment columns (Chemical and Drought). However, this would *only* be done *after* all data has been recorded in the spreadsheet, since the randomized order is the order in which the pots should ideally be processed. In sorting, the Unit, Bench and Pot columns must have their values sorted too, so that each measurement can be traced back to the particular bench and pot from which it derives.

Finally, although this spreadsheet has been constructed in Excel rather than in a spreadsheet facility of the statistical package to be used for analysis, this is solely because Excel is the package often used for storage and pre-processing of data. Its data analysis capabilities are somewhat limited though, so data can easily be imported from Excel into the statistics package. Obviously, a spreadsheet in the statistical package can be used directly for data entry, subsequent processing and analysis, if the user is more comfortable with this.

24.4. Analysis

Useful books on simple statistical analyses are Snedecor and Cochran (1989), Mead (1988) and Gomez and Gomez (1984). Having collected data from an experiment, the analysis can then be done. Initial data inspection should consider the variation of observations within treatments and what statistical distribution is appropriate for the data. Figure 5 shows a flow chart of statistical questions and procedures relevant for simple comparative experiments that may be encountered in nematology.

24.4.1. Initial Data Inspection

In statistical terms, data are either *continuous* or *discrete*. These terms relate to the type of statistical probability distribution which may be assigned. In the first case, the data are on a continuous numerical scale, e.g., dimensional measurements of individual nematodes, nematode eggs or cysts. In the second case, data are counts in categories, e.g., numbers of nematodes dead or alive, or numbers of live juvenile nematodes in each of the age categories J1 (first-stage juvenile), J2 (second-stage juvenile), J3 and J4 (third- and fourth-stage juvenile, respectively). Sometimes, discrete data can sensibly approximate a continuous distribution if the observations are seen to have a great enough range and sufficient variation. The shape and extent of variance in data for any particular variable measured can be seen by calculating summary statistics, per treatment and overall. Here, particular interest is focused on the maximum, minimum, mean, median and variance, to get a feel for the overall nature of the data set.

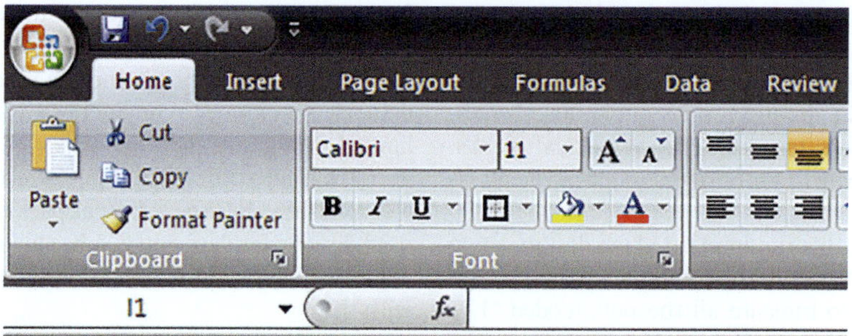

	A	B	C	D	E	F
1	Unit	Bench	Pot	Chemical	Drought	Count
2	11	1	1	250	Stressed	
3	12	1	2	0	Stressed	
4	13	1	3	50	Water	
5	14	1	4	250	Water	
6	15	1	5	50	Stressed	
7	16	1	6	100	Stressed	
8	17	1	7	0	Water	
9	18	1	8	100	Water	
10	21	2	1	250	Water	
11	22	2	2	100	Stressed	
12	23	2	3	0	Stressed	
13	24	2	4	50	Water	
14	25	2	5	250	Stressed	
15	26	2	6	0	Water	
16	27	2	7	50	Stressed	
17	28	2	8	100	Water	
18	31	3	1	250	Stressed	
19	32	3	2	0	Stressed	
20	33	3	3	250	Water	
21	34	3	4	100	Water	
22	35	3	5	0	Water	
23	36	3	6	50	Water	
24	37	3	7	50	Stressed	
25	38	3	8	100	Stressed	
26						
27						

Fig. 24.4. An example of how to lay out a spreadsheet for recording data from an experiment. Each row in the spreadsheet represents an experimental unit and the rows are in the order of the layout of the experiment itself, i.e., in the randomized order. The Unit, Bench and Pot columns describe the layout of the design while the Chemical and Drought columns describe the treatments. The Count column is ready for data entry, one observation per pot, in the sequence of the design layout. Note that the treatment columns have actual labels here, whereas in the design, codes (as in Figs 24.1-24.3) may have been used for efficiency.

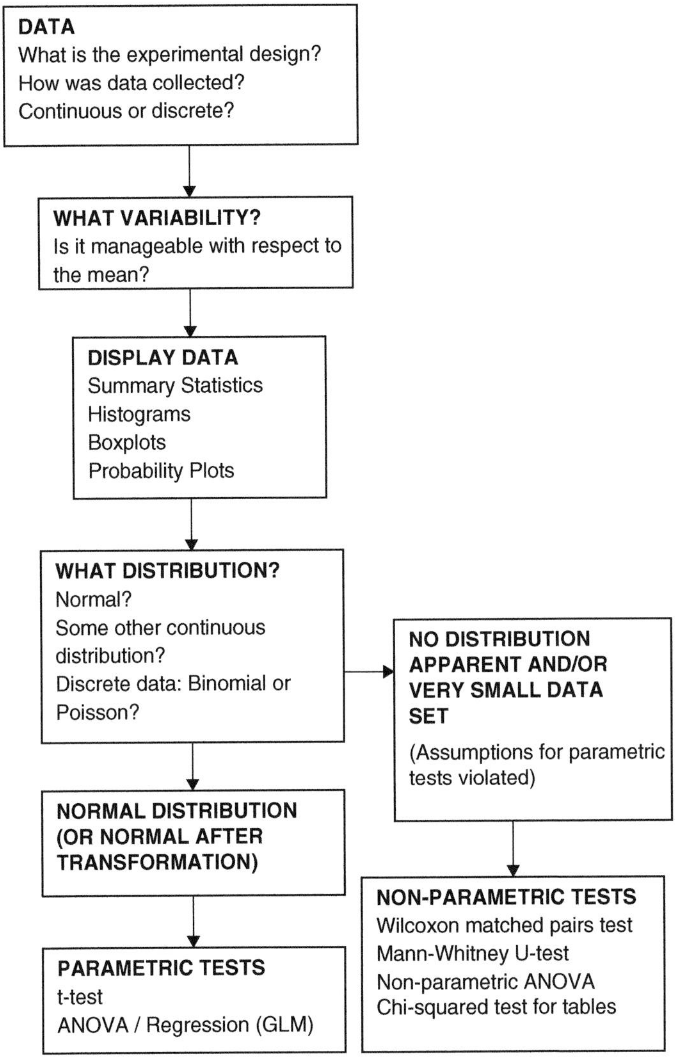

Fig. 24.5. Flow chart of initial data inspection and statistical testing procedure.

A histogram of the data is used to show its shape and spread across the possible values that the measurement may take, at least within the current experiment. In particular, it is the shape of the histogram that reveals what probability distribution may be appropriate. If the histogram is approximately symmetrical and bell-shaped with only one prominent peak, then it is likely that a Normal distribution will be appropriate. However, a skewed distribution will indicate the need for a transformation prior to analysis if the data are continuous, and similarly if they are discrete but, as mentioned above, have a great enough range and sufficient variation. For example, if the histogram of the data shows a positively skewed shape, then a log transformation is appropriate, or perhaps a square-root transformation if the log transformation is too strong and results in a negatively skewed histogram for the data on the log scale. A log or square-root transformation can also be appropriate for count data, in order to assume a Normal distribution for the data on the transformed scale. The logit is preferred in the case of proportions requiring transformation. This is calculated as the log-odds ratio, i.e., $\text{logit}(p) = \log(p/(1 - p))$, for proportion data p. It provides a transformed variable on a continuous scale and ensures additivity of effects, both of which criteria are not provided by the once-favoured arcsine transformation.

Certain discrete data, such as counts with a narrow range of values, are appropriate for analysis using the Poisson distribution. Proportion data where the Binomial totals are small, such that the possible values the proportions can take are few, should be analysed assuming the Binomial distribution. Probability plots can also be helpful in checking which probability distribution is appropriate for data.

The box plot is another useful tool for data inspection. It is simply a picture of the data on the number line, with a box (rectangle) forming a region indicating where 75% of the data lie. The edges of the box are plotted at the first (Q1) and third (Q3) quartiles of the data, these being $Q1 = 0.25 \times (n + 1)^{th}$ and $Q3 = 0.75 \times (n + 1)^{th}$ values in the ordered data. Lines then extend from each of these edges to the most extreme data observations within $1.5 \times (Q3 - Q1)$ beyond Q1 and Q3. Observations respectively less, and greater, than these observations are worth inspection, and those beyond $3.0 \times (Q3 - Q1)$ are outliers (very rare observations), and should be scrutinized for their accuracy.

Having decided on the distribution appropriate for the data, the method of testing can then be selected. For data that can be assumed Normal, t-tests, ANOVA, regression and correl ation can then be used, taking account of the design structure used for the experiment and the treatment structure. For data distributed as Poisson or Binomial, the methods of GLM should be invoked.

24.4.2. t-Tests and ANOVA

In a comparative experiment with only two treatments, the t-test is sufficient for comparing the means, under the assumption of a Normal distribution and constant variance across the data from the two treatments. For designed experiments with a treatment structure however, ANOVA should be used, to assess the overall significance of main effects and interactions between treatment factors using F-tests, having accounted for variation due to design features. The F-test considers the size of the ratio of variance due to a treatment factor or interaction to the residual variance. This ratio is distributed as an F-distribution, on the degrees of freedom for the respective variances in the ratio. So, a large F-value with a small probability (p-value) of occurrence given the corresponding theoretical F-distribution indicates statistical differences between means for that treatment factor or interaction. Rather than include specific data sets giving full analyses and p-values as results of F-tests, Table 1 shows the skeleton ANOVAs for the completely randomized design (Fig. 1), the randomized block design (Fig. 2) and the split-plot design (Fig. 3).

Underlying an ANOVA is a linear model with estimated parameters for the effects of different levels of the design and treatment factors, and interactions. For example, in the case of the randomized block design in Figure 2 with skeleton ANOVA in Table 1B, the model is:

$$y_{ijk} = \mu + \text{Block}_i + \text{Species}_j + \text{Media}_k + (\text{Species.Media})_{jk} + \varepsilon_{ijk}$$

where y_{ijk} is a data observation from a Petri dish modelled by μ, the grand mean; Block_i for i = 1, 2 or 3, the effect for each block; Species_j for j = 1, 2 or 3, the main effect for each nematode species; Media_k for k = 1, 2, 3, 4 or 5, the main effect for each media and $(\text{Species.Media})_{jk}$, the interaction (synergy) effect for the species by media combinations. The final term in the model is the residual (error), ε_{ijk}, for modelling the observation from the Petri dish in the ith block with the jth species and the kth media.

The residuals, i.e. estimates of ε given the linear model, from ANOVAs should be inspected for any systematic trend, evidence of a non-Normal distribution, and for non-constant variance. These problems may be accounted for by transformation of the data prior to the ANOVA.

Experiments with more complex treatment structures, perhaps involving different control treatments, may require specific ANOVAs involving *contrasts* (see, e.g., Mead, 1988) and/or *nesting* of some treatment factors. For example, it is often useful to 'take out' variance due to a control treatment compared to all the other treatments, thus giving rise to a so-called *control plus factorial* treatment structure. For the analysis of more complex experiments, it is often advisable to seek help from a qualified statistician.

Table 24.1. Skeleton ANOVAs for the designs shown in Figures 24.1-24.3 as provided by the Genstat statistical package. In each case (A–C), the ANOVA shows the degrees of freedom (d.f.) for the variance of design and treatment factors and indicates the stratum at which factors are tested (using the F-test). The d.f. for the single residual (underlying variation) term is shown in A and B and the two underlying residual terms at the main plot and the split plot level for C.

Source of variation	d.f.
A	
Dishes stratum	
Species	2
Media	4
Species.Media	8
Residual	30
Total	44
B	
Blocks stratum	2
Blocks.Plots stratum	
Species	2
Media	4
Species.Media	8
Residual	28
Total	44
C	
Blocks stratum	2
Blocks.MainPlots stratum	
Variety	5
Residual	10
Blocks.MainPlots.SplitPlot stratum	
Nematicide	2
Variety. Nematicide	10
Residual	24
Total	53

It is important that the implications of the results of the F-test are heeded. For example, if the interaction between two treatment factors is not significant, but the two main effects of those factors are significant, then the two independent main effects means tables should be interpreted, and *not* the full interaction means table. Similarly, if only one of the treatment factors is significant then only the means for this factor should be inspected for statistical comparison.

Having thus discovered which treatment factors or interactions are statistically significant, usually with $p < 0.05$ given the result of the F-test, the appropriate treatment means can be inspected given the relevant SED value on the degrees of freedom for the residual variance. Using this value, a *small* number of comparisons of treatment means can then be made, using the least significant difference (LSD), at a chosen level of significance (usually 5%). The LSD is calculated by multiplying the SED by the critical t-value on the degrees of freedom for the residual variance at the chosen level of significance. The comparisons to make should ideally have been decided on as being appropriate *a priori* the experiment, i.e. at the design stage. In a designed experiment with a small number of treatment combinations, making a small number of biologically sensible comparisons (tests) under the LSD is more appropriate than invoking one of the many multiple comparison tests (such as Duncan's multiple range, Student-Newman-Keuls and Tukey) that result in all comparisons of means being made. The issue here is: who really wants to make *all* the comparisons anyway? Arguably they should not all be made because the questions being asked in the experiment should be more specific than that. To make all comparisons even suggests laziness on the part of the researcher (and the statistician, if one is involved!), in that they did not arrive at a small number of *a priori* or *post-hoc* comparisons that should be made.

24.4.3. Repeated Measures

Repeated measures (of the same response), taken from all the experimental units over time, require special treatment if the effect of time is to be included in the analysis, see, for example, Hand and Crowder (1996). Data collected in such a time series are not independent and so analysis needs to take account of this aspect. In some cases, a so-called *split-plot in time ANOVA* may be sufficient, so that, in a design sense, it is assumed that the experimental unit is split by the number of time points. However, this approach assumes that there is no overall autocorrelation over consecutive time points for the experimental units. The correlation structure for the time points can be checked and appropriate adjustments to tests made by employing a *repeated measures ANOVA*, or by directly modelling the data using a linear mixed model incorporating a parameter for the autocorrelation. Alternatively, a two-stage modelling approach can be used if the data for each experimental unit has a very clear linear, or non-linear, relationship with time. In this case, if the same, biologically sensible, model can be fitted to the data for each experimental unit separately as the first stage, the sets of estimated parameters from the model can then be analysed by ANOVA, using the design and treatment structure of the experiment (e.g., Esteves *et al.*, 2009). The assistance of a trained statistician can be very helpful in the analysis of repeated measures data.

24.4.4. Correlation and Regression

It may be useful to consider whether two or more sets of measurements collected from experimental units are correlated, i.e. linearly associated. For example, it may be of interest to query whether root fresh weight is negatively correlated with the number of nematodes that invaded the roots. Clearly this is only appropriate if sufficient pairs of observations per treatment are available, so that in plotting all the pairs against each other the presence or absence of any trend can be assessed. Assuming a Normal distribution for the data, the strength of a linear association can be assessed by calculating *Pearson's correlation coefficient*, r, the significance of which is tested using an F-test. Statistical regression modelling can also be employed to investigate whether a single line is sufficient to explain the variance in a given variable (the *response*) with respect to an *explanatory* variable. More complex regression models can be developed to test, for example, whether lines with different slopes and intercepts are required in the presence of different treatments (i.e., a *parallel lines analysis*), or to test which of a number of explanatory variables are required to give a satisfactory explanation of a given response (i.e., *multiple regression*).

24.4.5. Generalized Linear Models

In many nematology experiments, the data collected are discrete, i.e. counts. These counts may be just small numbers of observations with little variation, in which case they should be analysed assuming a Poisson distribution, or they may be observations that fall into a number of categories. In this case the counts pertain to a Multinomial distribution, forming a contingency table. They pertain to a Binomial distribution if there are only two categories (e.g., dead or alive). All such data deriving from comparative experiments can be approached using GLM (McCullagh and Nelder, 1989; Dobson, 1993), as applied, for example, in Dutta *et al.* (2011) and Wishart *et al.* (2004). GLMs allow the Poisson or Binomial distribution to be assumed, rather than having to invoke a data transformation in order to assume a Normal distribution for such data in an ANOVA. The importance of the correct application of GLM rather than ANOVA in these situations must be emphasized.

In GLM, a linear model is fitted using a chosen *link function*, which is usually log for Poisson and Multinomial data, and logit for Binomial data, although other link functions and indeed other distributions may be appropriate for particular data sets and situations. The model is then fitted to estimate effects (parameters) for blocking terms, treatments and interactions using the method of maximum likelihood (see, e.g., DeGroot, 1986), given the particular distribution chosen for the data. This allows the size of the variance, or rather *deviance*, with respect to the different model terms to be assessed. Deviance is asymptotically distributed as Chi-squared on the relevant degrees of freedom, so the statistical significance of treatment effects and interactions can be considered in an accumulated analysis of deviance. However, in the presence of excessive residual variation, or *overdispersion*, for the model, i.e. beyond that which would be expected for the chosen distribution, approximate F-tests are invoked rather than Chi-squared tests, so that a more conservative result is output. With similarity to ANOVA, predicted means for the significant treatment terms in the model can then be displayed for inspection and comparison.

24.4.6. Non-parametric Tests

For data exhibiting particularly high variance and for which a distribution cannot be assumed, even after transformation, a non-parametric test may still be used in order to compare treatments (see, e.g., Siegel and Castellan, 1988; Sprent, 1993). The term 'non-parametric' implies that no distributional assumptions are made. The tests are usually interested in the location of data on the number line for the different treatments, and usually ask questions about the median rather than the mean of the data for each treatment. Therefore, in many non-parametric tests the data are ordered, and sums of ranks are used in the test statistics rather than absolute differences between observations. This implies that the variance of observations is not included in the assessment of difference between treatments. Although non-parametric tests are more robust in the sense of not having to rely on assumptions that need to be met, parametric tests are preferred because they are more powerful – but only, of course, if the assumptions hold. In the situation of two treatments, for paired data the Wilcoxon matched pairs test can be used, whereas with unpaired data the Mann-Whitney U-test can be applied. Non-parametric ANOVAs are also available, such as the Kruskal-Wallis one-way ANOVA and Friedman's non-parametric ANOVA. Spearman's rank correlation and robust regression are further methods for considering associations between pairs of observations deriving from variables exhibiting non-parametric characteristics. Chi-squared tests can be applied to simple two-way contingency tables, or in a goodness-of-fit test to see if frequencies in categories are close to those expected from a theory.

24.4.7. Multivariate Analyses

In nematological studies where multiple variables are measured, for example on each individual nematode found in a soil-sampling experiment, or from each plant (experimental unit) in a comparative experiment, multivariate analysis (MVA) can be useful (see, e.g., Digby and Kempton, 1987; Krzanowski, 2000). In particular, analyses which allow grouping and classification of nematodes, such as hierarchical cluster analysis (as used in Ye and Robbins, 2004) and Principal Coordinates Analysis (PCO) (also known as metric scaling), can be used to provide a summary of the data and thence to extract groups of interest for further inspection. Using the variables, these analyses are concerned with quantifying the similarity between any pair of individuals (units) as a first step, to thus construct a similarity matrix. In hierarchical cluster analysis, the second step is to use a clustering method (such as nearest neighbour, or group average) to group individuals together as the similarity between all the different pairs of individuals gradually decreases. In PCO, the second step is to employ a dimensionality reduction routine to a Euclidean distance matrix derived from the similarity matrix, in order to visualize the individuals in a low-dimensional space (preferably two or three dimensions).

When there are *known* groups in the data, for example in the situation of multivariate data from a comparative experiment with a manageable number of groups and a large number of variates, Canonical Variates Analysis (CVA) may be useful. This method is directly related to linear Discriminant Analysis (LDA) and seeks linear combinations of the measured variables that maximize the ratio of the between- to within-treatments variation given all the measured variables.

Another approach, Principal Components Analysis (PCA) extracts orthogonal linear combinations of variables that account for as much variation in the data set as possible. This allows visualization of the data in a low-dimensional space, but cannot be used considering the statistical separation of the units.

24.5. Statistical Dos and Don'ts

This final section of the chapter takes the form of a list of frequently asked questions, or else questions that should be asked, given the author's experience of being a statistical editor of biological manuscripts. The answers to the questions expand somewhat, to include other related information.

24.5.1. Do I Present Figures or Tables?

Both a table and a figure are not required to show the same results of an analysis, and certainly not in a paper, where space is often at a premium in the journal. It needs to be decided which of them makes the results clear and easy to interpret. For example, if one of the classifying treatment factors in a two-way experiment is quantitative with more than two levels, then a graph of the treatment means is likely the better option, with the x-axis displaying, to scale, the levels of the quantitative factor. Lines can be drawn between points to show trends *only* when the x-axis of a graph is quantitative, but bear in mind that this still does not demonstrate the exact relationship. In particular, a linear or non-linear model may be better fitted to such data, so that the fitted line or curve can then be added to the figure. Results from experiments with only two treatments, where the means were compared using a t-test, do not really require a table or a figure, but can be included in the text.

24.5.2. When Do I Report the SE of Individual Treatment Means?

A SE for a particular treatment mean is:

$$\frac{s_i}{\sqrt{n_i}}$$

where s_i is the *standard deviation* of the data for treatment i, and where replication is n_i for each of T treatments, i, with i = 1, 2, ..., T. These individual SE can be useful prior to analysis, to investigate the extent of variability across the means for the treatments. However, such SE *should not be used for statistical comparisons in the situation of a designed experiment* where there were two or more treatments, and certainly not after a t-test or ANOVA has been applied. After such analysis, individual SE should be modified according to the replication, n_i, as:

$$\frac{s}{\sqrt{n_i}}$$

where s is the *estimate of standard deviation* for all the data, i.e. the square root of the pooled variance from a t-test, or the square root of the residual variance, s^2, from ANOVA. Therefore, if really wanting (see next question!) to put SE bars on bar charts of means *after a statistical analysis*, then this is the SE that should be used.

24.5.3. When Do I Report the SED?

After a statistical analysis such as a t-test or ANOVA, a SED for the comparison of each pair of treatment means, and the corresponding degrees of freedom, should be output. The SED is the measure of variability that must be used to compare means, rather than either of the two SEs described in the section above.

Far too often bar charts and line graphs of means with SE bars on the means are presented after a statistical analysis has been applied. The results of this analysis allow such SE bars to be dispensed with and the SED (or LSD) to be presented instead. It is the difference between the treatment means that the reader wishes to assess given the results of the statistical analysis, and this is done using the SED, not the SE.

24.5.4. What Significance Level Should I Use?

The 5% level of significance ($p = 0.050$) is often used as a cut-off when performing tests. This is only a guide though, and should not be a hard and fast rule. The choice of p-value to use as a cut-off may depend on how much trust there should be in the results. So, setting a cut-off at 1% ($p = 0.010$) rather than 5% imposes more stringency. However, a better option is to quote all p-values in full unless they are less than 0.001 (0.1%) in which case '$p < 0.001$' can be used – any test result with $p < 0.001$ can simply be classified as 'highly significant'. A statement should be made (in the materials and methods) to say that significance was determined at the x% level, for the x of choice.

Clearly, if a p-value of between 0.051 and 0.100 is obtained, it can still be said that there is 'something going on', statistically significant at the 10% level. How much emphasis is placed on this depends on how biologically important the size of the difference being tested is. Similarly, a p-value of between 0.010 and 0.050 is not indicating overwhelming significance, and may be termed 'weak evidence', particularly when the test was based on a large number of degrees of freedom.

24.5.5. Writing up Statistical Methods: How Do I Say What I Have Done?

Obviously, the amount of detail included about a statistical analysis will depend on what is being written (paper, report or thesis), the complexity of the analysis done, and how much importance needs to be placed on the statistical methods used given the expected readership. In the situation of a simple t-test or non-parametric test, it will often suffice to state the nature of the design, the number of samples (or replication) and the name of the test applied. In the context of t-tests, it is usually requisite to assume that a 'two-tailed test' has been applied. This is the usual situation. If, *a priori* the test, it is proposed to test whether one treatment gives a greater (or lesser) mean than another it should be specifically stated that a 'one-tailed test' is to be used. So, the choice of a one-tailed or two-tailed test is made *a priori*, given some prior knowledge about the treatments. Generally, two-tailed tests are performed because it is not obvious beforehand in which direction a response might go in the presence of some treatment, and, in any case, two-tailed tests are more conservative.

Following on from this, another important point is to make sure that what is said matches the statistical test done! For example, a common mistake is to write up the results of a two-tailed t-test or from using the LSD as though a one-tailed test had been applied (the LSD is invoking a two-tailed test). The phrases 'significantly more', 'significantly greater', 'significantly lower' and 'significantly less' (along with the test details and p-values) should only be used in the context of having done a one-tailed test. The phrase 'significantly different' should be used when doing two-tailed tests and using the LSD.

Usually, there is no need to reference basic statistical procedures such as t-tests and ANOVA, although, in reports and theses, a general statistics book, or statistical software manual, can be referenced. For more complicated statistical procedures, specific books and papers will need to be referenced as appropriate. When a statistical computing package has been used in the analysis then a reference to that package must be made, including the edition or version number, the makers and their address.

24.5.6. Writing up Statistical Results: How Do I Say What the Results Are?

In a simple case, suppose a (two-tailed) t-test has been done to compare two treatments. A typical statement that includes all the required information might be: The difference between the means of treatments A and B was significant (SED = 0.67; $t = 10.5$; 6 df; $p < 0.001$).

This sentence includes the all-important SED, the calculated test statistic (t), the degrees of freedom (for the pooled variance) and finally the significance of the test (p-value). At present, what goes in the brackets in many cases (particularly in papers) is usually just the p-value, but it is preferable for all the information to be included about a test result, so that readers can see what was done and decide for themselves what evidence the application of statistics gives to the claims being made by the authors. Alternatively, the information can be presented in tables or figures and referenced.

At this point it is worthwhile highlighting the problem of using the word 'significant' in any other context than a statistical one in a piece of scientific writing where statistical tests have been applied. Moreover, in all the situations where it is used in the statistical sense it should always be qualified with at least a p-value in brackets, and preferably with the information about the test. In any case, the word 'significant' is most usually redundant in the non-statistical sense: For if what is being reported is not significant, why mention it?

Whatever the written work (paper, report, thesis etc.) may be, after reporting the result of a test, all the relevant information, such as the SED and the means, should be presented somewhere, in a table or in a figure, so that the correctness of the p-value reported can be seen and comparisons made by readers if they so desire.

A further issue is that a statistical test can only be used to give its verdict based on the evidence, and the assumptions made in applying the test. So, in the language of statistical testing, a 'null hypothesis' (that nothing changes and there is no difference) is never rejected but rather the phrase 'there is evidence in favour of the alternative hypothesis' is used, if a test produces a statistically significant result. This apparent lack of excitement about the result is because it is based only on the evidence acquired and used in the test. Also, the data are from a single experiment, and often the result and conclusions are based on a small number of observations. In statistical terms, other data could give other results, so there is always *evidence* but not *proof*. Similarly, it could also be said that a non-significant result does not mean that a null hypothesis is necessarily true, but just that we do not have enough evidence in favour of the alternative hypothesis given the data.

Finally, statistical results should always be related to biological meaning. After all, there is no point in listing statistical results and then having no discussion about what those results may imply.

24.6. Conclusion

Nematologists should have a clear understanding of the principles of statistics governing the planning of experiments and the analysis and interpretation of experimental data. There should be an understanding of why certain statistical methods should, or should not, be used. The purpose of an experiment should be thought about carefully so that available resources are used efficiently in its design and execution. It should then be possible to extract all relevant information from the data collected, by application of statistical tests and modelling.

24.7. References

Cochran, W.G. and Cox, G.M. (1957) *Experimental Designs*, 2nd edn. Wiley and Sons, New York, New York.

DeGroot, M.H. (1986) *Probability and Statistics*. Addison-Wesley Publishing, Wokingham, UK.

Digby, P.G.N. and Kempton, R.A. (1987) *Multivariate Analysis of Biological Communities*. Chapman and Hall, London, UK.

Dobson, A.J. (1993) *An Introduction to Generalized Linear Models*. Chapman and Hall, London, UK.

Dutta, T.K., Powers, S.J., Kerry, B.R., Gaur, H.R. and Curtis, H.C. (2011) Comparison of host recognition, invasion, development and reproduction of *Meloidogyne graminicola* and *M. incognita* on rice and tomato. *Nematology* 13, 509–520. DOI: 10.1163/138855410X528262.

Esteves, I., Peteira, B., Powers, S., Magan, N. and Kerry, B.R. (2009) Effects of osmotic and matric potential on growth and accumulation of endogenous reserves in three biotypes of *Pochonia chlamydosporia*. *Biocontrol Science and Technology* 19, 185–199.

Gomez, K.A. and Gomez, A.A. (1984) *Statistical Procedures for Agricultural Research*, 2nd edn. J. Wiley and Sons, New York, New York.

Hand, D.J. and Crowder, M.J. (1996) *Practical Longitudinal Data Analysis*. Chapman and Hall, London, UK.

Krzanowski, W.J. (2000) *Principles of Multivariate Analysis: A User's Perspective*. Revised edition, Oxford University Press, Oxford, UK.

McCullagh, P. and Nelder, J.A. (1989) *Generalized Linear Models*, 2nd edn. Chapman and Hall, London, UK.

Mead, R. (1988) *The Design of Experiments: Statistical Principles for Practical Applications*. Cambridge University Press, Cambridge, UK.

Mead, R., Curnow, R. and Hasted, A.M. (2003) *Statistical Methods in Agriculture and Experimental Biology*, 3rd edn. Chapman and Hall, Boca Raton, Florida.

Siegel, S. and Castellan, N.J. (1988) *Nonparametric Statistics for the Behavioural Sciences*. McGraw-Hill, New York, New York.

Snedecor, G.W. and Cochran, W.G. (1989) *Statistical Methods*. Iowa State University Press, Ames, Iowa.

Sprent, P. (1993) *Applied Nonparametric Statistical Methods*, 2nd edn. Chapman and Hall, London, UK.

Wishart, J., Blok, V.C., Phillips, M.S. and Davies, K.G. (2004) *Pasteuria penetrans* and *P. nishizawae* attachment to *Meloidogyne chitwoodi, M. fallax* and *M. hapla*. *Nematology* 6, 507–510.

Ye, W. and Robbins, R.T. (2004) Cluster analysis of *Longidorus* species (Nematoda: Longidoridae), a new approach in species identification. *Journal of Nematology* 36, 207–219.

Index to General Terms and Major Crops

(terms in species descriptions, minor hosts, weed hosts and countries are not indexed)

Index to Major Citations of Bacteria, Fungi and Viruses

Index to Major Citations of Nematode Higher Taxa, Genera and Species

Printed and bound by CPI Group (UK) Ltd, Croydon, CR0 4YY
16/02/2026
14827393-0001